Fluid Mechanics

Robert A. Granger
Professor of Mechanical Engineering
U.S. Naval Academy, Annapolis, Maryland

HOLT, RINEHART AND WINSTON

New York Chicago San Francisco Philadelphia Montreal Toronto
London Sydney Tokyo Mexico City Rio de Janeiro Madrid

To Ruth, who gives and gives

Cover photograph: The "barber pole" pattern of turbulent spirals, a phenomenon discovered by Coles, was formed by rotating the inner cylinder from rest to $R_D = 7500$. Courtesy of H. Werlé and Office National d'Etudes et de Recherches Aerospatiales, France.

Address correspondence to:
383 Madison Avenue, New York, NY 10017

Library of Congress Cataloging in Publication Data

Granger, Robert Alan.
Fluid mechanics.

Bibliography: p.
Includes index.
1. Fluid mechanics. I. Title.
TA357.G714 1985 620.1′06 84-28964
ISBN 0-03-062951-9

Printed in the United States of America
Published simultaneously in Canada

5 6 7 8 038 9 8 7 6 5 4 3 2 1

CBS COLLEGE PUBLISHING
Holt, Rinehart and Winston
The Dryden Press
Saunders College Publishing

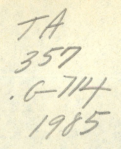

Contents

CHAPTER 14. BOUNDARY LAYER FLOWS 703

CHAPTER 15. ONE-DIMENSIONAL COMPRESSIBLE FLOW 801

Preface

Two remarks made years ago by my students typify the concerns that prompted the writing of this text. After spending a long and difficult night reviewing for her final, a perplexed senior told me, ''I think I understand each topic, but I can't get any sense of fluid mechanics as a whole. There must be some better way to organize all these equations!'' Her frustration is not uncommon. Encompassing as it does an enormous field of inquiry, touching nearly every phase of our physical existence, fluid mechanics too often seems to the student an amorphous mass of disparate formulas instead of the coherent body of theory and application it should be to the trained engineer. A primary purpose of this text, then, is to provide an orderly and structured introduction to fluid mechanics, one that will enable the student to see both the forest and the trees.

The second remark was a more plaintive one, but no less common than the first; ''My math is so weak I have trouble applying it to fluid mechanics problems.'' The student who said this wanted a course that would show him how to plug in a few formulas and come out with answers—whether or not he understood what was happening. This text does not seek to provide that kind of ''basic'' course: instead it seeks to train engineers even if they begin the course with less than desirable mathematical skills. It seeks, indeed, to improve those skills while teaching fundamentals of fluid mechanics. We will proceed on the premise that mathematical skills can be taught, used, and improved even as a student confronts new (sometimes esoteric) material.

As the above comments suggest, this text is written for the student rather than for the instructor or critic. The text presents a unified method of analysis that is unashamed of the detail devoted to posing fluid mechanics problems in precise mathematical language without becoming stiff or unnecessarily rigorous. This method involves three steps: First, the text carefully defines each problem so that the student knows what is given and what is missing. Second, each chapter treats the physical aspects of the problem so that the student can visualize how things work in the real world. Third, the text represents the physical model by appropriate mathematical symbols and operations, collects these into equations, and then solves them. Thus, the physics is never lost and the mathematical procedures are strengthened through repetition of examples. The text further illustrates each new topic with relevant engineering examples. By these means the text blends the principles of fluid sciences with problem-solving skills purposefully structured so as to strengthen the skills of students who lack confidence in their mathematical ability. The goal of this text is to train students to think like highly skilled engineers, not like mere mechanical ''plumbers'' or unnecessarily abstruse theoreticians.

The text covers ideal and real fluids in internal and external flows using fixed and moving frames of reference. Cartesian, cylindrical, and intrinsic coordinate systems are used, as well as vector and tensor notation. Some new material not covered in other fluid mechanics textbooks includes real vortex flows, detailed flow visualizations, new forms of differential linear momentum, certain energy loss expressions, free-surface effects, Prandtl's boundary layer equations, models in turbulent flow, boundary layer transition, and certain topics in drag.

The three-part organization reflects a "first the forest, then the trees" approach: Part I theoretically formulates fluid mechanics through separate applications (the formulation is partitioned into the differential approach and then the integral approach); Part II treats applications to solely incompressible flows, covering both internal and external flows of both real and ideal fluids; Part III covers applications to solely compressible flows.

To prepare the student for the fully elaborated topics in Part II and III, Part I ends with a summarizing chapter (Chapter 6) that develops detailed problem-solving methodologies (one involving the differential approach and one involving the integral approach). The instructor can select both or either approach in sequencing a one- or two-semester course without loss of continuity and without cumbersome jumping around in the text. To facilitate selection of material, each chapter begins with an outline of topics with appropriate statements of objectives. Each chapter contains numerous examples incorporating problem-solving techniques, demonstrations to illustrate topical material, study questions that act as chapter summaries or require the student to seek an answer from an external source, boxed equations of the significant results, appropriate references to supplementary materials, and a large number of exercises of varying degrees of difficulty using both SI and USCS units.

The perspective taken in this text is that fluid mechanics is an applied mathematics course. Without this perspective, fluid mechanics can become masked in mystery for the student, with results that seem more like magic than mathematics. Fluid mechanics, however, should not be reduced to a basic form that becomes a disservice in the teaching of outstanding engineers. This text will challenge the best and be a salvation to the weakest of students. As one student wrote after using this text, "This is the first time my calculus and differential equations make sense!"

The solutions manual contains a sample syllabus for both a short course and a longer, more complete one. Both have been used with success at the U.S. Naval Academy.

I am greatly indebted to Professor A. B. Strong of the University of Waterloo, Ontario, Canada and Allen R. Barbis of Auburn University and Visiting Professor at the U.S. Military Academy, West Point, N.Y. for reading and thoroughly criticizing the entire manuscript. Also to Sir G. I. Taylor, Professors J. Burgers, S. I. Pai, T. von Kármán, and M. G. Lighthill for their lectures, guidance, and inspiration. A special note of appreciation to Professor Stephen Ross of the English Department, USNA, for assistance in editing and offering numerous suggestions for improving the text. To John Beck, Editor at Holt, Rinehart and Winston, my appreciation in achieving the final product.

Finally, I want to thank my wife Ruth, who over three years not only endured my evening, weekend, and vacation hobby of writing this book, but accomplished all the typing flawlessly.

Any errors are my own fault, and I would greatly appreciate knowing about them along with any suggestions for improving the text.

R. A. Granger

1 Format and Fundamentals

1.1 Introduction: A Survey of Fluid Mechanics

Fluid mechanics is the branch of the physical sciences concerned with how fluids behave at rest or in motion. Its uses are limitless. We must understand fluid mechanics if we want to model the red spot on Jupiter, or measure the vorticity in a tornado, or design a transonic wing for an SST, or predict the behavior of subatomic particles in a betatron. To track the motions of fluids past objects or through objects, in oceans or in molecules, here on earth or in distant galaxies, fluid mechanics examines the behavior of liquids, gases, and plasma—of everything that is not solid. The theory of fluid mechanics is the foundation for literally dozens of fields within science and engineering: for meteorology, oceanography, and astronomy; for aerodynamics, propulsion, and combustion; for biofluids, acoustics, and particle physics.

Besides being one of the most important physical sciences in engineering, fluid mechanics is one of the oldest. As early as the 4th century B.C., Aristotle [1.1] was writing about density: ''Dense differs from rare in containing more matter in the same cubic area.'' Aristotle also conceived of uniform acceleration (an idea often credited to Galileo [1564]). The apple that fell on Sir Isaac Newton's head may have crowned him the founder of modern physics, but the following Aristotelian passages certainly have a familiar ring to our post-Newtonian ear:

1. ''Bodies which are at rest remain so owing to their resistance.''
2. ''When one is running fast it is difficult to divert the whole body from its impetus in one direction to some movement.''
3. ''The force of that which initiates movement must be made equal to the force of that which remains at rest. For there is a definite quantity of force or power by dint of which that which remains at rest does so, just as there is a force by dint of which that which initiates movement does so; and as there is a necessary proportion between opposite motions, so there is between absences of motion . . . For as the pusher pushes, so is the pushed pushed, and with equal force.''

It is, perhaps, an injustice of history that Newton's phraseology caught on, and Aristotle's lay dormant and nearly died.

Shortly after Aristotle another Greek philosopher-scientist, Archimedes (287–212 B.C.), offered mathematical formulations for those same physical problems that concerned Aristotle. Archimedes formulated, for example, the first quantitative law of the lever, which Aristotle had earlier described in words: ''As the weight moved is to the weight moving it, so inversely is the length of the arm bearing the weight to the length of the arm nearer the poser.'' Archimedes established many principles of fluid statics [1.2], among which the buoyancy principle is, in fact, named after him.

Then nothing significant happened for about 18 centuries, until Galileo started the modern inquiry into fluids [1.3]. Galileo used mathematics and deductive reasoning rather than direct experimental measurement to study bodies falling through fluids. His belief that mathematics was the language of motion and that change in nature was best described mathematically paved the way for the greatest awakening of scientific thought mankind had ever experienced. Though the calculus had not yet been invented, Galileo had the genius to realize that concepts of change were embedded somewhere in geometric arguments.

Then science erupted during the time of Newton (1642–1726). Rational mechanics* was conceived by James Bernoulli and Isaac Newton, the former developing ideas of continuous motion and of how dynamics was related to statics, and the latter describing motion as the result of external forces. Leibniz (1646–1716) was probably the first to develop theories of conservation of linear momentum to describe motion as energy. Another great Swiss mathematician, L. Euler (1707), is credited with modernizing mechanics by treating mechanics as analytical, and transforming Newton's physical concepts into mathematical equations.

The following gives some of the significant achievements of the scientists and mathematicians who initiated fluid mechanics during the period of Newton.

1. Hydrostatics—John Bernoulli [1.3, 112, 121–123]
2. Concept of pressure, integral form of linear momentum—Daniel Bernoulli [1.3, 11–112; 1.4, 14–15]
3. Equations of hydraulics for ideal fluids–John Bernoulli [1.4, 14–15]
4. Mechanics, concepts of force—Isaac Newton [1.5]
5. Concepts of energy, linear momentum—Gottfried Leibniz [1.6]
6. Theoretical hydrodynamics, theory of deformable bodies, concept of stress, kinematics, equations of motion of ideal fluids—Leonhard Euler [1.7]
7. Concept of stream function, velocity potential—Joseph-Louis Lagrange [1.8]
8. Potential theory, concept of vorticity—Augustin-Louis Cauchy [1.9]
9. Experimental hydrodynamics—Jean le Rond d'Alembert [1.10]

At first glance, this list might appear to be of negligible importance since it would seem to have no bearing on the technical aspect of the subject. Imbedded within the list, however, are little known but important human facts. Euler is known to have labored 20 years in developing some of the theories of fluid mechanics. In reading Laplace's statement that ''It is obvious'' we must translate it to mean (as his associates discovered when they queried him on the obviousness) that Laplace sometimes had to struggle for hours to duplicate a result. References are given throughout this book for a purpose. As Leibniz once said in the study of the history of science, the ''act of making discoveries should be extended by considering noteworthy examples of it.''

*''Rational mechanics is the science of the motions that results from any forces whatever, and of the forces that are required for any motions whatever, accurately set forth and proved.'' C. Truesdell, Newton Tricentennial Celebration, Univ. of Texas, 1966.

Next in the development of fluid mechanics came the age of 19th century theorists. This period witnessed mathematical developments in hydrodynamics and the birth of modern aerodynamics. The 19th century was a paradise for mathematicians, who had little or no regard for practical application. (There are exceptions, to be sure, such as practical work on lubrication, or on the filtration of liquids through sediments.) Following are some of the theories developed and the great thinkers who conceived them:

1. Complex variable method of two-dimensional potential flow to develop flow patterns—Gustave Kirchhoff [1.11] and Hermann Helmholtz [1.12]
2. Three-dimensional potential flow and viscosity—Siméon-Denis Poisson [1.13]
3. Method of singularities, gas dynamics, wave theory—William J. Rankine [1.14] and Lord Kelvin [1.15]
4. Rotational motion—H. Helmholtz [1.12] and Lord Kelvin [1.15]
5. Vortex flows, unsteady laminar flows—Gromeka [1.16]
6. Viscous flows—Claude L. M. H. Navier [1.17] and George G. Stokes [1.18]
7. Pipe flows—G. G. Stokes [1.18] and Jean L. M. Poiseuille [1.19]
8. Stability, turbulence—G. G. Stokes [1.18] and Osborne Reynolds [1.20]

Thinkers of a more practical bent, however, soon began to awaken to the applicability of these mathematical investigations. In the late 19th century engineers began to take over fluid mechanics, adapting it to designing machines of war, luxury, and production. Two remarkable theories started the age of the engineer: wing theory and air screw theory. With these concepts in hand, engineers raced to apply and extend the results; technical literature filled the shelves, and the aircraft was born. As more and more emphasis was placed on the behavior of bodies in a gas, scientists began building laboratory facilities to model the behavior of bodies moving through air. Tsiolkovskii (1896) [1.21] constructed one of the first wind tunnels to investigate drag on various bodies.* He postulated flight in all metal spacecraft into space, and he developed equations for the motion of a rocket-propelled spacecraft of variable mass.

At a time when one of the most adventurous ideas was to develop a body that a man could fit into which would behave like a fluid particle, it is no wonder that the age of the theorist gave way to the age of the engineer. While it would be impossible to list all the eminent contributors to fluid dynamics during this age, certain ones deserve mention. The list below is only a fraction of the major contributions.*

1. Shock waves—Doppler, Mach [1.22], and Reimann [1.23]
2. Wing theory—Lancaster [1.24], Zhukovskii [1.25], and Kutta [1.26]
3. Air screw theory—Zhukovskii [1.25] and Glauert [1.27]
4. Boundary layer concept—Rankine [1.28]
5. Boundary layer theory—Prandtl [1.29]
6. Turbulence—Burgers [1.30], Taylor [1.31], and von Kármán [1.32]

*It could also be argued that the first wind tunnel was built by Edme Mariotte (1620–1684) to measure the drag of a model held stationary in a moving stream: "Traite du mouvement des eaux" (1686).

*There are many fine surveys of fluid mechanics in the literature. C. Truesdell's [1.3] is recommended reading for the period between Aristotle and the 20th century. For the 20th century, S. Goldstein's brief survey paper [1.39] covers the period from 1900 to 1950.

7. High-speed aerodynamics—Rayleigh [1.33], Taylor [1.31], and Mach [1.22]
8. Magnetohydrodynamics—Hartman [1.34], Hoffman [1.35], and Alfvén [1.36]
9. Wave theory—Taylor [1.31] and Jeffreys [1.37]
10. Rarefied gases—Kantrowitz [1.38]

1.2 Format of This Text

1.2.1 The Subject of Fluid Mechanics

Fluid mechanics is one of the four branches of mechanics: elastic-body mechanics, fluid mechanics, relativistic mechanics, and quantum mechanics. The study of fluid mechanics subdivides into statics and dynamics which in turn divide into incompressible and compressible flow. Incompressible and compressible flow divide into real and ideal. Real divides into laminar and turbulent. And so on. Our subject grows out of a root system that seems tangled at first. Yet such elaborate classification is necessary in order to define accurately the types of fluid flow. Understanding fluid mechanics requires organization, summary, and recapitulation, all of which are integral to this text.

Fluid mechanics is based upon five great principles of physics:

1. Conservation of mass
2. Conservation of linear momentum
3. Conservation of angular momentum
4. Conservation of energy
5. Second law of thermodynamics

The first four principles are the same ones used in the other four branches of mechanics, and they are familiar to all. These principles govern the behavior of matter, whether it is solid or fluid; and it is the behavior of fluid matter that we seek to describe. If we can predict a fluid's behavior, regardless of the constraints, we can design, construct, and plan machines that use fluids: engines, ships, aircraft, pumps—the list is endless.

A fluid like oil or steam has stored energy. If we know how to predict the fluid's behavior, we can transport that energy from one location to another, or we can transform that stored energy into useful work. A fluid's energy can be converted into other forms of energy by combustion, mixing, impact reactions, and deflection. Blood is an excellent example of a fluid that transports energy in the form of nutrients, chemicals, and life-giving substances to our energy-burning cells. We are, after all, machines with pumps obeying the first and second laws of thermodynamics.

All fish, airplanes, ships, cars, and people are objects that move through fluids. And fluids move through objects, like blood through arteries, air in lungs, water in capillaries, oil in pipes, and gas in nozzles. Fluids deform as objects pass through them, sometimes translating, often rotating, expanding, compressing—all effortlessly with none of the failures, breaks, or fatigue that solids experience.

A fluid has the wonderful property of sometimes being a liquid and then again a gas. We can breathe the gas, but not the liquid, yet a fish does just the opposite. Some gases can quench a fire just as some liquids do, yet some gases are essential for fire just as some liquids are. In Chap. 2 we shall define the nature of a fluid, exploring its mysteries and its wonders.

1.2.2 The Structure of Fluid Mechanics

Structured in a manner similar to the other four branches of mechanics, as an organization of topics, fluid mechanics divides into subdivision after subdivision, each being necessary for one primary purpose: to classify *flow*, to give flow identity. Without such classification, we could not organize our techniques of problem solving. Figure 1.1 shows some of the classifications a fluid flow problem might possibly fit into. We can count 96 different combinations in just the categories given. For example, if we want to design a wing that would not flutter at transonic speeds, we would examine a *compressible gas* in *unsteady irrotational three-dimensional inviscid* flow. Each separate classification is important. The gas (air) will be *compressed* because we are studying velocities near the speed of sound. The flow is *unsteady* because gust loads

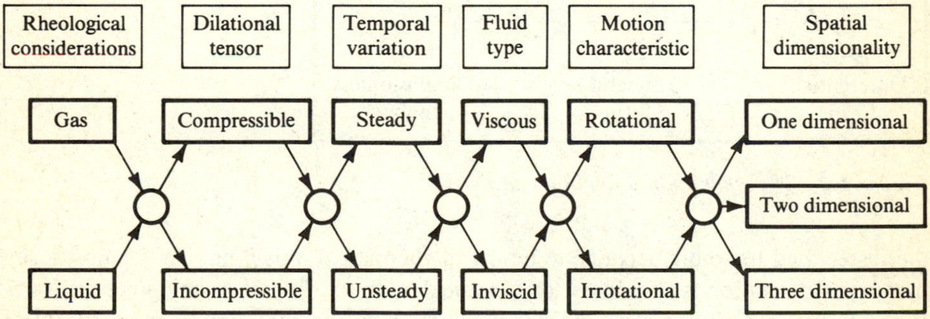

Figure 1.1 *Classification of fluid flows.*

and wind shear profiles continually deflect the elastic wing. The flow is *irrotational* because we shall assume that the pressure distribution over the wing is nearly the pressure distribution over the wing outside the boundary layer. The flow is *inviscid* because we shall assume that the frictional forces are small compared to the inertial and compressible forces and therefore neglect them. And the flow is *three-dimensional* because the wing has span length, chord width, and thickness, any variation of which can alter the flow. By classifying this problem as *unsteady, irrotational, inviscid, three-dimensional* flow, we can set up our problem quickly and select the appropriate form of our governing equations.

Once we have a problem defined, we seek its solution. Figure 1.2 shows a schematic we could use in solving a fluid flow problem. Any problem can be treated either by setting it up in a laboratory and measuring the desired unknowns, or by modeling the problem mathematically using symbols to represent physical quantities and operators to represent changes. Of the two choices, many might find the easier,

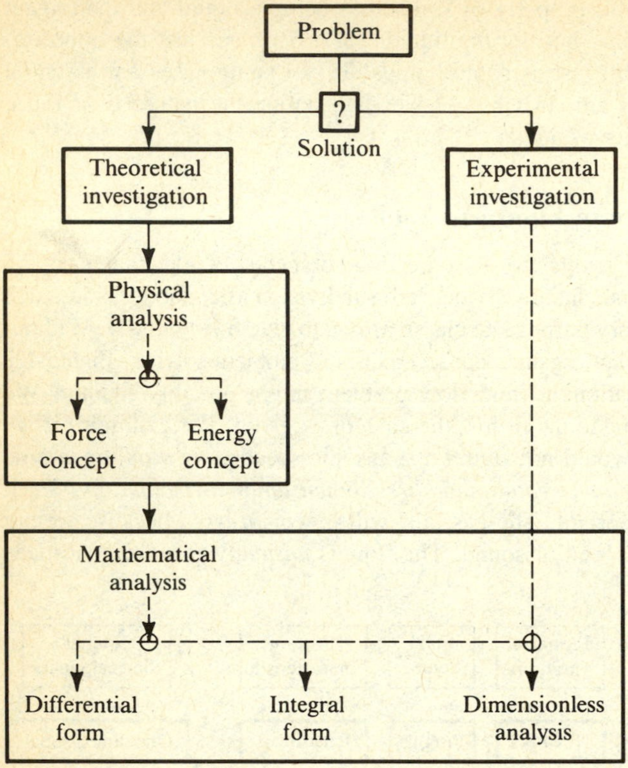

Figure 1.2 The solution of a fluid flow problem.

the faster, and the more accurate to be the mathematical modeling. This is not to say that experimentation is secondary to theoretical analysis. In fact, many practical problems in fluid mechanics are not yet mathematically tractable and must be solved in the lab. Good engineers always verify their theoretical predictions by testing their values experimentally. We would not expect 100% agreement since experimental errors are often found in instrumentation, in secondary fluid effects that had not been isolated from the experiment, and in measurement errors. But we would expect a close agreement, say within 93 or 94 percent accuracy, depending on the problem. Aircraft companies, ship builders, or high-quality industrial companies involved with machines that operate with fluids will usually perform a theoretical analysis first, then build a model or a prototype and test it to verify the results.

How is the theoretical analysis handled? Figure 1.2 shows two crucial steps that must be undertaken before any theoretical solution can be obtained. The first step is a physical analysis; the second step is a mathematical analysis.

A physical analysis is important in order to understand the physics of the problem. We are, after all, dealing with the real world, and thus nature affects that which is a part of it or moving through it. How do we understand the physical problem? Two concepts, *force* and *energy*, can be used to describe the manner in which the physical world behaves. We have a choice of using either or both to describe in a theoretical fashion a fluid itself, or the behavior of an object in a fluid. We often feel more comfortable working with the concept of force, as we can measure forces. Force is something we all have experienced. Energy, on the other hand, is a bit outside our experience. It is something most often calculated, not measured. But whichever we select, force or energy, we shall use it to describe the interaction between that which we are examining and its surroundings.

Forces have to act on something, just as energies have to be related to something. We need to define exactly what is acted upon. Either we may examine the effect of a force on a *system*, or we may examine its effect on a *control volume*. (Note that it is always possible for a system to occupy a control volume for an instant of time. We shall confine ourselves to systems and control volumes that are either independent of time or arbitrary in time.)

By definition, a *system* is a *predetermined identifiable quantity* of fluid. It could be a particle, or a collection of particles. The system may alter its geometric configuration, its location in space, its state variables such as pressure and temperature, but a system cannot alter its mass. An example of a system is shown in Fig. 1.3. Within the volume ∀, the system is the mass of gas. As the piston moves in the cylinder, the volume, pressure, and temperature of the system can change but the system, that is, the identity of mass, does not change. Thus, mass M remains constant, and the conservation of mass is automatically ensured.

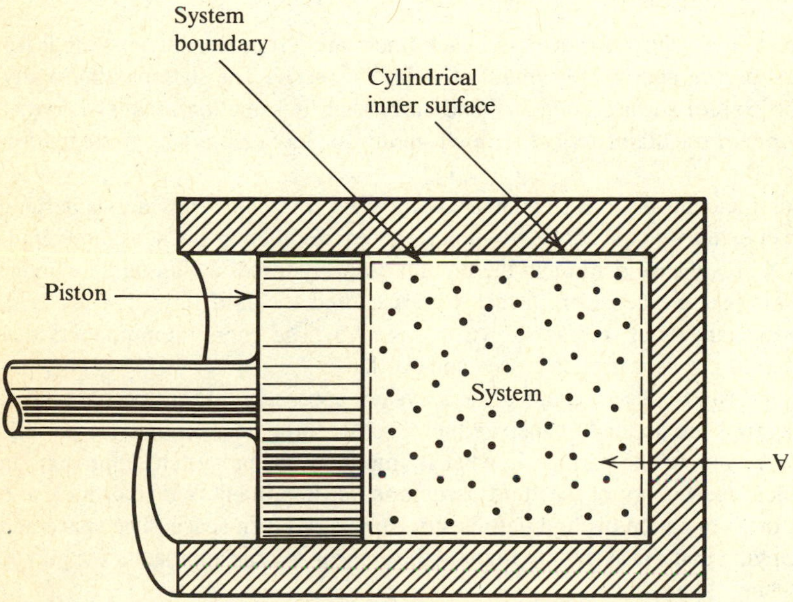

Figure 1.3 *Example of a system.*

In contrast to a system is the *control volume*. The control volume has a definite volume ∀, usually with a fixed shape for rigid boundaries and a variable shape for elastic or no boundaries. The thermodynamic term "closed system" corresponds to *system*, and the thermodynamic term "open system" corresponds to *control volume*.

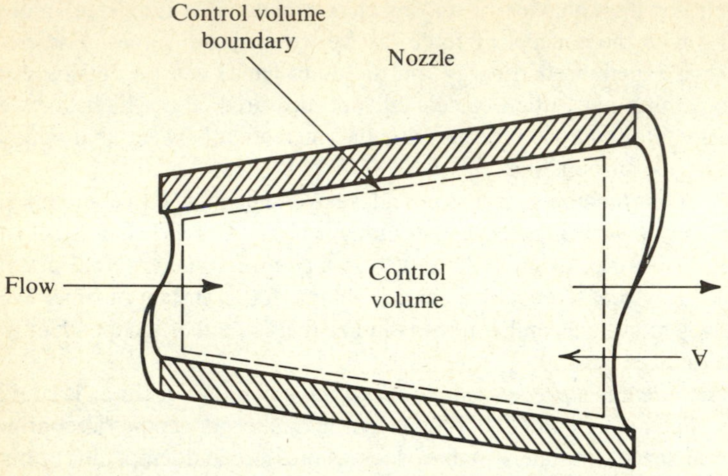

Figure 1.4 *Example of a control volume.*

A control volume allows mass to move through its *control surface*, which is the outer boundary of the control volume. An example of a control volume is shown in Fig. 1.4. The control surface boundary is customarily identified by dashed lines as illustrated in Fig. 1.4.

In addition to a control surface boundary, there are "stations." The symbol for a station is a subscript next to the symbol of a fluid property. The stations are usually located on the control surface boundary and carry such information as inlet i, outlet o, exit e, upstream (or high energy) 1, downstream (or low energy) 2, or stagnation (zero velocity) 0.

Acting on the control surface boundary will be the *environment*, which is defined as that which is exterior to the control volume. The environment influences the boundary either by energy transfers through the boundary, or by forces acting on the control volume. In this text we consider surface forces caused by fluid flow that are both normal, $\mathbf{F}_p$, and tangential, $\mathbf{F}_\tau$, as shown in Fig. 1.5. The energy components that can be transferred are heat friction losses and work.

Particle and rigid-body mechanics use a system approach, and the techniques of free-body diagrams are applied to analyze the dynamic force on each discrete particle or body. The *history* of the particle is what is important in the system approach. In fluid mechanics, the history of the fluid particle is *not* important. We shall look at a fluid particle or a collection of fluid particles occupying a given space. The space and the fluid occupying that space are important in the control volume approach. We pursue this idea in Chaps. 4 and 5.

Thus the distinction between the system approach and the control volume approach largely hinges on how the motion of the fluid is to be observed. We can observe the motion of flow particles from a fixed position in space, from which the velocities observed and measured are absolute velocities. By stationing ourselves, say, at the origin of an inertial frame of reference we are utilizing the control volume approach. A second way to observe the motion of fluid is to follow the collection of particles

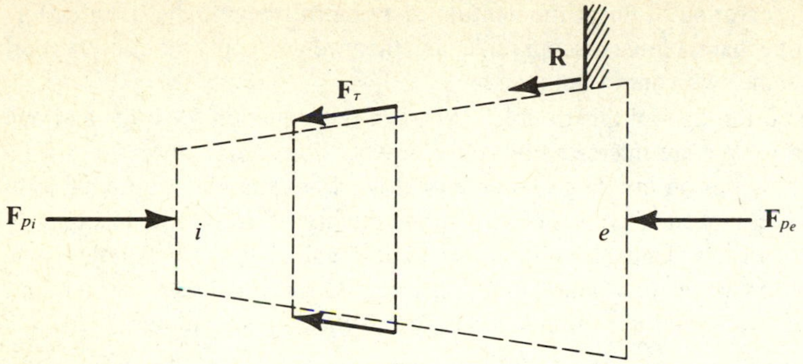

Figure 1.5 *Forces on control volume.*

or a single particle. This can be done by identifying the "system," attaching a coordinate system to its center of mass, and locating the observer at the origin of the moving frame of reference. The observer can then observe the resultant change in shape, position, and pressure. This is an example of the system approach.

As we will shortly discuss, the control volume represents an Eulerian point of view, where the dynamic behavior of a fluid motion is addressed at a fixed point in space. The system concept represents a Lagrangian point of view, where the dynamic behavior of a fluid particle is considered. From this latter point of view, one analyzes the *history* of the particle: where it has been, where it is, and where it will be. It predicts the future of the "system."

Let us summarize the two salient steps of the physical analysis:

1. Select the appropriate concept that can be used in describing a fluid motion: force or energy.
2. Select the appropriate control: the system (constant mass) or control volume (variable mass).

We shall use only the control volume in this text, and adopt both concepts of force and energy to describe fluid behavior.

Returning to Fig. 1.2, let us consider the next step in the theoretical analysis: the mathematical analysis. The choice of what mathematical tool to use in modeling the physical description of a fluid flow problem is usually dictated by the results of the physical analysis. If the control volume is of elemental size so that, in effect, a point in the flow field is being examined, we use the differential form of mathematical analysis. If the control volume is of macroscopic size (the size of machine components, sections of a river, part of a wing, and so on), we use the integral form of mathematical analysis. (It could be that we desire neither of these two forms. Suppose we are analyzing a physical model of a fluid flow problem. Then it would be appropriate to use an algebraic form of mathematical analysis called *dimensional analysis*. This is the topic of Chap. 7.) If we normally have a choice of two physical concepts (force and energy), and we have a choice of two mathematical forms (differential and integral), then there are four equations we can use. How do we obtain these four equations? We will apply the conservation of linear momentum to a control volume that is of differential size and to one that is of macroscopic size, and then we will apply the conservation of energy to the same two control volumes.

Why do we go through all this trouble? Why not one equation for force and one equation for energy? We are interested not only in what goes on at a point in a flow field, but also what goes on in a large volume of flow. For example, we might want to calculate the pressure at a point on an aircraft's wing at 15,000 m in order to calculate the speed of the plane. We might also want to calculate the lift on the same wing. To find the pressure we would use the differential form of linear momentum, and to find the lift we would use the integral form of linear momentum.

1.3 Fundamental Quantities, Units

There are five primary dimensions in fluid mechanics: length L, time T, temperature Θ, electric charge Q_e, mass M, or force F. In an engineering analysis, any equation relating physical quantities comprising these dimensions must be dimensionally homogeneous. This means that the dimensions of each term of the equation must be identical.

Units are arbitrary magnitudes that carry specific names for the primary quantities as well as combinations of dimensions. For example, the primary dimension of length might have units of yards, miles, inches, meters, Ångström, or furlongs. These units are related to one another by means of conversion factors. One way of writing unit conversions is

$$\frac{60 \text{ s}}{1 \text{ min}} = 1, \frac{1 \text{ m}}{3.2808 \text{ ft}} = 1, \frac{1 \text{ kg}}{0.06852 \text{ slugs}} = 1$$

Thus terms in equations can always be multiplied by unit conversions since it is always correct to multiply by one. Table 1.1 lists a number of useful unit conversions.

A large number of systems of units have been in use over the years. In this text, we shall use mostly two: the Systéme Internationale (SI) and the U.S. customary system (USCS).

1.3.1 Le Systéme International, (SI) System

The SI is the most widely used system of units in the world today. Table 1.2 lists the SI units for seven primary dimensions: length, time, temperature, mass or force, electric current, luminous intensity, and amount of a fluid substance. The SI unit of length is the meter (m) and is equal to the distance light travels through space in 1/299,792,458 of a second.

The unit of time is the second (s), defined as the period of 9,192,631,770 + 20 cycles of a transition of the cesium atom.

The unit of mass is the kilogram (kg) and is defined by the mass of a platinum-iridium cylinder in Sevres, France.

Table 1.1 *SI and USCS Conversion Units*

Ratio	Unit Conversion
Acceleration	$\frac{1\ \text{ft/s}^2}{0.3048\ \text{m/s}^2} = 1$
Angle	$\frac{1°}{0.0174533\ \text{rad}} = \frac{1\ \text{rev}}{2\pi\ \text{rad}} = 1$
Angular momentum	$\frac{1\ \text{lbm·ft}^2\text{/s}}{0.04214\ \text{kg·m}^2\text{/s}} = \frac{1\ \text{slug·ft}^2\text{/s}}{1.35582\ \text{kg·m}^2\text{/s}} = 1$
Angular speed	$\frac{1\ \text{rps}}{2\pi\ \text{rad/s}} = \frac{1\ \text{rpm}}{(2\pi/60)\ \text{rad/s}}$
Area	$\frac{1\ \text{ft}^2}{0.09290\ \text{m}^2} = \frac{1\ \text{in.}^2}{6.4516 \times 10^{-4}\ \text{m}^2} = \frac{1\ \text{mi}^2}{2.59 \times 10^6\ \text{m}^2} = 1$
Density	$\frac{1\ \text{slug/ft}^3}{515.379\ \text{kg/m}^3} = \frac{1\ \text{lbm/ft}^3}{16.0185\ \text{kg/m}^3} = \frac{1\ \text{lbm/in.}^3}{27.6799\ \text{mg/m}^3} = 1$
Energy	$\frac{1\ \text{J}}{9.4845 \times 10^{-4}\ \text{Btu}} = \frac{1\ \text{J}}{0.239\ \text{cal}} = \frac{1\ \text{J}}{1 \times 10^7\ \text{ergs}}$ $= \frac{1\ \text{J}}{0.7375\ \text{ft·lbf}} = \frac{1\ \text{J}}{0.102\ \text{kg·m}} = \frac{1\ \text{J}}{2.778 \times 10^{-4}\ \text{w·hr}} = \frac{1\ \text{J}}{1\ \text{N·m}}$ $= \frac{1\ \text{Btu}}{778\ \text{ft·lbf}} = \frac{1\ \text{J}}{\text{kg·m}^3\text{/s}^2} = 1$
Force	$\frac{1\ \text{N}}{0.2248\ \text{lbf}} = \frac{1\ \text{N}}{1 \times 10^5\ \text{dynes}} = \frac{1\ \text{kg}}{2.6792\ \text{lbm}} = \frac{1\ \text{kg}}{0.06852\ \text{slugs}}$ $= \frac{32.174\ \text{lbm·ft}}{\text{lbf·s}^2} = 1$
Heat capacity	$\frac{1\ \text{Btu/°F}}{1899.108\ \text{J/°C}} = 1$
Heat flow rate	$\frac{1\ \text{cal/s}}{4.1868\ \text{w}} = \frac{1\ \text{Btu/h}}{0.293071\ \text{w}} = 1$
Kinematic viscosity	$\frac{1\ \text{ft}^2\text{/s}}{0.092903\ \text{m}^2\text{/s}} = \frac{1\ \text{St}}{1 \times 10^{-4}\ \text{m}^2\text{/s}} = 1$
Length	$\frac{1\ \text{m}}{3.2808\ \text{ft}} = \frac{1\ \text{km}}{0.62137\ \text{mi}} = \frac{1\ \text{in.}}{25.4\ \text{mm}} = \frac{1\ \text{U.S. nautical mile}}{1.85325\ \text{km}} = 1$
Mass	$\frac{1\ \text{slug}}{14.5939\ \text{kg}} = \frac{1\ \text{lbm}}{0.45359\ \text{kg}} = 1$
Moment of inertia	$\frac{1\ \text{slug·ft}^2}{1.35582\ \text{kg·m}^2} = \frac{1\ \text{lbm·ft}^2}{0.04214\ \text{kg·m}^2} = 1$

Table 1.1 (Con't.)

Ratio	Unit Conversion
Momentum	$\frac{1\ \text{slug·ft/s}}{4.44822\ \text{kg·m/s}} = \frac{1\ \text{lbm·ft/s}}{0.13826\ \text{kg·m/s}} = 1$
Power	$\frac{1\ \text{hp}}{745.7\ \text{w}} = \frac{1\ \text{erg/s}}{1 \times 10^{-7}\ \text{w}} = \frac{1\ \text{ft·lbf/s}}{1.35582\ \text{w}} = 1$
Pressure and stress	$\frac{1\ \text{kg/m}^2}{9.6784 \times 10^{-5}\ \text{atm}} = \frac{1\ \text{kg/m}^2}{9.80665 \times 10^{-5}\ \text{bars}} = \frac{1\ \text{kg/m}^2}{0.1\ \text{g/cm}^2}$ $= \frac{1\ \text{kg/m}^2}{0.2048\ \text{lbf/ft}^2} = \frac{1\ \text{Pa}}{1.45 \times 10^{-4}\ \text{psi}} = \frac{1\ \text{Pa}}{1\ \text{N/m}^2}$ $= \frac{1\ \text{ft H}_2\text{O}}{2.98907\ \text{kN/m}^2} = \frac{1\ \text{in. Hg}}{3.38639\ \text{kN/m}^2} = \frac{1\ \text{mm Hg}}{133.322\ \text{N/m}^2} = 1$
Temperature ΔT	$\frac{\text{K}}{°\text{C}} = \frac{°\text{R}}{°\text{F}} = \frac{°\text{R}}{0.5556°\text{C}} = 1$
Velocity	$\frac{1\ \text{ft/s}}{0.3048\ \text{m/s}} = \frac{1\ \text{in./s}}{25.4\ \text{mm/s}} = \frac{1\ \text{mi/h}}{1.60934\ \text{km/h}} = \frac{1\ \text{knot}}{0.51479\ \text{m/s}} = 1$
Viscosity (dynamic)	$\frac{1\ \text{St}}{1\ \text{poise·cm}^3\text{/g}} = \frac{1\ \text{centipoise}}{1\ \text{g/cm·s}} = \frac{1\ \text{centipoise}}{1\ \text{lbf/ft·hr}}$ $= \frac{1\ \text{lbf·s/ft}^2}{47.8803\ \text{N·s/m}^2} = \frac{1\ \text{slug/ft·s}}{47.8803\ \text{N·s/m}^2} = 1$
Volume	$\frac{1\ \text{ft}^3}{2.83168 \times 10^{-2}\ \text{m}^3} = \frac{1\ \text{gal}}{3.785\ \ell} = 1$
Volume rate of flow	$\frac{1\ \text{gal/h}}{0.1337\ \text{ft}^3\text{/h}} = \frac{1\ \text{gal/h}}{3.7853\ \ell\text{/h}} = \frac{1\ \text{ft}^3\text{/s}}{0.02832\ \text{m}^3\text{/s}} = 1$

Table 1.2 *SI and USCS Units*

Primary Dimension	SI System Unit	USCS System Unit
Length L	meter (m)	foot (ft)
Time T	second (s)	second (s)
Temperature Θ	Kelvin (K)	Rankine (°R)
	Celsius (°C)	Fahrenheit (°F)
Electric current Q_e	ampere (A)	ampere (A)
Luminous intensity	candela (cd)	candle
Amount of substance	mole (mol)	
Mass M	kilogram (kg)	pound mass (lbm)
Force F	newton (N)	pound force (lbf)

The unit of temperature is Kelvin (K) for absolute temperature. Note that it is *not* degree Kelvin. Sometimes the temperature is measured in Celsius (°C), in which case the two temperatures are related by

$$\mathrm{T(K)} = \mathrm{T(^\circ C)} + 273.15\ \mathrm{K} \tag{1.1}$$

All other units in SI are secondary units and can always be expressed in terms of the seven primary units. Note that if mass is used as one of the primary dimensions, then force becomes a secondary dimension, since it can always be expressed in terms of the primary dimensions of M, L, and T. This is accomplished by using Newton's second law of motion:

$$F = MLT^{-2} \tag{1.2}$$

Conversely, if we designate force as one of the primary units, then from Newton's law mass can be expressed as

$$M = FT^2L^{-1} \tag{1.3}$$

The unit of a force is measured by the newton (N), so that from Eq. (1.2) a newton equals one kilogram of mass moving at one meter per second squared, such that

$$\frac{1\ \mathrm{N \cdot s^2}}{\mathrm{kg \cdot m}} = 1 \tag{1.4}$$

If we divide this force by an area, we get another secondary unit, that of pressure or stress:

$$\frac{1\ \mathrm{N}}{\mathrm{m^2}} = \frac{\mathrm{kg \cdot m}}{\mathrm{s^2}} \times \frac{1}{\mathrm{m^2}} = \frac{1\ \mathrm{kg}}{\mathrm{m \cdot s^2}} = 1\ \mathrm{Pa} \tag{1.5}$$

where a pascal (Pa) is defined as a kg mass divided by meter-second squared.

Table 1.3 presents a few secondary units that will be found in fluid mechanics. One of the quantities shown is energy. For example, kinetic energy is familiar as $MV^2/2$, and the units, ( ), would result in

$$MV^2 = (\mathrm{kg \cdot m^2/s^2})$$

Although $\mathrm{kg \cdot m^2/s^2}$ is correct as the unit of any energy quantity (including heat transfer), it is not acceptable. Using the SI conversion unit for energy in Table 1.1 and Eq. (1.4) we obtain

Table 1.3 Some Secondary Units in SI and USCS

Symbol	Quantity	Unit SI	Unit USCS	Dimensions MLT
a	Acceleration	m/s^2	ft/s^2	LT^{-2}
A	Area	m^2	ft^2	L^2
C_p, C_v	Specific heats	J/kg·K	ft·lbf/lbm°R	L^{2}T^{-2} Θ
g_c	Gravitational constant	kg·m/N·s^2	lbm·ft/lbf·s^2	
h	Head	m	ft	L
h	Enthalpy per unit mass	J/kg	ft·lbf/lbm	L^{2}T^{-2}
i	Specific internal energy	m·N/N	ft·lbf/lbf	L
K	Bulk modulus of elasticity	Pa	lbf/ft^2	ML^{-1}T^{-2}
L	Lift	N	lbf	MLT^{-2}
M	Mass	kg	lbm	M
p	Pressure	Pa	lbf/ft^2	ML^{-1}T^{-2}
Q	Discharge	m^3/s	ft^3/s	L^{3}T^{-1}
R	Gas constant	J/kg·K	ft·lbf/lbm°R	L^{2}T^{-2} Θ
S	Entropy	J/K	ft·lbf/°R	ML^{2}T^{-2} Θ
T	Temperature	°C	°F	Θ
T	Torque	N·m	lbf·ft	ML^{2}T^{-2}
V	Velocity	m/s	ft/s	LT^{-1}

$$MV^2 = (\text{kg}\cdot\text{m}^2/\text{s}^2)\left(\frac{1\ \text{J}}{1\ \text{N}\cdot\text{m}}\right)\left(\frac{1\ \text{N}\cdot\text{s}^2}{\text{kg}\cdot\text{m}}\right)$$

$$= (\text{J})$$

Thus, the unit of $MV^2/2$ is the joule.

1.3.2 The USCS System

The USCS system is one of several popular engineering systems still in use in the United States, though congressional impetus to convert all systems to SI is gradually eroding its popularity. Only a few technical journals accept papers using the USCS system. We include it in this text to aid in the transition toward the ultimate total conversion to SI.

The greatest difficulty in using USCS arises from the units for mass and force. The unit for mass is the pound-mass (lbm), and for force the pound-force (lbf). Using Newton's second law, we can see where the confusion lies:

$$\frac{F}{Ma} = \left(\frac{\text{lbf}\cdot\text{s}^2}{\text{lbm}\cdot\text{ft}}\right)$$

which does not equal unity unless one redefines a lbf or a lbm, whichever is considered the secondary unit. Consequently, we must modify Newton's second law whenever we use the USCS unit as

$$\frac{Fg_c}{Ma} = g_c \left(\frac{\text{lbf}\cdot\text{s}^2}{\text{lbm}\cdot\text{ft}}\right)$$

If we expect this ratio to be a unit conversion, that is, if we expect Newton's second law to be a homogeneous equation, then

$$g_c = \left(\frac{\text{lbm}\cdot\text{ft}}{\text{lbf}\cdot\text{s}^2}\right)$$

Mass can be either an inertial or a gravitational mass. If it is a gravitational mass, then a force of one lbf will accelerate a mass of one lbm at the rate of 32.174 ft/s^2 such that

$$g_c = 32.174 \left(\frac{\text{lbm}\cdot\text{ft}}{\text{lbf}\cdot\text{s}^2}\right) \tag{1.6}$$

Thus Newton's law for the USCS system becomes

$$F = \frac{Ma}{g_c} = \frac{Ma}{32.174} \tag{1.7}$$

In using Eq. (1.7), the mass M must be expressed in lbm and the acceleration in ft/s^2 so that the units of force will be lbf.

What then will our energy expression be? If we consider the kinetic energy ½MV^2, the units become

$$MV^2 = \left(\frac{\text{lbm}\cdot\text{ft}^2}{\text{s}^2}\right)$$

as was the case when we treated kinetic energy in SI units. We are correct as far as we have gone, except that the equation is not in acceptable form. From Eq. (1.7), we multiply the units of MV^2 by unity to obtain

$$MV^2 \times \frac{Ma}{32.174\ \text{F}} = \frac{MV^2}{32.174}\ (\text{ft}\cdot\text{lbf})$$

where now the units are ft·lbf, an acceptable form for kinetic energy.

There are other energy units besides ft·lbf, e.g., the N · m, J, or kg · m^2/s^2. The Btu (British thermal unit) is defined as 1.05506 kJ and is the energy required to raise 1 lbm of water 1°F. Another unit associated with heat and internal energy is the calorie (cal) and is the energy required to raise 1 g of water 1°C. The calorie is familiar to nutritionists and dieticians, who might prescribe a healthy energy input as being 2000 cal per day. Physicists frequently use the electron volt (eV) as a unit of energy, $1 \text{ eV} = 1.602 \times 10^{-19}$ J, and astronomers use ergs, $1 \text{ erg} = 10^{-7}$ J. Some typical energy levels range from lows of 10^{-16} ergs for the kinetic energy of a molecule, or 100 ergs for 1 gram of body tissue, to larger 10^8 ergs for a bullet's muzzle velocity, 10^{21} ergs for the daily output of a large hydroelectric dam, 10^{28} ergs for Mount St. Helen's eruption, or 10^{39} ergs for a daily solar output of energy.

1.3.3 Two Other Systems of Units

A third system currently used in many applications of fluid mechanics is the absolute engineering system of units. For this system, length, time, force, and temperature are the primary dimensions, and mass is a secondary dimension. The unit of force is the lbf and the unit of mass is the *slug*. One slug is defined as the mass of an object that is accelerated 1 ft/s^2 when acted upon by a 1 lbf force. Thus

$$1 \text{ slug} \equiv 1 \text{ lbf}\cdot\text{s}^2/\text{ft}$$

In this system, the universal gravitational constant g_c is unity and dimensionless. We shall use the absolute engineering system along with the USCS and SI systems.

A fourth system popular in many European countries is the absolute metric system. Here mass, length, time, and temperature are the primary dimensions, with force being a secondary dimension. The unit of mass is the gram, the unit of time is seconds, and the unit of length is centimeters (cm). The unit of force is the dyne, defined as that force required to accelerate a 1 gram mass at the rate of 1 centimeter per second squared:

$$1 \text{ dyne} \equiv 1 \text{ g}\cdot\text{cm/s}^2$$

For this system, the universal gravitational constant g_c is unity and dimensionless.

1.3.4 Secondary Dimensions

An amazing result of having identified the primary dimensions of fluid mechanics is that we can now express all other dimensions of all quantities involved in a fluid flow situation in terms of the primary dimensions. Table 1.3 contains a brief list of some of the quantities we shall encounter in this text. The reader should be familiar with both the SI and USCS units. The absolute engineering units will also appear at times. The conversion from one system to another is quite easy using the unit conversions shown in Table 1.1. Remembering that 1 ft = 0.3048 m and 1 slug = 14.5939 kg = 32.174 lbm, we can obtain the conversion factors of secondary dimensions.

1.4 Fundamental Idealizations

Theoretical fluid mechanics is an attempt to predict the behavior of real fluid motions by solving boundary value problems of either appropriate partial differential equations or integral equations. The roots from which these equations stem are the conservation principles of the physical sciences. It is assumed that the fluid (be it liquid or gas) is homogeneous and "continuum"* of matter, and that certain stresses caused by external forces or energies act on and within the fluid resulting in the fluid reacting either statically or dynamically.

In deriving the well-set boundary value equations, we postulate certain boundary and "inner" conditions which inevitably dictate the final form of the solution. With such a set of equations, we can solve few problems. Analytic solutions are impossible, numerical solutions are inappropriate, and nothing appears to work. Only the simplest fluid flow problems can be solved.

Therefore, we introduce idealizations into the problems. We might assume that the fluid is independent of time, reasoning that the disturbances are of secondary importance. We could assume that the fluid is ideal, when in fact no known fluid is ideal. But because the viscosity may be small, much smaller than, say, for water, the idealization will yield solutions that are acceptable. What else might we assume? The possibilities are endless. For example, we could assume the flow is (a) symmetric, (b) incompressible, (c) not rotating, (d) one-dimensional, (e) continuous, (f) isothermal, (g) isobaric, (h) adiabatic, (i) reversible, (j) homogeneous, etc. The flow of course, may be none of these, for all are idealizations.

Some physicists would reject the use of some of these in the solution of a real problem. Many mathematicians would question the appropriateness of applying some of these equations that we shall develop, such as Euler's equation to water flowing in a pipe, or Prandtl's boundary layer equation to the problem of fluids near the surface of moving bodies. But the engineer has to defend the idealizations he is imposing on the solution, and thus he must be part mathematician and part applied scientist. The engineer also has the added responsibility of ensuring that his design functions. It cannot work solely on paper. To make his design work, the engineer may have to relax some of the physical constraints of the real problem, or bypass some of the mathematical rigor, and, of course, whatever idealizations are chosen, they must be applied to any solution with considerable caution.

1.5 Fundamental Coordinates

We shall use three coordinate systems in this text: (a) rectangular Cartesian, (b) cylindrical, and (c) intrinsic. A coordinate system is necessary to identify the location of a point or object in the flow field. We will first consider the coordinate system as being fixed.

Let P denote a point in the flow field, such that the position vector from the origin of the fixed coordinate system to the point P is

*The continuum has been used as an idealization in earlier mechanics courses in the form of rigid body motion, behavior of elastic bodies and in thermodynamics. It is defined in Sec. 2.2.2.

$$\mathbf{r}_P = r\mathbf{e}_r + z\mathbf{k} \tag{1.8}$$

$$= x\mathbf{i} + y\mathbf{j} + z\mathbf{k} \tag{1.9}$$

Figure 1.6 shows the rectangular Cartesian and cylindrical coordinate systems with unit vectors **i**, **j**, **k** in the x, y, z directions, respectively, and $\mathbf{e}_r$, $\mathbf{e}_\theta$, **k** in the r, θ, z directions, respectively. A property of the former is that **i**, **j**, **k** are always constant, whereas $\mathbf{e}_r$ and $\mathbf{e}_\theta$ are functions of orientation. From the geometry of Fig. 1.6, we obtain

$$r = \sqrt{x^2 + y^2} \tag{1.10}$$

$$r_P = \sqrt{r^2 + z^2} \tag{1.11}$$

$$\theta = \tan^{-1}\frac{y}{x} \tag{1.12}$$

$$x = r\cos\theta \tag{1.13}$$

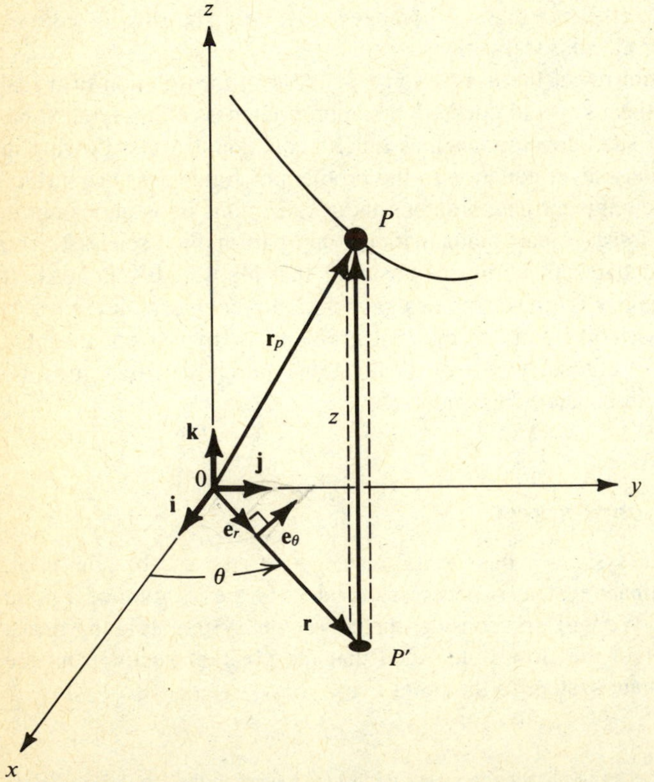

Figure 1.6 *Rectangular Cartesian and cylindrical coordinate systems.*

$$y = r \sin\theta \tag{1.14}$$

$$\mathbf{e}_r = \mathbf{i}\cos\theta + \mathbf{j}\sin\theta \tag{1.15}$$

$$\mathbf{e}_\theta = -\mathbf{i}\sin\theta + \mathbf{j}\cos\theta \tag{1.16}$$

Example 1.1

Show that

$$\frac{d}{dt}\mathbf{e}_r = \omega_z \mathbf{e}_\theta \text{ and } \frac{d}{dt}\mathbf{e}_\theta = -\omega_z \mathbf{e}_r$$

where $\omega_z = d\theta/dt$.

Solution

We start by differentiating Eq. (1.15):

$$d\mathbf{e}_r = -\sin\theta\,\mathbf{i}\,d\theta + \cos\theta\,\mathbf{j}\,d\theta \tag{i}$$

then compare it against Eq. (1.16) to obtain

$$d\mathbf{e}_r = \mathbf{e}_\theta\,d\theta \tag{ii}$$

Dividing through by dt, we find

$$\frac{d\mathbf{e}_r}{dt} = \frac{d\theta}{dt}\mathbf{e}_\theta = \omega_z\,\mathbf{e}_\theta \tag{iii}$$

We do the same thing with θ: differentiating Eq. (1.16),

$$d\mathbf{e}_\theta = -\mathbf{i}\cos\theta\,d\theta - \mathbf{j}\sin\theta\,d\theta \tag{iv}$$

and comparing it with Eq. (1.15), we find

$$d\mathbf{e}_\theta = -\mathbf{e}_r\,d\theta \tag{v}$$

so that

$$\frac{d\mathbf{e}_\theta}{dt} = -\frac{d\theta}{dt}\mathbf{e}_r = -\omega_z\,\mathbf{e}_r \tag{vi}$$

This completes the solution.

Consider an inertial reference xyz and a reference $x'y'z'$ moving in an arbitrary manner relative to xyz, as shown in Fig. 1.7. The origin O' of $x'y'z'$ is located in xyz by the position vector $\mathbf{r}_{O'}$, $\mathbf{r}_P$ is the absolute position vector of the point P with respect to the fixed origin 0, and $\mathbf{r}_{P/O'}$ is the relative position vector of the point P with respect to the moving origin O'. Examining the closed polygon in the figure, we obtain

$$\mathbf{r}_P = \mathbf{r}_{O'} + \mathbf{r}_{P/O'} \tag{1.17}$$

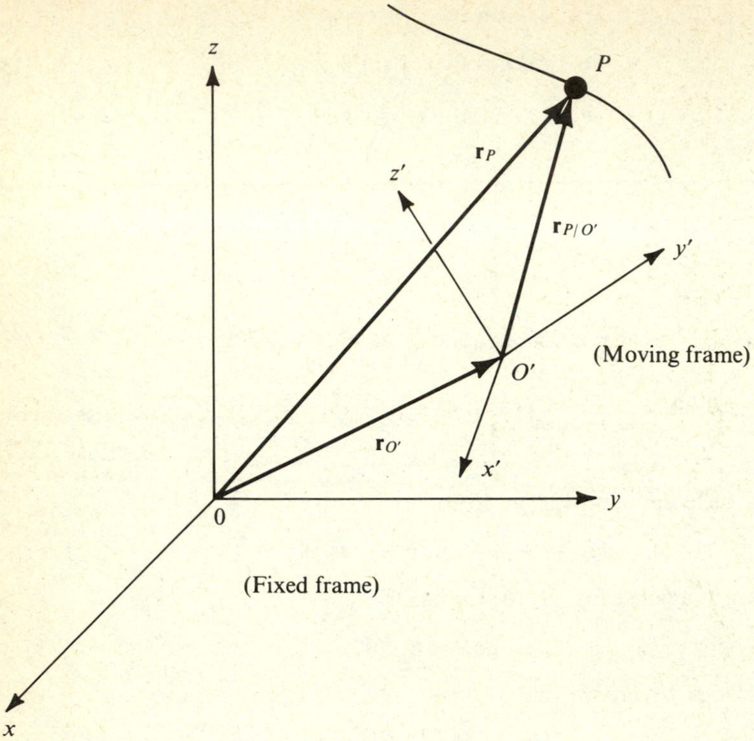

Figure 1.7 *Notation for moving frame.*

This equation will be of value in determining relative velocity and relative acceleration, which are extremely important quantities in turbomachinery, geostrophic flows, and other subjects that move with attached coordinate systems.

1.6 Fundamental Kinematic Field

One of the most important properties of a fluid flow is the velocity field **V**. It is a variable dependent upon location and real time. In a given flow situation, the determination of the velocity by theoretical analysis or by experimental measurement is one of the more important solutions to the problem. The kinematic field is composed of velocities and accelerations. We shall use three coordinate systems to describe kinematics.

1.6.1 Absolute Velocity

The velocity of a fluid particle at some vector position **r** at time t is defined as

$$\mathbf{V} = \frac{d\mathbf{r}}{dt} \tag{1.18}$$

and in intrinsic coordinates as

$$\mathbf{V} = V\mathbf{e}_t \tag{1.19}$$

where V is the magnitude of the velocity called *speed*, and $\mathbf{e}_t$ is the unit tangent vector defined as

$$\mathbf{e}_t = \frac{d\mathbf{r}}{|d\mathbf{r}|} \tag{1.20}$$

The velocity vector $\mathbf{V}$ has scalar components u, v, w in rectangular Cartesian coordinates, such that

$$\mathbf{V} = u\mathbf{i} + v\mathbf{j} + w\mathbf{k} \tag{1.21}$$

and are shown in Fig. 1.8 for a fluid particle at P and Q.

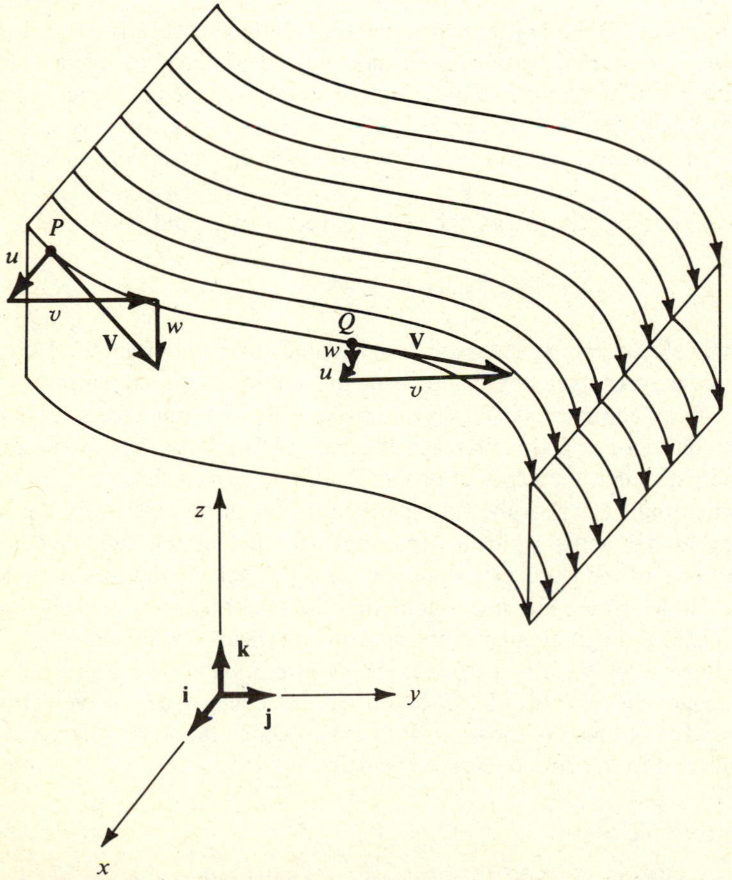

Figure 1.8 Velocity vector **V** *in Cartesian coordinates.*

At P, the velocity vector $\mathbf{V}$ is tangent to the path, that is, in the $\mathbf{e}_t$ direction. At Q, the direction of the velocity has changed, not only because of the change of the position vector $d\mathbf{r}$, but also because of the speed V. The advantage of the Cartesian unit vectors over the intrinsic and other coordinate systems is that the unit vectors $\mathbf{i}$, $\mathbf{j}$, $\mathbf{k}$ are always constant in any problem, whereas, for example, the unit vector $\mathbf{e}_t$ is often a variable.

In mechanics, the velocity components u, v, w are related to infinitessimal changes in the position of a fluid particle by

$$u = \frac{dx}{dt} \tag{1.22a}$$

$$v = \frac{dy}{dt} \tag{1.22b}$$

$$w = \frac{dz}{dt} \tag{1.22c}$$

and are the scalar forms of Eq. (1.18), provided that the position vector $\mathbf{r}$ is a function of time. In *mechanics*, $\mathbf{r} = \mathbf{r}(t)$, because the history of the object is of interest. In *fluid mechanics*, the position vector $\mathbf{r}$ may be independently selected. It is an independent variable just as time is an independent variable. In contrast, velocity is our fundamental dependent variable, just as x, y, z are the fundamental dependent variables for mechanics.

For a cylindrical coordinate system, the velocity vector of a fluid particle is

$$\mathbf{V} = v_r\mathbf{e}_r + v_\theta\mathbf{e}_\theta + w\mathbf{k} \tag{1.23}$$

where v_r is the radial velocity component, v_θ is the tangential or circumferential velocity component, and w is the axial velocity component, the axial direction being in the z direction. These velocity components are shown in Fig. 1.9. The unit vectors $\mathbf{e}_r$ and $\mathbf{e}_\theta$ are orthogonal to one another and lie in the horizontal xy-plane. The velocity vector $\mathbf{V}$ is tangent to the path, and is composed of three velocity components: one in the radial direction which indicates how the fluid particle moves in or away from the z-axis, one in the tangential direction which indicates how the fluid particle swirls around the centerline, and one in the axial or z direction which indicates how the particle moves up or down following the centerline. Such a description is necessary in analyzing flows around cylinders, through circular pipes, and in hurricanes or tornados.

Figure 1.10 shows a top view of a dye globule swirling in a vortex. Its radius is 0.35 in. and has a tangential velocity of 5 in./s. Notice the evidence of a radial velocity. A side view of the same globule is shown in Fig. 1.11. Notice the axial velocity, $w = 1.71$ in./s, compared to the circumferential velocity.

1.6.2 Relative Velocity

The relative velocity $\mathbf{V}_{P/O'}$, which is the velocity of the point P relative to the moving frame O', is obtained from Eq. (1.17) and (1.18) as being

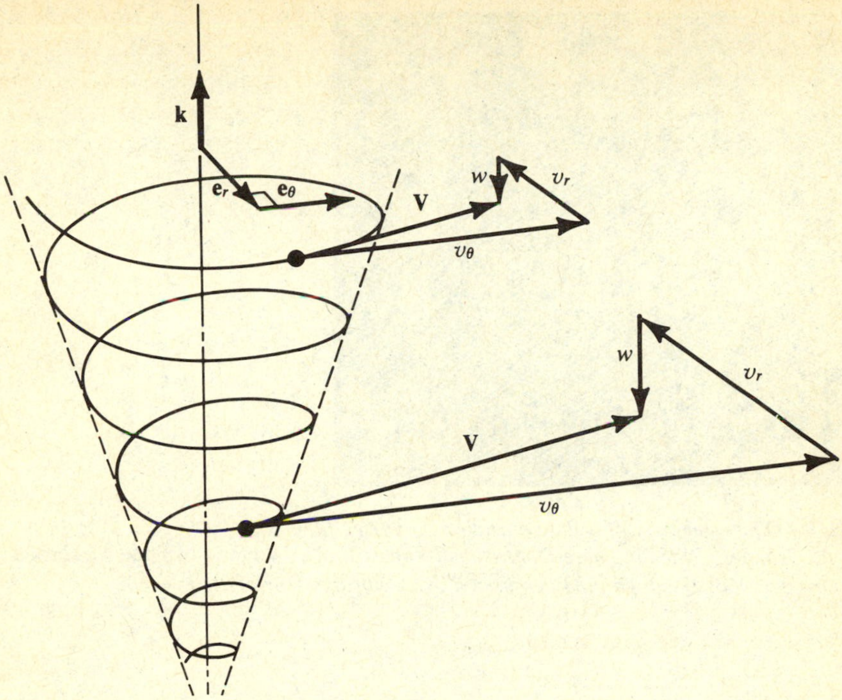

Figure 1.9 *Velocity vector* **V** *for a fluid particle in cylindrical coordinates.*

$$\mathbf{V}_{P/O'} = \mathbf{V}_P - \mathbf{V}_{O'} \tag{1.24}$$

From kinematics, we can express the above as

$$\mathbf{V}_{P/O'} = \dot{r}_{P/O'}\,\mathbf{e}_r + \boldsymbol{\omega} \times \mathbf{r}_{P/O'} \tag{1.25}$$

where $\boldsymbol{\omega}$ is the angular speed of the moving frame. If both points P and O' move together as a rigid body, then $\dot{r}_{P/O'}$ is zero, and we obtain

$$\mathbf{V}_P = \mathbf{V}_{O'} + \boldsymbol{\omega} \times \mathbf{r}_{P/O'} \tag{1.26}$$

1.6.3 Absolute Acceleration Field

The acceleration of a fluid particle is fundamental to fluid mechanics since it is closely connected to Newton's second law of motion. At some position **r** and at time t, the acceleration is defined as

$$\mathbf{a} \equiv \frac{d\mathbf{V}}{dt} \tag{1.27}$$

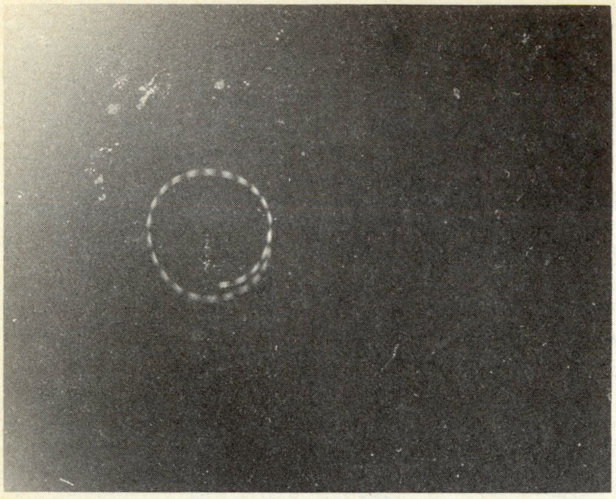

Figure 1.10 *Overhead view of a fluid particle in a vortex flow: v_θ = 5 in./s, r = 0.35 in. (Source: R. Granger, "Steady Three-Dimensional Vortex Flow,"* Journal of Fluid Mechanics, *vol. 25, part 3, 1966. Used with the permission of Cambridge University Press.)*

Figure 1.11 *Side view of Fig. 1.10; w = 1.71 in./s, v_θ = 5 in./s, r = 0.35 in. (Source: R. Granger, "Steady Three-Dimensional Vortex Flow,"* Journal of Fluid Mechanics, *vol. 25, part 3, 1966. Used with the permission of Cambridge University Press.)*

In intrinsic coordinates, the acceleration has components

$$\mathbf{a} = a_n\,\mathbf{e}_n + a_t\mathbf{e}_t + a_m\mathbf{e}_m \tag{1.28}$$

where a_t is the tangential acceleration in the $\mathbf{e}_t$ direction, a_n is the normal acceleration in the direction of the normal unit vector $\mathbf{e}_n$ (taken perpendicular to $\mathbf{e}_t$) and pointing toward the center of curvature, and a_m is the meridional acceleration in the direction of the unit vector $\mathbf{e}_m$ shown in Fig. 1.12. The meridional unit vector is always normal

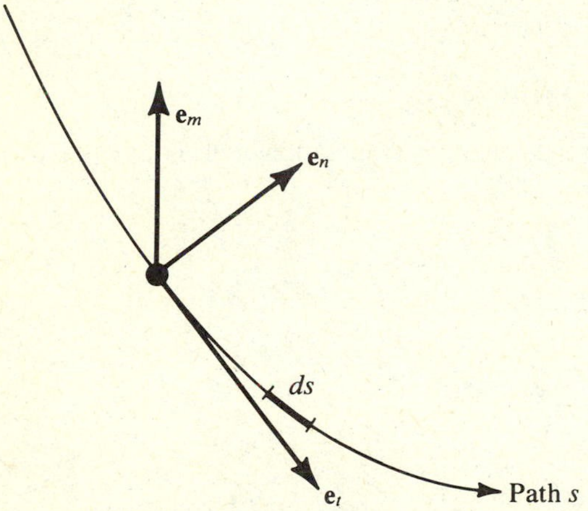

Figure 1.12 *Intrinsic coordinate system.*

to a surface generated by $\mathbf{e}_t$ and $\mathbf{e}_n$. Such a surface could be a lamina of fluid. For steady flows, the normal acceleration is given by

$$a_n = \frac{V^2}{R} \tag{1.29}$$

and is the fluid's centripetal acceleration. The radius of the path's curvature is denoted by R. From the kinematics, the tangential acceleration is expressed as

$$a_t = V\frac{dV}{ds} \tag{1.30}$$

Examining the accelerations of Eqs. (1.27) and (1.30), we observe that, in one expression, the velocity is only a function of time, and in the other expression, it is solely a function of space. What if the velocity were a function of both space and time? This question deals with the manner in which we *describe* velocity. It is the subject of Sec. 1.7.

The acceleration vector **a** expressed in Cartesian form is

$$\mathbf{a} = a_x\mathbf{i} + a_y\mathbf{j} + a_z\mathbf{k} \tag{1.31}$$

and in cylindrical form, the acceleration vector **a** is expressed as

$$\mathbf{a} = a_r\mathbf{e}_r + a_\theta\mathbf{e}_\theta + a_z\mathbf{k} \tag{1.32}$$

We shall obtain expressions for the scalar acceleration components in Eqs. (1.31) and (1.32) in Sec. 1.7.

1.6.4 Relative Acceleration Field

From the definitions of acceleration, the relative acceleration of the point P relative to a moving origin O' is

$$\mathbf{a}_{P/O'} = \mathbf{a}_P - \mathbf{a}_{O'} \tag{1.33}$$

From kinematics, we can show that

$$\begin{aligned} \mathbf{a}_P &= \frac{d\mathbf{V}_P}{dt} \\ &= \mathbf{a}_{O'} + \frac{d}{dt}(\dot{r}_{P/O'}\,\mathbf{e}_r) + \frac{d\boldsymbol{\omega}}{dt} \times \mathbf{r}_{P/O'} + \boldsymbol{\omega} \times \frac{d\mathbf{r}_{P/O'}}{dt} \end{aligned} \tag{1.34}$$

After a bit of manipulation, the acceleration can be written

$$\mathbf{a}_P = \mathbf{a}_{O'} + \boldsymbol{\alpha} \times \mathbf{r}_{P/O'} + \boldsymbol{\omega} \times (\boldsymbol{\omega} \times \mathbf{r}_{P/O}) + \mathbf{a}_{P/O'} + 2\boldsymbol{\omega} \times \mathbf{V}_{P/O'} \tag{1.35}$$

where $\boldsymbol{\alpha}$ is the angular acceleration of the moving frame of reference.

The term $\boldsymbol{\omega} \times (\boldsymbol{\omega} \times \mathbf{r}_{P/O})$ is recognized as the centripetal acceleration, and the term $2\boldsymbol{\omega} \times \mathbf{V}_{P/O'}$ is the familiar Coriolis acceleration. These accelerations are important in analyzing geostrophic flows and studying missile dynamics.

1.7 Fundamental Descriptions: Lagrange versus Euler Description

How are we going to *describe* the behavior of a fluid using either a system or control volume approach? If we wish to describe the behavior of a specific fixed quantity of fluid mass, then the mass becomes the system and we are interested in describing its *history*. On the other hand, if we wish to describe what takes place as fluid *passes* through a known volume of space, it would seem probable that the mathematical description of what transpires will be different from that of the fixed mass. Only when the fixed mass occupies the known volume of space would we suspect that the two mathematical approaches would yield identical results.

Figure 1.13 illustrates the essence of the two approaches. Let the fixed mass be the vibrating beam. The unknown, i.e., the dependent variable, is the position of a point on the beam. If we denote x to be that unknown position, then we inquire into the solution $x = x(\mathrm{t})$; that is, we independently select any time t we wish, and for that time we seek the beam's location. Hence time is the independent variable.

Things are different in fluid mechanics than in structural mechanics. Consider the river in Fig. 1.13b. The river might have the same shape as the beam for one special instant of time. The shape of the river is fixed (discounting erosion, etc.) whereas the beam's shape is not. The unknown in fluid mechanics is the fluid velocity at a point in the river. If we denote $\mathbf{V}$ to be the unknown velocity, then we seek the solution $\mathbf{V} = \mathbf{V}(\mathbf{r}, t)$. This means we independently select not only the time we wish, but any location $\mathbf{r}$ in the river. Thus space and time are independent variables.

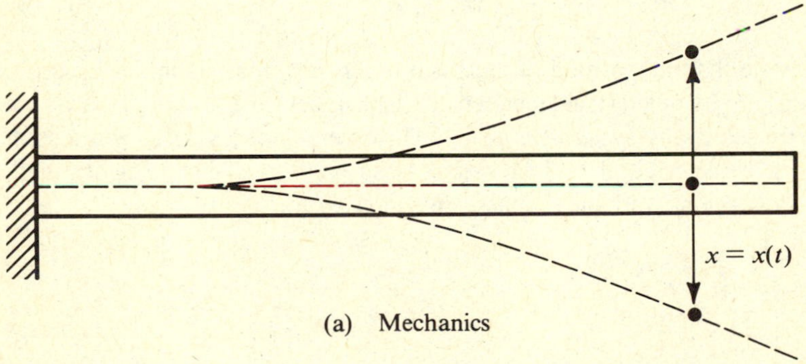

(a) Mechanics

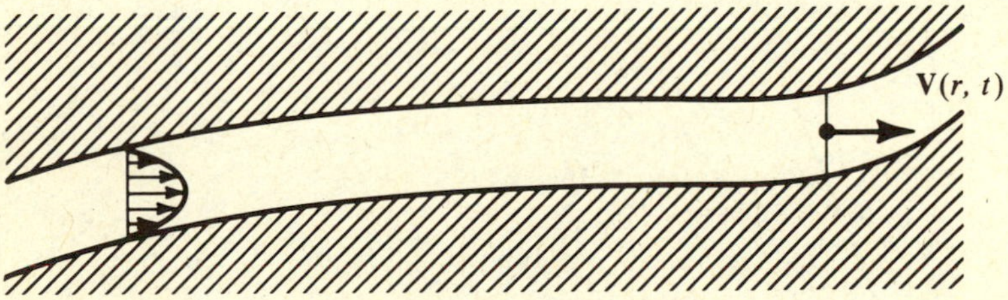

(b) Fluid mechanics

Subject	Independent variables	Dependent variables
Mechanics	Time; t	Location; x, y, z
Fluid mechanics	Time; t Location; x, y, z	Velocity; u, v, w

Figure 1.13 *Variables in mechanics and fluids. (a) Vibrating beam. (b) River.*

Because we have different independent variables, we can have different descriptions of motion. This will result in a different manner of describing how things change.

1.7.1 Lagrangian Description

Consider first the manner in which a system can be located. In Fig. 1.14, let P denote the position of the system, with respect to an inertial frame of reference, at an arbitrary instant of time t_0. If we reference the fluid continuum to a rectangular Cartesian coordinate system, the coordinates of the particle at P are x_0, y_0, z_0. On the other hand, we might desire a different system, Q at x_1, y_1, z_1 at initial time t_0. Thus, if we desire to locate the position of P for $t > t_0$,

$$x = x(x_0, y_0, z_0, t) \tag{1.36a}$$

$$y = y(x_0, y_0, z_0, t) \tag{1.36b}$$

$$z = z(x_0, y_0, z_0, t) \tag{1.36c}$$

These equations state that the position of the system depends upon its initial location. Note that x_0, y_0, z_0 are completely independent variables, selected at will. That is, we may independently select any system we desire. However, once a system is selected, i.e., once x_0, y_0, z_0 are selected, then the resultant position (x, y, z) can be determined. Thus, x, y, z are *dependent* variables, dependent upon x_0, y_0, z_0.

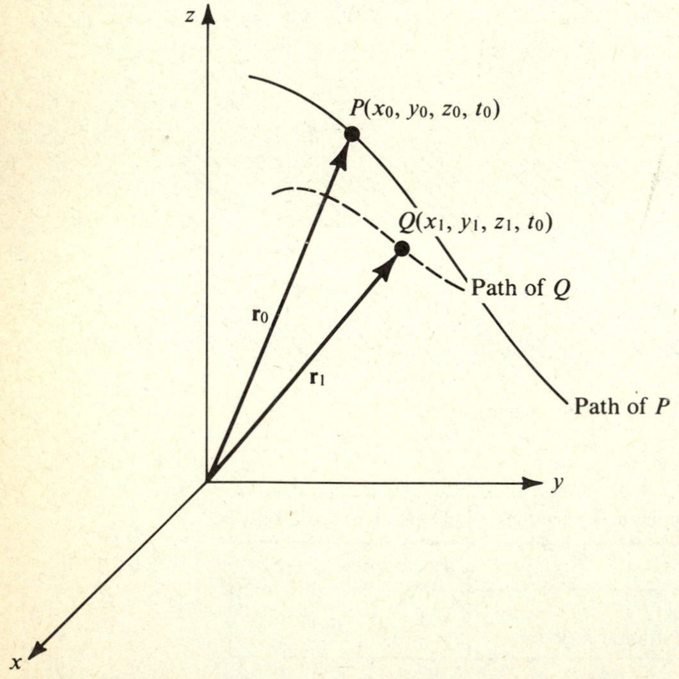

Figure 1.14 *Geometry for Lagrangian description at time* t_0.

This method of describing the position of a system is called the *Lagrangian description* and was devised by L. Euler. It describes the history of the particle exactly. The Lagrangian description allows us to study the paths of fluid particles of fixed identity. The velocity and acceleration components of the Lagrangian description in Cartesian form are

$$u = \frac{\partial x}{\partial t} \qquad a_x = \frac{\partial u}{\partial t} = \frac{\partial^2 x}{\partial t^2}$$
$$v = \frac{\partial y}{\partial t} \qquad \text{and} \qquad a_y = \frac{\partial v}{\partial t} = \frac{\partial^2 y}{\partial t^2} \tag{1.37}$$
$$w = \frac{\partial z}{\partial t} \qquad a_z = \frac{\partial w}{\partial t} = \frac{\partial^2 z}{\partial t^2}$$

The Lagrangian description is rarely used in fluid mechanics because of its significant mathematical complexities and experimental limitations, the former arising when the Lagrangian description is used in formulating the governing equations of motion. An example of this complexity occurs when we attempt to obtain directly the Lagrangian temperature of a fluid in a pipe. The Lagrangian description is used to describe the dynamic behavior of solids.

1.7.2 Eulerian Description

The Eulerian description is used to describe what is happening at a given spatial location $P(x, y, z)$ in the flow field at a given instant of time. In the Eulerian description, the spatial location $P(x, y, z)$ is independently chosen. The parameters dependent upon spatial variance are such tensors* as the pressure $p(x, y, z, t)$, the velocity field $\mathbf{V}(x, y, z, t)$, and the stress dyadic P.

By definition the acceleration **a** for a fluid particle is

$$\mathbf{a} = \frac{d\mathbf{V}}{dt} \tag{1.38}$$

for both Eulerian and Lagrangian descriptions. The acceleration gives the relationship connecting the two descriptions. This is done as follows. Since

$$u = f_1(x, y, z, t) \tag{1.39a}$$

$$v = f_2(x, y, z, t) \tag{1.39b}$$

*Scalars and vectors are special kinds of tensors. The scalar requires only a magnitude and is sometimes termed a zero-order tensor. A vector requires a specification of direction as well as magnitude, and is sometimes termed a first-order tensor. In addition, vectors applied at a point in space must follow the parallelogram law. There are also quantities that not only have both magnitude and direction, but also lie in certain planes. Since there can be three mutually orthogonal planes, and for each plane three possible directions, a complete description can require nine or more scalar components. Such quantities are called tensors.

$$w = f_3(x, y, z, t) \tag{1.39c}$$

for the Eulerian description in Cartesian coordinates, we integrate Eq. (1.38) for the spatial variables x, y, z of Eq. (1.36) using Eq. (1.39) with the initial conditions $t = t_0$, $x = x_0$, $y = y_0$, and $z = z_0$. The Eulerian and Lagrangian descriptions are related by the velocity relationships given by Eq. (1.22). We can obtain the Lagrangian Eq. (1.36) by solving Eq. (1.22) using the initial conditions at $t = t_0$ that $x = x_0$, $y = y_0$, and $z = z_0$. The solution of Eq. (1.22) would result in the parametric equations that describe the paths of the fluid particles. Hence, the Lagrangian description can always be derived through Eq. (1.38) and the Eulerian description.

1.7.3 Substantive Derivative *D/Dt*: The Stokes Derivative

It is convenient to derive the substantive derivative operator D/Dt by considering solely the x-component of velocity u. We could use any other continuum variable, such as temperature, pressure, density, stress, or vorticity.

At some spatial location x, y, z, consider the fluid velocity component u. Since

$$u = f_1(x, y, z, t) \tag{1.39a}$$

then at some small interval of time Δt later the particle will have translated to the position $(x + u\Delta t, y + v\Delta t, z + w\Delta t)$. It will possess a velocity $u + \Delta u$. Thus from the calculus,

$$\begin{aligned} u + \Delta u &= f_1(x + u\Delta t, y + v\Delta t, z + w\Delta t, t + \Delta t) \\ &= f_1(x, y, z, t) + \left(u\frac{\partial f_1}{\partial x} + v\frac{\partial f_1}{\partial y} + w\frac{\partial f_1}{\partial z} + \frac{\partial f_1}{\partial t} \right)\Delta t \end{aligned} \tag{1.40}$$

Since $u = f_1(x, y, z, t)$ and Δt is an arbitrary scalar, then

$$\Delta u/\Delta t = u\partial f_1/\partial x + v\partial f_1/\partial y + w\partial f_1/\partial z + \partial f_1/\partial t \tag{1.41}$$

We define the scalar component of acceleration a_x as

$$a_x \equiv \frac{Du}{Dt} = \lim_{\Delta t \to 0} \frac{\Delta u}{\Delta t} = \left(\frac{\partial}{\partial t} + u\frac{\partial}{\partial x} + v\frac{\partial}{\partial y} + w\frac{\partial}{\partial z} \right) u \tag{1.42}$$

where we adopt the symbol

$$\frac{D}{Dt} \equiv \left(\frac{\partial}{\partial t} + u\frac{\partial}{\partial x} + v\frac{\partial}{\partial y} + w\frac{\partial}{\partial z} \right) \tag{1.43}$$

$$= \left(\frac{\partial}{\partial t} + \mathbf{V}\cdot\boldsymbol{\nabla} \right) \tag{1.44}$$

calling it the substantive derivative, or the derivative following the fluid, or the material derivative. It is to be distinguished from the symbol d/dt, which is used as the Lagrangian rate of change.

Equation (1.44) gives the *local* rate of change $\partial/\partial t$ and the *convective* rate of change $(\mathbf{V}\cdot\nabla)$ following the path of the fluid particle. The convective rate of change represents the change in the continuum variable caused by the convection (spatial change) of a fluid particle from one field location to another.

The Acceleration in the Eulerian Description

The scalar acceleration component a_x has been derived in the previous section using the Eulerian description of fluid flow. We can derive all other scalar acceleration components in a similar manner. The results are given below in rectangular Cartesian form. The scalar components are

$$a_x = \frac{Du}{Dt} = \frac{\partial u}{\partial t} + u\frac{\partial u}{\partial x} + v\frac{\partial u}{\partial y} + w\frac{\partial u}{\partial z} \tag{1.45}$$

$$a_y = \frac{Dv}{Dt} = \frac{\partial v}{\partial t} + u\frac{\partial v}{\partial x} + v\frac{\partial v}{\partial y} + w\frac{\partial v}{\partial z} \tag{1.46}$$

$$a_z = \frac{Dw}{Dt} = \frac{\partial w}{\partial t} + u\frac{\partial w}{\partial x} + v\frac{\partial w}{\partial y} + w\frac{\partial w}{\partial z} \tag{1.47}$$

In cylindrical coordinates, the scalar acceleration components are

$$a_r = \frac{\partial v_r}{\partial t} + v_r\frac{\partial v_r}{\partial r} + \frac{v_\theta}{r}\frac{\partial v_r}{\partial \theta} - \frac{v_\theta^2}{r} + w\frac{\partial v_r}{\partial z} \tag{1.48}$$

$$a_\theta = \frac{\partial v_\theta}{\partial t} + v_r\frac{\partial v_\theta}{\partial r} + \frac{v_\theta}{r}\frac{\partial v_\theta}{\partial \theta} + \frac{v_r v_\theta}{r} + w\frac{\partial v_\theta}{\partial z} \tag{1.49}$$

$$a_z = \frac{\partial w}{\partial t} + v_r\frac{\partial w}{\partial r} + \frac{v_\theta}{r}\frac{\partial w}{\partial \theta} + w\frac{\partial w}{\partial z} \tag{1.50}$$

All of the above is related to an inertial frame of reference. For a moving frame of reference we have shown in Sec. 1.6.4 that there are additional terms for the total acceleration.

Example 1.2

Calculate the z-component of acceleration for a particle whose velocity vector is

$$\mathbf{V} = (3x - y)\,\mathbf{i} + yt^2\mathbf{j} + x^2zt\mathbf{k}, \text{ at } x = 2, y = 1, z = 3, \text{ and } t = 1$$

Solution

Since $u = (3x - y)$, $v = yt^2$, $w = x^2 zt$, we obtain from Eq. (1.47), $a_z = x^2z + (3x - y)(2xzt) + (yt^2)(0) + (x^2 zt)(x^2t)$

Thus the acceleration is evaluated as

$$\begin{aligned} a_z\,(2, 1, 3, 1) &= (3)(4) + (6 - 1)(12) + (12)(4) \\ &= 12 + 60 + 48 \\ &= 120 \end{aligned}$$

The units are implicit in the problem.

This completes the solution.

Example 1.3

Given $\mathbf{V} = 3r^2 \cos\theta\ \mathbf{e}_r - 2r \sin\theta\ \mathbf{e}_\theta$, what are (a) the radial and tangential velocity components, (b) the radial and tangential acceleration components, and (c) the material derivative of $\mathbf{V}$?

Solution:

(a) Since the flow is two dimensional, $\mathbf{V} = v_r\,\mathbf{e}_r + v_\theta\,\mathbf{e}_\theta$. We note $v_r = 3r^2 \cos\theta$ and $v_\theta = -2r \sin\theta$ by comparing the coefficients of the unit vectors of $\mathbf{e}_r$ and $\mathbf{e}_\theta$ with the given velocity vector expression.

(b) Using Eq. (1.48),

$$a_r = (3r^2 \cos\theta)(6r \cos\theta) + 6r^2 \sin^2\theta - 4r \sin^2\theta \qquad \text{(i)}$$

or

$$a_r = 2[r^2\,(9r \cos^2\theta + 3 \sin^2\theta) - 2r \sin^2\theta] \qquad \text{(ii)}$$

Using Eq. (1.49),

$$a_\theta = (3r^2 \cos\theta)(-2 \sin\theta) - \frac{2r \sin\theta}{r}(-2r \cos\theta) - \frac{6r^3 \sin\theta \cos\theta}{r} \qquad \text{(iii)}$$

or

Example 1.3 *(Con't.)*

$$a_\theta = 4r \sin \theta \cos \theta \,(1 - 3r) \tag{iv}$$

(c) The material derivative of **V** is simply

$$\mathbf{a} = a_r\mathbf{e}_r + a_\theta\mathbf{e}_\theta \tag{v}$$

Therefore,

$$\begin{aligned}\mathbf{a} = 2r[&(9r^2 \cos^2 \theta + 3r \sin^2 \theta - 2 \sin^2 \theta)\mathbf{e}_r \\ &+ 2 \sin \theta \cos \theta \,(1 - 3r)\mathbf{e}_\theta]\end{aligned} \tag{vi}$$

This completes the solution.

Example 1.4

Given

$$x = \frac{3x_0y_0t^2}{z_0}, \; y = \frac{5x_0z_0t}{y_0}, \; z = \frac{2y_0z_0t^3}{x_0}$$

find the velocity of the fluid particle and acceleration at

$$x_0 = 1 \text{ cm}, \; y_0 = 2 \text{ cm}, \; z_0 = 3 \text{ cm at } t = 2s.$$

Solution:

The velocity is given by

$$\begin{aligned}\mathbf{V} &= \frac{\partial x}{\partial t}\mathbf{i} + \frac{\partial y}{\partial t}\mathbf{j} + \frac{\partial z}{\partial t}\mathbf{k} \\ &= \frac{6x_0y_0}{z_0} t\mathbf{i} + \frac{5x_0z_0}{y_0}\mathbf{j} + \frac{6y_0z_0}{x_0} t^2\,\mathbf{k}\end{aligned} \tag{i}$$

Therefore,

$$\mathbf{V}(1, 2, 3, 2) = 8\,\mathbf{i} + 7.5\,\mathbf{j} + 144\,\mathbf{k} \tag{ii}$$

The acceleration is given by

$$\mathbf{a} = \frac{\partial^2 x}{\partial t^2}\mathbf{i} + \frac{\partial^2 y}{\partial t^2}\mathbf{j} + \frac{\partial^2 z}{\partial t^2}\mathbf{k} \tag{iii}$$

so that

$$\mathbf{a} = \frac{6x_0y_0}{z_0}\mathbf{i} + \frac{12y_0z_0}{x_0} t\,\mathbf{k} \tag{iv}$$

and thus

$$\mathbf{a}(1, 2, 3, 2) = 4\,\mathbf{i} + 144\,\mathbf{k} \tag{v}$$

This completes the solution.

References

1.1 Cooper, L., *Aristotle, Galileo, and the Leaning Tower of Pisa*, Cornell Univ. Press, Ithaca, N.Y., 1935.

1.2 Rouse, Hunter, and Ince, Simon, *History of Hydraulics*, Iowa Institute of Hydraulic Research, State University of Iowa, Iowa City, 1957.

1.3 Truesdell, C., *Essays in the History of Mechanics*, Archive for History of Exact Sciences, vol. 1, no. 1, 1960.

1.4 Robertson, James M., *Hydrodynamics in Theory and Application*, Prentice-Hall, Englewood Cliffs, N.J., 1965.

1.5 Newton, I., *Philosophiae Naturalis Principia Mathematica*, Book II, London, 1726.

1.6 Leibniz, G. W., "Sämtliche Schriften und Briefe," *Prussian Academy of Sciences*, Berlin, 1924.

1.7 Euler, L., "Déconvert d'un Nouveau Principe de Méchanique," *Mem. de l'Acad. des Sci. de Berlin*, 6:185–217, 1752.
— "Principes Généraux de l'Etat d'Equilibre des Fluides," *Mem. de l'Acad. des Sci. de Berlin*, 11:217–273, 1757.
— "Principes Généraux du Mouvement des Fluides," *Ibid.*, 11:274–315, 1757.
— "Continuation des Recherches sur la Theorie der Mouvement des Fluides," *Ibid.*, 316–361, 1757.

1.8 Lagrange, J. L., "Memoire sur la Theorie der Mouvement des Fluides," *Nonuv. Mem. de l'Acad. des Sci. de Berlin*, 151–198, 1781.
— "Méchanique Analytique," Paris, 1788.

1.9 Cauchy, A. L., "Recherches sur l'equilibre et le mouvement intérieur des corps solides or fluides, elastiques ou non elastiques," *Bull. Soc. Philomath*, 9–13, 1815.

1.10 d'Alembert, J., "Essai d'une Nouvelle Theorie de la Resistance des Fluides," Paris, 1752.

1.11 Kirchhoff, G., "On the Theory of Free Jets of Fluid," *J. Math.*, 70:289–298, 1869.

1.12 Helmholtz, H. von, "On Discontinuous Fluid Motion," Berlin, *Monatsber*, 215–228, 1868; *Phil. Mag.*, (4), 36, 337–346, 1868.

1.13 Poisson, S. D., "Theorie mathématique de la chaleur," Paris, Bachelier, 1835.
— "Memoir on the general equations of the equilibrium and motion of elastic solids and fluids," *J. École Polytech.*, 13:1–174, 1831.

1.14 Rankine, William John Macquorn, "On the Thermodynamic Theory of Waves of Finite Longitudinal Disturbance," *Philo. Trans. R. Soc. London* (A), 160:277, 1870.
— "Miscellaneous Scientific Papers," London, 1881.

1.15 Kelvin, Lord (Thomson, William), "On the Dynamical Theory of Heat," I–V, *Trans, R. Soc. Edinb.*, 20:261–293, 475–483, 1851; *Phil. Mag.*, (4) 4:8–20, 105–117, 168–176, 424–434, 1852.
— "Dynamical Problems Regarding Elastic Spheroidal Shells and Spheroids of Incompressible Liquid," *Phil. Trans. R. Soc. London* (A), 153:583–616, 1863.
— *Treatise on Natural Philosophy*, Pt. I, Cambridge, 1867.
— "On Vortex Motion," *Trans. R. Soc. Edinb.*, 25:217–260, 1869.

1.16 Loitsyanskii, L. G., *Mechanics of Liquids and Gases*, Pergamon, Oxford, 1966.

1.17 Navier, Claude-Louis-Marie-Henri, "Memorie sur les lois du mouvement des fluides," *Mem. Acad. Sci. Inst. France*, 6:289–440, 1827.

1.18 Stokes, George G., "On the Theories of the Internal Friction of Fluids in Motion, and of the Equilibrium and Motion of Elastic Solids," *Trans. Cambr. Phil. Soc.*, 8:287–319, 1845.
— "On the Effect of the Internal Friction of Fluids on the Motion of Pendulums," *Trans. Cambr. Phil. Soc.*, 9:8–106, 1850.

1.19 Poiseuille, Jean-Louis-Marie, "Recherches experimentales sur le mouvement des liquides dans les tubes de tres petits diameters," *Comptes Rendus d'Acadamie des Sciences*, 1841.

1.20 Reynolds, Osborne, "On the Dynamical Theory of Incompressible Fluids and the Determination of the Criterion," *Philo. Trans. of the R. Soc.*, vol. 186, 1894.

— "An Experimental Investigation of the Circumstances Which Determine Whether the Motion of Water Shall Be Direct or Sinuous and of the Law of Resistance in Parallel Channels," *Philo. Trans. of the R. Soc.*, vol. 174, 1883.

1.21 Tsiolkovskii, K. E., "Works on Rocket Technology," NASA TT F-243, 1965.

— Sobranie Sochinenii, Dirizhabli, Moscow, 1959.

1.22 Mach, E., and Salcher, P., "Photographische Fixierung der durch Projectile in der Luft engeleitenden Vorgänge," *Sitzungsberichte der Wien Akademie der Wissenschaften*, Abt. II, 95:764–780, 1887; Mach, E., and Mach, L., "Weitere ballistisch-photographische Versuche," *Ibid.*, Abt. IIa, 98:1310–1326, 1889; Reichenbach, H., "Contributions of Ernst Mach to Fluid Mechanics," *Ann. Rev. Fluid Mech.*, 15:1–28, 1983.

1.23 Reimann, B., "Gesammelte Werke," 1876.

1.24 Lancaster, F. W., "Aerodynamics," London, 1907; "Aerodonetics," London, 1908.

1.25 Zhukovskii, N., "On the Adjunct Vortices" (in Russian), Obshchestvo liubitelei estestvoznaniia, antropologii i etnografii, Moskva, *Izviestiia*, 112, *Transactions of the Physical Section*, 13:12–25, 1907; "De la chute dans l'air de corps légers de forme allongée, animés d'un mouvement rotatoire," *Bulletin de l'Institute Aérodynamique de Koutchino*, 1:51–65, 1912; "Über die Kontouren der Tragflächen der Drachenflieger," Zeitschrift für Flugtechnik und Motorluftschiffahrt, 1:281–284, 1910; 3:81–86, 1912; Aérodynamique, Paris, 1916 and 1931.

1.26 Kutta, M. W., "Auftriebskräfte in strömenden Flüssigkeiten," Illustrierte Aeronautische Mitteilungen, 6:133–135, 1902; "Über eine mit den Grundlagen des Flugproblems in Beziehung stehende zweidimensionale Strömung," *Sitzungsberichte der Bayerische Akademie der Wissenschaften, mathematisch-physikalische Klasse*, 1–58, 1910; "Über ebene Zirkulationsströmungen nebst flugtechnischen Anwendungen," *Ibid.*, 65–125, 1911.

1.27 Glauert, H., *Elements of Aerofoil and Airscrew Theory*, Cambridge Univ. Press, Cambridge, 1937.

1.28 Rankine, W. J. M., "On the Mechanical Principles of the Action of Propellers," *Trans. of the Inst. of Naval Arch.*, 6:13, 1865.

1.29 Prandtl, Ludwig, "Ueber Flüssigkeitsbewegung bei sehr kleiner Reibung," Verhandlungen des III. Internationalen Mathematikes Kongresses, Heidelberg, 1904, Leipzig, 1905.

— "The Generation of Vortices in Fluids of Small Viscosity," *J. R. Aeronautical Soc.*, 31:720–741, 1927.

— "Tragflügeltheorie," *Göttinger Nachrichten, mathematischphysikalische Klasse*, 451–477, 1918; 107–137, 1919 (reprinted by L. Prandtl and A. Betz in Vier Abhandlungen zur Hydrodynamik und Aerodynamik [Göttingen, 1927]); "Application of Modern Hydrodynamics to Aeronautics," N.A.C.A. Report No. 116, 1921.

1.30 Burgers, J. M., *Proc. Koninkl. Ned. Akad. Wetenschap*, 51:1073, 1948.

1.31 Taylor, G. I., *Proc. R. Soc. London*, 157A:537, 1936; *Proc. London Math. Soc.*, 20:196, 1921.

— "Applications to Aeronautics of Ackeret's Theory of Airfoils Moving at Speeds Greater than That of Sound," British A.R.C., R & M No. 1467, 1932.

1.32 von Kármán, T., *Proc. Natl. Acad. Sci. U.S.*, 34:530, 1948.

1.33 Rayleigh, J., *Proc. R. Soc. London*, 84:247, 1910.

1.34 Hartman, J., "Hg-dynamics" I. Kgl. Danske Videnski Selskab Math.-fys. Medd. XV:6, 1937.

1.35 de Hoffman, F., and Teller, E., "Magnetogasdynamic Shock," *Phys. Rev.*, 80 (4):692, 1950.

1.36 Alfvén, H., "On the Existence of Electromagnetic-Hydrodynamic Waves," *Arkiv F. mat. astr. o. fysik*, Bd. 29B, No. 2, 1942.

1.37 Jeffreys, Harold, "The Flow of Water in an Inclined Channel of Rectangular Section," *Philosophical Magazine*, Ser. 6, vol. 49:793, London, Edinburgh, and Dublin, 1925.

1.38 Kantrowitz, A., "Heat Capacity Lag in Gas Dynamics," *J. Chem. Phys.*, 14:150, 1946.

1.39 Goldstein, S., "Fluid Mechanics in the First Half of This Century," *Annual Reviews of Fluid Mech.*, 1(1), Ann. Rev., Inc., Calif., 1969.

Study Questions

1.1 When does a control volume become a system?

1.2 Express the unit vectors **i**, **j** in terms of the unit vectors $\mathbf{e}_r$, $\mathbf{e}_\theta$.

1.3 Transform cylindrical coordinates to rectangular coordinates.

1.4 When does a tensor become a scalar?

1.5 Explain the notation

$$\int_A F_x \, dA_x = \iint F_x \, dy \, dz$$

1.6 What is meant by "fluid parcel"?

1.7 Define and relate the operators D/Dt, $\partial/\partial t$, and d/dt.

1.8 Give the physical meaning of $D(\ )/Dt$, $\partial(\ )/\partial t$, $D/Dt \int (\ )\, d\forall$, $\partial/\partial t\ (\ )\, d\forall$.

1.9 Derive Eqs. (1.48) and (1.49). Where and how do the terms v_θ^2/r and $v_r v_\theta/r$ arise?

1.10 What are V, $\mathbf{V}$, $|\mathbf{V}|$?

Problems

Concepts and Definitions

1.1 Discuss the differences between a definition and a concept. Give some examples. What are some basic concepts of mathematics, physics, and engineering? What are some definitions of mathematics, physics, and engineering?

1.2 What is the composition of the universe? What is the fluid of space? Is the flow of electrons a fluid? Explain.

1.3 Galileo showed that uniform motion in a straight line is equivalent to a state of statics. Show this mathematically.

1.4 Prove that an object dropped from the masthead of a moving and heaving ship will always fall at the foot of the masthead despite the motion of the ship.

1.5 Newton conceived the earth as a rotating mass of fluid subject to its own gravitational field. Newton calculated the earth's shape as an oblate spheroid, and related the angular speed to the eccentricity of the earth. Outline the steps by which this is accomplished.

1.6 A convertible moves at high speed down the highway. Explain why the convertible top blows outward.

1.7 When opening the side window of a moving car, explain what causes the noise.

Units and Dimensions

1.8 Discuss the various considerations in establishing a system of units.

1.9 Discuss the various considerations in establishing the measure of a dimension of length and time.

1.10 Set up a system in which time is the only primary dimension. If second is the unit, determine the equivalence of a 1 lbm mass, 1 ft, and 1°F.

1.11 Set up a system in which time and length are the only primary dimensions. If seconds and feet are the units, determine the equivalence of 1 lbm, 1 lbf, and 1°F.

1.12 The daily output from Hoover dam is 10^{21} ergs of energy. (a) Compare this to the kinetic energy of a spacecraft of 10,000 kg moving at a velocity of 10,000 m/s with respect to the earth. (b) Compare it to a spherical iron ball 0.1 m in diameter shot out of a cannon with muzzle velocity 1000 m/s.

1.13 The basic dimension of electrical charge is determined from Coulomb's law

$$F = K\frac{Q_1 Q_2}{R^2}$$

Show the dimension of charge in the M, L, T system is

$$Q = \left(\frac{ML^3}{T^2}\right)^{1/2}$$

What is the dimension of charge in the F, L, T system?

1.14 Obtain the dimension of charge using the Biot-Sarvart law rather than the Coulomb law.

1.15 If charge Q is a primary quantity in the M, L, T system, show the constant of proportionality has the dimension of (ML^3/T^2Q^2).

1.16 Demonstrate the angular displacement θ is dimensionless.

1.17 If angular displacement has a unit of radians, show that the quantities s and r are not just lengths in the arc length relationship $s = r\theta$, but are in the same plane.

1.18 Repeat Prob. 1.17, using the dynamic equilibrium of moments $M = I\alpha$, and show α is orthogonal to the plane generated by M and I.

1.19 Repeat Prob. 1.17 using the equation for kinetic energy of a rotating body, $KE = \frac{1}{2}I\omega^2$. Using the USCS system, determine the meaning of (ft·lbf) = (ft·lbf·rad²), especially the significance of (radian × radian).

1.20 Using mixed units, show that the units of the following secondary quantities are as shown:

$$\text{Momentum} = (MV/g_c)$$

$$\text{Potential energy} = (Mgh/g_c)$$

$$\text{Kinetic energy} = (MV^2/2g_c)$$

$$\text{Kinetic energy of rotation} = \frac{1}{2}\left(\frac{I\omega^2}{g_c}\right)$$

where $g_c = 1.0$ for the lbm, ft, poundal; or the slug, ft, lbf; and $g_c = 32.174$ for the lbm, ft, lbf.

1.21 Discuss the two formulas for the velocity of sound: $c_1 = \sqrt{\gamma RT}$ and $c_2 = \sqrt{g\gamma RT}$. The first is a physical formula, the second is a numerical formula. Which requires only numbers, and which requires numbers and units?

1.22 Convert

$$\frac{0.4\ \text{Btu·in.}}{\text{ft}^2\text{·h·°F}}$$

to

$$k\frac{\text{cal·cm}}{\text{m}^2\text{·h·°C}}$$

What is the value of k?

1.23 Convert

$$Q = \left(0.305\frac{\text{ft}^3}{\text{min·in.}^{5/2}}\right)\text{H}^{5/2}$$

to

$$Q = \left(k\frac{\text{m}^3}{\text{h·cm}^{5/2}}\right)\text{H}^{5/2}$$

What is the value of k?

1.24 Students have difficulty with the difference between weight and mass and standard pound force as distinct from the weight of a pound anywhere (such as in a spacecraft). Show

$$\text{"lb-wt"} = \left(\frac{g}{g_c}\right)\text{lbf}$$

1.25 Show

$$\gamma = \frac{g}{g_c}(62.4)\frac{\text{lbf}}{\text{ft}^3}$$

for water.

1.26 If $g = 0.2g_c$ on the moon, what is the specific weight of water on the moon?

1.27 If power $P = b$ hp, torque $T = t$ lbf·ft, and speed $\omega = n$ rev/min, obtain the dimensionless number b, $b = ktn$. What is the value of k?

1.28 A girl of mass 50 kg falls from rest from a height of 50 ft. Express her energy in Btu, ft·lbf, J, erg, and eV.

1.29 An olympic star runs 15 mph. What is his kinetic energy in Btu, ft·lbf, J, and erg?

1.30 How many kilowatt-hours of energy are expended by a 100 hp motor operating for 8 hours?

1.31 What are the SI units of circulation Γ given

$$\Gamma = 5.6\left(\frac{dp}{dr}\right)^2 \frac{V}{\gamma^2}\sin(3x/l)$$

1.32 Verify the dimensions of dynamic viscosity μ by Newton's law

$$p_{xy} = \mu\left(\frac{\partial u}{\partial y} + \frac{\partial v}{\partial x}\right)$$

1.33 If the density $\rho = 10^3$ kg/m^3 and pressure $p = 1.013 \times 10^5$ N/m^2, at standard conditions, what are their values and units in USCS system?

1.34 Convert the following to SI units: (a) $g = 32.2$ ft/s^2, (b) $T = 32.2$°F, (c) $W =$ 350 ft·lbf, (d) $M = 3$ lbm, (e) $\rho = 2$ slugs/ft^3, (f) $v = 1 \times 10^{-5}$ ft^2/s, (g) 100 hp, (h) 5 gallons, (i) 2 liters.

1.35 Convert the following to USCS units: (a) $a = 3$ m/s^2, (b) $T = 125$ K, (c) $T =$ 21.3 N·m, (d) $M = 2$ g, (e) $\rho = 2$ lbm/m^3, (f) $v = 3 \times 10^{-3}$ m^2/s, (g) 8 w, (h) 5 liters, (i) 6 erg.

1.36 If $5xy + 2D^2\frac{z}{l}\sin(5y/r) = 0$ and

$$3 + \frac{t^2}{r}\frac{dV}{dt} = 0$$

why doesn't

$$5xy + \frac{2D^2z}{l}\sin\left(\frac{5y}{r}\right) = 3 + \frac{t^2}{r}\frac{dV}{dt}\,?$$

1.37 Is

$$p = \frac{Kx}{\nu}\frac{dx}{dt}$$

dimensionally homogeneous if p is pressure, K is bulk modulus of elasticity, and ν is viscosity?

1.38 What are the dimensions of dynamic viscosity? Show the relationship between SI and USCS units for dynamic viscosity. Express the unit of stoke in USCS system.

1.39 What must be the SI unit of k if the following equation is dimensionally homogeneous?

$$\mathbf{a} = \left(\frac{3x}{t^2} + k\nu\ddot{x}\right)\mathbf{k}$$

where ν is the kinematic viscosity.

Kinematics

1.40 A small object travels vertically through a fluid from A to B. If the particle is arrested at C by a magnetic force, then released under controlled fall such that $a_z = 5z$, determine (a) the velocity of the object when it reaches B ($\ell = 50$ cm), and (b) the time required to travel from C to B (Fig. P1.40).

1.41 An object moves vertically through a fluid such that its velocity is $3t$ m/s. Determine its position if the particle has an initial velocity of 1 m/s at $z = 0$, when $t = 5$ s.

1.42 A fluid particle moves along a straight-line path where its position is given by $\mathbf{r} = \mathbf{i}(3t^3 - 5t^2 + 6)$m. Determine (a) the distance the particle must travel from $t = 0$ until its velocity is zero, and (b) the time it travels until its acceleration is zero.

1.43 The motion of a fluid particle moves along the spiral path of a tornado like that

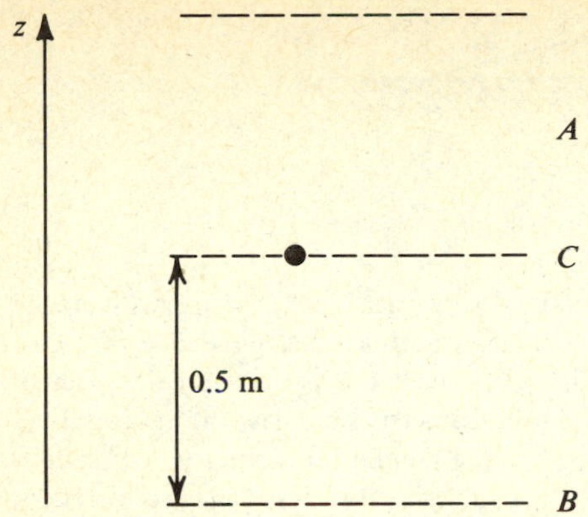

Figure P1.40

shown in Fig. P1.43. Given the position vector $\mathbf{r}_p = 2 \sin(3t)\mathbf{i} + 2 \cos(3t)\mathbf{j} - 5t\mathbf{k}$ (m), determine (a) the location of the fluid particle when $t = 1$ s, (b) the fluid particle velocity $\mathbf{V}$, and (c) the fluid particle acceleration $\mathbf{a}$.

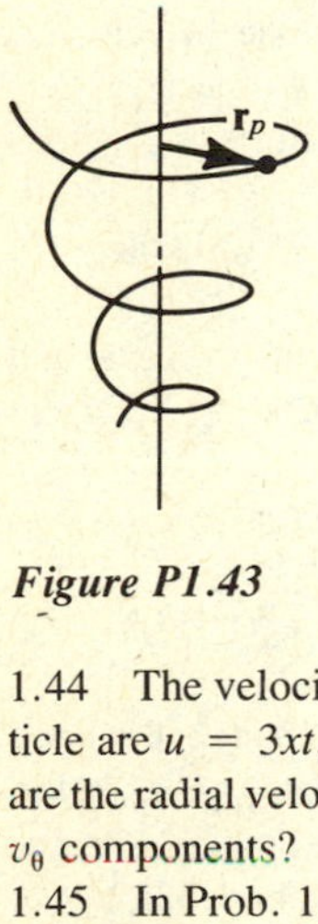

Figure P1.43

1.44 The velocity components of a fluid particle are $u = 3xt$, $v = 5xy^2$, $w = 7yz^3$. What are the radial velocity v_r and tangential velocity v_θ components?

1.45 In Prob. 1.44, what is the velocity vector $\mathbf{V}$ at time $t = 0$ and speed V?

1.46 In Prob. 1.44, what are the acceleration vector $\mathbf{a}$ and magnitude a?

1.47 In Prob. 1.44, what are the radial and tangential components of the acceleration?

1.48 The velocity components of a fluid particle are imagined to be $u = 3x \sin 2t$, $v = 2xz \cos 3t$, $w = xt^3$. What are the (a) vector velocity, (b) speed, (c) acceleration vector $\mathbf{a}$, (d) acceleration, (e) radial velocity, and (f) tangential velocity, where arguments for sine and cosine are given in radians (π radians $= 180°$)?

1.49 The curvilinear motion of a fluid particle is given as $x = 2t^2$, $y = 4t^3 + 5$, $z = 5t - 3$, where x, y, z are given in meters. Find the magnitude and directions of both the velocity and acceleration of the particle when $t = 3$ s.

1.50 The motion of a fluid particle is given as $\mathbf{r} = 22t^3\, \mathbf{e}_r + 5t\mathbf{k}$. Calculate the velocity $\mathbf{V}$ and acceleration $\mathbf{a}$.

1.51 The motion of a fluid particle is given as $\mathbf{r} = 3xyt^2\mathbf{i} - 2yz^2t\mathbf{j} + \dfrac{5}{z} \exp(2x)\mathbf{k}$, calculate (a) the velocity components u, v, w, (b) the velocity components v_r, v_θ, and (c) the acceleration a_r.

1.52 A hose discharges water at a velocity of 20 m/s. If the hose is held near the ground in a direction 45° with respect to the horizontal, calculate (a) the maximum height the water will rise, and (b) how far the water will go before striking the ground. (See Fig. P1.52.)

1.53 A fluid particle moves along the circular path of 100-m radius such that its position is given by $\theta = 5t^3$ (rad). Find (a) the velocity $\mathbf{V}$, (b) the acceleration $\mathbf{a}$, and (c) the speed V of the particle.

1.54 The position of a fluid particle is given by $x = 2t^2$, $y = 5$, $z = 3t - 2$, where x, y, z are measured in meters. Find (a) the velocity components u, v, w, (b) the acceleration components a_x, a_y, a_z, and (c) speed V when $t = 3$ s.

1.55 In Prob. 1.54, find (a) the velocity components v_r, v_θ, w, and (b) the acceleration components a_r, a_θ, a_z.

1.56 A fluid is moving along a circular path of radius 10 ft. Determine the magnitude of the fluid's acceleration if the fluid's speed is 7 ft/s and the rate of decrease due to viscosity is 1 m/s^2.

1.57 As Fig. P1.57 illustrates, water travels along a curved flume at a constant speed of 18 ft/s. Determine the acceleration of particle A when it is at $x = 3$ on the flume.

1.58 A small boat moves as a fluid parcel along the river whose shape is $y = x^2 + 5$ at a constant speed of 10 m/s. Determine the location where the acceleration is a maximum and calculate its value.

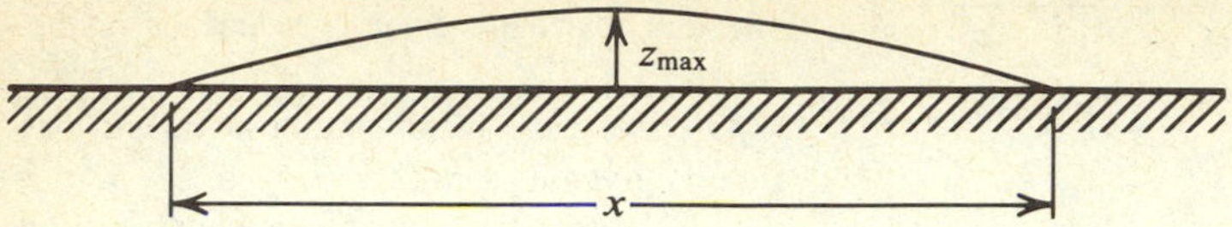

Figure P1.52

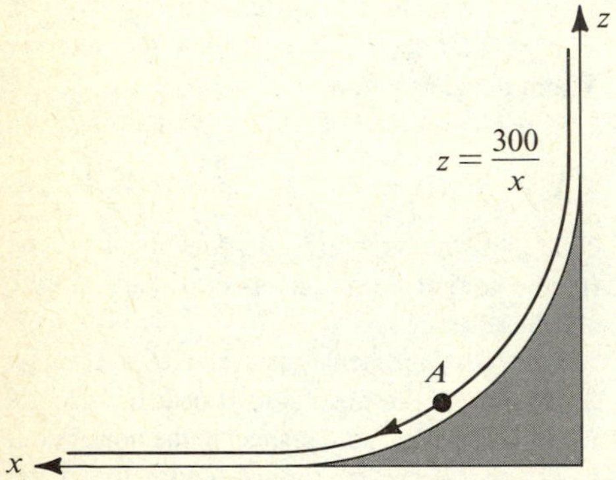

Figure P1.57

1.59 Two boats, A and B, are moving along the surface of two rivers, as shown in Fig. P1.59. If the velocity of A is 0.5 ft/s and that of B is 1.5 ft/s, and the angle between the river's centerline is 50°, determine the velocity of boat B relative to boat A.

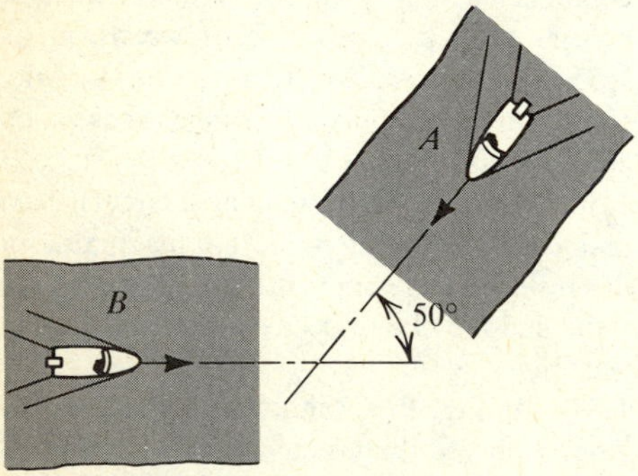

Figure P1.59

1.60 A very small block of wood swirls around a bathtub vortex at an angular rate of 1 cps. The block makes a perfect circular path of 5 cm in diameter. Calculate (a) the radial velocity, (b) tangential velocity, (c) speed, (d) normal velocity, (e) radial acceleration, (f) tangential acceleration, and (g) normal acceleration of the wooden block.

1.61 An object whose density is identical to air is placed in a wind tunnel and released. At a short time later it moves with the velocity of the air in the tunnel. At a certain point in the tunnel the object's velocity is measured and is found to be 20 m/s. Twenty meters downstream from that point the object is moving 40 m/s. Assuming a linear increase in velocity, what is the acceleration of the object?

1.62 Given the position vector

$$\mathbf{r} = (3t^2x)\mathbf{i} + (ty^3x)\mathbf{j} + (t^4xz^2)\mathbf{k}$$

find (a) the velocity components (u, v, w) and (b) the acceleration components (a_x, a_y, a_z) in the Eulerian description.

1.63 Given the position vector

$$\mathbf{r} = (3t^2x)\mathbf{i} + (ty^3x)\mathbf{j} + (t^4xz^2)\mathbf{k}$$

find (a) the velocity components (u, v, w) and (b) the acceleration components (a_x, a_y, a_z) in the Lagrangian description. Compare the solution with the solution of Prob. 1.62.

1.64 In Probs. 1.50 and 1.51, what is the local acceleration and what is the convective acceleration?

1.65 In Probs. 1.54 and 1.62, what is the local acceleration and what is the convective acceleration?

2 Description of Fluids

2.1 Introduction

We classify engineering into various disciplines, or as they are commonly called, various "fields." Civil, mechanical, chemical, aerospace are all familiar engineering fields. If we can use the metaphor of a "field" to distinguish types of engineering, perhaps we should compare fluid mechanics to a river or stream of knowledge that winds through virtually all the fields of engineering. Whatever kind of engineering we study, whether it is a familiar one or a lesser known (but equally important) one like naval architecture, geophysics, or biological fluid dynamics, fluid mechanics touches on its concerns. Just as every actual "field" needs water to be fruitful, so too does every engineering "field" need knowledge of fluid behavior to create the products that grow out of practical engineering. Turbines, pumps, aircraft, arterial circulatory systems—these all work because fluid flows through or around them.

Naturally, then, engineers must understand fluids themselves. Engineers must know all the properties of fluids, and how these properties affect a fluid's motion. In this chapter we will define and classify fluids and their properties. For our conceptual basis we turn first to thermodynamics, and then to other areas of physics and applied mathematics.

2.2 What Is a Fluid?

All that is material in the universe can be classified* as either solid, liquid or gas, or combinations thereof. We call both liquids and gases fluids because their dynamic behavior is similar. Both solids and fluids offer resistance to change of shape. The key distinction between a fluid and a solid lies in the *mode of resistance* to change of shape.

We can show, experimentally, that under a pure compressive force (which is a force applied normally to the boundary of matter) both solids and fluids deform proportionally to the compressive force. If we denote the deformation by the symbol ϵ (popularly called strain), and denote the compressive force per unit area by the symbol σ (popularly called a normal stress) then the mathematical equivalency of the experimental phenomenon is $\sigma = E\epsilon$, where E is the constant of proportionality called

*This classification is based on matter which follows the conservation laws of physics. For those phenomena that do not, such as the so-called bioplasma, aura, and corona discharge, other states may exist where the conservation laws may not hold. We do not consider such phenomena in this text.

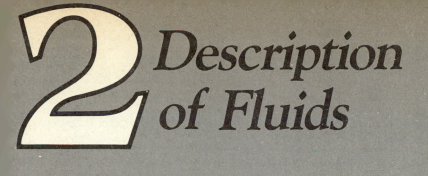

the modulus of elasticity. This equation is called Hooke's law and holds for solids up to the proportional limit of strain.

Consider next a tangential or shear force applied to matter. The fluid, unlike the solid, cannot sustain a finite deformation under the action of a shear force. Based on his experimental results, Newton noted that for most fluids the time rate of change of a fluid element's deformation is proportional to the shear force per unit area applied on the surface of the element. Mathematically, this statement can be expressed as $\tau \propto \dot{\epsilon}$. Newton wrote on this behavior in his "Principia":

> The resistance arising from the want of lubricity [slipperiness] in the parts of a fluid is, other things being equal, proportional to the velocity with which the parts of the fluid are separated from one another.

The velocity with which the parts are separated from one another is related to the rate of strain $\dot{\epsilon}$.

We have used the symbol τ, popular in engineering, to represent the shear force per unit area. Hereafter it will be more convenient to use the symbol p_{ij}, $i \neq j$, to represent the shear stress, since the shear stress has not only a direction of action (denoted by the subscript j), but also a particular plane in which it can lie (denoted by the subscript i). Use of p_{ij} also allows us to treat the normal stress, which are the components of p_{ij} when $i = j$. (We shall discuss this in detail in Chap. 4.)

The difference in behavior between a solid and a fluid then, is that under the action of a shear force, the solid undergoes a finite angular deformation proportional to the shear force, whereas the fluid undergoes a continuous angular deformation as long as the shear force is applied, offering no resistance to that force. The constants of proportionality are called the modulus of rigidity G for a solid, and the dynamic viscosity μ for a fluid.

Let us consider a three-dimensional fluid element of volume $dx \cdot dy \cdot dz$, as shown in Fig. 2.1. A shear stress p_{yx} acts on the elemental face $ABCD$. The stress will move

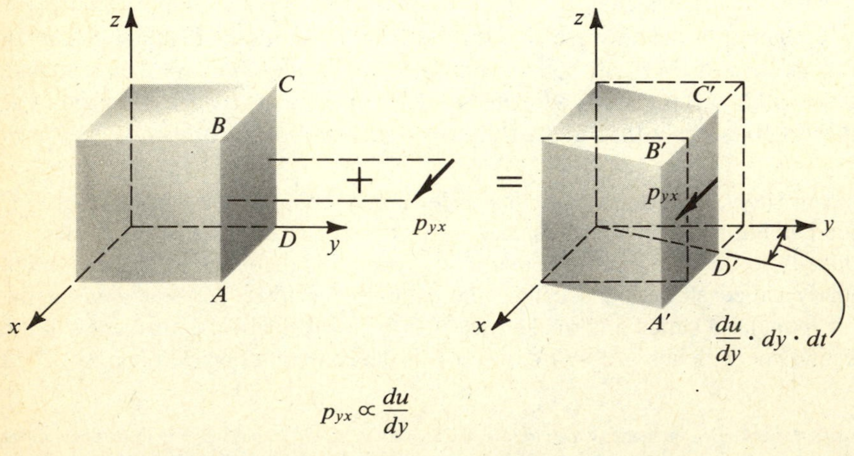

Figure 2.1 *Shear stress on a two-dimensional element, $p_{yx} \propto du/dy$.*

the face in the same direction the stress is being applied causing the fluid element to deform as shown. The distance the point C has moved from its initial position C is $(du/dy)\cdot dy\cdot dt$, where dy locates the point C, and dt denotes the interval of time C has moved to C'. The rate of angular deformation is denoted by $\dot{\epsilon}$.

From the mechanical foundations of elasticity and fluid dynamics [2.1], Hooke's law of shear is

$$p_{yx} = G\epsilon_{yx} \tag{2.1}$$

applied to a solid, and

$$\boxed{p_{yx} = 2\mu\dot{\epsilon}_{yx} = \mu\frac{du}{dy}} \tag{2.2}$$

applied to a fluid.*

Equation (2.2), the *mathematical definition of a fluid*, is sometimes called Newton's law of viscosity. It states what we have previously observed: when a shear stress is applied to any fluid, the fluid will deform continuously so long as the shear stress is applied. The viscosity μ is a measure of the resistance of the fluid to flow, and its behavior is discussed in Sec. 2.3.7.

In comparing Eq. (2.1) with Eq. (2.2), we note that the deformation for solids is due to displacement gradients, whereas the deformation for liquids is due to velocity gradients. Thus, in solid mechanics, displacements are the dependent variables (functions of time), whereas in fluid mechanics velocities are the dependent variables (functions of location and time). We mentioned this difference earlier in Chap. 1. We emphasized that the approach taken in fluids (the Eulerian approach) will be notably different than the approach used in solids (the Lagrangian approach). The approach used to set up the mathematical description of the physical problem is the key to the solution of problems in fluid dynamics. We shall first apply this difference in Chap. 4.

Since all fluids can be characterized by Eq. (2.2), we can plot the shear stress versus the rate of deformation for any fluid and obtain a rheological diagram that classifies nearly all fluids. Note that in Fig. 2.2 there is a linear relationship for Newtonian fluids between the shear stress p_{yx} and the rate of shearing strain $\dot{\epsilon}_{yx}$ where the dynamic viscosity μ of Eq. (2.2) is constant. For non-Newtonian fluids, which include pseudoplastics as well as plastics and rheopectics, the viscosity μ is a function of $\dot{\epsilon}_{yx}$ and thus a nonlinear relationship exists between shearing stress and rate of strain.

Examples of non-Newtonian fluids are molten, concentrated, or dilute polymers, grease, putty, blood, protein solutions, milk, colloidal suspensions of synthetic rubber in water, certain soap solutions, dyes, asphalts, honeys, marine glue, and starch suspensions. Tomato ketchup is an excellent example of a plastic fluid. If we shake the bottle lightly, no ketchup may flow. But if we shake it vigorously, the yield point is exceeded, the viscosity is reduced, and the ketchup flows.

A dilatant, such as printer's ink, paint, or gelatine, has a tendency to set when at rest, and to solidify at high rates of shear.

*Equation (2.2) can also be derived from a plane shear model.

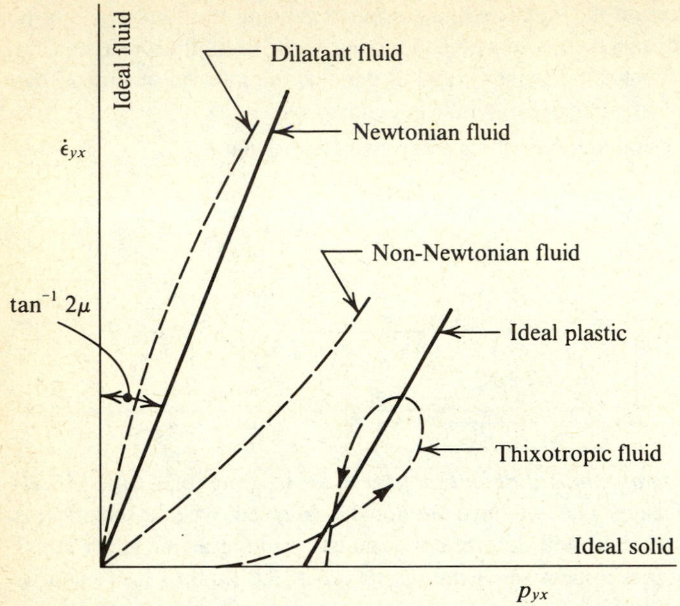

(a)

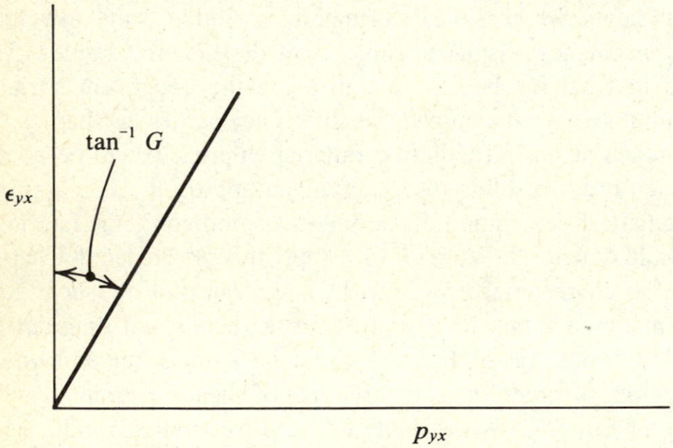

(b)

Figure 2.2 *Rheological diagram. (a) Fluids. (b) Solids.*

Certain fluids such as sludges and concentrated slurries behave like plastics, with an apparent yield stress. Most suspensions behave as pseudoplastics. Quicksand is an example of an inverted plastic called a thixotropic fluid, which becomes more and more fluid when agitated. Thus anyone caught in this water-sand mixture is more likely to survive by remaining motionless.

Plastics, pseudoplastics, and dilatant fluids are time-independent non-Newtonian fluids since their properties of flow are *independent* of time. The rate of shear stress at any point in the flow is a function of the shear stress at that point.

For thixotropic and rheopectic fluids, the viscosity depends not only on the rate of shear, but also on the interval of time during which shear has been applied. In drilling for oil, engineers pump a special mud with thixotropic properties down the hollow drill stem to force debris from drilling back to the surface. So long as the drill rotates, the mud remains a fluid. When drilling ceases, the mud gels and acts as a sealant.

DEMONSTRATION 2.1

Purpose:

This demonstration illustrates the behavior of a fluid when a shear force is applied.

An empty aquarium vessel is filled to within an inch of the top with clear glycerin. Dye made of glycerin and Sudan III vegetable dye is carefully injected by using an eye dropper so that a thin vertical line is clearly noticeable in the vessel as shown in Fig. 2.3. A smooth flat wooden block is then placed gently on the free-surface of the glycerin directly over the dye line and moved to some predetermined location. A nonlinear displacement of the dye line is observed.

- Why is the dye line nonlinear?
- What happens to the fluid particles in the vertical line as the block moves?
- How does this behavior differ from solids?

2.2.1 Concerning Water

By far the most abundant single substance in the biosphere is the familiar but baffling inorganic compound water. Our earth contains 1.5 billion cubic kilometers of water in one form or another. Its extraordinary physical properties endow it with a unique chemistry.

Table 2.1 shows some relevant properties of water for the condensed phase. Liquid and solid states are strikingly similar: note the small increase in volume, and the small difference between latent heats of fusion and evaporation. The similar specific heats indicate that water is highly condensed with strong intermolecular cohesive forces. None of the bulk properties has a unique range of values for a given state, except for surface tension. This suggests that liquid water and ice have a similar molecular arrangement, at least near the melting or freezing point. (Molecular arrangements are briefly discussed in Sec. 2.3.1.) But one difference between water and ice is crucial. Note in Table 2.1 that the density of water decreases as it freezes; because water expands upon freezing, ice is lighter than water. This is most fortunate, for if ice were heavier than liquid water, lakes and rivers would freeze from the bottom up. In colder climates ice at the deepest layers would not melt even in summer, thereby creating glacier-like conditions where fish and water flora could not survive.

Many equations have been proposed to represent the isothermal behavior of water, but none has been fully accepted. The best devised thus far is a single equation that represents the Helmholtz energy over both the liquid and vapor range, and possesses

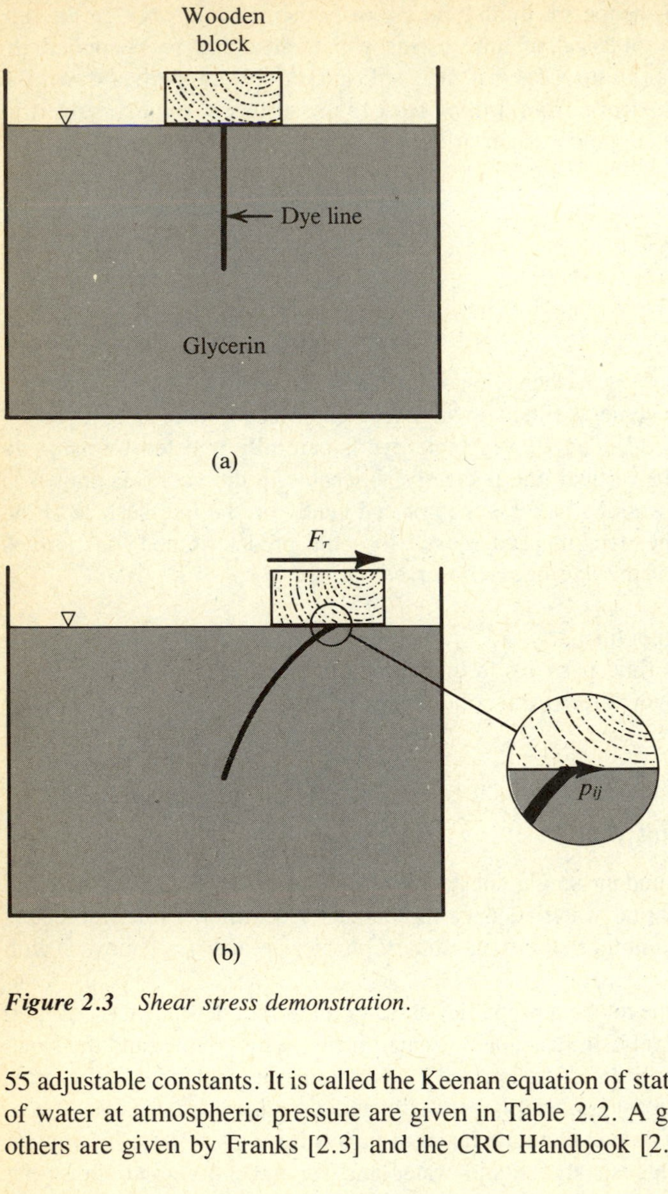

Figure 2.3 *Shear stress demonstration.*

55 adjustable constants. It is called the Keenan equation of state [2.2]. A few properties of water at atmospheric pressure are given in Table 2.2. A good source reference for others are given by Franks [2.3] and the CRC Handbook [2.4].

2.3 *Classification of Fluid Flows*

The vast and potentially confusing number of fluid flows can be classified into various types for more rational study. For example, if we wish to study the air flow around the Concord's wing as it flies through the sound barrier, we must know that we would be studying an *unsteady three-dimensional compressible irrotational inviscid* flow of a gas past a thick wing. This section will set forth the principles for classifying the types of fluids and fluid flows treated in this text.

Table 2.1 *Properties of Water for Condensed Phase*

	Solid		Liquid
Melting point (°C)		0	
Liquid range, deg.		100	
Density, kg/m^3	920		999.84
Viscosity, cP		1.7916	
Surface tension, dyne cm^{-1}		75.6	
Expansibility, per °F	0.00011		0.0003
Isothermal compressibility (at room temp. in % for each atm. change in pressure at pressure of 1 atm)	0.002		0.0046
Thermal conductivity (cal/s·cm.°C)	0.0052		3.471×10^{-5}
Heat capacity, cal(mol·deg)$^{-1}$	9		18
Latent heat of evaporation (Btu/lb)		970.3	
Latent heat of fusion (Btu/lb)		143.3	

Enthalpy and internal energy of saturated solid:

$$h_i = i_i = -334.6 + 1.96T \text{ (kJ/kg)}, \quad T \text{ in °C}$$

Specific volume of saturated solid:

$$\overline{v}_i = 1.091 \times 10^{-3} + 2.5 \times 10^{-6}T \text{ (m}^3\text{/kg)}, \quad T \text{ in °C}$$

Entropy of saturated solid:

$$s_i = -1.223 + 7.747 \times 10^{-3}T \text{ (kJ/kg·K)}, \quad T \text{ in °C}$$

Data abstracted from Refs. [2.2]–[2.4].

2.3.1 Gases versus Liquids

We stated previously that a fluid can be classified as either a liquid or a gas, or even a mixture of both. We can point to certain obvious differences between a liquid and a gas. A gas, for example, is far more compressible than a liquid; a gas does not possess a free-surface that divides it from its environment; a gas expands and occupies the entire space of its container; most gases have no wetting characteristics, while most liquids are wet.

But these common-sense differences do not take us very far. Ideally, to pinpoint the different characteristics of liquids and gases, we would need to peer through an electron microscope so powerful that we could compare the molecular structures of the two fluids. Short of this, however, we at least need theoretical models for molecular structure that enable us to predict the values of properties. Such a theory would yield

Table 2.2 *Thermodynamic Properties of Ordinary Fresh Water at Atmospheric Pressure*

T (°C)	T (°F)	Density (kg/m³)	Density (slug/ft³)	Dynamic Viscosity (cP)	Dynamic Viscosity (lbf·s/ft²)	Bulk Modulus (bar × 10⁴)	Bulk Modulus (psi × 10⁵)
0	32	999.84	1.94000	1.7916	3.74×10^{-5}	1.964	2.88855
5	41	999.96	1.94023	1.5192	3.17	2.034	2.98998
10	50	999.70	1.93973	1.3069	2.73	2.092	3.07524
15	59	999.10	1.93856	1.1382	2.38	2.140	3.14580
20	68	998.20	1.93682	1.0020	2.09	2.179	3.20313
25	77	997.05	1.93458	0.8903	1.86	2.210	3.24870
30	86	995.65	1.93187	0.7975	1.67	2.233	3.28251
35	95	994.03	1.92993	0.7195	1.50	2.250	3.30750
40	104	992.22	1.92521	0.6532	1.36	2.260	3.32220
45	113	990.21	1.92131	0.5963	1.25	2.265	3.32955
50	122	988.03	1.91708	0.5471	1.14	2.264	3.32808
55	131	985.69	1.91254	0.5042	1.05	2.257	3.31779
60	140	983.19	1.90769	0.4666	0.97	2.247	3.30309
65	149	980.55	1.90257	0.4334	0.91	2.232	3.28104
70	158	977.76	1.89716	0.4039	0.84	2.214	3.25458
75	167	974.84	1.89149	0.3775	0.79	2.192	3.22224
80	176	971.79	1.88557	0.3538	0.74	2.167	3.18549
85	185	968.61	1.87940	0.3323	0.69	2.138	3.14286
90	194	965.31	1.87300	0.3128	0.65	2.108	3.09876
95	203	961.89	1.86636	0.2949	0.62	2.076	3.05172
100	212	958.36	1.85951	0.2783	0.58	2.041	3.00027

Source: *Steam Tables,* by J. H. Keenan, et al., Copyright © 1969 John Wiley & Sons, Inc. Publishers. Reprinted by permission of John Wiley & Sons, Inc.

an equation of state relating the principal properties of pressure, temperature, and density.

Science has had less success defining the molecular structure of liquids than of gases. Indeed, the disordered state of the widely separated fast-moving molecules in a gas has been described for over a century, since the research of James Maxwell [2.5]. But nothing comparable exists for liquids. As we stated previously, there is no equation of state for liquids. Why is this? Liquids obviously do have molecular structure, but it is defined only for an instant. Liquids characteristically have very few molecules in a low-energy arrangement, and those that exist are only local. Since there is no time-invariant structure, then no fundamental theory can arise from which we can evaluate the properties. What we do know about liquids comes from empirical investigation. Scientists can accurately predict the properties of any liquid based on liquids already known.

Even if we have not yet fully defined the molecular structures of liquids, we do know a great deal about the behavior of molecules in fluids generally. We know something, for example, about the internal forces within liquids and gases. Consider the internal force between two neighboring molecules as a function of their separation distance d_0. At some small value of distance between the molecules, on the order of 10^{-8} cm, the mutual reaction is a strong force of quantum origin which may be either attractive or repulsive, depending upon the exchange of electron shells. When exchange is possible, the force is attractive and constitutes a chemical bond; if, on the other hand, exchange is not possible, the force is repulsive, and the force decreases rapidly as the distance between molecules increases. At large distances between molecules (on the order of 10^{-6} cm), the mutual reaction between the two molecules is a very weak attractive force. This cohesive force falls off as d^{-7} or d^{-8} when d is large. A sketch of the force exerted by one un-ionized molecule on another is shown in Fig. 2.4 as a function of the distance d between their centers.

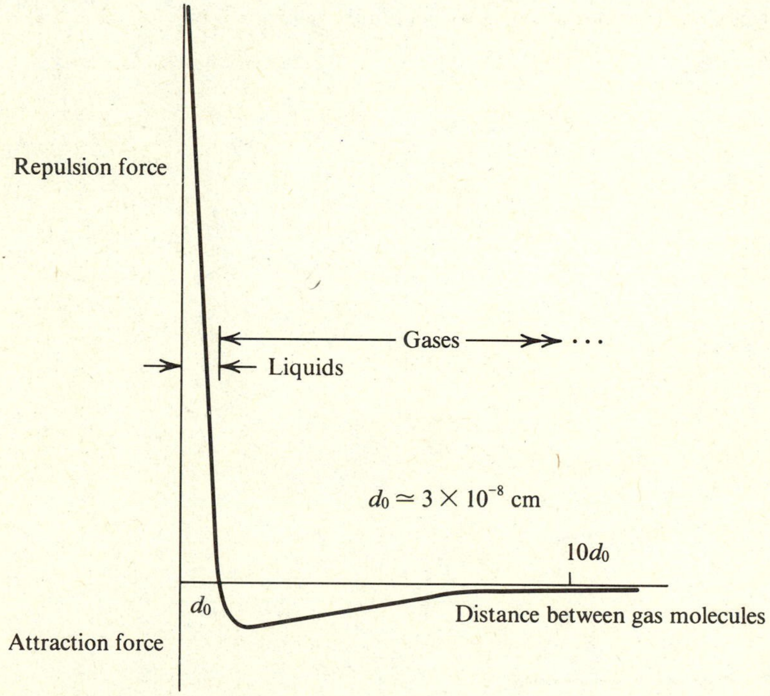

Figure 2.4 *Internal force on molecules.*

In gases, the average spacing between simple molecules at normal pressures and temperatures is of the order 10 d_0, whereas for liquids it is of the order d_0. The molecules are thus far apart for gases so that only weak cohesive forces act between molecules. In liquids, each molecule is within a strong force field and the molecules are packed as close together as the repulsive forces allow. The primary property of the liquid and solid phases of matter is that they are condensed phases when the molecules are within strong cohesive forces. The primary properties of the liquid and gaseous phases are the fluidity and the ability to change shape freely.

A fluid may first exist as a gas for some pressure and temperature, then as a liquid at a different pressure or temperature; this, of course, reflects changes in the intermolecular force and molecular spacing. We can understand this shift from gas to liquid by considering a gas that is being compressed isothermally. The kinetic energy of the molecule remains invariant as the distance between neighboring molecules decreases. When the specific volume of the gas becomes so small that the spacing between molecules is only a few times their diameter, attractive forces become significant. If the temperature of the gas is less than critical temperature, a further decrease in specific volume may prevent molecules from escaping the attractive forces of neighboring molecules, and thus clusters of molecules form. The formation of these clusters reduces the number density of the molecules still moving freely, so that a new equilibrium is established with an average proportion of the mass in the condensed (liquid) phase and the remainder in the gaseous phase. This is the mixed region shown in Fig. 2.5. Let us now consider a state that is entirely gaseous.

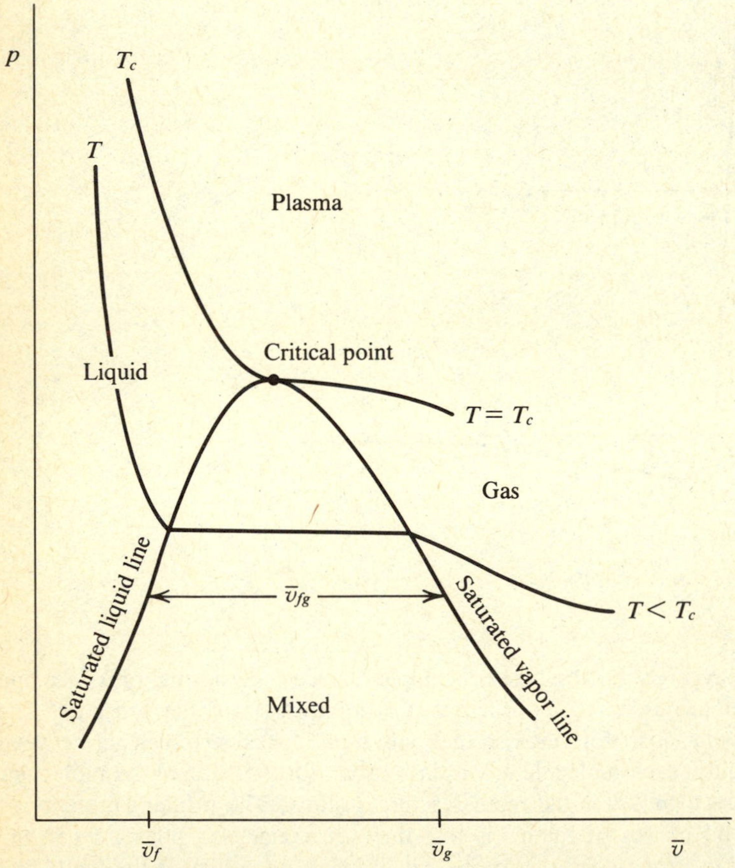

Figure 2.5 *$p-\overline{v}$ diagram for water.*

The equation of state of a perfect gas is well known if the pressures are low to moderate (i.e., less than 500 psi) and the gas is not near the condensation point:

$$\boxed{p\bar{v} = RT} \tag{2.3}$$

where $\bar{v}$ is the specific volume, T is absolute temperature, and R is the gas constant related to the universal gas constant R_u through the molar mass $\hat{M}$:

$$R \equiv \frac{R_u}{\hat{M}} \tag{2.4}$$

The values of R_u in several sets of units are

$$R_u = \begin{cases} 1.986 \text{ Btu/(lb·mol)(°R)}, \\ 1545 \text{ ft·lbf/(lb·mol)(°R)}, \\ 0.73 \text{ atm·ft}^3\text{/(lb·mol)(°R)}, \\ 8.315 \text{ kJ/(kg·mol)(K)}, \\ 0.08315 \text{ bar·m}^3\text{/(kg·mol)(K)}. \end{cases} \tag{2.5}$$

For example, if the average molar mass of air is taken as 29, then R for air in SI units is 0.287 J/(g·mol)(K).

Thermodynamics reveals that the compression and expansion of a perfect gas follow the Boyle-Mariotte and Guy-Lussac laws.

The Boyle-Mariotte law is for isothermal processes and is

$$p_1\,\bar{v}_1 = p_2\,\bar{v}_2 = \text{const.} \tag{2.6}$$

The Guy-Lussac law is for isobaric processes and is

$$\bar{v}_1 T_1 = \bar{v}_2 T_2 = \text{const.} \tag{2.7}$$

where $\bar{v}_1$ and $\bar{v}_2$ are specific volumes at a given pressure:

$$\bar{v} = v_0\,(1 + \alpha t) \tag{2.8}$$

where $\alpha = 0.00366$ is the expansion coefficient for all gases, and t is the relative gas temperature. Real gases do not obey the above two laws. An accepted relationship between pressure and volume of a real gas at constant temperature is the van der Waals' equation:

$$\left(p + \frac{a}{\bar{v}^2}\right)(\bar{v} - b) = RT \tag{2.9}$$

where a is a measure of attractive forces among gas particles, and b is the covolume of the particles, i.e., four times the volume of a gas molecule. There are other relationships, but van der Waals' equation is as good as any, within limitations.

If the gas volume is large in comparison with a and b, Eq. (2.9) reduces to Clapeyron's Eq. (2.3). The term $a/\overline{v}^2$ is the increased pressure resulting from the mutual attraction of fluid molecules. For water, it has a value of 11,000 atm, which indicates the high value of molecular pressure in water and to some extent explains its incompressibility.

If a gas changes reversibly and exchanges no heat with the surroundings, we can use the *isentropic* process equation

$$p\overline{v}^k = \text{const.} \tag{2.10}$$

where k is the ratio of constant specific heats

$$k = \frac{C_{p_0}}{C_{v_0}} \tag{2.11}$$

and where

$$C_{p_0} - C_{v_0} = R \tag{2.12}$$

Values of specific heats for air are given in Ref. 2.4 for a small temperature range. The value of k depends upon the molecular structure of the gas. For monatomic gases, $k = 1.66$, for diatomic gases (such as air), $k = 1.4$, and for polyatomic gases, $k = 1.33$.

We often consider isentropic processes in aerodynamics. For such a process forces due to friction do no work on the aircraft; that is, entropy remains constant during the change of state. This constant entropy will be important when we consider compressible flow in Chap. 15.

If the process is not isentropic, the entropy change may be expressed as

$$s_2 - s_1 = \int_{T_1}^{T_2} C_p \frac{dT}{T} - R \ln \frac{p_2}{p_1}, \tag{2.13}$$

where $C_p = C_p(T)$ and must be determined for the particular gas at a particular temperature.

2.3.2 Continuum versus Discrete Fluids

In this text we study only continuum fluid dynamics. A continuum is said to exist if in any given fluid volume $\forall$ (where the volume contains a sufficiently large number of molecules), the individual variational effects of molecules on the properties of density, temperature, and pressure of the fluid within the volume are negligible. From the continuum point of view, we consider properties at a point in space. By taking smaller and smaller sized volumes about the point, we approach a limit defined as the *property at the point*. Figure 2.6 shows how the specific energy e is defined at a point,

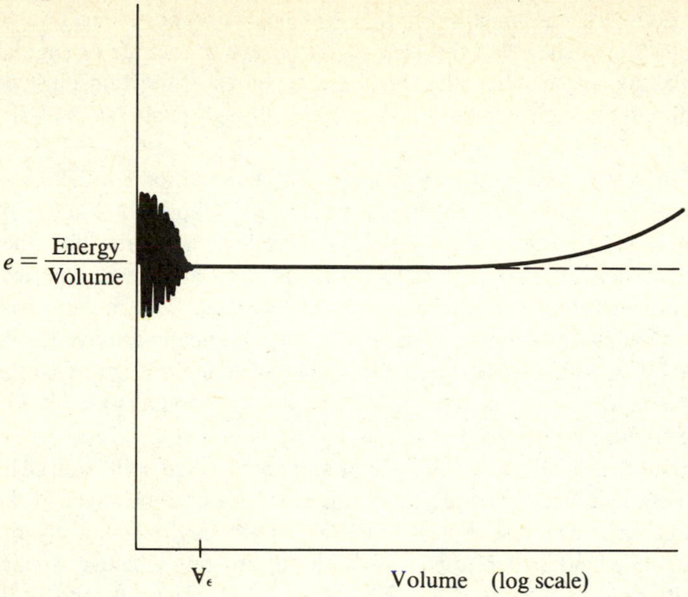

Figure 2.6 *Specific energy at a point.*

$$e = \lim_{\forall \to \forall_\epsilon} \frac{E}{\forall} \tag{2.14}$$

If the volume becomes too small ($\forall < \forall_\epsilon$), the ratio will deviate from the norm, since we must then consider the specific energy of each individual molecule. Thus, the properties of the fluid are based on averaged results in a continuum, and change smoothly with both time and space.

The concept of the continuum is quite arbitrary as it allows one to study the macroscopic behavior of the fluid rather than its microscopic behavior. Use of the continuum concept results in an analysis vastly less complicated than a discrete analysis. In the latter, a discrete or a particular molecule is identified. For each molecule the pressure and temperature must be determined because they are dependent on the probable state of the molecule. Hence, each molecule in the entire flow field has to be treated separately, since each molecule exhibits a behavior different from other molecules. A discrete analysis is sometimes required in high-energy fluid dynamics such as free molecule flow, or in the ionization and dissociation of gases, the latter two of which are popular subjects in re-entry and high-energy space physics. Fortunately, this text is confined to low energy fluid dynamics.

2.3.3 Perfect versus Real Fluids

A perfect fluid, according to L. Euler [2.6], who in 1768 wrote on the motion of ''fluida perfecta,'' does not sustain shear or the effects of compressive forces. To envision a fluid that does not sustain a shear p_{yx}, we must also posit one of two conditions: according to Eq. (2.2), either the viscosity μ is zero or the rate of angular deformation $\dot{\epsilon}_{yx}$ is zero. As we shall see shortly, for μ to be zero the fluid must be ideal. But in nature there is no such thing as an ideal fluid, though many (such as air) approach the ideal.

The second condition, i.e., $\dot{\epsilon}_{yx} = 0$, is a more plausible condition for the stress to be zero. What this latter condition means is that two adjacent horizontal layers of a perfect fluid can move at different velocities without one layer affecting the other layer through internal resistive stresses. We call such flows *slip flow*. Thus, a perfect fluid can slip by a solid wall, the wall acting only to redirect the flow moving past it. For perfect fluids, each layer of a moving fluid can be hypothetically removed from the flow and replaced by a solid boundary of the identical geometric shape as the removed layer, and this layer can be fixed or moving (it does not matter since it will in no way alter the resultant flow pattern).

A real fluid, on the other hand, is not allowed to slip past a fixed solid wall. The presence of the wall communicates dynamic information from one fluid layer to the next via the shear stress p_{yx}. The "stickiness" of fluid layers is controlled by the magnitude of the viscosity μ and is therefore responsible for creating whatever spatial variation of the velocity may exist in the flow field. Thus, at a stationary solid wall, the fluid lamina directly in contact with the wall has the velocity of the wall, provided that the mean free path of the molecule far from the wall is much *less* than the magnitude of a characteristic reference length.

If, however, the path of the molecule approaches the magnitude of the reference length—if, for example, the length of the mean free path of the molecule is much *greater* than the reference length—then the effect of the shear stress is totally negligible because of the ineffectiveness of the viscosity μ. Such a fluid can be considered perfect.* Free-molecule flow is where the path of the molecules is extremely large. It occurs at hypersonic speeds. In the problems we will be studying, the mean free path of the molecule is many orders of magnitude smaller than any reference length, so that the fluid lamina next to the fixed wall cannot slip past the wall.

Real fluid effects are apparent in a region called the *boundary layer*, a region very close to a body submerged in a fluid flow. The influence of viscosity μ spreads around the "wetted" surface of the body in a manner similar to the way heat would spread in that region. We shall devote Chap. 14 to boundary layer flows.

2.3.4 Newtonian versus Non-Newtonian Fluids

We previously stated that a Newtonian fluid is one in which the viscosity μ is a constant for a fixed temperature and pressure. A non-Newtonian fluid such as jello, ink, milk, therefore, would be a fluid in which the viscosity μ varies. We consider only Newtonian fluids in this text.

*The reader must exercise caution in describing a perfect fluid versus a perfect gas. They are quite different. See Chap. 15 for a definition of a perfect gas.

2.3.5 Compressible and Incompressible Fluids

Compressible fluids are fluids whose specific volume $\overline{v}$ is a function of pressure. Compressibility is not related to a fluid's ability to change shape, as is sometimes erroneously assumed. Conversely, an incompressible fluid is a fluid whose density is not changed by external forces acting on the fluid.

Almost everyone considers liquids as being incompressible. Figure 2.7 shows the effect of exceedingly large forces resulting from a sphere entering water at nearly 1.5 km/s. The pressure at the stagnation point (nose of the sphere) is approximately 6.9 GPa at the time of water impact. Note the absence of splash. One might wonder where the water has gone that was once in the cavity. It has gone into the outer hemispherical shock wave, giving it a greater thickness than a normal shock wave. Thus, for very high pressures, $p > 1$ GPa, water can be compressed. At such high pressures, water must be treated as a compressible fluid medium.

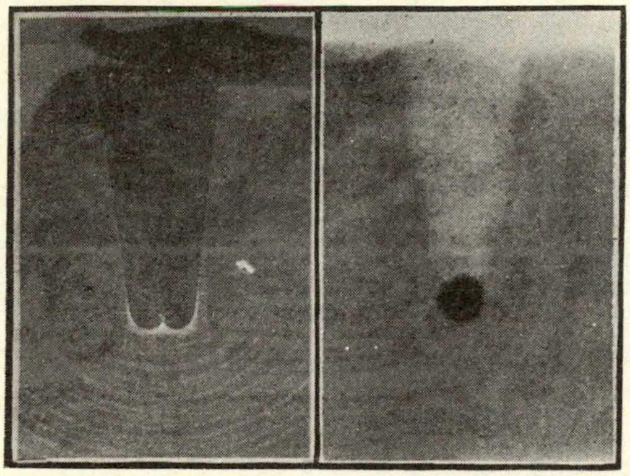

Figure 2.7 *High-speed water entry of a sphere into an incompressible fluid. (Source:* Journal of Applied Physics, *vol. 17, July 1946.)*

Hydrodynamics is the study of the behavior of incompressible fluids, whereas gas dynamics is the study of compressible fluids. The familiar Mach number M indicates the importance of the compressibility of gases in the dynamics of a fluid flow. The Mach number is defined as the ratio of the velocity of the fluid to the velocity of sound. Compressible fluids are subdivided into subsonic, transonic, supersonic and hypersonic compressible flows, meaning speeds less than, equal to, or greater than the speed of sound. In this text we treat both incompressible and compressible fluid flows.

2.3.6 Steady and Unsteady Fluid Flows

A steady fluid flow has properties and variables that are independent of real time. Mathematically, this can be stated as

$$\frac{\partial}{\partial t}(\quad) = 0 \tag{2.15}$$

Equation (2.15) states that none of the dependent variables change with time at any point in the flow. In unsteady flow or transient flow, however, the fluid exhibits variations at a fixed point in space with respect to time. Thus, we shall have to consider whether the flow through a nozzle is steady or unsteady or if the flow past a wing is steady or unsteady.

Consider steady and unsteady flows for two different real fluids. Let one fluid be laminar (well-behaved), and the other turbulent (random). These flows are shown in Fig. 2.8. A turbulent flow can be viewed as steady provided that its time *average* velocity is constant at a specific point in the flow. We are primarily concerned with steady fluid flow, although unsteady motions are also treated.

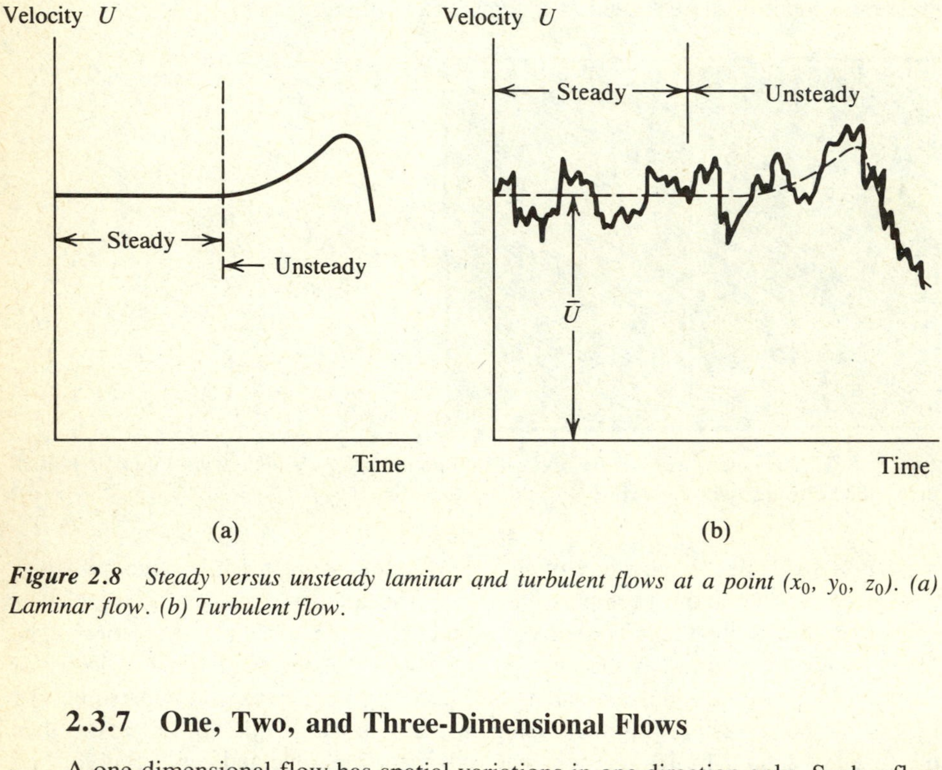

Figure 2.8 *Steady versus unsteady laminar and turbulent flows at a point* (x_0, y_0, z_0). *(a) Laminar flow. (b) Turbulent flow.*

2.3.7 One, Two, and Three-Dimensional Flows

A one-dimensional flow has spatial variations in one direction only. Such a flow is also said to be uniform at every cross section normal to the main direction of flow. Only one independent space variable is needed to describe the variation. We usually designate x to be that variable. Thus $f = f(x)$ is one-dimensional.

Similarly, a two-dimensional flow is one in which spatial variations exist in two directions, or variations exist along some planar surface. Two Cartesian independent space variables are needed to describe the variation. Thus $f = f(x,y)$ is two-dimensional.

A three-dimensional flow has spatial variations everywhere in the flow field. All turbulent fluid flows are three-dimensional. Thus $f = f(x,y,z)$ would be three-dimensional.

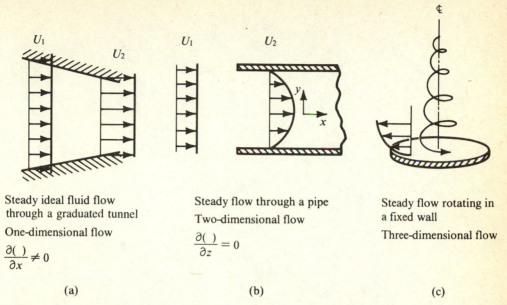

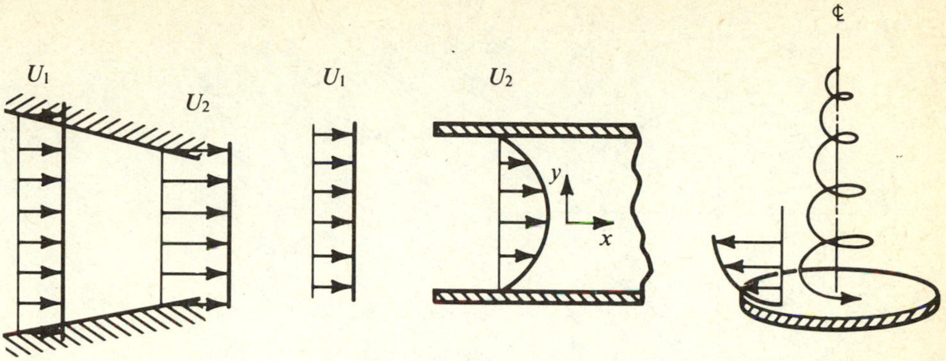

Steady ideal fluid flow through a graduated tunnel

One-dimensional flow

$\frac{\partial(\ )}{\partial x} \neq 0$

(a)

Steady flow through a pipe

Two-dimensional flow

$\frac{\partial(\ )}{\partial z} = 0$

(b)

Steady flow rotating in a fixed wall

Three-dimensional flow

(c)

Figure 2.9 *Steady ideal fluid flow through a graduated tunnel one-dimensional flow, $\partial(\ )/\partial x \neq 0$. (b) Steady flow through a pipe two-dimensional flow, $\partial(\ )/\partial z = 0$. (c) Steady flow rotating in a fixed wall three-dimensional flow.*

We shall be working with flows of all three dimensions, examples of which are shown in Fig. 2.9.

2.3.8 Rotational versus Irrotational Flow

A flow is irrotational if it exhibits no rate of angular deformation of any fluid particle. The converse holds for rotational flows. Irrotational flow means that the fluid masses may deform but cannot rotate. To recognize rotation we need to consider finite (though small) fluid masses called fluid parcels. Fluid particles, which are point masses, have no detectable rotation. To detect rotation, a coordinate system is attached to the fluid parcel, and if, as is shown in Fig. 2.10, the coordinate system rotates as the parcel moves along a path, then we say that the flow is rotational. We shall treat both rotational and irrotational flows in this text. We shall define a fluid flow that is irrotational as a *potential flow*. Potential flows are so important in both hydrodynamics and aerodynamics that we shall devote all of Chap. 12 to the subject.

2.4 Properties of Fluids

A property is an *observable* quality of a fluid. In an experimental measurement, the value of a property is always the same when the system is brought to the same conditions. The properties of a fluid system uniquely determine the state of that system.

All dynamic and energy properties in Newtonian mechanics derive from four fundamental properties: mass, length, time, and temperature. Of these four properties, only temperature is an "intensive property," meaning that it does not depend upon the extent of the system.

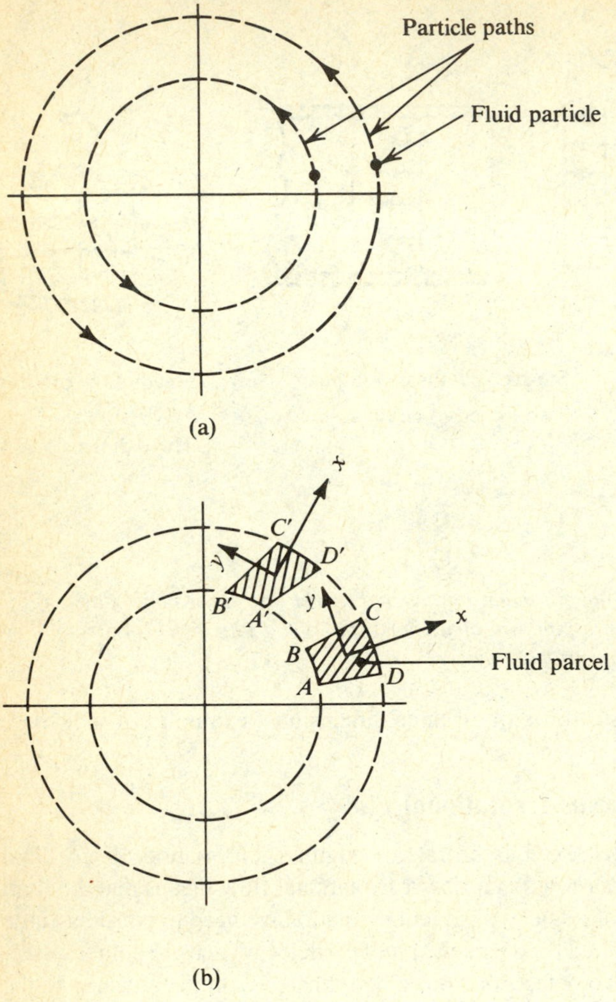

Figure 2.10 Rotational flow. (a) FLuid particles moving along circular paths. (b) Fluid parcel moving along a circular path. The cylindrical element ABCD distorts to A'B'C'D' since AB is moving faster than CD. The coordinate system rotates and thus indicates the flow is rotational flow.

2.4.1 Mass, *M*

We introduced mass in Chap. 1. It is a quantity of fluid, a function of the fluid's internal structure and dimensions. Though mass can neither be created nor destroyed, it can be transformed. This conservation of mass can be expressed mathematically as

$$\frac{dM}{dt} = 0 \tag{2.16}$$

The unit of mass is a kilogram in the SI system, and the lbm in the U.S. customary system.

2.4.2 Density, ρ

Density is the limiting value of mass per unit volume. The inverse of the density ρ is called the specific volume $\overline{v}$. Density, like viscosity, is a function of both pressure and temperature for fluids. Table 2.3 gives values of the density of water for a range

Table 2.3 (USCS) *Property of the Density[a] of Water versus Pressure and Temperature*

Temp. (°F)	Pressure 500 psi (467.13°F)	1000 psi (544.75°F)	1500 psi (596.39°F)	2000 psi (636.0°F)	3000 psi (695.52°F)	5000 psi
32	1.94329	1.94658	1.94999	1.95331	1.95983	1.97277
50	1.94281	1.94597	1.94914	1.95232	1.95847	1.97052
100	1.92978	1.93266	1.93555	1.93844	1.94414	1.95515
150	1.90471	1.90763	1.91056	1.91335	1.91905	1.93002
200	1.87145	1.87461	1.87755	1.88062	1.88644	1.89796
300	1.78462	1.78842	1.79214	1.79576	1.80284	1.81654
400	1.67095	1.67553	1.68069	1.68561	1.69527	1.71330
500	—	1.52657	1.53562	1.54325	1.55841	1.58552
Sat	1.57388	1.43953	1.32479	1.21178	0.90589	—

[a]Units of density are slug/ft^3.
Source: *Steam Tables,* by J. H. Keenan, et al., Copyright © 1969 John Wiley & Sons, Inc. Publishers. Reprinted by permission of John Wiley & Sons, Inc.

Table 2.3 (metric) *Property of the Density[a] of Water versus Pressure and Temperature*

Temp. °C	Pressure 25 bar (224°C)	50 bar (264°C)	75 bar (290.6°C)	100 bar (311.1°C)	150 bar (342.24°C)	200 bar (365.8°C)
20	994.0	1000.5	1001.6	1002.8	1005.0	1007.3
40	993.3	994.4	995.5	996.6	998.7	1000.8
80	972.8	973.9	975.0	976.1	978.3	980.5
120	944.3	945.5	946.8	948.0	950.4	952.7
160	908.6	910.1	911.6	913.0	915.9	918.7
200	865.4	867.3	869.2	871.1	874.7	878.1
220	840.5	842.7	845.0	847.1	851.2	855.2
260	—	784.4	787.6	790.8	796.8	802.4
300	—	—	—	715.7	726.2	735.5
Sat	835.2	777.7	731.1	688.5	603.1	491.2

[a]Units of density are in kg/m^3.
Source: *Steam Tables,* by J. H. Keenan, et al., Copyright © 1969 John Wiley & Sons, Inc. Publishers. Reprinted by permission of John Wiley & Sons, Inc.

of pressures from 25 to 340 bars with temperatures ranging from 20° to 300°C. Density can also be expressed as a function of altitude, and its values are presented in Table 2.4. The units of density are kilogram per cubic meter for the SI system, and slugs per cubic foot for the U.S. customary system.

The densities for a few other substances are presented in Table 2.5.

Table 2.4 *Altitude versus Density, Speed of Sound, and Viscosity of Air*

Altitude (ft)	Altitude (m)	*T* (°F)	*T* (°C)	p[a] (in. Hg)	ρ (slug/ft³)	ρ (kg/m³)	c (fps)	μ[b]
0	0	73.5	23.1	29.93	0.002378	1.192	1,132	1.228
1,000	305	70.7	21.4	28.89	0.002309	1.157	1,129	1.223
2,000	610	68.0	20.0	27.89	0.002242	1.123	1,126	1.218
3,000	915	65.2	18.4	26.91	0.002176	1.090	1,123	1.213
4,000	1,220	62.4	16.9	25.96	0.002112	1.057	1.120	1.209
5,000	1,525	59.7	15.4	25.05	0.002049	1.025	1,117	1.204
6,000	1,830	57.0	13.9	24.16	0.001987	.994	1,114	1.199
7,000	2,135	53.9	12.2	23.30	0.001928	.964	1,111	1.193
8,000	2,440	50.4	10.2	22.46	0.001869	.936	1,107	1.187
9,000	2,745	46.9	8.3	21.65	0.001812	.908	1,103	1.180
10,000	3,050	43.4	6.3	20.86	0.001756	.881	1,100	1.174
11,000	3,355	40.0	4.4	20.09	0.001701	.855	1,096	1.168
12,000	3,660	36.6	2.6	19.35	0.001648	.829	1,092	1.162
13,000	3,965	32.2	0.7	18.63	0.001596	.803	1,088	1.155
14,000	4,270	29.8	−1.2	17.94	0.001545	.779	1,085	1.149
15,000	4,575	26.5	−3.1	17.26	0.001496	.755	1,081	1.143
20,000	6,100	9.61	−12.5	14.19	0.001267	.643	1,062	1.111
25,000	7,625	−8.31	−22.4	11.58	0.001065	.545	1,042	1.077
30,000	9,150	−26.2	−32.3	9.38	0.000889	.460	1,021	1.043
40,000	12,200	−61.8	−52.1	5.98	0.000582	.319	978	0.972
50,000	15,250	−71.5	−57.5	3.70	0.000361	.203	966	0.952
75,000	22,875	−59.0	−50.6	1.13	0.000115	.0603	981	0.977
100,000	30,500	−37.3	−38.5	0.364	0.000035	.0183	1,008	1.021
150,000	45,750	29.3	−1.5	0.047	0.000004	.00206	1,084	1.148
200,000	61,000	−1.2	−18.4	0.0071	0.0000006	.000329	1,050	1.091
300,000	91,500	−156.6	−104.8	0.00004	—	.0000025	855	.770

[a]p, in. Hg. = pressure in inches of mercury. For atmospheres multiply by 0.0334210. For psia multiply by 0.491154.

[b]μ equals viscosity. For lbm/ft·s multiply by 10^{-5}. For poises multiply by 10^{-5} and 14.882.

Source: Compiled from: "U.S. Standard Atmosphere Supplements," U.S. Government Printing Office, 1966.

Example 2.1

Determine the density of water at standard pressure and 5°C in (a) SI units, (b) U.S. inconsistent units, (c) U.S. customary units, (d) cgs metric units, and (e) mks metric units.

Solution:

Step 1.

Find the property in the appropriate table and use the conversion Table 1.1.

(a) From Table 2.2,

$$\rho = 999.96 \text{ kg/m}^3$$

(b)

$$\rho = 999.96 \frac{\text{kg}}{\text{m}^3} \times \frac{1 \text{ slug}}{14.594 \text{ kg}} \times \left(\frac{0.3048 \text{ m}}{\text{ft}}\right)^3$$

$$= 1.94023 \text{ slug/ft}^3$$

which agrees with the tabulated value in Table 2.2.

(c) From Part (b),

$$\rho = 1.94023 \frac{\text{slug}}{\text{ft}^3}$$

$$= 1.94023 \frac{\text{lbf} \times \text{s}^2 \times 1 \text{ lbm}}{\text{ft}^4 \times \text{lbf}} \times 32.174 \frac{\text{ft}}{\text{s}^2}$$

$$= 62.425 \frac{\text{lbm}}{\text{ft}^3}$$

(d) From Part (a),

$$\rho = 999.96 \frac{\text{kg}}{\text{m}^3}$$

$$= \frac{999.96 \times 10^3 \text{ g}}{(10 \text{ cm})^3} = 0.99996 \text{ g/cm}$$

(e) From Part (a),

$$\rho = 999.96 \text{ kg/m}^3$$

This completes the solution.

2.4.3 Specific Weight, γ

Specific weight is weight per unit volume, which can also be expressed as

$$\gamma = \rho g \tag{2.17}$$

A typical value of γ for fresh water at sea level is 62.4 lbf/ft^3 at 60°F, and 9.388 kN/m^3 at 15°C.

The specific weight of any fluid is generally regarded as being a fluid property since it depends upon the density. Values of specific weight can be obtained for water at other pressures and temperatures using Eq. (2.17) and Table 2.3. If the fluid is air, the density can be calculated for any pressure and temperature using Eq. (2.3), provided that we assume that air is a perfect gas. Specific weight will be an important quantity in fluid statics; it will also appear in the famous Bernoulli equation and sometimes in the energy equation.

2.4.4 Specific Gravity, S

Specific gravity is the ratio of density to the density of pure water taken at a standard temperature of 15°C. Table 2.5 shows a few values of specific gravity for liquids. Specific gravity is primarily useful in identifying a particular fluid, and is also sometimes used to relate the property of one fluid to another fluid.

2.4.5 Pressure, p

Pressure is the normal stress in a fluid at rest (i.e., without relative motion between its parts), and is assumed to be positive if compressive. It is sometimes identified with thermodynamic pressure.* Pressure is strictly a "local" property, in that it must be defined at each point of the flow field just like velocity or shear stress. For a perfect gas in equilibrium, pressure is two-thirds of the total kinetic energy of translation of the molecules per unit volume, which is equivalent to the sum of the normal momentum flux per unit area and the resultant force between molecules on two sides of a fluid element:

$$p = \begin{pmatrix} \text{normal momentum flux} \\ \text{per unit area} \end{pmatrix} - a\rho^2 \tag{2.18}$$

*The thermodynamic definition of pressure is

$$p = \left(T \frac{\partial S}{\partial \forall}\right)_i$$

where the change in entropy is for a constant internal energy process.

Table 2.5 Specific Gravity and Kinematic Viscosity of Certain Liquids

Temp. (°F)	Specific Gravity, S					
	Water	*Med. Lub. Oil*	*Med. Fuel Oil*	*Heavy Fuel Oil*	*Regular Gasoline*	*Commercial Solvent*
40	1.0	0.905	0.865	0.918	0.738	0.728
50	1.0	0.900	0.861	0.915	0.733	0.725
60	0.999	0.896	0.858	0.912	0.728	0.721
70	0.998	0.891	0.854	0.908	0.724	0.717
80	0.997	0.888	0.851	0.905	0.719	0.713
90	0.995	0.885	0.847	0.902	0.715	0.709
100	0.993	0.882	0.843	0.899	0.710	0.705
110	0.991	0.874	0.840	0.895	0.706	0.702
120	0.990	0.866				
150	0.980	0.865				
	Kinematic Viscosity × 10^5, (ft^2/s)					
40	1.664	477	6.55	444	0.810	1.61
50	1.410	280	5.55	312	0.765	1.48
60	1.217	188	4.75	221	0.730	1.37
70	1.059	125	4.12	157	0.690	1.26
80	0.930	94	3.65	114	0.660	1.17
90	0.826	69	3.19	83.6	0.630	1.10
100	0.739	49.2	2.78	62.7	0.600	1.03
110	0.667	37.5	2.27	48.0	0.570	0.96

Source: *Fluid Mechanics and Hydraulics,* 2nd ed., by R. V. Giles, Copyright © 1962 McGraw-Hill Book Company. Used with the permission of McGraw-Hill Book Company.

where a is a function of the intermolecular force and is a constant for a given gas. (If the force is cohesive, then $a > 0$.) For a liquid, the contribution to pressure from the momentum flux is NkT, where N is the total number of molecules, k is Boltzmann's constant, and T is the absolute temperature.

If we assume an isentropic relationship for water, we can show that

$$\frac{p + 3000}{3001} = \left(\frac{\rho}{\rho_0}\right)^7 \tag{2.19}$$

where p is in atmospheres, and ρ_0 is the density of water at atmospheric pressure.

We shall use a variety of different pressures in this text: static pressure ($p = \gamma h$), stagnation pressure (p_0), dynamic pressure ($p = \frac{1}{2}\rho V^2$), total pressure ($p_T = \gamma h + \frac{1}{2}\rho V^2$), and others. Pressure, therefore, needs to be clarified each time it is used. Though the units are identical, magnitudes are not.

We shall deal with both relative and absolute pressures. Absolute pressure is used in all thermodynamic relationships, but not necessarily in all fluid relationships. Figure 2.11 depicts relationships among various pressures. A zero absolute pressure is the pressure in a perfect vacuum. A zero gauge pressure is the pressure of the local

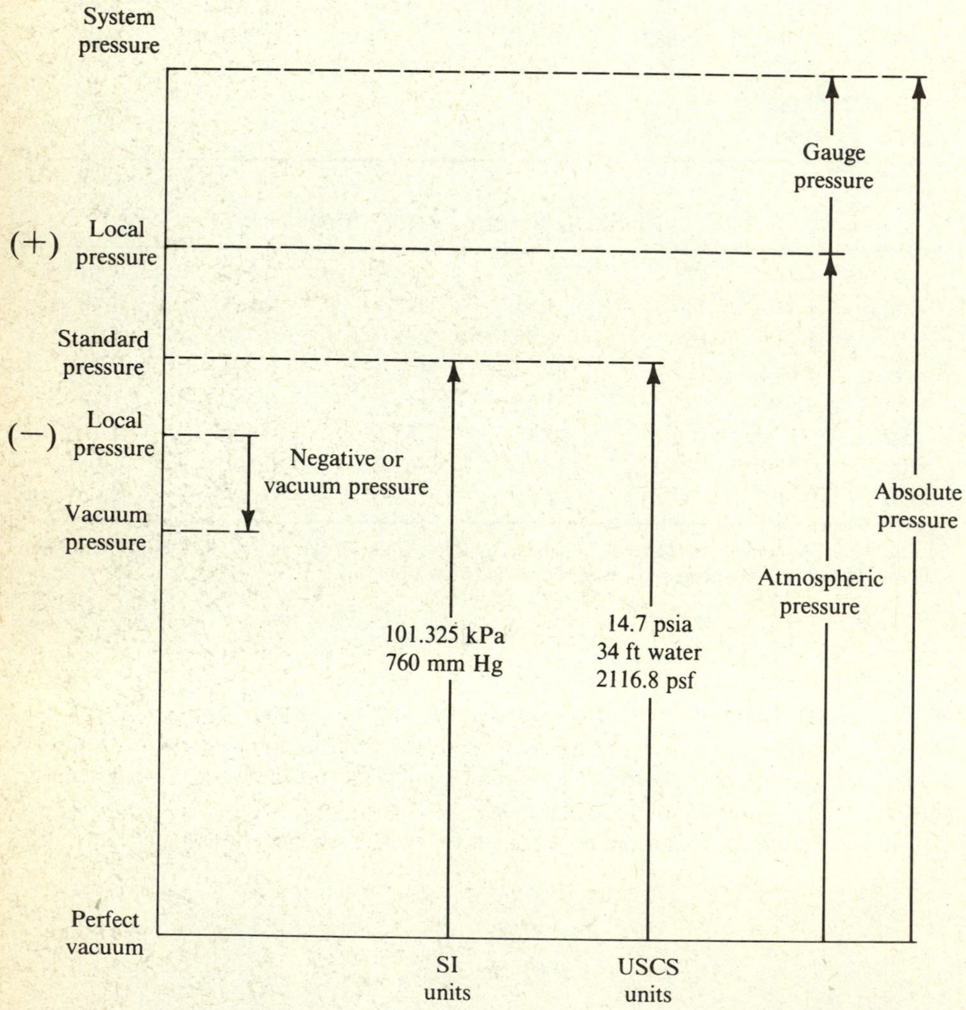

Figure 2.11 *Pressure scales.*

atmospheric pressure, or barometric pressure; its value is not fixed but varies with both time and location on earth. We sometimes measure a gauge pressure against a standard atmosphere, which is the pressure produced by a column of mercury exactly 760 mm high at 273.15 K or 33.91 ft of pure water. At these conditions, the absolute pressure is 14.7 psia, or 2116 lbf/ft^2, or 1 atmosphere or 101,325 pascals. If the gauge pressure is positive, the pressure is greater than the local atmospheric pressure; if the gauge pressure is negative, it is called suction pressure.

Some equivalent values in other units are:

$$1 \text{ standard atmosphere (atm)} = \begin{cases} 14.696 \text{ lbf/in.}^2 \\ 29.92 \text{ in. of Hg at } 0°\text{C} \\ 1.013 \text{ bar} \times 10^5 \text{ N/m}^2 \end{cases}$$

In the technical literature the letters *a* and *g* are often added to the abbreviations in order to indicate the difference between absolute and gauge pressures. For example, the absolute and gauge pressures in pounds per square inch are sometimes denoted by the symbols "psia" and "psig," respectively. In some circumstances the symbols "lbf/in.^{2} gauge" or "mbar gauge" may be used.

The standard pressure in SI is the pascal which is defined as 1 N/m^2. Although a kilopascal (kPa) is frequently employed as a convenient unit of pressure, in the present text we use the bar more frequently than the pascal as the SI pressure unit. As noted above, the bar is slightly smaller than 1 atmosphere.

Example 2.2

Using Eq. (2.19), determine the pressure *p* at the free-surface of water exposed to air at one atmosphere.

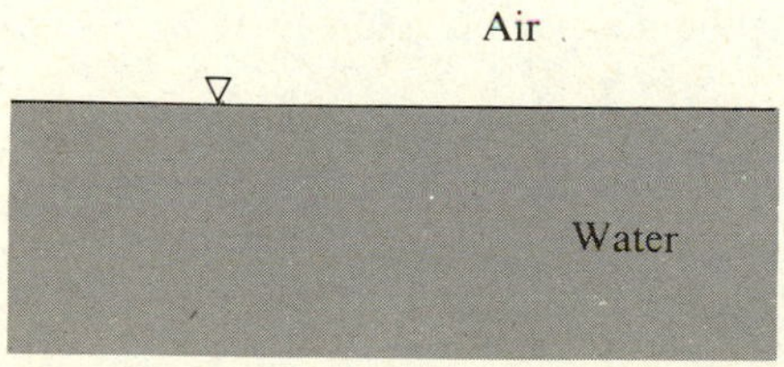

Figure E2.2

Example 2.2 *(Con't.)*

Solution:
From Eq. (2.19),

$$p + 3000 = (\rho/\rho_0)^7 \,(3001) \qquad \text{(i)}$$

Since we want p at the free-surface, then ρ is the density of water at the free-surface at one atmosphere, thus $\rho = \rho_0$. Hence,

$$p = 3001 - 3000 = 1 \text{ atm} \qquad \text{(ii)}$$

Thus, the pressure of water at the free-surface exposed to air at one atmosphere is one atmosphere.

This completes the solution.

2.4.6 Bulk Modulus of Elasticity, *K*

The bulk modulus of elasticity is used as a measure of the compressibility of liquids. We customarily view fluids as incompressible, but as we noted in Sec. 2.2.5, liquids can be compressed at very high pressures. If the pressure is increased isothermally by Δp, it will decrease in volume, $-\Delta\forall$, such that for any volume $\forall$ of liquid,

$$K = -\left(\frac{\forall\, dp}{d\forall}\right)_T \qquad (2.20)$$

Note the units of K are the same as for pressure.

Table 2.6a gives a few values of various liquid's bulk modulus of elasticity for two different atmospheric changes in pressure at around the pressure stated, and Table 2.6b gives the temperature variation of K for four different pressures in units of bars.

Example 2.3
Water is compressed isothermally from 15 psia to 1500 psia at 120°F in a piston cylinder arrangement. Estimate the work required in ft·lbf/slug, using ${}_1w_2 = -\int p\, d\bar{v}$ and (a) $\bar{v} = \bar{v}_f$ = saturated liquid specific volume, (b) $\bar{v} = \bar{v}(p)$, the equation of state for water.

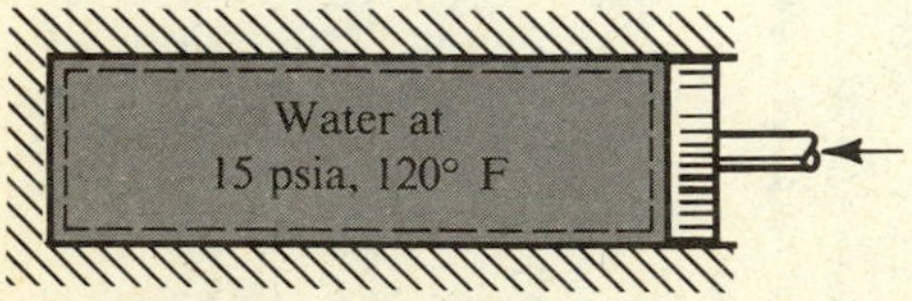

Figure E2.3

Example 2.3 *(Con't.)*

Solution:
The isothermal work of compression per unit mass is given by

$$ {}_1w_2 = -\int p\,d\overline{v} \tag{i} $$

Substituting the expression for the bulk modulus of elasticity K, Eq. (2.19), into Eq. (i) yields

$$ {}_1w_2 = \frac{1}{K}\int_{p_1}^{p_2} \overline{v}\,p\,dp \tag{ii} $$

(a) If $\overline{v}$ is assumed constant over the pressure change then

$$ {}_1w_2 = \frac{\overline{v}_f}{2K}(p_2^2 - p_1^2) \tag{iii} $$

where $\overline{v}_f$ is the saturation specific volume at the saturation temperature of 120°F, its value calculated from Table 2.2 ($\rho_f = 1.9179$ slug/ft³), and K is obtained from Table 2.6b ($K = 342{,}000$ psi). Thus, the work per unit mass assuming constant specific volume is

$$ \begin{aligned} {}_1w_2 &= \frac{(1500\ \text{lbf/in.}^2 \times 144\ \text{in.}^2/\text{ft}^2)^2 - (15\ \text{lbf/in.}^2 \times 144\ \text{in.}^2/\text{ft}^2)^2}{2 \times 1.9179\ \text{slug/ft}^3 \times 342{,}000\ \text{lbf/in.}^2 \times 144\ \text{in.}^2/\text{ft}^2} \\ &= 1.715\ \frac{\text{ft}\cdot\text{lbf}}{\text{slug}} \end{aligned} \tag{iv} $$

(b) If $\overline{v}$ varies as the pressure increases, we use the equation of state of water, Eq. (2.19), so that Eq. (ii) becomes

$$ {}_1w_2 = \frac{(3001)^{1/7}}{\rho_0 K}\int_{p_1}^{p_2} \frac{p\,dp}{(p + 3000)^{1/7}} \tag{v} $$

or upon integration

$$ \begin{aligned} {}_1w_2 = \frac{(3001)^{1/7}}{\rho_0 K}\Bigg\{&(p_2 + 3000)^{6/7}\left[\frac{7}{13}(p_2 + 3000) - 3500\right] \\ &(p_1 + 3000)^{6/7}\left[\frac{7}{13}(p_1 + 3000) - 3500\right]\Bigg\} \end{aligned} \tag{vi} $$

Since $p_1 = 1$ atm and $p_2 = 100$ atm, $\rho_0 = 1.9179$ slug/ft³ and $K = 342{,}000$ psi, the work per unit mass is

$$ \begin{aligned} {}_1w_2 &= \frac{1513.444\ \text{atm}^2}{1.9179\ \dfrac{\text{slug}}{\text{ft}^3} \times 342{,}000\ \text{psi}} \times \left(\frac{15\ \text{psi}}{1\ \text{atm}}\right)^2 \times \frac{144\ \text{in.}^2}{\text{ft}^2} \\ &= 74.76\ \frac{\text{ft}\cdot\text{lbf}}{\text{slug}} \end{aligned} \tag{vii} $$

Example 2.3 *(Con't.)*

which is over 40 times more work than the assumed constant specific volume case. Though the constant density case is easier to solve, it is not as realistic as the variable density case, and its estimated specific work is considerably different from what would be required to compress water up to 1500 psi.

This completes the solution.

Table 2.6a *Bulk Modulus of Elasticity of Liquids*[a]

Liquid	$K \times 10^6$, $(bar)^{-1}$	
	1 atm	*1000 atm*
Acetone	125	55
Aniline	45	30
Benzene	95	50
Carbon disulfide	95	50
Carbon tetrachloride	106	52
Chloroform	100	52
Ethanol	114	50
Glycerol	25	20
Mercury	4	4
Methanol	120	50
n-Octane	120	56
Petroleum oils	70	?
Vegetable oils	50	?
Water	46	35

[a]Compiled from several sources.

Table 2.6b *Bulk Modulus of Elasticity of Water, K* (bar)[a]

Pressure (bar)	Temperature, °C				
	0	*20*	*48.9*	*93.33*	*148.9*
1.034	20,127.7	22,057.7	22,884.9	21,230.5	
103.40	20,679.1	22,747.0	23,574.2	21,988.8	17,094.7
310.19	21,850.9	23,987.8	24,952.8	23,298.4	18,680.1
1033.95	26,193.5	28,261.4	29,364.3	27,916.8	24,125.6

[a]Compiled from several sources (1.013 bar = 14.696 psi).

Example 2.4

What pressure must be exerted on 1 cm^3 of water at 68°F to change the volume to 0.99 cm^3?

Solution:

Step 1.

Evaluate each quantity in the expression for the bulk modulus of elasticity.

From Eq. (2.20),

$$dp = p_2 - p_1 = p_2 \quad \text{(i)}$$

where p_1 is the initial pressure assumed zero gauge,

$$d\forall = \forall_2 - \forall_1 = -0.01 \text{ cm}^3 \quad \text{(ii)}$$

and

$$K = 22{,}057.7 \text{ bars} \quad \text{(iii)}$$

from Table 2.6b. Assuming an isothermal process and $\forall_1 = 1$ cm^3, Eq. (2.20) becomes

$$\frac{22{,}057.7\,(-0.01)}{1} = -p_2 \quad \text{(iv)}$$

Therefore, a pressure of 220.6 bars (2959 psi) must be exerted to compress the water 0.01 cm^3.

This completes the solution.

2.4.7 Absolute or Dynamic Viscosity, μ

Viscosity is perhaps the single most important property in fluid dynamics. All real fluids have a nonzero value of viscosity. All ideal fluids have a zero value for viscosity. In Fig. 2.2, one notes that the dynamic viscosity not only describes the nature of the fluid (whether it is Newtonian or not), but predicts the behavior of the shear stress with respect to the rate of the angular deformation of the fluid. We therefore know a great deal about the fluid if we know its viscosity.

The absolute viscosity is a measure of the internal resistance exhibited as one layer of a fluid is moved in relation to another layer (the layers need not be adjacent).

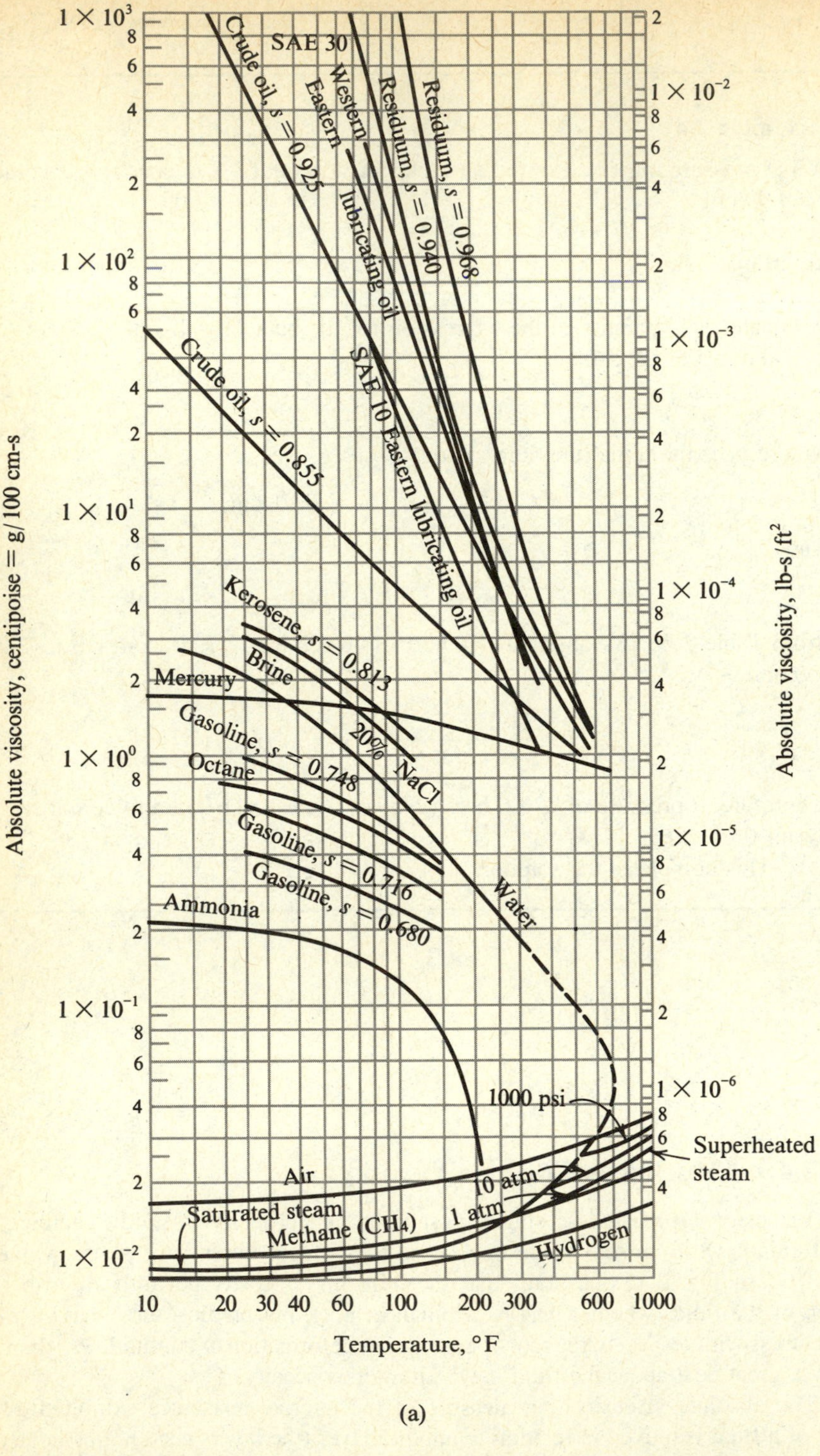

Figure 2.12 *(a) Absolute viscosity μ versus temperature for a variety of fluids.*

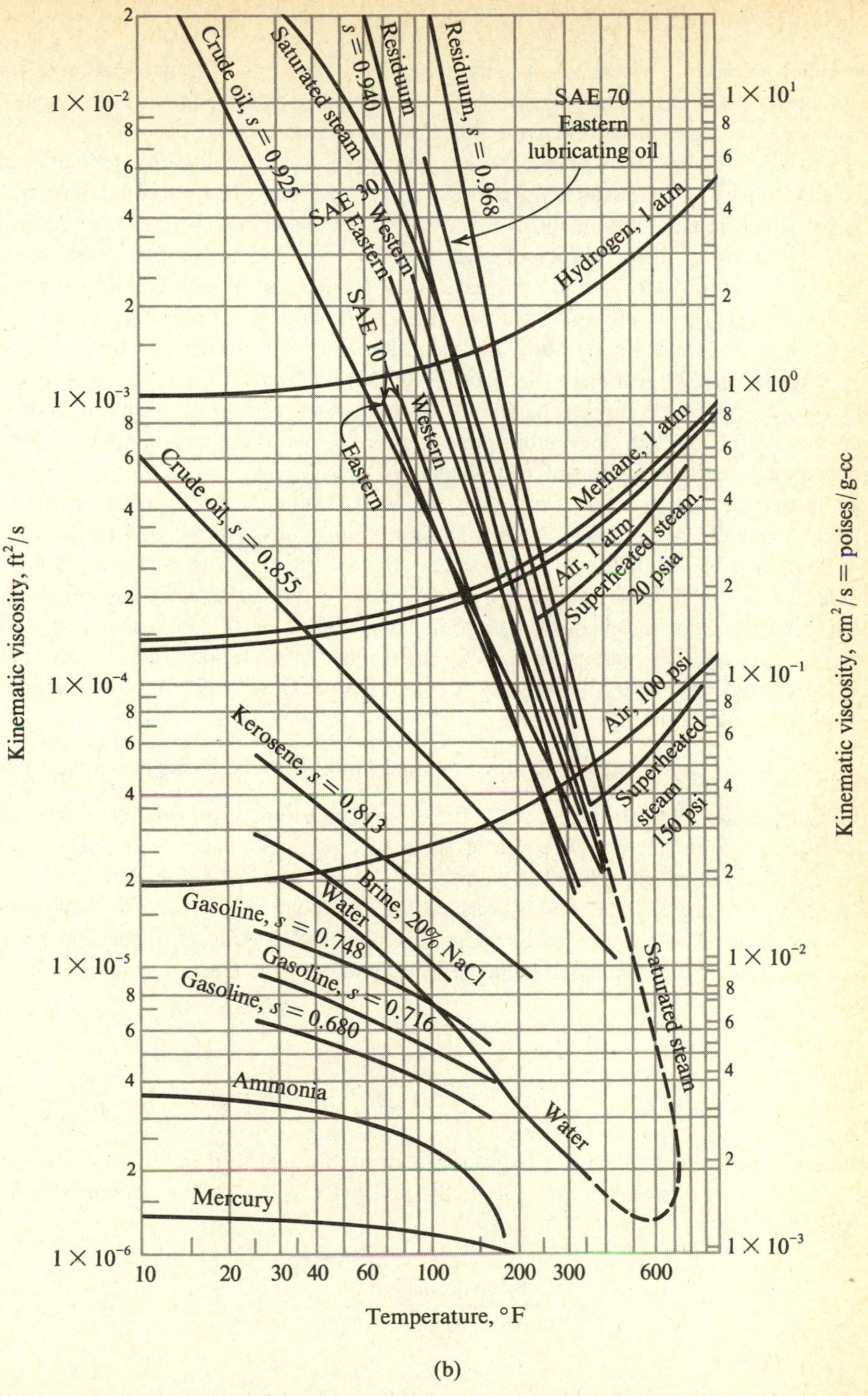

Figure 2.12 *(Con't.) (b) Kinematic viscosity v versus temperature for a variety of fluids. (Source: (a) and (b):* Handbook of Fluid Dynamics *by V. Streeter, Ed. Copyright © 1961. McGraw-Hill Book Company. Used with the permission of McGraw-Hill Book Company.)*

Two factors produce viscosity: (a) cohesion, and (b) the rate of transfer of molecular momentum. For liquids, the cohesive forces predominate over the inertial forces caused by changes in momentum. For gases, the reverse is true.

Temperature governs both cohesion and molecular activity, and so we would expect viscosity to depend on temperature. Figure 2.12a shows the variation of absolute viscosity versus temperature for a variety of fluids. Note that, because of increased molecular activity, the viscosity of a gas increases with an increase of temperature, whereas the viscosity of a liquid decreases with an increase of temperature. Viscosity is also a function of pressure, but we shall work largely with its dependency on temperature. Table 2.4 shows how the viscosity of air varies with altitude.

More than 200 years ago, Sir Isaac Newton [2.7] treated viscosity somewhat in the following manner. Suppose that a film of liquid, such as glycerin, is placed between two parallel plates with the bottom plate stationary, and the upper one moving with a constant velocity U by means of a shear force F_τ. Glycerin molecules between the two plates are visualized as small balls which roll along in layers between the flat plates. Since the glycerin will "wet" and cling tenaciously to the two surfaces, the bottommost layer of balls will not move at all, and the uppermost will move with a velocity U, with intermediate layers moving with velocities directly proportional to their distance from the stationary bottom plate. This orderly movement in parallel layers is known by a variety of names: streamline, laminar, or viscous flow. (The reader should compare the physics of the above flow with that described in Demonstration 2.1.)

The force per unit area $\dfrac{F_\tau}{A}$ required to impart motion to the layers has already been identified as the shear stress p_{yx}, while the movement of one layer relative to the other is the shear strain rate $\dot{\epsilon}_{yx}$. The rate of shear of a particular layer (sometimes called the velocity gradient) is defined as the ratio of its velocity to its perpendicular distance from the stationary surface, and is constant for each layer. Newton correctly deduced that the shear force F_τ required to maintain a constant velocity U of the upper plate is proportional to the area A and to the velocity gradient U/h. Thus, in a manner similar to Eq. (2.2)

$$F_\tau = \mu A \frac{U}{h} \tag{2.21}$$

The units of the dynamic viscosity μ are N·s/m^2 in SI, and lbf/ft·s in U.S. customary. A common unit of absolute viscosity is the poise (P), where 1 P = 1 dyne·s/cm^2 = 1 g/cm·s. Thus,

$$\frac{1 \text{ U.S. customary unit}}{47.9 \text{ SI unit}} = 1 \tag{2.22}$$

Newton further deduced that the viscosity of a given substance should be constant at any particular temperature and pressure, and independent of the rate of shear. In such "Newtonian fluids" we now know that shear stress is directly proportional to rate of strain.

Example 2.5

Calculate the absolute viscosity of fresh water at 30°C in units of (a) cP, (b) lbf·s/ft², (c) dyne·s/cm², (d) slug/ft·s, (e) kg/m·s.

Solution:

Step 1.

Find the appropriate table.

The absolute viscosity of fresh water is found in Table 2.2.

(a) From Table 2.2

$$\mu = 0.7975 \text{ cP}$$

(b) $\mu = 1.67 \text{ lbf}\cdot\text{s/ft}^2$

(c) Using conversions, and results from (a),

$$\mu = 7.975 \times 10^{-3} \text{ P}$$

$$= 7.975 \times 10^{-3} \frac{\text{dyne}\cdot\text{s}}{\text{cm}^2}$$

$$= 7.975 \times 10^{-3} \frac{\text{N}\cdot\text{s}}{\text{cm}^2} \times \frac{10^{-1}\text{ cm}^2}{\text{m}^2}$$

$$= 7.975 \times 10^{-4} \frac{\text{N}\cdot\text{s}}{\text{m}^2}$$

(d) Using conversions and result from (b)

$$\mu = 1.67 \frac{\text{lbf}\cdot\text{s}}{\text{ft}^2}$$

$$= 1.67 \frac{\text{lbf}\cdot\text{s}}{\text{ft}^2} \cdot \frac{\text{slug/ft}}{\text{lbf}\cdot\text{s}^2}$$

$$= 1.67 \text{ slug/ft}\cdot\text{s}$$

(e) Using conversions and result from (c)

$$\mu = 7.975 \times 10^{-3} \frac{\text{dyne}\cdot\text{s}}{\text{cm}^2}$$

$$= 7.975 \times 10^{-3} \frac{\text{g}}{\text{cm}\cdot\text{s}}$$

$$= 7.975 \times 10^{-3} \frac{\text{g}}{\text{cm}\cdot\text{s}} \times \frac{\text{kg}}{1000\text{ g}} \times \frac{100\text{ cm}}{\text{m}}$$

$$= 7.975 \times 10^{-4} \text{ kg/m}\cdot\text{s}$$

This completes the solution.

There are a number of empirical relationships for the viscosity of liquids and gases. For example, Bingham [2.8] obtained for fresh water at normal pressures

$$\mu = \{0.021482\,[T - 8.435 + (T^2 - 16.87T + 8149.5492)^{0.5}] - 1.20\}^{-1}, \text{ (cP)} \tag{2.23}$$

where T is the temperature in °C. The values obtained using the above equation are within 2% of those given in Table 2.2.

Not only does viscosity vary with temperature, but it also varies with pressure. Figure 2.13 and Table 2.7 show how pressure affects the viscosity of steam and water.

The viscosity of a gas is defined from the kinetic theory of gases by

$$\mu = 1/3\rho\tilde{V}\lambda$$

where $\tilde{V}$ is the average velocity of the molecular motion, and λ is the molecular mean free path. The above equation can also be expressed as

$$\mu = \frac{1}{\pi d_0}\frac{\hat{M}\,\tilde{V}}{3\sqrt{2}} \tag{2.24}$$

where d_0 is the molecular diameter and $\hat{M}$ is the molecular mass. One notes that the viscosity of a gas is independent of both density and pressure. Since the absolute temperature is proportional to the square of the root-mean-square velocity $\overline{V}$, the absolute viscosity of a gas is proportional to the square root of the absolute temperature.

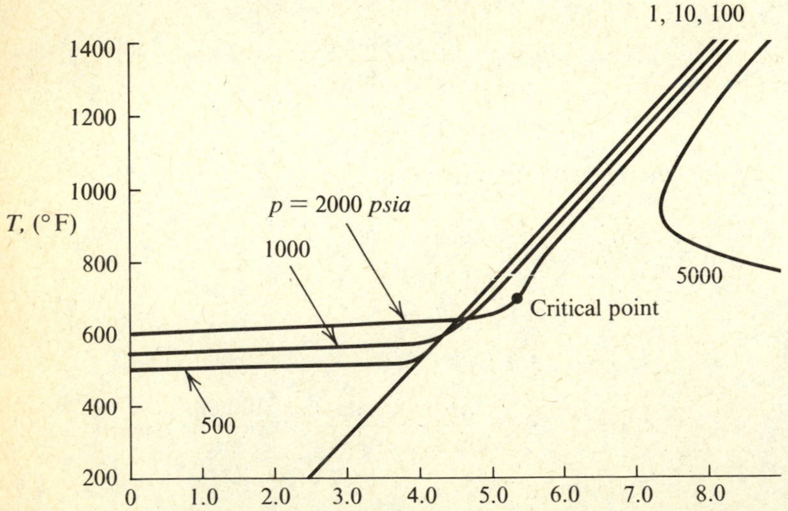

Figure 2.13 *Viscosity of steam and water for various pressures (psia).*

Table 2.7 *Dynamic Viscosity (micropoise) versus Pressure and Temperature for Steam*

Pressure		Temp.																		
		°C	0	50	100	150	200	250	300	350	375	400	425	450	475	500	550	600	650	700
(MPa)	lbf/in.^{2}	°F	32	122	212	302	392	482	572	662	707	752	797	842	887	932	1022	1112	1202	1292
0.1	14.504		17500	5440	121.1	141.5	161.8	182.2	202.5	223	233	243	253	264	274	284	304	325	345	365
0.5	72.52		17500	5440	2790	1810	160.2	181.4	202.3		234	244	254	264	274	284	305	325	345	366
1.0	145.04		17500	5440	2790	1810	158.5	180.6	202.2		234	244	255	265	275	285	305	326	346	366
2.5	362.6		17500	5440	2800	1820	1340	177.8	201.6		236	246	256	266	276	287	307	327	347	367
5.0	725.2		17500	5450	2800	1820	1350	1070	200.6		240	250	259	269	279	289	309	329	349	369
7.5	1087.8		17500	5450	2800	1830	1350	1080	199.2		244	253	263	273	282	292	312	332	352	372
10.0	1450.4		17500	5450	2810	1830	1360	1080	905		249	258	267	286	295	315	315	334	354	374
12.5	1813.0		17500	5460	2810	1840	1360	1090	911		254	263	271	280	289	299	318	337	357	376
15.0	2176		17400	5460	2820	1840	1370	1100	917		262	269	276	285	294	302	321	340	359	379
17.5	2538		17400	5460	2820	1850	1380	1100	924		273	276	282	290	298	307	324	343	362	381
20.0	2901		17400	5460	2830	1860	1380	1110	930	735	291	286	289	296	303	311	328	346	365	384
22.5	3263		17400	5460	2830	1860	1390	1120	936	747	491	299	298	302	309	316	332	350	368	386
25.0	3626		17400	5470	2840	1870	1390	1120	943	760	597	321	309	310	315	321	336	353	371	389
27.5	3989		17400	5470	2840	1870	1400	1130	949	772	633	367	324	320	322	327	341	357	374	392
30	4351		17400	5470	2850	1880	1400	1130	955	785	657	458	345	331	330	334	346	361	377	395
35	5076		17300	5480	2860	1890	1420	1150	968	805	693	573	416	363	351	349	357	369	385	401
40	5802		17300	5480	2870	1900	1430	1160	981	825	721	628	503	411	379	369	369	379	392	408
45	6527		17300	5490	2880	1910	1440	1170	993	837	743	664	565	468	415	393	383	389	401	415

Table 2.7 (Con't.)

Pressure		Temp.																		
		°C	0	50	100	150	200	250	300	350	375	400	425	450	475	500	550	600	650	700
(MPa)	lbf/in.^{2}	°F	32	122	212	302	392	482	572	662	707	752	797	842	887	932	1022	1112	1202	1292
50	7252		17200	5490	2890	1920	1450	1180	1010	850	762	693	609	521	456	421	400	401	410	423
55	7977		17200	5500	2900	1930	1460	1200	1020	860	780	716	643	564	497	453	418	414	420	431
60	8702		17200	5500	2910	1940	1480	1210	1030	870	795	736	670	600	534	485	439	428	430	439
65	9427		17200	5510	2920	1960	1490	1220	1040	882	809	754	693	629	567	516	460	442	441	448
70	10153		17100	5510	2930	1970	1500	1230	1060	895	822	770	713	654	596	545	482	458	453	458
75	10878		17100	5520	2940	1980	1510	1240	1070	905	835	784	732	676	621	572	504	474	466	468
80	11603		17100	5520	2950	1990	1520	1260	1080	915	846	798	748	695	644	596	526	491	478	478

Conversion Factors for Viscosity

$$10^6 \text{ micropoise} = 0.0020885 \text{ lbf} \times \text{s/ft}^2 = 241.91 \text{ lb/(hr} \times \text{ft)} = 0.1 \text{ kg/(m·s)}$$
$$= 0.067197 \text{ lb/(ft} \times \text{s)} = 0.58015 \times 10^{-6} \text{ lbf} \times \text{hr/ft}^2 = 0.1 \text{ N·s/m}^2$$

Source: *Steam Tables*, by J. H. Keenan, et al., Copyright © 1969 John Wiley & Sons, Inc. Publishers. Reprinted by permission of John Wiley & Sons, Inc.

In practice, we assume that the approximate formula for the viscosity of a gas is

$$\mu = \mu_0 (T/T_0)^n \tag{2.25}$$

where μ_0 is a referenced dynamic viscosity at the reference temperature T_0. The index of power n is different for different gases, and it also is a function of temperature. For air at ordinary temperature, $n = 0.76$. As the temperature increases, n decreases to 0.5.

DEMONSTRATION 2.2

Purpose:

To show how viscosity affects the motion of objects. Fixed volumes of varying viscous liquids are enclosed in tubes mounted in a frame that can rotate about a fixed axis, as shown in Fig. 2.14. Spherical balls of a diameter equal to ¼ the tube diameter (all of the same diameter and material) are placed in the tubes, and the tubes are then sealed so that no air is trapped in the tubes. Rotating the frame 180°, the balls are free to fall to the bottom of the tubes. The viscosity of the liquids in each tube can be calculated from Stoke's law

$$\mu = \frac{D^2(\gamma_s - \gamma_f)}{18l} t \tag{2.26}$$

where D is the diameter of the spheres, γ_s and γ_f are the specific weights of the spheres and fluid, respectively, l is the depth of fluid in the tube, and t is the time it took the sphere to fall from the top to the bottom.

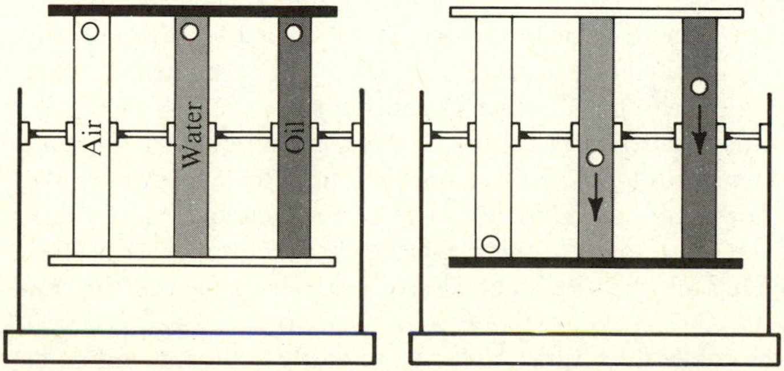

Figure 2.14

Discussion

- Verify Eq. (2.26) with experimental measurements.
- Give the physical explanation behind the fact that the viscosity depends upon two physical concepts: cohesion of molecules and the rate of transfer of molecular momentum. Which has the greater cohesive force in the fluids used in the demonstration?
- Explain how temperature affects the resistance to shear stress in air.

2.4.8 Kinematic Viscosity, ν

Kinematic viscosity is defined as the ratio of the absolute viscosity to density, i.e.,

$$\nu = \frac{\mu}{\rho} \tag{2.27}$$

Figure 2.12b shows the kinematic viscosity ν versus temperature T for a variety of fluids. Table 2.8 shows $\nu(p,T)$ for steam. The SI unit of kinematic viscosity is m^2/s, and the U.S. customary unit is ft^2/s. Units are often referred to as the stoke (St), where a stoke equals 1 cm^2/s.

Selection of Lubricants Based on Viscosity Consider the following factors in the specification for the viscosity of a fossilized fuel *lubricant*.

1. **Temperature**—as with all liquids, the viscosity of oil decreases as the oil temperature increases. Since viscosity is a measure of the internal friction of a fluid, the heat generated by the friction resulting from the rotating shaft in a bearing is roughly proportional to the viscosity of the lubricant.
2. **Shaft speed**—from (1) above it is seen that the faster the shaft rotation, the greater the friction created. Therefore the increased friction will raise the oil temperature, causing the oil's viscosity to fall.
3. **Bearing pressure**—the higher the viscosity of the oil, the greater its resistance to oil film rupture, since it is more difficult to squeeze out a high viscosity oil film than a low viscosity oil film. Therefore, a thicker oil (i.e., one of higher viscosity) can carry heavier loads, in the absence of other considerations.
4. **Bearing clearance**—the smaller the clearance, the greater the friction force created by the lubricant for a given viscosity oil. This can be seen from the approximation of Newton's law of fluid flow $F_\tau = \mu A(U/h)$ where U is the magnitude of the shaft speed at the surface of the shaft and h is the clearance. Larger clearances between bearing surfaces require high oil viscosity and cohesiveness to keep the oil film from rupturing.
5. **Starting torque**—the colder the oil, the higher its viscosity will be; therefore the greater the friction force will be. Thus for a given bearing, the friction force at starting may not exceed the starting force available at the bearing.

Table 2.8 *Kinematic Viscosity × 10^7 (m^2/s) versus Pressure and Temperature for Steam*

Pressure (MPa)	Pressure (lbf/in.²)	Temp. °C: 0	50	100	150	200	250	300	350	375	400	425	450	475	500	550	600	650	700
	°F:	32	122	212	302	392	482	572	662	707	752	797	842	887	932	1022	1112	1202	1292
0.1	14.504	17.5	5.51	205.4	273.9	351.4	438.3	534.4	640.2	695.9	754.0	814.4	880	945	1013	1154	1309	1469	1639
0.5	75.52	17.5	5.51	2.91	1.98	68.1	86.1	105.7		138.9	150.6	162.8	174	188	202	231	261	293	328
1.0	145.04	17.5	5.50	2.91	1.98	32.7	42.0	52.2		68.9	74.8	81.2	88	94	101	115	131	147	164
2.5	362.6	17.5	5.50	2.91	1.98	1.55	15.47	19.94		27.1	29.5	32.0	34.6	37.3	40.2	46.0	52.1	58.6	65.4
5.0	725.2	17.5	5.50	2.91	1.98	1.55	1.34	9.09		13.2	14.5	15.7	17.0	18.4	19.8	22.8	25.9	29.2	32.7
7.5	1087.8	17.4	5.50	2.92	1.99	1.56	1.34	5.32		8.5	9.3	10.3	11.2	12.1	13.1	15.1	17.2	19.4	21.8
10.0	1450.4	17.4	5.49	2.92	1.99	1.56	1.34	1.26		6.11	6.82	7.51	8.21	8.95	9.67	11.23	12.82	14.52	16.30
12.5	1813.0	17.4	5.49	2.92	1.99	1.56	1.34	1.26		4.46	5.26	5.84	6.44	7.03	7.66	8.91	10.21	11.60	13.01
15.0	2176	17.3	5.49	2.92	2.00	1.57	1.35	1.26		3.64	4.21	4.73	5.26	5.78	6.28	7.36	8.47	9.62	10.85
17.5	2538	17.3	5.48	2.92	2.00	1.57	1.36	1.27		2.88	3.44	3.93	4.40	4.86	5.33	6.25	7.23	8.23	9.27
20.0	2901	17.2	5.48	2.93	2.00	1.58	1.36	1.27	1.22	2.24	2.85	3.31	3.76	4.18	4.59	5.43	6.29	7.19	8.12
22.5	3263	17.2	5.48	2.93	2.00	1.58	1.36	1.27	1.22	1.23	2.35	2.83	3.25	3.65	4.03	4.79	5.58	6.38	7.20
25.0	3626	17.2	5.47	2.93	2.01	1.58	1.37	1.27	1.22	1.18	1.93	2.44	2.84	3.22	3.58	4.28	4.99	5.73	6.48
27.5	3989	17.1	5.47	2.93	2.01	1.59	1.37	1.27	1.22	1.18	1.54	2.11	2.51	2.87	3.20	3.86	4.52	5.19	5.89
30	4351	17.1	5.47	2.93	2.01	1.59	1.37	1.27	1.22	1.18	1.28	1.83	2.23	2.57	2.90	3.52	4.13	4.75	5.40
35	5076	17.0	5.46	2.94	2.02	1.60	1.38	1.28	1.22	1.18	1.21	1.43	1.80	2.13	2.42	2.98	3.52	4.07	4.63
40	5802	17.0	5.45	2.94	2.03	1.61	1.39	1.28	1.22	1.19	1.20	1.28	1.52	1.81	2.08	2.58	3.07	3.55	4.06
45	6527	16.9	5.45	2.94	2.03	1.61	1.40	1.29	1.23	1.19	1.20	1.24	1.37	1.59	1.82	2.27	2.72	3.17	3.62
50	7252	16.8	5.44	2.95	2.04	1.62	1.40	1.30	1.23	1.19	1.20	1.22	1.30	1.45	1.64	2.05	2.45	2.86	3.27

Conversion Factors for Kinematic Viscosity

$$10^{-7}\ m^2/s = 1.076 \times 10^{-6}\ ft^2/s = 3.875 \times 10^{-3}\ ft^2/hr = 3.6 \times 10^{-4}\ m^2/hr = 10^{-3}\ cm^2/s$$

Source: *Steam Tables*, by J. H. Keenan, et al., Copyright © 1969 John Wiley & Sons, Inc. Publishers. Reprinted by permission of John Wiley & Sons, Inc.

6. **Fuel dilution**—in internal combustion engines, fuel dilution (generally caused by ring leakage) reduces the viscosity of the lubricant. Fuel diluted lube oil cannot be renovated by centrifuge and must be discarded at 5% dilution. For diesel engine lubricating oil, 5% dilution is represented by 30% reduction in viscosity at 100°F.

In short, the higher the rubbing speed (i.e., surface speed) at a bearing surface, the lower the desirable oil viscosity. If an oil of higher viscosity is used, the higher internal friction will cause the oil to run at higher temperatures, resulting in a reduction in viscosity until some operating equilibrium is reached at a higher temperature. The higher the bearing pressure, the higher the viscosity of the oil needed to prevent oil film rupture. The amount of oil drawn into the clearance space between the journal and the bearing surfaces increases with rotating speed for the same lubricant.

Engine and Transmission Lubricants Engine and transmission lubricants are classified by viscosity according to standards established by the Society of Automotive Engineers (SAE). The allowable ranges of viscosity for several grades of oil are given in Table 2.9.

Table 2.9 *Allowable Viscosity Ranges for SAE Lubricant Classifications*

Lubricant Type	SAE Viscosity Number	Viscosity Range (cS) 0°F (−18°C) Minimum	0°F (−18°C) Maximum	210°F (99°C) Minimum	210°F (99°C) Maximum
Crankcase	5*W*		1,200	3.9	
	10*W*	1,200	2,400	3.9	
	20*W*	2,400	9,600	3.9	
	20			5.7	9.6
	30			9.6	12.9
	40			12.9	16.8
	50			16.8	22.7
Transmission and axle	75		15,000		
	80	15,000	100,000		
	90			75	120
	140			120	200
	250			200	
Automatic transmission fluid	Type A	39	43	7	8.5
Jet fuel	JP-5	1.9 at 100°F			

1 centistoke = 10^{-5} m²/s = 1.08×10^{-5} ft²/s.
Compiled from NAVSHIPS 0905-475-2010, ND Fuel Conversion Program.

Viscosity numbers with W (e.g., $20W$) are classified by viscosity at 0°F. Those without W are classified by viscosity at 210°F.

Multigrade oils (e.g., $10W$-40) with high-polymer additives are formulated to minimize the variation of viscosity with temperature. These high polymer additives are non-Newtonian fluids; i.e., their viscosity may change with shearing.

Example 2.6
SAE 10 Eastern motor oil is stored at a temperature of 75°F. At what temperature does the oil have to be heated to replace SAE 30 Eastern oil at 150°F?

Solution:
Step 1.
Find the appropriate table or figure.

For SAE 30 Eastern motor oil at 150°F, the kinematic viscosity ν is obtained from Fig. 2.12b:

$$\nu = 3.5 \times 10^{-4} \text{ ft}^2/\text{s} \tag{i}$$

Thus, for SAE 10 Eastern motor oil to have the same kinematic viscosity, the temperature must be increased from 75°F to 100°F.

This completes the solution.

Example 2.7
Calculate the kinematic viscosity of jet fuel JP-5 at 100°F. Find one other fluid whose viscosity is the same value and is within ±20° of 100°F.

Solution:
Step 1.
Find the appropriate table or figure.

From Table 2.9, the kinematic viscosity of JP-5 is

$$\nu = 2.052 \times 10^{-5} \text{ ft}^2/\text{s} \tag{i}$$

According to Fig. 2.12b, the only other fluid of a viscosity 2.052×10^{-5} ft^2/s that is within ±20° of 100°F is kerosene.

This completes the solution.

2.4.9 Surface Tension, σ

Surface tension is a phenomenon that occurs largely from the cohesion of molecules at the interface separating two different substances, such as air and water. Surface tension is therefore important in the behavior of bubbles, buoys, capillary flows, and osmosis.

The surface tension phenomenon is often studied by examining the behavior of a free-surface of liquid that acts like a stretched membrane. The student is referred to the well-known soapfilm experiments of the great mathematician, Richard Courant [2.9]. Owing to the action of surface tension, a film of liquid is in stable equilibrium only if its area is a minimum. Since fluids assume a minimal surface at a free-surface boundary, then a force exists that acts on the surface to prevent the surface from taking any other shape but the minimal. This force is called a surface tension force and is closely associated with the attractive force shown in Fig. 2.4. For a collection of molecules, the attractive (cohesive) force will be such that the average distance between molecules will be minimal at the free-surface boundary. When all external forces are zero the shape of the free-surface is that of a sphere.

Defining σ as the surface tension coefficient, the minimal principle can be stated

$$\sigma = \frac{dF}{dl} \tag{2.28}$$

where F is the cohesive force, and dl is an elemental length of the minimal surface.

The motion picture "Surface Tension in Fluid Mechanics" (L. M. Trefethen, film principal; No. 21609) of the NCFMF/EDC Series* is recommended as an outstanding supplement to surface tension.

For an air-seawater interface, the empirical relationship for surface tension

$$\sigma = 75.796 - 0.145t - 0.00024t^2 \tag{2.29}$$

works well for $10 \leq t \leq 60$ in degrees Celsius, where units of the surface tension σ are 10^{-4} lbf/ft. Salinity, density, air pollution, and particulates will alter any value of surface tension. Table 2.10 gives values of surface tension in an air-water interface for a range of temperature.

2.4.10 Capillary Rise or Depression, *h*

Capillarity is due either to cohesion or to adhesion, where adhesion enables a liquid to adhere to another body. It is a phenomenon of liquids that if cohesion is less effective than adhesion, the liquid will wet a solid surface with which it is in contact and rise at the point of contact. If cohesion is predominant over adhesion, the reverse occurs.

In order to understand capillary rise, let us consider how a drop of liquid rests on a smooth flat surface. Figure 2.15 shows two distinct shapes. Part (a) shows the droplet spread over the surface at a small angle-of-contact θ. We define such a liquid as a wet liquid. Part (b) shows a liquid that is quite round and produces a larger angle-of-contact than (a). It does not wet the surface.

In droplet (a) the resultant force **R** acting on a molecule at point P is due to a weak net molecular force of attraction **M** and a strong gravitational force **G**. The

*Available from Encyclopaedia Britannica Educational Corporation, 425 N. Michigan Ave., Chicago, Ill. 60611. See also "Illustrated Experiments in Fluid Mechanics," MIT Press, Cambridge, Mass., 1972.

Table 2.10 *Values of Surface Tension of Liquids in Contact with Air*

Liquid	Surface Tension (N/m)
Mercury	0.484
Water	0.072
Glycerin	0.063
Ethylene glycol	0.048
Bromine	0.042
Phenol	0.040
Carbon disulfide	0.032
Benzene	0.028
Alcohol	0.022
Octane	0.021

Temperature Variation of Water versus Surface Tension

Temp. (°C)	*Temp.* (°F)	*Surface Tension* σ *(N/m)*
−8	17.6	0.077
0	32	0.0756
10	50	0.0742
20	68	0.0728
30	86	0.0712
40	104	0.0696
50	122	0.0679
60	140	0.0662
70	158	0.0644
80	176	0.0626
100	212	0.0589

Source: Reprinted with permission from R. E. Bolz and G. L. Tuve, Eds., *Handbook of Tables for Applied Engineering Sciences,* © 1970. Copyright The Chemical Rubber Co., CRC Press, Inc.

resultant force at *P* is a surface force, and acts perpendicular to the liquid surface. The resultant forces produce the shape of the droplet.

In droplet (b) the resultant force **R** acting on a molecule at *Q* is due to a strong net molecular force **M** compared to a weaker gravitational force **G**. Like case (a), the resultant force is perpendicular to the surface and pulls the drop into the shape shown.

In capillary rise and fall a similar phenomenon exists. Figure 2.16a shows a small angle-of-contact fluid resulting in a large surface tension. The molecules of the fluid climb up the wall's inner surface with the main fluid moving up behind it creating the curved meniscus. In this figure, the molecular force **M** has the gravitational force included in it, and when combined with the force of attraction of the wall produces the resultant pull **R** that is perpendicular to the liquid surface.

Wetting agents (like soap) help water by making a small angle-of-contact between water and, say, cloth fibers or greasy dishes. The wetting agent must have a chemical

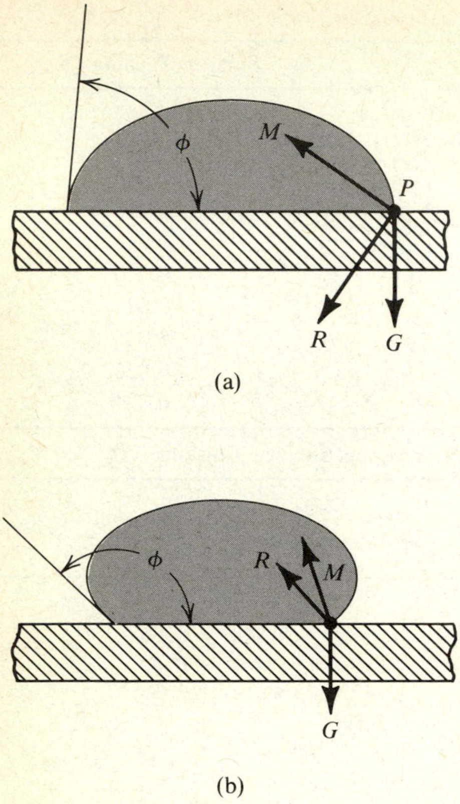

Figure 2.15 Forces on a liquid droplet.

composition such that its molecules are attracted to grease (or textile fibers), and also be attracted to water.

A quantity that occurs very often in the measurement of surface tension is the so-called capillary constant

$$a^2 = rh = \frac{2\sigma}{\gamma_1 - \gamma_2} \tag{2.30}$$

where r is the radius of curvature at the bottom (or top) of the meniscus, h is the capillary rise (or depression), γ_1 is the specific weight of the liquid below the meniscus, and γ_2 is the specific weight of the fluid above the meniscus. The product rh has the dimensions of area.

If the liquid is taken in contact with air, γ_2 can be omitted without serious loss of accuracy. Equation (2.30) also applies to the interface between two liquids, in which case γ_2 is the specific weight of the lighter fluid. Figure 2.17 shows the rise or fall in capillarity versus tube diameter for water and mercury at different temperatures. The curves are based on the application of Eq. (2.30) to water and mercury.

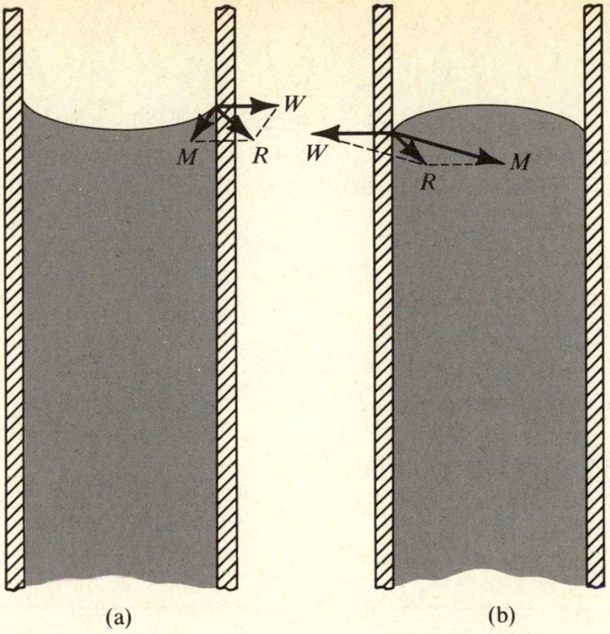

Figure 2.16 *Capillary rise and capillary depression.*

Example 2.8
Show that r in Eq. (2.30) is $R/\cos\theta$ where R is the radius of the tube.

Solution:
In order to derive Eq. (2.30), one must first be familiar with fluid statics. The student has already been exposed to the fact that the pressure at a point B is equal to the pressure at some point A above it, plus the specific weight of the fluid times the vertical distance between points A and B.

If we designate the pressure at point A by p_A, then at point B, the pressure is simply

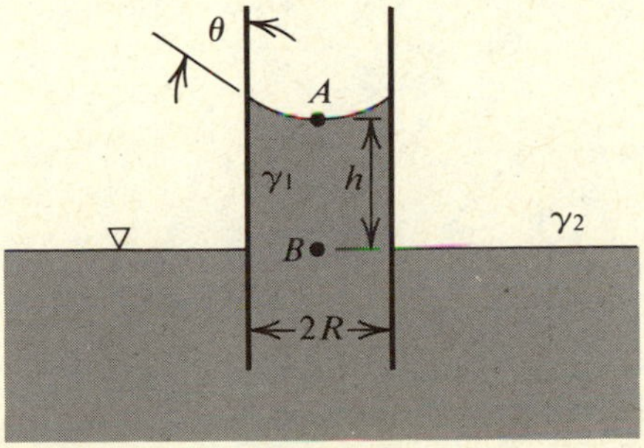

Figure E2.8

Example 2.8 *(Con't.)*

$$p_B = p_A + \gamma_2 h \tag{i}$$

where γ_2 is the specific weight of the surrounding medium. The upward force along the circumference $2\pi R$ is $2\pi R\sigma \cos\theta$, so that the upward pressure is $(2\pi R\sigma \cos\theta)/\pi R^2$. Applying the free-body diagram for the upward external forces yields

$$\frac{2\pi R\sigma \cos\theta}{\pi R^2} - \gamma_1 h_4 + p_B - p_A = 0 \tag{ii}$$

Substituting Eq. (i) into Eq. (ii) yields

$$h = \frac{2\sigma \cos\theta}{R(\gamma_2 - \gamma_1)} \tag{iii}$$

Comparing Eq. (iii) with Eq. (2.30) yields

$$r = R/\cos\theta \tag{iv}$$

This completes the solution.

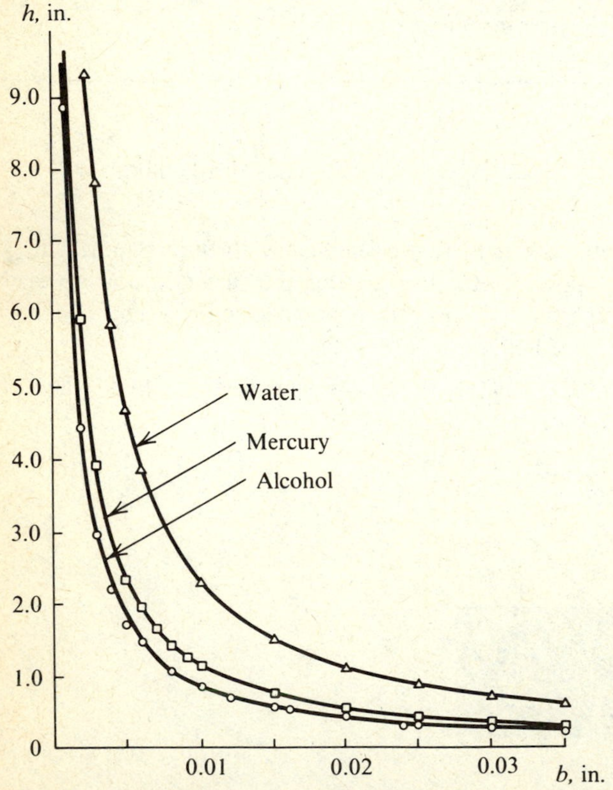

Figure 2.17 *Surface tension σ versus diameter b for $t = 68°F$, $h = 2\,\sigma/b(\gamma_1 - \gamma_2)$.*

References

2.1 Truesdell, C., *Continuum Mechanics I. The Mechanical Foundations of Elasticity and Fluid Dynamics*, Gordon and Breach Science Publishers, 1966.

2.2 Keenan, J. H., Keyes, F. G., Hill, P. G., and Moore, J. G., *Steam Tables* (English Units), Wiley, New York, 1969.
—*Steam Tables* (Metric Units), Wiley, New York, 1969.

2.3 Franks, Felix, *Water, A Comprehensive Treatise*, Plenum, New York, 1972.

2.4 Bolz, R. E., and Tuve, G. L., Eds., *Handbook of Tables for Applied Engineering Science*, The Chemical Rubber Co., Cleveland, Ohio, 1970.

2.5 Maxwell, J. C., "On the Dynamical Theory of Gases," *Phil. Trans. Roy. Soc. London*, (A), 157:49, 1866; *Phil. Mag.*, (4), 35:129–145, 185–217, 1868.

2.6 Euler, L. "Sectio prima de statu aequilibrii fluidorum," *Novi Comm. Petrop.*, 13:305, 1768.

2.7 Newton, Isaac, *Principia*, Book II, 1687.

2.8 Bingham, E. C., *Fluidity and Plasticity*, McGraw-Hill, New York, 1922.

2.9 Courant, R., and Robbins, H., *What Is Mathematics*, 5th Printing, Oxford Univ. Press, New York, 1941, pp. 386–397.

2.10 Fleming, R. H., and Revelle, R., "Symposium of Recent Marine Sediments," *Ann. Ass. Pet. Geol.*, P. D. Trask, Ed., 1939.

2.11 Giles, R. V., *Fluid Mechanics and Hydraulics*, Schaum, New York, 1962.

Study Questions

2.1 Define a fluid.

2.2 What distinguishes a liquid from a solid?

2.3 How would one express the shear stress of Eqs. (2.1) and (2.2) for other planes?

2.4 Discuss the continuum hypothesis and why it is important in fluid dynamics.

2.5 How can a fluid be considered a collection of lamina?

2.6 What are the requirements for fluid flow to be inviscid, ideal, perfect, potential, viscous, or real?

2.7 Suppose you were a salesman trying to sell fluid dynamics to a housewife. What would she buy and how could it change her life?

2.8 Think of two areas in fluid dynamics in which you believe no one is doing research.

2.9 Compare the number of fluid dynamics related journals in your technological library with the number of any other technological discipline's journals.

2.10 Why are you studying fluid dynamics? What role will it play in your planned vocation?

Problems

2.1 In Fig. 2.2, what is the slope of the shear stress versus rate of shearing strain equal to for a Newtonian fluid?

2.2 What is the relationship between the rate of shearing strain and the velocity gradient? What is the shear stress expression for a velocity gradient dv/dx, where v is the velocity component in the y direction?

2.3 What are the dimensions of p_{yx}, μ, $\dot{\epsilon}_{yx}$, and velocity gradient?

2.4 An ideal solid would have no velocity gradient under any loading condition. How would a real solid be plotted in Fig. 2.2 for a loading condition?

2.5 How would an ideal fluid be plotted in Fig. 2.2?

2.6 There are fluids called super fluids. Find what they are and how they would be plotted in Fig. 2.2.

2.7 What is a plasma, and how would it be plotted in Fig. 2.2?

2.8 Olive oil ($\mu = 3 \times 10^{-4}$ slug/ft·s, $\nu = 1.8 \times 10^{-4}$ ft²/s) flows over a wooden plank with a velocity profile given by $u = 53y - 1215y^2$ ft/s where y is the distance from the plank in feet. What is the shear stress at the surface of the plank, and where is it a maximum?

2.9 As Fig. P2.9 shows, air at 15°C flows over a smooth water surface with a velocity distribution $u = 1058y - y^3 \times 10^8$ m/s, where y is in meters. What is the shear stress at the smooth water surface? Discuss what would likely occur.

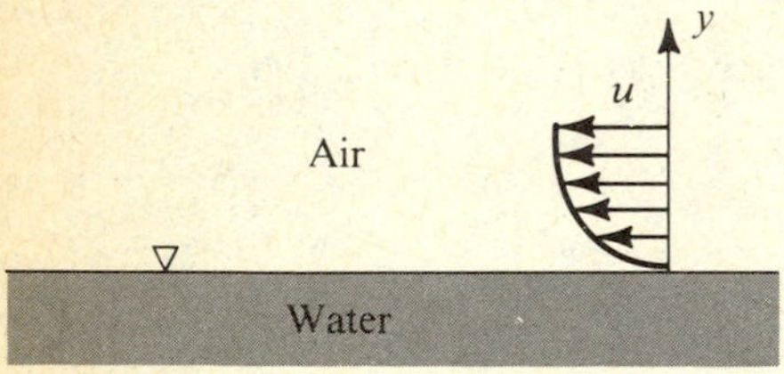

Figure P2.9

2.10 Obtain the velocity distribution $u = u(y)$ of air between two stationary plates of glass given the shear stress

$$p_{yx} = 2(p_{yx})_0 \frac{y}{h}$$

where $(p_{yx})_0$ is the shear stress on the glass ($y = \pm h/2$) (see Fig. P2.10).

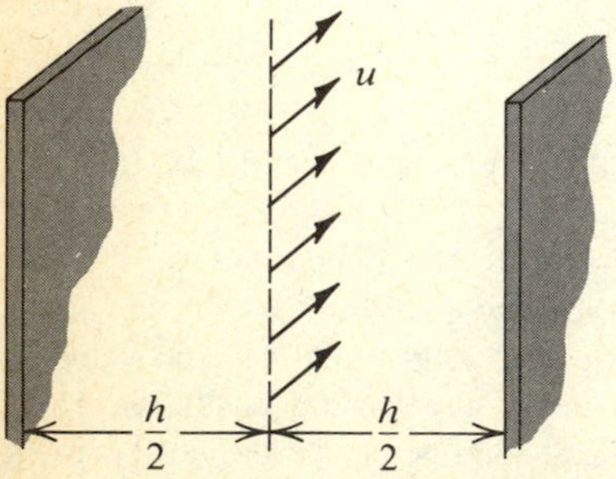

Figure P2.10

2.11 A liquid 1 cm thick moves past a stationary flat object at rest. The free surface of the liquid moves at a constant velocity of 1 cm/s. At 0.2 cm below the free surface a fluid particle moves steadily at a velocity of 0.8 cm/s. At 0.6 and 0.8 cm below the free surface, a fluid particle moves steadily at a velocity of 0.4 and 0.2 cm/s, respectively. (a) Calculate the angular rate of deformation $\dot{\epsilon}_{yx}$.

2.12 Obtain an expression for the shear stress of a Newtonian fluid if the velocity distribution measured normal to the flow is linear (Fig. P2.12).

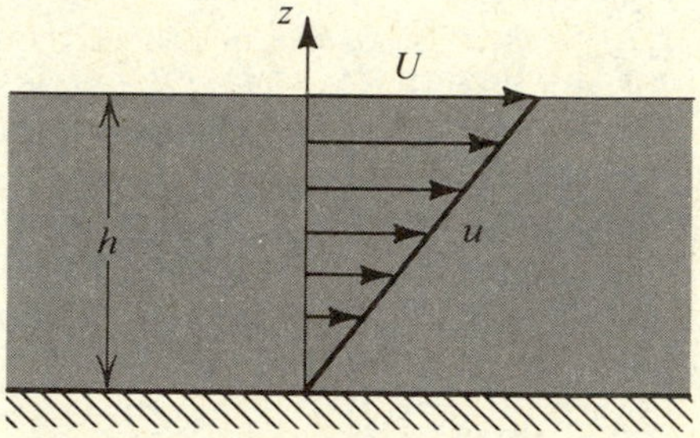

Figure P2.12

2.13 Consider a space between two large parallel sheets of smooth glass that is 0.25 in. apart, as shown in Fig. P2.13. Let the space be filled with an oil if viscosity $\mu = 0.003$ slug/ft·s. Calculate the force necessary to pull a 3 × 3-in. square piece of shim-stock through the oil at a velocity of 1 in./s if (a) the shim-stock is placed midway between the plates, and (b) if the shim-stock is placed 0.0625 in. from one of the plates. Assume a linear velocity distribution.

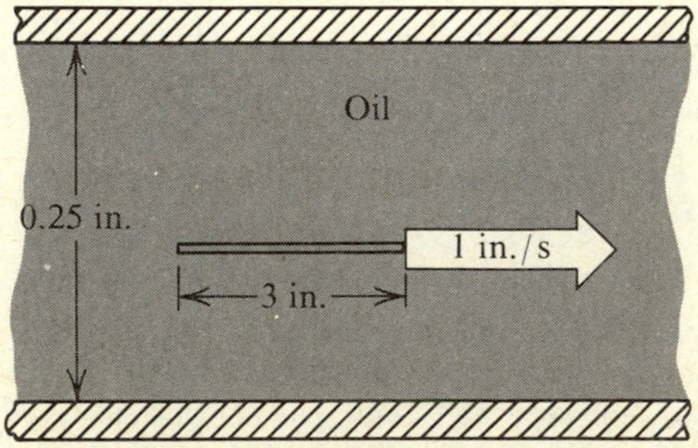

Figure P2.13

2.14 Air flows between two sheets of glass in a solar device. The distance between the two surfaces is 0.1 in. If the temperature of air is 120°F, calculate the maximum velocity if a stress of 0.05 psf exists on the plates, and a linear distribution of stress exists between the plates. (See Fig. P2.14.)

2.15 A viscometer is an apparatus that measures the viscosity of a fluid; it is shown in Fig. P2.15. It consists of an outer fixed cylinder with an inner rotating cylinder. What is the viscosity of the fluid which fills the annular region of both cylinders that are 9 in. long. The outer cylinder has a 6-in. radius, and the inner cylinder has a 5.7-in. radius. It takes a torque of 0.05 ft·lbf to maintain an angular speed of 50 rpm at a temperature of 40°F.

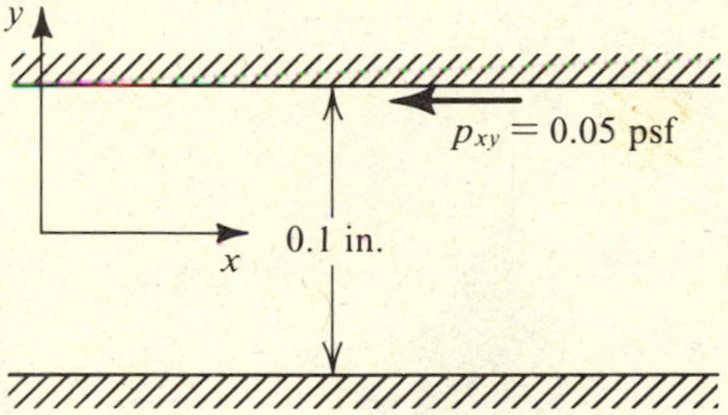

Figure P2.14

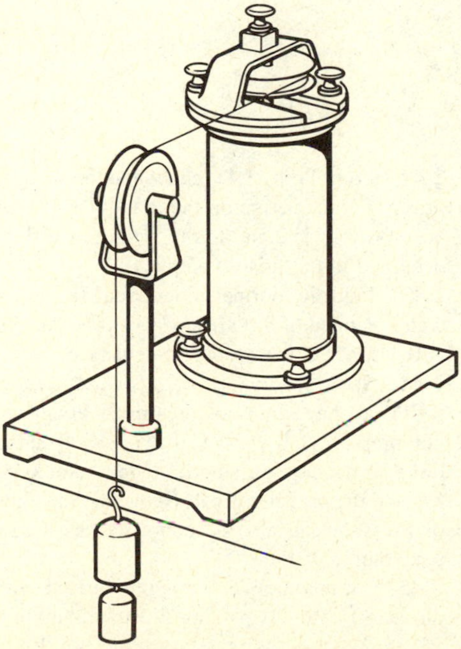

The coaxial, rotating-cylinder viscosimeter

Figure P2.15

2.16 Calculate the shear stress p_{yx} on the fixed flat surface shown in Fig. P2.16, given a linear distribution of velocity for a Newtonian fluid flow where the absolute viscosity is 0.008 lbf·s/ft². At the free-surface $y = 0.5$ in., the velocity is 10 in./s. Repeat the problem for a parabolic distribution.

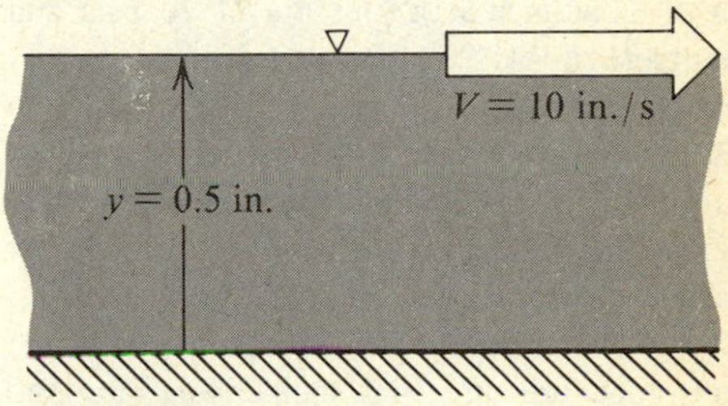

Figure P2.16

Density, Specific Weight, Mass, Entropy

2.17 If a barrel (55 gal) weighs 23 lbf, how much does the barrel weigh filled with heavy fuel oil at 70°F?

2.18 If 500 ft³ of a liquid weighs 25,000 lbf, calculate its density, specific weight, and specific gravity, and give a reasonable explanation of what the liquid could be along with its temperature.

2.19 Calculate the density, specific volume, and specific weight of air at 65°F and zero gauge pressure.

2.20 Calculate the density, specific volume and specific weight of air at 27°C and 1.4 bars.

2.21 A liquid has a density of 55 lbm/ft³. (a) If $g = 32.2$ ft/s², determine its specific volume, specific weights and density in slug/ft³. (b) If $g = 31.6$ ft/s², determine the three properties of (a).

2.22 What is the value of the universal gravitational constant g_c in a location where the fluid has a mass of 270 lbm and weighs 195 lbf?

2.23 What are the differential and integral relationships between mass M and density ρ?

2.24 What is the specific weight of air at (a) 50 psia, 100°F, (b) 2 bars, 15°C, and (c) 0.2 MPa, 25°C?

2.25 Given the specific gravity of a fluid as 0.8, determine the density in units of slug/ft^3 and kg/m^3.

2.26 The density of a fluid is 3 g/cm^3. What is its (a) specific volume, (b) specific weight, and (c) specific gravity in SI units and in U.S. customary units?

2.27 A gas of molecular weight 30 fills a 3 ft^3 container at a pressure of 20 psia and temperature 150°F. Find the specific gravity of the gas.

2.28 What is the density of air at 5000 ft altitude if the temperature is 41.2°F?

2.29 How much volume will be occupied by a 10 kg mass of helium at 15°C and 1 bar?

2.30 Calculate the density of a gas when the absolute pressure and temperature are 150 kN/m^2 and 30°C, respectively, if $R = 287$ J/kg·K. What is the gas?

2.31 The density of a certain liquid is 0.75 g/cm^3. Determine the specific weight in N/m^3 if the acceleration is (a) 3.2 m/s^2 and (b) 8.75 m/s^2.

2.32 The density of a certain liquid is 3 slug/ft^2. Determine the specific weight in lbf/ft^3 given (a) g is 5 ft/s^2 and (b) g is 36 ft/s^2.

2.33 The acceleration due to gravity on a distant planet is 1.5 m/s^2. If 1.5 kg of a fluid occupies a given volume of 1.1 m^3, determine (a) the specific weight (N/m^3), (b) the specific volume (m^3/kg), and (c) the density (kg/m^3).

2.34 Repeat Prob. 2.33 using U.S. customary units.

2.35 What is the weight of a 5 kg mass on the surface of the moon given the acceleration due to the moon's gravity is one-sixth that on earth?

2.36 In Prob. 2.35, if the mass occupies a volume of 0.2 ft^3, what is its density (a) on the earth's surface, and (b) on the moon's surface in both SI units and U.S. customary units?

2.37 In Prob. 2.36, what is the specific gravity of the 5-kg mass at a point where (a) $g = 5.36$ ft/s^2, (b) $g = 32.2$ ft/s^2?

2.38 Air is compressed from 15 psia, 100°F, to 50 psia, 300°F. Determine the entropy change given $C_p = C_{p0}(1 + aT)$, where a is constant.

2.39 Air is compressed from 1 bar, 27°C, to 3.5 bars, 127°C. Determine the entropy change given $C_p = C_{p0}$.

Pressure, Speed of Sound

2.40 In Prob. 2.28, what is the pressure and speed of sound at 5000-ft altitude?

2.41 Consider pressure forces acting on an elemental prism volume of an incompressible fluid at rest. Prove that the pressure in any horizontal plane through the prism is constant, but through any vertical plane is not. (Hint: Use a free-body diagram and the static equation $\Sigma F_x = 0$; $\Sigma F_y = 0$; $\Sigma F_z = 0$. Refer to Fig. P2.41.)

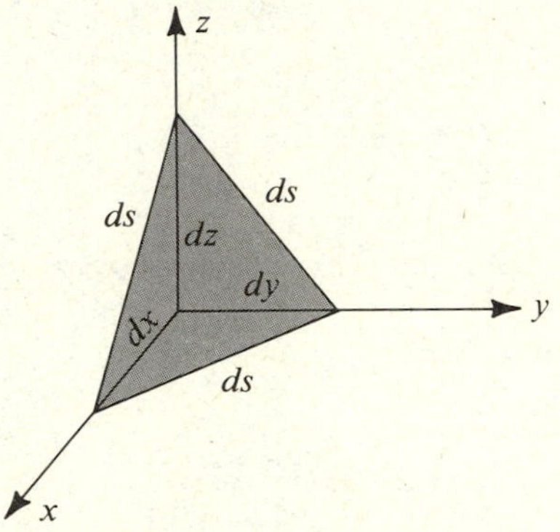

Figure P2.41

2.42 From Prob. 2.41, show that the vertical pressure force acting on the triangular face of area ½ $dx\ dy$ is equal to the weight of a fluid occupied in the prism volume.

2.43 Find the normal momentum flux on a water element of 0.0001 cm^2 area if its density is 0.1 times the atmospheric density ρ_0.

2.44 Show that γh, ρV^2, p_{xy}, F_τ/A, $M\mathbf{a}/A$, $\mu\nabla\cdot\mathbf{V}$ all have units of pressure. Which are not pressures? If $p = \frac{1}{2}\rho V^2$, show that the units of density are slug/ft^3 if the velocity is ft/s and the pressure is psf. Show that the units of pressure can also be expressed as energy per volume of fluid.

2.45 An automobile tire is pressurized with air at 80°F to 30 psi, as illustrated in Fig. P2.45. What is the pressure in the tire if it is (a) heated to 150°F, (b) cooled to 0°F, providing the volume of the tire does not change?

Figure P2.45

2.46 Determine the pressure equivalent of 1 bar in terms of (a) a vertical column of water in feet and in meters, (b) a vertical column of mercury in meters, (c) a vertical column of alcohol of specific gravity 0.6 in. meters.
2.47 If the barometric pressure is 31.2 in. Hg., convert (a) 28 psia to psi, (b) 15 psia to in. Hg. vacuum, (c) 25 in. Hg. gauge to psia.
2.48 An underwater evacuation system is evacuated until the gauge pressure indicates a system pressure of 22.6 in. Hg. If the local barometric pressure is 30.3 in. Hg., what is the absolute pressure within the tank (psia)?
2.49 5 kg of hydrogen gas at 7°C is confined in a diving bell of inner volume 2400 m^3. What is the pressure inside the diving bell?
2.50 Air at pressure 6.0×10^5 N/m^2 expands isentropically from 38°C to 2°C. (a) What is the final pressure? (b) What are the initial and final densities?
2.51 The pressure field in a step bearing is

$$p = \frac{6\mu Q}{\pi h_0^3} \ln \frac{R}{r}$$

where Q, h_0, and R are constants. Is the field one-, two-, or three-dimensional? Is it steady or unsteady?
2.52 The pressure field of an incompressible viscous core vortex is

$$p = -\frac{\rho\Gamma^2 a}{16\pi^2\nu} \cdot \frac{[1 - \exp(-x)]^2}{x}[1 + b\exp(-ct)]$$

where a, b, c, and Γ are constants. Is the field one-, two-, or three-dimensional? Is it steady or unsteady?
2.53 What is the relationship between p and T of a perfect gas with constant specific heats during an isentropic process?
2.54 A truck tire has an internal volume of 7500 in.^{3} and is inflated to a pressure of 42 psi at sea level and temperature T = 61°F. What is the pressure in the tire when it is driven to Paradise Inn, Washington at 9468 ft? What is the mass of air in the tire?
2.55 Determine the speed of sound in the following fluids at 70°F: water, air, glycerin, and mercury.
2.56 If the Mach number is defined as the ratio of the speed of an object to the local speed of sound, determine the Mach number of an airplane flying 600 mph at 50,000 ft where the temperature is −83°F.
2.57 Derive the following forms for the pressure of a perfect gas at low pressures

(a) $p = \rho RT$

(b) $p = \dfrac{MRT}{\forall}$

(c) $p = \dfrac{NR_uT}{\forall}$

(d) $p = \dfrac{N\hat{M}RT}{\forall}$

(e) $p = \dfrac{MR_uT}{\hat{M}\forall}$

(where N is the number of moles in the volume $\forall$).

Bulk Modulus of Elasticity

2.58 What are the density, specific weight, specific volume, specific heat ratios C_{p0} and C_{v0}, bulk modulus of elasticity, and velocity of sound of air at sea level for a temperature of 60°F?
2.59 A high-speed projectile enters water with a pressure at its nose of 500 psi. If the volume it displaces is 1 ft^3, find the change in volume for the water at 60°F.

2.60 Calculate the temperature of water if the following test data are obtained: at a pressure of 14.7 psia, the volume of fluid in the experiment was 1 ft^3; when the same fluid was subjected to a pressure of 10,000 psia, the volume was 0.9687 ft^3.

2.61 Suppose the bulk modulus of elasticity for a certain liquid is

$$K = 7\bar{v}(p + 3000)$$

where p is in atmospheres and $\bar{v}$ has units of cm^3/g. Determine the equation of state of this liquid, and compare it with the result given by Eq. (2.19).

2.62 Transform the expression for bulk modulus of elasticity in terms of pressure and density.

2.63 The equation of state of water is approximately

$$p = 3001\left(\frac{\rho}{\rho_0}\right)^7 - 3000$$

What is the bulk modulus of elasticity if the water temperature is 20°C and the pressure is 25 bars?

2.64 Show that the bulk modulus of elasticity of a perfect gas during an isentropic process is $K = kp$.

2.65 Show that the bulk modulus of elasticity of a perfect gas during an isothermal process is $K = p$.

2.66 If a fluid's bulk modulus of elasticity is linear with temperature, show that $\rho_2 \propto \rho_1 \exp(\Delta p/T)$.

2.67 What is the bulk modulus of elasticity of a liquid with a volume increase of 0.01 percent for a pressure increase of 100 kPa?

2.68 What is the bulk modulus of elasticity of (a) kerosene, (b) mercury, and (c) benzene at 1 atm and 1000 atm?

2.69 Using the definition of the work per unit mass of a simple compressible substance

$${}_1w_2 = -\int p \, d\bar{v}$$

and the bulk modulus of elasticity, show that the work necessary to compress a liquid isothermally is

$${}_1w_2 = \frac{\bar{v}}{2K}(p_2^2 - p_1^2)$$

2.70 Compute the work per unit mass to compress petroleum oils isothermally at room temperature from 1 to 100 bars.

2.71 Estimate the work (kJ) to compress 10 kg of water at 20°C and 1 bar isothermally to 1000 bars.

2.72 Nine kilometers below the surface of the ocean is where the pressure is 91.91 MN/m^2. Determine the specific weight of seawater at this depth if the specific weight at the surface is 10 kN/m^3 and the average bulk modulus of elasticity is 2.34 GN/m^2.

Viscosity

2.73 Calculate the conversion constant from $N{\cdot}s/m^2$ to (a) g/cm·s, (b) lbm/ft·s, (c) lbf·s/ft^2, (d) slug/ft·s, (e) dyne·s/cm^2, and (f) centipoise.

2.74 Calculate the conversion constant from m^2/s to (a) stoke, (b) ft^2/s.

2.75 What are the dynamic and kinematic viscosities of air, water, and alcohol at 80°F?

2.76 What is the kinematic viscosity of hydrogen at 1 atm and 150°F in stokes?

2.77 In the expression for the dynamic viscosity of a gas, Eq. (2.24), if $d_0 = 3 \times 10^{-8}$ cm, show using Fig. 2.12a that average velocity of the molecular motion is directly proportional to the square root of the temperature.

2.78 Verify Eq. (2.23) gives values shown in Table 2.2.

2.79 Obtain an expression for the kinematic viscosity in terms of the absolute viscosity, specific gravity, and the density of water.

2.80 What is the dynamic viscosity of fresh water at atmospheric pressure for a temperature of (a) 0°C, (b) 50°F, (c) 98.8°F, and (d) 100°C?

2.81 What is viscosity of steam at a pressure of 10,000 psia, and temperature of 1350°F?

2.82 Compare the values of the dynamic viscosity of fresh water at normal pressures in Table 2.2 with the results in Fig. 2.12a.

2.83 What is the viscosity range of 40 SAE crankcase oil at 210°F?

2.84 Figure P2.84 shows a hydraulic piston 100 mm in diameter and 1.1 mm long moving

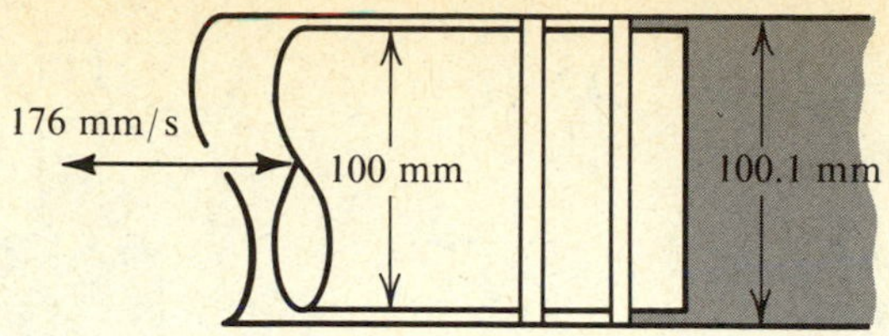

Figure P2.84

within a cylinder 100.1 mm in diameter. The annular region is filled with oil, $S = 0.85$, with kinematic viscosity 386 mm²/s. What is the viscous force F_τ resisting the motion when the piston moves at 176 mm/s?

2.85 A film of oil 0.17 mm thick separates two thin disks, each 170 mm in diameter and mounted co-axially. Calculate the moment necessary to rotate one disk relative to the other at a speed of 17 rpm if the oil has a viscosity of 0.17 N·s/m².

2.86 The volume between two flat parallel walls 17 mm apart is filled with a liquid of viscosity 0.67 N·s/m². Within the volume a flat piece of shim-stock 20 mm × 20 mm is towed at a velocity of 8 mm/s at a distance of 8 mm from one of the walls, the shim-stock and its movements being parallel to the wall. If a linear variation of velocity exists between the shim-stock and the walls, find the force F_τ exerted by the liquid on the shim-stock. (Refer to Fig. P2.86.)

Surface Tension

2.87 It is known that the smaller a droplet of water, the greater must be its inner pressure. Refer to Fig. P2.87 and derive the expression

$$\sigma = \frac{Dp}{4}$$

2.88 Using the sketch in Prob. 2.87, calculate the sphere's surface tension force given that the pressure inside the droplet is 0.1 greater than the pressure surrounding the droplet at a diameter 0.01 ft.

2.89 If the surface tension of a liquid is found to be related to the length l of its minimum surface by

$$\sigma = \sigma_0 \exp(-al)$$

find the cohesive force F.

2.90 What is the surface tension of water in contact with air at 59°F in (a) dynes/cm, (b) N/m, and (c) lbf/ft?

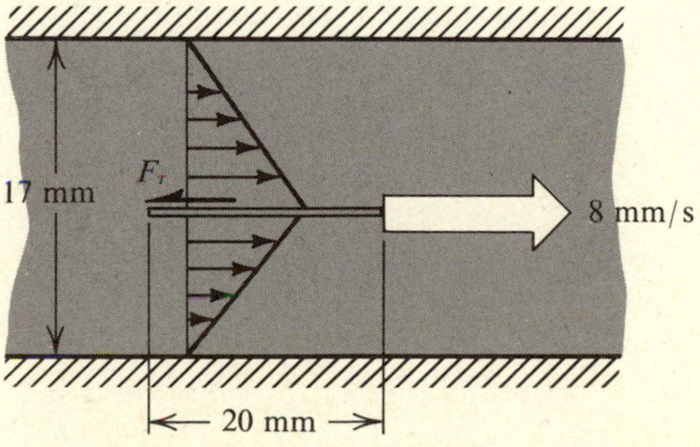

Figure P2.86

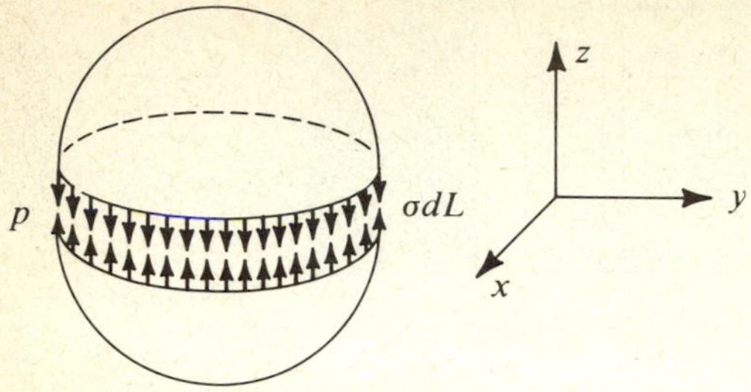

Figure P2.87

2.91 It has been suggested that there is a linear relationship among surface tension, density, and temperature. Investigate whether $(\sigma/\rho^{2/3})$ is linear in temperature for pure water in contact with air between 0 to 50°C.

2.92 A clean glass tube is placed vertically in water at 70°F. How much of the tube should be out of the water to prevent the water from reaching the top if the tube diameter is (a) 0.005 in., (b) 0.01 in., and (c) 0.1 in.? Express answers in centimeters.

2.93 Using Eq. (2.30), show that the capillary rise in a small tube can be approximated by

$$h = \frac{4\sigma}{\gamma D}$$

where γ is the specific weight of the tube in the capillary above which is air, and D is the tube diameter. What is the height water at 15°C will rise in a 2 cm diameter clean tube?

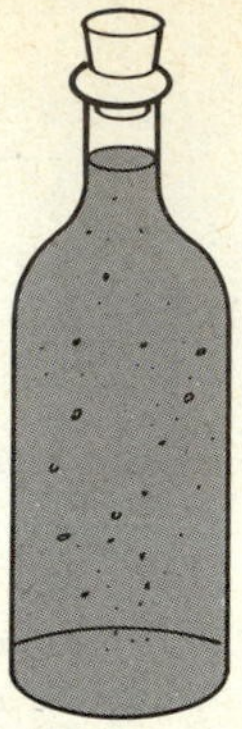

Figure P2.94

2.94 When a bottle of champagne is opened, gas bubbles quickly form due to a sudden reduction in pressure (Fig. P2.94). Assuming the wine is water at 34°F, what is the pressure difference between the inside and outside of a bubble if the bubble's diameter is 0.001 in.?

2.95 How much does the pressure in a cylindrical jet of water 3.2 mm in diameter exceed the pressure of the surrounding atmosphere, if the surface tension of pure water is based on 20°C at standard atmospheric pressure?

2.96 What is the approximate capillary rise of water at 20°C and standard atmospheric pressure in contact with air in a clean glass tube 10 mm in diameter?

2.97 What is the pressure increase over the atmospheric pressure in a drop of water 0.002 in. in diameter? What is the pressure excess for a spherical air bubble 0.002 in. diameter moving in water? Express answer in pascals.

3 Aerohydrostatics

3.1 Hydrostatics

The word *statics* comes from the Greek word *statikos*, meaning "causing to stand still." The word *equilibrium* has its roots in Latin meaning "equal balance." Equilibrium is a condition in which "a balance of all the external forces acting on the control volume" exists. In this chapter we are interested in the static equilibrium of fluids. The fluid may be either at rest or have constant velocity, so that if an observer were moving with it, the fluid would appear stationary.

Not only must the sum of all external forces acting on the fluid control volume be zero for static equilibrium, but the sum of all moments of these forces about an arbitrary space point O must also be zero. Mathematically, the statement of static equilibrium is given by Newton's familiar second law:

$$\Sigma \mathbf{F} = 0 \tag{3.1}$$

$$\Sigma \mathbf{M}_o = 0 \tag{3.2}$$

where the inertial force $M\mathbf{a}$ and moment $I\boldsymbol{\alpha}$ are nonexistent.

Let us consider a body at rest. Figure 3.1a shows a body of weight W acting at the center of mass G being directed down through the table. The table, being external to the body, acts *on* the body at the body's lower surface A by exerting a continuous pressure p over the entire area in such a manner

$$\int_A p dA = W \tag{3.3}$$

so that the body will not pass through the table. Notice that if $\int pdA$ is greater or smaller than W the body will move, and we will not have the static condition we desire.

We know that if the distribution of p is known, that is $p = p(A)$, we can evaluate the integral, calling it R. Since the body is homogeneous and the body is symmetrical, $R = pA$, such that $p = W/A = R/A$. This states that the pressure is the weight force per unit area, which is precisely our definition of pressure given in the preceding chapter.

Let us now go one step further. Suppose we consider the body to be an element of fluid with volume $dxdydz$ and weight W, where $W = \gamma dxdydz$. The elemental volume is surrounded by fluid. The external fluid will exert a pressure force on each

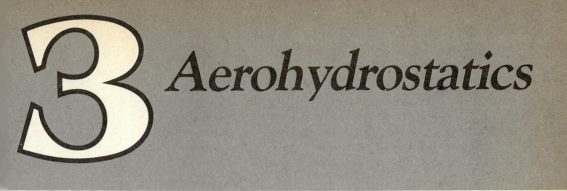

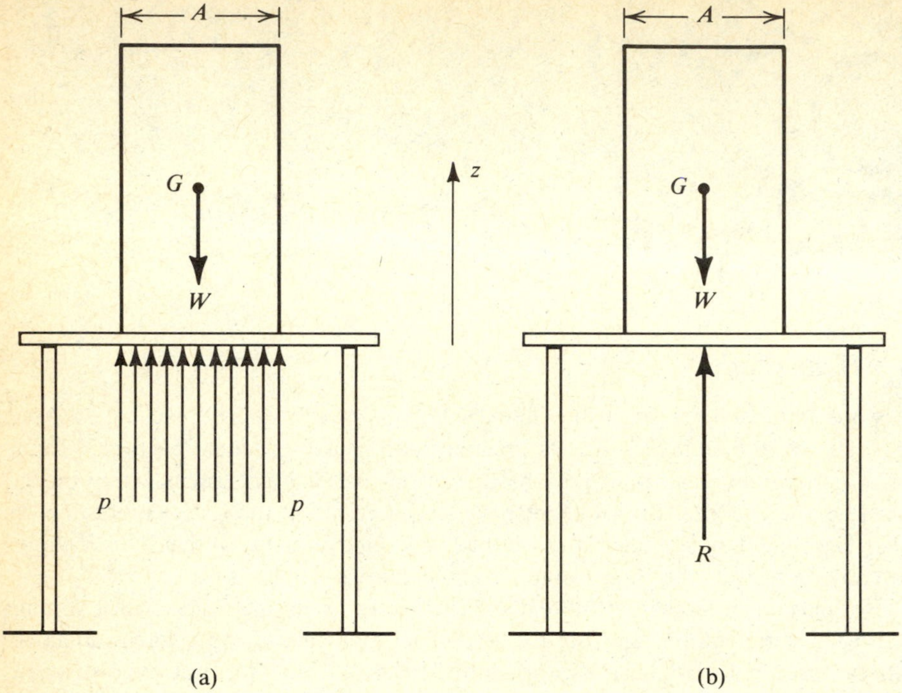

Figure 3.1 (a) $\Sigma F_z = 0$, $\int p dA - W = 0$. (b) $\Sigma F_z = 0$, $R - W = 0$.

of the element's six faces. In addition, we have to assume that each of these six pressure forces may be different, as shown in Fig. 3.2.

We use the symbol $\partial/\partial s$ to represent a change with respect to a change in the direction s. Thus $\partial p/\partial x$ means the pressure changes *only* in the x-direction. The term $\partial p/\partial x\ (dx)$ signifies that the pressure can change *only* in the x-direction through a distance dx. If the pressure force varies in three directions, we have three possible variations, shown in Fig. 3.2. Note that the z-axis is taken along the vertical.

Using Newton's second law for statics, Eq. (3.1), we obtain the vector differential equation of fluid statics:

$$\boxed{\frac{1}{\rho} \nabla p = \mathbf{g}} \tag{3.4}$$

Few simpler first degree linear partial differential equations than this exist. Assuming the gravitational potential for the earth as a homogeneous sphere, we obtain the scalar differential equation

$$\frac{\partial p}{\partial x} = \frac{\partial p}{\partial y} = 0 \tag{3.5}$$

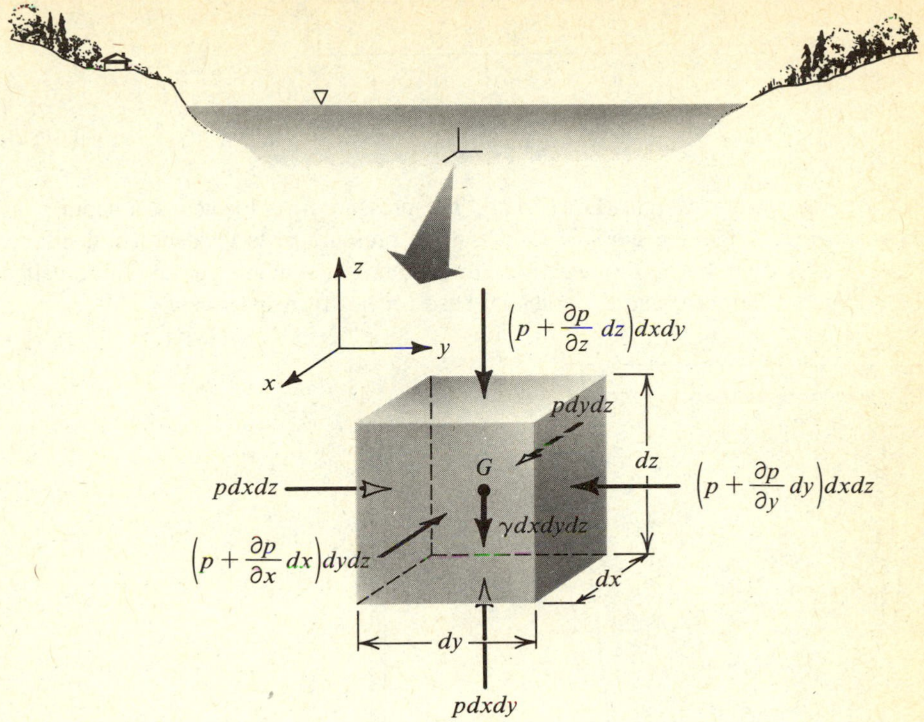

Figure 3.2 *Forces on a fluid element at rest or with constant velocity.*

and

$$\frac{1}{\rho}\frac{\partial p}{\partial z} = -g \tag{3.6}$$

What Eq. (3.5) tells us physically is that no pressure variation can exist in a horizontal plane, provided of course that we have a continuous fluid, one with no discontinuity in its field. Equation (3.6) says that all vertical pressure differences are constant for equal intervals of elevation, or depth. Since the pressure cannot vary in a horizontal plane, Eq. (3.6) is transformed to

$$\frac{dp}{dz} = -\rho g \tag{3.7}$$

and represents the scalar ordinary differential equation of fluid statics. The solution of Eq. (3.7) depends upon how the density is related to pressure p and/or height z. This brings us to the subjects of hydrostatics and aerostatics.

Hydrostatics is the science of the static equilibrium of *incompressible* fluids (one for which the density is constant, such as water). For incompressible fluids, a *definite* integration of Eq. (3.7) results in

$$p_2 - p_1 = \Delta p = \gamma h \tag{3.8}$$

where

$$h = z_1 - z_2 \tag{3.9}$$

and γ is the specific weight [Eq. (2.17)]. The pressure p_2 is located at a depth z_2 as shown in Fig. 3.3 in the glass of fluid and the pressure p_1 is located at a depth z_1. Thus, h is positive but in the *negative* z direction. The symbols Δp and h are called the pressure difference, and the potential head (or head), respectively.

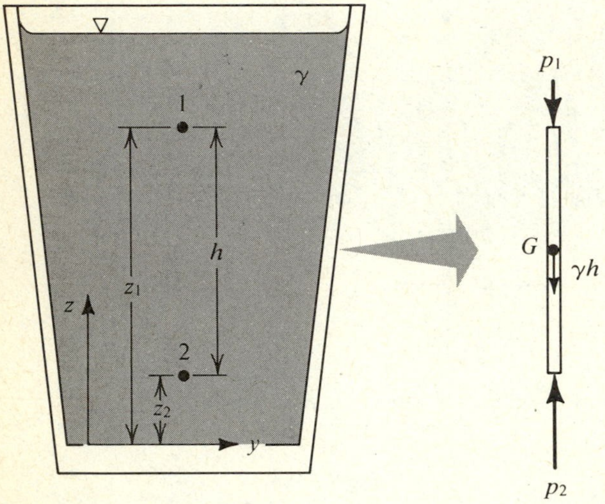

Figure 3.3 *Geometry for pressure difference Δp.*

From Eq. (3.8), or using Fig. 3.3, we see that the pressure at station 2 is equal to the pressure at station 1 plus the weight per unit area of the fluid between the two stations in the vertical z direction. This appears to be intuitively obvious, but it's nice to prove it.

An *indefinite* integration of the hydrostatic differential equation, Eq. (3.7), yields

$$p = -\gamma z + \text{const.} \tag{3.10}$$

which states that the pressure is distributed hydrostatically by increasing linearly with decreasing height.

The *limitations* on applying Eqs. (3.8) or (3.10) are that

1. The fluid is *homogeneous*, (meaning that it is of uniform chemical composition throughout), and *isotropic* (meaning that the properties have the same values when measured along axes in all directions at a point).
2. The fluid is *continuous* between any two stations.

Examining our differential equations of fluid statics, Eqs. (3.5) and (3.7), we note that geometric boundaries play a primary role in evaluating the pressure. For fluids satisfying the above limitations, all we need to solve for *any* problem in hydrostatics is one given value of pressure at a boundary or point in the fluid. This is illustrated in Fig. 3.4. If the specific weight of the fluid is the same for case (a), (b), and (c), and there are no obstructions in the manometer (c), then p_1 equals p_2 from Eq. (3.5), and p_3 equals p_1 plus γh from Eq. (3.7) for all three cases shown in Fig. 3.4. Hence, the somewhat complicated problem of case (c) is the same static situation as the very simple problem of case (a).

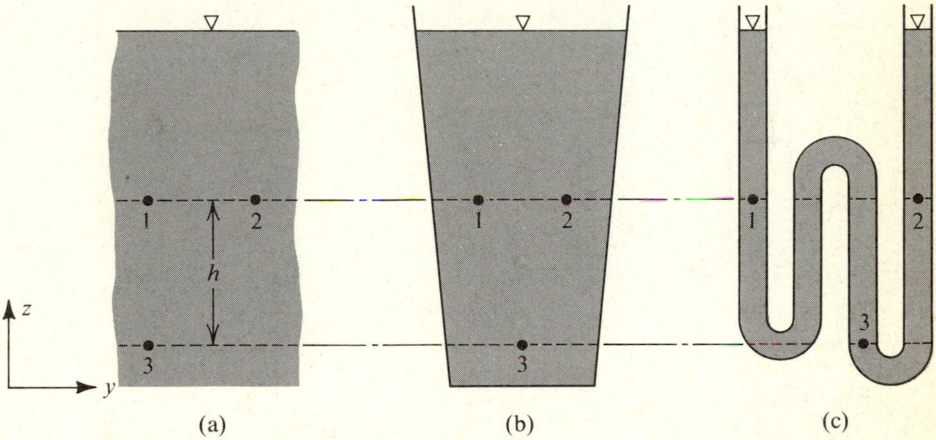

Figure 3.4 *Illustration of pressure variation.* $p_{1_a} = p_{2_a} = p_{1_b} = p_{1_c} = p_{2_c} \cdot p_{3_a} = p_{3_b} = p_{3c}.$

We can measure pressure by many methods and with many different apparatuses. They differ depending upon the range of pressure, the type of fluid being used, and the accuracy desired. We shall discuss some of the simpler devices.

3.1.1 Manometers

A manometer is a primary standard, meaning that we seldom have to calibrate it. We can calculate the pressure in a manometer directly if we know the specific gravity of the manometer fluid. Small errors can result from capillarity, unclean manometer glassware, or contaminants in the fluid, but these errors can be minimized with careful practice.

A manometer has its limitations. The range of pressure it can measure is necessarily limited by the height the manometer fluid can rise. It is not susceptible to electronic recording or to direct control of processes, and very small deflections of the manometer fluid in the capillary are difficult to read accurately (although accuracy can be improved by using slanted manometers).

A manometer measures the pressure in terms of the unbalanced height of the manometer fluid. The pressure to be measured is applied to the surface of the liquid, creating a force which must be balanced from the other side by the weight of the liquid column, as shown in Fig. 3.5. If the fluid in the reservoir is a gas at a pressure p, then from Eq. (3.8)

$$p = \gamma h \tag{3.11}$$

where γ is the specific weight of the manometer fluid.

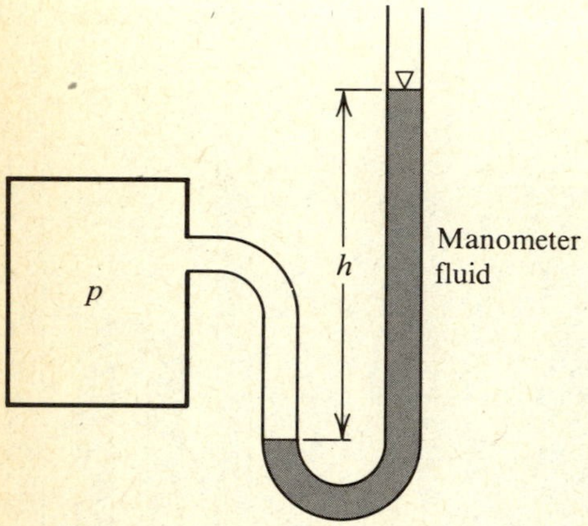

Figure 3.5 *A simple manometer.*

3.1.2 Barometer

A commercial mercury barometer of the well type is shown in Fig. 3.6. Stored in a reservoir open to the atmosphere A, the mercury rises in a sealed glass tube which has a vernier for accurate reading of the mercury column. The mercury level in the reservoir has to be adjusted to local changes in the barometric pressure before taking a reading.

3.1.3 U-Tube Manometer

A U-tube manometer indicates the pressure relative to the atmosphere. The manometer fluid may be colored water, alcohol, mercury, or any other liquid whose specific gravity is known. The manometer is attached to various apparatuses in which the fluid is moving, such as a water tunnel, wind tunnel, and pipe flow. It measures the difference in pressure, usually in units of millimeters or inches of that manometer fluid.

3.1.4 Inclined Manometer

We can often use an inclined manometer to increase our accuracy in reading a manometer. We incline one leg of the manometer at some small angle α with respect to the horizontal so that the fluid displacement is greater by approximately $1/\sin \alpha$. The other leg of the manometer is often the well.

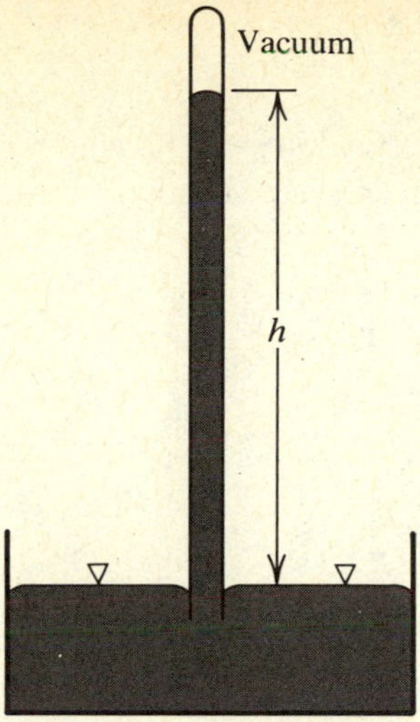

Figure 3.6 A barometer.

The zero setting of the manometer is important. Usually the scale is calibrated to account for the change in head of the vertical leg. The scale of some manometers shows directly the pressure in inches of water or millimeters of mercury, taking into account the specific gravity of the manometer fluid. Because of the high sensitivity required in this manometer, we shall use the lightest manometer fluid available. Example 3.1 illustrates an inclined manometer.

Example 3.1

The inclined manometer shown in Fig. E3.1 uses alcohol of specific weight $\gamma = 50$ lbf/ft^3. Determine the difference in pressure between points A and B.

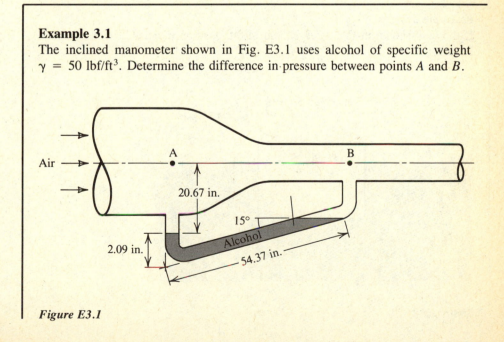

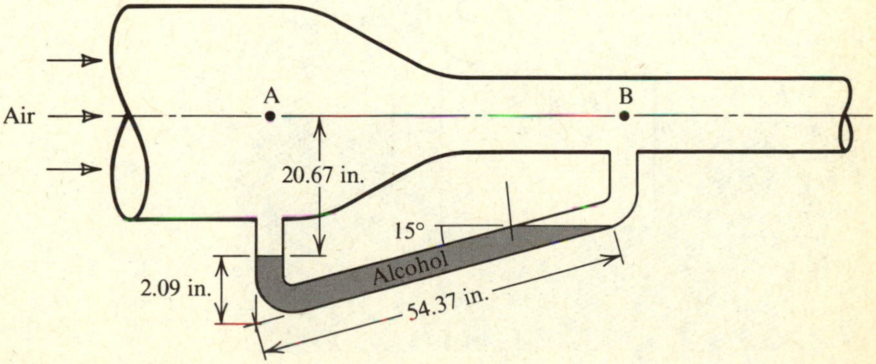

Figure E3.1

Solution:

Step 1.

Identify the characteristics of the fluid and flow field.

Although the fluid is flowing through the converging tunnel, it is not flowing through the inclined manometer, creating, therefore, a hydrostatic situation.

Step 2.

Write the appropriate form of the governing equation of flow.

From Eq. (3.8),

$$\Delta p = \gamma h$$

such that

$$p_A + \gamma_{\text{air}}\left(\frac{20.67}{12}\right) + \gamma_{\text{alch.}}\left(\frac{2.09}{12}\right) - \gamma_{\text{alch.}}\frac{54.37}{12}\sin 15^\circ - \gamma_{\text{air}}\left(\frac{22.76}{12} - \frac{54.37}{12}\sin 15^\circ\right) = p_B \tag{i}$$

Since $\gamma_{\text{air}} << \gamma_{\text{alch.}}$, one obtains

$$p_A - p_B = 0.347 \text{ psi} \tag{ii}$$

This completes the solution.

Example 3.2

If the tank shown in Fig. E3.2 is held in the position indicated, determine the length h.

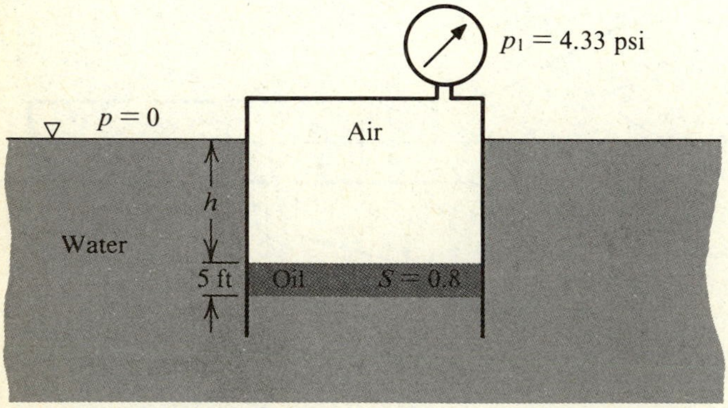

Figure E3.2

Solution:

Step 1.

Identify the characteristics of the fluid and flow field.

Since there is no motion, the situation is static.

Step 2.

Write the appropriate forms of the governing equation of flow.

The hydrostatic equation is

$$\Delta p = \gamma h \tag{i}$$

Neglecting the weight of air, one obtains

$$4.33 + 0.8 \times \frac{62.4 \times 5}{144} = \frac{62.4\,(h + 5)}{144} \tag{ii}$$

Solving for h in Eq. (ii)

$$h = \frac{4.33 \times 144}{62.4} + 4 - 5$$

or

$$h = 9 \text{ ft} \tag{iii}$$

This completes the solution.

3.2 Uniform Acceleration

We have uniform linear acceleration when the acceleration **a** has the same magnitude and direction for all particles. What does this mean? Let us examine our vector differential equation of statics: Eq. (3.4). What we are stating in Eq. (3.4) is that the pressure acceleration $(\nabla p)/\rho$ is not only equal to the gravitational acceleration **g**, but also that it must be constant. Thus, if we add another acceleration to the pressure acceleration, we will still have a static situation provided that the acceleration we add is also constant. Denoting **a** as the uniform acceleration, the vector differential equation of statics, Eq. (3.4), is transformed to the vector differential equation of quasistatics:

$$\frac{\nabla p}{\rho} = \mathbf{g} - \mathbf{a} \tag{3.12}$$

where **a** is constant.

An example of uniform acceleration is the centrifuge shown in Fig. 3.7. A fluid in a cylindrical container rotates at a constant angular speed ω. We notice that the free surface of the liquid is no longer horizontal. We need to find an expression for this free surface.

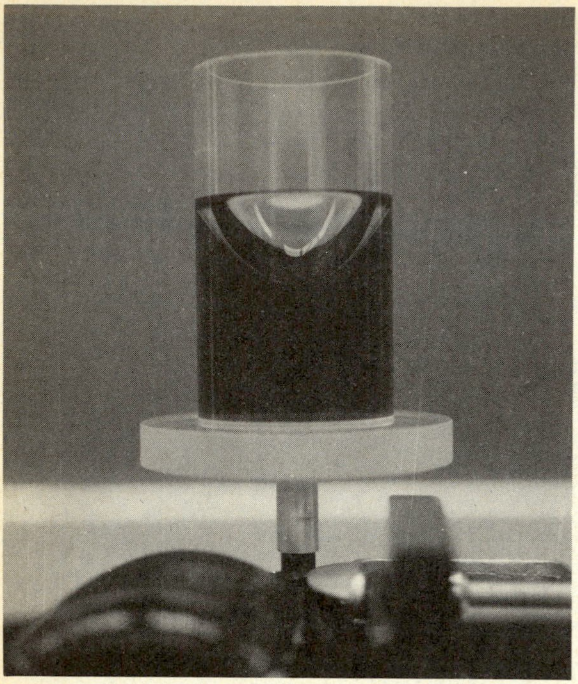

Figure 3.7 The free-surface shape of a centrifuge.

We first assume that the velocity field of the fluid in the cylinder consists solely of a swirl velocity v_θ:

$$\mathbf{V} = v_\theta \, \mathbf{e}_\theta \tag{3.13}$$

The outer boundary of the cylinder rotates with constant speed ω_z, the same speed as the turntable upon which the cylinder sits. The fluid rotates as solid body rotation:

$$v_\theta = \omega_z r \tag{3.14}$$

The acceleration of Eq. (3.12) is the normal acceleration, or radial acceleration, equal to the swirl velocity squared divided by the radius of curvature (or radius r). Transforming the vector differential equation of quasistatics to scalar forms yields

$$\frac{\partial p}{\partial z} = -\rho g \tag{3.15}$$

$$\frac{\partial p}{\partial r} = \rho \omega_z^2 r \tag{3.16}$$

Integrating Eq. (3.16) results in

$$p = \frac{\rho\omega_z^2 r^2}{2} + g(z) \tag{3.17}$$

To evaluate $g(z)$, let $r = 0$ be at the intersection of the centerline with the free-surface, where the pressure p is zero gauge pressure. Thus the pressure becomes

$$p = \frac{\rho\omega_z^2 r^2}{2} \tag{3.18}$$

where p is measured from station 1 of Fig. 3.8. Essentially we absorbed the constant of integration into the dependent variable.

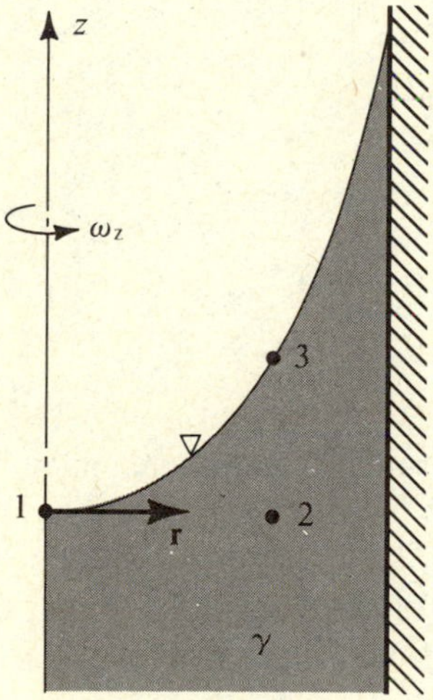

Figure 3.8 *Uniform acceleration.*

The free-surface shape z is found by integrating the second pressure expression, Eq. (3.15), from station 3 to station 2 of Fig. 3.8:

$$p_2 = \rho g z \tag{3.19}$$

The free-surface shape z is then obtained by substituting Eq. (3.19) into Eq. (3.18):

$$z = \frac{\omega_z^2 r^2}{2g} \tag{3.20}$$

which we recognize as the equation for a parabola. Thus, the free-surface shape of centrifuges, such as the familiar cream separator and a swirling cup of coffee, are parabolic surfaces like that shown in Fig. 3.7.

Example 3.3

The container shown in Fig. E3.3a is half full of water and open at the top. (a) What is the maximum acceleration in the y-direction without spilling water? (b) What is the pressure at point A when $a_z = 20$ ft/s^2 and $a_y = 0$?

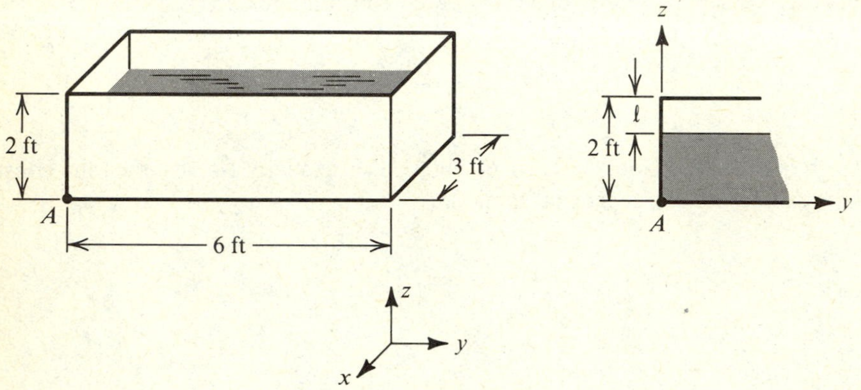

Figure E3.3a

Solution:

Step 1.

Identify the characteristics of the fluid and flow field.

We shall view this problem as quasihydrostatic, assuming that the maximum acceleration a_y and a_z are constants.

Step 2.

Write the appropriate form of the governing equation of flow.

The appropriate equation is the quasihydrostatic vector Eq. (3.12):

$$\frac{\nabla p}{\rho} = \mathbf{g} - \mathbf{a} \tag{i}$$

(a) The y-component quasistatic equation is

$$g \tan \theta = a_y \tag{ii}$$

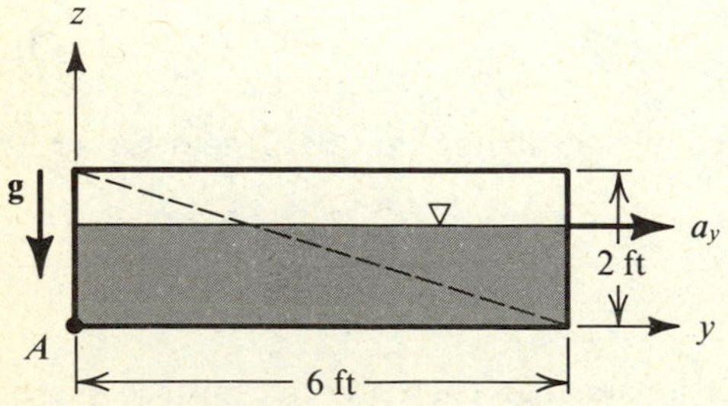

Figure E3.3b

Example 3.3 *(Con't.)*

From geometry,

$$g(2/6) = (a_y)_{\max}$$

or

$$(a_y)_{\max} = 10.72 \text{ ft/s}^2 \qquad \text{(iii)}$$

(b) From Eq. (3.12), the z-component quasistatic equation is

$$\frac{dp}{dz} = \gamma(1 + a_z/g) \qquad \text{(iv)}$$

Integrating Eq. (iv), where z goes from the free-surface location ($z = 0$) to the top of the container ($z = 1$), yields

$$\Delta p = \gamma l\,(1 + a_z/g) \qquad \text{(v)}$$

Thus the pressure at A is

$$p_A = 62.4 \times 1\left(1 + \frac{20}{32.2}\right)$$

or

$$p_A = 101.2 \text{ psf} \qquad \text{(vi)}$$

This completes the solution.

3.3 Aerostatics

Aerostatics is particularly valuable to meteorologists, because it allows them to make accurate predictions of the pressures and temperatures at different elevations. Aerostatics differs from hydrostatics in that the specific weight γ and/or density ρ is no longer considered constant. Since the density is a variable in air, or in any gas, the pressure in the atmosphere cannot be evaluated from the static equilibrium differential equation, Eq. (3.15), until the functional relationship between density and pressure is specified. Specific illustrations follow:

3.3.1 Halley's Law

The equation of state of air is governed by the perfect gas law relating density to pressure as

$$\rho = \frac{p}{RT} \qquad (3.21)$$

where the absolute temperature T is a prescribed function of altitude z. The density ρ of Eq. (3.21) is substituted into the static equilibrium Eq. (3.7) to yield

$$\frac{dp}{dz} = -\frac{pg}{RT(z)} \tag{3.22}$$

Equation (3.22) is integrable if an analytic expression of the temperature field is known. Let us assume it is isothermal at say the sea level temperature T_o. Integration of Eq. (3.22) then yields

$$p = p_o \exp\left(\frac{-gz}{RT_o}\right) \tag{3.23}$$

where p_o is the pressure at the earth's surface ($z = 0$). Equation (3.23) is called Halley's law and provides fairly accurate results to real problems where the pressure differences are approximately one atmosphere.

Example 3.4

Calculate the height z beneath which half of the earth's atmospheric weight is found, given that the temperature of the atmosphere is constant ($T_o = 518.7°R$). Assume that air is a perfect gas.

Solution:

Step 1.

Identify the characteristics of the fluid and flow field.

The fluid is compressible, and the flow is static.

Step 2.

Write the appropriate form of the governing equation of flow.

The local weight W of the earth's atmosphere is expressed as

$$W = \int_0^\infty p dz \tag{i}$$

where the pressure is given by Halley's law

$$p = p_o \exp(-gz/RT_o) \tag{3.23}$$

Substituting Eq. (3.23) into Eq. (i) and integrating results in

$$W = \frac{RT_o p_o}{g} \tag{ii}$$

But we are interested in the height z where one-half of the earth's atmospheric weight exists. Thus

$$\frac{W}{2} = \frac{RT_o p_o}{2g} = \int_0^z p dz \tag{iii}$$

From Eqs. (3.23) and (iii), we obtain

Example 3.4 *(Con't.)*

$$\frac{RT_op_o}{2g} = \int_0^z p_o \exp\left(\frac{-gz}{RT_o}\right)dz$$

$$= \frac{-RT_op_o}{g}\left[\exp\left(\frac{-gz}{RT_o}\right) - 1\right] \quad \text{(iv)}$$

or

$$\frac{1}{2} = 1 - \exp\left(-gz/RT_o\right) \quad \text{(v)}$$

Substituting the values $T_o = 518.7°R$, $g = 32.2$ ft/s^2, $R = 1718$ ft^2/(s$^2\cdot$°R) into Eq. (v), taking the logarithm of both sides, we obtain

$$z = \frac{0.691 \times 1718 \times 518.7}{32.2} = 19{,}100 \text{ ft} \quad \text{(vi)}$$

as the altitude.

This completes the solution.

3.3.2 Logarithmic Law

We can find a more realistic static pressure relationship for the earth's atmosphere than that given by Halley's law if we assume a linear temperature distribution for the earth's atmosphere. In particular, a linear variation of the temperature is valid from sea level to 45,000 ft altitude. Denoting the atmospheric temperature field as

$$T(z) = T_o - \alpha z \quad (3.24)$$

where T_o is the absolute temperature at z equals zero, and α is the *temperature lapse rate,* assumed constant, we find upon substituting Eq. (3.24) into Eq. (3.22) and integrating

$$p = p_o\left(1 - \frac{\alpha z}{T_o}\right)^{g/R\alpha} \quad (3.25)$$

which is called the logarithmic pressure law. A typical value of the temperature lapse rate α for the earth's atmosphere is

$$\alpha = 6.5\text{°C/km} \quad (3.26)$$

For a sea level temperature of 60°F, the normalized pressure distribution p/p_o of Eq. (3.25) is shown plotted in Fig. 3.9.

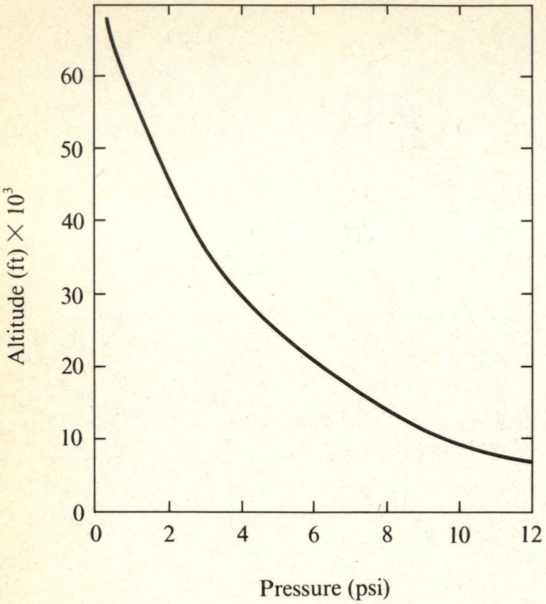

Figure 3.9 *Pressure variation in the atmosphere using the logarithmic law.*

Many other pressure distributions can be found for aerostatics if the equation of state and the thermodynamic process are known. The procedure is very simple, using Eq. (3.7) and integrating once the process is known.

Example 3.5

The process equation relating pressure to density of the earth's atmosphere is found to vary polytropically according to

$$\frac{p}{\rho^k} = \text{const.} \qquad \text{(i)}$$

where k is the polytropic expansion coefficient. For large values of altitude z, the gravitational acceleration g cannot be considered constant but varies inversely as the square of the distance from the center of the earth:

$$\mathbf{g} = -g(z)\mathbf{k} = -\frac{32.2\ \mathbf{k}}{(1 + z/R)^2} \qquad \text{(ii)}$$

where R is the radius of the earth having a value of approximately 4000 miles. Derive the expression for the altitude variation of pressure $p = p(z)$.

Solution:

Step 1.

Identify the characteristics of the fluid and flow field.

The fluid is air, and therefore compressible. The flow is static.

Example 3.5 *(Con't.)*

Step 2.
Write the appropriate form of the governing equation of flow.

Since the fluid is also static, we start with the basic governing Eq. (3.7), where the density can be evaluated from Eq. (i) as

$$\rho = \rho_o(p/p_o)^{1/k} \tag{iii}$$

where ρ_o is the density of air at sea level ($z = 0$). Substituting Eqs. (iii) and (ii) into Eq. (3.7) yields

$$\frac{dp}{dz} = -\frac{32.2\ \rho_o p^{1/k}}{p_o^{1/k}\ (1 + z/R)^2} \tag{iv}$$

Equation (iv) is separable and integrable:

$$\frac{k}{k-1}\left[p^{(k-1)/k} - p_o^{(k-1)/k}\right] = -\frac{32.2\rho_o}{p_o^{1/k}}\left[R - \frac{R}{1 + z/R}\right] \tag{v}$$

or from clearing the above equation, the altitude variation of the pressure becomes

$$p = \left[p_o^{(k-1)/k} - \frac{32.2\ (k-1)\ \rho_o z}{kp_o^{1/k}\ (1 + z/R)}\right]^{k/(k-1)} \tag{vi}$$

Though this pressure distribution is more complex than that given for the logarithmic law, it is no more difficult to obtain since the procedures are identical.

This completes the solution.

3.4 Forces on Planar Bodies

Often we must calculate the force on the wetted surface of a body in a static fluid environment. Let us first consider a body with a planar geometry.

3.4.1 Force on a Planar Body in a Horizontal Plane

Figure 3.10 shows two planar bodies with negligible thickness submerged in a static fluid. Each body has an area **A**, the vector area signifying not only the magnitude but the orientation of the body's surface. To emphasize the orientation of the area, we use

$$\mathbf{A} = \mathbf{e}_n A \tag{3.27}$$

where $\mathbf{e}_n$ is the *outward* unit normal vector. Thus, for the planar body lying in the horizontal *x-y* plane, $\mathbf{e}_n = \mathbf{k}$.

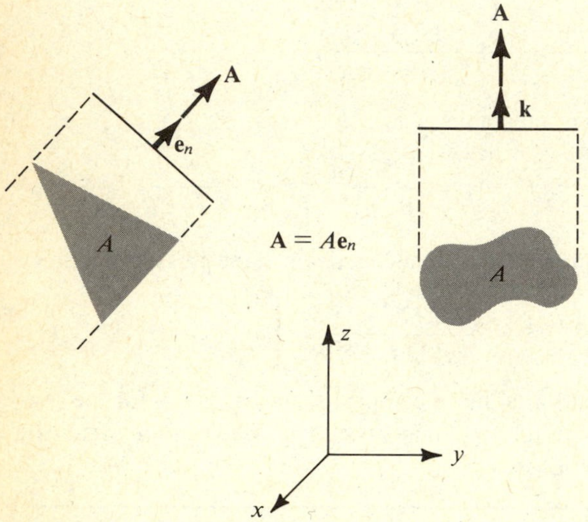

Figure 3.10 *The area vector* **A.**

The pressure force $\mathbf{F}_p$ is a *normal* surface force and is defined as

$$\mathbf{F}_p = -\int_A p d\mathbf{A} = -\mathbf{e}_n \int_A p dA \qquad (3.28)$$

where p is the hydrostatic or aerostatic pressure.

The pressure force should be familiar to the reader. Equation (3.28) states that the static pressure force is the sum of all static pressures over the body, and is in a direction normal to the surface. Substituting the hydrostatic pressure p of Eq. (3.10) into the integrand of the pressure force $\mathbf{F}_p$ of Eq. (3.28) and integrating yields

$$\mathbf{F}_p = -\gamma A h \mathbf{e}_n \qquad (3.29)$$

where h is measured with respect to the horizontal free-surface where the pressure is zero if the free-surface is exposed to the atmosphere at sea level (which is the case throughout this text unless otherwise indicated).

The pressure force $\mathbf{F}_p$ acts through a point C on the body. This point is called the center of pressure, and is shown in Fig. 3.11. The line through which the pressure force acts is called the line-of-action.

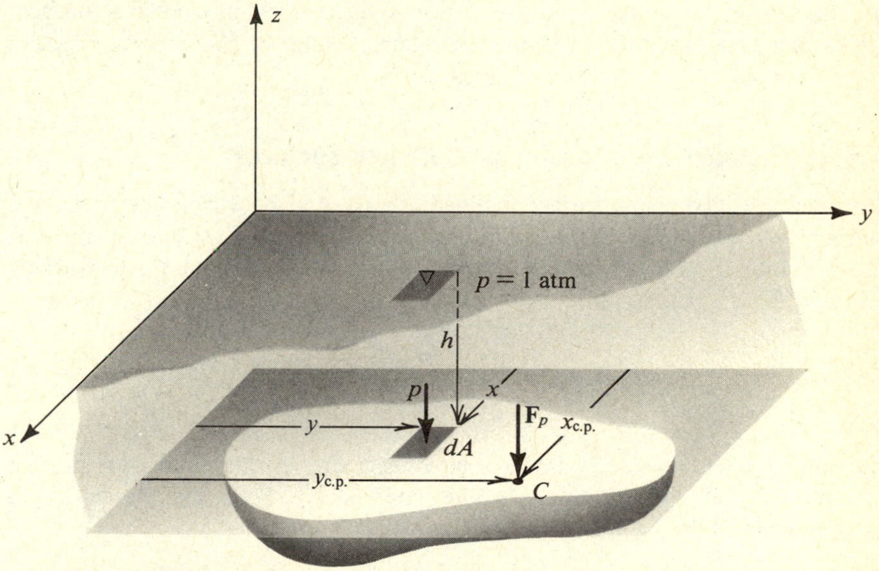

Figure 3.11 *Notation for determining line-of-action of the pressure force.*

To locate the coordinates of the line-of-action ($x_{c.p.}$, $y_{c.p.}$), we apply the second statement of static equilibrium: Eq. (3.2). What we want to state mathematically is the physical fact that the sum of the moments due to each pressure acting on an element dA of the planar body is to be balanced by the moment due to the equivalent pressure force times the moment arm measured in the y-direction from the x-axis. Thus

$$y_{c.p.} F_p = \int_A ypdA \tag{3.30}$$

The above equation can now be used to evaluate $y_{c.p.}$.

Using the relationship for the pressure of Eq. (3.30) (where station 1 is at the free-surface and station 2 is arbitrary) along with the pressure force expression given by Eq. (3.29), we solve for $y_{c.p.}$ in Eq. (3.30):

$$y_{c.p.} = \frac{1}{A}\int_A ydA = \bar{y} \tag{3.31}$$

in which $\bar{y}$ is the distance measured in the y-direction to the *centroid* of the planar surface.

In a similar fashion, we take the sum of moments about the y-axis and obtain

$$x_{\text{c.p.}} = \frac{1}{A}\int_A x\,dA = \bar{x} \tag{3.32}$$

which is the distance measured in the x-direction to the centroid of the planar surface. Thus, for a planar surface lying in a horizontal plane, the line-of-action of the pressure force passes through the centroid of the area.

3.4.2 Pressure Force on Inclined Planar Surfaces

Consider a wetted planar surface inclined at a fixed angle θ from the horizontal free-surface as shown in Fig. 3.14. To find the pressure force, we once again start with the definition given by Eq. (3.28). Substituting the expression for the hydrostatic pressure p of Eq. (3.10) into the integrand of Eq. (3.28), we obtain just as before

$$\mathbf{F}_p = -\mathbf{e}_n\gamma \int_A h\,dA \tag{3.33}$$

where h now becomes a variable, having rotated the Cartesian coordinate system from the horizontal position through an angle θ as shown in Fig. 3.12. The vertical depth h is easily related to the x-y plane through geometry:

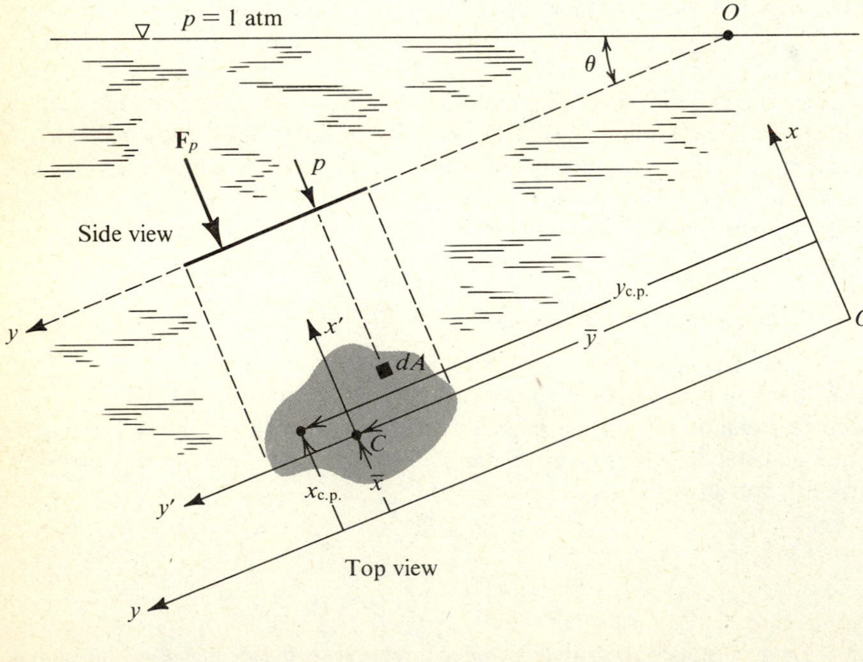

Figure 3.12 *Nomenclature for inclined planar surface.*

$$h = y \sin \theta \tag{3.34}$$

Substituting the expression for h into the integrand of the pressure force of Eq. (3.33) results in the pressure force for any inclined planar surface:

$$\mathbf{F}_p = -\mathbf{e}_n \, \gamma A \bar{y} \sin \theta \tag{3.35a}$$

$$= -\mathbf{e}_n \, \gamma \forall_p \tag{3.35b}$$

A geometric interpretation of the pressure force of Eq. (3.35b) is often extremely useful in calculating the magnitude of the force. The term $A\bar{y} \sin \theta$ of Eq. (3.35a) represents the volume of the pressure prism $\forall_p$ shown in Fig. 3.13.

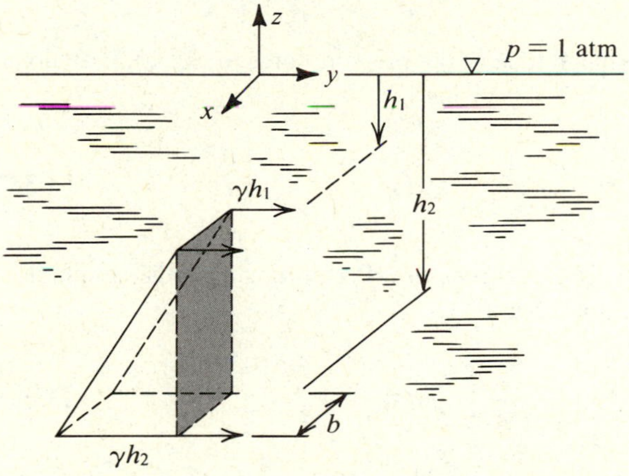

$$\gamma \forall_p = \gamma h_1 b(h_2 - h_1) + \tfrac{1}{2}\gamma(h_2 - h_1)(h_2 - h_1)b$$
$$= \frac{\gamma b}{2}(h_2 - h_1)(h_1 + h_2)$$
$$= \int p \, dA_y = F_{p_y}$$

Figure 3.13 *Geometric interpretation of magnitude of pressure force.*

To simplify the analysis, let us treat the planar body as lying in the vertical plane ($\theta = 90°$), as shown in Fig. 3.13. We know the pressure force $\mathbf{F}_p$ always acts perpendicular to the body. Thus $\mathbf{F}_p = F_{p_y} \mathbf{j}$, where F_{p_y} denotes the pressure force in the y-direction, since the outward normal vector $\mathbf{e}_n$ is in the y-direction.

Using the geometry of the pressure prism of Fig. 3.13,

$$F_{p_y} = \int p dA_y$$

$$= \frac{\gamma b}{2}(h_2 - h_1)(h_1 + h_2)$$

$$F_{p_y} = \gamma h_1 b(h_2 - h_1) + \tfrac{1}{2}\gamma(h_2 - h_1)(h_2 - h_1)b$$

$$= \gamma \forall_p$$

Thus for any inclined planar body

$$F_p = \gamma \forall_p \tag{3.36}$$

We obtain the line-of-action of the pressure force for an inclined planar surface in the same manner as for a planar surface in a horizontal plane. Summing moments about the y-axis gives

$$x_{\text{c.p.}} F_p = \int_A xp\,dA \tag{3.37}$$

Using the expression for the magnitude of the pressure force in Eq. (3.35a) along with the pressure expression of Eq. (3.10) yields

$$x_{\text{c.p.}} = \frac{1}{\bar{y}A} \int_A xy\,dA \tag{3.38}$$

and represents the x-location of the center of pressure. Performing a similar summation of moments about the x-axis, we can obtain

$$y_{\text{c.p.}} = \frac{1}{\bar{y}A} \int_A y^2\,dA \tag{3.39}$$

for the y-location of the center of pressure. The integral of Eq. (3.38) is called the product of inertia I_{xy} of the inclined area taken about the origin 0. Applying the parallel axis theorem, we transfer the product of inertia to the centroid of the inclined area:

$$I_{xy} = I_{x'y'} + A\bar{x}\bar{y} \tag{3.40}$$

where x', y' are axes parallel to x, y, respectively, passing through the centroid in Fig. 3.12. Substituting the product of inertia of Eq. (3.40) for the integral in Eq. (3.38) we obtain

$$x_{\text{c.p.}} = \frac{1}{\bar{y}A}(I_{x'y'} + A\bar{x}\bar{y}) \tag{3.41}$$

as the x-location of the center of pressure for inclined planar bodies.

Next, let us treat the y-location of the center of pressure. The integral in Eq. (3.39) is called the second moment of the area about the x-axis and is denoted by the symbol I_{xx}. Since the moment I_{xx} depends upon the depth of the inclined planar surface below the free-surface, it is convenient to express the second moment of the area about the parallel axis x' rather than the x-axis by use of the parallel axis theorem:

$$I_{xx} = I_{x'x'} + A\bar{y}^2 \tag{3.42}$$

Hence, the y-location of the center of pressure of Eq. (3.39) is easily expressed as

$$y_{\text{c.p.}} = \frac{1}{\bar{y}A}(I_{x'x'} + A\bar{y}^2) \tag{3.43}$$

A table of moments of inertia of some simple plane areas about their centroids is given in Table 3.1.

Table 3.1 Moments of Inertia of Plane Areas About Their Centroids

Rectangle	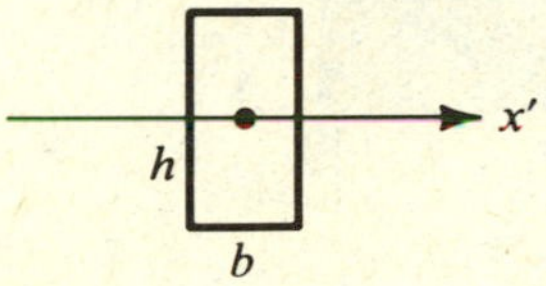	$I_{x'x'} = \dfrac{bh^3}{12}$
Circle	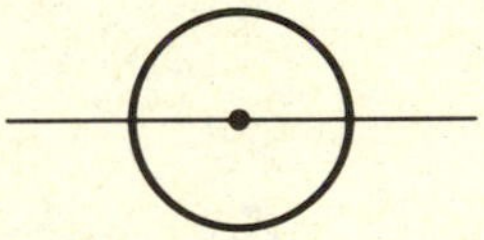	$I_{x'x'} = \dfrac{\pi D^4}{64}$
Triangle	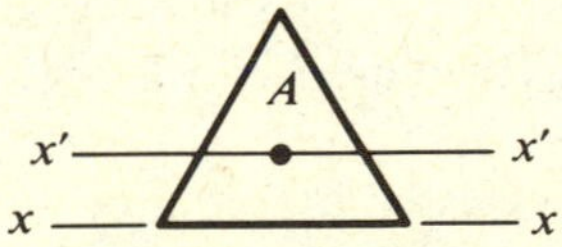	$I_{x'x'} = \dfrac{Ah^2}{18}$ $I_{xx} = \frac{1}{12}bh^3$
Semicircle	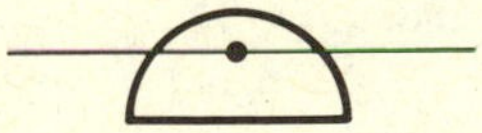	$I_{x'x'} = \dfrac{D^4(9\pi^2 - 64)}{1152\pi}$
Quadrant of a circle	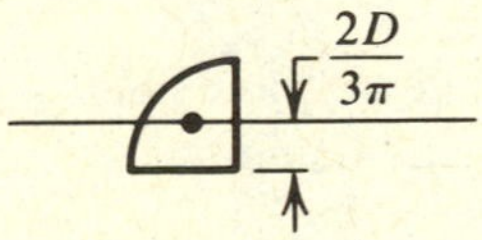	$I_{x'x'} = \dfrac{D^4(9\pi^2 - 64)}{2304\pi}$
Parabola	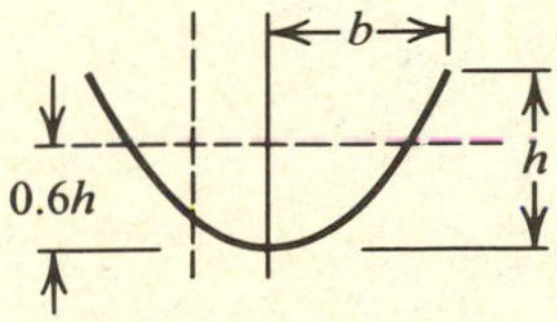	$I_{x'x'} = 0.0686\,Ah^2$
Ellipse	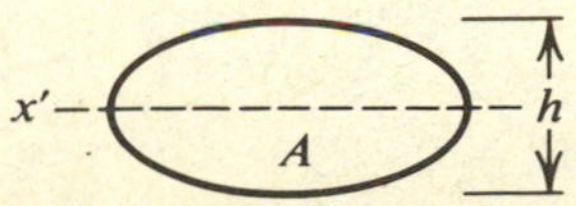	$I_{x'x'} = \dfrac{Ah^2}{16}$

Example 3.6

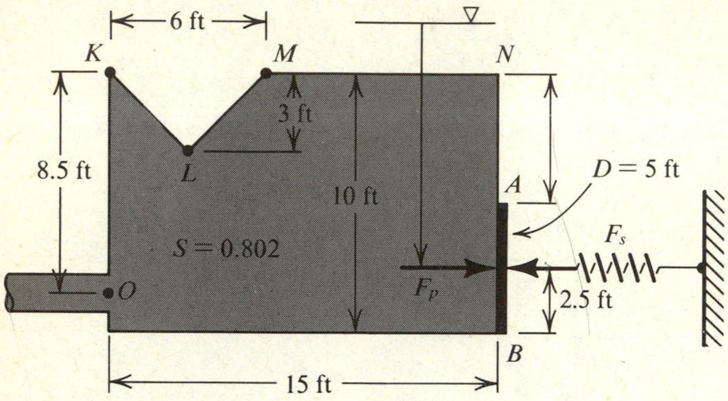

Figure E3.6

A tank containing oil has a pressure at point 0 of 525 lbf/ft^2 as shown in Fig. E3.6. The tank has *a* width (into the paper) of 10 ft. The oil is kept from draining out of the tank by the circular section *A-B*, hinged at point *B*, with a spring force (F_s) acting as shown. The dimension *y* is not to scale in Fig. E3.6. Find:

(a) The pressure at point *K* expressed in feet of oil.
(b) The force on the top surface of the tank (*K-L-M-N*).
(c) The force on surface *A-B* resulting from the oil pressure (F_p).
(d) The minimum spring force necessary to hold the gate *A-B* shut.

Solution:
Step 1.
Identify the characteristics of the fluid and flow field.
An inclined planar body in a static liquid.
Step 2.
Write the appropriate forms of the governing equations of flow:

$$\Delta p = \gamma h \quad \text{(i)}$$

$$F_p = \gamma \forall \quad \text{(ii)}$$

$$F_p = \gamma \bar{h} A \quad \text{(iii)}$$

$$y_{\text{c.p.}} = \bar{y} + \frac{I_{x'x'}}{\bar{y}A} \quad \text{(iv)}$$

$$\sum M_B = 0 \quad \text{(v)}$$

Example 3.6 *(Con't.)*

(a) From Eq. (i)

$$p_o = \gamma h_o = 525 \text{ lbf/ft}^2$$

$$\gamma = \gamma_w\,(.802) = 50 \text{ lbf/ft}^3$$

$$h_o = 525 \text{ lbf/ft}^2/50 \text{ lbf/ft}^3 = 10.5 \text{ ft}$$

$$h_K = h_o - 8.5 \text{ ft} = \boxed{2 \text{ ft of oil}}$$

(b) From Eq. (ii)

$$F_{KN} = \gamma \forall_p = \gamma\,[(2 \text{ ft})(15 \text{ ft})(10 \text{ ft}) + (3 \text{ ft})(3 \text{ ft})(10 \text{ ft})]$$

$$= 50 \text{ lbf/ft}^3 \cdot (300 + 90)\text{ft}^3$$

$$= \boxed{19{,}500 \text{ lbf}}$$

(c) From Eq. (iii)

$$F_p = \gamma \bar{h} A = 50 \text{ lbf/ft}^3\,(7.5 + 2)(\text{ft})[\pi(2.5 \text{ ft})^2]$$

$$= 9326.6 \text{ lbf}$$

(d) From Eq. (iv), and Table 3.1,

$$h_{\text{c.p.}} - \bar{y} = \frac{I_{x'x'}}{A\bar{y}} = \frac{Ad^2}{16A\bar{y}} = \frac{(5 \text{ ft})^2}{16\,(9.5 \text{ ft})} = 0.208 \text{ ft}$$

Utilizing the moment statement of static equilibrium, Eq. (v)

$$F_p(12 - y_{\text{c.p.}}) = F_s\,(2.5)$$

$$F_s = 9326.6 \text{ lbf}\,(2.34)/12.5 = 8730.0 \text{ lbf}$$

This completes the solution.

Example 3.7

A triangular weightless gate is hinged along $\overline{CD}$ as shown in Fig. E3.7. Find (a) the hydrostatic force on the gate by integration and by Eq. (3.35a), (b) the line-of-action of the force, and (c) the force P necessary to open the gate where the force P is normal to the gate and acts at point e.

Example 3.7 *(Con't.)*

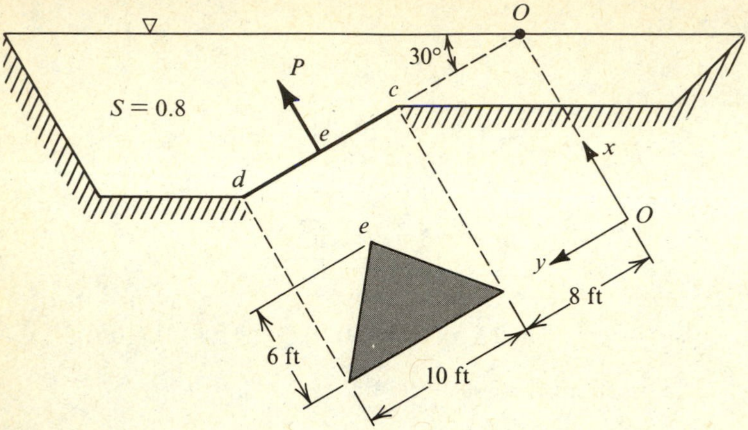

Figure E3.7

Solution:

Step 1.

Identify the characteristics of the fluid and flow field.

An inclined planar body in a static liquid.

Step 2.

Write the appropriate form of the governing equations of flow

$$F_p = \int p dA \tag{i}$$

$$F_p = \gamma A \bar{y} \sin \theta \tag{ii}$$

$$x_{\text{c.p.}} = \bar{x} + \frac{I_{xy}}{\bar{y}A} \tag{iii}$$

$$y_{\text{c.p.}} = \bar{y} + \frac{I_{x'x'}}{\bar{y}A} \tag{iv}$$

$$\Sigma M_{\overline{CD}} = 0 \tag{v}$$

(a) From Eq. (i), we integrate the pressure over the triangular gate to find the pressure force:

$$F_p = \int_A p dA = \int_A \gamma y \sin \theta \, x dy \tag{vi}$$

$$= \gamma \sin \theta \int_8^{13} x y dy + \gamma \sin \theta \int_{13}^{18} x y dy \tag{vii}$$

In order to evaluate the integral, the functional relationship between x and y is necessary.

Example 3.7 *(Con't.)*

At $y = 8$, $x = 0$; at $y = 13$, $x = 6$. But

$$x = ay + b \tag{viii}$$

Evaluating the constants a and b gives

$$0 = 8a + b;\ 6 = 13a + b$$

$$a = \frac{6}{5};\ b = -\frac{48}{5} \tag{ix}$$

$$x = \frac{6}{5}(y - 8),\ 13 \leq y \leq 8 \tag{x}$$

$$x = \frac{6}{5}(18 - y),\ 18 \leq y \leq 13 \tag{xi}$$

Therefore

$$F_p = \gamma \sin\theta \frac{6}{5}\left[\int_8^{13}(y - 8)y\,dy + \int_{13}^{18}(18 - y)y\,dy\right]$$

$$= \gamma \sin\theta \frac{6}{5}\left[\left(\frac{y^3}{3} - 4y^2\right)\Bigg|_8^{13} + \left(9y^2 - \frac{y^3}{3}\right)\Bigg|_{13}^{18}\right]$$

so that

$$F_p = 9734.4 \text{ lbf} \tag{xii}$$

We next evaluate the pressure force using Eq. (ii):

$$F_p = \gamma \sin\theta\, \bar{y}A$$

so that

$$F_p = 62.4 \text{ lbf/ft}^3\,(0.8)(0.5)(13 \text{ ft})(30 \text{ ft}^2) = 9734.4 \text{ lbf} \tag{xiii}$$

We note that the pressure force obtained by two different methods are identical. Which would you prefer using?

(b) From Eq. (iii), we evaluate the center of pressure:

$$x_{\text{c.p.}} = \frac{I_{xy}}{\bar{y}A} + \bar{x}$$

but $I_{xy} = 0$, so that

$$x_{\text{c.p.}} = \bar{x} = \boxed{2 \text{ ft}} \tag{xiv}$$

From Eq. (iv)

$$y_{\text{c.p.}} = \bar{y} + \frac{I_{x'x'}}{A\bar{y}} = 13 + [2(1/12)6(5)^3/30(13)]$$

$$= \boxed{13.32 \text{ ft}} \tag{xv}$$

Example 3.7 *(Con't.)*

(c) From Eq. (v), we sum the moments about $\overline{CD}$

$$P(6) - F_p(2) = 0 \quad \text{(xvi)}$$

Thus

$$P = 9734.4 \text{ lbf } (2)/6 = \boxed{3244.8 \text{ lbf}} \quad \text{(xvii)}$$

This completes the solution.

3.5 *Hydrostatic Forces on Curved Bodies*

When considering the hydrostatic forces on curved objects that are submerged in a static fluid environment, we resolve the hydrostatic force into horizontal and vertical components. We will discuss first the horizontal component of the pressure force and its line-of-action, and second the vertical component of the pressure force and its line-of-action. This order is necessary so that we can treat the buoyancy force and Archimedes principle, both of which are familiar topics in introductory physics courses.

3.5.1 Horizontal and Vertical Components of a Pressure Force

The horizontal component of the hydrostatic pressure force on the wetted surface of a curved body submerged in a static fluid environment is denoted by the symbol F_{p_y}. Its differential element of the force dF_{p_y} is the y-component of the pressure p_y as shown in Fig. 3.14 and acts perpendicular to the differential wetted area dA.

From the geometry shown in Fig. 3.14, we resolve the pressure p and differential area dA into their x- and y-components:

$$p_y = p \cos \theta \quad (3.44)$$

$$p_z = p \sin \theta \quad (3.45)$$

$$dA_y = dA \cos \theta \quad (3.46)$$

$$dA_z = dA \sin \theta \quad (3.47)$$

Using the definition of the pressure force $\mathbf{F}_p$ of Eq. (3.28), we express the y- and z-components of the pressure force as

$$F_{p_z} = \int_A p_z dA \quad (3.48)$$

and

$$F_{p_y} = \int_A p_y dA \quad (3.49)$$

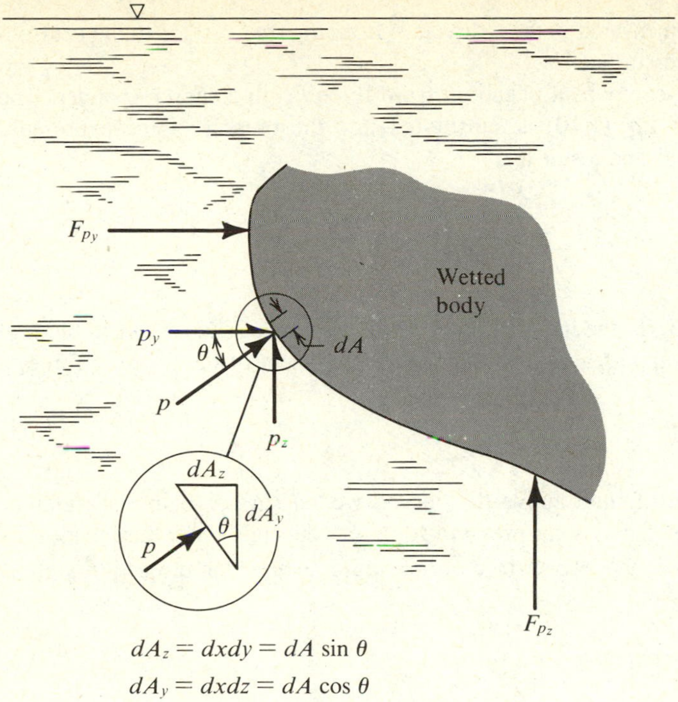

Figure 3.14 *Components of pressure and area.*

Substituting the pressure and area relationships of Eqs. (3.44)–(3.47) into the pressure force expressions of Eqs. (3.48) and (3.49), respectively, we obtain

$$F_{p_z} = \int_{A_z} p dA_z \tag{3.50}$$

for the vertical component, and

$$F_{p_y} = \int_{A_y} p dA_y \tag{3.51}$$

for the horizontal component of pressure forces.

Equation (3.51) is precisely the case presented in Fig. 3.13. Thus, the horizontal component of the pressure force can be evaluated by projecting the wetted curved surface onto a vertical plane, treating it in the manner shown in Sec. 3.4.2, with the result

$$F_{p_y} = \gamma A_y \bar{z} \tag{3.52}$$

where $\bar{z}$ is the distance measured in the z-direction from the free-surface to the centroid of the vertical projected surface.

We evaluate the z-component of the pressure force F_{p_z} in a similar manner. The hydrostatic pressure of Eq. (3.10) is substituted into the pressure force expression given by Eq. (3.50) with the result that

$$F_{p_z} = \gamma \int_{A_z} z dA_z \tag{3.53}$$

But as seen from Fig. 3.15, the integrand zdA_z is the volume of fluid between the free surface and the differential area dA_z, so that the z-component of the pressure force is

$$F_{p_z} = \gamma \forall \tag{3.54}$$

where $\forall$ is the volume of fluid above the curved wetted surface being considered. Thus, the magnitude of the vertical pressure force F_{p_z} depends upon the location of the wetted surface A from the free-surface of the liquid, even if the projection area A_z remains the same.

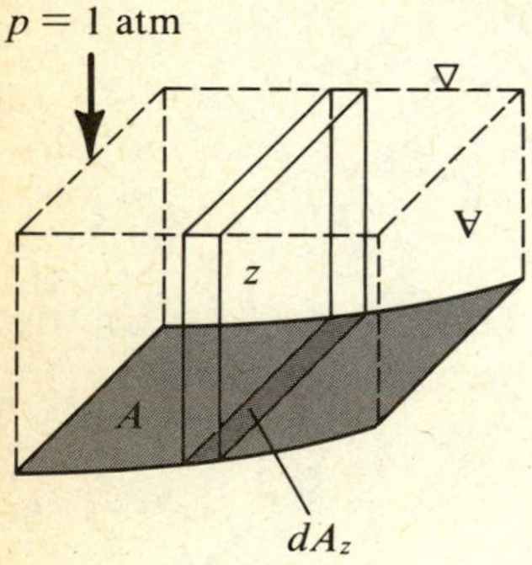

Figure 3.15 *Volume of fluid above wetted surface A.*

The location of the center-of-pressure ($x_{c.p.}$, $y_{c.p.}$, $z_{c.p.}$) is found in a manner identical to that used for locating the center of pressure of the pressure force on a planar surface. The results for the x, y, z locations of the center of pressure are

$$x_{c.p.} = \frac{\int_{A_y} xz dA_y}{\int_{A_y} z dA_y} = \frac{\int_{A_z} xz dA_z}{\int_{A_z} z dA_z}$$

$$x_{\text{c.p.}} = \frac{\int_{A_x} xzdA_x}{\int_{A_x} zdA_x} \tag{3.55}$$

$$y_{\text{c.p.}} = \frac{\int_{A_y} yzdA_y}{\int_{A_y} zdA_y} = \frac{\int_{A_x} yzdA_x}{\int_{A_x} zdA_x}$$

$$= \frac{\int_{A_z} yzdA_z}{\int_{A_z} zdA_z} \tag{3.56}$$

$$z_{\text{c.p.}} = \frac{\int_{A_y} z^2dAy}{\int_{A_y} zdA_y} = \frac{\int_{A_x} z^2dA_x}{\int_{A_x} zdA_x}$$

$$= \frac{\int_{A_z} z^2\, dA_z}{\int_{A_z} zdA_z} \tag{3.57}$$

respectively. The above integrals are not difficult to evaluate; however, we must exercise great care in defining the limits of integration. See the evaluation of the integral in Example 3.7.

Example 3.8

A depth charge has damaged the pressure hull of a submarine. Surfacing is deemed suicidal. The crew must shore from inside the hull to prevent rupture of the hull. Figure E3.8 gives pertinent dimensions and data. Given the specific gravity of seawater as 1.026, find (a) the magnitude of the hydrostatic force and (b) the vertical location of the center of pressure of the hydrostatic pressure force.

Example 3.8 *(Con't.)*

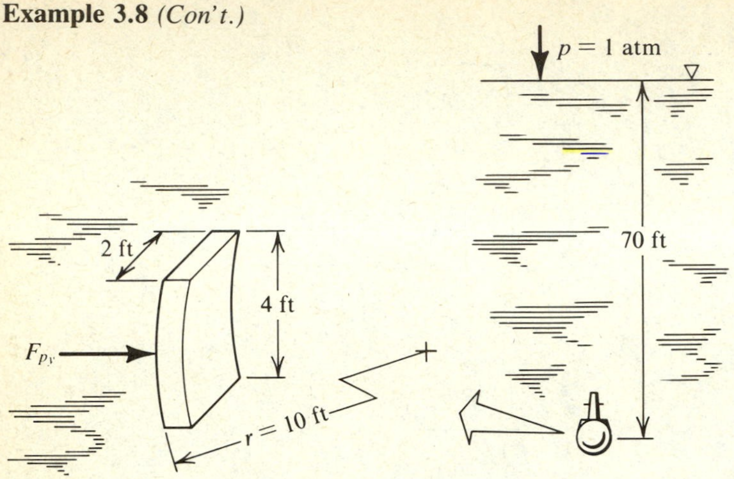

Figure E3.8

Solution:

Step 1.

Identify the characteristics of the fluid and flow field.

A curved section wetted by a static fluid on one side.

Step 2.

Write the appropriate form of the governing equations of flow:

$$F_{p_y} = \gamma \bar{z} A_y \qquad \text{(i)}$$

$$z_{\text{c.p.}} = \frac{1}{\forall} \int_{\forall} z \, d\forall \qquad \text{(ii)}$$

(a) The magnitude of the horizontal component of force is given by Eq. (i):

$$F_{p_y} = \gamma \bar{z} A_y$$

$$= 1.026 \times 62.4 \times 72 \times 8$$

or

$$F_{p_y} = 36{,}900 \text{ lbf} \qquad \text{(iii)}$$

(b) Since the center of pressure of the pressure prism of the projected area is the same center of pressure force for the curved surface, we can use Eqs. (ii) or (3.57):

$$z_{\text{c.p.}} = \frac{1}{\forall} \int_{\forall} z \, d\forall$$

$$= \frac{1}{\forall} \int_0^2 \left[\int_0^4 z \left(\int_0^{f(z)} dy \right) dz \right] dx \qquad \text{(iv)}$$

Example 3.8 *(Con't.)*

Since the volume of the pressure prism is

$$\forall = 2\gamma\left[\frac{1}{2}(4)(4) + 4(70)\right] = 36{,}000 \text{ ft}^3 \tag{v}$$

and the upper integration limit is

$$f(z) = \gamma(74 - z) \tag{vi}$$

substitution of Eqs. (v) and (vi) into Eq. (iv) yields

$$\begin{aligned} z_{\text{c.p.}} &= \frac{2}{36{,}000}\int_0^4 z(74 - z)\,dz \\ &= 1.98 \text{ ft} \end{aligned} \tag{vii}$$

Thus, the vertical distance from the free-surface to the center of pressure is

$$z_{\text{c.p.}} = 74 - 1.98 = 72.02 \text{ ft} \tag{viii}$$

This completes the solution.

3.6 Buoyant Forces on Submerged Bodies

Consider hydrostatic forces applied to the arbitrary body shown in Fig. 3.16. For both the body and the fluid to be in static equilibrium, we obtain from Newton's second law of motion, Eq. (3.1), the result

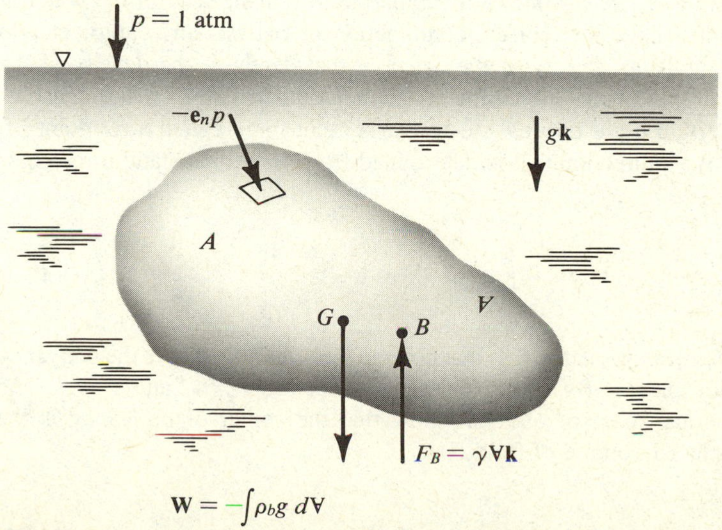

Figure 3.16 *Forces on a submerged body of area A.*

$$\int_{\forall} \rho_b \, \mathbf{g} \, d\forall - \oint_A \mathbf{e}_n \, p dA = 0 \tag{3.58}$$

where ρ_b is the density of the solid body shown in the figure. Applying Gauss's theorem, Eq. (B.37) to the pressure force of Eq. (3.58) results in

$$\int_{\forall} \rho_b \, \mathbf{g} \, d\forall - \int_{\forall} \nabla p \, d\forall = 0 \tag{3.59}$$

Noting that

$$\nabla(-\gamma \, z) = -\gamma \, \mathbf{k} \tag{3.60}$$

Equation (3.59) becomes

$$-\int_{\forall} \rho_b \, g \, \mathbf{k} \, d\forall + \mathbf{k} \int_{\forall} \rho g \, d\forall = 0 \tag{3.61}$$

The first term of Eq. (3.61) is simply the weight of the submerged body acting downward through the center of gravity G. The second term, opposite in direction to the weight, is called the buoyant force, $\mathbf{F}_b$. The buoyant force acts through a point B.

The interesting result of Eq. (3.61) is that the buoyant force is exactly the weight of the displaced fluid ($\gamma\forall$), and is identical to the weight of the body, ($\int \rho_b g d\forall$). This result is called the *principle of Archimedes*.

The conclusion reached through Eq. (3.61) is that for a body submerged in a fluid to be in static equilibrium, its density must equal the density of the surrounding fluid. For example, for a nuclear powered submarine to remain dead in the water at a fixed depth, it must have that correct combination of ballast, air, cargo, etc. to average its density with the density of the ocean at that depth. If the density of the ocean is lighter than that of the submarine, the submarine will sink. If the densities of the body ρ_b and fluid ρ are unequal, there exists a net force called a resultant lift force L in the positive z-direction that is determined from Newton's static force equilibrium law of Eq. (3.1):

$$\mathbf{L} = -\int_{\forall} g(\rho_b - \rho) \, \mathbf{k} \, d\forall \tag{3.62}$$

The *center of buoyancy* is defined as that point in the volume $\forall$ where the buoyancy force's line-of-action passes. By taking moments about the x-, y-, and z-axes, and setting the results equal to zero, one can show that the center of buoyancy is the centroid of the displaced volume of fluid, i.e.,

$$x_{\text{c.p.}} = \frac{1}{\forall} \int_{\forall} x \, d\forall \tag{3.63}$$

$$y_{c.p.} = \frac{1}{\forall}\int_{\forall} y\, d\forall \qquad (3.64)$$

$$z_{c.p.} = \frac{1}{\forall}\int_{\forall} z d\forall \qquad (3.65)$$

The only requirement that must be satisfied for the center of pressure to be the center of buoyancy is that the surface be a closed surface and completely wetted.

When a body is floating with two or more fluids wetting the body surface, the buoyancy force will not pass through the centroid of the body's volume. We have to calculate the centroids of the volumes of the displaced fluid.

Example 3.9

Determine the density, specific volume, specific weight, and volume of an object that weighs 3 lbf in water and 4 lbf in oil of specific gravity 0.83.

Solution:

Step 1.

Determine the characteristics of the fluid and the flow field.

Both fluids are incompressible and the solutions can be obtained assuming that the body rests in a static environment.

Step 2.

Write the equations governing the flow field.

$$F_b = \gamma\forall \qquad \text{(i)}$$

$$\gamma = \rho g \qquad \text{(ii)}$$

$$S = \frac{\gamma}{\gamma_w} \qquad \text{(iii)}$$

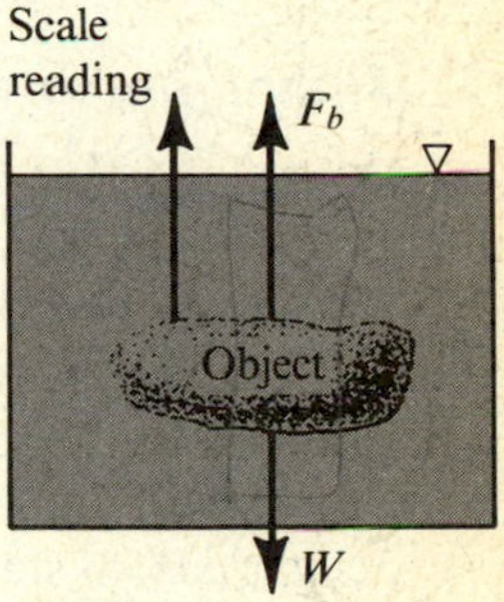

Figure E3.9

From Eq. (i), for water and oil,

$$F_b = \gamma_w\forall$$

$$F_b = \gamma_o\forall$$

respectively. From Eq. (3.1),

$$W = 4 + \gamma_o\forall \qquad \text{(iv)}$$

$$W = 3 + \gamma_w\forall \qquad \text{(v)}$$

for oil and water, respectively. Subtracting Eq. (v) from Eq. (iv) eliminates the object's weight:

$$0 = 1 + \forall\,(62.4)(0.83 - 1)$$

Example 3.9 *(Con't.)*

such that

$$\forall = 0.09426 \text{ ft}^3 \quad \text{(vi)}$$

Substituting the object's volume $\forall$ back into Eq. (v) yields an expression for the weight:

$$W = 3 + 62.4\,(0.09426) = 8.88 \text{ lbf} \quad \text{(vii)}$$

The density can be evaluated from Eq. (ii):

$$\rho = \frac{\gamma}{g} = \frac{W/\forall}{g} = 2.926 \text{ slug/ft}^3$$

The specific gravity is found from Eq. (iii):

$$S = \frac{\gamma}{\gamma_w} = \frac{(2.926)(32.2)}{62.4} = 1.51$$

This completes the solution.

3.7 Initial Stability of Floating and Submerged Ships

In order to understand completely the relationship between a ship's stability and equilibrium, it will be helpful to review some fundamental facts on physical equilibrium. Because the following study is one of balancing static forces, we shall deal with *static equilibrium*. While we realize that in a broader sense *equilibrium* refers to an overall balance of forces which involves no acceleration or deceleration, we will narrow our definition of static equilibrium as follows: *A body at rest is said to be in static equilibrium*. If this body is disturbed by an external force and returns to its original position when the force is removed, the body is said to be in *stable equilibrium*. Thus, a ship which is inclined from its normal upright position and tends to right or return to its normal upright position is said to be *stable*.

On the other hand, a body that continues to move in the same direction after it is originally set in motion by an external force is said to be in *unstable equilibrium*. A ship is said to be unstable if, after being inclined by a slight force (such as a wave or wind shear gust), it continues to incline, possibly until it capsizes. An initially unstable ship may sometimes incline until it reaches a point of stable equilibrium because of the change in underwater hull form.

A body is said to be in *neutral equilibrium* if it comes to rest at any position to which it is moved or displaced. A cylindrical homogeneous log floating in water would be in a state of neutral equilibrium.

Thus, there are three states of equilibrium: *stable, unstable*, and *neutral*.

A body in stable equilibrium tends to right itself when its aspect to the horizontal plane is changed. This tendency is called *statical stability* and is measurable. For a ship, this quantity is referred to as a ship's *initial stability*.

3.7.1 Relative Location of Reference Points

Before we look more closely at the initial stability of a ship, we first must become acquainted with certain standard reference points, identified in Fig. 3.17. A ship in normal position on the water surface has been chosen for convenience.

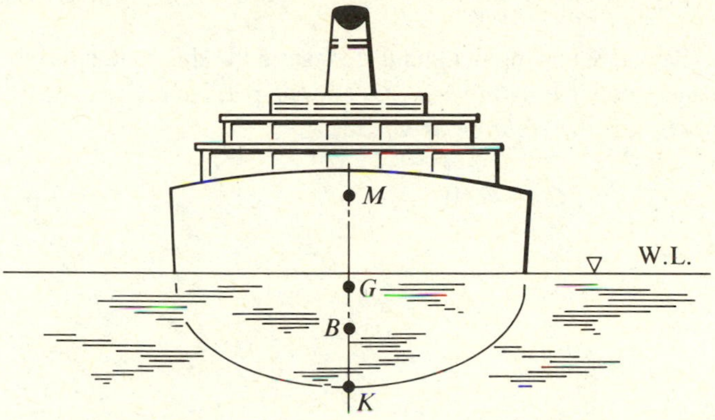

Figure 3.17 *Reference points for ship stability analysis.*

Point K refers to the *keel* of the ship. Vertical distances are normally measured from the keel.

Point B refers to the *center of buoyancy*. We showed in the previous section that it is the centroid of the underwater or displaced volume. It is the point through which the buoyant force acts vertically upward.

Point G refers to the *center of gravity*. The force of gravity acts vertically downward through this point.

Point M refers to the *metacenter*. It is the intersection of the buoyant force's line-of-action with the vertical centerline of the ship for very small angles of inclination. Its location is considered constant for angles up to 7° to 10° for most ships.

The following are standard lengths relating various reference points:

$$\overline{KM} = \text{height of metacenter above the keel}$$

$$\overline{GM} = \text{metacentric height}$$

$$\overline{BM} = \text{metacentric radius}$$

$$\overline{KB} = \text{height of center of buoyancy above the keel}$$

$$\overline{KG} = \text{height of center of gravity above the keel}$$

3.7.2 Initial Stability of a Surface Ship

Consider a ship floating upright on the surface of motionless water. In order to be at rest or in equilibrium, the ship can have no unbalanced forces or moments acting on it. This condition is satisfied if the forces of buoyancy and gravity are exactly equal and collinear.

When the ship is heeled by an external inclining force and the center of buoyancy has been moved from the centerline plane of the ship, gravity's line-of-action will usually diverge from buoyancy's line-of-action.

The separation of the lines-of-action of the two forces forms a couple whose magnitude is equal to the product of one of these forces and the distance separating them.

In the case of Fig. 3.18a where this moment tends to restore the ship to the upright position, the moment is called the *righting moment, C,* and the perpendicular distance between the two lines-of-action is the *righting arm* ($\overline{GZ}$):

$$C = \overline{GZ}W \tag{3.66}$$

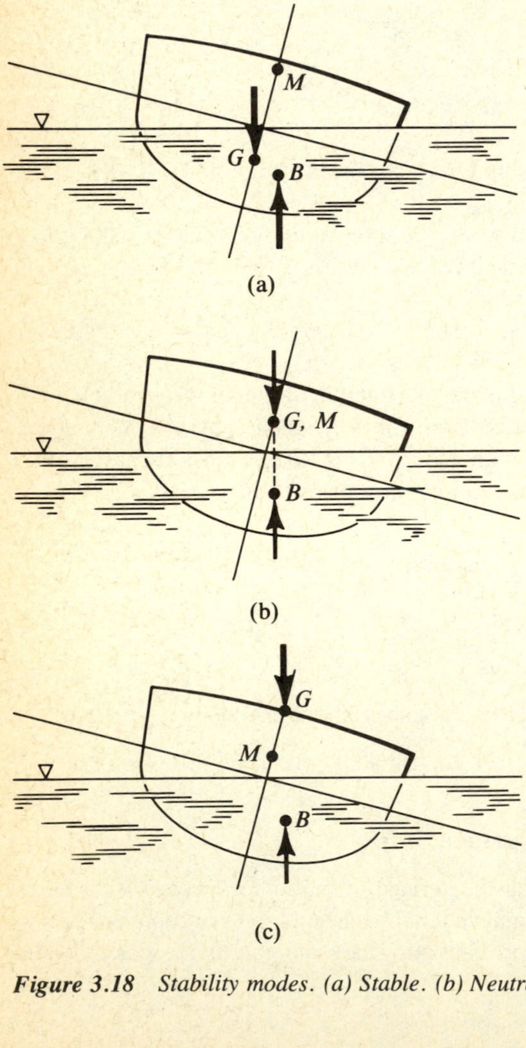

Figure 3.18 *Stability modes. (a) Stable. (b) Neutral. (c) Unstable.*

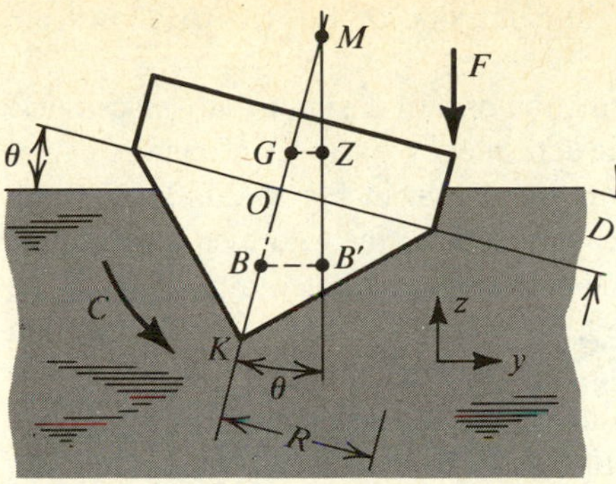

Figure 3.19 *Nomenclature for ship stability analysis.*

as shown in Fig. 3.19. The intersection of the lines-of-action of the old and new buoyancy forces locates the metacenter, M.

From the geometry of Fig. 3.19

$$\overline{GZ} = \overline{GM} \sin \theta \tag{3.67}$$

which is used to obtain a theoretical stability curve. Combining Eqs. (3.66) and (3.67) gives

$$C = W(\overline{GM}) \sin \theta \tag{3.68}$$

which states that the ship's righting moment depends upon the ship's weight, displacement, and angle-of-inclination.

From Fig. 3.19 we see that the direction of the righting couple determines the relative stability of the ship, and the position of the metacenter with respect to the center of gravity determines the direction of C. For the case shown, M is above G, and the couple C tends to oppose (or balance) the disturbance due to an external force F. This condition we have termed *stable* (see Fig. 3.18a). If the metacenter is below point G, however, then the moment due to C is in the same sense as the moment due to F and the ship is *unstable*. This is case (c) in Fig. 3.18. Finally, if M and G are coincident, the craft is marginally or *neutrally stable*. This is shown in Fig. 3.18b.

We can use Eq. (3.68) to compute the distance $\overline{GM}$ for a given vessel, its magnitude being a measure of the ship's stability. Summing the moments about the longitudinal x-axis at point 0 of Fig. 3.19 gives

$$C = FR \cos \theta + FD \sin \theta \tag{3.69}$$

Limiting ourselves to small angles, Eq. (3.69) is equated to Eq. (3.68)

$$\overline{GM} = \frac{FR}{W \tan \theta} \tag{3.70}$$

Thus, measuring all the quantities on the right-hand side of Eq. (3.70) allows the computation of the metacentric height, $\overline{GM}$.

The location of the metacenter, M, is a function of hull design and displacement. It is customary to represent such a functional dependency by a set of curves called the "curves of form." Typical curves of form are shown in Fig. 3.20. To find the distance from the keel to the metacenter, $\overline{KM}$, we first enter the figure on the ordinate at the mean draft of the ship (point 1). The mean draft is computed by averaging the fore and aft drafts of the ship. Next, we move horizontally to point 2, the intersection of curve *A-A* (called the $\overline{KM}$ diagonal). Moving vertically to curve *B-B* (the $\overline{KM}$ curve) and then horizontally back to the ordinate will then yield the distance $\overline{KM}$. Once we know the distance $\overline{KM}$, we can use the result of Eq. (3.70) and the geometric relation

$$\overline{KG} = \overline{KM} - \overline{GM} \tag{3.71}$$

to compute the location of the center of gravity of the ship.

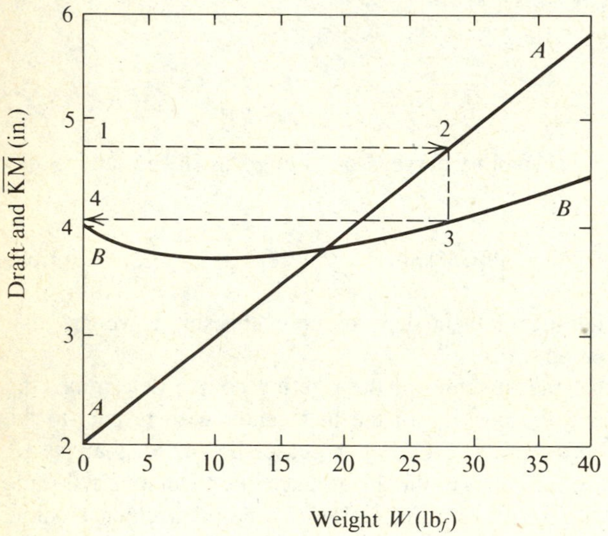

Figure 3.20 *Curves of form.*

Since the location of M relative to G dictates the stability of the ship, the length of the righting arm, $\overline{GZ}$, is a measure of the stability of a ship. A graph of $\overline{GZ}$ as a function of inclination θ is called the "static stability curve" of the ship. Such a curve is valid for a particular location of G. A family of such curves for a range of values of $\overline{KG}$ is termed a "general stability diagram." A brief look at the curve will determine whether the ship is a "tender" or "stiff" ship. Low values of the righting arm indicate a tendency for the ship to roll slowly back to its upright position which characterizes a tender ship and is desirable for gun platforms. A steep curve characterizes a stiff ship which has short, jerky rolls. The primary yardstick for ship safety is how large the maximum righting arm is and at what angle it occurs.

Once the static stability curve is generated, it can be translated into more useful forms for damage control onboard ship. For example, plates are made to depict the effects of flooding in each compartment of the ship. Consideration is given to list incurred, additional moments developed, and the overall weight added. The effect of a free-surface is to vertically raise and transversely shift the center of gravity. However, for small angles, the vertical shift is very small relative to the transverse shift. This effect decreases the righting couple and therefore decreases the stability of the ship.

For a floating ship of variable cross section, such as that shown in Fig. 3.21, we can develop a convenient formula for determining the metacentric radius (BM) of the ship. As the ship is inclined through a small angle, θ, the horizontal shift of the center of buoyancy, $\overline{BB_1}$ (Fig. 3.21b) is determined by the change in buoyant forces ΔF_B caused by the wedge WOW being submerged, which causes an upward force to the left, and by the other wedge LOL, decreasing the buoyant force by an equal amount ΔF_B on the right. The force system consisting of the original buoyant force at B and the couple $\Delta F_B xs$ due to the wedges, must have as resultant the equal buoyant force at B_1. Taking a moment about B, one obtains

$$C = \Delta F_B xs = WxBB_1 \tag{3.72}$$

where W is in this case the displacement of the ship.

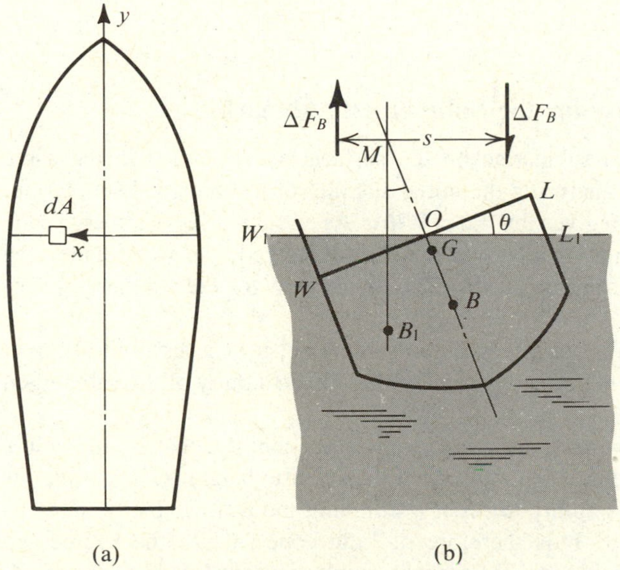

Figure 3.21 *Geometry for location of metacenter.*

The magnitude of the couple C may be determined by taking moments about O, the centerline of the body at the waterplane. For an element of area dA on the horizontal section through the body at the liquid surface an element of volume of the wedge is $x\theta dA$, in which θ is the small angle of heel (in radians). By integrating over the complete original waterplane area, the couple is determined to be

$$C = \Delta F_B xs = \Upsilon\theta \int_A x^2 dA = \Upsilon\theta I_{y-y} \tag{3.73}$$

in which I_{y-y} is the moment of inertia of the waterplane area about the y-axis.

Substituting Eq. (3.73) into Eq. (3.72) produces

$$C = \Delta F_B xs = \Upsilon\theta I_{y-y} = W(BB_1) = \Upsilon\forall(BB_1) \tag{3.74}$$

where $\forall$ is the underwater volume and where θ is assumed to be very small:

$$BM \sin\theta \; BM\,\theta = BB_1 \tag{3.75}$$

or

$$BM = \frac{I_{y-y}}{\forall} \tag{3.76}$$

It should be noted that the location of B and M depend upon the geometry of the ship. Their location is independent of the location of G. It should also be noted that the locations of B and M are functions of the ship's draft.

3.7.3 Initial Stability for the Submerged Submarine

The initial stability of a submarine on the surface is determined in the same manner as a surface ship. In analyzing the initial stability of a submerged submarine, first consider the expression given by Eq. (3.76). As a submarine submerges the waterplane area decreases and, finally, when fully submerged, the waterplane area becomes zero. Therefore, from Eq. (3.76), BM equals zero for the submerged submarine.

For positive initial stability of a submerged submarine, it is necessary to keep G below B (and M). Submariners normally speak of the initial stability of the submerged boat as BG rather than GM.

All previous discussions have dealt with transverse stability, that is, the ship's resistance to a heeling moment. The longitudinal metacentric height (GM'), which is a measure of a surface ship's ability to resist a trimming moment, is many times its transverse metacentric height. It is therefore of little concern. On the submerged submarine, however, the longitudinal metacentric height becomes equal to the transverse metacentric height. Since a fore and aft weight change on a submarine will probably produce a much greater trimming moment than an athwart-ships weight movement, because of the greater lever arms available, the longitudinal metacentric height is actually of greater concern to the operator of a submerged submarine.

3.7.4 Methods of Improving Initial Stability

We can analyze the principal alternatives available to the designer and the ship operator to improve a ship's initial stability. The methods available to the designer are:

1. Increase the beam of a surface ship. (Note that I and hence BM is approximately proportional to B_1^3.)
2. Place heavy weights low in the ship.
3. Use compartmentation and arrange compartments to minimize possible free-surface effects.

The methods available to the operator are:

1. Strictly control weights. Keep $\overline{KG}$ as low as possible and also keep items in excess of allowance off the ship.
2. Steam with a minimum of tanks other than 95% full or completely empty.
3. In event of damage, reduce free-surface effects as rapidly as possible.

Study Questions

3.1 Define the fluid static differential equation. What type of mathematical equation is it? What are the variables? What types of fluid are applicable?

3.2 Prove that the pressure is constant in a horizontal plane for a continuous homogeneous fluid.

3.3 What assumptions are necessary to impose if the pressure is the same in any direction at a point in a fluid?

3.4 Relate the gauge pressure of air at one atmosphere to absolute pressure, inches of mercury, and feet of water.

3.5 What is the boundary condition for fluid statics?

3.6 Where does the line-of-action of the horizontal component of the hydrostatic pressure force act?

3.7 Where does the line-of-action of the vertical component of the hydrostatic pressure force act?

3.8 What is the difference between the center of gravity, center of buoyancy, centroid, and center of pressure?

3.9 How does one calculate the vertical component of the hydrostatic pressure force if part of the body is exposed to air?

3.10 Define metacenter, metacentric height and how they are measured.

Problems

Pressure

3.1 On a day where $p_{atm} = 1.013 \times 10^2$ kPa, a pressure of -5 psi is noted in a particular system. If $g = 31.0$ ft/s^2, $\rho = 2.01$ slug/ft^3, and the specific gravity of mercury is 13.6, (a) convert this pressure to psia, (b) convert this pressure to feet of water, and (c) convert this pressure to inches of mercury vacuum.

3.2 A barometer in a laboratory shows 800 mm Hg. What is the pressure of the atmosphere in bars?

3.3 A pressure gauge is attached to a pressurized bottle of oxygen in a hospital's intensive care unit and reads 1.71 bars. The room where the bottle is stored has a barometer which reads 763 mm of mercury. Find the absolute pressure in the cylinder in bars.

3.4 A barometer in a home reads 31.4 in. of mercury. Convert this pressure to (a) bars, (b) atmospheres, (c) feet of water, (d) pascals, (e) MPa, (f) psi, (g) psia, and (h) N/m^3.

3.5 Knowing the density of mercury is based on the specific gravity 13.6, find the height of some column of liquid in meters that would be equivalent to 1 atmosphere.

3.6 A gauge pressure inside a manned spacecraft is equivalent to a height of 20 cm of a certain liquid having a specific gravity 0.8. Calculate the absolute pressure inside the spacecraft given the barometric pressure is 1 bar.

3.7 A research and salvage vehicle maneuvers to attach its hatch to another vehicle. The barometric pressure inside the salvage vehicle reads 700 mbar. Convert (a) a gauge reading of 1 atm to kPa, (b) 0.5 bar absolute to millibars, (c) an absolute pressure of 2 bar to a gauge reading in atmospheres.

3.8 The great mountain climber Messner notes that he starts climbing the Matterhorn where the barometric pressure is 1000 mbar. How high has he climbed when his barometer reads 750 mbar, assuming an average air density of 1.188 kg/m^3?

3.9 What is the pressure at the bottom of a barrel of oil 4 ft deep that has a specific gravity 0.8 (in psi)?

3.10 A glacier of specific weight 60 lbf/ft^3 moves slowly over a buried river. The glacier is approximately 500 ft thick. What is the pressure at a depth one foot in the river (psi)? (See Fig. P3.10.)

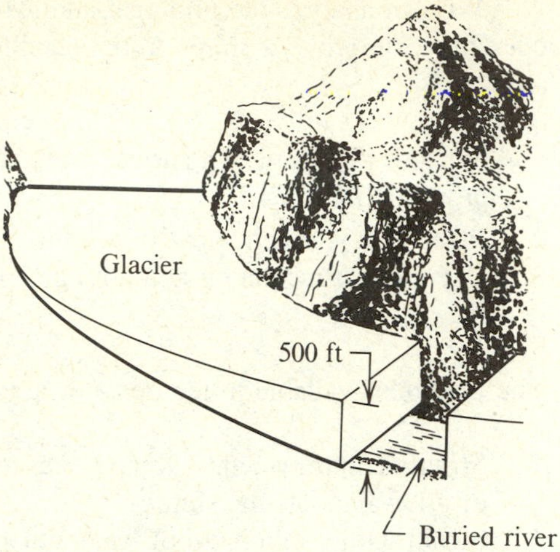

Figure P3.10

3.11 Convert 20 ft of water to (a) psi, (b) inches of Hg, and (c) ft of oil at $S = 0.8$.

3.12 The modulus of elasticity of a glass sphere is 10^4 psi. At what depth in water will the glass fracture (ft)?

3.13 Assuming that the density of the atmosphere $\rho = 0.002378$ slug/ft^3 does not change with altitude, what would be the height of the atmospheric column if the temperature and pressure are 60°F and 14.7 psi, respectively?

3.14 Calculate the temperature and pressure for the troposphere at 30,000 ft altitude, if $T_o = 68$°F.

3.15 Two psi suction is equivalent to a head of how many feet of water?

3.16 What vacuum in the mouth is necessary to draw water vertically upward 6 inches through a small diameter straw (neglect friction, psi)?

3.17 A barometer measures atmospheric pressure. If we use a liquid having a specific weight of 842 lbf/ft^3 and invert a tubeful of this material, as shown in Fig. P3.17, what is the value of h if the pressure at A is 3 psia?

3.18 A tank with a square cross section measuring 11 m by 11 m is floating in water as

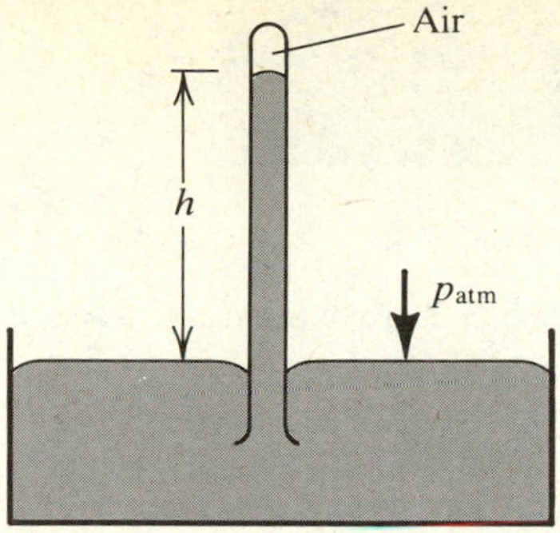

Figure P3.17

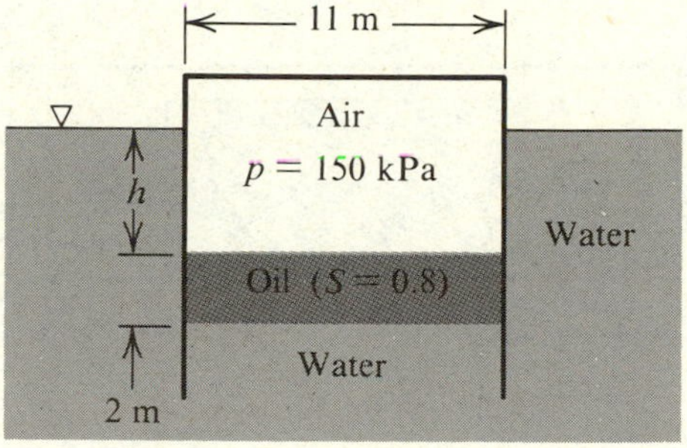

Figure P3.18

shown in Fig. P3.18. Assume the tank wall has a negligible thickness compared to the dimensions shown. Calculate the distance h (m).

3.19 Referring to Prob. 3.18, calculate the weight of the tank (N).

3.20 Derive an expression among pressure, density, and altitude, assuming air is isentropic, and $p = p_o$, $\rho = \rho_o$ at $z = 0$ (sea level).

3.21 A deep-sea diver is in a trench 1 mile below the ocean's surface. (a) What is the force exerted by the water on the diver's face mask which is 40 in.^{2}? (b) Repeat part (a) assuming water is compressible and has a bulk modulus of elasticity $K = 325{,}000$ psi? Assume the specific weight of seawater is 64 lbf/ft^3.

3.22 Nine kilometers below the surface of the ocean the pressure is 91.91 MN/m^2. Determine the specific weight of seawater at this depth if the specific weight at the surface is 10 kN/m^3 and the bulk modulus of elasticity is 2.34 GN/m^2.

3.23 A hydrogen-filled balloon is designed to expand to a near sphere 15 m in diameter at a height of 30 km where the absolute pressure is 1100 N/m^2 and the temperature is -38°C. If there is no stress in the fabric of the balloon, what volume of hydrogen must be added at ground level where the temperature is 22°C? (See Fig. P3.23.)

3.24 A liquid of specific gravity 0.75 is contained in a rectangular container 4 ft high, 6 ft long, and 1 ft wide. The container has a uniform acceleration of 16.1 ft/s^2 in the direction shown in Fig. P3.24. (a) Find the pressures at locations B and C if there is a hole at A. (b) Find the acceleration a_x for the case $p_B = 0$.

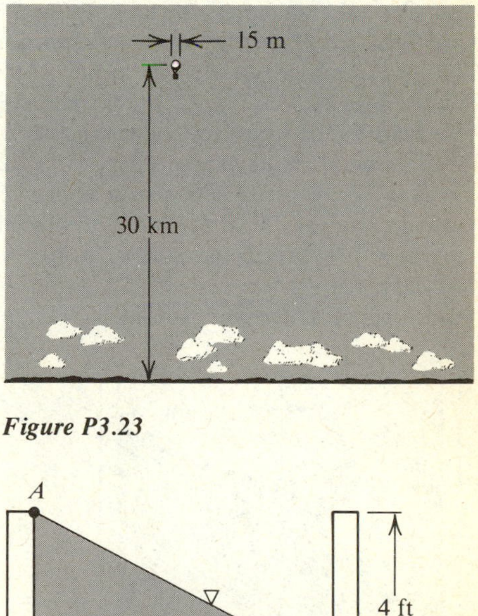

Figure P3.23

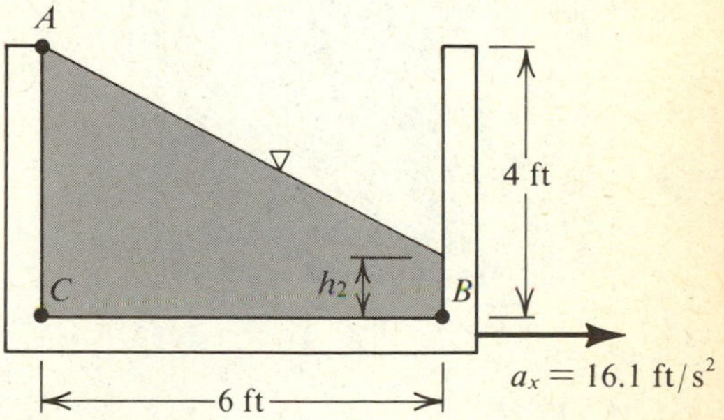

Figure P3.24

3.25 Mercury in the U-tube shown in Fig. P3.25 rotates about a vertical axis passing through the point A. Calculate the rotational speed ω when the difference between levels is 8 cm.

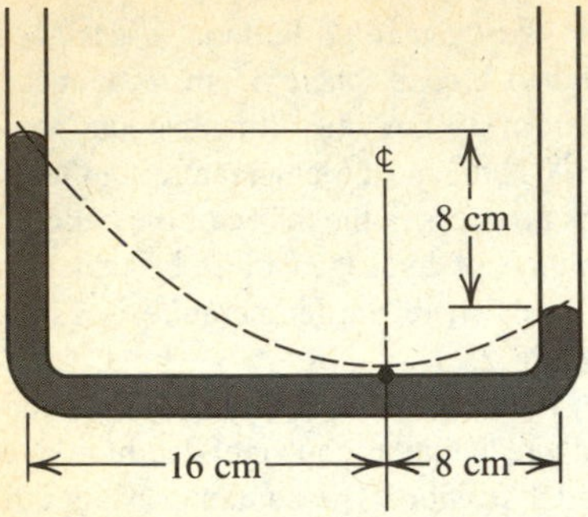

Figure P3.25

3.26 A body of fluid accelerates steadily with acceleration components a_x, a_y, and a_z. There is no relative motion between any of the fluid particles. (a) Find expressions for $\partial p/\partial x$, $\partial p/\partial y$, and $\partial p/\partial z$. (b) Find the pressure difference $(p_2 - p_1)$ between two points having coordinates (x_1, y_1, z_1) and (x_2, y_2, z_2). (c) Determine the shape of the surface of constant pressure.

3.27 Consider a liquid of specific gravity 1.2 rotating at a constant 200 rpm in a circular cylinder as shown in Fig. P3.27. At a radial location $r_A = 2$ ft, the pressure $p_A = 10$ psi. At the radial location $r_B = 3$ ft, the point B is 4 ft above A. Find the pressure p_B (psf).

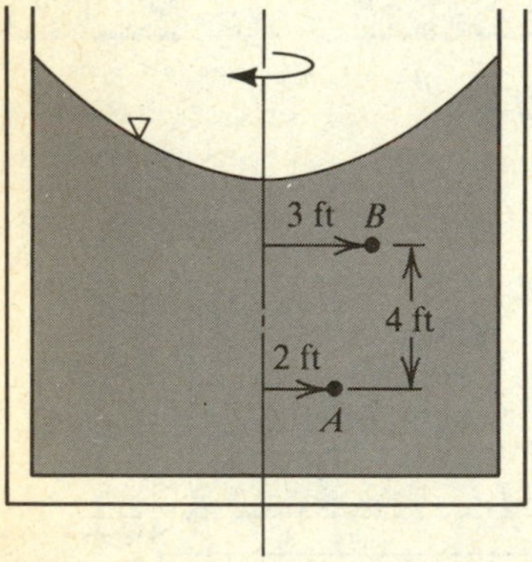

Figure P3.27

3.28 As Fig. P3.28 shows, compartments 1 and 2 are sealed compartments containing air. If the barometric pressure is 14.7 psi, the pressure recorded on gauge A is 8.0 psi vacuum, and the pressure recorded on gauge B is 10.0 psi, what is the pressure in compartments 1 and 2, and the pressure recorded by gauge C?

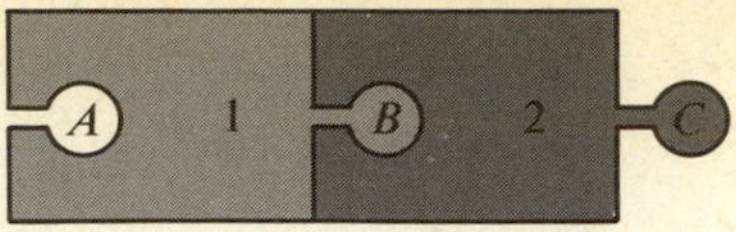

Figure P3.28

3.29 In Prob. 3.28, if compartment 2 is heated, what will be the effect on (a) the reading of gauge A, (b) the reading of gauge B, and (c) the reading of gauge C?

3.30 In the compartments shown below, each of which is sealed from the others, pressure gauge A reads 1.5 bar, gauge B reads 2.0 bar, and the pressure in D is 0.8 bar. What is the gauge pressure of C (bar)?

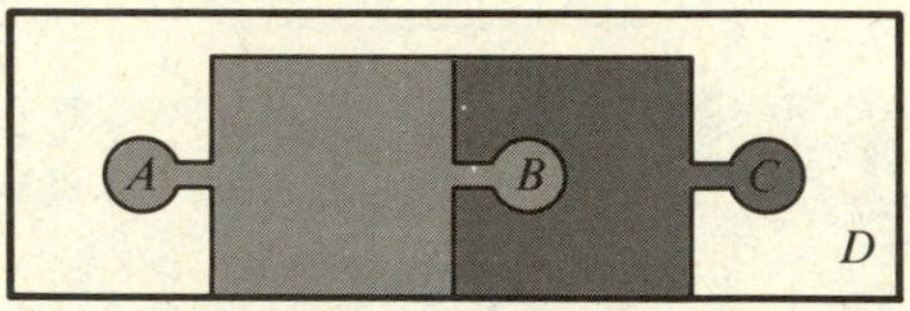

Figure P3.30

Manometers

3.31 A manometer reads 600 mbar of vacuum. Determine the absolute pressure in millibars and in kilopascals if the barometric pressure is 900 mbar.

3.32 In Fig. P3.32, if the specific gravities of A and B are $S_A = S_B = 1.0$, and the specific gravity of the liquid in the tank is 0.8, find $p_A - p_B$ (psi).

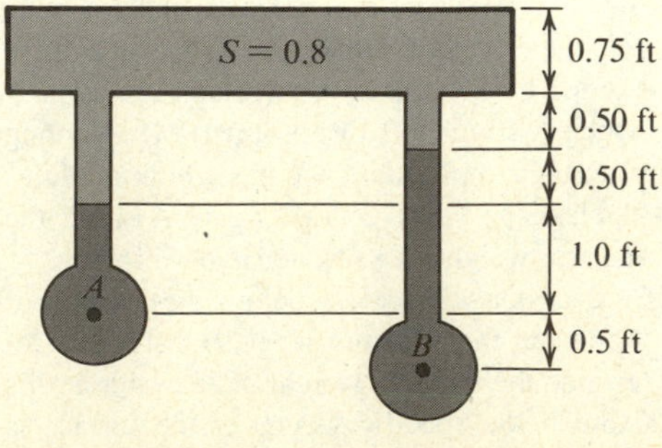

Figure P3.32

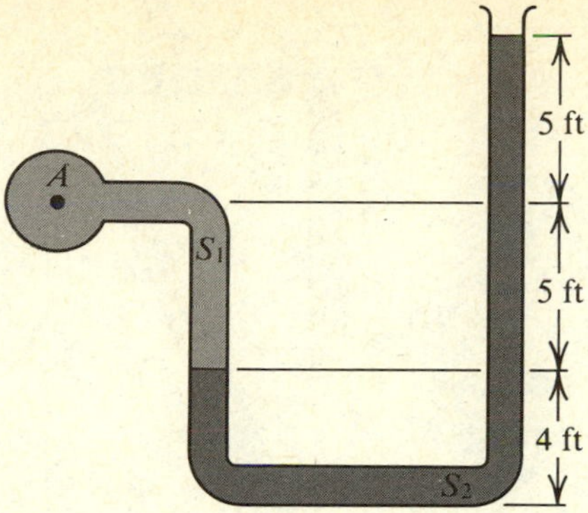

Figure P3.33

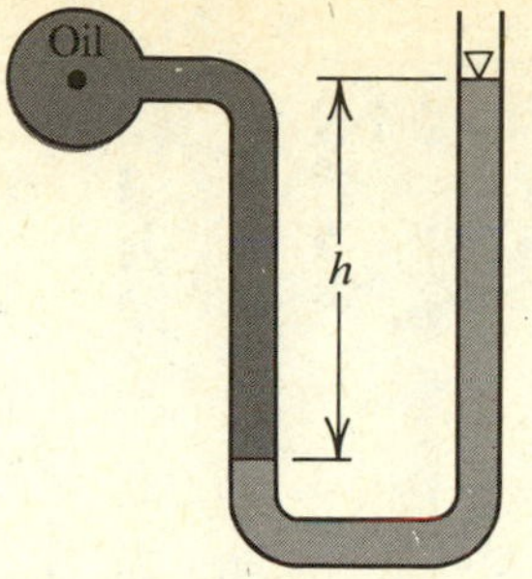

Figure P3.35

3.33 Given the manometer shown in Fig. P3.33, find the pressure p_A (psi) given $S_1 = 1.0$ and $S_2 = 13.55$.

3.34 In Fig. P3.34 gauge A reads 21.3 psi and p_{atm} is 14.7 psia. Compute the height h (ft). Assume that the pressure in the vacuum is 0 psia, the specific gravity of mercury and CCl_4 are 13.6 and 1.59, respectively.

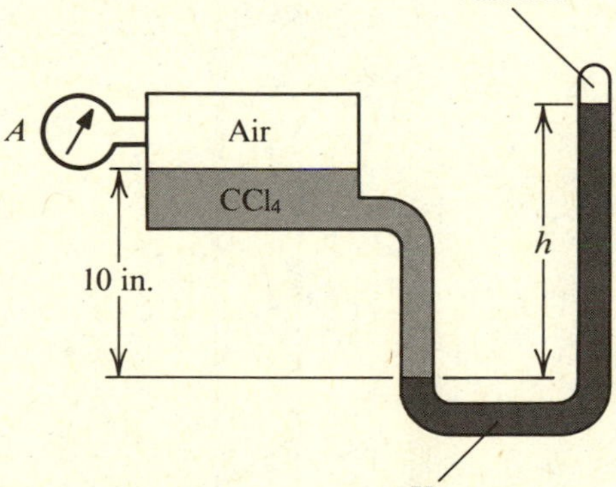

Figure P3.34

3.35 Given the elementary differential manometer shown in Fig. P3.35, if $p_{atm} = 14.7$ psia, $\gamma_{oil} = 50$ lbf/ft^3, and $h = 42$ in., (a) what is the density of the manometer fluid (slug/ft^3)? (b) What is the density of the manometer fluid (kg/m^3)? (c) What is the specific weight of the manometer fluid if $g = 30$ ft/s^2 (lbf/ft^3)? and (d) What is the specific gravity of the manometer fluid if $g = 30$ ft/s^2?

3.36 Repeat Prob. 3.35 if $p_{atm} = 1.013 \times 10^2$ kPa, and express parts (a), (c), and (d) in SI units.

3.37 The U-tube in Fig. P3.37 is open at both ends and is partially filled with mercury ($S = 13.6$). Water is then poured into the left leg until the water surface is at the top of the leg. What will be the height of the column of water (m)?

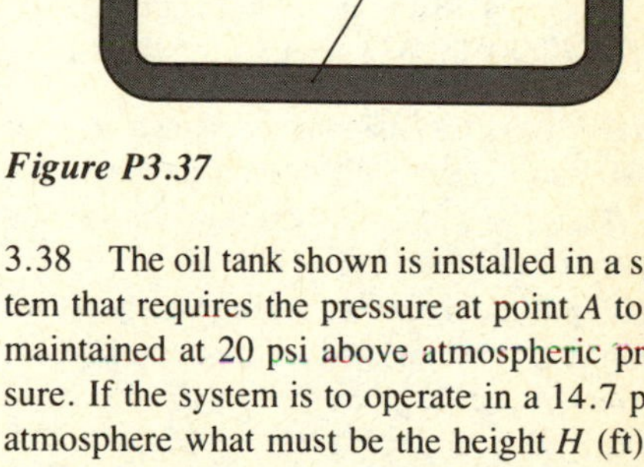

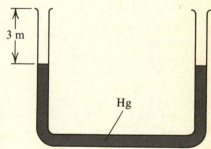

Figure P3.37

3.38 The oil tank shown is installed in a system that requires the pressure at point A to be maintained at 20 psi above atmospheric pressure. If the system is to operate in a 14.7 psia atmosphere what must be the height H (ft)?

3.39 Given: Liquids A and B are water; $S_2 = 0.8$, $h_1 = 1$ ft, $h_2 = 0.5$ ft, and $h_3 = 2$ ft. (a) Find $p_A - p_B$ in psi. (b) If $p_B = 10$ psia and the barometer reading is 29.5 in. of mercury, find the gauge pressure at A in lbf/ft^2.

$$\text{(a)}\quad h_A - h_1S_1 - h_2S_2 + h_3S_3 = h_B$$
$$h_A - 1(1) - 0.5(0.8) + 2(1) = h_B$$
$$h_A - h_B = 1 + 0.4 - 2$$
$$= -0.6 \text{ ft water}$$

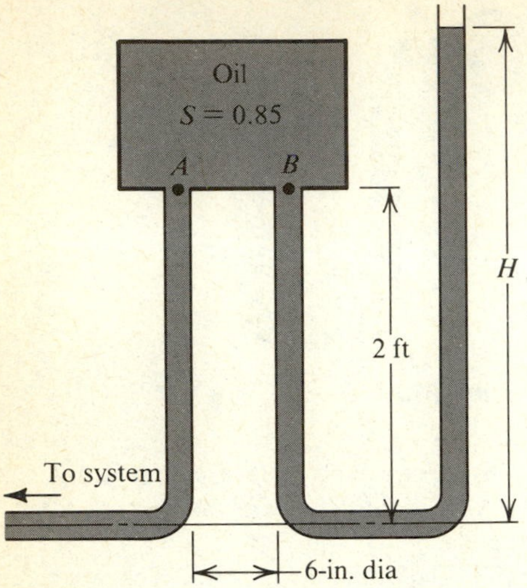

Figure P3.38

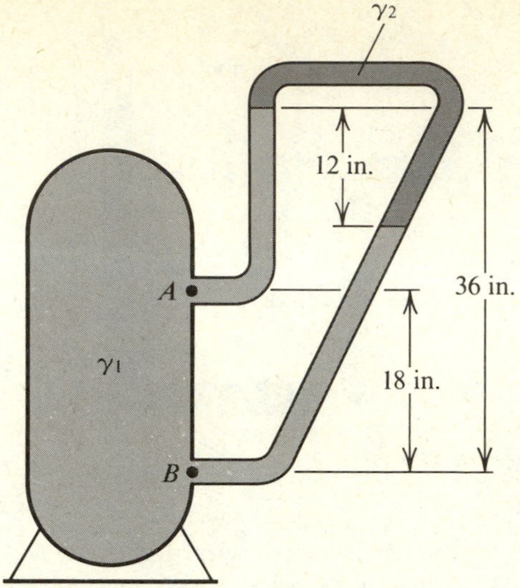

Figure P3.39

$$p_A - p_B = -0.6 \text{ ft water } (14.7 \text{ psi}/33.9 \text{ ft water}) = -0.26 \text{ psi}$$

(b) $p_A = p_B - 0.26$ psi $= 10 - 0.26 = 9.74$ psia

Atm pressure $= 29.5$ in. Hg $(14.7 \text{ psi}/30 \text{ in. Hg}) = 14.47$ psi

$$p_A = 9.74 - 14.47 = -4.73 \text{ psi} = -4.73 \text{ lbf/in.}^2 (144 \text{ in.}^2/\text{ft}^2) = 681 \text{ lbf/ft}^2 \text{ vacuum}$$

3.40 Given the manometer shown in Fig. P3.40 and the pressure difference between A and B is $p_A - p_B = 1.04$ psi, find the specific gravity of fluid labeled γ_2.

3.41 For the manometer arrangement shown in Fig. P3.41, calculate the pressure difference between A and B, given the specific gravity of mercury and turpentine are 13.6 and 0.873, respectively.

3.42 In Fig. P3.42, compartments 1 and 2 are large rigid compartments filled with air. If mercury is the manometer fluid, $S = 13.6$, the barometric pressure is 14.5 psia, and gauge A reads 30.0 psi, find (a) p_1 (psia), (b) p_2 (psia), and height X. If the atmospheric pressure were to increase, what would happen to the reading of gauge A and height X?

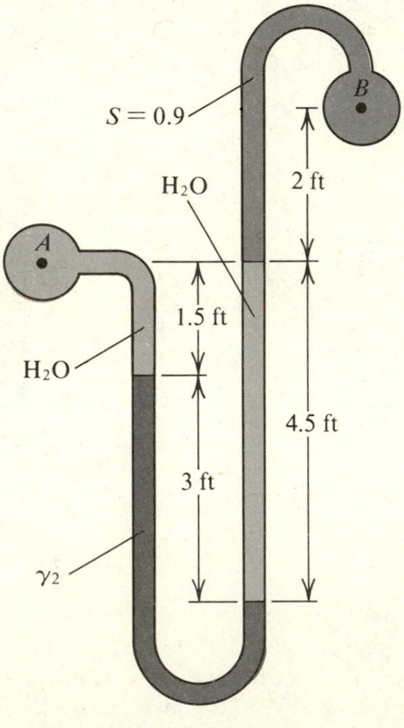

Figure P3.40

3.43 A differential manometer is shown in Fig. P3.43. The values of vertical lengths y are $y_1 = 18$ in., $y_2 = 2$ ft, $y_3 = 1.5$ ft, $y_4 =$

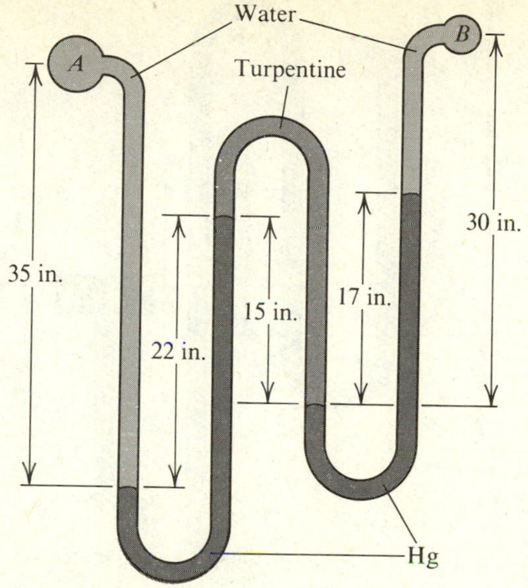

Figure P3.41

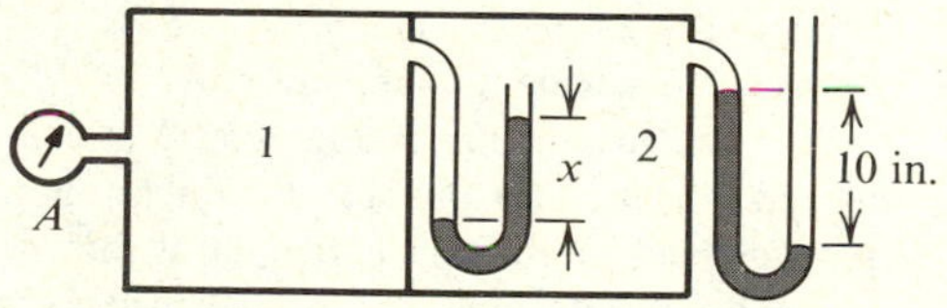

Figure P3.42

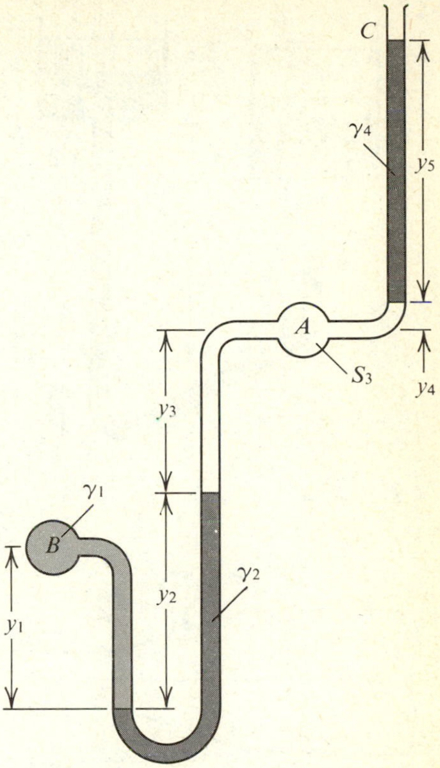

Figure P3.43

0.25 ft, and $y_5 = 2.4$ ft, and the specific weights of the manometer fluids are $\gamma_1 = 53.5$ lbf/ft^3, $\gamma_2 = 479$ lbf/ft^3, $\gamma_4 = 62.4$ lbf/ft^3, determine (a) the pressure p_A (psi), and (b) the pressure p_B if point C is open to the atmosphere and $S_3 = 1.59$.

3.44 The manometer in Fig. P3.44 is connected to an artery at two points, between which is a valve. If the manometer fluid is water, and the deflection of the water is 0.25 in., at its maximum reading, calculate the maximum difference in pressure in psi between A and B. Which way is the blood flowing?

3.45 Find the pressure at gauge A in a pressurized container shown in Fig. P3.45, given a piston force of 5 N acting on a piston of area 1 m^2 (kPa).

3.46 Air flows through the nozzle shown in Fig. P3.46 and causes the mercury to rise in the U-tube. Find the value of h if the pressure

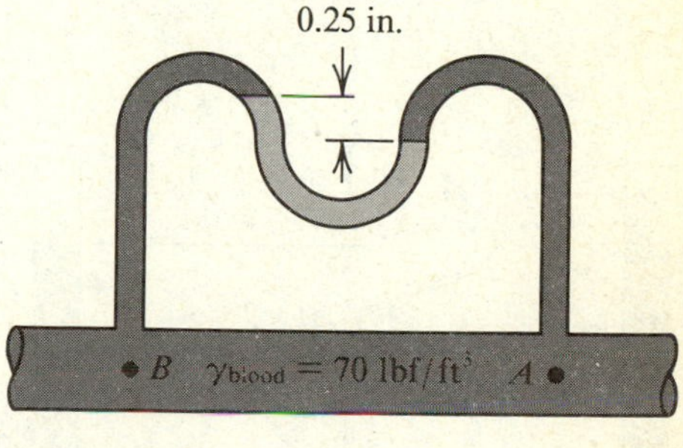

Figure P3.44

gauge at A reads 5 psia, and the manometer fluid is mercury, $S = 13.6$.

3.47 Find the specific gravity of the unknown liquid given both ends of the U-tube are open to the atmosphere (Fig. P3.47).

3.48 A single tube is filled with a variety of liquids of different densities. What is the difference in pressure at A and B for Fig. P3.48?

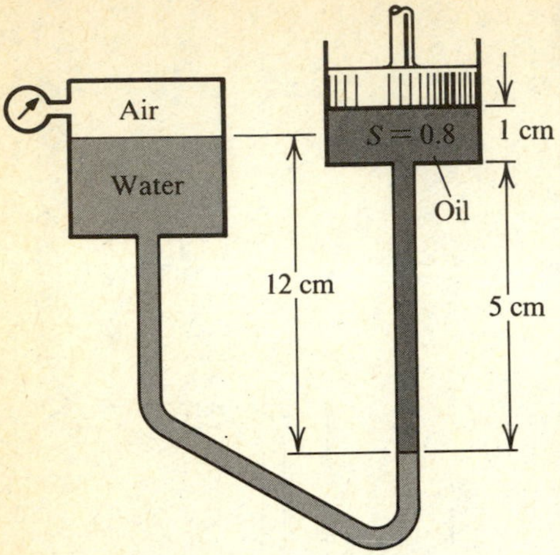

Figure P3.45

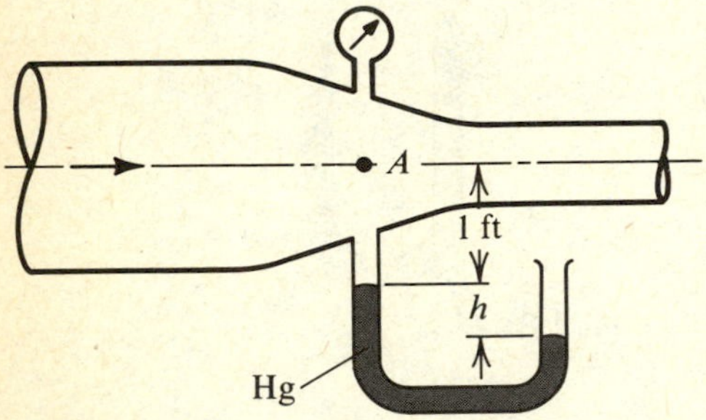

Figure P3.46

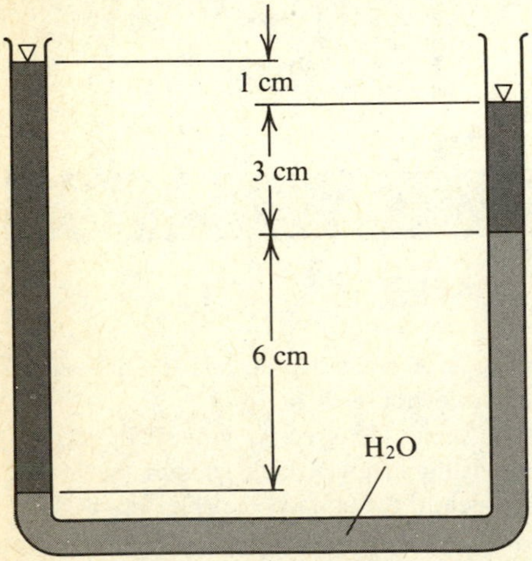

Figure P3.47

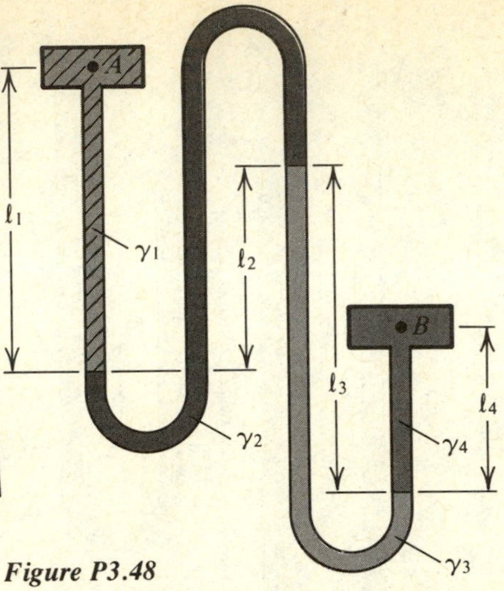

Figure P3.48

3.49 For the manometer shown in Prob. 3.48, if $p_B = 2$ kPa, $l_1 = 0.5$ m, $l_2 = 0.3$ m, $l_3 = 0.6$ m, $l_4 = 0.2$ m, and $\gamma_1 = 1 \times 10^4$ kg/m^2·s^2, $\gamma_2 = 0.5 \times 10^4$ kg/m^2·s^2, $\gamma_3 = 1.2 \times 10^4$ kg/m^2·s^2, and $\gamma_4 = 0.8 \times 10^4$ kg/m^2·s^2, calculate the pressure at A.

3.50 The manometer in Fig. P3.50 is connected to static pressure taps 10 ft apart on a pipe inclined 30° from the horizontal as shown. Water flows in the pipe. The difference in pressure at points 1 and 2 is to be calculated in psf.

3.51 Why would a differential manometer be unsuitable for measuring pressure drops in hydraulic systems aboard an orbiting spacecraft?

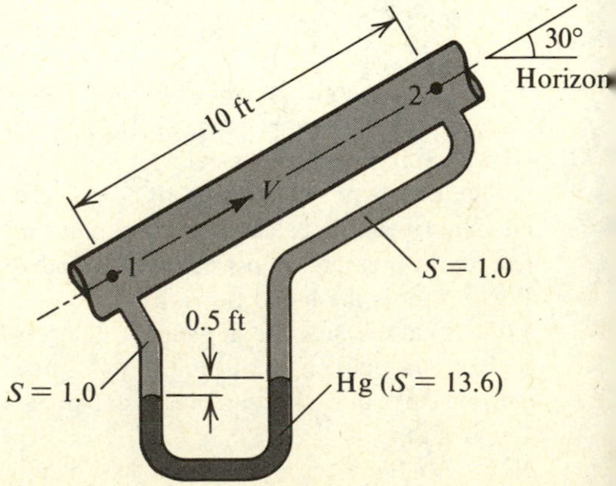

Figure P3.50

3.52 Hurricane winds of 60.8 m/s are blowing directly against a 2.78 m^2 window of the Empire State Building. Assuming air to be an ideal fluid having a specific weight of 12.566 N/m^3, what is the force exerted on the window if the stagnation pressure p_s is assumed to act over the entire window area if $p_s = \frac{1}{2}\rho V^2$?

3.53 From a manometer reading R and the depth L, compute the depth of water h that has risen in the diving bell shown in Fig. P3.53.

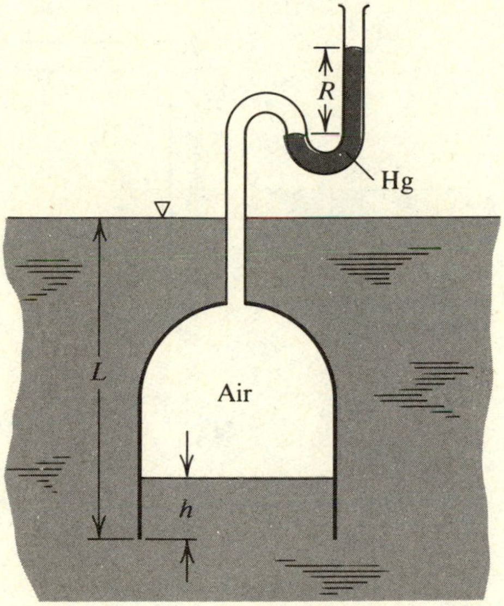

Figure P3.53

3.54 A force F of 20,000 lbf is to be developed in a hydraulic press as shown in Fig. P3.54. Find the required weight W to achieve this, where the area of the large piston is 100 ft^2 and that of the small one is 1 ft^2. The fluid is an oil of specific gravity 0.8 (lbf).

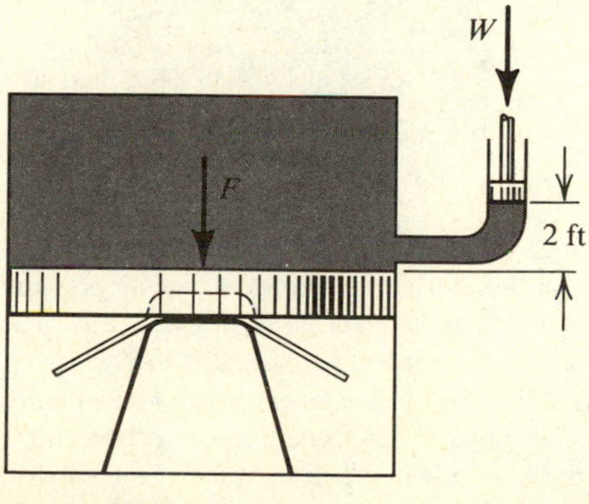

Figure P3.54

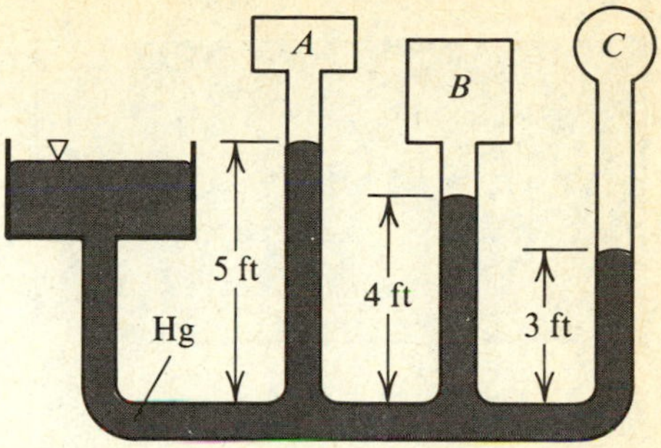

Figure P3.55

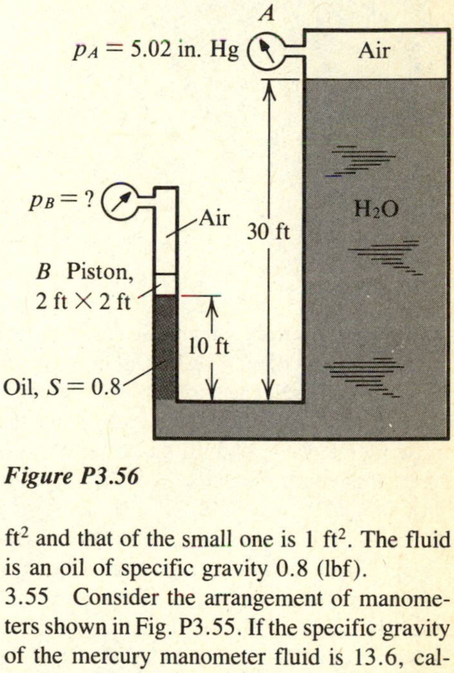

Figure P3.56

3.55 Consider the arrangement of manometers shown in Fig. P3.55. If the specific gravity of the mercury manometer fluid is 13.6, calculate the pressure difference $p_C - p_A$, $p_C - p_B$, $p_B - p_A$ (psi).

3.56 The air tank used to drive the tornado facility at the U.S. Naval Academy consists of three fluids: air, water, and oil. If the pressure at A is 5.02 in. Hg, what is the gauge pressure in the air tank shown in Fig. P3.56 to maintain the 2 ft by 2 ft square piston in equilibrium if friction and leakage are neglected?

Forces of Planar Objects

3.57 A hole in the side of a ship has been patched with a 4 foot square plate. The plane

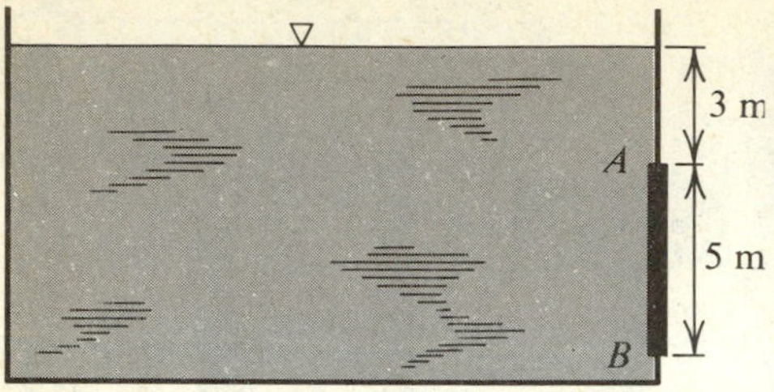

Figure P3.59

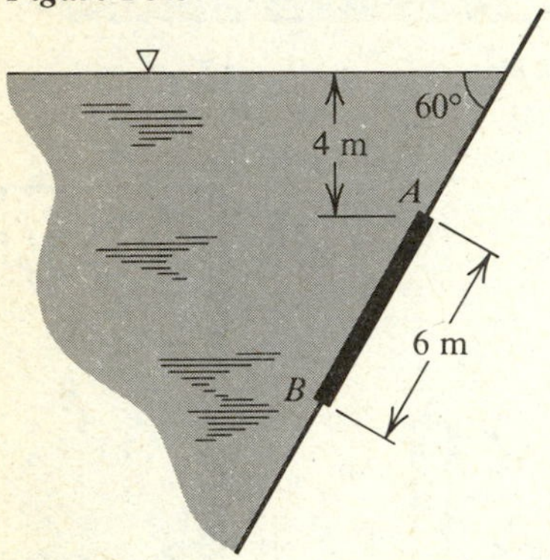

Figure P3.60

of the plate is vertical and its top edge is 3 feet below the waterline. The ship is in saltwater (S = 1.025). Neglecting atmospheric pressure, find (a) the total force exerted on the plate by the water, and (b) the center of pressure on the plate measured from the waterline (ft).

3.58 A rectangular plate 3 m by 4 m is submerged in water with its upper 3-m edge flush and horizontal to the surface. The plane of the plate is inclined 30° from the horizontal. Calculate (a) the force exerted on one side of the plate by the water (N), and (b) the location of the center of pressure (m).

3.59 Determine the hydrostatic pressure force F_p due to water acting on a 4 m by 5 m rectangular area *AB* shown in Fig. P3.59. Locate its line-of-action.

3.60 Determine the hydrostatic pressure force F_p due to water acting on a 4 m by 6 m triangular gate *AB* shown in Fig. P3.60, if its apex is at *A*. Locate the center of pressure of this force.

3.61 Determine the hydrostatic pressure force on side *ABC* in Fig. P3.61, which is 3 ft wide, given S_{oil} = 0.802.

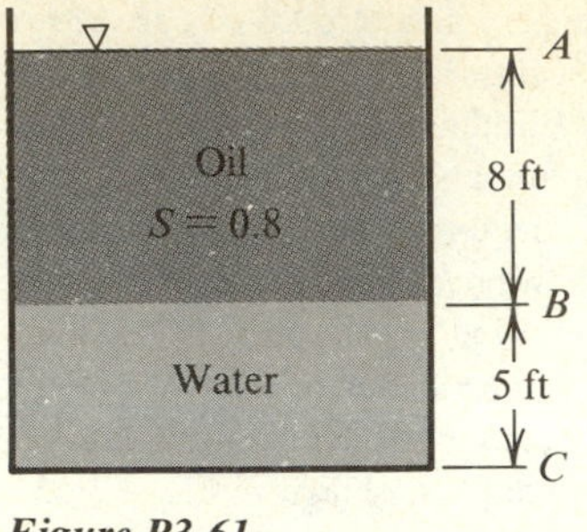

Figure P3.61

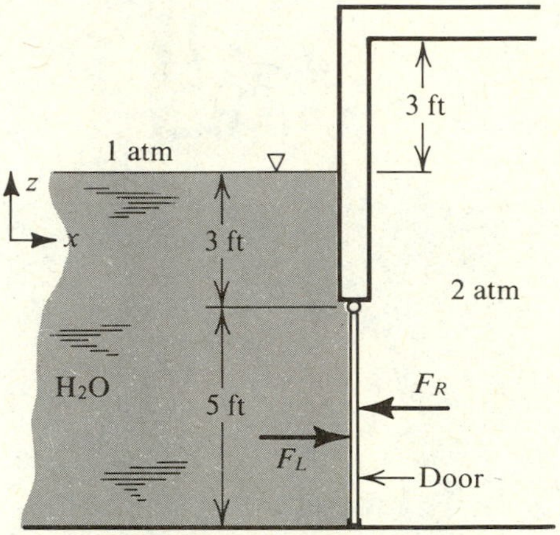

Figure P3.62

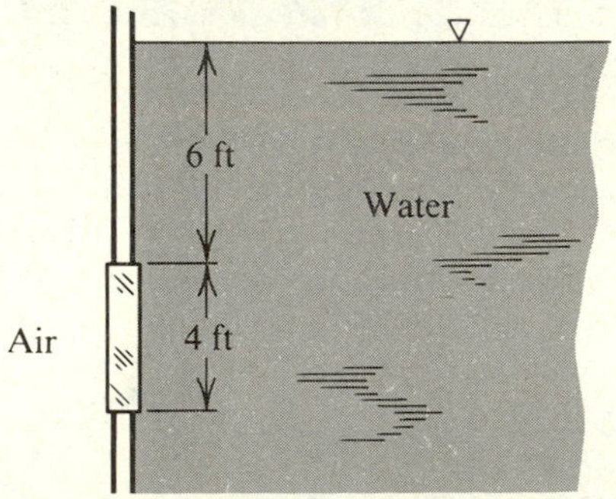

Figure P3.63

3.62 Determine the total force and its direction (left or right) due to water and air pressure in the surface shown in Fig. P3.62 (lbf). The door is hinged at the top and is 6 ft wide.

3.63 What is the magnitude of the hydrostatic pressure force and its location on a ship's porthole 4 ft in diameter whose center is 8 ft below the water surface (Fig. P3.63)?

3.64 Determine the depth of water h so that the resultant force on the gate shown in Fig. P3.64 is 6240 lbf acting to the right.

3.65 The rectangular gate in Fig. P3.65 AB is 2 m wide and 3 m high. The gate is hinged at A, separating water and oil of specific gravity 0.85. The water is confined in a tank with air pressured to 10 psi. Oil is confined in a second tank open to the atmosphere. What force must be exerted on the gate to keep it closed? Where should this force act?

3.66 The aquarium shown in Fig. P3.66 contains fresh water for studying aquatic life. An emergency access door 5 ft long and 10 ft wide has one edge located 10 ft below the water surface. Find the magnitude of the hydrostatic pressure force on the door AB by (a) formula and (b) integration.

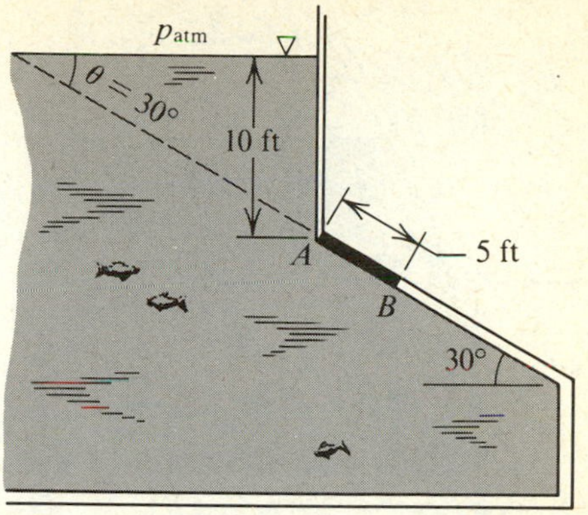

Figure P3.66

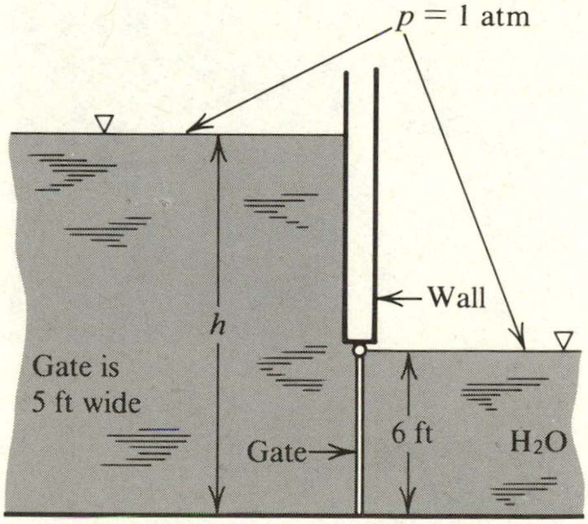

Figure P3.64

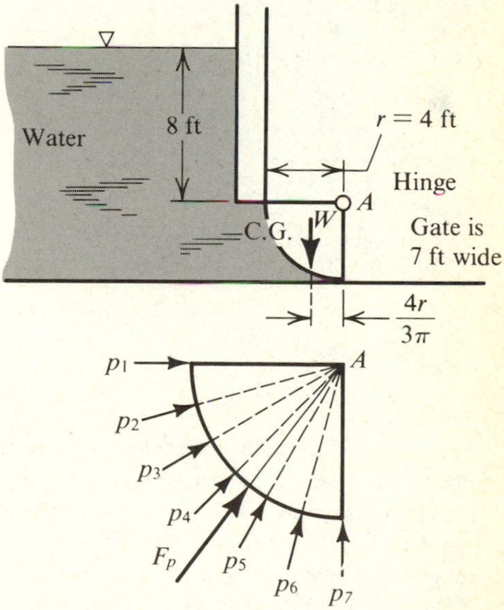

Figure P3.67

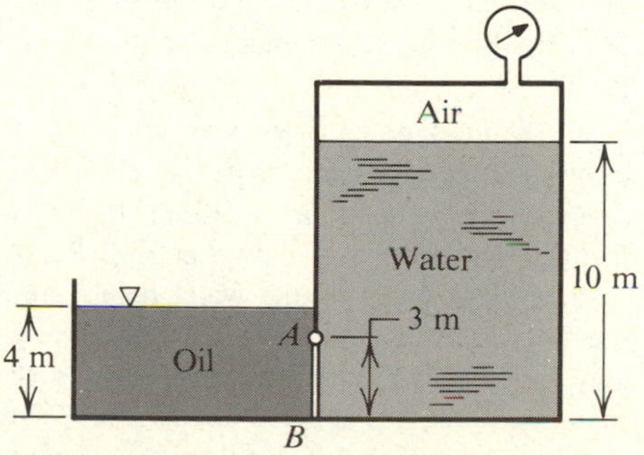

Figure P3.65

3.67 The curved gate (quarter of a circle) in Fig. P3.67, whose radius of curvature is 4 ft, is 7 ft wide and hinged at A. The top of the gate is 8 ft below the water surface. Find the moment required to open the gate if the gate weighs 300 lbf (ft-lbf).

3.68 The gate shown in Fig. P3.68 is in static equilibrium when the water level is 4 ft above the gate hinge and when $\theta = 60°$. Find the weight of the gate.

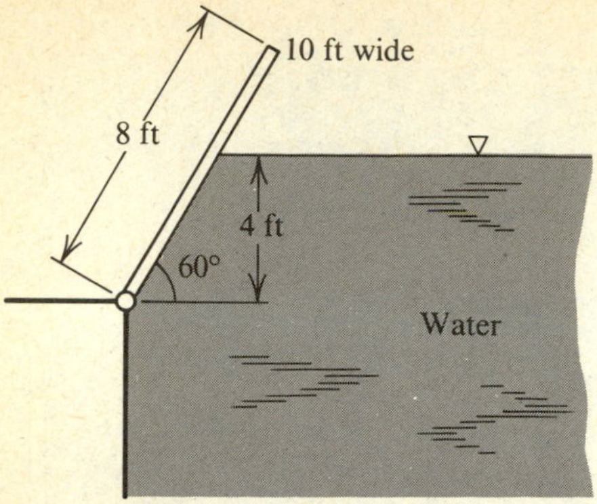

Figure P3.68

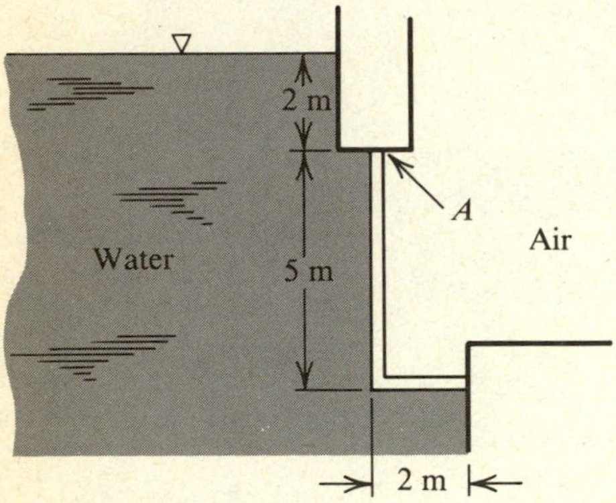

Figure P3.69

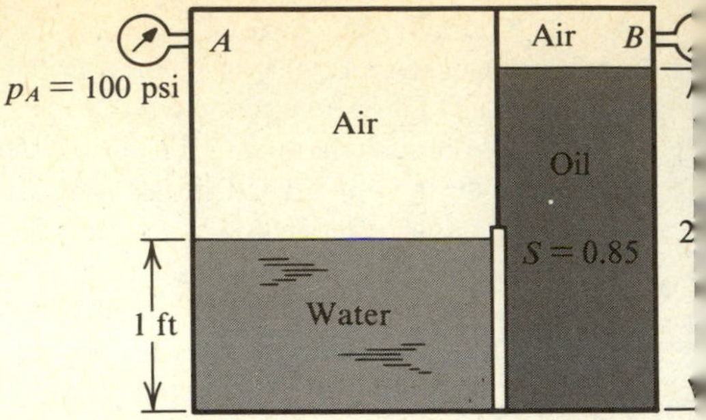

Figure P3.71

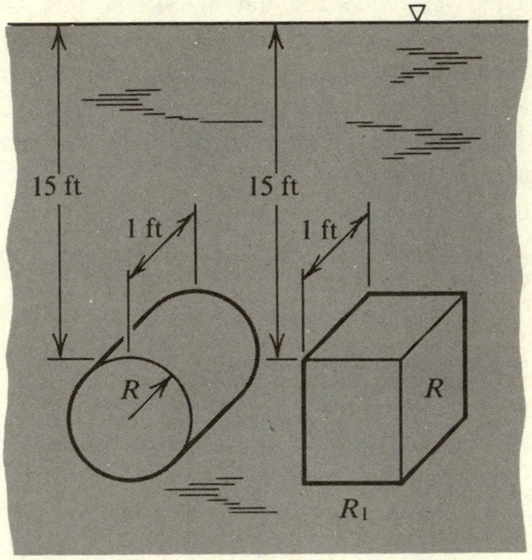

Figure P3.72

3.69 An L-shaped flat plate holds back a large quantity of water, as shown in Fig. P3.69. Find the moment about A due to (a) the water on the 5 m length and (b) the water on the 2 m length. The plate is 5 m wide.

3.70 With reference to Prob. 3.69, what is the magnitude of horizontal force necessary to open the gate if it were applied at the center of pressure along the 5 m length (N)?

3.71 The coffer dam shown in Fig. P3.71 supports a 1 ft head of water in a pressurized chamber on one side and 2 ft of oil on the other. (a) What is the force per foot of length on the water side of the dam if the chamber is pressurized to 100 psi? (b) What must the pressure be at B to keep the gate closed?

3.72 A triangular tank is filled with water, as shown in Fig. P3.72. (a) What is the total force on one end and one side of the tank? (b) What is the location of the force on one end?

3.73 Two check valves are located 15 ft below the water surface, as shown in Fig. P3.73. (a) Determine which valve has the greater total force. (b) In order to have equal total forces on both valves, what relationship must exist between R_1 and R, assuming R remains constant?

3.74 A pressure-sensitive deep submergence mine must be repeatedly tested at a *minimum* pressure of 40 psia in seawater (γ = 64 lbf/ft^3). The device is shown in Fig. P3.74. (a) How high must be the water height from the floor in order to achieve the desired pressure in the mine (in.)? (b) What will be the maximum pressure in the mine (psi)? (c) What is the total force on the longest sidewall of the mine (lbf)?

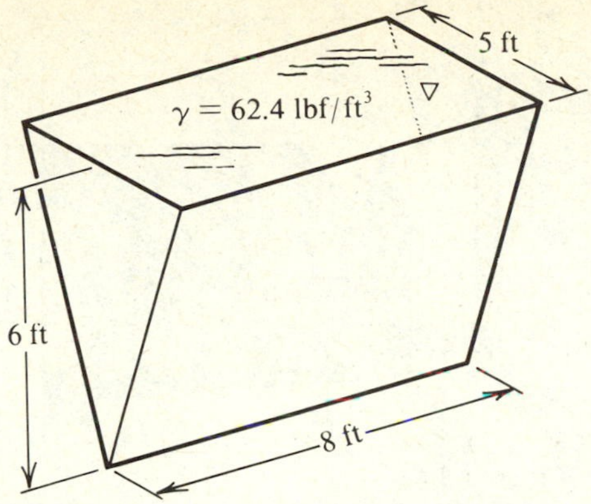

Figure P3.73

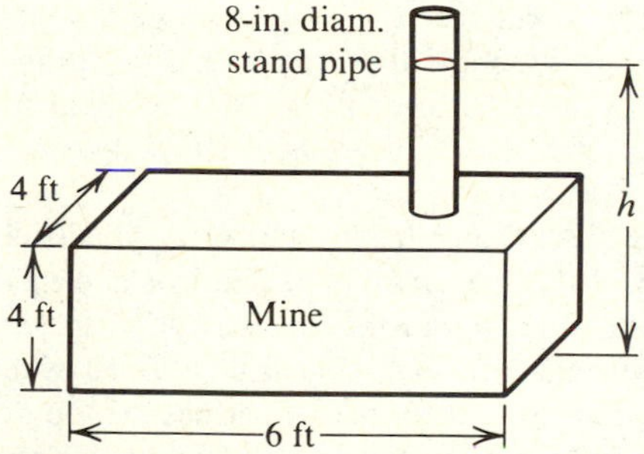

Figure P3.74

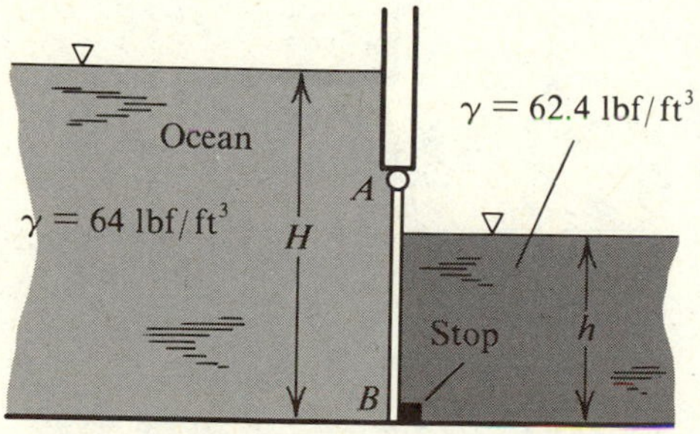

Figure P3.75

3.75 As shown in Fig. P3.75, a vertical gate *AB* 20 ft wide and 10 ft high is hinged at *A*. A mechanical stop is at *B*. The gate allows discharge water from a hydroelectric power plant to be dumped into the ocean. At what level *h* will the gate open (neglect the weight of the gate)?

3.76 Repeat Prob. 3.75 for a gate 10 ft wide.

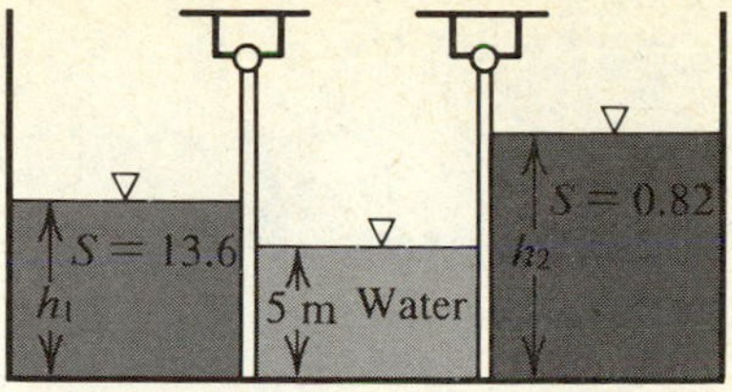

Figure P3.77

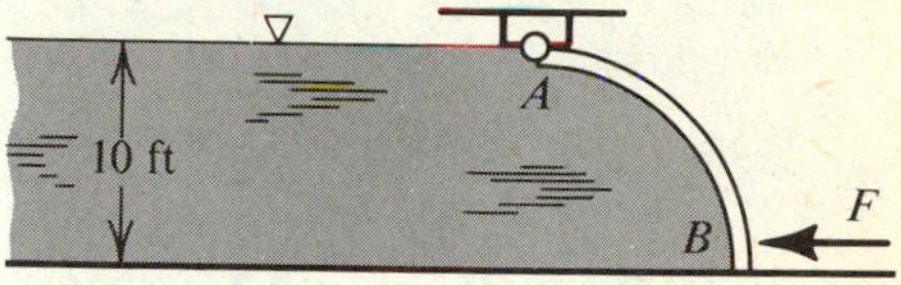

Figure P3.78

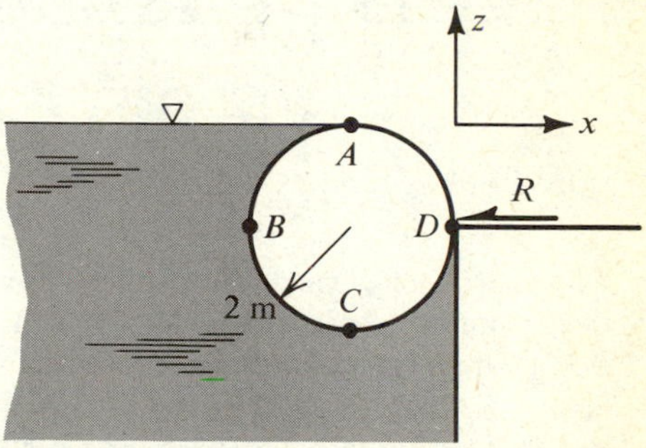

Figure P3.80

3.77 A series of gates keep various fluids separated. Consider the two gates separating three fluids shown in Fig. P3.77. If each gate is 1 m wide, calculate the heights h_1 and h_2 to keep the gates closed.

Forces on Curved Bodies, Buoyancy

3.78 The curved gate (a quarter of a circle) in Fig. P3.78 hinged at *A* confines water to a depth of 10 ft. Find the magnitude of the force *F* to keep the gate shut if the gate is 5 ft wide.

3.79 Repeat Prob. 3.78 except have the gate hinged at *B* and the force vertical at *A*.

3.80 A cylindrical beam 10 m long acts as a dam holding back water as shown in Fig. P3.80. Given that the friction between the wall and cylinder is zero, find (a) p_o (kPa), (b) F_{BCD} (N), (c) $(F_{ABC})_x$ (N), (d) $(F_{ABC})_z$ (N), (e) F_b (N), (f) reaction force *R* (N), and (g) cylindrical weight *W* (N).

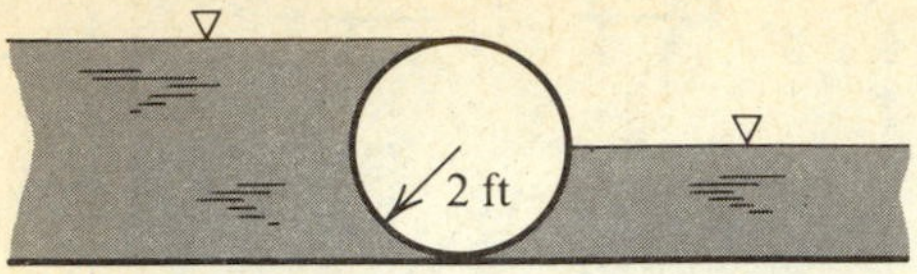

Figure P3.81

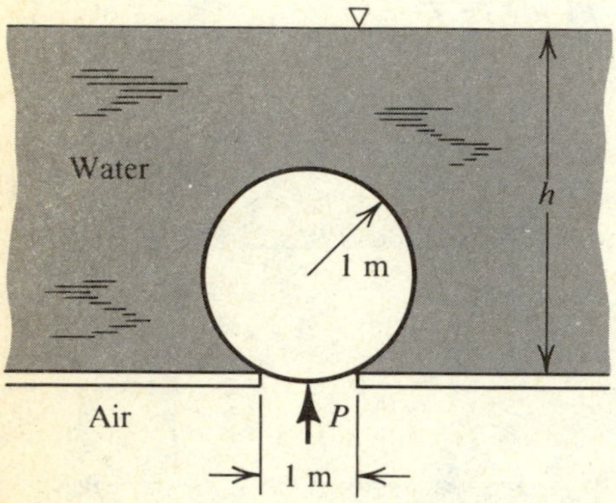

Figure P3.82

3.81 A 10 ft long cylindrical beam acts as a dam to partition water as shown in Fig. P3.81. Calculate the weight of the cylinder in its position.

3.82 A 2 m diameter cylindrical beam weighs 100 N and rests on a rectangular opening 1 m wide. If the opening is the same length as the beam (and no water escapes so that the water depth is a constant h feet), calculate the external force P to move the cylinder from the opening (see Fig. P3.82).

3.83 A spherical object weighs 20 N in air. A force of 200 N is required to keep the spherical object completely submerged in water. Calculate the radius of the object (m).

3.84 A large metal tank with an irregular shape is used to study predatory fish behavior patterns (Fig. P3.84). A hemispherical window, G-H, is used to view the fish. The tank has a width of 10 ft. The specific gravity of the seawater in the tank is 1.023. The gauge pressure at point A is 672 lbf/ft^2. Find (a) the pressure at point E expressed in feet of seawater, (b) the force on the top surface of tank (B-C-D-E-F), (c) the vertical force acting on the hemispherical surface G-H as a result of the water pressure, and (d) the horizontal force

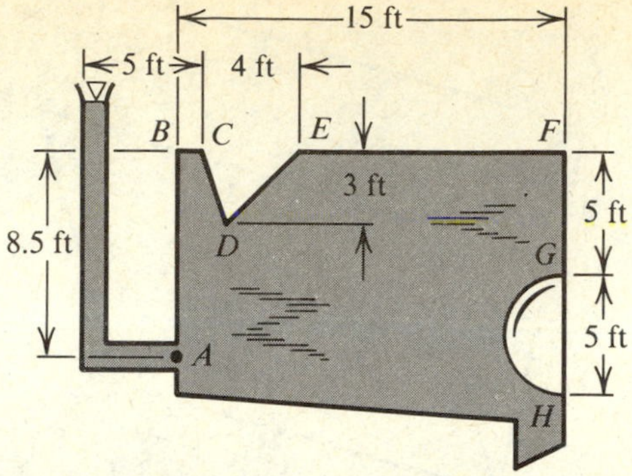

Figure P3.84

acting on the hemispherical surface G-H as a result of the water pressure.

3.85 A cylinder 3 m long floats on the surface of water. If the cylinder has a 1 m diameter, and a force of 13 N is required to submerge the cylinder, how much does the cylinder weigh in air?

3.86 As shown in Fig. P3.86, a cylindrical tank 2 ft in diameter lies in a horizontal plane. Located midway on the tank is a 1-in. diameter pipe vent 2 ft long. The tank is filled with gasoline $S = 0.63$ until it reaches the top of the vertical air vent pipe. What is the hydrostatic pressure force on one end of the tank?

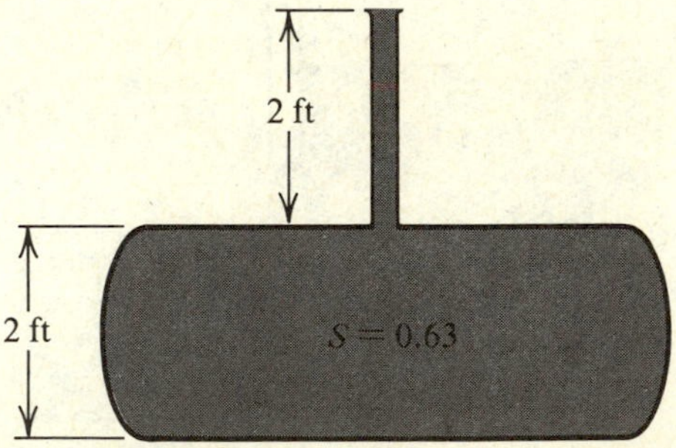

Figure P3.86

3.87 A spherical balloon is 5 m in diameter and is filled with a noncombustible gas where $R = 386$ ft-lbf/lbm°R. If the atmosphere is at 14.7 psia, and the gas in the balloon is at the same temperature as the atmosphere, 70°F, calculate the maximum load the balloon can lift (lbf).

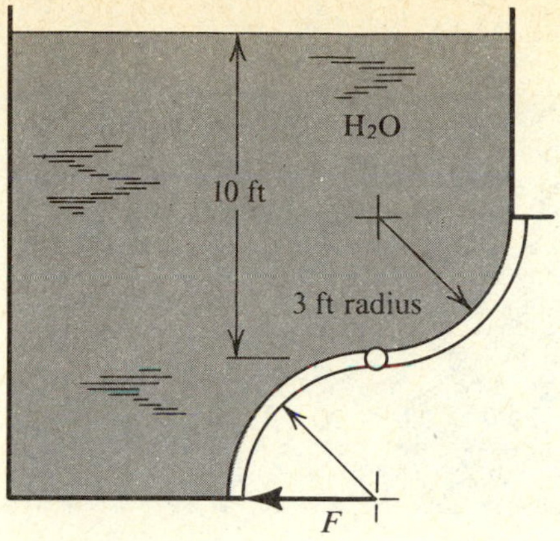

Figure P3.88

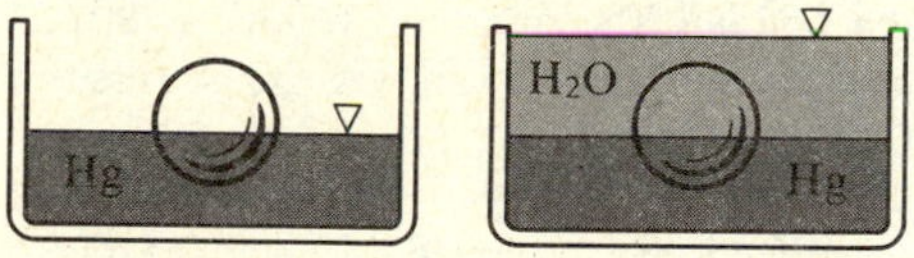

Figure P3.89

3.88 An S-shaped gate consisting of two quarter circles each of radius 3 ft, rotate about point *O* in Fig. P3.88. Calculate the horizontal and vertical force per unit width required to keep the gate shut (neglect any friction).

3.89 A hollow metal ball floats partly submerged in a dish of mercury, as shown in Fig. P3.89a. Water is then added until the ball is totally wetted, as shown in Fig. P3.89b. Does the ball sink deeper into the mercury because of the added weight or does it rise because of buoyancy? Explain why.

3.90 Repeat Prob. 3.89, using water rather than mercury, and oil rather than water.

3.91 A 30° sector gate of 2 m radius controls the flow of water in a flume (Fig. P3.91). Find the thrust per unit width on the gate (N/m).

3.92 A spherical air bubble rises in water. At a depth of 9 ft, its diameter is 0.01 in. What is its diameter just as it reaches the free-surface where the pressure is 1 atmosphere?

3.93 The profile of the inner face of a dam is described by a parabola having the form $x^2 = 3z$, where z is the height above the base and x is the horizontal distance of the face from

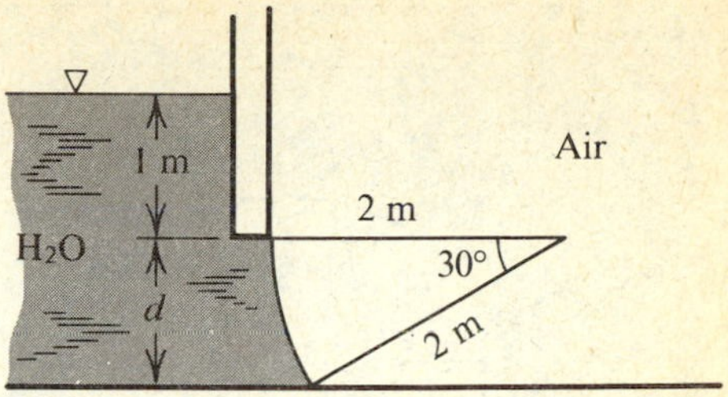

Figure P3.91

the vertical reference line. If the water level is 3 m above the base, find the thrust on the dam per unit width due to the water pressure, and its line-of-action.

3.94 A solid uniform cylinder of 2 m length and 0.25 m diameter is to float upright in water so that its upper surface is the free-surface of the water. Calculate its mass (kg).

3.95 A uniform wooden cylinder has a specific gravity of 0.65. Determine the diameter to length ratio of the cylinder so that it will just float upright in water.

3.96 A blob of unknown alloy weighs 6 lbf in air and 5 lbf in water. What is the volume of the blob, and what is its specific gravity?

3.97 A piece of ore weighing 7 lbf in air was found to weigh 5.6 lbf when submerged in water. Find (a) the volume of the ore, and (b) the specific gravity.

3.98 A sunken ship of original displacement of 5 tons is to be raised from 70 ft of seawater by means of a rubber flotation device. How many ft^3 of STD sea level air (60°F, 14.7 psia) must be pumped into the device to just lift the boat off the bottom?

3.99 A diving bell in equilibrium with local pressure is located 200 ft below sea level. An inclined manometer is connected to an air tank as shown in Fig. P3.99. How much air (lbm) is in the tank if the volume of the tank is 1.5 ft^3?

3.100 If the tank in Prob. 3.99 weighs 30 lbf (empty) how much does it weigh (filled) on a diver's back outside of the bell?

3.101 A round ball weighs 10 lbf in air, and a downward force of 160 lbf is required to keep it completely submerged in mercury ($S = 13.6$). Calculate the volume of the ball (ft^3).

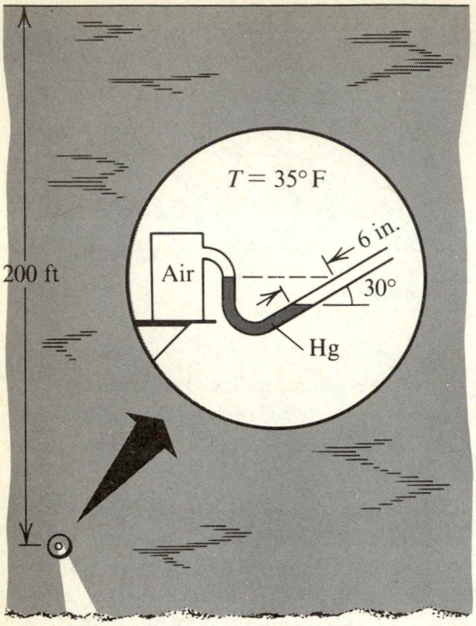

Figure P3.99

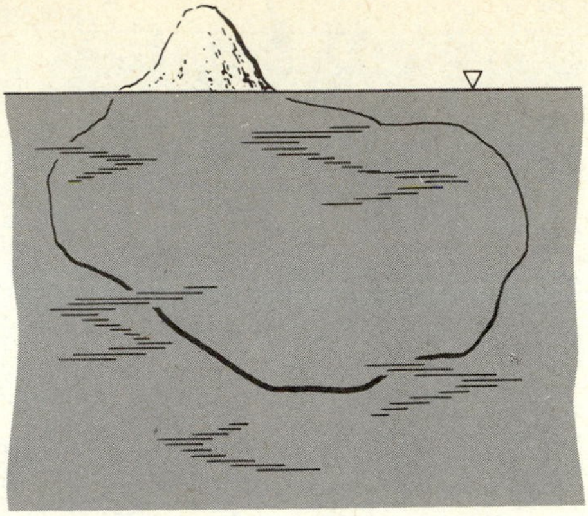

Figure P3.103

3.102 A piece of ore weighing 10 lbf in air was found to raise the level of water in a square tank (1 ft × 1 ft × 1 ft) by 0.03 ft when submerged in the tank. Find (a) the volume of the piece of ore (ft^3), (b) the buoyant force acting on the ore (lbf), and (c) the specific gravity (S) of the ore (neglect the buoyant force of air).

3.103 The iceberg shown in Fig. P3.103 has a density of 57.2 lbm/ft^3. What portion of its total volume will extend above the surface of the sea?

3.104 While at anchor in Sudha Bay, Crete, the passengers are allowed to have a beer bust/swim party on the ship's raft (see Fig. P3.104). The raft weighs 2000 lbf and is supported by 6 empty drums, each of which is 1.5 ft in diameter by 3.5 ft long and oriented vertically as shown below. Later it is discovered that the anchor is fouled on the bottom 50 ft below. A Tarzan type fills his lungs with air and using the anchor line hauls himself to the bottom. Tarzan's lung volume is 2.6 ft^3 and his specific gravity with lungs filled is 0.98. (a) How deep in the water do the barrels sink if the raft's weight is evenly distributed? (b) What is the absolute pressure on the diver? (c) If the diver exhales before returning to the surface will he move more rapidly? Why? (d) What is the magnitude of the force the diver must overcome to reach the anchor?

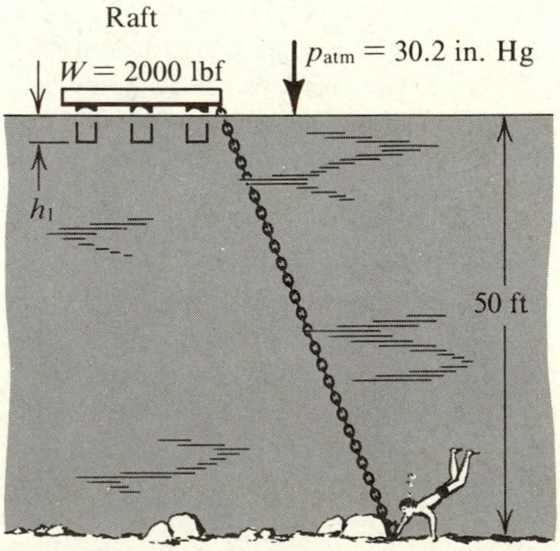

Figure P3.104

4 Differential Forms of Fluid Behavior

4.1 Introduction

Chapter 3 dealt with a special case of fluid flow: namely, where there is a fluid but no flow (except for the special case of constant acceleration). Even in this almost trivial case, we still have the problem of finding the distribution of each fluid property, such as pressure, density, and temperature. Such distributions have to be known in order to determine the loading on objects that were wetted by these fluids.

We now embark on the dynamics of fluid flows. We shall be interested in learning how the fluid's velocity is distributed as it moves over objects like ships, aircraft, and buildings, or through objects such as blades, channels, and pipes. In addition, we shall be concerned with the state of the fluid in space for any and all time.

Generally, we need to know the three scalar velocity components, the temperature, the pressure, and the density of the fluid—all six of which are functions of space and time. Since there are six primary unknowns, we must generate six mathematical relationships to evaluate them. All but one of these relationships are differential equations, i.e., rate equations expressing fluid behavior. They are:

1. *The equation of state*, which relates pressure, density, and temperature to each other. This was discussed in Chap. 2.
2. *The equation of continuity*, which is a statement of the conservation of mass.
3. *The equations of conservation of fluid momentum*, which are three scalar equations.
4. *The equations of conservation of fluid energy*.

We shall derive, discuss, and apply these equations first in differential form, and later in Chap. 5, in integrated form. The differential form permits us to evaluate the unknown dependent variables at any space point in the fluid flow for all time. The integrated form permits us to evaluate the quantities within a volume of fluid (such as fluid in a length of pipe). The method of describing the motion of the fluid will be in almost *every* case the Eulerian method described in Chap. 1. We shall use the Lagrangian descriptions to give some geometric insight into some fluid motions where necessary.

Of all the subjects in this text, the material in this chapter may prove the most challenging to the reader. The reasons are rather obvious. First, a fluid behaves in a more complicated manner when in motion than does the more familiar solid. Second, where multidimensions are used, any expression involving a number of different behaviors is bound to be fairly complicated when all variations in all directions for all variables are taken into account. In this chapter we have to face linear and nonlinear vector partial differential equations of first and second order. When we employ the scalar notation, these equations can unravel into rather long and tangled forms. We can package some expressions in tensors to keep them neater, since we know it is far easier to manipulate tensors than scalars. Other cases, however, will require generating quite a few terms.

4.2 General Property Balance

We presented in Chap. 1 two different ways to view the behavior of fluid flow, along with the appropriate mathematical operators that describe the rate of change of the various continuum variables encountered in fluid flow. We decided that focusing our attention at some predetermined field point in the fluid flow would be simpler than following a fluid particle. At this field point we can determine unique values of continuum variables such as pressure and velocity, values which characterize the flow's behavior. We suspect that, had we selected a different location, the unique value of any continuum variable might change. Or the property could change even in the same location. Such a condition we termed *unsteady*.

Suppose we let ϕ be the function that represents the *intensive* continuum property of the fluid that we are interested in determining (see Fig. 4.1). We sometimes speak of ϕ as the property per unit volume. Examples of intensive properties are density and specific weight. Let $\int \phi\ (x,y,z,t) d\forall$ represent the *extensive* continuum property in a volume $\forall$ of fluid and define

$$\Phi = \int_{\forall} \phi \, d\forall \tag{4.1}$$

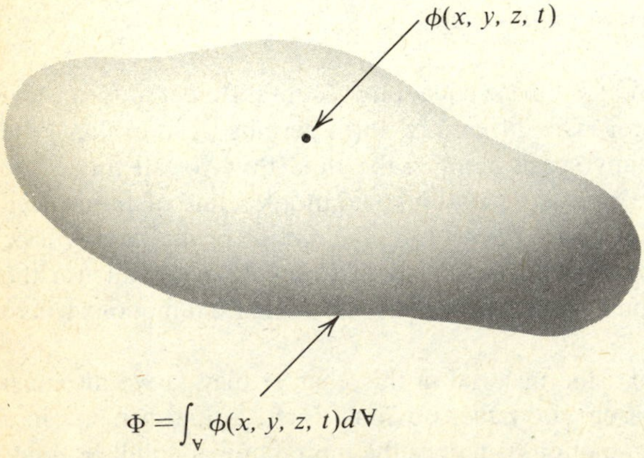

Figure 4.1 *Extensive versus intensive property of a fluid.*

Intensive properties can also be properties of mass, in which case we define

$$\Phi = \int_{\forall} \rho\overline{\phi}\, d\forall \tag{4.2}$$

where $\overline{\phi}$ is a property per unit mass. Examples of $\overline{\phi}$ are specific volume $\overline{v}$ and specific energy $\overline{e}$. If we wish to know how the intensive property $\phi\ (x,y,z,t)$ changes within the volume $\forall$, and across its surface, then we must operate on the property by either the ordinary differential d/dt using the Lagrangian description, or by the material derivative D/Dt using the Eulerian description. The following analysis allows us to transfer from the Lagrangian description to the Eulerian description.

Let us follow a *system* of fluid that is occupying a control volume $C{\cdot}\forall$ over an interval of time Δt. The control volume is indicated by the dashed line in Fig. 4.2, which also identifies the control surface. It is necessary to consider two other systems, labeled A and B, that are external to the control volume. The control surfaces at i (inlet) and e (exit) are open to the transfer of mass with intensive property ϕ_A and ϕ_B, respectively. At time t, the system is the sum of the mass within the control volume at that instant and the mass ΔM_A that is outside but adjacent to section i. At time $t + \Delta t$, the parcel A has moved into the control volume $C{\cdot}\forall$ and the mass ΔM_B has been transported out of the control volume through section e. Both the mass and volume of parcels A and B are assumed to be small compared to the mass and volume of the control volume. Thus, the change of the extensive property Φ for the time interval Δt is

$$\Phi_{C\cdot M} = \Phi_{C\cdot M,t+\Delta t} - \Phi_{C\cdot M,t} \tag{4.3}$$

Now we must transform Eq. (4.3), which is referenced to the control mass, to an equation referenced to the control volume of Fig. 4.2.

The property $\Phi_{C\cdot M,t}$ is the sum of the property of the system occupying the control volume at time t plus the property of the system A. That is, we state

$$\Phi_{C\cdot M,t} = \Phi_{C\cdot\forall,t} + \Delta\Phi_A \tag{4.4}$$

where $\Delta\Phi_A$ is the property Φ associated with the small parcel A which passes into the control volume. Similarly, the property of the control mass at time $t + \Delta t$ is

$$\Phi_{C\cdot M,t+\Delta t} = \Phi_{C\cdot\forall,t+\Delta t} + \Delta\Phi_B \tag{4.5}$$

where $\Delta\Phi_B$ is the property Φ associated with the small parcel B leaving the control volume. We substitute Eqs. (4.4) and (4.5) into Eq. (4.3) and divide through by an increment of time Δt to obtain

$$\frac{\Delta\Phi_{C\cdot M}}{\Delta t} = \frac{\Phi_{C\cdot\forall,t+\Delta t} - \Phi_{C\cdot\forall,t}}{\Delta t} + \frac{\Delta\Phi_B}{\Delta t} - \frac{\Delta\Phi_A}{\Delta t} \tag{4.6}$$

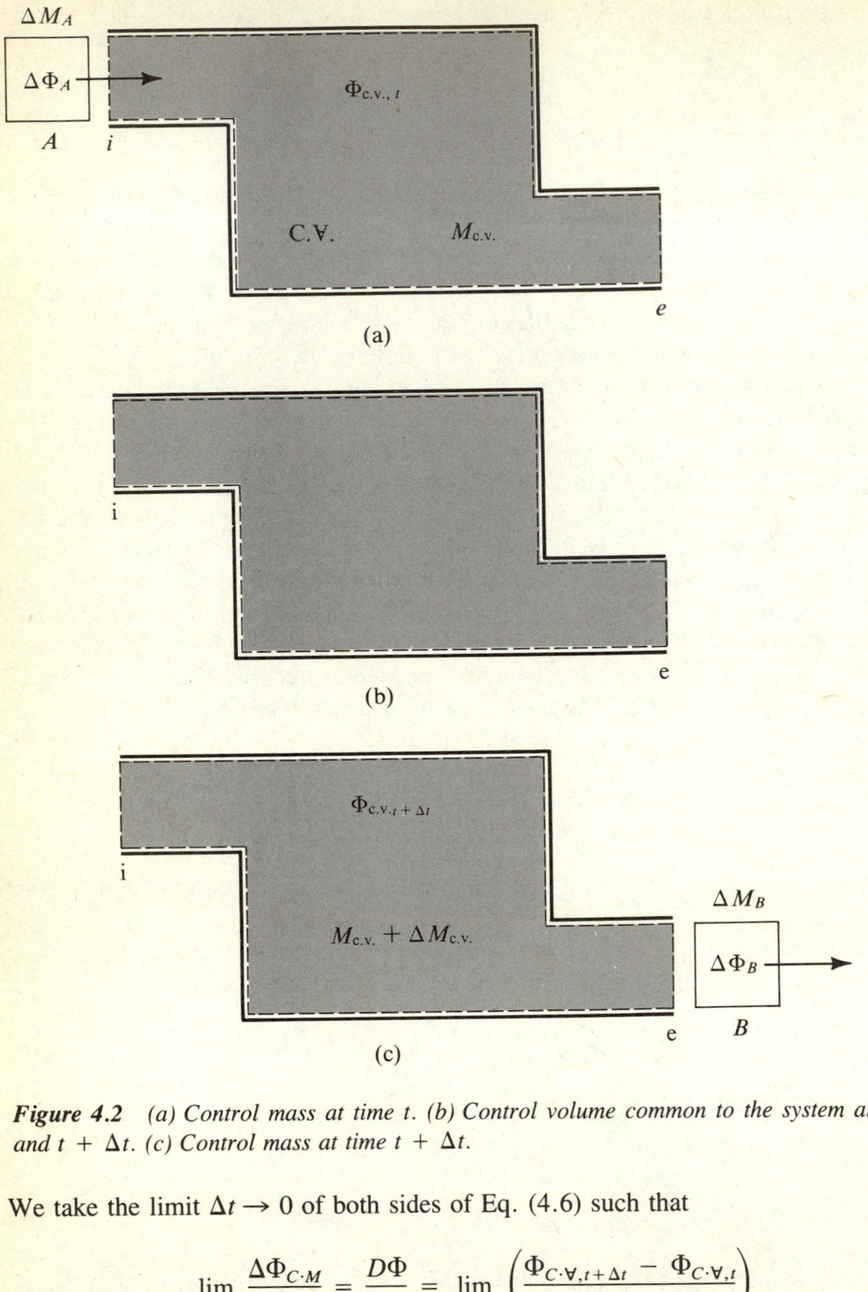

Figure 4.2 *(a) Control mass at time t. (b) Control volume common to the system at both t and t + Δt. (c) Control mass at time t + Δt.*

We take the limit $\Delta t \to 0$ of both sides of Eq. (4.6) such that

$$\lim_{\Delta t \to 0} \frac{\Delta \Phi_{C \cdot M}}{\Delta t} = \frac{D\Phi}{Dt} = \lim_{\Delta t \to 0} \left(\frac{\Phi_{C \cdot \forall, t + \Delta t} - \Phi_{C \cdot \forall, t}}{\Delta t} \right) + \lim_{\Delta t \to 0} \frac{\Delta \Phi_B}{\Delta t} - \lim_{\Delta t \to 0} \frac{\Delta \Phi_A}{\Delta t} \tag{4.7}$$

Substituting the expressions for the intensive property given by Eq. (4.1) into Eq. (4.7) results in

$$\frac{D\Phi}{Dt} = \lim_{\Delta t \to 0} \left(\frac{\int_{C \cdot \forall_{t+\Delta t}} \phi \, d\forall - \int_{C \cdot \forall_t} \phi \, d\forall}{\Delta t} \right) + \lim_{\Delta t \to 0} \frac{\Delta}{\Delta t} \int_{\forall_B} \phi_B \, d\forall - \lim_{\Delta t \to 0} \frac{\Delta}{\Delta t} \int_{\forall_A} \phi_A \, d\forall \tag{4.8}$$

The first term on the right-hand side of Eq. (4.8) involves the definition of a partial derivative:

$$\lim_{\Delta t \to 0} \frac{1}{\Delta t} \left(\int_{C \cdot \forall_{t+\Delta t}} \phi \, d\forall - \int_{C \cdot \forall_t} \phi \, d\forall \right) = \frac{\partial}{\partial t} \int_{\forall} \phi \, d\forall \tag{4.9}$$

where the volume $\forall$ is the volume of the control volume. The remaining two terms of Eq. (4.8) are easily transformed to area integrals by use of vector calculus:

$$\lim_{\Delta t \to 0} \frac{\Delta}{\Delta t} \int_{\forall_B} \phi_B \, d\forall = \frac{d}{dt} \int_{\forall_B} \phi_B \, d\forall = \int_{A_e} \phi_e \mathbf{V}_e \cdot d\mathbf{A} \tag{4.10}$$

It should be apparent that the volume of fluid crossing the exit section e into the region B in the interval of time dt is $d\forall_B = \mathbf{V}_B \cdot d\mathbf{A}dt = \mathbf{V}_e \cdot d\mathbf{A}dt$, since the elemental volume $d\forall_B$ can be expressed as the product of the exit area A_e and an incremental length ds which are, respectively, orthogonal and in the direction of the velocity of mass M_B. Similarly, we can obtain

$$\lim_{\Delta t \to 0} \frac{\Delta}{\Delta t} \int_{\forall_A} \phi_A \, d\forall = \int_{A_i} \phi_i \mathbf{V}_i \cdot d\mathbf{A} \tag{4.11}$$

Substituting Eqs. (4.9), (4.10), and (4.11) into Eq. (4.8) yields

$$\frac{D\Phi}{Dt} = \frac{\partial}{\partial t} \int_{\forall} \phi \, d\forall + \int_{A_e} \phi_e \mathbf{V}_e \cdot d\mathbf{A} - \int_{A_i} \phi_i \mathbf{V}_i \cdot d\mathbf{A} \tag{4.12}$$

or in more compact notation, we can state

$$\frac{D\Phi}{Dt} = \frac{\partial}{\partial t} \int_{\forall} \phi \, d\forall + \oint \phi \mathbf{V} \cdot d\mathbf{A} \tag{4.13}$$

which is the *integral form* of the general property balance.

Equation (4.13) states that the *total* change of any intensive property ϕ of the fluid in a volume $\forall$ fixed in space is equal to the sum of the local rate of change of that property momentarily contained in that volume and the *net flux* of that property *across* the surface of the volume. Or, to put it crudely, if we want to know how something is changing in the fluid flow, we must look at how it is changing in the control volume as well as at the boundaries. This important result is attributed to O.

Reynolds and is called the *Reynolds' transport theorem*. It allows us to transfer from a system approach (the Lagrangian description) to a control volume approach (the Eulerian description).

Figure 4.3 shows pictorially how a fluid property can change. The properties of pressure, density, mass, and temperature can change in the engine (denoted by control volume ∀), and at the surface of the boundary (the exhaust) where the fluid leaves the engine and enters the surroundings.

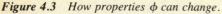

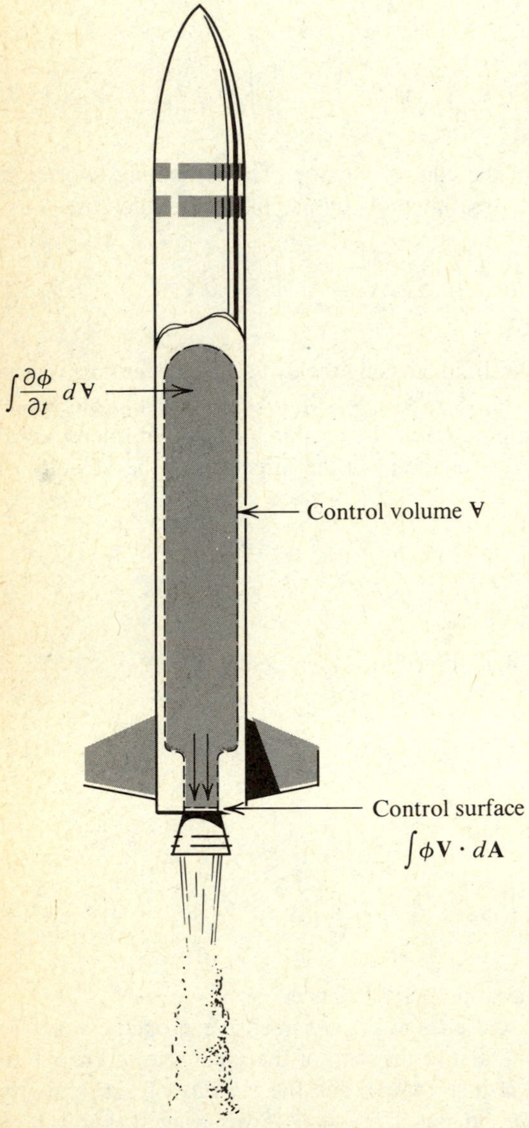

Figure 4.3 *How properties ϕ can change.*

If ϕ is independent of time, then the local change of the property vanishes, and Eq. (4.13) reduces to

$$\frac{D}{Dt}\int_{\forall}\phi \, d\forall = \oint_A \phi \mathbf{V}\cdot d\mathbf{A} \tag{4.14}$$

The differential form of the Reynolds' transport equation is a more useful form than the integral expression of Eq. (4.13) for describing the rate of change of a fluid property. We express the Reynolds' transport equation in terms of the intensive property ϕ by substituting Eq. (4.1) into Eq. (4.13), applying Gauss' law as given by Eq. (B.37) and rearranging terms to give us

$$\int_{\forall}\left[\left(\frac{D\phi}{Dt} - \frac{\partial\phi}{\partial t}\right) - \nabla\cdot\phi\mathbf{V}\right] d\forall = 0 \tag{4.15}$$

Since the property ϕ and velocity $\mathbf{V}$ are both continuous functions in the control volume $\forall$, the integrand exists at any arbitrary point in $\forall$ such that

$$\boxed{\frac{D\phi}{Dt} = \frac{\partial\phi}{\partial t} + \nabla\cdot(\phi\mathbf{V})} \tag{4.16}$$

Equation (4.16) is called the differential form (D.F.) of the general property balance.

Example 4.1

Express the transport Eq. (4.13) for (a) density ρ, and (b) energy E.

Solution:

(a) Density is related to the mass M by

$$\rho = \lim_{\Delta\forall\to\epsilon}\frac{\Delta M}{\Delta\forall}\text{*} \tag{i}$$

Substituting Eq. (i) into Eqs. (4.1) and (4.13) yields

$$\begin{aligned}\Phi &= \int \phi \, d\forall \\ &= \int \rho \, d\forall \\ &= \int dM = M\end{aligned} \tag{ii}$$

Thus, the transport equation for the continuum variable of density is

$$\frac{DM}{Dt} = \frac{\partial M}{\partial t} + \oint \rho\mathbf{V}\cdot d\mathbf{A} \tag{iii}$$

*The limiting volume ϵ is of the order of magnitude 10^{-9} mm^3 for fluids at atmospheric pressure (see Sec. 2.2.1).

Example 4.1 *(Con't.)*

(b) Energy E is related to the energy e per unit mass M by

$$\frac{dE}{dM} = e \tag{iv}$$

Substituting Eq. (i) into Eq. (iv) results in

$$dE = e d\forall \tag{v}$$

Thus,

$$\begin{aligned} \Phi &= \int \phi \, d\forall \\ &= \int \rho e \, d\forall \qquad \text{(vi)} \\ &= \int dE = E \end{aligned}$$

The transport equation for energy is

$$\frac{DE}{Dt} = \frac{\partial E}{\partial t} + \oint \rho e \mathbf{V} \cdot d\mathbf{A} \tag{vii}$$

If the term $\partial E/\partial t$ is zero, then the state of energy at each point in space is steady. This means that the energy contained within the control volume is invariant such that the only way the energy can change is across the control surface of the control volume. What we have now is the mathematical technique to express how fluid quantities can change in a given flow situation.

This completes the solution.

We shall now apply the differential form of the general property balance to the conservation of mass.

4.3 The Differential Form of the Conservation of Mass

The conservation principle of mass states that matter can neither be created nor destroyed. This is a well known, but not clearly understood principle. If we destroy mass, we have to take the volume of mass and in some fashion completely eliminate the mass within the confines of the control surface. We prohibit any transfer of any part of the mass through the control surface. To eliminate the mass requires that an implosion must take place with the result that nothing remains at the end of the implosion process. One need only consider nuclear physics to understand that mass can turn into energy. Here we have not destroyed mass. We have transformed it. This is allowed in the conservation principle. The reverse process of creating mass from energy is much more difficult to envision. It must be as if by magic that out of nothing comes something.

The conservation of mass states that the mass of the universe has always existed, and always shall exist. It can therefore only be manipulated. Thus matter can be transformed; e.g., gas to liquid, liquid to solid, solid to gas. What we want to state mathematically is that the total change of mass M cannot exist; that is,

$$\frac{DM}{Dt} = 0 \tag{4.17}$$

From the definition of the material derivative, Eq. (1.44), the total change of the mass consists of a local change plus a convective change:

$$\frac{\partial M}{\partial t} + (\mathbf{V}\cdot\nabla)M = 0 \tag{4.18}$$

the net flux of mass equals the accumulation or depletion of mass. If the mass does not vary with time, it cannot vary in space. Alternatively, it means that the mass per unit volume is a constant only for steady cases. Conversely, if the mass per unit volume is not a function of space, then it cannot vary locally with time.

It then follows from the definition of density and the conservation of mass, Eq. (4.17), that the total change of the density must be zero:

$$\frac{D\rho}{Dt} = 0 \tag{4.19}$$

Let the intensive continuum function ϕ be the fluid density ρ:

$$\phi(x,y,z,t) = \rho(x,y,z,t) \tag{4.20}$$

such that from the differential form of the general property balance

$$\boxed{\frac{\partial \rho}{\partial t} + \nabla\cdot(\rho\mathbf{V}) = 0} \tag{4.21}$$

The above equation represents the *continuity equation in differential form.*

Example 4.2

Express the continuity equation for compressible flow in (a) rectangular Cartesian coordinates and (b) cylindrical coordinates.

Solution:

Step 1.

Identify the characteristics of the fluid and flow field.

The fluid is compressible.

Example 4.2 *(Con't.)*

Step 2.
Write the appropriate form of the governing equations of flow.

The differential form of the continuity equation is

$$\frac{\partial \rho}{\partial t} + \nabla \cdot (\rho \mathbf{V}) = 0$$

(a) For rectangular Cartesian coordinates

$$\begin{aligned}
\frac{\partial \rho}{\partial t} &+ \nabla \cdot (\rho \mathbf{V}) \\
&= \frac{\partial \rho}{\partial t} + \left(\mathbf{i}\frac{\partial}{\partial x} + \mathbf{j}\frac{\partial}{\partial y} + \mathbf{k}\frac{\partial}{\partial z}\right) \cdot (\mathbf{i}\rho u + \mathbf{j}\rho v + \mathbf{k}\rho w) \\
&= \frac{\partial \rho}{\partial t} + \frac{\partial(\rho u)}{\partial x} + \frac{\partial(\rho v)}{\partial y} + \frac{\partial(\rho w)}{\partial z} \\
&= \frac{\partial \rho}{\partial t} + u\frac{\partial \rho}{\partial x} + v\frac{\partial \rho}{\partial y} + w\frac{\partial \rho}{\partial z} + \rho\left(\frac{\partial u}{\partial x} + \frac{\partial v}{\partial y} + \frac{\partial w}{\partial z}\right) = 0
\end{aligned} \tag{i}$$

(b) For cylindrical coordinates,

$$\begin{aligned}
\frac{\partial \rho}{\partial t} &+ \nabla \cdot (\rho \mathbf{V}) \\
&= \frac{\partial \rho}{\partial t} + \left(\mathbf{e}_r\frac{\partial}{\partial r} + \mathbf{e}_\theta\frac{1}{r}\frac{\partial}{\partial \theta} + \mathbf{k}\frac{\partial}{\partial z}\right) \cdot (\mathbf{e}_r\rho v_r + \mathbf{e}_\theta\rho v_\theta + \mathbf{k}\rho w) \\
&= \frac{\partial \rho}{\partial t} + \frac{\partial(\rho v_r)}{\partial r} + \frac{1}{r}\frac{\partial(\rho v_\theta)}{\partial \theta} + \frac{\rho v_r}{r} + \frac{\partial(\rho w)}{\partial z} \\
&= \frac{\partial \rho}{\partial t} + \frac{1}{r}\frac{\partial(\rho r v_r)}{\partial r} + \frac{1}{r}\frac{\partial(\rho v_\theta)}{\partial \theta} + \frac{\partial(\rho w)}{\partial z} \\
&= \frac{\partial \rho}{\partial t} + v_r\frac{\partial \rho}{\partial r} + \frac{v_\theta}{r}\frac{\partial \rho}{\partial \theta} + w\frac{\partial \rho}{\partial z} + \rho\left(\frac{1}{r}\frac{\partial(r v_r)}{\partial r} + \frac{\partial v_\theta}{r\partial \theta} + \frac{\partial w}{\partial z}\right) = 0
\end{aligned} \tag{ii}$$

In either case (a) or (b), we have a first degree partial differential equation with four unknowns. Obviously, the only way we can solve these equations is if we are given expressions for any three of the four variables.

This completes the solution.

If the flow is steady then the continuity equation, Eq. (4.21), becomes

$$\nabla \cdot (\rho \mathbf{V}) = \rho \nabla \cdot \mathbf{V} + \mathbf{V} \cdot \nabla \rho = 0 \tag{4.22}$$

Furthermore, if the fluid is incompressible, the continuity equation is expressed as

$$\nabla \cdot \mathbf{V} = D = 0 \tag{4.23}$$

where D is called the dilatation and is the divergence of the velocity vector. For any incompressible fluid, the dilatation is zero, which means that the fluid changes only its shape, not its volume. For such a condition the velocity field is called *solenoidal*.

The continuity equation for *incompressible flow*, Eq. (4.23) can be expressed in Cartesian coordinates with the aid of Eqs. (1.21) and (B.12) as

$$\boxed{\frac{\partial u}{\partial x} + \frac{\partial v}{\partial y} + \frac{\partial w}{\partial z} = 0} \tag{4.24}$$

The above equation is one of the most widely used equations in fluid mechanics. What it states is that if an incompressible fluid of elemental volume $dx \cdot dy \cdot dz$ is squashed as shown in Fig. 4.4, by pressing down vertically, the fluid will move toward the center at the rate $-\partial w/\partial z$. This in turn will cause the fluid to move outward in both the x- and y-directions at the rates $+\partial u/\partial x$ and $+\partial v/\partial y$. To see this movement, take a chalky blackboard eraser, place it on a table, and strike it. Lift up the eraser. The marks left by the chalk leaving the eraser will indicate that the chalk has flowed exactly as described here.

Example 4.3

Consider a two-dimensional flow of an incompressible fluid. If $u = 3 \sinh x$, what is the velocity field $\mathbf{V}$ if $v = \cosh x$ at $y = 0$?

Solution:

Step 1.

Identify the characteristics of the fluid and flow field.

The fluid is incompressible and the flow is steady ($\partial/\partial t = 0$) and two-dimensional ($\partial/\partial z = 0$).

Step 2.

Write the appropriate forms of the governing equations of flow.

The continuity equation for this flow is

$$\frac{\partial u}{\partial x} + \frac{\partial v}{\partial y} = 0 \tag{i}$$

Substituting $u = 3 \sinh x$ into Eq. (i) and integrating yields

$$v = -3y \cosh x + f(x) \tag{ii}$$

where $f(x)$ is an arbitrary function of integration. But at $y = 0$, $v = \cosh x$, then

$$f(x) = \cosh x \tag{iii}$$

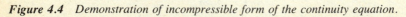

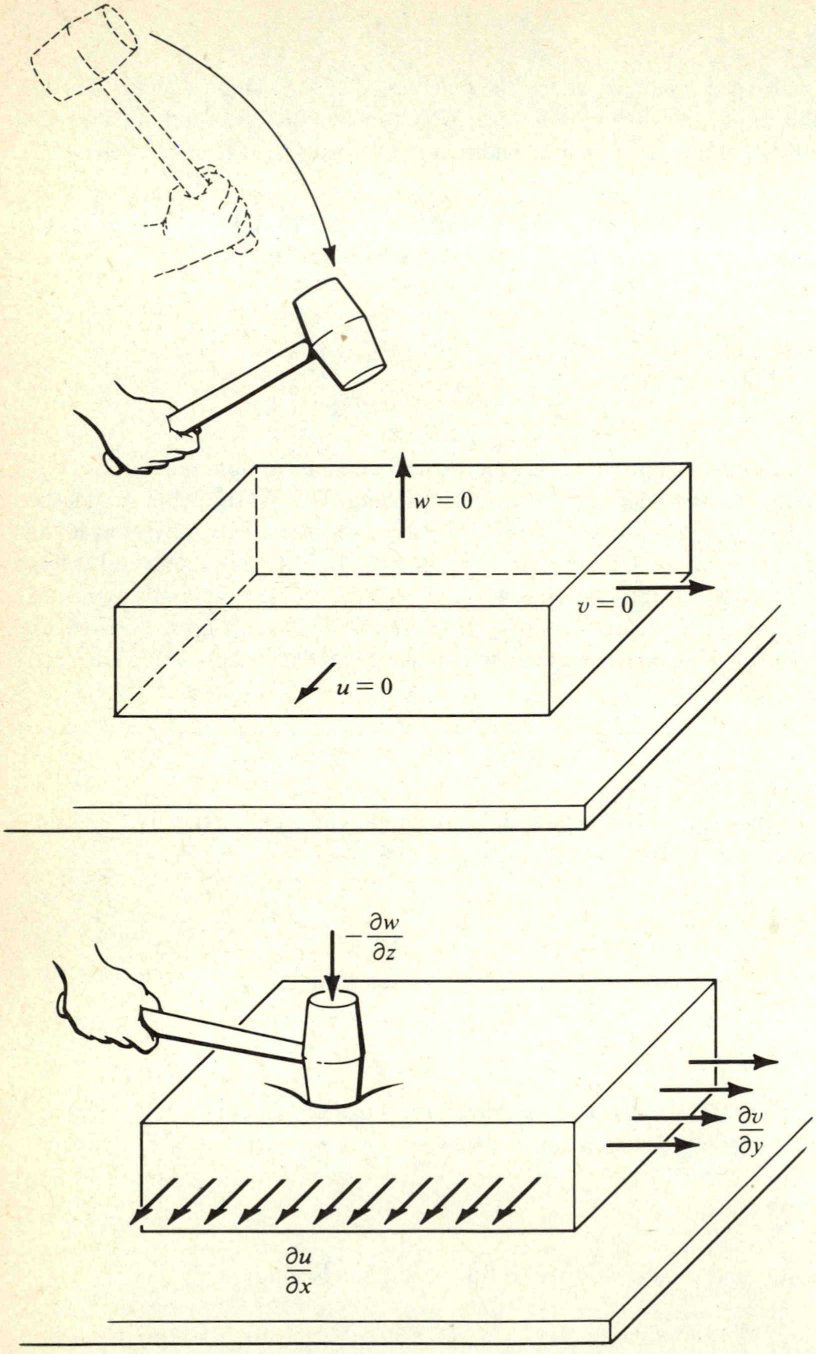

$$\frac{\partial u}{\partial x} + \frac{\partial v}{\partial y} = -\frac{\partial w}{\partial z}$$

Figure 4.4 *Demonstration of incompressible form of the continuity equation.*

Example 4.3 *(Con't.)*

The total solution for the velocity field is thus

$$\begin{aligned} \mathbf{V} &= u\mathbf{i} + v\mathbf{j} + w\mathbf{k} \qquad \text{(iv)} \\ &= 3 \sinh x\mathbf{i} + \cosh x(1 - 3y)\mathbf{j} \end{aligned}$$

This illustrates what we concluded in Example 4.2 and completes the solution.

The continuity equation in cylindrical coordinates is obtained from Eq. (4.23) using Eqs. (1.23) and (B.12) with the result

$$\boxed{\frac{\partial v_r}{\partial r} + \frac{v_r}{r} + \frac{1}{r}\frac{\partial v_\theta}{\partial \theta} + \frac{\partial w}{\partial z} = 0} \qquad (4.25)$$

The continuity equation in cylindrical coordinates for incompressible fluid flows informs us how the cylindrical velocity components must behave. For example, if the cylindrical flow is axisymmetric, like that shown in the tornado experiment of Fig. 4.5, the radial inflow governed by the expression $\frac{1}{r}\frac{\partial(rv_r)}{\partial r}$ is exactly balanced by the axial outflow $-\partial w/\partial v$ of Eq. (4.25). We can illustrate the three velocity components as follows. Take a ball on the end of a string and rotate it in a horizontal plane above your head. Keeping the string taut and rotating it with constant angular speed matches the case shown in Fig. 4.6a. Then, pulling the string with the other hand causes the swirling ball to move inward, as shown in Fig. 4.6b. The ball is still rotating in a horizontal plane (so there is no w-component of velocity), but now there is a radial velocity v_r and also a change in the tangential velocity v_θ. These velocities are governed by

$$\frac{\partial}{\partial r}(rv_r) = -\frac{\partial v_\theta}{\partial \theta}$$

Finally, with the ball rotating while the string is being pulled in, begin squatting. The act of squatting introduces the axial velocity w, so that the resultant behavior is shown as Fig. 4.6c.

We note in our two expressions for the conservation of mass, Eqs. (4.24) and (4.25), that if two of the three velocity components are given, the third can always be found by appropriate integration.

Example 4.4

Consider the three-dimensional incompressible vortex flow given by an axial velocity $w = 2az$, and a circumferential flow

$$v_\theta = \frac{A}{r}[1 - \exp(-ar)^2]$$

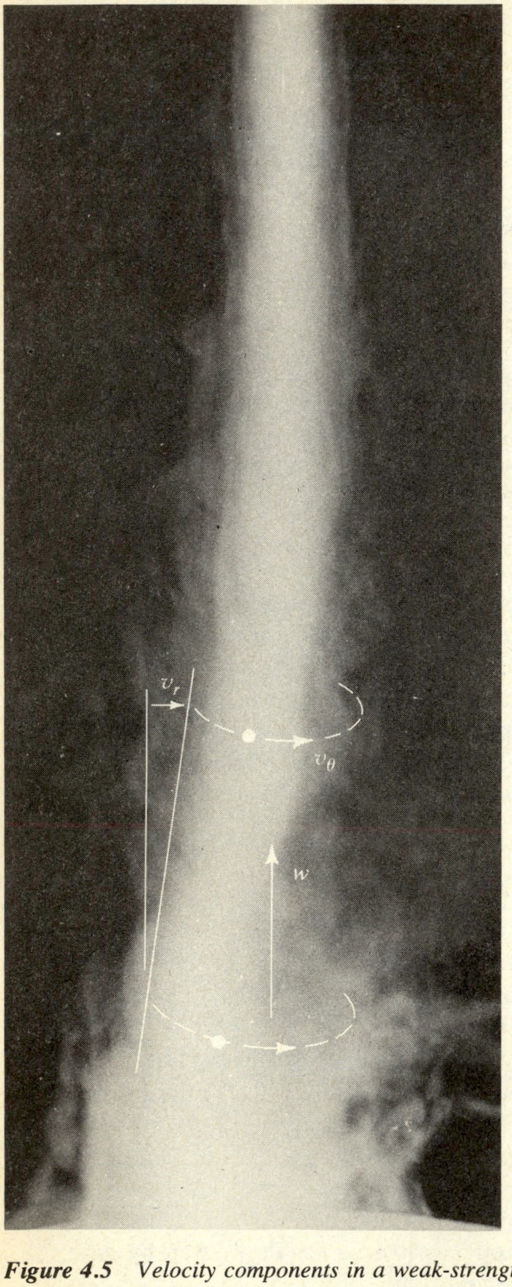

Figure 4.5 Velocity components in a weak-strength tornado.

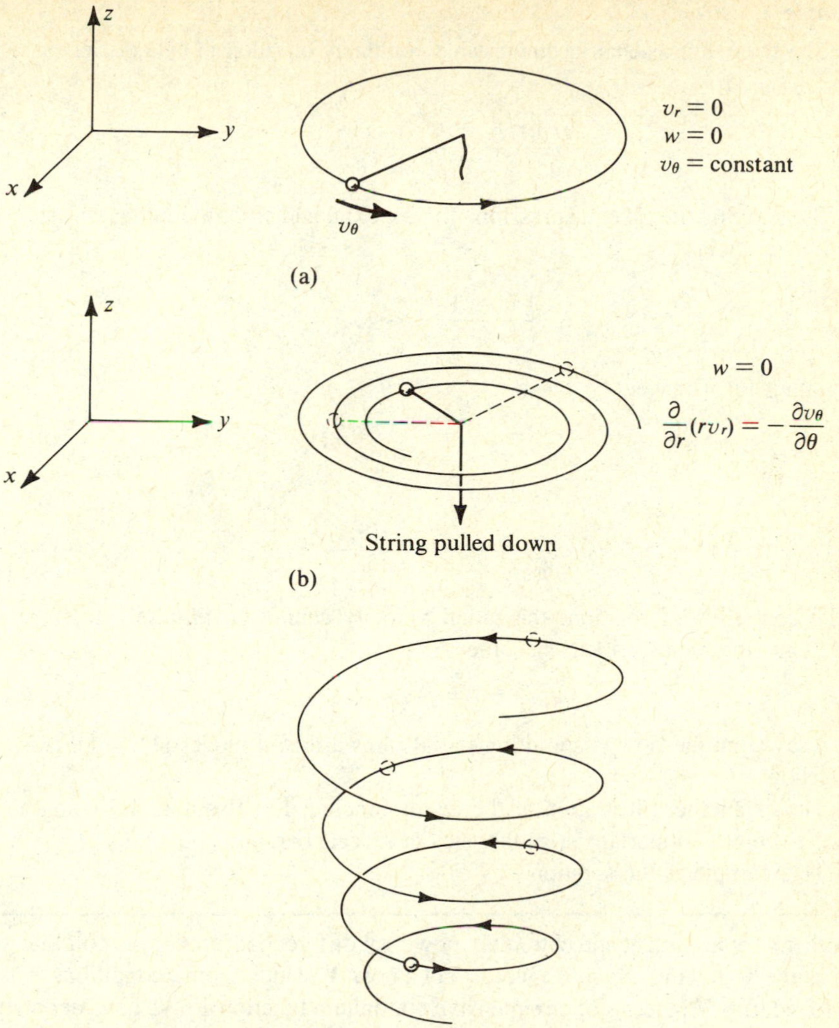

Figure 4.6 *(a) Pure circular motion in a horizontal plane. (b) Spiral motion in a horizontal plane. (c) Three-dimensional motion.*

Example 4.4 *(Con't.)*

Calculate the radial velocity v_r.

Solution:
Step 1.
Identify the characteristics of the fluid and the flow field.

The fluid is incompressible and the flow is steady and three-dimensional.

Step 2.
Write the appropriate forms of the governing equations of flow.

Example 4.4 *(Con't.)*

The three-dimensional incompressible continuity equation in cylindrical form is

$$\frac{1}{r}\frac{\partial(rv_r)}{\partial r} + \frac{1}{r}\frac{\partial v_\theta}{\partial \theta} + \frac{\partial w}{\partial z} = 0 \qquad \text{(i)}$$

Substituting the given expressions for the axial and circumferential velocity into Eq. (i) yields

$$\frac{1}{r}\frac{\partial(rv_r)}{\partial r} = -2a \qquad \text{(ii)}$$

Integrating Eq. (ii) yields

$$rv_r = -ar^2 + f(\theta,z) \qquad \text{(iii)}$$

or

$$v_r = -ar + \frac{f(\theta,z)}{r} \qquad \text{(iv)}$$

At the centerline of rotation, the radial velocity cannot be infinite, so $f(\theta,z) = 0$. Thus the radial velocity becomes

$$v_r = -ar \qquad \text{(v)}$$

which says that the flow is spiralling radially inward as it moves along the axis of rotation.

This is another illustration of the result concluded in Example 4.2 using a flow field that is important in geophysics and oceanography.

This completes the solution.

Another very important and extremely powerful differential form of the continuity equation is the following. Suppose we do not know $\mathbf{V}$. Under some conditions it is possible to redefine $\mathbf{V}$ in terms of our intensive continuum function ϕ. We have denoted ϕ as a scalar function. Since a vector is never equal to a scalar we need to make a vector relationship between the vector variable $\mathbf{V}$ and the scalar variable ϕ. This can be accomplished using operators. We multiply ϕ by the vector operator $\boldsymbol{\nabla}$, thus

$$\mathbf{V} = \boldsymbol{\nabla}\phi \qquad (4.26)$$

and adjust ϕ in such a way that Eq. (4.26) is satisfied. (We have simply redefined one unknown, $\mathbf{V}$, in terms of another unknown, $\boldsymbol{\nabla}\phi$.) Substituting Eq. (4.26) into the incompressible form of the continuity equation, Eq. (4.23), we obtain a rather interesting result:

$$\boldsymbol{\nabla}\cdot\boldsymbol{\nabla}\phi = \nabla^2\phi = 0 \qquad (4.27)$$

So, rather than solving the continuity equation as a first order partial differential equation with three unknown scalar velocity components [Eq. (4.24) or (4.25)], we express the continuity equation as a second order partial differential equation in *one* unknown: the scalar function ϕ. We can easily solve Eq. (4.27), whereas we cannot solve Eq. (4.23) unless two of the three unknown velocity components are specified. Additional boundary equations have to be generated to render the problem completely well posed. A word of caution: we have not mentioned it, but there is a restriction on using Eq. (4.27). We shall discuss this restriction in great detail later. For now, only the form matters.

We next turn to the second conservation law: the conservation of linear momentum. We will first address the behavior of fluid particles moving in an arbitrary manner.

4.4 The Differential Form of the Conservation of Linear Momentum

4.4.1 The Physics of the Problem

Consider a single fluid particle moving in a flow. We know that its behavior—that is, its velocity and its position—is affected by other particles in the flow. Just as people at rush hour jostle each other crowding through a gate on their way to the subway, so too do the particles in a flow push and turn and bump against one another (see Fig. 4.7). Particles deform as they move, and the *manner* in which each particle deforms helps determine the magnitude (or importance) of various forces. We must look first, then, at the analysis of fluid deformation.

In 1841, Cauchy [4.1] analyzed the deformation of solid bodies and derived his famous equation of motion, which properly predicts the behavior of solid particles. In 1845 Stokes [1.18] analyzed the deformation of viscous fluids and derived constitutive relations between stress components and rates of strain. And in 1858, Helmholtz [4.2] analyzed the deformation of ideal fluids. In the following progression, we employ Stokes' analysis.

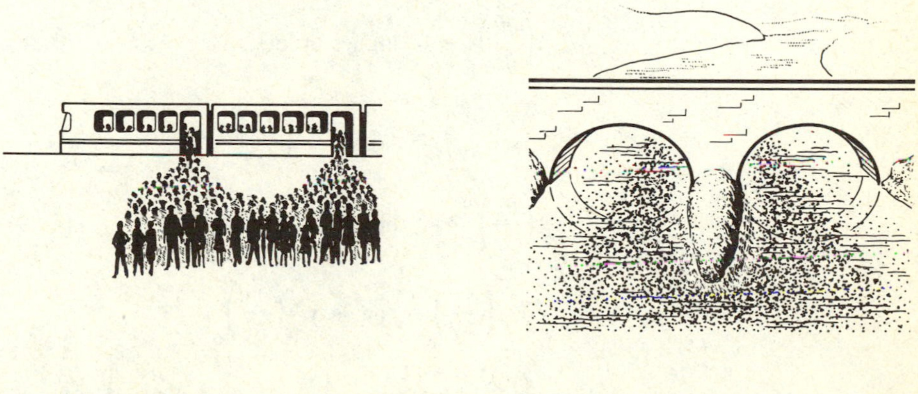

(a) (b)

Figure 4.7 *Particles in motion. (a) People particles. (b) Fluid particles.*

The reader has already learned in basic mechanics that the motion of particles is composed of

1. Pure translation
2. Pure rotation
3. Pure dilatation, sometimes referred to as linear strain
4. Pure shear, sometimes referred to as angular strain

We pointed out in the previous section that we are not concerned with the deformation of the fluid particle in itself, but rather with the *time rate of deformation*, or the velocity of the fluid particle. This is in essence the difference between the problem of solids and the problem of fluids. They have entirely different descriptions. The former we have identified as Lagrangian, and the latter as Eulerian. Since velocity is the principal variable of interest in fluids, it is appropriate we consider its composition in the Eulerian manner.

4.4.2 The Composition of Velocity

In studying the composition of velocity, we utilize the usual methods of mathematical analysis. Selecting the Eulerian description, we consider a fluid particle at some space point $P\ (x_0,y_0,z_0)$. In the following development all properties at this point will carry the suffix zero.

At some other point Q (see Fig. 4.8) close to P, the fluid has a velocity $\mathbf{V}$. We wish to know the velocity $\mathbf{V}$ in terms of what takes place at point P. To do this, it is convenient to develop the forthcoming relationships using scalar notation.

Using Taylor's series, we express the velocity components u,v,w of a particle at $Q\ (x,y,z)$ near $P\ (x_0,y_0,z_0)$ in series form:

$$u = u_0 + (x - x_0)\left(\frac{\partial u}{\partial x}\right)_0 + (y - y_0)\left(\frac{\partial u}{\partial y}\right)_0 + (z - z_0)\left(\frac{\partial u}{\partial z}\right)_0 + 0\ \text{(higher order)} \tag{4.28}$$

$$v = v_0 + (x - x_0)\left(\frac{\partial v}{\partial x}\right)_0 + (y - y_0)\left(\frac{\partial v}{\partial y}\right)_0 + (z - z_0)\left(\frac{\partial v}{\partial z}\right)_0 + 0\ \text{(higher order)} \tag{4.29}$$

$$w = w_0 + (x - x_0)\left(\frac{\partial w}{\partial x}\right)_0 + (y - y_0)\left(\frac{\partial w}{\partial y}\right)_0 + (z - z_0)\left(\frac{\partial w}{\partial z}\right)_0 + 0\ \text{(higher order)} \tag{4.30}$$

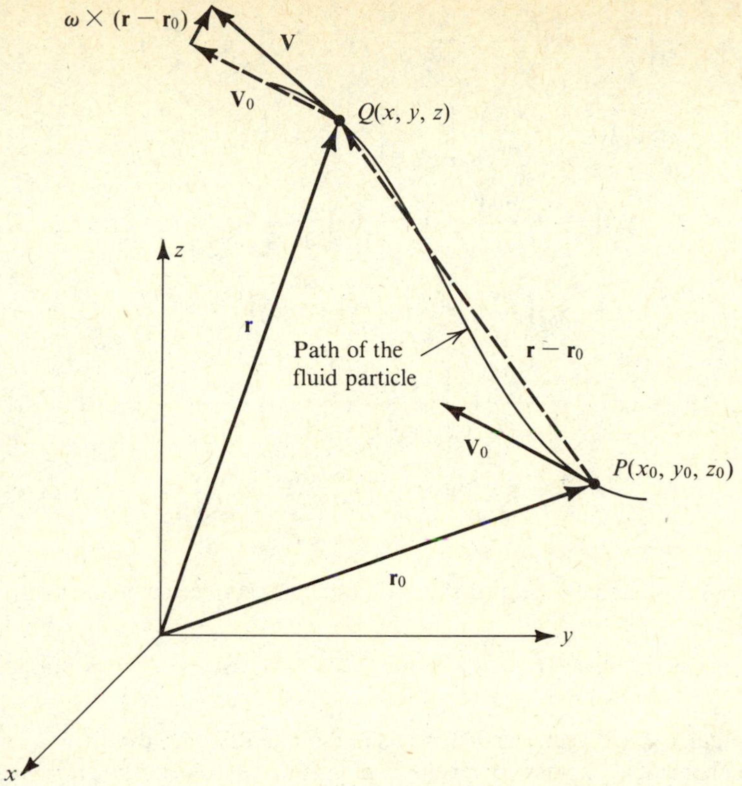

Figure 4.8 *Velocity of point Q with pure translation* $\mathbf{V}_0$ *and pure rotation* $(\mathbf{r} - \mathbf{r}_0) \times \boldsymbol{\omega}$.

The higher order terms involving terms $(\mathbf{r} - \mathbf{r}_0)^m$, $m \geq 2$, and higher order derivatives may be neglected only if $(\mathbf{r} - \mathbf{r}_0)$ is very small, i.e., if we consider velocity $\mathbf{V}$ very near $\mathbf{V}_0$. Thus, the difference between the velocity $\mathbf{V}_0$ at the reference point P and the velocity at some point near P is characterized by nine velocity derivatives. Furthermore, all values with the subscript zero are at most functions of time; whereas $\mathbf{V}$ is a *linear* function of the position vector $(\mathbf{r} - \mathbf{r}_0)$ which is usually designated $d\mathbf{r}$ in the limit $\mathbf{r} \to \mathbf{r}_0$. We shall neglect all higher order terms in Eqs. (4.28)–(4.30) as being small compared to $\mathbf{V}_0$ and first derivative terms.

Next, we add and subtract like differentials, collecting suitable terms together.

$$\begin{aligned} u = u_0 &+ \frac{1}{2}\left[(x - x_0)\left(\frac{\partial u}{\partial x} - \frac{\partial u}{\partial x}\right)_0 + (y - y_0)\left(\frac{\partial u}{\partial y} - \frac{\partial v}{\partial x}\right)_0\right. \\ &\left. + (z - z_0)\left(\frac{\partial u}{\partial z} - \frac{\partial w}{\partial x}\right)_0\right] + \frac{1}{2}\left[(x - x_0)\left(\frac{\partial u}{\partial x} + \frac{\partial u}{\partial x}\right)_0\right. \\ &\left. + (y - y_0)\left(\frac{\partial u}{\partial y} + \frac{\partial v}{\partial x}\right)_0 + (z - z_0)\left(\frac{\partial u}{\partial z} + \frac{\partial w}{\partial x}\right)_0\right] \end{aligned} \tag{4.31}$$

$$v = v_0 + \frac{1}{2}\left[(x - x_0)\left(\frac{\partial v}{\partial x} - \frac{\partial u}{\partial y}\right)_0 + (y - y_0)\left(\frac{\partial v}{\partial y} - \frac{\partial v}{\partial y}\right)_0\right.$$
$$\left. + (z - z_0)\left(\frac{\partial v}{\partial z} - \frac{\partial w}{\partial y}\right)_0\right] + \frac{1}{2}\left[(x - x_0)\left(\frac{\partial v}{\partial x} + \frac{\partial u}{\partial y}\right)_0\right.$$
$$\left. + (y - y_0)\left(\frac{\partial v}{\partial y} + \frac{\partial v}{\partial y}\right)_0 + (z - z_0)\left(\frac{\partial v}{\partial z} + \frac{\partial w}{\partial y}\right)_0\right] \tag{4.32}$$

$$w = w_0 + \frac{1}{2}\left[(x - x_0)\left(\frac{\partial w}{\partial x} - \frac{\partial u}{\partial z}\right)_0 + (y - y_0)\left(\frac{\partial w}{\partial y} - \frac{\partial v}{\partial z}\right)_0\right.$$
$$\left. + (z - z_0)\left(\frac{\partial w}{\partial z} - \frac{\partial w}{\partial z}\right)_0\right] + \frac{1}{2}\left[(x - x_0)\left(\frac{\partial w}{\partial z} + \frac{\partial u}{\partial z}\right)_0\right.$$
$$\left. + (y - y_0)\left(\frac{\partial w}{\partial y} + \frac{\partial v}{\partial z}\right) + (z - z_0)\left(\frac{\partial w}{\partial z} + \frac{\partial w}{\partial z}\right)\right] \tag{4.33}$$

Equations (4.31)–(4.33) can now be written in a more compact vector notation as

$$\mathbf{V} = \mathbf{V}_0 - (\mathbf{r} - \mathbf{r}_0) \times \boldsymbol{\omega} + (\mathbf{r} - \mathbf{r}_0)\cdot\dot{\mathbf{S}} \tag{4.34}$$

where $\boldsymbol{\omega}$ is the angular speed vector and $\dot{\mathbf{S}}$ is called the rate of strain dyadic.

From the mathematical expression for the composition of velocity given by the above equation, we obtain the *first theorem* of Helmholtz: "Any motion of a volume element of fluid can at a given moment be regarded as the sum of two motions: the motion as a solid [consisting of pure translation $\mathbf{V}_0$ and pure rotation $(\mathbf{r} - \mathbf{r}_0) \times \boldsymbol{\omega}$], and a deformation, $(\mathbf{r} - \mathbf{r}_0)\cdot\dot{\mathbf{S}}$."

The elemental components of the strain rate dyadic $\dot{\mathbf{S}}$ change from one field point of the fluid to another and can be regarded as functions of the spatial coordinates where the dyadic is given. This means that the *field* of dyadic $\dot{\mathbf{S}}$ is connected with the fluid stream. The strain rate dyadic characterizes the dilatation and angular deformation of fluid particles in the entire flow field. The strain rate dyadic $\dot{\mathbf{S}}$ can be conveniently expressed in matrix form:

$$\dot{\mathbf{S}} = \begin{bmatrix} \left(\dfrac{\partial u}{\partial x}\right)_0 & \dfrac{1}{2}\left(\dfrac{\partial u}{\partial y} + \dfrac{\partial v}{\partial x}\right)_0 & \dfrac{1}{2}\left(\dfrac{\partial u}{\partial z} + \dfrac{\partial w}{\partial x}\right)_0 \\ \dfrac{1}{2}\left(\dfrac{\partial v}{\partial x} + \dfrac{\partial u}{\partial y}\right)_0 & \left(\dfrac{\partial v}{\partial y}\right)_0 & \dfrac{1}{2}\left(\dfrac{\partial v}{\partial z} + \dfrac{\partial w}{\partial y}\right)_0 \\ \dfrac{1}{2}\left(\dfrac{\partial w}{\partial x} + \dfrac{\partial u}{\partial z}\right)_0 & \dfrac{1}{2}\left(\dfrac{\partial w}{\partial y} + \dfrac{\partial v}{\partial z}\right)_0 & \left(\dfrac{\partial w}{\partial z}\right)_0 \end{bmatrix} \tag{4.35}$$

The velocity $\mathbf{V}$ of Eq. (4.34) consists of a pure translational velocity $\mathbf{V}_0$ and a combined pure rotation about an axis through the point P along with a deformation of flow. The angular speed $\boldsymbol{\omega}$ is a property of the entire volume element. The angular speed $\boldsymbol{\omega}$ has components ω_x, ω_y, ω_z, respectively, i.e.,

$$\boldsymbol{\omega} = \omega_x\mathbf{i} + \omega_y\mathbf{j} + \omega_z\mathbf{k} \tag{4.36}$$

where

$$\omega_x = \frac{1}{2}\left(\frac{\partial w}{\partial y} - \frac{\partial v}{\partial z}\right)_0 \tag{4.37a}$$

$$\omega_y = \frac{1}{2}\left(\frac{\partial u}{\partial z} - \frac{\partial w}{\partial x}\right)_0 \tag{4.37b}$$

$$\omega_z = \frac{1}{2}\left(\frac{\partial v}{\partial x} - \frac{\partial u}{\partial y}\right)_0 \tag{4.37c}$$

The geometric interpretation of Eq. (4.34) is shown in Fig. 4.8. Note that the angular rotation $\boldsymbol{\omega}$ of Eq. (4.37) can be expressed in more convenient notation as

$$\boldsymbol{\omega} = \frac{1}{2}\nabla \times \mathbf{V} \tag{4.38}$$

In the kinematics of solid mechanics, there is of course no term involving the deformation $(\mathbf{r}_0 - \mathbf{r})\cdot\dot{\mathsf{S}}$, since solid particles are assumed rigid. Thus the velocity $\mathbf{V}$ for *solid* particles consists solely of pure translation and rotation.

4.4.3 The Strain Rate Dyadic $\dot{\mathsf{S}}$

The third component of the velocity composition $\mathbf{V}$ of Eq. (4.34) involves the dilatation and shearing strain of the fluid particle at P. Let ϵ_{ij}* denote the shear strain tensor. From Eq. (4.34), the components of the strain rate dyadic are defined as

$$\dot{\mathsf{S}} = \frac{\partial}{\partial t}\epsilon_{ij} = \begin{bmatrix} \dot\epsilon_{xx} & \dot\epsilon_{xy} & \dot\epsilon_{xz} \\ \dot\epsilon_{yx} & \dot\epsilon_{yy} & \dot\epsilon_{yz} \\ \dot\epsilon_{zx} & \dot\epsilon_{zy} & \dot\epsilon_{zz} \end{bmatrix}$$

$$= \begin{bmatrix} \left(\frac{\partial u}{\partial x}\right)_0 & \frac{1}{2}\left(\frac{\partial u}{\partial y} + \frac{\partial v}{\partial x}\right)_0 & \frac{1}{2}\left(\frac{\partial u}{\partial z} + \frac{\partial w}{\partial x}\right)_0 \\ \frac{1}{2}\left(\frac{\partial v}{\partial x} + \frac{\partial u}{\partial y}\right)_0 & \left(\frac{\partial v}{\partial y}\right)_0 & \frac{1}{2}\left(\frac{\partial v}{\partial z} + \frac{\partial w}{\partial y}\right)_0 \\ \frac{1}{2}\left(\frac{\partial w}{\partial x} + \frac{\partial u}{\partial z}\right)_0 & \frac{1}{2}\left(\frac{\partial w}{\partial y} + \frac{\partial v}{\partial z}\right)_0 & \left(\frac{\partial w}{\partial z}\right)_0 \end{bmatrix} \tag{4.39}$$

*The subscripts i and j may take on values x, y, or z. Note $\dot\epsilon_{xx} = (\partial\epsilon_{xx})/\partial t$, and so on.

The array in Eq. (4.39) is symmetric, that is,

$$\dot{\epsilon}_{ij} = \dot{\epsilon}_{ji} \tag{4.40}$$

The diagonal terms are the familiar normal strain rates, and their sum is the dilatation D of Eq. (4.23):

$$\begin{aligned} D &= \dot{\epsilon}_{xx} + \dot{\epsilon}_{yy} + \dot{\epsilon}_{zz} \\ &= \nabla \cdot \mathbf{V} \end{aligned} \tag{4.41}$$

Example 4.5

Consider a fluid particle with velocity

$$\mathbf{V} = 2x\mathbf{i} - 3y\mathbf{j} + z\mathbf{k}$$

Describe the deformation of the fluid particle.

Solution:

Given

$$\mathbf{V} = 2x\mathbf{i} - 3y\mathbf{j} + z\mathbf{k}$$

we see that the x-component of fluid velocity u varies only in the x-direction. The y-component of fluid velocity v varies only in the y-direction, and the z-component of fluid velocity w varies only in the z-direction. Intuitively we know there will be no angular deformation, only normal strains resulting in the fluid dilating.

For the angular rotation, we use Eq. (4.37) to obtain

$$\omega_x = \omega_y = \omega_z = 0 \tag{i}$$

since $u = u(x)$, $v = v(y)$, $w = w(z)$.

For the strain rate dyadic $\dot{\mathbf{S}}$, Eq. (4.35) can be written as

$$\dot{\mathbf{S}} = \begin{bmatrix} 2 & 0 & 0 \\ 0 & -3 & 0 \\ 0 & 0 & 1 \end{bmatrix} \tag{ii}$$

Equation (ii) indicates that the fluid is experiencing only dilation.

If for the moment we regard the fluid as being confined in a control volume, and allow the control surface to move with the y-z face having velocity u, and with the x-z face having velocity v, etc., then the fluid is expanding in the x-direction two units per unit length every second, one unit per unit length per second in the z-direction, and is contracting three units per unit length per second in the y-direction. For this problem, we see that

$$D = \nabla \cdot \mathbf{V} = 0 \tag{iii}$$

Example 4.5 *(Con't.)*
Thus, for a fixed control volume, the influx of fluid in the x, y, z directions is precisely equal to the efflux of fluid.

This completes the solution.

Example 4.6
A Newtonian flow field has the following velocity field:

$$\mathbf{V} = x^2y\mathbf{i} + 2xy^2z\mathbf{j} - yz^3\mathbf{k} \tag{i}$$

Calculate the normal and shearing rates of strain at the space location $x = -2$, $y = -1$, and $z = +2$. Also, determine if the fluid is rotating at that location, and about what axis.

Solution:
Equation (4.37) gives the angular speed as

$$\omega_x = \frac{1}{2}\left(\frac{\partial w}{\partial y} - \frac{\partial v}{\partial z}\right) = \frac{1}{2}(-z^3 - 2xy^2) = \frac{1}{2}(-8 + 4) = -4 \tag{ii}$$

$$\omega_y = \frac{1}{2}\left(\frac{\partial u}{\partial z} - \frac{\partial w}{\partial x}\right) = \frac{1}{2}(0 - 0) = 0 \tag{iii}$$

$$\omega_z = \frac{1}{2}\left(\frac{\partial v}{\partial x} - \frac{\partial u}{\partial y}\right) = \frac{1}{2}(2y^2z - x^2) = \frac{1}{2}(4 - 4) = 0 \tag{iv}$$

The fluid is rotating about the x-axis in a clockwise sense and has a magnitude of 4 rad/s.

The normal rates of shear are obtained from the diagonal terms of Eq. (4.39):

$$\dot{\epsilon}_{xx} = \frac{\partial u}{\partial x} = 2xy = 4 \tag{v}$$

$$\dot{\epsilon}_{yy} = \frac{\partial v}{\partial y} = 4xyz = 16 \tag{vi}$$

$$\dot{\epsilon}_{zz} = \frac{\partial w}{\partial z} = -3yz^2 = 12 \tag{vii}$$

The shearing rates of strain are the off-diagonal terms of Eq. (4.39) and are evaluated at the location $x = -2$, $y = -1$, and $z = +2$ as

$$\dot{\epsilon}_{xy} = \dot{\epsilon}_{yx} = \frac{1}{2}\left(\frac{\partial u}{\partial y} + \frac{\partial v}{\partial x}\right)$$

$$= \frac{1}{2}(x^2 + 2y^2z) = 4 \tag{viii}$$

Example 4.6 *(Con't.)*

$$\dot{\epsilon}_{xz} = \dot{\epsilon}_{zx} = \frac{1}{2}\left(\frac{\partial u}{\partial z} + \frac{\partial w}{\partial x}\right)$$

$$= \frac{1}{2}(0 + 0) = 0 \quad \text{(ix)}$$

$$\dot{\epsilon}_{yz} = \dot{\epsilon}_{zy} = \frac{1}{2}\left(\frac{\partial v}{\partial z} + \frac{\partial w}{\partial y}\right)$$

$$= \frac{1}{2}(2xy^2 - z^3) = -6 \quad \text{(x)}$$

Note how many different fluid quantities we have determined knowing solely the velocity of the flow. The angular speed and rates of shear will be shown to be important in calculating forces.

This completes the solution.

4.4.4 Geometric Interpretation of the Velocity Components

Because the expressions for the fluid particle velocity are so crucial, it is worthwhile to digress momentarily to interpret the results geometrically. We use a Lagrangian viewpoint to describe the motion.

Translation

Assume that a fluid element only translates. Locate a point $P(x, y)$ on the fluid element that is a distance $d\mathbf{r}$ from the center of the element, where

$$d\mathbf{r} = \mathbf{i}\,dx + \mathbf{j}\,dy \quad (4.42)$$

as shown in Fig. 4.9. Since the fluid element translates, then the velocity of any point in the element is

$$\frac{d\mathbf{r}}{dt} = \mathbf{i}u + \mathbf{j}v \quad (4.43)$$

where the velocity is positive for increasing coordinates. We note that the point P has moved to P' in an interval of time dt by the distance

$$\overline{PP'}\mathbf{e}_s = u\,dt\,\mathbf{i} + v\,dt\,\mathbf{j} \quad (4.44)$$

as evidenced by Eq. (4.43).

Dilatation

The dilatation is the condition of compression or expansion of a fluid element, and is shown in Fig. 4.10 for the case of an expansion in a typical cross-sectional area of a fluid element. If the point P has an x-component of velocity u, and if P' is downstream

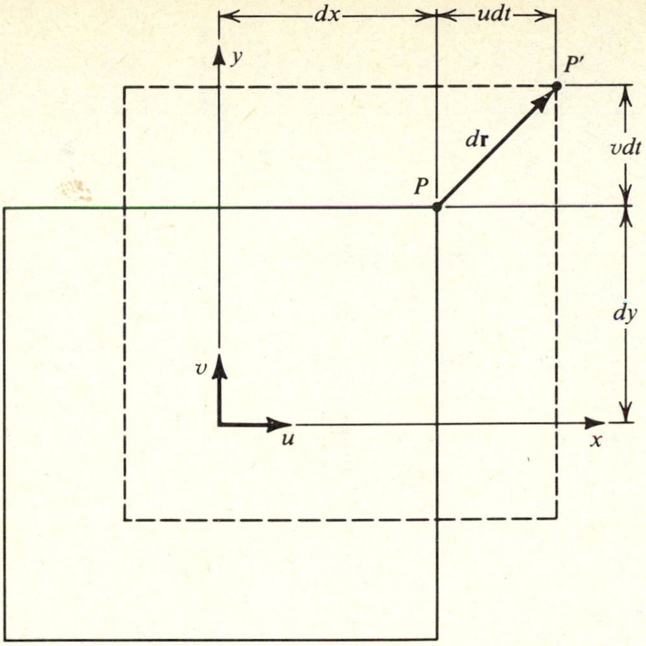

$d\mathbf{r} = udt\mathbf{i} + vdt\mathbf{j}$

Figure 4.9 *The displacement* $d\mathbf{r}$ of the point *P to P′ in time interval.*

of P, then P' has to have a velocity greater than P to have moved it downstream, or a velocity $u + (\partial u/\partial x)\,dx$. The change in the x-component of velocity is then $(\partial u/\partial x)\,dx$. The distance that point P' has moved relative to P in the x-direction is then $(\partial u/\partial x)\,dxdt$ in the interval of time dt. A similar explanation holds for the y-component of velocity v. The distance P' has moved in the y-direction is $\Delta y = (\partial v/\partial y)\,dydt$.

Thus, the two speeds of the dilatation are, from Eq. (4.39),

$$\dot{\epsilon}_{xx}\,dx = \frac{\partial u}{\partial x}\,dx \tag{4.45}$$

$$\dot{\epsilon}_{yy}\,dy = \frac{\partial v}{\partial y}\,dy \tag{4.46}$$

Rotation

In addition to translating and dilating, a fluid can rotate. A rotation is a turning motion about an axis perpendicular to the plane of rotation. An example of rotation is the familiar rigid body rotation where all points on the rigid body have the same angular speed ω. The rotation is *not* the same as an angular deformation. In Fig. 4.11, the element $\overline{OM} = dy$ has rotated with an angular speed equal to the velocity of M relative to O divided by the distance $\overline{OM}$.

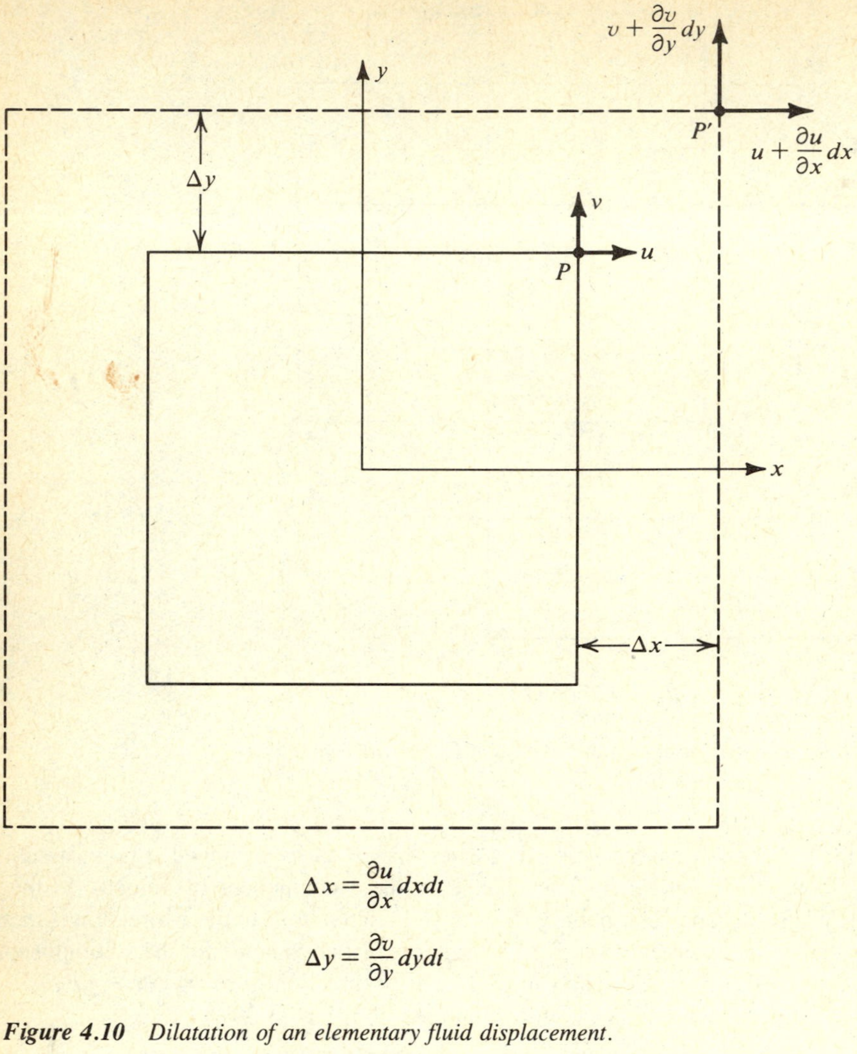

$$\Delta x = \frac{\partial u}{\partial x} dxdt$$

$$\Delta y = \frac{\partial v}{\partial y} dydt$$

Figure 4.10 *Dilatation of an elementary fluid displacement.*

If the x-component of velocity at O is u, then by Eq. (4.31) the velocity of M relative to O is $(\partial u/\partial y)\ (y - y_o)$, or $(\partial u/\partial y)\ dy$ as y approaches y_o. The distance M has traveled to M' is thus the product of the velocity times the interval of time or $-(\partial u/\partial y)\ dydt$. The distance is negative since it is in the direction opposite to increasing values of x. Similarly, the angular velocity of a vertical element can be found. The rotation is unaffected by the x-component of velocity u, and is found to be $\partial v/\partial x$.

In *rigid-body rotation*, the z-component of the angular speed of the element of Fig. 4.11 is given by Eq. (4.37). If

$$\frac{\partial v}{\partial x} = \frac{\partial u}{\partial y}$$

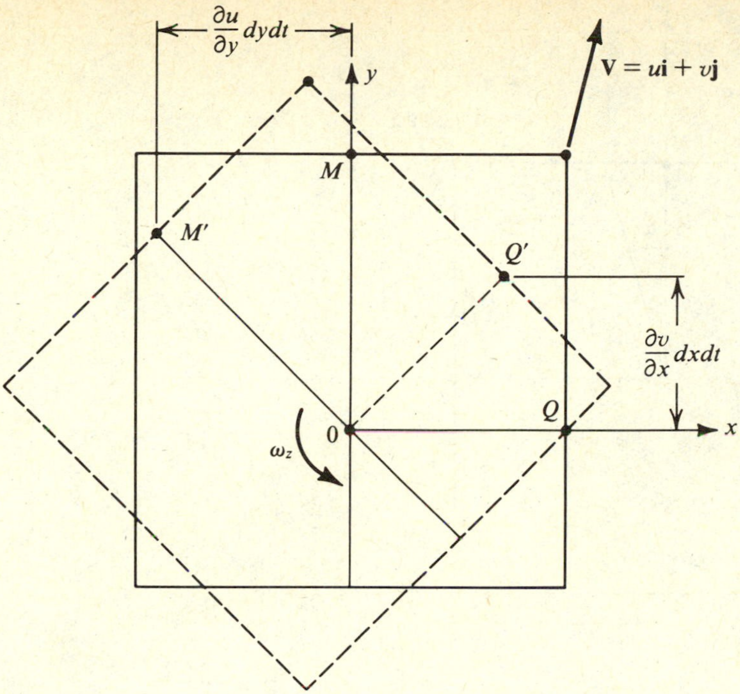

Figure 4.11 *Pure rotation of a fluid element.*

the position of lines remain the same. However, for *fluids*, the angular speed ω_z is defined as the *average* of the rotations of the horizontal and vertical line elements that were originally perpendicular.

A vortex can have a fluid parcel that translates in a circular path or it can rotate as it moves in a nearly circular path with an angular speed $\boldsymbol{\omega}$. In the latter case, the angular speed is not constant. This will be discussed further when we discuss vortex dynamics.

Angular Deformation

The only velocity components left that need a geometric interpretation are the off-diagonal terms in the strain rate tensor $\mathbf{S}$. In Fig. 4.12, let the horizontal and vertical faces rotate in *opposite* directions. This would occur when a fluid element is under a shear force on the top and side surfaces of the fluid element shown in the figure.

Using the results from Fig. 4.11, we find the angle α of Fig. 4.12 to be

$$\alpha \approx \tan \alpha = \frac{\dfrac{\partial u}{\partial y} dydt}{dy} = \frac{\partial u}{\partial y} dt \qquad (4.47)$$

provided that the distance $(\partial u/\partial y)\, dydt$ is very small compared to dy; that is, for very small deformations, the tangent of the angle is the angle itself. This means that the

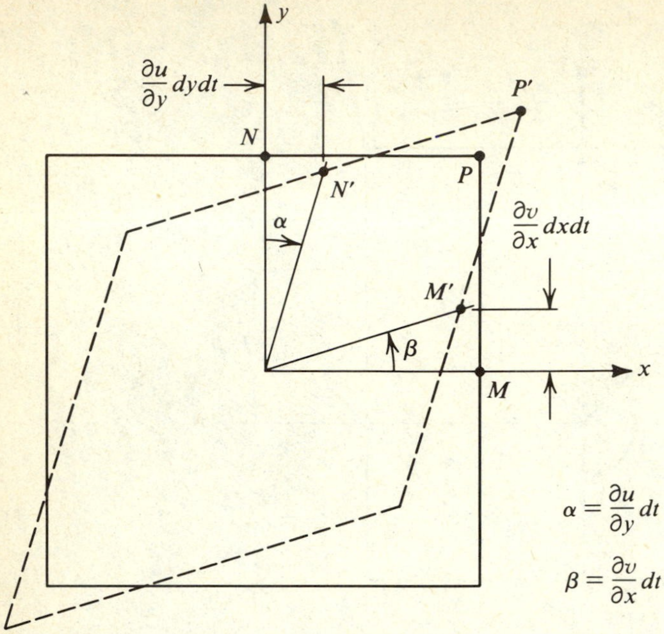

Figure 4.12 *Angular deformation of a fluid element.*

interval of time dt is exceedingly small, since $\partial u/\partial y$ can be very large. Similarly, the angle β is found from Figs. 4.11 and 4.12 as

$$\beta = \frac{\frac{\partial v}{\partial x} dxdt}{dx} = \frac{\partial v}{\partial x} dt \tag{4.48}$$

Comparing Eqs. (4.47) and (4.48) with Eq. (4.39), we note that the average value of the angular deformation, $(\alpha + \beta)/2$ of Fig. 4.12 is the shear strain rate $\dot{\epsilon}_{xy}$:

$$\dot{\epsilon}_{xy} = \dot{\epsilon}_{yx} = \frac{1}{2}(\dot{\alpha} + \dot{\beta}) \tag{4.49}$$

This then completes the geometric interpretation of all terms of the velocity component.

4.4.5 The Stress Dyadic, P

So far, we have been able to express velocity changes in terms of rate of strain. In this section, we will express rate of strain in terms of stress, and thus we will define a relationship between velocity changes and stress. This is the last formulation we need in order to obtain the rate equation which expresses how environmental stresses can alter the flow's velocity.

We shall confine ourselves exclusively to isotropic fluids. An isotropic fluid is a fluid that behaves in the same way for all directions. For such fluids, we can consider the most general form of a linear relation between a stress and rate of strain as

$$\mathbf{P} = a\dot{\mathbf{S}} + b\mathbf{I} \tag{4.50}$$

where $\mathbf{P}$ is called the stress dyadic with elements p_{ij};

$$\begin{aligned}\mathbf{P} = {} & \mathbf{ii}p_{xx} + \mathbf{ij}p_{xy} + \mathbf{ik}p_{xz} + \mathbf{ji}p_{yx} \\ & + \mathbf{jj}p_{yy} + \mathbf{jk}p_{yz} + \mathbf{ki}p_{zx} + \mathbf{kj}p_{zy} + \mathbf{kk}p_{zz}\end{aligned} \tag{4.51}$$

The outer subscript j denotes the direction of the stress, and the inner subscript i denotes the plane in which the stress acts. This is illustrated in Fig. 4.13. Only three faces are shown with the correct directions for the normal and shear stresses. These stresses create normal and tangential forces. To have a nonzero net force on the differential control surface, we must consider what stresses exist on the other three faces. We shall treat this in Sec. 4.4.6 dealing with surface forces. Meanwhile, let us return to the stress versus rate of strain expression.

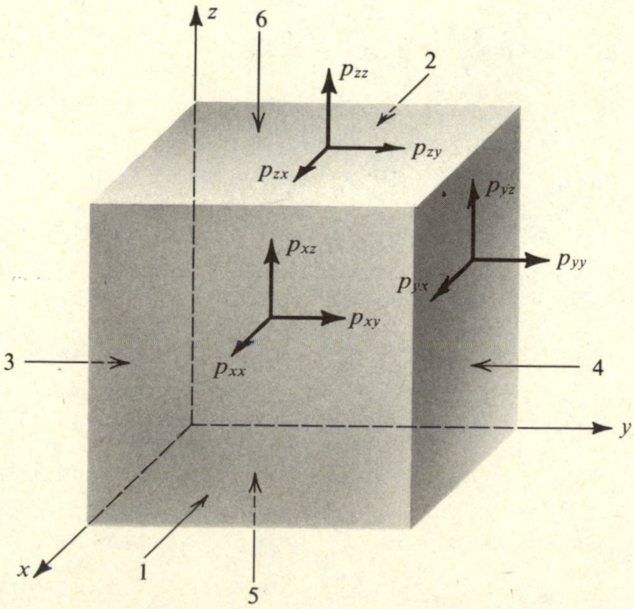

Figure 4.13 *Notation for stresses.*

From Eq. (4.50), a and b are constants associated with elements p_{ij} and the unit dyadic $\mathbf{I}$, defined as

$$\mathbf{I} = \begin{bmatrix} 1 & 0 & 0 \\ 0 & 1 & 0 \\ 0 & 0 & 1 \end{bmatrix} \tag{4.52}$$

Equation (4.50) is called the constitutive equation of fluid dynamics. It relates the stress components directly to the rate of strain, and is similar to Hooke's law in solids, which is a mathematical statement relating stress components directly to strain components. Equation (4.50) is a linear equation. Furthermore, in the absence of strain rates, the stress does not vanish because of the existence of the unit dyadic. Thus, we already suspect that stresses may exist for the case of no motion (i.e., no rate) of fluid. This is analogous to the prestressed conditions in certain solid materials like concrete, some plastics, and glass.

Examining Eq. (4.50), we note that there are 36 constants "a" (assuming symmetry), six for each strain rate for each stress equation. And there are three "b" constants: one for each normal stress equation. Thus we must evaluate a total of 39 constants [see Eq. (4.53)]. Fortunately, isotropy simplifies the problem tremendously. Let us experiment with the stress versus rate of strain Eq. (4.50) for a moment, and write a few of the scalar relationships for the stresses. We can write

$$\begin{aligned} p_{xx} &= a_{11}\dot{\epsilon}_{xx} + a_{12}\dot{\epsilon}_{xy} + a_{13}\dot{\epsilon}_{xz} \\ &\quad + a_{14}\dot{\epsilon}_{yy} + a_{15}\dot{\epsilon}_{yz} + a_{16}\dot{\epsilon}_{zz} + b_1 \end{aligned} \tag{4.53a}$$

$$\begin{aligned} p_{xy} &= a_{21}\dot{\epsilon}_{xx} + a_{22}\dot{\epsilon}_{xy} + a_{23}\dot{\epsilon}_{xz} + a_{24}\dot{\epsilon}_{yy} \\ &\quad + a_{25}\dot{\epsilon}_{yz} + a_{26}\dot{\epsilon}_{zz} \end{aligned} \tag{4.53b}$$

$$\begin{aligned} p_{xz} &= a_{31}\dot{\epsilon}_{xx} + a_{32}\dot{\epsilon}_{xy} + a_{33}\dot{\epsilon}_{xz} + a_{34}\dot{\epsilon}_{yy} \\ &\quad + a_{35}\dot{\epsilon}_{yz} + a_{36}\dot{\epsilon}_{zz} \end{aligned} \tag{4.53c}$$

$$\begin{aligned} p_{yy} &= a_{41}\dot{\epsilon}_{xx} + a_{42}\dot{\epsilon}_{xy} + a_{43}\dot{\epsilon}_{xz} + a_{44}\dot{\epsilon}_{yy} \\ &\quad + a_{45}\dot{\epsilon}_{yz} + a_{46}\dot{\epsilon}_{zz} + b_2 \end{aligned} \tag{4.53d}$$

$$\begin{aligned} p_{yz} &= a_{51}\dot{\epsilon}_{xx} + a_{52}\dot{\epsilon}_{xy} + a_{53}\dot{\epsilon}_{xz} + a_{54}\dot{\epsilon}_{yy} \\ &\quad + a_{55}\dot{\epsilon}_{yz} + a_{56}\dot{\epsilon}_{zz} \end{aligned} \tag{4.53e}$$

$$\begin{aligned} p_{zz} &= a_{61}\dot{\epsilon}_{xx} + a_{62}\dot{\epsilon}_{xy} + a_{63}\dot{\epsilon}_{xz} + a_{64}\dot{\epsilon}_{yy} \\ &\quad + a_{65}\dot{\epsilon}_{yz} + a_{66}\dot{\epsilon}_{zz} + b_3 \end{aligned} \tag{4.53f}$$

We count 36 unknown values of a and three unknown values of b.

From the assumption of isotropy (that there is no preferred direction), Stokes *proved* that

$$a_{11} = a_{44} = a_{66} \tag{4.54a}$$

$$a_{22} = a_{33} = a_{55} = \frac{1}{2}(a_{11} - a_{14}) \tag{4.54b}$$

$$b_1 = b_2 = b_3 \equiv b \tag{4.54c}$$

$$a_{14} = a_{41} = a_{46} = a_{61} = a_{64} \tag{4.54d}$$

and

$$\frac{1}{2}(a_{11} - a_{14}) = \mu \tag{4.54e}$$

All the other constants of a_{ij} were proven to be zero.

Consider the diagonal terms of the tensors **P** and $\dot{\mathbf{S}}$. From Eqs. (4.39) and (4.51) (and selecting $a_{11} = a_{44} = a_{66} \equiv a$), we obtain

$$p_{xx} + p_{yy} + p_{zz} = a(\dot{\epsilon}_{xx} + \dot{\epsilon}_{yy} + \dot{\epsilon}_{zz}) + 3b \tag{4.55a}$$

$$= a\nabla\cdot\mathbf{V} + 3b \tag{4.55b}$$

as the sum of the normal stresses.

The constant "a" represents a physical constant of the fluid medium. Stokes discovered that "a" was equal to 2μ. Thus, in Eq. (4.55b), the last remaining constant could be evaluated:

$$b = \frac{1}{3}(p_{xx} + p_{yy} + p_{zz}) - \frac{2}{3}\mu\nabla\cdot\mathbf{V} \tag{4.56}$$

By definition, the pressure p at a point in a fluid is the negative of the arithmetic mean of the three normal stresses p_{ii} acting on three mutually perpendicular elements of area:

$$p = -\frac{1}{3}(p_{xx} + p_{yy} + p_{zz}) \tag{4.57}$$

For an incompressible fluid, the divergence of the velocity vector $\nabla\cdot\mathbf{V}$ is identically zero, so that the pressure p is seen to be the negative of the mean bulk stress, and thus we can disregard the volume viscosity "a" of Eq. (4.55b).

Equation (4.57) approximates most liquids quite well, even those that are not especially incompressible. Even for a gas, which is highly compressible, Eq. (4.57) is valid since there is little energy interchange among the degrees of freedom of translation, rotation, spin, etc. of the atoms of ideal monatomic gases. (The correctness of Eq. (4.56) is justified only in practice.)

Thus, for an incompressible fluid, Eq. (4.56) becomes

$$b = -p \tag{4.58}$$

and the stress dyadic **P** becomes

$$\boxed{\mathbf{P} = 2\,\mu\,\dot{\mathbf{S}} - p\,\mathbf{I}} \tag{4.59}$$

and for a compressible fluid,

$$\mathbf{P} = 2\mu\dot{\mathbf{S}} - (p + \tfrac{2}{3}\mu\nabla \cdot \mathbf{V})\,\mathbf{I} \tag{4.60}$$

Comparing Eq. (4.59) with Eq. (4.39), we obtain the following expressions for the stresses:

$$p_{xx} = -p + 2\mu \frac{\partial u}{\partial x} \tag{4.61}$$

$$p_{yy} = -p + 2\mu \frac{\partial v}{\partial y} \tag{4.62}$$

$$p_{zz} = -p + 2\mu \frac{\partial w}{\partial z} \tag{4.63}$$

$$p_{xy} = \mu \left(\frac{\partial v}{\partial x} + \frac{\partial u}{\partial y}\right) \tag{4.64}$$

$$p_{xz} = \mu \left(\frac{\partial w}{\partial x} + \frac{\partial u}{\partial z}\right) \tag{4.65}$$

$$p_{yz} = \mu \left(\frac{\partial w}{\partial y} + \frac{\partial v}{\partial z}\right) \tag{4.66}$$

We can sometimes conveniently condense lengthy equations by using tensor notation. We denote the stress tensor p_{ij} as

$$p_{ij} = \begin{cases} \mu\left(\dfrac{\partial u_i}{\partial x_j} + \dfrac{\partial u_j}{\partial x_i}\right), & j \neq i \\[2ex] -p + 2\mu \dfrac{\partial u_i}{\partial x_i}, & j = i \end{cases} \tag{4.67}$$

where $x_1, x_2, x_3 = x, y, z$, respectively and $u_1, u_2, u_3 = u, v$, w, respectively. Though p is the mean of the normal pressure over three planes mutually at right angles (as given by Eq. (4.57), it is simply called the pressure. Some textbooks denote the shear stress $p_i;yj(i \neq j)$ as τ_{ij} and $p_{ij}(i = j)$ as σ_{ii}. The latter is the normal stress.

Comparing the stresses with the strain rate tensor **S** we easily determine

$$p_{xy} = 2\mu\, \dot{\epsilon}_{xy} \tag{4.68}$$

$$p_{xz} = 2\mu\, \dot{\epsilon}_{xz} \tag{4.69}$$

$$p_{yz} = 2\mu\, \dot{\epsilon}_{yz} \tag{4.70}$$

which are called Newton's viscosity postulates. Remember, they are based on the fluid being isotropic.

In cylindrical coordinates, the elements of the stress tensor are

$$p_{rr} = -p + 2\mu \frac{\partial v_r}{\partial r} \tag{4.71}$$

$$p_{\theta\theta} = -p + 2\mu \left(\frac{1}{r}\frac{\partial v_\theta}{\partial \theta} + \frac{v_r}{r} \right) \tag{4.72}$$

$$p_{zz} = -p + 2\mu \frac{\partial w}{\partial z} \tag{4.73}$$

$$p_{r\theta} = \mu \left[r \frac{\partial}{\partial r}\left(\frac{v_\theta}{r} \right) + \frac{1}{r}\frac{\partial v_r}{\partial \theta} \right] \tag{4.74}$$

$$p_{\theta z} = \mu \left(\frac{\partial v_\theta}{\partial z} + \frac{1}{r}\frac{\partial w}{\partial \theta} \right) \tag{4.75}$$

$$p_{rz} = \mu \left(\frac{\partial v_r}{\partial z} + \frac{\partial w}{\partial r} \right) \tag{4.76}$$

Equations (4.71)–(4.76) show the relationship between stresses and the fluid velocity components.

Example 4.7

Consider an incompressible steady axisymmetric vortex flow field given by

$$\mathbf{V} = -ar\mathbf{e}_r + \frac{b}{r}\mathbf{e}_\theta + 2az\,\mathbf{k}$$

Calculate all normal and shear stresses at the radial location $r = 2$ ft and axial station $z = 4$ ft if $a = 10^2$ rad/s, $b = 10^3$ ft²/s, $\mu = 1 \times 10^{-4}$ lbf·s/ft² and $p = 1 \times 10^{-2}$ psf.

Solution:

Step 1.

Identify the characteristics of the fluid and flow field.

The fluid is incompressible and the flow is a steady axisymmetric vortex.

Example 4.7 *(Con't.)*

Step 2.
Write the appropriate form of the governing equations of flow.

Since the flow is a vortex, we shall use the cylindrical form of the stress tensor elements. From Eqs. (4.71)–(4.76),

$$p_{rr} = -p + 2\mu \frac{\partial v_r}{\partial r}$$

$$= -p - 2\mu a \qquad \text{(i)}$$

$$p_{\theta\theta} = -p + 2\mu \left(\frac{1}{r}\frac{\partial v_\theta}{\partial \theta} + \frac{v_r}{r}\right)$$

$$= -p - 2\mu a \qquad \text{(ii)}$$

$$p_{zz} = -p + 2\mu \frac{\partial w}{\partial z}$$

$$= -p + 4\mu a \qquad \text{(iii)}$$

$$p_{r\theta} = \mu \left[r \frac{\partial}{\partial r}\left(\frac{v_\theta}{r}\right) + \frac{1}{r}\frac{\partial v_r}{\partial \theta}\right]$$

$$= \mu \left(-\frac{2b}{r^2}\right) \qquad \text{(iv)}$$

$$p_{\theta z} = \mu \left(\frac{\partial v_\theta}{\partial z} + \frac{1}{r}\frac{\partial w}{\partial \theta}\right)$$

$$= 0 \qquad \text{(v)}$$

$$p_{rz} = \mu \left(\frac{\partial v_r}{\partial z} + \frac{\partial w}{\partial r}\right)$$

$$= 0 \qquad \text{(vi)}$$

We check to see if the pressure p of Eq. (4.57) is satisfied:

$$p = -\tfrac{1}{3}(p_{rr} + p_{\theta\theta} + p_{zz}) \qquad \text{(vii)}$$

Substituting Eqs. (i)–(iii) into Eq. (vii) yields

$$p = -\tfrac{1}{3}(-p - 2\mu a - p - 2\mu a - p + 4\mu a)$$

$$= p \qquad \text{(viii)}$$

The normal stress components are constant everywhere in the flow field and are from Eqs. (i)–(iii),

$$p_{rr} = p_{\theta\theta} = -1 \times 10^{-2} - 2 \times 10^{-2}$$

$$= -3 \times 10^{-2} \text{ psf} \qquad \text{(ix)}$$

Example 4.7 *(Con't.)*

$$p_{zz} = -1 \times 10^{-2} + 4 \times 10^{-2}$$

$$= 3 \times 10^{-2} \text{ psf} \tag{x}$$

The shear stress in the $r\theta$ plane is a variable in the vortex flow, and at the location $r = 2$ ft, $z = 4$ ft,

$$p_{r\theta} = -2.5 \times 10^{-2} \text{ psf} \tag{xi}$$

Thus it is a rather simple task to calculate the stresses, once we are given the velocity field. The stresses will be important when we wish to calculate lift and drag.

This completes the solution.

4.4.6 The Surface Forces, $\mathbf{F}_s$

We are now in a position to discover what forces produce and then maintain the motion of the fluid. First we shall consider the surface force, a force that originates in the environment, and is applied on the surface of the control volume. Mathematically, the surface force $\mathbf{F}_s$ is the resultant, or integral of the stresses p_{ij} over the control surface. The surface force consists of forces that are normal, $\mathbf{F}_\sigma$, and forces that are tangential, $\mathbf{F}_\tau$. These are defined as

$$\mathbf{F}_\sigma = \int_A \mathbf{P} \cdot d\mathbf{A},\ i = j \tag{4.77}$$

and

$$\mathbf{F}_\tau = \int_A \mathbf{P} \cdot d\mathbf{A},\ i \neq j \tag{4.78}$$

Note that, although $\mathbf{P}$ is a tensor, its product with the elemental area $d\mathbf{A}$ results in an expression that will be a vector.*

The scalar surface forces are found by substituting the stress dyadic of Eq. (4.51) into the integrand of Eq. (4.77) and equating like coefficients of the unit vectors:

$$F_{\sigma_x} = \iint p_{xx}\, dydz \tag{4.79}$$

*Problem 4.45 is recommended to illustrate this fact.

$$F_{\sigma_y} = \iint p_{yy}\, dxdz \tag{4.80}$$

$$F_{\sigma_z} = \iint p_{zz}\, dxdy \tag{4.81}$$

for the normal surface forces, and

$$F_{\tau_x} = \iint p_{yx}\, dxdz + \iint p_{zx}\, dxdy \tag{4.82}$$

$$F_{\tau_y} = \iint p_{xy}\, dydz + \iint p_{zy}\, dxdy \tag{4.83}$$

$$F_{\tau_z} = \iint p_{xz}\, dydz + \iint p_{yz}\, dxdy \tag{4.84}$$

for the tangential surface forces.

We must keep in mind that in any volume there are always two opposite forces. In Fig. 4.13, face 1 lies in the x-z plane as well as face 2. Thus the normal surface force in the x-direction would be

$$F_{\sigma_x} = \iint_1 p_{xx}\, dydz + \iint_2 p_{xx}\, dydz \tag{4.85}$$

This equation will be important in applying the net force to real problems.

To find the net surface force, we require a net stress, as illustrated in Fig. 4.14. Notice that net stress is the difference, or gradient, of the stresses that generate the net force over the areas. We see, for example, that the normal force acting opposite to face 3 is $p_{xx}\, dydz$ and is opposed by the normal force on face 3

$$\left(p_{xx} + \frac{\partial p_{xx}}{\partial x}\, dx \right) dydz$$

such that the net normal force in the x-direction is simply $(\partial p_{xx}/\partial x) \cdot d\forall$.

Example 4.8

Consider an incompressible fluid flow with velocity vector

$$\mathbf{V} = 2xy\mathbf{i} + \frac{y^2}{2}\mathbf{j} - 3zy\mathbf{k}$$

Calculate the normal and tangential forces on face 1 of the control volume shown in Fig. E4.8 if the pressure $p = 2y$.

Example 4.8 *(Con't.)*

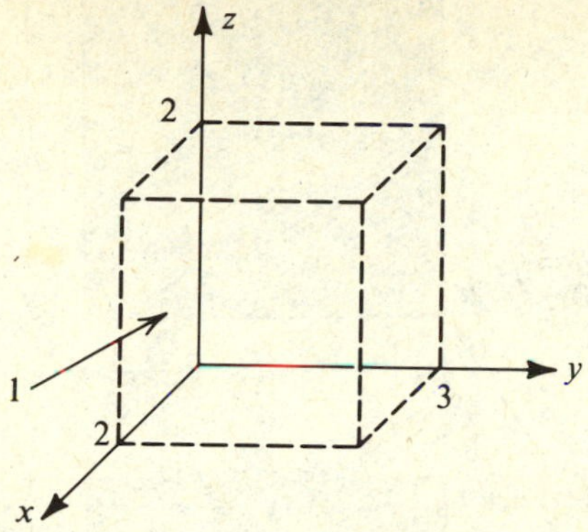

Figure E4.8

Solution:

Step 1.

Identify the characteristics of the fluid and flow field.

The fluid is assumed incompressible, and the flow is assumed steady and three-dimensional.

Step 2.

Write the appropriate form of the governing equations of flow.

$$\nabla \cdot \mathbf{V} = 0 \tag{4.23}$$

$$\mathbf{F}_\sigma = \int_A \mathbf{P} \cdot d\mathbf{A},\ i = j \tag{4.77}$$

$$\mathbf{F}_\tau = \int_A \mathbf{P} \cdot d\mathbf{A},\ i \neq j \tag{4.78}$$

First, continuity is satisfied. Second, we are interested in the normal force

$$F_{\sigma_1} = -\int_0^2\int_0^2 p_{yy}\,dx\,dz\,\bigg|_{y=0} \tag{i}$$

where p_{yy} is the normal stress in the y-direction acting on the face 1 ($y = 0$). From Eq. (4.62)

$$p_{yy} = -p + 2\mu\frac{\partial v}{\partial y}$$

$$= -2y + 2\mu\,(y) \tag{ii}$$

such that the normal stress on face 1 is

$$p_{yy}\bigg|_{y=0} = 0 \tag{iii}$$

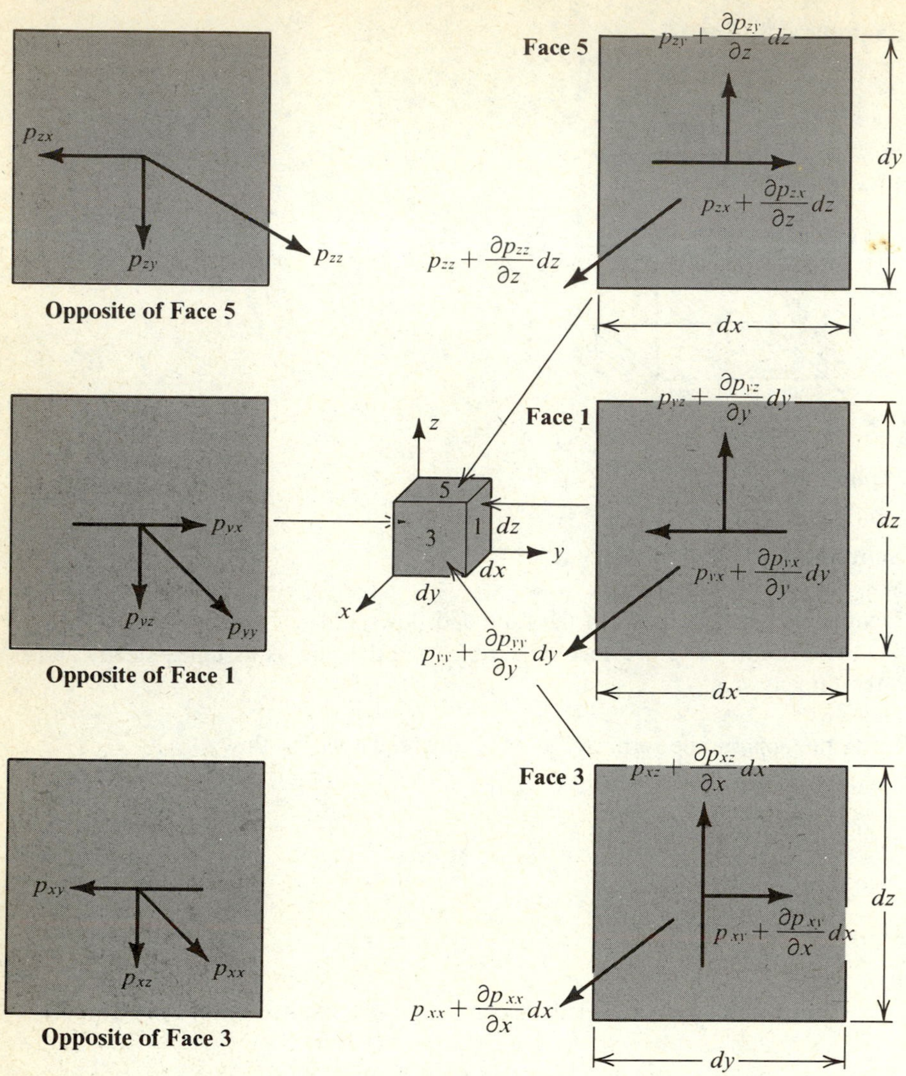

Figure 4.14 *Stresses acting on a fluid element.*

Example 4.8 *(Con't.)*

Substituting Eq. (iii) into Eq. (i) yields

$$\boxed{F_{\sigma_1} = 0} \tag{iv}$$

Two tangential stress forces act on face 1: one in the z direction, and one in the x direction. From Eq. (4.78)

$$F_{\tau_1}\mathbf{i} = \int_0^2\int_0^2 p_{yx}\mathbf{ji} \cdot dxdz\mathbf{j}\Bigg|_{y=0} \tag{v}$$

Example 4.8 *(Con't.)*

$$F_{\tau_1}\,\mathbf{k} = \int_0^2\int_0^2 p_{yz}\mathbf{jk}\cdot dxdz\mathbf{j}\Big|_{y=0} \qquad \text{(vi)}$$

Substituting Eqs. (4.64) and (4.66) into Eqs. (v) and (vi), respectively, yields

$$F_{\tau_1}\,\mathbf{i} = \mu\int_0^2\int_0^2 2xdxdz\mathbf{i}\Big|_{y=0} \qquad \text{(vii)}$$

$$F_{\tau_1}\,\mathbf{k} = -\mu\int_0^2\int_0^2 3zdxdz\mathbf{k}\Big|_{y=0} \qquad \text{(viii)}$$

Evaluating Eqs. (vii) and (viii) yields

$$\mathbf{F}_{\tau_1} = 2\,\mu\left(\frac{x^2}{2}\right)z\Big|_{\substack{x=2\\ z=2}}\mathbf{i} - 3\mu\left(\frac{z^2}{2}\right)x\Big|_{\substack{x=2\\ z=2}}\mathbf{k}$$

$$\mathbf{F}_{\tau_1} = 8\mu\mathbf{i} - 12\mu\mathbf{k} \qquad \text{(ix)}$$

We now have achieved one of our goals; i.e., we can calculate the resultant forces on an object placed in a flow whose velocity is known. This is an essential aspect of engineering design.

This completes the solution.

4.4.7 Vorticity, ζ

In a fluid motion, each small parcel of fluid may be viewed as a nonrigid body. Such a body may or may not be spinning. If it is spinning, it spins about an axis at a definite rate and thus its spin may be represented by an arrow. The arrow which points along the axis of spin has a length equal to the angular speed and is the *vorticity* (see Fig. 4.15). In a fluid motion, different fluid parcels may have different vortices, and the vorticity of a given parcel may change as the parcel moves. Vorticity has large values when the velocity changes significantly over a small spatial distance, as could be expected in a tornado. Vorticity is important near the surface of bodies immersed in a fluid; storms cannot exist without it, nor can aircraft fly with zero vorticity on the wing.

Vorticity $\boldsymbol{\zeta}$ is defined mathematically as the curl of the velocity vector $\mathbf{V}$:

$$\boxed{\boldsymbol{\zeta} = \nabla \times \mathbf{V}} \qquad (4.86)$$

and has scalar components ζ_x, ζ_y, ζ_z about the x, y, z axes, respectively:

$$\boldsymbol{\zeta} = \zeta_x\mathbf{i} + \zeta_y\mathbf{j} + \zeta_z\,\mathbf{k} \qquad (4.87)$$

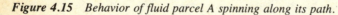

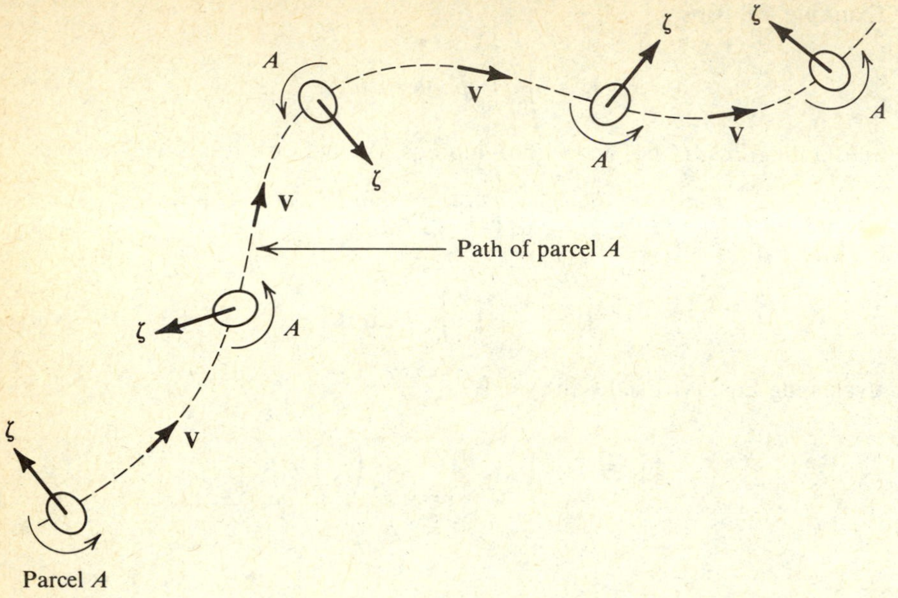

Figure 4.15 *Behavior of fluid parcel A spinning along its path.*

or has scalar components ζ_r, ζ_θ, ζ_z defined as

$$\boldsymbol{\zeta} = \zeta_r \mathbf{e}_r + \zeta_\theta \mathbf{e}_\theta + \zeta_z \mathbf{k} \tag{4.88}$$

Here ζ_r is the vorticity about an axis normal to the θ-z plane, ζ_θ is the vorticity about an axis normal to the r-z plane, and ζ_z is the vorticity about an axis normal to the polar or r-θ plane.

Comparing the vorticity of Eq. (4.86) with the angular speed $\boldsymbol{\omega}$ of Eq. (4.38), we obtain

$$\boldsymbol{\zeta} = 2\boldsymbol{\omega} \tag{4.89}$$

Thus, since $\boldsymbol{\omega}$ is a measure of the net angular rotation $\frac{1}{2}(\beta - \alpha)$ (see Fig. 4.12), vorticity measures the local spin of the fluid particle. Consequently, if $\boldsymbol{\omega} = 0$, then the flow is *irrotational* and the vorticity $\boldsymbol{\zeta}$ of the flow must be zero. This does *not* mean that a mass cannot revolve about a reference axis as described in Sec. 2.2.8. There we showed an example of irrotational motion following a circular path: the axis of the chair rotated with the wheel, but the chair rotated at the same rate but in the opposite sense about its own axis. Some other examples of various rotational motions are shown in Fig. 4.16.

In Fig. 4.16a the fluid particle is seen to translate along a curvilinear path without the fluid rotating. The shear stresses are such that the stress is zero everywhere in this flow:

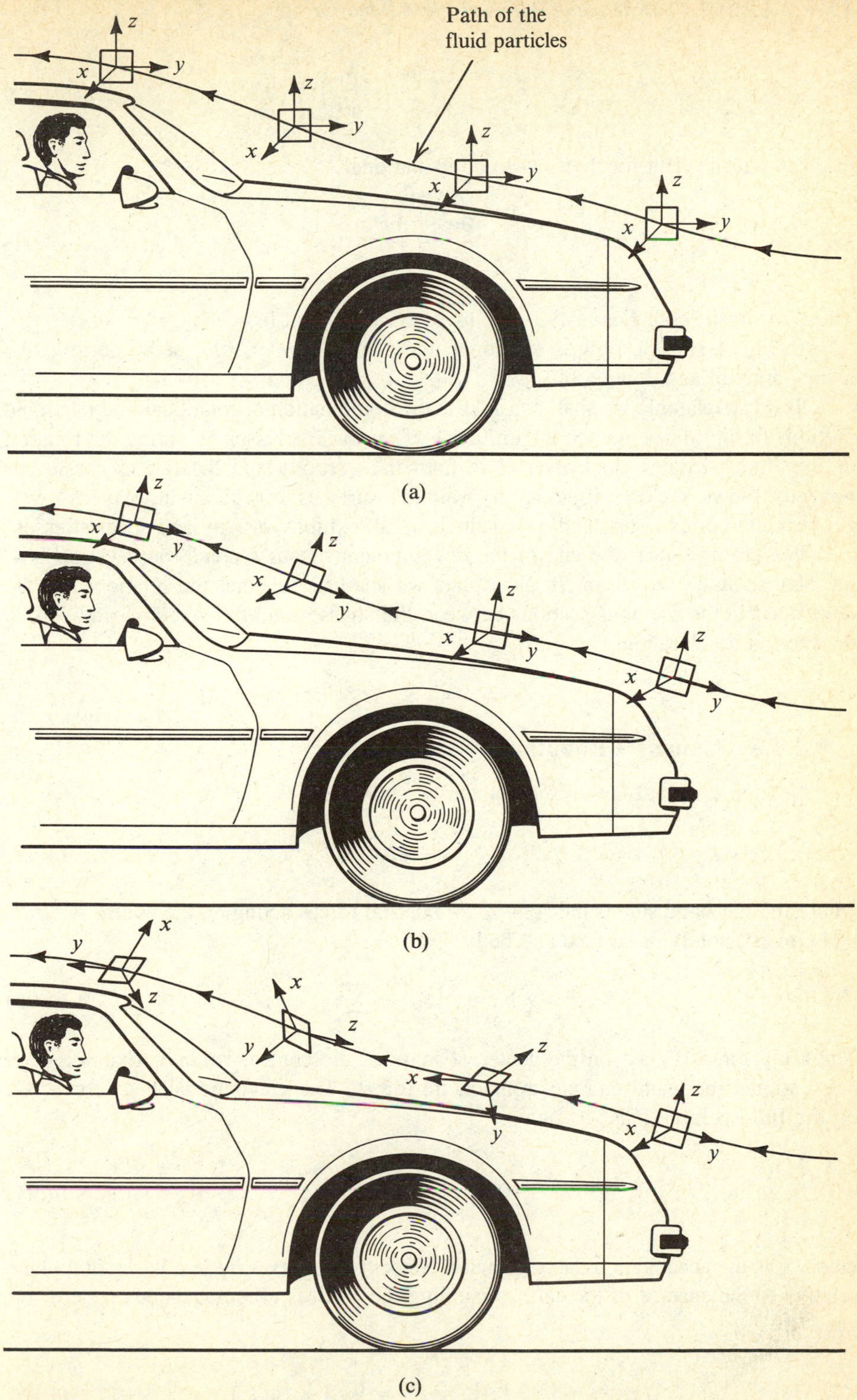

Figure 4.16 *(a) Irrotational frictionless motion. (b) Rotationless frictionless motion. (c) Rotational frictional motion.*

$$\frac{p_{xy}}{2\mu} = \dot{\epsilon}_{xy} = \frac{1}{2}\left(\frac{\partial u}{\partial y} + \frac{\partial v}{\partial x}\right) = 0 \tag{4.90}$$

in the x-y plane. But for the flow to be irrotational

$$\zeta_z = \frac{\partial v}{\partial x} - \frac{\partial u}{\partial y} = 0 \tag{4.91}$$

Thus, to satisfy both Eqs. (4.90) and (4.91), we see that $u = u(y)$, $v = v(x)$.

In Fig. 4.16b the flow is rotational, the classic case of solid body motion, like a car going up and down a hill.

The last example we shall consider is the combination of rotational and frictional motion. In Fig. 4.16c we see the influence of the shear stresses by noting the rotation of the fluid particles clockwise. Here both the vorticity and the rate of strain are nonzero. Flows in close proximity of ''wetted'' surfaces (e.g., boundary layer flows), geophysical vortex cores, and turbulent flows all exhibit vorticity and shear stresses.

We are now near the end of the development of our second conservation law: the Navier-Stokes equation. In particular, we shall tie together the relationships we developed in the previous sections between the stresses and the velocity gradients to the inertial acceleration.

4.4.8 Cauchy's Equation of Motion

Newton's second law of motion can be stated as

$$\sum \mathbf{F} = M\mathbf{a} \tag{4.92}$$

where the left-hand side is the sum of all external forces acting on the control volume and consists solely of surface and body forces:

$$\sum \mathbf{F} = \sum \mathbf{F}_s + \sum \mathbf{F}_b \tag{4.93}$$

The body forces $\mathbf{F}_b$ act on the center of mass of the control volume. Examples are gravitational, buoyant, and electromagnetic forces. The gravitational body force, $\mathbf{G}$, on the fluid volume $\forall$ is

$$\mathbf{G} = \int_{\forall} \rho \mathbf{g}\, d\forall \tag{4.94}$$

where $\mathbf{g}$ is the acceleration due to gravity. Since the x, y axes lie in a horizontal plane relative to the surface of the earth, assuming the earth is a homogeneous sphere, we can state

$$\mathbf{g} = -32.2\ \mathbf{k}\ (\text{ft/s}^2) = -9.81\ \mathbf{k}\ (\text{m/s}^2) \tag{4.95}$$

Using the relationship between mass M and density ρ and expressing the surface force in terms of the stress dyadic $\mathbf{P}$, we obtain

$$\int_{\forall} (\mathbf{g} - \mathbf{a})\, \rho\, d\forall + \int_{A} \mathbf{P} \cdot d\mathbf{A} = 0 \tag{4.96}$$

The area integral can be expressed in terms of a volume integral through Gauss's theorem, Eq. (B.37), such that Eq. (4.96) becomes

$$\int_{\forall} [(\mathbf{g} - \mathbf{a})\, \rho + \nabla \cdot \mathbf{P}]\, d\forall = 0 \tag{4.97}$$

Since the volume $\forall$ is completely arbitrary and the integrand is continuous throughout the volume, the integrand of Eq. (4.97) must also be zero:

$$(\mathbf{g} - \mathbf{a})\, \rho + \nabla \cdot \mathbf{P} = 0 \tag{4.98}$$

The above vector differential equation is called *Cauchy's equation of motion*. Physically, it represents a unique balance of body and surface forces with the inertial force. If the body forces do not nullify the surface forces a nonzero inertial force will cause the fluid to accelerate.

It is also important to state that Cauchy's equation of motion is valid for *both* incompressible and compressible fluids. The difference between these two fluids lies in the expressions for the stress. For an

- Incompressible fluid flow:

$$\mathbf{P} = 2\mu\dot{\mathbf{S}} - p\mathbf{I} \tag{4.59}$$

and for a

- Compressible fluid flow:

$$\mathbf{P} = 2\mu\, \dot{\mathbf{S}} - (p + \tfrac{2}{3}\mu\nabla \cdot \mathbf{V})\, \mathbf{I} \tag{4.60}$$

The equation of motion as given by Eq. (4.98) can easily be expressed in Cartesian form, by use of Eqs. (1.45)–(1.47) for the acceleration vector **a**, and Eq. (4.51) for the stress dyadic $\mathbf{P}$ to yield

1. x-component

$$\begin{aligned} &\frac{\partial u}{\partial t} + u\frac{\partial u}{\partial x} + v\frac{\partial u}{\partial y} + w\frac{\partial u}{\partial z} \\ &= g_x + \frac{1}{\rho}\left(\frac{\partial p_{xx}}{\partial x} + \frac{\partial p_{yx}}{\partial y} + \frac{\partial p_{zx}}{\partial z}\right) \end{aligned} \tag{4.99}$$

2. y-component

$$\frac{\partial v}{\partial t} + u\frac{\partial v}{\partial x} + v\frac{\partial v}{\partial y} + w\frac{\partial v}{\partial z} = g_y + \frac{1}{\rho}\left(\frac{\partial p_{xy}}{\partial x} + \frac{\partial p_{yy}}{\partial y} + \frac{\partial p_{zy}}{\partial z}\right) \tag{4.100}$$

3. z-component

$$\frac{\partial w}{\partial t} + u\frac{\partial w}{\partial x} + v\frac{\partial w}{\partial y} + w\frac{\partial w}{\partial z} = g_z + \frac{1}{\rho}\left(\frac{\partial p_{xz}}{\partial x} + \frac{\partial p_{yz}}{\partial y} + \frac{\partial p_{zz}}{\partial z}\right) \tag{4.101}$$

(recalling that $\mathbf{i}\cdot(\mathbf{ii}) = \mathbf{i}$, $\mathbf{i}\cdot(\mathbf{ij}) = \mathbf{j}$, $\mathbf{i}\cdot(\mathbf{ik}) = \mathbf{k}$ and any other dot product of $\mathbf{i}$ with the other unit dyads is zero).

A similar procedure is followed using cylindrical coordinates. The three scalar forms of Cauchy's equation of motion in cylindrical coordinates are

1. r-component

$$\frac{\partial v_r}{\partial t} + v_r\frac{\partial v_r}{\partial r} + \frac{v_\theta}{r}\frac{\partial v_r}{\partial \theta} - \frac{v_\theta^2}{r} + w\frac{\partial v_r}{\partial z} = g_r + \frac{1}{\rho}\left(\frac{1}{r}\frac{\partial (rp_{rr})}{\partial r} + \frac{1}{r}\frac{\partial p_{r\theta}}{\partial \theta} - \frac{p_{\theta\theta}}{r} + \frac{\partial p_{rz}}{\partial z}\right) \tag{4.102}$$

2. θ-component

$$\frac{\partial v_\theta}{\partial t} + v_r\frac{\partial v_\theta}{\partial r} + \frac{v_\theta}{r}\frac{\partial v_\theta}{\partial \theta} + \frac{v_r v_\theta}{r} + w\frac{\partial v_\theta}{\partial z} = g_\theta + \frac{1}{\rho}\left[\frac{1}{r^2}\frac{\partial (r^2 p_{r\theta})}{\partial r} + \frac{1}{r}\frac{\partial p_{\theta\theta}}{\partial \theta} + \frac{\partial p_{\theta z}}{\partial z}\right] \tag{4.103}$$

3. z-component

$$\frac{\partial w}{\partial t} + v_r\frac{\partial w}{\partial r} + \frac{v_\theta}{r}\frac{\partial w}{\partial \theta} + w\frac{\partial w}{\partial z} = g_z + \frac{1}{\rho}\left[\frac{1}{r}\frac{\partial (rp_{rz})}{\partial r} + \frac{1}{r}\frac{\partial (p_{\theta z})}{\partial \theta} + \frac{\partial p_{zz}}{\partial z}\right] \tag{4.104}$$

If the fluid is inviscid, all the shear stresses vanish because $\mu = 0$, resulting in the stress dyadic

$$\mathbf{P} = \mathbf{ii}p_{xx} + \mathbf{jj}p_{yy} + \mathbf{kk}p_{zz}$$
$$= \mathbf{e}_r\mathbf{e}_r p_{rr} + \mathbf{e}_\theta\mathbf{e}_\theta p_{\theta\theta} + \mathbf{kk}p_{zz}$$

Thus only normal stresses remain, resulting in

$$\nabla \cdot \mathbf{P} = -\nabla p \tag{4.105}$$

for an inviscid fluid.

One counts 13 dependent variables in the above equations: three velocity components, three components of the body force per unit mass, the density, and six stresses. We evaluate the dependent variables in terms of four independent variables, x, y, z, and t (or r, θ, z, and t). Thirteen dependent variables require 13 equations.

For an *incompressible* fluid, the density is constant. For a spherical potential field, the body force per unit mass $\mathbf{g}$ is known and is given by Eq. (4.95). Thus, the number of unknowns is reduced to nine. Section 4.4.5 discusses the stress dyadic $\mathbf{P}$ with rate of strain resulting in six additional equations: Eqs. (4.61)–(4.66). However, we also pick up an additional dependent variable: pressure. Thus, we have 10 unknowns and 10 equations, sufficient to solve most incompressible flow problems.

For a compressible fluid, the density is a variable, so we have 11 unknowns and 10 equations, and we need an additional equation, supplied by the fluid's equation of state.

The next section consolidates everything so that for the incompressible case, the number of unknowns is reduced to four, with four equations. The compressible case is also given. Three of these equations are called the Navier-Stokes equations. The other equations have already been developed. One is the continuity equation, and the other is the equation of state.

Example 4.9

Consider a one-dimensional flow that is incompressible, steady, and inviscid. Calculate the velocity of the flow in terms of the pressure given the pressure $p = \gamma x$ at $z = 0$, and $u = 0$ at $x = 0$.

Solution:

Step 1.

Identify the characteristics of the fluid and flow field.

The fluid is incompressible and inviscid; the flow is one-dimensional and steady.

Step 2.

Write the appropriate form of the governing equations of flow

Example 4.9 *(Con't.)*

$$\frac{\partial u}{\partial t} + u\frac{\partial u}{\partial x} + v\frac{\partial u}{\partial y} + w\frac{\partial u}{\partial z} = g_x - \frac{1}{\rho}\frac{\partial p}{\partial x} + \frac{\mu}{\rho}\left(\frac{\partial^2 u}{\partial x^2} + \frac{\partial^2 u}{\partial y^2} + \frac{\partial^2 u}{\partial z^2} + \frac{\partial D}{\partial x}\right) \tag{4.99}$$

$$\frac{\partial v}{\partial t} + u\frac{\partial v}{\partial x} + v\frac{\partial v}{\partial y} + w\frac{\partial v}{\partial z} = g_y - \frac{1}{\rho}\frac{\partial p}{\partial y} + \frac{\mu}{\rho}\left(\frac{\partial^2 v}{\partial x^2} + \frac{\partial^2 v}{\partial y^2} + \frac{\partial^2 v}{\partial z^2} + \frac{\partial D}{\partial y}\right) \tag{4.100}$$

$$\frac{\partial w}{\partial t} + u\frac{\partial w}{\partial x} + v\frac{\partial w}{\partial y} + w\frac{\partial w}{\partial z} = g_z - \frac{1}{\rho}\frac{\partial p}{\partial z} + \frac{\mu}{\rho}\left(\frac{\partial^2 w}{\partial x^2} + \frac{\partial^2 w}{\partial y^2} + \frac{\partial^2 w}{\partial z^2} + \frac{\partial D}{\partial z}\right) \tag{4.101}$$

$$p_{xx} = -p + 2\mu\frac{\partial u}{\partial x} \tag{4.61}$$

$$p_{yy} = -p + 2\mu\frac{\partial v}{\partial y} \tag{4.62}$$

$$p_{zz} = -p + 2\mu\frac{\partial w}{\partial z} \tag{4.63}$$

$$p_{xy} = \mu\left(\frac{\partial v}{\partial x} + \frac{\partial u}{\partial y}\right) \tag{4.64}$$

$$p_{xz} = \mu\left(\frac{\partial w}{\partial x} + \frac{\partial u}{\partial z}\right) \tag{4.65}$$

$$p_{yz} = \mu\left(\frac{\partial w}{\partial y} + \frac{\partial v}{\partial z}\right) \tag{4.66}$$

For inviscid steady one-dimensional flow, Eq. (4.99) in conjunction with Eqs. (4.61)–(4.66) becomes

$$u\frac{du}{dx} = -\frac{1}{\rho}\frac{\partial p}{\partial x} \tag{i}$$

since $\partial p/\partial y = 0$ and $\partial p/\partial z = -\rho g$. Integrating the result of Eq. (4.101) gives

$$p = -\rho g z + f(x) \tag{ii}$$

The function $f(x)$ is evaluated from the boundary condition $p = \gamma x$ at $z = 0$ with the result

$$p = -\rho g z + \gamma x \tag{iii}$$

Example 4.9 *(Con't.)*

Substituting Eq. (iii) into Eq. (i), then integrating, results in

$$u = \sqrt{2gx} \tag{iv}$$

We have just illustrated one way to obtain the velocity of the flow. Notice we had to carefully define what the fluid was, what the constraints were imposed on the flow, and what was taking place on the boundaries.

This completes the solution.

Example 4.10

Consider a steady inviscid two-dimensional flow of a fluid. Calculate the magnitude of the velocity $|\mathbf{V}|$ of the flow in terms of the pressure p and potential energy Ω, *given* that the slope of the path of the fluid particle is

$$\frac{dy}{dx} = \frac{v}{u} \tag{i}$$

and the gravitational acceleration $\mathbf{g}$ is the negative of the gradient of the potential energy

$$\mathbf{g} = -\nabla\Omega \tag{ii}$$

Solution:

The steps in the problem solution are identical to Example 4.19 except that this flow is now two-dimensional.

For inviscid steady two-dimensional flows, Eqs. (4.99) and (4.100) become with the help of Eqs. (4.61)–(4.66)

$$u\frac{\partial u}{\partial x} + v\frac{\partial u}{\partial y} = g_x - \frac{1}{\rho}\frac{\partial p}{\partial x} \tag{iii}$$

$$u\frac{\partial v}{\partial x} + v\frac{\partial v}{\partial y} = g_y - \frac{1}{\rho}\frac{\partial p}{\partial y} \tag{iv}$$

Substituting Eq. (ii) into Eqs. (iii) and (iv) results in

$$u\frac{\partial u}{\partial x} + v\frac{\partial u}{\partial y} = -\frac{\partial \Omega}{\partial x} - \frac{1}{\rho}\frac{\partial p}{\partial x} \tag{v}$$

$$u\frac{\partial v}{\partial x} + v\frac{\partial v}{\partial y} = -\frac{\partial \Omega}{\partial y} - \frac{1}{\rho}\frac{\partial p}{\partial y} \tag{vi}$$

Since the slope of the path of the fluid particle is given by Eq. (i), we substitute $v = u\dfrac{dy}{dx}$ into Eq. (v) and $u = v(dx/dy)$ into Eq. (vi) to obtain

Example 4.10 *(Con't.)*

$$u\frac{\partial u}{\partial x} + u\frac{dy}{dx}\frac{\partial u}{\partial y} = -\frac{\partial \Omega}{\partial x} - \frac{1}{\rho}\frac{\partial p}{\partial x} \qquad \text{(vii)}$$

$$v\frac{dx}{dy}\frac{\partial v}{\partial x} + v\frac{\partial v}{\partial y} = -\frac{\partial \Omega}{\partial y} - \frac{1}{\rho}\frac{\partial p}{\partial y} \qquad \text{(viii)}$$

Multiply Eq. (vii) by dx and Eq. (viii) by dy and add the resultant two equations:

$$u\frac{\partial u}{\partial x}dx + u\frac{\partial u}{\partial y}dy + v\frac{\partial v}{\partial x}dx + v\frac{\partial v}{\partial y}dy$$
$$= -\left(\frac{\partial \Omega}{\partial x}dx + \frac{\partial \Omega}{\partial y}dy\right) - \frac{1}{\rho}\left(\frac{\partial p}{\partial x}dx + \frac{\partial p}{\partial y}dy\right) \qquad \text{(ix)}$$

Since the flow is two-dimensional, $u = u(x, y)$, $v = v(x, y)$, $\Omega = \Omega(x, y)$, and $p = p(x, y)$ so that Eq. (ix) can be expressed in terms of exact differentials:

$$\frac{1}{2}d(u^2) + \frac{1}{2}d(v^2) = -d\Omega - \frac{1}{\rho}dp \qquad \text{(x)}$$

or it can be expressed in terms of the nabla operator ∇:

$$\nabla\left(\frac{p}{\rho} + \frac{V^2}{2} + \Omega\right) = 0 \qquad \text{(xi)}$$

Integrating Eq. (x) yields

$$\frac{1}{2}(u^2 + v^2) + \Omega + \frac{p}{\rho} = \text{const.} \qquad \text{(xii)}$$

Since the magnitude of the velocity $\mathbf{V}$ is defined as

$$|\mathbf{V}| = \sqrt{u^2 + v^2} \qquad \text{(xiii)}$$

we obtain

$$\boxed{|\mathbf{V}| = \sqrt{2}\left\{\text{const} - \frac{p}{\rho} - \Omega\right\}^{1/2}} \qquad \text{(xiv)}$$

This is the famous Bernoulli equation [Eq. (4.126)] which we shall apply to many different problems in Chap. 5.

This completes the solution.

Example 4.11

Oil is slowly flowing down the flat surface of a large rectangular tank because of gravity, as shown in Fig. E4.11. Its free-surface is exposed to air at 1 at-

Example 4.11 *(Con't.)*

mosphere. It has a thickness h. Let air be inviscid. If the flow of oil is steady and two-dimensional, determine (a) the stress $\mathbf{P}$ in terms of specific weight γ, thickness of oil h, and distance y, and (b) the velocity field $\mathbf{V}$. (c) Show that the flow is rotational.

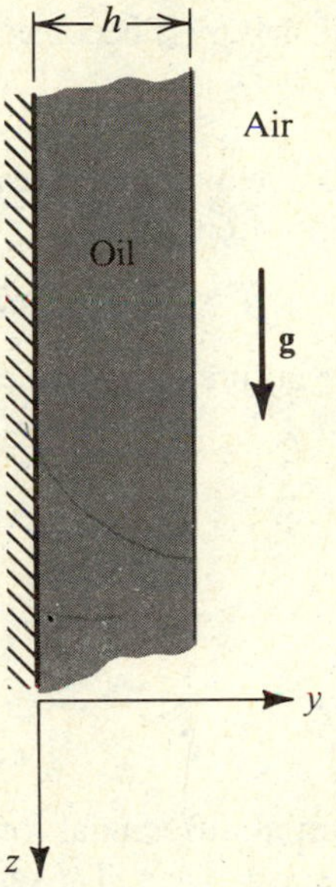

Figure E4.11

Solution:

Step 1.

Identify the characteristics of the fluid and flow field.

Oil is incompressible and viscous. The flow is steady and two-dimensional. The velocity components u, v are zero. Thus the velocity of the oil is in the z-direction and is expressed as

$$w = w(y, z) \tag{i}$$

Step 2.

Write the appropriate form of the governing equations.

From the conservation of mass, the continuity equation becomes

$$\cancelto{0}{\frac{\partial u}{\partial x}} + \cancelto{0}{\frac{\partial v}{\partial y}} + \frac{\partial w}{\partial z} = 0 \tag{ii}$$

Cauchy's equations of motion result in

$$\frac{Du}{Dt} = g_x + \frac{1}{\rho}\left(\frac{\partial p_{xx}}{\partial x} + \frac{\partial p_{yx}}{\partial y} + \frac{\partial p_{zx}}{\partial z}\right) \tag{iii}$$

Example 4.11 *(Con't.)*

$$\frac{Dv}{Dt} = g_y + \frac{1}{\rho}\left(\frac{\partial p_{xy}}{\partial x} + \frac{\partial p_{yy}}{\partial y} + \frac{\partial p_{zy}}{\partial z}\right) \quad \text{(iv)}$$

$$w\frac{\partial w}{\partial z} = g + \frac{1}{\rho}\left(\frac{\partial p_{xz}}{\partial x} + \frac{\partial p_{yz}}{\partial y} + \frac{\partial p_{zz}}{\partial z}\right) \quad \text{(v)}$$

The result shown in Eq. (ii) states that $w \neq w(z)$, such that Eq. (i) becomes

$$w = w(y) \quad \text{(vi)}$$

Since Cauchy's equations of motion are in terms of stresses, let us examine each separately. Since $u = v = 0$ and $w = w(y)$, Eqs. (4.61)–(4.66) reduce to

$$p_{xx} = p_{yy} = p_{zz} = -p = \text{const.} \quad \text{(vii)}$$

Note, so long as the oil has a free-surface, there will be no pressure gradient; i.e., $\nabla p = 0$. In addition to Eq. (vii), we have

$$p_{xy} = p_{xz} = 0 \quad \text{(viii)}$$

and

$$p_{yz} = \mu\frac{dw}{dy} \quad \text{(ix)}$$

It is important to point out that we have gone from the partial differential form of Eq. (4.66) to the total differential form of Eq. (ix) since $w = w(y)$. The only finite equation among Eqs. (iii)–(v) is Eq. (v):

$$0 = g + \frac{1}{\rho}\left(\frac{\partial p_{yz}}{\gamma y}\right) \quad \text{(x)}$$

Step 3.
Solve the differential equation. Since $w = w(y)$, the partial differential equation of Eq. (x) is transformed to a total differential. Integrating Eq. (x) yields

$$p_{yz} = -\gamma y + c_1 \quad \text{(xi)}$$

At $y = h$, the shear stress must vanish due to the stress boundary condition. Thus the stress becomes

$$p_{yz} = \gamma(h - y) \quad \text{(xii)}$$

a) The stress tensor $\mathbf{P}$ can now be evaluated using Eq. (4.51) and Eqs. (vii), (viii), and (xii):

$$\mathbf{P} = -p\mathbf{ii} - p\mathbf{jj} - p\mathbf{kk} + \gamma(h - y)\mathbf{jk} \quad \text{(xiii)}$$

Knowing the stress tensor enables us to evaluate such valuable engineering quantities as drag and normal forces on objects exposed to fluid flows (see Example 4.8).

Example 4.11 *(Con't.)*

(b) The velocity field **V** is obtained knowing the velocity components u, v, w. Since $u = v = 0$, we need only w. Equating the results of our shear stress p_{yz} [expressed by Eq. (xii) to Eq. (ix)] we integrate the result to obtain

$$w = \frac{\gamma}{\mu}(hy - y^2/2) + c_2 \qquad \text{(xiv)}$$

To evaluate the constant of integration c_2, we use the kinematic boundary condition of no slip; i.e., at $y = 0$, $w = 0$. Hence, the velocity field is

$$\mathbf{V} = \frac{\gamma y}{\mu}(h - y/2)\mathbf{k} \qquad \text{(xv)}$$

(c) To show the flow is rotational, we seek to evaluate the curl of the velocity vector **V**:

$$\nabla \times \mathbf{V} = \left(\frac{\partial w}{\partial y} - \frac{\partial v}{\partial z}\right)\mathbf{i}$$

$$= \frac{\gamma}{\mu}(h - y)\mathbf{i} \qquad \text{(xvi)}$$

We discover that the vorticity component ζ_x is proportional to the shear stress p_{yz}, that is,

$$\zeta_x = \frac{p_{yx}}{\mu} \qquad \text{(xvii)}$$

Thus if shear stresses exist in a flow we can expect fluid parcel rotations.

This completes the solution.

4.4.9 The Navier-Stokes Equations

The Navier-Stokes equations are used to solve many fluid flow problems. These equations represent the differential form of the conservation of linear momentum and are applicable in describing the motion of a fluid particle at an arbitrary location in the flow field at any instant of time.

We start with Cauchy's equation of motion:

$$(\mathbf{g} - \mathbf{a})\rho + \nabla \cdot \mathbf{P} = 0 \qquad (4.98)$$

The stress dyadic **P** is expressed in terms of the rate of strain dyadic $\dot{\mathbf{S}}$ and pressure p through Eqs. (4.59) or (4.60). The rate of strain tensor $\dot{\mathbf{S}}$ is related to the velocity gradients through Eq. (4.39). Substituting the stresses of Eqs. (4.61)–(4.66) into Eqs. (4.99)–(4.101) results in the governing equations of motion in Cartesian coordinates for *compressible* fluid flow.

1. *x*-component

$$\frac{\partial u}{\partial t} + u\frac{\partial u}{\partial x} + v\frac{\partial u}{\partial y} + w\frac{\partial u}{\partial z} = g_x - \frac{1}{\rho}\frac{\partial p}{\partial x} + \frac{\mu}{\rho} \cdot \left(\frac{\partial^2 u}{\partial x^2} + \frac{\partial^2 u}{\partial y^2} + \frac{\partial^2 u}{\partial z^2} + \frac{\partial D}{\partial x}\right) \tag{4.106}$$

2. *y*-component

$$\frac{\partial v}{\partial t} + u\frac{\partial v}{\partial x} + v\frac{\partial v}{\partial y} + w\frac{\partial v}{\partial z} = g_y - \frac{1}{\rho}\frac{\partial p}{\partial y} + \frac{\mu}{\rho} \cdot \left(\frac{\partial^2 v}{\partial x^2} + \frac{\partial^2 v}{\partial y^2} + \frac{\partial^2 v}{\partial z^2} + \frac{\partial D}{\partial y}\right) \tag{4.107}$$

3. *z*-component

$$\frac{\partial w}{\partial t} + u\frac{\partial w}{\partial x} + v\frac{\partial w}{\partial y} + w\frac{\partial w}{\partial z} = g_z - \frac{1}{\rho}\frac{\partial p}{\partial z} + \frac{\mu}{\rho} \cdot \left(\frac{\partial^2 w}{\partial x^2} + \frac{\partial^2 w}{\partial y^2} + \frac{\partial^2 w}{\partial z^2} + \frac{\partial D}{\partial z}\right) \tag{4.108}$$

We can easily express the above compressible equations in *incompressible form* by setting the dilation D equal to zero. One of the more convenient incompressible forms is the vector equation

$$\boxed{\underbrace{\frac{\partial \mathbf{V}}{\partial t}}_{ⓐ} + \underbrace{(\mathbf{V}\cdot\nabla)\mathbf{V}}_{ⓑ} = \underbrace{\mathbf{g}}_{ⓒ} \underbrace{- \frac{1}{\rho}\nabla p}_{ⓓ} + \underbrace{\nu\nabla^2\mathbf{V}}_{ⓔ}} \qquad M = 0 \tag{4.109}$$

The dimensions of each term are L/T^2; i.e., each term is an acceleration.

The term ⓐ, $\partial\mathbf{V}/\partial t$, represents the *local acceleration* of the fluid particle at a fixed point in space. For steady flow, this term is zero.

The term ⓑ, $(\mathbf{V}\cdot\nabla)\mathbf{V}$, is the *convective acceleration* of the fluid particle, and it predicts how the flow differs from one space location to the next at the same instant of time. Uniform flow has no convective acceleration, of course, since "uniform" means "of the same value."

The term ⓒ, $\mathbf{g}$, represents the acceleration due to gravity.

The term ⓓ, $-(1/\rho)\nabla p$, is the *pressure acceleration* due to the "pumping" action of the flow.

The last term, ⓔ, $\nu\nabla^2\mathbf{V}$, is the *viscous deceleration* due to the fluid's frictional resistance to objects moving through it.

The physical significance of the various terms of Eq. (4.109) can be illustrated to some extent by considering a few examples of fluid flow.

Consider the motion of an inviscid fluid. Such fluids are considered to be *ideal*, so that the Navier-Stokes equation for inviscid fluid flows would then include terms ⓐ–ⓓ of Eq. (4.109):

$$\frac{\partial \mathbf{V}}{\partial t} + (\mathbf{V}\cdot\nabla)\mathbf{V} = \mathbf{g} - \frac{1}{\rho}\nabla p \tag{4.110}$$

Popularly called *Euler's equation*, this is a first-order nonlinear partial differential equation and has some rather interesting solutions. Ideal fluid flows will be treated in succeeding chapters in much greater detail.

Another example of the Navier-Stokes equation is the case of very slow fluid motion. Such fluid flow problems are mathematically defined as those where the total acceleration $D\mathbf{V}/Dt$ [terms ⓐ and ⓑ of Eq. (4.109)] and the acceleration due to gravity, term c , are both zero. The Navier-Stokes equation for very slow motion is then

$$\nabla p = \mu\nabla^2\mathbf{V} \tag{4.111}$$

and is popularly called *Stokes flow* [4.3], or creep flow. It applies to the analysis of fluid behavior in lubrication mechanics, capillary flows, and certain molten metals.

Example 4.12. Very Slow Motion

Consider the two-dimensional motion of a cover plate moving with velocity V_∞ in the y-direction located a fixed distance h above a fixed flat plate. Let there be a fluid of density ρ and dynamic viscosity μ between the two plates. Assume no x- and z-component of velocity. Also assume the flow is steady. Calculate the pressure, velocity, and shear stress distributions (Fig. E4.12).

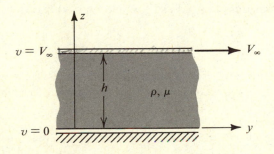

Figure E4.12

Solution:

Step 1.

Identify the characteristics of the fluid and flow field.

Assume the fluid is incompressible and real. The flow is steady and two-dimensional.

Example 4.12 *(Con't.)*

Step 2.
Write the appropriate form of the governing flow equations.

The continuity equation is

$$\frac{\partial u}{\partial x} + \frac{\partial v}{\partial y} + \frac{\partial w}{\partial z} = 0 \tag{4.24}$$

and the Navier-Stokes equation is

$$\frac{\partial \mathbf{V}}{\partial t} + (\mathbf{V}\cdot\nabla)\mathbf{V} = \mathbf{g} - \frac{1}{\rho}\nabla p + \nu\nabla^2\mathbf{V} \tag{4.109}$$

Using the assumptions given on the velocity field $u = w = 0$, we obtain from the continuity equation

$$\frac{\partial v}{\partial y} = 0 \tag{i}$$

The x-component scalar forms of the Navier-Stokes equations yields

$$\frac{\partial p}{\partial x} = 0 \tag{ii}$$

so that for this example

$$p = p(y) \tag{iii}$$

The y-component scalar form of the Navier-Stokes equation yields

$$\frac{\partial p}{\partial y} = \mu\left(\frac{\partial^2 v}{\partial y^2} + \frac{\partial^2 v}{\partial z^2}\right) \tag{iv}$$

Substituting the results obtained from Eqs. (i) and (iii) into Eq. (iv) results in

$$\frac{dp}{dy} = \mu\frac{d^2 v}{dz^2} \tag{v}$$

Since $p \neq p(x,z,t)$, and $v \neq v(x,y,t)$, Eq. (v) is readily integrable. Treating dp/dy as a constant, we have a simple second degree linear ordinary differential equation with the solution

$$v = \frac{1}{2\mu}\frac{dp}{dy}z^2 + c_1 z + c_2 \tag{vi}$$

From the given boundary conditions of the example

$$v = V_\infty \quad \text{at} \quad z = h \tag{vii}$$

and

$$v = 0 \quad \text{at} \quad z = 0 \tag{viii}$$

the coefficients c_1 and c_2 are evaluated with the result that the fluid flow velocity

Example 4.12 *(Con't.)*

$$v = \frac{1}{2\mu}\frac{dp}{dy}(z^2 - zh) + \frac{V_\infty}{h}z \tag{ix}$$

As shown by Eq. (iii), the pressure p varies only in the y-direction or

$$p = \mu\left(\frac{d^2v}{dz^2}\right)y + \text{const.} \tag{x}$$

where d^2v/dz^2, we recall, is a constant.

The shear stress $p_{xz} = p_{xy} = 0$, since the velocities $v = v(z)$ and $u = w = 0$. The only nonzero shear stress is

$$p_{zy} = \mu\frac{\partial v}{\partial z}$$

$$= \frac{1}{2}\frac{dp}{dy}(2z - h) + \frac{\mu V_\infty}{h} \tag{xi}$$

Note that the velocity field v is parabolic, and the stress field is linear. Note also that the stress p_{zy} is not zero at $h/2$ as we might suspect. From Eq. (x) we see that the shear stress is zero at the midpoint only if the upper plate is stationary (i.e., $V_\infty = 0$). This type of problem will reappear as one of the major topics in Chap. 12 on pipe flow.

This completes the solution.

Parallel flow constitutes that class of motion where all velocity components save one are zero. If that flow were in the x-direction, then terms ⓑ and ⓒ of Eq. (4.109) would be zero. The Navier-Stokes equation then becomes

$$\boxed{\frac{\partial u}{\partial t} = -\frac{1}{\rho}\frac{\partial p}{\partial x} + \nu\left(\frac{\partial^2 u}{\partial y^2} + \frac{\partial^2 u}{\partial z^2}\right)} \tag{4.112}$$

The above equation is a partial differential linear equation. One way to solve it is to try to transform it into an ordinary differential form. That will, of course, depend upon the particular flow problem.

The Navier-Stokes equation can also be expressed in other coordinate forms. Expressing the operator ∇ and D/Dt in cylindrical coordinates, Eq. (4.109) becomes

1. r-component

$$\frac{\partial v_r}{\partial t} + v_r\frac{\partial v_r}{\partial r} + \frac{v_\theta}{r}\frac{\partial v_r}{\partial \theta} - \frac{v_\theta^2}{r} + w\frac{\partial v_r}{\partial z}$$
$$= g_r - \frac{1}{\rho}\frac{\partial p}{\partial r} + \nu\left(\frac{\partial^2 v_r}{\partial r^2} + \frac{1}{r}\frac{\partial v_r}{\partial r} - \frac{v_r}{r^2}\right.$$
$$\left. + \frac{1}{r^2}\frac{\partial^2 v_r}{\partial \theta^2} - \frac{2}{r^2}\frac{\partial v_\theta}{\partial \theta} + \frac{\partial^2 v_r}{\partial z^2}\right) \tag{4.113}$$

2. θ-component

$$\frac{\partial v_\theta}{\partial t} + v_r\frac{\partial v_\theta}{\partial r} + \frac{v_\theta}{r}\frac{\partial v_\theta}{\partial \theta} + \frac{v_r v_\theta}{r} + w\frac{\partial v_\theta}{\partial z}$$

$$= g_\theta - \frac{1}{\rho r}\frac{\partial p}{\partial \theta} + \nu\left(\frac{\partial^2 v_\theta}{\partial r^2} + \frac{1}{r}\frac{\partial v_\theta}{\partial r} - \frac{v_\theta}{r^2}\right.$$

$$\left. + \frac{1}{r^2}\frac{\partial^2 v_\theta}{\partial \theta^2} + \frac{2}{r^2}\frac{\partial v_r}{\partial \theta} + \frac{\partial^2 v_\theta}{\partial z^2}\right) \tag{4.114}$$

3. z-component

$$\frac{\partial w}{\partial t} + v_r\frac{\partial w}{\partial r} + \frac{v_\theta}{r}\frac{\partial w}{\partial \theta} + w\frac{\partial w}{\partial z} = g_z - \frac{1}{\rho}\frac{\partial p}{\partial z}$$

$$+ \nu\left(\frac{\partial^2 w}{\partial r^2} + \frac{1}{r}\frac{\partial w}{\partial r} + \frac{1}{r^2}\frac{\partial^2 w}{\partial \theta^2} + \frac{\partial^2 w}{\partial z^2}\right) \tag{4.115}$$

The Navier-Stokes equation looks rather simple, compared to some equations found in engineering and physics—simple in form but not simple in solution. With our present technology, we cannot even imagine a computer capable of solving the Navier-Stokes equation. And yet it is the governing equation that correctly and completely describes the fundamental conservation law of linear momentum. There is nothing approximate in its formulation once we confirm that the fluid is Newtonian. Furthermore, it gives us important variables for dimensional analysis.

We have shown that by examining certain classes of flow, such as Stokes flow or inviscid flow, certain terms in the Navier-Stokes equation drop out and solutions become possible. Since the Navier-Stokes equation expresses the dynamic equilibrium condition between external forces and inertial forces, we must always ask which external forces may be large, and which may not exist or may be negligible compared to others. For example, *body* forces are important when there is a free-surface, when the density is stratified, or when buoyancy forces exist. *Surface* forces, which may be both normal and tangential, are reducible to normal forces if viscous effects can be neglected, or if certain velocity gradients are zero. We seek in every case, a way to simplify the Navier-Stokes equation, so that we can solve it—for *solving* problems is the essence of fluid dynamics. We always investigate what can be done to the assorted terms, what tricks can be introduced, and what transformations might exist to simplify the Navier-Stokes equation.

4.4.10 The Gromeka-Lamb Form of the Navier-Stokes Equation

The Navier-Stokes equation can be transformed into an alternate vector form, known as the Gromeka-Lamb equation by using the vector relationships of Eqs. (B.30) and (B.33). In reference to Eq. (B.30), we can show that

$$(\mathbf{V}\cdot\nabla)\mathbf{V} = \frac{1}{2}\nabla V^2 - \mathbf{V} \times (\nabla \times \mathbf{V}) \tag{4.116}$$

and in reference to Eq. (B.33), we can show that

$$\nabla^2\mathbf{V} = \nabla(\nabla\cdot\mathbf{V}) - \nabla \times (\nabla \times \mathbf{V}) \tag{4.117}$$

Making use of the differential forms of the incompressible continuity equation and the definition of a vorticity vector $\boldsymbol{\zeta}$, we transform Eqs. (4.116) and (4.117) into

$$(\mathbf{V}\cdot\nabla)\mathbf{V} = \frac{1}{2}\nabla V^2 - \mathbf{V} \times \boldsymbol{\zeta} \tag{4.118}$$

and

$$\nabla^2\mathbf{V} = -\nabla \times \boldsymbol{\zeta} \tag{4.119}$$

Substituting these two results into the Navier-Stokes equation results in a vector differential equation

$$\boxed{\frac{\partial \mathbf{V}}{\partial t} + \boldsymbol{\zeta} \times \mathbf{V} = \mathbf{g} - \nabla\left(\frac{p}{\rho} + \frac{V^2}{2}\right) - \nu(\nabla \times \boldsymbol{\zeta})} \tag{4.120}$$

which is called the *Gromeka-Lamb* form of the Navier-Stokes equation. It is just as exact as the latter.

The Gromeka-Lamb equation is particularly suitable for working with curvilinear coordinates, or when we are given some information about the vorticity $\boldsymbol{\zeta}$ for a particular flow. We note that for a flow that is both steady and irrotational, the equation has the elegant form

$$\nabla\left(\frac{p}{\rho} + \frac{V^2}{2}\right) = \mathbf{g} \tag{4.121}$$

which looks similar to the solution in Example 4.10. The integration of this equation is extremely simple—just as simple as the aerohydrostatic example—and will be discussed in Sec. 5.5.

Gromeka-Lamb Equations for Inviscid Incompressible Fluid Flow

An alternate form of Euler's equation can be obtained using the Gromeka-Lamb form of the Navier-Stokes equation. From Eq. (4.120), the *Gromeka-Lamb equation for inviscid fluid flow* is

$$\frac{\partial \mathbf{V}}{\partial t} + \boldsymbol{\zeta} \times \mathbf{V} = \mathbf{g} - \nabla\left(\frac{p}{\rho} + \frac{V^2}{2}\right) \tag{4.122}$$

Because $\mathbf{g}$ is the gravitational force per unit mass, it can be expressed in terms of the gradient of the potential energy Ω such that

$$\frac{\partial \mathbf{V}}{\partial t} + \boldsymbol{\zeta} \times \mathbf{V} = -\nabla\left(\frac{p}{\rho} + \frac{V^2}{2} + \Omega\right) \tag{4.123}$$

The above equation is the *fundamental* equation of equilibrium for *inviscid fluids*, and can be solved for a large number of vorticity cases.

Lamb's Equation

If the flow is steady, inviscid, and incompressible, Gromeka-Lamb's equation becomes

$$\mathbf{V} \times \boldsymbol{\zeta} = \nabla\left(\frac{p}{\rho} + \frac{V^2}{2} + \Omega\right) \tag{4.124}$$

and is called Crocco's or *Lamb's equation*. Lamb's equation gives us some very important geometric relationships between velocity vector $\mathbf{V}$ and vorticity vector $\boldsymbol{\zeta}$ and the quantity $(p/\rho) + (V^2/2) + \Omega$. Expressing the potential energy in terms of acceleration due to gravity, we obtain an alternate form of Lamb's equation:

$$\mathbf{V} \times \boldsymbol{\zeta} = \nabla\left(\frac{p}{\rho} + \frac{V^2}{2} + gz\right) \tag{4.125}$$

Figure 4.17 shows the geometric significance of Eq. (4.123). The surface $\mathbf{V} \times \boldsymbol{\zeta}$ contains the paths of the fluid flow. For steady flow, the fluid's path is called the streamline, so that streamlines lie in the surface $\mathbf{V} \times \boldsymbol{\zeta}$. From the properties of cross-products, the vector $\mathbf{V} \times \boldsymbol{\zeta}$ must be normal to the surface $(p/\rho) + (V^2/2) + gz$ equals constant. Thus, both the velocity $\mathbf{V}$ and vorticity $\boldsymbol{\zeta}$ must lie on the surface where

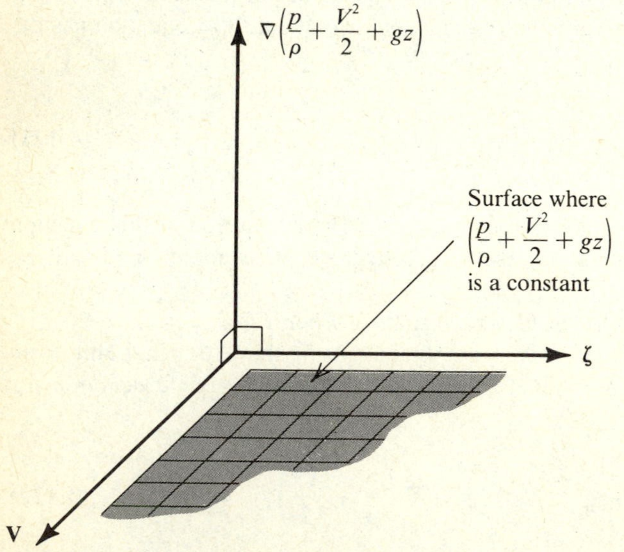

Figure 4.17 *Geometry of Lamb's equation.*

$(p/\rho) + (V^2/2) + gz$ equals constant. On another surface, the constant may be different and by an amount $\mathbf{V} \times \boldsymbol{\zeta}$. Since streamlines are tangent to $\mathbf{V}$, and vortex lines are tangent to $\boldsymbol{\zeta}$, the grid of streamlines and vortex lines must lie on the surface $(p/\rho) + (V^2/2) + gz$ equals constant. Thus, *along a streamline and a vortex line, the quantity* $(p/\rho) + (V^2/2) + gz$ *is constant*:

$$p/\rho + V^2/2 + gz = \text{const.} \tag{4.126}$$

This is the integrated form of Lamb's equation and is called Bernoulli's equation. It states that for a *steady incompressible inviscid flow*, the quantity $(p/\rho) + (V^2/2) + gz$ remains constant along the path of the fluid particle and along a vortex line. Note that Eq. (4.126) reduces to the hydrostatic equation when there is no change in the kinetic energy, $\Delta V^2 = 0$.

The quantity in parentheses in Eq. (4.126) is sometimes the *total head* per unit mass of fluid. Based upon Eq. (4.124), we can say that there will be vorticity in the flow field when there are gradients of total head; this means that if the total head changes along the flow, then there is vorticity, or alternatively, that the fluid particles have angular deformation.

The Gromeka-Lamb Equation for Irrotational Motion

For a flow to be irrotational, the angular rotation $\boldsymbol{\omega}$ of a fluid particle must be zero. In Sec. 4.4.7, the angular rotation $\boldsymbol{\omega}$ was shown to be equal to one-half the curl of the velocity vector. Thus as a consequence of irrotationality, we use the *necessary and sufficient condition for irrotational flow*, that is

$$\boxed{\nabla \times \mathbf{V} = 0} \tag{4.127}$$

We can satisfy the above relationship in three possible ways: (i) the velocity vector $\mathbf{V}$ is zero, (ii) the nabla operator ∇ is zero, or (iii) the velocity vector is defined as the gradient of a scalar potential function ϕ

$$\mathbf{V} = \nabla\phi \tag{4.128}$$

The first two ways are too restrictive, so we adopt the third way. In order for the velocity potential ϕ to exist, the curl of the velocity vector must vanish.

Since space and time variables are independent of each other, the order of the time variation can be changed with the differentiation of space if we take the time derivative of a gradient. The local acceleration $\partial\mathbf{V}/\partial t$ is then expressed as

$$\begin{aligned}\frac{\partial \mathbf{V}}{\partial t} &= \frac{\partial}{\partial t}\nabla\phi \\ &= \nabla\frac{\partial\phi}{\partial t}\end{aligned} \tag{4.129}$$

and is valid everywhere in the field of an *irrotational* flow. Substituting the local acceleration into the Gromeka-Lamb Eq. (4.123) we obtain

$$\nabla\left(\frac{p}{\rho} + \frac{V^2}{2} + gz + \frac{\partial\phi}{\partial t}\right) = 0 \tag{4.130}$$

for any irrotational motion of an incompressible inviscid fluid. Integration results in

$$\frac{p}{\rho} + \frac{V^2}{2} + gz + \frac{\partial\phi}{\partial t} = c(t) \tag{4.131}$$

and is applicable *everywhere* in the fluid. This is yet another form of Bernoulli's equation. Note that for steady flow, the term $\partial\phi/\partial t = 0$, and the term $c(t)$ becomes a constant. This is identical in *form* to Bernoulli's equation along a path line. The differences are that

$$\left(\frac{p}{\rho} + \frac{V^2}{2} + gz\right) = \text{const. (along streamline)} \tag{4.126}$$

is valid only along a streamline and vortex line for inviscid steady flows, and

$$\left(\frac{p}{\rho} + \frac{V^2}{2} + gz\right) = \text{const. (throughout flow)} \tag{4.132}$$

everywhere in the flow field for steady inviscid irrotational flows. The application of these integrated forms will be treated in Chap. 5.

Example 4.13

Consider a steady two-dimensional flow such that $\boldsymbol{\zeta} = \zeta_z\mathbf{k}$ and the velocity field is given by Fig. E4.13. Given that

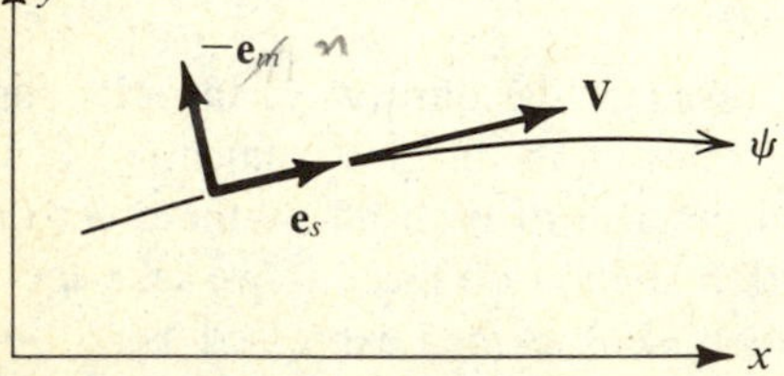

Figure E4.13

$$\frac{p}{\rho} + \frac{V^2}{2} + gz = c \tag{i}$$

where c is a constant on the path ψ, show that

$$\frac{\nabla c}{V} = -\zeta_z\mathbf{e}_n \tag{ii}$$

Example 4.13 *(Con't.)*

and that c therefore varies from path to path. Show also that if the flow is irrotational then the constant c is constant everywhere in the x-y plane.

Solution:

(This example does not require a step-by-step procedure.)

Substituting Eq. (i) into Eq. (4.125) produces

$$\mathbf{V} \times \boldsymbol{\zeta} = \nabla c \tag{iii}$$

Since the flow is two-dimensional,

$$\boldsymbol{\zeta} = \zeta_z \mathbf{k} \tag{iv}$$

$$\mathbf{V} = V\mathbf{e}_s \tag{v}$$

where $\mathbf{e}_s$ is the unit vector tangent to the path ψ where $\mathbf{V}$ is also tangent. Equation (iii) can be expressed as

$$\nabla c = -\mathbf{k} \times \mathbf{e}_s V\zeta_z \tag{vi}$$

using Eqs. (iv) and (v). Since the unit vectors $\mathbf{e}_n$ and $\mathbf{e}_s$ lie in the x-y plane, then from vector algebra

$$\mathbf{k} \times \mathbf{e}_s = \mathbf{e}_n \tag{vii}$$

Thus, the gradient of the constant c is in a direction $\mathbf{e}_n$, i.e.,

$$\nabla c = \frac{\partial c}{\partial n}\mathbf{e}_n = -V\zeta_z\mathbf{e}_n \tag{viii}$$

which is the result given by Eq. (ii).

Equation (viii) states that if c is constant along the path ψ, then it can vary only in the n direction, thus vary from path to path.

Equation (viii) also states that, for irrotational flow where ω_z equals zero, c is a constant everywhere in the x-y plane. These last two statements were summarized by Eqs. (4.126) and (4.132), respectively.

This completes the solution.

Beltrami Flow

The third solution of the Gromeka-Lamb equation, Eq. (4.123), is for *steady Beltrami flow*. A Beltrami flow is one in which the velocity vector $\mathbf{V}$ and the angular rotation vector $\boldsymbol{\omega}$ are parallel to each other. An example would be flow in a duct with vorticity (swirl) normal to the axial flow, such as flow from a fan blade. Examination of the angular rotation components reveals that such a flow can only take place in three dimensions. Thus, for steady three-dimensional motion where $\mathbf{V}$ and $\boldsymbol{\omega}$ are in the same direction, the Lamb equation becomes

$$\nabla\left(\frac{p}{\rho} + \frac{V^2}{2} + gz\right) = 0 \qquad (4.133a)$$

Integration results in

$$\frac{p}{\rho} + \frac{V^2}{2} + gz = \text{const.} \qquad (4.133b)$$

a result identical to the form given by Eq. (4.126).

4.4.11 Boundary Conditions

We have described a number of differential equations: the continuity equation, Navier-Stokes equation, Gromeka-Lamb equation, Euler's equation, Lamb's equation, Stokes equation, and others. In order to specify the mathematical problem for these equations completely, we must present boundary conditions. The set of equations that govern the type of fluid flow, coupled with the boundary conditions which define the particular flow, complete the fluid dynamic problem.

Any equation of the form

$$F(\mathbf{r},t) = 0 \qquad (4.134)$$

defines a surface in space such as an airplane wing, or a ship's hull. We note that the surface may continuously alter its shape like the interface between water and air (e.g., a wave), because Eq. (4.134) is a function of time. Such surfaces are called *material surfaces* if and only if the fluid particle that is on the surface remains on the surface for all values of time t. Thus a material surface always consists of the same fluid particles.

Furthermore, if a fluid particle located at position $\mathbf{r}_1$ at time t_1 is at position $\mathbf{r}_2$ a later time t_2, we obtain

$$\lim_{t_2 \to t_1} \frac{F(\mathbf{r}_2, t_2) - F(\mathbf{r}_1, t_1)}{t_2 - t_1} = \frac{DF}{Dt} = 0 \qquad (4.135)$$

since

$$F(\mathbf{r}_1, t_1) = 0 \quad \text{and} \quad F(\mathbf{r}_2, t_2) = 0$$

Thus, not only must Eq. (4.134) be satisfied, but its derivative, Eq. (4.135) must be satisfied *everywhere on the surface*.

Kinematic Boundary Condition

A material surface, which we'll now call a *boundary* of a fluid region, is characterized by the fact that every fluid particle remains on the boundary. Fluid particles adjacent to the material surface cannot penetrate nor cause a vacuum on the material surface.

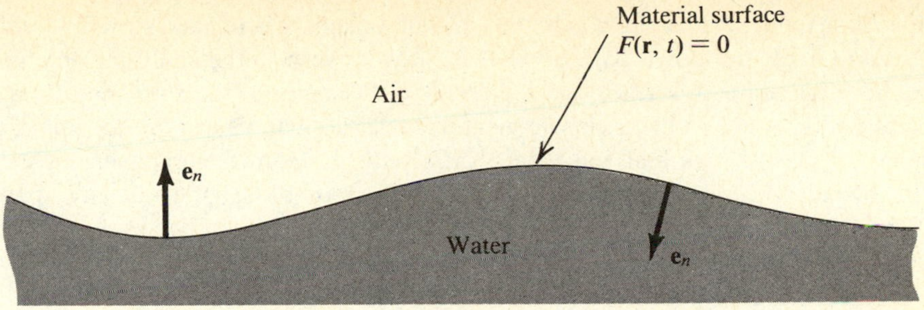

Figure 4.18 *The material surface F* (**r**, *t*) = 0.

In Fig. 4.18, let the normal unit vector $\mathbf{e}_n$ be perpendicular to the material surface $F(\mathbf{r},t) = 0$. The normal component of the velocity relative to the boundary must vanish:

$$\mathbf{V}\cdot\mathbf{e}_n = v_n = 0, \quad \text{at a boundary} \tag{4.136}$$

where v_n denotes the normal velocity component. If a material surface is made from solid material, and the *solid* material is moving with a velocity **U**, then the kinematic condition at the boundary of the material surface is adjusted so that

$$\mathbf{V}\cdot\mathbf{e}_n = \mathbf{U}\cdot\mathbf{e}_n \tag{4.137}$$

These mathematical statements are rather like scorekeeping in baseball. We must keep track of who's on first and who's on second. We must be certain that the fluid does not move through a solid boundary, and that the particles along the boundary or that comprise the boundary share the boundary's velocity.

Thus, for *real* fluids flowing past a *fixed* material surface the tangential velocity component of the fluid at the surface must vanish. This conclusion is based on the hypothesis that the fluid does not slip at the boundary. Our present experimental knowledge does not allow us to know precisely what takes place at a fixed boundary. We might say that slip, if any, is too small to be of any importance, or alternatively we might say that a quasisolid layer of fluid, if there ever is one, is too thin to be observed and therefore inconsequential. Using a microscope for turbulent flow of water through a pipe, two experimentalists (Fage and Townend) observed that the slowest observable particle of fluid near the pipe wall had a mean velocity of 0.006 ft/s when the average fluid velocity across the pipe section was 0.83 ft/s. Assuming no slip, they calculated that the particular particle was at a distance 2.5×10^{-5} in. from the pipe surface. Such an experimental result neither confirms nor denies the no-slip condition, but it does strongly suggest that it may exist.*

*The fact that the relative velocity of a fluid particle is zero next to a wall does not mean that molecules of the particle are at rest. There is considerable activity within the particle. It means that there is a 50% probability bouncing to the left and the rest going to the right such that they maintain a zero velocity within the particle.

For *ideal* fluids, the no-slip condition at a solid boundary is relaxed since no fluid shear stress retards the flow. To analyze these flows, we hypothetically replace the solid surface by a stream surface that represents the velocity of the streaming fluid. This substitution is one of the tricks we mentioned earlier that we can use to simplify a problem. Thus for *both* real and ideal fluids, only the normal component of the velocity boundary condition stated by Eq. (4.137) must be satisfied at any solid boundary.

Excluded from these general propositions are the boundary conditions in highly rarefied gases. Such fluids allow for slip: i.e., the slip velocity at the material surface is proportional to the rate of change of the tangential velocity in a direction normal to the boundary surface.

Figure 4.19 shows three different conditions at a boundary of a fluid flow. Profile (a) is a typical incompressible laminar flow profile of velocity that shows the maximum shear stress (i.e., the largest slope of the velocity) to exist at the boundary surface. Profile (b) is a typical slip flow velocity profile which occurs at high velocity compressible flows. The shear stresses are such that the fluid is slowed down by viscous dissipation. Because the flow's kinetic energy is extremely high, the dissipative energy cannot absorb the kinetic energy, so the fluid slips by the surface. Profile (c) is a typical free-molecule flow velocity profile. Free-molecule flow occurs at extremely high velocities where the effect of viscosity is negligible due to the extremely large value of the inertial force.

Stress Boundary Condition

In addition to the kinematic boundary condition, there is also a stress boundary condition. The stress boundary condition states that stress must be continuous across any

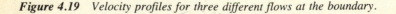

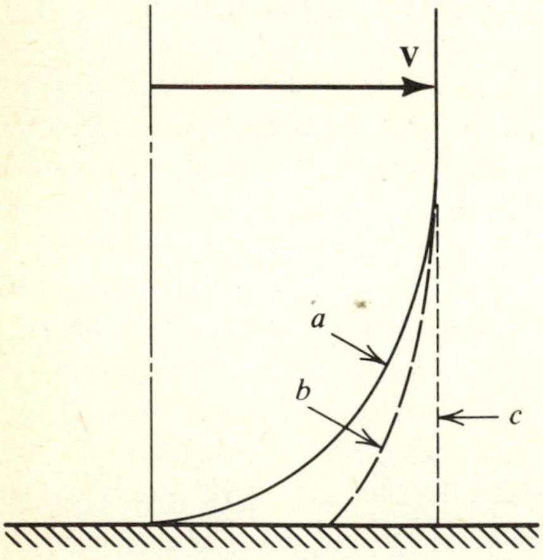

a Continuum flow "real flow"

b Slip flow

c Free molecule flow "inviscid flow"

Figure 4.19 *Velocity profiles for three different flows at the boundary.*

boundary. It is sometimes found that a natural boundary exists between two inmiscible fluids (such as oil and water, wind blowing over water, or a bubble moving through a liquid). If one fluid has a normal stress (p_{ii}) and a shear stress (p_{ij}) at the interface separating the two fluids, the second fluid must have identical normal and shear stresses in the region immediately adjacent to the first fluid. Thus,

$$(p_{ij})_{S_1} = (p_{ij})_{S_2} \tag{4.138}$$

at the interface $S_1 = S_2$, as shown in Fig. 4.20.

Where the boundary or the interface is to be determined in addition to the velocity and pressure fields, we have a *free-boundary problem*. A free-boundary problem occurs in unsteady fluid dynamic flows such as wave motion and fuel sloshing. If the free-surface amplitude is small we can often find the resultant shape of the free-surface or interface once the velocity field is known. The velocity field also depends upon the geometry of the boundary surface as well as on whether the amplitude of the free-surface is large or small compared to the fluid's depth.

At the bubble's interface
$(p_{ii})_{S_{water}} = (p_{ii})_{S_{air}}$
$(p_{ij})_{S_{water}} = (p_{ij})_{S_{air}}$

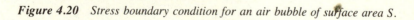

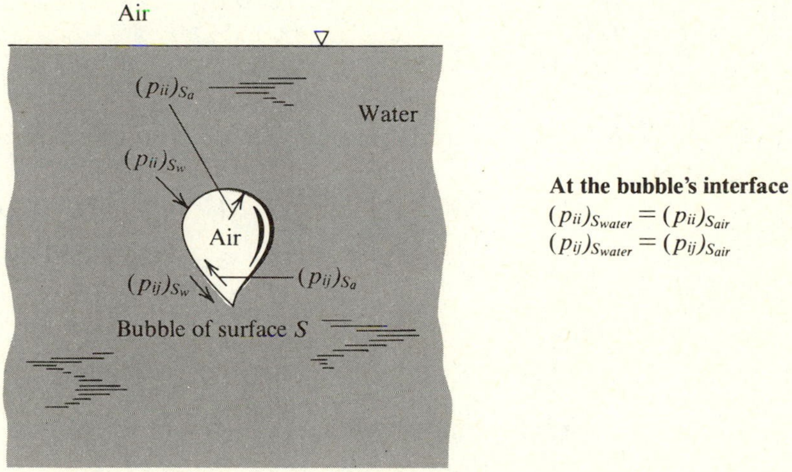

Figure 4.20 Stress boundary condition for an air bubble of surface area S.

4.5 The Differential Form of the Conservation of Energy*

For completeness, we treat the differential form of the conservation of energy. It is often neglected in incompressible fluid dynamics, but it is important when there are, for example, temperature variations in a flow. The conservation of energy principle may be applied to a control volume analysis by proper application of the differential forms of the general property balance, Eq. (4.16). We define the intensive property ϕ in terms of the specific energy $\bar{e}$ by

---

*This section may be omitted without loss of continuity.

$$\phi = \rho\bar{e} \tag{4.139}$$

where $\bar{e}$ is the *total* energy per unit mass and is composed of internal ($\bar{i}$), kinetic ($V^2/2$), potential (gz) energies plus others:

$$\bar{e} = \bar{i} + \frac{V^2}{2} + gz + \bar{e}_{\text{nuclear}} + \bar{e}_{\text{elect.}} + \bar{e}_{\text{magn.}} + \text{others} \tag{4.140}$$

In the present discussion, we shall neglect all energies except internal, kinetic, and potential, such that

$$\bar{e} = \bar{i} + \frac{V^2}{2} + gz \tag{4.141}$$

Substituting Eq. (4.139) into the (D.F.) general property balance, Eq. (4.16), results in

$$\frac{D(\rho\bar{e})}{Dt} = \frac{\partial(\rho\bar{e})}{\partial t} + \nabla\cdot(\bar{e}\rho\mathbf{V}) \tag{4.142}$$

The differential form of the first law of thermodynamics for a control volume analysis is

$$\frac{đq^*}{dt} + \frac{đw^*}{dt} = \frac{De}{Dt} \tag{4.143}$$

where $\frac{đq^*}{dt}$ is the *net* rate of heat transferred per unit volume from the surroundings and $\frac{đw^*}{dt}$ is the net power per unit volume transferred across the control surface from the surroundings into the differential fluid volume. Hence if there are no transfers of energy across the control volume, the total energy of the control volume is constant. Such a case would require the control volume to be an isolated system and is of no interest to us.

The rate of heat transfer per unit volume $\frac{đq^*}{dt}$ that we shall consider is due exclusively to conduction. The rate of heat transfer is related to the vector rate of heat flow by

$$\frac{đq^*}{dt} = -\nabla\cdot\mathbf{q} \tag{4.144}$$

where the vector rate of heat flow is related to temperature T and coefficient of thermal conductivity k by Fourier's law

$$\mathbf{q} = -k\nabla T \tag{4.145}$$

We can easily derive the relationship of Eq. (4.144) by examining how heat flows across an elemental Cartesian control volume.

The net power per unit volume $đw^*/dt$ results from three types of work. Work may be transferred across the control surface of our fluid control volume (see Fig. 4.21) by mechanical means using shafts or linkages, by viscous stresses, and by normal stresses (such as pressure):

$$\frac{đw^*}{dt} = \left(\frac{đw^*}{dt}\right)_{\text{mech}} + \left(\frac{đw^*}{dt}\right)_{v} + \left(\frac{đw^*}{dt}\right)_{p} \tag{4.146}$$

One way to express the loss of power due to viscous stresses is in terms of the strain rate dyadic $\mathbf{S}$ of Eq. (4.35):

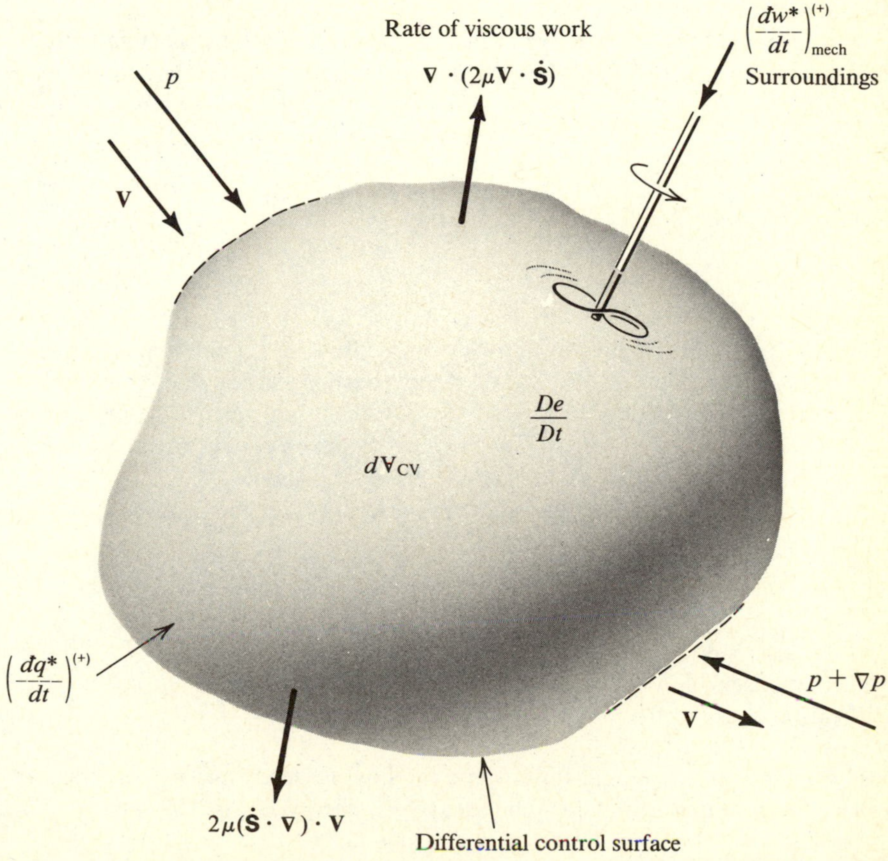

Figure 4.21 *Differential energy transfers.*

$$\left(\frac{đw^*}{dt}\right)_v = -2\mu\nabla\cdot(\mathbf{V}\cdot\dot{\mathbf{S}}) + 2\mu(\dot{\mathbf{S}}\cdot\nabla)\cdot\mathbf{V} \tag{4.147}$$

where the first term on the right is the power lost across the control surface due to the rate of viscous work per unit volume, and the second term is the power lost within the control volume, loss resulting from fluid friction being converted into heat. The derivation of Eq. (4.147) is left as an exercise. See Prob. 5.124. A second and more popular expression for the loss of power will be presented shortly.

The power per unit volume due to the normal stresses is also to be derived in Prob. 5.124. It is identified as

$$\left(\frac{đw^*}{dt}\right)_p = -\nabla\cdot(p\mathbf{V}) \tag{4.148}$$

This power expression is based on the possibility that a pumping action or a loss of pressure (from, say, friction and a turbine) may produce a difference in pressure between the inlet and exit sections.

The differential form of the energy equation can be obtained by substituting Eqs. (4.142) and (4.144)–(4.148) into the first law of thermodynamics, Eq. (4.143), to obtain

$$\left(\frac{đw^*}{dt}\right)_{\text{mech}} = \frac{\partial(\rho\bar{e})}{\partial t} + \nabla\cdot\left[\left(\bar{e} + \frac{p}{\rho}\right)\rho\mathbf{V} - k\nabla T + 2\mu\mathbf{V}\cdot\dot{\mathbf{S}}\right] - 2\mu(\dot{\mathbf{S}}\cdot\nabla)\cdot\mathbf{V} \tag{4.149}$$

The above equation applies to any Newtonian fluid in a field where the only transfer of heat is by conduction. The above energy equation is much more complicated than the Navier-Stokes equation, and, like the Navier-Stokes equation of linear momentum, it has no general solution. The number of unknowns require the Navier-Stokes equation for the general solution of the energy equation.

Certain assumptions can vastly simplify the energy equation. Some of these assumptions are given below.

- Steady flow: $\dfrac{\partial(\rho\bar{e})}{\partial t} = 0$
- No heat transfer: $\nabla T = 0$
- Inviscid flow: $\mu = 0$

Suppose we consider an inviscid fluid where the flow is steady and no heat transfer takes place. Substituting Eq. (4.141) into the (D.F.) energy Eq. (4.149) gives

$$\left(\frac{đw^*}{dt}\right)_{\text{mech}} = \nabla\cdot\left(\bar{h} + \frac{V^2}{2} + gz\right)\rho\mathbf{V} \tag{4.150}$$

where the specific enthalpy $\bar{h}$ is defined as

$$\bar{h} = \bar{i} + p/\rho \tag{4.151}$$

Equation (4.150) can be easily solved:

$$w_{\text{mech}} = \Delta\left(h + \frac{V^2}{2} + gz\right) \tag{4.152}$$

where w_{mech} is the mechancial work per unit mass.

Another important case where we can solve the (D.F.) energy Eq. (4.149) is for a fluid at rest or moving with negligible velocity and having no mechanical energy transfer. Equation (4.149) reduces to

$$\frac{\partial(\rho\bar{i})}{\partial t} = \nabla\cdot k\nabla T \tag{4.153}$$

If the fluid is a perfect gas, then the energy equation simplifies to

$$C_v\frac{\partial}{\partial t}(\rho T) = k\nabla^2 T \tag{4.154}$$

and is recognized as the unsteady heat conduction equation, solutions of which are found in Ref. [4.4] for a variety of boundary conditions. Equation (4.154) is one of the more important equations of heat transfer.

4.5.1 Boundary Conditions for the Energy Equation

The boundary conditions for the energy equation involve expressions for temperature and any of its derivatives on the boundaries confining or defining the flow. Across any surface separating a body and a fluid, or between any two fluids, the temperature and heat transfer must be continuous.

For the boundary condition of temperature at the surface of a body, the temperature of the fluid must equal the temperature of the body's surface. The fluid's normal temperature fluid gradient will equal the body's normal temperature gradient if both have the same thermal conductivity. Thus,

$$T = T_w \tag{4.155}$$

$$\frac{\partial T}{\partial n} = -\left(\frac{\dot{Q}}{k}\right)_w \tag{4.156}$$

where $\dot{Q}$ is the rate of heat transferred from the wall to the fluid.

At the edge of the boundary layer, the temperature of the fluid must be the free-stream temperature, T_∞, and the gradient of temperature must be zero, i.e.,

$$T = T_\infty, \qquad \text{at} \quad n = \delta \tag{4.157}$$

$$\frac{\partial T}{\partial n} = 0, \qquad \text{at} \quad n = \delta \tag{4.158}$$

where n is the coordinate normal to the boundary layer.

This presentation is purposefully brief. Detailed derivations and their explanations are presented in many heat transfer texts, such as Refs. [4.5] and [4.6].

4.6 Air as an Incompressible and/or Inviscid Fluid

The treatment of air, or any other gas, as an incompressible fluid is often said to be justified by postulating changes in density that are negligible. This justification is neither quite proper nor correct, because when we study compressibility effects of fluids, the changes in density are usually of the same order of magnitude as the change in pressure. Let us explore the real reason why Bernoulli's equation can be applied to gases.

The explanation was originally given by P. H. Oosthuizen [4.6]. Expressing the speed of sound c in terms of the temperature of the gas

$$c = \sqrt{C_p(k - 1)T} \tag{4.159}$$

where

$$k = C_{p_o}/C_{v_0} \tag{4.160}$$

and defining the local Mach number M as

$$M = \frac{U}{c} \tag{4.161}$$

Euler's equation can be expressed as

$$\frac{dp}{p} = -kM\frac{du}{u} \tag{4.162}$$

The energy equation for adiabatic flow can be written as

$$\frac{dT}{T} = -(k - 1)M^2\frac{du}{u} \tag{4.163}$$

and the equation of state for a perfect gas (in differential form) can be shown to be

$$\frac{d\rho}{\rho} = -M^2\frac{du}{u} \tag{4.164}$$

From thermodynamics, the ratio of specific heats lies in the range $1.3 \leq k \leq 1.7$. Hence any change in velocity du will result in very nearly the same changes in pressure, temperature and density according to $M^2 du/u$. In fact as one approaches the incompressible condition ($M \rightarrow 0$), the flow is isothermal and isobaric.

We need to integrate the one-dimensional Euler's equation to obtain

$$\begin{aligned} \frac{u_2^2 - u_1^2}{2} &= \int_{p_1}^{p_2} \frac{1}{\rho} dp \\ &= \frac{p_1}{\rho_1}\int_{p_1}^{p_2} \frac{dp}{p_1}\left(1 + \frac{d\rho}{\rho_1}\right) \end{aligned} \tag{4.165}$$

We have conditionally stated that dp/p_1 and $d\rho/\rho_1$ will both be small, so that higher order terms can be neglected. Thus Eq. (4.165) can be integrated to yield

$$\frac{p_i}{\gamma} + \frac{u_1^2}{2g} = \frac{p_2}{\gamma} + \frac{u_2^2}{2g} \tag{4.166}$$

which is the Bernoulli equation for incompressible flow.

Next, let us consider what is meant by frictionless motion. By definition, frictionless motion exists when the shear stress p_{ij} is zero. Two independent conditions can make the shear stress zero. First, the dynamic viscosity can be zero. Such flows are called inviscid flows and are characterized mathematically by $\mu = 0$. Second, all the components of the strain rate dyadic $\mathbf{S}$ can vanish: this condition means that no relative deformation occurs between points in the flow, as in fluid statics or rigid-body motion. Of these two conditions, the first, or inviscid, condition is the most *ideal* because there are *no inviscid* fluids in nature.

Inviscid flow often refers to the dynamic behavior of *perfect* fluids. The word *perfect* should *not* be mistaken with that of a perfect gas defined for thermodynamics. A perfect gas is one that is described by the familiar equation of state $p = \rho RT$. A perfect fluid, on the other hand, is one that obeys those kinematical laws that are void of vorticity. It is also incompressible.

Frictionless real fluids do exist. Again, *inviscid fluids* do not. Thus, the rate of strain is zero for frictionless fluids. We then ask where in a fluid flow can we assume a zero value of the shear stress? Though many have looked into this problem, L. Prandtl is credited with the concept that in a real fluid flow past a solid boundary, the entire flow field can be regarded as a boundary layer flow near a solid boundary and a ''free-stream'' flow far from the body.

The flow conceived as a ''free-stream'' flow is influenced solely by the *geometry of the boundary*. It is unaffected by shear stresses. Thus, Prandtl divided the flow into two regions: a region called boundary layer flow where shear stresses exist, and a frictionless region where no shear stresses exist.

References

4.1 Cauchy, A. L., "Memoire sur les dilations, les condensations et les rotations produits par un changement de forme dans un systéme de points materiels," *Ex. d'An Phys. Math.*, 2, 1841: Oeuvres (2)12, 343–377, 1841.

4.2 Helmholtz, H. von, "Über Integrale der hydrodynamischen Gleichungen, welche den Wirbelbewegungen entsprechen," *Journal für die reine und angewandte Mathematik*, 55, 1858, 22–55.

4.3 Lamb, H. *Hydrodynamics*, 1932, pp. 594–616.

4.4 Holman, J. P., Ed., *Heat Transfer*, McGraw-Hill, New York, 1966.

4.5 Thomas, L., *Fundamentals of Heat Transfer*, Prentice-Hall, Englewood Cliffs, N.J., 1980.

4.6 Oosthuizen, P. H., "A Note on the Meaning of Incompressible Gas Flow," *Bull. Mech. Engng. Educ.*, vol. 6, Pergamon, New York, 1967, pp. 369–370.

Study Questions

4.1 Transform the three-dimensional incompressible continuity equation in Cartesian coordinates directly to cylindrical coordinates.

4.2 Why is the continuity equation important? What kind of problems is it used in? Make up a problem where it is needed.

4.3 Give a geometric derivation of the continuity equation in polar coordinates.

4.4 What is the difference between angular deformation and fluid rotation? Give an example where the fluid can have angular deformation but no rotation, and vice versa.

4.5 Relate vorticity to velocity and to shear stress.

4.6 What is the expression for the Navier-Stokes equation for an irrotational flow?

4.7 Express the Navier-Stokes equation in terms of vorticity ζ rather than velocity $\mathbf{V}$.

4.8 What is a Beltrami flow? How does it differ from an irrotational flow?

4.9 What are both the mathematical and physical significance of $\mathbf{V} = \nabla\phi$ of Eq. (4.128)? What are the units of ϕ?

4.10 How can frictionless real fluids exist and inviscid fluids not exist?

Problems

4.1 Derive the relationship

$$\frac{D}{Dt}(d\forall) = (\nabla\cdot\mathbf{V})d\forall$$

using the geometry shown in Fig. P4.1 and $\Delta\forall = (\Delta x)(\Delta y)(\Delta z)$.

4.2 Express the general property balance Eq. (4.13) in terms of the enthalpy H and enthalpy per unit volume h.

4.3 Given $\phi = \rho V$, obtain the transport equation for linear momentum.

4.4 Write the continuity equation for steady two-dimensional compressible flow in (a) Cartesian coordinates and (b) polar coordinates.

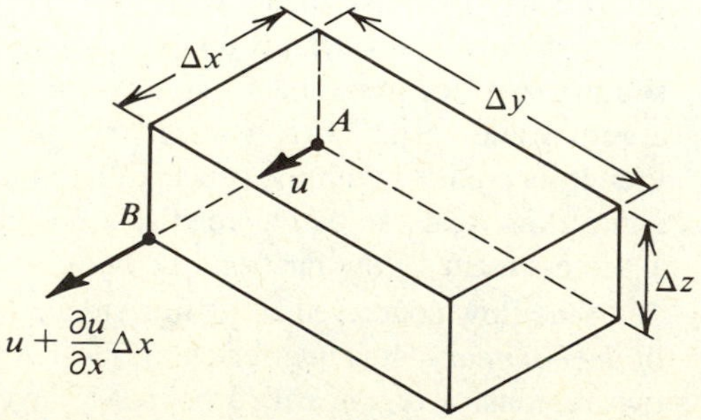

Figure P4.1

4.5 Write the continuity equation for steady two-dimensional incompressible flow in (a) Cartesian coordinates and (b) polar coordinates.

4.6 For one-dimensional incompressible flow, derive an expression for the average velocity through a duct of variable cross-sectional area $A = A(x)$.

4.7 Consider a steady compressible fluid flow where $\rho\mathbf{V} = ax\mathbf{i} + by\mathbf{j}$. What is the z-component of the momentum per unit volume?

4.8 Determine if the following expressions of the momentum per unit volume satisfy the steady compressible form of the continuity equation (a) $\rho\mathbf{V} = 3x^2\mathbf{i} - 3y^2\mathbf{j} + \mathbf{k}$; (b) $\rho\mathbf{V} = 3x\mathbf{i} - 3y\mathbf{j} + \mathbf{k}$; (c) $\rho\mathbf{V} = 3x^2\mathbf{i} - 3y^2\mathbf{j} + 6(x - y)z\mathbf{k}$; (d) $\rho\mathbf{V} = 3\mathbf{i} + 6\mathbf{j} + 1 \times 10^{12}\mathbf{k}$.

4.9 Derive the continuity equation for incompressible flow in polar coordinates by considering the momentum flux in and out of an elemental volume.

4.10 Which of the following velocity fields satisfies the continuity equation for incompressible flow? (a) $\mathbf{V} = ax\mathbf{i} + by\mathbf{j} + cz\mathbf{k}$; (b) $\mathbf{V} = ay\mathbf{i} + bx\mathbf{j} + cxy\mathbf{k}$; (c) $u = xt$, $v = -yt$; (d) $u = \sin x$, $v = \sin y$; (e) $u = x^2t^7$, $v = -2xyt^7$; (f) $v = y/z \sin t$, $w = (\ln z) \sin t$.

4.11 Given

$$\mathbf{V} = U\left(1 + \frac{x}{x^2 + y^2}\right)\mathbf{i} + \frac{Uy}{x^2 + y^2}\mathbf{j}$$

is the incompressible form of the continuity equation satisfied? What are the values of x and y for the flow to be stagnant?

4.12 Given

$$v_\theta = \frac{\Gamma}{2\pi r}[1 - \exp(-a^2 r)]$$

for a two-dimensional incompressible flow determine the radial velocity component v_r.

4.13 Given

$$v_\theta = -U \sin\theta\left(1 + \frac{b^2}{r^2}\right) + \frac{\Gamma_\infty}{2\pi r}$$

for a two-dimensional incompressible flow, determine the radial velocity component v_r.

4.14 Show that the velocity

$$\mathbf{V} = \frac{ax}{x^2 + y^2}\mathbf{i} + \frac{ay}{x^2 + y^2}\mathbf{j}$$

satisfies continuity everywhere except at the origin for incompressible flow.

4.15 If $\mathbf{V} = 3x\mathbf{i} + 4y\mathbf{j}$, perform the operation $\nabla \cdot \mathbf{V}$, and compare it with the operation $\mathbf{V} \cdot \nabla$. What are their similarities and dissimilarities?

4.16 If $u = 3x^2y$ and $v = -3xy^2$, for two-dimensional incompressible flow, what is a_x, a_y, v_r, v_θ, and V?

4.17 If $\mathbf{V} = 2xyz\mathbf{i} + y^2z\mathbf{j}$, what is $w(x,y,z)$, a_x, a_y, and a_z for steady incompressible flow?

4.18 Express the incompressible continuity equation in cylindrical coordinates, Eq. (4.25), using three terms rather than four. If the flow is two-dimensional and also axisymmetric, find the radial velocity, v_r. What type of flow is this?

4.19 Express Eq. (4.27) in terms of (a) Cartesian coordinates and (b) cylindrical coordinates.

4.20 Prove that (a) $\phi = \ln(x^2 + y^2)$; (b) $\phi = \tan^{-1} y/x$; (c) $\phi = Ax\cos\alpha + Ay\sin\alpha$; (d) $\phi = Ax/(x^2 + y^2)$. Satisfy Eq. (4.27).

4.21 Given

$$v_r = \left(a - \frac{b}{r^2}\right)\cos\theta$$

find v_θ for two-dimensional incompressible flow. What are the dimensions of a, b, and how does one evaluate them? Sketch the surface where v_r is zero everywhere.

4.22 A two-dimensional steady compressible flow has a velocity given by

$$\mathbf{V} = \frac{U}{\rho}\left(1 + \frac{x}{x^2 + y^2}\right)\mathbf{i} + \frac{Uy}{\rho(x^2 + y^2)}\mathbf{j}$$

Check to see if this flow satisfies continuity, and find the location where the flow is at rest.

4.23 The circumferential velocity of a cylindrical vortex is given by $v_\theta = A/r$. What is the radial distribution of the radial velocity? Assume the flow is incompressible and steady.

4.24 Given $\mathbf{V} = 3y\mathbf{i} + 6x\mathbf{j} + x^2\mathbf{k}$, calculate (a) the angular rotation, (b) the vorticity, and (c) the stress tensor.

4.25 If $u = x^2$, $v = y^2$, what must the z-component of velocity be for the dilation to be zero?

4.26 Given the velocity field $\mathbf{V} = x^2yz\mathbf{i} + xy^2z\mathbf{j} + xyz^2\mathbf{k}$, determine the velocities of (a) translation, (b) rotation, and (c) angular distortion.

4.27 Given

$$\mathbf{V} = -ar\mathbf{e}_r + \frac{b}{r}[1 - \exp(-ar)^2]\mathbf{e}_\theta + 2az\mathbf{k}$$

calculate all the stress tensors and the vorticity.

4.28 Obtain a relationship between the vorticity component ζ_z and the tangential velocity component v_θ if the flow is assumed axisymmetric.

4.29 Given the velocity field $\mathbf{V} = x^2yz\mathbf{i} - xy^2z\mathbf{j}$, calculate the normal and tangential forces on the face 1 of the body in Fig. P4.29 given that $p = 3z$.

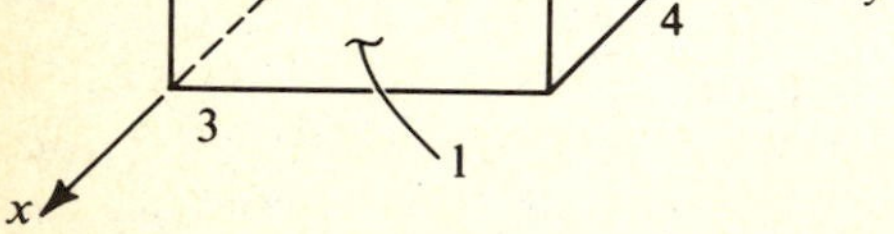

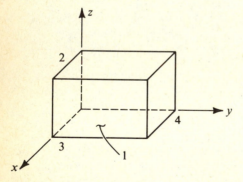

Figure P4.29

4.30 Given

$$\mathbf{V} = -ar\mathbf{e}_r + \frac{A}{r}$$

$$[1 - \exp(-ar)^2]\ \mathbf{e}_\theta + 2az\mathbf{k}$$

plot the shear stresses p_{rz}, $p_{r\theta}$, $p_{\theta z}$ for

$$0 \leq r \leq \frac{1}{a} \text{ at } z = 1.0$$

and

$$\mu = 1 \times 10^{-5}$$

Where are the stresses largest? What does this mean?

4.31 For Prob. 4.30, evaluate the tangential force on a cylinder of axial length 1.0 and radius R.

4.32 Given

$$\mathbf{V} = \left(a - \frac{b}{r^2}\right)\cos\theta\ \mathbf{e}_r - \left(a + \frac{b}{r^2}\right)\sin\theta\ \mathbf{e}_\theta$$

calculate (a) the angular rotation, (b) the vorticity $\boldsymbol{\zeta}$, and (c) the stress tensor.

4.33 The axial velocity for fully developed laminar flow in a pipe is $w = w_{\max}(1 - r^2/R^2)$. Assuming there is no tangential velocity v_θ, calculate the radial and axial distribution of the radial velocity $v_r(r,z)$ assuming the flow is incompressible.

4.34 From Prob. 4.33, determine the vorticity distribution $\boldsymbol{\zeta}(r,z)$.

4.35 From Prob. 4.33, determine all the elements of the stress tensor.

4.36 If an incompressible flow has velocity components $u = 3x$, $v = 3yz^2$, $w = 5x^3y^2$, calculate (a) the dilation D, (b) the strain rates $\dot{\epsilon}_{zx}$, $\dot{\epsilon}_{yy}$ and $\dot{\epsilon}_{yz}$, (c) the angular speed vector $\boldsymbol{\omega}$, and (d) the vorticity component ζ_y.

4.37 Expand the strain rate dyadic into a tensor of the form given by Eq. (4.51).

4.38 Find the divergence of the vorticity given

$$\mathbf{V} = -\frac{y}{x^2 + y^2}\mathbf{i} + \frac{x}{x^2 + y^2}\mathbf{j}$$

4.39 Find the divergence of the strain rate dyadic for the velocity field of Prob. 4.38.

4.40 Find the curl of the vorticity for the velocity field of Eq. (4.45).

4.41 Derive Eq. (4.55a) from Eqs. (4.53) and (4.54).

4.42 Given a velocity field $\mathbf{V} = 2Aye^{-x}\mathbf{i} - Ay^2e^{-x}\mathbf{j}$ of an incompressible fluid in a spiral centrifuge whose pressure distribution is $p = \cosh x + \gamma h$, calculate the stress tensor $p_{i,j}$ for $j = i$.

4.43 For the velocity field given by Prob. 4.32, calculate the elements of the stress tensor $p_{i,j}$ in cylindrical coordinates for a pressure distribution that is inversely proportional to the radial coordinate r.

4.44 For the velocity field given by Prob. 4.32, calculate the elements of the Cartesian stress tensor $p_{i,j}$ for a pressure distribution that is linear in y.

4.45 For the velocity field given by $\mathbf{V} = (x^2 - y^2)\mathbf{i} - 2xy\mathbf{j}$, calculate the normal surface force on each face of a 1-m cube if the pressure $p = -\gamma z$ (Fig. P4.45).

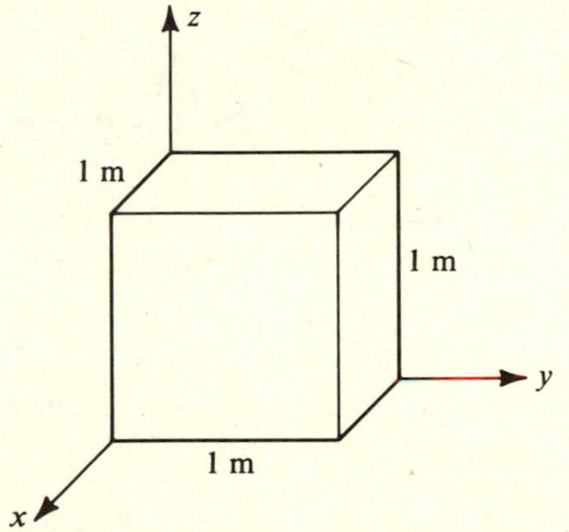

Figure P4.45

4.46 For the velocity field given by $\mathbf{V} = (x^2 - y^2)\mathbf{i} - 2xy\mathbf{j}$, calculate the tangential surface force on each face of a 1-m cube if the pressure $p = A \exp[-\cosh(\ln r^3)]$.

4.47 For the velocity field given by $\mathbf{V} = r^2 \cos\theta\, \mathbf{e}_r + r^3 \sin\theta\, \mathbf{e}_\theta$, calculate the normal and tangential surface forces on a cylinder 3 cm in diameter and 4 m long in the z-direction shown in Fig. P4.47.

4.48 If the velocity of a fluid is given by $\mathbf{V} = \nabla\phi$, where

$$\phi = \frac{x^3}{3} - xy^2$$

evaluate the vorticity ζ. Give a distribution for ϕ where the vorticity ζ does not vanish, and explain your answer.

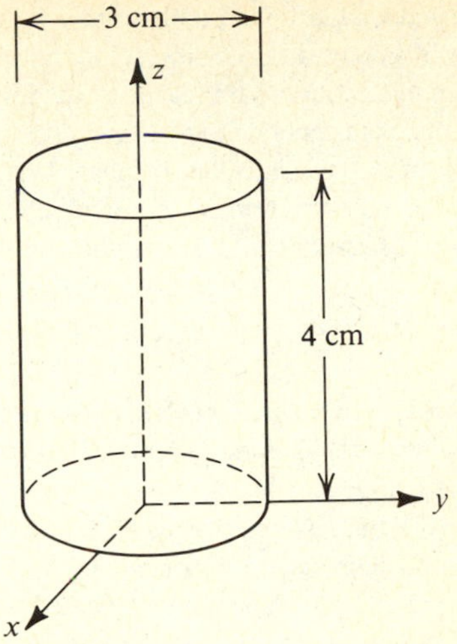

Figure P4.47

4.49 For the following velocity expressions, calculate the components of vorticity ζ_x, ζ_y, and ζ_z.

(a) $u = x^2y$, $v = -xy^2$

(b) $u = 2xy$, $v = a^2 + x^2 - y^2$

(c) $\mathbf{V} = \mathbf{i} \sinh a(x+y) - \mathbf{j} \sinh a(x+y)$

(d) $\mathbf{V} = \dfrac{\mathbf{i}}{y} \ln x - \dfrac{\mathbf{j}}{x} \ln y$

4.50 For the following velocity expressions, calculate the components of vorticity ζ_r, ζ_θ, and ζ_z.

(a) $v_r = \dfrac{1}{r}$; $v_\theta = r^3$; $v_z = 2r\cos\theta$

(b) $v_r = \left(1 - \dfrac{a^3}{r^3}\right)\cos\theta$;

$$v_\theta = -\tfrac{1}{2}\left(2 + \frac{a^3}{r^3}\right)\sin\theta$$

(c) $v_r = e^{-\theta}$: $v_\theta = e^{-i\theta}$

(d) $v_r = r\sin\theta$; $v_\theta = 2r\cos\theta$

4.51 Derive Eq. (4.106) from Eq. (4.98).

4.52 For two-dimensional steady incompressible flow, show Euler's equation can be expressed as

$$V dV + \frac{1}{\rho} dp = 0$$

where

$$V = \sqrt{u^2 + v^2}$$

4.53 Consider an infinite flat plate at the bottom of an infinitely deep ocean oscillating in linear harmonic motion parallel to itself. The motion of the plate generates a rectilinear flow, partially in phase, partially out of phase with the plate. The pressure remains constant. Plot the velocity distribution u/A above the oscillating plate in terms of $\sqrt{\frac{\omega}{2\nu}}\, z$ for (a) the plate at maximum displacement ($u = A$), and (b) plate at midcycle. (Note, A is the amplitude, such that $u = A \cos \omega t$ at $z = 0$).

4.54 Consider an incompressible fluid contained in a trough between two nonparallel walls shown in Fig. P4.54. A source is located at the vertex and has a volume output Q per unit length. Select the polar coordinates so that the walls correspond to $\theta = \pm\alpha$. The velocity components $v_r = v_\theta = w = 0$ at $\theta = \pm\alpha$, and

$$Q = \int_{-\alpha}^{+\alpha} r v_r \, d\theta$$

(a) Show

$$v_r = \frac{f(\theta)}{r}$$

(b) Obtain the ordinary differential equation for f. This is Hamel's problem of flow in a wedge-shaped region.

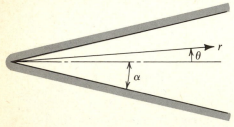

Figure P4.54

4.55 Suppose the spherical bubble of an inviscid gas shown in Fig. P4.55 is contained in an unlimited volume of liquid. Let the pressure of the gas p_g forming the bubble vary with time. This will cause the radius R of the bubble to vary with time. The bubble will generate a velocity field $\mathbf{V} = v_r(r,t)\mathbf{e}_r$ within the liquid creating a stress field $\mathbf{P}$. Obtain (a) the continuity equation and (b) the Navier-Stokes equation for this condition. (c) Show

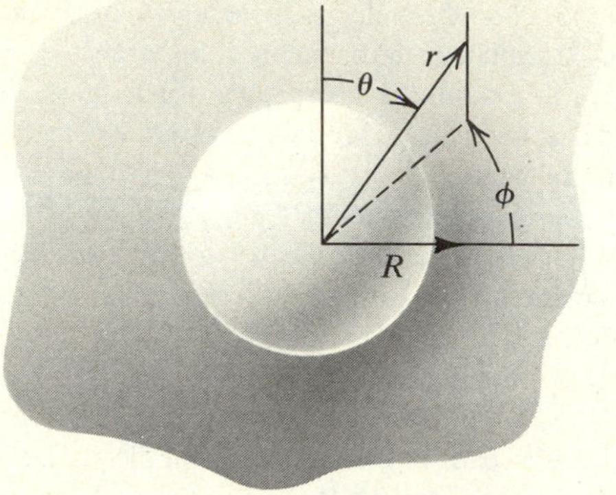

Figure P4.55

$$v_r = \frac{\dot{R}R^2}{r^2}$$

(d) Obtain the ordinary differential equation for the bubble radius as a function of time. (e) Obtain the components of stress p_{rr}, $p_{\theta\theta}$ in terms of pressure p, bubble radius R, coordinate r, and viscosity μ.

4.56 Consider the flow created by an infinite flat disc rotating in its own plane with constant angular velocity ω_d. Assume $v_r = rf(z)$, $v_\theta = r\omega(z)$, $w = w(z)$, $p = p(z)$. (a) Determine the form of the continuity equation. (b) Obtain three scalar forms of the Navier-Stokes equa-

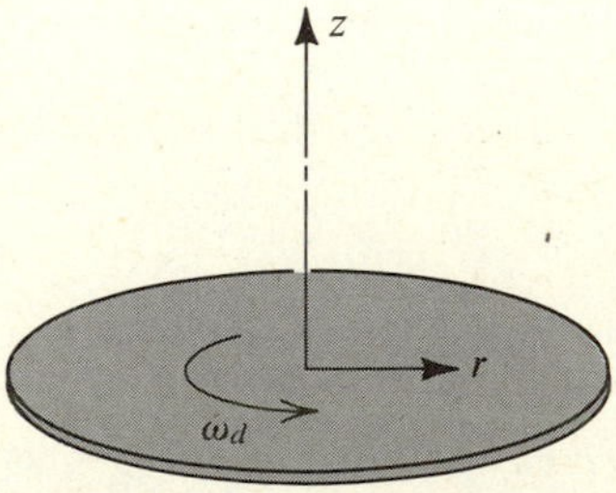

Figure P4.56

tion. (c) What are the boundary conditions for this problem? This problem was solved by von Kármán in 1934 (Fig. P4.56).

4.57 Consider fluid moving in the passsage of a turbine or a centrifugal pump that rotates with an angular speed ω about the vertical axis (Fig. P4.57). The fluid particles move with an acceleration $r\omega^2$, where $r^2 = x^2 + y^2$. Derive the Bernoulli equation for this problem:

$$\frac{p}{\gamma} + \frac{V^2}{2g} + z - \frac{r^2\omega^2}{2g} = \text{const.}$$

This equation is very important in hydraulic-turbine theory. It is called the turbine equation.

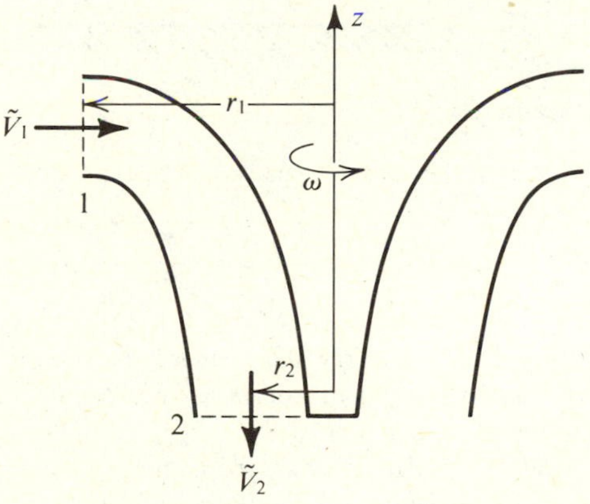

Figure P4.57

4.58 Show Euler's equation for a gas can be written as

$$\int \frac{dp}{\rho} + \frac{V^2}{2} = \text{const.}$$

4.59 Show Euler's equation for a gas can be expressed as

$$\frac{\partial \rho}{\partial x} = -\frac{\rho}{c^2}\left(u\frac{\partial u}{\partial x} + v\frac{\partial u}{\partial y}\right)$$

and

$$\frac{\partial \rho}{\partial y} = -\frac{\rho}{c^2}\left(u\frac{\partial v}{\partial x} + v\frac{\partial v}{\partial y}\right)$$

where c is the velocity of sound in the gas.

4.60 Consider two-dimensional steady uniform motion along the x-axis. (a) Show that the Navier-Stokes equations reduce to

$$\frac{1}{\rho}\frac{dp}{dx} = \nu \frac{d^2u}{dy^2}$$

(b) Express the above result in terms of the shear stress such that

$$\frac{dp}{dx} = \frac{dp_{xy}}{dy}$$

(c) Show that the shear stress is linear in y for all real fluids defined by the above characteristics of the flow. (d) Show that the velocity distribution is parabolic in y for all such flows. (e) Is pipe flow a good application of the above?

4.61 Test the following for rotationality.

(a) $u = \dfrac{x}{x^2 + y^2}, \; v = \dfrac{y}{x^2 + y^2}$

(b) $u = \dfrac{y^2 - x^2}{(x^2 + y^2)^2}, \; v = -\dfrac{2xy}{(x^2 + y^2)^2}$

(c) $u = x^3 - 3xy^2, \; v = y^3 - 3x^2y$

(d) $u = 2xy, \; v = -y^2$

4.62 Consider uniform flow in the positive x-direction such as water flowing in a river, or air moving in a valley. For such a flow we can show $\mathbf{V}$ = constant. (a) Is this flow irrotational? (b) If it is, what is the expression for the scalar potential function ϕ of Eq. (4.126)?

4.63 Suppose an incompressible fluid flowing past a 90° corner has velocity components $u = Ux$ and $v = -Uy$, as shown in Fig. P4.63. (a) Is the flow irrotational or rotational? (b) What is the fluid's acceleration $\mathbf{a}$? (c) What is the pressure distribution $p(x, y)$? (d) What are the flow's radial and tangential velocity components $v_r(r, \theta)$ and $v_\theta(r, \theta)$?

4.64 Derive Eq. (4.139) from Eq. (4.138).

4.65 A one-dimensional unsteady incompressible gas flows through a large duct in Fig. P4.65. Find the temperature distribution $T = T(x, t)$.

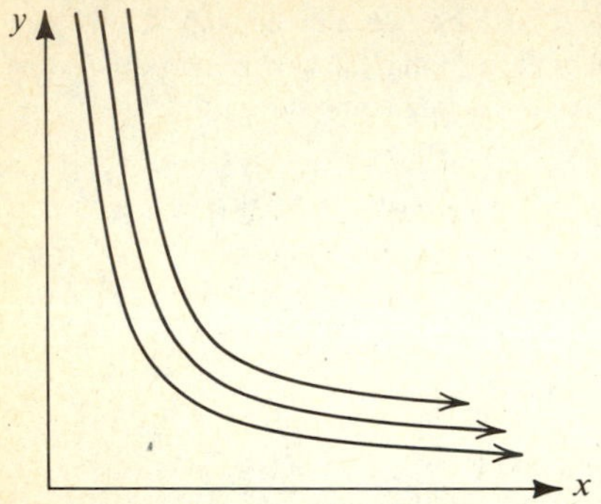

Figure P4.63

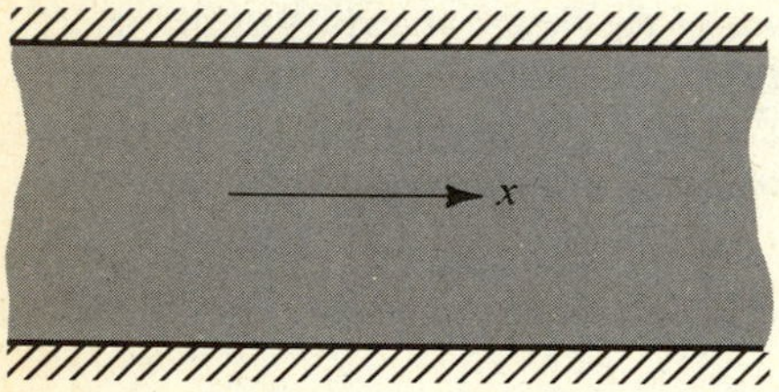

Figure P4.65

4.66 Express the energy equation for a perfect gas, Eq. (4.38), in terms of entropy, where, for the open system, $Tds = dh - dp/\rho$.

4.67 Consider laminar Couette flow of a very viscous fluid for the flow shown in Fig. P4.67 in which the temperature of both walls is maintained at T_o. Develop the temperature distribution in a region far downstream, given $u = Uy/h$.

4.68 Look at Fig. P4.68 and consider convection heat transfer associated with steady laminar boundary layer flow over a flat plate with uniform wall-flux heating maintained in the region $x \geqslant x_0$. (a) Show the continuity equation can be expressed as

$$\frac{\partial u}{\partial x} + \frac{\partial v}{\partial y} = 0$$

(b) Show the Navier-Stokes equation can be written as

$$u\,\frac{\partial u}{\partial x} + v\,\frac{\partial u}{\partial y} = \nu\,\frac{\partial^2 u}{\partial y^2}$$

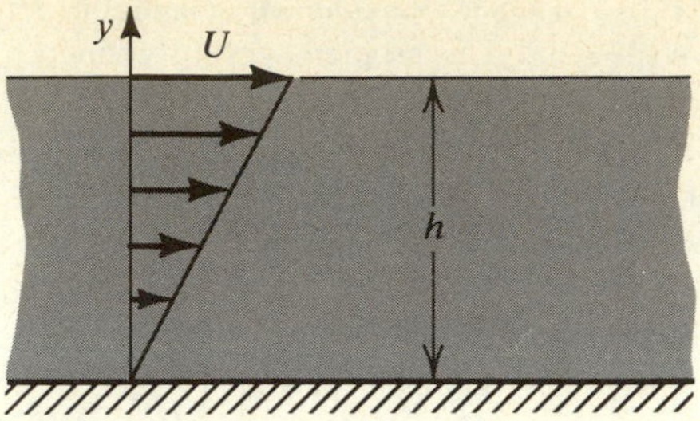

Figure P4.67

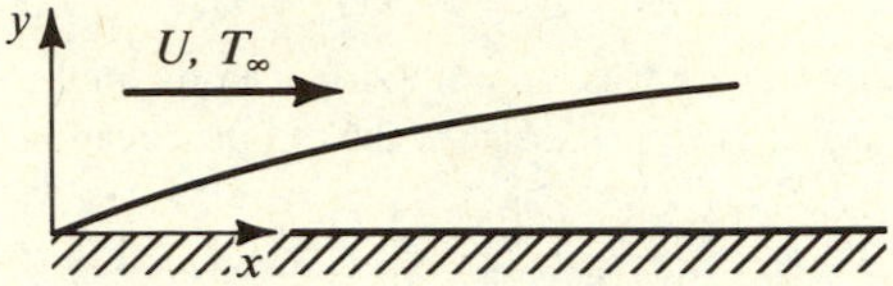

Figure P4.68

(c) Show the energy equation can be expressed as

$$u\,\frac{\partial T}{\partial x} + v\,\frac{\partial T}{\partial y} = \alpha\,\frac{\partial^2 T}{\partial y^2}$$

What is α? (d) Write the boundary condition for the velocity components and temperature at the plate and as $y \to \infty$.

4.69 Consider steady liquid laminar flow in a cylindrical tube of radius r_o with uniform wall-flux heating (Fig. P4.69). (a) Obtain the continuity equation for this flow. (b) Show the Navier-Stokes equation can be written as

$$\frac{\nu}{r}\frac{d}{dr}\left(r\,\frac{dw}{dr}\right) - \frac{1}{\rho}\frac{dp}{dz} = 0$$

What assumptions are necessary to obtain this form? (c) Show the energy equation for uniform properties can be expressed as

$$w\,\frac{\partial T}{\partial z} = \frac{\alpha\partial}{r\partial r}\left(r\,\frac{\partial T}{\partial r}\right)$$

What assumptions are necessary to obtain this form?

Navier-Stokes Problems

4.70 Consider a two-dimensional steady velocity field for which $u = ax$, $v = -ay$ of

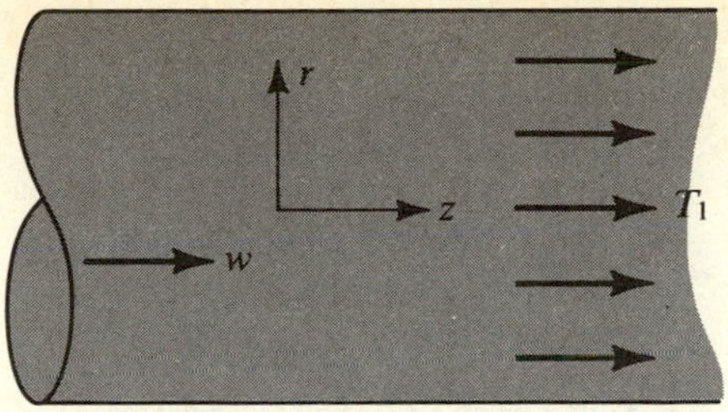

Figure P4.69

an incompressible fluid. (a) Show that continuity is satisfied. (b) Evaluate the pressure distribution $p(x, y)$. (c) Evaluate the stress components $p_{i,j}$. (d) Evaluate the vorticity $\boldsymbol{\zeta}$. (e) Is the flow irrotational?

4.71 Consider a three-dimensional steady velocity field for which $u = ax$, $v = ay$, $w = -2az$. (a) Show that continuity is satisfied. (b) Evaluate the pressure distribution $p(x, y, z)$. (c) Evaluate the stress components $p_{i,j}$. (d) Evaluate the vorticity $\boldsymbol{\zeta}$. (e) Is the flow irrotational?

4.72 For steady flow of a viscous flow past a solid flat boundary, the velocity is measured and found to be $\mathbf{V} = 5y\mathbf{i}/h$, as shown in Fig. P4.72. (a) What are the values of the stress components, and where are they maximum? (b) What is the shear force along the flat boundary of length l and width w? (c) What is the vorticity ζ, and where is it a maximum? (d) Plot the stress versus the rate of strain for this flow.

4.73 For steady flow of a viscous flow past a solid flat boundary, the velocity profile is $\mathbf{V} = 5\,x\,\mathbf{j}/l$ (a) Show continuity is satisfied. (b) Determine if the flow is rotational, and

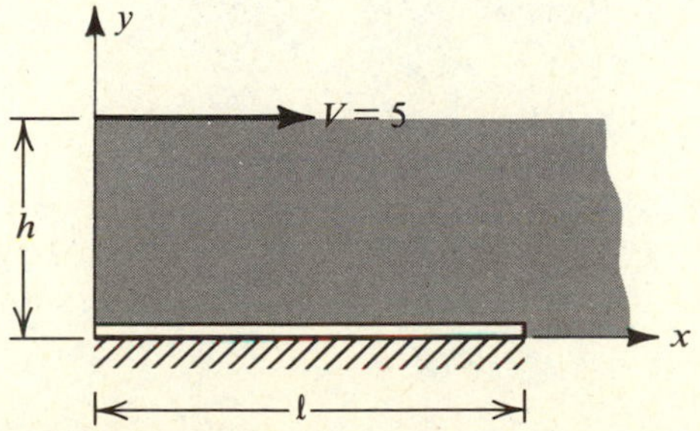

Figure P4.72

where it is a minimum. (c) Where is the flow stagnant, and what made it come to rest? (d) What is the value of the shear stress at this location? (e) Calculate the pressure distribution in the flow given $p = p_o$ at $x = l$.

4.74 The velocity distribution of a real fluid confined to move between two solid flat walls is given as $\mathbf{V} = V_{max}\,[1 - (y/h)^2] \cdot \mathbf{i}$, and is shown in Fig. P4.74. (a) Where is the flow stagnant? (b) Where is the flow maximum? (c) Determine the values of all the stresses $p_{i,j}$. (d) Where is the shear stress a maximum, and where is it a minimum? (e) What is the distribution of vorticity in the flow? (f) Determine the pressure distribution $p(x, y)$.

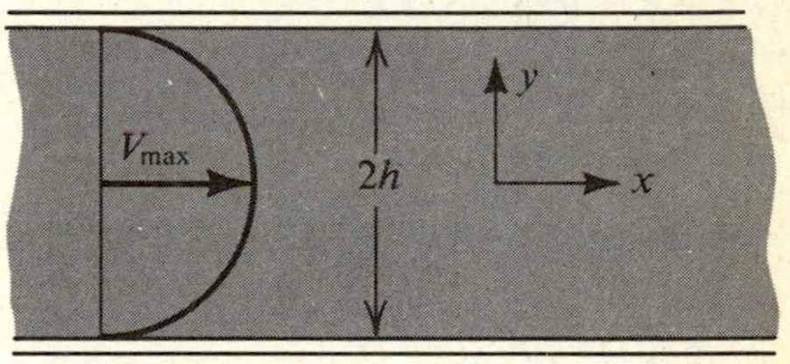

Figure P4.74

4.75 If the velocity field $\mathbf{V} = w_1(r)\exp(iat)\,\mathbf{k}$ discuss the solution of the Navier-Stokes equation given $p = p(z) \cdot \exp(iat)$. (Assume $\mathbf{g} = 0$.)

4.76 Consider the two-dimensional incompressible unsteady Navier-Stokes scalar equations. Express these two equations in terms of vorticity component ζ_z and show they can be combined to give

$$\frac{D\zeta_z}{Dt} = \nu\nabla^2\zeta_z$$

which is the vorticity transport equation. How does this equation differ in form to the Navier-Stokes equation in the absence of both pressure and body forces?

4.77 In Prob. 4.76, if the fluid is inviscid, what does the vorticity transport equation state about the vorticity? Compare the vorticity transport equation with the energy Eq. (4.139) in the absence of conduction. What can be said about vorticity and temperature?

4.78 Obtain the three-dimensional vorticity equation

$$\frac{D\boldsymbol{\zeta}}{Dt} = -(\boldsymbol{\zeta}\cdot\nabla)\mathbf{V} + \nu\nabla^2\boldsymbol{\zeta}$$

from the Navier-Stokes equation. Show that the stretching of the vorticity is the term $(\boldsymbol{\zeta}\cdot\nabla)\mathbf{V}$ (such as the whistling of air around a car's vented window).

4.79 In two-dimensional viscous flow with constant density, show that the vorticity vector is perpendicular to the plane of the motion and thus

$$(\boldsymbol{\zeta}\cdot\nabla)\mathbf{V} = 0$$

4.80 Show using the Navier-Stokes equation that in a constant density viscous motion with a conservative body force, the viscosity cannot produce vorticity.

4.81 As shown in Fig. P4.81, water is moving through a pipe. The velocity profile at a typical section is given as

$$u = \frac{B}{4\mu}\left(\frac{D^2}{4} - y^2\right)$$

where B = a constant, y = radial distance from centerline, and u = velocity at any position y.

(a) What is the shear stress, $p_{x,y}$, at the wall due to the water?
(b) What is the shear stress at $y = D/4$?
(c) If the profile persists a distance L along the pipe, what friction force is induced on the pipe by the water?

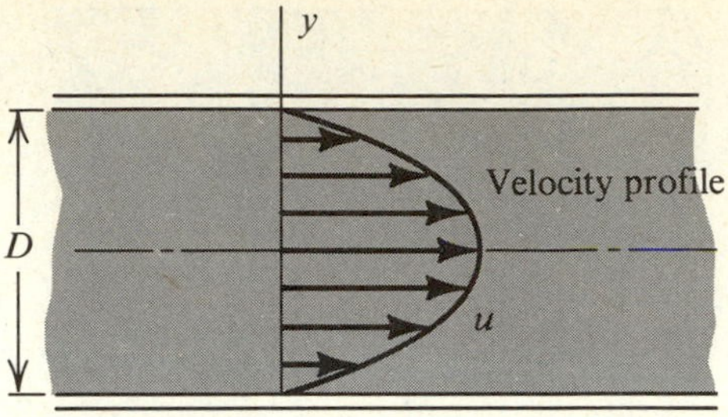

Figure P4.81

4.82 Flow occurs between two infinite, stationary, parallel flat plates, 1 ft apart. The velocity profile is given by $u(y) = -80y^2 + 80y$, where y is the distance from one plate in feet and u is in ft/s. The fluid is fresh water at standard conditions. (a) What is the maximum velocity and where does it occur? (b) What is the shear stress at the surface of one of the plates? (c) What is the frictional force per foot width exerted on the fluid by the plates in a distance of 100 ft? (d) What assumptions have you made?

4.83 Given a steady two-dimensional incompressible real fluid flow where $v = -10xy$, and $u = 0$ m/s on the y-axis, determine (a) the complete velocity field $\mathbf{V}(x, y)$, (b) the acceleration field $\mathbf{a}(x, y)$, (c) the strain rate field $\mathbf{S}(x, y)$, (d) the vorticity field $\boldsymbol{\zeta}(x, y)$, (e) the shearing stress field $\mathbf{P}(x, y)$, and (f) the gradient of the pressure field.

4.84 Repeat Prob. 4.83 given $u = -10xy$ and $v = 0$ m/s on the x-axis.

5 Integrated Forms of Fluid Behavior

5.1 Introduction

Chapter 4 applied the *differential form* to the conservation laws of physics to obtain rate equations expressing the behavior of fluid properties at a point in a flow field. In this chapter, we repeat the drill except we use a different approach: here we apply the *integral form* to the conservation laws of physics to obtain integrated equations expressing the behavior of fluid properties for finite control volumes in a flow field. All we are doing is integrating those differential equations developed in Chap. 4 from one point to another.

The problems we shall examine deal with almost every conceivable fluid situation, such as wind shear profiles past buildings, gust loads on aircraft wings, horsepower to a pump to move oil through a pipeline, or gas pressures exerted on an engine's piston. Each of the above situations involves either a volume or a surface of fluid flow. By applying the basic conservation laws of physics to our general property balance equation, which expresses how things can change, we can determine exactly how the fluid system interacts with its surroundings. This in turn enables us to design the fluid system or predict how an object placed in a fluid will behave or operate. The equations we will develop govern much of the analytical work in this text. Thus this chapter is one of the most important chapters in fluid mechanics.

The integral approach is straightforward. It consists of two major steps: (1) careful scrutiny of the physical problem to determine if the system is appropriate for a control volume analysis; and (2) examination of the behavior of the control volume. In the latter step, we must first ask whether the control volume is *fixed*, as in the case of water flowing through a pipe (Fig. 5.1a); or *moving*, as in the case of an airplane flying through air (Fig. 5.1b); or *elastic*, as in the case of the rising gas balloon (Fig. 5.1c).

In Fig. 5.1a the control volume is fixed, and the flow moving through the boundaries is based on an inertial reference frame. This is the simplest of the three cases shown, since the velocity components are absolute. In Fig. 5.1b the velocity components of the fluid are relative to a moving frame. Here noninertial coordinates should be used to describe the motion. In Fig. 5.1c the control volume deforms, and the motion at all boundaries is relative to a boundary whose shape keeps changing. Thus the different shapes of the control volume will have to be taken into consideration in the analysis. Problems in this category range from the difficult analysis of predicting the time dependent pressure distributions over an elastic wing (unsteady aerodynamics) to the simple problem shown in Example 4.2.

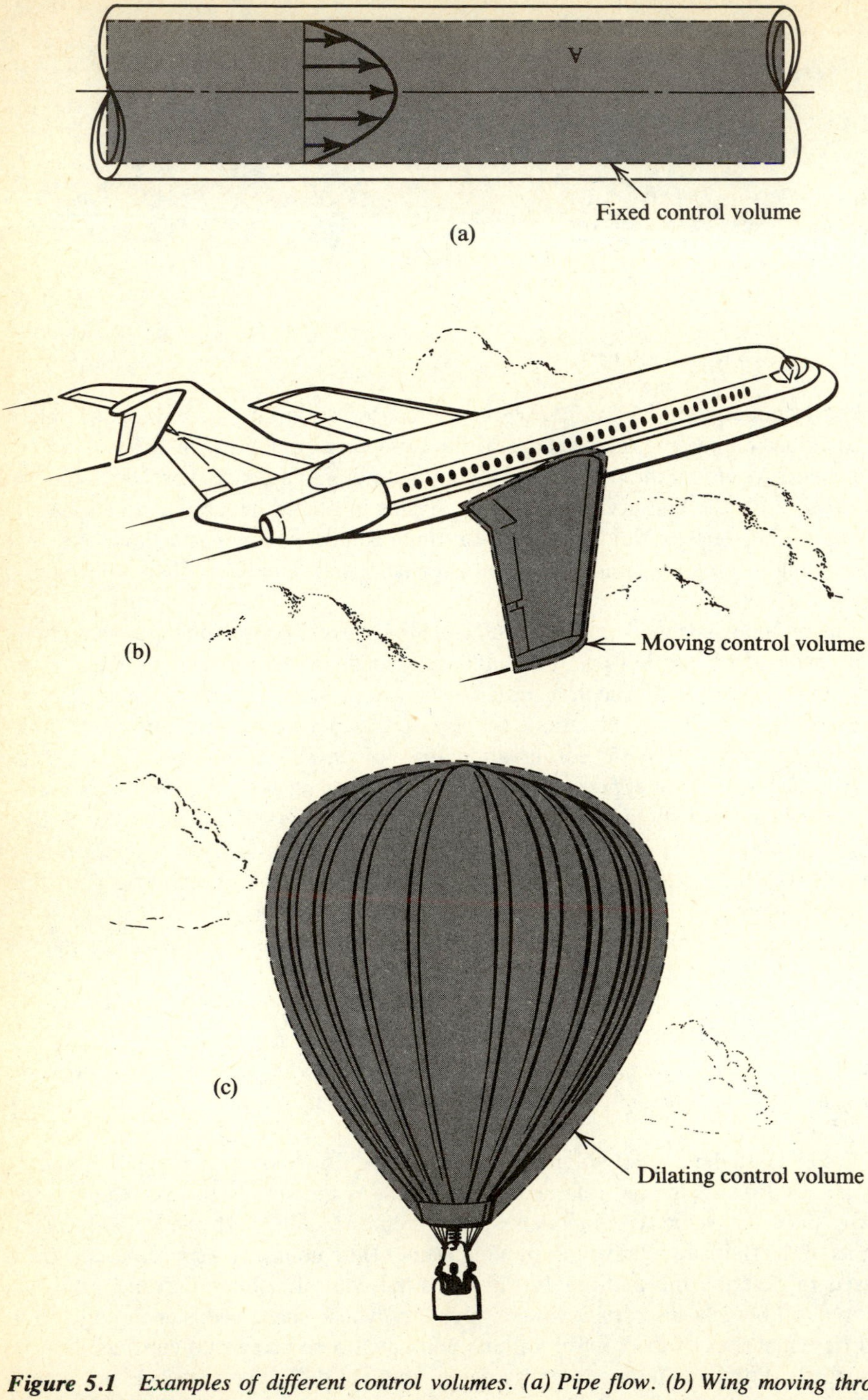

Figure 5.1 *Examples of different control volumes. (a) Pipe flow. (b) Wing moving through air. (c) Hot air balloon.*

We begin analyzing the integral forms of fluid behavior with the continuity equation. We shall obtain various expressions for the behavior of mass under various conditions. These conditions, or assumptions, should be carefully noted, as they state the strict limitations that must be observed if they are to be applied to any given flow condition.

5.2 The Integral Form of the Conservation of Mass

The conservation of mass is automatically preserved if we adopt the system method of analysis, since a system analysis has a fixed quantity of mass. However, since we are not interested in the history of a mass of fluid but rather we are interested in what takes place at a point or in a finite volume of fluid, we must use the control volume method. In using a control volume, mass is *not* identified. We must, therefore, exercise caution in applying the conservation of mass. In Chap. 4 we showed how to transform the system method of analysis to the control volume method through the use of the Reynolds transport, Eq. (4.13). The extensive property Φ is defined as the mass M of a fluid system, and the intensive property ϕ is the density ρ. This results in $\overline{\phi} = 1$ in Eq. (4.2). The Reynolds transport Eq. (4.13) can then be expressed as

$$\frac{D}{Dt}\int_{\forall} \rho \, d\forall = \int_{\forall} \frac{\partial \rho}{\partial t}\, d\forall + \oint_A \rho \mathbf{V}\cdot d\mathbf{A} \tag{5.1}$$

Applying the law of the conservation of mass, Eq. (4.19) (which states that the mass M of any system is constant), to Eq. (5.1) yields

$$\int_{\forall} \frac{\partial \rho}{\partial t}\, d\forall + \oint_A \rho \mathbf{V}\cdot d\mathbf{A} = 0 \tag{5.2}$$

This equation is identified as the *continuity equation* in *control volume form*, or alternatively the integral form of the continuity equation.

Let $\mathbf{A}_i$ be the area of the influx of the fluid flow and $\mathbf{A}_e$ designate the area of the efflux. The above continuity equation states that if mass per unit time accumulates in the control volume $\int_{\forall} (\partial\rho/\partial t)\, d\forall$, then more fluid is entering the control volume in that interval of time ($\int_{A_i} \rho_i \mathbf{V}_i\cdot d\mathbf{A}_i$) than is leaving ($\int_{A_e} \rho_e \mathbf{V}_e\cdot d\mathbf{A}_e$). Thus Eq. (5.2) is simply a balance sheet that accounts for how the mass can change in a period of time within and through the volume of fluid.

Similarly Fig. 5.2 depicts the situation in which fluid mass neither accumulates nor depletes in the control volume. Whatever mass comes into the control volume in an interval of time, $\int_{A_e} \rho_i \mathbf{V}_i\cdot d\mathbf{A}_i$, must leave the control volume in that same time interval, $\int_{A_e} \rho_e \mathbf{V}_e\cdot d\mathbf{A}_e$.

Thus for steady flow, Eq. (5.2) reduces to

$$\oint_A \rho \mathbf{V}\cdot d\mathbf{A} = 0 \tag{5.3}$$

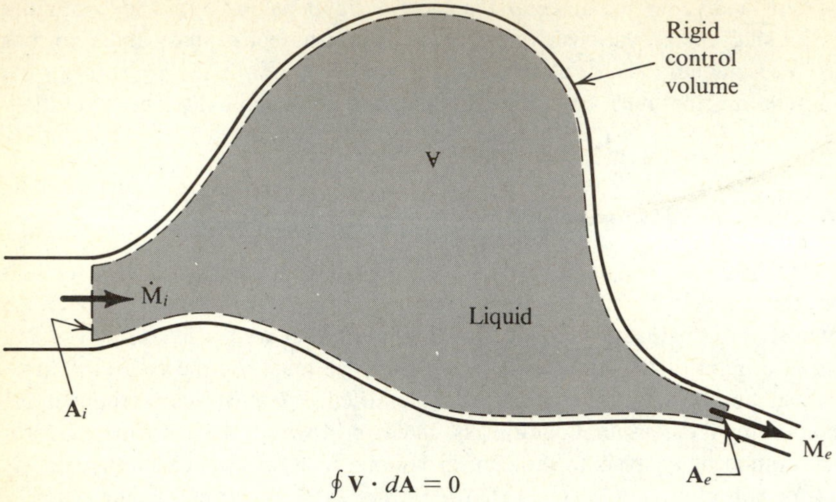

Figure 5.2 *The continuity equation in integral form for an incompressible fluid in a rigid control volume.*

which means that the *net* mass outflow through a fixed control volume for steady flow is zero. Equation (5.3) also implies that there are no sinks or sources within the control volume.

Let the control surface **A** that allows for influx and efflux of fluid flow be expressed in the form $d\mathbf{A} = \mathbf{e}_n dA$. Equation (5.3) can be further simplified as

$$\int_{A_i} \rho_i \mathbf{V}_i \cdot \mathbf{e}_{n_i}\, dA_i = \int_{A_e} \rho_e \mathbf{V}_e \cdot \mathbf{e}_{n_e} dA_e \tag{5.4}$$

The only way for these integrands to be nonzero is for the velocity vectors $\mathbf{V}_i$ and $\mathbf{V}_e$ to be those components which are orthogonal to the areas $\mathbf{A}_i$ and $\mathbf{A}_e$, respectively. An illustration is shown in Fig. 5.3. Note that both the area vector and velocity vector are expressed in terms of the normal unit vector components.

After we select the control surfaces $\mathbf{A}_i$ *and* $\mathbf{A}_e$ *so that they are perpendicular to the flow*, Eq. (5.4) becomes

$$\int_{A_i} \rho_i (V_i)\, dA_i = \int_{A_e} \rho_e (V_e)\, dA_e \tag{5.5}$$

Even this can be simplified by introducing the average density $\tilde{\rho}$ and mean velocity $\tilde{V}$, which are defined as

$$\tilde{\rho} = \frac{1}{A} \int_A \rho\, dA \tag{5.6}$$

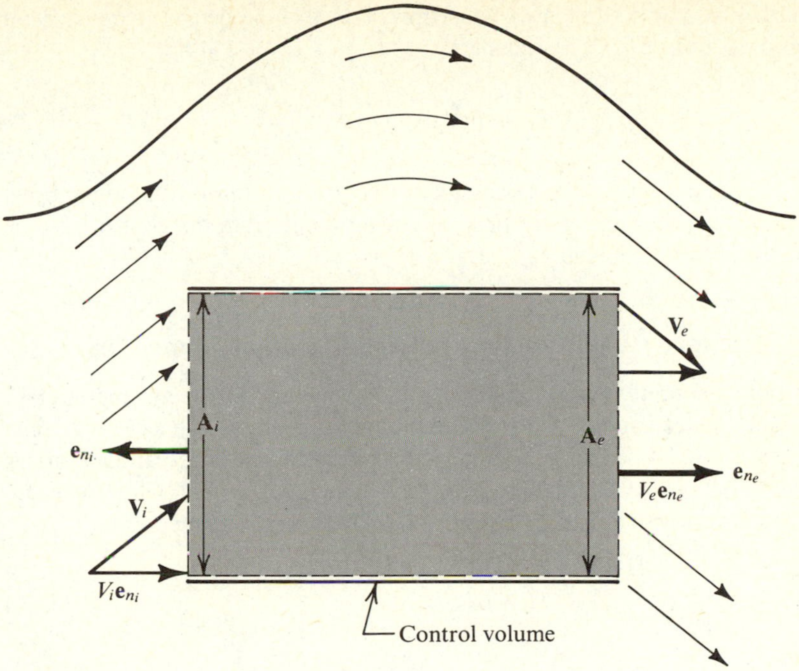

$$\int_{A_i} \rho_i V_i \mathbf{e}_{n_i} \cdot \mathbf{e}_{n_i}\, dA_i = \int_{A_e} \rho_e V_e \mathbf{e}_{n_e} \cdot \mathbf{e}_{n_e}\, dA_e$$

Figure 5.3 *Steady flow through a control volume.*

and

$$\tilde{V} = \frac{1}{A} \int_A \mathbf{V} \cdot d\mathbf{A} \tag{5.7}$$

respectively. For example, flow in a circular pipe of radius a has the average velocity

$$\tilde{V} = 2 \frac{\int_o^a \rho V r dr}{\tilde{\rho}\, a^2} \tag{5.8a}$$

Rectangular pipes have the average velocity

$$\tilde{V} = \frac{\int_{-a}^{+a} \int_{-b}^{+b} \rho w dx dy}{4\tilde{\rho} ab} \tag{5.8b}$$

where w is the velocity in the z (axial) direction.

If we further assume a one-dimensional flow, i.e., that the densities are constant across the cross section, Eq. (5.5) becomes

$$\tilde{\rho}_i \tilde{V}_i A_i = \tilde{\rho}_e \tilde{V}_e A_e = \dot{M} \tag{5.9}$$

which states that the mass rate $\dot{M}$ is a constant at any cross section normal to the flow; alternatively, we state that for steady flow, whatever comes into the control volume must go out. It cannot accumulate.

5.2.1 Incompressible Flow Form of the Continuity Equation

We can further simplify the continuity equation when the fluid is incompressible. (Almost all liquids are assumed to be incompressible, even though oceanographic scientists consider a 0.1% variation in water density significant. The variation arises out of temperature or salinity fluctuations.) For an incompressible fluid, whether steady or unsteady, our conservation law as given by Eq. (5.2) can be expressed as

$$\oint \mathbf{V} \cdot d\mathbf{A} = 0 \tag{5.10}$$

where we mean the surface area $\mathbf{A}$ to comprise inlet areas $\mathbf{A}_i$ and exit areas $\mathbf{A}_e$. We see from Eq. (5.10) that an incompressible flow cannot have just an inlet flow, or just an exit flow when the control volume is of fixed size. It must have both. It can have more than one inlet or exit; it can have as many as physically possible, provided

$$\sum_i \int_{A_i} \mathbf{V}_i \cdot d\mathbf{A}_i = \sum_e \int_{A_e} \mathbf{V}_e \cdot d\mathbf{A}_e \tag{5.11}$$

It is therefore convenient to define the volume rate of flow Q as

$$Q = \int_A \mathbf{V} \cdot d\mathbf{A} \tag{5.12}$$

such that the conservation law would require

$$\sum_{\text{in}} Q_i = \sum_{\text{exit}} Q_e \tag{5.13}$$

if several inlets and exits exist for the fluid passing through a control volume. This is illustrated in Fig. 5.4.

An alternate way of expressing the flow rate of Eq. (5.12) is to express it in terms of the average velocity $\tilde{V}$:

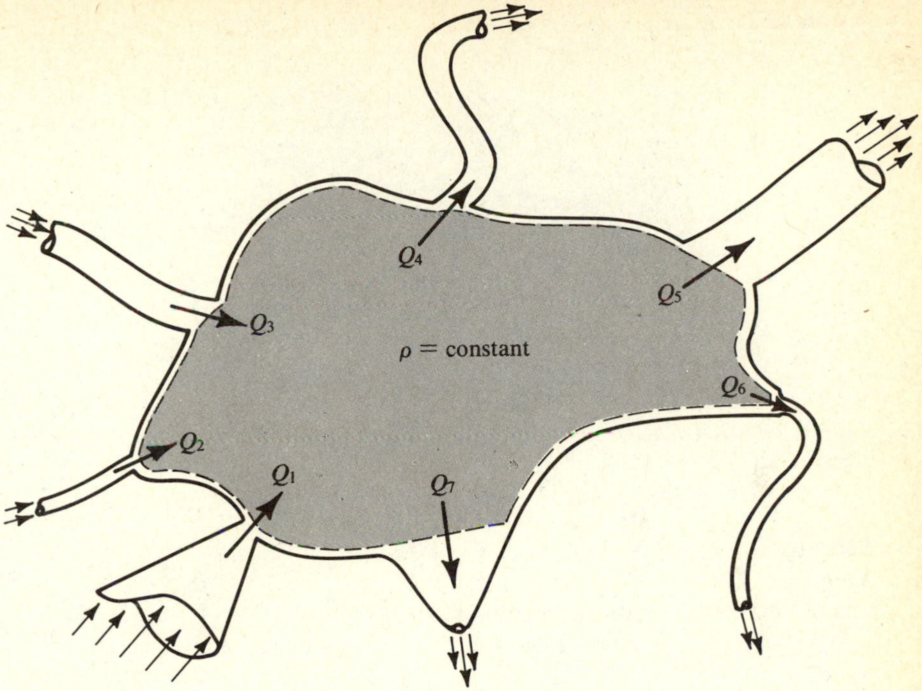

$$\Sigma Q_i = Q_1 + Q_2 + Q_3 = Q_4 + Q_5 + Q_6 + Q_7 = \Sigma Q_e$$

Figure 5.4 *Incompressible flow passing through a control volume.*

$$\tilde{V}_i A_i = \tilde{V}_e A_e = Q \tag{5.14}$$

which states that the volume rate of flow across any cross section normal to the flow is constant. The dimensions of Q are length cubed per unit time.

Equations (5.9) and (5.14) are the *continuity equations in control volume form* for steady flow; the former for compressible fluids, the latter for incompressible fluids. Both are for one-dimensional flow.

Example 5.1

Let $\dot{W}_s$ denote the weight of an incompressible fluid of density ρ removed per second at the sink located in the control volume shown in Fig. E5.1. Given the inlet and exit average velocities and the areas, what is the withdrawal rate $\dot{W}_s$ for such a flow?

Example 5.1 *(Con't.)*

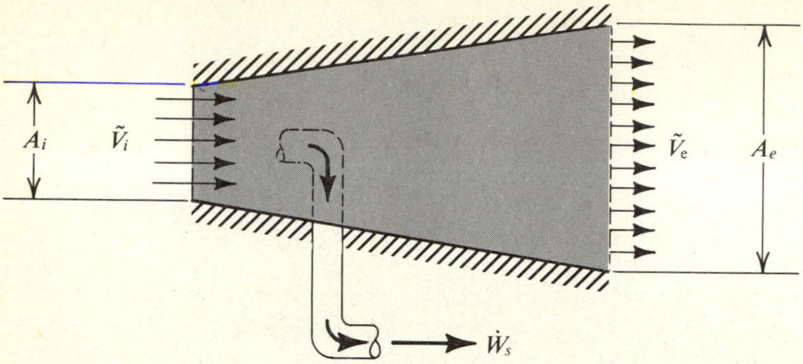

Figure E5.1

Solution:

Step 1.

Identify the characteristics of the fluid and flow.

The fluid is incompressible. The flow is one-dimensional.

Step 2.

Write the appropriate form of the governing equations of flow.

From Eq. (5.3), multiplying through by the acceleration due to gravity gives the continuity equation in terms of weight rate of flow:

$$\oint_A \gamma \mathbf{V} \cdot d\mathbf{A} = 0 \tag{i}$$

Adjusting Eq. (i) to allow for a weight rate of sink flow $\dot{W}_s$ results in

$$\oint_A \gamma \mathbf{V} \cdot d\mathbf{A} = \dot{W}_s \tag{ii}$$

Since the inlet and exit velocities are uniform at their respective stations, then Eq. (ii) becomes

$$\dot{W}_s = \gamma(V_i A_i - V_e A_e) \tag{iii}$$

This completes the solution.

Example 5.2

What is the average velocity at station 2 of air flowing through a circular pipe of radius a_1 at the inlet station 1 and radius a_2 at the exit station if the radial distribution of axial velocity at station 1 is $w = w_{\max_1}[1 - (r/a_1)^2]$? Assume that air is incompressible.

Example 5.2 *(Con't.)*

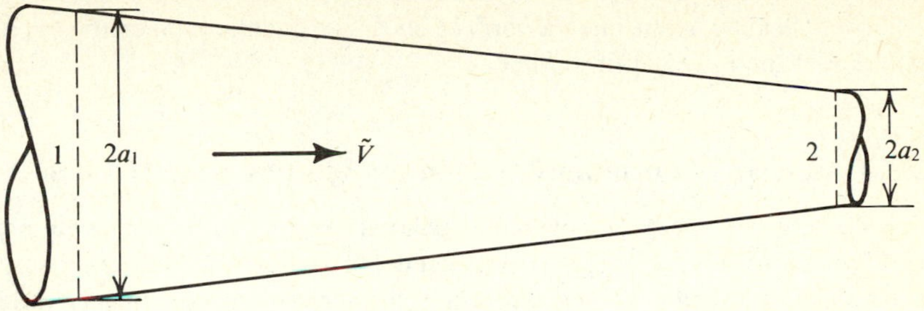

Figure E5.2

Solution:

The average velocity at the exit is related to the average velocity at the inlet for one-dimensional incompressible flow by

$$\tilde{V}_2 = \tilde{V}_1\left(\frac{A_1}{A_2}\right)$$

$$= \tilde{V}_1\left(\frac{a_1}{a_2}\right)^2 \tag{i}$$

The average value of the entering velocity $\tilde{V}_1$ is

$$\tilde{V}_1 = \frac{\iint V\, rdrd\theta}{\pi a_1^2} = \frac{2}{a_1^2}\int_o^{a_1} w_{\max_1}\,[1 - (r/a_1)^2]\, rdr$$

$$\tilde{V}_1 = \tfrac{1}{2}\, w_{\max_1} \tag{ii}$$

Substituting Eq. (ii) into Eq. (i) yields

$$\tilde{V}_2 = \frac{w_{\max_1}}{2}\left(\frac{a_1}{a_2}\right)^2 \tag{iii}$$

as the average value of the exiting velocity.

This completes the solution.

5.3 *The Integral Form of the Conservation of Linear Momentum*

The integral form of the linear momentum equation is extremely valuable when we wish to calculate the resultant force on objects that are placed in or along the surface of a flow field. Placing an object in a flow field may simply redirect the flow, as in the case of a fixed or moving blade, or it may transform the kinetic energy of the fluid into useful work, as in the case of moving blades converting fluid borne energy into shaft horsepower. A body shaped in the form of a convergent nozzle allows the transformation of the kinetic energy associated with the flow field into a force we call thrust. This is the basic idea behind propulsion. Subjects such as turbomachinery and propulsion rely heavily on the integral form of linear and angular momentum to predict the operation of these wonderful machines.

5.3.1 Linear Momentum Equation for Inertial Control Volume

In order to derive the basic governing equations expressing the momentum principles of fluid dynamics, we must perform a few fairly straightforward mathematical manipulations. Let us apply the integral form of the general property balance equation to the conservation of linear momentum as expressed by Newton's second law. Newton's second law can be expressed in terms of momentum $M\mathbf{V}$ rather than acceleration $\mathbf{a}$ by equating the inertial force $M\mathbf{a}$ to the material derivative of the momentum:

$$\sum \mathbf{F} = M\mathbf{a} = \frac{D}{Dt}(M\mathbf{V}) = \frac{D}{Dt}\int_{\forall} \rho \mathbf{V}\, dV \tag{5.15}$$

Using Reynolds transport, Eq. (4.13), and defining the intensive property variable ϕ as the momentum per unit volume

$$\phi = \rho \mathbf{V} \tag{5.16}$$

we quite readily obtain

$$\sum \mathbf{F} = \frac{D}{DT}\int_{\forall} \rho \mathbf{V}\, d\forall = \frac{\partial}{\partial t}\int_{\forall} \rho \mathbf{V} d\forall + \oint_{A} \rho \mathbf{V}(\mathbf{V}\cdot d\mathbf{A}) \tag{5.17}$$

which is called the *integral form* of the *linear momentum equation* or the *linear momentum principle* of fluid dynamics. Figure 5.5 illustrates the integral form of linear momentum. The total change of linear momentum consists of an accumulation or depletion of linear momentum within the control volume (denoted mathematically by $(\partial/\partial t) \int_{\forall} \rho \mathbf{V}\, d\forall$) due to a difference in momentum entering (denoted by $\int_{A_i} \rho_i \mathbf{V}_i (\mathbf{V}\cdot d\mathbf{A})_i$) and that leaving the control volume ($\int_{A_e} \rho_e \mathbf{V}_e (\mathbf{V}\cdot d\mathbf{A})_e$). If this total momentum is nonzero, then it is due to a resultant (or net) external force that may act exclusively on the surface (shown in Fig. 5.5 by $\Sigma\mathbf{F}$), or may act in concert with the body force. The direction of the net force must be in the same direction as the net change in linear momentum.

In terms of scalar components, we express Eq. (5.17) as

$$\sum F_x = \frac{\partial}{\partial t}\int_{\forall} \rho u\, d\forall + \oint_{A} \rho u(\mathbf{V}\cdot d\mathbf{A}) \tag{5.18}$$

$$\sum F_y = \frac{\partial}{\partial t}\int_{\forall} \rho v\, d\forall + \oint_{A} \rho v(\mathbf{V}\cdot d\mathbf{A}) \tag{5.19}$$

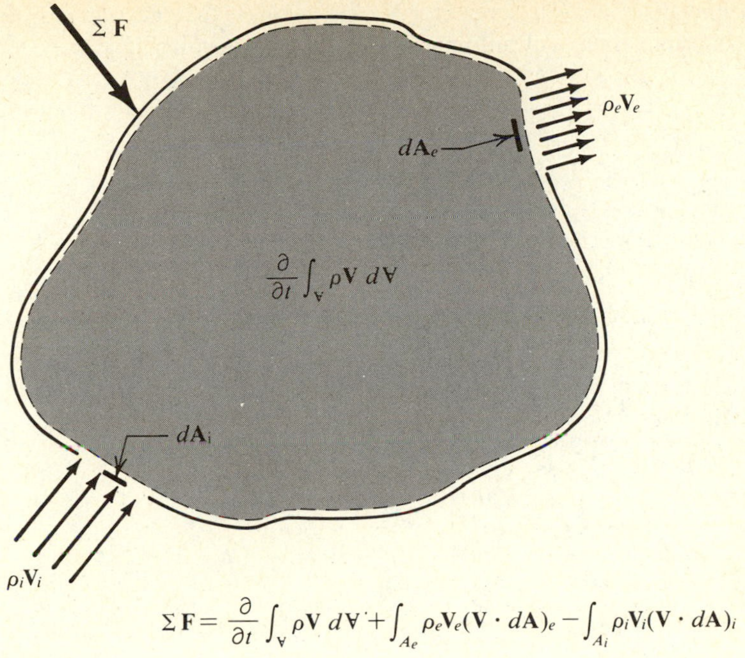
Σ F
$\rho_e \mathbf{V}_e$
$d\mathbf{A}_e$
$\frac{\partial}{\partial t}\int_{\forall} \rho \mathbf{V}\, d\forall$
$d\mathbf{A}_i$
$\rho_i \mathbf{V}_i$
$\Sigma \mathbf{F} = \frac{\partial}{\partial t}\int_{\forall} \rho \mathbf{V}\, d\forall + \int_{A_e} \rho_e \mathbf{V}_e(\mathbf{V}\cdot d\mathbf{A})_e - \int_{A_i} \rho_i \mathbf{V}_i(\mathbf{V}\cdot d\mathbf{A})_i$

Figure 5.5 *Integral form of conservation of linear momentum.*

$$\sum F_z = \frac{\partial}{\partial t}\int_A \rho w \, d\forall + \oint_A \rho w(\mathbf{V}\cdot d\mathbf{A}) \tag{5.20}$$

Thus at any time t, the net force can be found on any arbitrary volume of fluid by simply evaluating the rate momentum changes. Notice that these forces state that the momentum can change in just two ways: within a control volume and across the control surface.

In *steady* flow, momentum does not change within the control volume: with no accumulation or depletion of momentum, Eqs. (5.18)–(5.20) can be expressed in terms of direction cosines:

$$\sum F_x = \oint_A \rho \, Vu \cos \alpha_x \, dA \tag{5.21}$$

$$\sum F_y = \oint_A \rho \, Vv \cos \alpha_y \, dA \tag{5.22}$$

$$\sum F_z = \oint_A \rho \, Vw \cos \alpha_z \, dA \tag{5.23}$$

where V is the magnitude of the velocity vector $\mathbf{V}$, and $\cos \alpha_i$ is the direction cosine between the i coordinate and the outward unit normal e_n, as shown in Fig. 5.6.

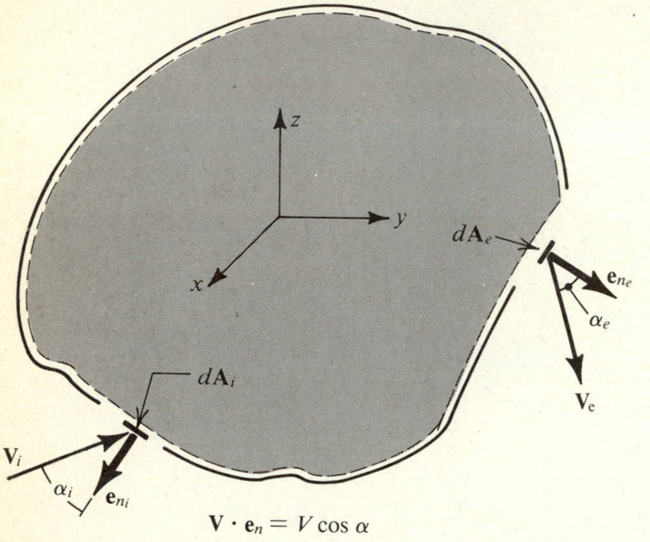

Figure 5.6 *Direction cosines.*

Simplified Steady State Forms

The integral form of the linear momentum Eqs. (5.21)–(5.23) can be simplified further. Suppose we consider a fluid entering a control volume through control surface 1, and leaving through control surface 2, as shown in Fig. 5.7. The boundaries A_1 and A_2 are selected so as to be normal to the velocity vectors $\mathbf{V}_1$ and $\mathbf{V}_2$, respectively. Applying Eqs. (5.21)–(5.23) to the flow illustrated by Fig. 5.7 results in

$$\sum F_x = \int_{A_2} \rho u \, dQ - \int_{A_1} \rho u \, dQ \tag{5.24}$$

$$\sum F_y = \int_{A_2} \rho v \, dQ - \int_{A_1} \rho v \, dQ \tag{5.25}$$

$$\sum F_z = \int_{A_2} \rho w \, dQ - \int_{A_1} \rho w \, dQ \tag{5.26}$$

where Q is the volume rate of flow defined by Eq. (5.12).

The net external force $\Sigma \mathbf{F}$ of Eqs. (5.24)–(5.26) is usually composed of body forces $\mathbf{F}_b$ due to gravitational and magnetic fields, shear forces $\mathbf{F}_\tau$ due to shear stress, forces due to pressure $\mathbf{F}_p$, plus a restoring force $\mathbf{R}$. When working with the integral forms of the momentum equation, always remember that the net external force consists of these singularly significant forces:

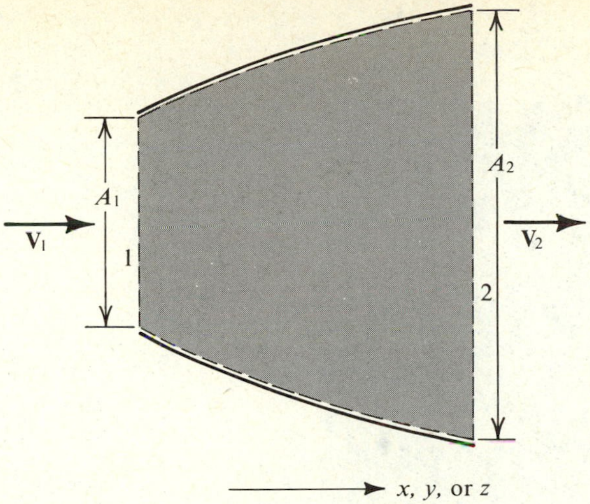

Figure 5.7 *Control volume for simplified analyses.*

$$\sum \mathbf{F} = \sum \mathbf{F}_b + \sum \mathbf{F}_\tau + \sum \mathbf{F}_p + \mathbf{R} \quad (5.27)$$

Also remember that each of the above forces can be resolved into components in the x, y, z directions.

One-Dimensional Steady Flow Case

As an additional simplification to the integral form of the momentum equation, we *assume* the velocity **V** and density ρ to be one-dimensional. The steady flow forms of the momentum equations are then

$$\sum F_x = (\tilde{V}_x \, \tilde{\rho} \, Q)_2 - (\tilde{V}_x \, \tilde{\rho} \, Q)_1 \quad (5.28)$$

$$\sum F_y = (\tilde{V}_y \, \tilde{\rho} \, Q)_2 - (\tilde{V}_y \, \tilde{\rho} \, Q)_1 \quad (5.29)$$

$$\sum F_z = (\tilde{V}_z \, \tilde{\rho} \, Q)_2 - (\tilde{V}_z \, \tilde{\rho} \, Q)_1 \quad (5.30)$$

When we use these equations, we must remember to use the mean values for density and velocity at each flow cross section. This is a good assumption as long as the flow is parallel at each station.

Free Jet Reaction

Consider a free jet of fluid impinging on a vertical wall. The pressure in a free jet is *always* atmospheric. Let its cross-sectional area be denoted by A_j. Let the jet impinge on a vertical wall, as shown in Fig. 5.8. From Eqs. (5.27) and (5.28) we obtain

$$R_x = \rho Q V_j \quad (5.31)$$

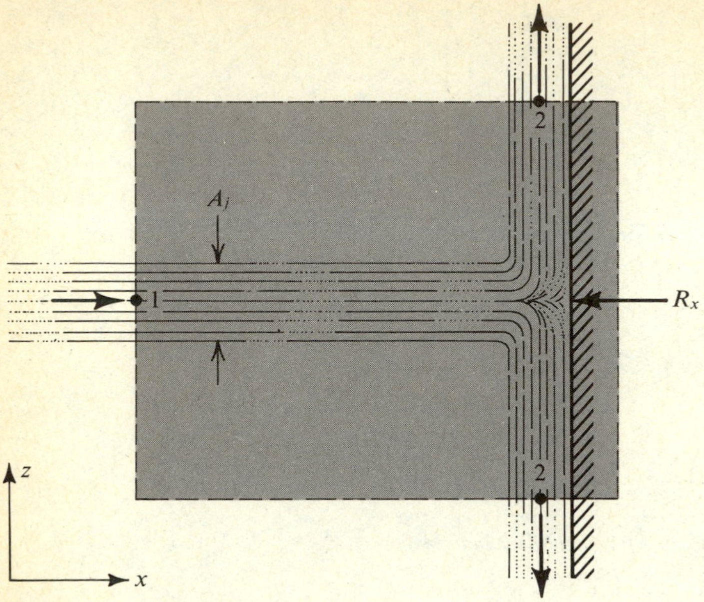

Figure 5.8 Control volume for a free-jet impinging on a vertical wall.

where the velocity $\tilde{V}_{1_x}$ is simply the velocity of the jet V_j. A restraining force R_x is necessary to keep the vertical wall stationary opposing the jet velocity V_j. Our equations give us quick solutions to problems that at first glance may appear difficult.

At this point we can now treat the case of a free jet impinging on an *inclined* wall—a slightly tougher problem than the preceding one, but just as easy to solve. Let α be the angle of inclination of a wall with the velocity vector $\mathbf{V}_j$, as shown in Fig. 5.9. Denoting R_n as the restraining force in the normal direction of the plate, we obtain from Eq. (5.31) the elegantly simple expression

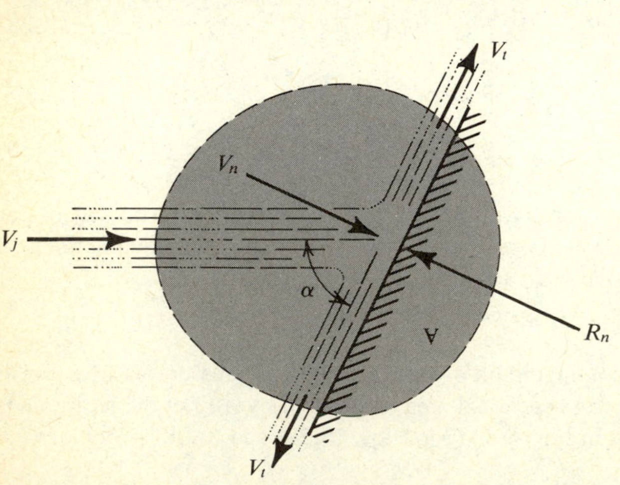

Figure 5.9 Jet reaction on an inclined wall.

$$R_n = \rho Q V_n = \rho Q V_j \sin \alpha \tag{5.32}$$

Note that we have neglected friction between the fluid and wall, just as we did in the previous case.

The momentum principle is particularly valuable when considering the forces acting on vanes or blades that serve solely to redirect the flow. Figure 5.10 shows a fluid jet moving in the positive x direction striking a vane that is stationary. The vane turns the jet through an angle θ in the horizontal plane.

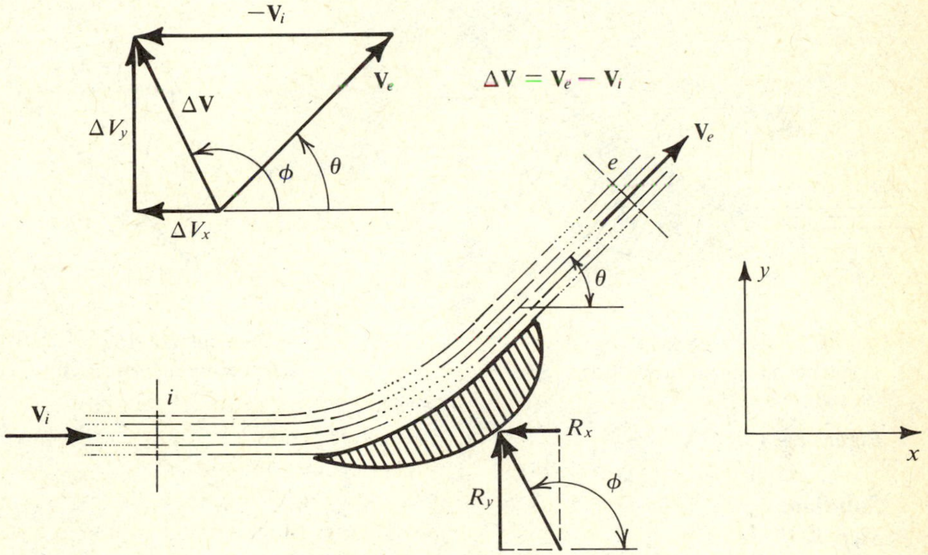

Figure 5.10 *Free jet past a fixed vane.*

Assume the fluid to be ideal. The magnitude of the exit velocity V_e equals the magnitude of the inlet velocity V_i, since the jet areas are identical.

It is sometimes convenient to sketch a velocity polygon for certain vane problems, as shown in Fig. 5.10. The resultant force **R** acting on the vane acts at an angle ϕ with respect to the horizontal. This angle is seen to be the same angle ϕ that defines the direction of the difference in velocities $\mathbf{\Delta V} = \mathbf{V}_e - \mathbf{V}_i$. Knowing the directions of the inlet and exit velocities, we can easily determine the angle ϕ.

For a free jet, the only net force in Eq. (5.27) is the reaction force **R** when $\mathbf{F}_T = 0$. Notice also that $\mathbf{F}_b = 0$ since the flow is in a horizontal plane, and that $\mathbf{F}_p = 0$ since the jet is exposed to the atmosphere.

The reaction force **R** on a fixed vane results in

$$\mathbf{R} = \rho Q \mathbf{\Delta V} \tag{5.33}$$

In Fig. 5.10, the change in velocity $\Delta\mathbf{V}$ may be easily computed by the law of sines. It is equally easy to solve for the x and y components of the resultant force using the scalar Cartesian components of Eq. (5.33) and working with ΔV_x and ΔV_y.

Example 5.3

Find the magnitude and direction of the force on a fixed vane which deflects a horizontal jet of water through a 45° angle. The velocity and cross-sectional area of the jet are 100 ft/s and 0.01 ft², respectively. Assume that the fluid is ideal.

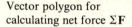

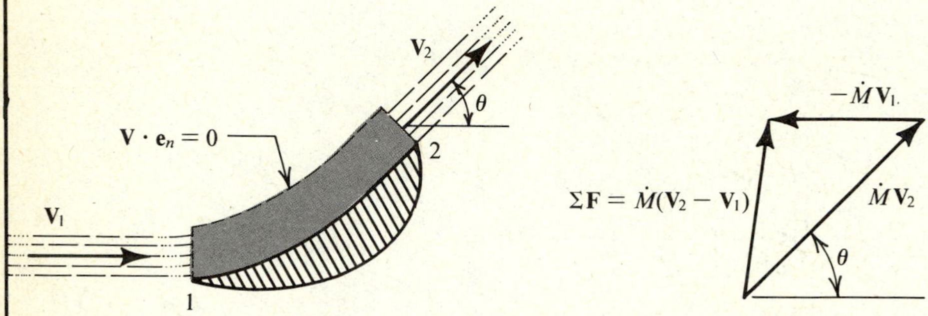

$\dot{M}_1 = \dot{M}_2 =$ constant
Stream tube in steady flow

Vector polygon for calculating net force $\Sigma\mathbf{F}$

Figure E5.3

Solution:

Step 1.

The fluid is both inviscid and incompressible. The flow is two-dimensional and steady.

Step 2.

The governing equations are

$$\tilde{V}_1 A_1 = \tilde{V}_2 A_2 \tag{i}$$

$$\mathbf{R} = \rho Q \Delta\mathbf{V} \tag{ii}$$

Using Fig. 5.10,

$$|\mathbf{V}_1| = |\mathbf{V}_2| = 100 \text{ ft/s}$$

$$\theta = 45^\circ$$

$$(180 - \phi) = 67.5^\circ$$

since the velocity vector triangle is isosceles. By the sine law,

$$\frac{|\Delta\mathbf{V}|}{\sin\theta} = \frac{|\mathbf{V}_2|}{\sin(180 - \phi)}$$

Example 5.3 *(Con't.)*
such that

$$|\Delta\mathbf{V}| = 100 \sin 45°/\sin 67.5°$$

or

$$|\Delta\mathbf{V}| = 76.5 \text{ ft/s} \tag{iii}$$

Solving for **R** in Eq. (ii),

$$\mathbf{R} = \rho Q(\Delta\mathbf{V}) = \rho A V_1 (\Delta\mathbf{V})$$

$$= 1.94\ (0.01 \times 100) \times 76.5$$

$$\mathbf{R} = 148.4 \text{ lbf at } 112.5°, \text{ (force on fluid)} \tag{iv}$$

$$\mathbf{R}_v = 148.4 \text{ lbf at } 292.5°, \text{ (force on vane)} \tag{v}$$

Or we could have worked this problem using scalar components. From Fig. 5.10

$$\Delta V_x = V_2 \cos\theta - V_1 = 100 \cos 45° - 100$$
$$= -29.3 \text{ ft/s} \tag{vi}$$

$$\Delta V_y = V_2 \sin\theta = 100 \sin 45° = 70.7 \text{ ft/s} \tag{vii}$$

Using Eq. (ii) in scalar form,

$$R_x = \rho Q(\Delta V_x) = 1.94\ (0.01 \times 100)(-29.3)$$
$$= -56.8 \text{ lbf} \tag{viii}$$

$$R_y = \rho Q(\Delta V_y) = 1.94\ (0.01 \times 100)(70.7)$$
$$= 137.1 \text{ lbf} \tag{ix}$$

so that the magnitude of the force R is

$$R = \sqrt{R_x^2 + R_y^2} = 148.4 \ lbf \tag{x}$$

which agrees with Eq. (iv). This force acts at an angle

$$(180 - \phi) = \tan^{-1}\frac{R_y}{R_x} = 67.50° \tag{xi}$$

This completes the solution.

Example 5.4
A jet of water strikes the plate as shown in Fig. E5.4a. Given the jet velocity 100 ft/s at station i, a stagnation pressure $p_o = 9700 \text{ lbf/ft}^2$, a volume rate of flow of 1 ft^3/s, and the areas of the jet and plate of 0.01 ft^2 and 1 ft^2, respectively, calculate the reaction force R_y necessary to keep the plate stationary.

Example 5.4 *(Con't.)*

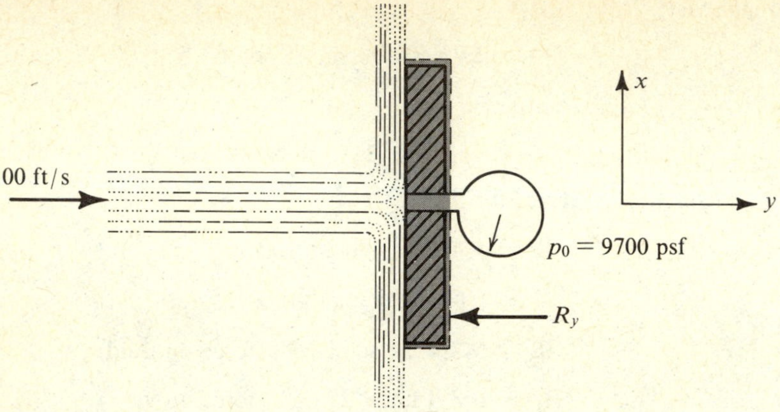

Figure E5.4a

Solution:
Step 1.
The fluid is incompressible. The flow is steady and two-dimensional.
Step 2.
The appropriate equations are either

$$R_y = F_p \tag{i}$$

or

$$-R_y = \rho Q(V_{e_y} - V_{i_y}) \tag{ii}$$

depending upon the choice of control volume.

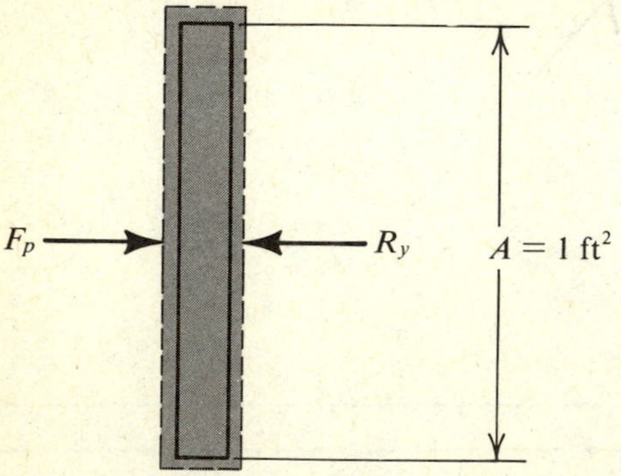

Figure E5.4b

If we select our control volume as we can find R_y from Eq. (i). Thus

$$R_y = \int_A p\,dA \tag{iii}$$

Example 5.4 *(Con't.)*

will give us the reaction force provided that the pressure distribution $p = p(A)$ is given. However, our problem states only that the stagnation pressure p_o is known. Since we suspect that the pressure distribution on the plate would be somewhat like that shown in Fig. E5.4c,

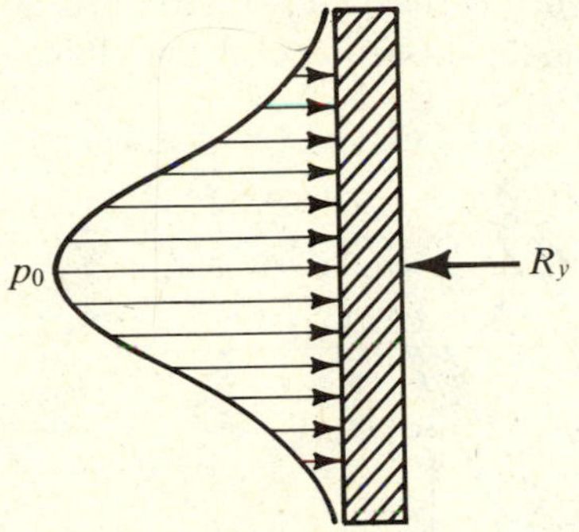

Figure E5.4c

we cannot use Eq. (iii), since the pressure distribution was not given.

Redefining our control volume as that shown in Fig. E5.4d,

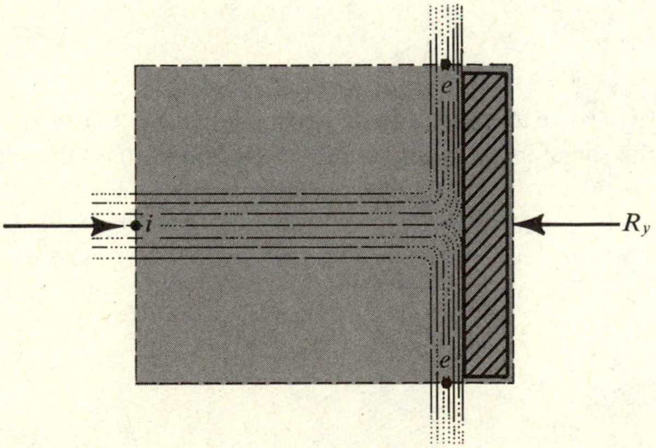

Figure E5.4d

we apply Eq. (ii) to obtain

$$-R_y = -\rho Q V_{i_y} \tag{iv}$$

$$= -1.935 \times (100 \times 0.01)(100)$$

Thus the y component of the restoring force is

$$R_y = 193.5 \text{ lbf} \tag{v}$$

This completes the solution.

Example 5.5

Water issues from a large tank attached to a cart. The jet strikes a stationary vane which turns the direction of flow through an angle of 60°, as shown in Fig. E5.5a. Assume that the velocity of the jet is 64.4 ft/s. (a) Calculate the thrust on the cart which is held stationary relative to the ground by the nonelastic cord, assuming the flow out of the tank orifice is steady and of cross-sectional area 1.2 ft². (b) Calculate the magnitude and line-of-action of the net force due to momentum change.

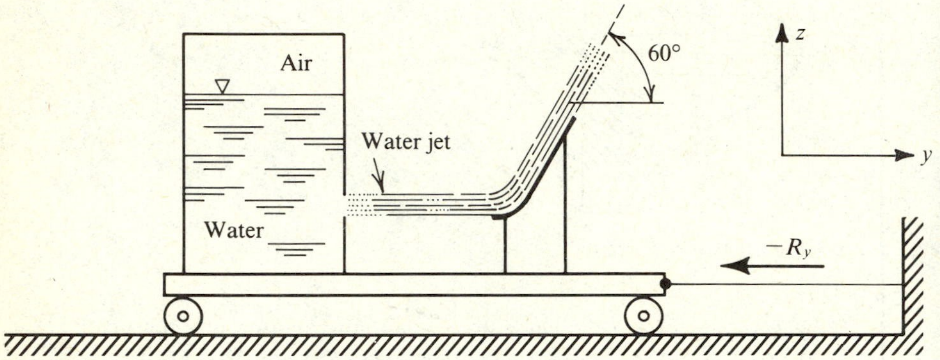

Figure E5.5a

Solution:

(a) There are two ways to solve this problem, depending upon the control volume used. Both result in the same answer, but one is vastly simpler than the other.

(i) First Way

Consider the problem where a force results from the jet leaving the tank, and a second force results from the redirection of the jet by the blade. The control volumes for this case are shown in Fig. E5.5b.

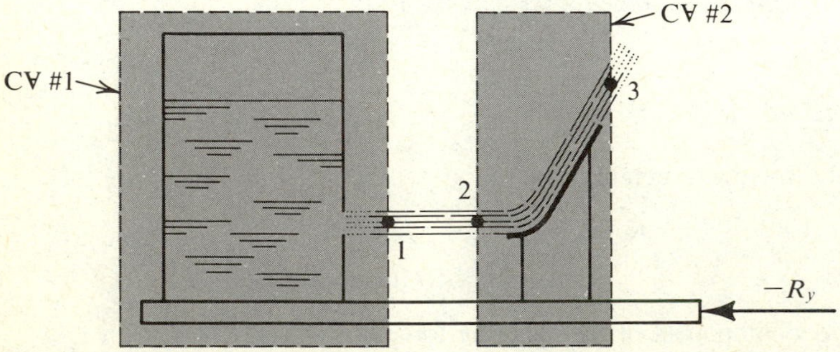

Figure E5.5b

Example 5.5 *(Con't.)*

From C.∀. #1, Eq. (5.29) produces

$$-R_{y_1} = \rho Q(V_{1_y}) \qquad \text{(i)}$$

$$= 1.935\ (64.4 \times 1.2)(64.4)$$

$$= 9630.2 \text{ lbf} \qquad \text{(ii)}$$

From C.∀. #2, Eq. (5.29) produces

$$-R_{y_2} = \rho Q(V_{3_y} - V_{2_y}) \qquad \text{(iii)}$$

$$= 1.935\ (64.4 \times 1.2)(64.4 \sin 30° - 64.4)$$

$$= -4815.1 \text{ lbf} \qquad \text{(iv)}$$

The total restoring force in the y direction is then

$$-R_y = -R_{y_1} - R_{y_2}$$

$$= 9630.2 - 4815.1$$

or

$$-R_y = 4815.1 \text{ lbf} \qquad \text{(v)}$$

(ii) Second Way

Let the control volume be the entire assembly.

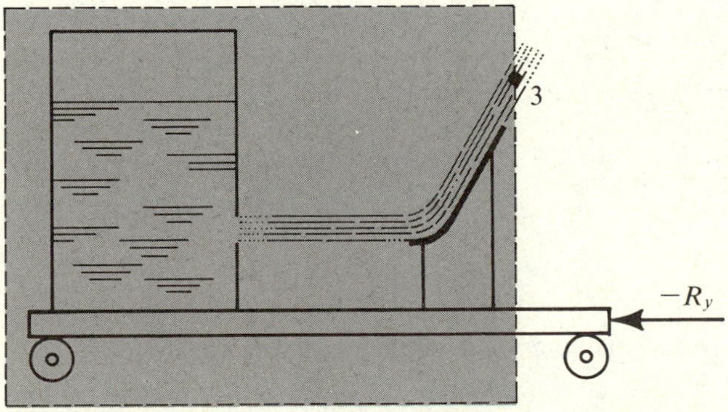

Figure E5.5c

Applying Eq. (5.29) to the above control volume gives

$$-R_y = \rho Q V_{3_y} \qquad \text{(vi)}$$

$$= 1.935\ (64.4 \times 1.2)(64.4 \sin 30°)$$

or

$$-R_y = 4815.1 \text{ lbf} \qquad \text{(vii)}$$

which agrees with Eq. (v).

Example 5.5 *(Con't.)*

This example illustrates the importance of the choice of control volume. Obviously whatever we select as our control volume will dictate the solution. Either control volume gave the correct answer, but the first one was more involved than the other. Try to select a control volume that minimizes the unknowns.

(b) The vector polygon can aid us in visualizing the various forces. Since the vector linear momentum equations can be expressed in terms of components given by Eqs. (5.28)–(5.30) as

$$\sum \mathbf{F} = \dot{M}\,(\mathbf{V}_2 - \mathbf{V}_1) \tag{viii}$$

we can resolve the vector diagram for the net force $\Sigma\mathbf{F}$ and momentum change $\dot{M}\Delta\mathbf{V}$ into an isosceles triangle as shown in Fig. E5.5d

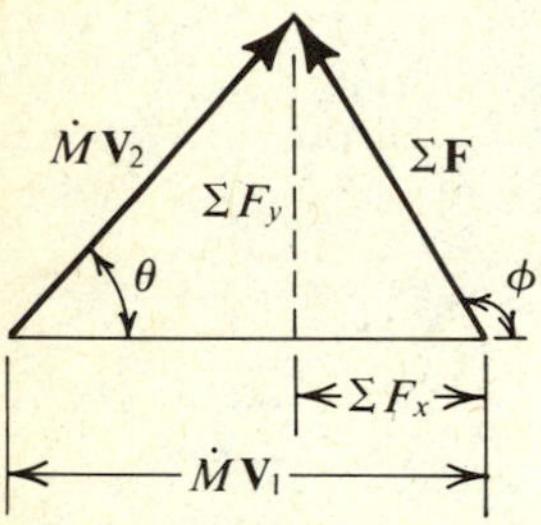

Figure E5.5d

where

$$\sum F_x = \rho V^2 A\,(\cos\theta - 1) \tag{ix}$$

$$\sum F_y = \rho V^2 A \sin\theta \tag{x}$$

since $\dot{M} = \rho VA$. Hence the *magnitude* of the net force $\Sigma\mathbf{F}$ is

$$\left|\sum \mathbf{F}\right| = \left[\left(\sum F_x\right)^2 + \left(\sum F_y\right)^2\right]^{1/2}$$

$$= \rho V^2 A\,(\cos^2\theta + \sin^2\theta + 1 - 2\cos\theta)^{1/2}$$

$$= 2\rho V^2 A \sin\frac{\theta}{2} \tag{xi}$$

Substituting the given values into Eq. (xi) results in

$$F = (2)(9630.2)(0.5) = 9630.2 \text{ lbf} \tag{xii}$$

Example 5.5 *(Con't.)*

The line-of-action of the net force can be obtained from the geometry of Fig. E5.5d:

$$\phi = 180° - \tan^{-1}\left(\frac{F_y}{F_x}\right)$$

$$= 90° + \frac{\theta}{2} \qquad \text{(xiii)}$$

Substituting the given values into Eq. (xiii) gives

$$\phi = 120°$$

This part of the solution reveals the advantages in using vector polygons.
This completes the solution.

Free Jet Reaction with Moving Wall

We have determined how a free jet of fluid impinges on a stationary wall. Let us now examine the more complex case of a moving wall. If the wall moves with a velocity u in say, a direction opposite to that of the jet, the flow becomes unsteady. But if an observer travels with the wall, then he sees a jet velocity $(V_j + u)$, and thus the resultant force R_n, from Eq. (5.32), is

$$R_n = \rho(V_j + u)^2 A_j$$

Note that in this problem the velocity of the plate is assumed to be constant and, therefore, both control volumes are observed by Newtonian observers. For the more general case of nonuniform motion of the plate, this type of problem involves the linear momentum principle for a noninertial control volume, which is the subject of Sec. 5.3.2.

Airfoil Forces in Plane Flow

Now we are ready to tackle a still more complex problem. Consider an airfoil of extremely long span placed in a moving stream of fluid as shown in Fig. 5.11. If the span is sufficiently long, the flow can be considered two-dimensional. Assume the body force due to the gravitational acceleration g and the shear stress force $\mathbf{F}_\tau$ to be negligible. Assume that the flow is steady. For the control volume, we choose that volume bounded by the airfoil and a large circle of radius r surrounding the airfoil. Remember, we can select any control surface we desire.

Suppose that we want to find the forces on the wing section using the linear momentum principle. The pressure force $\mathbf{F}_p$ for steady incompressible flow acting on the circle is obtained from Eq. (5.17) as

$$\mathbf{F}_p = \frac{\rho}{2}\int_A V^2\, d\mathbf{A} = \frac{\rho}{2}\int_A V^2\, \mathbf{e}_n\, dA \qquad (5.34)$$

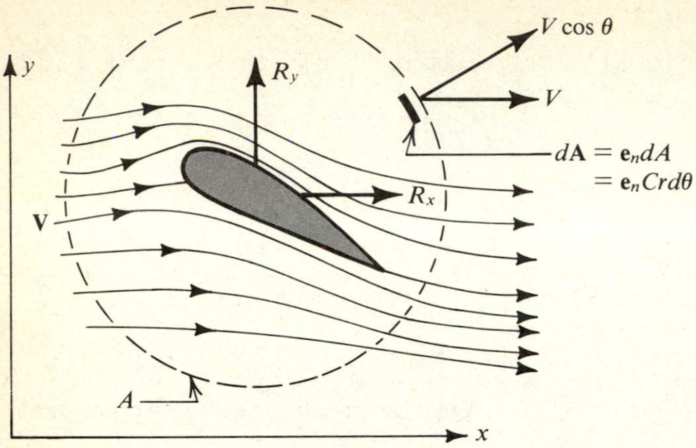

Figure 5.11 *Airfoil forces R_x and R_y.*

To obtain the scalar component forms of the force, we substitute Eq. (5.34) into Eqs. (5.21) and (5.22), which yields

$$R_x + \frac{\rho}{2}\int_A V^2 \cos\theta \, dA = \rho\int_A V^2 \cos\theta \, dA \tag{5.35}$$

$$R_y + \frac{\rho}{2}\int_A V^2 \sin\theta \, dA = 0 \tag{5.36}$$

The surface element dA in the above integrands is easily obtained from the geometry of Fig. 5.10 as $Crd\theta$, where C is a span or width (assumed constant) in a direction perpendicular to the xy plane. Equations (5.35) and (5.36) are evaluated to be

$$R_x = \rho\frac{V^2}{2}Cr\int_o^{2\pi} \cos\theta \, d\theta = 0 \tag{5.37}$$

and

$$R_y = \rho\frac{V^2}{2}Cr\int_o^{2\pi} \sin\theta \, d\theta = 0 \tag{5.38}$$

This exercise appears to be useless since our forces have been determined to be zero. How can we possibly have no forces existing in this problem? We see immediately that the reason lies in our assumption that the speed V is constant. If we select a large enough circle, we could easily be tempted to consider the velocity through the control surface a constant just as we did in obtaining the result of Eq. (5.38). The velocity variation at the control surface may be very small so small that it tempts us to assume it is constant. But we cannot do this when there is a thick body in the control volume. We shall see in Chap. 12 that this small variation in velocity contributes to lift. Because of symmetry the x-component of the force is zero, and so Eq. (5.37) is correct. The resultant zero force field for *ideal* fluid motions is known as d'Alembert's paradox and will be discussed in Chap. 12. For now we will tentatively accept the above results, keeping in the back of our minds that something is wrong in the formulation of the problem.

5.3.2 Integral Form of the Linear Momentum Equation for a Noninertial Control Volume

In many situations we may be required to analyze the behavior of a fluid flow with respect to an observer attached to a moving coordinate system. Moving coordinate systems are so bothersome that many students are reluctant to study the subject at all. This is unfortunate for we use noninertial coordinates in many areas (in which there are lucrative opportunities for research) including the study of oceanographic, geophysical, and meteorological flows, where ocean currents and huge air masses on the earth's surface could be accelerating or decelerating relative to fixed stellar masses. One of the more significant accelerations due to a noninertial coordinate system is the Coriolis acceleration. We also use noninertial coordinates in studying the dynamic behavior of rockets, spacecraft, and airplanes, where the coordinates are attached to the moving vehicle.

Although the resultant equations referring to a moving coordinate system are admittedly somewhat ghastly upon initial inspection, the student must realize that certain formulations of physical problems are necessarily difficult. We shall try to dodge some of the more difficult problems, but to expose a student to only simple problems would defeat one of the goals of education. Actually, the difficulty arises from the notation, not from the problem itself. Once the reader becomes familiar with the notation, results easily drop out of the formulation.

In describing the dynamics of the flow, we usually adopt Newton's laws of motion which are based entirely on an *inertial* frame of reference. It is convenient to develop the dynamics of the flow from an inertial frame resulting in the creation of a new force called the *apparent force*, consisting of such familiar forces as Coriolis and centripetal.

In developing the expression for the apparent force, we shall adopt the notation shown in Fig. 1.7, and place the observer on a moving frame, denoted by the subscript (m). At the origin O' of the moving (m) frame, we *observe* a control volume that is *fixed* relative to our observation platform as shown in Fig. 5.12. The position vector of a point P in the control volume relative to the moving point O' is denoted by the symbol $\mathbf{r}_{P/O'}$.

Consider the mathematical description of the flow through the control volume as *viewed from* the moving frame. Let $d\forall$ denote the fixed volume element of the space point P in the moving (m) coordinate system. The position vector $\mathbf{r}_P$ from the fixed frame is obtained from vector algebra

$$\mathbf{r}_P = \mathbf{r}_{O'} + \mathbf{r}_{P/O'} \tag{5.39}$$

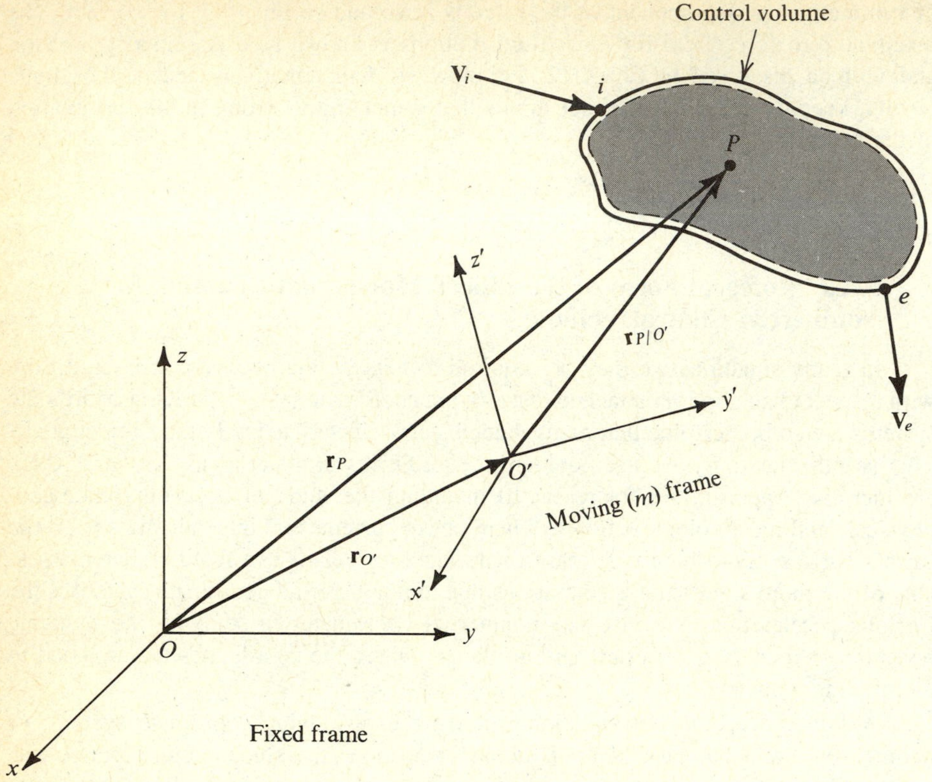

Figure 5.12 *The control volume in a noninertial frame of reference.*

At a fixed position $\mathbf{r}_{P/O'}$ we wish to determine the fluid particle velocity. We want the velocity with respect to the fixed frame as we need this expression in order to apply it to Newton's second law. Using Euler's description, the velocity of the flow with respect to the fixed frame is

$$\mathbf{V} = \frac{D\mathbf{r}_P}{Dt} \tag{5.40}$$

Substituting Eq. (1.17) into Eq. (5.40) yields the simple result

$$\frac{D\mathbf{r}_P}{Dt} = \frac{D\mathbf{r}_{O'}}{Dt} + \frac{D\mathbf{r}_{P/O'}}{Dt} \tag{5.41}$$

Note that we are still referenced to the fixed frame. In Eq. (5.41) we see that we have two different descriptions: one is Eulerian and the other is Lagrangian, since we have assumed that $\mathbf{r}_{o'}$ is *not a function of space but only of time*. Thus,

$$\frac{D\mathbf{r}_P}{Dt} = \mathbf{V}_{O'} + \boldsymbol{\omega}_{(m)} \times \mathbf{r}_{P/O'} + \frac{^{(m)}D\mathbf{r}_{P/O'}}{Dt} \tag{5.42}$$

where $\mathbf{V}_{O'}$ is the velocity of the moving (m) frame relative to the fixed frame, $\boldsymbol{\omega}_{(m)} \times \mathbf{r}_{P/O'}$ is the rotational velocity due to the rotating frame and

$$\frac{^{(m)}D\mathbf{r}_{P/O'}}{Dt}$$

is the relative velocity of the fluid flow relative to the origin of the moving frame (m). The term $\boldsymbol{\omega}_{(m)} \times \mathbf{r}_{P/O'}$ is rigid body rotation and the term $^{(m)}D\mathbf{r}_{P/O'}/(Dt)$ is the *relative velocity* of the flow through the control volume. Thus, the fixed frame material derivative D/Dt can be expressed as a collection of three separate and distinct operations:

$$\frac{D}{Dt} = \frac{d}{dt} + \boldsymbol{\omega}_{(m)} \times + \frac{^{(m)}D}{Dt} \tag{5.43}$$

In the above equation, d/dt is the Lagrangian ordinary differential operator, and the moving $^{(m)}D/Dt$ is treated in the manner of the fixed frame case; i.e., in Cartesian coordinates

$$\frac{^{(m)}D\mathbf{r}_{P/O'}}{Dt} = u\mathbf{i}' + v\mathbf{j}' + w\mathbf{k}' \tag{5.44}$$

where unit vectors $\mathbf{i}'$, $\mathbf{j}'$, $\mathbf{k}'$ of the moving frame need not necessarily be constant.

Next, consider the *dynamics* of the problem. Applying the summation of possible forces on the elemental volume $d\forall$ and integrating throughout the control volume yields our familiar

$$\sum \mathbf{F} = \int_{\forall} \rho \mathbf{g} \, d\forall + \int_{\forall} \nabla \cdot \mathbf{P} \, d\forall \tag{5.45}$$

where the weight of the flow in the control volumes is given by $\int_{\forall} \rho \mathbf{g} \, d\forall$, and the sum of *all external surface* forces is given by $\int_{\forall} \nabla \cdot \mathbf{P} \, d\forall$. Expressing Newton's second law of motion as

$$\sum \mathbf{F} = M\mathbf{a} \tag{5.46}$$

or in integral form

$$\sum \mathbf{F} = \int_{\forall} \rho \frac{D\mathbf{V}}{Dt} \, d\forall \tag{5.47}$$

we obtain with Eqs. (5.45) and (B.37)

$$\int_{\forall} \rho \mathbf{g}\, d\forall + \int_A \mathbf{P} \cdot d\mathbf{A} = \int_{\forall} \rho \frac{D\mathbf{V}}{Dt}\, d\forall \tag{5.48}$$

where the right-hand side represents the total linear momentum of the flow relative to the inertial frame of reference within the control volume.

An alternate expression for Eq. (5.47) is found using Eqs. (5.40) and (5.48) as

$$\int_{\forall} \rho \mathbf{g}\, d\forall + \int_A \mathbf{P} \cdot d\mathbf{A} = \frac{D^2}{Dt^2} \int_{\forall} \rho \mathbf{r}_P\, d\forall \tag{5.49}$$

where $\int_{\forall} \rho \mathbf{r}_P\, d\forall$ is the first moment of mass of the control volume's center of mass. Since the rotation $\boldsymbol{\omega}_{(m)}$ of the moving coordinate system and the acceleration of the origin O' do not depend upon the elemental volume $d\forall$, we can substitute the acceleration expression of Eq. (1.35) for the expression $D\mathbf{V}/Dt$ as

$$\mathbf{a}_P \equiv \frac{D\mathbf{V}}{Dt} \tag{5.50}$$

which results in the expression

$$\begin{aligned}\int_{\forall} \rho \mathbf{g}\, d\forall + \int_A \mathbf{P} \cdot d\mathbf{A} &= M\mathbf{a}_{O'} + \int_{\forall} \rho \boldsymbol{\alpha}_{(m)} \times \mathbf{r}_{P/O'}\, d\forall \\ &+ \int_{\forall} \rho \boldsymbol{\omega}_{(m)} \times (\boldsymbol{\omega}_{(m)} \times \mathbf{r}_{P/O'})\, d\forall + 2 \int_{\forall} \rho \boldsymbol{\omega}_{(m)} \\ &\times \mathbf{V}_{P/O'}\, d\forall + \int_{\forall} \rho \frac{D\mathbf{V}_{P/O'}}{Dt}\, d\forall\end{aligned} \tag{5.51}$$

Though rather long the above equation is nicely packaged. Let us interpret some of the terms. The first term we have identified as the body force, and the second term as our familiar surface forces. Next, $\mathbf{a}_{O'}$ is the acceleration of the moving frame relative to the fixed frame, $\boldsymbol{\omega}_{(m)} \times \mathbf{r}_{P/O'}$ is the tangential acceleration of a fixed point in the control volume relative to the fixed frame, $\boldsymbol{\omega}_{(m)} \times (\boldsymbol{\omega}_{(m)} \times \mathbf{r}_{P/O'})$ represents the normal acceleration of a fixed point in the control volume relative to the fixed frame, $2\boldsymbol{\omega} \times \mathbf{V}_{P/O'}$ is the Coriolis acceleration mentioned above, and $D\mathbf{V}_{P/O'}/Dt$ is the relative acceleration of a fluid particle relative to the moving frame.

Defining the body and surface forces as

$$\sum \mathbf{F}_b = \int_{\forall} \rho \mathbf{g}\, d\forall \tag{5.52}$$

$$\sum \mathbf{F}_s = \sum \mathbf{F}_\tau + \sum \mathbf{F}_p = \int_A \mathbf{P} \cdot d\mathbf{A} \tag{5.53}$$

respectively, we define the apparent force as

$$\sum \mathbf{F}_a = M\mathbf{a}_{O'} + \int_{\forall} \rho \boldsymbol{\alpha}_{(m)} \times \mathbf{r}_{P/O'}\, d\forall + \int_{\forall} \rho \boldsymbol{\omega}_{(m)} \times (\boldsymbol{\omega}_{(m)} \times \mathbf{r}_{P/O'})\, d\forall + 2\int_{\forall} \rho \boldsymbol{\omega}_{(m)} \times \mathbf{V}_{P/O'}\, d\forall \tag{5.54}$$

This gives us the linear momentum principle for a noninertial coordinate system; i.e., from Eqs. (5.51)–(5.54)

$$\boxed{\sum \mathbf{F}_b + \sum \mathbf{F}_\tau + \sum \mathbf{F}_P - \sum \mathbf{F}_a + \mathbf{R} = \mathbf{F}_{\text{ext}} = \frac{\partial}{\partial t}\int_{\forall} \rho \mathbf{V}_r\, d\forall + \int_A \rho \mathbf{V}_r\, (\mathbf{V}_r \cdot d\mathbf{A})} \tag{5.55}$$

Here the local momentum change is the first term on the right-hand side of Eq. (5.55), and the momentum flux across the control surface is the second term on the right-hand side.

It is customary and also more convenient to use the symbol $\mathbf{V}_r$ as the relative velocity rather than $\mathbf{V}_{P/O'}$. $\mathbf{V}_r$ represents the velocity of the fluid at point P relative to an observer attached to the origin of the noninertial platform. In addition, a restraining force $\mathbf{R}$ has been included in the external force as a force that may possibly exist in a given flow situation. Thus, *for a moving coordinate system, we have to worry about an additional force, namely the apparent force, given by Eq. (5.55)*. This is no more difficult than the kinematics of relative motion studied in rigid body mechanics.

Two cases are of particular interest to us.

1. *Case 1*: When the relative velocity is zero, i.e., $\mathbf{V}_r = 0$.

If the relative velocity $\mathbf{V}_r = 0$, then the fluid velocity relative to the moving frame is nonexistent, which in effect means that a static state exists between the flow and the moving coordinate system. For such a case, the apparent force $\Sigma\mathbf{F}_a$ of Eq. (5.54) becomes

$$\sum \mathbf{F}_a = M\mathbf{a}_{O'} + \int_{\forall} \rho \boldsymbol{\alpha}_{(m)} \times \mathbf{r}_{P/O'}\, d\forall + \int_{\forall} \rho \boldsymbol{\omega}_{(m)} \times (\boldsymbol{\omega}_{(m)} \times \mathbf{r}_{P/O'})\, d\forall \tag{5.56}$$

In addition, there is no relative acceleration $D\mathbf{V}_{P/O'}/Dt$. This means there is no net change of linear momentum locally (or convectively). The above equation can be simplified by evaluating the two integral expressions. Since $\mathbf{R}$, $\boldsymbol{\alpha}_{(m)}$, and $\boldsymbol{\omega}_{(m)}$ are independent of $d\forall$, then

$$\sum \mathbf{F}_a = M[\mathbf{a}_{O'} + \boldsymbol{\alpha}_{(m)} \times \mathbf{r}_{\text{C.M.}} + \boldsymbol{\omega}_{(m)} \times (\boldsymbol{\omega}_{(m)} \times \mathbf{r}_{\text{C.M.}})] \tag{5.57}$$

where $\mathbf{r}_{\text{C.M.}}$ is the location of the center of mass of the control volume.

We know from mechanics that $[\mathbf{a}_{O'} + \boldsymbol{\alpha} \times \mathbf{r}_{\text{C.M.}} + \boldsymbol{\omega}_{(m)} \times (\boldsymbol{\omega}_{(m)} \times \mathbf{r}_{\text{C.M.}})]$ is the total acceleration of the center of mass of the control volume. This means we have a *rigid body dynamics situation*:

$$\sum \mathbf{F}_a = M\mathbf{a}_{\text{C.M.}} \tag{5.58}$$

where the acceleration of the center of mass $\mathbf{a}_{\text{C.M.}}$ is

$$\mathbf{a}_{\text{C.M.}} = \mathbf{a}_{O'} + \boldsymbol{\alpha} \times \mathbf{r}_{\text{C.M.}} + \boldsymbol{\omega}_{(m)} \times (\boldsymbol{\omega}_{(m)} \times \mathbf{r}_{\text{C.M.}}) \tag{5.59}$$

2. *Case 2*: Where no rotation of the moving frame exists, i.e., $\boldsymbol{\alpha}_{(m)} = \boldsymbol{\omega}_{(m)} = 0$.

The moving frame translates only when $\boldsymbol{\alpha}_{(m)} = \boldsymbol{\omega}_{(m)} = 0$. In this instance, *the apparent force* is simply due to the acceleration of the origin of the moving frame O' relative to the fixed frame, i.e., from Eq. (5.54)

$$\sum \mathbf{F}_a = M\mathbf{a}_{O'} \tag{5.60}$$

so that the dynamics of the problem becomes from the above expression and Eq. (5.55)

$$\sum \mathbf{F}_b + \sum \mathbf{F}_\tau + \sum \mathbf{F}_p + \mathbf{R} = M\mathbf{a}_{O'} + \frac{\partial}{\partial t}\int_{\forall} \rho \mathbf{V}_r \, d\forall + \int_A \rho \mathbf{V}_r(\mathbf{V}_r \cdot d\mathbf{A}) \tag{5.61}$$

Example 5.6

Develop the equations of motion for a spacecraft accelerating from the surface of the earth vertically upward. The mass efflux of the exhaust gases $\dot{M}_g$, the relative exhaust velocity W_e, and the gravitational field potential are assumed to be constant. Denote the exhaust pressure p_e and let it be different from the ambient pressure of the air p_a. Let M_{s_o} denote the mass of the spacecraft at the prelaunch stage. Neglect the shear force due to air resistance. Find the absolute velocity W of the spacecraft as a function of time if $M_s = M_{x_o} - \dot{M}_g t$.

Example 5.6 *(Con't.)*

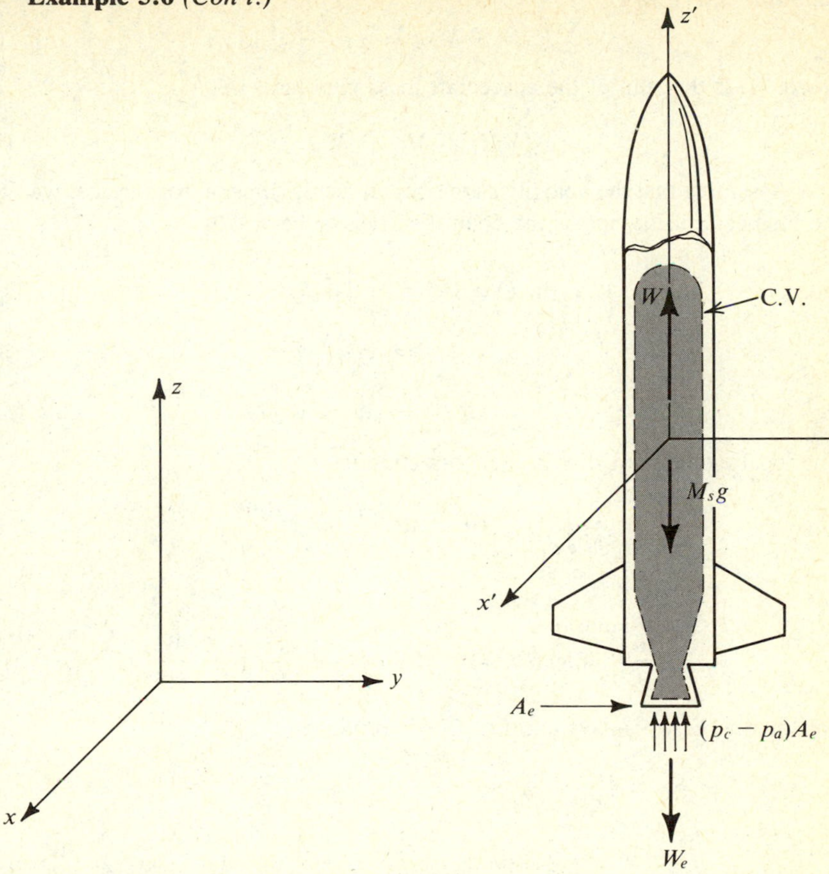

Figure E5.6

Solution:
Step 1.
The fluid is a gas. The flow is unsteady and one-dimensional.
Step 2.
The appropriate governing equation is

$$\sum \mathbf{F} = M\mathbf{a}_{O'} + \frac{\partial}{\partial t}\int_{\forall} \rho \mathbf{V}_r \, d\forall + \int_A \rho \mathbf{V}_r \, (\mathbf{V}_r \cdot d\mathbf{A}) \tag{5.61}$$

This problem illustrates Case 2. The control volume is the interior of the spacecraft, and since it is being accelerated vertically upward, the chosen control volume is noninertial.

The net external force in the vertical direction is composed of the body force and pressure force F_p:

Example 5.6 *(Con't.)*

$$\sum F_z = -M_s g + (p_e - p_a)A_e \tag{i}$$

where M_s is the sum of the spacecraft mass plus fuel, i.e.,

$$M_s = M_{s_o} - \dot{M}_g t \tag{ii}$$

Assuming that the velocities are constant across the control surface, we find the momentum flux across the control surface to be

$$\begin{aligned}\int \rho w_r(\mathbf{V}_r \cdot d\mathbf{A}) &= w_r \int \rho\, dQ \\ &= w_r \dot{M}_s \\ &= (W - W_e)\dot{M}_s\end{aligned} \tag{iii}$$

The local momentum flux in the z direction is

$$\frac{\partial}{\partial t}\int_{\forall} \rho w_r\, d\forall = W\frac{\partial M_s}{\partial t} + M_s \frac{\partial W}{\partial t}$$

or

$$\frac{\partial}{\partial t}\int_{\forall} \rho w_r\, d\forall = -\dot{M}_s W + (M_{s_o} - \dot{M}_g t)\frac{\partial W}{\partial t} \tag{iv}$$

Substituting Eqs. (i)–(iv) into Eq. (5.61) produces

$$-(M_{s_o} - \dot{M}_g t)g + (p_e - p_a)A_e$$
$$= (W - W_e)\dot{M}_s - \dot{M}_s W + (M_{s_o} - \dot{M}_g A)\frac{\partial W}{\partial t} \tag{v}$$

Rearranging and simplifying terms gives

$$\frac{\partial W}{\partial t} = \frac{\dot{M}_s W_e}{M_{s_o} - \dot{M}_g t} + \frac{(p_e - p_a)A_e}{M_{s_o} - \dot{M}_g t} - g \tag{vi}$$

Equation (vi) is easily integrated, resulting in

$$W = W_e \ln\left(\frac{M_{s_o}}{M_{s_o} - \dot{M}_g t}\right) + \frac{(p_e - p_a)A_e}{\dot{M}_g}\ln\left(\frac{M_{s_o}}{M_{s_o} - \dot{M}_g t}\right) - gt \tag{vii}$$

It is sometimes more convenient to identify the momentum efflux $W\dot{M}_g$ as $\rho W_e^2 A_e$.

This completes the solution.

Moving Vanes in Steady Nonaccelerating Flows

A moving vane differs from a fixed vane in that a portion of the total discharge from the nozzle overtakes the vane. An obvious advantage of the moving vane is that it generates useful power, whereas the fixed vane can generate no power because it is stationary.

Figure 5.13 presents the kinematics for the moving vane. Note first the change of notation for the fluid's relative velocity. We will emphasize the importance of the relative velocity *solely in moving vane problems* by setting $\mathbf{V}_r \equiv \mathbf{v}_r$. If the vane has a velocity $\mathbf{v}$ in the same direction as the jet velocity $\mathbf{V}_i$, and the jet enters the vane tangentially, then the fluid's velocity relative to the vane, $\mathbf{v}_{r_i}$, is equal to $\mathbf{V}_i - \mathbf{v}$. The volume rate of flow actually entering the vane is $\mathbf{A} \cdot \mathbf{v}_{r_i}$ and is less than the volume rate of flow upstream of the moving control volume, $\mathbf{A} \cdot \mathbf{V}_i$, the motion of the vane away from the jet. (A is the cross-sectional area of the jet.)

The best way to view this relative volume rate of flow is to imagine yourself seated at the lip of the moving vane and observe the jet as it moves past you. Measure

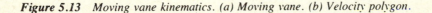

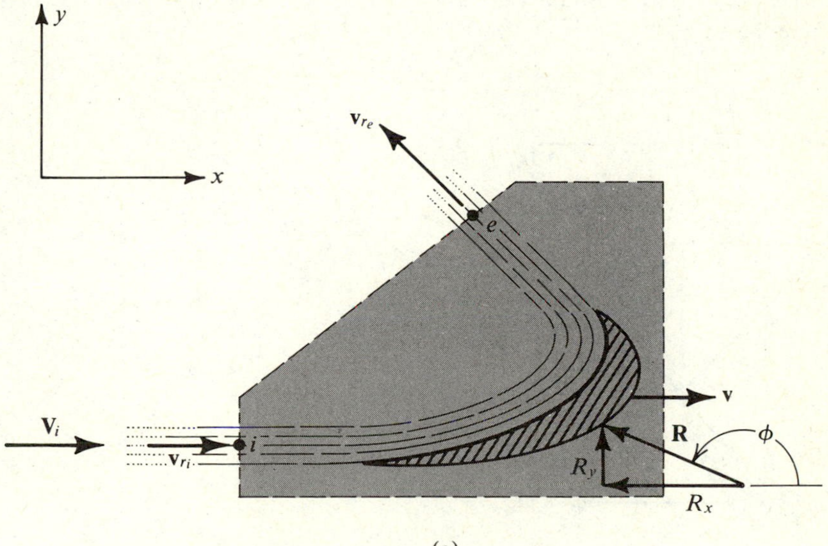

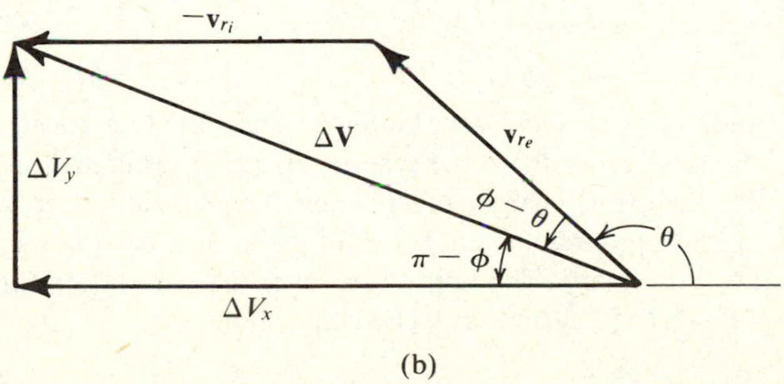

Figure 5.13 *Moving vane kinematics. (a) Moving vane. (b) Velocity polygon.*

the jet velocity $\mathbf{v}_r$ through the cross-sectional area of the jet. That quantity times the jet area is the relative volume rate of flow, i.e., the actual volume of fluid moving past you in an interval of time.

The resultant force of the vane acting on a steady flow of fluid in a nonaccelerating frame is obtained from Eq. (5.61) as

$$\mathbf{F}_{ext} = \rho Q \Delta \mathbf{v}_r$$
$$= \rho A v_{r_i} \Delta \mathbf{v}_r = \rho A v_{r_i} (\mathbf{v}_{r_e} - \mathbf{v}_{r_i}) \tag{5.62}$$

One small comment should be made regarding $\Delta \mathbf{v}_r$. The change in relative velocity is found by vector addition of relative velocities. As evident from the vector diagram in Fig. 5.14, the change in velocity using absolute velocities will yield the same results as that obtained using relative velocities. In many fluid flow problems, however, the use of relative velocities is desirable. (In Fig. 5.14 the vector $\mathbf{V}_e$ is the absolute exit velocity from the vane.)

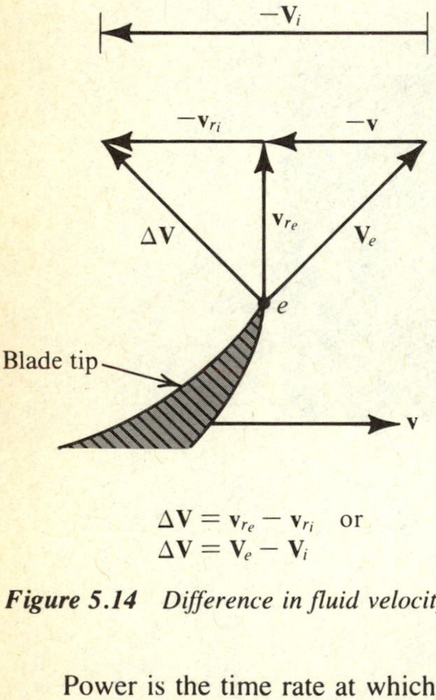

$\Delta \mathbf{V} = \mathbf{v}_{r_e} - \mathbf{v}_{r_i}$ or
$\Delta \mathbf{V} = \mathbf{V}_e - \mathbf{V}_i$

Figure 5.14 *Difference in fluid velocity* $\Delta \mathbf{V}$.

Power is the time rate at which work is done. Work due to a constant force, on the other hand, is the product of the force and the effective resultant change. Such a change might be displacement, charge, polarization, magnetization, area, or length. The unit of work is the appropriate energy, i.e., a joule, or a newton meter in SI, or ft-lbf, or Btu in U.S. customary. The unit of power is the watt or a joule/s in SI, or horsepower or ft-lbf/s in U.S. customary. The conversions are

$$1 \text{ J} = 1 \text{ N} \cdot \text{m} = 0.7376 \text{ ft-lbf} = 0.9478 \times 10^{-3} \text{ Btu} \tag{5.63}$$

conversely,

$$1 \text{ Btu} = 778 \text{ ft-lbf} = 1055 \text{ J} \tag{5.64}$$

for energy, and

$$1 \text{ W} = 1 \text{ J/s} = 0.7376 \text{ ft-lbf/s} = 1.341 \times 10^{-3} \text{ hp} \tag{5.65}$$

conversely,

$$1 \text{ hp} = 550 \text{ ft-lbf/s} = 0.7457 \text{ kW} \tag{5.66}$$

for power.

Therefore, the power obtained from a series of moving vanes designed to receive all of the jet rate of flow will be

$$P = \mathbf{F} \cdot \mathbf{v} = Q\rho \Delta v_{r_x} v \tag{5.67}$$

where Q is the total rate of flow (AV_1), ΔV_x is the change of velocity in the direction of the vane's motion, and v is the velocity of the vane.

From previous discussions of vector components, we can show the above equation for power to be

$$\begin{aligned} P &= \rho Q(v_{r_e} - v_{r_i})_x v \\ &= \rho Q(v_{r_i} \cos\theta - v_{r_i})v \\ &= \rho Q v_{r_i}(\cos\theta - 1)v \end{aligned} \tag{5.68}$$

$$P = \rho Q(V_i - v)(\cos\theta - 1)v \tag{5.69}$$

From this it is evident that no power is developed when (a) the vane is stopped ($\mathbf{v} = 0$), or (b) the vane's velocity equals that of the jet. When $\mathbf{v} = \mathbf{V}_i$, we say that the speed is the runaway speed.

With Q, ρ, V_i, and θ remaining constant, the relationship between power P and vane velocity v is parabolic. The symmetry of the parabola indicates that maximum output will be obtained when $\mathbf{v} = \mathbf{V}_1/2$. This results from the first derivative being zero at the maximum point. We then obtain

$$P_{\max} = Q\rho \frac{V_i^2}{4}(1 - \cos\theta) \tag{5.70}$$

For a blade angle $\theta = 180°$, the maximum power is

$$P_{\max} = \frac{Q\rho V_i^2}{2} = Q\gamma \frac{V_i^2}{2g} \tag{5.71}$$

This equation shows that if the blade angle is 180° and the peripheral speed is one-half the jet speed, all of the jet power of a frictionless fluid may be theoretically transferred to the machine.

Example 5.7

A single vane moves, as shown in Fig. E5.7a, with a velocity v = 50 m/s. Water impinges on the blade at an absolute velocity $\mathbf{V}_i = -150\mathbf{i}$ m/s. The cross-sectional area of the jet of water is 0.03 m². The blade angle is 60°. Calculate (a) the force on the vane in the x direction, (b) the power developed by the vane.

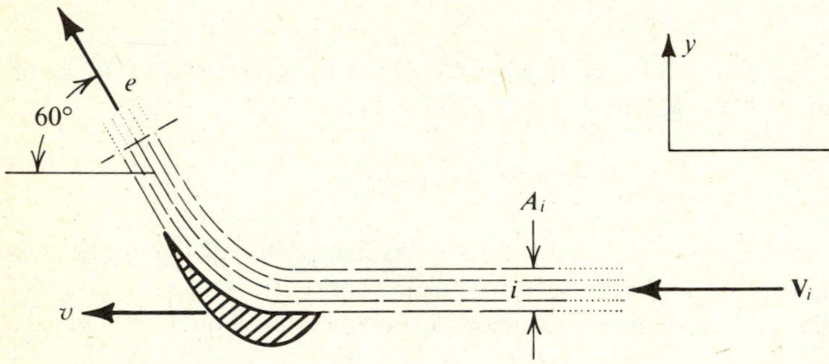

Figure E5.7a

Solution:

Step 1.

Assume that the fluid is incompressible. Assume that the flow is steady with no acceleration of the frame.

Step 2.

The appropriate equations are

$$\mathbf{F} = \rho Q \Delta \mathbf{v}_r \tag{5.62}$$

$$P = \mathbf{F} \cdot \mathbf{v} \tag{5.67}$$

The velocity polygon for the above flow is shown in Fig. E5.7b.

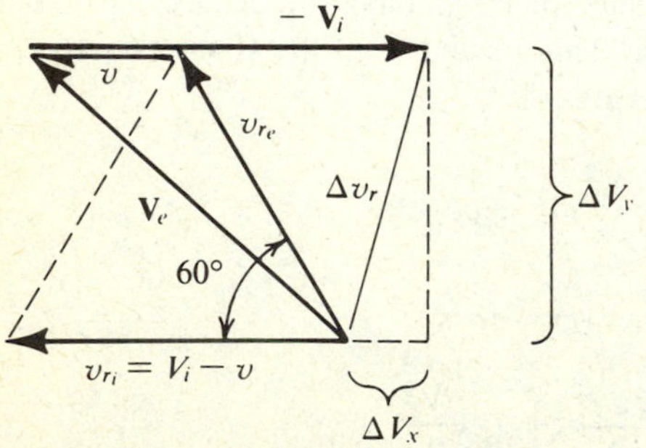

Figure E5.7b

Example 5.7 *(Con't.)*

(a) By geometry,

$$\Delta V_x = -v_r \cos 60^\circ + 100 = -50 + 100 = 50 \text{ m/s} \qquad \text{(i)}$$

$$\Delta V_y = v_r \sin 60^\circ = 100 \sin 60^\circ = 86.6 \text{ m/s} \qquad \text{(ii)}$$

Using Eq. (5.62), the x component of the force on the vane is

$$F_x = \rho v_{r_i} A_i (v_{r_e} - v_{r_i})_x$$

$$= 1000(150 - 50)0.03\ \Delta V_x \qquad \text{(iii)}$$

$$= 150 \text{ kN} \qquad \text{(iv)}$$

(b) The power developed by the vane can be found from Eq. (5.67) as

$$P = F_x v \qquad \text{(v)}$$

$$= 150 \text{ kN} \times 50 \text{ m/s}$$

$$= 7.5 \text{ MW} \qquad \text{(vi)}$$

This completes the solution.

Thus, the use of a moving coordinate system has introduced us to the concept of the apparent force, which includes centrifugal and Coriolis forces. These are very significant forces. For example, the Coriolis force causes clouds to rotate and it plays a key role in the origin of tornados and hurricanes.

5.4 *The Integral Form of the Conservation of Angular Momentum*

We have covered two conservation principles in integral form thus far in this chapter: the conservation of mass and the conservation of linear momentum. Just as linear momentum must be conserved, so must angular momentum. This is an important principle for any rotating system. Almost everywhere we look we see examples of rotating fluids: from the huge rotating stellar systems of a galaxy (see Fig. 11.7) to the small vortex eddies in a swirling cup of coffee.

Imagine a huge upright circular cylinder filled with water. The cylinder and the water are both at a state of rest. Suppose we are given the task of making the water in the tank rotate without moving the tank. To make the water rotate is to in effect swirl it, as when we stir coffee. We could rotate it using paddles (as we stir coffee with a spoon). But suppose we impose the condition the rotation must be introduced without using paddles or stirrers or any other mechanical device. One way might be to rotate the cylinder itself, so that the cylinder and the fluid move as a solid body, but we have already stipulated that the cylinder is fixed. Another way is to introduce small jets of water along the inner circumference of the cylinder, as shown in Fig. 5.15. The jets are directed to impart their tangential momentum to the stationary fluid and thereby commence the fluid rotating from the outer cylinder region inward. Each

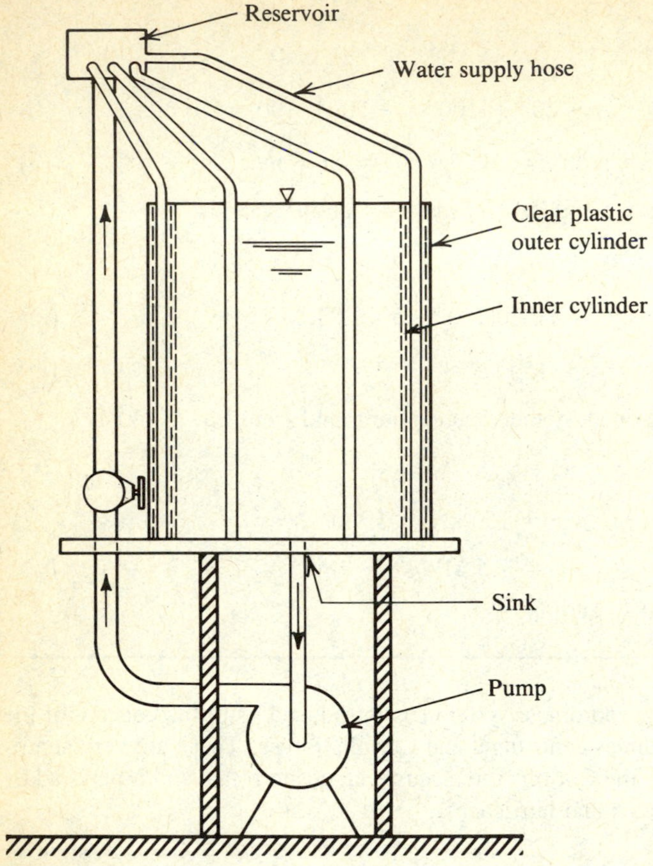

Figure 5.15 *Vortex generator. (Source: R. Granger, "Steady Three-Dimensional Vortex Flow,"* Journal of Fluid Mechanics, *vol. 25, part 3, 1966. Used with the permission of Cambridge University Press.)*

cylindrical laminae of fluid rotates at a different tangential velocity until they all reach the steady state condition of constant angular rotation ω_z. We should have a small drain centrally located at the bottom of the cylinder to allow the drained water to recirculate, as shown in Fig. 5.15, in such a way that it matches the volume rate of flow introduced at the jets. Thus the water level in the tank would be constant. At its steady state condition, we have a swirling mass of water, which could be the physical model for a water spout, or even a hurricane. If we now introduce paddles into the flow, the swirling energy of the fluid flow causes the paddles to rotate. And if the paddles are connected to a shaft, the shaft rotates, and we now have a crude turbine. Of if we want to swirl the water using the energy from rotating paddles, we can add energy to the fluid, the idea behind pumps.

The moment of momentum principle is Newton's second law of motion applied to rotating fluid masses. The statement of dynamic equilibrium of Newtonian mechanics is that the sum of all external moments must be in balance with the inertial moment. This section derives the mathematical statement of dynamic equilibrium for the *general case of a noninertial frame of reference*, then applies the results to two special cases. The first case is rigid body motion (where the relative fluid velocity is zero), and the second case is the steady flow expression of angular momentum for turbomachinery using the inertial coordinate system.

In Fig. 5.12, let O' be the point about which a moment will be taken. Let $\mathbf{r}$ designate the position vector from point O' on the moving frame (m) to a point P in the control volume where the fluid flow exists. A moment $\mathbf{M}_{O'}$ is defined as the cross product of the position vector $\mathbf{r}$ with a force $\mathbf{F}$ acting at P. The forces at P have already been derived and are given by Eq. (5.51), such that the statement of moment dynamic equilibrium is obtained by simply taking the cross product of each term with $\mathbf{r}$, thus:

$$\begin{aligned}\int_{\forall} \mathbf{r} \times \rho\mathbf{g}\, d\forall + \int_{A} \mathbf{r} \times \mathbf{P}\cdot d\mathbf{A} &= \int_{\forall} \mathbf{r} \times \rho\mathbf{a}_{O'}\, d\forall \\ &+ \int_{\forall} \mathbf{r} \times \rho\boldsymbol{\alpha}_{(m)} \times \mathbf{r}\, d\forall \\ &+ \int_{\forall} \mathbf{r} \times \left[\rho\boldsymbol{\omega}_{(m)} \times (\boldsymbol{\omega}_{(m)} \times \mathbf{r})\right] d\forall \\ &+ 2\int_{\forall} \mathbf{r} \times \left(\rho\boldsymbol{\omega}_{(m)} \times \mathbf{V}_r\right) d\forall + \int_{\forall} \rho\mathbf{r} \times \frac{D\mathbf{V}_r}{Dt}\, d\forall \end{aligned} \tag{5.72}$$

The two expressions on the left-hand side of Equation (5.72) represent the next *external* moment about O', i.e., the moments due to the body and surface forces:

$$\Sigma\mathbf{M}_{\text{ext.}O'} = \int_{\forall} \mathbf{r} \times \rho\mathbf{g}\, d\forall + \int_{A} \mathbf{r} \times \mathbf{P}\cdot d\mathbf{A} \tag{5.73}$$

$$\equiv \mathbf{r}_{\text{C.M.}} \times \mathbf{F}_b + \mathbf{r} \times \mathbf{F}_\tau + \mathbf{r} \times \mathbf{F}_p \tag{5.74}$$

where the distribution of elemental masses has been replaced by the body force $\mathbf{F}_b$ acting at the center of mass, $\mathbf{r}_{\text{C.M.}}$ is the position vector from the point O' to the center of mass, $\mathbf{F}_\tau$ and $\mathbf{F}_p$ are the resultant shear force and pressure force, respectively, and their position vector $\mathbf{r}$ is from O' to the point on the control volume $\forall$ through which the line-of-action passes.

The last term on the right-hand side of Eq. (5.72) can be simplified somewhat by considering how the operator D/Dt operates on the vector $(\mathbf{r} \times \rho\mathbf{V}_r\, d\forall)$. From the calculus,

$$\begin{aligned}\frac{D}{Dt}(\mathbf{r} \times \rho\mathbf{V}_r)\, d\forall &= \frac{D\mathbf{r}}{Dt} \times \rho\mathbf{V}_r\, d\forall \\ &+ \mathbf{r} \times \mathbf{V}_r \frac{D(\rho\, d\forall)}{Dt} + \mathbf{r} \times \rho\frac{D\mathbf{V}_r}{Dt} d\forall\end{aligned} \tag{5.75}$$

We can show that the expression

$$\frac{D\mathbf{r}}{Dt} \times \rho \mathbf{V}_r \, d\forall$$

vanishes, because

$$\frac{D\mathbf{r}}{Dt} \equiv \mathbf{V}_r$$

The expression

$$\mathbf{r} \times \mathbf{V}_r \frac{D(\rho \, d\forall)}{Dt}$$

can also be shown to be zero from the conservation of mass. Hence, Eq. (5.75) becomes

$$\mathbf{r} \times \rho \frac{D\mathbf{V}_r}{Dt} d\forall = \frac{D}{Dt}(\mathbf{r} \times \rho \mathbf{V}_r) \, d\forall \tag{5.76}$$

Substituting the results of Eqs. (5.74) and (5.76) into the moment of momentum, Eq. (5.72) yields the result we seek:

$$\begin{aligned} \Sigma \mathbf{M}_{\text{ext}O'} = & \int_\forall \mathbf{r} \times \rho \mathbf{a}_{O'} \, d\forall + \int_\forall \mathbf{r} \times (\rho \boldsymbol{\alpha}_{(m)} \times \mathbf{r}) \, d\forall \\ & + \int_\forall \mathbf{r} \times [\rho \boldsymbol{\omega}_{(m)} \times (\boldsymbol{\omega}_{(m)} \times \mathbf{r})] \, d\forall \\ & + 2\int_\forall \mathbf{r} \times (\rho \boldsymbol{\omega}_{(m)} \times \mathbf{V}_r) \, d\forall \\ & + \frac{D}{Dt} \int_\forall \mathbf{r} \times \rho \mathbf{V}_r \, d\forall \end{aligned} \tag{5.77}$$

This is the result of the moment of momentum principle. Notice we have returned to the notation $\mathbf{V}_r$ for the relative velocity. Again, if we apply our moment of momentum equation to moving blades, we shall set $\mathbf{V}_r$ equal to $\mathbf{v}_r$ to emphasize its importance. Next, we want to see how to use such an integral equation, so let us consider two special cases.

5.4.1 Case 1: $\mathbf{V}_r = 0$. Rigid Body Motion

The case where no fluid flows through the control volume $\forall$ with respect to the moving frame is equivalent to the frequently encountered situation where the fluid moves as a rigid body within the system. The moment of momentum principle for this case is obtained in a straightforward manner. From Eq. (5.77),

$$\Sigma \mathbf{M}_{\text{ext}_{O'}} = \int_{\forall} \mathbf{r} \times \rho \mathbf{a}_{O'} \, d\forall + \int_{\forall} \mathbf{r} \times (\rho \boldsymbol{\alpha}_{(m)} \times \mathbf{r}) \, d\forall$$
$$+ \int_{\forall} \mathbf{r} \times [\rho \boldsymbol{\omega}_{(m)} \times (\boldsymbol{\omega}_{(m)} \times \mathbf{r})] \, d\forall \tag{5.78}$$

Instead of five integrals to evaluate, we now have only three. Suppose we investigate an example of rigid body planar motion.

Example 5.8

To gain some insight into the application of Eq. (5.78), we investigate the planar motion

$$\left.\begin{aligned} \boldsymbol{\alpha} &= \alpha \mathbf{k} \\ \boldsymbol{\omega} &= \omega \mathbf{k} \\ \mathbf{r} &= r\mathbf{j} \\ \mathbf{a}_{O'} &= a_{O'}\mathbf{i} \end{aligned}\right\} \qquad \begin{aligned} a_{O'} &= \text{const.} \\ \alpha &= \text{const.} \end{aligned} \tag{i}$$

which is the case of a moving coordinate system (m) accelerating in translation in the **i** direction, rotating with angular acceleration α and velocity ω about the **k** axis. The motion of a river moving around a bend is a good illustration.

Substituting Eq. (i) into Eq. (5.78) yields

$$\Sigma \mathbf{M}_{\text{ext}_{O'}} = \left(-a_{O'} \int_{\forall} \rho r \, d\forall + \alpha \int_{\forall} \rho r^2 \, d\forall \right) \mathbf{k} \tag{ii}$$

The expressions given by $\int_{\forall} \rho r \, d\forall$ and $\int_{\forall} \rho r^2 \, d\forall$ are frequently termed inertial integrals I and are not difficult to evaluate given the control volume geometric configuration. Such integrals are frequently encountered in mechanics, the latter being the mass moments of inertia.

This completes the solution.

5.4.2 Case 2: Inertial Frame of Reference

The moment of momentum principle for an inertial frame is one where the coordinate system is not moving. For this case,

$$\mathbf{a}_{O'} = \boldsymbol{\alpha}_{(m)} = \boldsymbol{\omega}_{(m)} = 0 \tag{5.79}$$

such that Eq. (5.77) becomes

$$\Sigma \mathbf{M}_{\text{ext}_O} = \frac{\partial}{\partial t} \int_{\forall} \mathbf{r} \times \rho \mathbf{V} \, d\forall + \int_{A} \mathbf{r} \times \rho \mathbf{V} (\mathbf{V} \cdot d\mathbf{A}) \tag{5.80}$$

The first term on the right-hand side is the moment of momentum's net rate of change within the control volume at time t. The second term is the moment of momentum's net rate of efflux across the control surface at time t. The (m) frame is now the inertial frame, so the prime notation is no longer needed.

If we denote $\mathbf{T}_p$, $\mathbf{T}_\tau$, and $\mathbf{T}_b$ as the torques (i.e., moments) about the origin of the inertial frame of reference caused by pressure, shear stress, and gravitational potential, respectively, and if we take the components of the torques in Eq. (5.80) about the x-, y-, and z-axes, we obtain

$$\Sigma M_{o_x} = T_{p_x} + T_{\tau_x} + T_{b_x} = \frac{\partial}{\partial t}\int_{\forall} (rV\cos\alpha)_{yz}\rho\, d\forall + \int_A (rV\cos\alpha)_{yz}\rho\mathbf{V}\cdot d\mathbf{A} \tag{5.81}$$

$$\Sigma M_{o_y} = T_{p_y} + T_{\tau_y} + T_{b_y} = \frac{\partial}{\partial t}\int_{\forall} (rV\cos\alpha)_{xz}\rho\, d\forall + \int_A (rV\cos\alpha)_{xz}(\rho\mathbf{V}\cdot d\mathbf{A}) \tag{5.82}$$

and

$$\Sigma M_{o_z} = T_{p_z} + T_{b_z} + T_{\tau_z} = \frac{\partial}{\partial t}\int_{\forall} (rV\cos\alpha)_{xy}\rho\, d\forall + \int_A (rV\cos\alpha)_{xy}(\rho\mathbf{V}\cdot d\mathbf{A}) \tag{5.83}$$

In Eqs. (5.81)–(5.83), the term $(rV\cos\alpha)_{ij}$ has the following meaning: r_{ij} and V_{ij} are the components of the position vector $\mathbf{r}$ and velocity vector $\mathbf{V}$ that lie in the ij plane, where i and j can be x, y, or z. The angle α_{ij} is defined as the angle between V_{ij} and normal to r_{ij}. For example,

$$|\mathbf{r}\times\mathbf{V}|_{yz} = r_{yz}V_{yz}\sin\left(\frac{\pi}{2} - \alpha_{yz}\right) = (rV\cos\alpha)_{yz} \tag{5.84}$$

A geometric interpretation of one of the moment of momentum equations may be useful in applying the equations to specific problems. Consider Eq. (5.83). Here we are interested in moments about the z-axis, i.e., motion in the xy-plane. In Fig. 5.16, stations i and e denote the influx and efflux of flow, respectively. In Fig. 5.16 $(V_{xy})_i$ is orthogonal to the radius $(r_{xy})_i$, and $(V_{xy})_e$ is orthogonal to $(r_{xy})_e$. Thus, if the control surface is circular, the velocity $(V_{xy})_i$ and $(V_{xy})_e$ are the tangential velocities. This is illustrated next.

Moment of Momentum Equations Applied to Pumps and Turbines

Now we are in a position to discuss those marvelous machines that use the principle of angular momentum: pumps, turbines, compressors, blowers, and fans. Such machines are called turbomachines, since most have a rotating element, called a runner, whose rotating speed ω is constant. Fluid is permitted to flow between each pair of runners and the manner in which it flows designates the type of turbomachine.

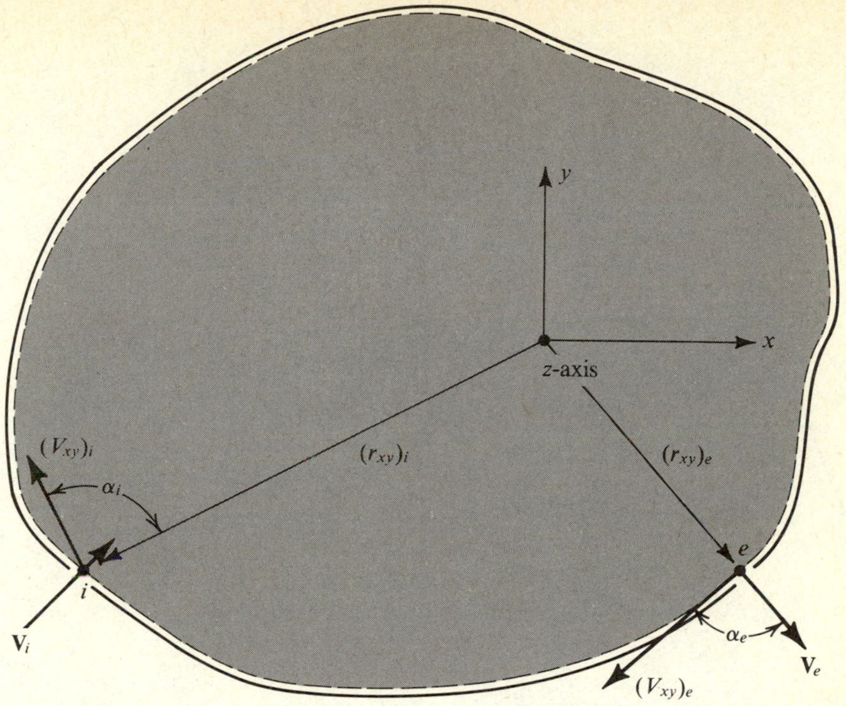

Figure 5.16 *Control volume in the xy-plane for integral form of moment of momentum.*

Rotodynamic machines like turbines and pumps differ from positive-displacement machines in that the former require *relative motion* between the fluid and the moving element of the machine. Rotodynamic machines consist of a rotor carrying a number of vanes. Here energy is transferred between the fluid and the rotor. Whether the fluid does work in the rotor (the turbine) or vice-versa (the pump), the machine is classified by the main direction of the fluid's path in the rotor. For radial flow machines, the path is in the plane of rotation, i.e., the fluid enters at one radius and exits at another, as in Fig. 5.17. The Francis turbine and centrifugal pump are examples. In an axial-flow machine the flow is mainly parallel to the axis of rotation; the Kaplan turbine and the propeller or axial flow pump are examples. If the machine has features of both radial and axial machines, it is called a mixed flow machine.

For any turbine the fluid's energy is initially in the form of high pressure, or temperature, or both. Thus, for a turbine in a hydroelectric complex, the high energy of the water comes from some high elevation, such as a dam. For steam turbines, the high pressure of the steam is produced by a pump and the high temperature from heat addition originates in boilers. For gas turbines, high-pressure energy comes from a chemical reaction of fuel and air in a combustion chamber. (These sources of energy and their transformations are studied in thermodynamics.)

In impulse turbines, one or more fixed nozzles convert the high pressure of the fluid to kinetic energy carried by a jet to impinge on the runner's moving blades. In

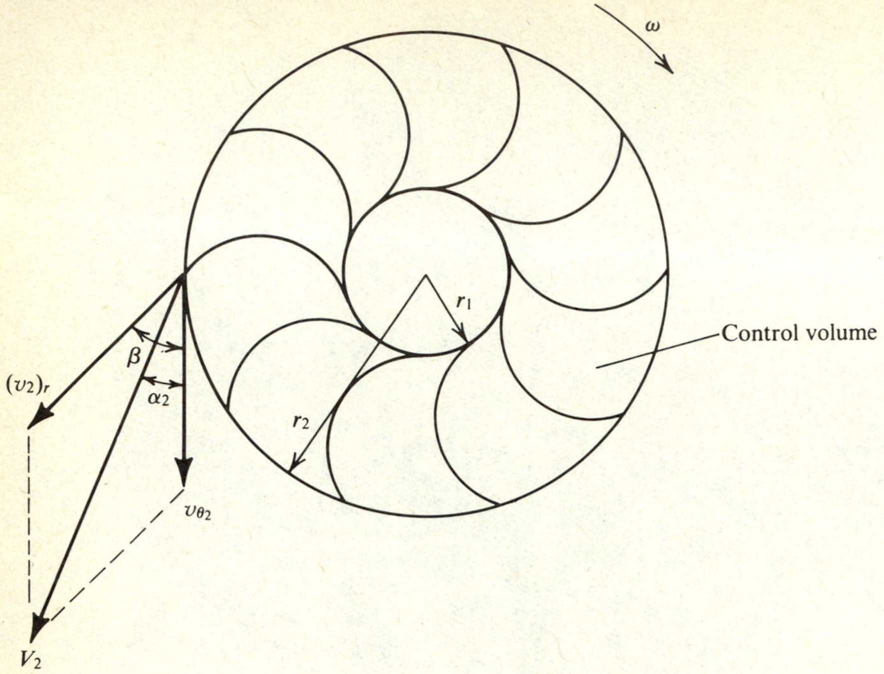

Figure 5.17 *Notation for radial flow pump.*

the course of moving through the blades, the kinetic energy is converted to work on the blades, leaving only a barely perceptible kinetic energy to move the fluid off the blades.

For such machines, it is convenient to consider the control volume shown in Fig. 5.17. Here β is the blade angle, v_θ is the tangential velocity equal to $r\omega$, and matching the tip speed of the impeller (V_e) is the absolute velocity at the exit station of the fluid leaving the blade, and (v_{r_e}) is the relative fluid velocity that is tangent to the blade angle β. If A_e, A_i denote the exit and inlet areas which enclose the runner, Eq. (5.83) becomes

$$\Sigma T_z = \int_{A_2} (r_2 V_2 \cos \alpha_2)\rho \, dQ - \int_{A_1} (r_1 V_1 \cos \alpha_1)\rho \, dQ \tag{5.85}$$

for steady flow, using the control volume shown in Fig. 5.17. If $\Sigma T_z > 0$, then the control volume is called a pump or compressor since energy is being added to the fluid. If $\Sigma T_z < 0$, then the control volume is called a turbine.

Most turbomachines do not have a constant absolute velocity across the control surface. If the absolute velocity V can be replaced by the average velocity $\tilde{V}$, Eq. (5.85) can be expressed as

$$\boxed{\Sigma T_z = \rho Q(r_2 \tilde{V}_2 \cos \alpha_2 - r_1 \tilde{V}_1 \cos \alpha_1)} \tag{5.86}$$

provided that the fluid is incompressible.

Equation (5.86), first developed by Euler, is popularly used as the one-dimensional steady moment of momentum equation. If the fluid spirals *freely* either radially inward or outward, then no torque exists, and our Eq. (5.86) becomes

$$r_2 \tilde{V}_2 \cos \alpha_2 = r_1 \tilde{V}_1 \cos \alpha_1 \tag{5.87}$$

If the absolute circumferential velocity components of the flow $\tilde{V}_2 \cos \alpha_2$, and $\tilde{V}_1 \cos \alpha_1$ are inversely proportional to the radius, then

$$r_2 \tilde{V}_2 \cos \alpha_2 = r_1 \tilde{V}_1 \cos \alpha_1 = r v_\theta = \text{const.} \tag{5.88}$$

and the resultant flow is the vortex flow shown in Fig. 5.18.

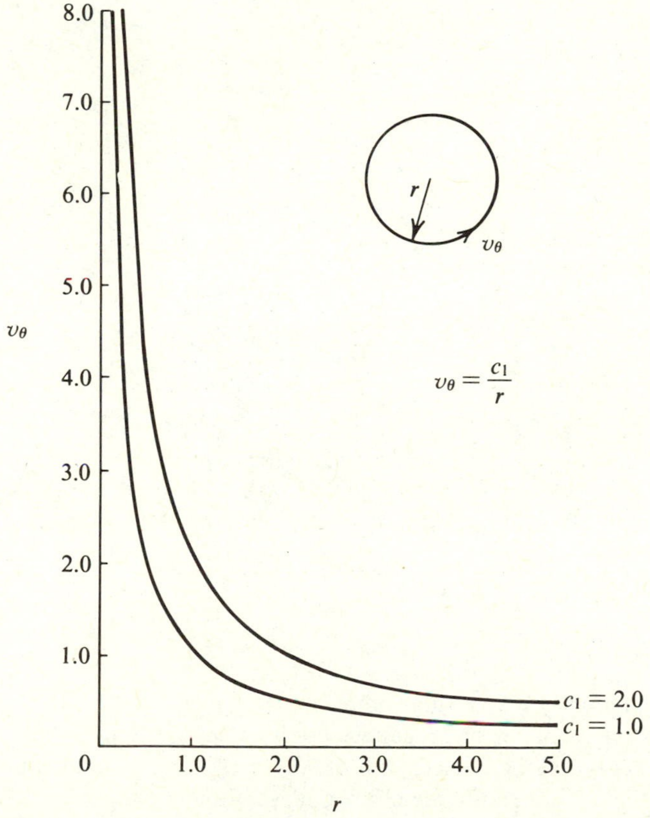

Figure 5.18 *Radial distribution of tangential velocity for one-dimensional steady moment of momentum.*

An Application: Centrifugal Pumps and Their Characteristics

A pump increases the energy of a fluid by applying mechanical energy at the pump shaft. Our discussion will be limited to centrifugal pumps which impart centrifugal forces to the fluid particles to transform mechanical to fluid energy.

The essential parts of a centrifugal pump are a rotating member called the impeller and a case surrounding it. The impeller can be driven by an electric motor, an I.C. engine, or by a steam turbine. As shown in Fig. 5.19, the fluid is led through the inlet pipe to the center or eye of the rotating impeller. In the volute type of centrifugal pump the impeller throws the fluid into the volute where it is led through the discharge diffuser to the discharge piping. The volute or spiral casing is so proportioned as to produce a gradual conversion of the fluid kinetic energy to pressure. Another type of centrifugal pump, the diffuser type, differs from the volute type only in the method of energy conversion. The impeller is surrounded by stationary guide vanes which form several gradually expanding passages that convert energy more efficiently than the volute. The diffuser type pump is not as common as the volute type because of its greater initial cost. Further classification designates the pump as single-suction if the fluid enters the eye from one side of the impeller, and double suction if entering from both sides.

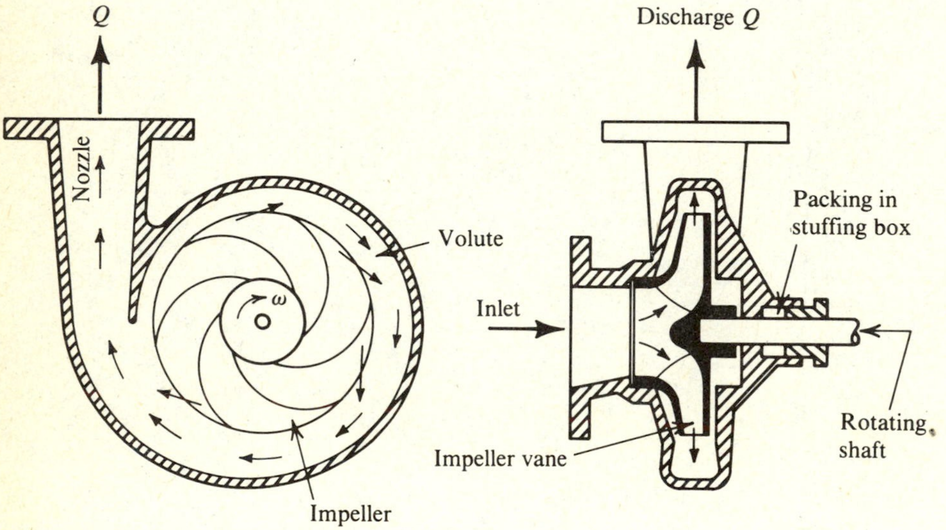

Figure 5.19 *The centrifugal pump schematic.*

The energy that can be produced by a single impeller pump is limited by the high rotational velocity or the large impeller size required. For high energy, staged pumps have two or more impellers on a single shaft with the discharge from one impeller leading to the suction of the next, creating the effect of several pumps in series.

To select the proper pump for a particular purpose, we must know the characteristics of the pump and of the system in which it is used. In order to discuss these characteristics, we should become familiar with the appropriate terms:

1. Head—Synonymous with specific energy (dimensions: length)
2. Capacity—Volume rate of fluid discharged (dimensions: length^3/time)
3. Rated or design capacity—Capacity at maximum pump efficiency

4. e_{p_s}—Specific energy input to pump shaft (dimensions: length)
5. Pump fluid losses (h_p)—Specific energy losses associated with the fluid within the pump (dimensions: length)
6. Total dynamic head (w_{mech})—Energy added to fluid by pump $w_{\text{mech}} = e_{p_s} - h_p$
7. Pump efficiency (η_p)—w_{mech}/e_{p_s}
8. Static suction lift—The vertical distance the surface of the liquid source is below the pump centerline
9. Static suction head—The vertical distance the surface of the liquid source is above the pump
10. Power delivered to the fluid—$P = \omega T = \rho Q(\omega_2 r_2 \tilde{V}_2 \cos \alpha_2 - \omega_1 r_1 \tilde{V}_1 \cos \alpha_1)$

Figure 5.20 illustrates the energy inputs and losses associated with a pump installation. The pump mechanical losses h_p are the energy losses due to friction within the pump bearings, packings, and so on. The pump fluid losses are the energy losses due to fluid friction within the pump impeller and casing. Energy transfers and losses will be presented in the following section.

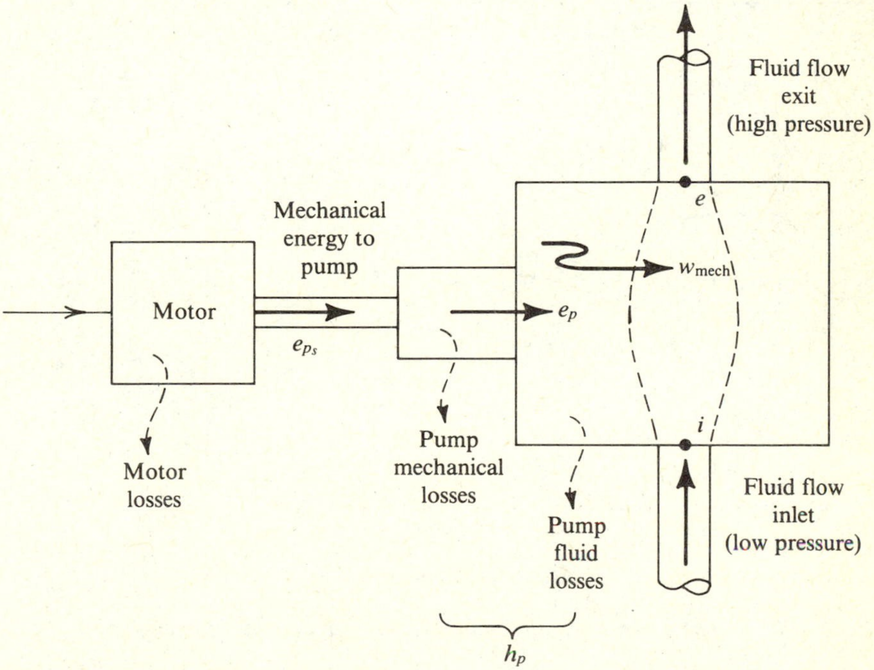

Figure 5.20 *Energy transfers in a pumping system.*

Pump Performance Analysis

Of particular importance in selecting a pump is the w_{mech} developed at various capacities. A curve of w_{mech} versus capacity is often referred to as the pump's characteristic.

As shown in Fig. 5.21, a centrifugal pump may have a rising, flat, or falling characteristic depending on the design. Other important performance characteristics are shown on the sample performance curve of Fig. 5.22.

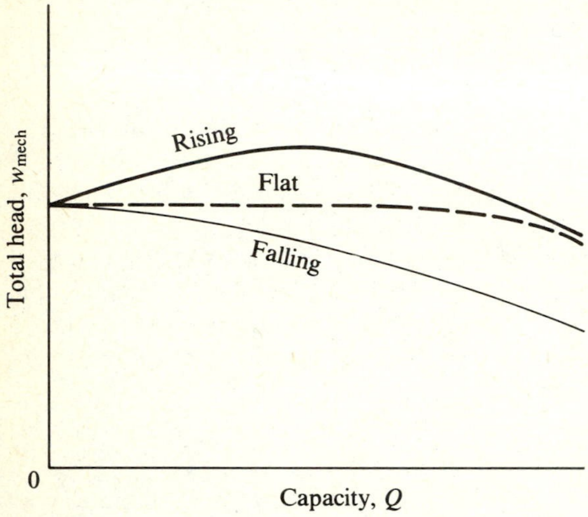

Figure 5.21 *Pump performance characteristics.*

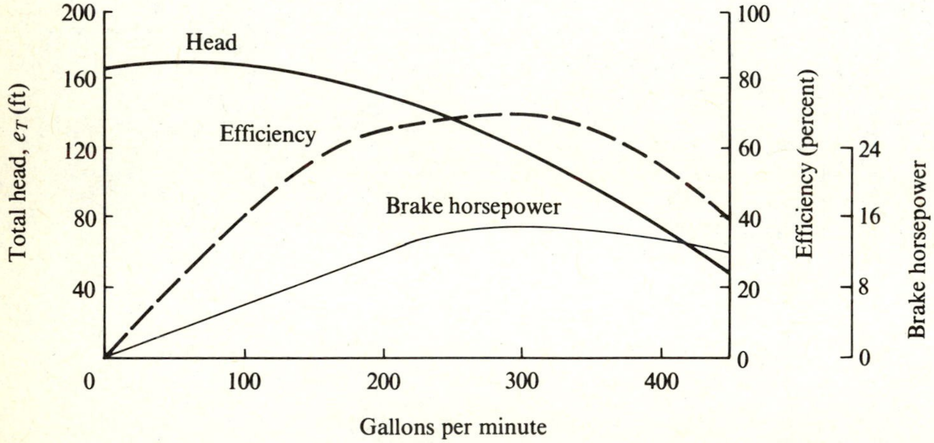

Figure 5.22 *Performance characteristics of a typical 3-inch centrifugal pump.*

Some Limiting Factors in Pump Operation

Because of the low pressures existing in the suction piping of a centrifugal pumping system with a static suction lift, the static fluid pressure frequently approaches the vapor pressure of the fluid being pumped. Extreme care should be exercised in designing the suction lift for a pump system if the fluids are volatile or if they are to be pumped at elevated temperatures. If the pressure in the suction line is below the vapor pressure of the fluid being pumped, the fluid will vaporize and the system will become "vapor-bound." The practical limit of the suction lift for water at 60°F is about 15 feet.

Cavitation is an undesirable fluid phenomenon associated with fluid machinery. The twofold undesirable effects manifest themselves in reduced hydraulic efficiency, and in pitting and erosion of metal surfaces exposed to cavitation. Cavitation means that, if at some point in the flow the existing fluid pressure equals the vapor pressure at the particular temperature, then the liquid will vaporize and a cavity or void will form. If the fluid pressure fluctuates slightly above and below the vapor pressure, vapor bubbles will form and collapse. This alternating collapse and formation of bubbles markedly lowers efficiency and pits the pump's metal parts, as can be seen in the German destroyer propeller of Fig. 5.23.

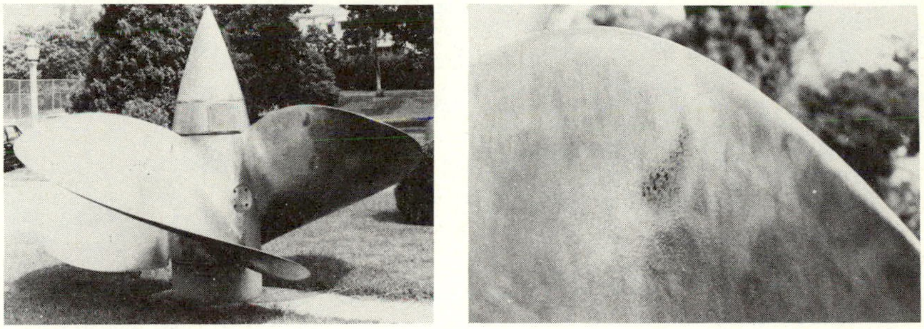

Figure 5.23 *Cavitation on a German destroyer's propeller.*

Combination of Pump and System

The pump is only one part of a fluid recirculating system. The engineer is often faced with the problem of selecting the best pump to match a given system. Pump characteristics are presented by the manufacturer in the form of curves as that in Fig. 5.22. The engineer must determine the head and capacity requirements of the system, and by comparing the characteristics of available pumps determine the one that best meets these needs.

As an example, consider the system of Fig. 5.24. By superimposing a curve of system characteristics on that of the pump we can see if a particular combination will satisfy our requirements. We can see that the intersection of the characteristic curves determines the capacity at which our system will operate. If it is desirable to operate at a lower capacity, the system can be throttled by use of the valves in Fig. 5.24 to give us the dashed curve. Note that our losses have increased.

Example 5.9

Consider flow through a propeller turbine, shown in Fig. E5.9a. At station 1, the angle α_1 between the absolute velocity $\vec{V}_1$ and the circumferential velocity $r_1\omega$ is 53° and is formed as a result of gates mounted at section 1. If the absolute

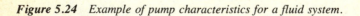

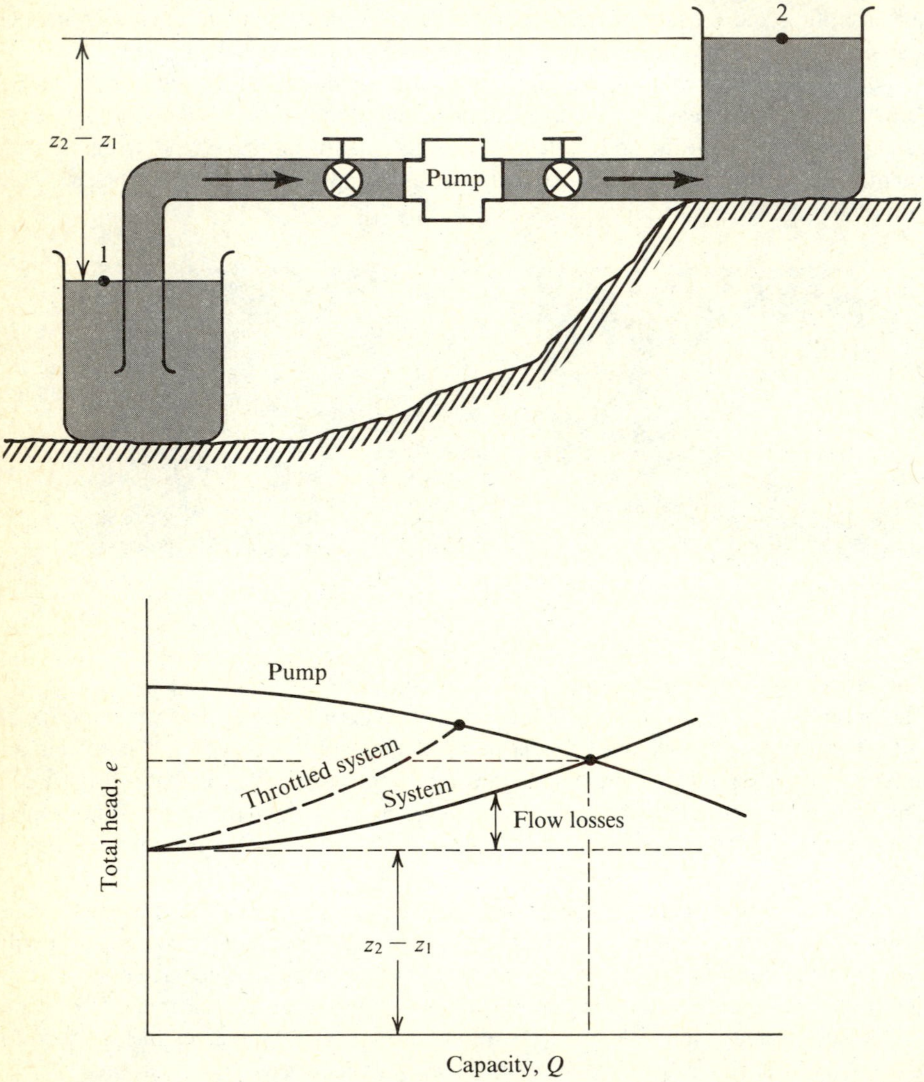

Figure 5.24 *Example of pump characteristics for a fluid system.*

Example 5.9 *(Con't.)*

velocity $\tilde{V}_1$ is 3 m/s, determine that part of the absolute velocity in the tangential direction ($\tilde{V}_2 \cos \alpha_2$) at station 2.

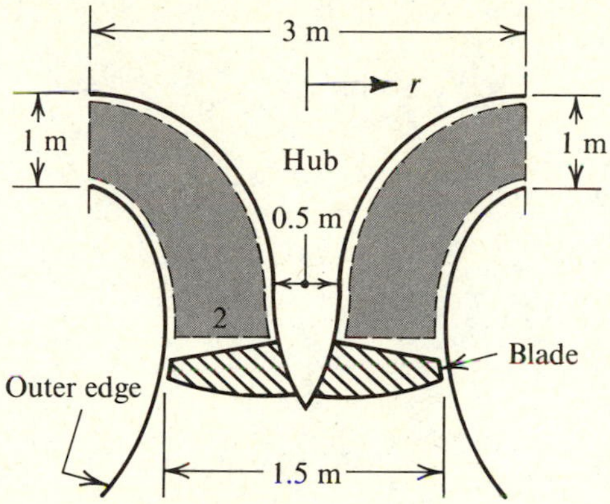

Figure E5.9a

Solution:

Step 1.

Assume that the fluid is incompressible and inviscid. Assume that the flow is steady and uniform at a section.

Step 2.

Since the blade is not within the control volume, there is no torque. We then have conditions necessary for a vortex flow

$$\rho Q(r_2\tilde{V}_2 \cos \alpha_2 - r_1\tilde{V}_1 \cos \alpha_1) = 0 \qquad \text{(i)}$$

This is the special case of a fluid moving in a spiral path. Notice that if $\alpha = 0$, the flow would be in a circular path. This is the free-vortex case.

It is convenient to sketch the velocity polygon at station 1 (Fig. E5.9b).

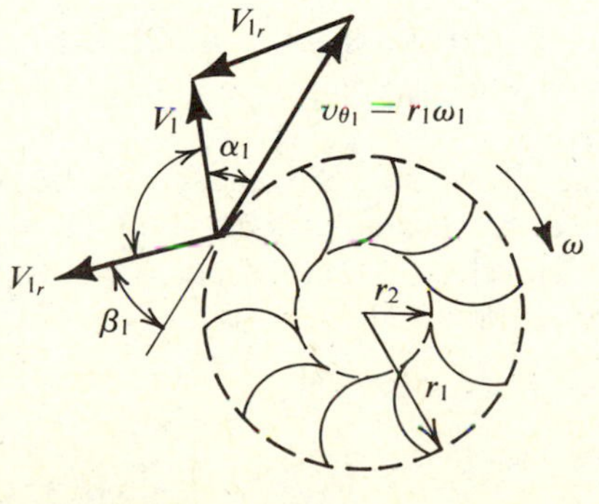

Figure E5.9b

Example 5.9 *(Con't.)*

At section 1 we have

$$r_1\tilde{V}_1 \cos \alpha_1 = (1.5 \text{ m})(3 \text{ m/s}) \cos 53° = 2.71 \text{ m}^2/\text{s} \tag{ii}$$

At section 2, Eq. (i) reveals that

$$\tilde{V}_2 \cos \alpha_2 = \frac{r_1\tilde{V}_1 \cos \alpha_1}{r_2} \tag{iii}$$

such that at the surface of the hub

$$\tilde{V}_2 \cos \alpha_2 = \frac{2.71 \text{ m}^2/\text{s}}{0.25 \text{ m}} = 10.84 \text{ m/s} \tag{iv}$$

and at the outer edge of the chamber

$$\tilde{V}_2 \cos \alpha_2 = \frac{2.71 \text{ m}^2/\text{s}}{0.75 \text{ m}} = 3.61 \text{ m/s} \tag{v}$$

This completes the solution.

Example 5.10

Consider the problem of Example 5.9. Assuming uniform axial velocity across section 2, and using the geometric data of the propeller turbine, find (a) the angle of the leading edge of the propeller at the radial distance $r = 0.25$ m and $r = 0.75$ m for a propeller speed of 200 rpm, (b) the magnitude of the axial velocity at section 2.

Solution:

The tangential velocity at $r = 0.25$ m is

$$r_2\omega = 0.25 \times 2\pi \times \frac{200}{60} = 5.24 \text{ m/s} \tag{i}$$

and represents the peripheral speed of the impeller at the base of the blade. In Example 5.9, the *fluid's* absolute tangential velocity component at this location was calculated as 10.86 m/s.

Before we can calculate the blade angle β we need the other component of flow velocity, the axial velocity V_z, so that the resultant velocity V of the flow can be found (assuming that there is no radial velocity). The discharge Q through the turbine at section 1 is all circumferential flow and is

$$Q = V_{\theta_1}A_1 = 1.81 \times 3\pi = 17.06 \text{ m}^3/\text{s} \tag{ii}$$

The flow through section 2 is all axial flow so that

$$Q = V_{z_2}A_2 \tag{iii}$$

Example 5.10 *(Con't.)*

so that

$$V_{z_2} = \frac{Q}{A_2} = \frac{17.06}{1.57} = 10.86 \text{ m/s} \tag{iv}$$

Next, we construct a velocity polygon to help evaluate the angle β (Fig. E5.10a).

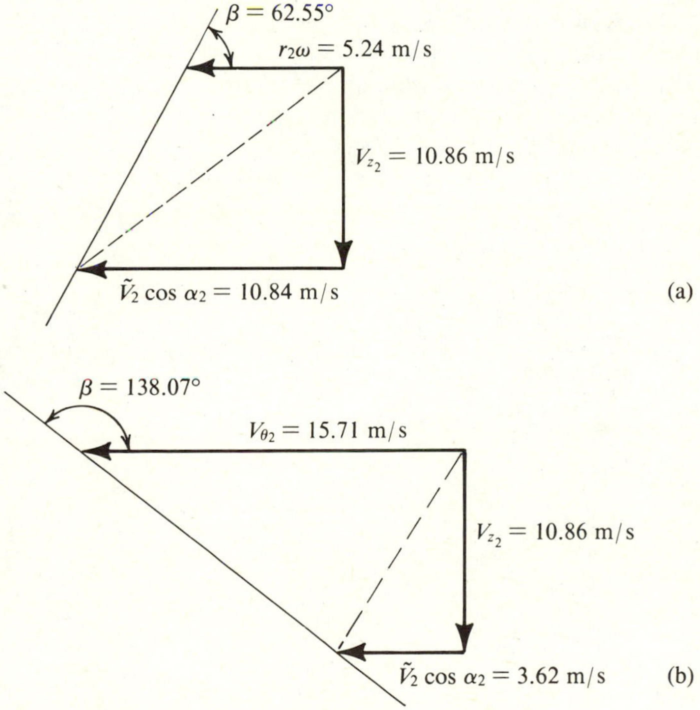

Figure E5.10

From the preceding geometry we find

$$\beta = \tan^{-1} \frac{10.86}{10.84 - 5.24} = 62.66°$$

Repeating the above procedure for $r = 0.75$ m, we find the tangential velocity of the impeller to be

$$r_2\omega = 0.75 \times 2\pi \times \frac{200}{60} = 15.71 \text{ m/s} \tag{v}$$

Example 5.9 calculated the flow's absolute tangential velocity component at $r = 0.75$ m as $\tilde{V}_2 \cos \alpha_2 = 3.61$ m/s. Since the flow rate Q is unchanged,

Example 5.10 *(Con't.)*

the axial velocity V_{z2} is unchanged at 10.86 m/s. Constructing a velocity polygon to evaluate the angle β for $r = 0.75$ m gives $\beta = 138.07°$ as the angle of the leading edge of a propeller turbine. See Fig. E5.10b.

This completes the solution.

Example 5.11

Consider the motion of a lawn sprinkler that discharges water from two nozzles as shown in Fig. E5.11. Let α denote the angle of the jet velocity with the normal to the radius r. Calculate (a) the torque required to keep the sprinkler from rotating and (b) the resultant circular speed for zero torque. Let the fluid enter the sprinkler rotary arms in the z-direction.

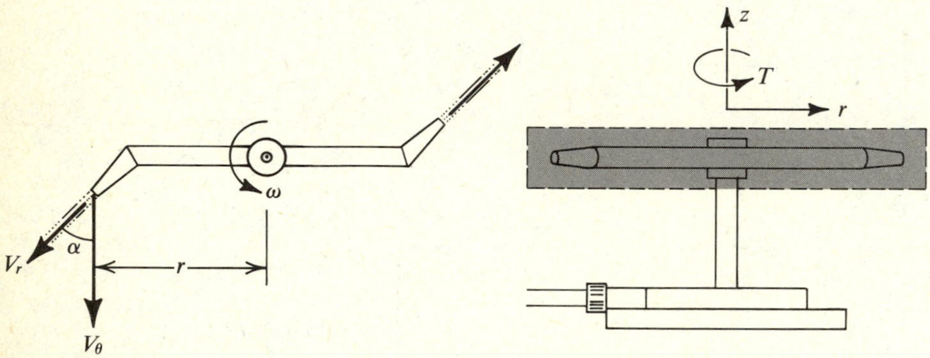

Figure E5.11

Solution:

Step 1.

The fluid is assumed to be incompressible. The flow is steady and two-dimensional.

Step 2.

The appropriate equation governing the flow is

$$\Sigma T_z = \rho Q r V \cos \alpha \tag{5.86}$$

The control volume is the sprinkler in the horizontal xy-plane. Since the fluid enters the control volume in the z-direction, there is no entering angular momentum. Thus the torque results solely from the efflux out of the two nozzles.

From Eq. (5.86) Q is the volume rate out of *both* nozzles, and V is the absolute velocity of the fluid. We express the absolute velocity V in terms of the relative velocity v_r of the flow (relative to the moving nozzle) and rotational velocity of the nozzle. The tangential velocity component V_θ of the absolute velocity V is

Example 5.11 *(Con't.)*

$$V_\theta = v_r \cos\alpha - \omega r \tag{i}$$

where $v_r \cos\alpha$ is the tangential component of the relative velocity. The tangential component of the absolute velocity V_θ is related to the absolute velocity V by

$$V_\theta = V\cos\alpha \tag{ii}$$

Substituting Eqs. (i) and (ii) into Eq. (5.86) yields

$$\Sigma T_z = \rho Q r(v_r \cos\alpha - \omega r) \tag{iii}$$

The volume rate of flow Q is found using the relative velocity v_r and diameter D of the nozzle:

$$Q = 2\left(\frac{\pi}{4}D^2 v_r\right) \tag{iv}$$

Substituting Eq. (iv) into Eq. (iii) gives

$$\Sigma T_z = \rho Q r\left(\frac{2Q}{\pi D^2}\cos\alpha - \omega r\right) \tag{v}$$

Thus, for a given density ρ, nozzle diameter D, nozzle setting α, sprinkler arm r, volume rate of flow Q, and speed of rotation ω, the torque can be calculated.

(a) The torque required to keep the sprinkler from rotating is found by setting the angular speed ω equal to zero in Eq. (v) such that

$$\Sigma T_z = \frac{2\rho Q^2 r\cos\alpha}{\pi D^2} \tag{vi}$$

(b) The rotational speed of the sprinkler for no torque (the case of maximum speed) is found by setting ΣT_z equal to zero in Eq. (v) and solving for ω:

$$\omega = \frac{2Q}{\pi D^2 r}\cos\alpha \tag{vii}$$

Thus, we obtain the maximum speed for a nozzle setting of zero degrees, large volume flow rate, very small nozzle diameter and sprinkler arm.

This completes the solution.

5.5 *The Integral Form of the Conservation of Energy*

Our fourth and final conservation principle is that of energy. The derivation of this extremely important and powerful equation has been presented in thermodynamics. Letting the intensive property ϕ of the fluid be the energy per unit volume e, the integral form of the general property balance, i.e., Reynolds transport Eq. (4.13) becomes

$$\frac{DE}{Dt} = \int_{\forall} \frac{\partial e}{\partial t}\, d\forall + \oint_A e\mathbf{V}\cdot d\mathbf{A} \tag{5.89}$$

where the total energy per unit volume is related to the energy per unit mass of internal, kinetic, potential, nuclear, electrical, magnetic, and other energies by Eq. (4.140). Equation (5.89) expresses some now familiar concepts. The first term on the left of the equality sign denotes the total change of energy of a *system*. However, we do not wish to use the system method of analysis since we are concerned with describing the behavior of energy for a finite volume of fluid. The total energy E is a point function, dependent upon spatial variables as well as time. The notation DE/Dt indicates that we are examining the *total* behavior of the system's total energy. In transforming to the control volume method of analysis, we make use of the Reynolds transport Eq. (4.13). On the right side of the equality sign are the changes of total energy applicable to a control volume $\forall$ with control surface $\mathbf{A}$. The first term on the right is the accumulation or depletion of energy *within* the control volume. It can change only with respect to time. The second term is the total energy being transported across the control surface by the fluid moving at velocity $\mathbf{V}$.

As stated in Sec. 4.5, energy can be transferred not only by the influx and efflux of fluid but in the form of heat and work. The first law of thermodynamics states how the energy is conserved:

The sum of the heat added to a system of masses and the work done on the system depends solely on the initial and final total energy of the system.

The first law of thermodynamics is expressed mathematically as

$$\begin{aligned}\frac{đQ}{dt} + \frac{đW}{dt} &= \frac{DE}{Dt} \\ &= \int_{\forall} \frac{\partial e}{\partial t}\, d\forall + \oint_A e\mathbf{V}\cdot d\mathbf{A}\end{aligned} \tag{5.90}$$

If no energy is transferred by heat or work between the fluid control volume and the environment, then the control volume is isolated and the total change of energy DE/Dt is zero. If the control volume has an adiabatic surface, then no energy can be transferred as heat, leaving work as the only means whereby energy can be transferred. We shall consider both heat ${}_iQ_e$ and work ${}_iW_e$ as possible energy transfers between the inlet and exit stations of a control volume.

To simplify the notation, we shall consider only internal, kinetic, potential, and flow energies in Eq. (5.90). Substituting the specific energy of Eq. (4.141) into the general property balance of Eq. (5.90) results in

$${}_i\dot{Q}_e + {}_i\dot{W}_e = \frac{\partial}{\partial t}\int_{\forall} \rho\left(\bar{i} + \frac{V^2}{2} + gz\right) d\forall + \oint_A \rho\left(\bar{i} + \frac{V^2}{2} + gz\right)\mathbf{V}\cdot d\mathbf{A} \tag{5.91}$$

where energy transfer is positive if it is transferred from the surroundings into the control volume.

5.5.1 Rate of Heat Transfer, $_i\dot{Q}_e$

The rate of heat transfer is composed of heat by conduction, convection, and radiation. As explained in Sec. 4.5, we shall not consider heat transferred by convection and radiation. According to Fourier's law, the quantity of heat per unit time $đQ/dt$ passing across a fluid control surface area $d\mathbf{A}$ is given by

$$ {}_i\dot{Q}_e = \frac{đQ_c}{dt} = \oint_A k\nabla T \cdot d\mathbf{A} = \oint_A k\frac{\partial T}{\partial n}\,dA \tag{5.92}$$

where k is the coefficient of thermal conduction, and where the derivative of the temperature T is taken in the direction $\mathbf{e}_n$ normal to the differential area. Given the temperature field $T\,(x,y,z)$ of the fluid, we can evaluate the integral expression of Eq. (5.92).

5.5.2 Fluid Power, $_i\dot{W}_e$

We shall consider two basic types of power: mechanical power transmitted by some mechanical means $\dot{W}_{\text{mech}}$, e.g., rotating shafts, blades, and linkages that transmit energy across the control surface between the finite control volume and the surroundings, and surface forces power $\dot{W}_s$:

$$ {}_i\dot{W}_e = \dot{W}_{\text{mech}} + \dot{W}_s \tag{5.93}$$

Mechanical power is either a given quantity or is the desired unknown to be calculated in a particular problem. So let us consider surface power $\dot{W}_s$.

The surface power $\dot{W}_s$ is defined in terms of the stresses p_{ns} as

$$\dot{W}_s = \frac{d}{dt}\int_{\forall} p_{ns}\,d\forall \tag{5.94}$$

Here it is convenient to use the intrinsic coordinate system and its notation. For example, if we wish to consider the normal stress p_{nn} for an inviscid flow, we obtain $p_{nn} = -p$.

Work Due to Shear Stresses, W_{s_τ}

The power lost by viscous shear stresses consists primarily of the product of a viscous stress and the respective velocity component.

$$\dot{W}_{s_\tau} = -2\mu\int_A \mathbf{e}_n \cdot (\mathbf{V}\cdot\dot{\mathbf{S}})\,dA + 2\mu\int_{\forall} (\dot{\mathbf{S}}\cdot\nabla)\mathbf{V}\,d\forall \tag{5.95}$$

The first integral is the power loss across the control surface *due to the rate of viscous work*, and the second integral is the power loss within the control volume *due to friction*. Equation (5.95) is derived in Prob. 5.124.

The viscous work term W_s is often awkward to evaluate. We therefore seek a control surface where this term may be negligible compared to other energies. Since shear stresses are large where velocity gradients are large, and small where the velocity gradients and viscosity are small, we shall seek to place our control volume along streaming surfaces where changes in velocity are negligible. Figure 5.25 illustrates a possible control surface along which shear stresses might be neglected. Using similar control surfaces in defining the control volume for a particular flow problem may help simplify computations.

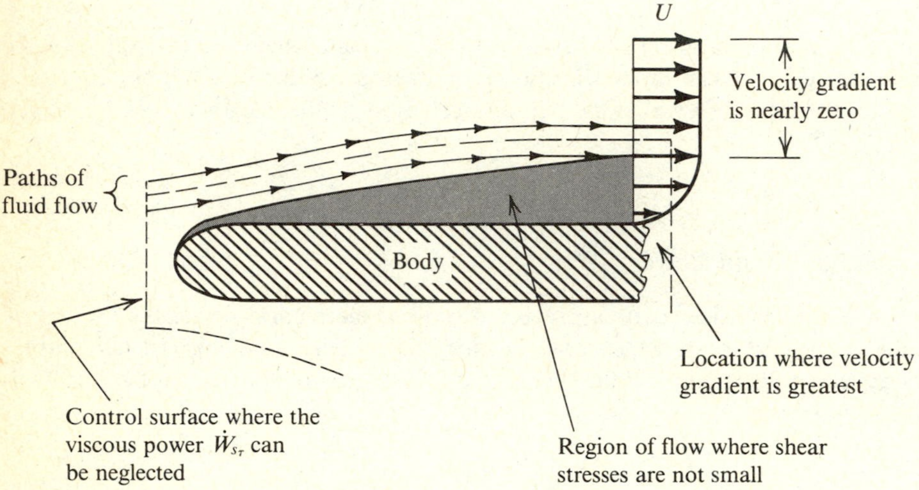

Figure 5.25 *Control surface for negligible viscous power;* $\dot{W}_{s_\tau} = 0$.

Work Due to Normal Stresses, W_{s_p}

Due to its lengthy derivation, the power done on the normal stresses has been left as an exercise (Prob. 5.124). Considering only the power done by pressure forces, it can be shown that

$$\dot{W}_{s_p} = - \oint \mathbf{e}_n \cdot \mathbf{V} p \, dA \tag{5.96}$$

5.5.3 Integral Form of the Energy Equation

The net result of the foregoing discussion is that the power term of Eq. (5.91) consists of

$$_i\dot{W}_e = \dot{W}_{\text{mech}} - \oint \mathbf{e}_n \cdot \mathbf{V} p \, dA + \dot{W}_{s_\tau} \tag{5.97}$$

where $\dot{W}_{s_\tau} = 0$ if the flow is inviscid, or if the velocity is uniform. Since fluid power is lost in regions where velocities are not negligible, it would be advisable to select a control volume that is outside this region so that $\dot{W}_{s_\tau}$ can be neglected.

Substituting the power expression of Eq. (5.97) and the rate of heat expression of Eq. (5.92) into the energy Eq. (5.91) results in

$$\dot{W}_{\text{mech}} = \frac{\partial}{\partial t}\int_{\forall} \rho\left(\frac{V^2}{2} + gz\right) d\forall + \oint_A \rho\left(\frac{p}{\rho} + \frac{V^2}{2} + gz\right)\mathbf{V}\cdot d\mathbf{A} + \left[\int_{\forall}\left(\frac{\partial}{\partial t} + \nabla\right)\rho\bar{i}\, d\forall - \dot{W}_{s_\tau} - \oint k\frac{\partial T}{\partial n} dA\right] \tag{5.98}$$

The terms have been purposely rearranged to emphasize that the terms in the brackets are involved with temperature and velocity gradients. Notice that both viscosity and the coefficient of heat conduction appear in some of the terms. Let us define these terms in brackets as

$$\dot{H}_f = \int_{\forall}\left(\frac{\partial}{\partial t} + \mathbf{V}\cdot\boldsymbol{\nabla}\right)\rho\bar{i}\, d\forall - \dot{W}_{s_\tau} - \oint k\frac{\partial T}{\partial n} dA \tag{5.99}$$

and define $\dot{H}_f$ as the rate of *friction loss*. All of the problems for incompressible fluid flow that we shall be treating in this text have a constant value of internal energy and no heat conduction. The rate of friction loss results from power lost from fluid flow shear stresses resulting in Eq. (5.99) reducing to

$$\dot{H}_f = -\dot{W}_{s_\tau}, \quad \text{for } \bar{i} = \text{const.}, \quad \frac{\partial T}{\partial n} = 0 \tag{5.100}$$

Equation (5.100) is generally not applicable to compressible fluid flows.

Expressions and values of friction loss will be investigated further in Chap. 12 when we investigate the behavior of fluid flow through pipes.

Steady State Form

The steady state form of the energy equation is

$$\dot{W}_{\text{mech}} = \oint_A \rho\left(\frac{p}{\rho} + \frac{V^2}{2} + gz\right)\mathbf{V}\cdot d\mathbf{A} + \dot{H}_f \tag{5.101}$$

The above expression can be rearranged with the aid of Eqs. (5.92) and (5.99) to obtain

$$\dot{W}_{\text{mech}} = \int_{A_e} \rho_e\left(h + \frac{V^2}{2} + gz\right)_e dQ_e - \int_{A_i} \rho_i\left(h + \frac{V^2}{2} + gz\right)_i dQ_i - \dot{W}_{s_\tau} - {}_i\dot{Q}_e \tag{5.102}$$

where h is the enthalpy defined by Eq. (4.151). Of the two expressions, we prefer Eq. (5.101) when dealing with incompressible fluids, and Eq. (5.102) is more useful for compressible fluid flow.

One-Dimensional Form

If the fluid properties vary at most in one direction, and if the control volume has a number of inlet and exit stations through which the fluid flows, then the energy equation can be expressed as

$$w_{\text{mech}} = \sum_e \left(\frac{p}{\gamma} + \frac{\tilde{V}^2}{2g} + z \right)_e - \sum_i \left(\frac{p}{\gamma} + \frac{\tilde{V}^2}{2g} + z \right)_i + h_f \tag{5.103}$$

where

$$h_f = \frac{\dot{H}_f}{\dot{M}g} \tag{5.104}$$

is called the friction head loss, and

$$w_{\text{mech}} = \frac{\dot{W}}{\dot{M}g} \tag{5.105}$$

is the mechanical work per weight of fluid flow. Equation (5.103) is the most widely used form of the steady flow one-dimensional energy equation.

Figure 5.26 illustrates the equation. The equation states that the energy per unit weight carried into and out of the control volume is exactly balanced by the energies per unit weight added and taken away by mechanical devices, heat transfer, and viscous work. If the flow is steady, we have the ultimate simple statement of conserving energy:

$$\text{Energy going in} = \text{energy going out} \tag{5.106}$$

The energy going in is carried by the fluid through the inlet stations plus any energy coming in by means of mechanical devices and heat. The energy going out is that carried by the fluid plus any energy transferred by mechanical devices and heat and any energy lost by other means. The energy of the fluid consists primarily of pressure, kinetic, and potential energies for incompressible fluids, while for compressible fluids the energy consists primarily of enthalpy, kinetic and potential energies.

Example 5.12

An ideal gas whose gas constant $R = 772$ ft-lbf/lbm°R is used as the propellant for a rocket engine in a space station. The rocket consists of a rigid tank with numerous solar panels that transfer heat. The gas in the tank is initially at 40 psia and 20°F. An automatic device opens a valve in a line from the tank permitting gas to escape through a nozzle. The valve throttles the gas to 20 psia, and the nozzle discharge state is 5 psia and −100°F. Assume that the specific enthalpies $h_1 = h_0$ are average values during a 2-second burst resulting in a

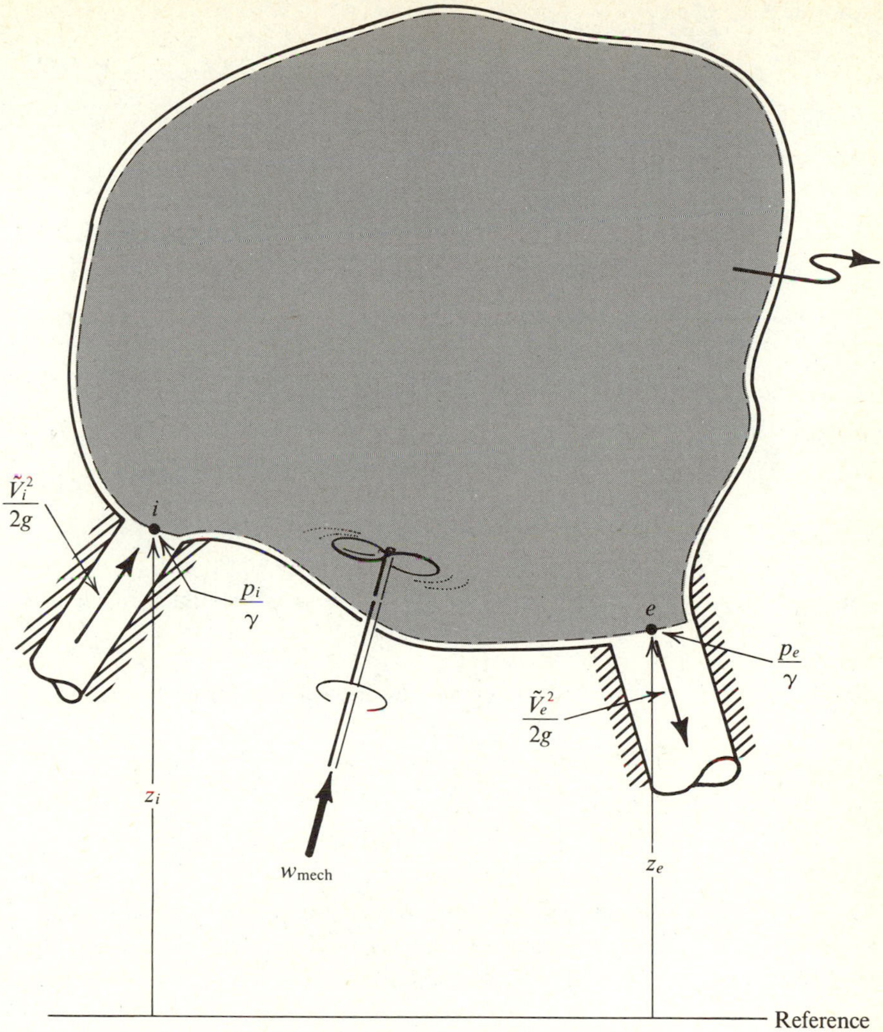

Figure 5.26 *Energy per unit weight transfer for one-dimensional steady flow.*

Example 5.12 *(Con't.)*

thrust of 100 lbf. Assume the average velocity $\tilde{V}_1$ is negligibly small compared to $\tilde{V}_3$, and that the flow through line 1–3 is steady and adiabatic. Calculate (a) the mass M of gas that flows in the 2-second interval, and (b) the heat transfer Q necessary to keep the pressure constant at 40 psia in the tank, given C_p = 3.1 Btu/lbm°R,

$$\Delta h = C_p \Delta T \tag{i}$$

$$\Delta i = C_v \Delta T \tag{ii}$$

Example 5.12 *(Con't.)*

$$\mathbf{F} = \dot{M}\Delta\mathbf{V} \tag{iii}$$

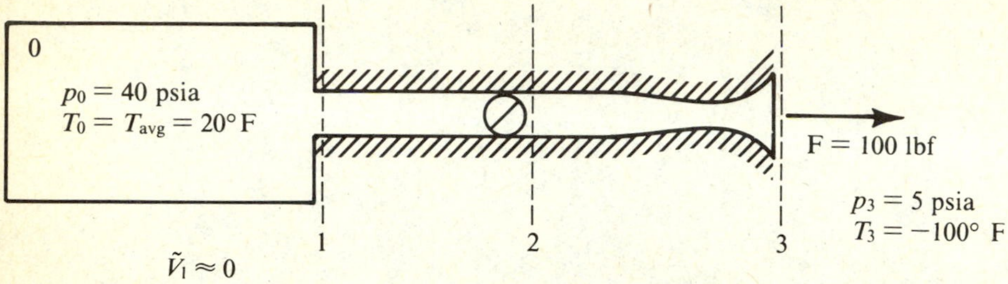

Figure E5.12a

Solution:

Step 1.

The gas is assumed a perfect gas. The flow is steady, one-dimensional, and inviscid from 1 to 3.

Step 2.

Since the fluid is a perfect gas, the enthalpies are evaluated from Eq. (i) as

$$h_0 = C_p T_0 = \left(\frac{3.1 \text{ Btu}}{\text{lbm°R}}\right)(480°\text{R}) = 1488 \text{ Btu/lbm} \tag{iv}$$

$$h_3 = C_p T_3 = \left(\frac{3.1 \text{ Btu}}{\text{lbm°R}}\right)(360°\text{R}) = 1116 \text{ Btu/lbm} \tag{v}$$

(a) Consider the pipe and nozzle as our control volume. Since the flow is adiabatic and no energy is transferred to the fluid by mechanical work, the steady state energy Eq. (5.102) is expressed as

$$\dot{M}_1 h_1 = \dot{M}_3\left(h_3 + \frac{\tilde{V}_3^2}{2}\right) \tag{vi}$$

or

$$\tilde{V}_3 = \sqrt{2 \times 32.2 \times (1488 - 1116) \times 778} = 4317.2 \text{ ft/s} \tag{vii}$$

The specific impulse SI is based on a difference in velocity $\tilde{V}_3 - \tilde{V}_1$ or

$$\text{SI} = \frac{\tilde{V}_3}{g_c} = \frac{4317.2 \text{ ft/s}}{32.2 \text{ ft}\cdot\text{lbm/lbf}\cdot\text{s}^2} = 134.1 \text{ lbf}\cdot\text{s/lbm} \tag{viii}$$

The mass rate $\dot{M}$ is related to thrust by Eq. (iii):

$$\dot{M} = \frac{F}{\text{SI}} = \frac{100 \text{ lbf}}{134.1 \text{ lbf}\cdot\text{s/lbm}} = 0.746 \text{ lbm/s} \tag{ix}$$

Example 5.12 *(Con't.)*

Thus for two seconds of thrust, tne mass of gas leaving the nozzle is

$$M = \dot{M}\Delta t = 0.746 \text{ lbm/s} \times 2 \text{ s} = 1.492 \text{ lbm} \qquad \text{(x)}$$

(b) Consider the tank as the control volume. This is an unsteady flow problem. The energy equation for this case becomes*

$$Q = \Delta I + M_1 h_1 \qquad \text{(xi)}$$

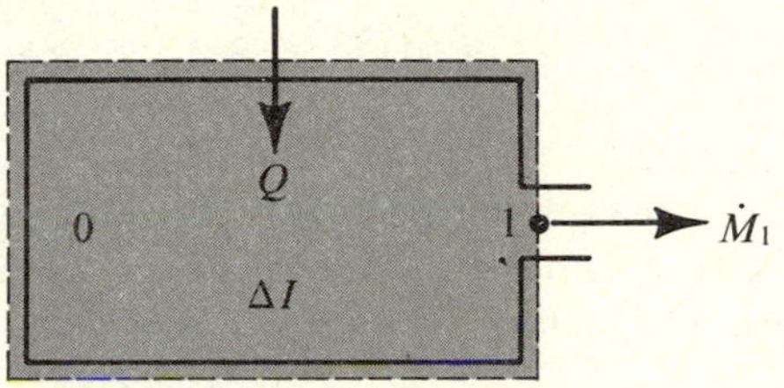

Figure E5.12b

The change in the total internal energy ΔI is expressed in terms of the difference in the final specific internal energy i_f and initial specific energy i_0:

$$Q = M_f i_f - M_0 i_0 + M_1 h_1 \qquad \text{(xii)}$$

where the final mass M_f and initial mass M_0 are expressed in terms of the density ρ, and the internal energies are expressed in terms of temperature T by Eq. (ii):

$$\begin{aligned} Q &= \forall C_v(\rho_f T_f - \rho_0 T_0) + M_1 h_1 \\ &= \frac{\forall C_v}{R}(p_f - p_0) + M_1 h_1 \qquad \text{(xiii)} \end{aligned}$$

But the pressure is constant in the tank during the 2-second flow such that the heat transferred to the tank is simply equal to the exiting enthalpy:

$$\begin{aligned} Q &= M_1 h_1 \\ &= (1.492 \text{ lbm})(1488 \text{ Btu/lbm}) \\ &= \boxed{2220.1 \text{ Btu}} \qquad \text{(xiv)} \end{aligned}$$

Thus 2220.1 Btu of heat must be supplied from the solar panels to the gas in order to maintain a constant pressure of 40 psia to obtain 100 lbf of thrust for a 2-second blast.

This completes the solution.

*The reader should derive Eq. (xi) as an exercise.

5.5.4 The Steady Flow Energy Equation versus Bernoulli's Equation

From our discussion of energy, we have seen that if the fluid is inviscid, there is no work transfer because there is no shear stress:

$$W_{s_\tau} = 0 \tag{5.107}$$

If in addition the fluid is incompressible, then the work due to compression is zero, leaving only one possible type of work transfer; displacement work:

$$W_d = M\left(\frac{1}{\rho}\int dp + \int \tilde{V}\,d\tilde{V} + g\int dz\right)$$

This type of work transfer is independent of the internal energy of the fluid. If the fluid flow takes place within an adiabatic control surface, the internal energy does not change. We then write the steady flow energy equation for no mechanical work as

$$\left(\frac{p}{\gamma} + \frac{\tilde{V}^2}{2g} + z\right)_1 = \left(\frac{p}{\gamma} + \frac{\tilde{V}^2}{2g} + z\right)_2 \tag{5.108}$$

which is exactly the same as Bernoulli's Eq. (4.126).

Bernoulli's equation does *not* account for energy transfers which take place when there is fluid shear stresses or compressibility. The effect of either a shear stress or compressibility is to change the internal energy of the fluid, which in turn usually results in a change in temperature. If the control volume process were isothermal, all of this energy would be transferred in the form of heat from the control volume. Thus, there could be a pressure drop in, say, a constant diameter conduit flow due to real fluid flow. The pressure at the outlet would be *lower* than that predicted by Bernoulli's equation, and the difference between that value predicted by Bernoulli's equation and the actual has been given the name "losses."

These losses are predicted by the energy equation and *not* by Bernoulli's equation. Thus, we cannot speak of Bernoulli's equation in the same context as the energy equation even though both can have identical form.

Section 5.5 has presented the energy balance for real fluid flow. Consider the one-dimensional form. The energy Eq. (5.103) can be approximated by letting the loss of energy h_f be proportional to the average kinetic energy at the efflux station for an isothermal and adiabatic process:

$$h_f \propto \frac{V_e^2}{2g} = K\frac{\tilde{V}_e^2}{2g} \tag{5.109}$$

where K is a constant of proportionality popularly called a *loss coefficient*. (This type of approximation is a favorite trick of some hydraulic and civil engineers.) Expressions for and discussion of the loss coefficient K will be presented in Chap. 12.

Substituting Eq. (5.109) into Eq. (5.103) (where the properties are assumed to be uniform normal to the flow direction), we obtain

$$w_{\text{mech}} = \left(\frac{p}{\gamma} + \frac{\tilde{V}^2}{2g} + z\right)_e - \left(\frac{p}{\gamma} + \frac{\tilde{V}^2}{2g} + z\right)_i + K\frac{\tilde{V}_e^2}{2g} \tag{5.110}$$

The above equation is referred to as the mechanical energy balance for isothermal conduit flow. The terms can be rearranged to give for no mechanical work

$$\frac{p}{\rho} + (1 + K)\frac{\tilde{V}^2}{2} + gz = \text{const.} \tag{5.111}$$

Examining the above energy equation, we see that it is identical to Bernoulli's equation if $K = 0$. But remember that the Bernoulli equation stems from the Gromeka-Lamb equation for steady incompressible inviscid flow, a linear momentum concept, whereas Eq. (5.111) stems from an energy concept. It too is for steady incompressible flow, but not necessarily inviscid; that there is no mechanical power and the net internal energy per unit time $\oint \rho \bar{i} \mathbf{V} \cdot d\mathbf{A}$ is equal and opposite to the viscous power $\dot{W}_{S_T}$.

5.5.5 Energy Grade Lines

It is often convenient to plot the various energy terms in the energy equation. Hydraulic engineers call the quantity

$$\frac{p}{\gamma} + (1 + K)\frac{\tilde{V}^2}{2g} + z$$

the *total head*. The quantity $p/\gamma + z$ is called the *piezometric head* or the *hydraulic grade line*. The quantity $\tilde{V}^2/2g$ is called the *velocity head* and the quantity $K(\tilde{V}^2/2g)$ is called the *friction head*. Figure 5.27 is a typical flow with energy losses. Using Eq. (5.110) and Fig. 5.27, the constant is evaluated as z_1 and represents the total upstream energy per weight of fluid, e. Applying the energy equation we obtain

$$\frac{p_2}{\gamma} + z_2 = z_1 = e_1$$

since the total energy at station 2 equals that at station 1. Applying Eq. (5.110) to station 3, we obtain

$$\frac{p_3}{\gamma} + \frac{\tilde{V}_3^2}{2g} + z_3 + K\frac{\tilde{V}_3^2}{2g} = z_1 \tag{5.112}$$

where a loss of energy $K(\tilde{V}_3^2/2g)$ is due to the pipe inlet. Applying Eq. (5.110) to station 4, we must add the energy w_{mech} to obtain

$$\frac{p_4}{\gamma} + \frac{\tilde{V}_4^2}{2g} + z_4 + w_{\text{mech}} = e_4 \tag{5.113}$$

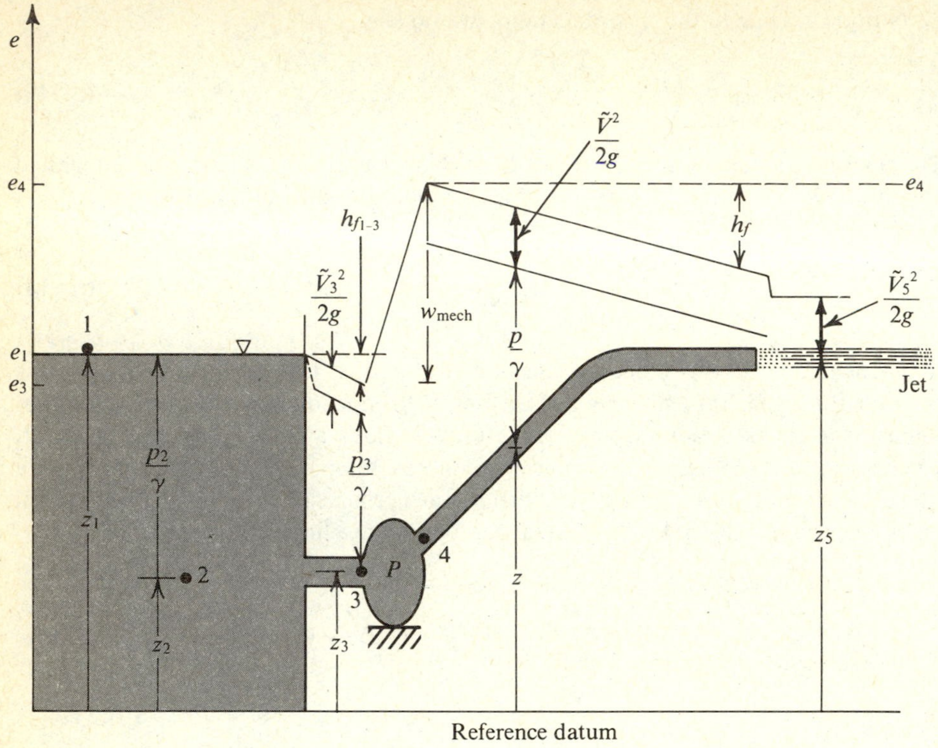

(I. F.) Energy Equations

$$e_1 = z_1 = e_2 = z_2 + \frac{p_2}{\gamma}$$

$$e_1 = e_3 + h_{f1\text{-}3} = z_3 + \frac{p_3}{\gamma} + \frac{\tilde{V}_3^2}{2g} + h_{f1\text{-}3}$$

$$e_1 = w_{mech} = e_4 = e - h_f = z_5 + \frac{\tilde{V}_5^2}{2g} + h_{f4\text{-}5}$$

Figure 5.27 *Geometric interpretation of energy levels.*

where from the conservation of energy

$$e_4 = e_3 + w_{mech} \tag{5.114}$$

is the energy per unit weight of fluid at station 4. Again, the total energy drops off as the fluid moves from station 4 to the nozzle owing to fluid shear stresses and eddy motion. At station 5 the pressure drops to zero gauge pressure and the velocity increases. The total specific energy at 5 is simply

$$e_5 = \frac{\tilde{V}_5^2}{2g} + z_5 \tag{5.115}$$

Both the energy and Bernoulli's equations enable us to learn how the energy components alter. Use of grade lines allows us to obtain a geometric interpretation of the various energy components, but has no practical value as a mathematical technique in solving a problem.

Example 5.13

For the flow pictured in Fig. E5.13, determine the energy loss per pound of fluid $K\tilde{V}^2/2g$.

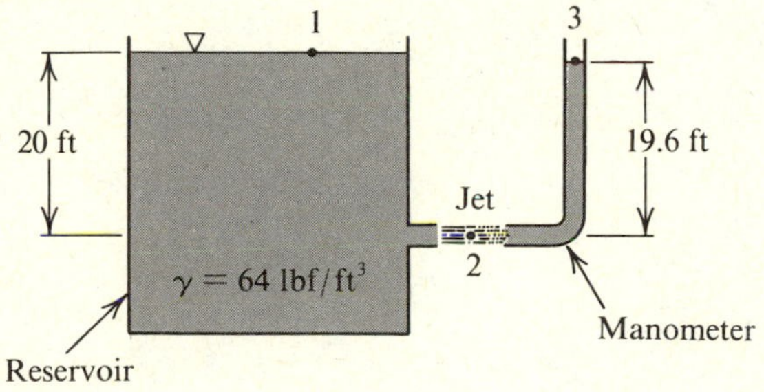

Figure E5.13

Solution:

Step 1.

The fluid is incompressible and viscous. The flow is steady and one-dimensional.

Step 2.

The appropriate equation governing the flow is therefore the energy Eq. (5.110):

$$\frac{K\tilde{V}^2}{2g} = \left(\frac{p}{\gamma} + \frac{\tilde{V}^2}{2g} + z\right)_1 - \left(\frac{p}{\gamma} + \frac{\tilde{V}^2}{2g} + z\right)_2 \qquad \text{(i)}$$

At the free-surface station, $p_1 = 0$, $\tilde{V}_1 = 0$, and $z_1 = 20$ ft. If station 2 is selected, the velocity $\tilde{V}_2$ is unknown. If station 3 is selected, then enough information is known to solve the problem: i.e., $p_3 = 0$, $\tilde{V}_3 = 0$, and $z_3 = 19.6$ ft. Substituting these values into Eq. (i) gives

$$h_f = \frac{K\tilde{V}^2}{2g} = 20 - 19.6 \text{ (ft)} \qquad \text{(ii)}$$

or energy loss per pound of fluid out of the orifice is

$$h_f = 0.4 \text{ (ft)} \qquad \text{(iii)}$$

This completes the solution.

Example 5.14
Water flows in a constant diameter pipe between sections A and B. The pressure head at A is 45 m greater than at B. In what direction is the flow? What is the head loss $K\tilde{V}^2/2g$?

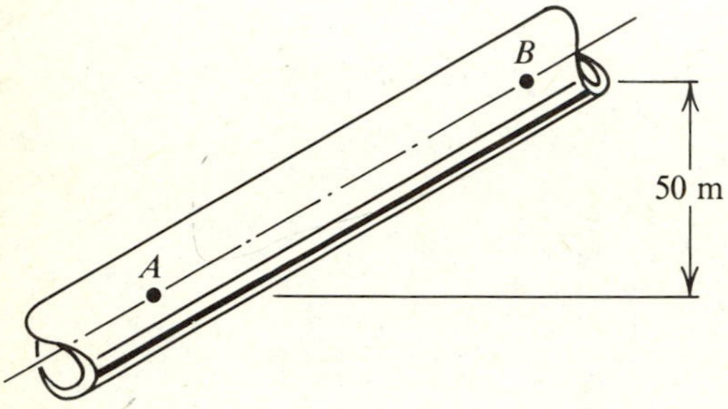

Figure E5.14

Solution:
Step 1.
The fluid is incompressible and viscous. The flow is steady and one-dimensional.
Step 2.
The appropriate equation governing the flow is therefore the energy Eq. (5.110).

Assume the flow goes in the direction B to A. Applying the energy equation yields

$$\left(\frac{p}{\gamma} + z\right)_B = \left(\frac{p}{\gamma} + z\right)_A + K\frac{\tilde{V}^2}{2g} \tag{i}$$

Using the above values for elevation and pressure difference gives

$$\begin{aligned} K\frac{\tilde{V}^2}{2g} &= (z_B - z_A) - \left(\frac{p_A}{\gamma} - \frac{p_B}{\gamma}\right) \\ &= 50 - 45 = 5 \text{ m} \end{aligned} \tag{ii}$$

Since $K(\tilde{V}^2/2g)$ is positive, the flow does indeed move from B to A.

This completes the solution.

Example 5.15
The head extracted by turbine CR in Fig. E5.15 is 200 ft and the pressure at T is 72.7 psi. Losses of $2(\tilde{V}^2/2g)$ through the 24-in. diameter pipe exist between W and R and losses of $3(\tilde{V}^2/2g)$ through the 12-in. diameter pipe exist between

Example 5.15 *(Con't.)*

C and T. Calculate (a) how much water is flowing? (ft³/s) and (b) the pressure head at R (ft). Assume that the fluid is real.

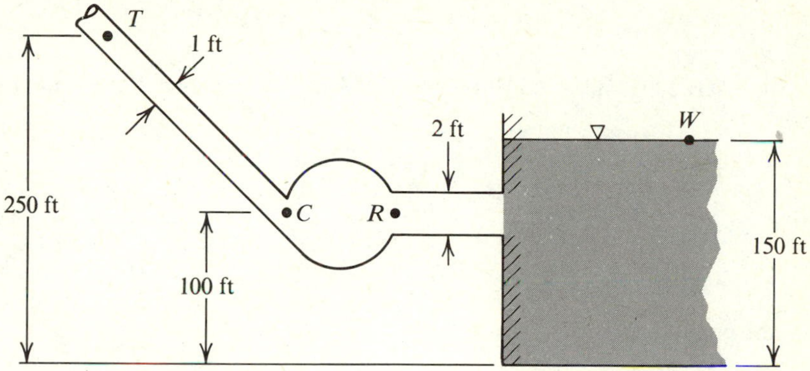

Figure E5.15

Solution:

Step 1.

The fluid is real. The flow is steady and one-dimensional. Assume that the flow goes from T to W.

Step 2.

The appropriate equations governing the flow are the continuity Eq. (5.14) and the energy Eq. (5.110).

(a) Write the energy equation between stations T and W:

$$\frac{\tilde{V}_T^2}{2g} + \frac{72.7 \times 144}{62.4} + 250 - 200 = 150 + h_f \qquad \text{(i)}$$

where

$$h_f = \frac{2\tilde{V}_R^2}{2g} + \frac{3\tilde{V}_T^2}{2g} \qquad \text{(ii)}$$

The continuity equation gives $(\tilde{V}A)_T = (\tilde{V}A)_R$ or

$$\tilde{V}_R = \frac{\tilde{V}_T}{4} \qquad \text{(iii)}$$

Therefore

$$\frac{\tilde{V}_T^2}{2g} - \frac{3\tilde{V}_T^2}{2g} - \frac{2}{16}\frac{\tilde{V}_T^2}{2g} = 150 + 200 - 250 - 168$$

or

Example 5.15 *(Con't.)*

$$\tilde{V}_T = 45.4 \text{ ft/s} \tag{iv}$$

Thus the volume rate of flow

$$Q = V_T A_T = 45.4 \times \frac{\pi \times 1}{4} = 35.6 \text{ ft}^3\text{/s} \tag{v}$$

(b) Write the energy equation between R and W and solve for the pressure head:

$$\begin{aligned}\frac{p_R}{\gamma} &= (z_w - z_R) + \frac{2V_{24\text{ in.}}^2}{2g} - \frac{V_R^2}{2g} \\ &= 50 + \frac{2 \times 11.32 \times 11.32}{2 \times 32.2} - \frac{(11.32)^2}{64.4}\end{aligned} \tag{vi}$$

Thus the pressure head is

$$\frac{p_R}{\gamma} = 51.99 \text{ ft} \tag{vii}$$

This completes the solution.

Example 5.16

Find the energy loss for steady one-dimensional flow of an incompressible liquid flowing through the geometry shown in Fig. E5.16.

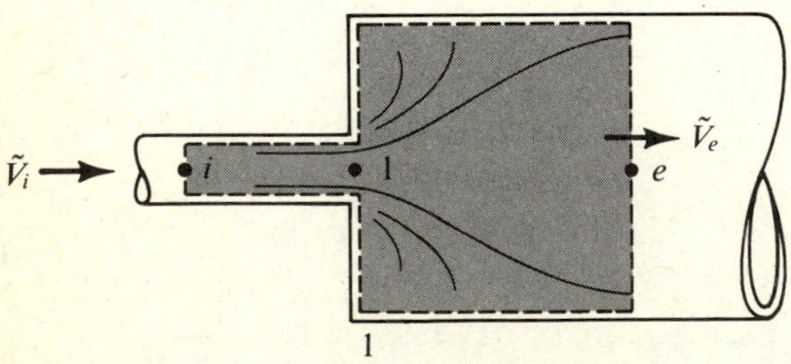

Figure E5.16

Example 5.16 *(Con't.)*

Solution:
For steady one-dimensional flow of an incompressible fluid, the energy Eq. (5.110) becomes

$$\frac{1}{2}(\tilde{V}_i^2 - \tilde{V}_e^2) + \frac{(p_i - p_e)}{\rho} = gh_f \quad \text{(i)}$$

Using Fig. E5.16, we assume that

$$\tilde{V}_i = \tilde{V}_1, \quad p_i = p_1 \quad \text{(ii)}$$

The momentum integral equation is applied to the dotted control volume shown in the figure to produce

$$p_1A_1 - p_eA_e = \rho Q(\tilde{V}_e - \tilde{V}_1) \quad \text{(iii)}$$

Using the equivalent for the mass flow rate, we modify Eq. (iii) to read

$$\tilde{V}_e(\tilde{V}_1 - \tilde{V}_e) + \frac{(p_1 - p_e)}{\rho} = 0 \quad \text{(iv)}$$

Comparison of Eqs. (i), (ii) with (iv) results in

$$h_f = \frac{1}{2g}(\tilde{V}_i - \tilde{V}_e)^2 \quad \text{(v)}$$

and represents the loss of energy due to sudden expansion. This will be an important minor loss expression in pipe flow. See Sec. 10.5.4.

This completes the solution.

Example 5.17
Consider the flow of an ideal liquid of specific weight γ through the circular tube shown in Fig. E5.17a. Calculate

(a) The difference in pressure between points 1 and 2 as a function of flow rate Q, geometry, and density.
(b) The difference in pressure between points 1 and 2 as a function of a manometer fluid's specific weight γ_m, and elevation h.
(c) The velocity of the jet V_j in terms of conditions at 1.
(d) The y-component of the restraining force R_y in terms of velocities $\tilde{V}_1$ and $\tilde{V}_j$, diameters D_1 and D_j, and the density of the fluid ρ.

Assume that the velocities and pressures are uniform over the areas A_1 and A_2.

Example 5.17 *(Con't.)*

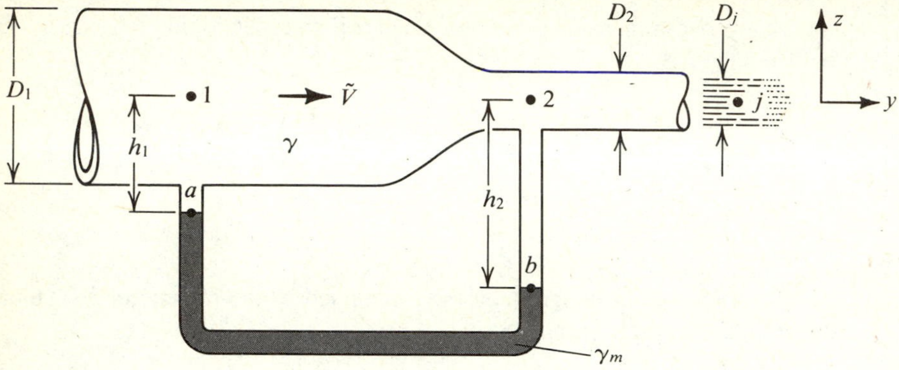

Figure E5.17a

Solution:

Step 1.

The fluid is ideal. The flow is steady and one-dimensional.

Step 2.

The appropriate equations governing this flow are the continuity Eq. (5.14), the linear momentum equation (5.29), and Bernoulli's equation (4.126).

(a) Applying Bernoulli's equation between stations 1 and 2 gives

$$\left(\frac{p}{\gamma} + \frac{\tilde{V}^2}{2g} + z\right)_1 = \left(\frac{p}{\gamma} + \frac{\tilde{V}^2}{2g} + z\right)_2 \tag{i}$$

Since stations 1 and 2 are at the same elevation, $z_1 = z_2$ the velocity $\tilde{V}$ is expressed in terms of flow rate Q and geometry by the continuity equation

$$\tilde{V} = \frac{Q}{A} = \frac{4Q}{\pi D^2} \tag{ii}$$

Substituting Eq. (ii) into Eq. (i) and rearranging terms yields the pressure difference

$$p_1 - p_2 = 8\frac{\rho Q^2}{\pi^2}\left(\frac{1}{D_2^4} - \frac{1}{D_1^4}\right) \tag{iii}$$

expressed in terms of flow rate, density, and geometry.

(b) To calculate the pressure difference between points 1 and 2 as a function of manometer fluid's specific weight γ_m and elevation h, we consider points 1 and a and apply the hydrostatic equation

$$p_a = p_1 + \gamma h_1 \tag{iv}$$

Between a and b we obtain

$$p_b = p_a + \gamma_m(h_2 - h_1) \tag{v}$$

Example 5.17 *(Con't.)*

Between b and 2 we obtain

$$p_b = p_2 + \gamma h_2 \tag{vi}$$

Combining results of Eqs. (iv)–(vi) produces

$$p_1 - p_2 = (\gamma - \gamma_m)(h_2 - h_1) \tag{vii}$$

Note from part (a) that $D_1 > D_2$ so that $p_1 > p_2$ as given by Eq. (iii). From Fig. E5.17a, $h_2 > h_1$, so that from Eq. (vii) we know something about the fluids, i.e., $\gamma > \gamma_m$.

(c) Applying Bernoulli's equation between stations 1 and j gives

$$\frac{p_1}{\gamma} + \frac{\tilde{V}_1^2}{2g} = \frac{\tilde{V}_j^2}{2g} \tag{viii}$$

since the pressure in a jet is atmospheric. Solving for the velocity of the jet, we obtain

$$\tilde{V}_j = \sqrt{\tilde{V}_1^2 + \frac{2p_1}{\rho}} \tag{ix}$$

(d) The restraining force on the system is evaluated using the control volume shown in Fig. E5.17b.

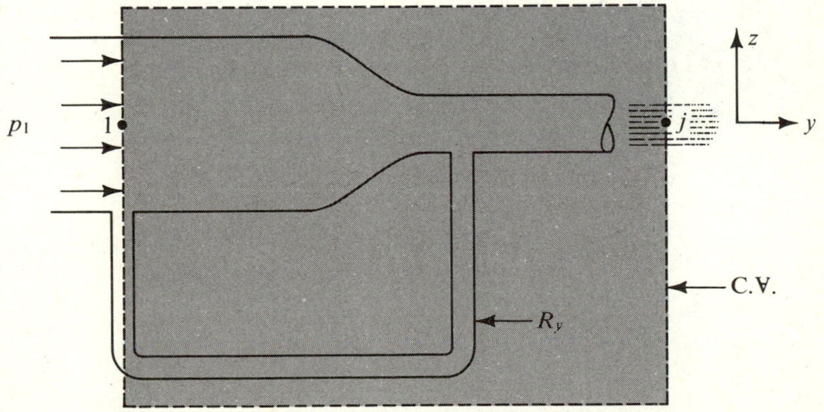

Figure E5.17b

Applying the integral form of the linear momentum equation in the y-direction yields

$$\frac{\pi}{4} p_1 D_1^2 - R_y = \rho Q(V_j - V_1) \tag{x}$$

Example 5.17 *(Con't.)*

Substituting Eqs. (viii) and (ii) into Eq. (x) yields

$$R_y = \frac{\pi\rho}{4}\left[\frac{\tilde{V}_1^2 D_1^2}{2} + \tilde{V}_j^2\left(\frac{D_1^2}{2} - D_j^2\right)\right] \qquad \text{(xi)}$$

This completes the solution.

Thus far all the examples we have been treating in this section have been largely for liquids. Though all the previous examples can be adjusted for gases at high velocity to accommodate compressibility effects (with a subsequent alteration of the appropriate equations in step 2 of the solution), it is worthwhile to consider the following example to show the similarity of results be the fluid gas or liquid.

Example 5.18

Consider air flowing through an adiabatic convergent nozzle. At the inlet, the velocity is 20 m/s at a temperature of 27°C. At the exit, the velocity is 100 m/s at a temperature of 127°C. Assuming no change in potential energy, calculate the change in internal energy and enthalpy of the air.

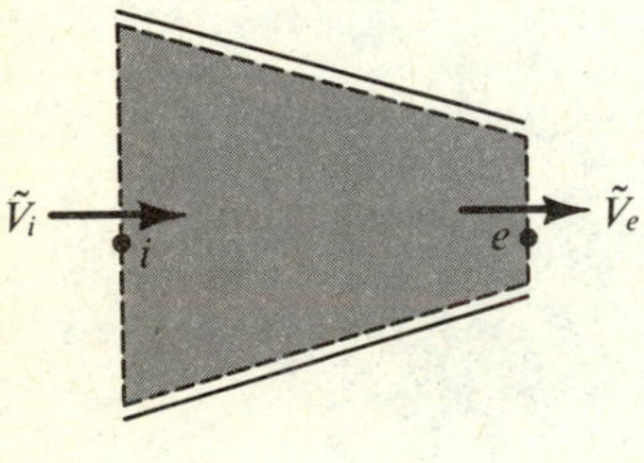

Figure E5.18

Solution:

There is no transfer of energy so the energy equation becomes

$$h_i + \frac{\tilde{V}_i^2}{2} = h_e + \frac{\tilde{V}_e^2}{2} \qquad \text{(i)}$$

The change of enthalpy $\Delta h = h_i - h_e$ is

$$\Delta h = \frac{1}{2}(\tilde{V}_e^2 - \tilde{V}_i^2) = \frac{100^2 - 20^2}{2} = 4800 \text{ m}^2/\text{s}^2$$
$$= 4.8 \text{ J/g} \qquad \text{(ii)}$$

and represents the increase in enthalpy of the fluid. The enthalpy is related to internal energy by

Example 5.18 *(Con't.)*

$$h = \bar{i} + \frac{p}{\rho} \qquad \text{(iii)}$$

Therefore

$$\Delta\bar{i} = \Delta h - \Delta\left(\frac{p}{\rho}\right) = \Delta h - R\Delta T$$

$$= 4.8 - 0.2867(27 - 127)$$

$$= 33.47 \text{ kJ/kg}$$

which shows a significant increase in internal energy of the air.

This completes the solution.

Study Questions

5.1 What is the integral form of the continuity equation for compressible flow with two sources and one sink?

5.2 What is the integral form of the continuity equation with an elastic control volume?

5.3 What is the physical significance of

(a) $\rho\mathbf{V}$

(b) $\rho\mathbf{V}\, d\forall$

(c) $\dfrac{D}{Dt}\displaystyle\int_{\forall} \rho\mathbf{V}\, d\forall$

(d) $\dfrac{\partial}{\partial t}\displaystyle\int_{\forall} \rho\mathbf{V}\, d\forall$

(e) $\rho\mathbf{V}(\mathbf{V}\cdot d\mathbf{A})$

(f) $\displaystyle\int \rho\mathbf{V}(\mathbf{V}\cdot d\mathbf{A})$

5.4 Give the physical significance of

(a) $M\mathbf{a}_{O'}$

(b) $\boldsymbol{\alpha}_{(m)} \times \mathbf{r}_{P/O'}$

(c) $\boldsymbol{\omega}_{(m)} \times (\boldsymbol{\omega}_{(m)} \times \mathbf{r}_{P/O'})$

(d) $\boldsymbol{\omega} \times \mathbf{v}_{P/O'}$

(e) $\dfrac{D\mathbf{V}_{P/O'}}{Dt}$

5.5 Explain the difference between v_θ, V_θ, and V_{θ_r}.

5.6 What is work, and how does one measure it? Demonstrate that in moving a frictionless object from point *A* to point *B* the shortest distance is less than moving it along a circuitous route. Justify the results mathematically.

5.7 How can the temperature of a fluid be increased without heat transfer? Give an example. How can the temperature remain constant when heat transfer takes place?

5.8 Give a physical meaning for the following terms in the energy and Bernoulli equations

$$\frac{\bar{V}_2^2 - \bar{V}_1^2}{2g}, \quad \frac{p_2 - p_1}{\gamma}, \quad (z_2 - z_1), \quad h_f, \quad \dot{H}_f, \quad w_{\text{mech}}$$

as energy terms; as dynamic terms.

5.9 Show that power is given by

$$\gamma Q w_{\text{mech}}, \quad Q\Delta p, \quad \frac{\rho Q \Delta \tilde{V}^2}{2}, \quad \dot{M}\Delta z$$

What is the physical meaning of each of the above expressions?

5.10 What is the mathematical and physical difference between Bernoulli's equation and the energy equation?

Problems

Mass Flow

5.1 Water is moving through a nozzle at a volume rate of flow of 3 m^3/s. A pinhole leak in the nozzle exists. If the velocity at a downstream area of 8 cm^2 is 12 m/s, how much fluid is lost every 10 seconds?

5.2 Calculate the average velocity of a real fluid flowing through a pipe of area 3 m^2 if the volume rate of flow of the water is 6 m^3/s.

5.3 An isentropic gas moves through a horizontal pipe of constant cross-sectional area. If the inlet pressure is 12 psi with a velocity of 70 ft/s, and the exit pressure is 1 psi, calculate the exit velocity of gas, given the ratio of specific heats $k = 1.65$.

5.4 Gasoline (specific gravity 0.7) is being dumped into a pink tank at a rate of 720,000 lbf/hr through a 4-in. diameter nozzle. The temperature of gasoline is 95°F. Barometric pressure is 28.5 in. Hg. What is the force on the bottom of the tank? (See Fig. P5.4.)

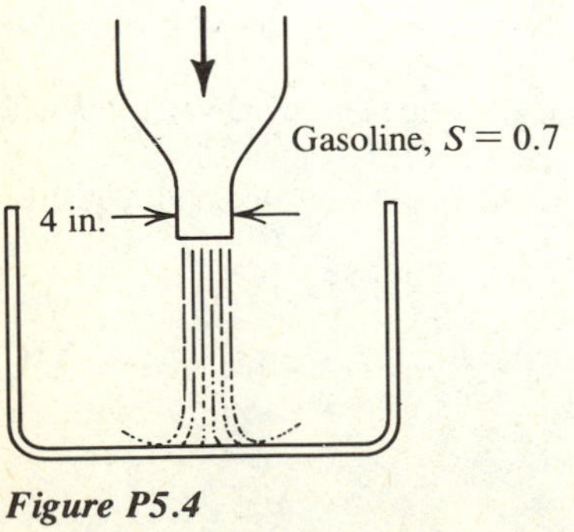

Figure P5.4

5.5 Air at 60 psia and 140°F flows through a 12-in. diameter pipe at 0.5 slug/s. What is the average velocity of the air?

5.6 Your oceanographic ship lands on an exotic Pacific Island which contains a stream 24 in. deep. The stream makes a bend formed by concentric arcs 10 ft and 20 ft. Wading to the center of the stream you measure the velocity there to be 8 ft/s. Wading ashore you assume two-dimensional frictionless flow and calculate the volume flow rate to be what?

5.7 Water is forced into the device shown below at the rate of 5 ft^3/s through pipe A, while oil of specific gravity 0.8 is forced in at the rate of 1 ft^3/s through pipe B. If both liquids are incompressible and form a homogeneous mixture of oil and water leaving pipe C, find (a) the average velocity at C, and (b) the average density. If the piston at D is 12 in. in diameter and moves at 1 ft/s to the left, find (c) the average velocity leaving pipe C. (See Fig. P5.7.)

Conservation of Mass

5.8 Consider water at 68°F flowing through the divergent tunnel. If the inlet area A_i is 2 ft in diameter, calculate the exit area A_e if (a) $\tilde{V}_i$ = 25 ft/s, (b) $u_{\max}$ = 25 ft/s. (See Fig. P5.8.)

5.9 Obtain the differential equation of compressible mass flow

$$\frac{dA}{A} + \frac{d\tilde{V}}{\tilde{V}} + \frac{d\rho}{\rho} = 0$$

What assumptions are necessary in deriving it?

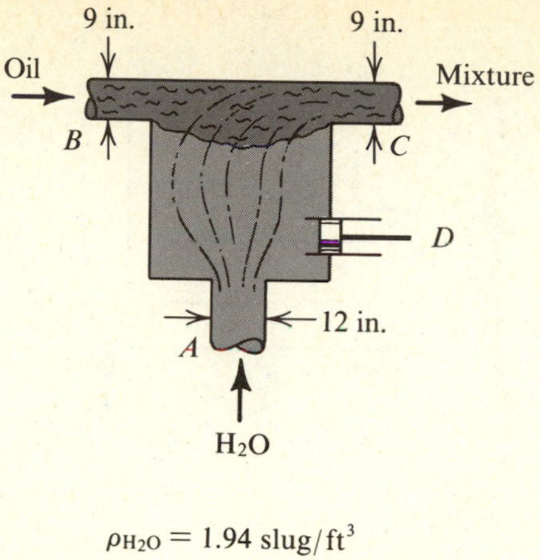

Figure P5.7

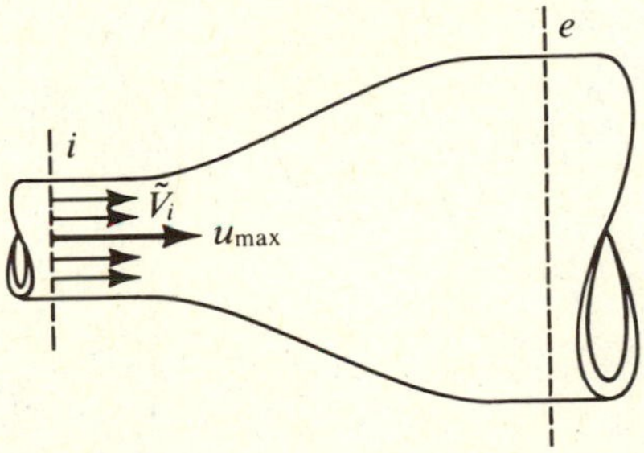

Figure P5.8

5.10 Calculate the average velocity $\bar{V}$ given the velocity distribution $u/U = (y/h)^{1/3}$ for the flow shown in Fig. P5.10.

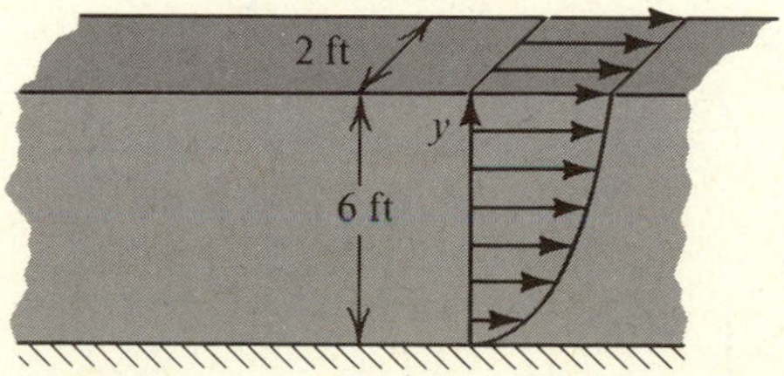

Figure P5.10

5.11 A rocket sled weighs 3 tons (6000 lbf) including 1 ton of fuel and rests on a level section of ground. At time $t = 0$, the solid fuel of the rocket is ignited and burns at the rate of 150 lbm/s. The exit velocity of the exhaust gas relative to the rocket is 3500 ft/s. Neglecting friction and air resistance, what is the velocity of the sled at the instant at which all the fuel is burned? Be sure to state all assumptions clearly.

5.12 Water flows steadily through the nozzle shown in Fig. P5.12 with a volume rate $Q = 3\ \text{m}^3/\text{s}$. Determine (a) the velocity at the exit, and (b) the acceleration at the position x.

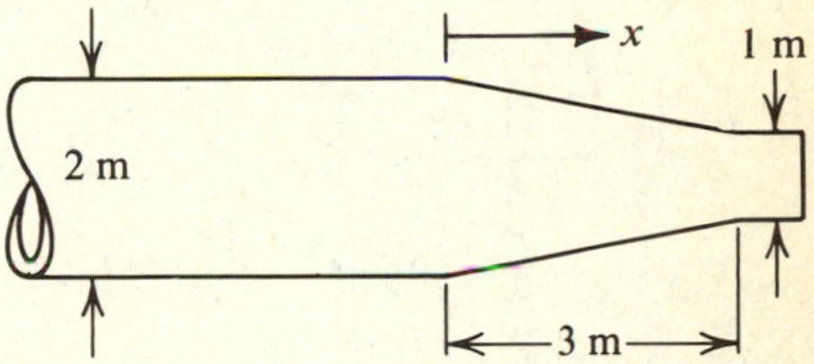

Figure P5.12

Momentum

5.13 Kerosene (specific gravity 0.76) is being dumped into a large tank where it hits a splash plate at a 45° angle. Kerosene is being dumped at a rate of 1000 gal/min through a 3 in. diameter nozzle. What is the total force on the plate? (See Fig. P5.13.)

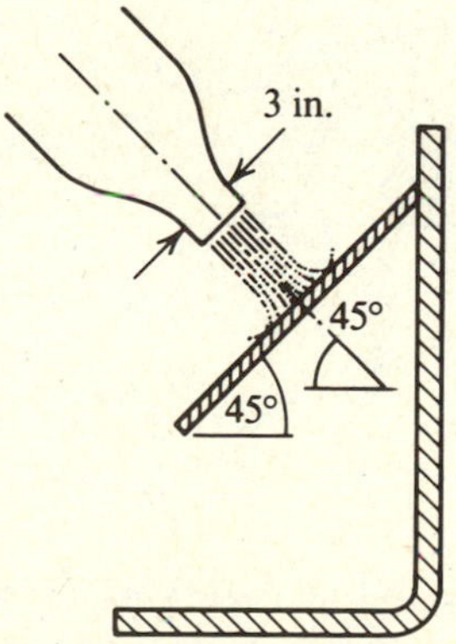

Figure P5.13

5.14 Consider a water jet with exit velocity 100 ft/s, exit area 0.05 ft² impinging on the blade of a turbine, $\theta = 120°$. A set of reduction gears are used to drive a pulley with a weight attached. What is the maximum weight that can be lifted with this apparatus? (See Fig.

P5.14.) Assume the force acts on the center of the blades, which are 2 ft from the center shaft. (Assume massless system—zero inertia or friction.)

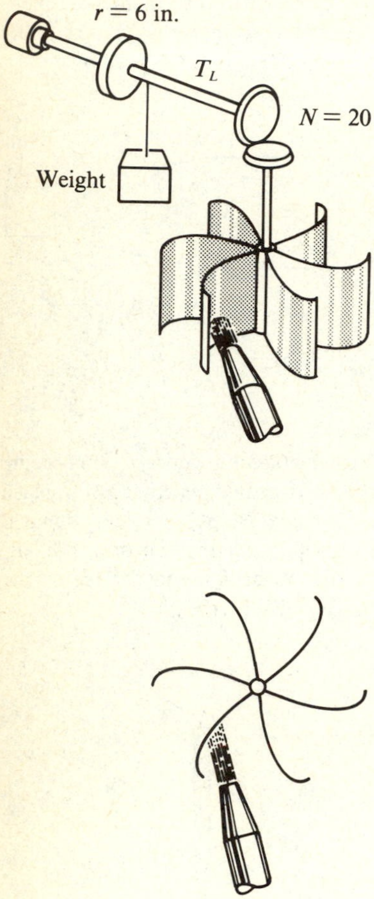

Figure P5.14

5.15 Water flows steadily through the constant area pipe shown below at a velocity of 8 ft/s. The pipe makes a 90° bend in the horizontal plane. The pressure at the entrance 1 is 14 psi. Calculate the force in the x-direction on the bend, if the fluid is ideal (Fig. P5.15).

5.16 Sugarcane syrup (specific gravity 1.35) flows through a pipe into the beginning of a large manifold, as shown in Fig. P5.16. Since the pressure rise between these two sections is a direct measure of the mass flow, a pressure

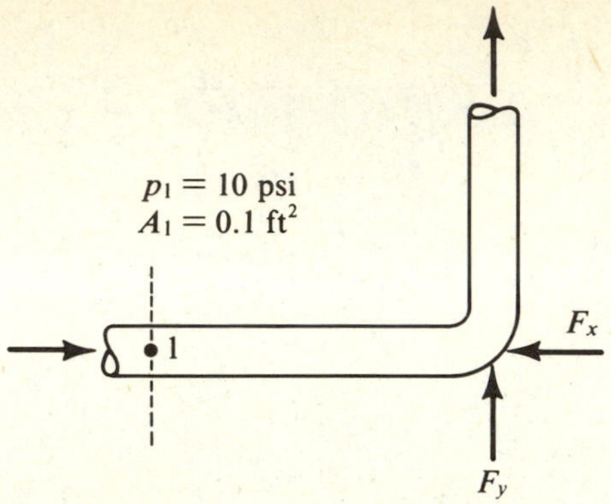

Figure P5.15

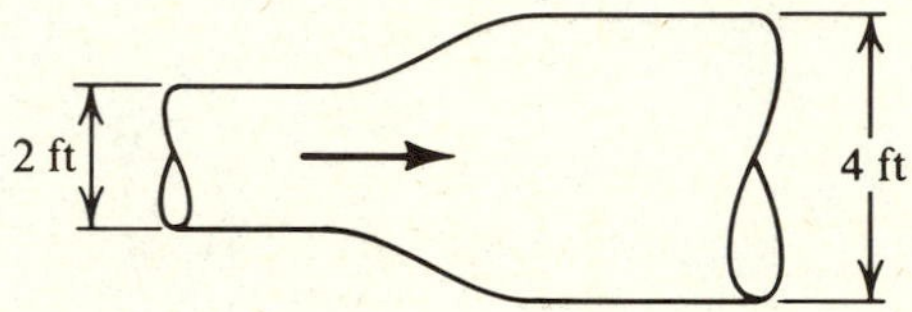

Figure P5.16

rise of 1 psi is measured. Is this sufficient to indicate a mass flow of 4000 lbm/s?

5.17 A water tunnel, as shown in Fig. P5.17, is used to test propellers. The area ratio is 8, the flow rate has been measured to be 750 lbm/s, the pressure gauge at station 1 reads 7 psi, and the inlet velocity is 3 ft/s. Find (a) the pressure at the throat, and (b) the throat area.

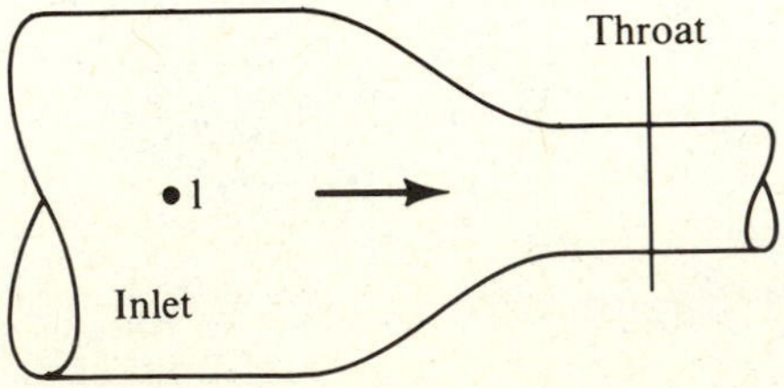

Figure P5.17

5.18 The vertical reducing section shown in Fig. P5.18 weighs 200 lbf. It contains oil (S = 0.86) flowing *upward* at 15 ft^3/s. The pressure in the larger section is 20 psia. Neglecting losses, determine the force on the contraction due to the flow. Disregard the weight of the fluid in the contraction in your calculations.

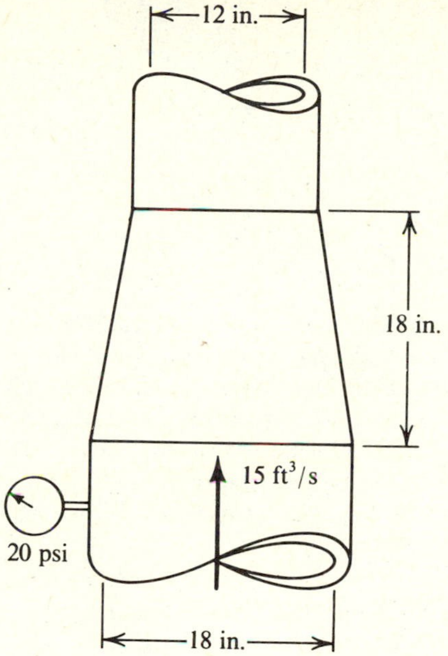

Figure P5.18

5.19 A nozzle is to discharge fresh water to the open atmosphere. Considering frictionless flow, if a pressure p_1 is at the nozzle inlet of 10 lbf/in.^{2} and a velocity V_1 is at the nozzle inlet of 30 ft/s, what is the discharge velocity V_2 from the nozzle (Fig. P5.19)?

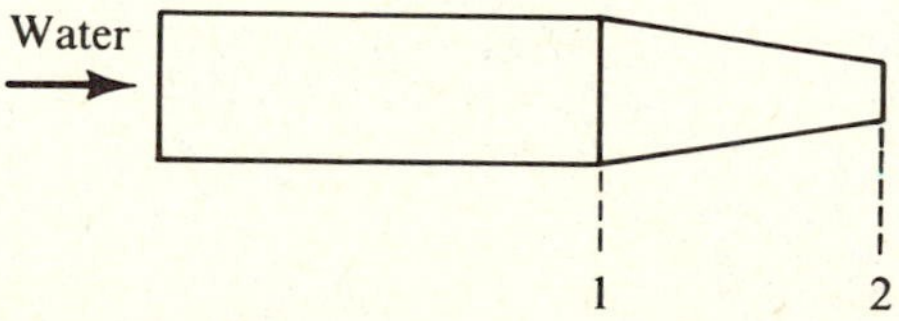

Figure P5.19

5.20 Fresh water is flowing through a converging nozzle section under the following conditions: $p_1A_1 = 250$ lbf, $p_2A_2 = 10$ lbf, $V_1 =$ 10 ft/s, $V_2 = 110$ ft/s, $Q = 1.0$ ft^3/s. Calculate the force of the water on the nozzle (Fig. P5.20).

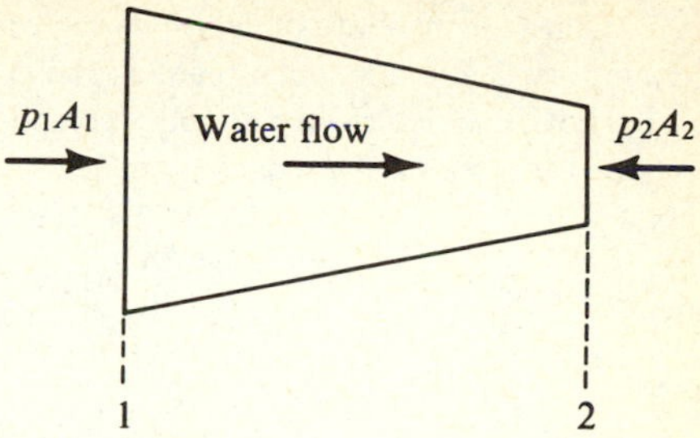

Figure P5.20

5.21 The rate of change of momentum between sections 1 and 2 in Fig. P5.21 for frictionless steady flow of water through a pipeline of expanding cross-sectional area shown below is -29 lbf. (a) What is the pressure at section 2? (b) What is the magnitude of the force F_w of the water on the expanding section? (c) What direction is the force F_w of the water on the expanding section?

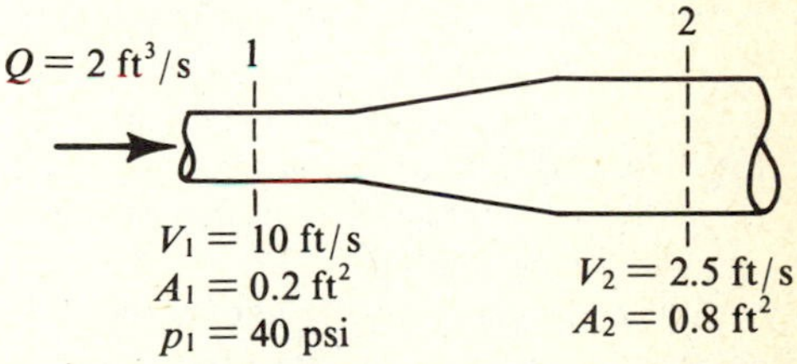

Figure P5.21

5.22 As shown in Fig. P5.22, water enters the pipe elbow at a velocity of 20 ft/s. The elbow lies in the horizontal plane where there is a negligible friction loss between A and B. Calculate the x-component of the restraining force.

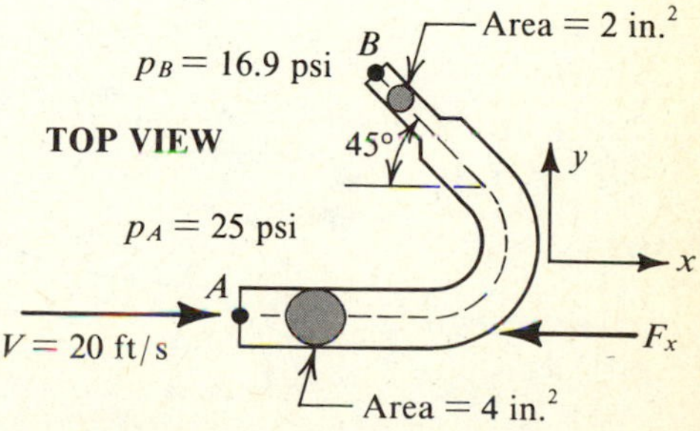

Figure P5.22

5.23 Consider the flow of water through a reducing bend in a horizontal pipe, as shown in Fig. P5.23. Calculate the x- and y-components of the force on the reducing bend.

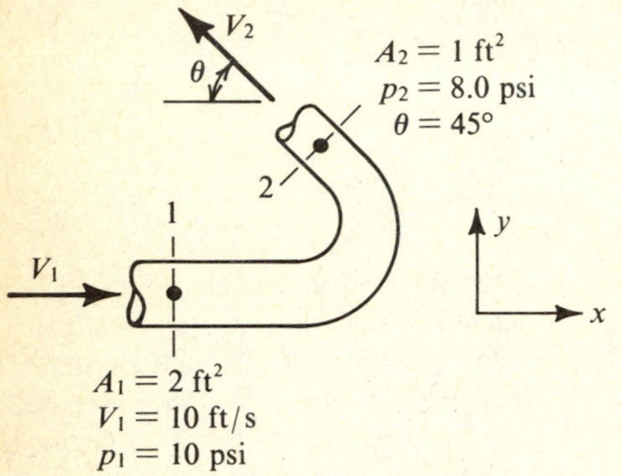

Figure P5.23

5.24 Water flows from a horizontal pipe into a 90° elbow as shown below and exits at atmospheric pressure (Fig. P5.24). Neglecting friction and all losses, calculate (a) the velocity at the exit, (b) the horizontal component of the force that the fluid exerts on the elbow, and (c) the vertical component of the force that the elbow exerts on the fluid.

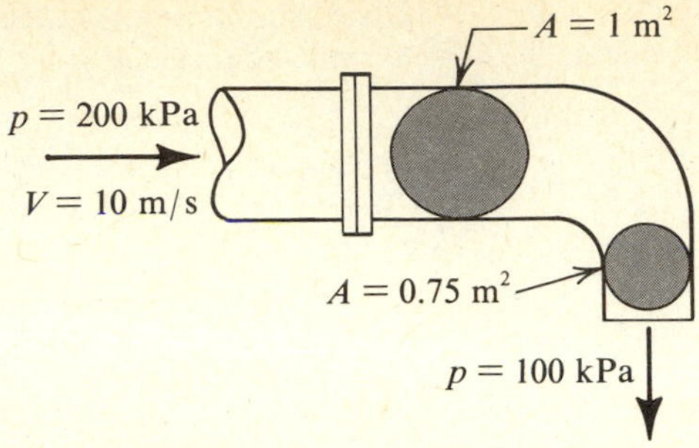

Figure P5.24

5.25 An incompressible fluid of density 1.9 slug/ft³ flows at a steady rate through a horizontally mounted circular pipe (Fig. P5.25). If the magnitude of the y-direction component of force is 325.53 lbf, what is the magnitude of the resultant force that must be applied to hold the pipe in equilibrium? Neglect frictional losses through the pipe bend.

5.26 Water flows steadily through a stationary fire hose and nozzle which have inside diameters of 3 in. and 1 in., respectively. The water pressure in the hose is 75 psi and the stream exiting the nozzle is uniform. The exit velocity and pressure are 106 ft/s and 1 atm, respectively. Calculate the force transmitted by the coupling between the nozzle and hose. Indicate whether the coupling is in tension or compression.

Conservation of Linear Momentum

5.27 A free jet of water which has a cross-sectional area of 0.04 ft², and a velocity of 25 ft/s is deflected 180° by a fixed vane. If frictional losses are neglected, calculate the force exerted on the vane by the fluid.

5.28 Fresh water at standard temperature flows in the device shown in Fig. P5.28 with Q = 10 ft³/s. A reaction force **R** is needed to anchor the device which weighs 250 lbf with water in it. If p_1 = 65 psia, p_2 = 47 psia, and p_{atm} = 29.92 in. Hg, what are (a) average velocities $\tilde{V}_1$ and $\tilde{V}_2$, (b) the mass rate $\dot{M}$, (c) the horizontal and vertical components of the reaction force, R_x, R_y, respectively?

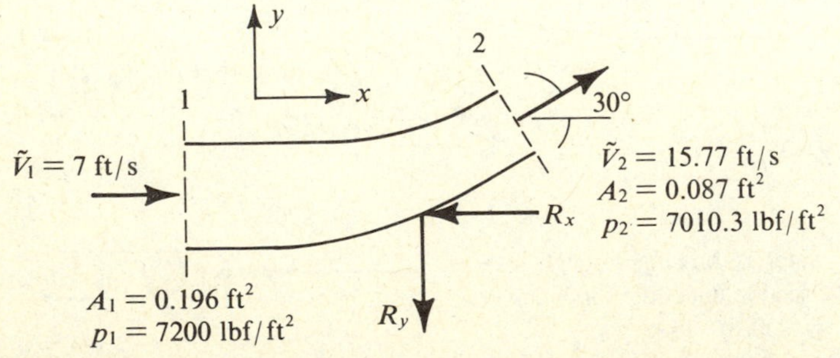

Figure P5.25

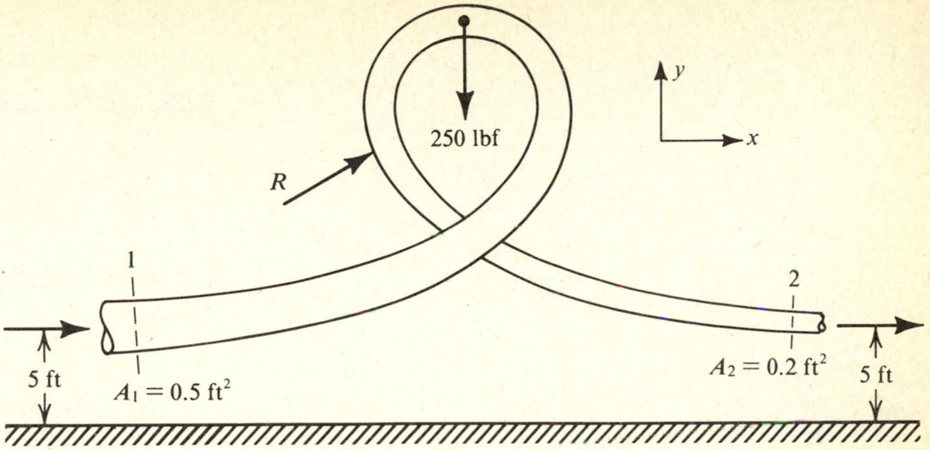

Figure P5.28

5.29 A closed tank on frictionless wheels is held in place by a restraining cable, as shown in Fig. P5.29. The pressure above the water in the tank is 5.0 psi, and the nozzle has a cross-sectional area of 3 in.^{2}. (a) What is the exit velocity of the water leaving the nozzle? (b) What is the mass flow rate of water leaving the nozzle? (c) What is the tension in the cable?

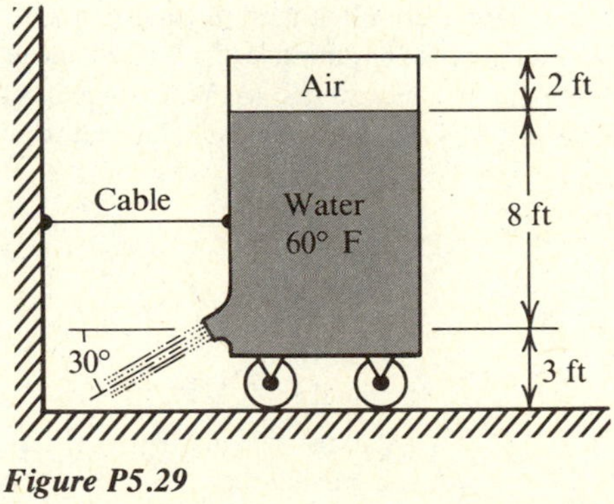

Figure P5.29

5.30 A stationary curved vane deflects a 20-mm diameter jet of water through 150°. Because of friction over the vane, the exiting water has only 83% of its original velocity. Calculate the rate of flow necessary to produce a force of 3000 N on the vane.

5.31 Figure P5.31 shows a test assembly for measuring the thrust of a liquid-oxygen hydrocarbon rocket. The liquid oxygen enters at 1 with a velocity of 35 ft/s, at a pressure of 600 psia, and with a mass flow rate of 11 lbm/s. The inlet pipe for the lox is elastic, so that the force it exerts in the combustion chamber can be ignored. In the exit plane of the jet nozzle 2, it is known that the pressure is 18 psia and the exit plane area is 13 in.^{2}. The force exerted in the scale when the silencer is *not* used is 300 lbf. Atmospheric pressure is 15 psia. Estimate the velocity of the jet in the exit plane of the nozzle when the silencer is not used.

5.32 A steady jet of water, as seen in Fig. P5.32, flows smoothly onto a moving curved vane which turns the jet through 60°. The diameter of the jet is 35 mm and the velocity is 30 m/s. The velocity of the water leaving the surface is 25 m/s. Neglecting friction and gravitational effects, calculate the velocity and direction of the vane if the force on the fluid in the y-direction is 2000 N, and in the x-direction is -1500 N.

5.33 A 45° reducing pipe-bend (in a horizontal plane) tapers from a 500 mm diameter at the inlet to 250 mm diameter at the outlet. The pressure at the inlet is 200 kN/m^2 and the rate of flow of water through the bend is 0.5 m^3/s. If the pipe moves with a constant velocity, $v = 1.5$ m/s, calculate the resultant force exerted by the water on the bend (neglect friction).

5.34 A 420 is sailing downwind with just its spinnaker up. If the wind speed is 25 mph and the boat is going 10 mph, what is the forward force generated in the spinnaker? Assume the

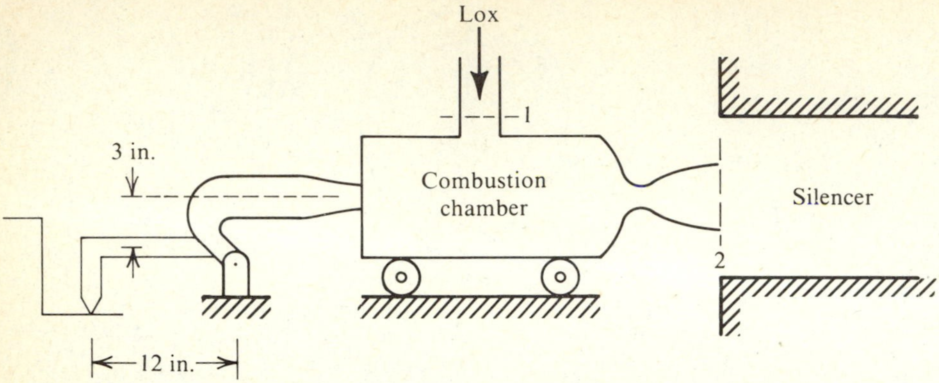

Figure P5.31

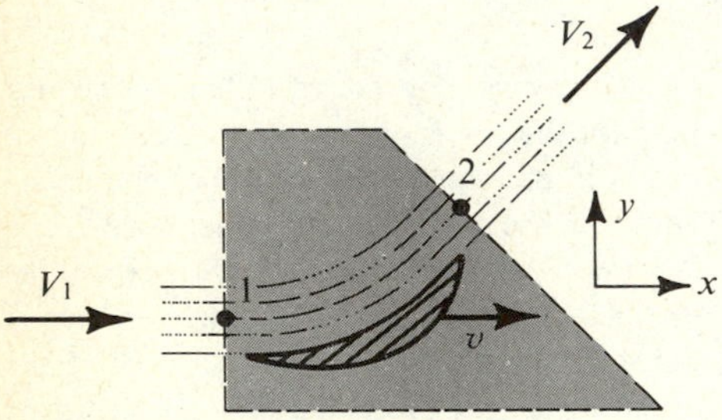

Figure P5.32

hull drag is 400 lbf at 10 mph. (Refer to Fig. P5.34.)

5.35 Neglecting losses what is the power of the sail if V_{boat} = 10 mph, V_{wind} = 30 mph, A_{wind} = 1 ft^2 given γ = 0.08 lbf/ft^3? (Refer to Fig. P5.35.)

5.36 What force **F** is required to hold the plate shown in Fig. P5.36 in place? The fluid is water at standard conditions, the jet area is 1 m^2, and the jet velocity is 7 m/s.

5.37 Water at 68°F flows through a 180° horizontal bend in a piping system at a steady rate of 6 ft^3/s. What is the force required to hold this bend in place if the inlet pressure is 175 psi and exit pressure is 162 psi (Fig. P5.37)?

5.38 Hoses are often used to control crowds such as in riots. Assume use of a 2-in. diameter hose with velocity 50 ft/s. (a) What force must be exerted by the rioter to prevent himself from

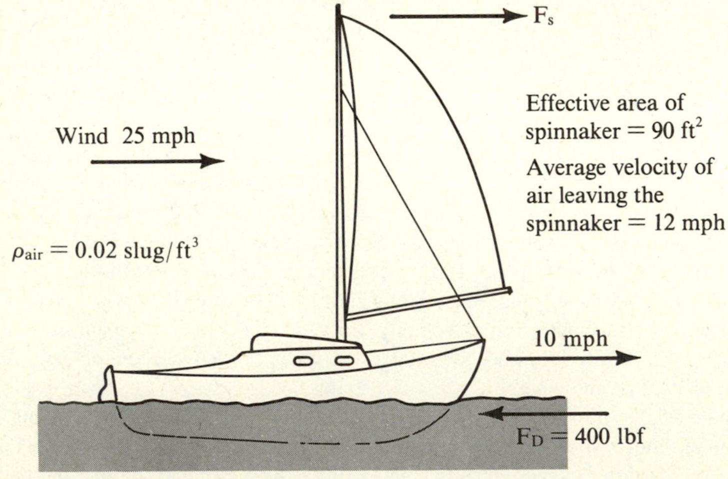

Figure P5.34

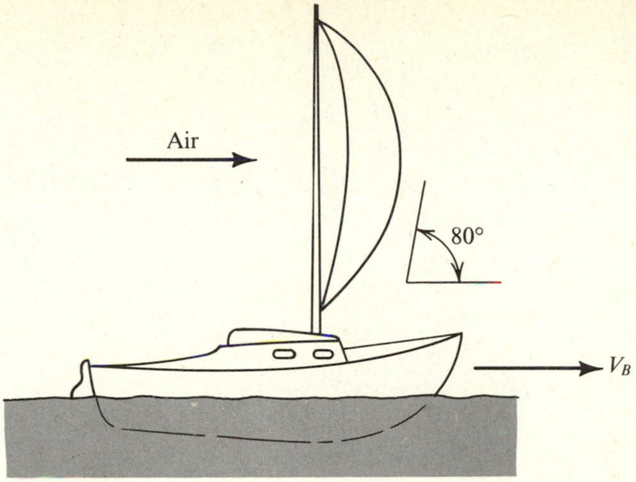

Figure P5.35

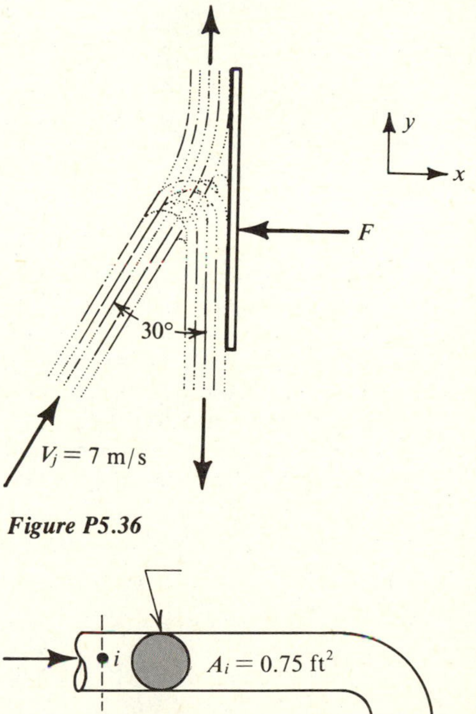

Figure P5.36

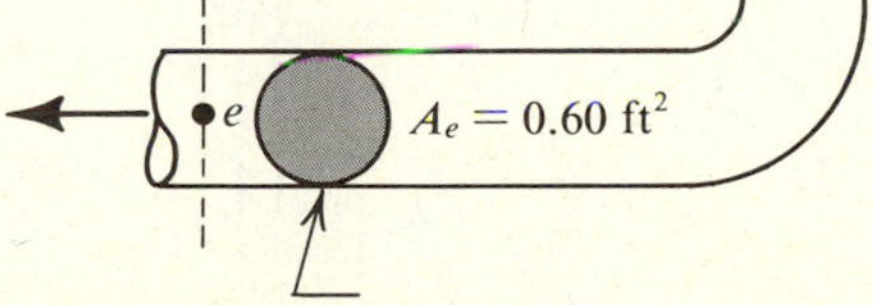

Figure P5.37

being knocked over? (b) If he leans forward at a 10° angle, what is the force? (See Fig. P5.38.)

5.39 A free jet of water which has a cross-sectional area of 0.01 m^2 and a velocity of 3 m/s is deflected 180° by a fixed vane. If frictional losses are neglected, calculate the force exerted on the vane by the fluid.

5.40 A spacecraft's fuel consumption of 5 slugs/s is discharged at 3000 ft/s relative to the rocket at ambient pressure. If the rocket has a constant velocity, calculate the propulsive force on the rocket.

5.41 Referring to Fig. P5.41, calculate the resultant force of water flowing at a velocity of 3 m/s and having a cross-sectional area of 0.1 m^2.

5.42 A jet of water with a velocity of 11 ft/s produces a force of 80 lbf on a stationary plate normal to the jet. If the plate is moved away from the jet at 4 ft/s, calculate the force on the plate due to the water.

5.43 A square plate of uniform thickness and length on each side of 30 cm hangs vertically from hinges at the top edge. When a horizontal jet strikes the plate at its center, the plate is deflected and comes to rest at an angle of 30° to the vertical as shown in Fig. P5.43. The jet velocity is 6 m/s and the jet diameter is 25 mm. Calculate the mass (kg) of the plate.

5.44 A water jet 0.2 m in diameter strikes a fixed vane and is deflected 90° from its original

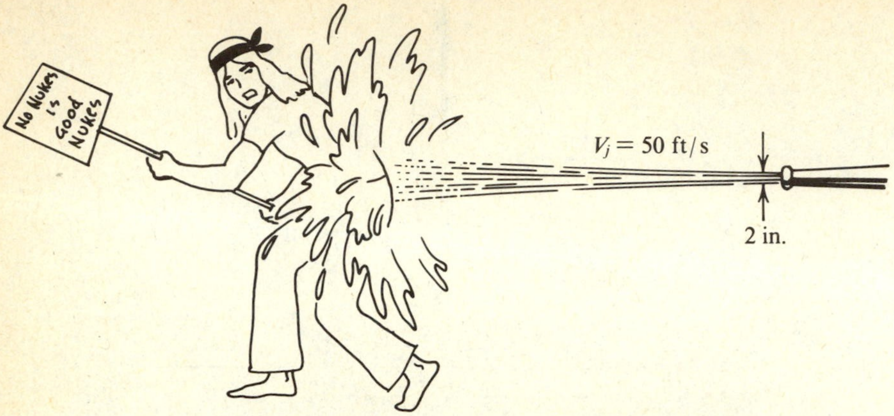

Figure P5.38

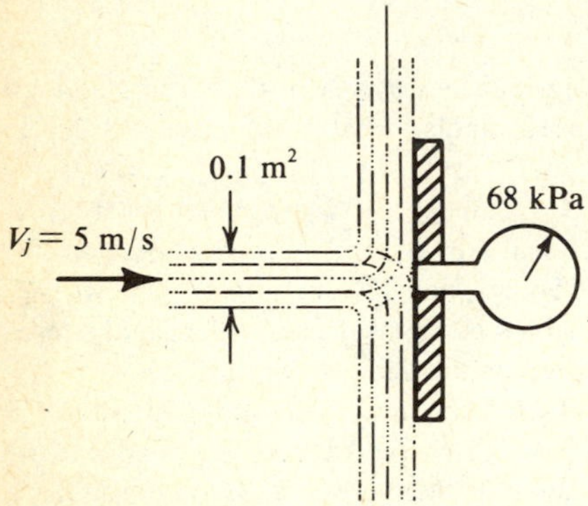

Figure P5.41

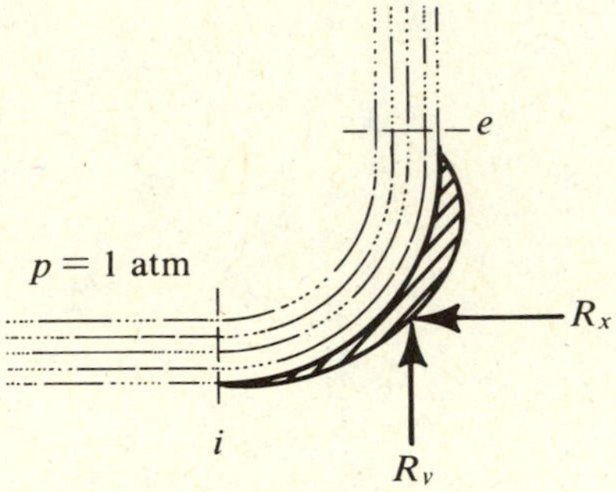

Figure P5.44

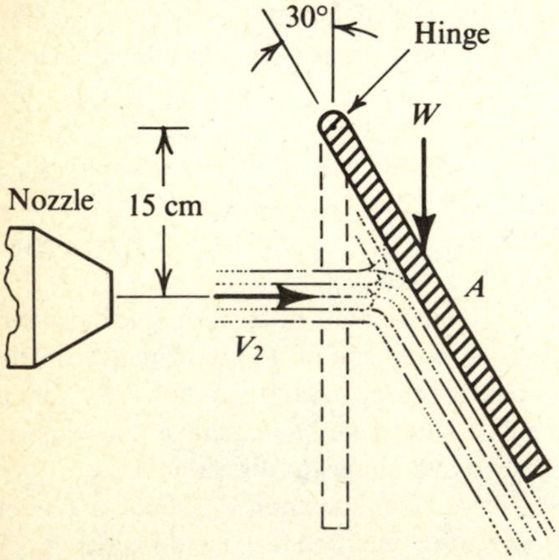

Figure P5.43

direction. Determine the forces of reaction R_x and R_y acting on the vane given an entrance velocity of 3 m/s (Fig. P5.44).

5.45 A firehose directs a horizontal stream of water against a flat object. If the volume rate of flow is 2 m³/s, and the velocity of the jet is 15 m/s, calculate the force required to hold the plate in position if the density of water is 1000 kg/m³ (Fig. P5.45).

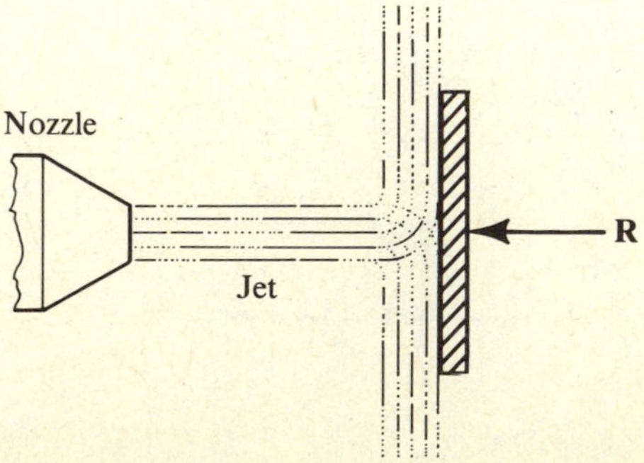

Figure P5.45

5.46 A jet hits a plane moving with a velocity v. Find the maximum height h the water jet will reach and show that this h is not a function of g (Fig. P5.46).

5.47 A brilliant child has built a frictionless self-propelled go-cart using a large tank of water a hose, an adjustable board, and his very own little red wagon. He has figured out that by moving the board through θ, as shown in Fig. P5.47, he can vary the speed (and direction) that he moves. At what angle θ must he hold the board in order to be given no resultant force on his racing machine? (Assume the height of the tank water to be constant.)

5.48 A stream of water strikes a plate on a cart. The tank rests on the same cart. The frictionless cart is moving to the right with a constant velocity 16 m/s. The height of water in the tank is 16 m above the jet. What is the force propelling the cart and the discharge from the jet (Fig. P5.48)?

5.49 As Fig. P5.49 shows, in an impulse turbine, water flows steadily through a frictionless nozzle and exits to the atmosphere as a free jet that impinges on a turbine blade. The jet velocity is measured at 120 ft/s, and the velocity at piont 1, where the diameter is 1 ft, is measured at 10 ft/s. Flow over the blade is also frictionless. (a) What is the gauge reading at point 1 in psia? (b) What is the cross-sectional area of the jet? (c) What is the horizontal force on the blade, if the blade is held stationary? (d) What is the horizontal force on the blade, if the blade is moving to the right at 80 ft/s?

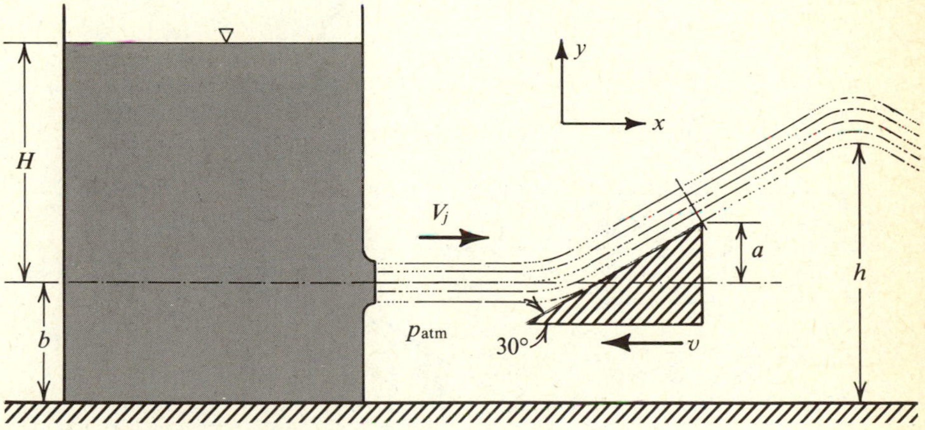

Figure P5.46

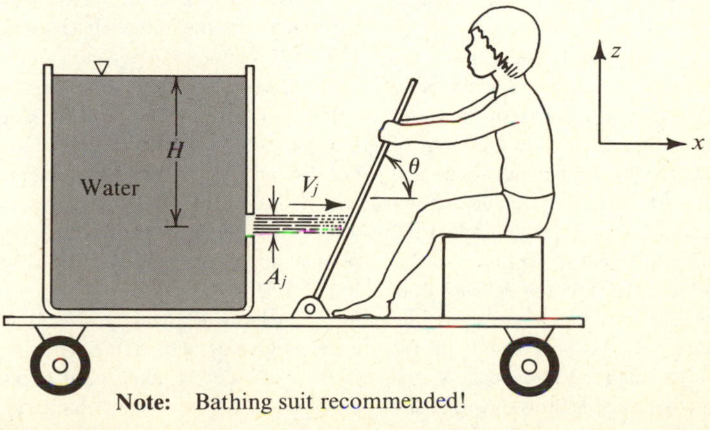

Figure P5.47

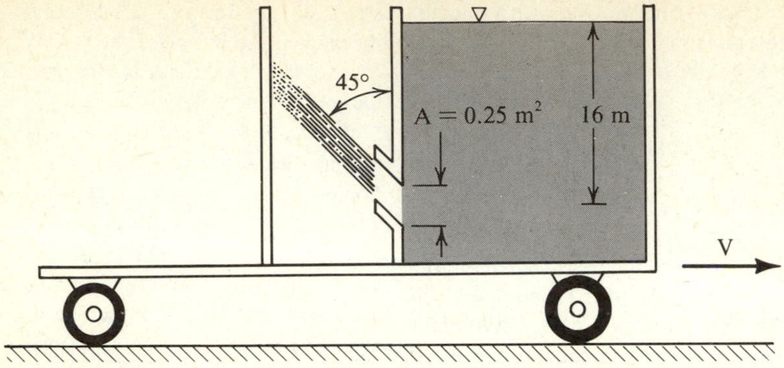

Figure P5.48

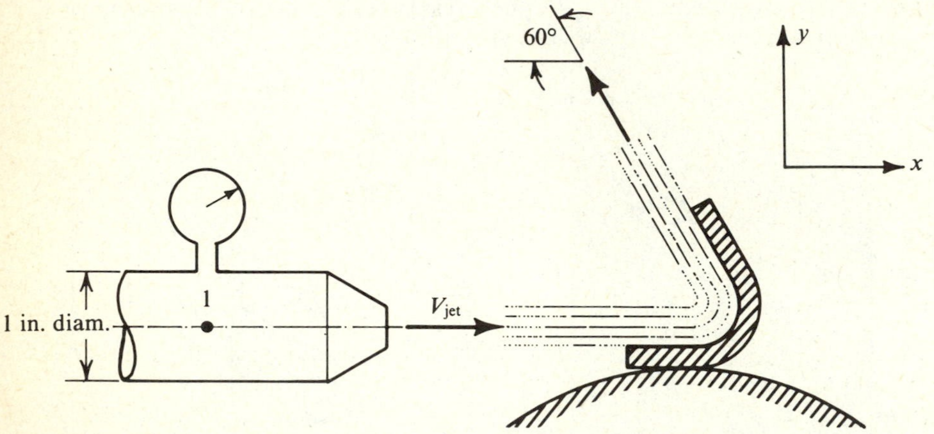

Figure P5.49

5.50 Water flows steadily through a frictionless nozzle and exits as a free jet. The measured velocity and static pressure at state 1 are $\bar{V}_1 = 20$ ft/s and $p_1 = 80$ psi. The jet strikes the cart as shown in Fig. P5.50. Neglect body forces. (a) Calculate the velocity of the water jet. (b) Calculate the cross-sectional area of the jet. (c) Calculate the horizontal component of force on the cart, if the cart is held stationary. (d) Calculate the force on the cart, if the cart is moving to the right at 80 ft/s.

5.51 A jet-propelled boat draws in water from under the bow and expels it at the stern. Let U be the velocity of the boat and $\bar{V}$ be the velocity of the jet relative to the boat. A_1 and A_2 are the cross-sectional areas of the inlet and outlet, respectively. The boat operates in a lake where the water is static. Write mathematical expressions for the following terms: (a) absolute velocity of incoming water; (b) absolute velocity of expelled water; (c) change in absolute velocity; (d) mass rate of water flow through the boat; (e) propulsion force; (f) work done per second by jet; (g) kinetic energy per s of water at inlet; (h) kinetic energy per s of water at outlet; (i) energy per s supplied by internal pumps; and (j) efficiency of power unit. (See Fig. P5.51.)

5.52 A jet boat travels upstream in a river at a constant velocity of 30 ft/s. The river velocity is 5 ft/s. Water is taken in at 0 psi, and the exhaust jet is 0 psi. The volumetric flow rate

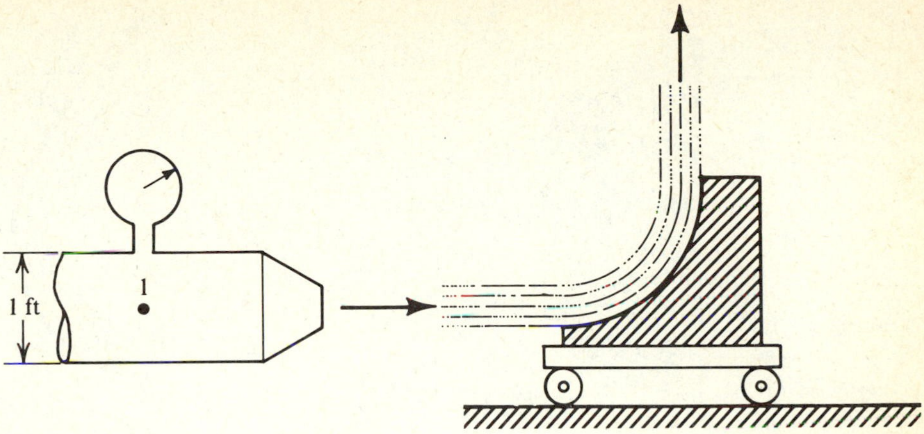

Figure P5.50

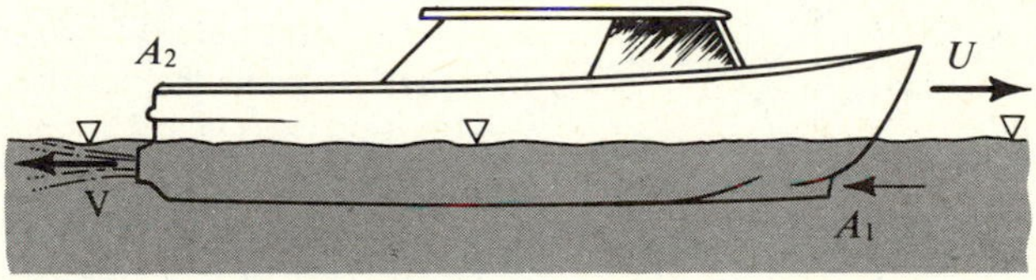

Figure P5.51

of water through the boat and out the jet is 5 ft^3/s, and the jet velocity, relative to the boat, is 60 ft/s. Choose a control volume that moves with the boat and contains the water inside the boat. Find the net force per unit mass rate acting on this control volume.

5.53 In a river flowing at 1.0 ft/s, a motorboat travels upstream at 10 ft/s, relative to the land. The boat is powered by a jet propulsion unit which takes in water at the bow and discharges it beneath the water surface at the stern. The discharge velocity is 20 ft/s relative to the boat, and the flow rate is 0.20 ft^3/s, producing 23 hp. Estimate the propulsive force.

5.54 A toy balloon of 100 g mass is filled with air of density 1.030 kg/m^3, by a small hose 5 mm in diameter oriented in a vertical direction. The balloon is released, letting air escape through the hose. Neglecting friction, calculate the rate at which air escapes if the initial acceleration of the balloon is zero.

5.55 A vane moves with steady speed v, from a jet stream of water of velocity V. Assuming frictionless flow, show that the work done against the restraining force R_x is a maximum when $v/V = 1/3$.

5.56 A jet of water is directed against a vane, as shown in Fig. P5.56. The vane moves to the right at 25 ft/s and the inside surface makes a 180° angle with the positive x-direction. If the jet velocity is 50 ft/s and the volumetric flow rate is 5 ft^3/s, find (a) area of water jet, (b) velocity of jet relative to the vane, (c) mass flow rate, and (d) reaction (and direction) of the vane on water.

5.57 Figure P5.57 shows water at 26.7 psia pressure entering a 60° horizontal reducing elbow with a velocity of 15 ft/s, and leaving against a pressure of 5.8 psi. The diameters of the entrance and exit sections are 12 in. and 8 in., respectively. Assuming incompressible

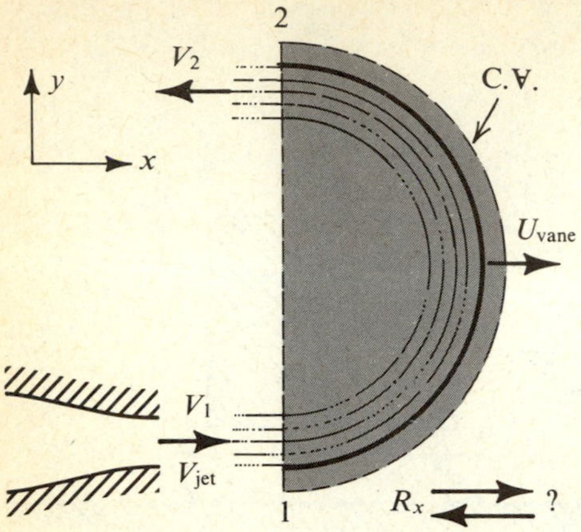

Figure P5.56

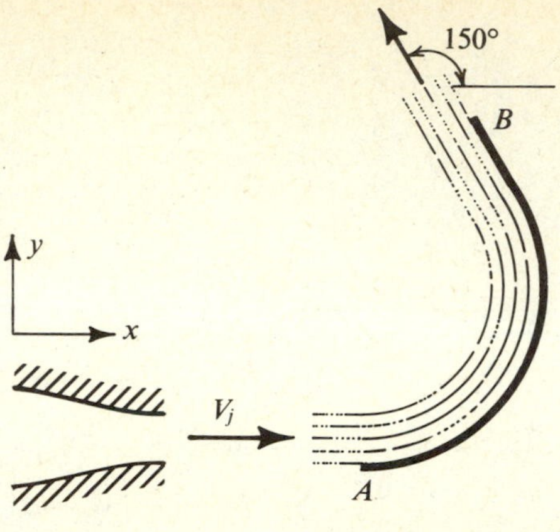

Figure P5.58

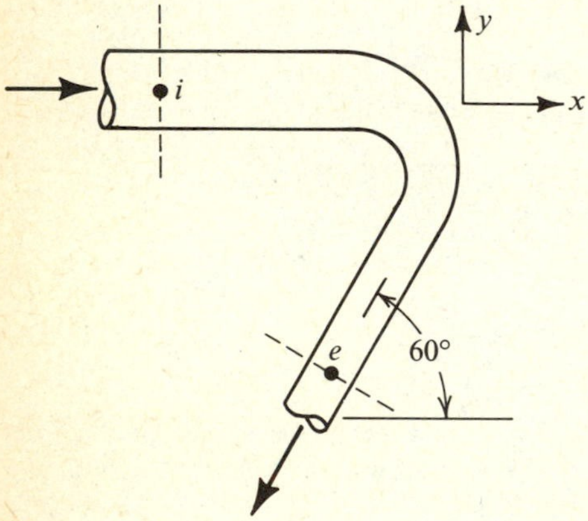

Figure P5.57

steady flow, determine the resultant force of the water on the elbow.

5.58 A jet of water impinges on a stationary horizontal vane, as shown in Fig. P5.58. The stream velocity is 40 ft/s, and the volumetric flow rate is 3 ft^3/s. (a) Determine the x-component of the force exerted on the vane by the water. (b) Determine the y-component of the force exerted on the vane by the water. (c) If the vane is moving to the right at a velocity of 20 ft/s, determine the new volumetric flow rate. (d) If the vane is moving to the right at a velocity of 20 ft/s determine the x-component of the force of the water on the vane. (The density of water is 1.94 slug/ft^3.)

5.59 The ducted propeller system shown in Fig. P5.59 is proposed for use in propelling a surface ship. For a ship speed of 20 ft/s and a depth of operation of 10 ft, determine (a) the maximum flow rate through the propeller before cavitation sets in just upstream of the propeller (station 1 in Fig. P5.59), p_{atm} = 14.7 psia and vapor pressure p_v = 0.36 psia; (b) the thrust produced at this flow rate; and (c) the horsepower required to drive the propeller at this condition.

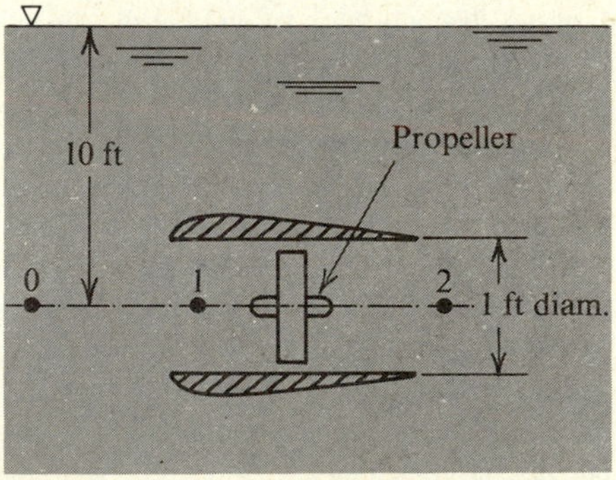

Figure P5.59

5.60 In a surface effect ship it may be assumed that the "air cushion" supporting the ship is contained within a canvas-like skirt fastened around the periphery of the ship. Air, supplied by a compressor to the cushion, escapes through the clearance between the end of the skirt and the sea. The skirt is stretched in a rectangular shape 42 ft × 82 ft. The weight of the ship is 200,000 lbf and the average ground clearance is 2 in. The volume of the cushion is sufficiently large so that the velocity of air in this region is negligible. The air flow may be treated as incompressible (γ_{air} = 0.078 lbf/ft^3). Determine (a) the mass flow rate of air required to maintain the cushion, and (b) the required horsepower of the compressor (Fig. P5.60).

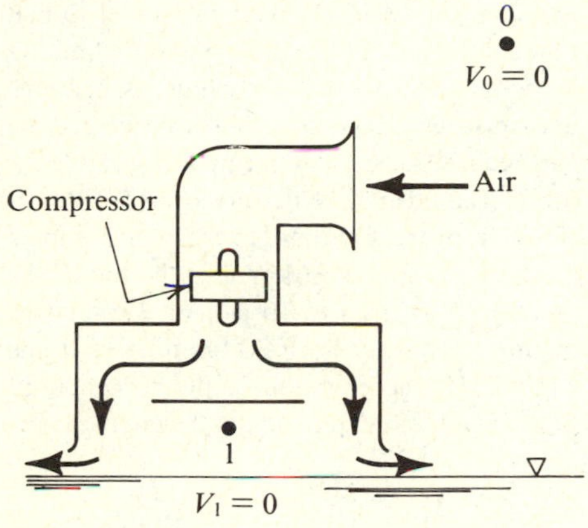

Figure P5.60

5.61 As shown in Fig. P5.61, an aircraft jet engine consumes 200 lbf/s air-fuel mixture of density ρ = 0.0030 slug/ft^3 (measured at exhaust). The air-fuel mixture is exhausted through a 12-in. diameter duct into thrust reversers when landing. The thrust reversers turn the exhaust through an angle of 150°. What is the braking thrust for such a situation? Assume the engine is symmetrical with respect to the centerline, that all flow is frictionless, and that the exhaust is a free stream.

5.62 An incompressible, *frictionless* fluid of height h_1 and velocity u_1 flows through a sluice gate, as shown in Fig. P5.62. The downstream depth of the fluid is h_2. Assume that the flow is irrotational, steady, and the velocity profiles are parallel and uniform. The pressure distribution, however, need not be uniform. (a) Show that the absolute pressures at 1 and 2 are given by $p_1 = \rho g(h_1 - y) + p_A$, and $p_2 = \rho g(h_2 - y) + p_A$, where p_A is the atmospheric pressure. (b) Find the upstream velocity $\bar{V}_1$ in terms of g, h, and h_2. (c) Find the force F required to hold the gate in equilibrium in terms of ρ, g, h_1, and h_2.

5.63 In an experiment to determine drag, a circular cylinder of diameter d was immersed in a steady two-dimensional incompressible flow. Measurements of velocity and pressure were made at the boundaries of the control surface shown. The pressure was found to be uniform over the entire control surface. The x-component of velocity at the control surface boundary was approximately as indicated by Fig. P5.63. From the measured data, calculate the drag force per unit length of the cylinder, based on diameter d and the free stream dynamic head ($\frac{1}{2}\rho V_\infty^2$).

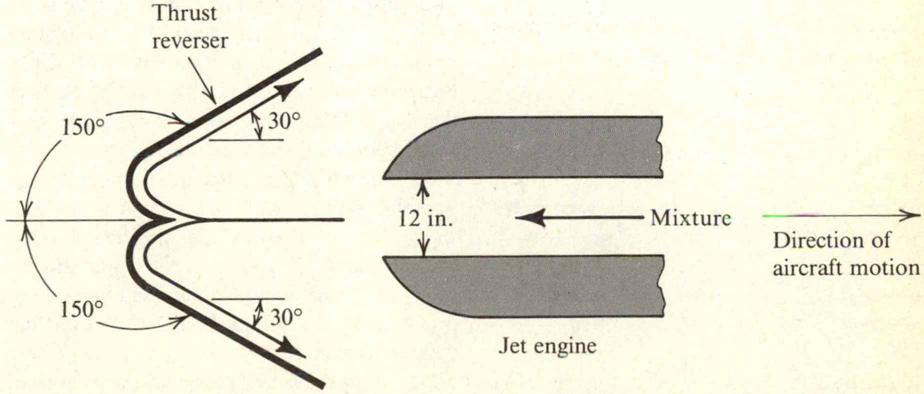

Figure P5.61

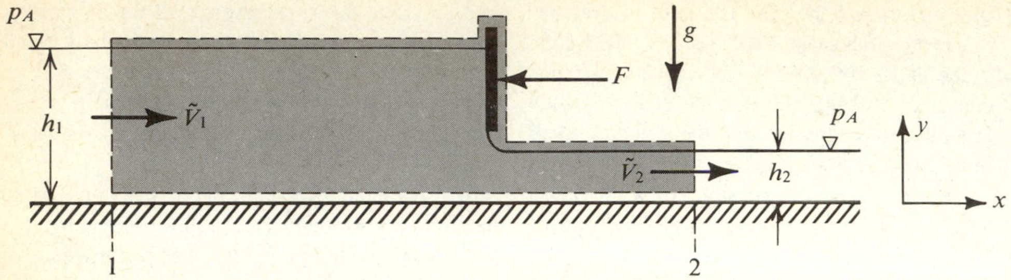

Figure P5.62

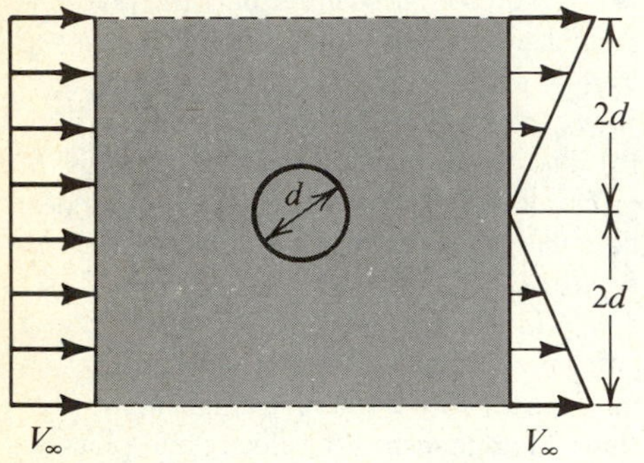

Figure P5.63

5.64 As Fig. P5.64 shows, water flows over a flat plate. The velocity profile above the leading edge is uniform. The x-component of the velocity above the trailing edge is Vy/h for $y < h$, and V for $y \geqslant h$. If the momentum flux per unit width of plates for flow across $\overline{ab}$ is $-\rho h V^2$, what is the x-component of the momentum flux per unit width for the flow across area $\overline{cd}$? Is this an x-component of momentum flux for area $\overline{bc}$? Explain your answer.

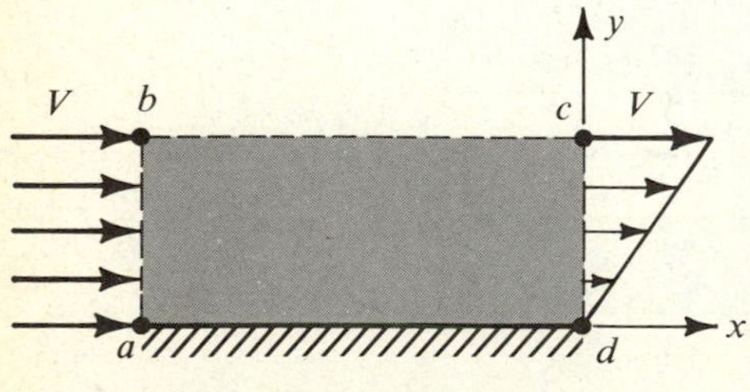

Figure P5.64

5.65 The wind is blowing at a constant velocity V_o. There is a cart with a sail area A_o which is aimed in a direction parallel to that of the direction of the wind. The cart's limiting velocity is V_o assuming frictionless bearings and surfaces. How long will it be before the cart reaches speed V_o (cart mass $= M_o$)?

5.66 Liquid fuel of density $\rho = 50$ lbm/ft^3 flows from the fuel tank to the rocket engine as shown in Fig. P5.66. Air at the top of the tank is pressurized to 10 psi. Fuel enters the engine at velocity $\tilde{V} = 20$ ft/s relative to the engine. The acceleration of the rocket is 50 ft/s^2. Find the fuel pressure at the engine inlet.

Energy Equation and Bernoulli's Equation

5.67 A jet of water leaving a nozzle is directed vertically upward. The jet leaves the nozzle with a speed of 30 ft/s. Neglect thermal and friction effects. How far above the nozzle will the water travel?

5.68 A circular jet of water leaves a nozzle in a vertical upward direction with a velocity of 20 ft/s. The jet diameter is 1 inch. A large circular disk weighing 2 pounds is held in a horizontal position above the nozzle. Neglect thermal and friction effects. What is the distance between nozzle and disk?

5.69 On a beautiful standard day your oceanographic ship sights a water spout in the east Indian Ocean. You fix your position as 6000 ft from the spout and record a wind velocity of 10 ft/s. What wind velocity and barometric pressure does a fishing boat 600 ft from the spout experience?

5.70 A hydrofoil is being tested in a wind tunnel (air), where the static pressure is 1990 lbf/ft^2 and the temperature is 125°F, at 80 knots. What is the maximum pressure on this airfoil? Assume steady inviscid incompressible flow.

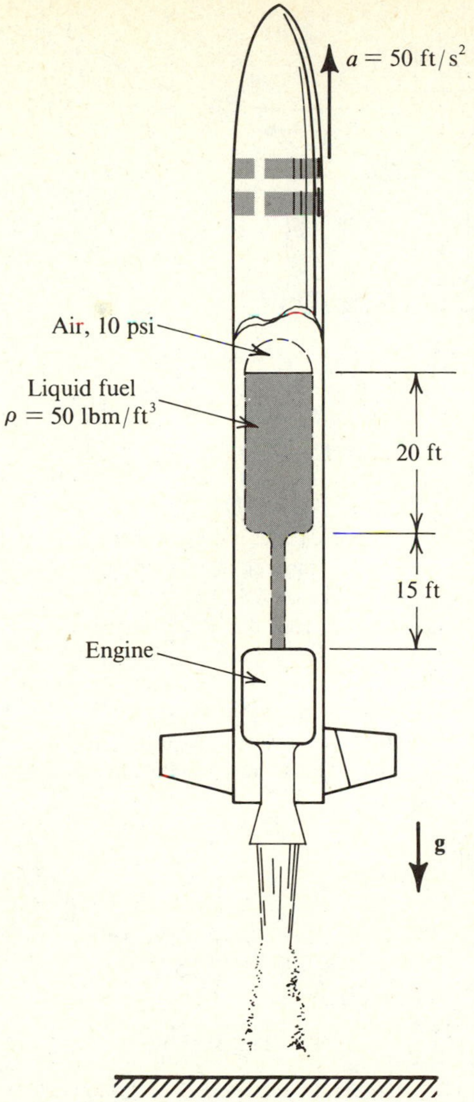

Figure P5.66

5.71 The speed of sea lions has been measured to be as great as 25 mph. At that speed, what is the pressure, above that of the ambient seawater, on the nose of the sea lion?

5.72 A combined pitot tube is used to measure the flow of ammonia gas in a 10 in. diameter pipe where the pressure and temperature are 30 psia and 150°F, respectively. If the pitot tube is hooked up to a water manometer which reads 7 in., what is the velocity in the pipe?

5.73 Just upstream of a fan in the ship's ventilation system is installed a pitot static tube and a static tube connected to water manometers. The pitot tube is located at the center of a 9 in. × 12 in. rectangular duct. The air stream temperature is 70°F. (a) What is the static pressure of the stream? (b) What is the density of the air stream? (c) What is the value of the velocity indicated by the pitot static tube? (d) Is this the average or the maximum velocity? (e) How would the mass flow rate in the duct be determined? (See Fig. P5.73.) Assume ideal fluid in steady incompressible flow.

5.74 Water is flowing through the venturi tube as shown in Fig. P5.74. If, because of the flow, the manometer shown deflects 4 ft, what is the velocity at section 2?

5.75 A rocket-propelled craft flies at a speed of 1700 ft/s in a westerly direction. The combustion gases exhaust aft of the craft through an area 1 ft² at 15 psia. Ambient pressure is 11 psi. The exhaust gases leave the aircraft at a speed of 300 ft/s *relative to the ground* in an easterly direction. (a) Determine the velocity of the exhaust gases relative to the aircraft. (b) Calculate the thrust of the gases on the aircraft.

5.76 The fire-fighting system installed at a Naval Station is sketched in Fig. P5.76. (a) If

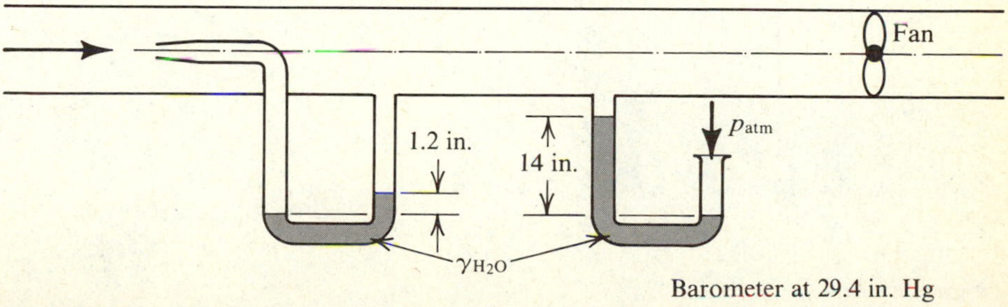

Figure P5.73

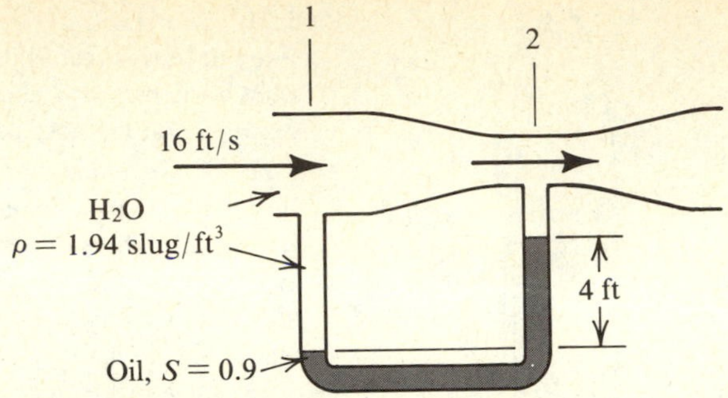

Figure P5.74

the flow rate Q is 0.4 ft³/s, what is the pressure at A? (b) For $Q = 0.4$ ft³/s, what are the water velocities at points B, C, and D? (c) If the nozzle has an inside diameter of ¾ in., what is the velocity of the water leaving the nozzle? ($Q = 0.4$ ft³/s.) (d) What would be the force on a fireman holding the nozzle for the conditions of question (c)? (e) How high will the water jet go? (f) What will be the force on a vertical wall if the jet hits it ($\theta = 0°$)? (g) What is the force tending to pull the nozzle off of the end of the hose if the pressure inside the hose is 100 psi?

5.77 Find the velocity of the water flowing from the faucet if $p_1 = 75$ psia, $p_2 = 14.7$ psia, $D_1 = 20$ ft, $D_2 = 0.5$ in., $z_1 = 120$ ft, and $z_2 = 15$ ft (Fig. P5.77).

5.78 An ideal gas flows compressibly and isothermally through a converging duct, as shown in Fig. P5.78. For inviscid adiabatic steady flow, find the exit pressure if $p_i = 75$ psia, $A_e = 0.5\,A_i$, $V_i = 10$ ft/s, and $V_e = 27$ ft/s.

5.79 In the jet pump shown in Fig. P5.79, a high-velocity stream of water ($\rho = 1.94$ slug/ft³) entrains a low-velocity stream at sec-

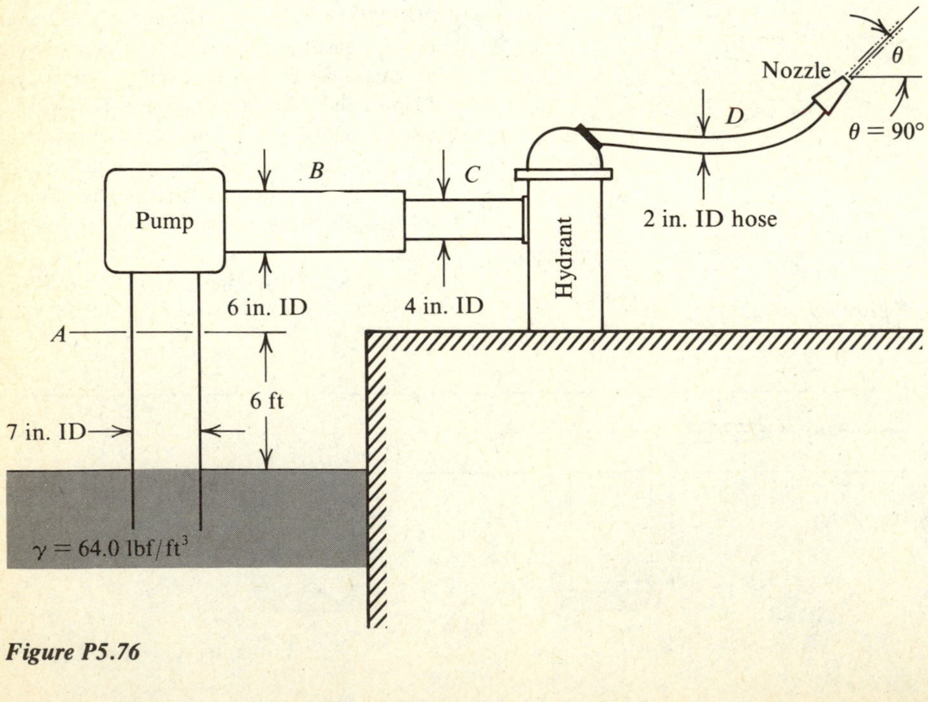

Figure P5.76

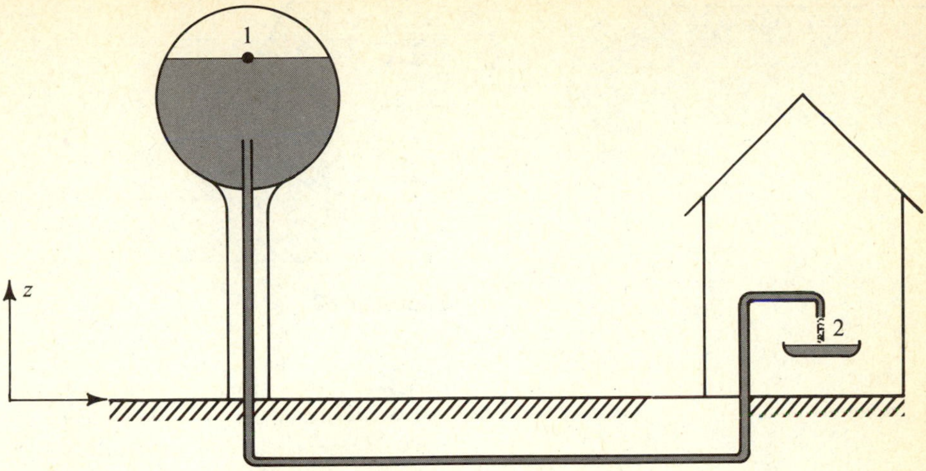

Figure P5.77

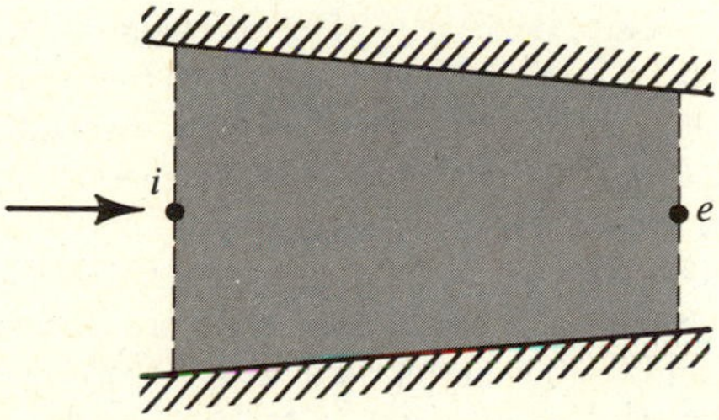

Figure P5.78

tion A. At the end of a constant diameter mixing-tube, section B, the streams are completely mixed as a result of friction between the streams. Assuming that at A both streams have the same pressure, that the velocities are uniform across the section at A and B, and that shearing stress at the tube walls is negligible, find (a) the velocity at B, $\bar{V}_3$ (ft/s), and (b) the pressure difference, $p_B - p_A$ (psi).

5.80 Water flows from a tap at a constant rate Q, and A is the cross-sectional area of the stream. (a) Show that the velocity of the stream is $V = \sqrt{2gy}$. (b) What is the time rate of change of the area DA/Dt in terms of Q and y (Fig. P5.80)? Assume inviscid fluid in steady flow.

5.81 Water flows without losses through the inclined reducer shown in Fig. P5.81. Given a flow rate of 1 m³/s, an inlet diameter of 0.25 m, an exit diameter of 0.1 m, and an inlet pressure of 200 kPa, find (a) the inlet and exit average velocities, and (b) the exit pressure.

5.82 As Fig. P5.82 shows, a turbine located 100 ft below a water supply delivers 50 ft³/s of water of density 1.937 slug/ft³. Assuming

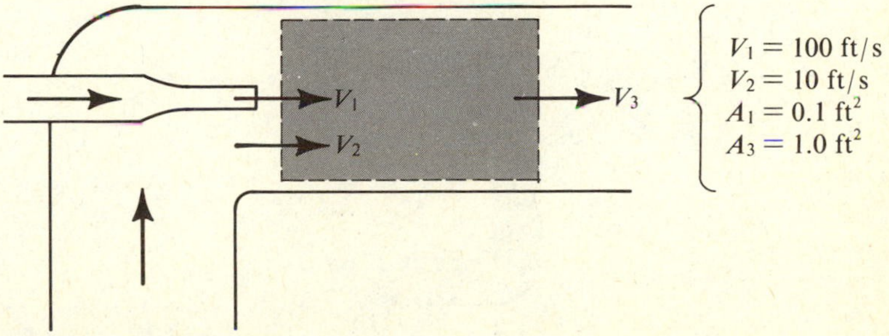

Figure P5.79

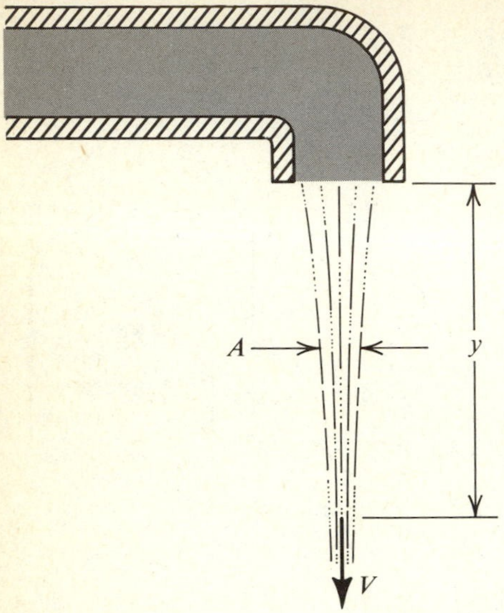

Figure P5.80

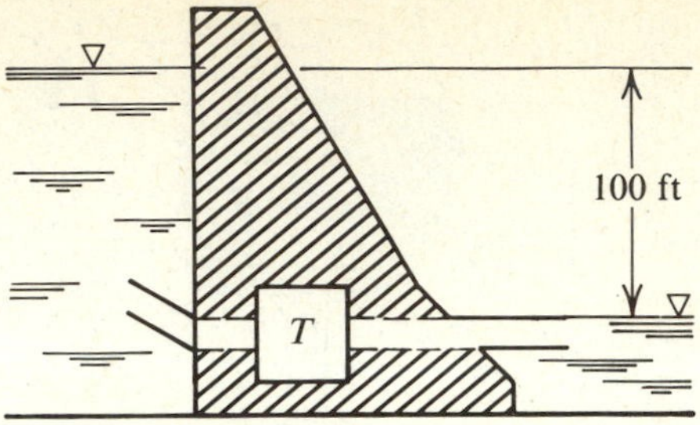

Figure P5.82

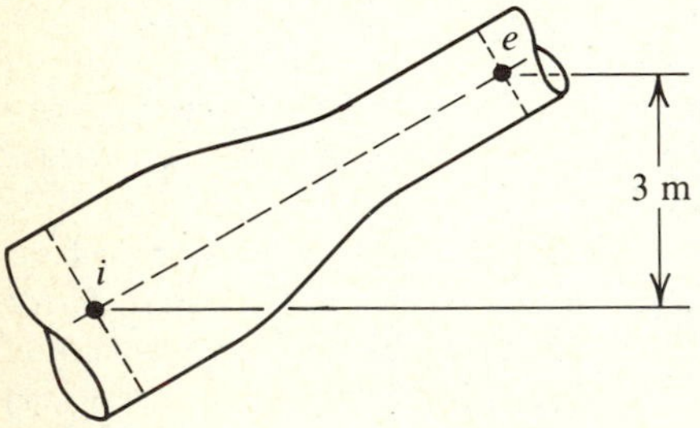

Figure P5.81

no power losses, calculate (a) the mass rate, and (b) the maximum available horsepower.

5.83 A submarine moves steadily through water at a depth of 350 m and a speed of 5 m/s. (a) Find the static pressure p_∞ far away from the submarine at the 350 m depth (kPa). (b) Find the pressure on the nose of the submarine (kPa).

5.84 Given the flow of water out of the tank in Fig. P5.84, find the pressure p_e assuming the flow is steady and the fluid incompressible and inviscid if (a) $p_o = 0$, and (b) $p_o = 50$ kPa.

5.85 The vertical reducing section and elbow shown in Fig. P5.85 contains water at standard conditions flowing upward at a rate of 7.85 ft^3/s. The pressure at 1 is 34.7 psia, and the water exits at 4 to standard atmospheric conditions. Assume ideal flow. (a) What is the velocity at 2? (b) What is the gauge pressure at 2? (c) Determine the force on the pipe due to the contraction (ignore effect of exit 4 momentum). (d) Determine pressure at 3. (e) What is the horizontal component force exerted on the 90° elbow by the water?

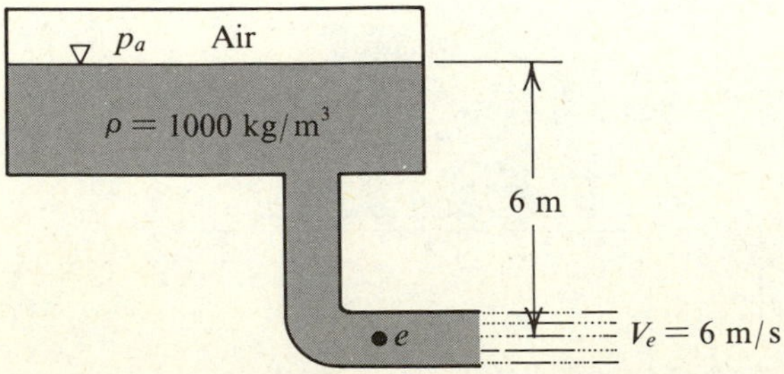

Figure P5.84

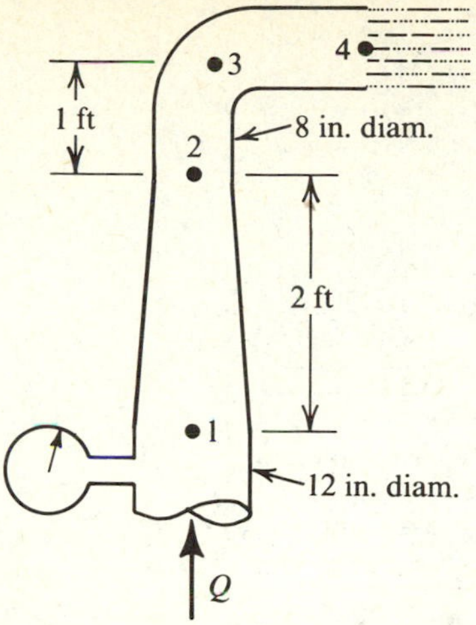

Figure P5.85

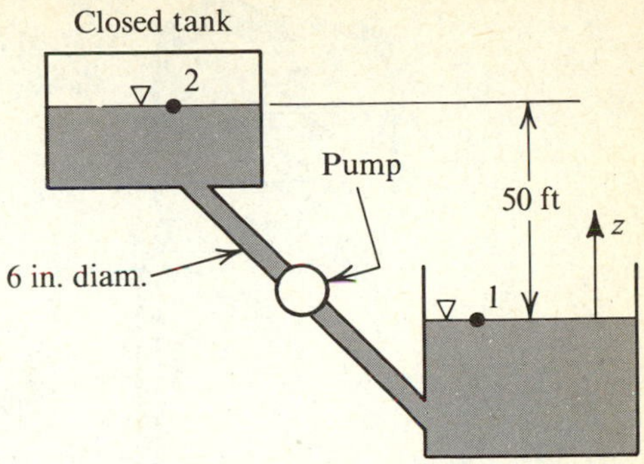

Figure P5.87

5.86 (a) Write the general integral forms of the continuity, linear momentum, and energy equations. (b) Consider steady, incompressible, 1-D, adiabatic, inviscid, isothermal flow through a horizontal, converging nozzle section of the piping system shown in Fig. P5.86. Reduce the three equations in part *a* to apply to a control volume between sections *A* and *B* of the nozzle.

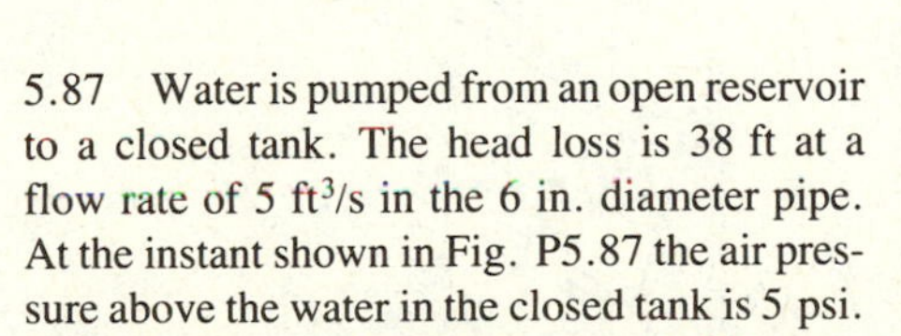

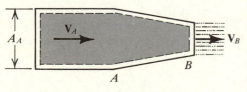

Figure P5.86

5.87 Water is pumped from an open reservoir to a closed tank. The head loss is 38 ft at a flow rate of 5 ft^3/s in the 6 in. diameter pipe. At the instant shown in Fig. P5.87 the air pressure above the water in the closed tank is 5 psi. (a) Determine the pump work. (b) Compute the pump horsepower. (c) The head loss through the pump and pipe is 68 ft at a flow rate of 7 ft^3/s when water flows back through the system from the tank to the reservoir. Compute the gauge pressure required in the tank in order to obtain a flow rate of 7 ft^3/s from the tank to the reservoir.

5.88 Air enters a horizontal pipe 14 in. in diameter at a pressure of 14.8 psia and a temperature of 60°F. At the exit the pressure is 14.5 psia; entrance velocity is 35 ft/s; and exit velocity is 50 ft/s. What is the heat added or extracted?

5.89 A rocket travels through air at 650 mph. The temperature of the air at rest far from the rocket is 35°F. Assume an adiabatic compression process and that potential energy changes can be neglected. What is the air temperature at the stagnation point of the rocket?

5.90 Your wife decides that she needs a heart-shaped Jacuzzi in the bedroom of your three-story house and you plan to pump the water from your large heated swimming pool out back. If 2 in. diameter pipe is used, the water must be delivered at 0.2 ft^3/s, and the pumps advertised claim to be 70% efficient. Determine the horsepower needed. You guess from taking a college friend's course that the losses will be $(25\,\bar{V}^2)/2g$ due to friction in the pipe. Assume pressure is 0 at both elevations. (See Fig. P5.90.)

5.91 As demonstrated by Fig. P5.91, in order to control the head on beer exiting the tap the exit velocity must be less than 0.01 m/s. If one stroke of the pump adds 100 Pa to the keg, what is the maximum number of pumps that can be added and still have a suitable head? The initial pressure in the keg is 0 Pa. Assume line losses of 0.5 (N·m)/N.

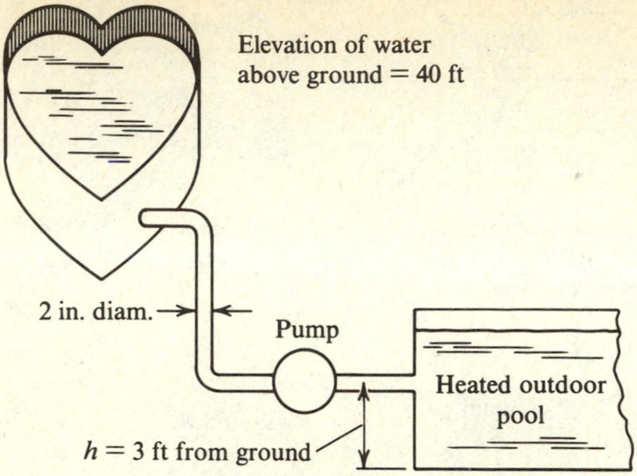

Figure P5.90

5.92 Beer is flowing from a keg at the rate of 0.01 m^3/s. The diameter of the tap is 2 cm. If the level of beer in the keg is 2 m above the tap, what is the pressure at the free-surface? (See Fig. P5.92.)

5.93 Water is pumped from the lower to upper reservoir, as shown in Fig. P5.93. The pump delivers 104.5 ft of work; $Q = 2.5$ ft^3/s. What are the losses?

5.94 Consider beer jetting out of an enormous vat and impinging on turbine blades that have a 140° curve, as shown in Fig. P5.94. Find the horsepower that can be obtained if the beer's free-surface is 200 ft above the jet, the pipe diameter is 6 in., the losses are 3 $\bar{V}_2^2/2g$, and the beer's jet is 20 ft/s.

5.95 As shown in Fig. P5.95, a hydroelectric plant has a difference in elevation from head water to tail water H and a flow $Q = 5$ m^3/s of water through the turbine. Output of the generator is 2100 kW. Determine (a) the minimum difference in water levels (assuming the system is frictionless); (b) the difference in water levels if the power out is ½ the power into the system (i.e., the lost power = 2100 kW); (c) the power loss in the T-G set when H is that of (b); and (d) the system's efficiency.

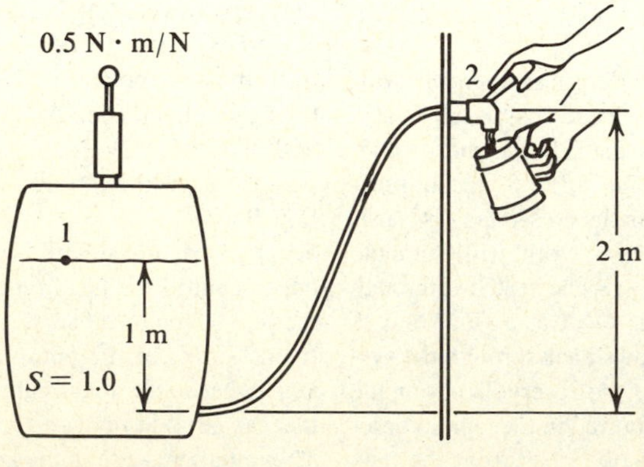

Figure P5.91

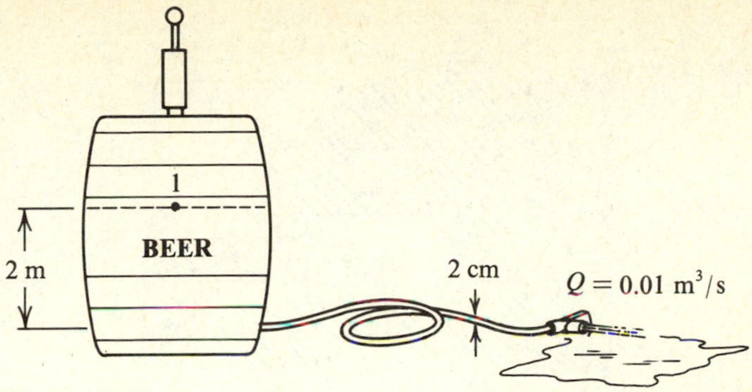

Figure P5.92

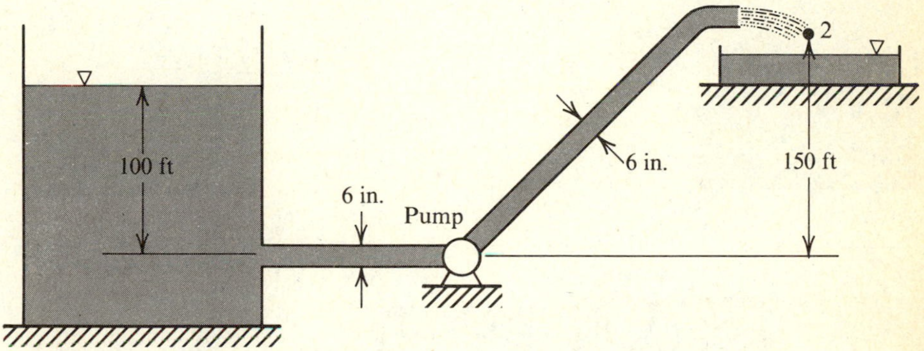

Figure P5.93

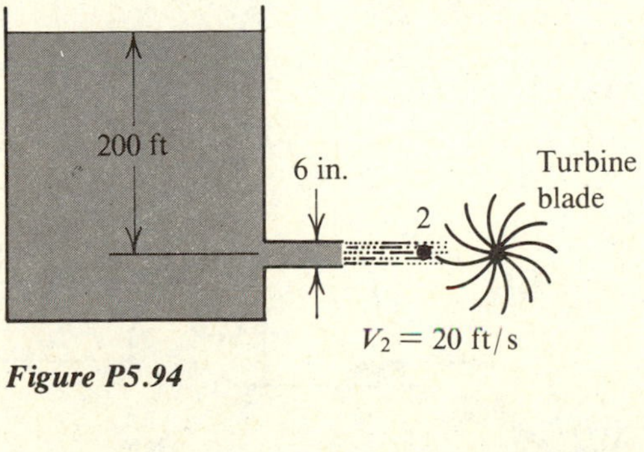

Figure P5.94

5.96 How much work must the pump do to draw water into the upper reservoir at 150 l/s? How much power does the pump use? (Refer to Fig. P5.96.)

5.97 What is the maximum height achievable by the water if $Q = 1\ \text{ft}^3/\text{s}$, and what are the total losses and pressure required in the reservoir (Fig. P5.97)?

5.98 A main feed booster pump has a capacity of 480 gpm. It takes a suction at 20 psi and discharges at a pressure of 85 psi. A temperature rise of 0.1°F is measured between the suction side of the pump and the discharge side. Changes in kinetic and potential energies are negligible. There is no heat transfer through the pump casing. (a) Compute the minimum

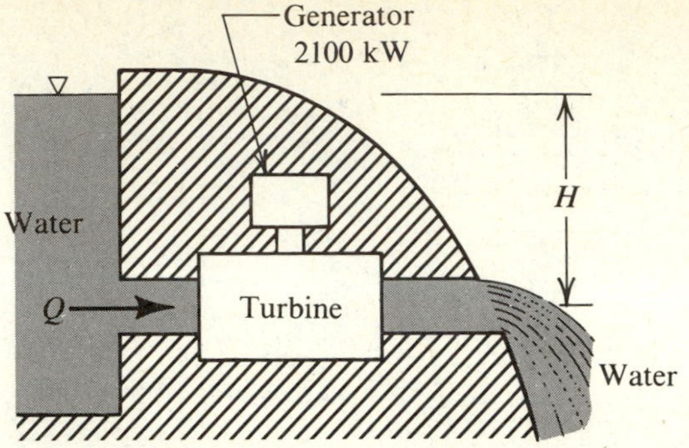

Figure P5.95

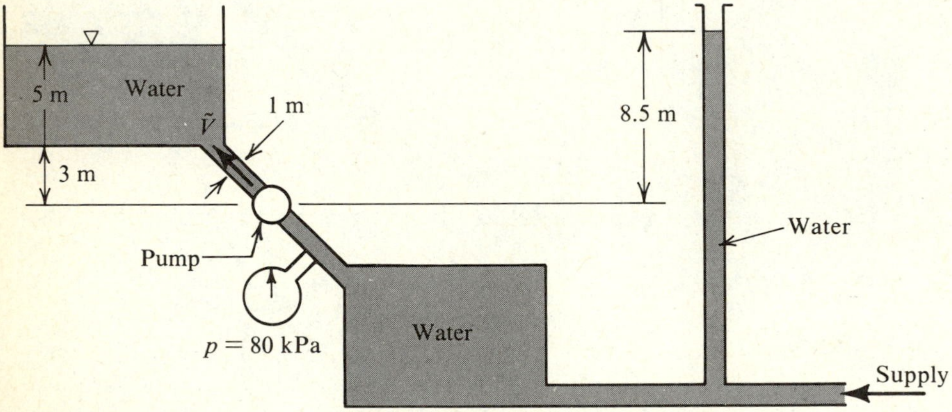

Figure P5.96

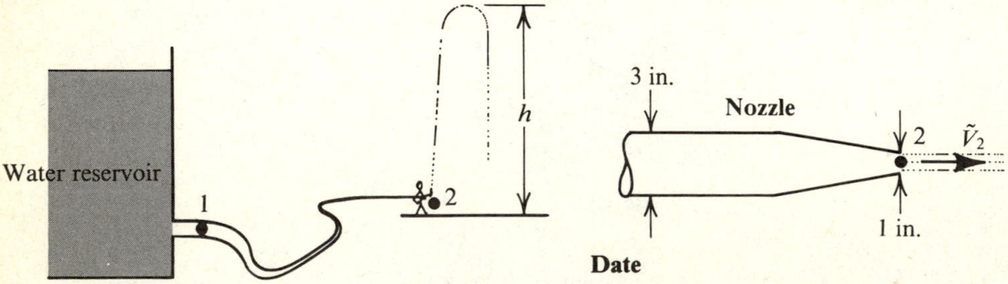

Date

Hose diameter = 3 in.

Hose length = 150 ft

Losses in nozzle = $0.01 \dfrac{\tilde{V}_2^2}{2g}$

Losses in hose = $\dfrac{5\tilde{V}_1^2}{2g}$

Figure P5.97

horsepower required of a motor to operate this pump, and (b) what proportion of the horsepower computed in (a) is used in overcoming friction if the specific heat of water is 1 Btu/lbm°R.

5.99 The "principal expansion loop" of a main steam system involves the turning of the steam flow through 270° prior to admitting the steam to a steam turbine. For steam at a pressure of 1200 psi (and constant density of 0.050 slug/ft^3) compute the magnitude and direction of the force necessary to hold this "expansion loop" in place for a flow rate of 2500 slug/h of steam in a 6-in.-diameter piping system. Assume frictionless, incompressible steady flow—no heat transfer or changes in elevation. (See Fig. P5.99.)

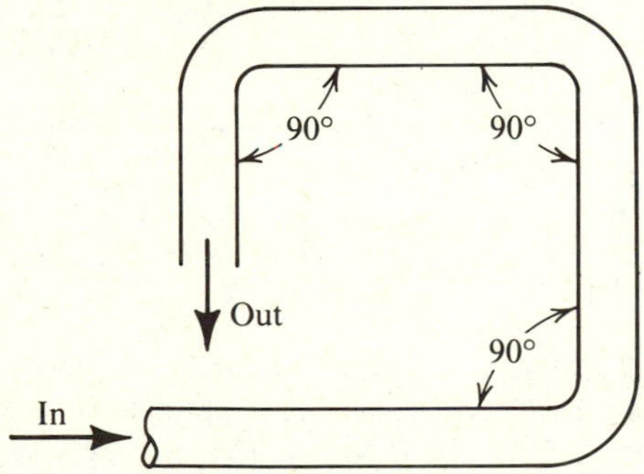

Figure P5.99

5.100 In a certain horizontal piping installation the flow is to be reversed 180° with a return bend ($K = 2.2$), and accelerated by a sudden reduction in pipe area. The pipe size can be reduced either in the beginning of the 180° reversal or at the end of it. The flow rate is 200 gpm of fresh water and pressure at inlet is 20 psi. (a) Which installation experiences the greater loss in pressure due to fittings? (b) Which installation will require less force to hold it in place? Why? (Words, numbers, or equations.) (c) For this installation, what is the magnitude and direction of the force to hold it in place? (d) If operating costs are more important than initial cost (as represented by the cost of material, piping supports, etc.), which installation should be chosen and why? (Refer to Fig. P5.100.)

5.101 Water moves steadily through the turbine shown in Fig. P5.101 at 8 ft^3/s. The pressures at 1 and 2 are 25 psi and 6.1 in. Hg vacuum, respectively. Neglect heat transfer. (a) Find the inlet velocity. (b) Find the exit velocity. (c) What is the ideal shaft horsepower of the turbine? (d) Since the turbine is assumed to be adiabatic, if the turbine is less than 100% efficient, how does the energy equation account for the "lost" energy? (e) What is the actual horsepower of the turbine if it is 90% efficient?

5.102 As Fig. P5.102 shows, an aircraft powered by a Pratt & Whitney J57 turbojet is flying at an air speed of 560 knots (932 ft/s)

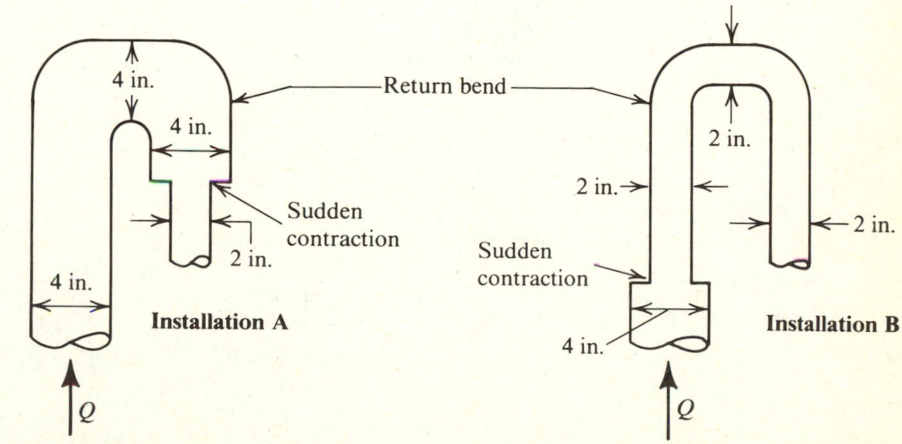

Figure P5.100

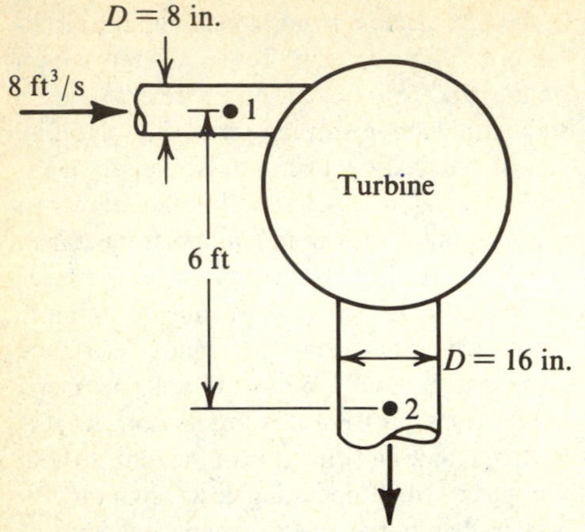

Figure P5.101

in standard air (59°F, 14.7 psia). At this speed the engine burns 0.0863 slug/s of fuel. The fuel is introduced into the jet engine at zero velocity relative to the engine and the area of the fuel inlet piping is negligible. At exhaust from the jet the air-fuel mixture has a jet velocity of 1500 ft/s relative to the engine. The air-fuel mixture is underexpanded and has a static pressure of 17.2 psi at the exit. The cross-sectional area of the jet engine is assumed constant at 2.38 ft^2. Compute the thrust developed by the engine.

5.103 The discharge from a fire hose nozzle of exit area 0.006 ft^2 is 0.45 ft^3/s. The pumper maintains a static pressure of 100 psi on the hose of cross-sectional area 0.05 ft^2 that supplies the nozzle. (a) What is the loss of energy in the hose and nozzle? (b) What is the mass rate of flow at the pumper discharge?

5.104 Air enters a 10-cm horizontal pipe at 550 kPa and 20°C with a velocity of 50 m/s. At the pipe outlet, the pressure is 412 kPa and the temperature is 20°C. Calculate the loss of energy (m).

5.105 Air flows isothermally through a 6 in. diameter inclined pipe with fluid losses (Fig. P5.105). Given a mass rate of 1 lbm/s, air temperature of 40°F, an inlet pressure of 20 psia, and an exit pressure of 19 psia, (a) find the inlet and exit velocities (ft/s), and (b) the head loss h_f. Do not assume air is incompressible.

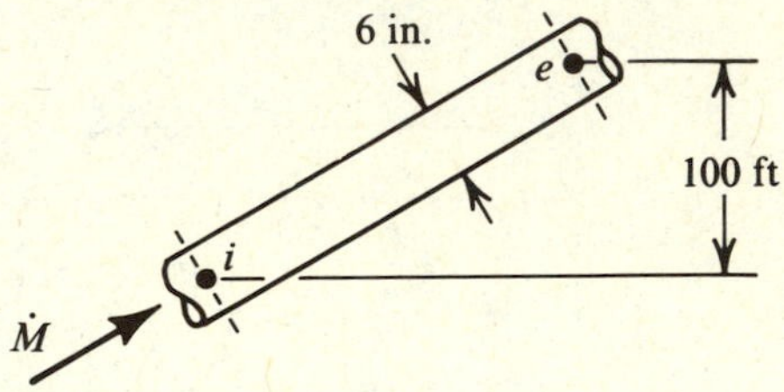

Figure P5.105

5.106 In an experiment designed to measure the head loss of a pipe, 7.07 kg of water flowed through the pipe in 60 s. (a) Calculate the mean

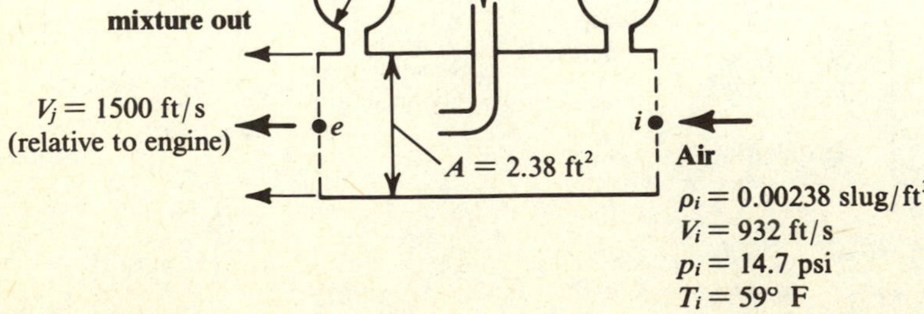

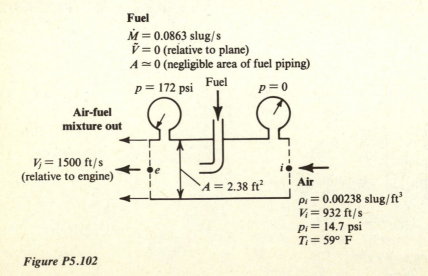

Figure P5.102

flow velocity. (b) Calculate the power loss if the process is steady, isothermal, and adiabatic (Fig. P5.106).

5.107 A steam turbine generates 254.5 Btu/lbm of shaft work. Given $\tilde{V}_i = 200$ ft/s, $\tilde{V}_e = 900$ ft/s, $h_i = 1200$ Btu/lbm, and $h_e = 900$ Btu/lbm, find the heat transfer per pound mass absorbed by or rejected by the fluid, if Δz and viscous losses can be neglected (Fig. P5.107).

5.108 The flow between parallel flat plates is obstructed by a half-cylinder attached to the lower plate. Assume the velocity profiles are uniform and the fluid is incompressible. Given the data in Fig. P5.108, find $\tilde{V}_2$ for an inviscid isothermeal steady flow.

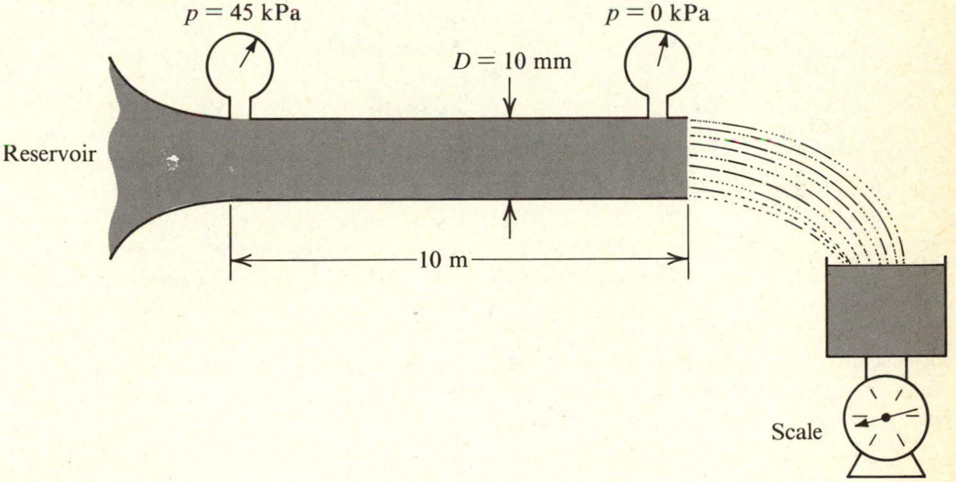

Figure P5.106

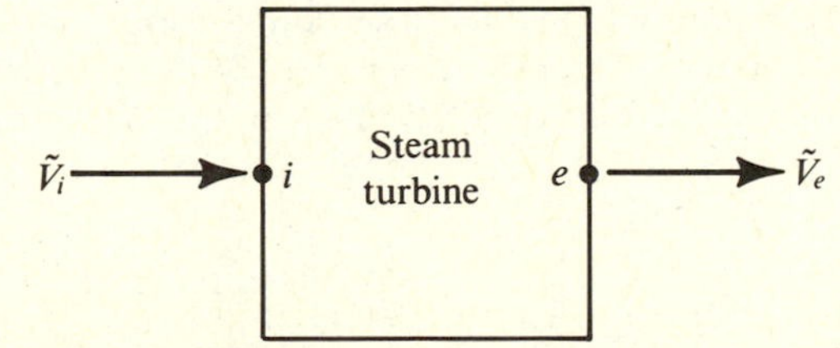

Figure P5.107

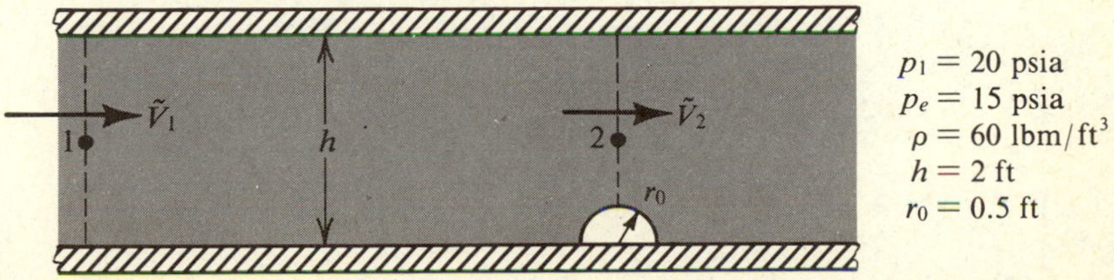

Figure P5.108

5.109 The pipe shown in Fig. P5.109 is 0.5 m in diameter and carries water at standard conditions at a steady rate of 0.5 m^3/s. What power must be supplied to the flow by the pump if the gauge pressure at 2 is to be 350 kPa, and the head loss h_f in the pipe is given as 50 $(\tilde{V}^2/2g)$.

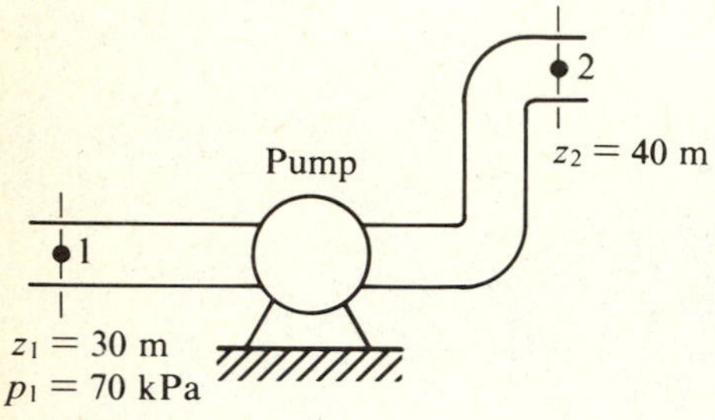

Figure P5.109

5.110 A horizontal pipe carries cooling water at standard conditions for a thermal power plant from the reservoir shown in Fig. P5.110. If the head loss h_f in the pipe can be expressed as

$$h_f = 0.002 \left(\frac{L}{D}\right) \frac{\tilde{V}^2}{2g}$$

where L is the length of pipe from inlet to exit station, and D is the inner diameter of the pipe, what is the pressure in a pipe 6 in. in diameter at the location $L = 4000$ ft?

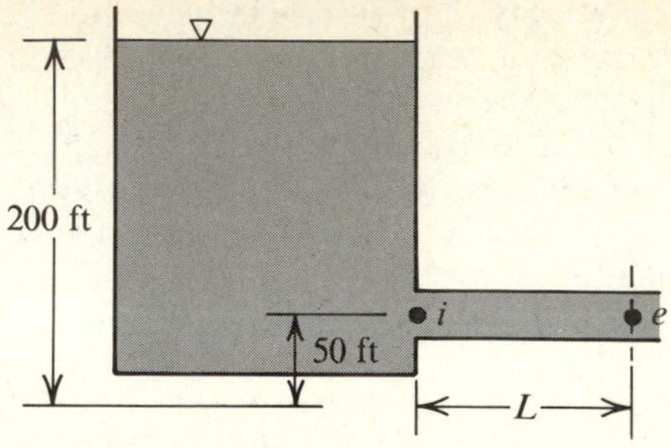

Figure P5.110

5.111 Consider the system in Fig. P5.111, which has been set up to determine the power of a hydraulic turbine. Assume that the system is 100% efficient, i.e., no friction or other losses. For a reading of 100 lbf on the scale, determine: (a) the volume flow rate Q through the system, and (b) the horsepower of the turbine.

5.112 In a steady flow device air is compressed adiabatically. Given the data in Fig. P5.112, determine the required shaft work per pound mass of air where $\dot{M} = 0.5$ slug/s. (Use gas tables for air.)

5.113 Water flows steadily through a pipe as shown in Fig. P5.113. The discharge point at 3 is open to atmospheric pressure. A heater around the pipe adds 100 Btu/lbm to the flow. The flow is steady, frictionless, and incompressible. The specific heat of water is C = 1 Btu/lbm°R. Assume $\Delta z = 0$. Find (a) $\dot{M}$, (b) the temperature of the water at 2 in °F.

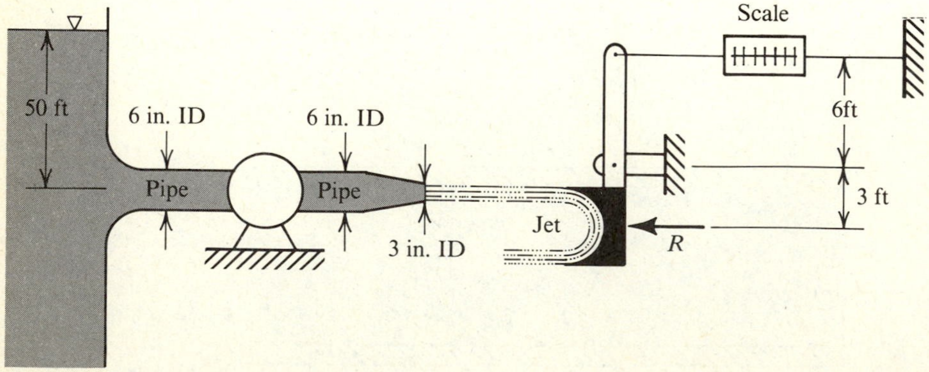

Figure P5.111

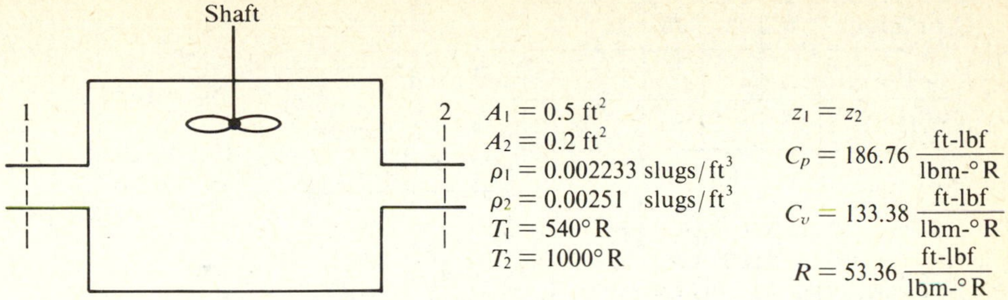

Shaft

$A_1 = 0.5 \text{ ft}^2$
$A_2 = 0.2 \text{ ft}^2$
$\rho_1 = 0.002233 \text{ slugs/ft}^3$
$\rho_2 = 0.00251 \text{ slugs/ft}^3$
$T_1 = 540° \text{R}$
$T_2 = 1000° \text{R}$

$z_1 = z_2$

$C_p = 186.76 \dfrac{\text{ft-lbf}}{\text{lbm-°R}}$

$C_v = 133.38 \dfrac{\text{ft-lbf}}{\text{lbm-°R}}$

$R = 53.36 \dfrac{\text{ft-lbf}}{\text{lbm-°R}}$

Figure P5.112

Heater

$\tilde{V}_1 = \dfrac{16}{\pi} \text{ ft/s}$
$T_1 = 40° \text{ F}$
$D_1 = 6 \text{ in.}$

$D_2 = 6 \text{ in.}$

$p_3 = p_{\text{atm}}$
$D_3 = 3 \text{ in.}$

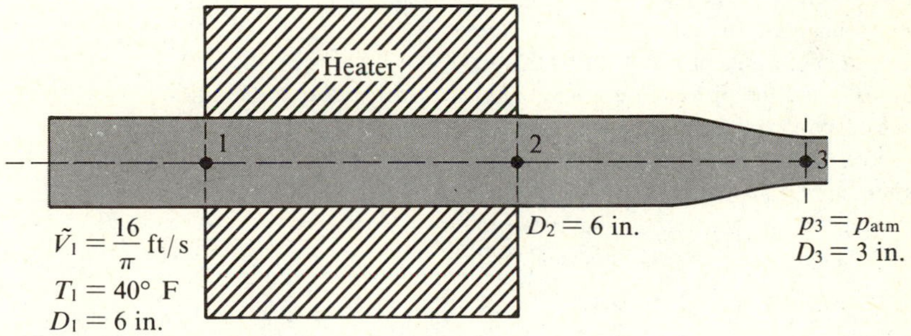

Figure P5.113

5.114 Air flows adiabatically through a supersonic wind tunnel as shown in Fig. P5.114. At the inlet to the tunnel a large reservoir exists. Therefore, $\tilde{V}_1 \cong 0$. The air temperature and pressure in the reservoir of the tunnel are $T_1 = 1000°\text{R}$ and $p_1 = 10$ atm. The throat temperature is $T_2 = 800°\text{R}$. Calculate the velocity at the throat.

5.115 Fresh water flows through a piping system. The head required by the system and the head available from the pump (at 1800 rpm) are given by the following equations: $H_{\text{reqd}} \cong 100 + 1.335 \ Q^2$, $H_{\text{avail}} \cong 500 - 0.26 \ Q^2$ where H is in feet and Q is in 10^3 gpm. (a) Fill in the chart for each value of Q. (b) Plot both curves on the grid. (c) Estimate Q at the operating point. (d) What head is required by the pump? (e) Calculate the specific speed, N_s. (f) If the pump is operating at its maximum efficiency (1800 rpm) what type of pump is this? (g) Estimate the horsepower required to drive this pump. Refer to Fig. P5.115.

Throat

p_1
T_1
$V_1 \approx 0$

• 1

Large
reservoir

$C_v = 0.171 \text{ Btu/lb-°R}$
$C_p = 0.24 \text{ Btu/lb-°R}$

2

$T_2 = 800° \text{R}$

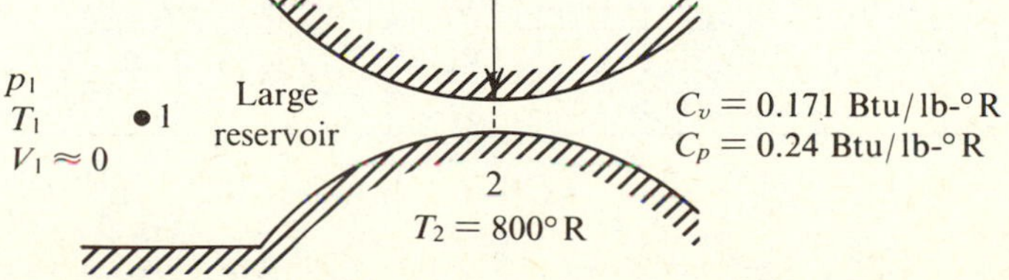

Figure P5.114

$Q[10^3$ gpm]	0	5	10	15	20
H_{avail} [ft]					
H_{reqd} [ft]					

5.116 An 8.95 cm outer diameter by 9.0 cm long piston moves downward in a 9.0 cm inner diameter cylinder. The downstroke velocity is 40 m/s. SAE 30 oil at 100°C provides lubrication between cylinder wall and piston. If the oil velocity profile is assumed to be linear, and the piston weight is negligible, find (a) the shear stress on the piston, (b) the force required to maintain the piston speed, (c) the energy equation for this problem. (d) If the oil were at 10°C and the piston 12 cm long, find the force required for the same speed.

5.117 Consider steady uniform flow in the convergent nozzle shown in Fig. P5.117. Obtain an expression for the heat transfer between stations 1 and 2 if the flow is isothermal.

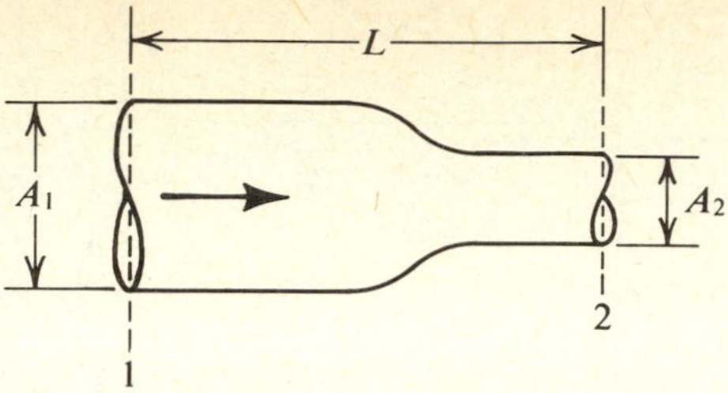

Figure P5.117

5.118 Consider a fluid flow whose velocity field is

$$\mathbf{V} = \frac{Vy}{h}\mathbf{i}$$

a temperature distribution

$$\frac{T - T_w}{T_\infty - T_w} = \frac{y}{h} + \frac{\mu V^2}{2k(T_\infty - T_w)}\left(\frac{y}{h}\right)(1 - y/h)$$

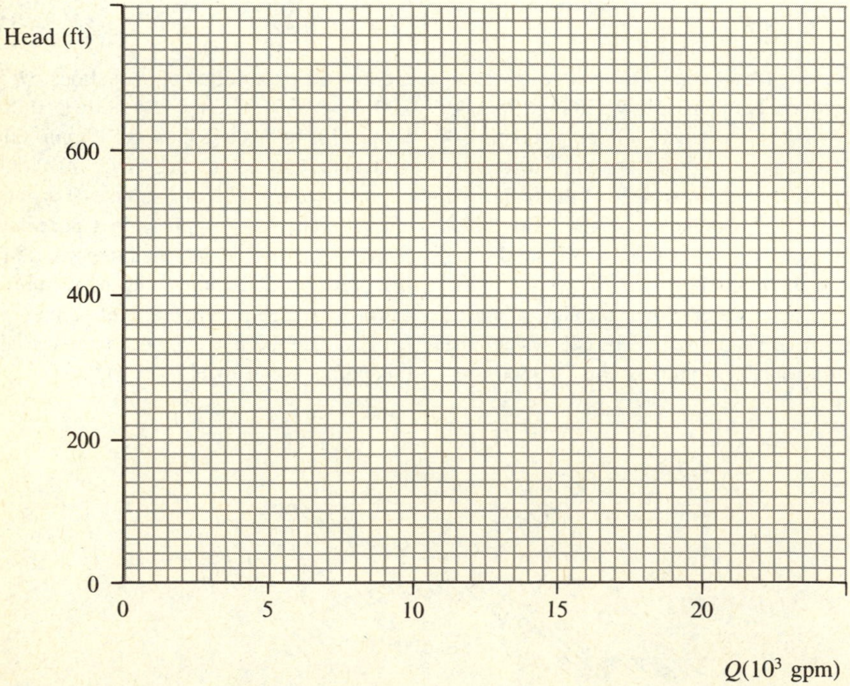

Figure P5.115

and a pressure field that is constant everywhere, where T_w and T_∞ are wall and ambient temperatures considered constant and μ, h, and k are constants. Verify that such a flow exists and satisfies the energy equation.

5.119 Given that the velocity and temperature distributions are

$$\mathbf{V} = V\left[1 - \left(\frac{2y}{h}\right)^2\right]\mathbf{i}$$

$$T = T_\infty + \mu\frac{V^2}{3k}\left[1 - \left(\frac{2y}{h}\right)^4\right]$$

where T_∞, μ, V, and k are constant, show that the energy equation is satisfied.

5.120 Figure P5.120 shows the free-surface of a whirlpool at distances not too close to the center. For this reason, it is known that the paths of fluid particles are concentric circles and that the velocity varies with radius according to $Vr = k$. Neglecting friction, determine the shape of the free-surface in terms of h, r, k, and fluid properties, where h is the distance measured downward from the level of the liquid at great distance from the center.

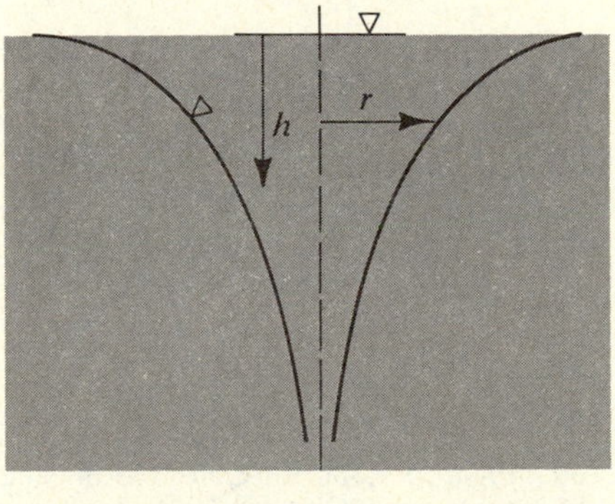

Figure P5.120

5.121 A point source of fluid is an imaginary point from which fluid flows radially outward in all directions with complete symmetry about the point. Consider the ejection of water from a point source with an instantaneous volume rate Q and with an instantaneous rate of change of volume rate $\dot{Q}$. Calculate the acceleration of a fluid particle which is instantaneously at the radius r.

5.122 To eliminate propellers on submarines, the propulsive duct illustrated in Fig. P5.122 has been suggested. Water is taken into the submarine at the nose, is raised in pressure by a pump, and then discharged through a nozzle at the rear of section 2. It is desired to investigate this system with the following: (a) pump increases the head by 100 ft (no losses); (b) pressure at discharge 2 is the same as at 1; (c) submarine travels at a steady speed of 20 ft/s; (d) external drag of the submarine is 2380 lbf. Calculate (a) pump volume rate of flow capacity and (b) discharge area A_2.

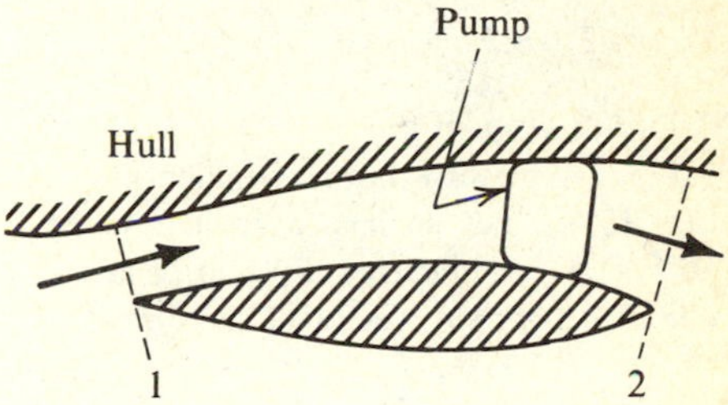

Figure P5.122

5.123 Consider a bubble of high pressure gas expanding in water in a spherical manner, as shown in Fig. P5.123. The gas is not soluble in water and water does not evaporate into the gas. At any instant, R is the radius of the bubble, dR/dt is the velocity of the interface, p_g is the gas pressure, u is the water velocity at the radius r, and p_∞ is the water pressure at a large distance from the bubble. Neglecting gravity, show (a) that at any instant

$$u = (dR/dt)\frac{R^2}{r^2}$$

(b) that the rate of bubble growth is

$$R\frac{d^2R}{dt^2} + \frac{3}{2}\left(\frac{dR}{dt}\right)^2 + \frac{2\sigma}{\rho R} = \frac{p_g - p_\infty}{\rho}$$

where σ is the surface tension at the gas-water interface.

5.124 Derive the power loss term $\dot{W}_\tau$ of Eq. (5.95), commencing with Cauchy's equation of motion.

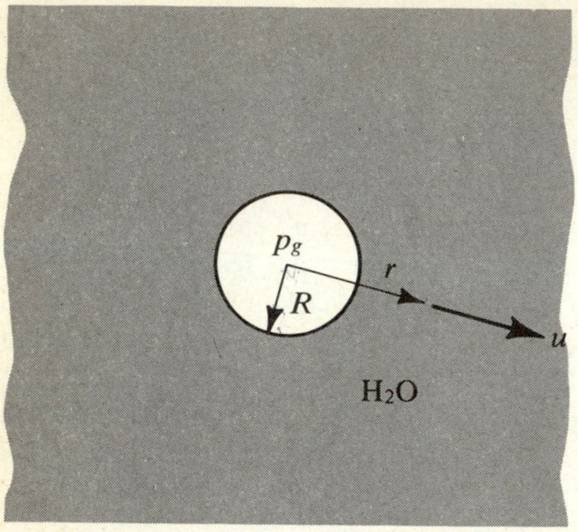

Figure P5.123

Angular Momentum

5.125 A rotating lawn sprinkler shown in Fig. P5.125 has an inlet average velocity of 50 ft/s, an exit diameter of 0.1 in., and a mass rate of 0.1 lbf·s/ft. Calculate the torque T.

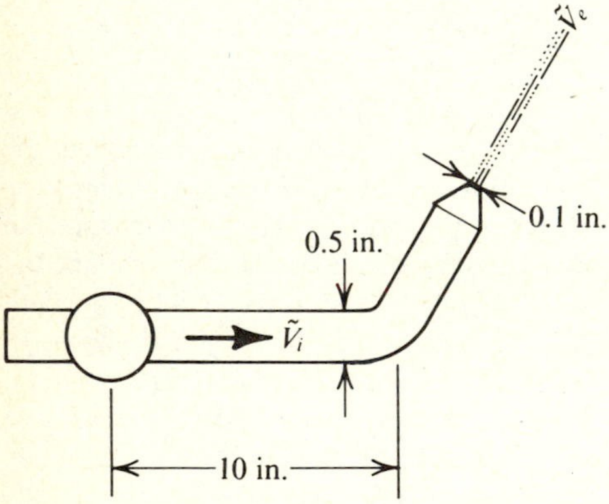

Figure P5.125

5.126 A disc rotates near a fixed surface. The radius of the disc is R, and the space between the disc and the surface is filled with a fluid of viscosity μ. The spacing between disc and surface is h feet, and the disc rotates at angular velocity ω. Find the dependence between the torque on the disc, T, and the other variables.

5.127 For a free-vortex flow, plot the velocity versus radius and compare it with the velocity versus radius for rigid body motion given that constant equals ω_1 in Eq. (5.88). What condition needs to be imposed for the velocities to be equal?

5.128 A fluid particle leaves the outer perimeter of a centrifugal impeller at a peripheral speed of 100 ft/s and a radial speed of 20 ft/s. What will be the peripheral velocity of this particle after it has traveled a radial distance of 0.5 in. from the perimeter of the impeller without any interference? Assume an incompressible inviscid fluid in steady flow.

5.129 Find the torque required to maintain a rotation of the sprinkler arms at 850 rpm, given a steady flow of water at the rate of 75 lbm/ft^3 and volume rate of 3.2 ft^3/s where the volume rate is equally divided between the two sprinkler arms. The two areas are A_1 = 0.5 in.^{2} and A_2 = 0.25 in.^{2}. What is the direction of rotation? (Refer to Fig. P5.129.)

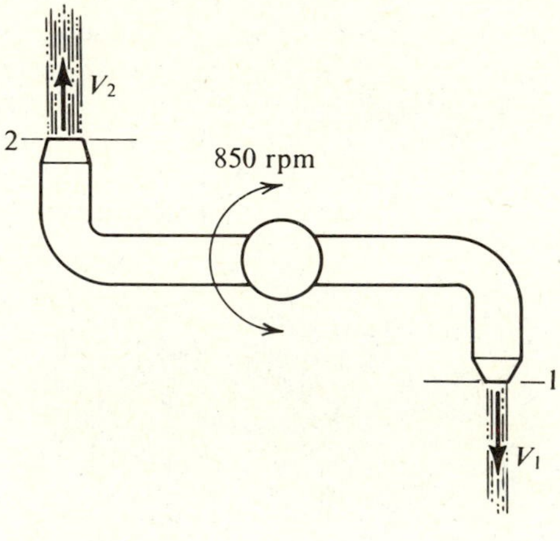

Figure P5.129

5.130 Figure P5.130 illustrates the impeller wheel of a turbine with an outside radius of 2 m. The blades are designed to deflect the relative flow by an angle of 172°. Flow strikes the blades with a velocity of 50 m/s from three equally spaced nozzles. Each nozzle delivers a jet of water 10 cm in diameter. If the turbine delivers 75 hp, find the rotational speed ω.

5.131 A stern-wheeler is a romantic rotational thruster used on river steamboats for transporting cargo and people in no hurry to get anywhere. The device consists of a paddle

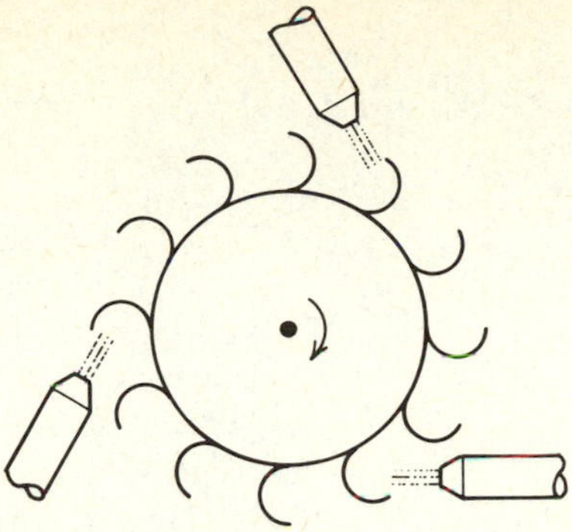

Figure P5.130

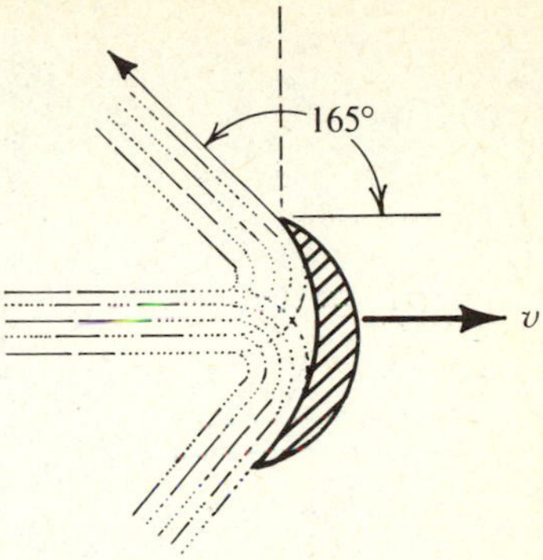

wheel that is turned by a steam engine through a crank and linkage assembly. If a torque T turns the wheel of diameter d and blade length l at ω rad/s, estimate the force per unit width of the wheel in a fluid of density ρ which moves at a velocity V toward the wheel. (See Fig. P5.131.)

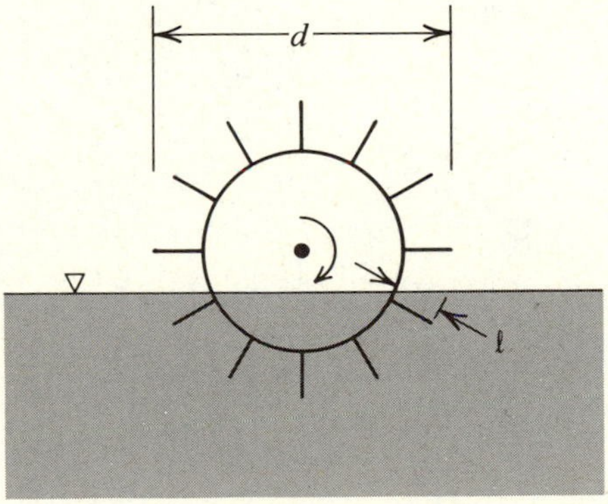

Figure P5.131

5.132 Consider fluid flowing over the blade shown in Fig. P5.132. The blade rotates at 600 rpm from water entering the blade radially at a velocity of 150 ft/s. If the distance between blades is 4 in., calculate the torque developed.

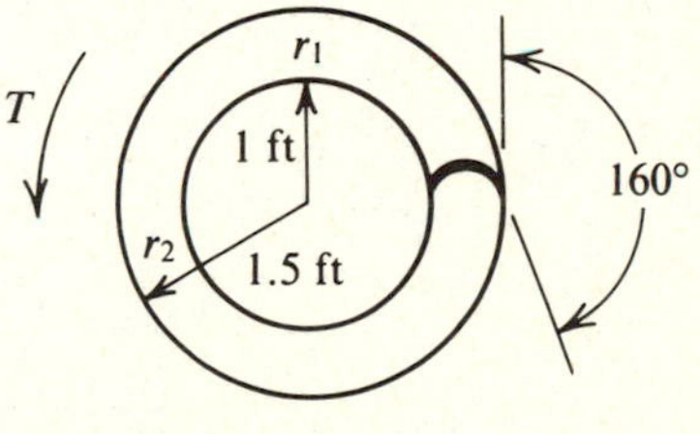

Figure P5.132

5.133 For the turbine shown in Fig. P5.133, derive the expression that the torque exerted on the wheel is $T = \rho Q r\,(V_1 - \omega r)(1 + \cos\theta)$.

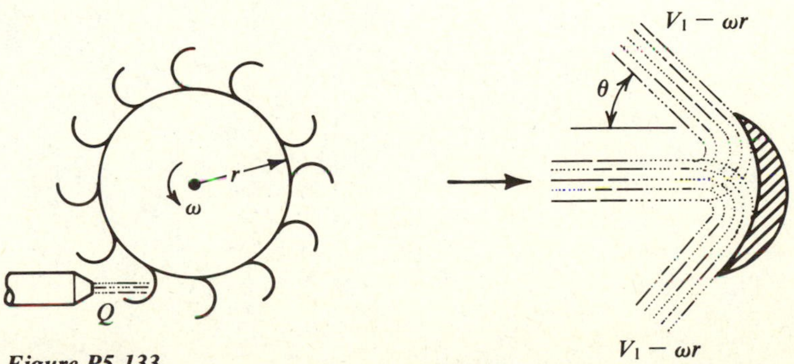

Figure P5.133

6 Recapitulation

6.1 Summary

Figure 6.1 schematizes the results we have reached thus far in our first five chapters. Note the dividing line in Fig. 6.1. In Chap. 1 we stated that any problem has two fundamental aspects to its solution: (1) we must select the correct mathematical tool to describe the functions and their changes, and (b) we must incorporate the appropriate physical laws that define the problem fully.

Three methods govern the mathematical modeling of any problem: methods of analysis, description, and approach. When we *analyze* a problem, we must first select the appropriate control: we can choose either a system (where we identify a fixed mass), or a control volume (where we identify a volume of space). Once we have selected the control, we then determine the method that *describes* the fluid's behavior for that control. If we have chosen a system, we select the Lagrangian description, because we want to study the history of a fixed mass. If we have chosen a control volume, we select the Eulerian description, because mass can enter, leave, accumulate, or deplete with respect to the control. Finally, we must decide how to *approach* the description of the control. Two forms of approach are available: the integral form (I.F), or the differential form (D.F.). The former is used for large volumes or quantities of mass, the latter for differential or elemental sizes.

Referring again to Fig. 6.1, we note that whatever choices we have made within the three methods of modeling, we will eventually have to consider the (I.F.) general property balance Eq. (4.13), or the (D.F.) general property balance Eq. (4.16). In Chap. 1 we applied the conservation principles of physics to these equations. Applying the conservation of mass, we derived a mass balance equation that predicts the behavior of a parcel of mass or mass through a continuous volume of space. Applying the conservation of linear and angular momentum, we produced a similar result for the behavior of the fluid flow linear and angular momentum. Applying the conservation of energy, we found an energy balance for either an elemental volume or a macroscopic one. The results of applying conservation laws to fluid flow are summarized as follows.

1. The (D.F.) equation for the conservation of mass is

$$\frac{\partial \rho}{\partial t} + \nabla \cdot \rho \mathbf{V} = 0 \tag{6.1}$$

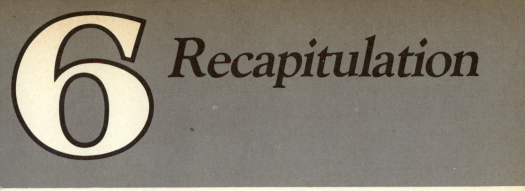

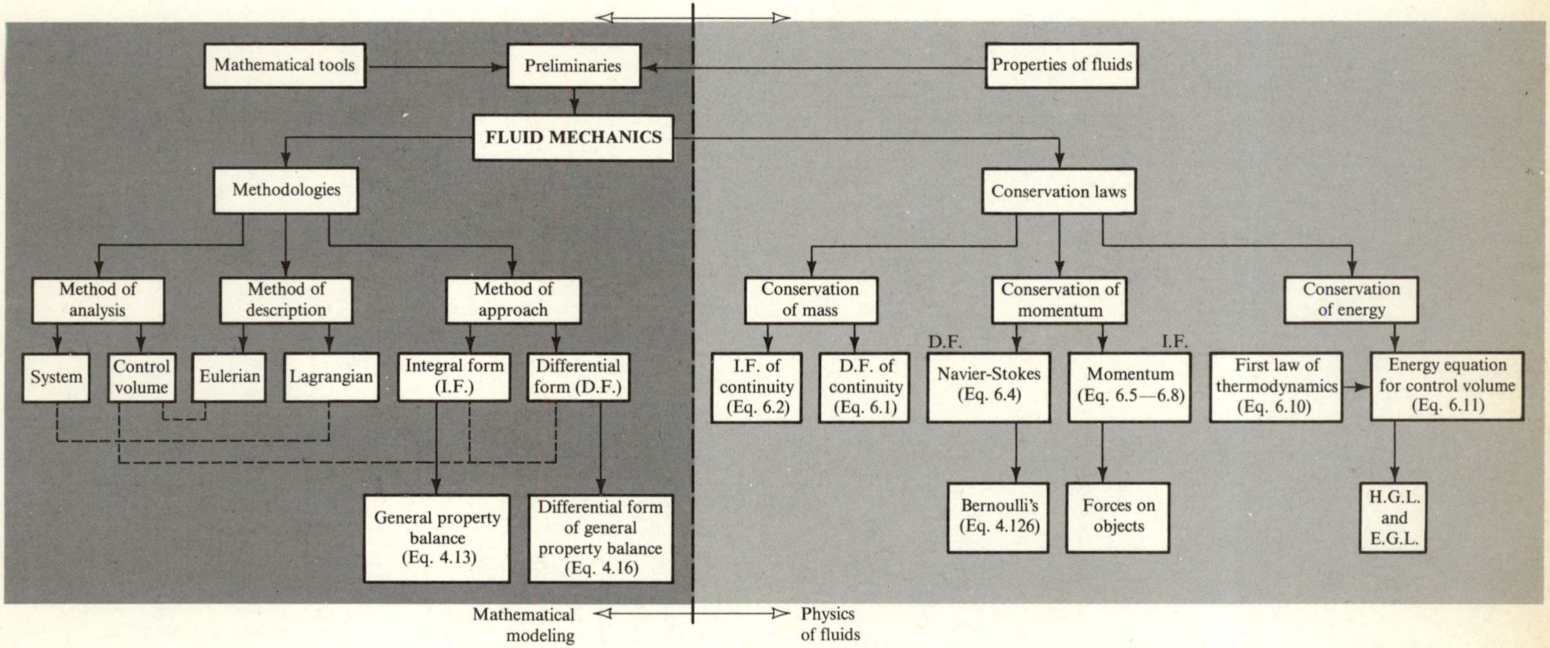

Figure 6.1 *Schematic of introduction to the theory of fluid mechanics.*

2. The (I.F.) equation for the conservation of mass is

$$\frac{\partial}{\partial t}\int_{\forall} \rho d\forall + \oint_A \rho \mathbf{V}\cdot d\mathbf{A} = 0 \tag{6.2}$$

The one-dimensional incompressible form is

$$Q = \tilde{V}A \tag{6.3}$$

3. The (D.F.) equation for the conservation of linear momentum is

$$\frac{\partial \mathbf{V}}{\partial t} + (\mathbf{V}\cdot\nabla)\mathbf{V} = \frac{1}{\rho}\nabla p + \mathbf{g} + \nu\nabla^2\mathbf{V} + \frac{1}{3}\nu\nabla(\nabla\cdot\mathbf{V}) \tag{6.4}$$

For incompressible flow, $\nabla\cdot\mathbf{V} = 0$, and the last term vanishes.

4. The (I.F.) equation for the conservation of linear momentum is

$$\frac{\partial}{\partial t}\int_{\forall} \rho \mathbf{V}\, d\forall + \oint_A \rho \mathbf{V}\,(\mathbf{V}\cdot d\mathbf{A}) = \Sigma\mathbf{F} \tag{6.5}$$

For incompressible steady one-dimensional flow, the equation reduces to

$$\Sigma\mathbf{F} = \rho Q\Delta\tilde{\mathbf{V}} \tag{6.6}$$

5. The (I.F.) equation for the conservation of angular momentum is

$$\frac{\partial}{\partial t}\int_{\forall} \rho\mathbf{r} \times \mathbf{V}\, dA + \oint_A \rho\mathbf{r} \times \mathbf{V}(\mathbf{V}\cdot d\mathbf{A}) = \Sigma\mathbf{r} \times \mathbf{F} \tag{6.7}$$

For incompressible steady one-dimensional flow, the equation reduces to

$$\Sigma T = \rho Q(r_e\tilde{V}_e \cos\alpha_e - r_i\tilde{V}_i \cos\alpha_i) \tag{6.8}$$

6. The (D.F.) equation for the conservation of energy is

$$\left(\frac{\text{đ}w}{dt}\right)_{\text{mech}} = \frac{\partial}{\partial t}(\rho\bar{e}) + \nabla\cdot[(\bar{e} + p/\rho)\rho\mathbf{V} - k\nabla T + 2\mu\mathbf{V}\cdot\dot{\mathbf{S}}] - 2\mu(\dot{\mathbf{S}}\cdot\nabla)\cdot\mathbf{V} \tag{6.9}$$

7. The (I.F.) equation for the conservation of energy is

$$ {}_i\dot{Q}_e + {}_i\dot{W}_e = \frac{\partial}{\partial t}\int_{\forall} \rho\left(\bar{i} + \frac{V^2}{2} + gz\right) d\forall + \oint_A \rho\left(\bar{i} + \frac{V^2}{2} + gz\right) \mathbf{V}\cdot d\mathbf{A} \quad (6.10) $$

For incompressible, steady, one-dimensional flow, Eq. (6.10) can be expressed as

$$ ({}_iw_e)_{\text{mech}} + \left(\frac{p}{\gamma} + \frac{\tilde{V}^2}{2g} + z\right)_i - \left(\frac{p}{\gamma} + \frac{\tilde{V}^2}{2g} + z\right)_e = h_f \quad (6.11) $$

The foregoing set of equations represents the governing equations of motion for fluid flow. We shall next summarize some special forms of the governing equations.

6.2 Special Forms of the Governing Equations

Special forms of some of the governing equations of motion are of value when we deal with certain types of fluid flow. For example, whenever the flow is

1. Steady:

$$ \frac{\partial}{\partial t}(\qquad\qquad) = 0 \quad (6.12) $$

2. Inviscid:

$$ \nu(\qquad\qquad) = 0 \quad (6.13) $$

3. Two-dimensional:

$$ \frac{\partial}{\partial z}(\qquad\qquad) = 0 \quad (6.14) $$

   where x and y are the principle direction of flow.
4. One-dimensional:

$$ \frac{\partial}{\partial y}(\qquad\qquad) = \frac{\partial}{\partial z}(\qquad\qquad) = 0 \quad (6.15) $$

   where x is the principle direction of flow.
5. Incompressible:

$$ \nabla\cdot\mathbf{V} = 0 \quad (6.16) $$

6. Irrotational:

$$\nabla \times \mathbf{V} = 0 \tag{6.17}$$

7. Definition of a real average:

$$(\qquad\qquad)\tilde{\mathbf{A}} = \int_{\mathbf{A}} (\qquad\qquad) d\mathbf{A} \tag{6.18}$$

Substitution of these results in any of the governing equations helps reduce the complexity of the equations. Special forms of the equations can now be constructed. Certain (D.F.) equations of linear momentum are presented in Table 6.1. So important are these forms in fluid dynamics that we will devote entire chapters to them. For example, Laplace's equation is the governing equation for potential theory (the topic of Chap. 12). The differential equation for the accelerating flat plate is important as it introduces us to the concept of the boundary layer (the topic of Chap. 14). Poiseuille's equation is important in describing laminar flow in a pipe (the subject of Chap. 10). A solution of Euler's equation yields one of the most widely used equations in fluid dynamics, namely, Bernoulli's equation.

All the equations shown in Table 6.1 have exact solutions. Though the list is far from complete, it does illustrate that, for certain fluid flows, the Navier-Stokes equations can be reduced to a form where we can obtain closed form analytic solutions. This will be pursued in Chap. 9.

Figure 6.2 shows how we might take one of the two forms of approach (the D.F.) and apply it to a general class of incompressible fluid flows. We start by identifying the two governing equations: the continuity equation and the Navier-Stokes equation.

The types of flows we can treat depend upon the relative significance of the various acceleration terms in the Navier-Stokes equation. Referring to Fig. 6.2, we can identify the following terms:

1. ⓐ The local acceleration, which is important only if the fluid properties change at a point in the flow field.
2. ⓑ The convective acceleration, which would be zero if the flow is uniform because there is no change in the velocities in space.
3. ⓒ The pressure acceleration, which would vanish because, for a fluid with a free-surface, the pressure cannot change (except hydrostatically).
4. ⓓ The body force per unit mass. This deceleration would vanish for two-dimensional planar flow.
5. ⓔ The viscous deceleration, which vanishes for an inviscid fluid.

All incompressible flows using the (D.F.) in describing the fluid motion must satisfy this Navier-Stokes equation. We have already studied some special cases:

- Hydrostatics: ⓐ = ⓑ = ⓔ = 0, in Chap. 3.
- Inviscid flows: ⓔ = 0, in Chaps. 4 and 5.

Table 6.1 *Equations of Fluid Flow That Have Exact Solutions*

Assumptions	Origin (Equation)		Mathematical Expression	Coord. System	Name	References (equation)
ρ = constant		(i)	$\frac{\partial V}{\partial t} + V\frac{\partial V}{\partial s} = -\frac{1}{\rho}\frac{\partial p}{\partial s}$	Intrinsic		(4.110)
$\nu = 0$	(4.109)	(ii)	$\frac{\partial u}{\partial t} + u\frac{\partial u}{\partial x} + v\frac{\partial u}{\partial y} = -\frac{1}{\rho}\frac{\partial p}{\partial x}$	Rectangular	Euler's equation	
		(iii)	$\frac{\partial v}{\partial t} + u\frac{\partial v}{\partial x} + v\frac{\partial v}{\partial y} = -\frac{1}{\rho}\frac{\partial p}{\partial y}$			
$\frac{\partial}{\partial \theta} = 0$		(iv)	$\frac{\partial v_r}{\partial t} + v_r\frac{\partial v_r}{\partial r} - \frac{v_\theta^2}{r} = -\frac{1}{\rho}\frac{\partial p}{\partial r}$	Polar		
$\frac{\partial}{\partial z} = 0$		(v)	$\frac{\partial v_\theta}{\partial t} + v_r\frac{\partial v_\theta}{\partial r} + \frac{v_r v_\theta}{r} = 0$			
ρ = constant		(vi)	$\frac{\partial^2\phi}{\partial x^2} + \frac{\partial^2\phi}{\partial y^2} + \frac{\partial^2\phi}{\partial z^2} = 0$	Rectangular		
$\nabla \times \mathbf{V} = 0$	(4.24)				Laplace's equation	(12.2)
(Potential flow)		(vii)	$\frac{\partial^2\phi}{\partial r^2} + \frac{1}{r}\frac{\partial\phi}{\partial\theta} + \frac{1}{r^2}\frac{\partial^2\phi}{\partial\theta^2} + \frac{\partial^2\phi}{\partial z^2} = 0$	Cylindrical		
ρ = const.						
$\frac{\partial(\ )}{\partial t} = 0$	(4.109)	(viii)	$\frac{d^2u}{dy^2} = \frac{1}{\mu}\frac{dp}{dx}$	Rectangular	Poiseuille's equation	(9.6)
$\mathbf{g} = 0$						
$v = w = 0$						

Table 6.1 (*Con't.*)

Assumptions	Origin (Equation)	Mathematical Expression	Coord. System	Name	References (equation)
$\rho = \text{const.}$, $\frac{\partial(\ )}{\partial t} = 0$, $\mathbf{g} = 0$, $v_r = w = 0$, $\nu = 0$	(4.109)	(ix) $v_\theta^2 = \frac{r}{\rho}\frac{dp}{dr}$; (x) $\frac{d^2 v_\theta}{dr^2} + \frac{d}{dr}\left(\frac{v_\theta}{r}\right) = 0$	Polar	Axisymmetric; Inviscid circular vortex flow	
$\rho = \text{const.}$, $\frac{\partial(\ )}{\partial t} = 0$, $\mathbf{g} = 0$, $\frac{\partial}{\partial z} = 0$, $v_r = v_\theta = 0$	(4.109)	(xi) $\frac{d^2 w}{dr^2} + \frac{1}{r}\frac{dw}{dr} = \frac{1}{\mu}\frac{dp}{dz}$	Cylindrical	Hagen-Poiseuille equation	Prob. 4.69
$\rho = \text{const.}$, $\frac{\partial}{\partial t} = 0$, $\frac{\partial}{\partial z} = 0$, $\mathbf{g} = 0$, $v_\theta = w = 0$	(4.109)	(xii) $v_r \frac{dv_r}{dr} = -\frac{1}{\rho}\frac{dp}{dr} + \nu\left(\frac{\partial^2 v_r}{\partial r^2} + \frac{1}{r}\frac{\partial v_r}{dr} + \frac{1}{r^2}\frac{\partial^2 v_r}{\partial \theta^2} - \frac{v_r}{r^2}\right)$; (xiii) $0 = -\frac{\partial p}{\partial \theta} + \frac{2\mu}{r^2}\frac{\partial v_r}{\partial \theta}$	Cylindrical	Hamel's equation	

Table 6.1 (Con't.)

Assumptions	Origin (Equation)	Mathematical Expression	Coord. System	Name	References (equation)
ρ = const. $\mathbf{g} = 0$ p = const. $v = w = 0$	(4.109)	(xiv) $\frac{\partial u}{\partial t} = \nu \frac{\partial^2 u}{\partial y^2}$	Rectangular	Accelerating flat plate	(9.24)
ρ = const. $\mathbf{g} = 0$ $v_\theta = w = 0$ Axisymmetric	(4.109)	(xv) $\frac{\partial v_r}{\partial t} + v_r \frac{\partial v_r}{\partial r} = -\frac{1}{\rho}\frac{\partial p}{\partial r} + \nu\left(\frac{\partial^2 v_r}{\partial r^2} + \frac{2}{r}\frac{\partial v_r}{\partial r} - \frac{2v_r}{r^2}\right)$	Polar	Bubble dynamics	

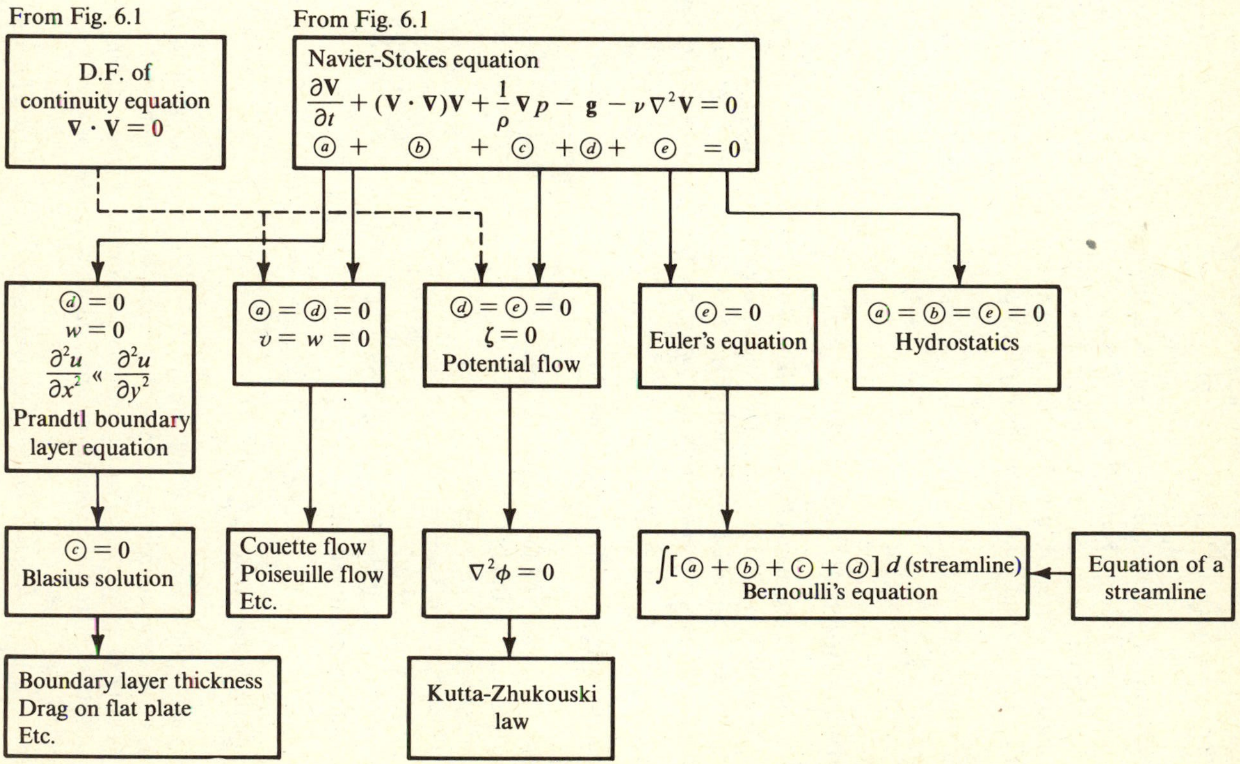

Figure 6.2 *Continuation of schematic of introduction to the theory of fluid mechanics.*

In the chapter to follow, we shall study

- Poiseuille flows: ⓐ = ⓓ = 0, in Chaps. 9 and 10.
- Potential flows: ⓓ = ⓔ = 0, in Chaps. 12 and 13.
- Boundary layer flows: ⓓ = 0, in Chap. 14.

In particular, we shall examine a variety of applications. For example, we shall study the principle of flight (Kutta-Zhukouski law) which arises out of potential flow theory, or the friction drag on a ship's hull, which arises out of the Blasius solution of the boundary layer flow theory. We shall learn about pipe flow and free channel flow. Not all branches of fluid dynamics can be treated in this text, but many shall be.

6.3 Problem-Solving Technique

How do we solve fluid dynamics problems? There is no mutually agreed upon single way to set up solutions. The choice of problem-solving technique may depend on nothing more than the familiarity and confidence of the user. For ease and clarity, especially for beginners, we suggest the following problem-solving methodology:

Step 1.
Identify the characteristics of the fluid and flow field.

1. List the data given and the results required.
2. Check to see if the fluid is
   (a) incompressible or compressible
   (b) inviscid or real
   (c) steady or unsteady
   (d) one-, two-, or three-dimensional flow
   (e) rotational or irrotational

Step 2.
Select the method of approach.

1. Examine the problem to see if quantities are to be solved at a point in the flow (D.F.), or at some boundary, or on some volume of a large control (I.F.).
2. Draw and label any necessary diagrams needed for the solution. Examine what forces or stresses are acting on the control and what transfer of energies through the control may exist.
3. Determine if the description takes place with respect to a *fixed* reference frame or a *moving* reference frame. The former utilizes absolute velocities and accelerations; the latter utilizes relative velocities and accelerations.
   (a) Select the appropriate coordinate system that best describes the motion.
   (b) Let one of the coordinates be oriented in the direction of the flow.

Step 3.
Write the appropriate form of the governing equations of flow (see the summary of equations in Sec. 6.1).

1. Apply the results of Step 1 to reduce the equations to their simplest form.
2. Examine all conditions given at a point or boundary.
   (a) Identify all kinematic conditions.
   (b) Identify all stress conditions.
   (c) Identify all "inner" conditions.
   (d) Identify all initial conditions.

Step 4.
Solve the problem.

1. Think about the problem in terms of its physical significance. Each term of the reduced governing equation and its boundary conditions have a physical interpretation. The dimensions of each term must be identical and correct.
2. Check to see if there are sufficient equations to solve for the unknowns. If not, then additional equations will have to be sought, e.g., the equation of state and stress-strain relationships.
3. Select a method for solving the reduced governing equations. The following two sections offer a few aids in selecting an appropriate method.

6.4 Examples of Problem-Solving Technique

We are now at a position to apply our problem-solving technique to a flow problem. We have all the necessary tools to commence building a solution: the governing equations of fluid flow; the various methods of analysis, description, and approach to treat a problem; ways to fit the equations to the given problem and simplify their expressions; the mathematical symbols and operators necessary to describe the behavior of the flow and the methods to solve them; and the step-by-step procedure to bring all this together. Two problems will be analyzed in detail. The first will involve the (D.F.) approach, and the second will involve the (I.F.) approach.

Example 6.1. The (D.F.) Approach
Consider a steady two-dimensional incompressible viscous flow with no body forces, where the pressure gradient $\partial p/\partial x$ is zero and the y-component of velocity v is equal in magnitude to the kinematic viscosity ν and is a constant. At the location y equals zero, the shear stress p_{xy} is unity. (a) Find an expression for the shear stress p_{xy} and vorticity component ζ_z about the z-axis that is valid everywhere in the flow field. (b) Find the distribution of velocity given it is zero at y equal zero.

Example 6.1 *(Con't.)*

Solution:

Step 1.

Identify the characteristics of the fluid and flow field:

- The flow is steady:

$$\frac{\partial(\quad)}{\partial t} = 0$$

- The flow is two-dimensional:

$$\frac{\partial(\quad)}{\partial z} = 0$$

- The flow is incompressible:

$$\rho = \text{const.}$$

- The flow is real:

$$\nu \neq 0$$

- There are no body forces:

$$\mathbf{g} = 0$$

- The pressure gradient is zero:

$$\frac{\partial p}{\partial x} = 0$$

- The velocity component v is given:

$$v = \nu = \text{dynamic viscosity}$$

- A boundary condition is given:

$$p_{xy}\Big|_{y=0} = 1$$

- Find $p_{xy}\,(x, y, z, t)$ and $\zeta_z\,(x, y, z, t)$

Step 2.

Select the method of approach:

- (D.F.)
- No external forces, no energy transfers
- Fixed reference frame (Cartesian)

Example 6.1 *(Con't.)*

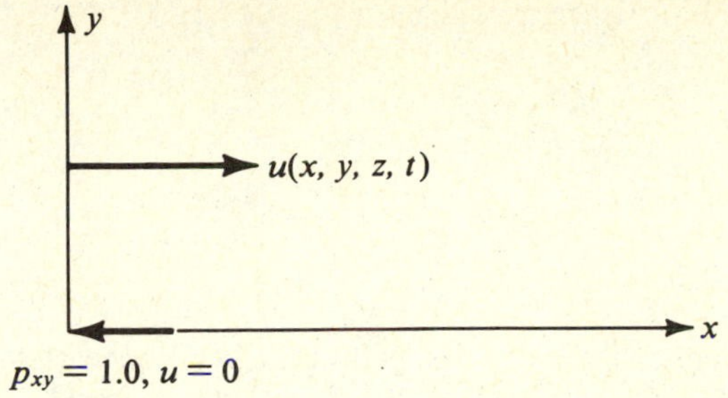

Figure E6.1

Step 3.
Write the *appropriate* form of the governing equations of flow:

- Continuity equation:

$$\frac{\partial u}{\partial x} + \cancelto{0}{\frac{\partial v}{\partial y}} = 0 \tag{i}$$

- Navier-Stokes equations:

$$u\frac{\partial u}{\partial x} + v\frac{\partial u}{\partial y} = -\frac{1}{\rho}\cancelto{0}{\frac{\partial p}{\partial x}} + \nu\left(\frac{\partial^2 u}{\partial x^2} + \frac{\partial^2 u}{\partial y^2}\right) \tag{ii}$$

$$u\cancelto{0}{\frac{\partial v}{\partial x}} + v\cancelto{0}{\frac{\partial v}{\partial y}} = -\frac{1}{\rho}\frac{\partial p}{\partial y} + \nu\left(\cancelto{0}{\frac{\partial^2 v}{\partial x^2}} + \cancelto{0}{\frac{\partial^2 v}{\partial y^2}}\right) \tag{iii}$$

since we were given $\partial p/\partial x = 0$ and $v = \nu$.

Equation (i) becomes

$$\frac{\partial u}{\partial x} = 0 \tag{iv}$$

which means that since the velocity component u is not a function of x, z, and t, it is at most a function only of y.

In a similar fashion, from Eq. (iii),

$$\frac{\partial p}{\partial y} = 0 \tag{v}$$

Thus, the pressure everywhere in the flow field is constant.

Example 6.1 *(Con't.)*

The Navier-Stokes equation, Eq. (ii), reveals that

$$\frac{d^2u}{dy^2} = \frac{du}{dy} \tag{vi}$$

Step 4.
Solve the problem.

We can solve this problem analytically. Integrating Eq. (vi) once yields

$$\frac{du}{dy} = c_1 e^y \tag{vii}$$

where c_1 is an arbitrary constant of integration. To evaluate it, we use the boundary condition that

$$p_{xy} = \mu \left. \frac{du}{dy} \right|_{y=0} = 1 \tag{viii}$$

From Eqs. (vii) and (viii), we obtain

$$c_1 = \frac{1}{\mu} \tag{ix}$$

(a) The shear stress distribution is from Eqs. (vii) and (ix),

$$p_{xy} = e^y \tag{x}$$

The vorticity component ζ_z is now easily obtained as

$$\zeta_z = -\frac{du}{dy} \tag{xi}$$

or

$$\zeta_z = -\frac{e^y}{\mu} \tag{xii}$$

The quantities p_{xy} and $\mu\zeta_z$ are shown plotted in Fig. 6.3 for values of negative y.

(b) If we impose the condition that the velocity u is zero at y equals zero, we obtain

Example 6.1 *(Con't.)*

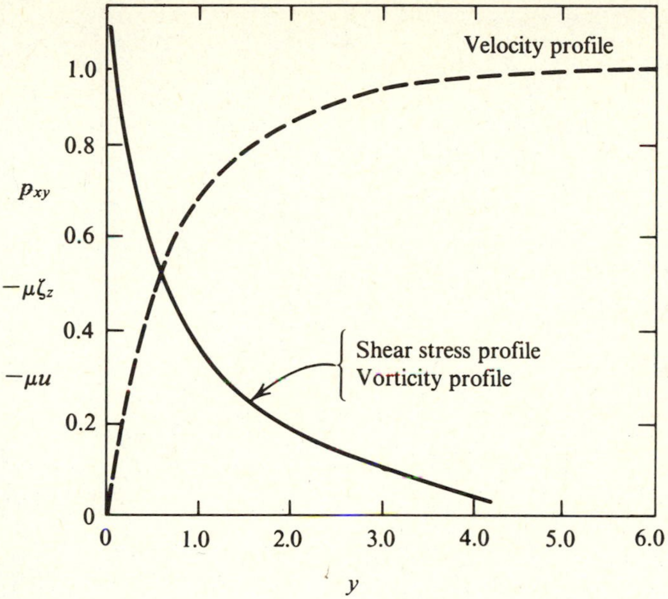

Figure 6.3 *Example of the (D.F.) problem.*

$$u = \frac{1}{\mu}(e^y - 1) \qquad \text{(xiii)}$$

as the velocity profile. The distribution is shown plotted in Fig. 6.3 above.

This completes the solution.

Example 6.2. The (I.F.) Approach

Water enters a 60° horizontal reducing elbow with a velocity of 15 ft/s at a pressure of 12 psi. Calculate the x- and y-components of the reaction force acting on the elbow. The entering diameter is 12 in., and the exit diameter is 8 in. Neglect viscous effects.

Solution:

Step 1.

Identify the characteristics of the fluid and flow field:

- The flow is steady:

$$\frac{\partial}{\partial t}(\qquad) = 0$$

Example 6.2 *(Con't.)*

- The flow is one-dimensional:

$$\mathbf{V} = \tilde{V}\mathbf{e}_s$$

- The flow is incompressible:

$$\rho = \text{const.}$$

- The flow is ideal:

$$\nu = 0$$

- There are no body forces:

$$\mathbf{W} = 0$$

- Given data:

$$\tilde{V}_i = 15 \text{ ft/s}$$

$$p_i = 12 \text{ psi}$$

$$D_i = 12 \text{ in.}$$

$$D_e = 8 \text{ in.}$$

$$\theta = 60°$$

- Find: p_e, $\tilde{V}_e$, R_x, R_y

Step 2.
Select the method of approach:
- (I.F.)
- External forces are pressure forces; no energy transfers
- Fixed reference frame (Cartesian)

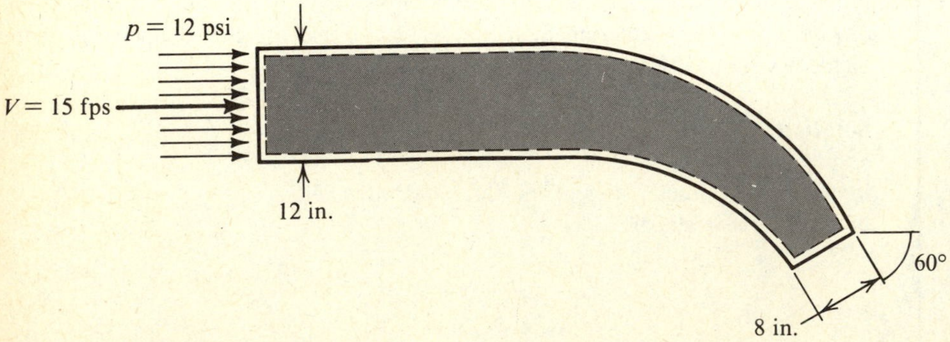

Figure E6.2

Example 6.2 *(Con't.)*

Step 3.
Write the appropriate form of the governing equations of flow:

- Continuity equation:

$$Q = (\tilde{V}A)_i = (\tilde{V}A)_e \tag{i}$$

- Linear momentum:

$$\Sigma F_x = \rho Q(\tilde{V}_e - \tilde{V}_i)_x \tag{ii}$$

$$\Sigma F_y = \rho Q(\tilde{V}_e - \tilde{V}_i)_y \tag{iii}$$

- Energy:

$$\left(\frac{p}{\gamma} + \frac{\tilde{V}^2}{2g} + z\right)_i = \left(\frac{p}{\gamma} + \frac{\tilde{V}^2}{2g} + z\right)_e \tag{iv}$$

Step 4.
Solve the problem:
From continuity equation (i),

$$15 \times \frac{\pi}{4} \times 12^2 = \tilde{V}_2 \times \frac{\pi}{4} \times 8^2$$

or

$$\tilde{V}_2 = \frac{9}{4} \times 15 = 33.75 \text{ fps} \tag{v}$$

From Bernoulli equation (iv),

$$\frac{p_1}{\gamma} + \frac{\tilde{V}_1^2}{2g} = \frac{p_2}{\gamma} + \frac{\tilde{V}_2^2}{2g}$$

$$\frac{p_2}{\gamma} = \frac{p_1}{\gamma} + \frac{\tilde{V}_1^2}{2g} - \frac{\tilde{V}_2^2}{2g}$$

or

$$\frac{p_2}{\gamma} = \frac{12 \times 144}{62.4} + \frac{1}{64.4}(15^2 - 33.75^2)$$

$$p_2 = 5.85 \text{ psi} \tag{vi}$$

From the momentum equation (ii),

$$\Sigma F_x = -R_x + p_1A_1 - p_2A_2 \cos 60°$$

$$= \rho Q(\tilde{V}_1 \cos\theta - \tilde{V}_1)$$

$$\rho Q\tilde{V}_2 = 1.94 \times 11.8 \times 33.75 = 773$$

Example 6.2 *(Con't.)*

$$\rho Q\tilde{V}_1 = 1.94 \times 11.8 \times 15 = 343$$

$$p_2A_2 = 5.85\left(\frac{\pi}{4}\right)8^2 = 294$$

$$p_1A_1 = 12\left(\frac{\pi}{4}\right)12^2 = 1357$$

Therefore

$$R_x = -773(.5) + 343 - 294(.5) + 1357$$

or

$$\boxed{R_x = 1167 \text{ lbf}} \tag{vii}$$

From the momentum Eq. (iii),

$$\Sigma F_y = -R_y + p_2A_2 \sin 60° = \rho Q(\tilde{V}_2 \sin 60° - 0)$$

therefore,

$$R_y = 294(.866) + 773(.866) = 254.5 + 670$$

or

$$\boxed{R_y = 924.5 \text{ lbf}} \tag{viii}$$

This completes the solution.

7 Dimensional Analysis and Similitude

7.1 *Introduction*

In 1942 a V-2 rocket (see Fig. 7.1) flew across a London residential area at around Mach 5. The world had seen nothing like this terrifying new weapon. It almost resulted in Britain's capitulation. When the allies captured the Nazi rocket test facility at Peenemünde, they were astounded to find not only a Mach 5 wind tunnel and scaled models of the V-2 rocket, but also plans for a Mach 10 tunnel and models of missiles

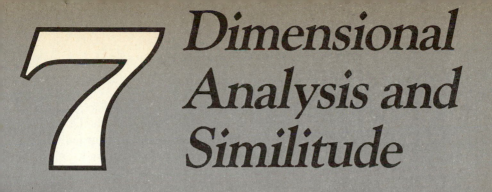

Figure 7.1 *A V-2 rocket being prepared for launching. (Source: Official U.S. Navy photograph.)*

targeted against the United States. German engineers, working in haste and secrecy, understood the immense value of models.

Today no airplane is built that has not first been tested in a wind tunnel, and no large ship that has not been modeled and run in a tow tank. The space shuttle went through 100,000 hours of wind tunnel testing. The Boeing company spent millions testing their 727, 737, 747, and 767 commercial jets before releasing them for mass production. By building a model carefully, engineers can reduce an object, or a whole system like a galaxy, to a manageable form; better yet, they can tinker with it in a controlled environment. Engineers use wind tunnels (Fig. 7.2), tow tanks, and water tunnels (Fig. 7.3) to study the behavior of models for missiles, planes, and submarines, for cars, boats, and spacecraft, for tents, skyscrapers, whole towns, and even birds—for anything surrounded by a moving fluid.

We create experimental problems and build models not because we lack confidence in theoretical predictions, but because some problems in fluid dynamics are so complex and unwieldy that it is not possible to solve them directly. This is true on both a theoretical and a practical level: devising valid experiments requires both a careful formulation of an experimental problem and the precise construction of physical models.

Figure 7.2 *Advanced wind tunnel: Photograph of Ames 80 ft × 120 ft test section, 345 mph maximum wind speed. Ames wind tunnel facility. (Source:* Wind Tunnels of NASA, *D. D. Boals and W. R. Corliss, NASA SP-440. Courtesy of NASA.)*

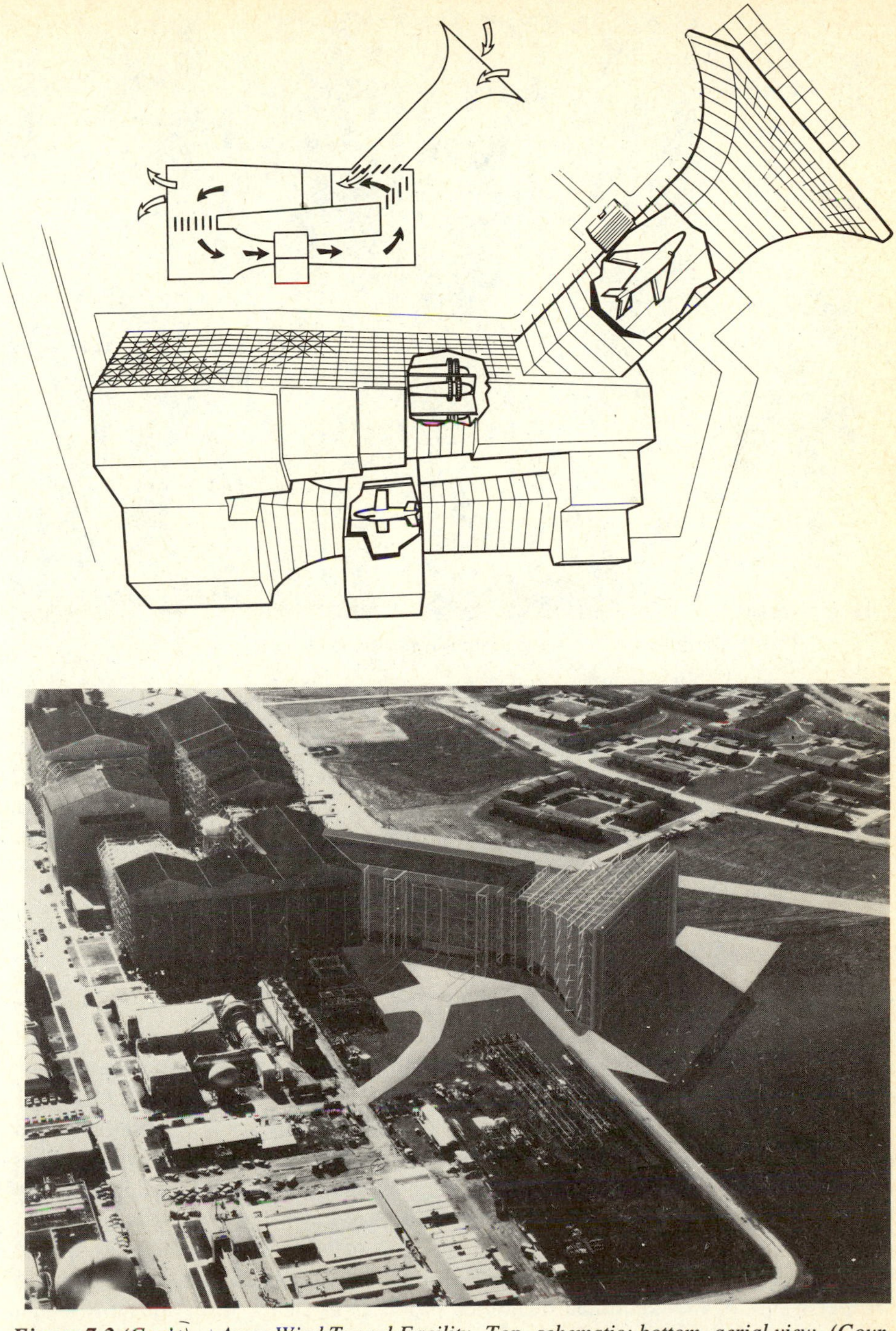

Figure 7.2 (Con't). Ames Wind Tunnel Facility. Top, schematic; bottom, aerial view. (Courtesy of NASA.)

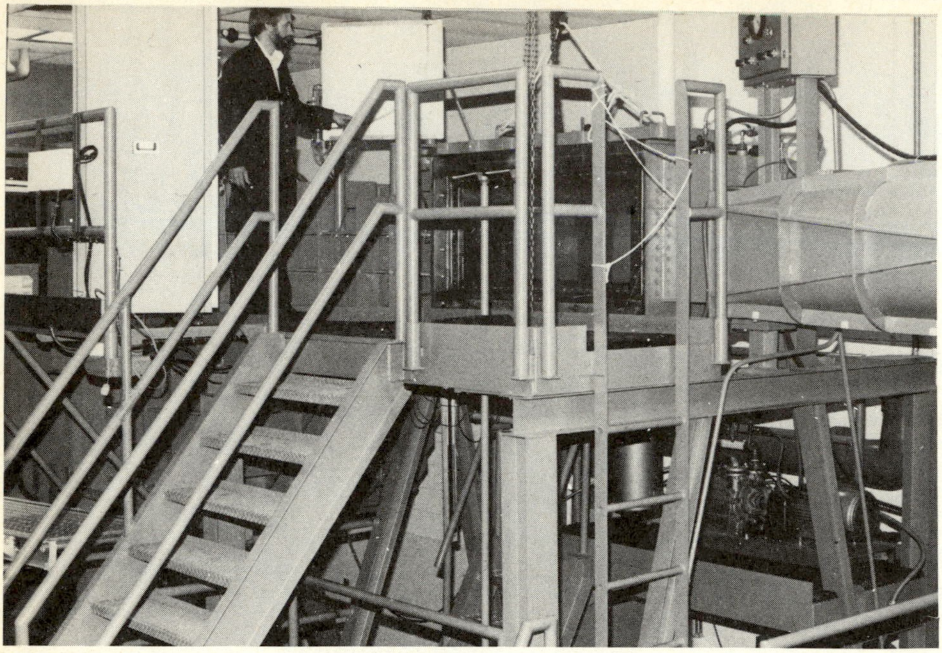

Figure 7.3 *Photograph of USNA recirculating water tunnel.*

In this chapter we will consider the concepts necessary in creating useful models for studying fluid flow. First, we will discuss *dimensional analysis*, the mathematical techniques by which we determine how an experiment should be set up, i.e., what forces and energies are significant, and what can be ignored. Second, we will study *similitude*, or those sets of laws through which we can construct models sufficiently similar to actual phenomena so as to produce experimental results that can be validly applied to actual situations.

7.2 Dimensional Analysis

No experimental problem or model is useful unless it behaves in a manner similar to the actual phenomenon being studied. To model the actual problem properly, we must devise a problem with similar conditions that can be studied exactly. The quantitative conditions for similarity of behavior are based on a particular technique called "dimensional analysis" (formerly called "the method of dimensions"). We call physical quantities "dimensional" if their numerical value depends upon whatever scale is utilized to measure them.* Nondimensional quantities therefore have values that are independent of units. The principal assumption of a dimensional analysis is that the physical quantity being investigated is a function of other known quantities. These other quantities are identified, in a carefully controlled experiment, by noting if a change in one of the controlled quantities produces a change in the quantity being measured.

*An exception to this definition is angles. Angles are dimensionless, yet their numerical values depend upon the scale being used.

Consider, for example, uniform potential flow of velocity U past a two-dimensional cylinder of radius a. From "potential theory" (which we shall discuss in Chap. 12), we can show that the velocity components are

$$\frac{u}{U} = \left(\frac{a}{r}\right)^2 \cos 2\theta \tag{7.1}$$

$$\frac{v}{U} = \left(\frac{a}{r}\right)^2 \sin 2\theta \tag{7.2}$$

Here, u and v are fixed fractions of the main flow U. Equations (7.1) and (7.2) are examples of a similarity law for an ideal flow, i.e., when a body of given shape advances steadily through an otherwise stationary fluid at a given fixed altitude and velocity. The velocity components for a fixed point in the fluid, whose coordinates relative to the body bear fixed ratios to the body geometry, are in fixed ratio to U.

Now, if p is that fluid pressure which depends upon U, then $p/\rho U^2$ is constant for a point whose coordinates relative to the body bear fixed ratios to the body geometry.

Suppose now that the fluid is not irrotational, but has vorticity. Certain forces now exist which were absent in potential flows. As we will show in detail later, *similarity of flow is maintained only when some new condition is satisfied, in addition to those required for potential flow*. The new condition is that the dimensionless quantity

$$R_L = \frac{\rho U L}{\mu} \tag{7.3}$$

shall have the identical numerical value for the bodies being considered. The symbol R is a nondimensional parameter and is called the Reynolds number. The quantities ρ and μ being properties of the fluid, can in a given experiment be controlled. The reference length L is a description of the body, and being one variable of design, it may also be controlled. The velocity U is defined as a reference velocity: the velocity of the body or of the flow. It is a kinematic variable.

In some cases, an experiment cannot be treated directly. Then we have to conduct experimental measurements on a reduced or perhaps a magnified scale in order to handle the problem. Regardless, it is imperative that the reduced or magnified scale be similar in all respects to the actual phenomena. To accomplish this, it is essential we select certain *nondimensional parameters* correctly. These parameters reflect the fundamental effect of the actual phenomena on the scaled phenomena.

Dimensional analysis assists the experimenter in selecting the appropriate dimensionless parameters. This mathematical tool is not a theoretical solution to the problem. At most, it shows only macroscopic results. It provides some insight into how the experiment should be set up, and in how to scale the actual physical problem. In scaling, we must determine the proper geometric size of the models that possess proper inertial and elastic properties. Next we must have fluid flow patterns that are similar to that of a prototype. Then we must find a way to correlate the empirical results based on the final system. The purpose of this chapter is to develop some rules to conduct a dimensional analysis where variables are known but equations are not. We shall present two techniques: the Buckingham pi theorem and the Rayleigh method, which enable us to obtain empirical relationships between known variables and fluid properties. We shall also discuss some important dimensionless parameters, and some restrictions on how we may use them.

Before proceeding, a word of caution. The empirical relationships obtained by a dimensional analysis should never be interpreted as empirical laws. A *law requires both theoretical and experimental verification* within specified limitations. The end product of a dimensional analysis may be properly called an empirical relationship, but never a law.

7.2.1 The Principle of Dimensional Homogeneity

The very basis of dimensional analysis is the *principle of dimensional homogeneity*, which can be stated as follows: "Every term in a complete physical equation has the same measure formula." Thus, an equation is complete when every essential factor is explicitly represented.

Let there exist n quantities Q ($Q_1, Q_2, \ldots, Q_n$), that are involved in an experiment, and postulate that they are related by a dimensionally *homogeneous* equation

$$f(Q_1, Q_2, \ldots, Q_n) = 0 \tag{7.4}$$

Since Eq. (7.4) is dimensionally homogeneous, we can express it in dimensionless form by normalizing each term by the dimensions of any other term in Eq. (7.4). For example, consider the dimensionally homogeneous algebraic equation

$$F(x, y, z, l) = x^3y + \frac{xy^5}{z^2} + z^4 \sin\frac{x}{l} = 0 \tag{7.5}$$

The dimension of Eq. (7.5) is L^4. The equation can be made dimensionless by dividing each term by some reference length l^4, which must be a constant:

$$F\left(\frac{x}{l}, \frac{y}{l}, \frac{z}{l}\right) = \frac{x^3y}{l^4} + \frac{xy^5}{z^2l^4} + \left(\frac{z}{l}\right)^4 \sin\frac{x}{l} = 0 \tag{7.6}$$

We have expressed a relationship among four quantities (x, y, z, l) in terms of three dimensionless quantities x/l, y/l, z/l. In general, Eq. (7.4) can be expressed in terms of dimensionless quantities $\Pi_1, \Pi_2, \Pi_3, \ldots$ formed with the quantities $Q_1, Q_2, \ldots, Q_n$. The usefulness of this is quite evident, in that the number of quantities that are involved can be reduced not necessarily by one, but perhaps two or more. This is a considerable simplification of the problem.

What remains now is to determine how many dimensionless quantities there are in a given functional relationship, as expressed by Eq. (7.4). We can determine this by the Buckingham pi theorem, which is a general theorem of dimensional analysis.

7.3 Buckingham Pi Theorem

Never has a method of dimensional analysis, or perhaps even any mathematical analysis, evoked so much heated controversy as the Buckingham pi theorem. Protagonists of the method use it to solve dimensional analysis problems ranging from the difficult to the impossible. Antagonists expound the dangers of using the method under any condition. The central issue arises from the inability of the Buckingham pi theorem to provide either a unique or a complete solution to a problem. The theorem requires experiments involving models that provide solutions which can never be free from inaccuracies or procedural errors. Thus, complete similarity is impossible to achieve.

Theoretically the Buckingham pi theorem is mathematically correct though not necessarily accurate. For the user, it has the advantage of speed and a certain ease. If an accidental slip occurs in the analysis, it can easily be corrected by examining the dimensionless quantity. Its dangerous weakness is that it can be learned by following some simple steps but without grasping the essence of modeling or similitude.

We will present the rule first and then step-by-step procedures for applying it; we will include suggestions for making those correct choices that render the analysis most effective.

We stated in Chap. 1 that a basic dimensional system used in mechanics is M, L, and T, i.e., mass, length, and time in a physical system. Thus, each mechanical quantity, say Q, dimensionally denoted by $[Q]$, can be represented as a power product of the basic physical dimensions

$$[Q] = M^a L^b T^c \tag{7.7}$$

Or, we can represent the mechanical quantity $[Q]$ as a power product of the technical system dimensions

$$[Q] = F^a L^b T^c \tag{7.8}$$

Let there exist n quantities, $Q_1, Q_2, \ldots, Q_n$, involved in an experimental fluid dynamics problem. We wish to determine the dimensions of each quantity Q such that there are m dimensions involved in the n quantities. We denote the fundamental dimensions by D (e.g., $D_1 = F$, $D_2 = L$, $D_3 = T$, using the technical system in mechanics), and then construct a dimensional formula for each quantity Q in terms of dimensions D:

$$[Q_1] = D_1^{a_1} D_2^{b_1} \ldots D_m^{m_1} \tag{7.9}$$

$$[Q_2] = D_1^{a_2} D_2^{b_2} \ldots D_m^{m_2} \tag{7.10}$$

$$\vdots$$

$$[Q_n] = D_1^{a_n} D_2^{b_n} \ldots D_m^{m_n} \tag{7.11}$$

where the exponents of Eqs. (7.9)–(7.11) are known values of a dimension D.

The question arises if there exist dimensionless terms of the form

$$\Pi = Q_1^{k_1} Q_2^{k_2} \ldots Q_n^{k_n} \tag{7.12}$$

and if they do exist, how many are there? Since Π is dimensionless, it can be expressed in the form

$$[\Pi] = D_1^0 D_2^0 \ldots D_m^0 \tag{7.13}$$

Substituting Eqs. (7.13) and (7.9)–(7.11) into Eq. (7.12) yields

$$\begin{aligned} D_1^0 D_2^0 \ldots D_m^0 &= (D_1^{a_1} D_2^{b_1} \ldots D_m^{m_1})^{k_1} \\ &\times (D_1^{a_2} D_2^{b_2} \ldots D_m^{m_2})^{k_2} \ldots \\ &\times (D_1^{a_n} D_2^{b_n} \ldots D_m^{m_n})^{k_n} \end{aligned} \tag{7.14}$$

Equating like powers of the dimensions $D_1, D_2, \ldots, D_m$, we obtain the following set of linear algebraic equations:

$$\text{for } D_1 : 0 = a_1k_1 + a_2k_2 + \ldots a_nk_n \tag{7.15}$$

$$\text{for } D_2 : 0 = b_1k_1 + b_2k_2 + \ldots b_nk_n \tag{7.16}$$

$$\vdots \quad \vdots \quad \vdots \quad \vdots \quad \vdots$$

$$\text{for } D_m : 0 = m_1k_1 + m_2k_2 + \ldots m_nk_n \tag{7.17}$$

where the coefficients $a_n, b_n, \ldots, m_n$ are known, and k_n is to be determined.

Equations (7.15)–(7.17) form a set of homogeneous linear system of m equations in n unknown variables $k_1, k_2, \ldots, k_n$. According to Cramer's rule, "If the determinant $\mathcal{D}$

$$\mathcal{D} = \begin{vmatrix} a_1 & a_2 & a_3 & \ldots & a_n \\ b_1 & b_2 & b_3 & \ldots & b_n \\ \vdots & & & & \\ m_1 & m_2 & m_3 & \ldots & m_n \end{vmatrix} \tag{7.18}$$

does not vanish, the simultaneous equations of Eqs. (7.15)–(7.17) are satisfied by one and only one set of values of the unknowns k_n. The theory of algebraic equations tells us that a homogeneous linear system of equations in n unknown variables whose matrix of the coefficients exhibits the rank r, has exactly $(n - r)$ linearly independent solutions. So, given n physical quantities $Q_1, Q_2, \ldots, Q_n$ with a relation among them, there exist exactly $(n - r)$ independent dimensionless Π terms, with $r \leq m \leq n$ being the rank of the dimensional matrix. We can then mathematically pose this theorem as

$$F(\Pi_1, \Pi_2, \ldots, \Pi_{n-r}) = 0 \tag{7.19}$$

calling it the *Buckingham pi theorem*. In words, the Buckingham pi theorem states that "*any* complete physical relationship can be represented as one subsisting between a set of independent nondimensional product combinations of the physical quantities concerned." *The least possible number of independent nondimensional quantities which appear in the relationship is equal to the number of related physical quantities less the number of the fundamental units.*

7.3.1 Applications of the Buckingham Pi Theorem

To solve problems using the Buckingham pi therorem, follow these steps:

Step i

Isolate the physical quantities in the given problem. Identify the quantities Q_n, and the number n.

Step ii

Select the M, L, T, or F, L, T system. For each quantity Q, select the appropriate dimensions, and determine the number m.

Step iii

Construct the dimensional matrix and evaluate the rank r (this step is sometimes omitted for $r = m$).

Step iv

Evaluate the $(n - r)$ dimensionless Π coefficients.

Step v

Apply Eq. (7.19) to obtain the desired empirical relationship governing the physical problem. A few examples will now be given to illustrate applications of the Buckingham pi theorem.

Example 7.1

Consider a cylindrical pipe of uniform cross section, so that the geometry of the pipe is completely defined by the inner diameter D of the pipe and its axial length l, which is sufficiently large to neglect end effects. Let the fluid motion be steady, and consider the inertia and viscosity as represented by the fluid density ρ and dynamic viscosity μ. In addition, the experimental measurements show that the composition of the inner surface affects the flow, particularly the pressure drop along the axis of the pipe. Let the composition of the inner surface be denoted by an absolute roughness ϵ. Last, the average fluid velocity $\bar{V}$ over the pipe cross section can vary in the experiment. Using the Buckingham pi theorem, find the empirical relationship for the experimental problem.

Solution:

The following subheadings conform to the numbered steps discussed in the problem solution technique.

Example 7.1 *(Con't.)*

Step i

Q_1	Q_2	Q_3	Q_4	Q_5	Q_6	Q_7
D	l	$\tilde{V}$	g	ϵ	μ	ρ

$n = 7$

Step ii

D_1	D_2	D_3
F	L	T

$m = 3$

$$[D] = L$$

$$[l] = L$$

$$[\tilde{V}] = LT^{-1}$$

$$[g] = LT^{-2}$$

$$[\epsilon] = L$$

$$[\mu] = FL^{-2}T$$

$$[\rho] = FL^{-4}T^2$$

Step iii

	D	l	$\tilde{V}$	g	ϵ	μ	ρ
F	0	0	0	0	0	1	1
L	1	1	1	1	1	−2	−4
T	0	0	−1	−2	0	1	2

$r = 3$

Step iv

Thus, there are four Π terms, and we have some freedom of choice in how to solve the four equations for the seven unknowns. So let us arbitrarily select $\tilde{V}$, D, and μ as the repeating variables.

$$\Pi_1 = \tilde{V}^{k_1} D^{k_2} \mu^{k_3} \rho \tag{i}$$

$$\Pi_2 = \tilde{V}^{k_4} D^{k_5} \mu^{k_6} g \tag{ii}$$

$$\Pi_3 = \tilde{V}^{k_7} D^{k_8} \mu^{k_9} \epsilon \tag{iii}$$

$$\Pi_4 = \tilde{V}^{k_{10}} D^{k_{11}} \mu^{k_{12}} l \tag{iv}$$

From Eq. (i)

$$\Pi_1 = F^0 L^0 T^0 = (LT^{-1})^{k_1} (L)^{k_2} (FL^{-2}T)^{k_3} (FL^{-4}T^2)^1 \tag{v}$$

Example 7.1 *(Con't.)*

Equating exponents of like dimensions, we obtain

$$
\begin{aligned}
F:&\quad 0 = k_3 + 1 \\
L:&\quad 0 = k_1 + k_2 - 2k_3 - 4 \\
T:&\quad 0 = -k_1 + k_3 + 2
\end{aligned}
\tag{vi}
$$

Solving for the exponents results in

$$k_1 = k_2 = -k_3 = 1 \tag{vii}$$

so that substituting the results of Eq. (vii) into Eq. (i) gives

$$\Pi_1 = \frac{\rho \tilde{V} D}{\mu} \tag{viii}$$

From Eq. (ii), we work on the second dimensionless parameter

$$\Pi_2 = F^0 L^0 T^0 = (LT^{-1})^{k_4} (L)^{k_5} (FL^{-2}T)^{k_6} (LT^{-2}) \tag{ix}$$

Equating exponents of like dimensions

$$
\begin{aligned}
F:&\quad 0 = k_6 \\
L:&\quad 0 = k_4 + k_5 - 2k_6 + 1 \\
T:&\quad 0 = -k_4 + k_6 - 2
\end{aligned}
\tag{x}
$$

or

$$
\begin{aligned}
k_6 &= 0 \\
k_5 &= 1 \\
k_4 &= -2
\end{aligned}
\tag{xi}
$$

so that substituting the results of Eq. (xi) into Eq. (ii) gives

$$\Pi_2 = \frac{Dg}{\tilde{V}^2} \tag{xii}$$

From Eq. (iii), we work on the third pi parameter.

$$\Pi_3 = F^0 L^0 T^0 = (LT^{-1})^{k_7} (L)^{k_8} (FL^{-2}T)^{k_9} (L) \tag{xiii}$$

Equating exponents of like dimensions

$$
\begin{aligned}
F:&\quad 0 = k_9 \\
L:&\quad 0 = k_7 + k_8 - 2k_9 + 1 \\
T:&\quad 0 = -k_7 + k_9
\end{aligned}
\tag{xiv}
$$

Example 7.1 *(Con't.)*

or

$$k_7 = k_9 = 0$$
$$k_8 = -1 \quad \text{(xv)}$$

so that substituting the results of Eq. (xv) into Eq. (iii) gives

$$\Pi_3 = \epsilon/D \quad \text{(xvi)}$$

From Eq. (iv), we find for the fourth parameter

$$\Pi_4 = \frac{l}{D} \quad \text{(xvii)}$$

Step v

Finally, we apply the results of Eqs. (viii), (xii), (xvi), and (xvii) to Eq. (7.19) and obtain

$$F\left(\frac{\rho\tilde{V}D}{\mu}, \frac{Dg}{\tilde{V}^2}, \frac{\epsilon}{D}, \frac{l}{D}\right) = 0 \quad \text{(xviii)}$$

which represents the solution of the problem. We shall show in Chap. 10 that a dimensionless friction factor f is defined as

$$f = f\left(\frac{\rho\tilde{V}D}{\mu}, \frac{\epsilon}{D}\right) \quad \text{(xix)}$$

so that Eq. (xviii) can be abbreviated as

$$F\left(f, \frac{\tilde{V}^2}{Dg}, \frac{l}{D}\right) = 0 \quad \text{(xx)}$$

A typical empirical equation that is represented by Eq. (xx) is Darcy's equation

$$h_f = f\frac{l}{D}\frac{\tilde{V}^2}{2g} \quad \text{(xxi)}$$

which will be discussed in Chap. 10.

This completes the solution.

Example 7.1 showed the application of the Buckingham pi theorem, and it appears to work very nicely. The student might well ask, can we arbitrarily select any quantity as the repeating variable if $n - r > 1$? The answer is yes, but the results may yield an indeterminant solution or even a trivial solution, such as one with all unknown exponents of zero. Worse, arbitrarily selecting any grouping of variables that we desire could result in valid dimensionless pi parameters having no physical meaning—or even physical meaning that prove incorrect when modeled and applied.

So there is more to consider. We must remember this rule of procedure:

> Of all the quantities involved in a problem, choose any three such that all three basic dimensions are included among them and then with each of the remaining quantities perform the analysis.

The difficulty here lies in choosing the right three quantities. Experience helps, but the novice is going to experience serious problems.

In all fluid flow problems, the quantities describing the situation can be divided into three categories. One category should describe the device or facility through or around which the flow takes place. Typically, these quantities express the geometry of the device, with the result that the pi parameter will express the geometric similarity desired. The second category should account for the properties describing the flow. This category includes density, viscosity, specific weight, etc. The third and last category consists of those quantities describing the kinematics of the fluid and device. This includes velocity, vorticity, acceleration, and so on.

To have significance, therefore, the fluid flow phenomenon should include quantities from each of the three categories, at least one quantity from each, if possible. Which one? That question can't be answered, and shows a weakness of the pi theorem. But having a dimensionless parameter contain one quantity describing the flow device, a second describing a fluid property, and a third describing the relative motion between the flow and a boundary helps achieve some useful form in trying to simulate the flow in the laboratory. It will definitely give the solution physical meaning.

Example 7.2 illustrates this procedure.

Example 7.2

Consider a vertical plate with a cutout having a notch of angle ϕ cut into the top of it and placed across an open channel containing water. The plate backs up the water in the channel until it flows through the notch. The volume rate of flow Q is some function of the elevation H of upstream liquid surface above the bottom of the notch. In addition, the discharge depends upon gravity and upon the velocity of approach $\tilde{V}$ to the vertical plate. Determine the form of discharge equation.

Example 7.2 *(Con't.)*

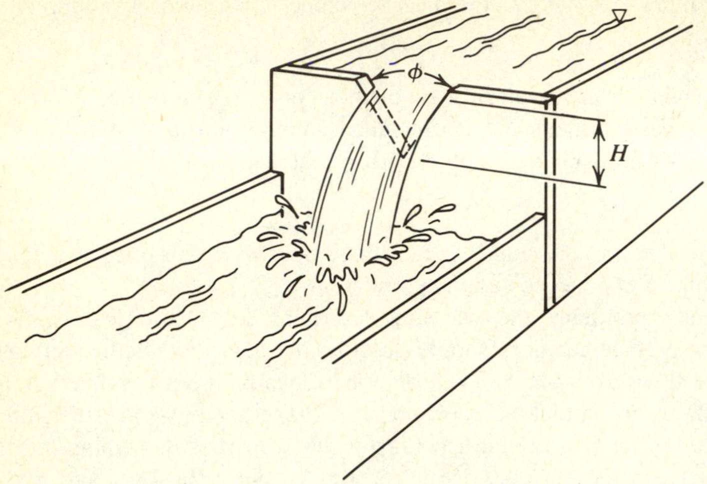

Figure E7.2

Solution:

Step 1.

Identify the quantities in the problem. A functional relationship

$$F(Q, H, g, \tilde{V}, \phi) = 0 \tag{i}$$

is to be grouped into dimensionless parameters; ϕ is considered dimensionless, hence it is one of the π parameters.

Step 2.

Determine the number of dimensions in the problem. Only two dimensions are needed, L and T.

Step 3.

Select the repeating variables. We need one geometric quantity, so we will choose elevation H. For the second category, we need a fluid property. We notice that no fluid property is given: no density, viscosity, or surface tension. So we advance to the third category and select a kinematic property. We have Q, g, and V to select from. Any would be appropriate, so let us choose acceleration g. Thus our repeating quantities will be H and g.

Step 4.

Follow Steps i–v. As in Example 7.1, we construct the solution.

Step i:

Q_1	Q_2	Q_3	Q_4
Q	H	g	$\tilde{V}$
n = 4			

Example 7.2 *(Con't.)*

Step ii:

D_1	D_2
L	T

$m = 2$

$$[Q] = L^3T^{-1}$$

$$[H] = L$$

$$[g] = LT^{-2}$$

$$[\tilde{V}] = LT^{-1}$$

Step iii:

	Q	H	g	$\tilde{V}$
L	3	1	1	1
T	−1	0	−2	−1

$r = 2$

Step iv:

Thus there are two pi terms, and we shall use H and g as the repeating variables:

$$\Pi_1 = H^{k_1} g^{k_2} Q \qquad \text{(ii)}$$

Therefore

$$[\Pi_1] = L^{k_1}(LT^{-2})^{k_2}(L^3T^{-1})^1 \qquad \text{(iii)}$$

Similarly,

$$\Pi_2 = H^{k_3} g^{k_4} \tilde{V} \qquad \text{(iv)}$$

Therefore,

$$[\Pi_2] = L^{k_3}(LT^{-2})^{k_4}(LT^{-1})^1 \qquad \text{(v)}$$

Following the procedure of Example 7.1,

$$k_1 + k_2 + 3 = 0 \qquad \text{(vi)}$$

$$-2k_2 - 1 = 0 \qquad \text{(vii)}$$

and

$$k_1 = -\frac{5}{2}, \quad k_2 = -\frac{1}{2} \qquad \text{(viii)}$$

Therefore from Eq. (ii)

$$\Pi_1 = \frac{Q}{\sqrt{g}H^{5/2}} \qquad \text{(ix)}$$

Example 7.2 *(Con't.)*

Similarly from $[\pi_2]$, we find

$$k_3 + k_4 + 1 = 0 \tag{x}$$

$$-2k_4 - 1 = 0 \tag{xi}$$

thus

$$k_3 = -\frac{1}{2} \qquad k_4 = -\frac{1}{2} \tag{xii}$$

so that

$$\Pi_2 = \frac{\tilde{V}}{\sqrt{gH}} \tag{xiii}$$

We have stated that ϕ is dimensionless, so

$$\Pi_3 = \phi \tag{xiv}$$

Substituting Eqs. (ix), (xiii), and (xiv) into Eq. (7.19) gives

$$F\left(\frac{Q}{\sqrt{g}H^{5/2}}, \frac{\tilde{V}}{\sqrt{gH}}, \phi\right) = 0 \tag{xv}$$

or an alternate form

$$\frac{Q}{\sqrt{g}H^{5/2}} = f\left(\frac{\tilde{V}}{\sqrt{gH}}, \phi\right) \tag{xvi}$$

Solving for the volume rate of flow Q,

$$Q = \sqrt{g}H^{5/2} f\left(\frac{\tilde{V}}{\sqrt{gH}}, \phi\right) \tag{xvii}$$

An experiment or analysis is needed to yield additional information as to how $\tilde{V}/\sqrt{gH}$ and ϕ are related.

This completes the solution.

Before we finish with the Buckingham pi theorem, we need to mention a number of small but important points, points that might be regarded as clues to solving problems in dimensional analysis.

Points

1. If a significant quantity is left out of the original listing of all the quantities, one clue may be that a primary dimension appears only in the dimensions of one

variable. There is no way to cancel this dimension. But this is not infallible. The only guarantee that all significant quantities have been considered is that a functional relationship results where pi parameters are plotted as a function of one another.

2. If an extraneous variable is included as a repeating parameter then the value of its exponent is zero. If an extraneous variable is included as a nonrepeating parameter, then one or more of the resulting equations for the values of the unknown exponents results in a mathematical contradiction (such as one equals zero).
3. Dependent variables should appear in only one dimensionless group each. Guarantee this by *not* selecting a dependent variable as a repeating quantity.
4. If two variables have the same dimensions, one pi term is their ratio.
5. Dimensional analysis gives no information on the functional relationship between dimensionless groups. This is found by experiment.
6. Dimensional analysis seldom helps if we do not have a good idea of the variables involved in the first place.
7. Primary dimensions must be truly independent. For example, it is possible that mass and time might always occur as the ratio mass/time when the dimensions of the variables are written out. In this case there is really only one primary dimension.

Besides the Buckingham pi theorem, a variety of other techniques can be used to perform a dimensional analysis. They are all similar in principle to the pi theorem. The next section treats the Rayleigh method.

7.4 The Rayleigh Method

In 1899 Lord Rayleigh [7.1] proposed an easy method for analyzing the behavior of fluid motion. He proposed that the dimensions of any term in a homogeneous equation be the same. Let f be a function of properties Q_n as in Eq. (7.4). Since the equation is to be homogeneous, we can arrange Eq. (7.4) as

$$f \propto Q_1^a Q_2^b Q_3^c \ldots Q_n^r \tag{7.20}$$

where $a, b, c, \ldots, r$ are to be determined from the fact that the arrangement of quantities Q_n must be reducible to the dimensions of f.

The Rayleigh method requires no formal rules as in the Buckingham pi theorem, and is best understood by studying an example.

Example 7.3

Using the Rayleigh method, determine an expression for the drag on a missile in supersonic flow. Consider the primary quantities in the problem as the density, dynamic viscosity, bulk modulus, reference length l, and flight speed V.

Example 7.3 *(Con't.)*

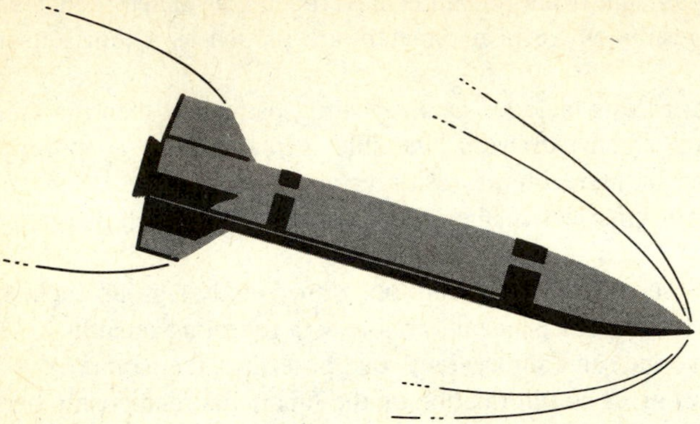

Figure E7.3

Solution:

Step 1.

Express the problem mathematically. Using Eq. (7.20), the drag force D is expressed as

$$D = c\rho^a \mu^b K^c l^d V^e \qquad \text{(i)}$$

where both sides of the equation have the units of force.

Step 2.

Select a primary system, and express the dimensions of each quantity. Choose the FLT system. Equation (i) is written

$$F = (FT^2L^{-4})^a(FTL^{-2})^b(FL^{-2})^c(L)^d(LT^{-1})^e \qquad \text{(ii)}$$

Step 3.

Equate exponents of like dimensions and solve for their value. From Eq. (ii)

$$F: \quad 1 = a + b + c$$

$$L: \quad 0 = -4a - 2b - 2c + d + e$$

$$T: \quad 0 = 2a + b - e$$

or

$$a = 1 - b - c$$

$$e = 2 - b - 2c \qquad \text{(iii)}$$

$$d = 2 - b$$

Step 4.

Evaluate the desired quantity. Substituting Eq. (iii) into Eq. (i) yields

$$D = c\rho^{1-b-c}\mu^b K^c l^{2-b} V^{2-b-2c} \qquad \text{(iv)}$$

Example 7.3 *(Con't.)*

Step 5.
Collect all terms with common exponents. From Eq. (iv), we can group the various quantities as

$$D = c(\rho l^2 V^2)\left(\frac{\mu}{\rho l V}\right)^b \left(\frac{K}{\rho V^2}\right)^c \tag{v}$$

where each term in parentheses with unknown exponent b and c is dimensionless, and the term in parentheses to the one power has the dimensions of the drag force D.

Equation (v) can be rearranged to read

$$D = c(\rho l^2 V^2)\left(\frac{1}{R_l}\right)^b \left(\frac{1}{M^2}\right)^c \tag{vi}$$

or

$$D = \rho l^2 V^2 g(M, R_l) \tag{vii}$$

From Eq. (vii), we can make this dimensionless and obtain

$$C_D = C_D(M, R_l) \tag{viii}$$

This completes the solution.

7.4.1 A Critique of the Two Methods

What experience can we draw upon to dictate the best choice of the primary quantites with which to form the nondimensional parameter? Certainly the student does not have the experience, and therefore sees the choice as being arbitrary, and cannot appreciate the generality of the dynamic model laws. Some authors use Newton's second law to derive the force ratio for each nondimensional group, and others use dimensional transformation. The different approaches to dimensional analysis are not comprehensive, and the student may wish to read the article by B. W. Imrie [7.2] to gain an appreciation for the problem.

Most fluid dynamicists believe that the Buckingham pi theorem is the most effective means of dimensional analysis; the Rayleigh method is an alternative, but is generally avoided. It probably should be just the opposite. There is great confusion concerning the pi method, as opposed to the pi theorem, as well as about the rule concerning the number of dimensionless parameters. Buckingham's proof is not a proof, nor is it rigorous, though many superior treatises have been written on both subjects. The Rayleigh method might be considered as existing without proof. It is certainly easier than the pi theorem. Both methods can treat eight and more quantities without significant difficulty.

7.5 Dimensionless Parameters

We will initially be interested in seven dimensionless parameters in the flow of fluids. The seven dimensionless parameters that are of interest are Reynolds, Froude, Euler, Mach, Weber, and Cauchy numbers, plus the pressure coefficient. These dimensionless parameters are extremely valuable in that they describe the *nature* of the flow. The behavior of the fluid flow can be described by specifying a value to these dimensionless parameters. For example if the Reynolds number were extremely large, it might indicate very high-speed flow, or perhaps a nearly inviscid low-speed flow. The Navier-Stokes equation can be expressed in dimensionless form, with some terms involving these dimensionless parameters. By inspecting the dimensionless form of the Navier-Stokes equation, we can determine what accelerations may or may not be important in a particular physical flow problem, thereby possibly simplifying the problem so it can be solved. That is one of the goals we seek, i.e., to find ways to simplify the Navier-Stokes equation so we can solve it theoretically.

7.5.1 Dimensionless Navier-Stokes Equation

It is convenient to make the Navier-Stokes equation (4.109) dimensionless. We define the *dimensionless Euler (or outer) variables* as

$$\mathbf{r}^* \doteq \frac{\mathbf{r}}{L} \tag{7.21}$$

or in scalar Cartesian form

$$x^* \doteq \frac{x}{L}; \quad y^* \doteq \frac{y}{L}; \quad z^* \doteq \frac{z}{L} \tag{7.22}$$

where L is a characteristic length and is assumed to be a constant. We also define a dimensionless velocity

$$\mathbf{V}^* \doteq \frac{\mathbf{V}}{U} \tag{7.23}$$

or expressed in scalar Cartesian form as

$$u^* \doteq \frac{u}{U}; \quad v^* \doteq \frac{v}{U}; \quad w^* \doteq \frac{w}{U} \tag{7.24}$$

where U is a characteristic velocity and is assumed to be a constant. The dimensionless pressure and nabla operator are defined as

$$p^* \doteq \frac{p}{\rho U^2}; \quad \boldsymbol{\nabla}^* \doteq L\boldsymbol{\nabla} \tag{7.25}$$

Multiplying each term of the Navier-Stokes equation (4.109) by L/U^2 and using the above definitions yields

$$\frac{\partial \mathbf{V}^*}{\partial \tau} + (\mathbf{V}^* \cdot \nabla^*)\mathbf{V}^* = -\nabla^* p^* - \frac{\mathbf{k}}{F_r^2} + \frac{1}{R_L}\nabla^{*2}\mathbf{V}^* \tag{7.26}$$

where

$$\tau = \frac{Ut}{L} \tag{7.27}$$

$$\boxed{F_r = \frac{U}{\sqrt{gL}}} \tag{7.28}$$

$$\boxed{R_L = \frac{UL}{\nu}} \tag{7.29}$$

all of which are dimensionless quantities.

We assume that an exact solution of Eq. (7.26) exists satisfying the no-slip condition on the boundary

$$\mathbf{V}^* = 0 \tag{7.30}$$

with accompanying posed conditions at infinity. We consider bodies which extend downstream to infinity, and which have a *non*decreasing body thickness in the cross-stream direction so that no wake forms and the boundary layer does not separate.

We have defined R_L of Eq. (7.29) as the Reynolds number:

$$R_L = \frac{\rho UL}{\mu} \tag{7.3}$$

which represents the ratio of the inertial force (which has the characteristic quantity of ρL^2U^2) to the viscous force (which has the characteristic quantity of μUL). In this text we shall consider numerous expressions for the Reynolds number, each depending upon what is being used as the reference velocity U and reference length L. We shall adopt the following notations for the Reynolds number:

$$R_x = \frac{Ux}{\nu} \tag{7.3a}$$

x being a coordinate in the direction of flow velocity U

$$R_l = \frac{Ul}{\nu} \tag{7.3b}$$

l being in the direction of the flow velocity U

$$R_D = \frac{UD}{\nu} \tag{7.3c}$$

D being a diameter

$$R_a = \frac{Ua}{\nu} \tag{7.3d}$$

a being a radius

We define F_r as the Froude number, which represents the square root of the ratio of the inertial force to body force due to a gravitational potential (which has the characteristic quantity $g\rho L^3$).

Thus, to find the flow at high Reynolds number (high velocities or very low viscosities) the idea is to take the Euler limit, that is, to let $R_L \to \infty$ for fixed Euler variables. *Thus, the Euler limit of the dimensionless Navier-Stokes equation for incompressible flow is the Euler equation for motion of an inviscid fluid, namely*

$$\frac{\partial \mathbf{V}^*}{\partial \tau} + (\mathbf{V}^* \cdot \nabla^*)\mathbf{V}^* = -\nabla^* p^* - \frac{\mathbf{k}}{F_r^2} \tag{7.31}$$

Equation (7.31) is the dimensionless Euler equation. We can absorb the constant term $\mathbf{k}/F_r^2$ in Eq. (7.31) by using the gravitational potential Ω:

$$\mathbf{g} = -\nabla\Omega \tag{7.31a}$$

so that Eq. (7.31) reads

$$\boxed{\frac{\partial \mathbf{V}^*}{\partial \tau} + (\mathbf{V}^* \cdot \nabla^*)\mathbf{V}^* = -\nabla^*(\tilde{p}^*)} \tag{7.32}$$

where we have defined a new dimensionless total pressure

$$\tilde{p}^* = p^* + \frac{\Omega}{U^2} \tag{7.33}$$

To complete the picture, we need to consider the continuity equation. It is quite easy to show that the dimensionless continuity equation is

$$\boxed{\nabla^* \cdot \mathbf{V}^* = 0} \tag{7.34}$$

It is well known that the solution of Eqs. (7.32) and (7.34) *cannot* approximate the exact solution of the Navier-Stokes equations even for large R_L throughout the flow field. The trouble lies in neglecting the term $\dfrac{\nabla^{*2}\mathbf{V}^*}{R_L}$ in Eq. (7.26). Note that the order of the differential equations is lowered from second to first order. To find any solution of Euler's equation (7.32), it is necessary to omit one of the boundary conditions on the velocity $\mathbf{V}^*$. The boundary condition that we require in solving Eqs. (7.32) and (7.34) is that the normal velocity component vanish on the boundary,

$$\mathbf{V}^* \cdot \mathbf{e}_n = 0 \tag{7.35}$$

on the boundary. The boundary condition we relax is the tangential velocity on the boundary. Thus, *Euler's equations are not applicable in the near vicinity of the body where no-slip effects are evident.*

7.5.2 Scaling Rules

By use of the Euler (or outer) variables, we have assumed that the fluid flow region can be characterized by one geometric characteristic length L and one characteristic velocity U. The length L could represent any reference length (such as the diameter of a circular channel, the length of a span, or the length of a chord). From Eq. (7.27), we see there is an inherent time scale (L/U), which is interpreted as the time it takes a fluid particle to traverse a distance L moving at a speed U.

Similarity rules, or scaling rules, allow us to relate the solutions of two different flow problems, each satisfying the same Navier-Stokes equation (7.26), but pertaining to configurations which are different yet *geometrically similar*. By making the governing differential equations dimensionless, we in effect *make* the equation fit a very general class of bodies, so that the dimensionless solution of the equations *can then be fitted to a specific case by making the solution dimensional*. This is done by selecting a value for U and L. Before we illustrate this technique (which will be treated in Chap. 8), we need to discuss the dimensionless parameters.

7.5.3 Reynolds Number, R_L

In 1908 Sommerfeld suggested that O. Reynolds be credited with the dimensionless quantity

$$R_L = \frac{UL}{\nu} \tag{7.29}$$

It was Reynolds who in 1883 used this quantity as the principal hydrodynamic property that characterized the problem of stability in the transition from laminar to turbulent flow in pipes. Much later, in 1963, Oswatitsch showed that the Reynolds number could just as easily be defined as the ratio of the momentum flux to the shearing stress, showing the interesting result that the inertial force is *not necessarily* the decisive quantity. Other plausible ratio definitions may exist, just as long as the final result is that given by Eq. (7.29).

The Reynolds number measures the relative importance of the fluid's inertia and viscosity. Thus, if the viscous forces play a predominant role, as in the case of a flow very near a body, then R_L is small. If, on the other hand, inertia effects are predominant, then R_L is very large. Problems where R_L is usually numerically very large or infinite, are

1. Turbulent flows
2. Inviscid flows
3. Potential flows
4. Flows far removed from boundaries

Cases where R_L is numerically very small are

1. Creeping flows
2. Laminar flows
3. Stokes flow and lubrication theory
4. Bubble flows
5. Flows very close to a boundary

Reynolds number is thus used to characterize the speed and/or viscous properties of a flow field. We say that similarity rules can be established between two geometrically similar problems only if their Reynolds numbers are the same for a body wetted by a single fluid.

7.5.4 Froude Number, F_r

Froude number, F_r, is defined by Eq. (7.28). M. Weber, in 1919, suggested the name Froude number be used for the relationship, as it was Froude in 1869 who was the first to determine the drag of ships by using laws of similarity. When gravitational potential forces are one of the major forces that govern the flow, the Froude number becomes an important factor in describing the flow. Usually the length characteristic L is the length of a ship if one is dealing with water gravity waves, or in problems dealing with an open channel flow, L is the water depth. If $F_r < 1$, open channel flow indicates a subcritical condition, and if $F_r > 1$, we have supercritical flow corresponding to tranquil flow or rapid flow in a channel. This will be discussed in Chap. 13. We say that similarity rules can be established between two geometrically similar problems only if their Froude numbers are the same for a body or interface that is wetted by two or more fluids, such as air and water.

7.5.5 Mach Number M and Cauchy Number C

Mach number M is a measure of the compressibility of a fluid. Mach number is the ratio of the speed of the fluid flow U to the speed of sound c in the fluid:

$$\boxed{M = \frac{U}{c}} \tag{7.36}$$

where

$$c = \sqrt{dp/d\rho} \tag{7.37}$$

J. Ackeret, in 1928, was the first to suggest the name Mach for this dimensionless parameter in honor of the great German engineer Ernst Mach.

For an isentropic process, the pressure p is a function of density ρ only so that we can show from Eq. (7.37) that

$$c = \sqrt{\frac{kp}{\rho}} \tag{7.38}$$

where k is a measure of the internal complexities of the molecules and is related to the ratio of two constant specific heats:

$$k = \frac{C_p}{C_v} \tag{7.39}$$

We now need the equation of state of the fluid to simplify Eq. (7.38). For instance, for a perfect gas, Eq. (7.38) can be written as

$$c = \sqrt{kg_c RT} \tag{7.40}$$

where T is the absolute temperature, g_c is a universal constant, and R is the gas constant.*

For very small Mach numbers, the variation of density (which measures the compressibility effect) due to the variation of the flow field is negligible and the fluid may be considered incompressible. For large Mach numbers, $M \geqslant 0.3$, the effect of compressibility must be considered. When $0.3 < M < 1$, the flow is termed subsonic flow, and when $M > 1$, the flow is termed supersonic. The field of gas dynamics, or compressible flow, covers the behavior of fluid particles moving at speeds where the gas compresses (see Chap. 15).

The Mach number can be viewed as the ratio of the inertial force to the compressibility force which is proportional to KL^2, where K is the bulk modulus of compressibility.

The Cauchy number C is defined as the ratio of the compressibility force to the inertial force, or

$$C = \frac{K}{\rho U^2} \tag{7.41}$$

Comparing Eq. (7.41) with the Mach number M, we obtain

$$M = \frac{1}{\sqrt{C}} \tag{7.42}$$

so that $M = 0$, or $C = \infty$, defines incompressible flow.

*Some authors combine g_c and R, calling the result a gas constant.

7.5.6 Weber Number, *W*

There is a large class of fluid flows that deals with a free-surface where surface tension forces exist and are important. The surface tension, denoted by σ, is the ratio of the inertial force to the surface tension force (proportional to σL), giving

$$W \doteq \frac{U^2 L}{\sigma} \tag{7.43}$$

Large values of the Weber number indicate that surface tension is relatively unimportant, compared to the inertial force.

7.5.7 Euler Number *E*, and the Pressure Coefficient C_p

The Euler number E is defined as the ratio of the pressure force of the ambient flow to the inertial force, with the result that

$$E \doteq \frac{p_\infty}{\rho U^2} \tag{7.44}$$

The pressure coefficient $\tilde{C}_p$, on the other hand, is the ratio of a difference in pressure Δp to the dynamic pressure ρU^2. Whenever it is used, we should be aware of how it is defined, since it can have two different forms

$$\tilde{C}_p \doteq \frac{\Delta p}{\rho U^2} \tag{7.45}$$

or

$$C_p \doteq \frac{\Delta p}{\frac{1}{2}\rho U^2} \tag{7.46}$$

where Δp may be a pressure difference in a streamwise direction or a pressure difference in a crosswise direction, depending upon the problem.

7.6 Similitude

Similarity laws permit results from a test using one set of conditions to be easily applied to another set of conditions without having to repeat the experimental procedure. The basic idea behind these laws is as follows: the behavior of a fluid (or a body in a fluid) in one set of conditions is related to the behavior of the same (or another) fluid (or body in a fluid) in another set of conditions. This comparison is usually (but not necessarily) made between the full-scale apparatus (or fluid) and the modeled apparatus (or fluid). Certain conditions of similarity between the model and the prototype are necessary to make the model testing truly representative of what can be expected in the prototype. These conditions are (1) geometric similarity of the physical boundaries, and (2) dynamic similarity between the flow fields.

We need to discuss what is called similar solutions of dimensionless differential equations. Like the laws of similarity between prototype and model, there exists a similarity law for this *theoretical* problem. We shall also treat similarity solutions, geometric and dynamic similitude, modeling, similarity from fluid resistance, and how to achieve similar lift for similar bodies.

7.7 Similarity Solutions and Transformations

By similar solutions, we mean those solutions whose longitudinal component of velocity u has the property that $u(x, y)$ differs at two different points x, say, x_1 and x_2, only by a scale factor for the velocity and y coordinate. Thus, similar solutions coincide for all points x:

$$\frac{u\{x_1, [y/g(x_1)]\}}{U(x_1)} = \frac{u\{x_2, [y/g(x_2)]\}}{U(x_2)} \tag{7.47}$$

when $y_1/g(x_1) = y_2/g(x_2)$. Similarity of profile shapes occurs along coordinate lines: hence, similarity solutions will depend on the coordinate system used. Linear partial differential equations may admit similarity solutions which can be superimposed to form new solutions.

The similarity solutions involve one independent variable, say η, where η is of the form shown in the brackets of Eq. (7.47). The dependent variable u is then a function of this new independent variable (also called a coordinate) along with x, according to Eq. (7.47).

Two methods are used for the analysis. They differ only on what they emphasize: the dependent or independent variables. The first, called the *free-parameter or similarity method*, is a method where the dependent variable is first specified in terms of the similarity variable η *without specifying what* η *is*. We apply boundary conditions in terms of the dependent variable. The second, called *the separation of variable method*, has η *initially specified*. Consider the following example of the similarity method.

Example 7.4
Consider the problem of calculating the distribution of the axial velocity of a real fluid's circular jet. The jet flows through a small circular orifice in a wall, and the resultant fluid motion is axisymmetric about the axial axis of the jet, as shown in Fig. E7.4. Let r denote the normal to the axis. The appropriate equation of linear momentum is

$$w\frac{\partial w}{\partial z} + v_r\frac{\partial w}{\partial r} = \frac{\nu}{r}\frac{\partial}{\partial r}\left(r\frac{\partial w}{\partial r}\right) \tag{i}$$

Example 7.4 *(Con't.)*

with boundary conditions

$$\frac{\partial w}{\partial r} = 0, \quad \text{at} \quad r = 0 \tag{ii}$$

$$v_r = 0, \quad \text{at} \quad r = 0 \tag{iii}$$

$$v_r = 0, \quad \text{at} \quad r = \infty \tag{iv}$$

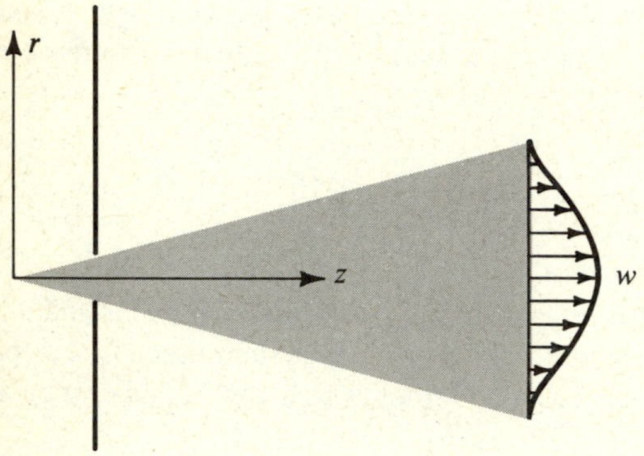

Figure E7.4

Let us define a function $\psi(r, z)$ that is uniquely related to the velocity components v_r and w so that it satisfies the (D.F.) continuity equation.

$$w = \frac{1}{r}\frac{\partial \psi}{\partial r} \tag{v}$$

$$v_r = -\frac{1}{r}\frac{\partial \psi}{\partial z} \tag{vi}$$

We introduce a similarity variable η as

$$\eta = \frac{1}{\sqrt{\nu}}\left(\frac{r}{z}\right) \tag{vii}$$

that transforms our two independent variables into a single independent variable. We now wish to transform our partial differential Eq. (i) into an ordinary differential equation. We investigate

$$\psi = \nu z f(\eta) \tag{viii}$$

and seek to find expressions for the velocity components v_r and w if the rate of flow of momentum $\dot{M}$ across any cross section of the jet is constant, where

$$\dot{M} = 2\pi\rho \int_0^\infty rw2dr = \text{const.} \tag{ix}$$

Example 7.4 *(Con't.)*

Solution:

Step 1.

Express the velocity components in terms of the similarity functions.

Substituting Eq. (viii) into the expressions for the velocity components w and v_r, Eqs. (v) and (vi), respectively, gives

$$w = \frac{f'}{\eta z} \tag{x}$$

$$v_r = \frac{\sqrt{\nu}}{z}\left(f' - \frac{f}{\eta}\right) \tag{xi}$$

Step 2.

Transform the partial differential equation of linear momentum into an ordinary differential equation.

We note in Eqs. (x) and (xi) that both the independent similarity variable η and independent variable z exist, which we cannot allow in the differential equation. So, to check this, we substitute the results of Eq. (x) and (xi) into the equation of linear momentum (i), with the result

$$\frac{d}{d\eta}\left(f'' - \frac{f'}{\eta}\right) + \frac{d}{d\eta}(ff'/\eta) = 0 \tag{xii}$$

along with the boundary conditions from Eqs. (ii)–(iv)

$$f' - \frac{f}{\eta} = 0, \qquad \eta = 0 \tag{xiii}$$

$$\frac{d}{d\eta}\left(\frac{f'}{\eta}\right) = 0, \qquad \eta = 0 \tag{xiv}$$

From Eq. (xii), we obtain

$$\eta f'' - f' + ff' = 0 \tag{xv}$$

We now have our ordinary differential equation in terms of η.

Step 3.

Obtain a solution of the ordinary differential equation of linear momentum that satisfies boundary conditions.

Equation (xv) has the solution

$$f = \frac{\xi^2}{1 + \xi^2/4} \tag{xvi}$$

where

$$\xi = a\eta \tag{xvii}$$

and a is an arbitrary constant.

Example 7.4 *(Con't.)*

Step 4.
Evaluate the velocity components.
From Eq. (x),

$$w = \frac{2a^2}{z(1 + \xi^2/4)^2} \qquad \text{(xviii)}$$

so that the integral expression for the rate of momentum $\dot{M}$ of Eq. (ix) becomes

$$\dot{M} = \frac{16}{3}\pi\mu a^2 \qquad \text{(xix)}$$

Solving for the value of a in Eq. (xix), we find

$$w = \frac{3\dot{M}}{8\pi\mu z}(1 + \xi^2/4)^2 \qquad \text{(xx)}$$

$$v_r = \sqrt{\frac{3\dot{M}}{16\pi\rho}}\,\frac{\xi(1 - \xi^2/4)}{z(1 + \xi^2/4)^2} \qquad \text{(xxi)}$$

where

$$\xi = \sqrt{\frac{3\dot{M}}{16\pi\rho\nu^2}}\left(\frac{r}{z}\right) \qquad \text{(xxii)}$$

This problem was initially solved by Schlichting in 1933.

What was formerly a complicated mathematical model for a simple jet flow, Eq. (i) has been transformed into an easy differential equation, Eq. (xv) through the use of similarity transformations.

This completes the solution.

We now turn to integrated results of a similarity analysis, popularly called *similitude*. Here we use the results of similar solutions for similar flow conditions, as expressed by Eq. (7.47).

7.8 Geometric and Dynamic Similitude

At the outset, we cannot *define* the meaning of similitude, as too many different "laws" can be grouped under the heading.

Similar flow fields can be defined mathematically using Eq. (7.47). Though this is necessary, it is not sufficient in most instances. Other similarity requirements might also be required in a given fluid flow problem. These requirements are expressed in equivalent mathematical form like Eq. (7.47). In this section we shall examine solely the results of the similarity analysis. These results are expressed in terms of similarity parameters, much like the pi parameters in Sec. 7.3.

Flow fields are *geometrically similar* when the geometry of one flow is a scale factor of the other throughout the entire flow field, that is, there is a similarity of shape. This, of course, requires both the flow and fluid boundaries to possess the same geometric scale. For *geometric similitude*, the ratio of characteristic lengths is

$$\frac{\mathbf{r}_2}{\mathbf{r}_1} = \text{const.} \tag{7.48}$$

In dynamic similitude, the ratio of corresponding forces $\mathbf{F}_2/\mathbf{F}_1$ in both flow fields is the *same*, or there is a similarity of forces. For complete similitude to exist in two flow fields (one involving the model, and the other involving the prototype), the ratio of forces of the *same nature* must be the same in both flow fields. Some of the forces that influence fluid behavior are inertial, pressure, body or gravitational, magnetic and electrical, viscous, surface tension, and compressible forces. Similarity of forces is necessary because the direction taken by any fluid particle is determined by the resultant force acting on it. Thus, complete similarity of two flows can be achieved only where corresponding particles are acted on by forces that have the same direction and are in a fixed ratio of magnitude.

In *kinematic similitude* the ratio of corresponding velocities $\mathbf{V}_2/\mathbf{V}_1$ is a constant for all field points of the two systems, or there is a similarity of streamlines.

Consider two different flows. Let the subscript 1 denote the model system and the subscript 2 denote the prototype. For geometric similarity

$$\frac{\mathbf{r}_2}{\mathbf{r}_1} = c_1 \tag{7.49}$$

where $\mathbf{r}_2$ and $\mathbf{r}_1$ are two position vectors to the same characteristic point in the flow, and c_1 is a constant. The geometries of both flow fields are identical if all space dimensions in flow 2 are made dimensionless with $\mathbf{r}_2$ and all space dimensions in flow 1 are made dimensionless with $\mathbf{r}_1$.

Now that the linear dimensions have been made similar, real time also possesses similarity: from kinematic similitude,

$$\frac{\mathbf{V}_2}{\mathbf{V}_1} = c_2 \tag{7.50}$$

so that from Eq. (7.49) a similarity in time is

$$\frac{T_2}{T_1} = c_3 = \frac{c_1}{c_2} \tag{7.51}$$

In the case of density, there is a similarity between all homogeneous Newtonian fluids (as evidenced by specific gravity); hence

$$\frac{\rho_2}{\rho_1} = c_4 \tag{7.52}$$

Similar ratios hold for pressure:

$$\frac{p_2}{p_1} = c_5 \tag{7.53}$$

The scale of acceleration **a** stems from kinematic similitude, so that from Eqs. (7.50) and (7.51) (along with the definition of acceleration) we obtain

$$\frac{\mathbf{a}_2}{\mathbf{a}_1} = \frac{c_2}{c_3} = \frac{c_2^2}{c_1} \tag{7.54}$$

The scale for kinematic viscosity ν is obtained from Newton's law of friction, so that from Eqs. (7.49)–(7.51) we find

$$\frac{\nu_2}{\nu_1} = \frac{c_1^2}{c_3} = c_1 c_2 \tag{7.55}$$

and so on.

Consider now two flows around similar bodies moving in a real fluid. For the flows to be dynamically similar, the forces must be proportional. From Eq. (7.26)

$$\frac{\partial \mathbf{V}_1^*}{\partial \tau_1} + (\mathbf{V}_1^* \cdot \boldsymbol{\nabla}_1^*)\mathbf{V}_1^* = -\boldsymbol{\nabla}_1^* \tilde{p}_1^* + \frac{1}{R_{L_1}} \nabla_1^{*2} \mathbf{V}_1^* \tag{7.56}$$

representing the dimensionless Navier-Stokes equation for the model and

$$\frac{\partial \mathbf{V}_2^*}{\partial \tau_2} + (\mathbf{V}_2^* \cdot \boldsymbol{\nabla}_2^*)\mathbf{V}_2^* = -\boldsymbol{\nabla}_2^* \tilde{p}_2^* + \frac{1}{R_{L_2}} \nabla_2^{*2} \mathbf{V}_2^* \tag{7.57}$$

representing the dimensionless Navier-Stokes equation for the prototype. For these two equations to yield similar solutions, the two equations must differ by a proportionality constant. Thus we can show, by ratioing the force per unit mass of each term of Eq. (7.56) with that of Eq. (7.57), that

$$\begin{aligned} \frac{\partial \mathbf{V}_1^*/\partial \tau_1}{\partial \mathbf{V}_2^*/\partial \tau_2} &= \frac{(\mathbf{V}_1^* \cdot \boldsymbol{\nabla}_1^*)V_1^*}{(\mathbf{V}_2^* \cdot \boldsymbol{\nabla}_2^*)V_2^*} = \frac{\boldsymbol{\nabla}_1^* \,\tilde{p}_1^*}{\boldsymbol{\nabla}_2^* \,\tilde{p}_2^*} \\ &= \left(\frac{\nabla_1^{*2}\mathbf{V}_1^*}{R_{L_1}}\right)\left(\frac{R_{L_2}}{\nabla_2^{*2}\mathbf{V}_2^*}\right) = 1 \end{aligned} \tag{7.58}$$

with the result that *each force ratio is equal to the same constant, and the flows of the prototype and model are dynamically and geometrically similar.*

7.9 Modeling

Modeling is a human instinct, utterly necessary to understanding and manipulating the environment. As children, we continually create models of the world around us, models that we can control and experiment with. We build sand castles and block towers—and learn about erosion and unstable structures. We play with dolls, toy soldiers, and stuffed animals—and fashion models of the social world we must grow into. Modeling is critical to a child's development and allows the creative imagination to discover and expand. As adults, engineers of all persuasions still use models as professional tools to understand how and why something may or may not work. No longer a game when an engineer says, "Let's build a model," or "Let's play with that idea," modeling is an activity that is at once serious and still tinged with magic and wonder. The ideal model is the essence of a theory or a problem, from which all trivia has been stripped. The great trick, of course, is knowing what is trivia.

Figure 7.4a shows a wooden model of an ice-breaker's hull. The bow is a radical departure from conventional designs and must be tested in a tow tank before it can be approved for construction. A naval architect designed the ship's hull theoretically using a hydrodynamic analysis. Satisfied with his theoretical predictions of the ship's

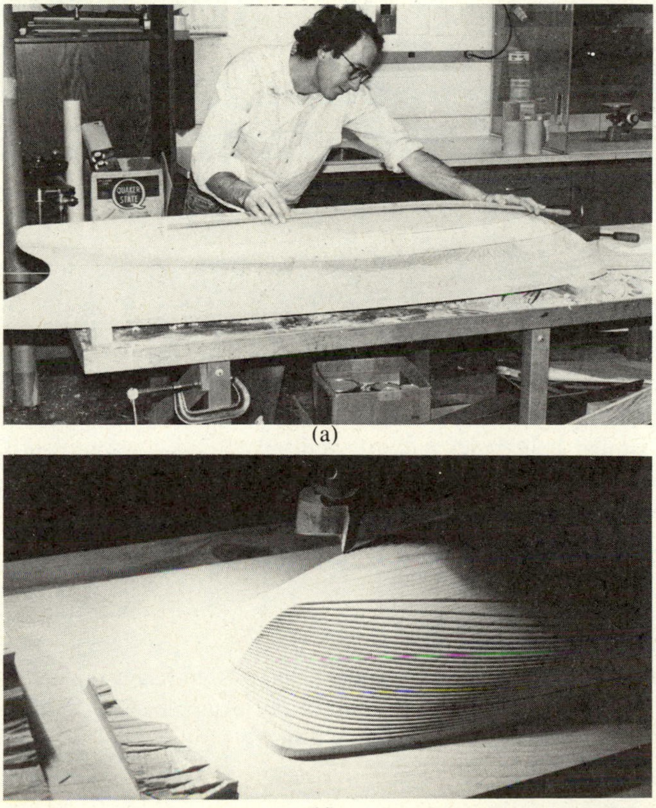

(a)

(b)

Figure 7.4 *(a) Model maker working on model icebreaker hull. (b) Computer-graphic controlled milling machine cutting out wooden ship hull.*

performance, the architect fed the equations for the hull's geometry into a computer graphic program which instructed a milling machine to reproduce the desired hull shape. Figure 7.4b shows the result. The model was then carefully refined by a model maker, who smoothed out the imperfections by the machine.

Often we have to modify the model to create flow phenomena such as turbulence. For example, to scale the drag of water to the surface of a model ship's hull, experimenters attach small sections of sandpaper or a line distribution of brad nails near the bow below the waterline, as shown in Fig. 7.5, to generate small vortices akin to desired levels of turbulence. To simulate supersonic flight, gas in a supersonic wind tunnel is subcooled to −350°F in order to match a viscosity for turbulence at supersonic flows. We can thus study the model's behavior under controlled conditions and make predictions as to how the prototype will behave.

Let us now consider some of the criteria necessary to achieve similitude.

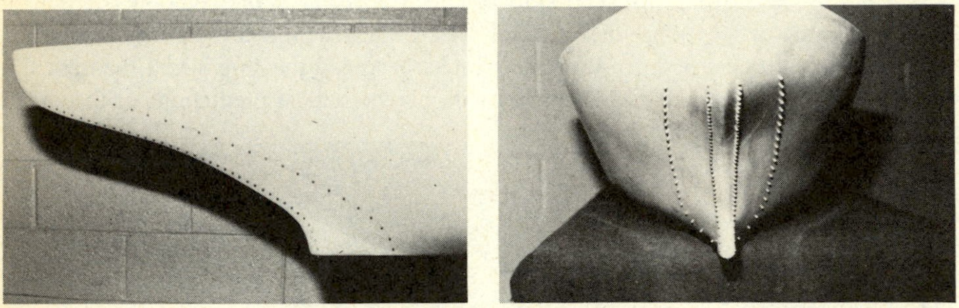

Figure 7.5 Turbulence generators on model's ship bow. (a) Side view. (b) Front view.

7.9.1 Reynolds Number Modeling

For two dynamically similar flows where shear stresses play an important role in the fluid flow, the Reynolds numbers must be equal. For dynamic similarity of two flows past geometrically similar boundaries and affected only by viscous, pressure, and inertial forces, the magnitude ratio of inertial and viscous forces at corresponding points must be the same:

$$\left(\frac{UL}{\nu}\right)_1 = \left(\frac{UL}{\nu}\right)_2 \tag{7.59}$$

The length L in Eq. (7.59) may be any reference length that is significant in characterizing the flow. The length is to be defined where used. For example, Eq. (7.59) is important in any flow of a single fluid past or through objects.

7.9.2 Froude Number Modeling

For two dynamically similar flows where gravity forces play a significant role, the Froude numbers must be equal:

$$\left(\frac{U^2}{gL}\right)_1 = \left(\frac{U^2}{gL}\right)_2 \tag{7.60}$$

The condition for dynamic similarity of flows of this type is that the magnitude ratio of inertial to gravity forces be the same at corresponding points in geometrically similar systems. The pressure forces are taken care of by the requirement of zero resultant force. Gravity forces are important in any flow with a free-surface. Thus, the Froude number is a significant parameter in determining what part of a ship's resistance is caused by surface waves. Also, Froude number is important when there is an interface between two immiscible fluids.

It is sometimes impossible to match certain dimensionless parameters. For instance, suppose we desire to match both Reynolds number and Froude number, which is an important consideration in studying the behavior of surface ships. Let us consider the fluid for the prototype to be the same for the model:

$$\nu_1 = \nu_2 \tag{7.61}$$

and let

$$g_1 = g_2 \tag{7.62}$$

From Reynolds number modeling, Eq. (7.59) gives

$$\frac{U_2}{U_1} = \frac{L_1}{L_2} \tag{7.63}$$

and from Froude number modeling, Eq. (7.60) gives

$$\frac{U_2}{U_1} = \sqrt{\frac{L_2}{L_1}} \tag{7.64}$$

Clearly, by comparing Eqs. (7.63) and (7.64) we see that we cannot match the two force ratios for the *same fluid environment*. The model would have to be tested in a fluid whose viscosity was different from the fluid for the prototype, according to

$$\nu_1 = \nu_2 \left(\frac{L_1}{L_2}\right)^3 \tag{7.65}$$

if $R_{L_1} = R_{L_2}$ and $(F_r)_1 = (F_r)_2$. For example, if a model is 1/10 the size of its prototype, the prototype would have to be immersed in residium or oil if the model were in water.

In actual practice, ship models are tested in nearly the same liquid for the prototype: water. To find the ship's drag from model tests, two separate tests are made. The ship drag is assumed to be composed of wave drag and friction drag. The wave drag is based upon a free-surface phenomenon caused by gravity waves. So, to calculate the wave drag, a model of the ship is built and tested for Froude number similitude. We do not consider Reynolds number for this test.

A drag coefficient C_D is defined as

$$\boxed{C_D = 2D/\rho U^2 A} \tag{7.66}$$

so that the drag coefficient for wave drag is

$$C_{D_W} \equiv \frac{2(D)_w}{\rho U^2 A} \tag{7.67}$$

The wave drag of the model is scaled up to the wave drag of the prototype by dynamic similitude:

$$\left(\frac{D_w}{\rho U^2 L^2}\right)_2 = \left(\frac{D_w}{\rho U^2 L^2}\right)_1 \tag{7.68}$$

Substituting Eqs. (7.60) and (7.67) into Eq. (7.68) we obtain

$$(D_w)_2 = (D_w)_1 \left(\frac{L_2}{L_1}\right)^3 \tag{7.69}$$

which states the prototype's wave drag is increased as the cube of the length ratio c_1.

The friction drag of the model is scaled up to the friction drag of the prototype by

$$\left(\frac{D_f}{\rho U^2 L^2}\right)_2 = \left(\frac{D_f}{\rho U^2 L^2}\right)_1 \tag{7.70}$$

Substituting Eqs. (7.59) and (7.61) into Eq. (7.70) yields

$$(D_f)_2 = (D_f)_1 \tag{7.71}$$

which means one can directly measure the friction drag for the model directly and equate it to the prototype frictional drag.

Example 7.5

Air at standard conditions of 15.56°C flows at an average velocity U of 6.1 m/s through a circular pipe of diameter 0.762 m. A model of this flow is desired using water. (a) What is the average velocity of water at standard conditions of 15.56°C flowing in a 7.62 cm diameter pipe if the flow is to be similar to the prototype? (b) If the pressure drop Δp in the prototype is 3.447 kPa, what is the pressure drop in the flow of water?

Solution:

Step 1.

Determine appropriate dimensionless parameters to match, and then evaluate them.

Example 7.5 *(Con't.)*

(a) For complete dynamic similarity in this problem of a single fluid that has no interface with a second fluid, the Reynolds numbers of the prototype and model must be identical. Thus, from Eq. (7.59), taking the characteristic length as the diameter of the pipe,

$$\left(\frac{\rho UD}{\mu}\right)_1 = \left(\frac{\rho UD}{\mu}\right)_2 \tag{i}$$

From Table 2.4, the density of air at sea level is $\rho_2 = 1.192\ \text{kg/m}^3$, and the density of water at 15.56°C is $\rho_1 = 1000\ \text{kg/m}^3$. From Fig. 2.12, the values of viscosity of air and water at 15.56°C are $\mu_2 = 0.018$ cP and $\mu_1 = 1.1$ cP. Using these values of density and viscosity, the average velocity U_1 of water becomes

$$\begin{aligned} U_1 &= U_2\left(\frac{\rho_2}{\rho_1}\right)\left(\frac{\mu_1}{\mu_2}\right)\left(\frac{L_2}{L_1}\right) \\ &= 6.1\left(\frac{1.192}{1000}\right)\left(\frac{1.1}{0.018}\right)(10) \\ &= 4.44\ \text{m/s} \end{aligned} \tag{ii}$$

Thus, to obtain dynamic similarity, the average velocity of the water must be 4.44 m/s.

Step 2.

Match the pressure coefficients.

(b) To find the pressure drop in water, we equate the pressure coefficient $\tilde{C}_p$ of Eq. (7.45) of the prototype with the pressure coefficient of the model

$$(\tilde{C}_p)_2 = (\tilde{C}_p)_1 \tag{iii}$$

or

$$\left(\frac{\Delta p}{\rho U^2}\right)_2 = \left(\frac{\Delta p}{\rho U^2}\right)_1 \tag{iv}$$

Solving for the pressure drop in water gives

$$\Delta p_1 = \Delta p_2\left(\frac{\rho_1}{\rho_2}\right)\left(\frac{U_1}{U_2}\right)^2 \tag{v}$$

Substituting the given and calculated values,

$$\Delta p_1 = 3.447\left(\frac{1000}{1.192}\right)\left(\frac{4.44}{6.1}\right)^2$$

or

$$\Delta p_1 = 1.53\ \text{MPa}$$

Thus, we find that a pressure drop of 1.53 MPa will occcur in the flow of water.

This completes the solution.

Example 7.6

A rectangular platform juts out in a stream. The object is 1.5 m × 3 m in area exposed to the air surface. The stream is 2.5 m deep. A model of the flow is to be constructed at a scale 1:20. The average velocity of the free stream is 1.5 m/s, and a pressure force of 22 kN acts on the platform because of the stream. (a) Calculate the velocity of the flow and force on the model for dynamic similarity if the water is used for both fluids in the prototype and the model. (b) If a standing wave is 5 cm high in the model flow, what height of a wave can be expected in the prototype flow?

Solution:

Step 1.

Determine appropriate dimensionless parameters to match, and evaluate.

(a) In this problem, water has an interface with air, so that the predominant forces are gravity and inertial. From Eq. (7.60) the matching of the Froude numbers gives

$$\left(\frac{U^2}{gL}\right)_1 = \left(\frac{U^2}{gL}\right)_2 \tag{i}$$

where 1 and 2 denote model and prototype, respectively. Solving for the average velocity of the model U_1 gives

$$U_1 = U_2\sqrt{\frac{L_1}{L_2}} \tag{ii}$$

and substituting in given values of velocity and geometric scale, we obtain, from Eq. (ii),

$$U_1 = 1.5\sqrt{\frac{1}{20}}$$

$$= 0.335 \text{ m/s} \tag{iii}$$

as the velocity of the flow in the model.

To calculate the pressure force on the model, we now shift our emphasis to where inertial and pressure forces are the predominant forces in the flow field. The ratio of the pressure forces is

$$\frac{F_{p_1}}{F_{p_2}} = \left(\frac{\gamma_1}{\gamma_2}\right)\left(\frac{L_1}{L_2}\right)^3 \tag{iv}$$

Since the fluid is the same for both model and prototype, the pressure force on the model platform becomes (using the given values for the prototype's pressure force and scale factor)

Example 7.6 *(Con't.)*

$$F_{p_1} = F_{p_2}\left(\frac{L_1}{L_2}\right)^3 \qquad \text{(v)}$$

$$= 22\ \text{kN}\left(\frac{1}{20}\right)^3$$

$$= 2.75\ \text{N} \qquad \text{(vi)}$$

(b) To obtain the height of the wave in the prototype flow, we again make use of the Froude number model law for kinematic similitude Eq. (ii). Expressing the characteristic length L as the wave height h,

$$h_2 = h_1\left(\frac{U_2}{U_1}\right)^2 \qquad \text{(vii)}$$

$$= 5\left(\frac{1.5}{0.335}\right)$$

$$= 1\ \text{m} \qquad \text{(viii)}$$

Thus, a wave height of 1 m would occur in the prototype flow if dynamic similarity existed.

This completes the solution.

7.10 Drag

Having gained some appreciation for the composition of drag for surface ships, we can now reiterate the salient aspects of drag and apply them to a few other shapes in assorted fluid environments. We will, of course, have much more to say about drag later: we will discuss the calculation of friction drag after delving into boundary layer flows in Chap. 14. Later, we will devote an entire section to alleviating this important force.

Drag over a wetted body is caused by two stresses: the tangential or shearing stresses, and the normal or pressure stresses. The former are the result of both the fluid's viscosity and of the spatial gradients of the velocity components. Since velocity gradients increase in magnitude as the boundary of the body is approached, the stresses reach a maximum at the surface. At the boundary the shear stresses must maintain the fluid flow to equal the velocity of the boundary. Outside of this boundary the fluid layers can slip, but at the boundary the lamina next to the surface may not slip.

The other aspect of drag is caused by pressure in the direction of drag. For a fully wetted body immersed in a fluid flow, the vector sum of the normal and tangential surface stresses integrated over the wetted surface in the direction of the relative velocity past the body is the total drag force, D_T. That component perpendicular to the relative velocity is called the lift force. The total drag on the wetted body of Fig. 7.6 is therefore

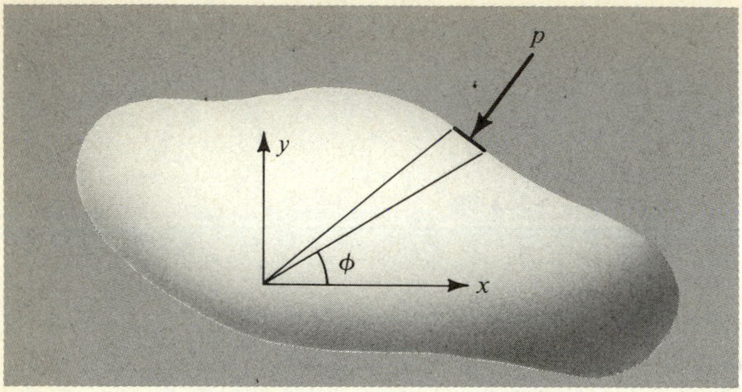

Figure 7.6 Pressure on an arbitrary body immersed in a fluid.

$$D_T = D_f + D_p \tag{7.72}$$

with the components

$$D_f = \text{frictional drag} = \int_S (p_{xy})_o \sin \phi \, dS \tag{7.73}$$

$$D_p = \text{pressure drag} = -\int_S (p \cos \phi)_o \, dS \tag{7.74}$$

where S is the total surface area, and ϕ is the angle between the normal to the surface element and the absolute flow direction. Other names for the frictional drag are skin friction drag, or surface resistance drag. Another name for pressure drag is form drag. Pressure drag would predominate if the body were blunt and the surface very smooth. Friction drag would predominate for streamlined bodies and a rough surface. Thus, airfoils have small pressure drag and large frictional drag, whereas spheres and cars have large pressure drag relative to frictional drag.

Friction and pressure drag are often expressed in terms of the dimensionless drag coefficients in the form of Eq. (7.66):

$$D_f = C_{D_f} \left(\frac{1}{2} \rho U^2 A_f \right) \tag{7.75}$$

$$D_p = C_{D_p} \left(\frac{1}{2} \rho U^2 A_p \right) \tag{7.76}$$

where A_f is usually the wetted area, and A_p is the frontal or projected area. Thus, the total drag coefficient on wetted bodies is

$$\boxed{C_{D_T} = C_{D_f} + C_{D_p}} \tag{7.77}$$

The total drag coefficient for a few simple shapes is presented in Fig. 7.7 for a fairly large range of Reynolds number. Note how the reference length L of the Reynolds number is defined for the various bodies. The area A_p is taken to be the projected area normal to the flow. The results in Fig. 7.7 show that streamlining significantly reduces the total drag coefficient C_{D_T} owing to a significant reduction of pressure drag. Note that, by comparing results in Table 7.1, we see that three-dimensional flow results in a lower drag than two-dimensional flow because pressure drag is relieved. All the shapes have smooth surfaces. If they had rough surfaces, the total drag coefficients would be greater than that shown. This is discussed in Chap. 14.

We must turn to boundary layer theory to gain an appreciation of the manner in which the flow about a body may be only slightly influenced by shear stresses, and yet be subjected to a significant viscous or friction drag force. The boundary layer concept, conceived by L. Prandtl in 1904, is limited to a very thin region near the body surface. The larger the Reynolds number, the less the viscous effect, so the layer is thinner. But within this layer, the shear stresses decelerate the fluid from the relative free-stream velocity value as given by ideal fluid flow theory at the outer edge of the layer, to zero relative velocity at the surface of the body. Thus, outside of the boundary layer, the real fluid flow is, for all intensive purposes, ideal. We will analyze ideal fluid flow by techniques we call potential theory. The region of the boundary layer is taken into account by creating a slightly larger body to accommodate the boundary layer thickness. This will be discussed in Chap. 12.

Table 7.1 *Drag Coefficients for Different Shapes at* $R_D \sim 10^5$

2-D Shapes	C_D	3-D Shapes		C_D
U		U		
	1.16	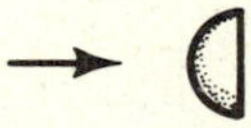		0.38
	1.17		(sphere)	0.47
	1.20		(60° cone)	0.50
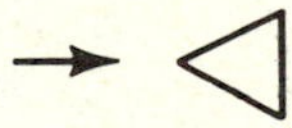	1.55			
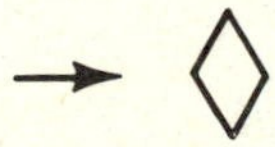	1.55	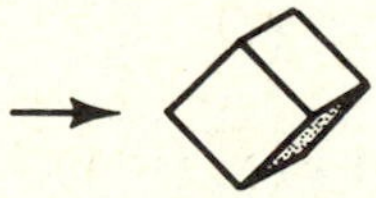	(cube)	0.80
	1.98			
	2.00	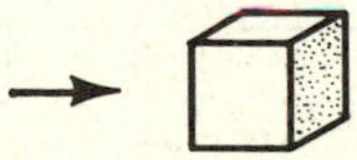	(cube)	1.05
	2.20		(hemisphere)	1.17
	2.30		(cup)	1.42
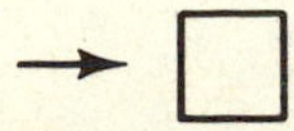	2.05			

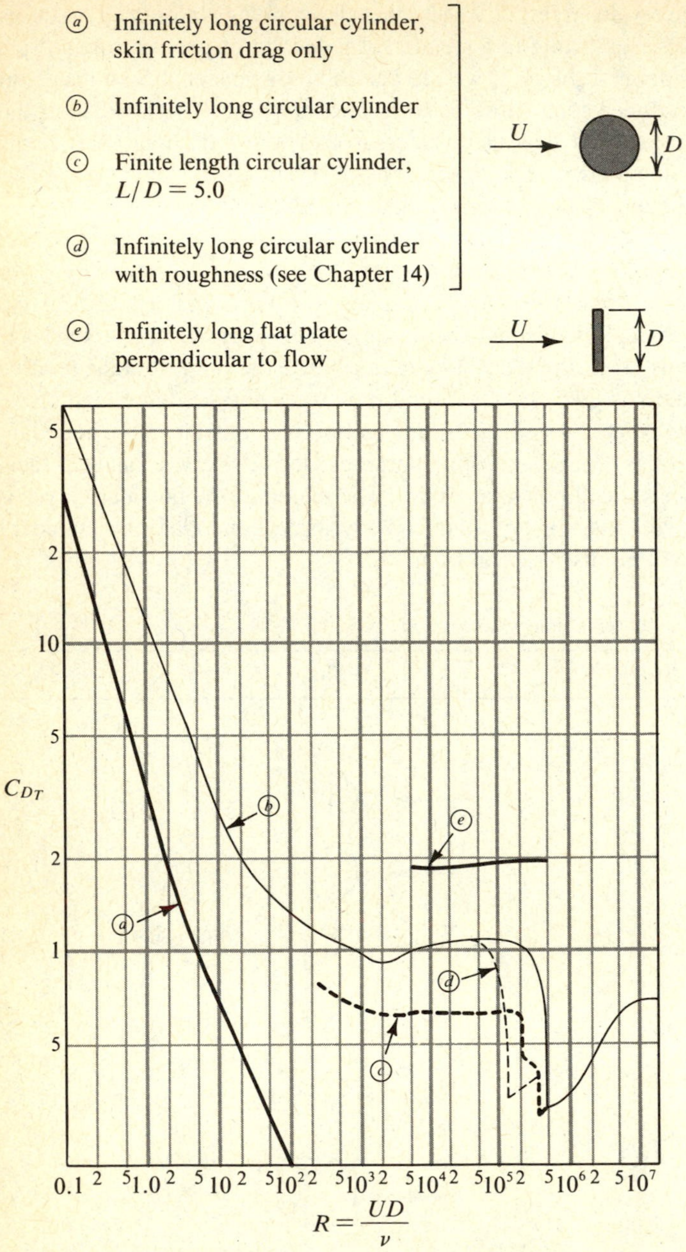

***Figure** 7.7 (a) Drag coefficients for sphere, cylinder, and disk versus Reynolds number.*

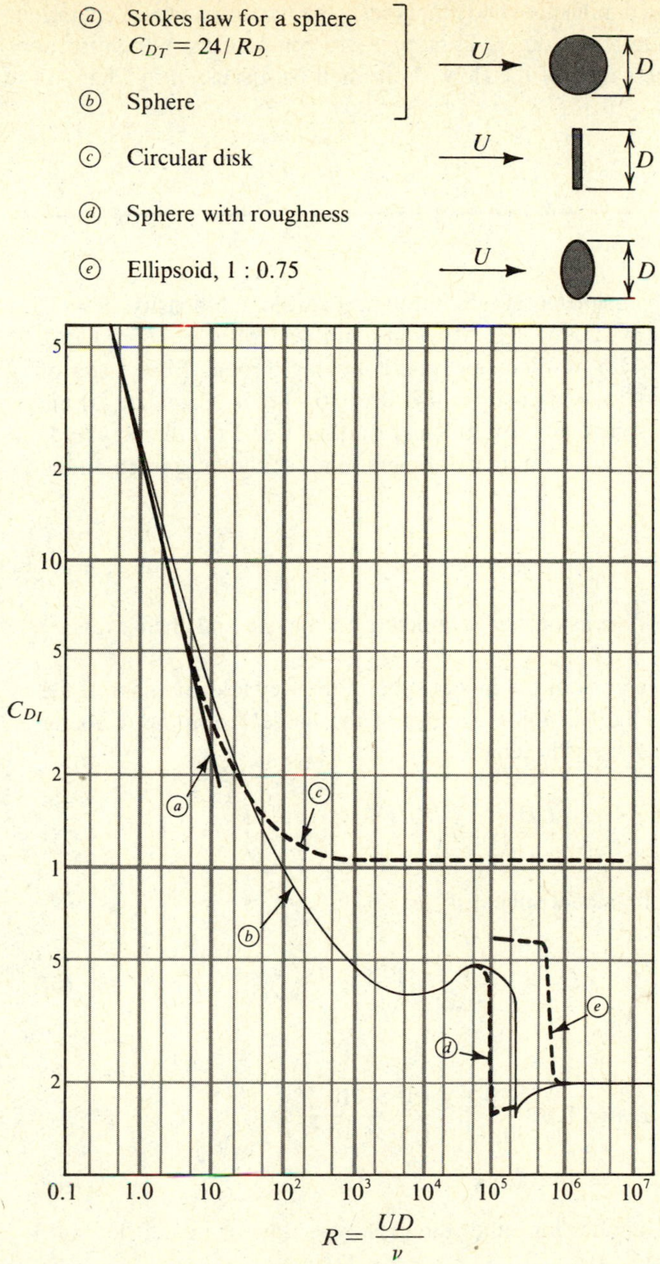

Figure 7.7 *(b) Drag coefficients for sphere, cylinder, and disk versus Reynolds number.*

For bodies of finite and diminishing body thickness, the flow separates. Separation not only produces large drag forces due to large pressure components in the direction of drag, but it modifies the nature of the flow. This shall be discussed in Chap. 14.

Example 7.7

A bathysphere 10 ft in diameter moves through seawater of density $\rho = 2$ slug/ft^3 and viscosity $\nu = 1.5 \times 10^{-5}$ ft^2/s at a uniform velocity $U = 1$ ft/s, and is to be modeled by a 6-in.-diameter sphere moving through fresh water of density $\rho = 1.935$ slug/ft^3 and viscosity $\nu = 1.0 \times 10^{-5}$ ft^2/s. Calculate (a) the velocity of the model sphere for dynamically similar flow, (b) the Reynolds number for the flow, (c) the total drag coefficient, and (d) the drag force on the prototype sphere.

Solution:

Step 1.

Determine appropriate dimensionless parameters to match, and then evaluate them.

(a) To obtain the velocity of the model sphere, the Reynolds number of the prototype is equated to the Reynolds number of the model using the diameter of the sphere as the characteristic length

$$\left(\frac{UD}{\nu}\right)_1 = \left(\frac{UD}{\nu}\right)_2 \tag{i}$$

so that the velocity of the model sphere is

$$U_1 = U_2 \left(\frac{D_2}{D_1}\right)\left(\frac{\nu_1}{\nu_2}\right) \tag{ii}$$

such that

$$U_1 = 1.0 \left(\frac{10}{0.5}\right)\left(\frac{1.0 \times 10^{-5}}{1.5 \times 10^{-5}}\right)$$

$$= 13.33 \text{ fps} \tag{iii}$$

(b) The Reynolds number for either the prototype flow or model flow is

$$R_D = \frac{UD}{\nu}$$

$$= \frac{1 \times 10}{1.5 \times 10^{-5}}$$

$$= 6.67 \times 10^5 \tag{iv}$$

Example 7.7 *(Con't.)*

Step 2.
Evaluate the total drag coefficient either by Eq. (7.66) or from Fig. 7.7.

(c) The total drag coefficient C_{D_T} is obtained from Fig. 7.7b as

$$C_{D_T} = 0.2 \tag{v}$$

(d) The drag force on the prototype sphere is obtained from Eq. (7.66) as

$$D_{T_2} = \frac{1}{2}(\rho U^2 A\, C_{D_T})_2 \tag{vi}$$

Putting in the given and calculated data we obtain

$$D_T = \frac{1}{2} \times 2 \times 1 \times \pi(25) \times 0.2$$

$$= 15.7 \text{ lbf} \tag{vii}$$

as the drag force on the sphere moving through seawater at 1 ft/s.

This completes the solution.

7.11 Lift

The lift component of the resultant force need not be separated into frictional and pressure components. The lift is the other pressure component of Eq. (7.74):

$$L = -\int (p \sin \phi)_o dS \tag{7.78}$$

or in terms of the dimensionless lift coefficient, C_L,

$$\boxed{L = C_L \frac{1}{2} \rho U^2 A} \tag{7.79}$$

where A may be either the projected area normal to U, or the largest projected area of the body (such as the planar area of a wing). The lift coefficient C_L is a function of angle-of-attack and Reynolds number, as well as the geometric shape (i.e., the thickness distribution) of the body. The Reynolds number can be based on any length (span, chord, thickness) as long as it is consistently chosen. Figure 7.8 presents the lift coefficient as a function of angle-of-attack for various values of the Reynolds number for two specific airfoils, whose cross-sectional geometry is shown. Note that lift is relatively insensitive to Reynolds number. Figure 7.8b is similar to Fig. 7.8a except that the thinner airfoil does not have as much lift as the thick airfoil for the same angle-of-attack, although the thinner airfoil would have less drag. So we have to consider trade-offs in selecting a particular wing geometry. Note that, in Fig. 7.8, the lift drops off at a moderately high angle-of-attack. This is caused by stall.

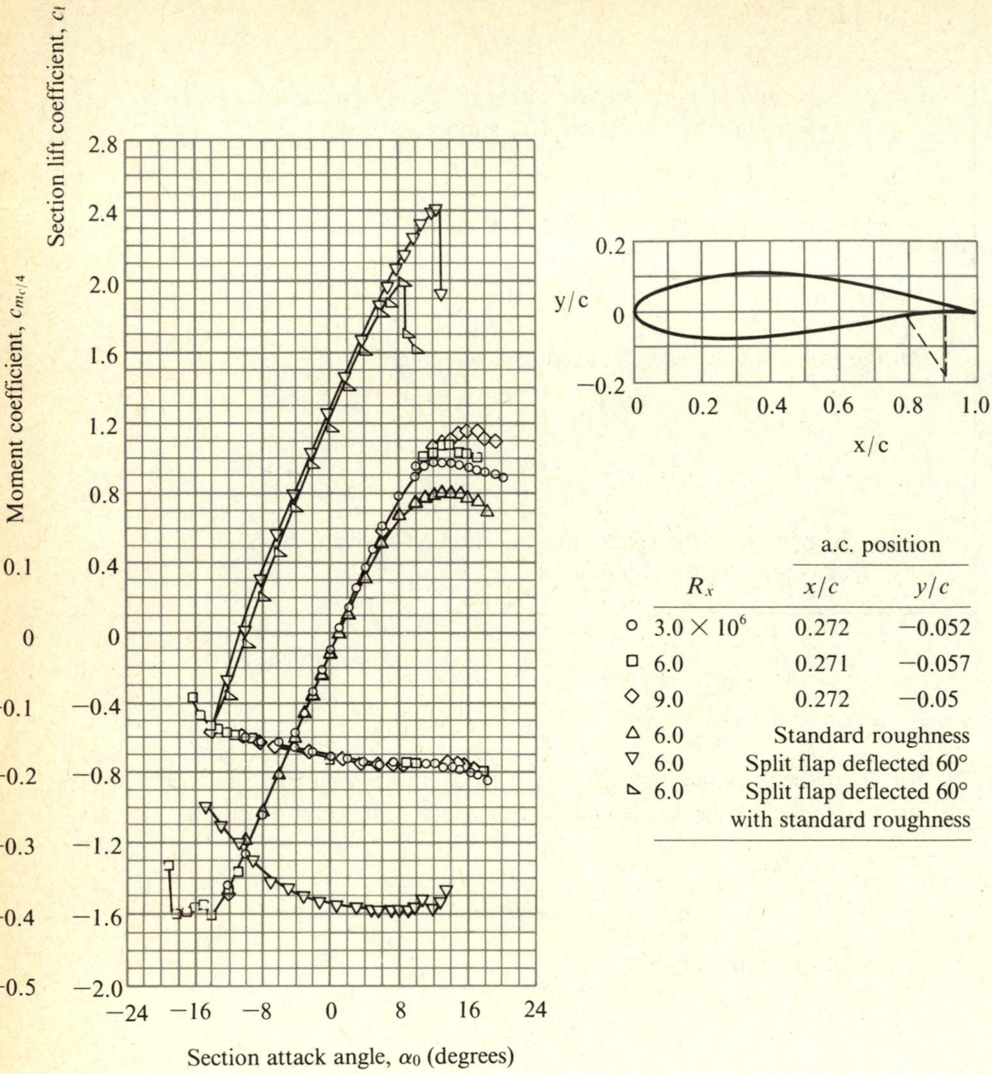

Figure 7.8 *(a) Lift coefficient for NACA 63-418 airfoil.*

In the designation of NACA 64-110, the number sequence is significant. The 6 is the series designation. The 4 denotes the chordwise position of the minimum pressure in tenths of the chord behind the leading edge for the basic symmetrical section at zero lift. The 1 following the dash gives the design lift coefficient in tenths, and the last two digits (10) indicate the thickness of the wing section in percent of the chord.

These figures contain so many results that it would be rather exhausting to discuss them all. Note the effect of (1) thickness, (2) Reynolds number, (3) roughness, (4) deflected flap, and (5) angle-of-attack on the lift and moment coefficients.

Section lift coefficient, c_l

Moment coefficient, $c_{m_{c/4}}$

	R_x	a.c. position	
		x/c	y/c
○	3.0×10^6	0.26	-0.013
□	6.0	0.26	-0.018
◇	9.0	0.261	-0.022
△	6.0	Standard roughness	
▽	6.0	Split flap deflected 60°	
◁	6.0	Split flap deflected 60° with standard roughness	

Section angle of attack, α_0 (degrees)

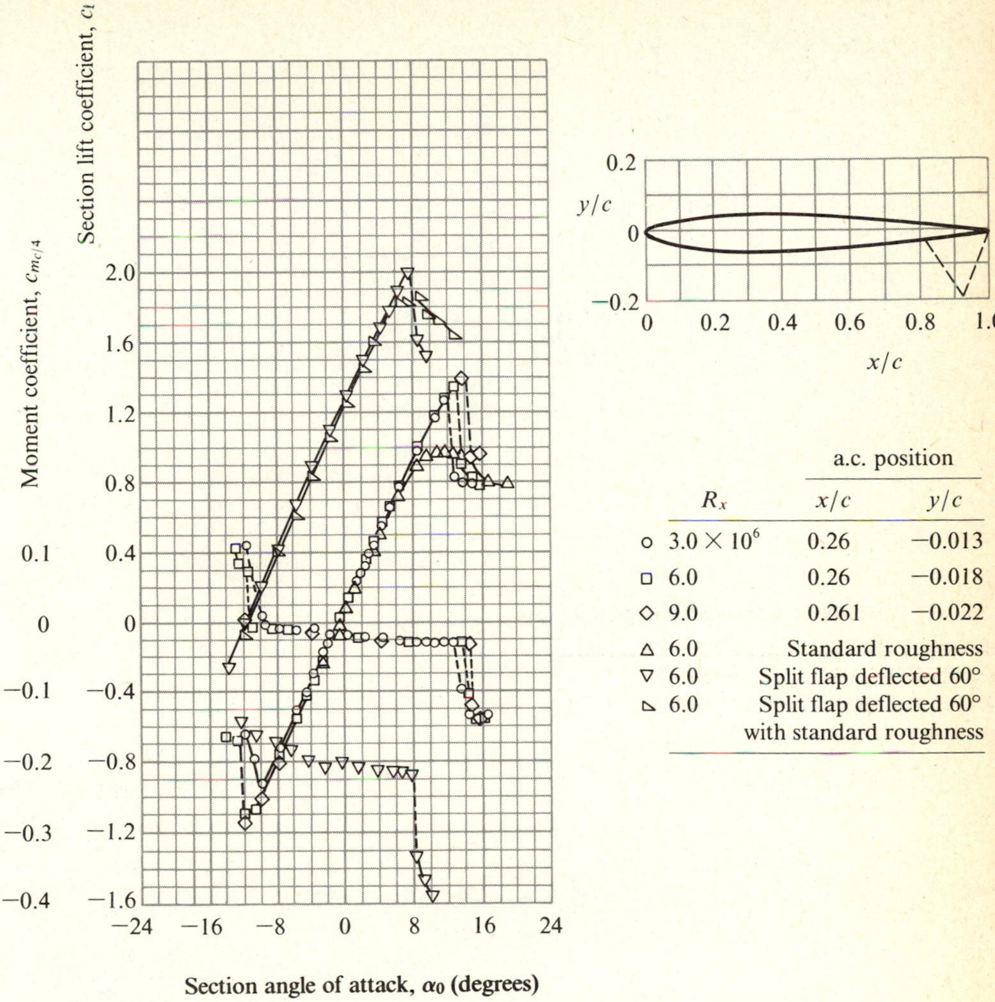

Figure 7.8 (b) *Lift coefficient for NACA 64-110 airfoil. (Source: I. Abbott and A. E. von Doenhoff, Theory of Wing Sections, © 1949, Dover Publications, Inc. Used with the permission of the publishers.)*

7.12 *Vorticity Effect in Lift and Drag*

Consider a finite thin flat wing moving with velocity U. Let the pressure on the upper side of the wing be less than that of the lower side, resulting in a pressure difference and subsequently a lift. For this case, the fluid beneath the wing near the wing tips will move around the tips toward the lower energy side, as shown in Fig. 7.9. Each vortex will have a circulation Γ given by

$$\Gamma = \oint \mathbf{V} \cdot d\mathbf{r} \qquad (7.80)$$

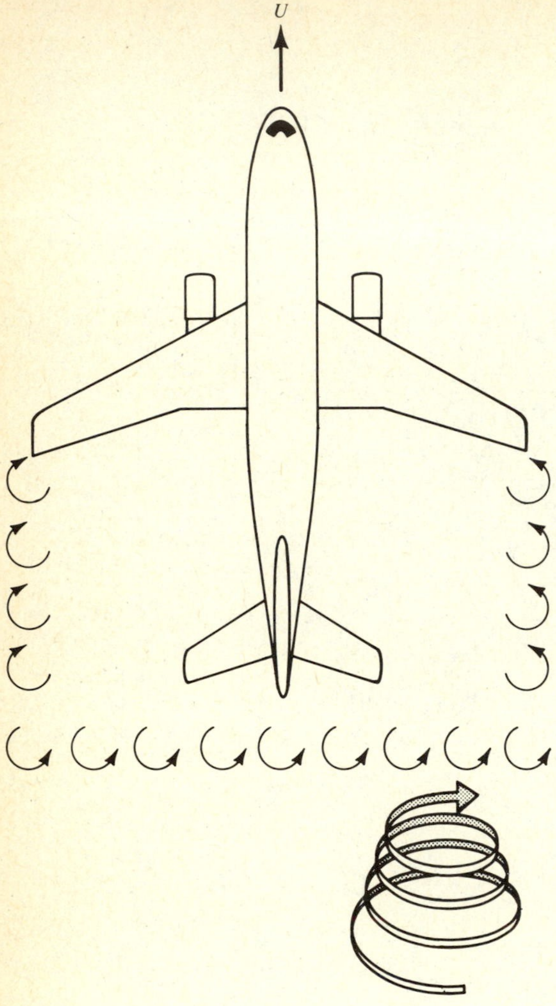

Figure 7.9 Vorticity distribution on a finite wing.

which is nonzero. Furthermore, the vortex tube must obey Kelvin's theorem that it cannot end in the fluid, but must attach itself to the boundary (the wing) or else extend to infinity. This means that lift is not divorced from vorticity. Furthermore, because of the presence of vorticity and the fact the fluid is real, a drag force exists. The energy expended by the drag D is DU. Since the kinetic energy of the flow does not change, the drag experienced by the body is zero for an irrotational flow. Yet, when vorticity is present, it lies in the wake in a length, say Ut. If we denote the mean kinetic energy of the fluid per unit length by K, the vorticity in the wake contains kinetic energy which increases with time at a rate KU. But this increase in kinetic energy has to stem from the power done by the drag force of the wing. Thus, a nonzero drag force (which is identified as *induced drag*)* is always present where there is lift on any *finite* body, and its magnitude must equal K. Thus, vorticity is very closely related to both lift and drag and plays a significant role in the design of aircraft.

References

7.1 Lord Rayleigh, "On the Viscosity of Argon as Affected by Temperature," *Proc. Roy. Soc.* (London), 66:68–94 (1899).

7.2 Imrie, B. W., "The Dimensionless Efficiency of Teaching Dimensional Analysis," *Bull. Mech. Engr. Ed.*, 7, 227–235.

Study Questions

7.1 What is the difference between units and dimensions?
7.2 Why is it important to differentiate between lbm and lbf, lbm and slug?
7.3 Explain an important use of dimensional analysis in experimental fluid dynamics.
7.4 State the Buckingham pi theorem. Explain how it works.
7.5 What does one do with the dimensionless pi term once it is evaluated?
7.6 Arrange Bernoulli's equation in dimensionless form and identify the dimensionless terms.
7.7 Arrange the continuity equation in differential and dimensionless form.
7.8 How is a dimensionless parameter derived?
7.9 Explain the words "characteristic length" and "characteristic velocity." What makes a quantity characteristic?
7.10 What is the advantage of the dimensionless form of the Navier-Stokes equation (7.26) over the dimensional form of the Navier-Stokes equation (6.4)?

Problems

7.1 Express the dimensions of force in the physical system, and the basic SI, metric, and USCS system.

7.2 Obtain the magnitude and units of the universal gravitational constant g_c for the SI, metric, and absolute engineering system.

7.3 Verify the dimensions of dynamic viscosity by using Newton's definition of viscosity.

7.4 Show that the incompressible continuity equation combined with the Navier-Stokes equation can be properly written as

$$\rho\frac{D\mathbf{V}}{Dt} + \nabla\cdot\mathbf{V} = -\nabla p + \gamma + \mu\nabla^2\mathbf{V}$$

but is not homogeneous.

7.5 Arrange the following into dimensionless parameters: (a) a, V; (b) F, γ, A, D; (c) K, σ, L; and (d) Δp, a, M, A, where a is acceleration and A is area.

7.6 Arrange the following into dimensionless parameters: (a) Q, g, ν; (b) Δp, ρ, V; (c) V, a, L; and (d) K, γ, A, where a is acceleration and A is area.

*A superb description of the drag due to lift is given in Chap. VII of *Fluid Dynamics Drag*, S. Hoerner, published by the author, Midland Park, N.J., 1965.

7.7 Give the dimensions of shear stress, circulation, vorticity, power, momentum, lift, and angles in terms of (a) M, L, T, and (b) F, L, T.

7.8 Determine the form of Newton's second law of motion using the M, L, T system, and compare results using the F, L, T system.

7.9 Determine Newton's law of viscosity by dimensional analysis given that the shear stress depends upon viscosity and rate of deformation for laminar flows.

7.10 Using the Buckingham pi theorem, obtain a relationship for a pump's power in terms of flow rate Q, pressure rise Δp, density, efficiency, dynamic viscosity, and pipe diameter D.

7.11 Repeat Prob. 7.4 using the Rayleigh method.

7.12 Construct a dimensionless arrangement of shear force F_τ, pressure drop Δp, surface tension σ, and real time t, using the Buckingham pi theorem. Compare results using the Rayleigh method.

7.13 A parachutist falls through the air with an acceleration g. If his body area and mass are A and M, respectively, and the wind resistance is D, determine the pi parameters.

7.14 Natural gas is flowing through a 6 in. pipe at 2 ft/s. The gas is characterized by its density and dynamic viscosity. The gas company wishes to model this flow, and needs an expression for the speed. Find an expression for the speed V using the pi theorem.

7.15 Water moves up a straw placed in a glass with a velocity V (Fig. P7.15). The properties of the water are density and kinematic viscosity, such that a change in either one will give different results. If the surface tension is σ in a diameter D, obtain pi parameters characterizing the problem.

7.16 The drag D on a rock falling from a cliff depends upon g, rock diameter D, density of air, and the mass of the rock. Using the Buckingham pi theorem, develop an expression for the drag. Compare the solution using the Rayleigh method.

7.17 Show that Bernoulli's equation can be written in nondimensional form

$$\frac{F_r^2}{2} + \frac{p}{\gamma z_o} + \frac{z}{z_o} = \text{const.}$$

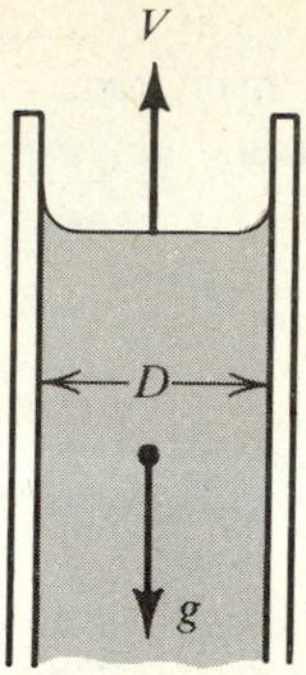

Figure P7.15

where z_o is a reference height. Determine what F_r should be.

7.18 The mean free path λ of gas molecules depends upon the molecular diameter d_o, molecular mass M, and density ρ of the gas. Find the dimensionless pi parameters descriptive of the problem.

7.19 Find the pi dimensionless parameters when the following quantities are important in an experiment.

Power, P
Length, l
Number of revolutions n in unit time
Linear velocity, U
Density of fluid, ρ
Viscosity of fluid, μ

7.20 Obtain the pi dimensionless parameters for an experiment involving

Moment of the force, M
Reference length, l
Uniform velocity, U
Acceleration due to gravity, g
Density, ρ
Speed of sound, c

7.21 In a vortex dynamics experiment, it was found that a surge moving with a velocity W depended upon the radius of the vortex R, viscosity of the fluid ν, and circulation of the swirl Γ. Find a relationship for the velocity w in terms of R, ν, and Γ. (See Fig. P7.21.)

7.22 In a particular fluid dynamics experiment, the fluid pressure depends upon a linear dimension l, density ρ of the fluid, and viscosity μ. Find an expression for the pressure p, using the Rayleigh method.

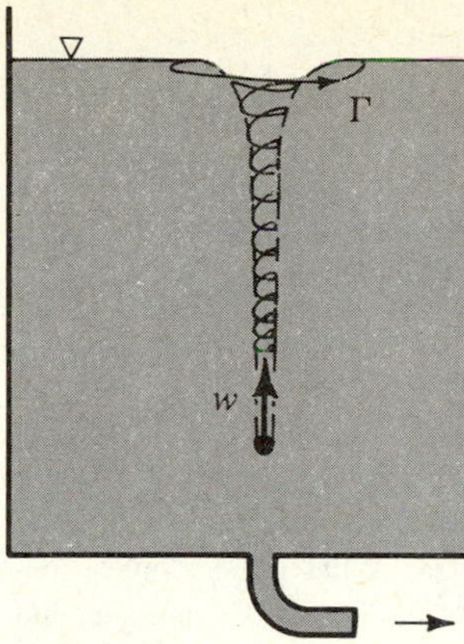

Figure P7.21

7.23 A seaplane lands on the water surface and a hydrodynamic pitching moment T results (Fig. P7.23). The following are the variables that play a significant role in the pitching moment:

- L, length of seaplane hull
- M, mass of seaplane
- ρ, density of water
- g, acceleration due to gravity
- R, radius of gyration about pitching axis
- α, flight path angle

Determine expressions for the pi dimensionless numbers.

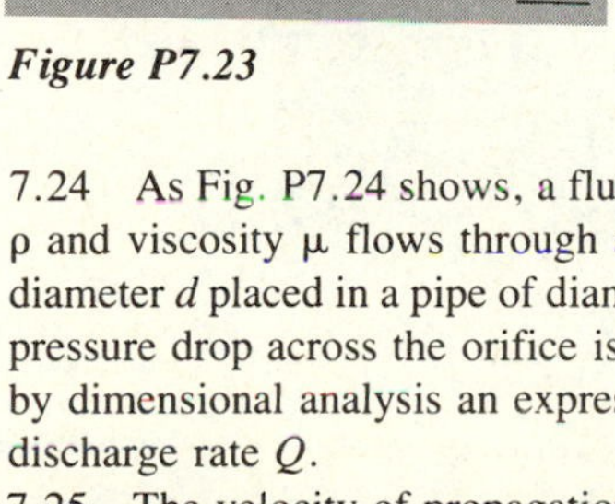

Figure P7.23

7.24 As Fig. P7.24 shows, a fluid of density ρ and viscosity μ flows through an orifice of diameter d placed in a pipe of diameter D. The pressure drop across the orifice is Δp. Derive by dimensional analysis an expression for the discharge rate Q.

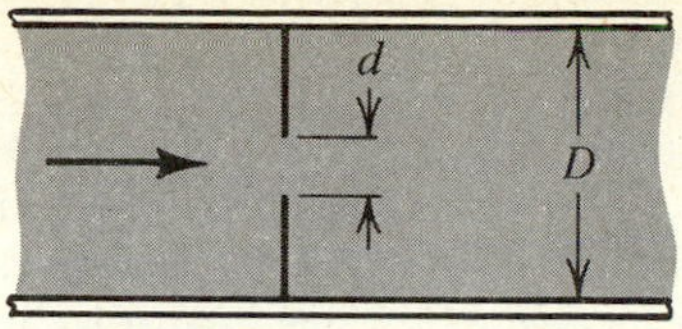

Figure P7.24

7.25 The velocity of propagation c of a surface wave in deep water is assumed to depend upon the wavelength λ, the density ρ, the acceleration of gravity g, and surface tension σ. (a) Make a dimensional analysis of this problem. Ignore viscous forces. (b) Also, make a dimensional analysis assuming that surface tension forces are negligible compared with gravity forces.

7.26 A metal ball falls at steady speed in a large tank containing a viscous liquid. The ball falls so slowly that it is known that inertia forces may be ignored in the equation of motion compared with the viscous forces. Make a dimensional analysis of this problem using the M, L, T system, with the aim of relating the speed of fall V to the diameter of the ball D, the density of the ball ρ_b, the density of liquid ρ_c, and any other variables that play a significant role.

7.27 Observations indicate that the resistance R which the air offers to an airplane wing depends mainly on some characteristic length L, speed V of the wing, and the density and viscosity of the air. Find a conventional set of pi parameters.

7.28 Assume the speed of sound c in a gas depends upon the gas density ρ, pressure p, and dynamic viscosity μ. Find a functional relationship for the speed of sound.

7.29 Given a uniform flow velocity of 10 m/s of salt water of kinematic viscosity 5.6×10^{-7} m^2/s, plot the distribution of Reynolds number over 2 m of a flat plate at zero angle-of-attack in water.

7.30 A circular cylinder 1 m in diameter has air flowing past it at a uniform velocity of 30 m/s. The length of the cylinder is 3 m. If the temperature of air is 40°C, calculate the Reynolds number for the flow.

7.31 Derive Eqs. (7.32) and (7.34).

7.32 What is the differential equation of linear momentum when one takes the divergence of each term of Eq. (7.32)? What is a possible solution of the differential equation?

7.33 Derive an expression for the Froude number F_r. What is the Froude number for a ship cruising at 20 knots with a 20 ft waterline, a 100 ft beam, and 575 ft long? What is the ship's speed for the flow to be critical?

7.34 What conditions must exist for the Froude number to equal the Reynolds number?

7.35 Consider the motion of an ocean liner through the high seas. The principle quantities in the design and performance of the ship are D (drag), V (ship speed in still water), g (acceleration potential), σ (surface tension), dynamic viscosity, density of the water, L (length of ship), b (beam of ship), and d (draft of ship). Choosing μ, σ, and ρ as the repeating variables, obtain the result

$$\frac{D\mu^{1/2}}{\sigma^{3/2}} = f(\Pi_2, \Pi_3, \Pi_4, \Pi_5, \Pi_6)$$

Discuss these results.

7.36 An underwater kite is to be simulated in fresh water at 70°F using a scaled-down model of 1:10. If the prototype kite is to be towed at 20 knots in salt water at 50°F, what velocity must the model be towed for dynamic similarity (Fig. P7.36)?

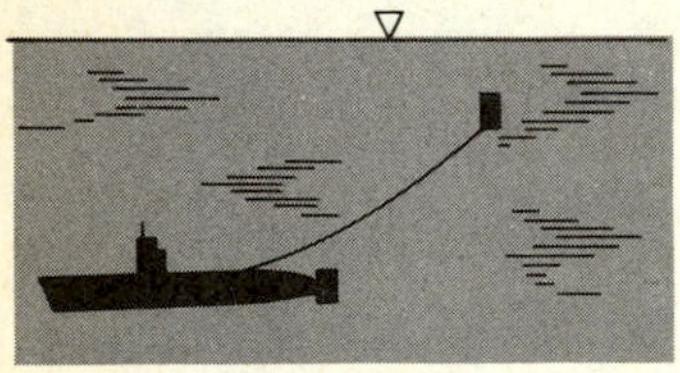

Figure P7.36

7.37 For Prob. 7.36, what is the magnitude of the viscous force, and what is the magnitude of the inertia force, given L_p = 10 ft?

7.38 A model of a salt water harbor is desired. If the actual harbor is 3000 ft wide, and 50 ft deep with 5 ft waves to be simulated, what are the dimensions of the model harbor if 3 in. waves are the maximum possible (Fig. P7.38)?

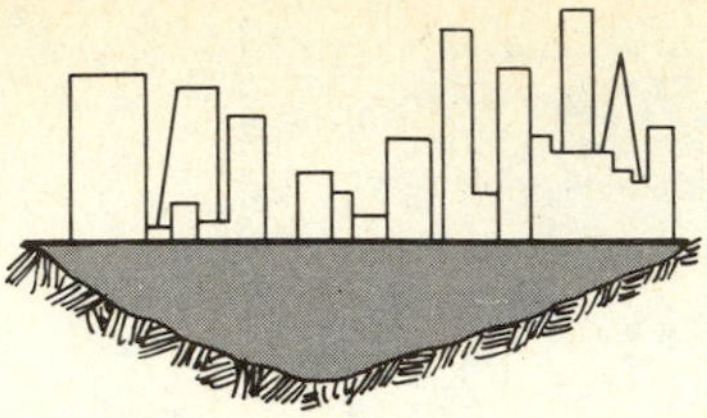

Figure P7.38

7.39 A submarine moves below the water surface at 10 knots. The model is 1/100 scale and is to be tested. What dimensionless parameters should be equated between model and prototype if the drag on the prototype is desired to be calculated? What is the speed the model should be towed using the same water as for the prototype (Fig. P7.39)?

Figure P7.39

7.40 A 10 ft diameter bathysphere is towed through fresh water of density 2 slug/ft^3 at constant speed (see Fig. P7.40). Calculate the drag force on the sphere if a 1/10 scaled model is towed at 2 ft/s.

Figure P7.40

7.41 Oil is flowing through the Alaskan pipeline that has an inner diameter of 18 in. What is the velocity of fresh water at 70°F if we want to have similitude of the oil of viscosity $\nu = 5.0 \times 10^{-5}$ ft²/s? Estimate the drag force per running length on the inside of the pipe, given that $\bar{V}_o$ is 1 in./s (see Fig. P7.41).

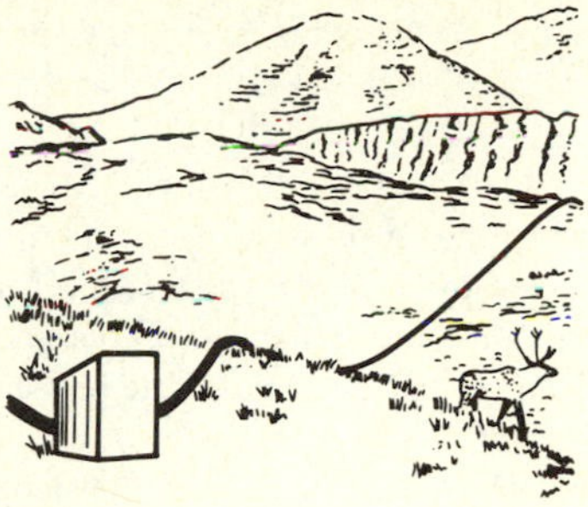

Figure P7.41

7.42 For compressible fluids, show that

$$\frac{U_2}{U_1} = \left[\left(\frac{K}{\rho}\right)_2\left(\frac{\rho}{K}\right)_1\right]^{0.5}$$

and

$$\frac{T_2}{T_1} = \left(\frac{L_2}{L_1}\right)\left[\left(\frac{K}{\rho}\right)_2\left(\frac{\rho}{K}\right)_1\right]^{0.5}$$

7.43 For fluid flows where surface tension is a predominant force per unit length, show

$$\frac{U_2}{U_1} = \left[\left(\frac{\sigma}{L\rho}\right)_2\left(\frac{L\rho}{\sigma}\right)_1\right]^{0.5}$$

and

$$\frac{T_2}{T_1} = \left(\frac{L_2}{L_1}\right)^{3/2}\left[\left(\frac{\rho}{\sigma}\right)_2\left(\frac{\sigma}{\rho}\right)_1\right]^{0.5}$$

7.44 Solve Example 7.4 using the method of separation of variables given $v_r(r, z) = 0$.

7.45 In Example 7.4, plot $w(r, z)$ for various values of r and z using Eqs. (xvi), (xvii), (vii), and (x).

7.46 It is desired to test the water entry of ballistic missiles by studying the behavior of models in laboratory-conducted experiments. Suppose a prototype missile 12 m long enters water at 900 m/s. A 1⁄20 scale model is built to represent the prototype. Neglecting the compressibility effect of water entry and assuming the flow is unaffected by changes in Reynolds number, determine (a) the water entry velocity of the model, and (b) the dynamic force F_2 on the prototype if the model experiences an impact force F_1 of 100 N at water entry.

7.47 The Tucumcari hydrofoil boat has a lift force of 57.5 tons for a Reynolds number of 2×10^8 (based on a reference length of 75 ft at a water temperature of 60°F). What is the lift force on a model hydrofoil of reference length 3 ft if the Reynolds number is the same? What velocity is the hydrofoil operating for the condition given above if the planar area is 1000 ft²? (Refer to Fig. P7.47.) Discuss what is wrong with these assumptions particularly if $\rho_m = \rho_v$, $\nu_m = \nu_p$.

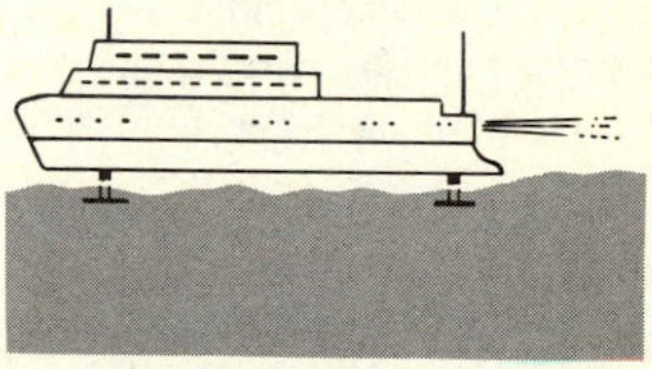

Figure P7.47

7.48 A special water siphon is to be built and tested. The siphon will draw water at one atmosphere and raise it over a 50 ft obstacle. The model is to be ⅛ full size. The vapor pressure of water is 0.44 psia. What is the ambient pressure for the model siphon for dynamic similitude if viscous effects can be neglected and equality of flow rates exists between model and prototype? (Refer to Fig. P7.48.)

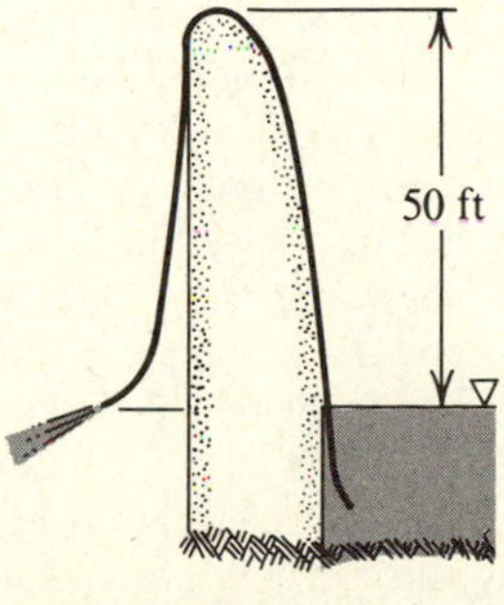

Figure P7.48

7.49 An airplane is to be designed that will fly 650 km/h at an altitude where the density $\rho = 0.655$ kg/m^3 and the kinematic viscosity is 2×10^{-5} m^2/s. A 1/15 scale model is built to determine the drag on the prototype, using a wind tunnel speed of 650 km/h. (a) Assuming the air at the test section of the tunnel is 55°C and that viscosity is independent of pressure, determine the test section pressure so that the model data are useful in designing the prototype. (b) What is the relation between the drag of the prototype and that of the model?

7.50 Looking at Fig. P7.50, see that a missile is designed to reach a velocity of 10 m/s in the ocean whose temperature is 15°C. A model of the missile is tested in a tow tank at a velocity of 40 m/s. (a) Determine the scale c_l that has been used. (b) If the model had been tested in a wind tunnel whose test section pressure was 2 MPa and temperature was 24°C, what would the test section wind velocity be? Let the tow tank have the same fluid as the ocean.

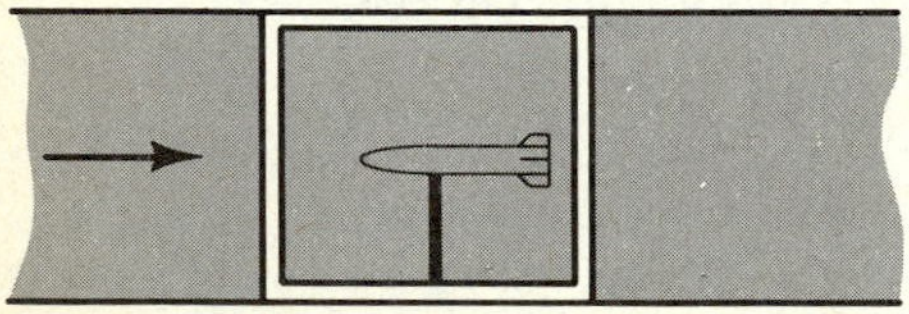

Figure P7.50

7.51 A DGL 51 was once designed to have a 100 m waterline hull length, and travel at 15 m/s. (a) Calculate the Froude number. (b) What velocity should a 1/50 model be towed through water?

7.52 A model of a harbor is made of a prototype using the geometric ratio 1:280. Storm waves of 5 ft amplitude and 30 ft/s velocity occur on the breakwater of the prototype harbor. (a) Neglecting friction, what should be the size and speed of the waves in the model? (b) If the time between tides in the prototype is 12 hours, what should be the tidal period in the model in hours?

7.53 A 6 ft ship model is towed at a velocity of 5 ft/s with a drag force of 8 lbf. (a) What is the velocity of a 150 ft ship? (b) What thrust is necessary to move the ship at this speed through fresh water? (c) Repeat parts (a) and (b) for a submarine.

7.54 A glider whose wing chord is 1 m and span is 20 m is to glide at 15 m/s. A model whose chord is 10 cm is tested in a low-speed wind tunnel at 17 m/s and 20°C. What is the wind tunnel pressure?

7.55 Consider a two-dimensional circular cylinder which has a pressure distribution on the body given by $p = a + b \sin^2 \phi$, where a and b are constants. Calculate (a) the pressure drag D_p, (b) the lift, and (c) the pressure drag coefficient, given $p_\infty = 0$.

7.56 A 0.0168 m^2 gas pipeline is designed to carry 5624.5 kg/h of hydrogen at 689.5 kPa and 30°C. We wish to model this using air at 30°C with a pressure loss of 34.5 kPa. What pressure loss can be expected for the system using hydrogen if the geometric scale ratio $c_l = 1$?

7.57 The pressure drop in a waterline is assumed to be a function of its length L, diameter D, average velocity U, density ρ, and viscosity of the fluid μ. (a) How many dimensionless parameters will be necessary to describe the problem and why? (b) What are the variables and their dimensions? (c) What nonrepeating parameter is mandatory based on the problem statement? (d) Solve the problem, referring to Fig. P7.57, and express the answer in the form $\pi_1 = f(\pi_2, \pi_3, \ldots, \pi_{n-m})$, using D, ρ, U as repeating variables.

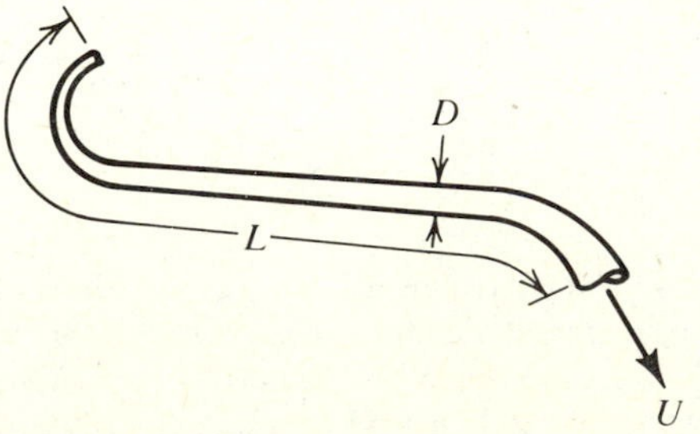

Figure P7.57

7.58 The torque T of an axial flow fan is believed to be a function of the blade diameter D, its angular speed ω, the fluid properties ρ,

μ, and the average flow-through velocity U. (a) How many dimensionless parameters are necessary to describe the problem, and why? (b) What are the variables and their dimensions? (c) What nonrepeating parameter is mandatory? (d) Using ω, ρ, D as the repeating parameters, find the π parameters, and express the solution in the form $\pi_1 = f(\pi_2, \pi_3, \ldots, \pi_{n-m})$.

7.59 The radiator fan in an auto is a source of considerable noise. The sound power P of the fan depends on the fan diameter D, angular speed ω, fluid density ρ, and speed of sound in air c. (a) How many dimensionless parameters are necessary to describe the problem? (b) What are the variables and their dimensions? (c) Selecting ρ, ω, D as repeating variables, determine the dependency of sound power P (energy per unit time) on the angular speed ω using dimensionless analysis (Fig. P7.59). Use the M, L, T system.

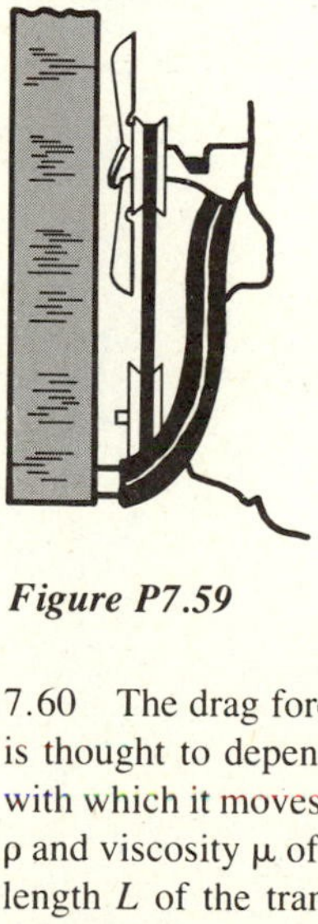

Figure P7.59

7.60 The drag force D on a sonar transducer is thought to depend on the towed velocity $\tilde{V}$ with which it moves through water, the density ρ and viscosity μ of water, and a characteristic length L of the transducer. (a) Find a convenient set of dimensionless parameters about which to organize experimental test data. (b) What is the significance of each dimensionless term? (c) A one-half scale model of the transducer is tested in a wind tunnel using air at standard temperature and pressure. What must the speed of air in the wind tunnel be for a dynamically similar flow in salt water at 20 ft/s, $\gamma = 64$ lbf/ft^3, and $\nu = 1.4 \times 10^{-5}$ ft^2/s?

7.61 In Prob. 7.60, if the model registers a drag force of 5.58 lbf in the wind tunnel at the speed computed in part (c), what will be the prototype's drag force in salt water at 20 ft/s?

7.62 The drag force on a submarine hull is thought to depend on the density ρ, viscosity μ of the fluid, the speed of the submarine **V**, the hull diameter D, and ship's length L. A 1/10 scale model is tested in the same fluid the prototype will operate in. What is the drag force predicted for the prototype in terms of the drag force measured on the 1/10 scale model (neglect cavitation considerations)?

7.63 The power P consumed by a journal bearing is thought to depend on the bearing clearance c (a linear dimension), the bearing diameter D, the rotational speed N, the viscosity of the lubricant μ, and the weight of the load carried by the bearing. A 1/5 scale model of the bearing is to be tested using the same lubricant and carrying the same weight as the prototype. Determine the power required by the prototype bearing in terms of the power consumed by the 1/5 scale model bearing under these circumstances.

7.64 The drag on a whale is to be predicted based on wind tunnel testing of a model. If the actual whale is 10 m long and moves at a velocity of 0.5 m/s in water of viscosity $\nu = 1.233 \times 10^{-6}$ m^2/s, calculate the drag on the whale if the drag on the model is 0.558 N during the test. The length of the model whale is 5 m. Refer to Fig. P7.64.

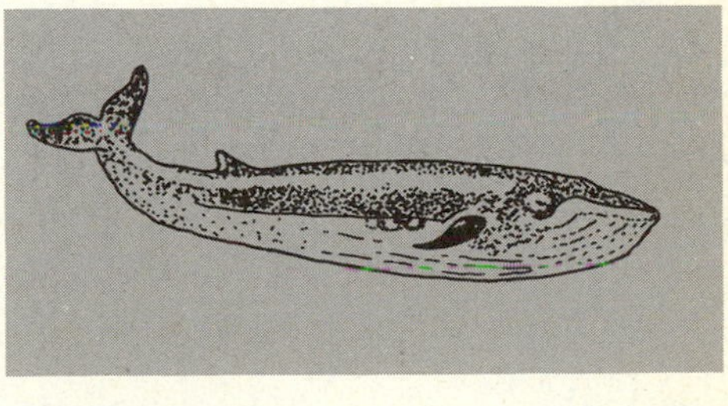

Figure P7.64

7.65 A model of Lake Okeechobee is to be constructed to study the effects of hurricane winds on the lake. The scale ratio of 1/150 is to be used. Assuming that fresh water at 45°C flows in the prototype, what must the kinematic viscosity of the water in the model be in order to maintain similarity between the two if both Reynolds and Froude numbers are satisfied?

7.66 Obtain the expression for the Weber number if we observe that the motion of a spherical bubble of diameter D moves with a velocity V through a fluid of density ρ. The size of the bubble is governed by the surface tension σ.

8 Flow Visualization

8.1 Introduction

We often hear the saying, "one picture is worth a thousand words." This is particularly true in fluid mechanics. Often teachers use some simple demonstration to illustrate a fundamental fluid mechanics phenomenon. For example, to demonstrate the transition from laminar to turbulent flow, we need only to light a wad of paper, then extinguish the flame and observe the behavior of the filament of smoke, as shown in Fig. 8.1. Notice that the smoke rises nearly vertically like a ribbon of moving particles, each well behaved, each on a predictable path. Then without warning there occurs a burst, followed by an unpredictable motion of smoke. The well-behaved lamina of smoke illustrates laminar flow, the burst is transition, and the disordered motion is turbulence. We do not need smoke to visualize this. In fact, if the transition region were of major importance to us, we might see it using water and dyes. We tried this and the result is shown in Fig. 8.2. Various colored dyes judiciously inserted into the flow color

Figure 8.1 *Smoke visualization.*

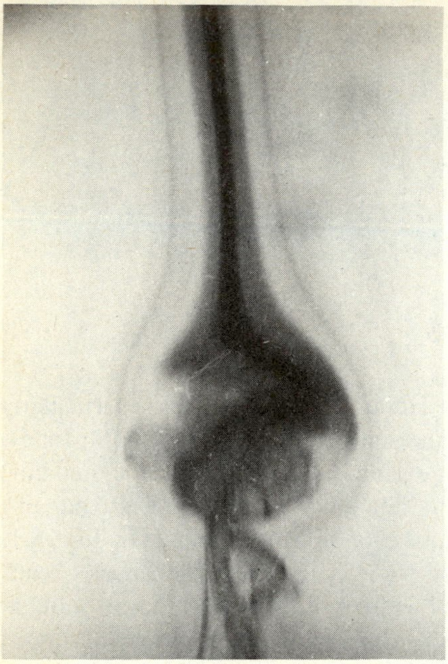

Figure 8.2 *Flow visualization by dyes (diameter of vortex breakdown is 0.5 in.). (Source: R. Granger, "Speed on a Surge in a Bathtub Vortex,"* Journal of Fluid Mechanics, *vol. 34, part 4, 1968. Used with the permission of the Cambridge University Press.)*

different regions of transition. Upstream of the burst we notice laminar flow, and downstream we recognize the onset of turbulence. The photograph was taken at high speed using a telescopic lens. The burst is only ¼ inch wide.

Some excellent films* show a variety of flow visualization methods. For example, we have mentioned injection of a stream of dye into a flow as one way to visualize fluid behavior. By studying the path the dye takes, we can determine some useful properties of the flow, such as velocity. When certain conditions prevail, we call these paths streamlines. The word "streamline" means to shape an object in such a way that the flow moves easily past it with as little resistance as possible. Streamlining occurs constantly in nature: rough rocks become smooth when exposed to wind and water currents; the wings of birds and the shape of a downhill skier both conform to a configuration that minimizes drag. Streamlining is a natural phenomenon that proves that nature abhors resistance.

To understand streamlining, we need to know what a streamline is and what type of a function describes it best. In steady flow the function that denotes the path of a fluid particle is the stream function. The *stream function exists by virtue of the equation of continuity*. The stream function not only reduces the complexity of the analysis by reducing the number of variables by one, but it is also extremely valuable in helping to visualize what takes place in the flow field. The French scientist d'Alembert is

*Table 8.1 lists films that illustrate flow phenomenon.

*Table 8.1 Films of Flow Visualization Method**

Film Type	Author(s), Affiliation, and/or Location	Date	Avail. From	Film Type	Running Time (min)
Acoustic Tripping of the Flow Around a Sphere	F. N. M. Brown, U. of Notre Dame	—	A	3	20
Waves in Fluids	A. E. Bryson	1964	B	3	33
Flow Visualization using Water as the Working Medium (Hydrogen Bubble)	D. W. Clutter et al., Douglas Aircraft	1959	A	1	24
Flow Visualization Studies of Free Convection Transition (On a Vertical Flat Plate)	E. R. G. Eckert et al., U. of Minn.	1958	A	1	10
Transition from Laminar to Turbulent Flow in a Thin Film of Water Flowing Down an Inclined Glass Plate	H. W. Emmons et al., Harvard U.	1952	A	1	14
Some Phenomena of Open Channel Flow	St. Anthony Falls Hydraulic Lab.	1947	F	4	33
Flow in Culverts	St. Anthony Falls Hydraulic Lab.	1948	F	4	20
Surface Waves	St. Anthony Falls Hydraulic Lab.	1952	F	4	24
Surface Tension in Fluid Mechanics	L. M. Trefethen, Tufts U.	1964	B	5	29
Secondary Flow	E. S. Taylor, MIT	1964	B	3	30
Visual Cavitation Studies of Mixed Flow Pump Impellers	G. M. Wood, Pratt and Whitney Aircraft	1963	A	3	22

*Source: A complete list of fluid mechanics films is found in the ASME Film Catalog, *Journal of Fluids Engineering*, June 1976, pp. 151–155.

Table 8.1 (*Con't.*) *Films of Flow Visualization Method*

Film Type	Author(s), Affiliation, and/or Location	Date	Avail. From	Film Type	Running Time (min)
Growth of Hydrogen Bubbles During Electrolysis	J. W. Westwater and D. E. Westerheide, U. of Ill.	1960	A	1	19
Boundary Layer Control	D. Hazen	1965	B	3	25
Some Preliminary Results of Visual Studies of the Flow Model of the Wall Layers of the Turbulent Boundary Layer	S. J. Kline and P. W. Runstadler, Stanford U.	1958	A	4	13
Flow Visualization	S. J. Kline, Stanford U.	1963	B	3	30
Smoke Study of Nozzle Secondary Flows in a Low-Speed Turbine	M. G. Kofsky, and H. W. Allen, NASA, Lewis	1954	C	3	20
A Visual Study of Turbulent Energy Production Mechanism Near Wall Region in Fully Developed Turbulent Boundary Layer in a Flat Plate	H. T. Kim et al., Stanford U.	1967	A	1	8
Visualization Studies by Smoke	A. M. Lippisch, Collins Radio Corp.	1959	A	1	30
Deformation of Continuous Media	J. L. Lumley, Penn State U.	1963	B	3	30
Stratified Flow	R. R. Long, Johns Hopkins U.	1969	B	5	26
The Structure and Stability of Turbulent Wall Layers in Rotating Channel Flow	D. K. Lezius and J. P. Johnston, Stanford U.	1971	A	1	15

Table 8.1 (*Con't.*)

Film Type	Author(s), Affiliation, and/or Location	Date	Avail. From	Film Type	Running Time (min)
Flow of Water in a Glass Pipe	P. E. Mohn, U. of Buffalo	—	A	1	16
Later Stages of Laminar-Turbulent Transition on a Flat Plate	K. A. Meyer and S. J. Kline, Stanford U.	1961	A	4	24
Rheological Behavior of Fluids	H. Markovitz	1965	B	3	22
Flow Instabilities	E. L. Mollo-Christensen, MIT	1969	B	5	27
A Visual Study of Turbulent Shear Flow	S. Nychas et al., Ohio State U.	1974	A	3	6
High Speed Color Schlieren Studies Over Two-Dimensional Training-Edge Flap Type Control (L-341)	NASA, Langley	1958	D	4	6
Hypersonic Flow Visualization by Interferential Schlieren	ONERA, France	1959	A	5	11
Flow Separations and Formation of Vortices	ONERA, France	1960	A	5	16
Boundary Layer Separation Along a Wall	ONERA, France	1960	A	5	8
Physical Study of Vortex Phenomena	ONERA, France	1961	A	5	18
On Flow Patterns Observed with Doubly Refracting Solutions of Milling Yellow	F. N. Peebles, U. of Tennessee	1953	A	4	12
Production of Vortices by Bodies Travelling in Water	L. Prandtl and F. Ahlborn, Gottingen, Germany	1920	A	1	15
Separation Studies	R. H. Page, et al., Rutgers U.	1966	A	4	15
Flow Structure in the Fully Developed Turbulent	P. W. Runstadler et al.,				

Table 8.1 (Con't.) Films of Flow Visualization Method

Film Type	Author(s), Affiliation, and/or Location	Date	Avail. From	Film Type	Running Time (min)
Boundary Layer of a Flat Plate	Stanford U.	1962	A	4	22
Fundamental Principles of Flow	H. Rouse and L. M. Brush, U. of Iowa	1962	E	5	23
Characteristics of Laminar and Turbulent Flow	H. Rouse and E. M. O'Loughlin, U. of Iowa	1964	E	5	26
Flow Visualization in Combustion Systems	Shell Research Ltd.	—	A	3	12
Smoke Visualization of Rotating Stall in Axial Flow Compressors	G. Sovran, General Motors Res. Lab.	1959	A	1	12
Vorticity (Parts I and II)	A. H. Shapiro, MIT	1959	B	3	45
The Periodic Breakdown of Vortices	T. Sarpkaya, U.S. Naval Postgraduate School	1965	A	3	8
Visual Observations of Flow Through a Radial-Bladed Impeller	R. F. Soltis and M. J. Miller, NASA, Lewis	1968	C	5	22
A Visual Study of Velocity and Buoyancy Effects on Boiling Nitrogen	R. J. Simoneau and F. F. Simon, NASA, Lewis	1966	A	5	16
Some Visual Observations of Cavitation in Rotating Machinery	R. F. Soltis, NASA, Lewis	1965	C	3	17
Flow Phenomena Associated with Underwater Launching of Missiles	F. O. Ringleb, Naval Air Material Center	1958	A	2	26
Diffusion of Smoke and Gas by Wind	H. Rouse, U. of Iowa	1956	A	1	26
Vortex Motion and Fluids Turbulence	H. Rouse, U. of Iowa	1959	A	4	3

Table 8.1 (Con't.)

Index of Film Locations

	Address	*Rental Fee*
A.	Engineering Societies Library 345 East 47th St. New York, NY 10017	Yes
B.	Encyclopedia Brittanica Educational Corp. 425 North Michigan Ave. Chicago, IL 60611	Yes
C.	National Aeronautics and Space Administration Lewis Research Center, Cleveland, OH 44135	No
D.	National Aeronautics and Space Administration Scientific and Technical Information Prog. Langley Research Center Hampton, VA 23365	No
E.	University of Iowa Media Library, Audiovisual Center Iowa City, IA 52242	Yes
F.	St. Anthony Falls Hydraulic Laboratory Mississippi River at 3rd Ave. S.E. Minneapolis, MN 55414	Yes

Key to Film Types

All films are 16 mm. Film speeds indicated are the standard running speeds although some of them contain "slow motion" sequences shot at higher speeds.

Film Type Code	B&W or Color	Silent or Sound	Running Speed Frames/s
1.	B&W	silent	16
2.	B&W	silent	24
3.	B&W	sound	24
4.	color	silent	16
5.	color	sound	24

The films listed above can be obtained from Engineering Societies Library, United Engineering Center, 345 East 47th St., New York, NY 10017 credited with first introducing the concept of the stream function. It is used primarily in two-dimensional fluid flow, and therefore it must be defined carefully for *plane flow* as well as for *axisymmetric flow*. If the flow is axisymmetric, the stream function is more popularly termed the *Stokes stream function*.

8.2 Equation of a Streamline

A *streamline* is defined as a line whose tangent at any point is in the direction of the velocity at that point. A *pathline*, on the other hand, is the trajectory of a single particle of fluid. A *streakline* is a line joining the instantaneous positions of a succession of particles which have issued from one source or passed through one point. For steady flow, streamlines, pathlines, and streaklines coincide. In flow visualization we usually see streaklines. In some unsteady flows, however, the three lines can be distinguished, as shown in Fig. 8.3.

Figure 8.3a shows a uniform flow of smoke moving past an autorotating flat plate. Upstream of the plate the intervals between the traces of smoke are equidistant indicating the flow is uniform. As the flow approaches the plate, the smoke lines near the center of the plate are wide apart, indicating that the flow has slowed down. Near the edges of the plate, the smoke lines are crowded together, indicating that the flow has speeded up. Behind the plate we notice a large swirling flow of smoke near the top which is an attached vortex; at the bottom we notice a wave-like pattern and far downstream a similar swirling flow of smoke like the top vortex. Directly behind the plate is little smoke, indicating a region nearly devoid of fluid motion.

The different flow lines are explained in Fig. 8.3b. For the configuration shown in Fig. 8.3a, we identify a common point from which a fluid particle may originate. The *streakline* is the picture of one of the lines of smoke. The *pathline* cannot be seen in the photograph, but is sketched in Fig. 8.3b to represent the time exposure of one particular particle (that particle originating from the common point). The dashed trail is the path that the particle took in an interval of time.

Streamlines are trajectories that at an instant of time are tangent to the direction of flow at each and every point in the flow field. If the velocity is a function of time, then the shape of the streamlines may vary from instant to instant.

The aggregate of all streamlines is called the *flow net*. Since by definition the streamline is that line where the velocity vector **V** is tangent, we can derive a mathematical expression for it. Consider the steady two-dimensional (planar) flow pattern of Fig. 8.4. At point $P(x, y)$ one and only one streamline can pass. By definition the streamline is tangent to the velocity vector **V** at P. (This can also be seen in Eq. (1.19).) using Cartesian coordinates, we obtain from geometry

$$\frac{v}{u} = \tan\theta = \frac{dy}{dx} \tag{8.1}$$

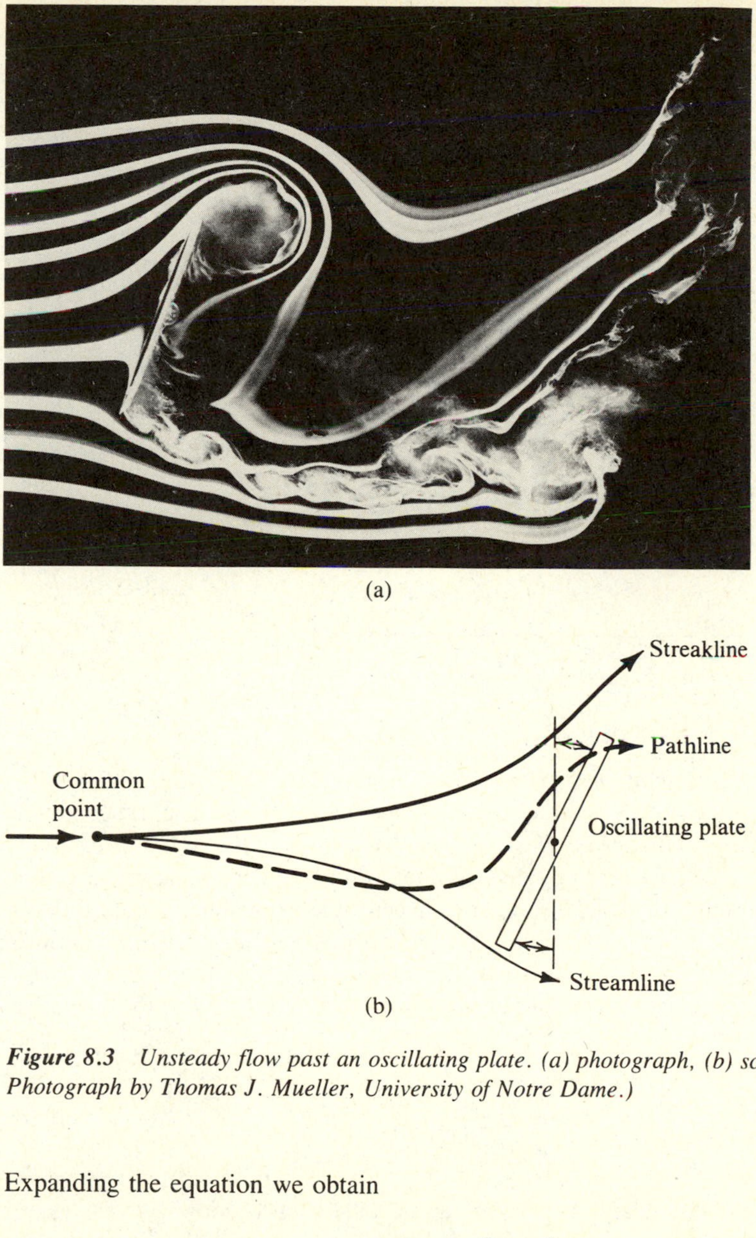

Figure 8.3 Unsteady flow past an oscillating plate. (a) photograph, (b) schematic. (Source: Photograph by Thomas J. Mueller, University of Notre Dame.)

Expanding the equation we obtain

$$u dy - v dx = 0 \tag{8.2}$$

For three-dimensional flow, we then find

$$\mathbf{V} \times d\mathbf{r} = 0 \tag{8.3}$$

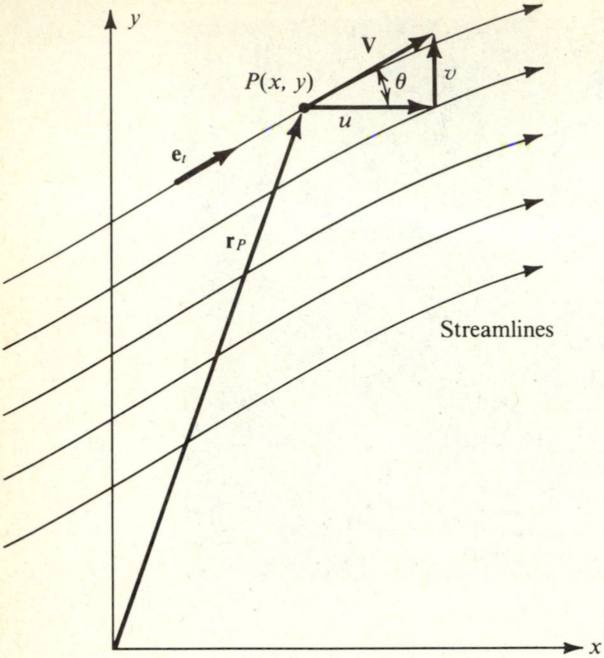

Figure 8.4 *Velocity components at a point P(x, y).*

which is known as the *equation of a streamline*. Note that to satisfy Eq. (8.3), either $\mathbf{V} = 0$, $d\mathbf{r} = 0$, or the flow **V** must be tangent to the streamline $d\mathbf{r}$. The last condition is the most favorable. *This means then that no flow can pass through a streamline.* Thus, streamlines are *viewed*, physically, as being solid boundaries that move with a velocity **V**, possessing the precise geometric shape of the streamline. It then follows that a surface across which no flow passes is called a *stream surface*. If the surface has the geometric form of a tube, it is called a *stream tube*. (See Fig. 8.5.)

We have shown that the streamlines of a flow are defined by Eq. (8.3). Notice that we can also define the equation by

$$\frac{dx}{\rho u} = \frac{dy}{\rho v} = \frac{dz}{\rho w} \tag{8.4}$$

These equations are well determined except at the following locations:

- Where the velocity **V** is infinite (which might be streamline sources)
- At infinity, if there is a streamline source at infinity
- At boundaries of an obstacle's analytic domain
- Where ρu, ρv, and ρw are simultaneously zero

We should always keep these considerations in mind when treating fluid flows.

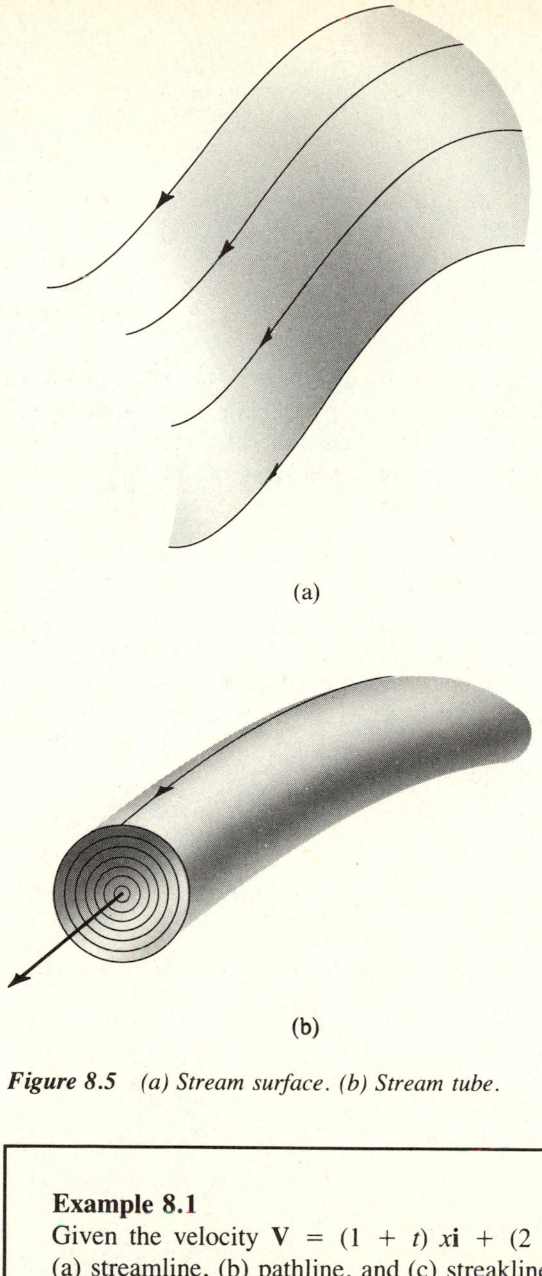

Figure 8.5 *(a) Stream surface. (b) Stream tube.*

Example 8.1

Given the velocity $\mathbf{V} = (1 + t)\, x\mathbf{i} + (2 + t)\, y\mathbf{j}$, find the equation of the (a) streamline, (b) pathline, and (c) streakline, given that the common point for all three is $x = 1$, $y = 2$, and $z = 0$ at $t = 0$.

Solution:

Step 1.

Identify the characteristics of the fluid and flow field. The flow is unsteady and two-dimensional.

Example 8.1 *(Con't.)*

Step 3.

Write the appropriate governing equations of flow.

- Continuity:

$$\frac{\partial u}{\partial x} + \frac{\partial v}{\partial y} = 0 \tag{i}$$

- Streamline:

$$\frac{dx}{u} = \frac{dy}{v} \tag{ii}$$

The continuity equation $\nabla \cdot \mathbf{V} = 0$ is *not* satisfied. This does not mean that it is not a real fluid flow; rather it means that it is not a possible incompressible flow. We shall assume that the problem is strictly a mathematical exercise.

(a) To find the equation of the streamline, we substitute the appropriate velocity components into Eq. (ii) and integrate:

$$\frac{dx}{(1 + t)x} = \frac{dy}{(2 + t)y} \tag{iii}$$

$$(2 + t)\ln x = (1 + t)\ln y + \ln c \tag{iv}$$

To evaluate the constant, we use the given initial condition:

$$0 = \ln 2 + \ln c \tag{v}$$

Rearranging terms and taking antilogs gives

$$\ln x^{2+t} = \ln y^{1+t} - \ln 2 \tag{vi}$$

$$\frac{y^{1+t}}{x^{2+t}} = 2 \tag{vii}$$

or

$$y = (2x^{2+t})^{1/(1+t)} \tag{viii}$$

which is plotted in Fig. E8.1.

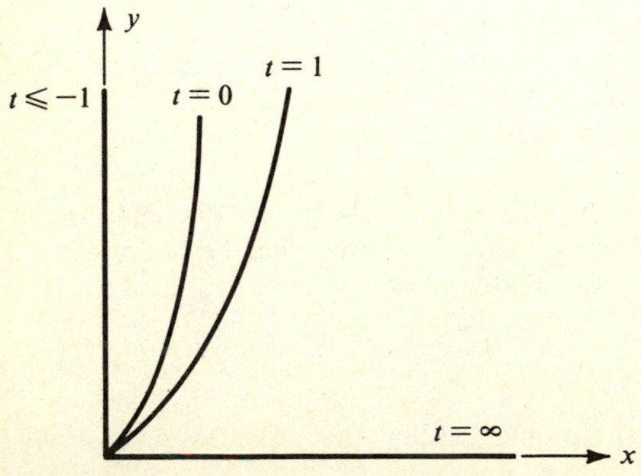

Figure E8.1

Example 8.1 *(Con't.)*

(b) We obtain the equation for the pathlines using the definition of the velocity:

$$u = \frac{dx}{dt} = (1 + t)x \quad \text{(ix)}$$

$$v = \frac{dy}{dt} = (2 + t)y \quad \text{(x)}$$

Integrating the preceding two equations gives

$$x = c_1 \exp (t + t^2/2) \quad \text{(xi)}$$

$$y = c_2 \exp (2t + t^2/2) \quad \text{(xii)}$$

Using the given initial conditions, we obtain

$$c_1 = 1, c_2 = 2 \quad \text{(xiii)}$$

such that Eqs. (xi) and (xii) become

$$x = \exp (t + t^2/2) \quad \text{(xiv)}$$

$$y = 2 \exp (2t + t^2/2) \quad \text{(xv)}$$

To find the equation of the pathline, we eliminate time t from Eqs. (xiv) and (xv). This results in

$$2 \ln x - [\ln (y/2x) + 1]^2 + 1 = 0 \quad \text{(xvi)}$$

which will allow a pathline through (1, 2) at time $t = 0$. It is plotted in Fig. 8.6. Note the pathline is not similar to the streamline at time $t = 0$.

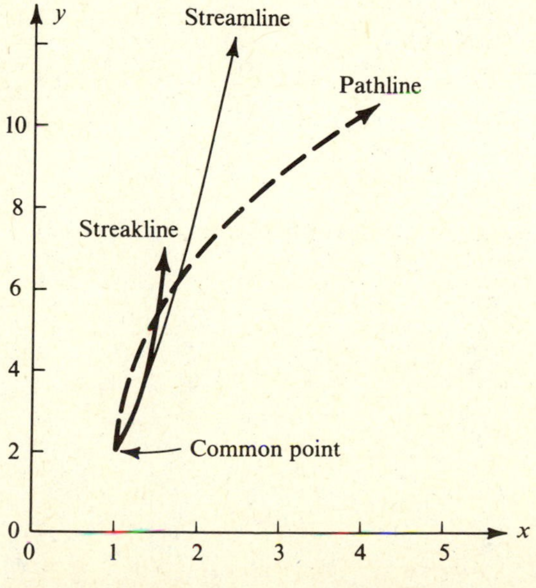

Figure 8.6 *Flow lines for $t = 0$, Example 9.2.*

Example 8.1 *(Con't.)*

(c) To find the equation of the streaklines, we shall investigate that streakline solely at time $t = 0$ since it is less cumbersome to calculate.

We begin with Eqs. (xi) and (xii). For $x = 1$, $y = 2$, we solve for c_1 and c_2 holding time constant: let $k = t$.

$$c_1 = \exp(-k - k^2/2), \; c_2 = 2\exp(-2k - k^2/2) \tag{xvii}$$

Substituting these expressions back into Eqs. (xi) and (xii) gives

$$x = \exp(-k - k^2/2), \; y = 2\exp(-2k - k^2/2) \tag{xviii}$$

Next, we eliminate k from the above two expressions. This results in

$$2 \ln x + \left[\ln\left(\frac{y}{2x}\right) - 1\right]^2 - 1 = 0 \tag{xix}$$

Let us compare the streakline of Eq. (xix) with that of the pathline, Eq. (xvi) and the streamline, Eq. (viii). They are all different expressions. Figure 8.6 shows this difference and the condition they all pass through the same point at time $t = 0$.

This completes the solution.

8.3 Stream Function, ψ

Consider an arbitrary scalar field function $\psi(x, y)$, and its total differential $d\psi$. If the function depends upon the location x, y,

$$\psi = \psi(x, y) \tag{8.5}$$

then the differential can be written as

$$d\psi = \frac{\partial \psi}{\partial x} dx + \frac{\partial \psi}{\partial y} dy \tag{8.6}$$

Compare Eq. (8.6) with the equation of the streamline given by Eq. (8.2). After equating similar terms, we find that if

$$u = \frac{\partial \psi}{\partial y} \tag{8.7}$$

$$v = -\frac{\partial \psi}{\partial x} \tag{8.8}$$

then

$$d\psi = 0 \tag{8.9}$$

Comparison of Eq. (8.6) along with the relationships given by Eqs. (8.7) and (8.8) with Eq. (8.2) shows that they are identical. *Thus, Eq. (8.9) also represents the equation of a streamline.* We can easily integrate Eq. (8.9) and obtain

$$\boxed{\psi = \text{const.}} \tag{8.10}$$

for an alternate form of the *equation of a streamline*. Thus, each streamline can be represented by Eq. (8.10) using, however, a different value of the (plus or minus) constant for each streamline. We call ψ the Lagrange stream function of a two-dimensional fluid flow.

A requirement of the velocities given by Eqs. (8.7) and (8.8) is that they must satisfy the differential form of the continuity equation. To accomplish this, we substitute Eqs. (8.7) and (8.8) into Eq. (4.24) to obtain

$$\frac{\partial^2\psi}{\partial x\partial y} - \frac{\partial^2\psi}{\partial y\partial x} \stackrel{?}{=} 0 \tag{8.11}$$

Since the order of the differentiation is inconsequential, the above equation shows that continuity is automatically satisfied.

We can now easily generate some interesting two-dimensional cases using various two-dimensional coordinate systems:

1. Steady *compressible* two-dimensional flow in the x, y plane:

$$u = \frac{1}{\rho}\frac{\partial\psi}{\partial y} \tag{8.12}$$

$$v = -\frac{1}{\rho}\frac{\partial\psi}{\partial x} \tag{8.13}$$

2. Unsteady incompressible axisymmetric flow in the r, z plane (the case for flow around a body of revolution with the free stream flows in the direction of the axis of symmetry):

$$v_r = \frac{1}{r}\frac{\partial\psi}{\partial z} \tag{8.14}$$

$$v_z = -\frac{1}{r}\frac{\partial\psi}{\partial r} \tag{8.15}$$

Equations (8.14) and (8.15) define a Stokes stream function.

3. Steady *compressible* axisymmetric flow in the r, z plane:

$$v_r = \frac{1}{\rho r}\frac{\partial\psi}{\partial z} \tag{8.16}$$

$$v_z = -\frac{1}{\rho r}\frac{\partial\psi}{\partial r} \tag{8.17}$$

Equations (8.16) and (8.17) also define a Stokes stream function.

4. Incompressible flow in the r, θ plane:

$$v_r = \frac{1}{r}\frac{\partial\psi}{\partial\theta} \tag{8.18}$$

$$v_\theta = -\frac{\partial\psi}{\partial r} \tag{8.19}$$

Equations (8.18) and (8.19) define a Lagrangian stream function. We should verify that the above expressions for the four special flows do satisfy continuity.

Let us consider two streamlines, $\psi = c_1$ and $\psi = c_2$, as shown in Fig. 8.7. We know that no fluid can flow *through* the two streamlines. Hence, these two streamlines act as moving boundaries. What can we say about the flow between these two boundaries? Obviously there exists a continuous collection of other streamlines between them, and their sum will be shown to be the volume rate per unit width. We evaluate he volume rate of flow Q in terms of the stream function by using the geometry shown in Fig. 8.7. We see that the elemental area $d\mathbf{A}$ can be expressed as

$$d\mathbf{A} = \mathbf{i}\,dydz - \mathbf{j}\,dxdz \tag{8.20}$$

Using the integrated form of the continuity equation

$$Q = \int dQ = \int_A^B \mathbf{V}\cdot d\mathbf{A} \tag{8.21}$$

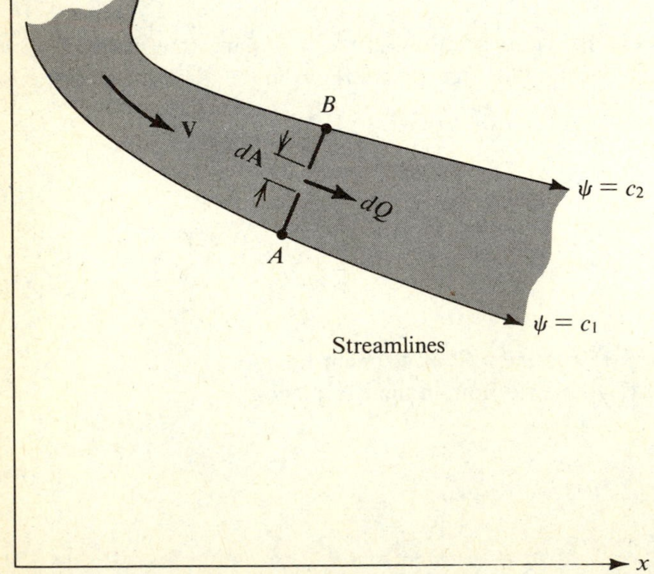

Figure 8.7 Geometry for volume rate of flow, Q.

we obtain after substitution of the elemental area of Eq. (8.20)

$$Q = \int_A^B (u\mathbf{i} + v\mathbf{j}) \cdot (\mathbf{i}\, dydz - \mathbf{j}\, dxdz) \tag{8.22}$$

If we let the depth be unity, then the volume rate of flow per unit depth becomes

$$\frac{Q}{\text{unit depth}} = \int_A^B (udy - vdx) \tag{8.23}$$

The integrand of Eq. (8.23) represents the change of the stream function $d\psi$ (as given by Eq. 8.6)) or

$$\frac{Q}{\text{unit depth}} = \int_A^B d\psi \tag{8.24}$$

Performing the indicated definite integration results in

$$\boxed{\frac{Q}{\text{unit depth}} = \psi_B - \psi_A} \tag{8.25}$$

Thus the flow rate Q per unit depth between any two streamlines in a two-dimensional incompressible flow is numerically equal to the difference in the values of the stream function ψ.

Example 8.2

Find the stream function of the two-dimensional incompressible flow

$$u = U\left[\left(\frac{y}{l}\right)^2 - \left(\frac{y}{l}\right)\right]$$

$$v = w = 0$$

shown in Fig. E8.2.

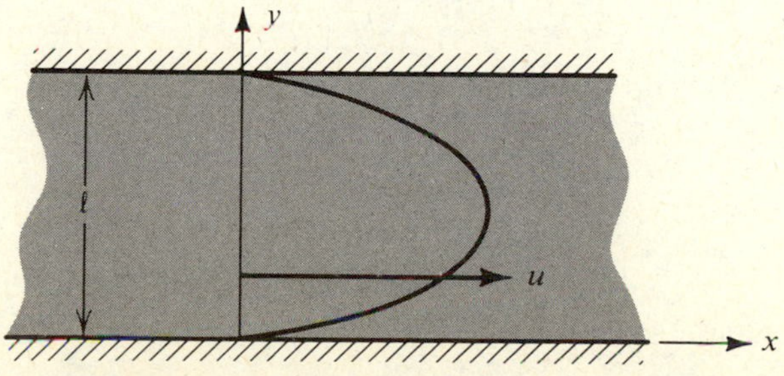

Figure E8.2

Example 8.2 *(Con't.)*

Solution:

Step 1.

Identify the characteristics of the fluid and flow field.

The fluid is incompressible, and the flow is steady and two-dimensional.

Step 3.

Write the appropriate forms of the governing equations of flow.

- Continuity:

$$\frac{\partial u}{\partial x} + \frac{\partial v}{\partial y} = 0 \tag{i}$$

- Stream functions:

$$u = \frac{\partial \psi}{\partial y}, \qquad v = -\frac{\partial \psi}{\partial x} \tag{ii}$$

Substituting the given velocity components into Eqs. (i) and (ii) results in

$$u = \frac{\partial \psi}{\partial y} = U\left[\left(\frac{y}{l}\right)^2 - \left(\frac{y}{l}\right)\right] \tag{iii}$$

$$v = -\frac{\partial \psi}{\partial x} = 0 \tag{iv}$$

Integrating Eq. (iii) gives

$$\psi = U\left[\frac{y^3}{3l^2} - \frac{y^2}{2l} + f_1(x)\right] \tag{v}$$

and integration of Eq. (iv) gives

$$\psi = f_2(y) \tag{vi}$$

The stream function ψ *must satisfy* both of the requirements as stated by Eqs. (v) and (vi) such that comparing Eq. (v) with Eq. (vi) we obtain

$$f_1(x) = \text{const.} \tag{vii}$$

$$f_2(y) = U\left(\frac{y^3}{3l^2} - \frac{y^2}{2l}\right) \tag{viii}$$

Thus using either Eqs. (v) and (vii) or Eqs. (vi) and (viii), we find the stream function ψ to be

$$\psi = Ul\left[\frac{1}{3}\left(\frac{y}{l}\right)^3 - \frac{1}{2}\left(\frac{y}{l}\right)^2\right] \tag{ix}$$

Note that we could also add a constant c to both Eqs. (v) and (vi), so that Eq. (ix) reads

Example 8.2 *(Con't.)*

$$\psi = Ul\left[\frac{1}{3}\left(\frac{y}{l}\right)^3 - \frac{1}{2}\left(\frac{y}{l}\right)^2\right] + c \qquad \text{(x)}$$

The value of c is immaterial and depends upon that choice of the streamline which one wishes to designate $\psi = 0$. It is convenient to select the streamline along the bottom of the channel for $\psi = 0$. Then c must be zero so that $\psi = 0$ at $y = 0$, which is the case given by Eq. (ix).

This completes the solution.

Example 8.3

Find the stream function for the vortex whose flow field is given by

$$v_r = w = 0$$

$$v_\theta = \omega_r$$

where ω is the angular speed and is a constant. Let the $\psi = 0$ streamline be at $r = 0$.

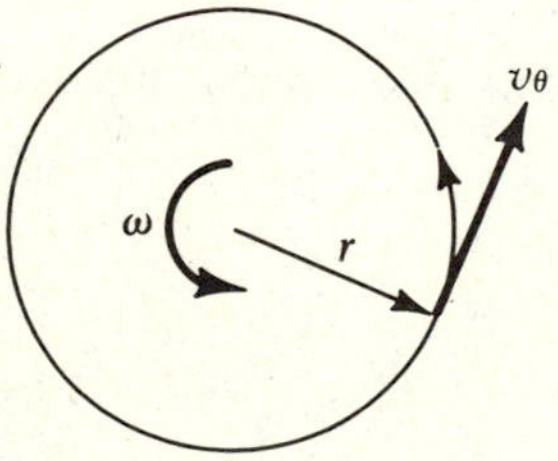

Figure E8.3

Solution:

Step 1.

Identify the characteristics of the fluid and flow field.

Assume that the fluid is incompressible. The flow is planar and axisymmetric.

Step 3.

Write the appropriate governing equations of flow:

- Continuity:

$$\frac{\partial v_r}{\partial r} + \frac{v_r}{r} + \frac{1}{r}\frac{\partial v_\theta}{\partial \theta} = 0 \qquad \text{(i)}$$

Example 8.3 *(Con't.)*

- Stream function:

$$v_r = \frac{1}{r}\frac{\partial\psi}{\partial\theta} \tag{ii}$$

$$v_\theta = -\frac{\partial\psi}{\partial r} \tag{iii}$$

Substituting the given velocity components into Eq. (ii) and integrating results in

$$\psi = f_1(r) + c \tag{iv}$$

Substituting the velocities into Eq. (iii) and integrating results in

$$\psi = -\frac{\omega r^2}{2} + f_2(\theta) + c \tag{v}$$

The stream function ψ must satisfy both requirements as given by Eqs. (iv) and (v) such that when comparing the results we obtain

$$f_1(r) = -\frac{\omega r^2}{2} \tag{vi}$$

$$f_2(\theta) = \text{const.} \tag{vii}$$

Using either Eqs. (iv) and (vi) or Eqs. (v) and (vii), the stream function ψ becomes

$$\psi = -\frac{\omega r^2}{2} + c \tag{viii}$$

Denoting $\psi = 0$ at $r = 0$ gives

$$c = 0 \tag{ix}$$

so that

$$\psi = -\frac{\omega r^2}{2} \tag{x}$$

is the stream function for this vortex flow. Note that at a particular radius, the stream function is constant, which means that the path is circular. This was rather obvious since we were given no radial inflow or outflow, only a circumferential velocity which signifies that the fluid particles are traveling in circular paths.

We could also obtain the equation of the streamline by using a different approach. Since the equation of a streamline is given by

$$\mathbf{V} \times d\mathbf{r} = 0 \tag{8.3}$$

Example 8.3 *(Con't.)*

where

$$\mathbf{V} = \omega r \, \mathbf{e}_\theta \tag{xi}$$

and

$$\mathbf{r} = r\mathbf{e}_r \tag{xii}$$

so that

$$d\mathbf{r} = dr \, \mathbf{e}_r + r \, d\theta \, \mathbf{e}_\theta \tag{xiii}$$

we find, after substituting Eqs. (xi) and (xiii) into Eq. (8.3), that

$$\omega r \, \mathbf{e}_\theta \times (dr \, \mathbf{e}_r + r \, d\theta \, \mathbf{e}_\theta) = 0 \tag{xiv}$$

or

$$-\omega r \, dr \, \mathbf{k} = 0 \tag{xv}$$

Integrating Eq. (xv) results in

$$-\frac{\omega r^2}{2} = \text{const.} \tag{xvi}$$

which is precisely what we obtained by using Eq. (x).

This example shows a flow that is both axisymmetric and two-dimensional so that either the Lagrange or Stokes stream function can be used.

This completes the solution.

The only restriction we must keep in mind when using the stream function is that the flow must be two-dimensional. It can be unsteady or viscous, compressible or incompressible, but it must be planar or axisymmetric.

We have learned that the stream function is a mathematical tool that can be used successfully in describing the direction of a flow. If we wished to see what the flow looks like, we should examine not only streamlines, but lines that are orthogonal to the streamlines. The next two sections treat the manner in which flow nets can be calculated.

8.3.1 Cauchy-Riemann Conditions

Consider an irrotational two-dimensional fluid flow. For a flow to be irrotational, we have shown that a necessary and sufficient condition is that the curl of the velocity vector vanish, such that a scalar potential function ϕ exists [as given by Eq. (4.128)]:

$$\mathbf{V} = \nabla \phi \tag{8.26}$$

It is now possible to find a relationship between the velocity potential ϕ and the stream function ψ. For instance, using Eqs. (8.26), (8.7), and (8.8), we find the following relationships between ϕ and ψ:

$$u = \frac{\partial \phi}{\partial x} = \frac{\partial \psi}{\partial y} \tag{8.27}$$

$$v = \frac{\partial \phi}{\partial y} = -\frac{\partial \psi}{\partial x} \tag{8.28}$$

Using Eqs. (8.26), (8.18), and (8.19) produces another set of relationships between ϕ and ψ:

$$v_r = \frac{\partial \phi}{\partial r} = \frac{1}{r}\frac{\partial \psi}{\partial \theta} \tag{8.29}$$

$$v_\theta = \frac{1}{r}\frac{\partial \phi}{\partial \theta} = -\frac{\partial \psi}{\partial r} \tag{8.30}$$

The above relationships are called the Cauchy-Reimann conditions, and always exist so long as the flow is two-dimensional and irrotational.

Example 8.4
Given $\phi = x^2 - 2y - y^2$ for a two-dimensional irrotational incompressible flow field, (a) find the stream function ψ, and (b) identify the type of flow.

Solution:
Step 1.
Identify the characteristics of the fluid and flow field.

The fluid is incompressible, and the flow is steady, two-dimensional, and irrotational.

Step 3.
Write the appropriate forms of the governing equations of flow:

$$\frac{\partial \phi}{\partial x} = \frac{\partial \psi}{\partial y}$$

$$\frac{\partial \phi}{\partial y} = -\frac{\partial \psi}{\partial x}$$

(a) Substituting the given ϕ function into the preceding equations results in

$$\frac{\partial \psi}{\partial y} = 2x \tag{i}$$

$$\frac{\partial \psi}{\partial x} = 2(1 + y) \tag{ii}$$

Example 8.4 *(Con't.)*

respectively. Integration of these equations shows that

$$\psi = 2xy + f_1(x) \tag{iii}$$

$$\psi = 2x + 2xy + f_2(y) \tag{iv}$$

Comparing Eqs. (iii) and (iv) enables us to evaluate

$$f_1(x) = 2x \tag{v}$$

$$f_2(y) = 0 \tag{vi}$$

with the result

$$\psi = 2x(1 + y) \tag{vii}$$

as the expression for the stream function.

(b) To find what the flow looks like, we first set $\psi = 0$, and obtain

$$x = 0 \tag{viii}$$

$$y = -1 \tag{ix}$$

which represent the zero streamline.

To obtain the relationship for $\psi = 2$ streamline, we construct a data table.

$\psi = \pm 2$

x	y
± 1	$\pm 0, -2$
± 2	$-\frac{1}{2}, -\frac{3}{2}$
± 3	$-\frac{2}{3}, -\frac{4}{3}$
$\pm \infty$	-1
$\pm \frac{1}{2}$	$1, -3$
$\pm \frac{1}{3}$	$2, -4$

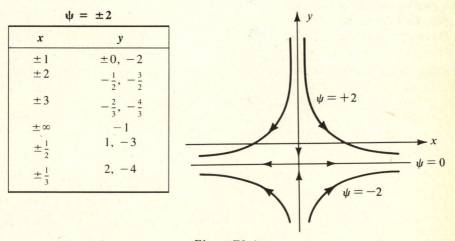

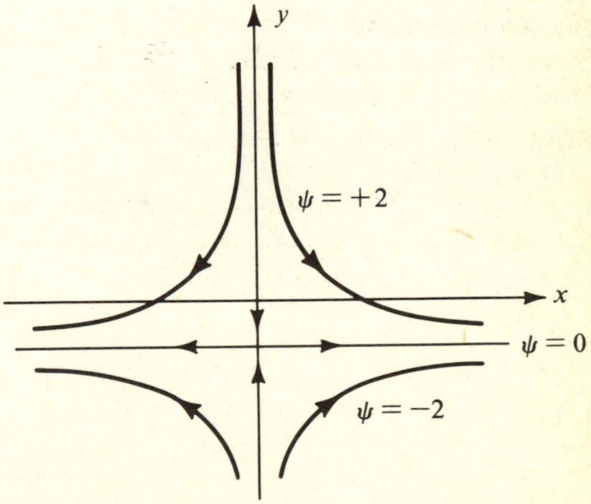

Figure E8.4

The direction of the flow can be found by noting the sign of the x, y components of velocity, since

$$u = \frac{\partial \psi}{\partial y} = 2x \tag{x}$$

$$v = -\frac{\partial \psi}{\partial x} = -2(1 + y) \tag{xi}$$

Example 8.4 *(Con't.)*

Then ($x > 0$, $y > 0$) in the first quadrant such that $u > 0$, $v < 0$. Thus the slope of the streamline is negative in the first quadrant. This flow is called stagnation point flow, or flow in a corner.

This completes the solution.

8.3.2 Orthogonality of ϕ and ψ

Since the stream function and velocity potential function are related mathematically through the Cauchy-Reimann conditions, they possess orthogonality properties.

Consider the velocity potential in planar Cartesian coordinates:

$$\phi = \phi(x, y) \tag{8.31}$$

The total change of ϕ expressed in terms of convective changes is

$$d\phi = \frac{\partial \phi}{\partial x} dx + \frac{\partial \phi}{\partial y} dy \tag{8.32}$$

Substituting the scalar velocities of Eq. (4.128) into the expression for change of potential results in

$$d\phi = udx + vdy \tag{8.33}$$

If we define an equipotential as having velocity potential ϕ as constant, then $d\phi = 0$ such that, from Eq. (8.33),

$$\left.\frac{dy}{dx}\right|_{\phi = \text{const.}} = -\frac{u}{v} \tag{8.34}$$

A similar operation is performed on the stream function. We express the total change of the stream function as

$$d\psi = udy - vdx \tag{8.35}$$

Thus, the slope of a streamline is found to be the ratio of the velocities v/u, i.e.,

$$\left.\frac{dy}{dx}\right|_{\psi = \text{const.}} = \frac{v}{u} \tag{8.36}$$

We could have obtained this result more directly by using the equation of a streamline: Eq. (8.3). Comparing Eq. (8.34) with Eq. (8.36), we see that the slopes of equipotential lines are orthogonal to the slopes of streamlines. This, of course, was already obvious from Eq. (4.128); however, the above explanation is perhaps more enlightening.

Velocity Potential as a Fluid Property

We have given both a physical and a mathematical explanation of the stream function ψ. We have shown it to be the volume rate of flow per unit depth of a fluid surface passing through a point. This makes the stream function a real property of the flow. Since the stream function is mathematically related to the velocity potential function through the Cauchy-Riemann equations, we could ask if the velocity potential ϕ is also a flow property.

To answer this, let us consider a flow field which is irrotational. The velocity field $\mathbf{V}$ can then be defined in terms of ϕ by

$$\mathbf{V} = \nabla\phi \tag{4.128}$$

Taking the dot product of an elemental length of a streamline $d\mathbf{r}$ and integrating from a point A to a point B gives

$$\int_A^B \mathbf{V}\cdot d\mathbf{r} = \int_A^B \nabla\phi\cdot d\mathbf{r} \tag{8.37}$$

But the integrand $\nabla\phi\cdot d\mathbf{r}$ is just the total change $d\phi$, such that

$$\int_A^B \mathbf{V}\cdot d\mathbf{r} = \int_A^B d\phi \tag{8.38}$$

Integrating, we obtain

$$\phi_B - \phi_A = \int_A^B \mathbf{V}\cdot d\mathbf{r} \tag{8.39}$$

Equation (8.39) states that the difference in potential is independent of path. Hence, in a closed path, the velocity potential does not change. This is one of Lagrange's postulates, showing that the velocity potential is a flow property: "If a velocity potential exists at some instant for a portion of fluid in irrotational motion, then a velocity potential exists for this portion of the fluid at all times." This is because the difference in velocity potential $\Delta\phi$ depends upon "the portion" or the path, and its magnitude depends upon time as well as space. Only in a closed path will the net velocity potential be zero for an irrotational flow.

8.4 Visualization Techniques

Flow visualization is an extremely powerful experimental technique for gaining insight into the fluid flow. Often it can save valuable time in designing aerodynamic and hydrodynamic vehicles. Engineers use it with high-speed computers to interpret test data, and in setting up various devices that measure flow velocities. If the visualization technique permits the motion of discrete particles that do not diffuse in the fluid, photographs* can be taken that allow one to make quantitative flow velocity measurements.

Texts (e.g., Merzkirch [8.1]) cover most techniques used in visualizing a flow. This section shall only encapsulate essentials of flow visualization.

8.4.1 Methods for Visualizing Flows of Liquids and Gases

Most of the methods used in visualizing streaklines require the experimenter to inject some foreign material into the flow that makes the particle, or path, or surface visible. Though the manner of injection is extremely important, it shall not be discussed here because it depends upon how experienced the experimenter is in flow visualization. Here experience is the best teacher. The one major requirement an experimenter must keep in mind is that the material injected should reach the flow velocity as quickly as possible.

We shall next discuss and illustrate the various materials.

Dyes

Dyes are popular in flow visualization because they are easy to handle. The dye must be thick enough to minimize diffusion in the working substance. It should possess the same density as the working substance as well as necessary visibility for observation or photography.

Table 2.1 of Ref. 8.1 presents a number of materials popularly used for marking filament lines. One of the materials mentioned, often used in water, is Sudan III. This dye is especially good in rotational flows as it does not diffuse rapidly. Figure 8.8 shows the use of Sudan III in water to visualize a particular region of historical value: the region of shear flow near a corner. The original sketch of what might take place was offered by da Vinci (see Fig. 8.8a) who did not use dyes. Da Vinci envisioned a rolling apart of vortex filaments that formed around a blunt body placed in a uniform flow. To validate da Vinci's concept, an actual experiment (Fig. 8.8b) shows that the vortex does indeed break down much the way da Vinci conceived it. The explanation of the various regions is shown in Figure 8.8c, each region of which is made clearly visible by the dye lines. Thus dye lines visualize the physical aspects of fluid flow quite nicely.

Another dye from Table 2.1 of Ref. 8.1 that mixes well with water is the rhodamine fluorescent. This dye is also well suited for visualizing vortex flows, as shown in Fig. 8.9. The dye is injected out of the curved hypodermic tube. Surprisingly, when we used the dye we saw that the flow was *not* moving toward the sink (where the orifice is pictured), but away from it. This occurred only at a unique radius, which dye injection helped find.

Some salient aspects important in the use of dye are:

- Injection: largely through hypodermic tubes or walls of submerged bodies.
- Region: good for uniform and rotational flows, especially good in laminar regions, poor in turbulent zones.

*Beautiful and important illustrations are in "An Album of Fluid Motion" by M. Van Dyke, (1983) Parabolic Press, P.O. Box 3032, Stanford, Ca. 94305-0030. Students can gain valuable insight into types of fluid motion upon examining this pictorial essay.

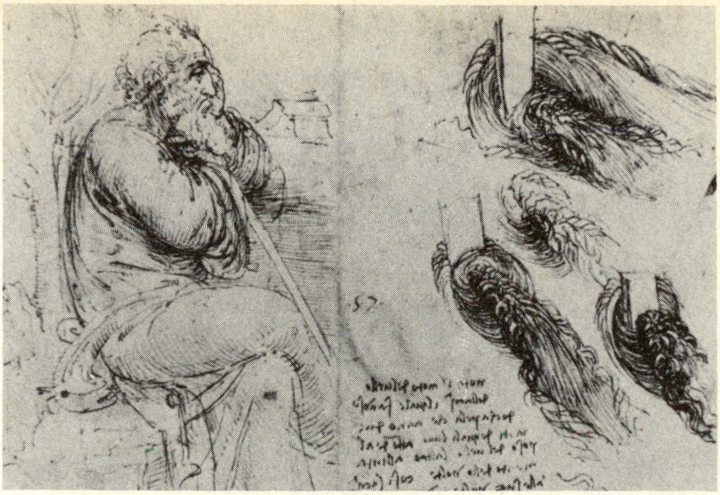

(a)

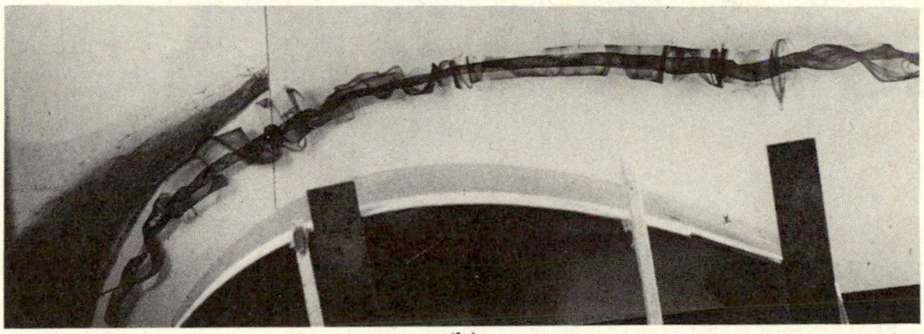

(b)

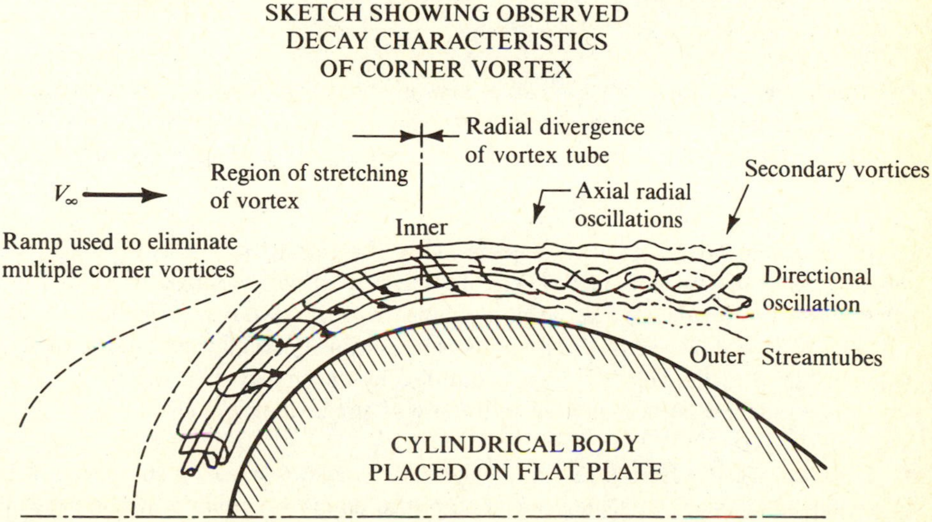

(c)

Figure 8.8 *(a) The horseshoe vortex envisioned by Leonardo da Vinci. (b) The horseshoe vortex in an experiment. (c) Regions of flow in Fig. 8.8b. (Source: (a) Royal Library, Windsor Castle. Copyright reserved. Reproduced by gracious permission of Her Majesty Queen Elizabeth II.)*

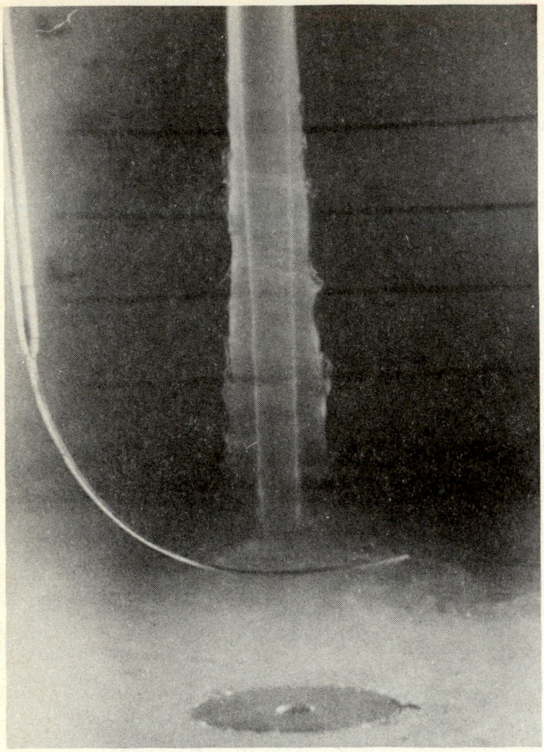

Figure 8.9 *Light-sensitive dye in a vortex flow (sink is orifice at bottom, dye streamtube shows flow moving away from sink. Within inner streamtube, flow is toward sink).*

- Advantages: nontoxic, noncorrosive, neutral buoyancy, good reflective properties for photography and photographic contrast.
- Disadvantages: diffuse easily—cannot be used in recirculating flows.

Smoke

Smoke is employed for gas flows, particularly for visualizing the streamlines past airfoil sections in a wind tunnel. There are a variety of ways to make smoke. Table 8.2 shows a few chemicals that when combined with air give dense smoke.

Figure 8.10 shows smoke in a vortex flow confined in an upright glass cylinder. The breakdown of the vortex is clearly outlined by the smoke streaklines.

Some salient aspects important in the use of smoke are:

- Injection: usually from a reservoir where the smoke is made, then bled into the test area. Titanium tetrachloride can be placed on local regions. When exposed to air, it turns to smoke.
- Region: good for low speed uniform flows, poor in turbulent regions.
- Advantages: inexpensive and easy to make, neutral buoyancy, good reflective properties.

Table 8.2 Smoke Used for Marking Filament Lines

Chemicals	Technique	Reference
Mineral oil	Boiled and mixed	8.5
Oils	Hot plate	8.6
Bromonaphythalene and titanium tetrachloride	Evaporation	8.7
White pine	Burned	8.8
Wheat straw	Coking	8.9

Source: *Flow Visualization,* by Wolfgang Merzkirch, Academic Press, Inc., © 1974, used with permission of the author and the publisher.

- Disadvantages: highly diffusive, can be toxic and corrosive, cannot be used in recirculating flows.

Sometimes flames can be used as a flow visualization technique, particularly if the working substance is a combustible gas. Figure 8.11 shows the same fluid flow

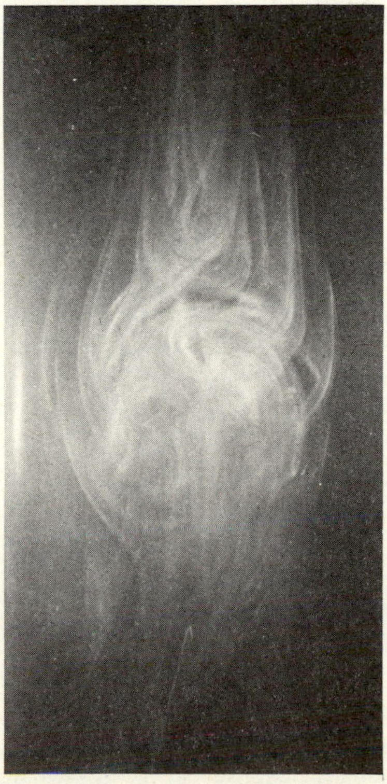

Figure 8.10 *Smoke in cylindrical vortex; visualizing vortex breakdown.*

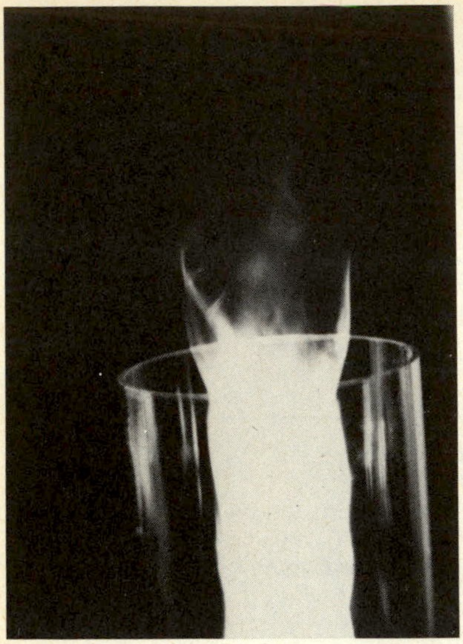

Figure 8.11 Flow visualization by flame: cylindrical vortex flow.

phenomenon as in Fig. 8.10, except that air was replaced by butane gas. Flames especially help visualize flame dynamics, fire whirls, and the onset of turbulence.

Tufts

Tufts (usually wool, tufted nylon, or cotton) are often used to discover the *direction* of the flow. They are used to identify the region of the wake, separation line, instability, and transition. The yarn should not be any longer than 2 cm. Tufts work best in a velocity range of 1 m/s $\leqslant V \leqslant$ 30 m/s.

Small Particles

The injection of small neutrally buoyant particles is ideal if we wish to study the behavior of a fluid particle flow. While we cannot say that these small particles follow the flow precisely (every particle has a drag and thus its velocity never equals the fluid particle velocity), they can come close. The particles can be solid, liquid, or gas.

Solids

Solids are popular tracers since they are inexpensive, expendible, nearly neutrally buoyant, stable, and have good photographic qualities. Some solid particles can be quite large. Figure 8.12 shows three different flow examples, each using a different solid particle. Figure 8.13a shows a vorticity meter. The span was made out of aluminum foil glued to a length of cotton twine dipped in hot wax and allowed to cool, thus possessing the characteristics of a tiny airplane. By placing the meter in a glass of water, we can note its behavior. If it falls, we can trim the aluminum until it is neutrally buoyant; if it rises, we can trim the tail. We then place the meter on the centerline of a vortex, release it, and, as it falls, make a time exposure as shown in Fig. 8.13b. Knowing the time exposure and counting the exposed surfaces, we can evaluate the vorticity.

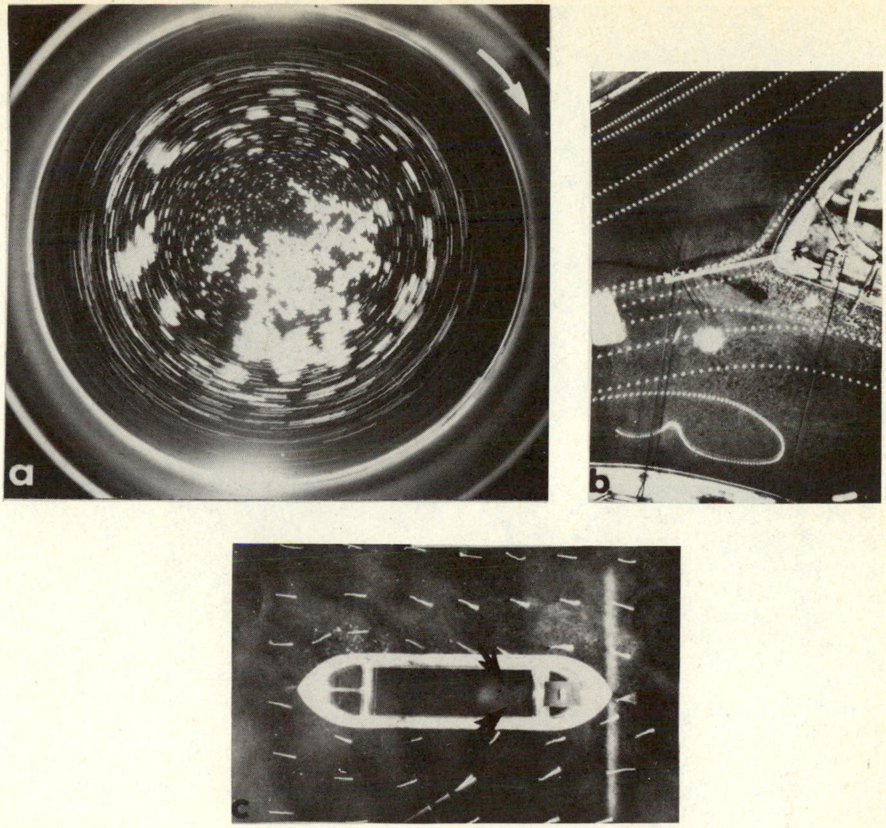

Figure 8.12 Flow visualization using solid particles. (a) Rotation initiation of a fluid in a rotating cylinder. $N = 6$ rpm, diameter $= 94$ mm, $t = 75$ s after start. Visualization with hostaflon powder placed on the free-surface of the fluid. (b) Flow at the surface of a Y-shaped channel (multiple-exposure photograph of floating agents). (c) Flow around an obstacle with suction (textile filaments). (Source: H. Werlé, Hydrodynamic Flow Visualization, reproduced, with permission, from the Annual Review of Fluid Mechanics, Volume 5. © 1973 by Annual Reviews Inc. Reproduced with permission from ONERA.)

Liquids

Liquid particles are especially effective in water flows. Figure 8.14 shows dye globules translating in a vortex flow. The dye globule is made of paraffin oil and bromonaphthalene combined in proper amounts to achieve the density of water. This combination makes an imiscible fluid, which, when combined with Sudan III dye, has excellent reflective qualities.

Gas

Use of gas bubbles in water flows and helium bubbles in gas flows have developed into a fine art. For gas bubbles in water flows, the density of the tracer particulate is less than that of the working substance, creating serious buoyancy problems. References 8.2 and 8.3 discuss the optical problems of using gas bubbles.

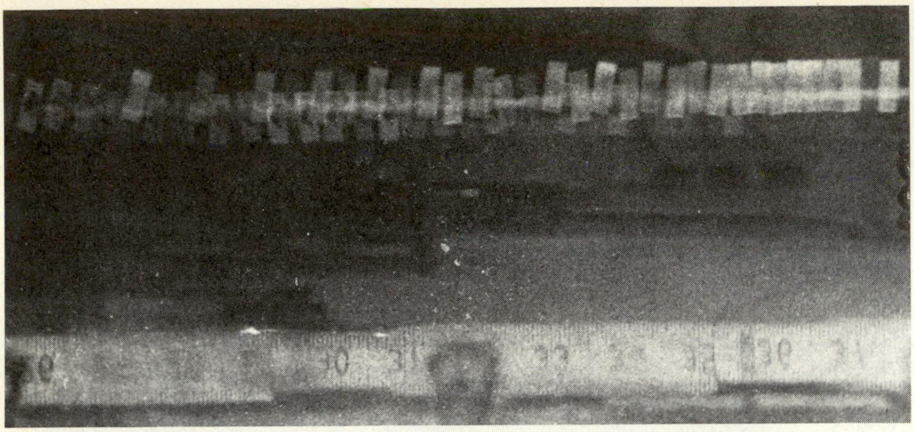

(a)

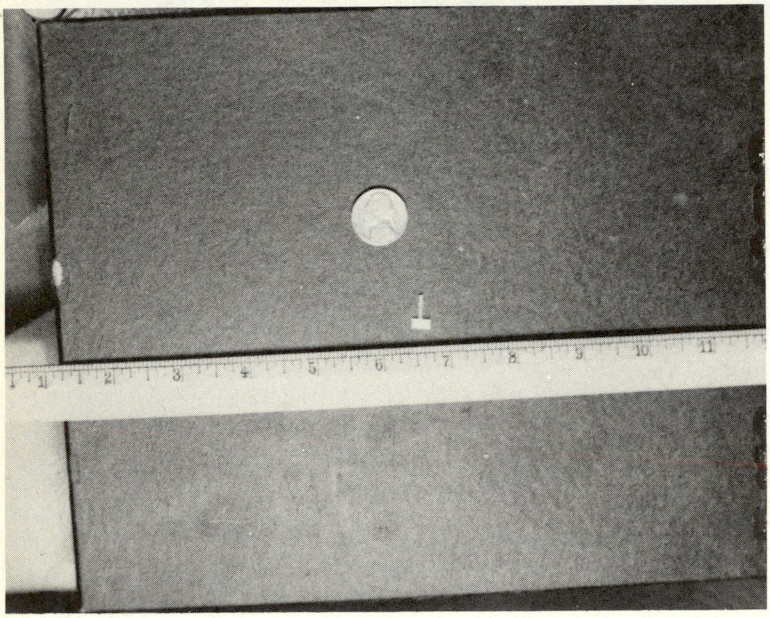

(b)

Figure 8.13 *(a) Measurement of centerline vorticity by vorticity meter. (b) The vorticity meter (aluminum foil wing, wax cotton string body). (Source: (a) R. Granger, ''Steady Three-Dimensional Vortex Flow,''* Journal of Fluid Mechanics, *vol. 25, part 3, 1966. Used with the permission of Cambridge University Press.*

The hydrogen-bubble technique was developed by Clutter and Smith (1961) [8.2], and involves placing a small diameter wire (0.01 mm) of platinum (commonly called Wollaston wire) in a region of the flow. The bare wire acts as a cathode of a dc circuit for electrolyzing the water (though glycerin and other fluids have been used). The anode is placed at some other point in the flow. With the current on, hydrogen bubbles

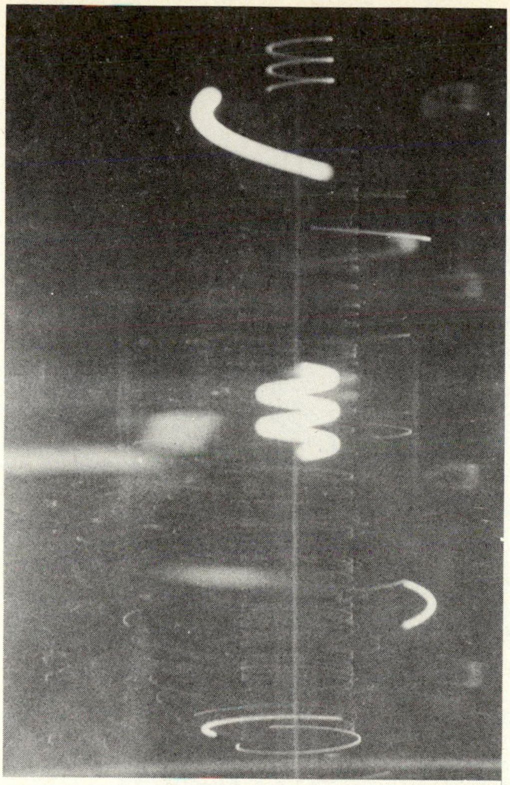

Figure 8.14 Four different size dye globules at different radial locations in a liquid vortex flow showing regions of large tangential velocity and axial velocity.

form at the exposed portion of the wire, and are carried away by the flow shown in Fig. 8.15. The circuit can also be pulsed, resulting in sheets of bubbles as shown in Fig. 8.15b. These sheets of bubbles are useful in studying the angular deformation of the flow.

The helium bubble technique of Ref. 8.4 is a clever way to create a tracer for high speed air flows. Helium is injected into a liquid soap bath that makes fairly uniform bubbles. The mixture is then injected into air flows through specially designed nozzles. The bubbles are neutrally buoyant and can trace airflow patterns as high as transonic flow and as low as 3 ft/s. With an uninterrupted light source, the motion of each bubble appears as a broken streak on a photograph (see Fig. 8.16). These streaks can allow quite accurate measurements of local velocities.

Some other techniques using air bubbles for tracers are shown in Fig. 8.17. When compared to flow visualization by colored emission or an iodine-starch reaction, there is a diffusion in the dyes, and lack of diffusion in the air bubbles.

Table 2.3 of Ref. 8.1 is a short compilation of various tracer particle/fluid combinations used for velocity measurements. Though the list is incomplete it does show great variety.

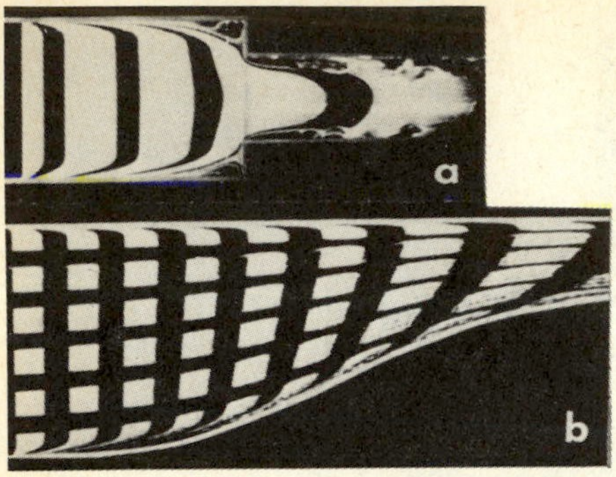

Figure 8.15 Flow visualization by hydrogen bubbles. Visualization by timed lines of hydrogen bubbles (intermittent electrolysis) of the flow in a plane convergent section. (a) Model with wall step. (b) Half-model with continuous wall. (Source: H. Werlé, Hydrodynamic Flow Visualization, reproduced, with permission, from the Annual Review of Fluid Mechanics, Volume 5. © 1973 by Annual Reviews Inc. Reproduced with permission from ONERA.)

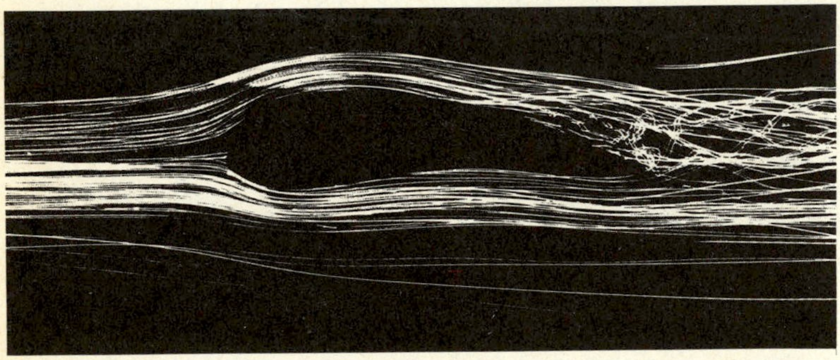

Figure 8.16 Flow visualization past an airfoil using helium bubbles and time photography. (Source: Photograph taken by SAGE ACTION, Inc., Ithaca, New York, under Office of Naval Research Contract N00014-68-C-0434.)

Optical Set-Ups

Reference 8.1 and Table 8.3 present four optical set-ups popularly used in visualizing flows in air. Though these devices are not used to obtain quantitative measurements of pressure and velocity, they are valuable in determining certain flow characteristics.

The Schlieren and shadowgraph optical systems are used most often in experimental compressible flows. They are based on a relatively simple concept. Figure 8.18 shows one simple type of Schlieren set-up. Light from a concentrated source (mercury arc lamp) at *A* is at the focal point of the convex lens *B*. Between the two convex lenses *B* and *E*, the light is straight and parallel, yet orthogonal to the direction of flow. The flow is confined in a high speed wind tunnel where two opposite walls *CD* are made of glass allowing the light to pass through. The light rays passing through lens *E* are then focused at *F* and pass to an image screen at *G*. If the gas through which the light rays pass is of uniform density, the light rays will not bend or refract,

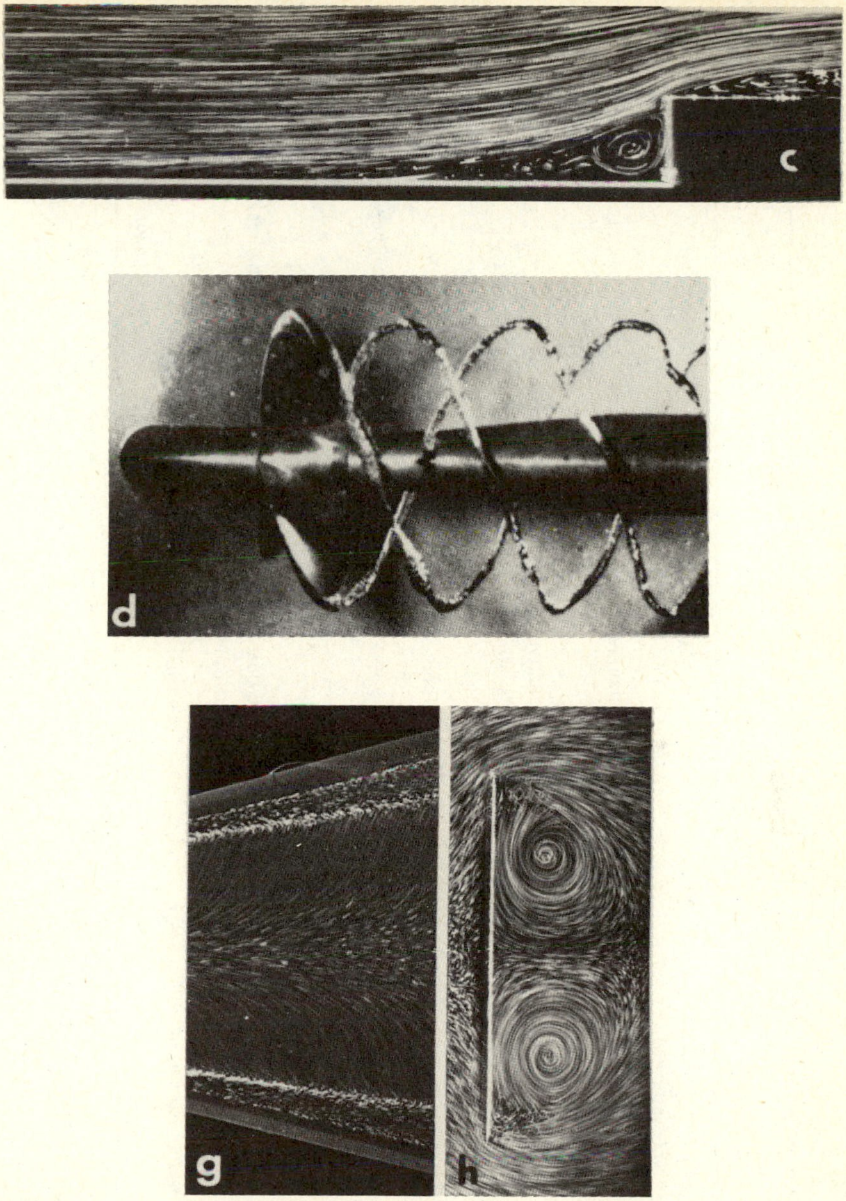

Figure 8.17 *Flow visualization by air bubbles. (c) Flow along a flat wall with a step (visualization by air bubbles—short time exposure). (d) Cavitation on a ship propeller. The air bubbles are concentrated along the helicoidal axis of the vortices issued from the blades of the model. (g, h) Slender delta wing ($\Lambda = 75°$, $\alpha = 15°$). Visualization by air bubbles. (g) Flow near the wall (upper surface). (h) Transverse section at the trailing edge. (Source: H. Werlé, Hydrodynamic Flow Visualization, reproduced with permission from the Annual Review of Fluid Mechanics, Volume 5. © 1973 by Annual Reviews Inc. Reproduced with permission from ONERA.)*

Table 8.3 *Other Flow Visualization Techniques*

Technique	Medium	Working Fluid	Advantages	Disadvantages	Reference
1. Electrolytic Dye Production	• Thymol blue dye (pH indicator)	Water	Neutrally buoyant dye good for low velocities	Large diffusion for high velocity	8.10
	• Tellurium wire	Water	Good for laminar low-speed flows	Resultant dye lines not clear	8.11
2. Surface Flow Patterns	• Oil flow • Paints	Air and water	• Only qualitative • Good for studying flow pattern	Repeatability ease	8.12
	• Soot	Air	Good for high-speed flows	Qualitative only	8.13
3. Shadowgraph	• Optical set-up, film	Air	Quick way to get a flow pattern	No quantitative measurement	8.14
4. Schlieren	• Optical set-up	Air	High degree of resolution	No quantitative measurement	8.15
5. Mach–Zehnder Interferometer	• Optical set-up	Air	Can make quantitative density measurements	High degree of mechanical precision	8.16
6. Schlieren Interferometer	• Optical set-up	Air	Good in visualizing weak shock waves	Not appropriate for continuous density changes	8.17

Source: *Flow Visualization*, by Wolfgang Merzkirch, Academic Press, Inc., © 1974, used with permission of the author and the publisher.

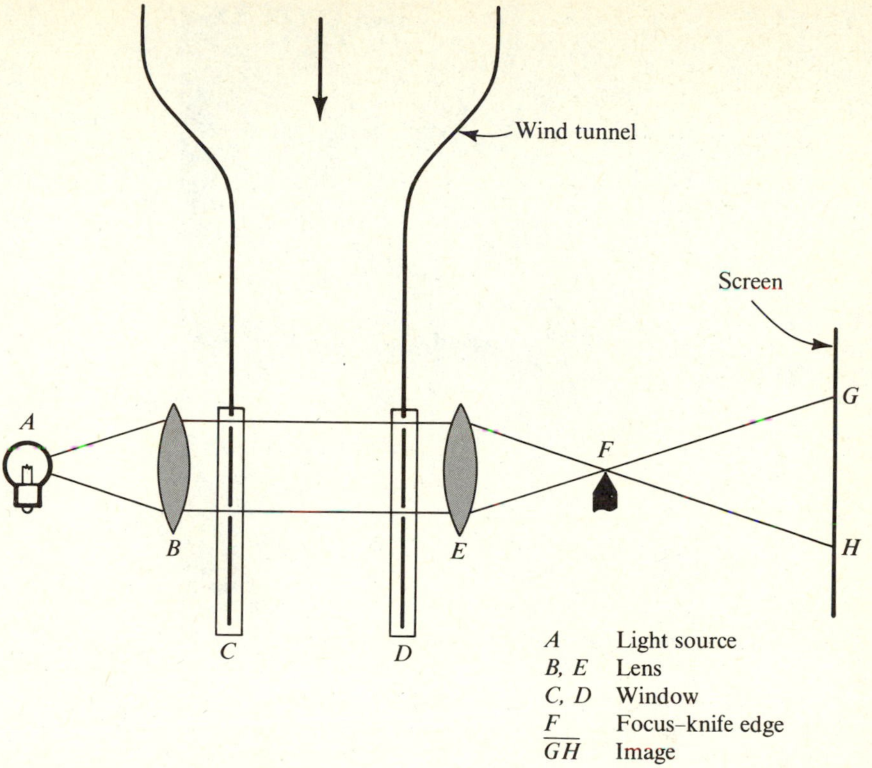

Figure 8.18 *Schlieren set-up.*

and the image on the screen will be uniform. If, however, there are any density changes in the gas, the light beams will be refracted going through one density to another, as shown in Fig. 8.19. Definite changes in shadows or illumination indicate density variation.

There are three types of systems: (i) the shadow, (ii) black and white Schlieren, and (iii) colored Schlieren. The shadow utilizes no knife edge at *F*. For the black and white Schlieren, a razor blade makes a near perfect knife edge that acts as an optical filter. Contrast can be intensified by judicious positioning of the razor blade. For the colored Schlieren, a series of colored bands are placed at *F*. Table 8.3 presents the advantages and disadvantages of four optical set-ups. As mentioned earlier, such optical devices do not measure local velocities of the fluid flow. To achieve such measurements, we often select a particle tracking technique. Table 8.4 presents a collection of different systems for measuring fluid velocities by particle tracking. The references presented alongside each subclass of measurement technique give detailed explanations of the technique.

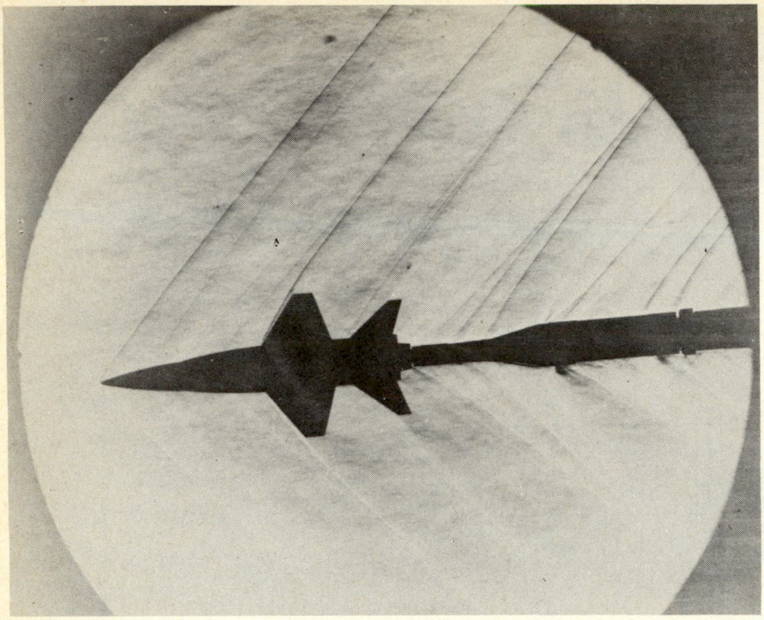

Figure 8.19 Schlieren photograph. (Source: Courtesy NASA.)

Table 8.4 Observation Systems for Measurement of Fluid Velocities by Particle Tracking

Class	Subclass	Velocity Range (mm/s)	References
Visual	Timing	0–2	8.18
	Moving graticule	0–1	8.19
	Moving spot	0.02–2	8.20
	Streak image	0–16	8.20
	Point image	0–3000	8.21
	Stroboscope	0–300	8.22
Photographic	Interrupted illumination	0–5000	8.3
	Multiple frame movie	0–2000	8.23
	Streak image	0–16	8.4
Integrated photoelectronic	Moving grating	0–5	8.24
	slit image		8.20
	Flying spot		8.25
	Laser doppler	$0–1.5 \times 10^4$	8.26
	Electro-optical tracker	0–500	8.27

Source: Reproduced with permission © Instrument Society of America 1981, from *Flow: Its Measurement and Control in Science and Industry,* Vol. 2, R. B. Dowdell, Ed.

References

8.1 Merzkirch, W., *Flow Visualization*, Academic, New York, 1974.

8.2 Clutter, D. W., and Smith, A. M. O., "Flow Visualization by Electrolysis of Water," *Aerosp. Eng.*, 20:24–27; 74–76, 1961.

8.3 Birkhoff, G., and Caywood, J., *J. Appl. Phys.*, 20:646, 1949.

8.4 Hale, R. W., Tan, P., Stowell, R. C., and Ordway, D. E., "Development of an Integrated System for Flow Visualization in Air Using Neutrally Buoyant Bubbles," SAI RR 7107, 1971.

8.5 Maltby, R. L., and Keating, R. F. A., "Smoke Technique for Use in Low Speed Wind Tunnels," *AGARDograph*, No. 70, pp. 87–109, 1962.

8.6 Goddard, V. P., McLaughlin, J. A., and Brown, F. N. M., "Visual Supersonic Flow Patterns by Means of Smoke Lines," *J. Aerosp. Sci.*, 26:761–762, 1959.

8.7 Dewey, J. M., "The Properties of a Blast Wave Obtained from an Analysis of the Particles Trajectories," *Proc. Roy. Soc.*, Ser. 324A:275–299, 1976.

8.8 Yu, J. P., Sparrow, E. M., and Echert, E. R. G., "A Smoke Generator for Use in Fluid Flow Visualization," *Int. J. Heat Mass Transfer*, 15:557–558, 1972.

8.9 Brown, F. N. M., "A Photographic Technique for the Mensuration and Evaluation of Aerodynamic Patterns," *Photog. Eng.*, 4:146–156, 1953.

8.10 Baker, D. J., "A Technique for the Precise Measurement of Small Fluid Velocities," *J. Fluid Mech.*, 26:573–575, 1966.

8.11 Wortman, F. X., "Eine Methode zur Beobachtung und Messung von Wasserströmungen mit Tellur," *Z. Angew. Phys.*, 5:201–206, 1953.

8.12 Maltby, R. L., and Keating, R. F. A., "The Surface Oil Flow Technique for Use in Low Speed Wind Tunnels," *AGARDograph*, No. 70, pp. 29–38, 1962.

8.13 Strehlow, R. A., Maurer, R. E., and Rajan, S., "Transverse Waves in Detonations: I Spacing in the Hydrogen-Oxygen System," *AIAA J.*, 7:323–328, 1969.

8.14 Dvorak, V., "Über eine neue einfache Art der Schlierenbeobachtung," *Ann. Phys. Chem.*, 9:502–512, 1880.

8.15 Schardin, H., "Das Toeplersche Schlierenverfahren," *VDI Forschungsh*, No. 367.

8.16 Ladenburg, R., and Bushader, D. *Interferometry in Physical Measurements in Gas Dynamics and Combustion*, Princeton Univ. Press, Princeton, N.J., pp. 47–78.

8.17 Francon, M., "Interférométrie par double réfraction en lumiére blanche," *Rev. Opt.*, No. 31, pp. 65–80, 1952.

8.18 Malkus, W. V. R., *Proc. Roy. Soc. London*, 225A:185, 1954.

8.19 Bennett, L., *Biorheology*, 5:253, 1968.

8.20 Monro, P. A. G., *Advances in Optical and Electron Microscopy*, Academic, New York, 1966, p. 1.

8.21 Fage, A., and Townsend, H. C. H., "An Examination of Turbulent Flow with an Ultramicroscope," *Proc. Roy. Soc. London*, 135A:656–677, 1932.

8.22 Brookes, A. M. P., and Monro, P. A. G., "High Speed Photography," *Proc. 3rd International Congress* (Butterworths, London, 1956), p. 351.

8.23 Kalinske, A. A., *Trans. Am. Soc. Civ. Engrs.*, 111:355, 1946.

8.24 Gaster, M., "A New Technique for the Measurement of Low Fluid Velocities," *J. Fluid Mech.* 20:183–192, 1964.

8.25 Lamport, H., *Nature*, 206:132, 1965.

8.26 Morse, H. L., et al., Rept. SUDAAR No. 365, v. II, Stanford University, Aeronautics and Astronautics Dept., 1968.

8.27 Somerscales, E. F. C., *Phys. Fluids*, 13:1866, 1970.

Study Questions

8.1 Define streamline, streakline, and pathline. Give an illustration of each.

8.2 How does one evaluate the volume rate of flow given (a) the velocity field, and (b) the equation of the streamlines?

8.3 Express the equation of streamlines in polar coordinates and cylindrical coordinates.

8.4 Why can't stream functions be used for three-dimensional flows?

8.5 What must the stream function satisfy?

8.6 Express vorticity in terms of the stream function.

8.7 What is the significance of the velocity potential? What is its physical interpretation?

8.8 Equation (8.1) is the slope of a streamline in Cartesian coordinates. Find the slope of a streamline in polar coordinates using v_r and v_θ.

8.9 Supposing the streamlines in Fig. 8.7 had positive slopes. Obtain an expression for $d\mathbf{A}$ similar to Eq. (8.20) and evaluate the volume rate of flow Q/unit depth.

8.10 If $\psi = 2x$, plot ϕ and show it is everywhere orthogonal to ψ.

Problems

8.1 Given the stream function $\psi = 3x^2 + 2y^2$, calculate the velocity field and draw the streamlines $\psi = 0$, $\psi = 1$, and $\psi = -1$.

8.2 Given the stream function $\psi = e^x$, calculate the velocity field and draw what this flow looks like.

8.3 Given the stream function $\psi = r \sin \theta$, calculate the radial and tangential velocity components, and sketch a few of the streamlines.

8.4 Given the stream function $\psi = rz$, calculate the velocity field and sketch a few of the streamlines.

8.5 Given $u = 2x$, $v = -2y$, determine the stream function $\psi = \psi(x, y)$.

8.6 Given $u = 5$, determine the stream function. What type of flow is this?

8.7 Given $v_\theta = r \sin \theta$ calculate v_r and the stream function.

8.8 For Prob. 8.5, determine the volume rate of flow Q per unit depth between the points (1, 2) and (−1, −2).

8.9 Given $u = x^2$, $v = -2xy$, determine the equation of the streamline.

8.10 If the velocity potential $\phi = x^2 - y^2$, what is the stream function $\psi = \psi(x, y)$? Draw a few equipotential lines and streamlines for this flow. What kind of a flow can you imagine this to be?

8.11 If the stream function $\psi = 2xy$, determine $\phi = \phi(r, \theta)$.

8.12 Given $u = 3xy$, find an expression for the equation of a streamline if the flow is steady and two-dimensional.

8.13 In Example 8.1, show that the equations for the streamline, streakline, and pathline are one in the same.

8.14 Find the equations of the streamline, the pathline, and the streakline at time t given a flow field

$$\mathbf{V} = \frac{x}{1+t}\mathbf{i} + \frac{y}{t}\mathbf{j}$$

8.15 Given $\psi = 2x$, find (a) $\phi(x, y)$, (b) $\mathbf{V}(x, y)$, (c) $\mathbf{a}(x, y)$, (d) $\zeta_z(x, y)$, and (e) $p(x, y)$.

8.16 Given $\phi = x^2 - y^2$, find (a) $\psi(x, y)$, (b) $\mathbf{V}(x, y)$, (c) $\mathbf{a}(x, y)$, (d) $\zeta_z(x, y)$, and (e) $p(x, y)$.

8.17 Given $\phi = 4xy$, find (a) $\phi(r, \theta)$, (b) $\psi(r, \theta)$, (c) v_r, (d) v_θ, (e) a_r, and (f) a_θ.

8.18 For the steady incompressible flow shown in Fig. P8.18, find the equation of the streamline passing the point (1, 2, 3) if $u = 3ax$, $v = 4ay$, and $w = -7az$.

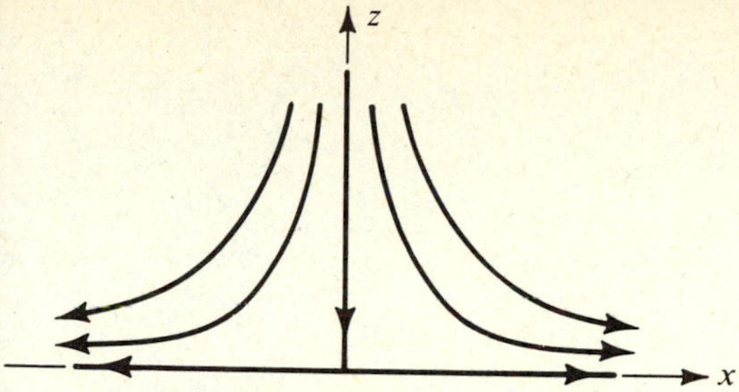

Figure P8.18

8.19 The stream function of a two-dimensional incompressible flow is given by

$$\psi = x^3 - 3y$$

(a) Determine the magnitude of the velocity V at the point (2, 3). (b) Determine the magnitude of the acceleration at the point (2, 3).

8.20 Express the vorticity ζ in terms of the stream function ψ in both Cartesian and polar coordinates.

9 Viscous Fluid Flows

9.1 Introduction

In this chapter, we shall discuss the behavior of a real fluid flowing through objects that are confined in a particular manner. The confinement may be at some finite geometric domain, or it may be at some infinite or semi-infinite location. For the former, we will be treating viscous flows between two flat or curved boundaries (such as wind blowing between two skyscrapers, or through a pipe), and for the latter we will be treating various types of rotational flows (such as models for tornados). In all cases, we shall consider the fluid flow to be well behaved (laminar) and incompressible.

We have shown the Navier-Stokes equations to be the governing partial differential equations describing the dynamics of fluid flow. Because of the nonlinear nature of these equations there are very few cases, as was shown in Chap. 4, for which the equations can be solved exactly without introducing certain approximations. It can be shown mathematically (and observed physically) that the flow as predicted by these equations becomes unstable when the specific momentum becomes too high, or the dynamic viscosity becomes too small, or the dimensions of the containing boundary becomes too great, or any combination of these phenomena. For example, the flow becomes unstable (turbulent) when the inertial force becomes large compared to the viscous force. So we shall confine ourselves to the case where this does *not* occur. *Rectilinear flow between parallel flat plates* as well as flow induced by a *suddenly accelerated flat plate* immersed in a semi-infinite fluid are two very important examples of real fluid flow where the velocity distribution can be obtained analytically in closed form. This kind of flow also has a wide range of applications, making it extremely worthwhile to study.

9.2 Rectilinear Flow Between Parallel Plates

Consider the incompressible flow contained between two infinite parallel plates that move steadily in their own plane, such that we have rectilinear flow in the fully developed sense; i.e., we neglect inlet effects, etc. In order to observe and describe the fluid flow between the two plates we choose an inertial rectangular Cartesian system with the z-axis normal to the plates and the xy-plane lying midway between the plates so that the plates are located at $z = \pm h$ as shown in Fig. 9.1.

Based on symmetry, the flow is assumed to be steady and parallel to the plate. The boundary conditions for no-slip would then be

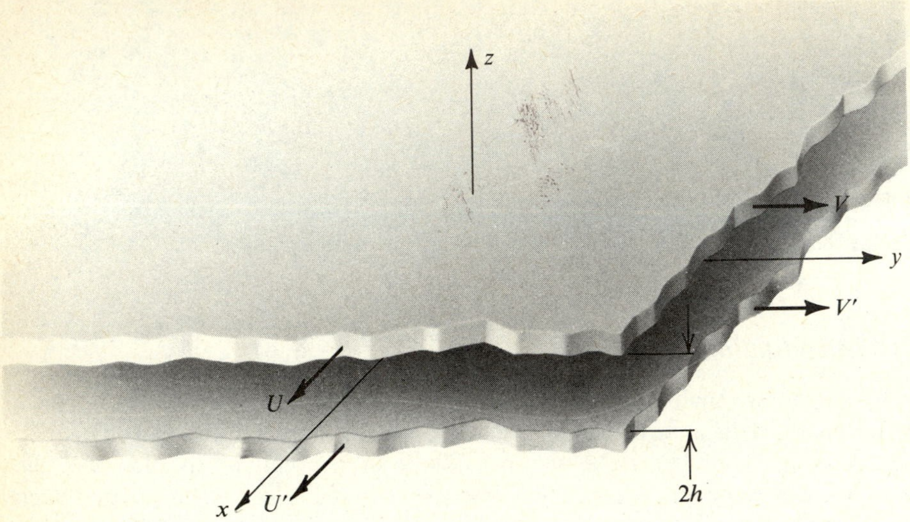

Figure 9.1 Flow between infinite parallel plates.

$$\left.\begin{aligned} u &= U \\ v &= V \end{aligned}\right\}, \quad z = h$$
$$\left.\begin{aligned} u &= U' \\ v &= V' \end{aligned}\right\}, \quad z = -h \tag{9.1}$$

where the velocity of the upper plate is $U\mathbf{i} + V\mathbf{j}$ and the velocity of the lower plate is $U'\mathbf{i} + V'\mathbf{j}$. We seek solutions for the velocity in the form

$$\mathbf{V} = u(z)\mathbf{i} + v(z)\mathbf{j} \tag{9.2}$$

The velocity field given by Eq. (9.2) satisfies the (D.F.) continuity equation since the z-component of velocity w does not exist, and the other velocity components are solely functions of z.

By substituting the assumed velocity field $\mathbf{V}$ of Eq. (9.2) into the Navier-Stokes equation (4.109), we obtain

$$\left(-\frac{1}{\rho}\frac{\partial p}{\partial x} + \nu\frac{d^2u}{dz^2}\right)\mathbf{i} + \nu\frac{d^2v}{dz^2}\mathbf{j} + \left(-g - \frac{1}{\rho}\frac{\partial p}{\partial z}\right)\mathbf{k} = 0 \tag{9.3}$$

The boundary conditions as given by Eq. (9.1) are

$$\mathbf{V}(x, y, h) = U\mathbf{i} + V\mathbf{i} \tag{9.4a}$$

$$\mathbf{V}\,(x, y, -h) = U'\mathbf{i} + V'\mathbf{i} \tag{9.4b}$$

The z-component of linear momentum of Eq. (9.3) can be integrated in a straightforward manner to give

$$p = -\rho g z + f_1(x, y) \tag{9.5}$$

Using the above results for the pressure, we can write the x-component of linear momentum of Eq. (9.3) as

$$\mu \frac{d^2u}{dz^2} = \frac{\partial p}{\partial x} = \frac{\partial f_1}{\partial x} \tag{9.6}$$

Since the left-hand side of this last expression is a function of z only and the right side a function of x and y only, both sides must be equal to a constant G. Thus,

$$G = \frac{\partial f_1}{\partial x} \tag{9.7a}$$

or

$$G = \mu \frac{d^2u}{dz^2} \tag{9.7b}$$

or

$$G = \frac{\partial p}{\partial x} \tag{9.7c}$$

Integrating the pressure relationship of Eq. (9.7c) gives

$$p = Gx + f_2(z) \tag{9.8}$$

Combining the two pressure expressions of Eqs. (9.5) and (9.8) results in

$$p = Gx - \rho g z + p_o \tag{9.9}$$

where p_o is taken to be the pressure at the origin of our coordinate system. Integration of the velocity expression of Eq. (9.7b) gives

$$u = \frac{G}{2\mu} z^2 + c_1 z + c_2 \tag{9.10}$$

where c_1 and c_2 are arbitrary constants of integration.

The y-component of linear momentum of Eq. (9.3) can be integrated to give

$$v = c_3 z + c_4 \tag{9.11}$$

where c_3 and c_4 are arbitrary constants of integration.

The kinematic boundary conditions as given by Eq. (9.4) are substituted into the velocity expression for u and v with the result

$$u = \frac{Gz^2}{2\mu} + \frac{U - U'}{2h} z + \frac{U + U'}{2} - \frac{Gh^2}{2\mu} \tag{9.12}$$

and

$$v = \frac{V - V'}{2h} z + \frac{V + V'}{2} \tag{9.13}$$

The velocity components as given by Eqs. (9.12) and (9.13) and the pressure field as given by Eq. (9.9) are a complete closed form analytic solution of the Navier-Stokes equation. Having thus determined the velocity and pressure fields, we leave it to the reader to determine the location and magnitude of the maximum velocity, the stress field, the vorticity field, and the viscous dissipation function.

Several physical and mathematical results are important to consider. First, notice that the form of the velocity field given in Eq. (9.2) resulted in the convective acceleration vanishing. This linearized the Navier-Stokes equation as well as decoupled the three resultant scalar equations which allowed a simple integration of each scalar form of the Navier-Stokes equation. Physically, the decoupling of the variables means that the flow along the x-axis does not influence the flow along the y-axis and vice versa. Hence, we can look at the velocity profiles independently.

Figure 9.2 depicts the velocity profile, shear stress distribution, and vorticity distribution in the xz plane for the case where the upper plate is the free-surface moving at velocity U and the lower plate moves at velocity U'. When the flow along the x-axis has no pressure gradient, we call the flow uniform shear flow or *Couette flow*. Note that not only is the normal stress distribution p_{zx} constant throughout the flow, but so is the vorticity ζ_y. Notice that the velocity distribution is linear, which has to result in the shear stress and the vorticity being constant everywhere in the flow.

Next, consider a flow that has no free-surface, shown in Fig. 9.3. Here we are viewing the velocity profile, shear stress distribution, and vorticity distribution in the xz plane for $U > U' > 0$. Notice that the velocity profile in this case is parabolic, owing to the existence of the pressure gradient. The shear stress and the vorticity distribution across the flow are linear, since for laminar flows they are always one order lower than the velocity distribution. By varying the parameters of G, U, and U', many important special cases can be studied. Figure 9.4 depicts four special cases.

In Fig. 9.4a the flow is symmetric about the x-axis, because both upper and lower plates are at rest. The characteristic of this flow is that the pressure decreases along

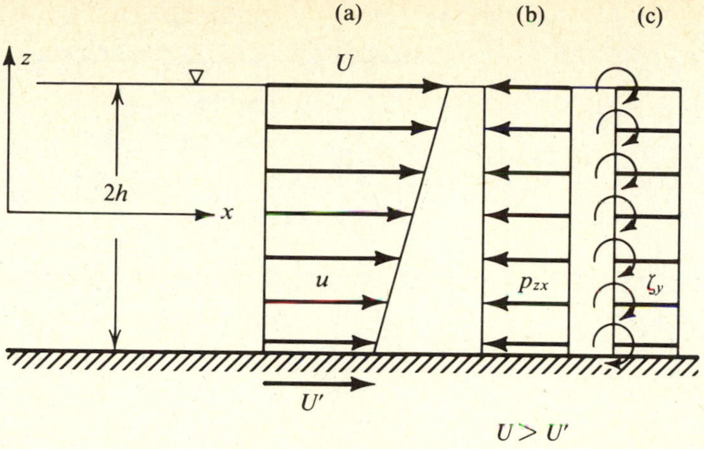

Figure 9.2 *Couette flow, $dp/dx = 0$. (a) Velocity distribution. (b) Shear stress distribution. (c) Vorticity distribution.*

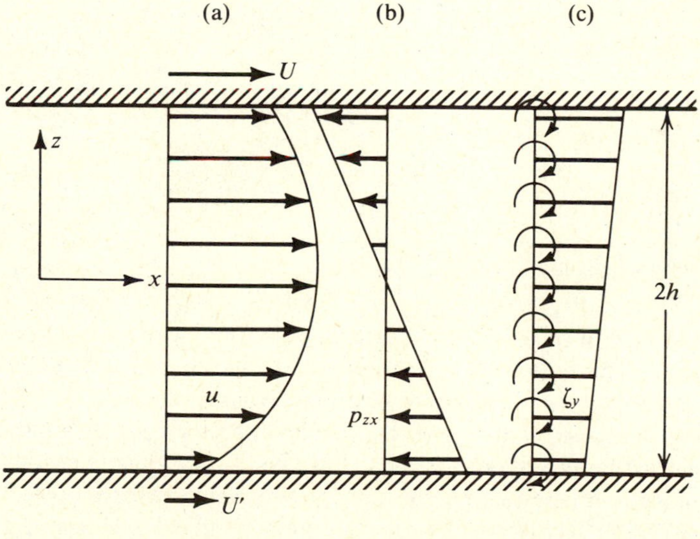

Figure 9.3 *Poiseuille flow, $dp/dx \neq 0$. (a) Velocity distribution. (b) Shear stress distribution. (c) Vorticity distribution.*

the direction of flow. Without this pressure drop, the flow would not exist. Also the maximum velocity is at the center of the channel.

In Fig. 9.4b the lower plate is at rest and the upper plate is moving at uniform velocity U. The characteristic of this flow is that the pressure decreases along the direction of flow with the specific value $dp/dx = -(U\mu/2h^2)$. For this flow, the maximum velocity occurs at the top plate.

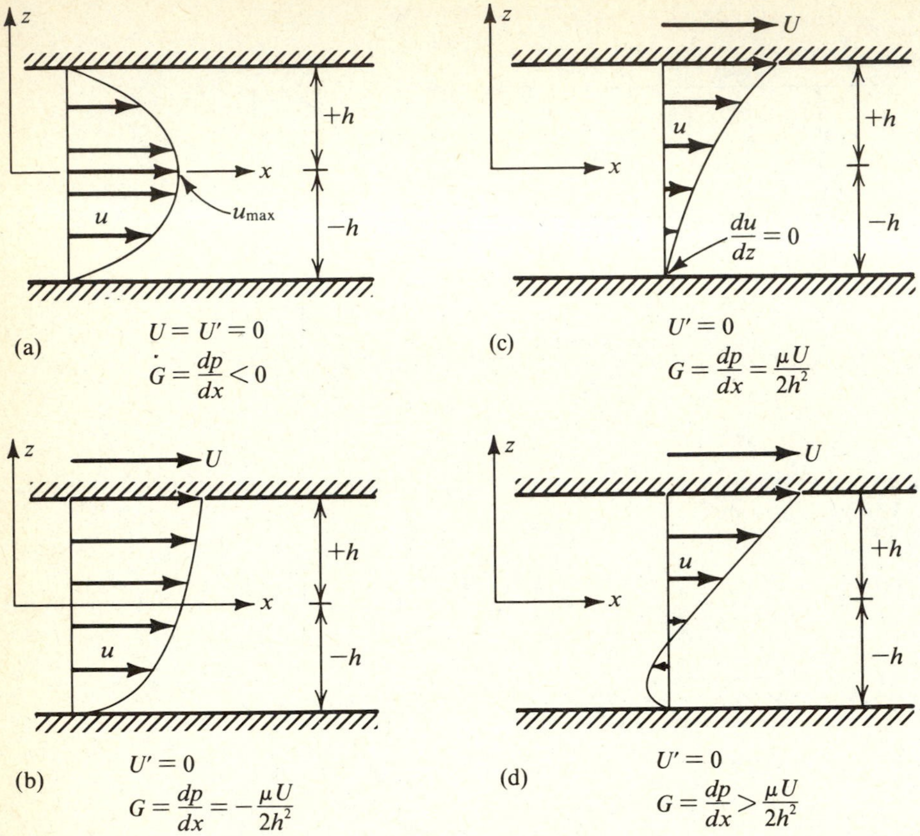

Figure 9.4 *Types of Poiseuille flow.*

In Fig. 9.4c the pressure increases along the direction of flow with the pressure gradient $dp/dx = U\mu/2h^2$. This unique value of the pressure gradient causes both the velocity u and its gradient du/dz to vanish at the lower stationary plate.

In Fig. 9.4d the pressure increases along the direction of flow causing a back flow, like a wake, near the lower stationary plate. This type of back flow is characteristic of flows that move over tapered bodies or through divergent channels—where the flow decelerates. We call such a condition of the pressure an *adverse pressure gradient*.

Example 9.1

Consider two fixed parallel flat plates that are a distance $2a$ apart. Let the plates be inclined at an angle θ, as shown in Fig. E9.1, and have a fluid flowing between them of viscosity μ at a sufficiently low velocity so that the flow can be assumed laminar. Express the angle θ of the slope in terms of viscosity, flow rate Q/L, specific weight γ, and distance a, if the pressure gradient $\partial p/\partial s$ is zero.

Example 9.1 *(Con't.)*

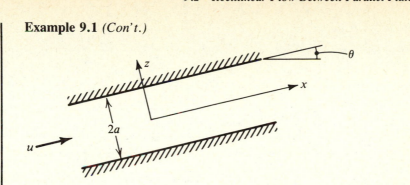

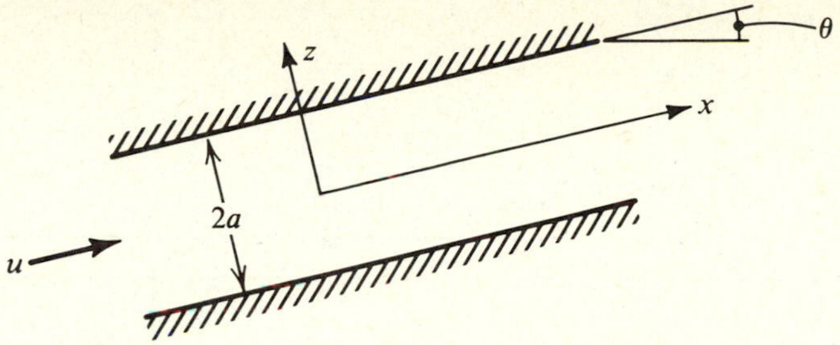

Figure E9.1

Solution:
Step 1.
The fluid is viscous and incompressible. The flow is steady, laminar, and one-dimensional.
Step 3.
The appropriate governing equations are

$$u = \frac{Gz^2}{2\mu} + \frac{U - U'}{2h}z + \frac{U + U'}{2} - \frac{Gh^2}{2\mu} \tag{i}$$

and

$$Q = \int \mathbf{V} \cdot d\mathbf{A} \tag{ii}$$

Applying the assumptions used in Fig. 9.4a, we find the velocity distribution u for this flow to be

$$u = \frac{G}{2\mu}(z^2 - a^2) \tag{iii}$$

where

$$G = -\frac{\partial(p + \gamma z)}{\partial s} \tag{iv}$$

The volume rate of flow per unit length L is

$$\frac{Q}{L} = \int_{-a}^{a} u dz \tag{v}$$

$$= \frac{1}{2\mu}\left[-\frac{\partial(p + \gamma z)}{\partial s}\right]\frac{4}{3}a^3 \tag{vi}$$

If the pressure gradient $\partial p/\partial s$ is zero, then

$$-\frac{\partial(\gamma z)}{\partial s} = \frac{3\mu Q}{2a^3 L} \tag{vii}$$

Example 9.1 *(Con't.)*

such that the slope θ becomes

$$\theta = -\frac{\partial z}{\partial s} = \frac{3\mu Q}{2\gamma a^3 L} \tag{viii}$$

This completes the solution.

Example 9.2

Two parallel flat plates are a distance $2h$ apart. The top plate is moving with a velocity U to the right and the lower plate is moving with a velocity U' to the left and V' in the positive y-direction, as shown in Fig. E9.2. (a) Calculate the location and value of the maximum velocity of the flow between the two plates, and (b) calculate the distribution of vorticity ζ.

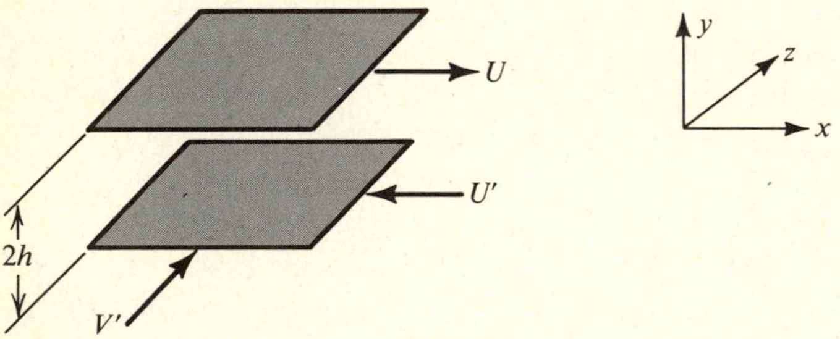

Figure E9.2

Solution:

Step 1.

The fluid is viscous and incompressible. The flow is steady, laminar, and two-dimensional.

Step 3.

The appropriate governing equations are

$$u = \frac{G}{2\mu}(z^2 - h^2) + \frac{U + U'}{2h}z + \frac{U - U'}{2} \tag{i}$$

$$v = -\frac{V'}{2h}z + \frac{V'}{2} \tag{ii}$$

(a) The maximum velocity u_{max} is found by taking the derivative of Eq. (i) and setting the result equal to zero:

$$\frac{\partial u}{\partial z} = \frac{Gz}{\mu} + \frac{U + U'}{2h} = 0 \tag{iii}$$

Example 9.2 *(Con't.)*

so that solving for z gives

$$z = -\mu\left(\frac{U + U'}{2hG}\right) \tag{iv}$$

which is where the velocity component u is a maximum. Substituting Eq. (iv) into the velocity expression for u gives

$$u_{\max} = \frac{G}{2\mu}\left[\mu^2\left(\frac{U + U'}{2hG}\right)^2 - h^2\right] - \frac{\mu}{G}\left(\frac{U + U'}{2h}\right)^2 + \frac{U - U'}{2} \tag{v}$$

as the expression for the maximum velocity component in the x-direction.

The maximum value of the y-component of velocity v occurs at $z = -h$, and thus

$$v_{\max} = V' \tag{vi}$$

(b) The vorticity ζ_y is quite easy to obtain:

$$\begin{aligned} \zeta_y &= \frac{\partial u}{\partial z} \\ &= \frac{Gz}{\mu} + \frac{U + U'}{2h} \end{aligned} \tag{vii}$$

The vorticity ζ_x is

$$\begin{aligned} \zeta_x &= -\frac{\partial v}{\partial z} \\ &= \frac{V'}{2h} \end{aligned} \tag{viii}$$

Hence, the vorticity distribution is

$$\boldsymbol{\zeta} = \frac{V'}{2h}\mathbf{i} + \left(\frac{Gz}{\mu} + \frac{U + U'}{2h}\right)\mathbf{j} \tag{ix}$$

This completes the solution.

Intriguing examples of Couette and Poiseuille flows include the wind moving along the ground, oil pumped along the Alaskan pipeline, and the motion of a pulley belt moving through water. The important issue is whether the particular flow being studied has a pressure gradient so that the proper distribution of velocity, shear stress, and pressure can be made. Though other fluid dynamicists may have different interpretations as to what comprises Couette flow and Poiseuille flow, we define Couette flow as any real fluid flow that has no pressure gradient ($G = 0$), with all other cases being classified as Poiseuille flow ($G \neq 0$).

9.2.1 Temperature Distributions for Couette and Poiseuille Flows

Once we know the velocity distribution for Couette or Poiseuille flow, the temperature distribution may be calculated using the (D.F.) energy Eq. (4.149). If we have no heat transfer, and there are constant values of thermal conductivity and specific heats for steady incompressible flow, then we can express the energy Eq. (4.149) as

$$\rho C_v \frac{DT}{Dt} = k\nabla^2 T + \Phi \tag{9.14}$$

If in addition we confine our flow to one dimension, then Eq. (9.14) becomes

$$\rho C_v u \frac{\partial T}{\partial x} = k\left(\frac{\partial^2 T}{\partial x^2} + \frac{\partial^2 T}{\partial z^2}\right) + \mu\left(\frac{\partial u}{\partial z}\right)^2 \tag{9.15}$$

Further, if we assume that the boundary plates maintain constant temperature, then Eq. (9.15) reduces to

$$k\frac{\partial^2 T}{\partial z^2} + \mu\left(\frac{\partial u}{\partial z}\right)^2 = 0 \tag{9.16}$$

since $\partial T/\partial x = 0$. Thus if we know the velocity distribution of u, we can find the temperature distribution $T(z)$. Let us consider Couette and Poiseuille flows using the above assumptions.

Couette Flow Temperature Distribution

Couette flow is defined as $G = 0$ such that substituting the velocity distribution Eq. (9.12) into the one-dimensional steady energy Eq. (9.16) results in

$$k\frac{d^2T}{dz^2} = -\mu\frac{(U - U')^2}{4h^2} \tag{9.17}$$

with boundary conditions $T = T_w$ at $z = -h$, $T = T_1$ at $z = +h$, when T_w and T_1 are the temperatures of the lower and upper plates, respectively. The solution of Eq. (9.17) is

$$T = \frac{T_w + T_1}{2} + \mu\frac{(U - U')^2}{8h^2k} + \frac{(T_1 - T_w)}{2h}z - \mu\frac{(U - U')^2}{8h^2k}z^2 \tag{9.18}$$

It is sometimes worthwhile to investigate whether there is a cooling or heating of the flow, which can be easily checked by examining the sign of the temperature gradient dT/dz:

$$\frac{dT}{dz} = \frac{(T_1 - T_w)}{2h} - \mu\frac{(U - U')^2}{4h^2k}z \tag{9.19}$$

The sign of the heat transfer depends on whether T_1 is greater or less than T_w, as well as whether $(T_1 - T_w)$ is greater than

$$\mu \frac{(U - U')^2}{2hk} z$$

Remember that z may be either positive or negative, depending upon which plate has heat transfer.

Poiseuille Flow Temperature Distribution

Poiseuille flow is defined as $G \neq 0$. Substituting the velocity distribution of Eq. (9.12) into Eq. (9.16) results in

$$k\frac{d^2T}{dz^2} = -\frac{G^2z^2}{\mu} - G(U - U')\frac{z}{h} - \mu\frac{(U - U')^2}{4h^2} \tag{9.20}$$

Integrating Eq. (9.20) twice to obtain the temperature distribution yields

$$T = -\frac{G^2z^4}{12\mu k} - \frac{G(U - U')}{6kh}z^3 - \frac{(U - U')^2}{8kh^2}z^2 + c_1 z + c_2 \tag{9.21}$$

where

$$c_1 = \frac{T_1 - T_w}{2} + \frac{G(U - U')}{6k}h^2 \tag{9.22}$$

$$c_2 = \frac{T_1 + T_w}{2} + \frac{G^2h^4}{12\mu k} + \mu\frac{(U - U')^2}{8k} \tag{9.23}$$

9.3 Suddenly Accelerated Flat Plate in a Viscous Fluid

Consider a semi-infinite flat plat immersed in a viscous fluid. The inertial coordinate system is shown in Fig. 9.5. At time $t < 0$ the plate is stationary. At $t = 0$ the plate

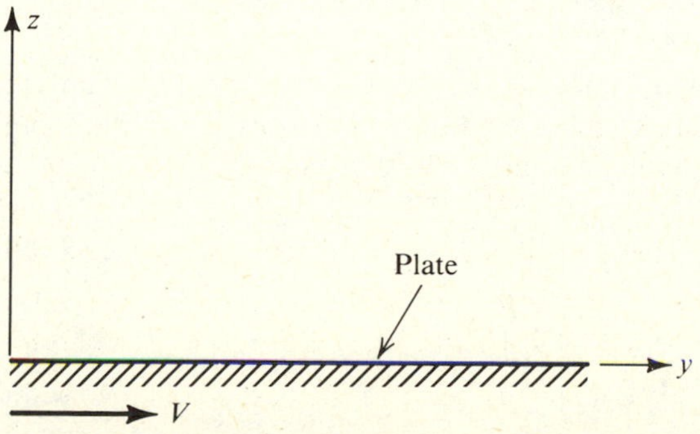

Figure 9.5 *Semi-infinite flat plate immersed in a viscous fluid.*

is suddenly given a velocity V in the positive y-direction. We seek the resultant velocity and pressure field in the fluid as a result of this sudden acceleration of the flat plate.

Clearly, the plate induces a velocity only in the y-direction, since the plate is assumed to have no appreciable thickness. Hence, the velocity components $u = w = 0$ and $v = v(z, t)$. Such a flow is seen to satisfy the (D.F.) continuity equation. The Navier-Stokes equations reduce to the simple form

$$\frac{\partial v}{\partial t} = \nu \frac{\partial^2 v}{\partial z^2} \tag{9.24}$$

$$\frac{dp}{dz} = -\rho g \tag{9.25}$$

Our pressure is thus the hydrostatic pressure.

The boundary conditions for our problem are

$$v = V \quad \text{at} \quad z = 0, \quad \text{for } t \geqslant 0 \tag{9.26}$$

$$v \to 0 \quad \text{as} \quad z \to \infty, \quad \text{for all } t \tag{9.27}$$

Equation (9.24) is a well-known linear partial differential equation called the *diffusion equation*. For example the diffusion equation governs the heat transfer of conduction through solids. It also describes the behavior of a gas passing from one cell into another, and the diffusion of neutrons through matter.

The diffusion equation differs from the wave equation by having a first time derivative instead of a second time derivative. Correspondingly, diffusion is an irreversible process analogous to fluid friction through the shear stress term $\mu(\partial v/\partial z)$, where energy is lost; whereas wave motion diffusion is reversible, having no viscous forces to contend with.*

The diffusion Eq. (9.24) is a parabolic partial differential equation, and a number of analytic methods enable us to solve this equation. We suspect that the velocity component v will be some function of a combination of location z and real time t. Thus, we seek a solution of the form

$$v = v(z, \nu, t) = Vf(z^{k_1}\nu^{k_2}t^{k_3}) = Vf(\eta) \tag{9.28}$$

where η is a nondimensional variable called the similarity variable. To obtain the expression for the similarity variable, we must determine the values k_1, k_2, k_3. From dimensional analysis, we consider finding a dimensionless parameter

$$\Pi_1 = z^{k_1}\nu^{k_2}t^{k_3} \tag{9.29}$$

such that

$$[\Pi_1] = L^{k_1}(L^2T^{-1})^{k_2}(T)^{k_3} \tag{9.30}$$

*This is the idealized situation. Actually, apparent stresses arise from the turbulent action of the flow, but these are independent of viscous effects, yet irreversible.

Following the procedure of Sec. 7.3, we evaluate the exponents as

$$k_1 + 2k_2 = 0$$

$$k_2 = -\frac{1}{2k_1}$$

$$-k_2 + k_3 = 0$$

or

$$k_3 = k_2 = -\frac{1}{2k_1} \tag{9.31}$$

Substituting the values of the exponents given by Eq. (9.31) into Eq. (9.28), we get

$$\eta = A\left(\frac{z}{\sqrt{\nu t}}\right)^{k_3} \tag{9.32}$$

Of course k_3 can be absorbed into the unknown function η. For convenience we take k_3 equal to unity and set A equal to one-half:

$$\eta = \frac{1}{2}\frac{z}{\sqrt{\nu t}} \tag{9.33}$$

Substituting the results for the similarity variable η of Eq. (9.33) and the velocity component v of Eq. (9.28) into the linear momentum Eq. (9.24) yields after some algebraic manipulations

$$\frac{d^2f}{d\eta^2} + 2\eta\frac{df}{d\eta} = 0 \tag{9.34}$$

Observe that we have transformed the partial differential equation into an ordinary differential equation by using the single independent similarity variable η, and thus our differential equation is of a form which we now can solve. The boundary conditions for our problem need to be transformed into conditions for f. These are found from the kinematic conditions stated by Eqs. (9.26) and (9.27) along with the similarity relationships of Eqs. (9.28) and (9.33) such that

$$f = 1 \quad \text{at} \quad \eta = 0$$

$$f \to 0 \quad \text{as} \quad \eta \to \infty \tag{9.35}$$

To solve the differential Eq. (9.34), we set $g(\eta) = df/d\eta$ to obtain a reduced first order linear differential equation:

$$\frac{dg}{d\eta} + 2\eta g = 0 \tag{9.36}$$

This equation is homogeneous and separable. Integration yields

$$g = c_1 e^{-\eta^2} = \frac{df}{d\eta} \tag{9.37}$$

Integrating again this time for f gives

$$f(\eta) = c_1 \int_o^\eta e^{-\eta^2} d\eta + c_2 \tag{9.38}$$

We evaluate the constants of integration c_1 and c_2 using the boundary conditions of Eq. (9.35) to give

$$c_2 = 1$$

and

$$c_1 = \frac{-1}{\displaystyle\int_o^\infty e^{-\eta^2} d\eta} \tag{9.39}$$

Hence, the solution for the similarity function $f(\eta)$ is

$$f(\eta) = 1 - \frac{\displaystyle\int_o^\eta e^{-\eta^2} d\eta}{\displaystyle\int_o^\infty e^{-\eta^2} d\eta} = 1 - \frac{2}{\sqrt{\pi}} \int_o^\eta e^{-\eta^2} d\eta \tag{9.40}$$

The ratio of the two integrals shown above is the well-known error function, erf η, found in probability theory.* The velocity distribution for v can be obtained by substituting the similarity function $f(\eta)$ into the velocity expression of Eq. (9.28) to obtain the result which we desire:

$$v = V\left(1 - \operatorname{erf} \frac{z}{\sqrt{4\nu t}}\right) \tag{9.41}$$

This is shown in Fig. 9.6. We see from the figure that the velocity v rapidly decreases as one moves further away from the plate. This decrease is attributed to diffusion. Figure 9.7 is a graphical representation of the meaning of the error function. It is the ratio of the cross-hatched area to the total area of the plot of the exponential $e^{-\eta^2}$ versus η. When $\eta = 2$ the error function is approximately 0.99. Thus, at this value

*$\operatorname{erf} \eta = \frac{2}{\sqrt{\pi}} \int_o^\eta \exp(-\eta)^2 \, d\eta.$

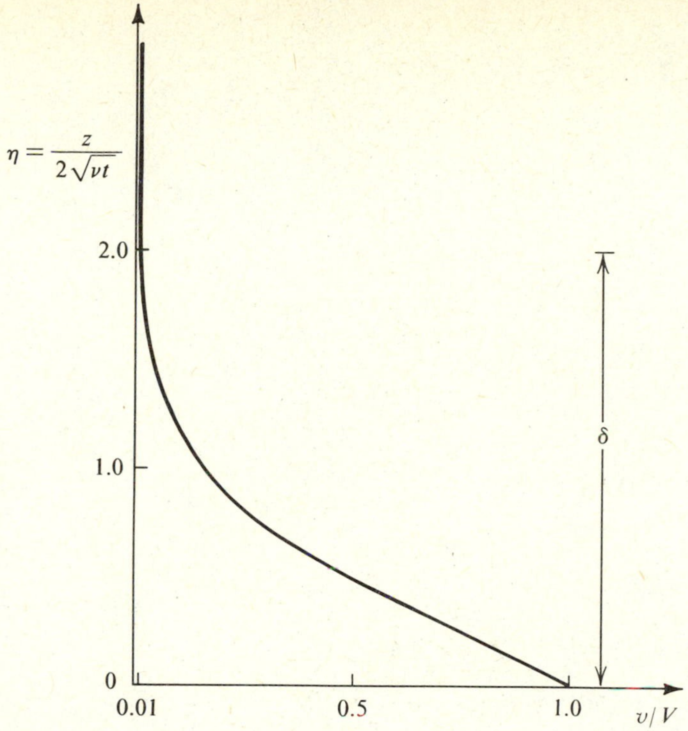

Figure 9.6 Velocity profile above a suddenly accelerated flat plate.

we observe (from Fig. 9.6) that the fluid velocity v is about $0.01V$. Let us define a layer of thickness δ as the distance from the plate's boundary at which the fluid velocity is approximately 99% of the undisturbed velocity V. Then, substituting the value $\eta = 2$ into the expression for η given by Eq. (9.33) results in

$$\eta(\delta, t) = 2 = \frac{\delta}{2\sqrt{\nu t}} \tag{9.42}$$

Solving for the thickness δ, we obtain

$$\delta = 4\sqrt{\nu t} \tag{9.43}$$

Now we ask the question: How far past the plate will the fluid be influenced by the movement of the plate? We can answer this question by setting $t = L/V$, which is the time it takes a fluid particle to go a distance L at a velocity V. Thus

$$\delta = 4L\sqrt{\frac{\nu}{VL}}$$

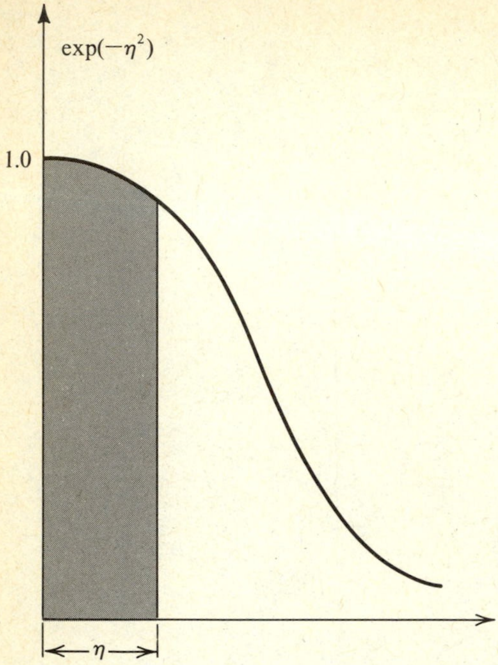

Figure 9.7 Graphical representation of the error function.

or

$$\delta = \frac{4L}{\sqrt{R_L}} \tag{9.44}$$

where R_L is the Reynolds number based on the distance moved by the plate. Therefore, we might expect a laminar boundary layer to have a thickness inversely proportional to the square root of some Reynolds number based on some characteristic length in the direction of flow. This is an approximate way to determine the laminar ''influence'' thickness. We shall find a more elegant way to express this ''influence'' thickness when we discuss laminar boundary layers, so as to refine the value of the constant in Eq. (9.44).

9.4 Rotational Viscous Flows

Vortex flows are all around us: from the large-scale typhoon to the small-scale eddy. We can treat many of them as being laminar, and even steady, but the nonlinearity of some of those Navier-Stokes equation terms cannot be thrown away and can be a nuisance.

In this section we want to expose the reader to a few problems faced in treating vortex flows. By carefully selecting topics, we have simplified the mathematical analysis and at the same time illustrated a powerful mathematical tool: the method of similarity transformations.

We shall first transform the incompressible form of the Navier-Stokes equation to a dimensionless form in terms of two unknowns: the stream function and the circulation. This is much better than working with four unknowns (three velocity components and the pressure). Using these equations, we shall treat a variety of vortex flows, each describing a different physical vortex situation.

9.4.1 Equations of Motion

The Navier-Stokes equations for axisymmetric flow of an incompressible fluid medium, expressed in cylindrical coordinates, are

$$\rho\left(\frac{\partial v_r}{\partial t} + v_r\frac{\partial v_r}{\partial r} + w\frac{\partial v_r}{\partial z} - \frac{v_\theta^2}{r}\right) = -\frac{\partial p}{\partial r} + \mu\left(\frac{\partial^2 v_r}{\partial r^2} + \frac{1}{r}\frac{\partial v_r}{\partial r} - \frac{v_r}{r^2} + \frac{\partial^2 v_r}{\partial z^2}\right) \tag{9.45}$$

$$\rho\left(\frac{\partial w}{\partial t} + v_r\frac{\partial w}{\partial r} + w\frac{\partial w}{\partial z}\right) = -\frac{\partial p}{\partial z} + \mu\left(\frac{\partial^2 w}{\partial r^2} + \frac{1}{r}\frac{\partial w}{\partial r} + \frac{\partial^2 w}{\partial z^2}\right) \tag{9.46}$$

$$\rho\left(\frac{\partial v_\theta}{\partial t} + v_r\frac{\partial v_\theta}{\partial r} + w\frac{\partial v_\theta}{\partial z} + \frac{v_r v_\theta}{r}\right) = \mu\left(\frac{\partial^2 v_\theta}{\partial r^2} + \frac{1}{r}\frac{\partial v_\theta}{\partial r} - \frac{v_\theta}{r^2} + \frac{\partial^2 v_\theta}{\partial z^2}\right) \tag{9.47}$$

The continuity equation is

$$(1/r)(\partial(rv_r)/\partial r) + \partial w/\partial z = 0 \tag{9.48}$$

It is important to note at this point that, because a fixed nonrotating coordinate system has been chosen, the Coriolis terms do not enter the equations of motion.

We simplify the equations of motion by reducing the number of unknowns. The first and second momentum Eqs. (9.45) and (9.46) are combined to eliminate the pressure. We express the radial and axial velocity components in terms of the stream function ψ, and the circumferential velocity component in terms of the circulation Γ by defining

$$v_r = (Q_s/r_0 l\eta^{1/2})(\partial\bar{\psi}/\partial\xi) \tag{9.49}$$

$$w = (-2Q_s/r_0^2)(\partial\bar{\psi}/\partial\eta) \tag{9.50}$$

$$v_\theta = \Gamma_\infty\bar{\Gamma}/2\pi r_0\eta^{1/2} \tag{9.51}$$

where the dimensionless dependent variables are defined as

$$\bar{\Gamma} = \Gamma/\Gamma_\infty, \quad \bar{\psi} = \psi/Q_s \tag{9.52}$$

Flow parameters that are constant are the sink or source volume rate of flow Q_s, the potential circulation Γ_∞, a reference radial length r_0, and a reference axial length l.

The dimensionless independent variables η, ξ, and τ are defined as

$$\eta = (r/r_0)^2, \quad \xi = z/l, \quad \tau = 4\nu t/r_0^2 \tag{9.53}$$

If we define

$$\alpha = (r_0/l)^2, \quad N = Q_s/\nu l, \quad R_0 = (2\pi Q_s/r_0\Gamma_\infty) \tag{9.54}$$

as a dimensionless characteristic length, the radial Reynolds number and the Rossby number, respectively, then the equations of motion produced by substituting Eqs. (9.49)–(9.54) into the momentum equations resulting from elimination of the pressure are, after algebraic manipulations,

$$\begin{aligned}
\frac{1}{8}NR_0^{-2}\eta^{-2}\bar{\Gamma}\frac{\partial\bar{\Gamma}}{\partial\xi} = {} & \frac{1}{2}N\left(\frac{\partial\bar{\psi}}{\partial\xi}\frac{\partial^3\bar{\psi}}{\partial\eta^3} - \frac{\partial\bar{\psi}}{\partial\eta}\frac{\partial^3\bar{\psi}}{\partial\xi\,\partial\eta^2}\right) \\
& + \frac{1}{8}\alpha N\eta^{-2}\left(\eta\frac{\partial\bar{\psi}}{\partial\xi}\frac{\partial^3\bar{\psi}}{\partial\eta\,\partial\xi^2} - \frac{\partial\bar{\psi}}{\partial\xi}\frac{\partial^2\bar{\psi}}{\partial\xi^2} - \eta\frac{\partial\bar{\psi}}{\partial\eta}\frac{\partial^3\bar{\psi}}{\partial\xi^3}\right) \\
& - \eta\frac{\partial^4\bar{\psi}}{\partial\eta^4} - 2\frac{\partial^3\bar{\psi}}{\partial\eta^3} - \frac{\alpha}{2}\frac{\partial^4\bar{\psi}}{\partial\eta^2\,\partial\xi^2} - \frac{1}{16}\alpha^2\eta^{-1}\frac{\partial^4\bar{\psi}}{\partial\xi^4} \\
& + \frac{\partial^3\bar{\psi}}{\partial\tau\,\partial\eta^2} + \frac{1}{4}\alpha\eta^{-1}\frac{\partial^3\bar{\psi}}{\partial\tau\,\partial\xi^2}
\end{aligned} \tag{9.55}$$

and

$$\frac{\partial\bar{\Gamma}}{\partial\tau} + \frac{N}{2}\left(\frac{\partial\bar{\psi}}{\partial\xi}\frac{\partial\bar{\Gamma}}{\partial\eta} - \frac{\partial\bar{\psi}}{\partial\eta}\frac{\partial\bar{\Gamma}}{\partial\xi}\right) = \eta\frac{\partial^2\bar{\Gamma}}{\partial\eta^2} + \frac{\alpha}{4}\frac{\partial^2\bar{\Gamma}}{\partial\xi^2} \tag{9.56}$$

from Eq. (9.47).

Let us consider real vortex flows of large axial extent compared to radial extent. For $\alpha \ll 1$, we compare the individual terms of Eqs. (9.55) and (9.56) so that Eq. (9.55) reduces to

$$\eta^2\left[\frac{\partial^2}{\partial\eta^2}\left(\frac{\partial\bar{\psi}}{\partial\tau} - \eta\frac{\partial^2\bar{\psi}}{\partial\eta^2}\right) + \frac{N}{2}\left(\frac{\partial\bar{\psi}}{\partial\xi}\frac{\partial^3\bar{\psi}}{\partial\eta^3} - \frac{\partial\bar{\psi}}{\partial\eta}\frac{\partial^3\bar{\psi}}{\partial\xi\,\partial\eta^2}\right)\right] = \frac{1}{8}NR_0^{-2}\bar{\Gamma}\frac{\partial\bar{\Gamma}}{\partial\xi} \tag{9.57}$$

and Eq. (9.56) reduces to

$$\frac{\partial\bar{\Gamma}}{\partial\tau} + \frac{N}{2}\left(\frac{\partial\bar{\psi}}{\partial\xi}\frac{\partial\bar{\Gamma}}{\partial\eta} - \frac{\partial\bar{\psi}}{\partial\eta}\frac{\partial\bar{\Gamma}}{\partial\xi}\right) = \eta\frac{\partial^2\bar{\Gamma}}{\partial\eta^2} \tag{9.58}$$

The above equations account for two phenomena of three-dimensional rotational flows: vortex diffusion and vortex stretching. Equation (9.57) accounts for vortex stretching through the swirl term $\frac{1}{8}NR_0^{-2}\bar{\Gamma}(\partial\bar{\Gamma}/\partial\xi)$ and is absolutely essential for vortex breakdown. The swirl term gives the crucial interaction between the axial and rotational components of motion. Equation (9.58) is called the general circulation theorem.

9.4.2 Some Exact Solutions

Though some might balk at pursuing a solution to our two governing equations of unsteady rotational motion, we should not be too intimidated by their unfamiliar form. After all we have two equations with two unknowns: one is fourth order and nonlinear, the other is second order linear but coupled. Physically, they represent long thin vortex flows, as in tornados, the bathtub vortex, and circular vortices. We shall examine two closed form analytic solutions.

Oseen's Solution

Oseen [9.1] assumed that the circulation is a function of the stream function. He found that

$$\bar{\psi} = -\eta/k\tau, \qquad \bar{\Gamma} = (a_1/k)\exp(-\eta/\tau) - 1 \tag{9.59}$$

satisfied Eqs. (9.57) and (9.58). The constants k and a_1 are evaluated from initial and boundary conditions, whatever they may happen to be for the type of vortex. The vortex lines corresponding to this vortex flow are helices around the axis of the vortex, as shown in Fig. 9.8.

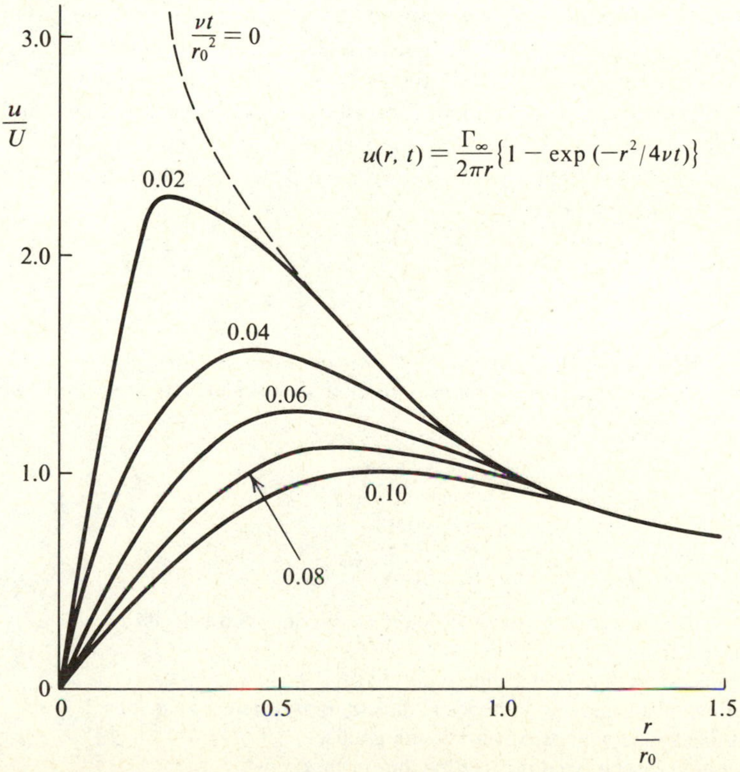

Figure 9.8 *Oseen's vortex.*

A Decaying Vortex

Another interesting solution of our governing equations of vortex motion, Eqs. (9.57) and (9.58), can be found by the method of separation of variables and by assuming the solutions to be harmonic in time. From the definitions of $\overline{\psi}$ and $\overline{\Gamma}$, it is obvious that in order to keep velocities finite on the axis, allowing no steps or kinks in the profile and having a nonrotating coordinate system, it is necessary to have

$$\left.\frac{\partial \overline{\psi}}{\partial \xi}\right|_{\eta=0} = \overline{\Gamma}|_{\eta=0} = \left.\frac{\partial w}{\partial r}\right|_{r=0} = 0 \tag{9.60}$$

Fundamental solutions of Eqs. (9.57) and (9.58) satisfying the "inner" boundary conditions of Eq. (9.60) are

$$\overline{\psi} = 2(\beta\eta)^{1/2} f(\xi) J_1[2(\beta\eta)^{1/2}] \exp(-\beta\tau) \tag{9.61}$$

$$\overline{\Gamma} = \tfrac{1}{2}\beta^{-1/2}\overline{\psi} \tag{9.62}$$

where $\beta = \frac{1}{4}R_o$ is the characterizing frequency and $f(\xi)$ is based upon physically admissible expressions at, say, the outer boundary, where r is equal to r_0. For example, if $f(\xi) = A\xi + B$, the vortex is identified as a Beltrami vortex flow, since the cross product of the vorticity vector with the velocity vector is zero. If $f(\xi)$ = constant, we obtain the results of Caldonazzo [9.2], who considered a vortex as having no radial flow. The latter solution, given by Eq. (9.62), easily leads to the relation that the ratio of the axial velocity to the vorticity about the axis of rotation is constant. Thus streamlines and vortex lines can be coincident for unsteady viscous vortex flows, and the flow is not circulation preserving. Note that these are exact solutions of the Navier-Stokes equations; this is a significant point for we have stated that there are only a few exact solutions of these equations.

References

9.1 Oseen, C. W., "Hydromechanik," *Ark. f. Math. Astron. och Fys.*, 7:82, 1911.

9.2 Caldonazzo, B., "Un osservazione a proposito di moti viscosi simmetrici rispetto ad un asse," *Rend. Acc. Lincei*, 6:152, 1927.

Study Questions

9.1 Sketch a velocity profile for laminar steady flow

1. Between a fixed lower plate and a free-surface moving with velocity U
2. Between two fixed plates
3. In a circular constant diameter pipe
4. Between two plates moving in opposite directions at different velocities
5. Along a fixed plate with negative pressure gradient
6. Down an incline with fixed inclined bottom surface

9.2 Sketch the vorticity distribution, shear stress distribution, and pressure gradient for laminar steady flow between two fixed flat plates.
9.3 Find the volume rate of flow Q per unit length of steady laminar flow between two fixed flat plates. Explain why the sign is negative. What does this say the pressure gradient must be?
9.4 Evaluate the error function whose arguments are 0, 1, ∞, $-\infty$. What is the derivative of an error function (d/dx) (erf x)?
9.5 Go to the library and research other cases of incompressible fluid flow that have exact solutions of the Navier-Stokes equations.
9.6 Explain how δ varies with viscosity and time, and what is its significance at $t \to \infty$.
9.7 Use separation of variables technique to solve Eq. (9.24).
9.8 What is the justification in setting A equal to 0.5 in Eq. (9.32)?
9.9 What is meant by circulation? How is the tangential velocity related to it?
9.10 How is vorticity related to circulation? How is vorticity related to tangential velocity?

Problems

9.1 Oil flows between two parallel fixed flat plates that are 1 cm apart. Assuming the oil temperature to be 40°C and the volume rate of flow to be 4 m^3 per min per meter width, find (a) the drag per meter of width exerted on a 3 m length of channel assuming the flow fully developed, and (b) the pressure drop over 3 m of channel.

9.2 Show that the velocity profile for flow between two stationary flat plates a distance $2h$ apart can be written $u = u_{max}(1 - z^2/h^2)$. Determine the shear stress at the upper plate. What are the normal stresses?

9.3 Two different fluids flow down an inclined plane at an angle θ to the horizontal. A fluid of depth d_2, viscosity μ_2, and density ρ_2 is below an upper fluid of depth d_1, viscosity μ_1, and density ρ_1. The upper fluid has a free-surface. Find the velocity and shear stress distribution of both fluids.

9.4 Glycerin whose viscosity is 0.24 N·s/m^2 is flowing between two parallel fixed plates 10 cm apart (Fig. P9.4). If the shear stress on the plate is 10 kPa and the flow is laminar, calculate the velocity of the flow at a point 0.2 cm from the plane of the plate.

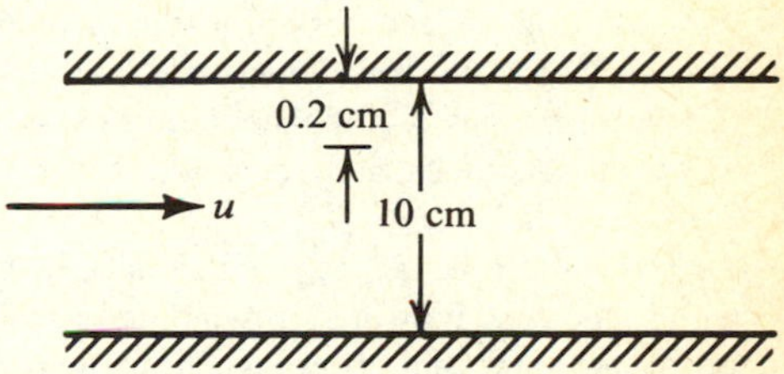

Figure P9.4

9.5 A 20° roof with respect to the horizontal is to be covered with a 0.25 in. layer of liquid roofing asphalt. Find the minimum effective kinematic viscosity of the asphalt in a 6 hr period if the top surface of the layer is not to move more than 0.00100 in. relative to the bottom surface, and the width of the covered surface is 30 ft.

9.6 Consider the flow in Fig. P9.6. If the pressure difference is 10 psi over a length L, and $U = 2U' = 10$ ft/s when $2h = 0.06$ in. and $\mu = 1 \times 10^{-4}$ lbf·s/ft^2, find the shear stress on each plate.

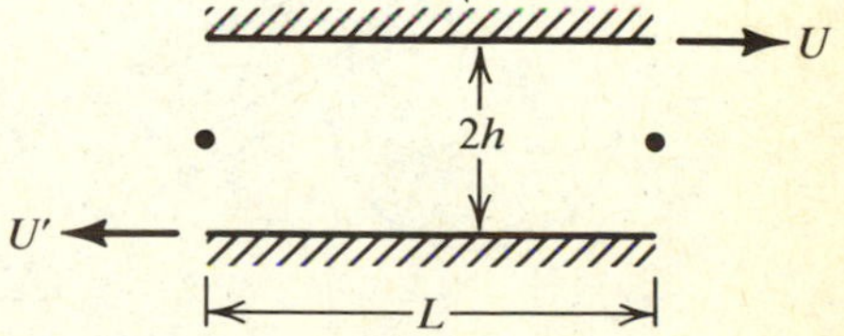

Figure P9.6

9.7 Show that the volume rate of flow q per unit width for laminar flow down an inclined plane of angle θ with respect to the horizontal and having a free-surface is

$$q = \frac{gh^3}{3\nu} \sin \theta$$

where h is the depth of the liquid.

9.8 An endless belt, one end of which is immersed in a glycerin tank, is used to transport small amounts of glycerin to an upper tank. The thickness of the fluid is 0.04 ft. The belt surface is completely smooth, the flow is laminar, and the specific weight of the fluid is 60 lbf/ft^3. Referring to Fig. P9.8, assume the viscosity to be 0.002 lbf·s/ft^2, and the belt velocity to be 1 ft/s. Determine the flow rate discharged into the upper tank per unit ft of the belt, and sketch the velocity profile across a plane normal to the belt.

9.9 What is the thickness of a film of fluid moving at constant speed down an infinite vertical wall in terms of volume rate of flow per unit width? Assume that the shear stress is zero at the free-surface.

9.10 Consider the generalized Poiseuille flow with suction and blowing for two-dimensional steady incompressible flow, where the upper plate ($y = h$) moves at constant velocity U. (a) Show the Navier-Stokes equations reduce to

$$v_o \frac{du}{dy} = -\frac{1}{\rho}\frac{\partial p}{\partial x} + \nu \frac{d^2u}{dy^2}$$

$$v_o \frac{du}{dy} + \beta = \nu \frac{d^2u}{dy^2} \qquad \text{(i)}$$

where $\beta = 1/\rho\ (dp/dx)$ and v_o is the velocity at $y = 0$ and $y = h$. (b) For impermeable walls, show that the solution of Eq. (i) is

$$u = \frac{3Uh - 6Q}{h^3}(y^2 - hy) + U\frac{y}{h} \qquad \text{(ii)}$$

where the flow is between two parallel flat plates at a fixed distance h from each other. (c) Show for the case of uniform suction and blowing that

$$u = \left(U + \frac{\beta h}{v_o}\right)\frac{\exp(\alpha y) - 1}{\exp(\alpha h) - 1} - \frac{\beta}{v_o}y \qquad \text{(iii)}$$

where $\alpha = v_o/v$.

9.11 If the radial velocity v_r is zero everywhere in the flow field, the vortex flow is said to rotate as a solid body at an angular speed ω, such that $\zeta_z = 2\omega$. In addition, the local radial Reynolds number N is also zero. Prove that the centerline axial velocity w_o cannot vary in the axial direction.

9.12 A free vortex is characterized by two conditions: (i) a linear radial distribution of radial velocity, (ii) zero effects of viscosity. Show that from (i) this results in a linear axial distribution of the axial velocity, and from (ii) the tangential velocity $v_\theta = \Gamma_\infty/2\pi r$.

9.13 Consider a vortex whose vorticity is

$$\zeta_z(r) = \frac{a\Gamma_\infty}{2\pi\nu}\exp(-ar^2/2\nu)$$

What is the tangential velocity for such a flow?

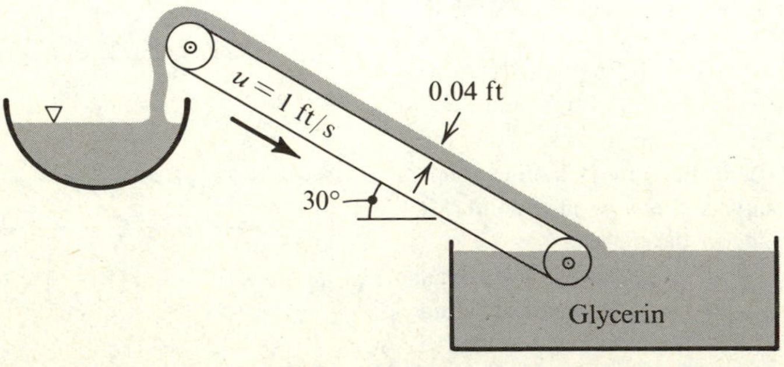

Figure P9.8

9.14 Using the mathematical definition for the vortex core radius r_c, show that for vortex motions of the Oseen form that (a) the core radius is inversely proportional to the square root of the vorticity along the centerline, and (b) the core radius is proportional to the core radius of a Rankine vortex by a factor of 1.12.

9.15 If the radial velocity $v_r = v_r(r)$, the circumferential velocity $v_\theta = v_\theta(r)$, and the axial velocity $w = z\overline{w}(r)$ for a steady incompressible axisymmetric vortex, show that the Navier-Stokes equations become

$$v_r \frac{d\overline{w}}{dr} + \overline{w}^2 - \frac{\nu}{r}\frac{d}{dr}\left(r\frac{d\overline{w}}{dr}\right) = C$$

and

$$\frac{\partial p}{\partial z} = c\rho z$$

9.16 In Prob. 9.15, if the stream function $\psi = \nu z f(x)$, where $x = r/z$, show that the velocity components can be expressed as

$$v_r = -\frac{\nu f}{r} + \frac{\nu f'}{z}$$

$$v_\theta = \frac{\nu}{r} m(x)$$

$$w = \frac{\nu f'}{r}$$

where $f' = df/dx$.

9.17 In Prob. 9.16, if the pressure p can be expressed as $p = -(\nu^2/z^2)h(x)$, show that the three Navier-Stokes equations can be expressed as

$$-f^2 + ff'x - m^2x^3h' - x^4f''$$

$$- x^2f'' - zx^4h + f'x - x^5h' = 0$$

$$f''x(1 + x^2) - f'(1 - f) - x^3h = 0$$

$$m''x(1 + x^2) - m'(1 - 2x^2 - f) = 0$$

9.18 Using Euler's equations of motion in cylindrical coordinates, show that the equations reduce to

$$\frac{\partial^2\psi}{\partial z^2} + \frac{\partial^2\psi}{\partial r^2} - \frac{1}{r}\frac{\partial\psi}{\partial r} = r^2F(\psi)$$

for a steady axisymmetric incompressible vortex, where F is an arbitrary function of the stream function ψ.

9.19 In Prob. 9.18, show that $\psi = (A/2)r^2(a^2 - z^2 - r^2)$ is a solution for the vortex, where $F(\psi) = 5A$ and $z^2 + r^2 = a^2$, where $\psi = 0$. What is the vorticity inside the surface $\psi = 0$?

9.20 Consider a two-dimensional unsteady vortex in a viscous incompressible fluid. Using polar coordinates, transform the Navier-Stokes equations in terms of vorticity $\zeta_z(r, \theta)$ and the stream function $\psi(r, \theta)$.

9.21 In Prob. 9.20, find the value of the stream function $\psi(r, t)$ for an axisymmetric, unsteady, two-dimensional vortex where the vorticity ζ_z is constant ζ_o.

10 Laminar Pipe Flow

10.1 Introduction

In this chapter we shall apply the energy and continuity equations to a large class of internal flow problems of enormous interest to engineers. These are problems of transferring energy from one location to another by means of fluid flow through pipes. We cannot always package energy as in a battery, place it in our car, and carry it to where it shall be used; instead we must move energy through conduits such as pipes. Because fluids (such as steam and oil) can store enormous levels of energy, transporting fluids is a crucial fluid mechanics problem requiring careful design of pipe circuits and fluid machinery. We shall first treat the (D.F.) of pipe flow equations that help us calculate how pressure and velocity are distributed, then the (I.F.) pipe flow equations that will give us the Hagen-Poiseuille result, which expresses the head loss representing loss of energy in a pipe.

We start by considering a fundamental problem. What horsepower would a pump require to move a fluid through a very long pipe? The value for the horsepower can be determined using the energy equation. Some of the horsepower will be used to compensate for energy lost in the pressure drop caused by the friction of the fluid flowing past the pipe's wetted surface. This energy loss is called a *major pipe "head" loss* and is calculated from a value of a term we call the friction factor f, length and diameter of pipe, and velocity of the flow. The friction factor f can be a function of both Reynolds number and the relative roughness of the pipe. This chapter will develop an expression for the energy loss which the pump will compensate, and will describe ways to evaluate the friction factor for a variety of pipe flows. We will also examine a method of analysis to accomplish all this.

10.2 Description of the Physical Phenomenon

When fluid moves through a pipe, its behavior can be described by the Reynolds number

$$R_D = \frac{\tilde{V}D}{\nu} \tag{10.1}$$

where the characteristic length is the inner diameter D of the pipe, $\tilde{V}$ is the mean velocity of the flow, and ν is the kinematic viscosity of the fluid. Three basic types of flow are possible in the pipe, each possessing different characteristics of behavior. One of the easiest ways to observe these three basic types of flow is to light a small roll of paper, extinguish the flame, and observe the behavior of the filament of smoke, as shown in Fig. 10.1. The first type is called *laminar flow*, where the fluid flows in a well-behaved fashion in smooth laminae. In laminar flow, a fluid particle stays in the laminae layer as long as the flow remains laminar. The precise definition of laminar flow will be presented later, but for the present, it is sufficient to say that a laminar flow is one where the trajectory of a fluid particle is predictable, not random or unstable. A typical velocity distribution for laminar flow in a pipe is shown in Fig. 10.2, and is seen to be a paraboloid, with maximum velocity at the pipe centerline and zero velocity along the wall.

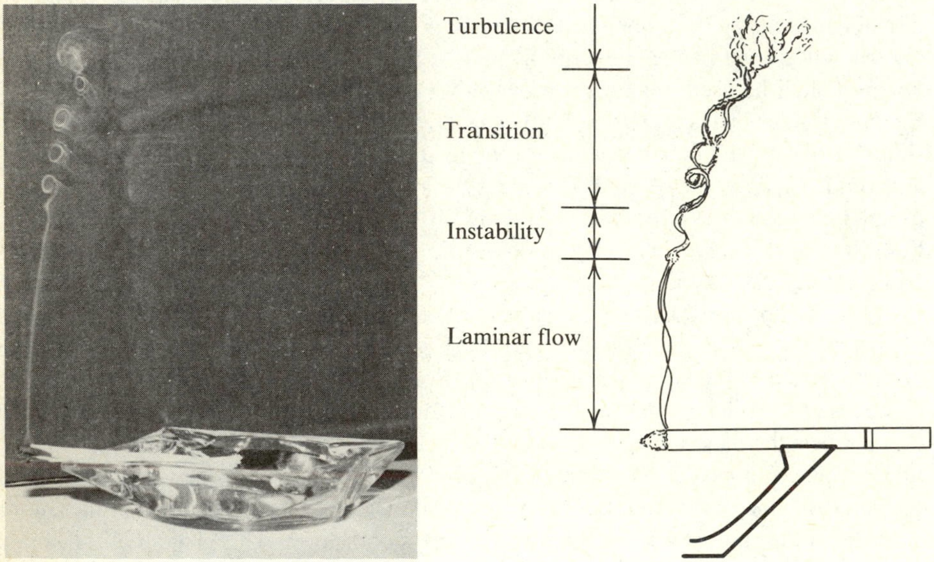

Figure 10.1 *Regions of fluid flow.*

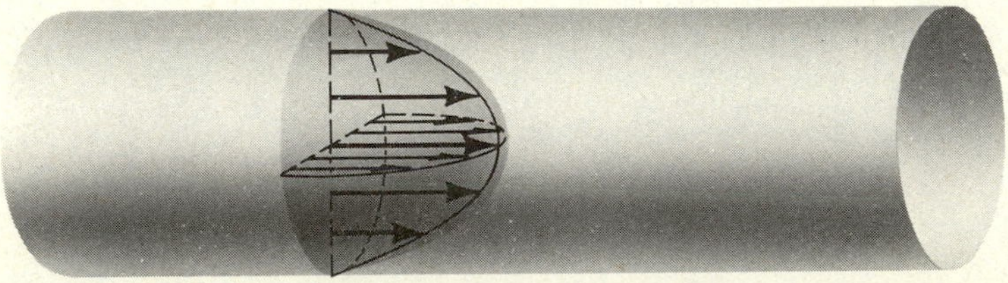

Figure 10.2 *Velocity distribution of laminar flow in a pipe.*

The second type of flow is called *transitional flow*. During transitional flow, the fluid particle moves from its well-behaved laminae into other adjacent layers in a somewhat oscillatory manner that grows in amplitude and quickly produces an unstable pattern. The spatial zone of transitional fluid flow is exceedingly small compared to the spatial zones of the other two types of flow. We shall treat transition in Chaps. 11 and 14. Figure 11.6 shows a typical streakline in the transitional range for water flowing through a pipe. There the burst is not as well-defined as in Fig. 10.1.

The third and last type of flow is called *turbulent flow*. The fluid particles move in all three directions most often in an irregular fashion. A typical cross-sectional view of the velocity distribution through a pipe is shown in Fig. 10.3. The front of the profile appears serrated owing to the axial velocity fluctuations, but near the wall it is smooth and laminar. The mean flow is more nearly uniform in the central portion of the pipe for turbulent flow than for laminar flow, the latter being parabolic.

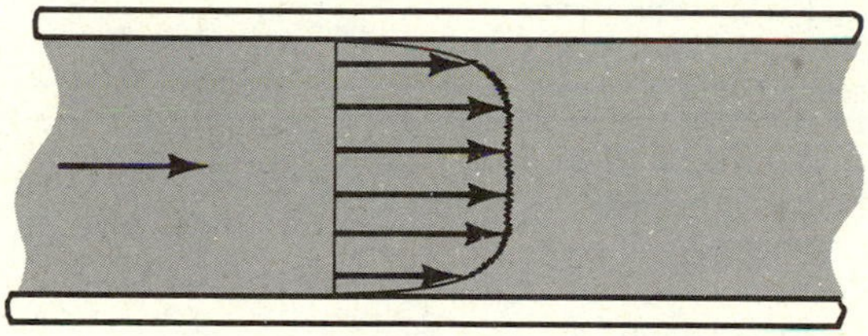

Figure 10.3 *Velocity distribution of turbulent flow in a pipe.*

Under ordinary conditions, laminar flow exists in a pipe for a range of Reynolds number R_D:

$$R_D \leq 2300, \quad \text{laminar flow} \tag{10.2}$$

although laminar flow has been observed for values of Reynolds number up to 40,000. As we shall see in Chap. 11, *fully turbulent flow* (where the flow is everywhere turbulent) in a pipe depends strongly on the pipe's inlet conditions, the inherent turbulence level of the inflowing fluid, and on the roughness of the pipe's wetted inner surface. For very rough pipes, turbulent flow may exist at a Reynolds number of 10^4, and for a very smooth pipe, turbulent flow may exist at a Reynolds number of 10^8. Though roughness can be controlled experimentally, it plays a major role in the behavior of real fluids flowing in pipes along with entrance effects and the inherent turbulence level of the upstream fluid.

In the following discussion, we assume that the fluid flow is *fully established*, i.e., that the flow (whether laminar or turbulent) completely fills the cross section of the pipe. Thus a fully established flow is one in which the velocity profile does not change with respect to the streamwise direction. For a fully established flow we consider stations at large distances downstream of any entrance to a pipe. In the fully established region of a pipe (or a channel) of uniform cross section, there is no streamwise acceleration; that is, $Du/Dt = 0$, since by definition $\partial u/\partial x \rightarrow 0$ and $v \rightarrow 0$. Figure 10.4 shows a development region where $a_x \neq 0$, and a fully established region. At the entrance, the flow is stagnant along the edge of the pipe. Along this edge the shear stress is a maximum. As the flow moves into the pipe, a region is created where there is large velocity gradient $\partial u/\partial y$. This region is called the boundary layer. Since $\partial u/\partial y$ is large in this region, both the shear stress p_{xy} and the vorticity ζ_z are finite. The two boundary layers meet at an axial distance L, at which place we say the flow is fully established.

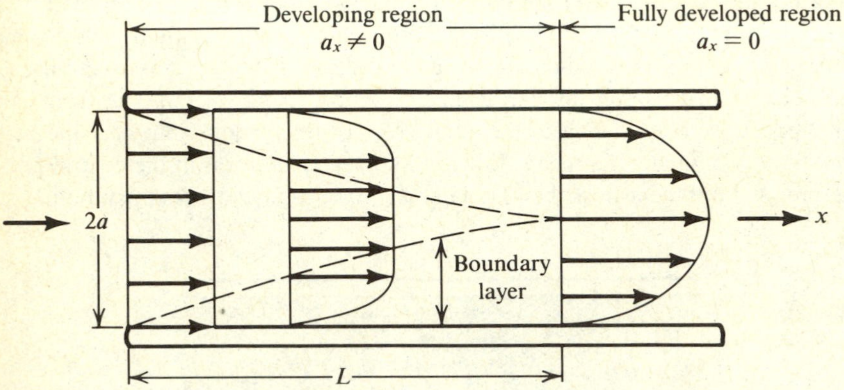

Figure 10.4 *Development region in pipe flow.*

For a pipe of uniform radius a, an accepted correlation for the length of the developing region L is

$$L \cong 0.23 \frac{a^2}{\nu} \tilde{V} \tag{10.3}$$

whereas for a channel of width $2a$, the length is

$$L = 0.16 \frac{a^2}{\nu} \tilde{V} \tag{10.4}$$

For very low Reynolds numbers, a better correlation is given by

$$L = \frac{0.32a^2}{\nu} \tilde{V} + 1.4a \tag{10.4a}$$

Thus, for a Reynolds number of 2000, the development length L is 115 pipe diameters before the flow can be considered fully established. It is important to mention that the flow may become turbulent long before it becomes fully developed.

10.3 Equations of Motion for Laminar Flow in a Pipe

This section is devoted to finding analytic expressions for the velocity distribution of *laminar flow* in a pipe and for the pressure drop due to energy loss from friction, and to defining a quantity called the friction factor f.

One of the most important *exact* solutions of the Navier-Stokes equations is that for a fully established steady flow of a viscous fluid through a straight round pipe. Let us consider a pipe of inside radius a located in a gravitational field, as shown in Fig. 10.5. The centerline of the pipe lies in the $y'z'$ plane. We assume that fluid completely fills the pipe, allowing no cavities or free-surface in the pipe. Since the outer boundary of the flow is cylindrical, we will use cylindrical coordinates to describe the flow. A boundary condition of the flow is that there is no slip at the pipe wall:

$$v_r = v_\theta = w = 0, \quad \text{at} \quad r = a \tag{10.5}$$

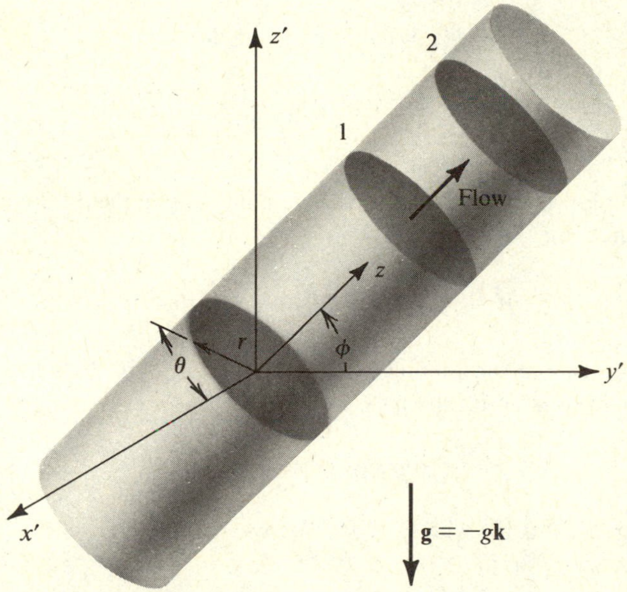

Figure 10.5 *Geometry for pipe flow.*

The flow is assumed axisymmetric, fully developed, and steady. Thus we seek solutions of the form

$$\begin{aligned} v_\theta &= 0 \\ w &= w(r) \end{aligned} \tag{10.6}$$

It is obvious that this flow field satisfies the continuity equation as given by Eq. (4.25). The Navier-Stokes equations written in cylindrical coordinates are given by Eqs. (4.113), (4.114), and (4.115).

For the case under consideration we can show that the governing dynamical equations reduce to

$$r\text{ - component:} \quad 0 = g_r - \frac{1}{\rho}\frac{\partial p}{\partial r} \tag{10.7}$$

$$\theta \text{ - component:} \quad 0 = g_\theta - \frac{1}{\rho r}\frac{\partial p}{\partial \theta} \tag{10.8}$$

$$z \text{ - component:} \quad 0 = g_z - \frac{1}{\rho}\frac{\partial p}{\partial z} + \frac{\mu}{\rho}\frac{1}{r}\frac{\partial}{\partial r}\left(r\frac{\partial w}{\partial r}\right) \tag{10.9}$$

where $\mathbf{g} = g_r\mathbf{e}_r + g_\theta\mathbf{e}_\theta + g_z\mathbf{k}$. Resolving $\mathbf{g}$ into scalar components, and using the geometry given in Fig. 10.5, we get

$$g_r = -g \cos\phi \sin\theta \tag{10.10}$$

$$g_\theta = -g \cos\phi \cos\theta \tag{10.11}$$

$$g_z = -\frac{\partial}{\partial z} g \sin\phi \, z \tag{10.12}$$

Substituting Eq. (10.10) into Eq. (10.7) and integrating the differential equation, we obtain the pressure

$$p = -\rho g r \cos\phi \sin\theta + f_1(\theta, z) \tag{10.13}$$

In a similar fashion, substituting Eq. (10.11) into Eq. (10.8) and integrating, we obtain a similar expression for the pressure

$$p = -\rho g r \cos\phi \sin\theta + f_2(r, z) \tag{10.14}$$

Comparing Eq. (10.13) with Eq. (10.14), we see that

$$p = -\rho g r \cos\phi \sin\theta + f(z) \tag{10.15}$$

Substituting Eqs. (10.12) and (10.15) into Eq. (10.9) yields

$$\frac{\mu}{r}\frac{\partial}{\partial r}\left(r\frac{\partial w}{\partial r}\right) = -G \tag{10.16}$$

where

$$G = -\rho g \sin\phi - \frac{df}{dz} \tag{10.17}$$

and is a constant (see Eq. (10.21)).

Integration of Eq. (10.16) results in the axial velocity expression

$$w = -\frac{Gr^2}{4\mu} + c_1 \ln r + c_2 \tag{10.18}$$

where the arbitrary function of integration is

$$f(z) = -Gz - \rho g \sin\phi\, z + c_3 \tag{10.19}$$

Note that c_1 must be zero; otherwise the velocity w would become infinite at the center of the pipe. The constant of integration c_2 can be determined from the boundary condition $w(a) = 0$, yielding the result

$$w = \frac{Ga^2}{4\mu}\left(1 - \frac{r^2}{a^2}\right) \tag{10.20}$$

Equation (10.20) shows the velocity distribution to be parabolic in radius r with w_{max} occurring at $r = 0$. Thus, setting $r = 0$, $w = w_{max}$, and solving Eq. (10.20) for G gives

$$G = \frac{4\mu w_{max}}{a^2} = \frac{16\mu w_{max}}{D^2} \tag{10.21}$$

The flow rate Q is evaluated as

$$Q = A\tilde{w} = 2\pi\int_o^R \frac{G}{4\mu}(a^2 - r^2)r\,dr = \frac{\pi G a^4}{8\mu} \tag{10.22}$$

resulting in a relationship between the average velocity $\tilde{w}$ and maximum velocity w_{max}:

$$\tilde{w} = \frac{w_{max}}{2} \tag{10.23}$$

Next let us discuss the pressure distribution. Substituting Eq. (10.19) into Eq. (10.15) yields the pressure

$$p = -\rho g r \cos\phi \sin\theta - Gz - \rho g z \sin\phi + c_3 \tag{10.24}$$

Consider the change in pressure between two points along a pipe. We see from Eqs. (10.16) and (10.19) that

$$\frac{\partial}{\partial z}(\rho g z \sin\phi + p) = -G \tag{10.25}$$

Integrating the above equation along the axial extent of the pipe from station 1 to 2 results in

$$\rho g \sin\phi\,(z_2 - z_1) + p_2 - p_1 = -G(z_2 - z_1) \tag{10.26}$$

or

$$(p_1 + \rho g \sin \phi \, z_1) - (p_2 + \rho g \sin \phi \, z_2) = GL \tag{10.27}$$

where we now define $L = z_2 - z_1$ to be the length of pipe between the stations. Using Eqs. (10.21) and (10.23), and noting from the geometry of Fig. 10.5 that $z' = z \sin \phi$, we can write Eq. (10.27) in more useful form as

$$(p_1 + \gamma z_1') - (p_2 + \gamma z_2') = \frac{32\mu\tilde{w}L}{D^2} \tag{10.28}$$

This result is the famous *Hagen-Poiseuille* expression for the pressure drop in a pipe due to viscosity. It is, however, applicable only for laminar flow. Note that there is no obstruction in the pipe between stations 1 and 2 in the equation. The equation also assumes that the pipe is of constant diameter.

The Hagen-Poiseuille equation is a simple statement that the loss in the energy per unit volume h_f between stations 1 and 2 is given by

$$h_f = \frac{32\mu\tilde{w}L}{D^2\gamma} \tag{10.29}$$

This can be verified by comparing Eq. (10.28) with Eq. (5.103).

Many engineers working pipe flow problems express the loss of mechanical energy per unit volume h_f in terms of the kinetic energy:

$$h_f = \frac{32\nu\tilde{V}L}{D^2 g} = f\frac{L}{D}\frac{\tilde{V}^2}{2g} \tag{10.30}$$

where f is called the *friction factor*. Solving this last expression for f we get

$$f = \frac{64\mu}{\tilde{w}D\rho} = \frac{64}{\dfrac{\rho\tilde{w}D}{\mu}} = \frac{64}{R_D} \tag{10.31}$$

where R_D is the Reynolds number based on the characteristic length being the diameter of pipe and $\tilde{w}$ being the mean velocity in the pipe (see Eq. (10.1)). We have stated previously that if the value of the Reynolds number is below a value of 2300, the flow in the pipe will be laminar. If the Reynolds number is larger than 2300, the flow *could* be turbulent, and the friction factor f as given by Eq. (10.31) is not valid. The parabolic velocity profile and the expression for the frictional pressure loss in round pipes are in excellent agreement with experimental results for fully established flow.

Example 10.1
Consider fluid of kinematic viscosity ν between two circular pipes shown in Fig. E10.1.

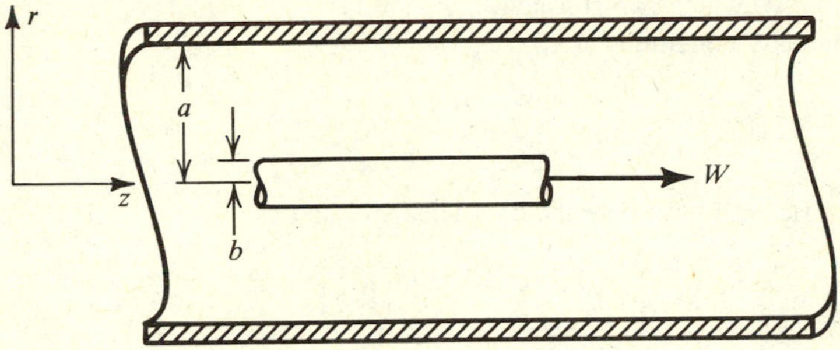

Figure E10.1

Let the outer cylinder of radius a be stationary, and the inner cylinder of radius b move in the positive x-direction with uniform velocity W. Assume the flow to be steady, fully established, and axisymmetric. Determine (a) the radial distribution of axial velocity component w; (b) the shear stress at the outer wall $r = a$, and the shear stress at the inner wall $r = b$; (c) the average velocity $\tilde{W}$; and (d) the maximum velocity w_{max}.

Solution:
Step 1.
The fluid is viscous and incompressible. The flow is steady and fully established.
Step 2.
The (D.F.) governing equations will be used.
Step 3.
The appropriate form of the governing equation is:

- Linear momentum:

$$\frac{1}{r}\frac{d}{dr}\left(r\frac{dw}{dr}\right) = \frac{1}{\mu}\frac{dp}{dz} \tag{i}$$

This problem is precisely like that presented in Sec. 10.3 except for different boundary conditions.

(a) Integrating Eq. (i) yields the radial distribution of velocity w as given by Eq. (10.18):

$$w = -\frac{Gr^2}{4\mu} + c_1 \ln r + c_2 \tag{ii}$$

Example 10.1 *(Con't.)*

where

$$G = -\frac{dp}{dz} \tag{iii}$$

from Eq. (10.25), which is a constant.

The boundary conditions are

$$w = W, \quad \text{at} \quad r = b \tag{iv}$$

$$w = 0\,, \quad \text{at} \quad r = a \tag{v}$$

which are sufficient to evaluate the coefficients c_1 and c_2:

$$c_1 = \frac{1}{\ln \dfrac{a}{b}}\left[\mu W - \frac{1}{4}\frac{\partial p}{\partial z}(a^2 - b^2)\right] \tag{vi}$$

$$c_2 = -\frac{b^2}{4}\frac{\partial p}{\partial z} - \frac{\ln b}{\ln \dfrac{a}{b}}\left[\mu W - \frac{1}{4}\frac{\partial p}{\partial z}(a^2 - b^2)\right] \tag{vii}$$

Substituting Eqs. (vi) and (vii) into Eq. (ii) gives

$$w = \frac{1}{4}\frac{\partial p}{\partial z}\left[\frac{r^2}{\mu} - \frac{(a^2 - b^2)\ln r}{\ln \dfrac{a}{b}} - b^2 + \frac{(a^2 - b^2)\ln b}{\ln \dfrac{a}{b}}\right] + \frac{\mu W}{\ln \dfrac{a}{b}} \ln \frac{r}{b} \tag{viii}$$

(b) We evaluate shear stress p_{rz} at the outer wall $r = a$ using

$$p_{rz}\Big|_{r=a} = \mu \left.\frac{\partial w}{\partial r}\right|_{r=a} \tag{ix}$$

such that substituting Eq. (ii) into Eq. (ix) results in

$$p_{rz}\Big|_{r=a} = \frac{c_1}{a} + \frac{a}{2}\frac{\partial p}{\partial z} \tag{x}$$

where c_1 is given by Eq. (vi). We evaluate the shear stress at the inner wall $r = b$ using

$$p_{rz}\Big|_{r=b} = \mu \left.\frac{\partial w}{\partial r}\right|_{r=b} \tag{xi}$$

Example 10.1 *(Con't.)*

such that

$$p_{rz}\Big|_{r=b} = \frac{c_1}{b} + \frac{b}{2}\frac{\partial p}{\partial z} \qquad \text{(xii)}$$

(c) The one-dimensional (I.F.) continuity equation is

$$Q = \tilde{V}A = \int_o^{2\pi}\int_a^b wr\, dr\, d\theta \qquad \text{(xiii)}$$

Substituting the velocity distribution for w of Eq. (ii) into Eq. (xiii) and integrating results in

$$\tilde{V} = -\frac{1}{8\mu}\frac{\partial p}{\partial z}\left[a^2 + b^2 + \frac{(a^2 - b^2)}{\ln \dfrac{b}{a}}\right] \qquad \text{(xiv)}$$

(d) The maximum velocity is where $\partial w/\partial r$ is zero. Differentiating Eq. (ii) with respect to the radius r and setting the result equal to zero yields the radius

$$r = \sqrt{\frac{2c_1}{\mu(\partial p/\partial z)}} \qquad \text{(xv)}$$

which is where the axial velocity w is a maximum. Substituting the value of the radius of Eq. (xv) back into the velocity expression of Eq. (ii) gives

$$w_{\max} = \frac{c_1}{2\mu}\left[1 + \ln 2c_1 - \ln \mu\left(\frac{\partial p}{\partial z}\right)\right] \qquad \text{(xvi)}$$

where c_1 is given by Eq. (vi).

This completes the solution.

Example 10.2

Consider fluid confined between two cylinders as shown in Fig. E10.2.

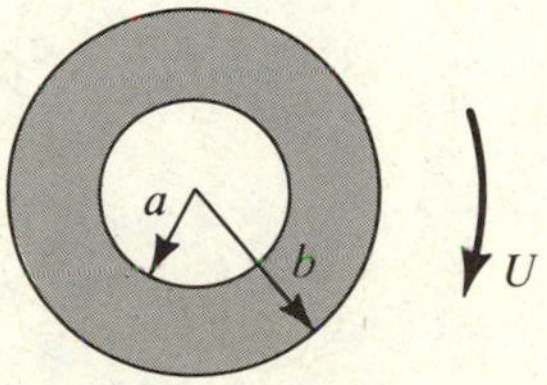

Figure E10.2

The radius of the inner stationary cylinder is a, and the radius of the outer cylinder is b. The outer cylinder rotates at a uniform circumferential velocity U.

Example 10.2 *(Con't.)*

Let the fluid between the two cylinders be steady, axisymmetric, and invariant in the axial direction, so that the fluid motion is purely circular. Determine (a) the velocity components v_r, v_θ, and w, and (b) the shear stresses on the inner and outer walls.

Solution:

Step 1.

The fluid is viscous and incompressible. The flow is steady, axisymmetric, and fully established.

Step 2.

The (D.F.) governing equation in cylindrical coordinates will be used.

Step 3.

The appropriate form of the governing equations are

- Linear momentum:

$$\frac{\partial w}{\partial t} + v_r \frac{\partial w}{\partial r} + \frac{v_\theta}{r}\frac{\partial w}{\partial \theta} + w\frac{\partial w}{\partial z} = g_z - \frac{1}{\rho}\frac{\partial p}{\partial z} + \nu\left(\frac{\partial^2 w}{\partial r^2} + \frac{1}{r}\frac{\partial w}{\partial r} + \frac{1}{r^2}\frac{\partial^2 w}{\partial \theta^2} + \frac{\partial^2 w}{\partial z^2}\right) \quad \text{(i)}$$

$$\frac{\partial v_\theta}{\partial t} + v_r \frac{\partial v_\theta}{\partial r} + \frac{v_\theta}{r}\frac{\partial v_\theta}{\partial \theta} + \frac{v_r v_\theta}{r} + w\frac{\partial v_\theta}{\partial z} = g_\theta - \frac{1}{\rho r}\frac{\partial p}{\partial \theta} + \nu\left(\frac{\partial^2 v_\theta}{\partial r^2} + \frac{1}{r}\frac{\partial v_\theta}{\partial r} - \frac{v_\theta}{r^2} + \frac{1}{r^2}\frac{\partial^2 v_\theta}{\partial \theta^2} + \frac{2}{r^2}\frac{\partial v_r}{\partial \theta} + \frac{\partial^2 v_\theta}{\partial z^2}\right) \quad \text{(ii)}$$

- Continuity:

$$\frac{\partial v_r}{\partial r} + \frac{v_r}{r} + \frac{1}{r}\frac{\partial v_\theta}{\partial \theta} + \frac{\partial w}{\partial z} = 0 \quad \text{(iii)}$$

(a) For a flow to be invariant in the axial direction

$$\frac{\partial}{\partial z} = 0 \quad \text{(iv)}$$

and for a flow to be axisymmetric

$$\frac{\partial}{\partial \theta} = 0 \quad \text{(v)}$$

The differential form of the continuity equation of Eq. (iii) then becomes

$$\frac{\partial (r v_r)}{\partial r} = 0 \quad \text{(vi)}$$

Example 10.2 *(Con't.)*

so that upon integration

$$v_r = \frac{c_1}{r} \tag{vii}$$

But the boundary condition states that at the outer radius $r = b$ the radial velocity is zero since the cylinder wall is solid. Thus the value of c_1 is zero, resulting in

$$v_r = 0 \tag{viii}$$

Substituting Eq. (iv), (v), and (viii) into the axial component of the Navier-Stokes equation (i) gives

$$\frac{\partial}{\partial r}\left(r\frac{\partial w}{\partial r}\right) = 0 \tag{ix}$$

Integrating Eq. (ix) twice for the axial velocity component produces

$$w = c_2 \ln r + c_3 \tag{x}$$

The boundary conditions on the axial velocity component are

$$w = 0, \quad \text{at} \quad r = a \tag{xi}$$

$$w = 0, \quad \text{at} \quad r = b \tag{xii}$$

Evaluating the constants c_2 and c_3 yields

$$c_1 \ln \frac{a}{b} = 0 \tag{xiii}$$

Since $a \neq b$, then c_1 is zero. Also, c_2 is zero. Thus, the axial velocity w for this flow is

$$w = 0 \tag{xiv}$$

The problem thus reduces to one dependent variable: v_θ.

From the θ-component form of the Navier-Stokes equation (ii), we have

$$\frac{1}{r}\frac{\partial}{\partial r}\left(r\frac{\partial v_\theta}{\partial r}\right) - \frac{v_\theta}{r^2} = 0 \tag{xv}$$

The solution of this equation is straightforward and results in

$$v_\theta = \frac{c_4}{r} + c_5 r \tag{xvi}$$

The boundary conditions on the tangential velocity are

$$v_\theta = 0\,, \quad \text{at} \quad r = a \tag{xvii}$$

$$v_\theta = U, \quad \text{at} \quad r = b \tag{xviii}$$

Example 10.2 *(Con't.)*

so that substituting Eqs. (xvii) and (xviii) into Eq. (xvi) produces

$$c_4 = \frac{U(a^2 b)}{(a^2 - b^2)} \tag{xix}$$

and

$$c_5 = -\frac{bU}{a^2 - b^2} \tag{xx}$$

Hence the tangential velocity v_θ becomes

$$v_\theta = \frac{Ub}{a^2 - b^2}\left(\frac{a^2}{r} - r\right) \tag{xxi}$$

(b) The shear stress $p_{r\theta}$ is found using Eq. (4.74). For axisymmetric flow

$$p_{r\theta} = \mu r \frac{\partial}{\partial r}\left(\frac{v_\theta}{r}\right) \tag{xxii}$$

For steady motion, the total moment on any annular element must vanish, so that

$$(p_{r\theta})_{\text{wall}} a^2 = (p_{r\theta}) r^2$$

or

$$p_{r\theta} = \left(\frac{a}{r}\right)^2 (p_{r\theta})_{\text{wall}} \tag{xxiii}$$

Substituting Eq. (xxi) for the velocity into the shear stress expression of Eq. (xxiii) results in

$$(p_{r\theta})_{\text{wall}} = -\frac{2b\mu U}{a^2 - b^2} \tag{xxiv}$$

for the inner wall, and

$$\begin{aligned}(p_{r\theta})_{\text{wall}} &= \frac{a^2}{b^2}\left(-\frac{2b\mu U}{a^2 - b^2}\right) \\ &= -\frac{2a^2\mu U}{b(a^2 - b^2)}\end{aligned} \tag{xxv}$$

for the shear stress on the outer wall.

This completes the solution.

10.4 The Moody Diagram

The friction factor f has been defined in terms of energy loss per unit volume h_f by Eq. (10.30). In particular, for laminar flow, Eq. (10.31) showed that the friction factor was inversely proportional to the Reynolds number. For laminar flow, the energy loss is considered to be independent of the roughness of the pipe because very little of the flow comes in contact with the protuberances at the pipe surface: only those laminae of the total thickness of the height of the protuberances are affected, and the eddies resulting from the protuberances are quickly damped out by extremely large viscous dissipation forces. At higher Reynolds numbers where the flow may be turbulent, the higher velocities bring in fluid particles coming into contact with the protuberances so that the protuberances do affect the flow, causing a greater loss of energy than for laminar flow.

For many years, experimentalists have tried to express the friction factor f in terms of (1) roughness, and (2) Reynolds number for various Reynolds number regimes. So far they have found only empirical relationships. Some of these will be discussed in Chap. 11 on turbulent pipe flow. But for now, we shall turn to the Moody diagram to aid us in determining the correct friction factor f for a specific flow in a particular kind of pipe.

Figure 10.6 is a plot of the friction factor f versus Reynolds number R_D for a variety of dimensionless roughness ϵ/D. The figure, one of the more widely used in pipe flow, is called the *Moody diagram* after the American engineer, L. F. Moody, who obtained the necessary data to develop the figure. On the right-hand side of the figure is the *relative roughness* ϵ/D, obtained by taking a typical value of *absolute roughness* ϵ* and dividing it by the inner diameter of the pipe, D.

Inspecting the Moody diagram of Fig. 10.6, we note four flow regimes: laminar flow, critical zone flow, transitional flow, and complete turbulent flow. If $R_D \leqslant 2300$, we know the flow is laminar, and the friction factor is inversely proportional to Reynolds number. Its value is given by Eq. (10.31). The critical zone is that regime where $2000 \leqslant R_D \leqslant 4500$. Here the flow can either be laminar [and therefore obey the relationship given by Eq. (10.31)], or it can be transitional, where the effect of the protuberance produces in a greater loss of energy than in laminar flow for the same Reynolds number. In the transitional zone is where the friction factor f is dependent upon both Reynolds number and relative roughness. Note that the lowest curve in the family of curves in the transitional zone represents the friction factor for a smooth pipe. The last zone in the figure is the completely turbulent zone where the friction factor is seen to be independent of Reynolds number and dependent solely upon relative roughness.

Figure 10.6 offers three expressions for the friction factor f if the pipe flow is turbulent. The first expression is the Colebrook interpolation formula that yields good results for the region between smooth walls and fully rough conditions. The other two relations will be derived in Chap. 11 for turbulent pipe flow. They exist for either a smooth pipe or fully turbulent flow. Applicable to both circular and noncircular pipes, as well as to open channel flow, the Moody diagram is fairly accurate: up to $\pm 15\%$ for engineering applications.

For pipes that are noncircular, and are of low aspect ratio geometries, there is good correlation between theory and experiment for *completely developed flow* using

*Table 10.1 is a list of various roughness heights ϵ representative of a variety of materials.

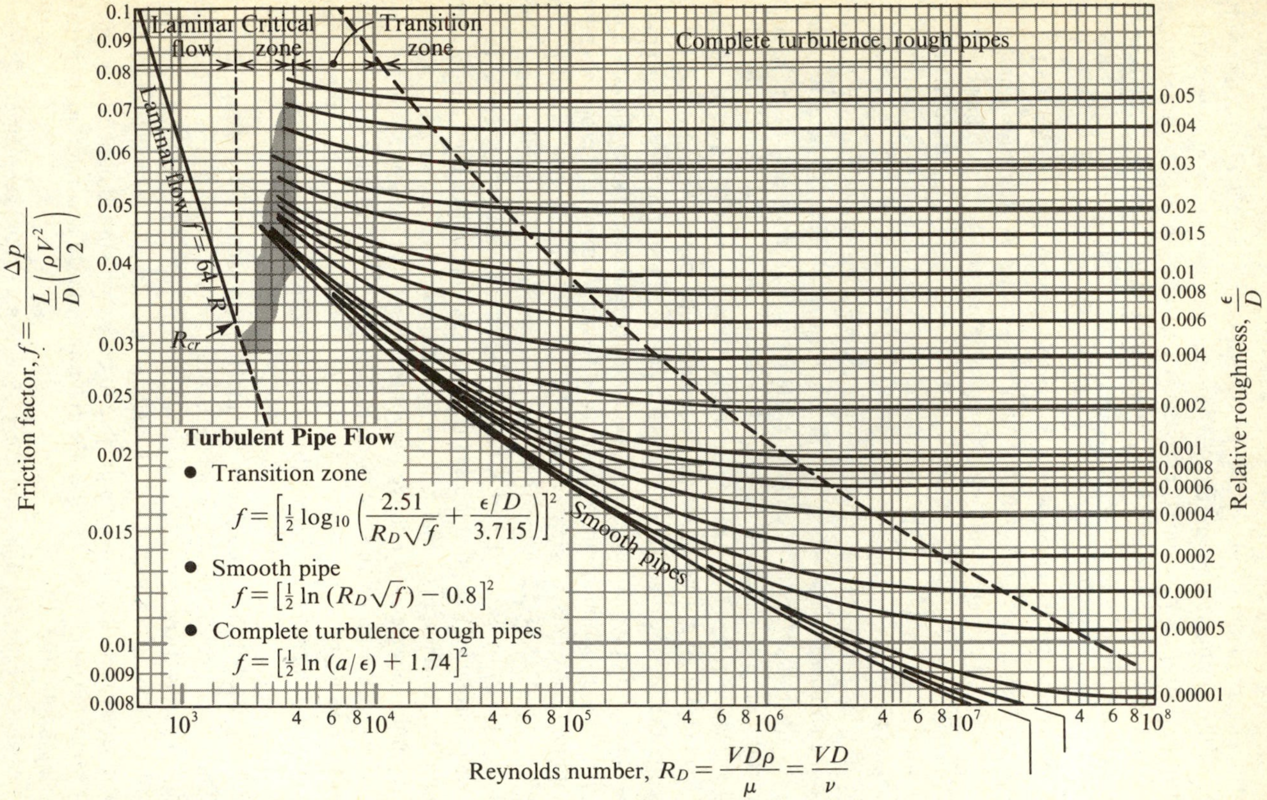

Figure 10.6 *Friction factors for commercial pipe. (Source: Reproduced with the permission of the American Society of Mechanical Engineers, from L. F. Moody, "Friction Factors for Pipe Flow,"* Trans. ASME, *vol. 66, Nov. 1944.)*

Table 10.1 *Table of Absolute Roughness, ϵ (for pipes and channels)*

Type	(ft)	(mm)
Glass	Smooth, < 0.000001	< 0.0003
Drawn tubing	0.000005	0.0015
Wrought iron	0.00015	0.046
Asphalted cast iron	0.0004	0.122
Galvanized iron	0.0005	0.152
Cast iron	0.00085	0.259
Asphalted steel	0.0001	0.030
Welded steel	0.0003	0.091
Riveted steel	0.0025	0.762
Rusted steel	0.005	1.524
Vibrated concrete	0.0002	0.061
Smooth concrete	0.0006	0.183
Cement plaster	0.0015	0.457
Unfinished concrete	0.01	0.305
Old concrete	0.05	15.24
Planed wood	0.001	0.030
Rough wood	0.002	0.610
Old wood	0.005	1.524
Brick	0.002	0.610
Smooth earth	0.02	6.096
Gravel	0.07	21.336
Coarse gravel	0.2	60.96
Stones	0.4	121.92

the Moody diagram if we replace D in the Reynolds number Eq. (10.1) with the hydraulic diameter D_h, which is defined as

$$D_h = \frac{4\text{ (cross-sectional area)}}{\text{(wetted circumference)}}$$

$$= \frac{4wd}{w + 2d} \tag{10.32}$$

where lengths d and w are specified in the geometry of Fig. 10.7. Note that we must also replace D with D_h in the relative roughness expression ϵ/D. For a pipe we must use the entire cross-sectional area and circumference, whereas for free-surface channels, we use the area occupied by the fluid. Also, the circumference is only the wetted circumference. The area must not be too narrow. Engineers sometimes use as a rule-of-thumb that

$$A \leqslant 4\, D_h \tag{10.33}$$

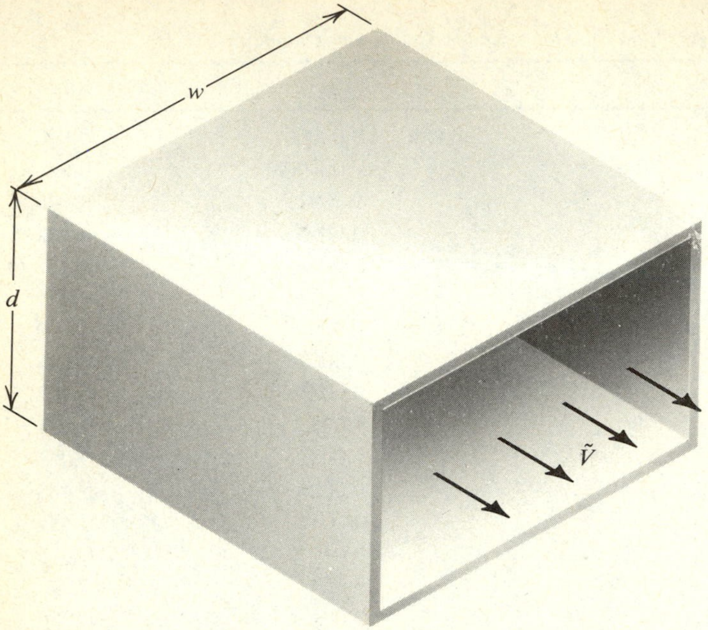

Figure 10.7 *Flow through a rectangular channel.*

Example 10.3

Calculate the friction factor f for the following fluid flows through a 1 ft diameter pipe: (a) $R_D = 0$, (b) $R_D = 10$, (c) $R_D = 1000$, $\epsilon = 0.0267$ ft, (d) $R_D = 20{,}000$, $\epsilon = 0.002$ ft, (e) $R_D = 10{,}000$, material is smooth concrete pipe, (f) $R_D = 10^7$, $\epsilon/D = 0.004$, and (g) $R_D = 1 \times 10^{12}$, $\epsilon = 0.005$ ft.

Solution:

We use the Moody diagram of Fig. 10.6 for the solution of these problems.

(a) For $R_D = 0$, the flow is static and we cannot calculate the friction factor.

(b) For $R_D = 10$, the flow is laminar, and we can use the friction factor expression of Eq. (10.31) to find

$$f = \frac{64}{R_D} = \frac{64}{10} = 6.4 \qquad \text{(i)}$$

(c) For $R_D = 1000$, the flow is laminar, and we need not concern ourselves with what value the roughness ϵ has, so that

$$f = \frac{64}{1000} = 0.064 \qquad \text{(ii)}$$

Example 10.3 *(Con't.)*

(d) For $R_D = 20{,}000$, the flow is in the region that is neither laminar nor completely turbulent. The relative roughness is $\epsilon/D = 0.002$ and is the right-hand ordinate of Fig. 10.6 with $R_D = 20{,}000$ as the abscissa giving

$$f = 0.0297 \tag{iii}$$

(e) For a smooth concrete pipe, Table 10.1 gives an absolute roughness ϵ of 0.0006 ft, so that the relative roughness is $\epsilon/D = 0.006$. With $R_D = 20{,}000$, we find that

$$f = 0.027 \tag{iv}$$

(f) For $R_D = 10^7$ and $\epsilon/D = 0.004$, a direct reading gives

$$f = 0.0282 \tag{v}$$

(g) For $R_D = 1 \times 10^{12}$ and $\epsilon = 0.005$ ft, we note that the flow is completely turbulent, so that the value of the friction factor is independent of Reynolds number. Thus, using the relative roughness $\epsilon/D = 0.005$ on the ordinate,

$$f = 0.03 \tag{vii}$$

This completes the solution.

10.4.1 Other Ways to Use the Moody Diagram

We have seen that given the pipe geometry, the flow velocity, the viscosity of the fluid, and the material of the pipe, one can directly calculate the friction factor f, and thus the pressure drop Δp. Suppose these particular parameters are not given. Table 10.2 presents three of the more popular cases found in pipe flow. The pipe flows of case *A* are the easiest to solve as we merely follow the procedure described on the next page.

Table 10.2 *Fundamental Description of Pipe Flow Problems*

	GIVEN								
	Geometry		*Fluid*		*Flow Rate*	*Pipe Material*	*Head Loss*		
CASE	D	L	ρ	μ	Q	ϵ	h_f	**FIND**
A	x	x	x	x	x	x		h_f
B	x	x	x	x		x	x	Q
C		x	x	x	x	x	x	D

x means given information.

Step 1.
Calculate the Reynolds number:

$$R_D = \frac{4\rho Q}{\pi D \mu} \qquad (10.34)$$

Step 2.
Calculate the relative roughness ϵ/D.
Step 3.
Read the value of f from Fig. 10.6, the Moody diagram.

Step 4.
Calculate the head loss h_f

$$h_f = f\frac{L}{D}\frac{\tilde{V}^2}{2g} \qquad (10.30)$$

For case B, White [10.1] develops a slick cross-plot of the Moody diagram. He defines the abscissa of Fig. 10.8 as a ratio of the pressure force over a viscous force

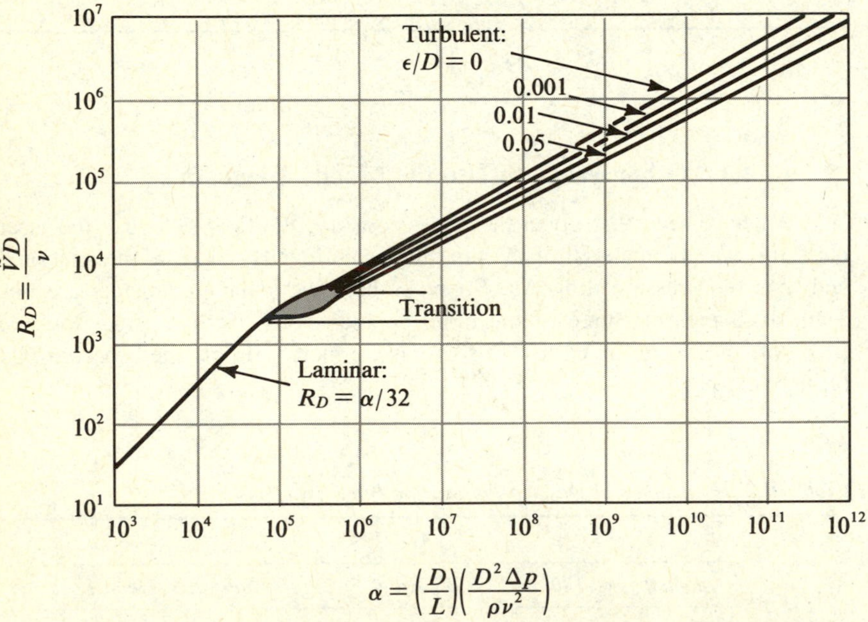

Figure 10.8 *Modified Moody diagram for finding pipe flow rate Q. (Source:* Fluid Mechanics *by F. H. White. Copyright © 1979. McGraw-Hill Book Company. Used with the permission of McGraw-Hill Book Company.)*

times a fineness ratio:

$$\alpha = \left(\frac{D}{L}\right)\left(\frac{D^2\Delta p}{\rho\nu^2}\right) \tag{10.35}$$

and the ordinate as the Reynolds number. In case B we are given the head loss h_f, so we can easily evaluate the pressure drop Δp for a horizontal pipe of constant cross-sectional area using the energy Eq. (5.103):

$$h_f = \frac{\Delta p}{\gamma}$$

Notice that flow is laminar when

$$R_D = \frac{\alpha}{32} \tag{10.36}$$

which is found using Eqs. (10.30), (10.31), and (10.35).

The transition region for pipe flow occurs when $\alpha \simeq 10^5$. (The transition region for flow over a flat plate is when the Reynolds number is of the order 10^5.) For turbulent flow, a smooth pipe is now the upper curve in Fig. 10.8, whereas it was the lowest curve in Fig. 10.6. The procedure for solving case B problems is straightforward.

Step 1.
Calculate the White number α:

$$\alpha = \left(\frac{D}{L}\right)\left(\frac{D^2\,\Delta p}{\rho\nu^2}\right) = \frac{D^3 g h_f}{L\nu^2}$$

Step 2.
Calculate the relative roughness ϵ/D.

Step 3.
Using Fig. 10.8, find the Reynolds number.

Step 4.
Calculate the flow rate Q:

$$Q = \frac{\pi D \mu R_D}{4\rho}$$

For pipe flows of case C, White [10.1] cross-plots the Moody diagram of Fig. 10.6 so that we can easily find the appropriate pipe diameter D. If a new dimensionless parameter β is defined as

$$\beta = R_D\sqrt{2R_D\alpha} \tag{10.37}$$

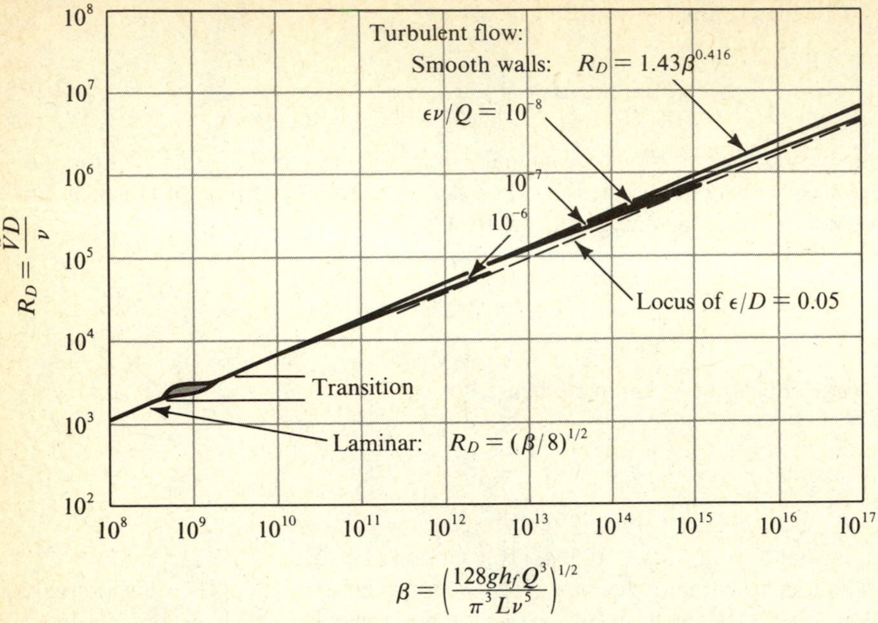

Figure 10.9 *Modified Moody diagram for finding pipe diameter D. (Source:* Fluid Mechanics *by F. H. White. Copyright © 1979. McGraw-Hill Book Company. Used with the permission of McGraw-Hill Book Company.)*

then the pipe diameter can be canceled so that our parameter is in terms of given values:

$$\beta = \left(\frac{128gh_fQ^3}{\pi^3 L\nu^5}\right)^{1/2} \tag{10.38}$$

which is the abscissa of Fig. 10.9. The family of relative roughness curves of the Moody diagram of Fig. 10.6 are transformed to a new dimensionless roughness parameter $\epsilon\nu/Q$, and is one of the coordinates shown in Fig. 10.9. The recommended procedure for solving case C problems is as follows:

Step 1.
Calculate the value of the dimensionless parameter β.

Step 2.
Calculate the value of the dimensionless parameter $\epsilon\nu/Q$.

Step 3.
Using Fig. 10.9, find the appropriate value of the Reynolds number.

Step 4.
Determine the value of pipe diameter D using Eq. (10.34):

$$D = \frac{4\rho Q}{\pi\mu R_D}$$

Example 10.4

Gasoline of density $\rho = 1.32$ slug/ft^3 and viscosity $\mu = 0.61 \times 10^{-5}$ lbf·s/ft^2 flows through a rubber hose of absolute roughness 0.0004 ft. Find the volume rate of flow Q given $D = 1$ in., L = 10 ft, and a pressure drop of 1 psi. (See Fig. E10.4.)

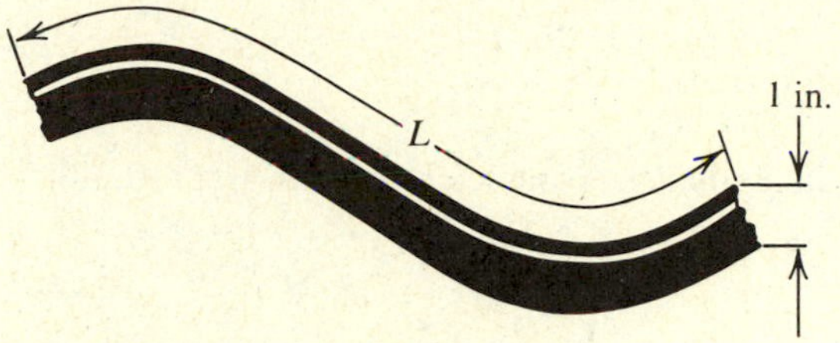

Figure E10.4

Solution:

This is a case B pipe flow problem.

Step 1.

Calculate α. From Eq. (10.35):

$$\alpha = \left(\frac{1/12}{10}\right)\left[\frac{(1/12)^2\,(1 \times 144)}{1.32 \times 0.61 \times 10^{-5}}\right] = 1.035 \times 10^3$$

Step 2.

Calculate relative roughness:

$$\frac{\epsilon}{D} = \frac{0.0004}{1/12} = 0.0048$$

Step 3.

Using Fig. 10.8, find the Reynolds number. Since we are in the laminar flow regime,

$$R_D = \frac{\alpha}{32} = \frac{1.035 \times 10^3}{32} = 32.343$$

Step 4.

Calculate the flow rate Q:

$$Q = \frac{\pi D \mu R_D}{4\rho} = \frac{\pi(\tfrac{1}{2})(0.61 \times 10^{-5})(32.343)}{4(1.32)}$$

$$= 0.978 \times 10^{-5} \text{ ft}^3/\text{s}$$

This completes the solution.

Example 10.5

Water is to be pumped through 100 miles of the Rub' al Khali desert in a smooth concrete pipe at a flow rate of 5 ft^3/s. Given the density $\rho = 1.908$ slug/ft^3, and viscosity $\mu = 0.975 \times 10^{-5}$ lbf·s/ft^2, what size pipe is required if the pumps are designed to handle a head loss of only 200 ft?

Solution:

This is a case C pipe flow problem.

Step 1.

Calculate the value of the dimensionless parameter β. From Eq. (10.38):

$$\beta = \left(\frac{128g\, h_f\, Q^3}{\pi^3\, L\nu^5}\right)^{1/2}$$

$$= \left[\frac{128(32.2)(200)(5)^3}{(\pi)^3\,(528{,}000)(0.511 \times 10^{-5})^5}\right]^{1/2}$$

$$= 4.25021 \times 10^{13}$$

Step 2.

Calculate the value of the dimensionless parameter $\epsilon\nu/Q$: From Table 10.1, $\epsilon = 0.0006$ ft, such that

$$\frac{\epsilon\nu}{Q} = \frac{(0.0006)(0.511 \times 10^{-5})}{5} = 0.6132 \times 10^{-9}$$

Step 3.

Using Fig. 10.9, find the appropriate value of the Reynolds number.

The results from Steps 1 and 2 indicate an approximate value of

$$R_D \cong 1.43\beta^{0.416} = 6.68 \times 10^5$$

Step 4.

Determine the value of the pipe diameter using Eq. (10.34):

$$D = \frac{4Q}{\pi\nu R_D} = \frac{4 \times 5}{\pi \times 0.511 \times 10^{-5} \times 6.68 \times 10^5} = 1.87 \text{ ft}$$

Because of the value of the Reynolds number being approximate, a 1.9-ft-diameter smooth concrete pipe is appropriate.

This completes the solution.

10.5 Minor Losses

When fluids flow through a pipe, other losses in addition to the friction loss are apparent. These additional losses are attributed to the flow through valves and fittings, the abrupt enlargement or contraction of the pipe, obstructions, and losses at the entrance and exit of the pipe.

Losses of energy due to sudden or gradual changes in the channel geometry and due to fittings placed in the channel are smaller than the major loss of energy h_f due to friction, and are therefore called *minor losses*. In some instances, the minor losses can be ignored when a length of 1000 diameters separates minor loss items.

The minor loss of energy, which is evidenced by a pressure drop, can be treated as some equivalent fractional loss of the kinetic energy ($\frac{1}{2}\rho\tilde{V}^2$). The energy drop per weight of fluid is expressed as

$$\frac{\Delta p}{\gamma} = (h_f)_{\text{minor}} = +k\left(\frac{\tilde{V}^2}{2g}\right) \tag{10.39}$$

where the value of k is determined in most cases by experimental measurements.

10.5.1 Fittings and Obstructions

The value of the constant k in Eq. (10.39) for various fittings placed in the pipe can be obtained from experimental measurements. These fittings introduce serious obstructions to the flow field which in turn create disturbances that convert a part of the flow energy into thermal energy that is subsequently transferred out of the piping control volume through the pipe walls. The loss of energy can be easily detected from the drop in pressure across the fitting. The next section presents a few typical values of k for various valves.

Valves

Essential in any piping system, valves control the flow of fluid. They throttle it, they direct it, they prevent back flow—but mostly they turn the flow on or off. Valves must be selected carefully for the particular flow requirement. If we want to be able to turn a flow on or off, then we may wish to use a gate, plug, or ball valve. If we want to carefully regulate or throttle the flow, then a globe, needle, or butterfly valve might be our choice. If we need to guarantee no back flow, then our choice is a lift or swing check valve.

The factors that govern the selection of a valve are

- Is the fluid a gas or liquid?
- Is the fluid abrasive?
- Is temperature an important factor?
- Is volume rate critical?
- At what pressures will the fluid flow?
- What pressure drop can the system tolerate?
- Must the system be leak proof?

For excellent information on factors governing the selection of valves, see Ref. 10.2.

A few valves and their cutaways are presented in Fig. 10.10. Those shown in Figs. 10.10d–f are special valves used in steam engineering, and show the variety that can be adapted to special flow situations. The valves in *d* are pressure relief valves: the relief valve is designed to relieve *all* the boiler pressure, the sentinel valve has a

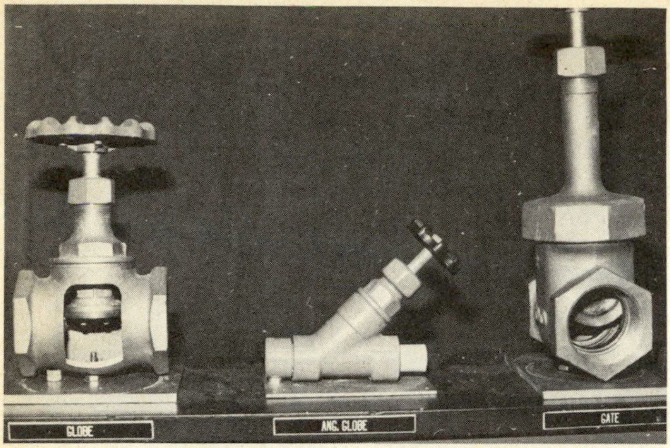

(a)

(b)

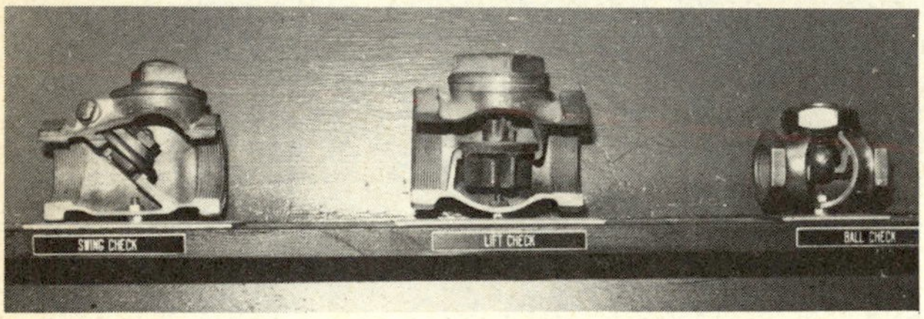

(c)

Figure 10.10 *Cutaways of some valves.*

whistle, and the pressure reducer lowers high pressure to a desired low pressure (for example, it can reduce 500 psi steam to 5 psi).

The next three valves in Fig. 10.10e are steam traps. The bellows is a thermostatic steam trap that has a diaphragm which expands when hot, closing the steam orifice. Cold condensate on the diaphragm shrinks it and thus opens the valve. In the impulse steam trap pressure rather than temperature moves the diaphragm (two sizes are shown).

The last three valves in Fig. 10.10 are also steam traps. The ball float allows condensate to form on the ball's surface, giving it added weight and closing off the inlet. The bucket works in much the same way but is more efficient than the ball float.

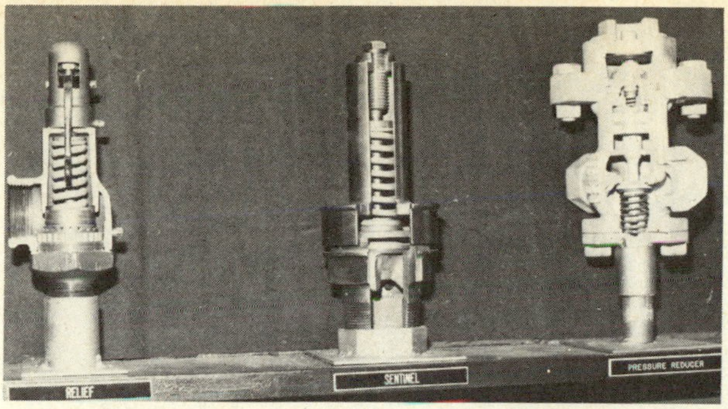

(d)

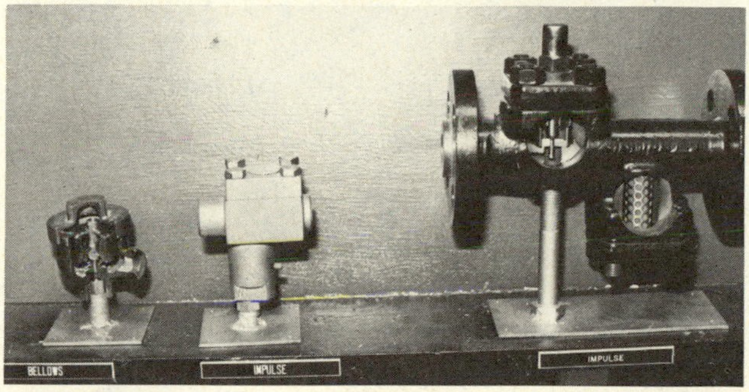

(e)

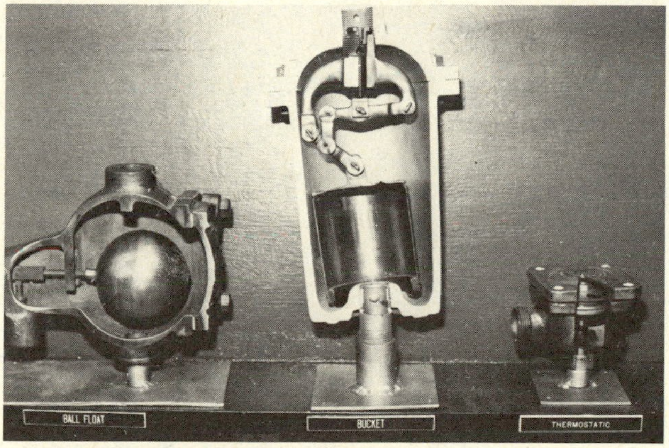

(f)

Figure 10.10 *(Con't.)*

The thermostatic valve works like the bellows but with a different diaphragm. An excellent reference on steam traps is Ref. 10.3.

While the valves in Fig. 10.10 are small, that shown in Fig. 10.11 is quite large. Some are manually operated as in Fig. 10.11, while others are motor driven. The marine valve in Fig. 10.11 is a main steam stop valve, similar to a globe valve. It has special seal, corrosion, watertight integrity and pressure requirements that must conform to marine engineering regulations that meet the requirements of the American

Bureau of Shipping, Lloyd's Register of Shipping, and the U.S. Coast Guard. Such a valve can work above 225 psi at temperatures above 350°F, handling everything from water, petroleum products, and molten sulphur to liquefied flammable gases and poisons. Loss coefficients for these valves depend on size opening and nominal pipe inlet diameter.

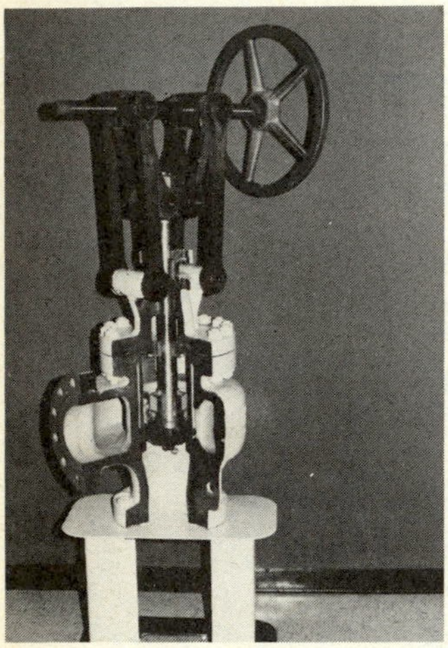

Figure 10.11 *Main steam stop valve.*

Loss Coefficient k for a Few Valves

1. Gate valve

The gate valve shown in Fig. 10.10a has a loss coefficient that is a function of the degree of opening w/D. Figure 10.12 presents a few values of k for four settings of w/D.

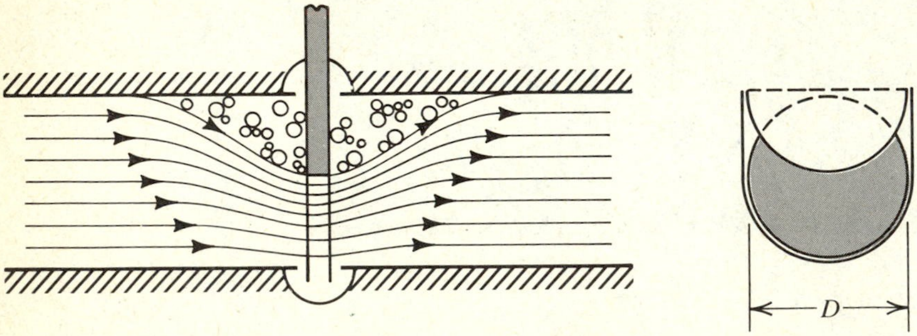

w/D	0	0.25	0.50	0.75
k	0.2	0.9	4.5	24.0

Figure 10.12 *Degree of opening w/D and loss coefficient k for a gate valve.*

2. Test tap

The loss coefficient for a test tap is given in Fig. 10.13 as a function of the angle-of-opening α.

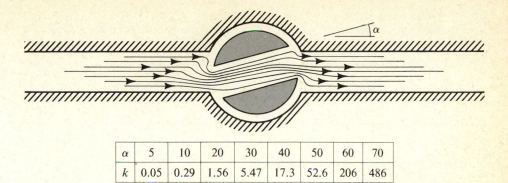

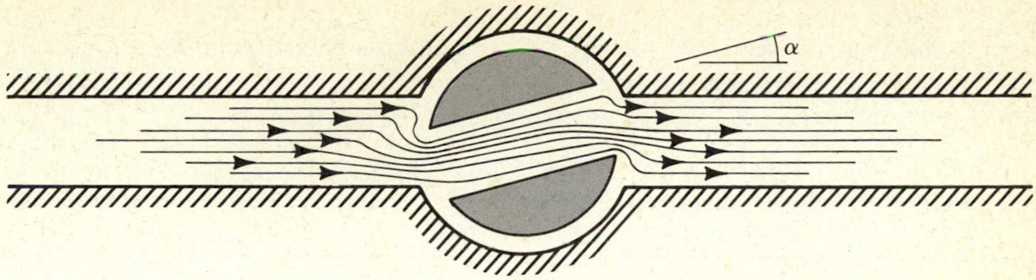

α	5	10	20	30	40	50	60	70
k	0.05	0.29	1.56	5.47	17.3	52.6	206	486

Figure 10.13 *Loss coefficient for a test tap.*

3. Throttle valve

A throttle valve is sometimes called a needle valve (see Fig. 10.10b). We can determine the loss coefficient by Abelev's formula

$$k = \frac{1000}{\exp(5.57\ \beta)} \tag{10.40}$$

where $\beta = \pi\alpha/180$ is the angle of revolutions in radians. Some typical values are:

α	20°	30°	40°	50°	60°	70°	80°	90°
k	142.2	55.21	20.23	7.85	2.88	1.11	0.41	0.11

4. Hinged valve and back valve

A hinged valve, like the butterfly valve, is shown in Fig. 10.10b. Values of the loss coefficient k are given in Fig. 10.14.

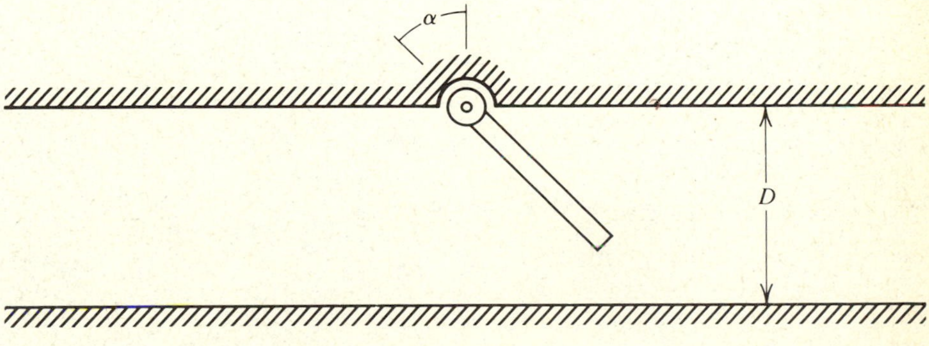

α	15	20	25	30	35	40	45	50	55	60	65	70
k	90	62	42	30	20	14	9.5	6.6	4.6	3.2	2.3	1.7

Figure 10.14 *Loss coefficient for hinged valve.*

5. Globe valve

A globe valve is one of the most popular ones used in marine engineering, because its loss coefficient is relatively small. The loss coefficient k for a fully opened globe valve is given below for a variety of pipe inner diameters*:

D (in.)	0.5	0.75	1.0	1.5	2	3	4	5	6
k	9.2	8.5	7.8	7.1	6.5	6.1	5.8	5.4	5.1

10.5.2 Elbows, Tees, and Such

Many pipe fittings, when installed in a line, reduce pressure, thereby requiring additional energy from the pump. A number of fittings and their loss coefficients are presented as follows.

1. Elbows and bends in pipes

The loss coefficient for a rounded elbow such as that shown in Fig. 10.15 is given by the empirical relationship

$$k = 0.13 + 1.83\left(\frac{r}{a}\right)^{3.5} \tag{10.41}$$

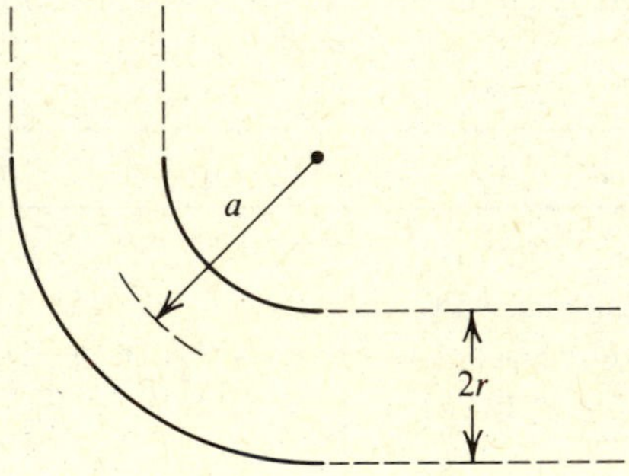

r/a	0.1	0.2	0.3	0.4	0.5	0.6	0.7	0.8	0.9	1.0
k	0.131	0.138	0.158	0.206	0.294	0.440	0.661	0.977	1.408	1.978

Figure 10.15 *Loss coefficient for rounded elbow (90°).*

If no value of (r/a) is given, then we can use

$$k_{45°} = 0.35 \tag{10.42}$$

$$k_{90°} = 0.75 \tag{10.43}$$

for standard elbows.

*Data from "Flow of Fluids," Tech. Rpt. 410, Crane Co., 1979.

2. Couplings, unions, and tees

Some other accepted loss coefficients are

$$k_{\text{coupling}} = 0.04 \tag{10.44}$$

$$k_{\text{union}} = 0.04 \tag{10.45}$$

$$k_{\text{tee}} = 0.04 \tag{10.46}$$

10.5.3 Sudden Contractions

In sudden contraction, the loss of energy is somewhat complicated. As the flow enters the small area, it accelerates, resulting in an effective jet. Here the flow is "pinched" to a minimum area: the *vena contracta*. Some energy is lost around the entrance where a reversed vortical flow exists. Typical values of the loss coefficient are given in Fig. 10.16. In using $(h_f)_{\text{minor}}$ of Eq. (10.39) for sudden contractions, the average velocity $\tilde{V}$ is taken downstream of the vena contracta, i.e., at the smaller area A_2.

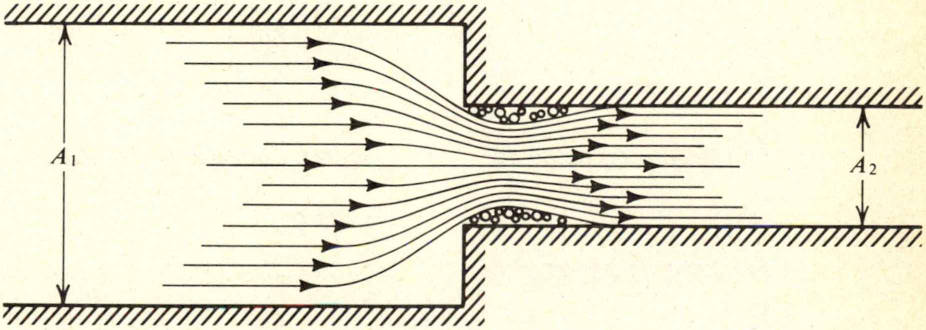

A_2/A_1	0.1	0.2	0.3	0.4	0.5	0.6	0.7	0.8	0.9	1.0
k	0.37	0.35	0.32	0.27	0.22	0.17	0.10	0.06	0.02	0

(Data obtained from several sources)

Figure 10.16 *Loss coefficient for sudden contraction.*

If the flow is from a reservoir into a circular pipe, then the shape of the pipe at the entrance is important. Figure 10.17 shows three types of entrances and the appropriate loss coefficient for each. If the pipe is not perpendicular to the reservoir, but is inclined at an angle α, as shown in Fig. 10.18, then we use

$$k = 0.5 + 0.3 \sin \alpha + 0.226 \sin^2 \alpha \tag{10.47}$$

to estimate the loss coefficient.

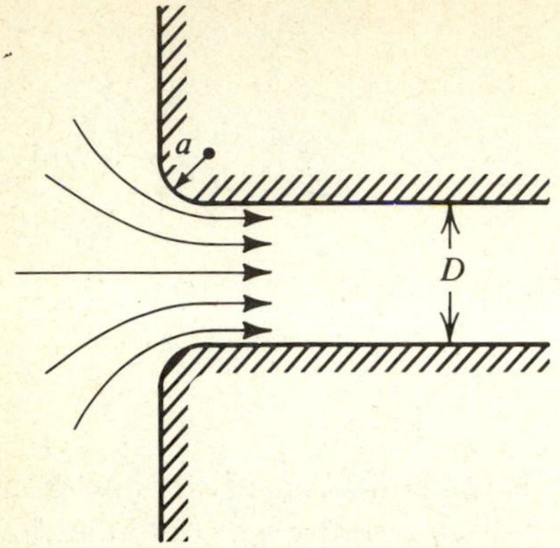

a/D	0.02	0.04	0.06	0.10	>0.15
k	0.28	0.24	0.15	0.09	0.04

(a)

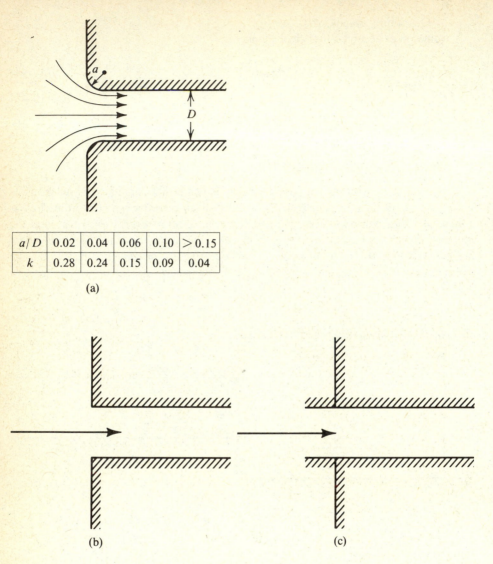

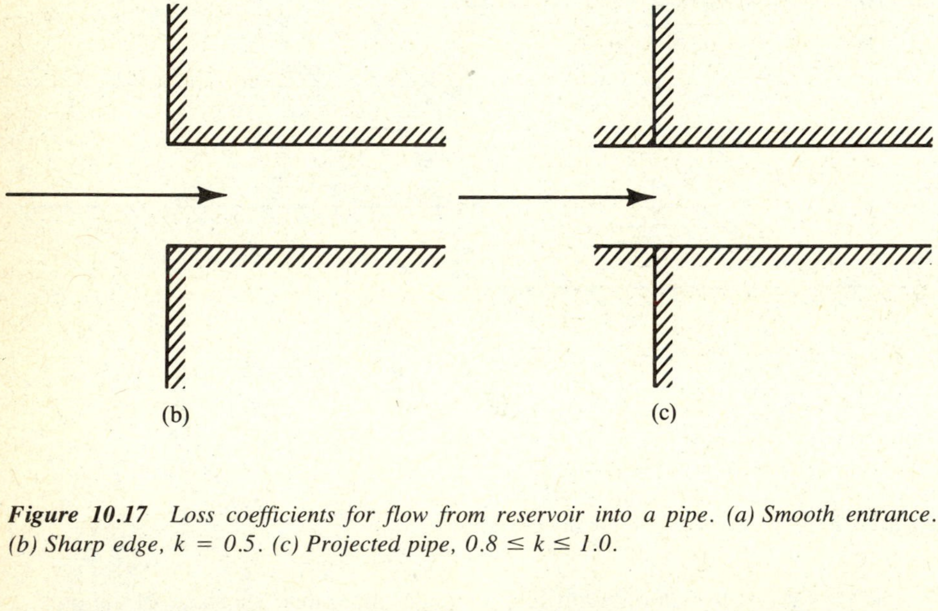

(b) (c)

Figure 10.17 *Loss coefficients for flow from reservoir into a pipe. (a) Smooth entrance. (b) Sharp edge,* $k = 0.5$. *(c) Projected pipe,* $0.8 \leq k \leq 1.0$.

10.5.4 Sudden Expansion

The energy lost per pound of fluid flow from a small cross-sectional area pipe to a large cross-sectional area pipe was given by Eq. (v) in Example 5.16:

$$(h_f)_{\text{minor}} = \frac{1}{2g}(\tilde{V}_{\text{i}} - \tilde{V}_e)^2 \tag{10.48}$$

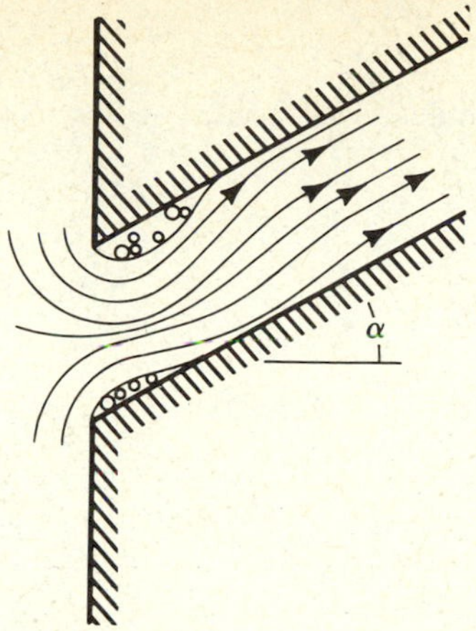

Figure 10.18 Flow into an inclined pipe.

From the (I.F.) continuity equation, Eq. (10.48) becomes

$$(h_f)_{\text{minor}} = \frac{\tilde{V}_i^2}{2g}\left(1 - \frac{A_i}{A_e}\right)^2 \tag{10.49}$$

In the extreme case of a sudden expansion from a pipe into a reservoir, we use the minor loss expression of Eq. (10.39). For the flow from a pipe into a reservoir (see Fig. 10.19), the loss coefficient k is unity. For this type of flow, it makes no difference if the outlet is streamlined, square, or the pipe projects into the reservoir: the loss coefficient k is unity for any geometry.

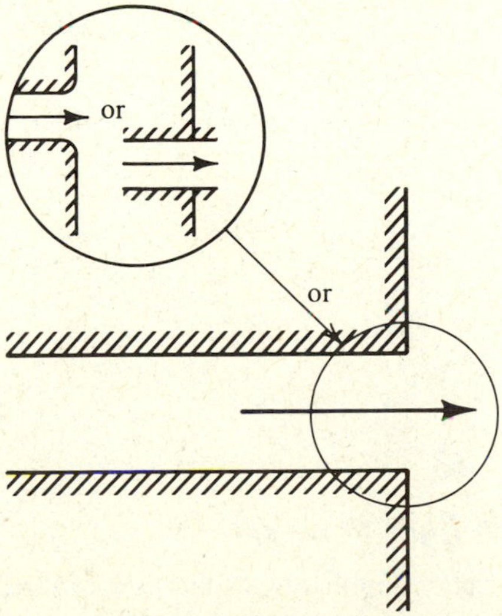

Figure 10.19 Flow from a pipe into a reservoir.

10.5.5 Gradual Expansion

The loss of energy per pound of fluid through a gradual short conical diffuser is given by

$$(h_f)_{\text{minor}} = k\,\frac{(\tilde{V}_i - \tilde{V}_e)^2}{2g} \tag{10.50}$$

where k is given for three angles α in Fig. 10.20. A fair approximation for the loss coefficient of short conical diffusers is

$$k \simeq \sin 2\,\alpha$$

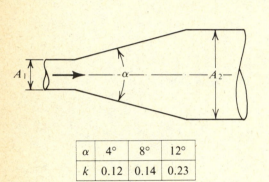

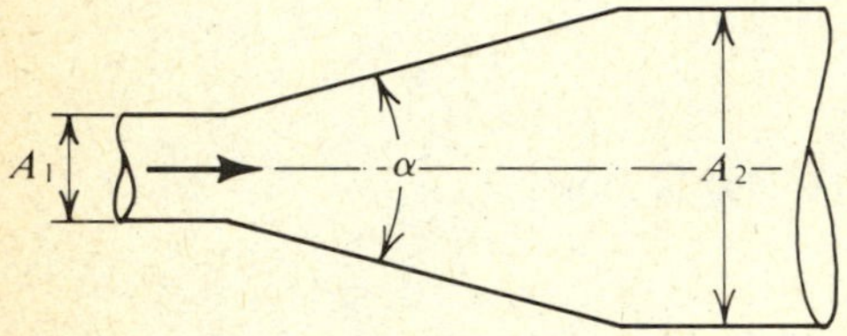

α	4°	8°	12°
k	0.12	0.14	0.23

Figure 10.20 *Loss coefficient for short diffuser.*

for $\alpha/2 < 25°$, and

$$k \simeq 1$$

for $\alpha/2 > 25°$. For a long conical diffuser, the loss coefficient must provide for friction losses. For long diffusers,

$$k = \frac{f}{2\tan\frac{\alpha}{4}}\left[\left(\frac{A_2}{A_1}\right)^2 - 1\right] + \sin\frac{\alpha}{2}\left(\frac{A_2}{A_1} - 1\right)^2 \tag{10.51}$$

gives fairly good results.

10.6 Energy Equation for Real Fluid Flow in a Pipe

Bernoulli's equation has been shown to be a relationship that treats the flow of ideal fluids. We have shown that when considering real systems with real fluids, we must insert a "head loss" term to account for the losses between the entrance and exit of the system. This head-loss term accounts for friction between the wetted surface and the fluid, for losses at the entrance and exit, and for losses due to abrupt turns and enlargements or contractions in the path of the fluid, or due to valves or other obstructions.

For those cases where losses occur, we can no longer consider Bernoulli's equation as being valid. Instead, we use the energy equation. The head-loss term may be a loss due exclusively to friction, in which case the pressure loss is directly proportional to the length, the roughness of the pipe, the viscosity and velocity of the fluid, and is inversely proportional to the pipe diameter. For laminar flows, the pressure drop is linear in velocity and nonlinear in pipe diameter by $1/D^2$. In turbulent flows, the pressure drop is nonlinear in velocity by $\tilde{V}^2$.

The energy equation as given by Eq. (5.103) is expressed in terms of the major and minor losses of energy by substituting Eqs. (10.30) and (10.39) into Eq. (5.103), so that

$$\left(\frac{p}{\gamma} + z + \frac{\tilde{V}^2}{2g}\right)_1 + ({}_1w_2)_{\text{mech}} = \left(\frac{p}{\gamma} + z + \frac{\tilde{V}^2}{2g}\right)_2 + f\frac{L}{D}\frac{\tilde{V}^2}{2g} + \Sigma k\frac{\tilde{V}^2}{2g} \qquad (10.52)$$

In Eq. (10.52), 1 is the upstream or high-energy station, $({}_1w_2)_{\text{mech}}$ is the shaft mechanical energy transfer (positive if added to the system), 2 is the downstream or low-energy station, $f(L/D)/(\tilde{V}^2/2g)$ is the major head loss due to pipe friction, and $\Sigma k(\tilde{V}^2/2g)$ is the sum of all minor losses between 1 and 2.

Example 10.6

Oil is flowing through 100 ft of a 1-ft inner diameter smooth pipe at an average velocity of 8 ft/s. The kinematic viscosity of the oil is 4×10^{-4} ft^2/s. Calculate the head loss h_f.

Solution:

Step 1.

The fluid is viscous and incompressible. The flow is steady, fully established, and axisymmetric.

Step 2.

Use the (I.F.) governing equations.

Step 3.

The appropriate governing equation is

- Energy equation:

$$\frac{\Delta p}{\gamma} + \Delta z + \frac{\Delta\tilde{V}^2}{2g} + ({}_1w_2)_{\text{mech}} = h_f \qquad \text{(i)}$$

The Reynolds number of the pipe flow is

$$R_D = \frac{\tilde{V}D}{\nu} \qquad \text{(ii)}$$

Example 10.6 *(Con't.)*

$$R_D = \frac{8.00 \times 1}{4 \times 10^{-4}}$$

$$= 2 \times 10^4 \qquad \text{(iii)}$$

which is in a zone neither laminar nor completely turbulent. From the Moody diagram of Fig. 10.6, the friction factor f is found to be

$$f = 0.0253 \qquad \text{(iv)}$$

The head loss h_f is calculated from the energy Eq. (i). But we do not know the pressure drop Δp or change of elevation Δz, so we must use a different way to calculate head loss. Since we know f, L, D, and $\tilde{V}$, we can use Eq. (10.30):

$$h_f = f \frac{L}{D} \frac{\tilde{V}^2}{2g}$$

$$= 0.0253 \times \frac{100}{1} \times \frac{64}{64.4}$$

$$= 2.514 \text{ ft} \qquad \text{(v)}$$

This completes the solution.

Example 10.7

Consider a recirculating flow in the power cycle shown in Fig. E10.7. Let the fluid flowing in 100 ft of smooth pipe have an average specific weight of 50 lbf/ft^3 and average dynamic viscosity of 5×10^{-4} lbf·s/ft^2. Suppose the head loss in the nuclear reactor is 75 ft·lbf/lbf of fluid, and the head loss in the steam generator is 15 ft. The pipe inside diameter is 6 in., and the volume rate of flow of fluid through the closed pipe system is 22.2 ft^3/s. There are four 90° standard elbows, one throttle valve of 50° revolution and a fully opened globe valve. Find the horsepower to circulate the flow.

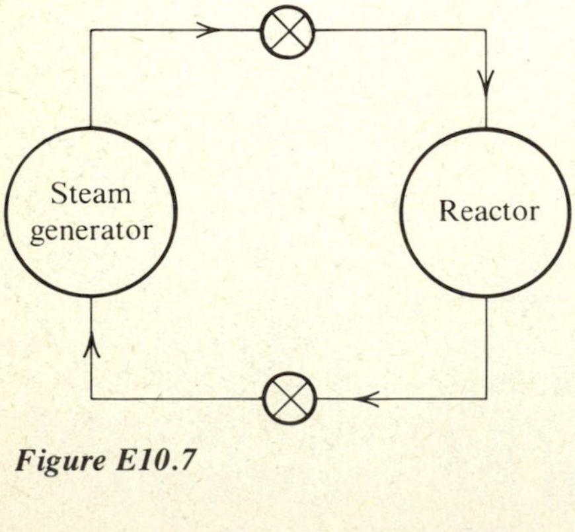

Figure E10.7

Example 10.7 *(Con't.)*

Solution:

Step 1.

The fluid is viscous and incompressible. The flow is steady, one-dimensional, and fully established.

Step 2.

Use the (I.F.) governing equations.

Step 3.

The appropriate governing equations are:

- Continuity

$$Q = \tilde{V}A \tag{i}$$

- Energy:

$$\frac{\Delta p}{\gamma} + \Delta z + \frac{\Delta \tilde{V}^2}{2g} + ({}_1w_2)_{\text{mech}} = f\frac{L}{D}\frac{\tilde{V}^2}{2g} + \Sigma k\frac{\tilde{V}^2}{2g} \tag{ii}$$

The one-dimensional continuity equation gives

$$\tilde{V} = \frac{Q}{A} \tag{iii}$$

such that the given values of flow rate Q and inner diameter of the pipe results in an average velocity

$$\tilde{V} = \frac{22.2}{\pi/16} = 113.2 \text{ ft/s} \tag{iv}$$

The Reynolds number R_D for this flow is

$$R_D = \frac{\gamma \tilde{V} D}{g\mu} \tag{v}$$

$$= \frac{50 \times 113.2 \times 0.5}{32.2 \times 5 \times 10^{-4}}$$

$$= 1.758 \times 10^5 \tag{vi}$$

which from Fig. 10.6 indicates the flow is neither in the laminar nor completely turbulent regime. For a smooth pipe, we use the lower curve in the Moody diagram that results in

$$f = 0.016 \tag{vii}$$

The power P required to maintain the flow is simply the power required to overcome the head losses due to major and minor pipe losses plus any other losses such as those due to the reactor and generator:

Example 10.7 *(Con't.)*

$$P = \frac{\gamma Q h_f}{550} \tag{viii}$$

where the head loss h_f is obtained from Eq. (ii):

$$h_f = \Sigma k \frac{\tilde{V}^2}{2g} + f \frac{L}{D} \frac{\tilde{V}^2}{2g} + w_\tau(\text{reactor}) + w_\tau(\text{generator})$$

$$= (4 \times 0.75 + 5.1 + 7.85) \frac{(113.2)^2}{64.4} + 0.016 \times \frac{100}{0.5} \tag{ix}$$

$$\times \frac{(113.2)^2}{64.4} + 75 + 15$$

$$= 3900.4 \text{ ft}$$

Substituting the head loss of Eq. (ix) into the expression for power of Eq. (viii) gives

$$P = \frac{50 \times 22.2 \times 3900.4}{550} = 7872 \text{ hp} \tag{x}$$

This completes the solution.

Example 10.8

Water of kinematic viscosity $\nu = 1 \times 10^{-5}$ ft²/s flows out of a tank at an average velocity of 10 ft/s through a 2 in. galvanized iron pipe of 100 ft length as shown in Fig. E10.8. Calculate the depth of water h in the tank given that the pressure head at 2 is negligible.

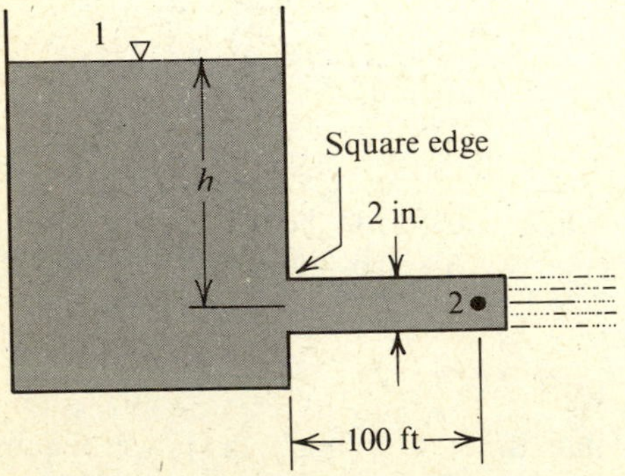

Figure E10.8

Example 10.8 *(Con't.)*

Solution:

Step 1.

The fluid is viscous and incompressible. The flow is steady, one-dimensional, and fully established.

Step 2.

Use the (I.F.) governing equations.

Step 3.

The appropriate form of the governing equation is:

- Energy:

$$\frac{\Delta p}{\gamma} + \Delta z + \frac{\Delta \tilde{V}^2}{2g} + ({}_1w_2)_{\text{mech}} = f\frac{L}{D}\frac{\tilde{V}^2}{2g} + \Sigma k\,\frac{\tilde{V}^2}{2g} \quad \text{(i)}$$

The Reynolds number R_D for the flow in the pipe is

$$R_D = \frac{\tilde{V}D}{\nu} \quad \text{(ii)}$$

$$= \frac{(10)(2/12)}{1 \times 10^{-5}} = 1.67 \times 10^5 \quad \text{(iii)}$$

which, according to the Moody diagram of Fig. 10.6, is neither a laminar nor a completely turbulent flow. From Table 10.1, the absolute roughness ϵ of galvanized iron is 0.0005 ft. Thus the relative roughness ϵ/D is

$$\frac{\epsilon}{D} = \frac{0.0005}{2/12} = 0.003 \quad \text{(iv)}$$

Using the relative roughness of Eq. (iv) and the Reynolds number R_D of Eq. (iii), we find the friction factor

$$f = 0.0265 \quad \text{(v)}$$

using Fig. 10.6.

The major loss $(h_f)_{\text{major}}$ due to pipe flow is

$$(h_f)_{\text{major}} = f\frac{L}{D}\frac{\tilde{V}^2}{2g} \quad \text{(vi)}$$

$$= 0.0265 \times \frac{100}{2/12} \times \frac{100}{64.4}$$

$$= 24.7 \text{ ft} \quad \text{(vii)}$$

The minor loss $(h_f)_{\text{minor}}$ is due to the square edge orifice at the pipe entrance. The loss coefficient k for this configuration is given by Fig. 10.17 as 0.5. Thus the minor loss $(h_f)_{\text{minor}}$ is found from Eq. (10.39) as

Example 10.8 *(Con't.)*

$$(h_f)_{\text{minor}} = 0.5\,\frac{100}{64.4} = 0.78 \text{ ft.} \tag{viii}$$

Applying the energy Eq. (i) and substituting in the calculated and given data results in

$$z_1 = h = \frac{\tilde{V}_2^2}{2g} + (h_f)_{\text{major}} + (h_f)_{\text{minor}} \tag{ix}$$

$$= \frac{100}{64.4} + 24.7 + 0.78$$

$$= 27.03 \text{ ft} \tag{x}$$

Thus, the depth of water is 27.03 ft to obtain an average velocity of 10 ft/s through 100 ft of 2 in. galvanized iron pipe.

This completes the solution.

10.7 Examples of Pipe Flow

It is appropriate to devote one section to the methodology of solving problems of fluid flow in closed conduits. The problems that we shall consider in this section apply solely to steady flows. The illustrations to be presented are (1) the siphon, (2) flow in a pipe series, and (3) flow in parallel pipes.

In the previous sections we learned how to evaluate the head loss h_f for steady flow in a constant diameter pipe. In this section, we shall complicate matters and combine pipes in assorted fashion, the desired goal being once again to calculate the head loss. Little will be new: it is principally a matter of bookkeeping.

10.7.1 The Siphon

A siphon is an elementary pipe flow shown in Fig. 10.21. The purpose of a siphon is to move fluid from one level to a second lower level by use of the gravitational potential force. Certain problems can occur at the summit s of the siphon because of low pressures in the line. The expression for the pressure at the summit of the siphon is obtained by applying the energy Eq. (10.52) to the flow between the free-surface station s_1 and the summit station s, with the result

$$p_s = -z_s - \frac{\tilde{V}^2}{2g}\left(f\frac{L}{D} + 2\right) \tag{10.53}$$

Note first of all that $p_s < 0$ in Eq. (10.53), and hence we must exercise care in the magnitude of the elevation z_s, as well as the velocity head so that the summit pressure

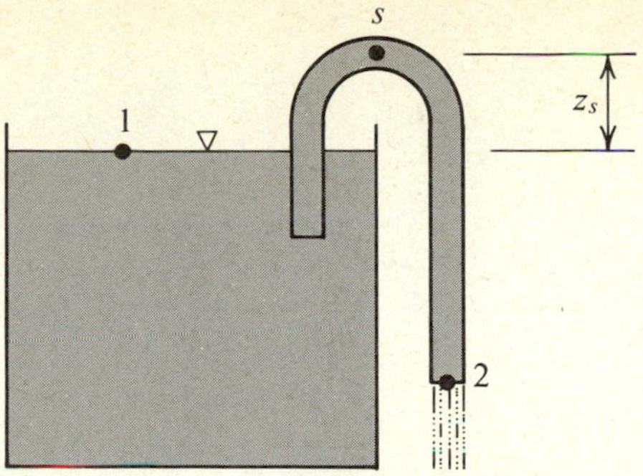

Figure 10.21 *The siphon.*

p_s does not equal or become less than the vapor pressure p_v of the liquid.* If this occurs, then cavitation results. If $p_s \leq p_v$, then the energy Eq. (10.53) is not applicable to the flow since it can no longer be considered incompressible and homogeneous. Furthermore, a siphon will not perform satisfactorily when the summit pressure p_s is *even close* to the vapor pressure. Any gas which may be entrained in the fluid will collect at the summit, and affect the length of the fluid in the right-hand side of the siphon column which in turn governs the pressure p_s. A practical solution, if this occurs, is to install a vacuum pump that can remove the collected gases at the summit.

10.7.2 Pipes in Series

Pipes in series require the sum of energy *losses* h_f, and the *equality of flow rates*. If pipes a, b, c are in series, then the total head loss h_{f_T} is

$$h_{f_a} + h_{f_b} + h_{f_c} = h_{f_T} \tag{10.54}$$

and the flow rates are

$$Q_a = Q_b = Q_c \tag{10.55}$$

We see that pipe flow in series is like electrical circuits in series; that is, we add resistances and equate current.

We can use three general types of analysis to treat fluid flow in pipes that are connected in series. They involve (1) calculation of the pressure drop $\Delta p/\gamma$, (2) calculation of the volume rate of flow Q for a given pipe, and (3) calculation of the size of the pipe. These three analyses are discussed below.

1. Calculation of pressure drop $\Delta p/\gamma$.

Consider the piping configuration of Fig. 10.22. The pressure drop $\Delta p/\gamma$ for the flow shown in the figure is calculated from the energy Eq. (10.52) in a manner following case A of Sec. 10.4.1. The appropriate energy equation is

*The vapor pressure of water at 68°F is 0.773 ft of water, absolute.

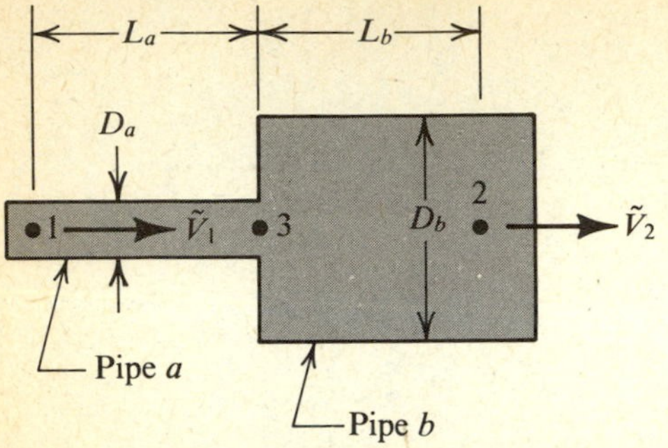

Figure 10.22 Pipes in series.

$$\frac{\Delta p}{\gamma} = \frac{p_1 - p_2}{\gamma} = (z_2 - z_1) + \frac{\tilde{V}_2^2}{2g}\left(2 + f_b \frac{L_b}{D_b}\right) + f_a \frac{L_a}{D_a}\frac{\tilde{V}_1^2}{2g} - \frac{\tilde{V}_1 \tilde{V}_2}{g} \tag{10.56}$$

where subscripts a and b refer to pipes a and b of Fig. 10.22, respectively, and subscripts 1 and 2 refer to stations in pipes a and b, respectively. Given the volume rate of flow Q, geometric properties D_a, D_b, L_a, L_b and the construction material of the pipe, it is a straightforward procedure to calculate the relative roughness ϵ/D and Reynolds number R_D for each pipe, as well as the friction factors f_a and f_b using the Moody diagram of Fig. 10.6. This is perhaps the simplest analysis one can make for pipes in series.

2. Calculation of the discharge Q for a given pipe.

Let us introduce two methods that might be used to calculate the volume rate of flow through a series of pipes.

(a) The first method involves assuming a value of the friction factor f given the pressure drop Δp in the pipe. Consider the flow in the straight pipe a of Fig. 10.22, and a step-by-step procedure for calculating Q:

Step 1.
Assume a value of friction factor f_a.

Step 2.
Solve for the average velocity $\tilde{V}_1$. From Eq. (10.56)

$$\frac{\Delta p}{\gamma} = \frac{p_1 - p_3}{\gamma} = f_a \frac{L_a}{D_a}\frac{\tilde{V}_1^2}{2g} \tag{10.57}$$

we solve for the average velocity $\tilde{V}_1$ given values of L_a, D_a, and $\Delta p/\gamma$.

Step 3.
Calculate the Reynolds number R_D.

Step 4.
Recalculate the friction factor f_a.

Using the given relative roughness $(\epsilon/D)_a$, we obtain an improved friction factor from the Moody diagram of Fig. 10.6.

Step 5.
Repeat Steps 2–4.

We can repeat the procedure until we obtain a satisfactory value of velocity such that we can find the flow rate Q using

$$Q = \tilde{V}_1 \frac{\pi}{4} D_a^2 \tag{10.58}$$

Example 10.9

Consider the two pipes in series shown in Fig. E10.9. Let pipe 1 be 500 ft long, with an inner diameter $D_1 = 2$ ft and an absolute roughness $\epsilon_1 = 0.003$ ft. Let pipe 2 be 1000 ft long with an inner diameter $D_2 = 3$ ft and an absolute roughness $\epsilon_2 = 0.001$ ft. Let the fluid have kinematic viscosity $\nu = 0.000010$ ft²/s. Given a potential head of 50 ft, calculate the volume rate of flow Q through the pipe system.

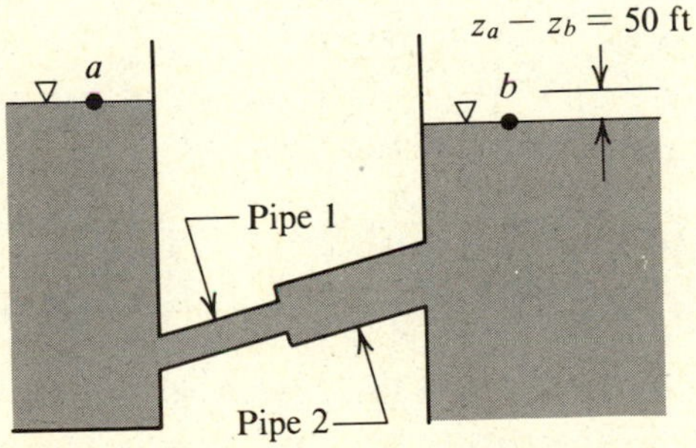

Figure E10.9

Solution:

Applying the energy Eq. (10.52), or Eq. (10.56), to the above flow gives

$$\left(\frac{p}{\gamma} + z + \frac{\tilde{V}^2}{2g}\right)_a = \left(\frac{p}{\gamma} + z + \frac{\tilde{V}^2}{2g}\right)_b + 0.5\frac{\tilde{V}_1^2}{2g} + \left(f\frac{L}{D}\frac{\tilde{V}^2}{2g}\right)_1 + \frac{(\tilde{V}_1 - \tilde{V}_2)^2}{2g} + \left(f\frac{L}{D}\frac{\tilde{V}^2}{2g}\right)_2 + \frac{\tilde{V}_2^2}{2g} \tag{i}$$

Example 10.9 *(Con't.)*

Substituting the one-dimensional continuity equation

$$\tilde{V}_2 = \tilde{V}_1 \left(\frac{D_1}{D_2}\right)^2 \tag{ii}$$

into Eq. (i) and letting

$$p_a = p_b = \tilde{V}_a = \tilde{V}_b = 0,\ z_a - z_b = 50 \text{ ft}$$

gives

$$50 = \frac{\tilde{V}_1^2}{2g}\left\{0.5 + f_1\frac{500}{2} + \left[1 - \left(\frac{2}{3}\right)^2\right]^2 + f_2\frac{1000}{3}\left(\frac{2}{3}\right)^4 + \left(\frac{2}{3}\right)^4\right\} \tag{iii}$$

Simplifying Eq. (iii) gives

$$50 = \frac{\tilde{V}_1^2}{2g}(1.006 + 250 f_1 + 65.84 f_2) \tag{iv}$$

Since the relative roughness

$$\left(\frac{\epsilon}{D}\right)_1 = 0.0005,\ \left(\frac{\epsilon}{D}\right)_2 = 0.001 \tag{v}$$

we first assume an f_1 and f_2 in the completely turbulent range: let

$$f_1 = 0.020 \tag{vi}$$

$$f_2 = 0.020 \tag{vii}$$

From the Moody diagram of Fig. 10.6, and using the expressions of Eqs. (v) and (vi), we obtain trial values of the Reynolds number:

$$R_{D_1} = 8 \times 10^4 \tag{viii}$$

$$R_{D_2} = 8 \times 10^6 \tag{ix}$$

Substituting Eqs. (vi) and (vii) into Eq. (iv), we evaluate $\tilde{V}_1$ to be

$$\tilde{V}_1 = 20.97 \text{ ft/s} \tag{x}$$

so that

$$\tilde{V}_2 = 9.32 \text{ ft/s} \tag{xi}$$

using Eqs. (ii) and (x).

With these values of velocity we calculate the Reynolds numbers as

Example 10.9 *(Con't.)*

$$R_{D_1} = \left(\frac{\tilde{V}D}{\nu}\right)_1 = 4.19 \times 10^6 \quad \text{(xii)}$$

and

$$R_{D_2} = \left(\frac{\tilde{V}D}{\nu}\right)_2 = 2.8 \times 10^6 \quad \text{(xiii)}$$

Using the results given by Eqs. (xii), (xiii), and (v) in Moody's diagram, we compute a second estimate of friction factor f

$$f_1 = 0.016 \quad \text{(xiv)}$$

$$f_2 = 0.02 \quad \text{(xv)}$$

Substituting the new values of friction factor into Eq. (iv) gives

$$\tilde{V}_1 = 22.32 \text{ ft/s} \quad \text{(xvi)}$$

and

$$\tilde{V}_2 = 9.92 \text{ ft/s} \quad \text{(xvii)}$$

From Eq. (ii) we calculate a second set of Reynolds numbers to be

$$R_{D_1} = \frac{22.32 \times 2}{1 \times 10^{-5}} = 4.46 \times 10^6 \quad \text{(xviii)}$$

$$R_{D_2} = \frac{9.92 \times 3}{1 \times 10^{-5}} = 2.97 \times 10^6 \quad \text{(xix)}$$

Again, we use the Moody diagram to find a third value of friction factor

$$f_1 = 0.016 \quad \text{(xx)}$$

$$f_2 = 0.02 \quad \text{(xxi)}$$

which agree with the second estimates given by Eqs. (xiv) and (xv), respectively.

Thus, the volume rate of flow Q is

$$Q_1 = \tilde{V}_1 \left(\frac{\pi}{4}\right) D_1^2 = 22.32 \times \frac{\pi}{4} \times 4 = 70.13 \text{ ft}^3\text{/s} \quad \text{(xxii)}$$

Example 10.9 *(Con't.)*

and using the results of Eq. (xvii),

$$Q_2 = \tilde{V}_2 \left(\frac{\pi}{4}\right) D_2^2$$

$$= 9.92 \times \frac{\pi}{4} \times 9 \qquad \text{(xxiii)}$$

$$= 70.14 \text{ ft}^3/\text{s}$$

We want $Q_1 = Q_2$, and the above results indicate that they may be of acceptable value.

This completes the solution.

(b) The second method is to assume a velocity instead of a value of the friction factor. It works just as fast as method (a). Recall that we are given $\Delta p/\gamma$.

Step 1.
Assume an average velocity $\tilde{V}$.
Step 2.
Calculate the Reynolds number R_D.
Step 3.
Evaluate the friction factor f using Fig. 10.6.
Step 4.
Evaluate the pressure drop $\Delta p/\gamma$ using Eq. (10.57).
Step 5.
Compare $\Delta p/\gamma$ of Step 4 with the given value and evaluate a second $\tilde{V}$.
Step 6.
Repeat Steps 1–6 until an appropriate value of $\tilde{V}$ is found.

Example 10.10

Consider three circular pipes in series of lengths 200 m, 300 m, and 100 m, as shown in Fig. E10.10. Let their inner diameter be 1 m, 2 m, and 0.5 m, respectively, and their relative roughness 0.003, 0.004, and 0.005, respectively. If the total pressure drop through the three pipes is 200 kPa and the change in elevation from inlet to exit of the three pipes is 10 m, calculate the volume rate of flow Q through the system if $\rho = 1000 \text{ kg/m}^3$, and $\nu = 1.0 \times 10^{-6} \text{ m}^2/\text{s}$.

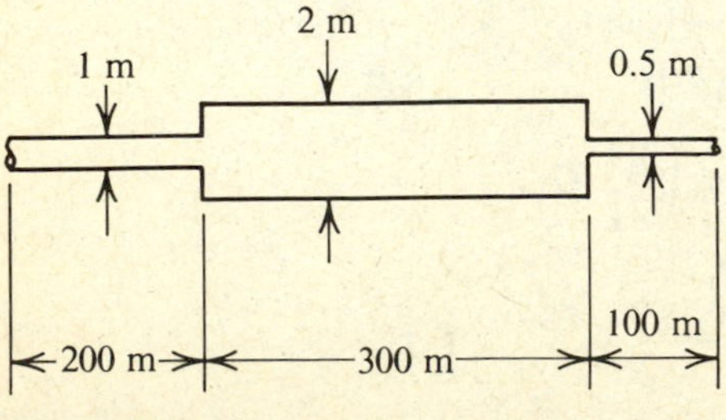

Figure E10.10

Example 10.10 *(Con't.)*

Solution:
The energy Eq. (10.52) gives

$$\frac{\Delta p}{\gamma} + \Delta z = h_f \tag{i}$$

or

$$h_f = \frac{200{,}000}{(1000)(9.81)} + 10 = 30.39 \text{ m} \tag{ii}$$

Thus for all three pipes in series

$$h_f = \left(f\frac{L}{D}\frac{\tilde{V}^2}{2g}\right)_a + \left(f\frac{L}{D}\frac{\tilde{V}^2}{2g}\right)_b + \left(f\frac{L}{D}\frac{\tilde{V}^2}{2g}\right)_c \tag{iii}$$

Since

$$\tilde{V}_b = \left(\frac{D_a}{D_b}\right)^2 \tilde{V}_a = (\tfrac{1}{2})^2 \tilde{V}_a \tag{iv}$$

$$\tilde{V}_c = \left(\frac{D_a}{D_c}\right)^2 \tilde{V}_a = \left(\frac{1}{0.5}\right)^2 \tilde{V}_a \tag{v}$$

we can express Eq. (iii) as

$$h_f = \frac{\tilde{V}_a^2}{2g}\left[(f_a)\left(\frac{200}{1}\right) + (f_b)\left(\frac{300}{2}\right)\cdot\left(\frac{1}{4}\right) + (f_c)\left(\frac{100}{0.5}\right)\left(\frac{1}{0.5}\right)^2\right] \tag{vi}$$

Step 1.
Assume a value of the friction factors.

Let us select the three friction factors in the fully rough region:

$$f_a = 0.026, \quad f_b = 0.024, \quad f_c = 0.03 \tag{vii}$$

Step 2.
Evaluate the velocity.

Substituting Eqs. (iii) and (vii) into our energy Eq. (vi) for series pipe flow gives

$$30.39 = \frac{\tilde{V}_a^2}{2g}(5.2 + 0.9 + 24) \qquad \text{(viii)}$$
$$\tilde{V}_a = 4.45 \text{ m/s}$$

Step 3.
Evaluate the Reynolds number.

Example 10.10 *(Con't.)*

Using our first estimate of the velocity $\tilde{V}_a$ from Eq. (viii)

$$\left.\begin{aligned} D_a &= 4.45 \times 10^6 \\ D_b &= \tfrac{1}{2}R_a = 2.225 \times 10^6 \\ D_c &= 8.9 \times 10^6 \end{aligned}\right\} \qquad \text{(ix)}$$

Step 4.
Use the Moody diagram to find a second estimate of the friction factors:
From Fig. 10.6,

$$f_a = 0.0255, \quad f_b = 0.027, \quad f_c = 0.03 \qquad \text{(x)}$$

Substituting these second estimates of the friction factor into Eq. (vi) gives

$$\tilde{V}_a = 4.45 \text{ m/s} \qquad \text{(xi)}$$

which agrees with the first estimate of Eq. (viii). Thus the flow rate Q is

$$Q = 4.45 \times \frac{\pi}{4} \times (1)^2 = 3.495 \text{ m}^3\text{/s}$$

This completes the solution.

10.7.3 Flow in Parallel Pipes

When two or more pipes are connected in parallel, the energy loss through one pipe must be the same as the energy loss for each of the others. Thus, if there are pipes 1, 2, and 3 connected in parallel, as shown in Fig. 10.23, then

$$h_{f_1} = h_{f_2} = h_{f_3} \qquad (10.59)$$

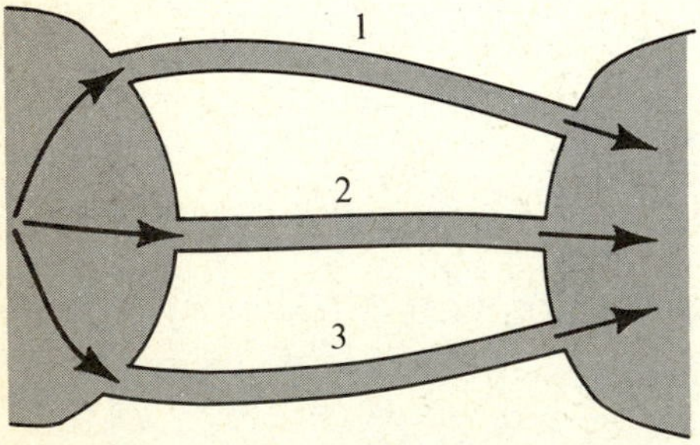

Figure 10.23 *Pipes in parallel.*

In addition, the total flow rate Q for the parallel pipe system is the algebraic sum of the flow rates through each pipe:

$$Q_{total} = Q_1 + Q_2 + Q_3 \tag{10.60}$$

The results shown here are the opposite to the relations shown for pipes in series.

Assuming the geometric and material properties of the pipes are known and the loss coefficients k for each pipe are given, we find from Eq. (10.59) that the velocity ratios are

$$\frac{\tilde{V}_1}{\tilde{V}_2} = \frac{\tilde{V}_1}{\tilde{V}_3} = \sqrt{\frac{f_2(L_2/D_2) + \Sigma k_2}{f_1(L_1/D_1) + \Sigma k_1}} = \sqrt{\frac{f_3(L_3/D_3) + \Sigma k_3}{f_1(L_1/D_1) + \Sigma k_1}}$$
$$= \sqrt{\frac{f_3(L_3/D_3) + \Sigma k_3}{f_2(L_2/D_2) + \Sigma k_2}} \tag{10.61}$$

Example 10.11

Consider two pipes connected in parallel between two reservoirs shown in Fig. E10.11. The properties of pipe 1 are an inner diameter $D_1 = 1$ ft, length $L_1 =$ 1000 ft, and absolute roughness $\epsilon_1 = 0.002$ ft. The properties of pipe 2 are an inner diameter of 18 in., length of 1000 ft, and absolute roughness $\epsilon_2 = 0.003$ ft. The change in the surface elevations of the two reservoirs is 50 ft. Calculate the total flow of water at 73°F flowing through the two pipes.

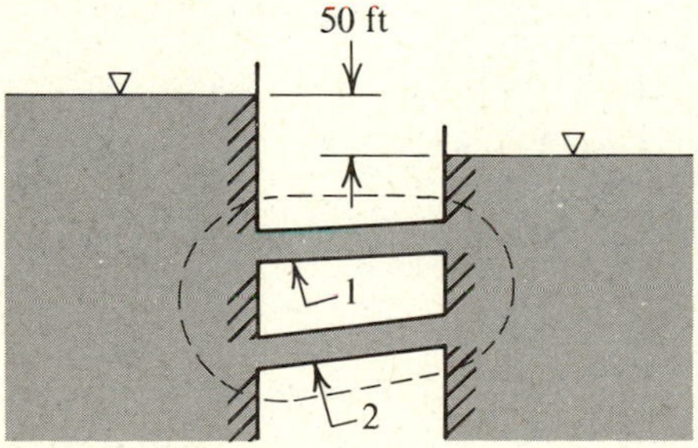

Figure E10.11

Solution:

We neglect the minor losses since the pipe lengths are sufficiently large that their effect on the flow is negligible compared to the major loss. From the energy Eq. (10.52), we find for pipe 1

$$50 = f_1\left(\frac{1000}{1}\right)\frac{\tilde{V}_1^2}{2g} \tag{i}$$

An iteration is once again necessary. Assume

$$f_1 = 0.023 \tag{ii}$$

Example 10.11 *(Con't.)*

and substitute it into Eq. (i). The velocity $\tilde{V}_1$ is calculated to be

$$\tilde{V}_1 = 12 \text{ ft/s} \tag{iii}$$

which gives a Reynolds number

$$R_{D_1} = \left(\frac{\tilde{V}D}{\nu}\right)_1 = \frac{12 \times 1}{1 \times 10^{-5}} = 1.2 \times 10^6 \tag{iv}$$

For a Reynolds number of 1.2×10^6 and relative roughness ϵ/D of 0.002, the friction factor is found to be 0.023, which agrees with that given by Eq. (ii).

The flow rate Q is evaluated as

$$Q_1 = (\tilde{V}A)_1 = 12 \times \frac{\pi}{4} = 9.43 \text{ ft}^3\text{/s} \tag{v}$$

For pipe 2, we repeat the procedure.

$$50 = f_2\left(\frac{2000}{1.5}\right)\frac{\tilde{V}_2^2}{2g} \tag{vi}$$

Assume a trial value of friction factor:

$$f_2 = 0.023 \tag{vi}$$

and substitute it into Eq. (vi). The velocity $\tilde{V}_2$ is calculated to be

$$\tilde{V}_2 = 10.25 \text{ ft/s} \tag{vii}$$

which gives a Reynolds number

$$R_{D_2} = \left(\frac{VD}{\nu}\right)_2 = \frac{10.25 \times 1.5}{1 \times 10^{-5}} = 1.54 \times 10^6 \tag{viii}$$

For a Reynolds number of 1.54×10^6 and relative roughness ϵ/D of 0.002, a second value of the friction factor f_2 is found to be 0.023 which agrees with that given by Eq. (vi).

The flow rate Q going through pipe 2 is evaluated as

$$Q_2 = (\tilde{V}A)_2 = 10.25 \times \frac{\pi}{4}(1.5)^2 = 18.11 \text{ ft}^3\text{/s} \tag{ix}$$

The total flow rate for the parallel pipe is found using Eq. (10.60):

$$Q_{\text{total}} = Q_1 + Q_2 = 9.43 + 18.1 = 27.54 \text{ ft}^3\text{/s} \tag{x}$$

This completes the solution.

Example 10.12

Two pipes are in parallel. Pipe 1 has a pipe cross-sectional area of 1 ft^2, length of 2000 ft, and friction factor $f_1 = 0.020$, and allows gasoline to flow through it at a volume rate of 10 ft^3/s. Pipe 2 has a pipe cross-sectional area of 2 ft^2, a length of 3000 ft, and friction factor $f_2 = 0.025$. Find the average velocity of the gasoline through the second pipe.

Solution:

The average velocity of the gasoline through the first pipe is

$$\tilde{V}_1 = \frac{Q_1}{A_1} = \frac{10}{1} = 10 \text{ ft/s} \tag{i}$$

Using Eq. (10.61), the average velocity in the second pipe $\tilde{V}_2$ is related to the average velocity in the first pipe $\tilde{V}_1$ by

$$\tilde{V}_2 = \tilde{V}_1 \sqrt{\frac{f_1 L_1 D_2}{f_2 L_2 D_1}} \tag{ii}$$

Substituting the given values into Eq. (ii) gives

$$\begin{aligned} \tilde{V}_2 &= 10 \sqrt{\frac{0.020 \times 2000 \times 1.414}{0.025 \times 3000}} \\ &= 8.68 \text{ ft/s} \end{aligned} \tag{iii}$$

The gasoline will flow through the second parallel pipe with an average velocity of $\tilde{V}_2 = 8.68$ ft/s.

This completes the solution.

Hence, the analysis of pipes in series or parallel may require some initial estimate of one of the parameters, but this estimate can quickly be refined to an accurate value by iteration techniques. An excellent source for more complex systems is given by H. Cross [10.4].

References

10.1 White, Frank H., *Fluid Mechanics*, McGraw-Hill, New York, 1979.

10.2 Walworth Valves, Catalog 130, One Decker Square, Bala Cynwyd, Pa. 19004.

10.3 Pollard, R. S., "Industrial Steam Trapping Course," Yarway, Yarway Corp., Philadelphia, 1963.

10.4 Cross, H., "Analysis of Flow in Networks of Conduits or Conductors," U. of Il. Engr. Expt. Sta., Bull. 286, 1936.

Study Questions

10.1 What is the Reynolds number range for laminar, transitional, and turbulent flow in a pipe?
10.2 What is the axial velocity profile, and shear stress profile of laminar flow in a pipe?
10.3 What does G physically represent in Eq. (10.17)?
10.4 What proportion of the maximum axial velocity of laminar flow in a pipe is the average velocity?
10.5 How does one determine the friction factor for pipes that are other than circular?
10.6 Where is $\tilde{V}$ measured for the minor loss expression? Upstream or downstream of the obstruction, and why?
10.7 What is the minor loss coefficient for pipe flow of water into the atmosphere?
10.8 Plot k versus α for $0 \leq \alpha \leq 90°$ for a conical diffuser.
10.9 What is the maximum height that point s of Fig. 10.21 could be above the reservoir free surface without stopping discharge of the siphon?
10.10 In solving a pipe system problem in series for the volume rate of flow Q_1, explain how the energy equation and the continuity equation are used to obtain an expression that contains the kinetic energy $\tilde{V}^2/2g$ and f_1, f_2.

Problems

10.1 Medium lubricating oil, $S = 0.86$, is pumped through 1000 ft of horizontal 2 in. diameter pipe at the rate of 0.0436 ft³/s. If the pressure drop is 30 psi, what is the viscosity of the oil?

10.2 Calculate the loss of energy of water at 20°C flowing through 200 m of 100 mm diameter galvanized steel pipe at a flow rate of 50 l/s?

10.3 Water flows through a smooth square pipe with an average velocity of 10 ft/s. The pipe's internal cross section is 6 in. × 6 in. The pipe makes an angle of 60° with the vertical and the flow is up. Determine the pressure drop in an 800 ft length of pipe given $\rho = 1.937$ slug/ft³, $\mu = 20.92 \times 10^{-6}$ lbf·s/ft².

10.4 Determine the pressure drop in a smooth pipe 100 m long and 50 mm in diameter if oil of density 800 kg/m³ and viscosity 0.039 N·s/m² flows at a rate of 3 m³/s.

10.5 A horizontal 4 in. pipe is 200 ft from the discharge side of a pump. Determine the outlet pressure of the pump if the water discharges to the atmosphere, and the pipe has the following fittings: two globe valves fully open and four 90° elbows. The average velocity of the water in the pipe is 7 ft/s, and the density and viscosity are 1.937 slug/ft³ and 20.29×10^{-6} lbf·s/ft². Assume a smooth pipe.

10.6 Oil of viscosity 5×10^{-6} ft²/s and density 57 lbm/ft³ flows through a 30 ft long tube that is 2 in. in diameter. Calculate the pressure drop if the pipe is (a) wrought iron, and (b) smooth if $R_D = 1.5 \times 10^6$ and $Q = 100$ ft³/min.

10.7 Air flows at 100°F in a pipe at the rate of 0.2 lbm/s. What must be the pipe diameter just to keep the flow laminar in the pipe?

10.8 Water at 150°F flows in a 6 in. diameter smooth pipe at a velocity of 80 ft/s as shown in Fig. P10.8. If the pressure at the exit is 20 psia, what is the pressure at A, assuming that the 90° elbow adds 16 ft effective length to the pipe length according to an elbow loss formula $L/D = 32$.

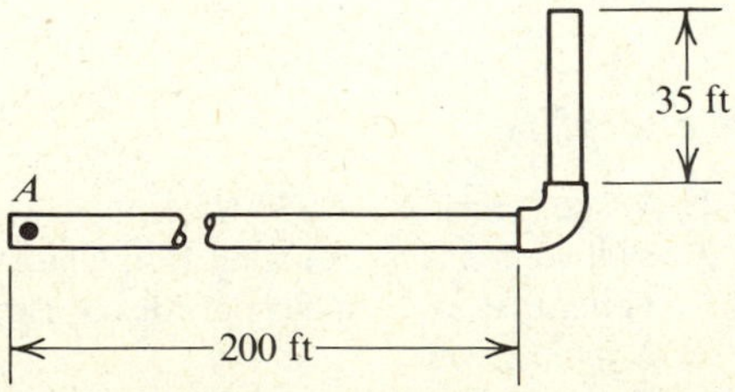

Figure P10.8

10.9 Calculate the flow rate of fuel oil of viscosity 1×10^{-5} m^2/s in a horizontal commercial steel pipe 200 mm in diameter and 200 m long if the pressure drop across the 200 m length is 5 m.

10.10 On board the British liner HMS Pipeflow, the junior engine officer receives a calibrated section of smooth pipe. An attached tag is torn, and all that remains is a statement $f = 0.0200$ in turbulent flow at a Reynolds number _____. What is the value of the missing Reynolds number?

10.11 The tag in Prob. 10.10 also says $f = 0.0600$ in laminar flow at $R =$ _____. What is the value of the missing Reynolds number?

10.12 In a test with the pipe of Prob. 10.10, let $R_D = 10^5$ and $f = 0.018$. The pipe is 10 ft long and 1 in. in diameter. If seawater ($\gamma = 64$ lbf/ft^3, $\mu = 2.391 \times 10^{-5}$ slug/ft·s) is flowing through the pipe, calculate the flow rate.

10.13 In Prob. 10.12, what is the pressure loss (psi)? What horsepower is required to pump the seawater through a similar pipe 100 times as long as the test pipe?

10.14 Water at 15°C moves through 200 m of cast iron pipe at a rate of 100 liter/s. What diameter pipe is required if the head loss is to be 10 m?

10.15 Calculate the loss of energy for (a) sudden expansion from 50 mm diameter pipe to a 100 mm diameter pipe, and (b) sudden contraction from 100 mm diameter pipe to 50 mm diameter pipe, given a flow rate of 0.05 m^3/s.

10.16 Water at 59°F flows through a horizontal 6 in. diameter asphalted cast iron pipe with a velocity of 4 ft/s. You find the pressure drop is 0.7 psi in a length of 120 ft. If this pipe is placed in a vertical position and the flow is up at the same speed, what is the pressure drop?

10.17 A uniform circular pipe of radius a has a fixed coaxial circular rigid core of radius b. Discuss the case where $h = a - b$ is small and compare the axial velocity distribution and rate of discharge Q with flow between two parallel plates.

10.18 Consider a fluid moving through a pipe whose cross section is a narrow rectangle of depth h. Assume that the pressure difference, length, and viscosity are fixed. Find the rate of discharge Q per unit width.

10.19 Show that the force of resistance F on the wetted surface of a pipe is given by $F = CD\bar{p}_m$, where C is the perimeter of the pipe, D is the diameter, and $\bar{p}_m$ is the mean value of the frictional stress on the wall.

10.20 Calculate the head loss for fresh water at 70°F flowing at 5 ft/s through a concrete pipe of diameter 2 ft and length 1200 ft. If no minor losses occur, what horsepower pump is required to move the fluid along 1200 ft of horizontal pipe?

10.21 Repeat Prob. 10.20 except that the last 200 ft of the pipe moves the water up a 200-ft high hill to a reservoir. Neglect minor losses.

10.22 The kinematic viscosity of water flowing through a 100 mm diameter pipe is 1.05×10^{-6} m^2/s. Find the head loss (m) across 900 m of pipe if the discharge is 0.05 m^3/s and the relative roughness is 0.005.

10.23 A pipe carrying fresh water has a 75 mm diameter. When the discharge rate is 3.7×10^{-2} m^3/s, the head loss between pressure taps 60 m apart is 90 m. Find the friction factor f.

10.24 Water of viscosity 1.14×10^{-6} m^2/s flows through a 610 m long 200 mm diameter pipe. The major loss is 15.5 m for a discharge of 0.0713 m^3/s. Estimate what the material of the pipe is.

10.25 Oil is being pumped from a truck to a tank 10 ft higher than the truck through a 2 in. galvanized pipeline 75 ft long. If the pressure at the discharge side of the pump is 15 psi, at what rate (ft^3/s) is oil flowing through the pipe? The oil has a kinematic viscosity of 0.001 ft^2/s and $\gamma = 0.92$ lbf/ft^3.

10.26 How much oil is flowing through a venturi meter with a 4 in. throat which is placed in a smooth pipeline 6 in. in diameter? A mercury/oil manometer connected between the upstream section and the throat shows a difference of 8 in. The specific gravity of oil is 0.95.

10.27 An ice-skating rink is flooded by means of a 2 in. diameter rubber hose 75 ft long with a nozzle of 0.75 in. diameter attached at one end. If the nozzle is held level 3 ft from and above the ice, the jet strikes the ice 15 ft away from the nozzle orifice. Find the flow from the nozzle and the pressure required at the valve at the upstream end of the hose. The rubber hose may be assumed similar to welded steel pipe. The kinematic viscosity of water is 1.2×10^{-5} ft²/s. Let the absolute roughness of the rubber hose be 0.0003 in.

10.28 A pipe 2000 m long and 0.5 m in diameter connects two reservoirs having a difference in level of 100 m. Calculate the discharge in liters per day if $f = 0.0047$ (consider major friction losses only). Assume laminar flow.

10.29 The roughness ratio of a 50 mm diameter pipe is 0.002. Estimate the kinematic viscosity of the fluid flowing through the pipe if the head loss is 900 mm over 20 m of pipe, and the discharge rate is 3.0×10^{-3} m³/s.

10.30 Water is to be pumped through a city's sewage concrete pipelines to purge the system. If the desired flow rate is 0.7 ft³/s, and the pressure drop is 150 ft over 200 miles of pipe, what is the diameter of the pipe given $\rho = 1.935$ slug/ft³ and $\mu = 1 \times 10^{-5}$ lbf·s/ft²?

10.31 The QE2 wishes to test its freshwater lines, so it sections off a 40 ft length of 1 in. drawn tubing and installs pressure gauges at both ends. If the maximum pressure drop that the gauges can detect is 50 psi, what is the maximum flow rate of freshwater at 15°C?

10.32 A 0.4 mm pipe is taken off the line in a nuclear power cooling line, and after testing, it is found that $f = 0.027$ for $R_D = 7 \times 10^5$. (a) If the head loss for the pipe is 7 m, calculate the head loss if the pipe length were halved. (b) What is the value of the absolute roughness? (c) What is the probable material of the pipe?

10.33 Water at 15°C flows through a 0.4-m diameter pipe that is 50 m long and horizontal. If the average velocity of the water is 5 m/s, what is the pressure difference (kPa)? Assume a smooth pipe.

10.34 Water at 15°C flows through a horizontal pipe made of wood. If the pipe is 50 mm in diameter and is 20 m long, determine the friction factor f and flow rate if the average velocity is 5 m/s.

10.35 Water at 15°C flows through a pipe that is 3 m in diameter and 10 m long. If $\epsilon = 0.13$ m and the energy loss is 10 m, what is the volume rate of flow of the water?

10.36 The flow rate of water at 15°C through a brick pipe that is 1 m in diameter and 300 m long is 0.4 m³/s. Find the value of the friction factor.

10.37 A ship's captain stateroom is 85 ft above the ship's fresh-water pump. There is 250 linear feet of smooth ½ in. pipe between the pump and the captain's shower. The sum of the loss coefficients for all fittings between the pump and the shower is 45. The captain takes Hollywood showers using 7 gal/min of water. (a) What is the pressure loss due to turbulence and friction between the pump and the shower? (b) What is the total change in pressure between the pump discharge and the shower? (c) How low can the discharge pressure of the pump be before there is no water pressure in the captain's shower?

10.38 In Prob. 10.37, suppose the pipe is insulated. The pump's efficiency is 60%, and the pump takes suction at 0 psi. Suppose there is no significant change in kinetic energy of the fluid as it passes through the pump. Let the suction pressure be measured at the same elevation as the pump inlet. If 10 psi pressure must be maintained in the captain's shower, what is the minimum horsepower motor which can be used to drive the pump?

10.39 It is desired to determine the loss coefficient for a new valve, illustrated in Fig. P10.39. The pressure drop across the valve is measured with a manometer using glycerin. The flow rate through the valve is measured with a sharp edge orifice (orifice diameter = 0.8 in., flow discharge coefficient = 0.61). The orifice has vena contracta taps connected to a mercury filled manometer. The fluid flowing through the pipe is fresh water and the mercury manometer reading is 4 in. of mercury and the glycerin manometer reading is 5 in. of glycerin. The piping system and the valve have a 2 in. diameter. (a) What is the average velocity of the fresh water? (b) What is the loss coefficient for the new valve?

10.40 The surface blowdown piping from a ship's boiler leads from the steam drum of the boiler to a sea chest (i.e., opening to the sea) that is 20 ft below the center of the steam drum.

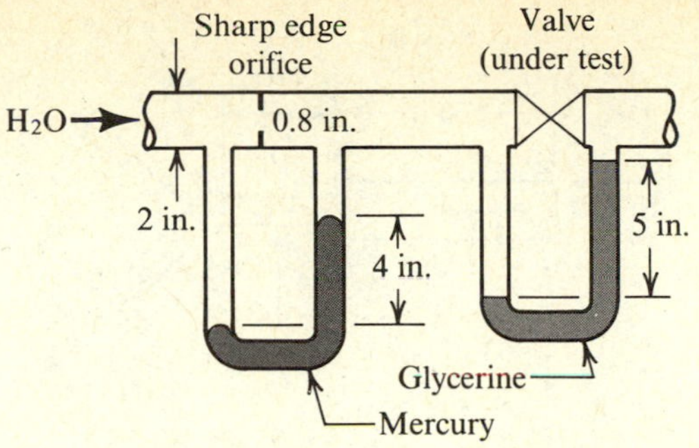

Figure P10.39

The sea chest is 10 ft below the waterline of the ship. The required average velocity for a satisfactory blowdown is 15 ft/s. The fluid in the blowdown piping is feedwater (γ = 62.4 lbf/ft^3) at 200°F. The blowdown system consists of the following pipe and fittings: two std. tees, 100 ft of 1.5 in. commercial steel pipe, two globe valves (wide open when in use), one angle valve (wide open when in use), one gate valve (wide open when in use), 16 std. elbows, and four 45° elbows. What is the minimum pressure at point 1 (the center of the steam drum) for a satisfactory blowdown? (Refer to Fig. P10.40.)

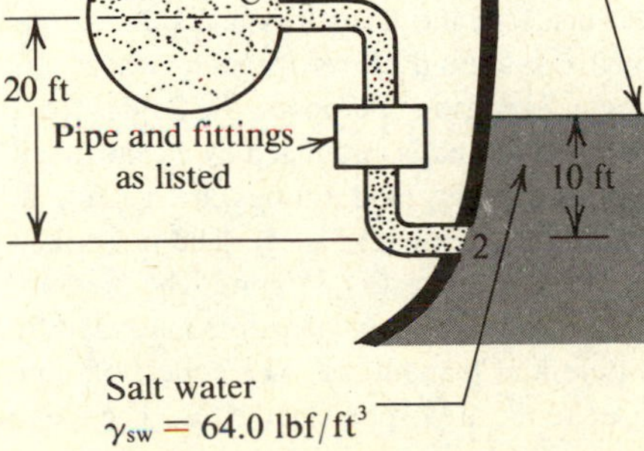

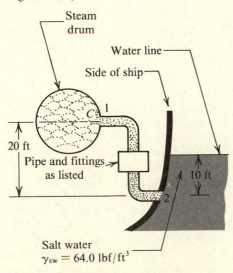

Figure P10.40

10.41 Water flows through an 8 × 12 in.^{2} section of galvanized pipe 500 ft long at an average velocity of 14 ft/s. What is the pressure difference between the inlet and outlet?

10.42 As shown in Fig. P10.42, the volumetric flow rate of water at standard conditions in the smooth pipeline shown is 1.96 ft^3/s. The pipe suddenly contracts to 6 in. diameter, turns a 45° bend and suddenly expands back to a 12 in. diameter section. Find (a) Reynolds number in the 500 ft section, (b) head loss in the 500 ft section, (c) head loss due to the sudden contraction, (d) head loss in the 6 in. pipe section (f = 0.0138), (e) head loss due to sudden expansion, and (f) pressure drop over entire pipeline.

10.43 A motor-driven centrifugal pump delivers water at 60°F from an open reservoir to an elevated tank, vented to the atmosphere. The water level in the tank is 100 ft above the level in the reservoir. The supply line from the pump is 10,000 ft long, 12 in. diameter cast iron pipe, with a friction factor f = 0.02. A venturi meter, installed in a horizontal portion of the pipeline, is in Fig. P10.43. (a) As a result of losses, the actual velocity through the venturi meter is 98% of the ideal velocity. Calculate the flow rate through this pipeline. (b) If the pump efficiency is 60%, calculate the horsepower output of the motor.

10.44 Fresh water at 60°F flows from a large reservoir through a constant diameter 6 in. commercial steel pipe. The water discharges into the atmosphere. (a) Find the velocity $\bar{V}$ in the pipe assuming an ideal fluid. (b) Determine the relative roughness. (c) Estimate an initial friction factor. (d) What is the velocity based on this value of f? (e) What is the Reynolds

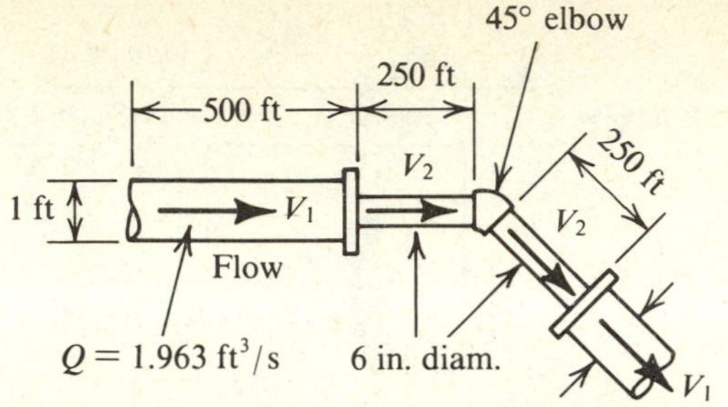

Figure P10.42

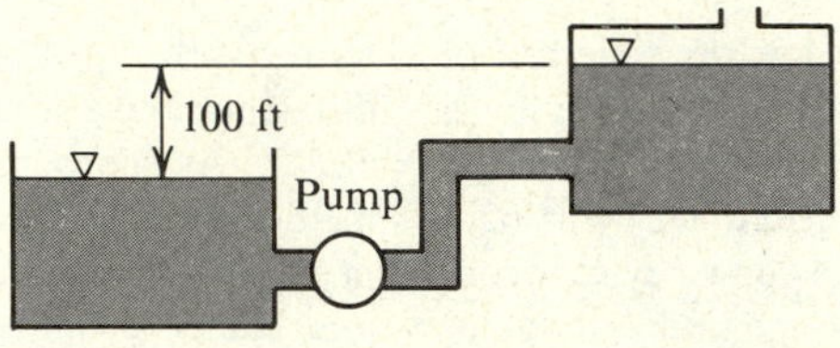

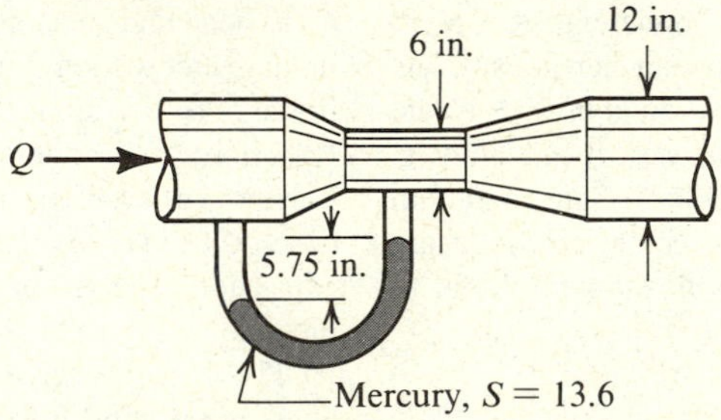

Figure P10.43

number based on this value of $\tilde{V}$? (f) Iterate for the next value of f. (g) What is the flow rate based on your velocity in part (d)? (See Fig. P10.44.)

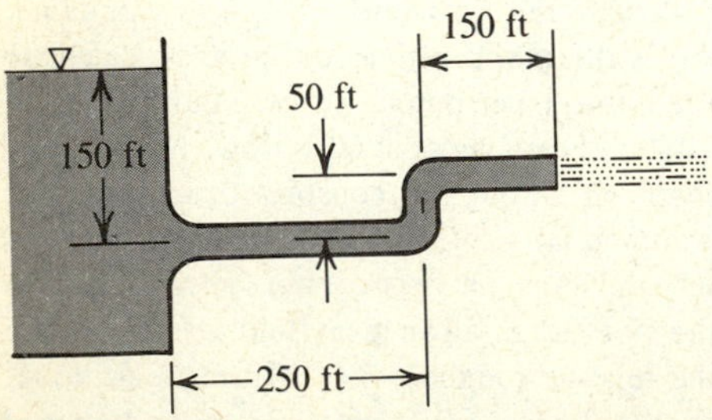

Figure P10.44

10.45 The system shown in Fig. P10.45 is an installation for evaluating the pressure drop (and therefore the head loss and power consumed) as a fluid flows through a heat exchanger. For testing purposes, the fluid is fresh water at 70°F and is circulated by a centrifugal pump. All piping and fittings are 1.0 in. in diameter of commercial steel. The prescribed flow rate for the test is 35 gpm. The pressure drop across the heat exchanger is measured by a differential manometer. The collection tank is open to the atmosphere at the top. The globe valve will be wide open when the system is in use. The entry from the tank to the pipe is a sharp edge entry. (a) What is the average velocity of the fluid in the pipe? (b) What are

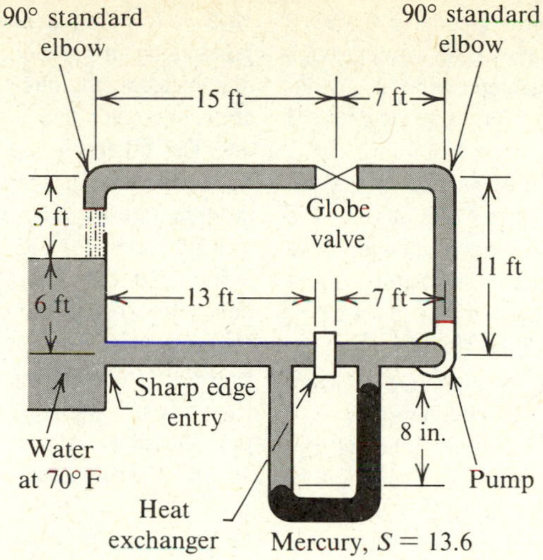

Fluid is water at 70° F
Pipe diam. = 0.0874 ft
Flow area = 0.006 ft^2
Flow rate = 35 gpm
Atm. pressure = 15.0 psia

Figure P10.45

the values of the pipe Reynolds number, relative roughness, and friction factor? (c) What is the pressure at the suction side of the pump? (d) What is the power added to the fluid by the pump? (e) The manufacturer guarantees its pump to have a minimum efficiency of 60%. Motors are available in 0.5 hp, 1.0 hp, 1.5 hp, 2.0 hp, and 2.5 hp sizes. Which size motor is required?

10.46 As a frigate is moving through rough seas, an unsecured coffee pot falls from its position through a hatch and into the bilge. At the bilge it strikes and shears a pipe connected to the sea at a point where the pipe makes a transition from 4 in. to 2 in. diameter. The open section is 18 ft below the surface. On one side of the break there is a 150 ft of 4 in. pipe with six elbows, two open gate valves, and a nozzle-like transition to a 2 in. pipe. On the other side of the break there is 300 ft of 2 in. pipe with two closed gate valves. The roughness ϵ of all piping is 0.004 in. Seawater has a specific weight of 64 lbf/ft^3, and a kinematic viscosity of 10^{-5} ft^2/s. (a) What is the leakage in gpm through the 4 in. line? (b) A pump is moved to the bilge and rigged with 100 ft of 3 in. diameter hose of roughness ϵ = 0.009 in. The pump discharges 15 ft above sea level, and has an efficiency of 70%. What pump power is needed to deliver twice the leakage (hp)? (Refer to Fig. P10.46.)

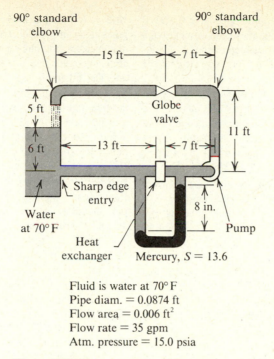

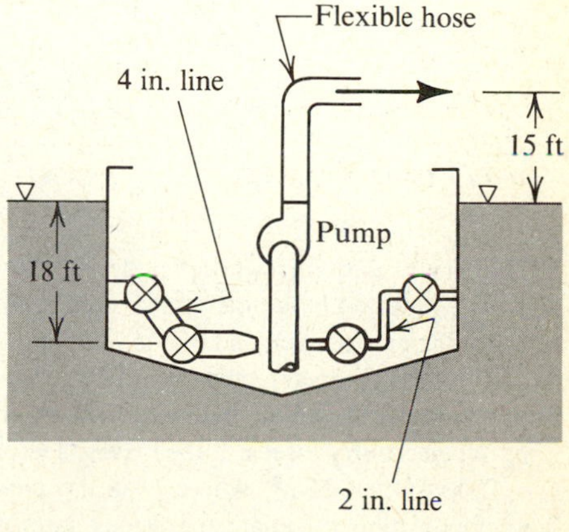

Figure P10.46

10.47 Oil of dynamic viscosity 0.00210 lbf·s/ft^2 and specific gravity 0.85 flows through 10,000 ft of 12 in. diameter pipe at a volumetric flow rate of 1.57 ft^3/s. Calculate the head loss in the pipe.

10.48 Water at 20°C flows in a 3 in. 250 psi cast iron pipe with a velocity of 6 m/s. For the following data determine the pressure drop (kPa) in 20 m of this pipe: $\rho_w = 998$ kg/m^3, $\mu_w = 10.02 \times 10^{-4}$ Pa·s, Internal pipe diameter = 0.085 m, $\epsilon/D = 3.072 \times 10^{-3}$.

10.49 A schematic of a water tunnel is shown in Fig. P10.49. The combined length $l_1 + l_2 + 2l_3 + l_4 + l_5 = 125$ ft, and all the material in the pipe is commercial steel. The length of the cylindrical test section is $l_6 = 4$ ft, and material is also commercial steel. The minor loss coefficient of the section between l_1 and l_2 is $k = 0.13$ (use the velocity $\bar{V}_6$ in l_6 for the kinetic energy). If the velocity $\bar{V}_6$ in the l_6 section is 100 ft/s, find the head to be delivered by the pump.

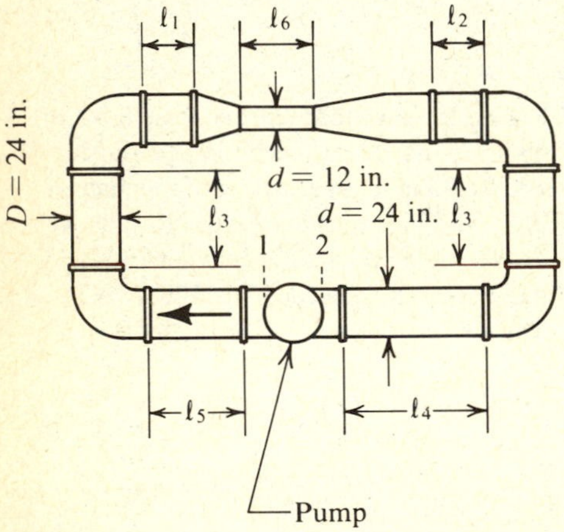

Figure P10.49

10.50 Two reservoirs whose levels differ by 2 m are connected by a pipe system consisting of a sloping pipe at each end that is 7.5 m long and 75 mm in diameter, and joined by a horizontal pipe 300 mm in diameter and 60 m long. Neglect entry losses. If the friction factor $f = 0.008\,(1 + 25/D)$ where D is the pipe diameter in mm, calculate the steady rate of flow through the pipe.

10.51 A pump delivers water through two pipes lying in parallel. One pipe is 50 mm in diameter, 50 m long, and discharges to the atmosphere at a level 10 m above a pump outlet. The other pipe is 200 mm in diameter, 80 m long, and discharges at a level 8 m above the pump outlet. The two pipes are connected to a junction near the pump. Both pipes have a friction factor $f = 0.008$. The inlet to the pump is 500 mm below the level of the outlet. Using the reference point as that of the inlet, find the total level at the pump outlet if the flow rate is 0.04 m^3/s (neglecting losses at the pipe junction).

10.52 Consider the flow in a pipe shown in Fig. 10.22. Let pipe *a* be 1000 ft long, and have an inner diameter of 4 ft. The pressure drop in the pipe is 625 psf. The fluid flowing in the pipe has a kinematic viscosity $\nu = 1 \times 10^{-5}$ ft^2/s and density $\rho = 2$ slug/ft^3. The pipe's relative roughness is 0.001. Calculate the average velocity $\bar{V}$ of the fluid flow.

10.53 Calculate the diameter of 400 ft of smooth pipe that contains oil at kinematic viscosity $\nu = 2.75 \times 10^{-5}$ ft^2/s flowing at a volume rate of flow of 0.178 ft^3/s under a pressure head of 10 ft. Refer to Fig. P10.53.

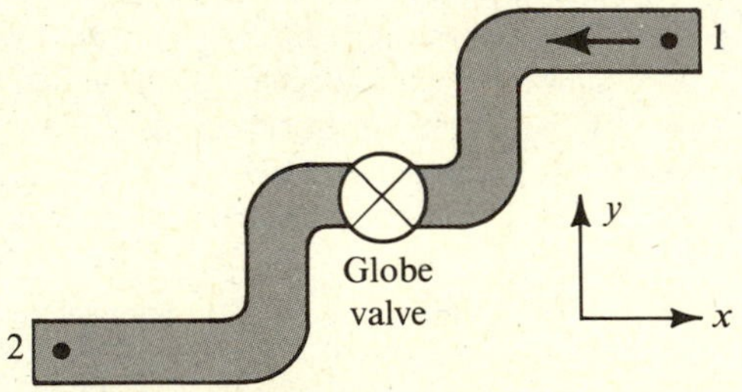

Figure P10.53

10.54 Consider the pipe system shown in Fig. P10.54. The length of pipe in 1 is 2000 ft with an inner diameter of 1 ft, whereas the length of pipe in 2 is 1500 ft with a 1.5 ft inner diameter. The pipe is made of cast iron, and allows 20 ft^3/s of water to flow from point A at 20 ft elevation to point B at 50 ft elevation. If the pressure is 100 psi at A, what is the pressure at B, neglecting all *minor* losses?

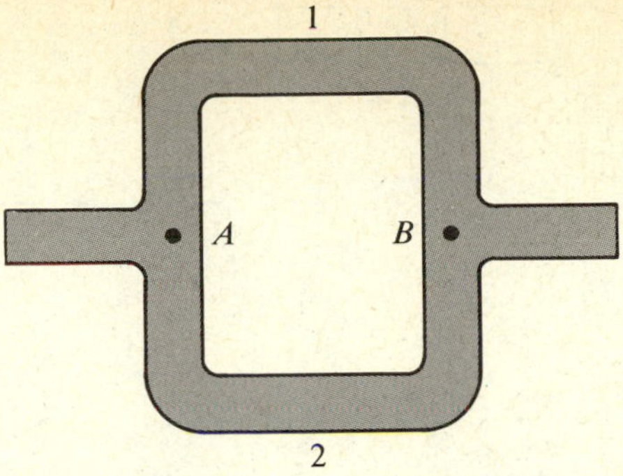

Figure P10.54

10.55 Consider the commercial steel piping system shown in Fig. P10.55. If 2 ft^3/s of water flows into the system at *A* at a pressure of 50 psi, what is the pressure at *B*, neglecting minor losses?

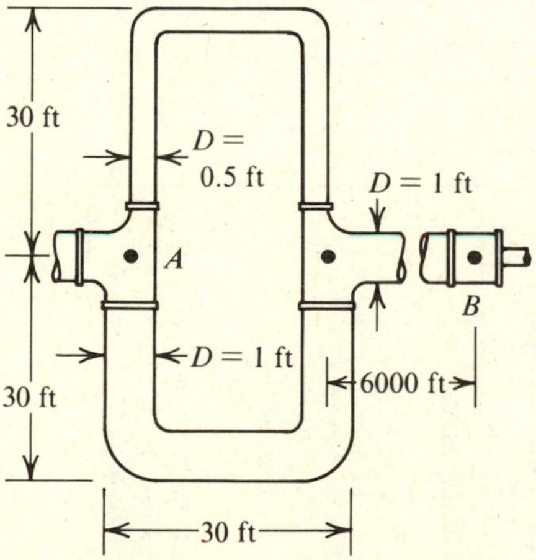

Figure P10.55

10.56 Determine the pump horsepower to pump 0.05 ft^3/s of water through the system shown in Fig. P10.56 where p_B = 20 psia. The pipes are constructed of cast iron. Neglecting minor losses, what are the head losses for the 2 ft pipe and the 3 in. pipe?

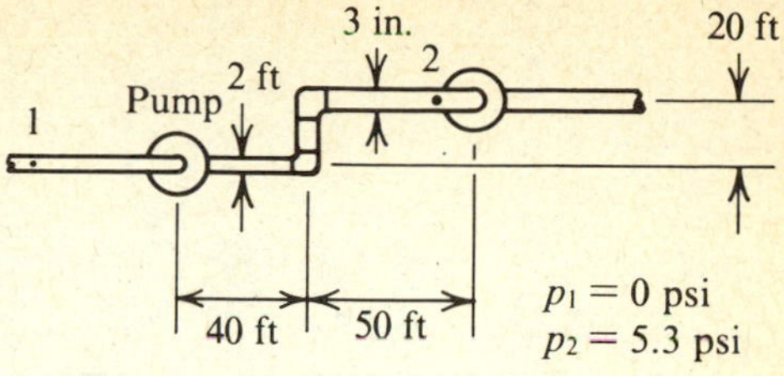

Figure P10.56

10.57 Consider the siphon in Fig. P10.57. Neglecting frictional effects, what is the velocity of the water jet at *C*? Also, what are pressures at points *A* and *B*?

10.58 Referring to Fig. P10.57, if the vapor pressure of water at 70°F is 6 ft of water, tell how high point *B* can be above point *A* before the siphon loses its action?

10.59 A pump discharges water (ν = 1.217 $\times$ 10^{-5} ft^2/s) at a constant rate through a 6 in. diameter pipe. The total length of pipe is 600 ft. The two elbows are medium sweep elbows (k = 0.75) and the valve is a wide open globe valve. The pump discharges through a 2 in. nozzle into the atmosphere. Assume the nozzle loss to be 10% of the exit velocity head. The discharge pipe of the pump contains a 2 in. venturi meter across which is connected a differential U-tube manometer with mercury (S = 13.6) as the gauge fluid and a deflection

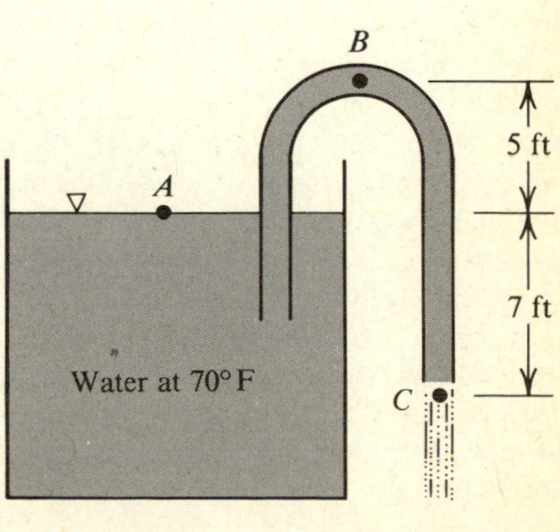

Figure P10.57

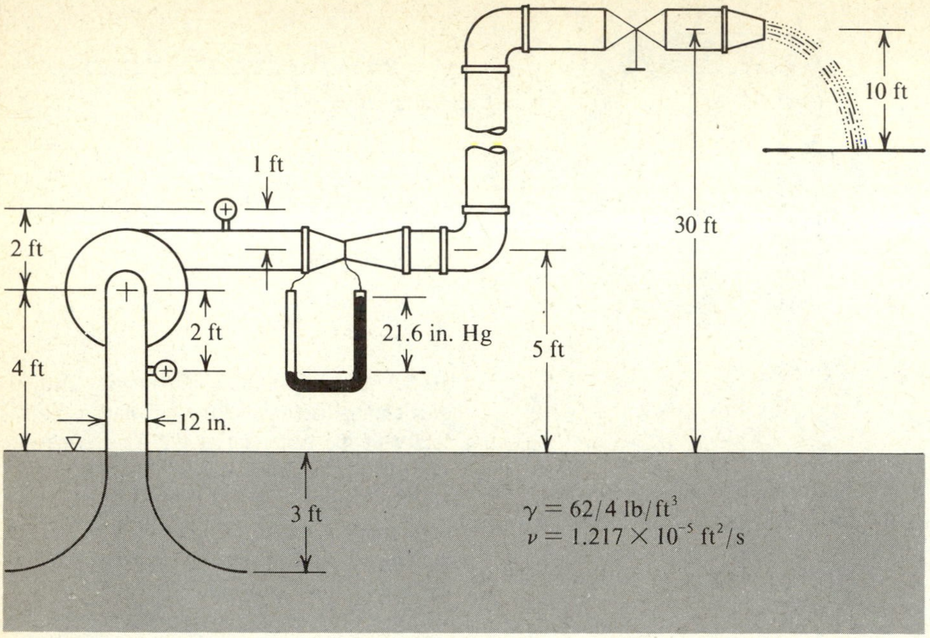

Figure P10.59

of 21.6 in. is noted. The barometer pressure is 29.85 in.Hg and the ambient temperature is 75°F. Assume all $\epsilon = 0.0001$ ft. For this arrangement, assuming negligible losses across the venturi and at the entrance to the system, calculate (a) the rate of flow through the system, (b) the horsepower delivered to the system, (c) the fraction of the total power used to overcome losses, and (d) the horizontal length of travel of the fluid leaving the nozzle. (See Fig. P10.59.)

10.60 By opening valve D in the system shown in Fig. P10.60, water is supplied to the manifold and pipes. What minimum pressure would gauge p have to read before flow would occur?

10.61 In Fig. P10.60, find the flow rate in each pipe for water at 60°F, $\epsilon = 0.0001$ ft, and total flow rate of 0.25 ft³/s. Neglect losses between supply manifold and pressure tap in the pipe. (*Hint*: Start analysis with flow through largest pipe.)

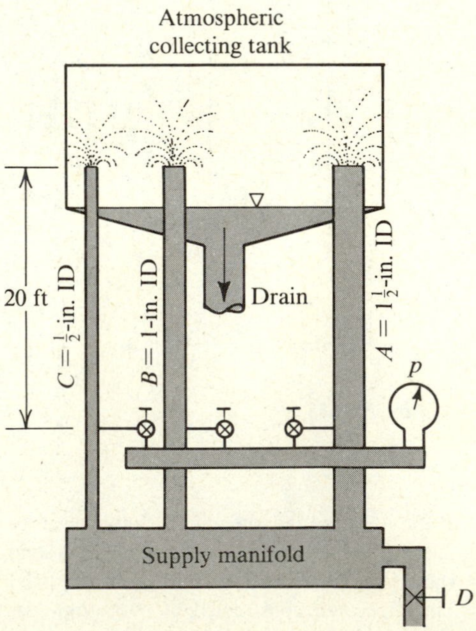

Figure P10.60

11 Turbulent Pipe Flow

11.1 Introduction

Much of our natural universe swirls and eddies in turbulent motion, as if disorder were the rule and order only an occasional respite from chaos. From the unsteady motion in interstellar dust clouds to the motion of blood in minute capillaries, the universe abounds with examples of the turbulent transport of matter, momentum, and heat. Physicists, geophysicists, oceanographers, engineers of many kinds must understand turbulence, though it is without doubt the most complex kind of fluid motion. "Even though the Navier-Stokes equation describes turbulence for Newtonian fluids, expressing the simplest principle that viscous stress is directly proportional to rate of strain, some of the possible solutions to this equation for the simplest of flow geometries are too complicated for the human mind to understand."* With such testimony from professional researchers, it is no wonder that students have often been mystified when introduced to concepts of turbulence. Like turbulent motion itself, the subject of turbulence seems at first to be mysterious and chaotic; but just as theorists today believe that turbulent motion actually possesses a reasonable degree of order, so can we order the study of turbulence for reasonable comprehension.

One symptom of the subject's complexity can be seen in how drastically views of turbulence have changed over the last 40 years or so. In the 1920s to 1930s, theorists viewed turbulence as essentially a stochastic phenomenon having a well-defined mean that was superimposed on a randomly fluctuating velocity field. The fluid motion was characterized by a range of scales dictated by the dimensions of the flow. Such a viewpoint led Prandtl [11.2] and Taylor [11.3] to turbulence models where the convective stresses were related to the mean flow by an effective eddy viscosity or mixing length. Though still popularly used, their models are somewhat restrictive. The greatest obstacle facing Prandtl and Taylor was that they lacked information on the physical structure of turbulence. It is remarkable that their inventive genius conjured up a model that worked fairly well.

In the 1940s turbulence was treated by statistical theories (see Kolmogorov (1914) [11.4]). A significant breakthrough in understanding turbulence was the hypothesis that if a turbulent shear flow is driven by a relatively slow motion of large eddies, an interface exists that is intermittently turbulent along with an *outer region* that is unsteady irrotational potential flow. The new model was based on the contribution of Corrsin (1943), (1955) [11.5], Klebanoff (1954) [11.6], and Townsend (1956) [11.7].

*P. Bradshaw [11.1].

In the 1960s to the present, the view of turbulence changed to the realization that the transport properties of turbulent shear flow are governed by vortex motions that are *not* random. Today we know that turbulence possesses a remarkable degree of order, even though it is characterized by an infinitely variable state. Turbulence still remains a major unsolved problem in classical physics. Part of this difficulty stems from the enormous number of discrete points required to solve the numerical equations of motion. Assuming that the average size of an eddy responsible for the decay of turbulence is 0.1 mm in diameter, we would need approximately 10^5 points, covering a volume of fluid 1 cm^3, scarcely large enough to get significant information. Working solely with that fine a mesh, we have already exceeded the storage capacity of many high-speed computers.

Fortunately, engineers can evade such troublesome details by time-averaging values of the flow. Time averaging produces statistical correlations that involve fluctuating velocities, pressures, and temperatures. These fluctuating properties are the unknown variables that appear in the governing equations of turbulent motion. Since no general solution of these fluctuating properties are known to exist, we are forced to model them in terms of properties we do know.

Four references offer particularly useful review of work in turbulence: Batchelor [11.8] discusses work up to 1950, Orszag [11.9] reviews the problems to 1970, and Cantwell [11.10] presents a thorough survey up to the 1980s. An excellent literature review is given by Monin and Yaglom [11.11].

11.2 Detecting Turbulence

We can detect turbulence with such instruments as the piezoelectric pressure transducer (Fig. 11.1), the hot-wire anemometer (Fig. 11.2), and the LDV (Fig. 11.3). If a flow is fully turbulent, the velocity will appear in a form similar to that shown in Fig. 11.4, where the frequency can range from 1 to 10,000 Hz with an equally wide range of wavelength from 0.01 to 400 cm when the Reynolds number* is large. If we are dealing with free-surface flows, turbulence can be detected visually. Figure 11.5 shows a close-up view of the surface of the sun, where both large- and small-scale turbulence is evident. Figure 11.6 shows three regimes of viscous flow: laminar flow at low Reynolds number, transition at intermediate Reynolds number, and turbulent flow at large Reynolds number. Close examination of the dye lines in Fig. 11.6 reveals the growth of a complicated motion typical of turbulence: very spotty on a small scale, changing rapidly at every point.

Another type of turbulent motion is the galaxy. The giant whirlpool galaxy of Fig. 11.7, seen from a distance of millions of light-years, appears as a huge eddy possessing the characteristics of turbulence. On a much smaller scale is another familiar eddy: the tropical cyclone and hurricane (see Fig. 11.8). Each example of turbulence can be *structured* differently. Each behaves differently. Each is either a solitary eddy or composed of smaller eddies. Each arises from a different mechanism.

*Reynolds number depends upon the reference length and reference velocity.

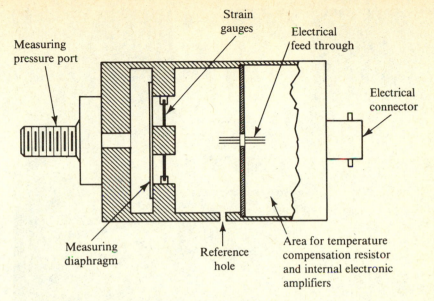

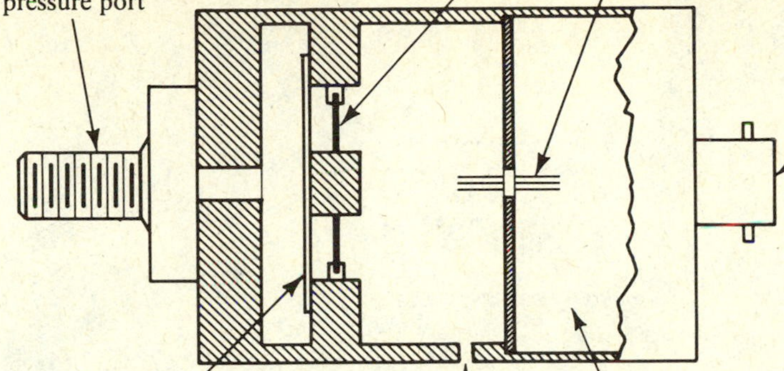

Figure 11.1 *Standard gauge pressure transducer.*

11.3 On the Origin of Turbulence

Consider laminar flow in a constant diameter pipe. We can view the motion of fluid as consisting of steadily advancing layers of fluid. If we increase the flow rate, which results in an increase of Reynolds number, we would observe the flow becoming unsteady with slugs of fluid superimposed on the main flow moving in a rather chaotic fashion. The flow is now turbulent. An overly simplified explanation that might have been sufficient in the 1940s is that near the surface of the pipe the shear stresses slow the fluid down, creating large velocity gradients. If the velocity gradient is large enough, the local surface friction transforms the fluid into small eddies. Some of these eddies are transported toward the center of the pipe where they combine with other eddies to form larger ones. Figure 11.9 is a schematic representation of how this might work. Today this explanation is *incomplete*. There is a great deal more to be said.

The inherent requirement for the existence of turbulence is shear (whether the flow is in pipes, ducts, boundary layers, wakes behind grids, or in more complex flows involving buoyancy and curvature). Turbulence is the result of viscosity but in a subtle way. Though viscosity *dampens* out turbulence and is one of the factors that makes a flow well behaved, it cannot by itself make turbulence. Turbulence is *made by surface friction*, which is a local phenomenon. It is not spread out like laminar friction. Experiments have been conducted using a wide variety of fluids in pipes of various diameters that confirm Reynolds number characterizes the velocity of the flow wherein laminar flow breaks down into turbulent flow. At $R_D \leqslant 2000$, the flow in a smooth surfaced pipe is always laminar. For $2000 < R_D < 4000$, there is a gradual change to turbulence which is called transition. For $R_D > 4000$, the flow is customarily considered turbulent, though portions of the flow may still remain laminar. If there

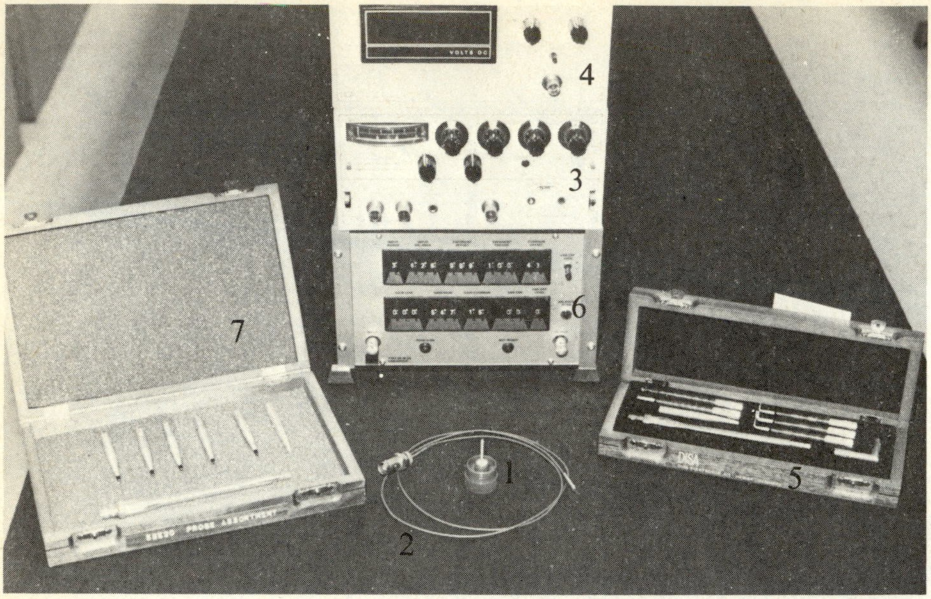

(a)

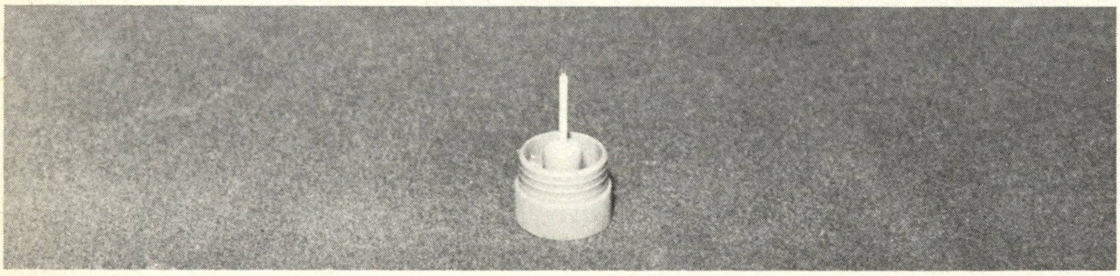

(b)

Figure 11.2 *Hot-wire and hot-film anemometry equipment.* ① *Hot wire probe DISA P11.* ② *DISA probe support and connector.* ③ *Anemometer, DISA type 55M10 (main unit plus bridge).* ④ *Digital volt meter, DISA type 55D31.* ⑤ *Assorted hot-film probes—wedge shape, 90°, flush mounted, straight wedge, adapters and wedge support; DISA.* ⑥ *Linearizer, DISA type 55M25.* ⑦ *Assorted hot-wire probes plus support, DISA. (Source: Reproduced with the permission of DISA Electronics.)*

are protrusions in the flow, it may trip into turbulence at a lower Reynolds number. We shall discuss the Reynolds number requirement for turbulent flow later in this chapter.

Objects placed in a laminar flow can induce turbulence at a Reynolds number much less than 4000. As will be shown in Chap. 14, eddies form in the region of the wake behind a cylinder at a Reynolds number between 3 and 40, and 1 for a sphere.

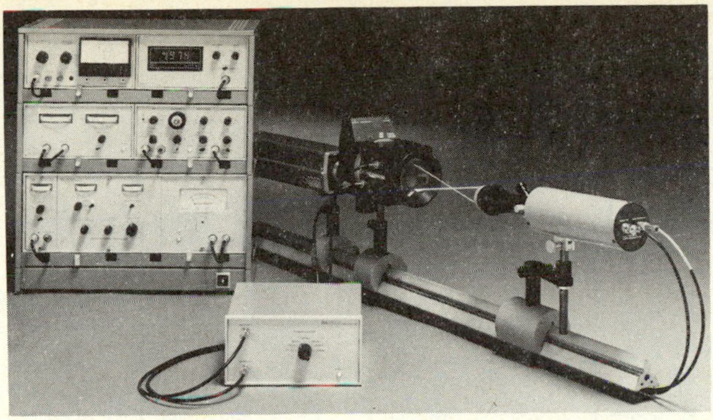

(a)

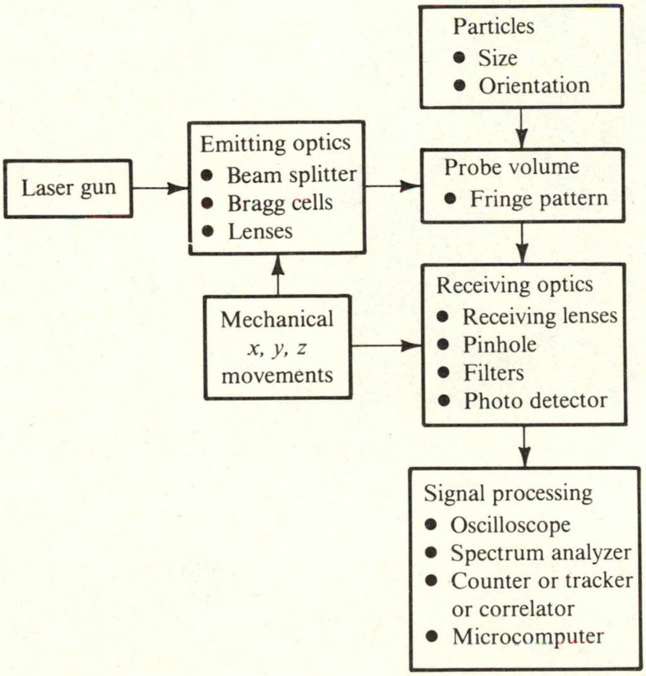

(b)

Figure 11.3 *(a) DISA type 55L laser Doppler anemometer. (b) Schematic of an LDV system. (Source: Reproduced with the permission of DISA Electronics.)*

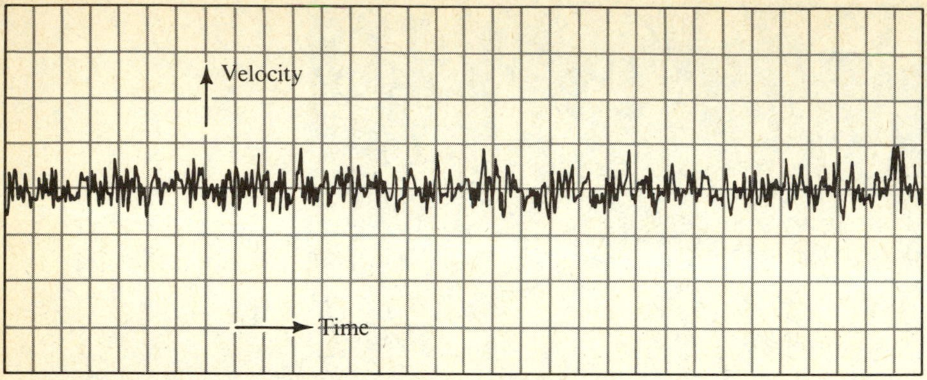

Figure 11.4 *Typical chart recording of turbulence.*

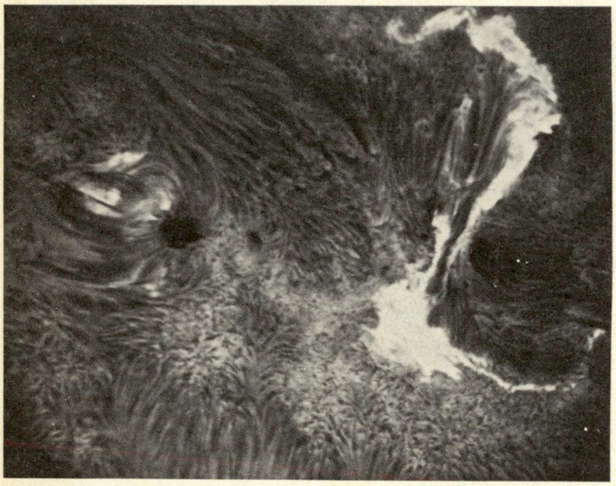

Figure 11.5 *A "microscopic" view of the sun. (Source:* A New Sun *by John H. Eddy, NASA SP-402. Courtesy of NASA.)*

11.3.1 The Role of Vorticity in the Origin of Turbulence

The origin of turbulence consists of four basic stages:

1. The development of an unstable shear layer
2. The growth of two-dimensional disturbances with periodic fluctuations of vorticity
3. The transformation of the disturbance to three-dimensionality leading to a spectral broadening by vortex filament interaction
4. The development of a random behavior wherein the vortex filaments become complicated, and vorticity is transferred across the spectrum to smaller and smaller scale

We shall briefly discuss these stages and show how one leads to the next.

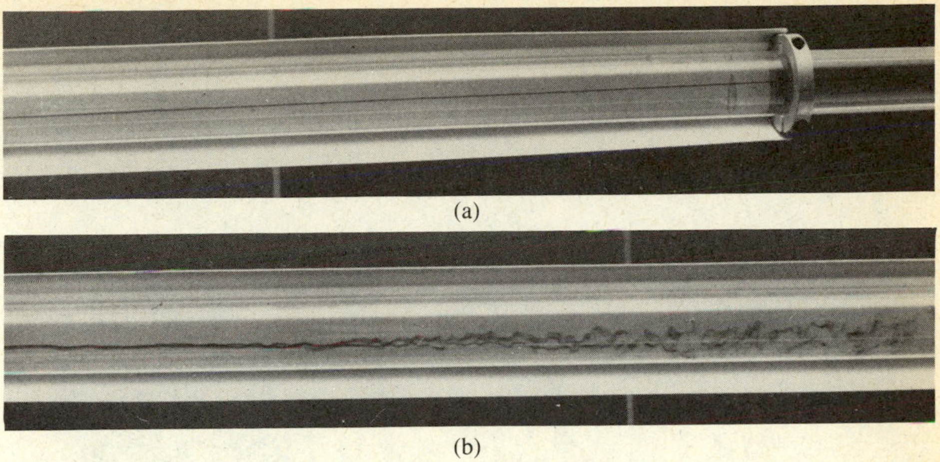

Figure 11.6 Reynold's pipe flow experiment. (a) Laminar flow. (b) Instability, transition and turbulent flow.

Figure 11.7 A turbulent spiral galaxy. (Source: © Association of Universities for Research in Astronomy, Sacramento Peak Observatory.)

A flow with a total pressure gradient normal to the streamlines can be destabilized by small disturbances. The unstable shear flow gives rise to some complicated mathematics. The key to both the development of the instabilities and the maintenance of turbulent flow lies in the vorticity $\boldsymbol{\zeta}$, defined in Chap. 4 as

$$\boldsymbol{\zeta} = \boldsymbol{\nabla} \times \mathbf{V} \tag{4.86}$$

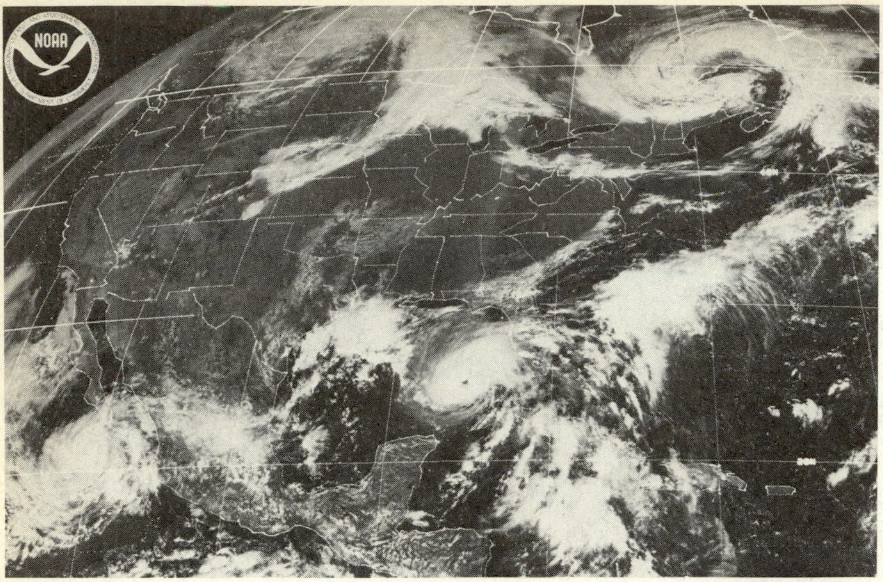

Figure 11.8 NOAA satellite photograph of three giant storms. (Source: Weatherwise, vol. 33, p. 111, June 1980. Reproduced with the permission of National Oceanic and Atmospheric Administration.)

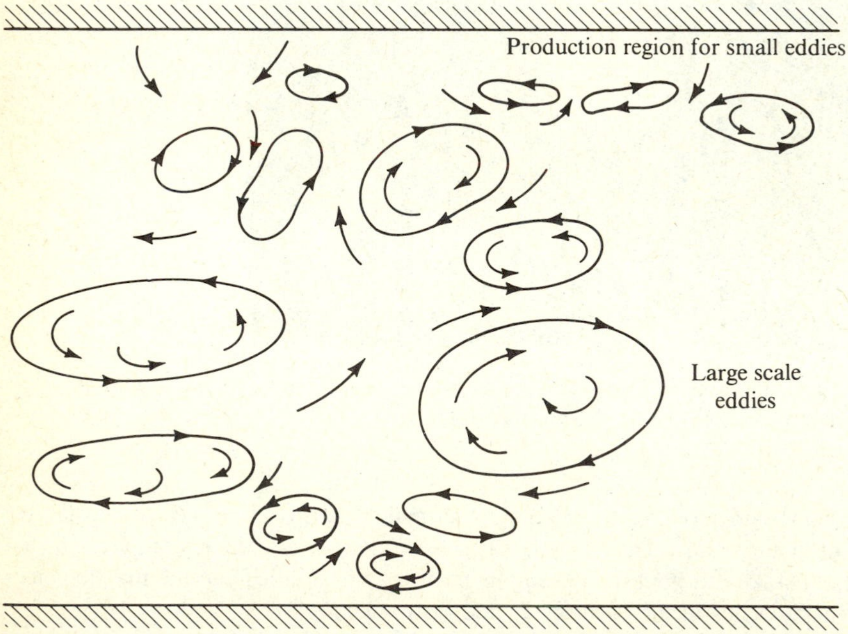

Figure 11.9 Production of eddies in turbulent shear flow.

The transport equation for vorticity is obtained by taking the curl of the Navier-Stokes Eq. (4.109). For incompressible flow,

$$\boxed{\frac{\partial \boldsymbol{\zeta}}{\partial t} + (\mathbf{V}\cdot\nabla)\boldsymbol{\zeta} = (\boldsymbol{\zeta}\cdot\nabla)\,\mathbf{V} + \nu\nabla^2\boldsymbol{\zeta}} \tag{11.1}$$

Comparing the vorticity transport equation (11.1) with the Navier-Stokes equation (4.109), we note that the pressure term has disappeared and that a new term $(\boldsymbol{\zeta}\cdot\nabla)\,\mathbf{V}$ has appeared which is the vorticity/velocity gradient interaction. This nonlinear term has two important effects, depending upon the direction of the flow. Figure 11.10 shows the effect of a velocity gradient on a particular vortex line $\boldsymbol{\zeta}$. The velocity gradient $\partial u/\partial y$ tilts a vortex filament, whereas the velocity gradient $\partial v/\partial y$ stretches a vortex filament. The vortex filament will conserve its angular momentum provided that we neglect viscous diffusion and the cross section of a vortex filament is circular. Thus if its cross-sectional area decreases because of vortex stretching (accelerated flow), its vorticity must increase. If we include viscous diffusion, this simple flow pattern transforms into an immensely complicated distribution of velocity and vorticity, what we call *turbulence*.

In Fig. 11.10, we see a simple flow pattern. The shear flow that is shown is two-dimensional, the velocity vector is everywhere parallel to the xy-plane. We know from the definition of vorticity $\boldsymbol{\zeta}$, Eq. (4.86), that the vorticity vector $\boldsymbol{\zeta}$ would be normal to that plane such that the term $(\boldsymbol{\zeta}\cdot\nabla)\,\mathbf{V}$ would be zero. Hence in a two-dimensional shear flow, a disturbance is a two-dimensional traveling wave. Bradshaw [11.1] states that "although the most unstable infinitesimal disturbance in a steady two-dimensional shear flow is a two-dimensional traveling wave, amplified disturbances of sufficient amplitude (which can be regarded as packets of vortex lines with spanwise axes) are themselves unstable to infinitesimal three-dimensional perturbations." Imagine, then, our vortex filament becoming disturbed by the traveling two-dimensional flow, somewhat in the manner shown in Fig. 11.11. If the viscous diffusion is sufficiently small, the distortion of the vortex filament will continue indefinitely. We can now see the development of a very complicated vortex pattern caused by the tilting and stretching of all the vortex filaments in a shear flow. As the motion becomes increasingly more complicated, the effects become random.

The distance between any two randomly perturbed particles will, on the average increase. Applying this fact to two particles at the ends of a vortex filament in a flow field of random disturbances, the length of the vortex filament will, on the average, increase and thus its area will decrease causing the vorticity to increase. This is a simplified explanation of the physical model presently accepted for turbulence. Turbulence is viewed as a maze of vortex filaments tangled together, interacting one with the other in a random manner, transferring vorticity to smaller and smaller length scales.

When the vortex stretches, its rotational kinetic energy increases. The increase of rotational kinetic energy comes from the mean flow through the vortex motion. The kinetic energy is converted into thermal energy as a result of the work done against the viscous stresses, provided that a mean strain rate exists. If there is no mean strain rate, then the loss of power $\dot{W}_{s_T}$ decays. Hence there is energy transfer in turbulence.

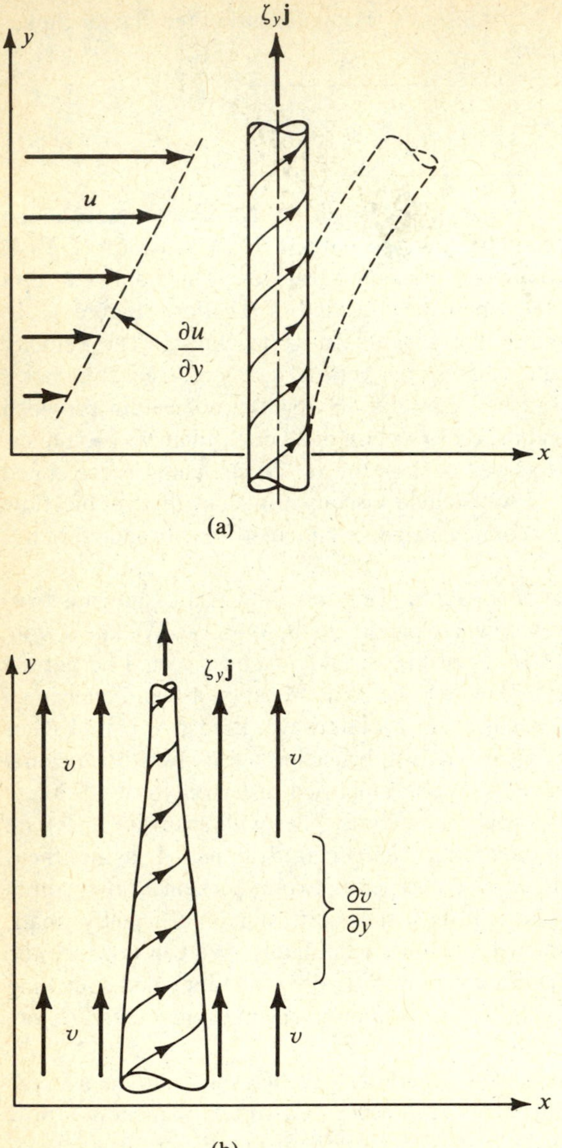

Figure 11.10 *Velocity gradients affecting a vortex filament. (a) Tilting the vortex filament ζ_y. (b) Stretching the vortex filament ζ_y.*

It is transferred to smaller and smaller scales just as vorticity, and is independent of viscosity until the final stages of transfer. The viscosity can cause viscous dissipation of energy and is proportional to the mean square of the rate of strain.

Certain questions inevitably arise about the origin of turbulence. When does turbulence begin? There is no exact point at which we can say the motion is now turbulent. Is there a single model universally accepted that we can use to study turbulence? There are three widely used models one may select to study turbulence: the

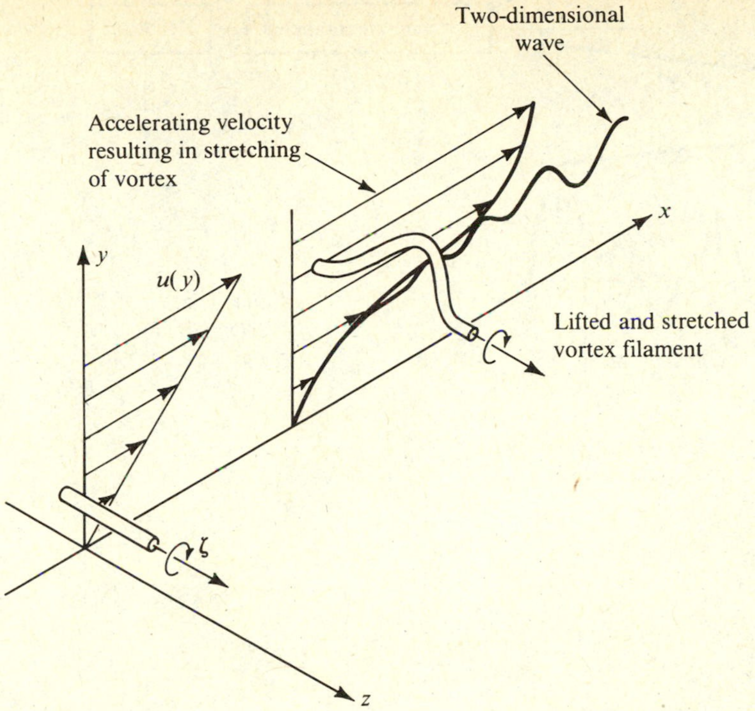

Figure 11.11 *The development of the kink in a vortex filament for two-dimensional sheer flow. (Source: Adapted from* Topics in Applied Physics, *vol. 12, "Turbulence," Springer-Verlag 1976. Used with the permission of the publisher and the author, Peter Bradshaw.)*

vortex filament which we have used, the vortex sheet, and the eddy. Is there one characteristic of turbulence that is uniformly accepted? Turbulence is a three-dimensional nonlinear phenomenon having the essential characteristic of transferring energy to smaller spatial scales across a continuous wave number spectrum. The next section examines a way to mathematically describe turbulent motion.

11.4 Definitions of Various Velocity Terms

Turbulence involves the random motion of fluid particles. At any point in a three-dimensional turbulent fluid, the velocity of the particles varies with respect to both space and time. At some instant of time, the difference between the instantaneous velocities (u, v, w) and the mean velocities ($\overline{u}$, $\overline{v}$, $\overline{w}$), shown in Fig. 11.12a, represents the magnitude of fluctuation of the instantaneous quantity from the average quantity. Thus,

$$\begin{aligned} u'(x, y, z, t) &= u(x, y, z, t) - \overline{u}(x, y, z, t) \\ v'(x, y, z, t) &= v(x, y, z, t) - \overline{v}(x, y, z, t) \\ w'(x, y, z, t) &= w(x, y, z, t) - \overline{w}(x, y, z, t) \\ p'(x, y, z, t) &= p(x, y, z, t) - \overline{p}(x, y, z, t) \end{aligned} \tag{11.2}$$

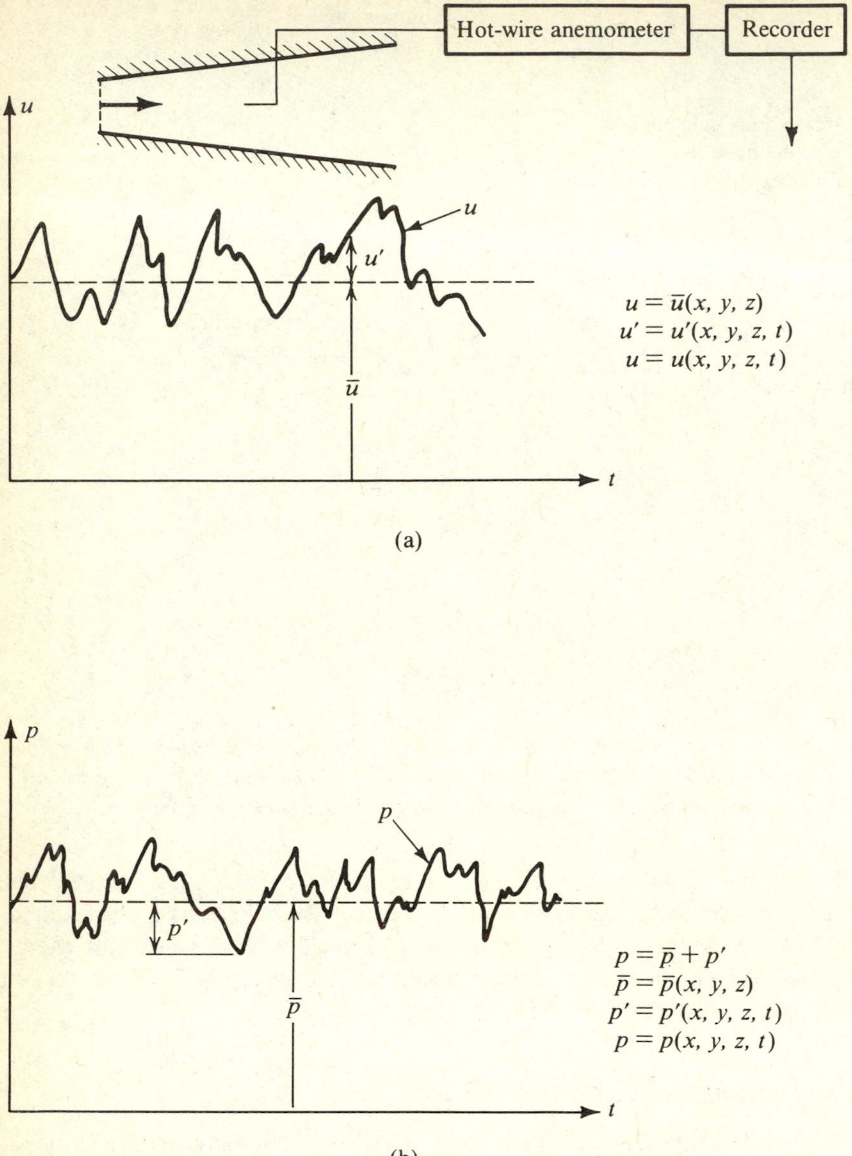

Figure 11.12 *Turbulent velocity and pressure. (a) Example of typical recorder readout of velocity component u for steady turbulent flow. (b) Example of pressure for steady turbulent flow.*

Since the velocity and pressure at a point in the flow field are random, we evaluate them using statistical averaging. The simplest form of statistical average is the mean, with respect to time, at a single point. This is useful provided that the mean is independent of the time at which the averaging process is started. From the definition of average, we obtain

$$\frac{1}{T}\int_t^{t+T} u'(\mathbf{r}, t)\, dt = \overline{u'} = 0$$
$$\frac{1}{T}\int_t^{t+T} u(\mathbf{r}, t)\, dt = \overline{u} \tag{11.3}$$

with similar expressions involving v', $\overline{v}$, w', $\overline{w}$, etc. The symbol T represents an averaging period that is greater than any significant period of the typical fluctuation. A typical magnitude of the averaging period is about 5 seconds for flows of air and water. In a similar manner we can define the mean square of the velocity fluctuations as

$$\frac{1}{T}\int_t^{t+T} u'^2\,(\mathbf{r}, t)\, dt = \overline{u'^2}$$
$$\frac{1}{T}\int_t^{t+T} v'^2\,(\mathbf{r}, t)\, dt = \overline{v'^2} \tag{11.4}$$
$$\frac{1}{T}\int_t^{t+T} w'^2\,(\mathbf{r}, t)\, dt = \overline{w'^2}, \qquad \frac{1}{T}\int_t^{t+T} p'^2\,(\mathbf{r}, t)\, dt = \overline{p'^2}$$

Thus, the root-mean-square (RMS) of the fluctuations are denoted by $(\overline{u'^2})^{1/2}$, $(\overline{v'^2})^{1/2}$, $(\overline{w'^2})^{1/2}$, $(\overline{p'^2})^{1/2}$.

It would then follow that

$$\frac{1}{T}\int_t^{t+T} u'(\mathbf{r}, t)\, v'(\mathbf{r}, t)\, dt = \overline{u'v'}$$
$$\frac{1}{T}\int_t^{t+T} u'(\mathbf{r}, t)\, w'(\mathbf{r}, t)\, dt = \overline{u'w'} \tag{11.5}$$
$$\frac{1}{T}\int_t^{t+T} v'(\mathbf{r}, t)\, w'(\mathbf{r}, t)\, dt = \overline{v'w'}$$

where the quantities on the right-hand side are nonzero. It is significant to point out that when the mean flow is in the direction of the x-axis, the quantities $\overline{u'^2}$, $\overline{v'^2}$, $\overline{u'v'}$, $\overline{v'w'}$, $\overline{u'w'}$, and $\overline{p'^2}$ are all finite.

11.4.1 The Equations of Motion for Turbulent Flow

It should be noted that the velocities and pressure as defined by Eq. (11.2) are of a macroscopic nature. This means that the velocity fluctuations we are defining at fixed points in space are of fluid masses in chunks, globules, or lumps as opposed to microscopic motion of the molecules and atoms of the fluid. This is important because it allows us to use the Navier-Stokes equations to describe the dynamics of turbulent flow. That, along with the continuity equation, assures us that the fluid flows as a continuum. Restricting the analysis to incompressible mean flow, the Navier-Stokes equation and the continuity equation are written in the form

$$\frac{\partial u}{\partial t} + \frac{\partial(u^2)}{\partial x} + \frac{\partial(uv)}{\partial y} + \frac{\partial(uw)}{\partial z} = -\frac{1}{\rho}\frac{\partial p}{\partial x} + \nu\nabla^2 u + g_x \tag{11.6}$$

$$\frac{\partial v}{\partial t} + \frac{\partial(vu)}{\partial x} + \frac{\partial(v^2)}{\partial y} + \frac{\partial(vw)}{\partial z} = -\frac{1}{\rho}\frac{\partial p}{\partial y} + \nu\nabla^2 v + g_y \tag{11.7}$$

$$\frac{\partial w}{\partial t} + \frac{\partial(wu)}{\partial x} + \frac{\partial(wv)}{\partial y} + \frac{\partial(w^2)}{\partial z} = -\frac{1}{\rho}\frac{\partial p}{\partial z} + \nu\nabla^2 w + g_z \tag{11.8}$$

and

$$\frac{\partial u}{\partial x} + \frac{\partial v}{\partial y} + \frac{\partial w}{\partial z} = 0 \tag{11.9}$$

We derive Eqs. (11.6)–(11.8) from Eqs. (4.106)–(4.108) by using the continuity equation. Following our hypothesis that the dependent variables $\{u, v, w, p\}$ can be decomposed into time-averaged plus fluctuating components as given by Eq. (11.2), we can generate the equations which describe the time-averaged properties of the flow. To do this we substitute Eqs. (11.2) into Eqs. (11.6)–(11.8) and perform the time-averaging process of each term in each equation on a term-by-term basis. Certain operations are useful for finding typical time-averaged values. For example, using Eq. (11.3)

$$\overline{\frac{\partial u}{\partial x}} = \frac{1}{T}\int_t^{t+T} \frac{\partial u}{\partial x}\, dt = \frac{\partial}{\partial x}\left[\frac{1}{T}\int_t^{t+T} (\bar{u} + u')\, dt\right] = \frac{\partial \bar{u}}{\partial x} \tag{11.10}$$

and by definition

$$\overline{\frac{\partial u'}{\partial t}} = \frac{\partial \overline{u'}}{\partial t} = 0$$

Similarly,

$$\overline{\frac{\partial v}{\partial y}} = \frac{\partial \bar{v}}{\partial y}, \qquad \overline{\frac{\partial w}{\partial z}} = \frac{\partial \bar{w}}{\partial z} \tag{11.11}$$

If we substitute the velocity components as given by Eq. (11.2) into the continuity equation we obtain

$$\frac{\partial \bar{u}}{\partial x} + \frac{\partial \bar{v}}{\partial y} + \frac{\partial \bar{w}}{\partial z} + \frac{\partial u'}{\partial x} + \frac{\partial v'}{\partial y} + \frac{\partial w'}{\partial z} = 0 \tag{11.12}$$

But if we take the time-averaged value of Eq. (11.12) and use the results of Eqs. (11.3) and (11.10) we get

$$\frac{\partial \bar{u}}{\partial x} + \frac{\partial \bar{v}}{\partial y} + \frac{\partial \bar{w}}{\partial z} = 0 \tag{11.13}$$

indicating that the mean flow satisfies continuity. Comparing Eqs. (11.12) and (11.13), we readily see that the fluctuating components also satisfy continuity.

To carry out the time-averaging process for the Navier-Stokes equations we need operations of the following type:

$$\begin{aligned}
\overline{\frac{\partial(u^2)}{\partial x}} &= \frac{\partial}{\partial x}\left[\frac{1}{T}\int_t^{t+T} (\bar{u} + u')^2\, dt\right] \\
&= \frac{\partial}{\partial x}\left[\frac{1}{T}\int_t^{t+T} (\bar{u}^2 + 2\bar{u}u' + u'^2)\, dt\right] \\
&= \frac{\partial(\bar{u}^2)}{\partial x} + \frac{\partial\overline{(u')^2}}{\partial x} \\
\overline{\frac{\partial(uv)}{\partial y}} &= \frac{\partial}{\partial y}\left[\frac{1}{T}\int_t^{t+T} (\overline{uv} + \bar{u}v' + \bar{v}u' + u'v')\, dt\right] \\
&= \frac{\partial}{\partial y}\,(\overline{uv}) + \frac{\partial}{\partial y}\,(\overline{u'v'}) \\
\overline{\nabla^2 u} &= \nabla^2\left[\frac{1}{T}\int_t^{t+T} (\bar{u} + u')\, dt\right] = \nabla^2\bar{u}
\end{aligned} \tag{11.14}$$

and

$$\bar{g}_x = \frac{1}{T}\int_t^{t+T} g_x\, dt = g_x \tag{11.15}$$

In arriving at the results above, we made use of Eq. (11.3) and similar relationships.

Substituting Eqs. (11.2) into the Navier-Stokes equations (11.6)–(11.8), performing the time-averaging process, we obtain

$$\rho\left(\frac{\partial \bar{u}}{\partial t} + \bar{u}\,\frac{\partial \bar{u}}{\partial x} + \bar{v}\,\frac{\partial \bar{u}}{\partial y} + \bar{w}\,\frac{\partial \bar{u}}{\partial z}\right) = -\frac{\partial \bar{p}}{\partial x} + \mu\nabla^2\bar{u}$$
$$- \rho\left[\frac{\partial(\overline{u'^2})}{\partial x} + \frac{\partial\overline{u'v'}}{\partial y} + \frac{\partial\overline{u'w'}}{\partial z}\right] + \rho g_x$$

$$\rho\left(\frac{\partial \bar{v}}{\partial t} + \bar{u}\frac{\partial \bar{v}}{\partial x} + \bar{v}\frac{\partial \bar{v}}{\partial y} + \bar{w}\frac{\partial \bar{v}}{\partial z}\right) = -\frac{\partial \bar{p}}{\partial y} + \mu\nabla^2\bar{v} - \rho\left[\frac{\partial \overline{u'v'}}{\partial x} + \frac{\partial(\overline{v'^2})}{\partial y} + \frac{\partial \overline{v'w'}}{\partial z}\right] + \rho g_y \tag{11.16}$$

$$\rho\left(\frac{\partial \bar{w}}{\partial t} + \bar{u}\frac{\partial \bar{w}}{\partial x} + \bar{v}\frac{\partial \bar{w}}{\partial y} + \bar{w}\frac{\partial \bar{w}}{\partial z}\right) = -\frac{\partial \bar{p}}{\partial z} + \mu\nabla^2\bar{w} - \rho\left[\frac{\partial \overline{u'w'}}{\partial x} + \frac{\partial \overline{v'w'}}{\partial y} + \frac{\partial(\overline{w'^2})}{\partial z}\right] + \rho g_z$$

The quadratic terms in the fluctuating velocity components have been transferred to the right-hand side of Eqs. (11.16) for reasons which are apparent. Close examination reveals that all the terms in Eqs. (11.16) with the exception of the terms in brackets are identical in form to the equation for laminar flow. The terms on the left-hand side of the equation are inertia forces per unit volume of the mean flow. The first and second terms on the right-hand side are the mean pressure and viscous forces per unit volume, respectively. The terms in the brackets involving the fluctuating velocity components are forces per unit volume caused by the turbulent stresses in the flow field.

In fact, Eq. (11.16) can be written in a form analogous to Cauchy's equation of motion, i.e.,

$$\mathbf{a} = \nabla\cdot\mathbf{P} + \mathbf{g} \tag{4.98}$$

We modify the constitutive Eq. (4.98) to include effects of turbulence:

$$\mathbf{P} = -p\mathbf{I} + 2\mu\mathbf{S} + \mathbf{P}' \tag{11.17}$$

where the components of the turbulent stress dyadic $\mathbf{P}'$ are given by

$$\begin{bmatrix} p'_{xx} & p'_{xy} & p'_{xz} \\ p'_{yx} & p'_{yy} & p'_{yz} \\ p'_{zx} & p'_{zy} & p'_{zz} \end{bmatrix} = -\rho\begin{bmatrix} \overline{u'^2} & \overline{u'v'} & \overline{u'w'} \\ \overline{u'v'} & \overline{v'^2} & \overline{v'w'} \\ \overline{u'w'} & \overline{v'w'} & \overline{w'^2} \end{bmatrix} \tag{11.18}$$

These additional stresses are called *apparent, turbulent,* or *Reynolds stresses.* They are caused by turbulent fluctuations and are given by the time-averaged values of the quadratic terms in the turbulent components. Since these stresses are added to the ordinary viscous terms in laminar flow and have similar influence on the course of the flow, it is sometimes said that they are caused by *eddy viscosity*. In many practical cases of turbulent flow the apparent or Reynolds stresses far outweigh their viscous counterpart, with the result that the viscous stresses can be neglected with very little loss of accuracy.

Inspection of the three equations given by Eq. (11.16) reveals the fairly obvious fact that these equations are not solvable. The number of unknowns exceeds the number of equations. The incompressible instantaneous Navier-Stokes equations plus the continuity equation are a closed soluble set, but the process of averaging loses information. This lost information must be gained back somehow by empirical results. We say such a problem is a *closure problem*. Solutions of closure problems are called *turbulence models*. There are five turbulence models, only one of which we shall discuss.* The five models are

- Zero-equation models: models using only the instantaneous Navier-Stokes equation and *no* turbulence differential equations. They are based on the eddy viscosity and mixing length concept.
- One-equation models: models using turbulent kinetic energy.
- Two-equation models: models using one differential equation involving the turbulence velocity scale and another differential equation involving a turbulence length scale.
- Stress-equation models: models involving differenital equations for all components of the Reynolds stress tensor and a length scale.
- Large-eddy simulation: models involving calculation of the three-dimensional eddy structure plus a model for small-scale turbulence.

Due to the very complicated form of these models, most of the closure work pertains to *homogeneous turbulence*, which is defined as turbulent flow possessing fluctuation components u', v', w', p' that are independent of position in space. In practice, it is difficult to produce homogeneous turbulence except over short distances. Figure 11.13 provides a useful insight into homogeneous turbulence behind a grid. Downstream of this homogeneous region, we see a different pattern, illustrative of *isotropic turbulence*.

In order to complete the definition of the fluid dynamics problem, we must specify the boundary conditions. Here we must be extremely careful. In this chapter, we shall be investigating internal turbulent flows, whereas in Chap. 14 we shall be investigating external turbulent flows. The simplest example of internal turbulent flows is a fully developed turbulent flow in a circular pipe. This is a flow that is independent, in the time-mean, of the streamwise position along the pipe, that is independent of either an inlet or outlet state, and that has conditions statistically identical at each axial position. Before we can discuss this problem, we must identify the various zones where different conditions exist.

There are four basic zones for a steady flow entering a constant area duct as shown in Fig. 11.14:

- Zone 1: Converging section
- Zone 2: Displacement interaction zone
- Zone 3: Shear layer interaction zone
- Zone 4: Fully developed flow

*An excellent treatise on turbulence models is given by W. C. Reynolds and T. Cebeci [11.1].

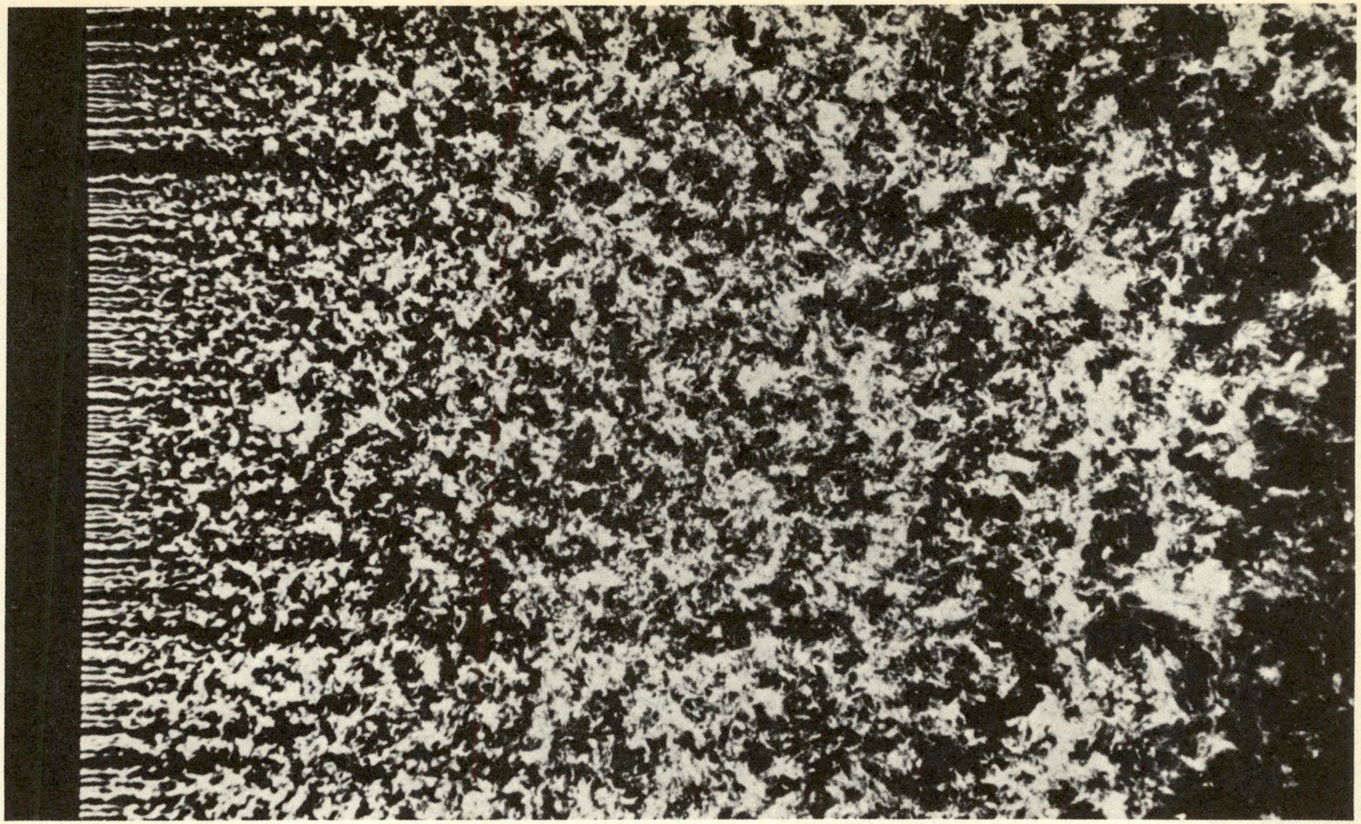

Figure 11.13 *Homogeneous turbulence behind a grid. Behind a finer grid than above, the merging unstable wakes quickly form a homogeneous field. As it decays downstream, it provides a useful approximation to the idealization of isotropic turbulence. (Source: Photograph by Thomas Corke and Hasson Nagib, from* An Album of Fluid Motion, *M. Van Dyke, Parabolic Press.)*

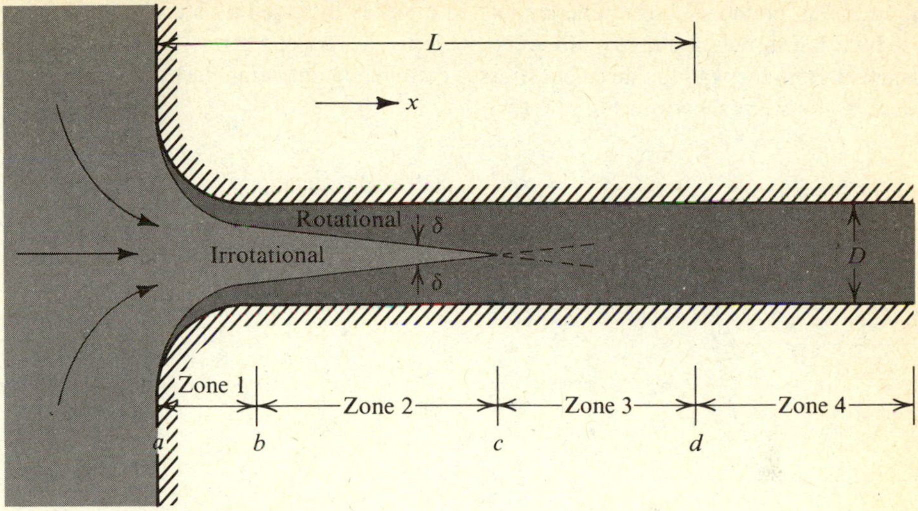

Figure 11.14 *Zones for turbulent flow in a circular pipe. (Source: Adapted from* Topics in Applied Physics, *vol. 12, "Turbulence," Springer-Verlag 1976. Used with the permission of the publisher and the author, Peter Bradshaw.)*

Zone 1 was treated in Chap. 10. The flow accelerates into the pipe where the interior of the flow is largely void of shear stresses. A thin region, close to the wall, is where significant vorticity and shear stress exist. In Zone 2, the effect of the blocking of the flow by the increased thickness of the shear stress region affects the flow's velocity in the central region by causing deviations of predicted static pressure changes between sections b and c. Note that the flows in Zones 1 and 2 are physically similar. In Zone 3, the regions of shear flow meet and overlap. This is a particularly difficult region to analyze because of the complexity of the interactions between the two regions. (J. P. Johnston [11.1] has a nice discussion of the problems in this zone.) Zone 4 is the only zone we shall treat in this chapter.

The boundary conditions for the velocity components in Eq. (11.16) are the same as for laminar flow, namely at a boundary the fluid has the same velocity as the boundary (no slip condition). This means that at the boundary the time-averaged velocity components have the same velocity as the boundary, and the fluctuating component vanishes at solid boundaries. Since the fluctuating components vanish at the solid boundary, so do the turbulent stresses. Furthermore, in the immediate neighborhood of a solid boundary, the turbulent stresses are small compared to the viscous stresses, and it follows that in every turbulent flow there exists a very thin laminar-like layer next to the solid boundary. This thin layer where viscous forces dominate is sometimes called the *laminar sublayer*, and its velocities are so small that the viscous forces dominate over the inertial forces. It does not have a definable thickness, so the term "layer" should be taken as a conceptual idea. It is a useful descriptive concept, but not quantitatively true. Experimental studies indicate the "laminar sublayer" consists of a complex vortex structure. The best that can be said is that it is neither definable nor measurable. We shall discuss this in depth in Chap. 14, where we treat the turbulent boundary layer. The laminar sublayer is followed by a transitional layer in which the turbulent stresses are comparable to the viscous stresses. At still larger distances from the wall the turbulent stresses completely outweigh the viscous stresses. These regions are shown in Fig. 11.15.

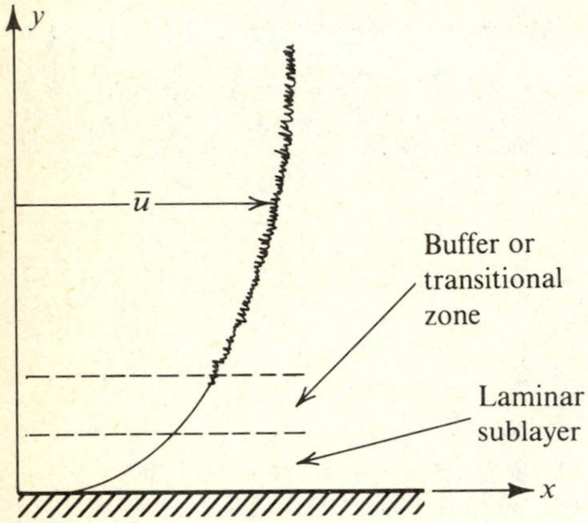

Figure 11.15 *Turbulent flow near a fixed solid boundary.*

Example 11.1

Given a velocity and pressure field with components described in terms of cylindrical coordinates by

$$v_r(r, \theta, z, t) = \overline{v}_r(r, \theta, z, t) + v_r'(r, \theta, z, t) \tag{i}$$

$$v_\theta(r, \theta, z, t) = \overline{v}_\theta(r, \theta, z, t) + v_\theta'(r, \theta, z, t) \tag{ii}$$

$$w(r, \theta, z, t) = \overline{w}(r, \theta, z, t) + w'(r, \theta, z, t) \tag{iii}$$

$$p(r, \theta, z, t) = \overline{p}(r, \theta, z, t) + p'(r, \theta, z, t) \tag{iv}$$

(a) Express the Navier-Stokes equations in terms of the average velocity components and velocity component fluctuations. (b) Derive the Reynolds stresses for cylindrical coordinates.

Solution:

Step 1.

Write the governing equations of motion.

The equations for the conservation of mass are

$$\frac{\partial \overline{v}_r}{\partial r} + \frac{\overline{v}_r}{r} + \frac{1}{r}\frac{\partial \overline{v}_\theta}{\partial \theta} + \frac{\partial \overline{w}}{\partial z} = 0 \tag{v}$$

$$\frac{\partial v_r'}{\partial r} + \frac{v_r'}{r} + \frac{1}{r}\frac{\partial v_\theta'}{\partial \theta} + \frac{\partial w'}{\partial z} = 0 \tag{vi}$$

Example 11.1 *(Con't.)*

Since the equations of linear momentum are

$$\rho\left(\frac{Dv_r}{Dt} - \frac{v_\theta^2}{r}\right) = -\frac{\partial p}{\partial r} + \mu\left(\nabla^2 v_r + \frac{v_r}{r^2} - \frac{2}{r^2}\frac{\partial v_\theta}{\partial \theta}\right) + \rho g_r \qquad \text{(vii)}$$

$$\rho\left(\frac{Dv_\theta}{Dt} + \frac{v_r v_\theta}{r}\right) = -\frac{1}{r}\frac{\partial p}{\partial \theta} + \mu\left(\nabla^2 v_\theta - \frac{v_\theta}{r^2} + \frac{2}{r^2}\frac{\partial v_r}{\partial \theta}\right) + \rho g_\theta \qquad \text{(viii)}$$

$$\rho\frac{Dw}{Dt} = -\frac{\partial p}{\partial z} + \mu\nabla^2 w + \rho g_z \qquad \text{(ix)}$$

where

$$\frac{D}{Dt} = \frac{\partial}{\partial t} + v_r\frac{\partial}{\partial r} + \frac{v_\theta}{r}\frac{\partial}{\partial \theta} + w\frac{\partial}{\partial z} \qquad \text{(x)}$$

and

$$\nabla^2 = \frac{\partial^2}{\partial r^2} + \frac{1}{r}\frac{\partial}{\partial r} + \frac{1}{r^2}\frac{\partial^2}{\partial \theta^2} + \frac{\partial^2}{\partial z^2} \qquad \text{(xi)}$$

we substitute the velocity and pressure expressions of Eqs. (i)–(iv) into the Navier-Stokes equations (vii)–(ix), putting the terms with turbulence fluctuations on the right-hand side of the equations, resulting in

$$\rho\left(\frac{D\bar{v}_r}{Dt} - \frac{\bar{v}_\theta^2}{r}\right) = -\frac{\partial \bar{p}}{\partial r} + \mu\left(\nabla^2\bar{v}_r + \frac{\bar{v}_r}{r^2} - \frac{2}{r^2}\frac{\partial \bar{v}_\theta}{\partial \theta}\right) - \rho\left[\frac{1}{r}\frac{\partial}{\partial r}(r\overline{v_r'^2}) + \frac{1}{r}\frac{\partial}{\partial \theta}(\overline{v_r' v_\theta'}) + \frac{\partial}{\partial z}(\overline{v_r' w'}) - \frac{\overline{v_\theta'^2}}{r}\right] + \rho g_r \qquad \text{(xii)}$$

$$\rho\left(\frac{D\bar{v}_\theta}{Dt} + \frac{2\bar{v}_r\bar{v}_\theta}{r}\right) = -\frac{1}{r}\frac{\partial \bar{p}}{\partial \theta} + \mu\left(\nabla^2\bar{v}_\theta - \frac{\bar{v}_\theta}{r^2} + \frac{2}{r^2}\frac{\partial \bar{v}_r}{\partial \theta}\right) - \rho\left[\frac{1}{r}\frac{\partial}{\partial \theta}(\overline{v_\theta'^2}) + \frac{\partial}{\partial r}(\overline{v_\theta' v_r'}) + \frac{\partial}{\partial z}(\overline{v_\theta' w'}) + \frac{2\overline{v_\theta' v_r'}}{r}\right] + \rho g_\theta \qquad \text{(xiii)}$$

$$\rho\frac{D\bar{w}}{Dt} = -\frac{\partial \bar{p}}{\partial z} + \mu\nabla^2\bar{w} - \rho\left[\frac{\partial \overline{w'^2}}{\partial z} + \frac{1}{r}\frac{\partial}{\partial r}(r\overline{v_r' w'}) + \frac{1}{r}\frac{\partial}{\partial \theta}(\overline{v_\theta' w'})\right] + \rho g_z \qquad \text{(xiv)}$$

Step 2.

Evaluate the stress components.

Example 11.1 *(Con't.)*

The stress components are

$$p_{rr} = -p + 2\mu \frac{\partial v_r}{\partial r} \tag{xv}$$

$$p_{\theta\theta} = -p + 2\mu \left(\frac{1}{r}\frac{\partial v_\theta}{\partial \theta} + \frac{v_r}{r}\right) \tag{xvi}$$

$$p_{zz} = -p + 2\mu \frac{\partial w}{\partial z} \tag{xvii}$$

$$p_{r\theta} = \mu \left(\frac{\partial v_\theta}{\partial r} + \frac{1}{r}\frac{\partial v_r}{\partial \theta} - \frac{v_\theta}{r}\right) \tag{xviii}$$

$$p_{\theta z} = \mu \left(\frac{1}{r}\frac{\partial w}{\partial \theta} + \frac{\partial v_\theta}{\partial z}\right) \tag{xix}$$

$$p_{zr} = \mu \left(\frac{\partial v_r}{\partial z} + \frac{\partial w}{\partial r}\right) \tag{xx}$$

Substituting the velocity and pressure expressions of Eqs. (i)–(iv) into the stress equations of Eqs. (xv)–(xx) gives

$$p_{rr} = -\overline{p} + 2\mu \frac{\partial \overline{v}_r}{\partial r} - \rho\, \overline{v_r'^2} \tag{xxi}$$

$$p_{\theta\theta} = -\overline{p} + 2\mu \left(\frac{1}{r}\frac{\partial \overline{v}_\theta}{\partial \theta} + \frac{\overline{v}_r}{r}\right) - \rho\overline{v_\theta'^2} \tag{xxii}$$

$$p_{zz} = -p + 2\mu \frac{\partial \overline{w}}{\partial z} - \rho\overline{w'^2} \tag{xxiii}$$

$$p_{r\theta} = \mu \left(\frac{\partial \overline{v}_\theta}{\partial r} + \frac{1}{r}\frac{\partial \overline{v}_r}{\partial \theta} - \frac{\overline{v}_\theta}{r}\right) - \rho\overline{v_r' v_\theta'} \tag{xxiv}$$

$$p_{zr} = \mu \left(\frac{\partial \overline{v}_r}{\partial z} + \frac{\partial \overline{w}}{\partial r}\right) - \rho\overline{v_r' w'} \tag{xxv}$$

$$p_{\theta z} = \mu \left(\frac{1}{r}\frac{\partial \overline{w}}{\partial \theta} + \frac{\partial \overline{v}_\theta}{\partial z}\right) - \rho\overline{v_\theta' w'} \tag{xxvi}$$

Using the constitutive equation for turbulent flow, Eq. (11.17), the components of the turbulence stress dyadic $\mathbf{P}'$ can be expressed as

$$\begin{bmatrix} p'_{rr} \; p'_{r\theta} \; p'_{rz} \\ p'_{\theta r} \; p'_{\theta\theta} \; p'_{\theta z} \\ p'_{zr} \; p'_{z\theta} \; p'_{zz} \end{bmatrix} = -\rho \begin{bmatrix} \overline{v_r'^2} & \overline{v_r' v_\theta'} & \overline{v_r' w'} \\ \overline{v_\theta' v_r'} & \overline{v_\theta'^2} & \overline{v_\theta' w'} \\ \overline{w' v_r'} & \overline{w' v_\theta'} & \overline{w'^2} \end{bmatrix} \tag{xxvii}$$

This completes the solution.

We have developed the basic governing dynamic relationship, Eq. (11.16), for the mathematical treatment of turbulent flow problems, or more precisely, for the calculation of the time averages of the magnitudes of the dependent variables which describe the flow. We observed that the time-averaged values of the quadratic fluctuating velocity components can be interpreted as the components of a stress dyadic, but it must be kept in mind that such an interpretation does not itself lead to quantified results. The equations we have developed do no more than point out how turbulence affects various stresses. We still have the identical number of equations we had for laminar flow, but now for turbulent flow we have four additional unknowns: u', v', w', p'.

Future theorists may provide us with those additional four equations that will enable us to have a deterministic set of equations upon which to solve turbulence problems. In the meantime, we must rely on empirical relationships, intuitive arguments, similarities, and analogs to improve upon our understanding of turbulence. The next section discusses some empirical concepts.

11.5 Zero-Equation Model for Fully Turbulent Flow

We have stated that we have no direct way to evaluate the properties (u, v, w, p, u', v', w', p') of turbulent flow. We must therefore approximate or model them in terms of properties we can determine. We shall examine the manner in which the shear stress is calculated by zero-equation turbulent models.

Over 100 years ago, Boussinesq (1877) [11.12] recommended that the effective turbulent shear stress p'_{xy} that arose from the cross correlation of fluctuating velocities $-\overline{u'v'}$ could be replaced with the product of the mean velocity gradient $\partial\bar{u}/\partial y$ and a quantity that he termed the eddy viscosity η. From Eq. (11.18), we identify

$$p'_{xy} = -\rho\overline{u'v'} \tag{11.19}$$

$$= \eta\frac{\partial\bar{u}}{\partial y} \tag{11.20}$$

The eddy viscosity η is unlike dynamic viscosity μ. The eddy viscosity is not a property of the fluid. It can vary from point to point in the flow. The introduction of η provides a framework for constructing the zero-equation turbulence model. By itself, it is not a model, since its functional dependency is unknown at this stage of the development. We shall consider how various engineers have evaluated it.

11.5.1 The Mixing Length Hypothesis (MLH)

One of the earliest approaches to estimating the Reynolds stresses was Prandtl's [11.2]. Prandtl tried to apply the notion of the mean free path (which is used in the kinetic theory of gas dynamics) to the theory of turbulence. In gas dynamics, the mean free path is easy to calculate since the particles are molecules and well defined. But in turbulence, eddies are not well defined. Prandtl introduced a path of convection, or mixing length, into the turbulent mixing picture, and let the experimentalist determine its magnitude.

Von Kármán [11.13] and Boussinesq treated the mixing length differently, and it is their concept which is usually adopted. They assumed that turbulent flow patterns are similar in the neighborhood of any two points in the flow. They differ only in their length and time scales.

One idea is based on the correlation of the velocity fluctuations in simple shear flow where $\bar{v} = 0$. Consider the location $y = y_o$ in a simple turbulent flow having fluctuation u', v' superimposed over the mean flow $\bar{u}(y)$. At $y = y_o$, let the fluid particles experience the fluctuation $v' > 0$, and have a velocity $u = \bar{u}(y)$ in the flow direction as shown in Fig. 11.16. Using the figure, if a particle of fluid moves from the y_o layer to the y layer, then an exchange of linear momentum between the two layers results. The slower particle enters the layer of the faster particles with a vertical velocity v' and acts as a drag on it. Similarly, faster particles from the y layer can enter the slower layer and tend to speed up the particles. The result is the equivalent of a shear force between the two layers. From the geometry of Fig. 11.16, we show that

$$u' = \overline{u(y)} - \overline{u(y_o)} = (y - y_o)\frac{d\bar{u}}{dy} = l\frac{d\bar{u}}{dy} \tag{11.21}$$

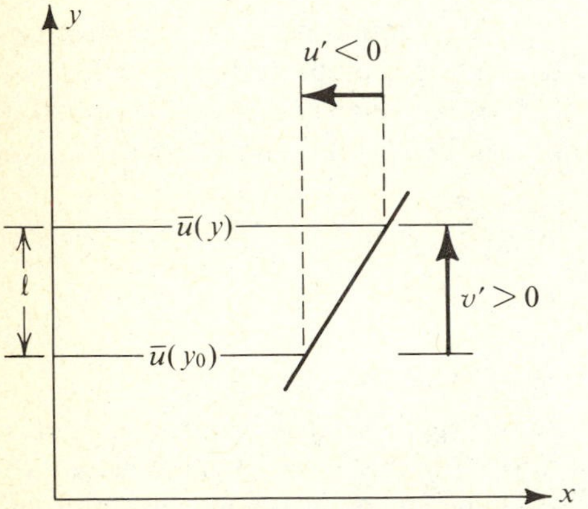

Figure 11.16 Turbulent velocity fluctuations, u', v'.

if the displacement $(y - y_o)$ is sufficiently small that a Taylor series expansion in $(y - y_o)$ can be truncated. If $u' < 0$, then $(y_o - y)(d\bar{u}/dy) < 0$; for $d\bar{u}/dy > 0$, then $u'v' < 0$. Similarly, if the fluid particle is displaced in the opposite direction, i.e., $v' < 0$, the same conclusion can be drawn since $u' > 0$ and $u'v' < 0$ due to the excess of mean velocity of the fluid particles carried into the region $y < y_o$ where the velocity $\bar{u}$ is less than $\bar{u}(y_o)$. Thus, in summary, a lump of fluid comes from a layer $y_o - l$ and has a velocity $\bar{u}(y_o - l)$. The fluid is displaced a transverse distance l

with velocity $v' > 0$. As the lump of fluid retains its original momentum, its velocity in the new lamina at y_o is smaller than the velocity prevailing there, the difference being u', given by Eq. (11.21). Similarly, a lump of fluid which arrives at y_o from the lamina $(y_o + l)$ possesses a velocity which exceeds that around it, the difference being

$$u' = \overline{u}(y_o + l) - \overline{u}(y_o) \tag{11.22}$$

The velocity differences caused by the transverse motion can be regarded as the turbulent velocity component at y_o, thus

$$\overline{|u'|} = l\left|\frac{\partial \overline{u}}{\partial y}\right| \tag{11.23}$$

where l is called the mixing length. The absolute sign is necessary because the sign for u' depends on both $\partial\overline{u}/\partial y$ and v'. A partial derivative is necessary since $\overline{u}$ may vary spatially in several directions. The term $\overline{|u'|}$ was identified by Prandtl as a random velocity that varied from place to place in a flow and was related to the eddy viscosity η through Newton's viscosity law. He reasoned that Newton's law was just as applicable to turbulent flow as it was to laminar flow, both fluids being real and coexisting in a flow. Adopting Boussinesq's idea that p'_{xy} and $\partial\overline{u}/\partial y$ might vanish together, Prandtl presumed, on the basis of the kinetic theory of gases, that the eddy viscosity η could be formed from a product of the density ρ, the mixing length l, and the random velocity $\overline{|u'|}$:

$$\eta = \rho l\overline{|u'|} \tag{11.24}$$

Substituting Eqs. (11.23) and (11.24) into the turbulent shear stress Eq. (11.20) results in the Reynolds stress

$$p'_{xy} = \rho l^2 \left|\frac{\partial \overline{u}}{\partial y}\right| \frac{\partial \overline{u}}{\partial y} \tag{11.25}$$

The sign of the Reynolds stress is positive when the shear $\partial\overline{u}/\partial y > 0$. Equation (11.25) is called the *Prandtl mixing length theory*.

Boussinesq defined the turbulent shear stress p'_{xy} in a manner similar to the Newtonian shear stress for laminar flow, i.e.,

$$(p_{xy})_{\text{total}} = (p_{xy})_{\text{lam}} + p'_{xy} \tag{11.26}$$

$$= \mu \frac{d\overline{u}}{dy} + \eta \frac{d\overline{u}}{dy}$$

$$= (\mu + \eta) \frac{d\overline{u}}{dy} \tag{11.27}$$

In applying Eq. (11.26), we should remember that the influence of viscosity is restricted to a narrow layer near a wall where $(p_{xy})_{\text{lam}}$ is significant, while beyond this layer only the turbulent shear stress p'_{xy} is of practical importance. Figure 11.17 shows the distribution of $(p_{xy})_{\text{lam}}$ and p'_{xy} in turbulent flow near a wall. In the linear laminar sublayer, laminar shear predominates. In the buffer layer, both turbulent and laminar shear are important, and in the outer layer the turbulent shear dominates. Some prefer to consider solely two layers: a wall layer where

$$0 < \frac{y}{\nu}\sqrt{\frac{(p_{xy})_o}{\rho}} < 100$$

which includes the linear sublayer and the buffer layer, all the rest being the outer layer. The velocity in the outer layer, therefore, is considered to be independent of molecular viscosity. Comparing Eqs. (11.23) and (11.24), the eddy viscosity of Boussinesq is related to Prandtl's mixing length l by

$$\boxed{\eta = \rho l^2 \left|\frac{d\bar{u}}{dy}\right|} \tag{11.28}$$

Unfortunately we still face a great difficulty in applying the Reynolds stress of Eq. (11.25). We still have no information on the functional relationship between mixing length l and position y. Von Kármán [11.13] found certain similarity conditions for the Prandtl mixing length. He stated that the mixing length magnitude is

$$l = l(y) = \kappa \frac{d\bar{u}/dy}{d^2\bar{u}/dy^2}\,^* \tag{11.29}$$

where κ is an empirical constant, is independent of the properties of a fluid, and has a value very nearly 0.4. Thus, von Kármán's model for the mixing length is that it is a point function and depends only on the velocity distribution in the vicinity of a particular point.

Another concept frequently used for the mixing length is

$$l = l(y) = \kappa y \tag{11.30}$$

which was adopted by Prandtl and used only for turbulent flow near a wall. The value of Prandtl's κ is the same as for the von Kármán expression, i.e., 0.4.

*This result can be obtained from dimensional analysis by assuming η to be a function of $d\bar{u}/dy$ and $d^2\bar{u}/dy^2$.

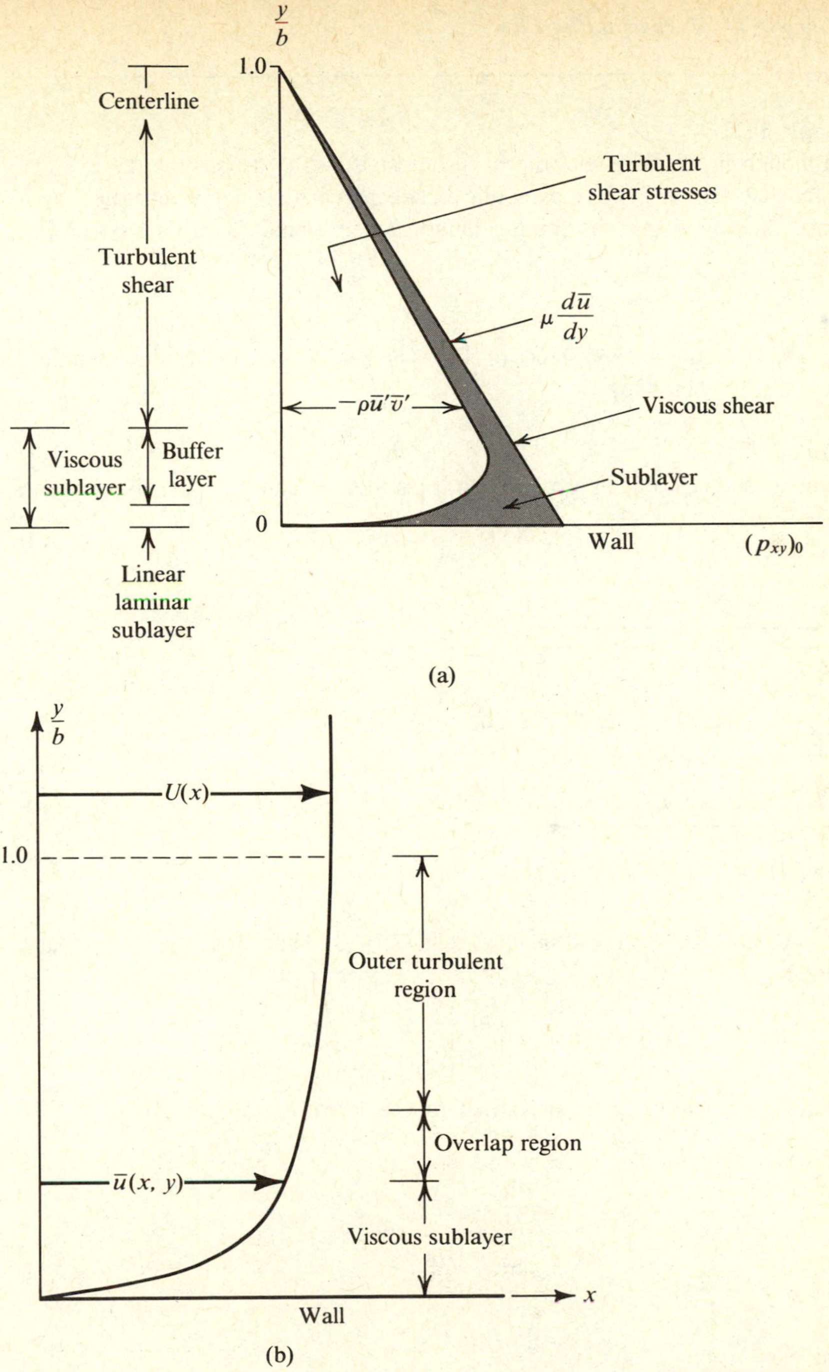

Figure 11.17 *(a) Shear stress distribution for turbulent flow past a wall. (b) Typical velocity distribution for turbulent flow past a wall. The viscous sublayer region is where viscous shear dominates, the overlap region is where both types of shear are important, and the outer turbulent region is where turbulent shear dominates. (Source: From lectures on viscous flows by T. Sarpkaya. Reproduced here by written permission of Professor T. Sarpkaya.)*

Example 11.2
Even though it can never be true in turbulent flow, let us assume the average velocity distribution $\overline{u}(y)$ is parabolic between two flat plates separated by a distance $2h$. Show that the mixing length $l(y)$ is Prandtl's expression of Eq. (11.30)

$$l = \kappa y \tag{i}$$

using von Kármán's expression of Eq. (11.28). Compare results assuming $\overline{u}(y) = u_{max} \cos(\pi y/2h)$.

Solution:
Assume as a very crude approximation a parabolic mean velocity distribution

$$\overline{u}(y) = ay^2 + b \tag{ii}$$

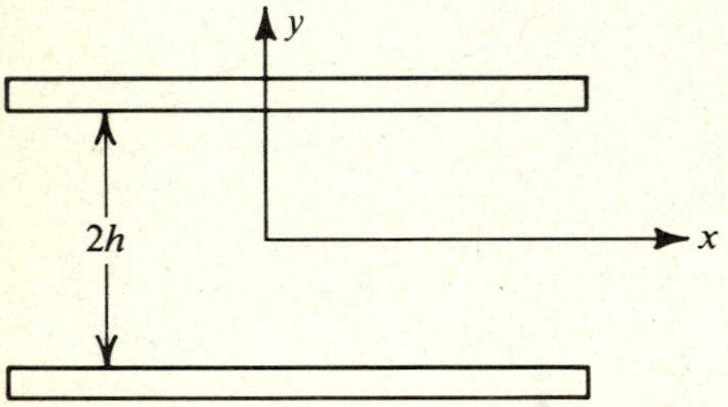

Figure E11.2a

At $y = 0$, $\overline{u}(0) = \overline{u}_{max}$, and at $y = \pm h$, $\overline{u} = 0$. Thus from these boundary conditions

$$\overline{u}(y) = \overline{u}_{max}\left[1 - \left(\frac{y}{h}\right)^2\right] \tag{iii}$$

Substituting Eq. (iii) into the von Kármán mixing length expression of Eq. (11.29) gives

$$\frac{d\overline{u}}{dy} = 2\overline{u}_{max}\frac{y}{h^2}$$

$$\frac{d^2\overline{u}}{dy^2} = \frac{2\overline{u}_{max}}{h^2}$$

so that

$$l(y) = \kappa\frac{d\overline{u}/dy}{d^2\overline{u}/dy^2} = \kappa\left[\frac{2\overline{u}_{max}\,y/h^2}{2\overline{u}_{max}/h^2}\right] \tag{iv}$$

Example 11.2 *(Con't.)*

Therefore

$$l = \kappa y \tag{i}$$

is the Prandtl mixing length expression. For

$$\bar{u} = u_{\max} \cos \frac{\pi y}{2h} \tag{v}$$

we find from Eqs. (v) and (iv)

$$l(y) = \kappa 2h \tan \frac{\pi y}{2h} \tag{vi}$$

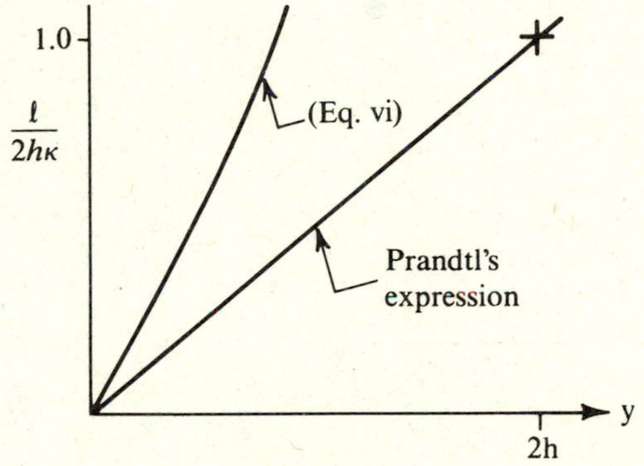

Figure E11.2b

This completes the solution.

Example 11.3

Using the same average velocity of Example 11.2, calculate the Reynolds stress for pipe flow assuming $\bar{u} = \bar{u}_{\max}$ at the centerline, and plot the dimensionless Reynolds stress $(p'_{xy}/\rho\bar{u}^2_{\max})$ versus (y/h), where h is the pipe radius.

Solution:

From Example 11.2

$$\bar{u} = u_{\max}[1 - (y/h)^2] \tag{i}$$

The mixing length was calculated to be

$$l = 0.4y \tag{ii}$$

Example 11.3 *(Con't.)*

The turbulent shear stress p'_{xy} is given by Eq. (11.25) as

$$p'_{xy} = \rho l^2 \left| \frac{d\bar{u}}{dy} \right| \frac{d\bar{u}}{dy} \tag{iii}$$

Substituting Eqs. (i) and (ii) into (iii), and simplifying gives

$$\frac{p'_{xy}}{\rho \bar{u}^2_{max}} = 0.64 \left(\frac{y}{h} \right)^4$$

and is plotted in Fig. E11.3.

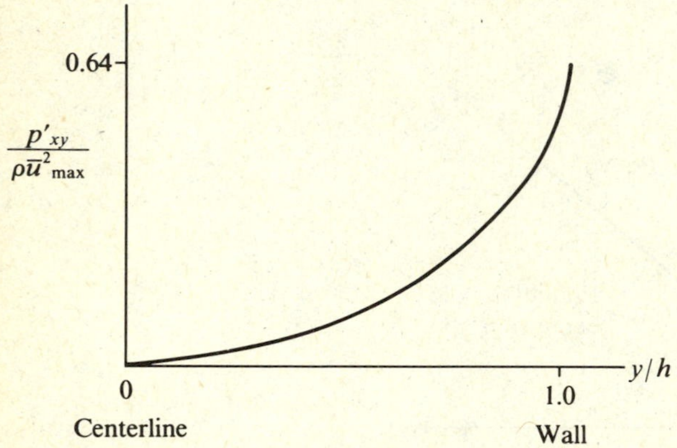

Figure E11.3

The result is a fair representation how the apparent stress behaves throughout the radial extent of the pipe. Near the centerline, the turbulent stresses are negligible and rapidly increase as one moves toward the wall. Experimental data of Sandborn* and Laufer** using hot-wire anemometry techniques show the stress decreasing in value once $y \geqslant 0.9h$.

This completes the solution.

11.5.2 Experimental Determination of Mixing Length

Using a velocity distribution obtained by experimental measurements, Nikuradse [11.14] calculated the mixing length l for flow through a smooth pipe. Using Eq. (11.25), we show the results in Fig. 11.18 for a small range of Reynolds number. In these figures, a is the inner radius of the pipe, so that $y = a - r$ denotes the distance from the wall.

*V. A. Sandborn, *NACA TN*, 3266, 1955.

**J. Laufer, *NACA TR*, 1174, 1954.

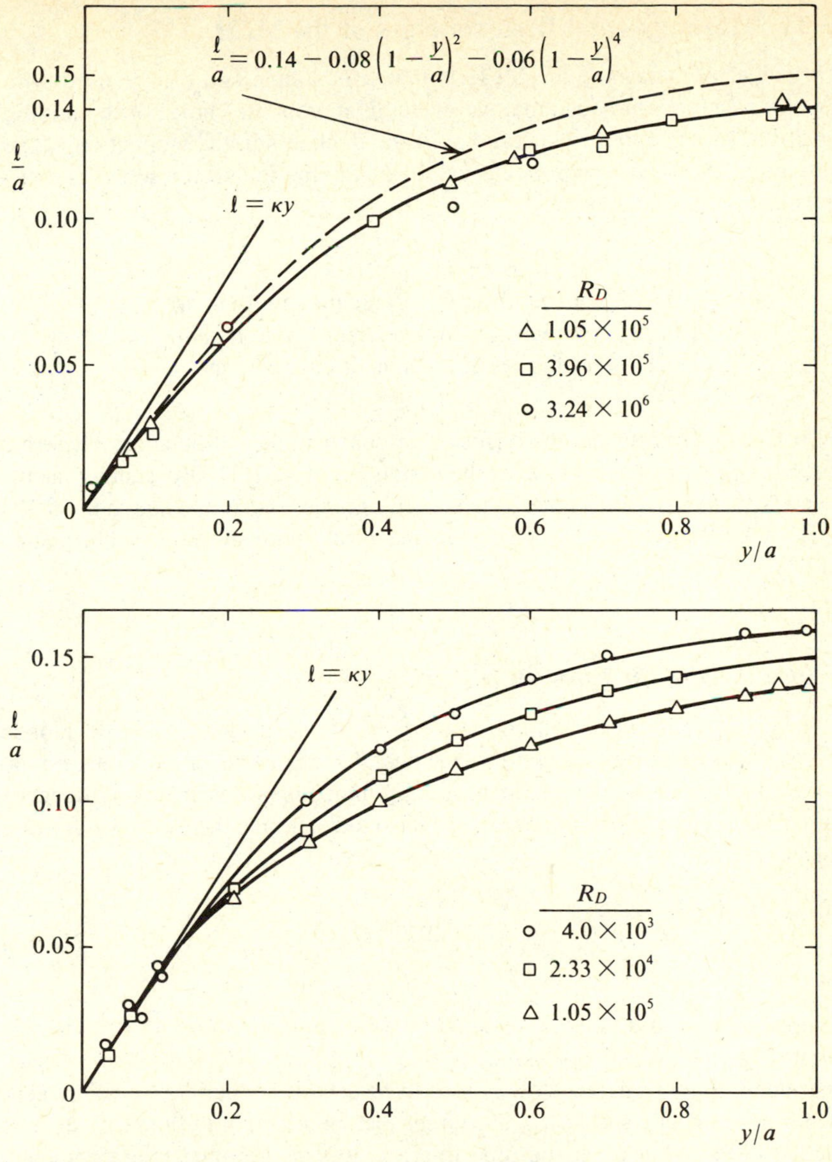

Figure 11.18 *Experimental value of mixing length.*

In obtaining these curves, Nikuradse had to differentiate numerically the velocity measurements in order to obtain l. The value of l could, perhaps, be better obtained using either Prandtl's Eq. (11.30) or von Kármán's Eq. (11.29), so that the velocity distribution is obtained by one or two integrations. Near the pipe wall, the slope of the curve in Fig. 11.18 has a value of 0.4. We have to exercise caution when we treat the region close to the wall. For example, we must modify the mixing length to account for laminar effects.

11.5.3 Advantages and Disadvantages of the MLH

The principal advantage of the MLH is that it is simple and can be used with some degree of accuracy, provided that we make an appropriate choice of the mixing length $l = l(y)$. For example, if we define δ as the thickness in the y-direction of the turbulence region, then some appropriate values of mixing length for some *free* turbulent flows are*

$$\frac{l}{\delta} = \begin{cases} 0.07, & \text{for plane mixing layer} \\ 0.075, & \text{for round jet in a stagnant environment} \\ 0.09, & \text{for plane jet in a stagnant environment} \\ 0.125, & \text{for fan jet in a stagnant environment} \end{cases}$$

The MLH does not consider the diffusion or convection of turbulence. These are two important aspects in any turbulent phenomenon. The MLH also states that the effective viscosity does not exist where the velocity gradient is zero. Though the MLH has its uses, it is destined to be replaced in the near future by more sophisticated analyses.

11.6 *Fully Turbulent Flow in a Pipe*

We must be able to know when and under what conditions a state of fully developed turbulent flow exists in a pipe or a two-dimensional duct. Because of the controversial nature of the structure of turbulence, there are no unique answers. It largely depends upon what we are seeking. For example, if we wish to determine the local skin friction coefficient c_f,

$$c_f = \frac{2(p_{xy})_o}{\rho \bar{V}^2} \tag{11.31}$$

fully developed flow is said to exist when $R_D > 3000$ and $20 \leq L/D \leq 40$, where L is defined in Fig. 11.14 and discussed in Sec. 10.2. For turbulent flow in two-dimensional ducts, fully turbulent flow is said to be achieved when $R_h = 10^5$ (where h is the depth of the duct), and $L/h > 80$. Other factors affecting turbulence, such as the turbulence level and scale at the pipe inlet, as well as the pipe inlet geometry, make the problem of determining the exact conditions extremely difficult.

Consider turbulent flow in a smooth circular pipe. Let us use Prandtl's expression for the mixing length, Eq. (11.30), in the expression for the turbulent shear stress given by Eq. (11.25), with the result

$$p'_{xy} = \rho \kappa^2 y^2 \left| \frac{d\bar{u}}{dy} \right| \frac{d\bar{u}}{dy} \tag{11.32}$$

*Launder [11.15].

We assume that the turbulent shearing stress is constant (which does not mean the total shear stress is constant). Define

$$u_* = \sqrt{\frac{p'_{xy}}{\rho}} \tag{11.33}$$

as a friction velocity. (We are treating turbulent flows at a distance y above a solid surface.)

Substituting Eq. (11.33) into Eq. (11.32) yields for $d\bar{u}/dy > 0$

$$\frac{1}{u_*}\frac{d\bar{u}}{dy} = \frac{1}{\kappa y} \tag{11.34}$$

Integrating Eq. (11.34) yields

$$u^+ \equiv \frac{\bar{u}}{u_*} = \frac{1}{\kappa}\ln y + c \tag{11.35}$$

where

$$c = a - \frac{1}{\kappa}\ln\left(\frac{a\mu}{\rho u_*}\right) \tag{11.36}$$

and is an empirical constant. Thus, for smooth pipes, the mean velocity distribution for turbulent flow is

$$u^+ = \left[G + \frac{1}{\kappa}\ln(y^+)\right] \tag{11.37}$$

where

$$y^+ = u_* y/\nu \tag{11.38}$$

and

$$G = a - \frac{1}{\kappa}\ln(a) \tag{11.39}$$

If the flow is laminar, Eq. (11.37) does not hold. For such a flow, i.e., for flow very close to the solid wall, u^+ and y^+ are linearly related to each other. Prandtl showed that

$$u^+ = y^+ \tag{11.40}$$

Equation (11.40) is applicable in a region from the wall to the limit of the viscous sublayer, $0 \leqslant y^+ \leqslant 5$. Outside this layer, in a region $5 \leqslant y^+ \leqslant 15$, bursts of fluid move outward from the wall. Experimental evidence has shown that the bursts move into regions of large velocity and serve as part of the mechanism for turbulence generation (which shall be treated in Chap. 14). The production of these bursts is a maximum in a region $11 \leqslant y^+ \leqslant 14$. The relationship between u^+ and y^+ that can be applied with less than 10% error is the logarithmic law of Eq. (11.37).

In the region $y^+ > 70$, we say that the flow is completely turbulent. This region is often called the *defect-law region*, or the *log-law region*. Here our logarithmic law of Eq. (11.37) holds very well, for $70 \leqslant y^+ \leqslant 200$. In the outer region,* $y^+ > 200$, effects of the "wake" become significant. The outer region is where the turbulent shear stress dominates. Figure 11.19 summarizes the various wall-layer regions.

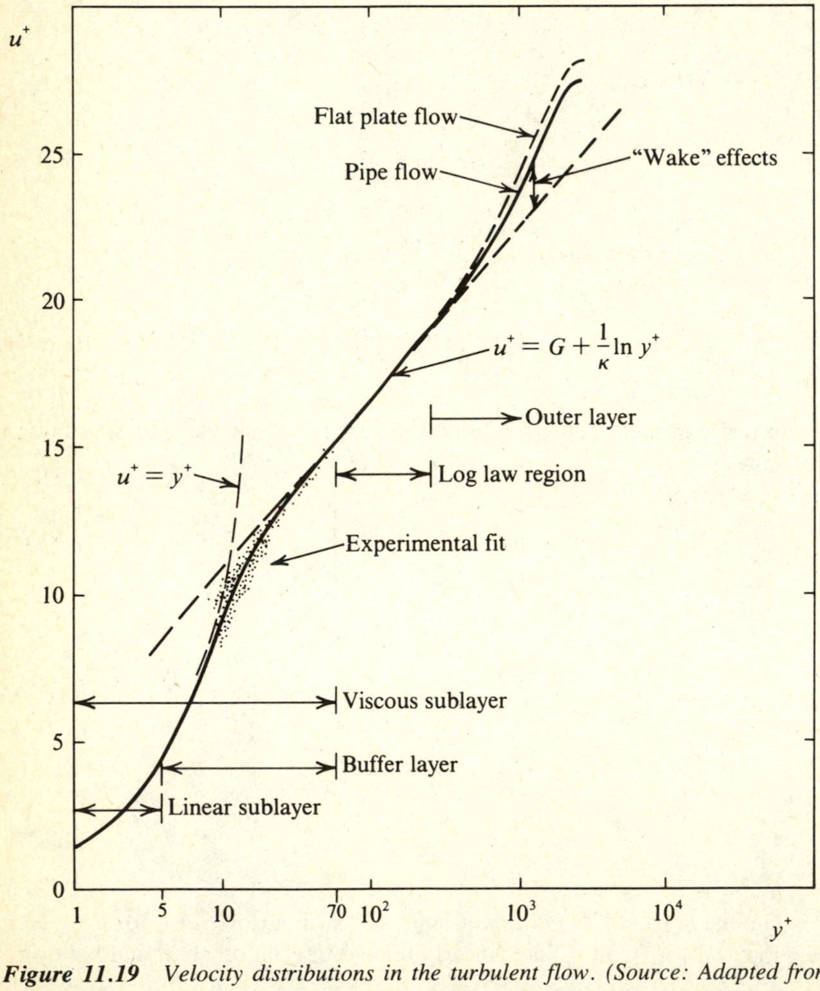

Figure 11.19 *Velocity distributions in the turbulent flow. (Source: Adapted from* Topics in Applied Physics, *vol. 12, "Turbulence," Springer-Verlag 1976. Used with the permission of the publisher and the author, Peter Bradshaw.)*

*Controversy exists as to the lower limit for y^+. Cantwell [11.10] suggests 100, and Bradshaw [11.1] suggests 300.

For turbulent pipe flow over the full range of the radius, we use the approximate values** $\kappa = 0.406$ and $G = 5.67$. (For turbulent flow over a flat plate, $\kappa = 0.42$, and $G = 5.8$.)

The foregoing analysis has been based on a smooth pipe. To properly describe wall roughness, we need only the distribution of the absolute roughness ϵ of the protrubances. What is important is the size ϵ compared to the thickness of the laminar sublayer. A pipe is considered smooth if $\rho u_* \epsilon/\mu < 5$ so that all protuberances are within the laminar sublayer, and therefore do not affect the flow. The surface is considered rough if $\rho u_* \epsilon/\mu > 70$. For this case, the protuberances are in the completely turbulent region. Hence, form drag caused by the geometric size of the protuberances in turn causes flow resistance. Viscosity plays no significant role in this area.

For rough pipes, the mean flow velocity as given by Eq. (11.37) is slightly modified to read

$$\bar{u} = u_*[A + 5.75 \ln(y/\epsilon)] \tag{11.41}$$

where A equals 8.5. Figure 11.20 plots the dimensionless mean flow velocity u/u_* versus $\rho u_* \epsilon/\mu$ for both smooth and rough pipes.

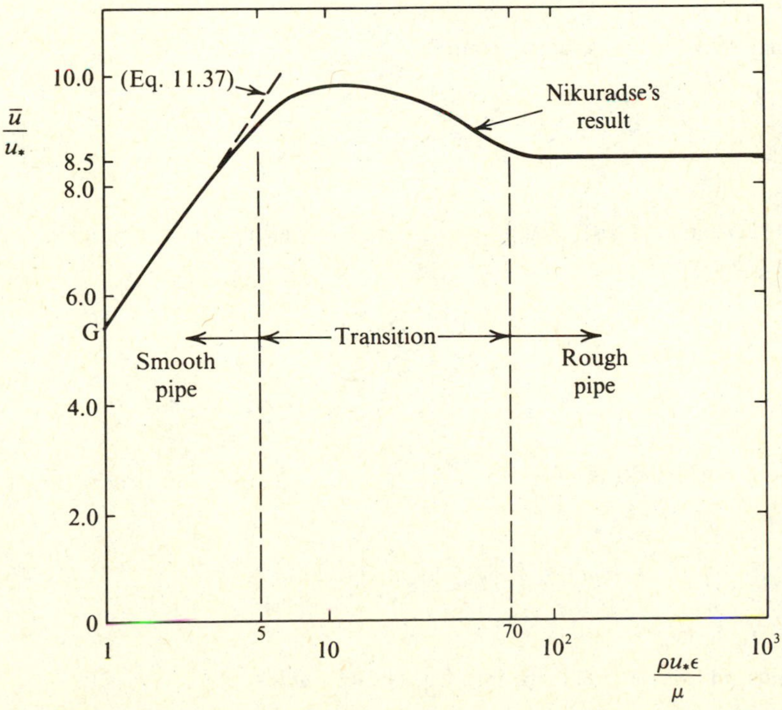

Figure 11.20 *Velocity distribution for smooth and rough pipes.*

**According to A. J. Reynolds [11.16], the best log-law fit over the entire radius (neglecting the viscous sublayer) is when $\kappa = 0.406$ and $G = 5.67$ for the tested Reynolds number range.

The friction factor f has been defined for a pipe in horizontal flow as

$$\frac{\Delta p}{\gamma} = w_\tau = f\frac{L}{D}\frac{\tilde{V}^2}{2g}$$

or in differential form

$$\frac{\partial p}{\partial x} = \frac{f}{2D}\rho\tilde{V}^2 \tag{11.42}$$

where x is in the direction of flow in the pipe. But

$$(p_{xy})_o = -\frac{D}{4}\frac{\partial p}{\partial x}$$

so that from Eq. (11.42)

$$f = \frac{8(p_{xy})_o}{\rho\tilde{V}^2} = \frac{8u_*^2}{\tilde{V}^2} \tag{11.43}$$

But the average velocity $\tilde{V}$ is simply, from Eq. (5.12),

$$\tilde{V} = \frac{1}{\pi a^2}\int_o^a \overline{u}(2\pi)(a - y)\,dy \tag{11.44}$$

Using Eqs. (11.37) and (11.39), changing to a base-10 logarithm and rearranging, we obtain for smooth pipes

$$\frac{\tilde{V}}{u_*} = 1.75 + 2.5\log\left(\frac{\rho u_* a}{\mu}\right)\Bigg|_{\text{smooth pipe}} \tag{11.45}$$

and from Eqs. (11.41) and (11.44)

$$\frac{\tilde{V}}{u_*} = 4.75 + 5.75\log\left(\frac{a}{\epsilon}\right)\Bigg|_{\text{rough pipe}} \tag{11.46}$$

Substituting Eqs. (11.45) and (11.46) into Eq. (11.43) yields

$$\frac{1}{\sqrt{f}} = 2.0\log\left(\frac{\rho\tilde{V}D}{\mu}\sqrt{f}\right) - 0.8 \tag{11.47}$$

for smooth pipes, and

$$\frac{1}{\sqrt{f}} = 2.0 \log\left(\frac{a}{\epsilon}\right) + 1.74 \tag{11.48}$$

for rough pipes. These expressions are plotted in Fig. 10.6 (the Moody diagram) where the constants of Eqs. (11.47) and (11.48) have been used to conform to experimental measurements, as 0.8 and 1.74, respectively. Equation (11.47) is frequently referred to as the Kármán-Prandtl universal friction factor equation for turbulent flow in smooth pipes. Unfortunately, Eq. (11.47) is implicit, that is, f appears in two places in the transcendental equation so that one must use either iteration or work with the Moody diagram to find f. Haaland [11.18] suggests an explicit equation

$$\frac{1}{\sqrt{f}} = 1.8 \log\left(\frac{R_D}{6{\cdot}9}\right) \tag{11.49}$$

for smooth pipes and

$$\frac{1}{\sqrt{f}} = -1{\cdot}8 \log\left[\frac{6{\cdot}9}{R_D} + \left(\frac{\epsilon}{7{\cdot}4a}\right)^{1.11}\right] \tag{11.50}$$

for the fully rough regime. Equation (11.49) approximates Eq. (11.47) within 1% for $5 \times 10^3 \leqslant R_D \leqslant 5 = 10^7$ and Eq. (11.50) reduces to the smooth pipe formula when $\epsilon/a \to 0$.

Example 11.4
Using

$$p_{xy} = (p_{xy})_o \frac{r}{a}$$

as the radial distribution of shear stress in a pipe, find the radial distribution of the mean flow velocity $\overline{u}$, where $(p_{xy})_o$ is the shear stress at the wall ($r = a$) and a is the pipe inner radius.

Solution:
The shear stress p_{xy} is expressed in terms of von Kármán's mixing length expression

$$p_{xy} = \frac{\rho\kappa^2(d\overline{u}/dr)^4}{(d^2\overline{u}/dr^2)^2} = (p_{xy})_o \frac{r}{a} \tag{i}$$

or

$$\frac{d^2\overline{u}/dr^2}{(d\overline{u}/dr)^2} = \frac{\kappa}{u_*}\sqrt{\frac{a}{r}} \tag{ii}$$

Example 11.4 *(Con't.)*

Integrating Eq. (ii) once results in

$$\frac{d\bar{u}}{dr} = \frac{1}{(2k/u_*)\sqrt{ar} + c_1} \quad \text{(iii)}$$

Integrating Eq. (iii) yields

$$\bar{u} = c_1\sqrt{r} + c_2 \ln (A\sqrt{r} + B) + c_3 \quad \text{(iv)}$$

as the mean velocity of the turbulent flow. The boundary conditions are

$$\bar{u} = \bar{u}_{max} \quad \text{at} \quad r = 0 \quad \text{(v)}$$

$$\bar{u} = 0 \quad \text{at} \quad r = a \quad \text{(vi)}$$

As an exercise, evaluate c_1, c_2, c_3, A, and B in terms of r, a, $\bar{u}_{max}$, k, and u_*.

This completes the solution.

Example 11.5

Find the relative roughness of a pipe given that $\bar{u}$ is 4.5 m/s for y/a equals 0.25, and $\bar{u}$ is 5.2 m/s for y/a equals 0.5.

Solution:

From Eq. (11.41)

$$\frac{\bar{u}}{u_*} = 5.75 \log \frac{y}{\epsilon} + 8.5 \quad \text{(i)}$$

Letting $\bar{u}_1 = 5.2$ m/s, $y_1 = 0.5a$, and $\bar{u}_2 = 4.5$ m/s, $y_2 = 0.25a$, then Eq. (i) can be expressed as

$$\frac{\bar{u}_1 - \bar{u}_2}{u_*} = 5.75 \log \frac{y_1}{y_2} \quad \text{(ii)}$$

or

$$\frac{5.2 - 4.5}{u_*} = 5.75 \log 2 = 1.7309 \quad \text{(iii)}$$

Solving for u_* gives

$$u_* = 0.404 \text{ m/s} \quad \text{(iv)}$$

Example 11.5 *(Con't.)*

Substituting the expression for u_* back into Eq. (i), we find

$$\log \frac{0.5a}{\epsilon} = \frac{1}{5.75}\left(\frac{5.2}{0.404} - 8.5\right) \quad \text{(v)}$$

or

$$\log \frac{0.5a}{\epsilon} = 0.76 \quad \text{(vi)}$$

Taking antilogs of both sides of Eq. (vi) gives

$$0.5\frac{a}{\epsilon} = 5.76 \quad \text{(vii)}$$

or the relative roughness becomes

$$\frac{\epsilon}{a} = 0.087 \quad \text{(viii)}$$

This completes the solution.

A simple piece of apparatus demonstrates a typical turbulent pipe flow problem. Head loss characteristics are studied through both the laminar and turbulent ranges. At the same time, a visual indication of the type of flow is presented.

The apparatus is from J. A. Charlton [11.17]. It follows very closely Reynolds' original apparatus that is used to visualize flow regimes. Those interested in an excellent yet simple demonstration that enables the student to plot the friction factor against Reynolds number, and to see the flow as well, are encouraged to read Charlton's article.

References

11.1 Bradshaw, P., "Turbulence," *Topics in Applied Physics*, vol. 12, Springer-Verlag, New York, 1976.

11.2 Prandtl, L., "Bericht über Untersuchungen zur ausgebildeten Turbulenz," *Z. Angew. Math. Mech.*, 5:136, 1925.

11.3 Taylor, G. I., "Eddy Motions in the Atmosphere," *Philos. Trans. R. Soc. London*, Ser. A, 215:1, 1915.
— "The Statistical Theory of Turbulence," Parts I–IV, *Proc. R. Soc. London*, Ser. A, 151:421, 1935.

11.4 Kolmogorov, A. N., "The Local Structure of Turbulence in Incompressible Flow for Very Large Reynolds Number," C.R. Acad. Sci., U.S.S.R. 30:301, 1941.

11.5 Corrsin, S., "Investigations of Flow in an Axially Symmetric Heated Jet of Air," *NACA Adv. Cond.*, Rep. 3123, 1943.
— "The Free Stream Boundaries of Turbulent Flows," *NACA TR*, 1244, 1955.
11.6 Klebanoff, P. S., "Characteristics of Turbulence in a Boundary Layer with Zero Pressure Gradient," *NACA TN*, 3278, 1954.
11.7 Townsend, A. A., *The Structure of Turbulent Shear Flow*, 1st ed., Cambridge Univ. Press, 1976.
11.8 Batchelor, G. K., *An Introduction to Fluid Dynamics*, Cambridge Univ. Press, 1967.
11.9 Orszag, S. A., and Israeli, Moshe, "Numerical Solution of Viscous Incompressible Flows," *Ann. Rev. Fluid Mech.*, 6:281, 1970.
11.10 Cantwell, B. J., "Organized Motion in Turbulent Flow," *Ann. Rev. Fluid Mech.*, 13:457, 1981.
11.11 Monin, A. S., and Yaglom, A. M., *Statistical Fluid Mechanics*, 2 vols., MIT Press, Cambridge, Mass., 1971, 1975.
11.12 Boussinesq, J., "Théorie de l'écoulement tourbillant," *Mem. Pre. par. div. Sav.*, 23, Paris, 1877.
11.13 von Kármán, T., "Mechanische Ahnlichkeit und Turbulenz," *Proc. 3rd Int. Congress Appl. Mech.*, Stockholm, pt. 1, 85, 1930.
11.14 Nikuradse, J., "Untersuchungen über die Greschwindigkeitsverteilung in Turbulenten Strönungen," *VDI-Forschungsheft*, 281, Berlin, 1926.
— "Gestzmassigkeit der Turbulenten Strömung in glatten Rohren," *VDI-Forschungsheft*, 356, 1932.
11.15 Launder, B. E., and Spalding, D. B., *Mathematical Models of Turbulence*, Academic, New York, 1972.
11.16 Reynolds, A. J., *Turbulent Flows in Engineering*, Wiley, London, 1974.
11.17 Charlton, S. A., *Bulletin Mech. Engr. Educ.*, 6:181–184, 1967.
11.18 Haaland, S. E., "Simple and Explicit Formulas for the Friction Factor in Turbulent Pipe Flow," *J. Fluids Eng.*, 105:89, 1983.

Study Questions

11.1 Define the mixing length. What is its physical description? How can one measure it? Why is it important to turbulence?

11.2 If

$$|\overline{u}'| = l\left|\frac{d\overline{u}}{dy}\right|$$

then explain how

$$|\overline{v}'| = cl\,\frac{d\overline{u}}{dy}$$

What is c?

11.3 Explain

$$\overline{u'v'} = -cl^2\left|\frac{d\overline{u}}{dy}\right|\frac{d\overline{u}}{dy}$$

Why is the absolute sign used on only one velocity gradient?

11.4 If the turbulent stresses p'_{xy} are not dependent upon viscosity, explain then how they are dependent upon $\bar{u}$ by Eq. (11.25) and $\bar{u}$ is dependent upon viscosity.

11.5 What is the significance of eddy viscosity? How does it arise, both physically and mathematically?

11.6 What kind of velocity profile is assumed if Prandtl's expression of the mixing length of Eq. (11.30) is used? What limitations are required for its use? Is it independent of Reynolds number? Is it valid near the centerline of pipe flow?

11.7 What is meant by the friction velocity u_*? Is it constant?

11.8 What is the "law of the wall"? How is the mathematical expression used, and in what flow regime is it applicable?

11.9 Discuss Eq. (11.37) and the values of G and κ.

11.10 Derive Eq. (11.30). If the friction factor f is 16 times the value of the dimensionless wall shear stress

$$\frac{(p_{xy})_o}{\frac{1}{2}\rho\bar{V}^2}$$

then how does the dependency on relative roughness ϵ/a enter the picture?

Problems

11.1 Given a 0.4 m³/s flow of gasoline through a 500 mm diameter smooth hose, calculate the friction factor f given $\mu = 0.018$ N·s/m². (Refer to Fig. P11.1.)

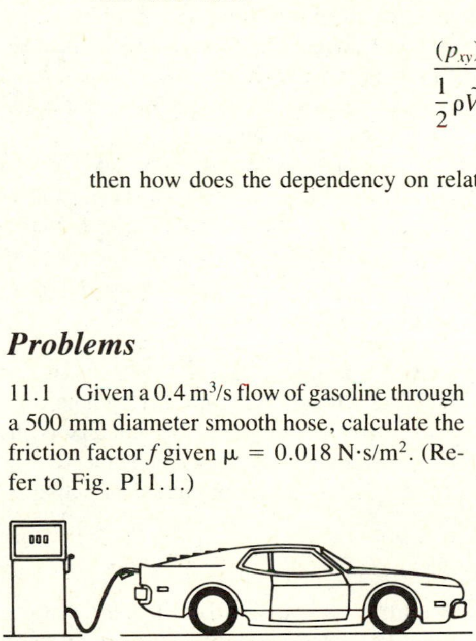

Figure P11.1

11.2 Consider kerosene flowing through a 6 in. diameter cast iron pipe (Fig. P11.2). What is the distribution of velocity in the pipe for a volume rate of flow of 10 ft³/s? What is the friction drag produced by the flow on 100 ft of pipe, if the kinematic viscosity of kerosene is 5×10^{-6} ft²/s, and the density is 1.5 slug/ft³?

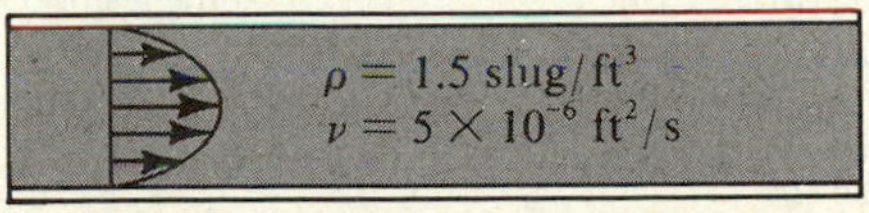

Figure P11.2

11.3 Obtain curves of various turbulent velocity profiles using $\bar{u} = \bar{u}_{max}(y/h)^{1/n}$ for n = 6, 7, 8, and 9. Discuss results.

11.4 Twenty cubic feet per second of water at 65°F is flowing through a 12 in. diameter pipe in a dam (Fig. P11.4). The pipe is assumed completely smooth. Estimate the shear stress on the pipe wetted surface.

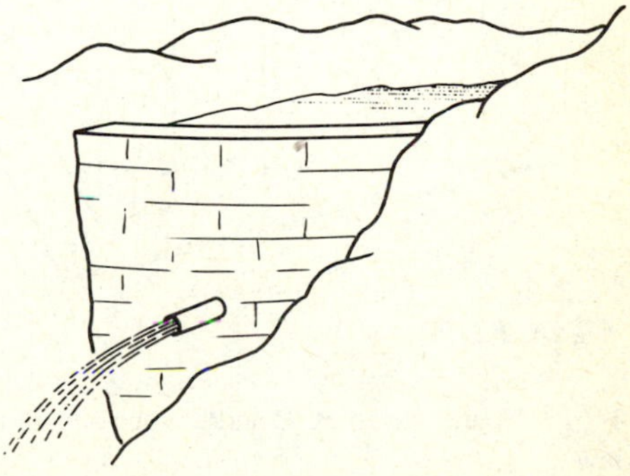

Figure P11.4

11.5 Referring to Fig. P11.5, estimate the difference between the pressure at a pipe's centerline and that at the wall if air has a maximum velocity of 200 ft/s in a tube 20 in. in diameter. Compare the result with the longitudinal pressure drop in the same distance. Let air be at 75°F and 1 atm.

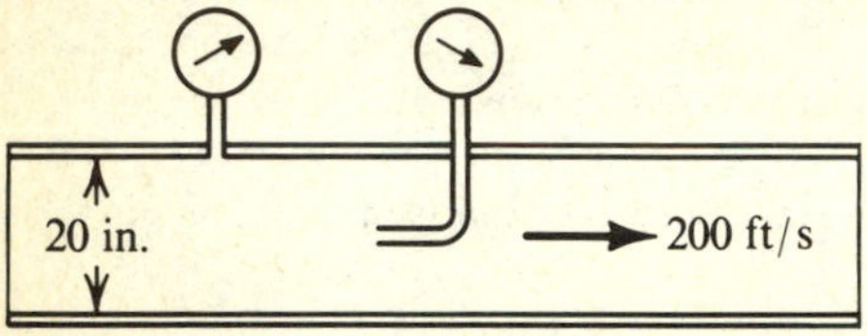

Figure P11.5

11.6 What is the shear stress distribution for fully turbulent flow between two flat surfaces given

$$\bar{u} = \bar{u}_{max}\left[1 - \left(\frac{y}{h}\right)^2\right]$$

Refer to Fig. P11.6.

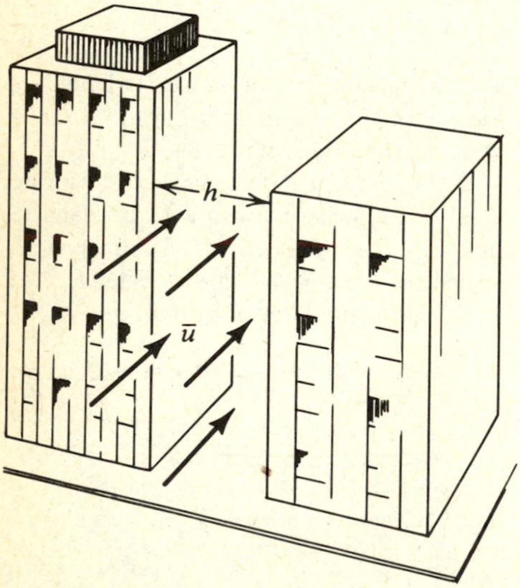

Figure P11.6

11.7 If one assumes Prandtl's one-seventh law

$$\frac{\bar{u}}{u_{max}} = \left(\frac{r}{a}\right)^{1/7}$$

determine the radial distribution of the mixing length $l = l(r)$.

11.8 When cooling water flows steadily at the rate of 16.0 l/s through 20 m of a 50 mm diameter tubing, the head loss is 0.9 m. Estimate the shear stress on the tube's wall, the relative roughness and the centerline velocity of the flow in the tube. Assume $\nu = 1.2$ mm²/s.

11.9 If the velocity of a flow is the average velocity $\bar{V}$, determine r/a in a pipe of constant diameter for turbulent flow.

11.10 Assume a mean velocity distribution $\bar{u}(y)$ that is of the form $\bar{u}(y) = ay^n$. Let the flow be through a concrete conduit of diameter R. Obtain a value of the mixing length l and compare the results with empirical results using various values for n.

11.11 Plot the eddy viscosity η versus y/h for turbulent flow between two sheets of glass using Prandtl's linear distribution of mixing length and the logarithmic velocity distribution of Eq. (11.41). (Refer to Fig. P11.11.)

11.12 Calculate the friction velocity u_* versus y/ϵ for flow in a rough pipe.

11.13 Consider turbulent flow through 1 m diameter wrought iron pipe at very high pressure. Above what Reynold's number is the flow independent of Reynold's number?

11.14 Given the shear stress distribution

$$p_{xy} = (p_{xy})_o(r/a)$$

in a tube of radius a, calculate the radial distribution of the mean velocity $\bar{u}$ in terms of $\bar{u}_{max}$, u_*, κ, r, and a.

11.15 An office cooling system moves 10,000 ft³/min of air at 70°F through 100 ft of 1 ft diameter galvanized pipe (Fig. P11.15). What is the pressure drop (ft of water)?

11.16 A pipe 100 m long and 50 mm in diameter carries water at a rate of 9 l/s from the basement of an office building to an office 50 m above the main. If the pressure at the main is 2 MPa, what is the pressure of the water at the office? (See Fig. P11.16.) (Assume a smooth pipe.)

11.17 Determine the absolute roughness in a pipe given that $\bar{u} = 15$ ft/s for y/a equal 0.38

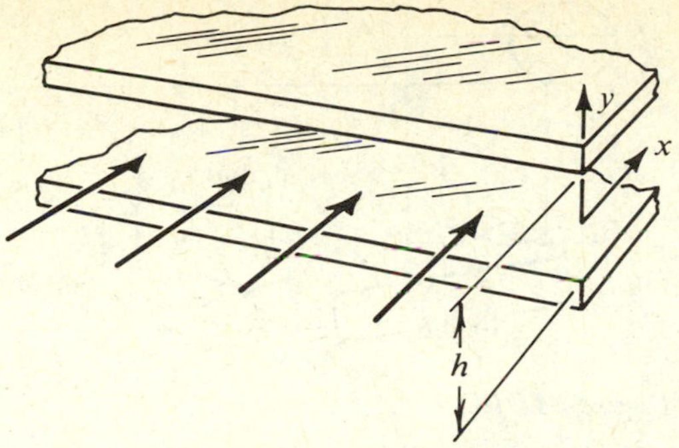

Figure P11.11

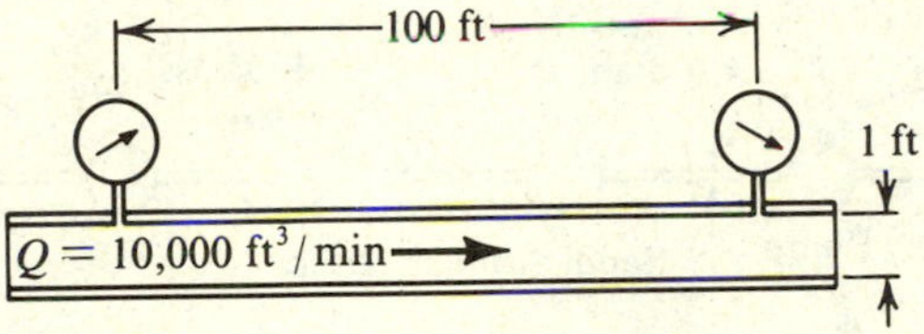

Figure P11.15

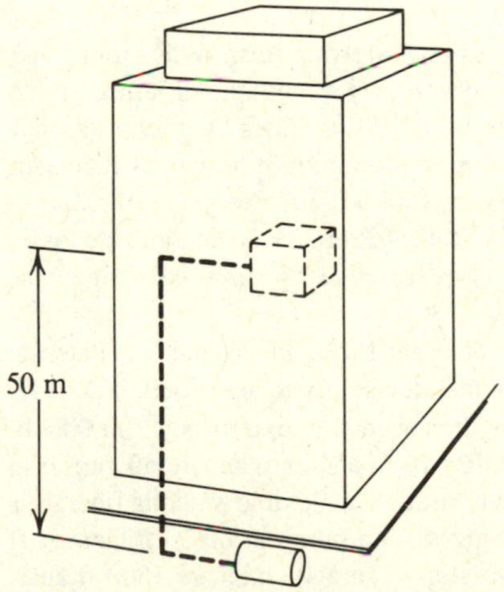

Figure P11.16

and $\overline{u}$ = 17 ft/s for y/a equal 0.76, if the pipe diameter is 1 ft.

11.18 A city's water reservoir is situated 300 ft above the town. What is the weight rate of flow for the main water line shown in Fig. P11.18? Assume a smooth pipe.

11.19 Air at a pressure of 1 bar and 23°C moves through 200 m of a 3 cm diameter smooth coil of tubing in Johns Hopkins Hospital's cardiovascular unit at a rate of 10 m^3/s. Calculate the pressure drop in the line.

11.20 Crude oil at μ = 0.158P, γ = 55 lbf/ft^3 is pumped from Ras Tanura to Yanbu through a 4 ft diameter pipe of fiberglass at the rate of 4 ft^3/s. If each pump produces 43.5 psi, how far apart should the pumps be placed? Let ϵ/D = 0.00005. (See Fig. P11.20.)

11.21 As shown in Fig. P11.21, drinking water at 25°C is being pumped through 2 km of 100 mm diameter commercial steel pipe with a flow rate of 50 l/s. Determine the pump work (m) required to maintain steady flow.

11.22 An energy loss of 100 mm of water exists through a 0.75 m diameter circular channel of length 300 m. If the channel is constructed of concrete, calculate the flow rate.

11.23 Oxygen flows steadily through a 0.2 in. diameter aluminum pipe from a gas bottle to a main at 100 psia and 85°F. What is the flow rate in 4 ft of pipe if the pressure drop is 4 in. of water (slug/min)?

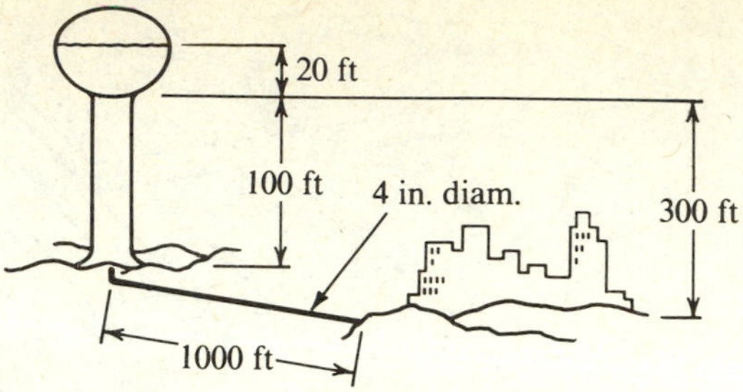

Figure P11.18

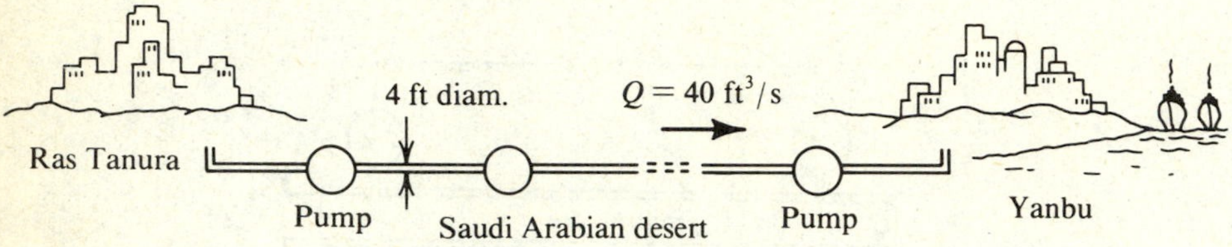

Figure P11.20

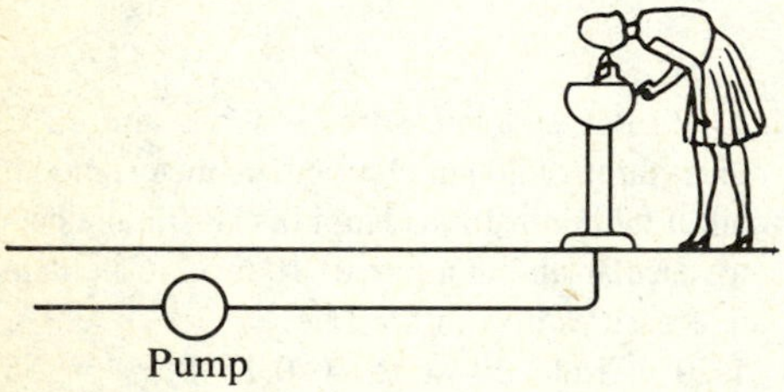

Figure P11.21

11.24 Miners deep in a coal mine require 300 lbm/min of air to breathe. If the air moves through 4650 ft of 8 in. diameter wrought iron pipe, what power is required for a blower to maintain the flow, given a pressure of 1 atm and temperature of 100°F? (See Fig. P11.24.)

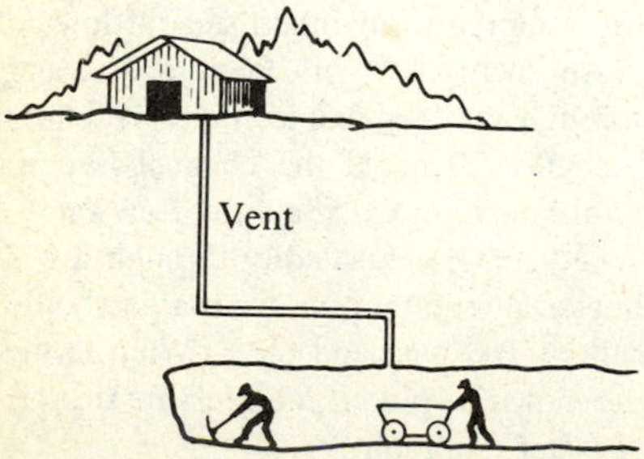

Figure P11.24

11.25 In Prob. 11.18, determine the weight rate of flow (N/s) if μ = 5 cP and ϵ = 1.2 mm.

11.26 In transferring fresh water from a lake to a reservoir, what must the difference in elevation be if 100 l/s flows by gravity through 100 m of an aluminum pipe 1 m in diameter, given ϵ = 0.00001 m? (See Fig. P11.26.)

11.27 Repeat Prob. 11.26 and include losses from a square edge opening and exit in the pipe.

11.28 Repeat Prob. 11.26 and calculate the energy lost due solely to skin friction (kJ/kg).

11.29 It is necessary to transport 100,000 lbf of gasoline from a ship to an aircraft carrier in an hour through a flexible smooth fiberglass line using a 2 hp pump at 60% efficiency. If the two ship's storage tanks are 1500 ft apart during the transfer and both tanks are at the same elevation, what size diameter line is necessary? (Refer to Fig. P11.29.)

11.30 If a 8 kW pump adds 70 kPa to a closed circuit of oil (S = 0.8, μ = 6.9 cP), what is the volume rate of flow (m^3/s)? (See Fig. P11.30.) Does one need to know whether the fluid is oil or not?

11.31 Cooling water to a nuclear reactor shows a pressure drop of 6 kPa in a 10 m section of

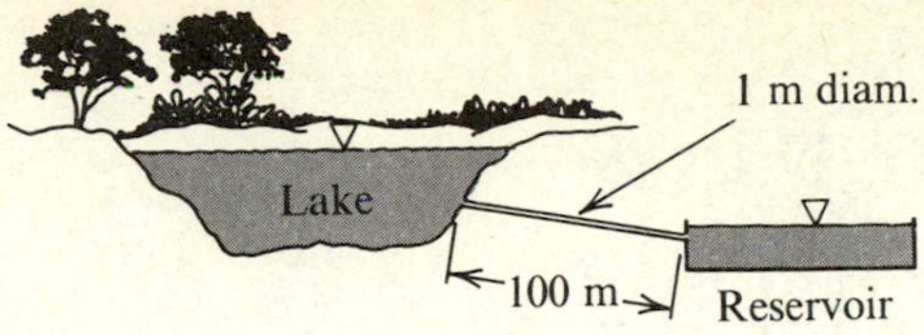

Figure P11.26

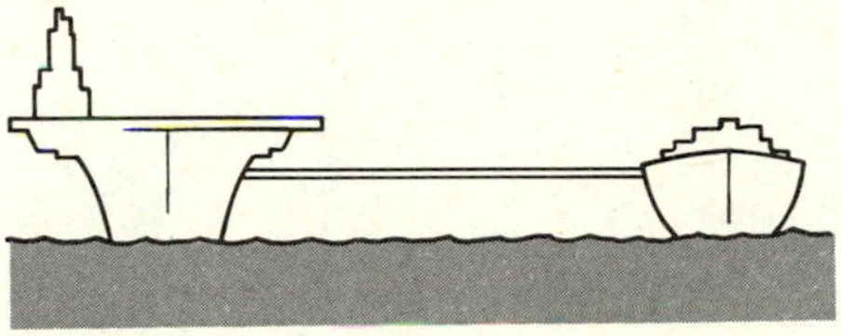

Figure P11.29

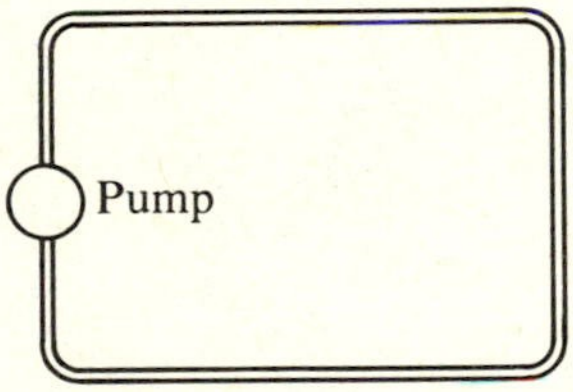

Figure P11.30

stainless pipe of constant diameter. Calculate the size diameter of the pipe and determine if there must be an obstruction in the flow. (Refer to Fig. P11.31.)

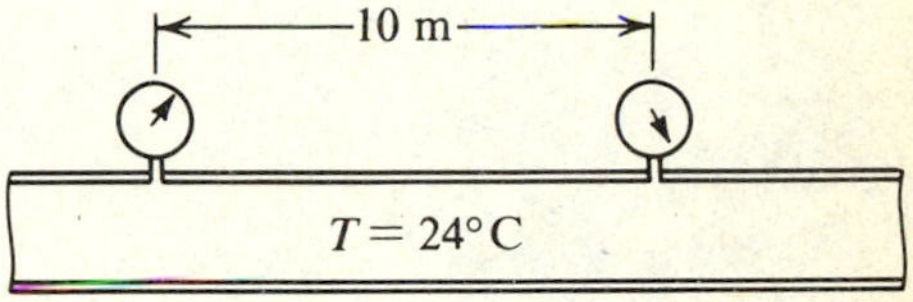

Figure P11.31

11.32 Using Eq. (11.48), plot the friction factor f versus Reynolds number for a relative roughness $\epsilon/D = 0.001$, and compare the result with the Moody diagram. Change ln to log.

11.33 Using Eq. (11.48), what is the value of relative roughness ϵ/D to generate the smooth pipe curve of Fig. 11.20?

11.34 Compare the smooth pipe curve of Fig. 11.20 with Eq. (11.48) for $R_D = 10^5$, 10^6, 10^7, and 10^9.

11.35 In Fig. P11.35, find the height h and the average velocity through the pipe entering the reservoir at station 3, given a flow rate of 2 ft^3/s of water at 40°F through cast iron pipes.

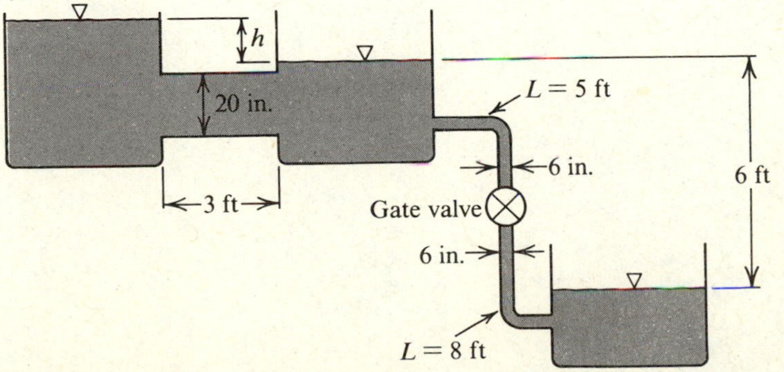

Figure P11.35

11.36 Show that

$$\rho\overline{\frac{D\mathbf{V}}{Dt}} = \rho\frac{\partial\overline{\mathbf{V}}}{\partial t} + \rho(\overline{\mathbf{V}}\cdot\nabla)\overline{\mathbf{V}} + \boldsymbol{\nabla}\cdot\rho\overline{\mathbf{V}'\mathbf{V}'}$$

11.37 Derive Reynolds equation for turbulence

$$\rho\frac{D\overline{\mathbf{V}}}{Dt} = -\nabla\overline{p} + \mu\nabla^2\overline{\mathbf{V}} - \boldsymbol{\nabla}\cdot\rho\overline{\mathbf{V}'\mathbf{V}'}$$

12 Potential Flow

12.1 Introduction

In the following sections we are going to confine ourselves to flows that are incompressible and irrotational. We will show that in using the scalar field function ϕ (which has been identified as the *velocity potential*), we can transform the differential form of the continuity equation into a differential equation that can *easily* be solved. The continuity equation, as we developed in Chap. 4, is a first-order partial differential equation in three unknowns, having three unknown velocity components. Use of the velocity potential transforms the continuity equation into a *second*-order linear partial differential equation with only one unknown: the velocity potential ϕ. This second-order partial differential equation has many solutions, each representing a different type of fluid flow. Some of the solutions are elementary in their mathematical form; others are fairly complex. For each solution, however, there is an equivalent idealized physical flow that closely duplicates a real fluid flow situation. The use of the velocity potential to describe the fluid flow is a popular technique called potential flow* theory. Currently used to describe the behavior of airplanes, boats, and automobiles moving through their respective fluids, potential flow theory is one of the most powerful vehicle design tools in the fluid dynamicist's arsenal. Part of its popularity lies in the fact it is extremely simple to use.

12.2 Laplace's Equation

The continuity equation in differential form for an incompressible fluid is given by

$$\nabla \cdot \mathbf{V} = 0 \tag{4.23}$$

If we impose the condition of irrotationality, then

$$\nabla \times \mathbf{V} = 0 \tag{12.1}$$

A solution of this differential equation is the velocity potential ϕ given by

$$\mathbf{V} = \nabla\phi \tag{4.26}$$

*Fluid dynamicists differ as to what they call potential flow, or potential theory. Some call it irrotational ideal fluid flow, some may call it incompressible irrotational fluid flow, but the proper definition is that it is the flow described by Laplace's equation, i.e., strictly irrotational motion. Potential flow theory is thus applicable to compressible flows also.

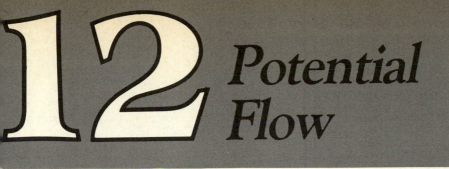

which, when substituted into Eq. (4.23), results in

$$\nabla \cdot \nabla\phi = \nabla^2\phi = 0 \tag{12.2}$$

This is called *Laplace's equation*. If we wish to express Laplace's equation in Cartesian coordinates, then Eq. (12.2) becomes

$$\frac{\partial^2\phi}{\partial x^2} + \frac{\partial^2\phi}{\partial y^2} + \frac{\partial^2\phi}{\partial z^2} = 0 \tag{12.3}$$

Using cylindrical coordinates, we express Eq. (12.2) as

$$\frac{\partial^2\phi}{\partial r^2} + \frac{1}{r}\frac{\partial\phi}{\partial r} + \frac{1}{r^2}\frac{\partial^2\phi}{\partial\theta^2} + \frac{\partial^2\phi}{\partial z^2} = 0 \tag{12.4}$$

The velocity potential ϕ for every possible irrotational motion of incompressible fluid flow *must* satisfy the Laplace Eq. (12.2). Most ideal fluid motions are irrotational (see Sec. 4.4.7 and page 656). We recall that it is the angular deformations created by shear stresses that set fluid particles rotating; since an ideal fluid has no shear stresses, then no particles can originate in rotation.

In 1752, L. Euler [1.7] first suggested the form $\nabla^2\phi = 0$ in a fluid dynamics treatise. This form was later used by Laplace, the great French mathematician, when he discussed the utilization of Newtonian potentials in polar coordinate form. Four years later, Laplace studied Cartesian coordinates, and actually treated the solutions of Eq. (12.3). So the equation has a long history.

Scalar field functions that *satisfy* Laplace's equation are termed *harmonic* functions. Hence, any harmonic function represents a possible irrotational fluid motion. We know, however, that a general solution requires satisfying not only the differential equation but also specific *boundary conditions*. This latter requirement often makes the general solution of Laplace's equation difficult.

Example 12.1

Show that $\phi = x^3t + 2y^2t - 3txz^2 - 2z^2t$ is a possible velocity potential for a three-dimensional incompressible potential fluid flow.

Solution:

Step 1.

The fluid flow is three-dimensional, unsteady, incompressible, and irrotational.

Step 3.

The appropriate governing equation of motion is Laplace's equation:

$$\nabla^2\phi = 0 \tag{i}$$

Substituting

$$\phi = x^3t + 2y^2t - 3txz^2 - 2z^2t \tag{ii}$$

Example 12.1 *(Con't.)*

into Eq. (i) gives

$$\frac{\partial \phi}{\partial x} = 3x^2t - 3tz^2$$

$$\frac{\partial^2 \phi}{\partial x^2} = 6xt \tag{iii}$$

$$\frac{\partial \phi}{\partial y} = 4yt$$

$$\frac{\partial^2 \phi}{\partial y^2} = 4t \tag{iv}$$

$$\frac{\partial \phi}{\partial z} = -6txz - 4zt$$

$$\frac{\partial^2 \phi}{\partial z^2} = -6xt - 4t \tag{v}$$

Substituting Eqs. (iii)–(v) into Eq. (i) gives

$$\frac{\partial^2 \phi}{\partial x^2} + \frac{\partial^2 \phi}{\partial y^2} + \frac{\partial^2 \phi}{\partial z^2} = 6xt + 4t - 6xt - 4t = 0 \tag{vi}$$

Thus, Laplace's equation is satisfied, and the flow is irrotational.

This completes the solution.

Example 12.2

Given the velocity field $\mathbf{V} = 6\mathbf{i} + 8\mathbf{j} - 10\mathbf{k}$, (a) show that the flow is irrotational, (b) find the velocity potential ϕ.

Solution:

Step 1.

The fluid flow is three-dimensional, steady, and incompressible.

Step 3.

The appropriate equation is Laplace's equation:

$$\nabla^2 \phi = 0 \tag{i}$$

- Definition of velocity potential:

$$\mathbf{V} = \boldsymbol{\nabla}\phi \tag{ii}$$

(a) To show that the flow is indeed irrotational, we must show that the vorticity ζ vanishes; that is,

Example 12.2 *(Con't.)*

$$\boldsymbol{\zeta} = \nabla \times \mathbf{V} \tag{iii}$$

$$= \nabla \times (6\mathbf{i} + 8\mathbf{j} - 10\mathbf{k})$$

$$= \left(\mathbf{i}\frac{\partial}{\partial x} + \mathbf{j}\frac{\partial}{\partial y} + \mathbf{k}\frac{\partial}{\partial z}\right) \times (6\mathbf{i} + 8\mathbf{j} - 10\mathbf{k}) \tag{iv}$$

Since the velocity field is independent of spatial variation, the curl of a constant vector field is zero. Thus,

$$\boldsymbol{\zeta} = 0 \tag{v}$$

(b) Given

$$\mathbf{V} = 6\mathbf{i} + 8\mathbf{j} - 10\mathbf{k} \tag{vi}$$

we find, using Eq. (ii), the velocity components

$$u = 6 = \frac{\partial \phi}{\partial x} \tag{vii}$$

$$v = 8 = \frac{\partial \phi}{\partial y} \tag{viii}$$

$$w = -10 = \frac{\partial \phi}{\partial z} \tag{ix}$$

Integrating Eqs. (vii)–(ix), we obtain

$$\phi = 6x + f_1(y) + f_2(z) \tag{x}$$

$$\phi = 8y + f_3(x) + f_4(z) \tag{xi}$$

$$\phi = -10z + f_5(x) + f_6(y) \tag{xii}$$

respectively. Comparing Eqs. (x)–(xii) with one another, we note

$$f_1(y) = f_6(y) = 8y \tag{xiii}$$

$$f_2(z) = f_4(z) = -10z \tag{xiv}$$

$$f_3(x) = f_5(x) = 6x \tag{xv}$$

such that we obtain

$$\phi = 6x + 8y - 10z \tag{xvi}$$

as the velocity potential that will produce the given velocity field. (A constant may or may not be included in Eq. (xvi).)

This completes the solution.

12.2.1 Methods of Solving Laplace's Equation

Laplace's equation can be solved in various ways:

- Direct Method. We directly integrate the first-order partial differential equation given by Eq. (4.26): $\mathbf{V} = \nabla\phi$. This method is not used except in special academic cases; for instance, given the velocity $\mathbf{V}$, find the potential ϕ.
- Indirect Method. Using trial and error we combine analytic expressions so as to satisfy both Laplace's equation and boundary conditions. Most of the fundamental solutions (such as the source, vortex, and doublet which, we will consider in the following sections) appear to have been found using this method. We seek to find some function ϕ that satisfies Laplace's equation, and then we ask what boundary conditions it can satisfy. This method will be used in Sec. 12.5.
- Combinations of Elementary Flows. This method is based on the superposition principle that fundamental solutions are combined to represent more complex flow solutions. This will be illustrated in Sec. 12.10.
- Analytic Solutions. Many mathematical techniques exist that enable us to collapse or transform a linear partial differential equation into ordinary differential equations. Besides familiar techniques such as separation of variables, others might involve similarity transformations, where we express two or three variables in terms of some new variable.
- Conformal Transformations. Conformal transformations involve changing variables using a particular mapping technique. We map *known* simple flow solutions in a unique fashion onto geometries for which the potential is not known. For instance, suppose we know the potential function relationship for flow past a cylinder, and suppose we want to know the potential function relationship for flow past a particular wing. Using a particular mathematical transformation equation, we map all points lying on the wing surface to points on the circumference of a circular cylinder. Conformal mapping will be discussed later in this chapter. A good treatment of conformal mapping can be found in Chap. 9 of Ref. 12.1.
- Numerical Solution. A number of high-speed computing programs enable us to solve Laplace's equation. The more popular ones involve finite differencing and finite element numerical schemes.* We shall not discuss these in this text. For those interested in pursuing this aspect, the text by Roache [12.2] is superb.

12.3 The Complex Potential, Ω

In Chap. 8 the intimate relationship between the stream function ψ and velocity potential ϕ was given for planar incompressible flows. We showed for instance, that they were orthogonal to each other. Since a mathematical relationship ties the two functions together (the Cauchy-Riemann condition), then it is mathematically conceivable that a new function takes advantage of such a relationship. We can accomplish this using

*An excellent description of the use of the finite element method in potential fluid flow is given by A. T. Sayers, I. Mech. E. & UMIST, *International J. Mech. Engrg. Ed.*, vol. 4, 4 pp. 309–318, 1976.

analytic functions of a complex variable. We are taking this tack because the advantages of working with complex variables will reward us with a simplified technique for evaluating the hydrodynamic forces on solid bodies in potential flows.

One of the more significant relationships in potential theory is defined by the Cauchy-Riemann equations developed in Chap. 8. We showed that if the potential function ϕ and stream function ψ are real, single-valued functions of space, with their four first partial derivatives continuous throughout a fluid region R, then they are related to each other by

$$\frac{\partial \phi}{\partial x} = \frac{\partial \psi}{\partial y} \quad \text{and} \quad \frac{\partial \phi}{\partial y} = -\frac{\partial \psi}{\partial x} \tag{12.5}$$

in Cartesian coordinate form, and

$$\frac{\partial \phi}{\partial r} = \frac{1}{r}\frac{\partial \psi}{\partial \theta} \quad \text{and} \quad \frac{1}{r}\frac{\partial \phi}{\partial \theta} = -\frac{\partial \psi}{\partial r} \tag{12.6}$$

in polar coordinate form.

We define a new function, a complex function

$$\Omega = \phi + i\psi \tag{12.7}$$

Since ϕ and ψ are analytic in R, then Ω must be analytic in R. We call Ω the *complex velocity potential*. The real part of the complex potential is therefore the velocity potential ϕ, and the imaginary part is the stream function ψ. If we let the abscissa be the velocity potential, and the ordinate the stream function, then this complex potential Ω generates a *flow net* of the fluid flow like that shown in Fig. 12.1. Let us look at some illustrations.

Consider the complex velocity potential to be of the form

$$\Omega = Uz \tag{12.8}$$

where U is the average velocity and z is the independent complex variable

$$z = x + iy \tag{12.9}$$

in the Cartesian frame, and

$$z = re^{i\theta} \tag{12.10}$$

in the polar frame. The real and imaginary parts of the complex velocity potential are easily identified as

$$\phi = Ux, \quad \psi = Uy \tag{12.11}$$

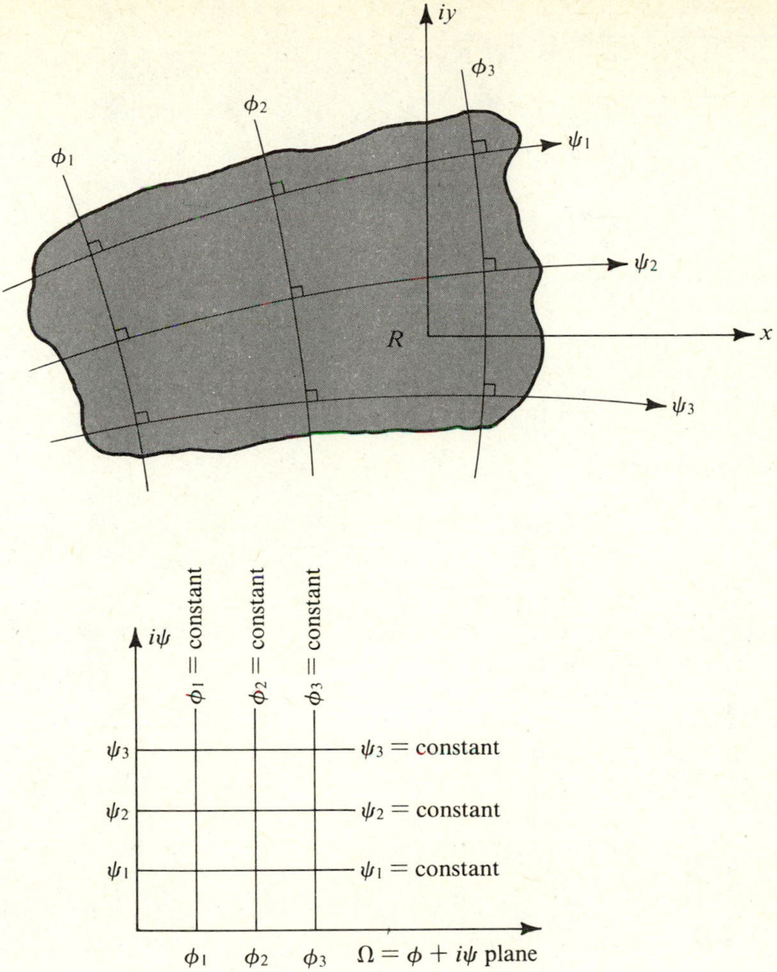

Figure 12.1 *Potential lines and streamlines in a typical inviscid flow net.*

The results of the streamlines for this flow are shown in Fig. 12.2. Since by definition no fluid crosses ψ = const., then any ψ = const. locus can be taken as the boundary of a solid object.

Consider a second illustration. Suppose our complex velocity potential is

$$\Omega = \frac{Ua^2}{z} \tag{12.12}$$

where a is some length. Once again, we identify the real and imaginary parts of the complex velocity potential so that the velocity potential ϕ is

$$\phi = \frac{Ua^2}{r} \cos \theta = \frac{Ua^2 x}{x^2 + y^2} \tag{12.13}$$

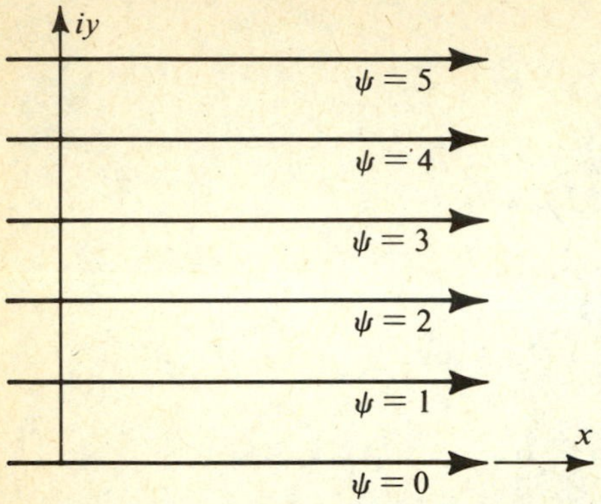

Figure 12.2 *Streamlines for $\Omega = Uz$.*

and the stream function ψ is

$$\psi = -\frac{Ua^2}{r}\sin\theta = -\frac{Ua^2y}{x^2 + y^2} \tag{12.14}$$

Setting ψ = const., we see from Eq. (12.14) that the streamlines are circles that touch the x-axis at the origin, as shown in Fig. 12.3.

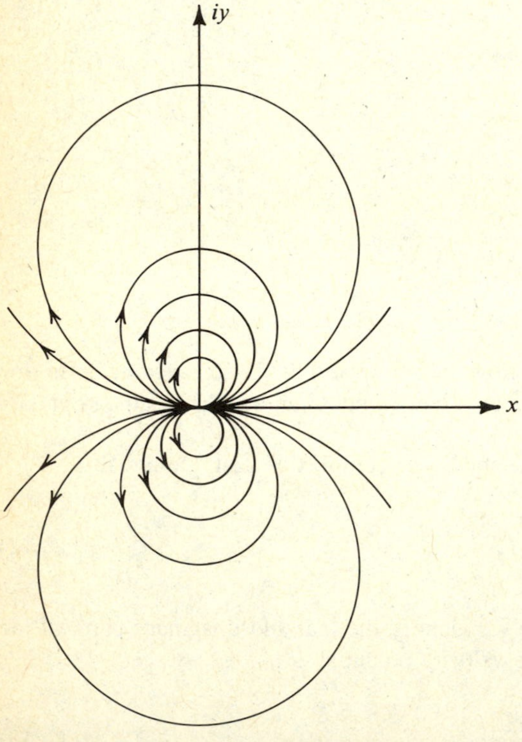

Figure 12.3 *Streamlines for $\Omega = Ua^2/z$.*

12.4 The Complex Velocity, $d\Omega/dz$

The complex velocity $d\Omega/dz$ can be easily found from the complex potential Ω in the following manner. From Eq. (12.7), consider the total change of the complex potential Ω:

$$d\Omega = d\phi + i d\psi \tag{12.15}$$

where the total change of the velocity potential is

$$\begin{aligned} d\phi &= \frac{\partial \phi}{\partial x} dx + \frac{\partial \phi}{\partial y} dy \\ &= u dx + v dy \end{aligned} \tag{12.16}$$

and the total change of the stream function is

$$\begin{aligned} d\psi &= \frac{\partial \psi}{\partial x} dx + \frac{\partial \psi}{\partial y} dy \\ &= -v dx + u dy \end{aligned} \tag{12.17}$$

Substituting Eqs. (12.16) and (12.17) into Eq. (12.15) yields

$$\begin{aligned} d\Omega &= u dx + v dy + i(-v dx + u dy) \\ &= (u - iv)(dx + i dy) \\ &= (u - iv) dz \end{aligned} \tag{12.18}$$

Thus, the total change of Ω is expressed in terms of the velocity field as

$$\boxed{\frac{d\Omega}{dz} = u - iv} \tag{12.19}$$

where the real part of $d\Omega/dz$ is the x component of velocity u, and the imaginary part of $d\Omega/dz$ is the negative of the y component of velocity v.

12.4.1 Stagnation Points

At a stagnation point, the velocity of the flow is zero. Starting with the expression for the complex velocity field, Eq. (12.19), we can find the stagnation points by setting the complex velocity equal to zero and evaluating the values of the independent variables:

$$\boxed{\frac{d\Omega}{dz} = 0} \tag{12.20}$$

12.4.2 The Speed

The speed V is given by

$$V = \sqrt{u^2 + v^2} = \left|\frac{d\Omega}{dz}\right| \tag{12.21}$$

Thus,

$$V^2 = u^2 + v^2 = (u - iv)(u + iv)$$

or

$$\boxed{V^2 = \left(\frac{d\Omega}{dz}\right)\left(\frac{d\overline{\Omega}}{d\bar{z}}\right)} \tag{12.22}$$

Example 12.3
Consider an irrotational incompressible flow field where the complex velocity potential is given as

$$\Omega = z + \ln z \tag{i}$$

(a) Calculate the velocity potential $\phi(r, \theta)$, $\phi(x, y)$, and the stream function $\psi(r, \theta)$, $\psi(x, y)$.
(b) Show that the complex potential $\Omega(z)$ is analytic.
(c) Calculate the radial and tangential velocity components $v_r(r, \theta)$ and $v_\theta(r, \theta)$, respectively.
(d) Find the (x, y) location of the stagnation points.
(e) What is the expression for the speed $V(x, y)$?

Solution:
Step 1.
The flow is incompressible potential flow.
Step 3.
The appropriate equations are:

- Definition of velocity potential:

$$\mathbf{V} = \nabla\phi \tag{ii}$$

- Definition of complex velocity potential:

$$\Omega = \phi + i\psi \tag{iii}$$

- Definition of stagnation:

$$\frac{d\Omega}{dz} = 0 \tag{iv}$$

Example 12.3 *(Con't.)*

(a) From Eqs. (i) and (iii)

$$\Omega = \phi + i\psi = z + \ln z \tag{v}$$

Substituting Eqs. (12.11) and (12.12) into Eq. (v) gives

$$\phi + i\psi = x + iy + \ln r + i\theta \tag{vi}$$

$$= r\cos\theta + \ln r + i(r\sin\theta + \theta) \tag{vii}$$

Equating the real parts of Eq. (vii) gives

$$\phi(r, \theta) = r\cos\theta + \ln r \tag{viii}$$

and

$$\phi(x, y) = x + \tfrac{1}{2}\ln(x^2 + y^2) \tag{ix}$$

Equating the imaginary parts of Eq. (vii) gives

$$\psi(r, \theta) = r\sin\theta + \theta \tag{x}$$

and

$$\psi(x, y) = y + \tan^{-1}\frac{y}{x} \tag{xi}$$

(b) For $\Omega(z)$ to be analytic, it must satisfy the Cauchy-Riemann conditions. From Eqs. (8.29), (viii), and (x)

$$\frac{\partial\phi}{\partial r} = \cos\theta + \frac{1}{r} \stackrel{?}{=} \frac{1}{r}\frac{\partial\psi}{\partial\theta} = \frac{r\cos\theta}{r} + \frac{1}{r} \tag{xii}$$

and from Eqs. (8.30), (viii), and (x)

$$\frac{1}{r}\frac{\partial\phi}{\partial\theta} = -\frac{r\sin\theta}{r} \stackrel{?}{=} -\frac{\partial\psi}{\partial r} = \sin\theta \tag{xiii}$$

Thus $\Omega(z)$ is analytic.

(c) The radial component of velocity $v_r(r, \theta)$ is found using Eq. (xii):

$$v_r(r, \theta) = \frac{\partial\phi}{\partial r} = \cos\theta + \frac{1}{r} \tag{xiv}$$

The tangential velocity component $v_\theta(r, \theta)$ is obtained from Eq. (xiii):

$$v_\theta(r, \theta) = \frac{1}{r}\frac{\partial\phi}{\partial\theta} = -\sin\theta \tag{xv}$$

(d) We calculate the stagnation point using Eqs. (iv) and (i):

$$\frac{d\Omega}{dz} = 1 + \frac{1}{z} = 0 \tag{xvi}$$

Example 12.3 *(Con't.)*

Solving for z gives

$$z = -1 \tag{xvii}$$

Substituting the definition of the complex variable z of Eq. (12.11) into Eq. (xvii) gives

$$\begin{aligned} x &= -1 \\ y &= 0 \end{aligned} \tag{xviii}$$

which is the (x, y) location of the stagnation point.

(e) The expression for the speed is given by Eq. (12.21):

$$\begin{aligned} V &= \sqrt{v_r^2 + v_\theta^2} \\ &= \sqrt{\left(\cos\theta + \frac{1}{r}\right)^2 + (\sin\theta)^2} \qquad \text{(xix)} \\ &= \sqrt{1 + \frac{2\cos\theta}{r} + \frac{1}{r^2}} \end{aligned}$$

Note that at $r = 1$, $\theta = \pi$, the speed V is zero, which is our stagnation point.

This completes the solution.

12.5 Complex Potential for Fundamental Flows

We will investigate some fundamental solutions of Laplace's equation and then examine some boundary conditions relating the solutions to equivalent physical flows. This method of solving Laplace's equation is called the indirect method, as explained in Sec. 12.2.1. We shall investigate solutions for the following elementary flows: uniform flow, sources and sinks, a vortex, and a doublet. Though there are an inexhaustible number of flows we could consider, these few will allow us to grasp the general idea of potential flow theory. All of what follows is not simply a mathematical exercise. When we realize that all the giant aircraft companies use potential flow theory to design their commercial and military aircraft, then the topic should lose some of its purely academic flavor, and gain the practical engineer's respect.

12.5.1 Uniform Flow

Consider a uniform flow of magnitude U inclined at an angle α with respect to the x-axis, as shown in Fig. 12.4. The x- and y-components of fluid velocity are related to U using the geometry of Fig. 12.4 as

$$u = U\cos\alpha \tag{12.23}$$

$$v = U\sin\alpha \tag{12.24}$$

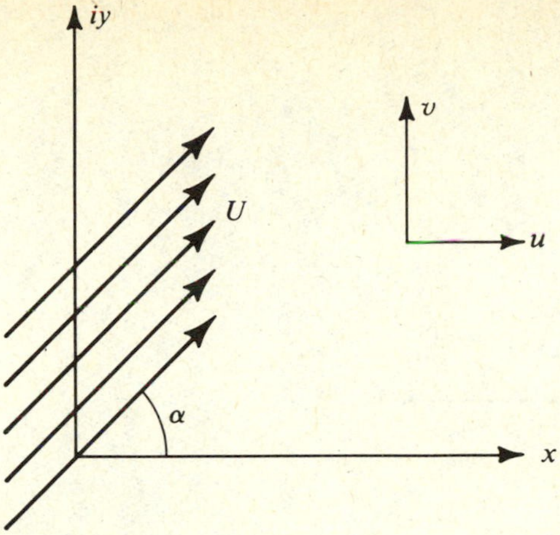

Figure 12.4 *Uniform flow.*

In order to find the complex velocity potential Ω, we substitute Eqs. (12.23) and (12.24) into the complex velocity field expression of Eq. (12.19),

$$\frac{d\Omega}{dz} = Ue^{-i\alpha} \tag{12.25}$$

and integrate to obtain

$$\boxed{\Omega(z) = Uze^{-i\alpha}} \tag{12.26}$$

where the constant of integration is absorbed into $\Omega(z)$. The reader should compare Eq. (12.26) with that given by Fig. 12.2, and verify that it satisfies Laplace's equation.

12.5.2 Sources and Sinks

Conceptually, a source is a *point* at which fluid is flowing at a uniform rate and from which the flow is moving radially away in a uniform fashion in all directions. Such a flow is shown in Fig. 12.5.

At some radial location r, the volume rate of flow q per unit width is evaluated using the net volume flow rate expression of Eq. (5.12) along with the geometry shown in Fig. 12.5. We find

$$q = \int_o^{2\pi} v_r r \, d\theta \tag{12.27}$$

Since the radial velocity component $v_r \neq v_r(\theta)$, we evaluate the integral of Eq. (12.27) as

$$q = 2\pi r v_r \tag{12.28}$$

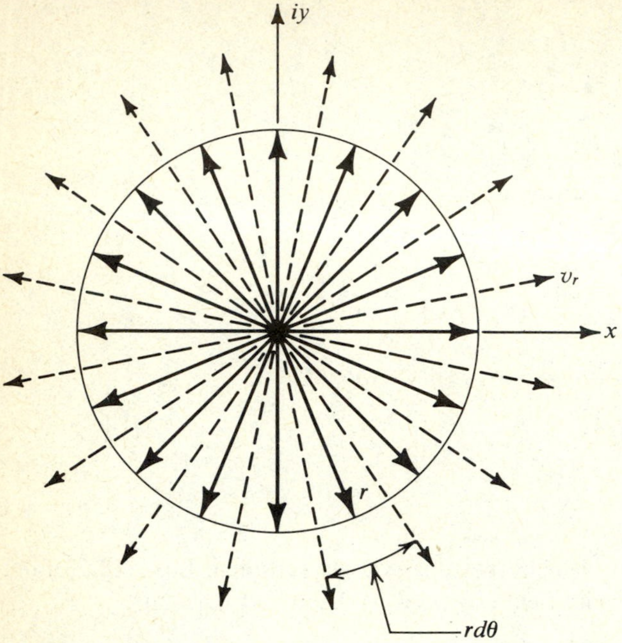

Figure 12.5 *Two-dimensional source flow.*

where q is constant. Substituting the velocity potential ϕ and stream function ψ relationships of Eqs. (8.29) and (8.30) into Eq. (12.28) results in

$$v_r = \frac{\partial \phi}{\partial r} = \frac{1}{r}\frac{\partial \psi}{\partial \theta} = \frac{q}{2\pi r} \tag{12.29}$$

Also, we find

$$v_\theta = \frac{1}{r}\frac{\partial \phi}{\partial \theta} = -\frac{\partial \psi}{\partial r} = 0 \tag{12.30}$$

since no source or sink flows have circumferential velocity component v_θ. It can easily be shown that the complex velocity field $d\Omega/dz$ in polar coordinates is

$$\frac{d\Omega}{dz} = (v_r - iv_\theta)e^{-i\theta} \tag{12.31}$$

using a technique similar to that given in Sec. 12.4. Substituting the velocity components of Eqs. (12.29) and (12.30) into Eq. (12.31) and integrating gives

$$\Omega(z) = \frac{q}{2\pi} \ln z \tag{12.32}$$

where the constant of integration is absorbed in $\Omega(z)$.

When the center of the source is located at some space point z_o different from the origin of the coordinate system, the complex potential is simply

$$\Omega(z) = \frac{q}{2\pi} \ln (z - z_o) \tag{12.33}$$

Equation (12.33) therefore represents the complex potential for a source located at z_o with a strength q, length^2/s.

The singular nature of the source is found by integrating the flux through a closed path. This is equivalent to determining the change in the stream function ψ. The result is nonzero for a closed circuit enclosing the source.

A *sink* is but a negative source, where once again the radial velocity is uniform in all directions. In this case, however, the flow is reversed and moves toward the origin. The complex potential Ω for a sink is simply the negative of the complex potential for a source:

$$\Omega(z) = -\frac{q}{2\pi} \ln (z - z_o) \tag{12.34}$$

Example 12.4

Given a source located at $a + ib$, (a) what is the value of the z- and y-components of velocity u, v, respectively, at the origin? (b) What strength q is necessary to have a speed $V = 10$ m/s at the origin? (See Fig. E12.4.)

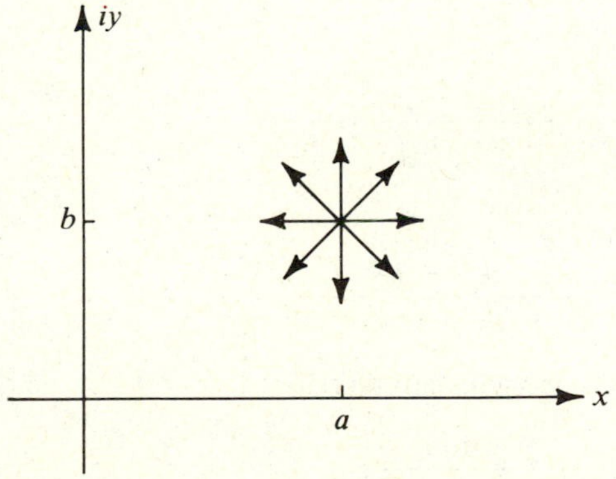

Figure E12.4

Solution:

Step 1.

The flow is incompressible source potential flow.

Step 3.

The appropriate equations are:

- Definition of the complex velocity potential:

$$\Omega = \phi + i\psi \tag{i}$$

Example 12.4 *(Con't.)*

- Definition of the complex velocity:

$$\frac{d\Omega}{dz} = u - iv \tag{ii}$$

- Complex potential for a source:

$$\Omega = \frac{q}{2\pi} \ln (z - z_o) \tag{iii}$$

(a) The complex potential for a source comes from Eqs. (i) and (iii):

$$\Omega(z) = \phi + i\psi = \frac{q}{2\pi} \ln (z - z_o) \tag{iv}$$

We find the complex velocity using Eq. (ii):

$$u - iv = \frac{q}{2\pi(z - z_o)} \tag{v}$$

Evaluating the real and imaginary parts of Eq. (v) results in

$$\begin{aligned} u - iv &= \frac{q[(x - a) - i(y - b)]}{2\pi[(x - a) + i(y - b)][(x - a) - i(y - b)]} \\ &= \frac{q}{2\pi}\left\{\frac{x - a}{(x - a)^2 + (y - b)^2} - \frac{i(y - b)}{(x - a)^2 + (y - b)^2}\right\} \end{aligned} \tag{vi}$$

so that the x-component of velocity is

$$u = \frac{q}{2\pi} \frac{(x - a)}{(x - a)^2 + (y - b)^2} \tag{vii}$$

and the y-component of velocity is

$$v = \frac{q}{2\pi} \frac{(y - b)}{(x - a)^2 + (y - b)^2} \tag{viii}$$

At the origin, the velocity components are therefore

$$u = \frac{-qa}{2\pi(a^2 + b^2)} \tag{ix}$$

and

$$v = \frac{-qb}{2\pi(a^2 + b^2)} \tag{x}$$

Note that the above two expressions state that there is a finite velocity at the origin of the *coordinate* system due to a source *not* located at the origin. If the source were at the origin, then our velocity components would be undefined at the origin.

Example 12.4 *(Cont'd.)*

(b) For a speed $V = 10$ m/s at the origin, we use Eq. (12.21) to calculate the value of the strength q:

$$V^2 = 100 = \left[\frac{q}{2\pi(a^2 + b^2)}\right]^2 (a^2 + b^2) \tag{vii}$$

or

$$q = 20\pi(a^2 + b^2)^{1/2} \text{ m}^2/\text{s} \tag{viii}$$

This is precisely what is given by Eq. (12.28), where $v_r = V$ and $r^2 = a^2 + b^2$.
This completes the solution.

12.5.3 Vortex Motions

This section presents a few aspects of a two-dimensional vortex motion. Only a z-component of vorticity ζ_z is perpendicular to the xy plane of motion. The thickness of the fluid is assumed to be unity, meaning that the vortex is confined between two horizontal planes separated by a unit thickness. The vortex lines are straight and parallel, and all vortex tubes are concentric cylinders. Such vortices are called *rectilinear vortices*.

We shall confine ourselves to a *Rankine vortex*, a rectilinear vortex that has two distinct regions of flow. Figure 12.6 shows the geometry for this vortex. Two regions of the vortex have to be considered: an inner region $r \leqslant r_c$ (where r_c denotes the core radius), where the vorticity exists, and an outer region $r > r_c$ where no vorticity exists. The Rankine vortex has a constant value of vorticity in the inner region. Thus,

$$\zeta_z = \text{const.} \quad 0 \leqslant r \leqslant r_c \tag{12.35}$$

$$\zeta_z = 0 \qquad r > r_c \tag{12.36}$$

Using the vorticity-velocity relationship of Eq. (4.86) and Stokes theorem, we find

$$\oint \mathbf{V} \cdot d\mathbf{r} = \int_A \zeta_z \, dA \tag{12.37}$$

Since the only velocity is a tangential velocity v_θ

$$\mathbf{V} = v_\theta \mathbf{e}_\theta \tag{12.38}$$

and the incremental path of the fluid is

$$d\mathbf{r} = r \, d\theta \, \mathbf{e}_\theta \tag{12.39}$$

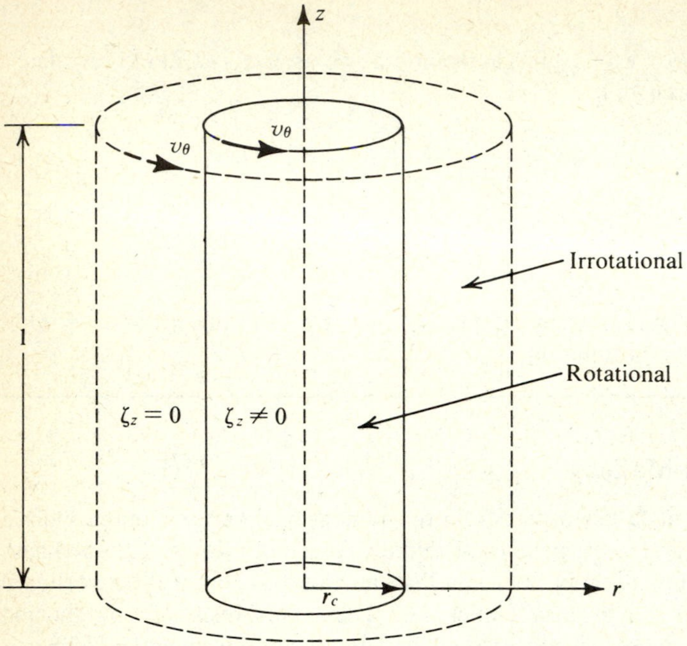

Figure 12.6 *Rankine vortex geometry.*

we can express Eq. (12.37) as

$$\oint v_\theta r\, d\theta = \zeta_z \pi r^2 \tag{12.40}$$

Since v_θ and r are both independent of angular variation θ, for both the rotational and irrotational regions of Fig. 12.6, Eq. (12.40) becomes

$$v_\theta = \frac{r\zeta_z}{2} \tag{12.41}$$

Thus, for a *Rankine vortex*, the vorticity is twice the angular rotation ω, so that we obtain the familiar result

$$v_\theta = \omega r, \qquad r \leqslant r_c \tag{12.42}$$

for the inner region. This we recognize as the expression for solid body rotation. Note that at the centerline the circumferential velocity is zero, even though the fluid particles have vorticity (i.e., are spinning at the centerline r equals zero).

Equation (4.86) gives the relationship between vorticity and velocity. In cylindrical coordinates, the vorticity about the x-axis is expressed as

$$\zeta_z = \frac{1}{r}\frac{\partial(r v_\theta)}{\partial r} - \frac{1}{r}\frac{\partial v_r}{\partial \theta} \tag{12.43}$$

For axisymmetric motion, we note that $\partial v_r/\partial\theta = 0$, so that for axisymmetric two-dimensional vortex flow

$$\zeta_z = \frac{v_\theta}{r} + \frac{dv}{dr} \tag{12.44}$$

Equation (12.44) is the proper expression for the vorticity expressed in terms of the tangential velocity. Note that both Eqs. (12.41) and (12.42) satisfy Eq. (12.44), because for a *Rankine vortex* both the vorticity ζ_z and the angular speed ω are constant. Equation (12.44), on the other hand, holds for *all* axisymmetric vortex flows of which the Rankine vortex is but one example.

The terms "free" and/or "forced" are sometimes used in describing irrotational and solid body rotations. These words can be misleading, however. Consider the following.*

Why have shear stresses been neglected in the analysis of irrotational vortex motions? There can be but three possible explanations: (1) the vortex motion can only be produced in an ideal fluid, (2) the motion might have originated in a real fluid, where certain rates of shear strain are zero, or (3) the net moment on the fluid parcel is zero.

Explanation (2) is possible only if the vortex motion is solid body rotation (not a forced vortex), and (3) is possible only if the vortex is not free, but irrotational. Let us consider our two regions of the Rankine vortex: first, the solid body rotation region.

Solid body rotation is given by Eq. (12.42) along with $v_r = w = 0$. This type of vortex motion can be produced when a cylindrical tank containing a fluid is rotated on a table at constant angular speed ω. Here, we see that viscosity plays no role in the existence of the shear stress $p_{r\theta}$:

$$p_{r\theta} = \mu\left[r\frac{\partial}{\partial r}\left(\frac{v_\theta}{r}\right) + \frac{1}{r}\frac{\partial v_r}{\partial\theta}\right] \tag{12.45}$$

Substituting Eq. (12.42) into Eq. (12.45) results in $p_{r\theta} = 0$, which clearly shows that the viscosity of the fluid plays no role in this vortex. This is the reason why the shear stress need not be taken into account for circular vortex flow.

We now come to the issue of whether this solid body rotation can be properly called "forced." The phrase "forced vortex" arises largely from hydraulic engineering. Some hydraulic engineers state that a constant external torque *must be maintained* on the fluid so as to keep it rotating like a solid body, and because of this, they call it "forced" vortex flow. But at steady state no torque is necessary to rotate a cylinder filled with a liquid, assuming that the system is the vortex and does not include anything external to the vortex such as resistance by the atmosphere and friction bearings.

We can treat a forced vortex in two ways. One is to impose a sink or source on the rotation. But then we no longer have solid body rotation, and the angular speed is a variable, a function of radius r and no longer a constant. The second way where

*From P. K. Kundu in *Bull. Mech. Engrg. Educ.*, vol. 7, pp. 361–368. Used with the permission of Pergamon Press Ltd.

"forced" may be justified is for unsteady motion. Here, an external torque would be necessary to create vorticity from an irrotational fluid to rotate as a solid body.

We should be aware that solid body rotation is preferable to rotational vortex, since we can have rotational motion, such as $v_\theta = \omega_1 r + \omega_2/r$, $v_r = w = 0$, but not solid body unless $\omega_2 = 0$.

Next let us consider an incompressible viscous fluid surrounded by a rotating cylinder of radius a and with surface speed V_o. For axisymmetric steady flow, the equations of motion yield

$$v_r = 0 \tag{12.46}$$

from the (D.F.) continuity Eq. (4.25), such that

$$\frac{\partial p}{\partial r} = \frac{v_\theta^2}{r} \tag{12.47}$$

and

$$\mu \frac{d}{dr}\left(\frac{dv_\theta}{dr} + \frac{v_\theta}{r}\right) = 0 \tag{12.48}$$

from the Navier-Stokes equation (4.114). If $\mu \neq 0$, two integrations of the θ-momentum Eq. (12.48) yield

$$v_\theta = \frac{c_1 r}{2} + \frac{c_2}{r} \tag{12.49}$$

The boundary conditions for our problem are

$$v_\theta(a) = V_o \tag{12.50}$$

and

$$v_\theta(\infty) \to 0 \tag{12.51}$$

so that the circumferential velocity v_θ becomes

$$v_\theta = V_o \frac{a}{r} \tag{12.52}$$

Thus, the circumferential velocity given by Eq. (12.52) satisfies the Navier-Stokes equation for a viscous fluid. Looking at a fluid parcel in the form of a small cylindrical wedge of fluid, we find that (a) the net tangential shear force on the parcel is zero (just as in Couette flow) and (b) the net moment on the parcel is zero (unlike Couette flow). Condition (b) occurs because the flow is irrotational in that the *net moment* on a fluid parcel is zero. It *seems* justifiable to ignore the shear stresses in the formulation of the equations of motion of vortex flow—but only because the net moment on the fluid parcels is zero. Thus, an irrotational vortex is *not "free"* because shear stresses *do* exist.

Circulation, Γ

Circulation has frequently been viewed as solely a mathematical concept: as the line integral of the velocity around a closed fluid particle path (see Eq. (12.53)). Actually, it is a real physical quantity that can be calculated from measured data. It is a measure of the *swirl* of the fluid flow, and represents the *net vorticity* in an area bounded by any closed path of a fluid particle. The path need not be circular, as will be illustrated shortly.

Circulation is *defined mathematically* as

$$\Gamma = \oint \mathbf{V} \cdot d\mathbf{r} \tag{12.53}$$

where $d\mathbf{r}$ is an incremental arc length of a closed curve C (see Fig. 12.7). For example, for a fluid particle moving in a closed circular path of radius a, with a velocity v_θ given by Eq. (12.49), the circulation is

$$\Gamma = \oint v_\theta r \, d\theta = 2\pi \left(\frac{c_1 a^2}{2} + c_2 \right) \tag{12.54}$$

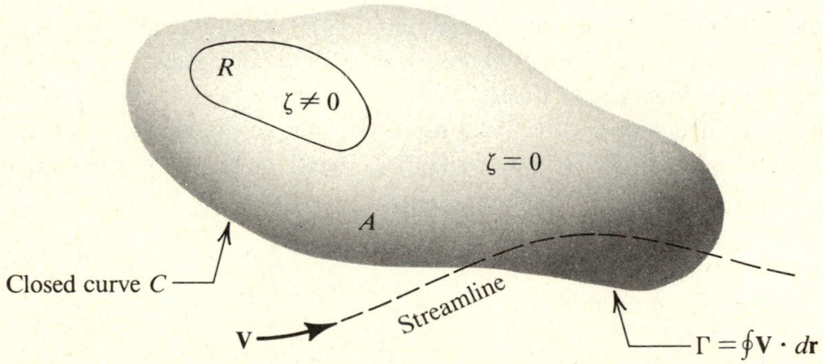

Figure 12.7 *A rotational region within an irrotational region.*

In addition, for an *irrotational* flow the circulation can be shown to be a constant (or zero), since the vorticity is locally constant (or zero). Thus, from Eq. (12.54)

$$\Gamma_\infty = 2\pi c_2 \tag{12.55}$$

where Γ_∞ denotes the *potential circulation* and is a constant in any irrotational region, its value depending upon the boundary condition of a given problem.

Another way to look at circulation is to use the physical definition

$$\Gamma = \int_A \boldsymbol{\zeta} \cdot d\mathbf{A} \tag{12.56}$$

which is derived from using Eq. (12.53) and Stokes theorem. This relationship is useful because it gives a physical interpretation for circulation; that is, circulation is the sum of all vorticity in an area A. Thus, if A has no vorticity, then the circulation is zero, which is the case for an irrotational region. But suppose in a region A there exists a subregion R, as shown in Fig. 12.7. Then from Eq. (12.56), we can evaluate the circulation of region R if we know the analytic expression for vorticity $\boldsymbol{\zeta}$. Since there is no vorticity $\boldsymbol{\zeta}$ in region A, the circulation of region A must be the circulation of region R, since R is a subregion of A enclosed by the outer curve of region A.

In discussing the Rankine vortex, we stated that there was vorticity in the inner region $r \leqslant r_c$ and no vorticity outside the region $r > r_c$. Thus

$$\Gamma_\infty = 2\pi\omega r_c^2 \tag{12.57}$$

for the Rankine vortex. Outside of the inner region, $r > r_c$, the circumferential velocity can be found from Eq. (12.55), or preferably from Eq. (12.56):

$$v_\theta = \frac{\omega r_c^2}{r} \tag{12.58}$$

Figure 12.8 shows the radial distribution of circumferential velocity, vorticity, and circulation for the Rankine vortex. Note the abscissa for the circulation.

Complex Potential for an Irrotational Vortex

Consider the case of an irrotational vortex that has zero value of the vorticity outside of the radius r_c. From Eqs. (12.58), (12.48), and (12.7), we find the complex potential to be

$$\Omega(z) = -i\omega r_c^2 \ln(z - z_o) \tag{12.59}$$

or

$$= -\frac{i\Gamma_\infty}{2\pi} \ln(z - z_o) \tag{12.60}$$

These equations are valid for a vortex *whose center is located at z_o and in a direction that is counterclockwise.*

Example 12.5

Consider a vortex whose tangential velocity $v_\theta(r)$ is given as

$$v_\theta = \frac{\Gamma_\infty}{2\pi r}[1 - \exp(-ar^2)] \tag{i}$$

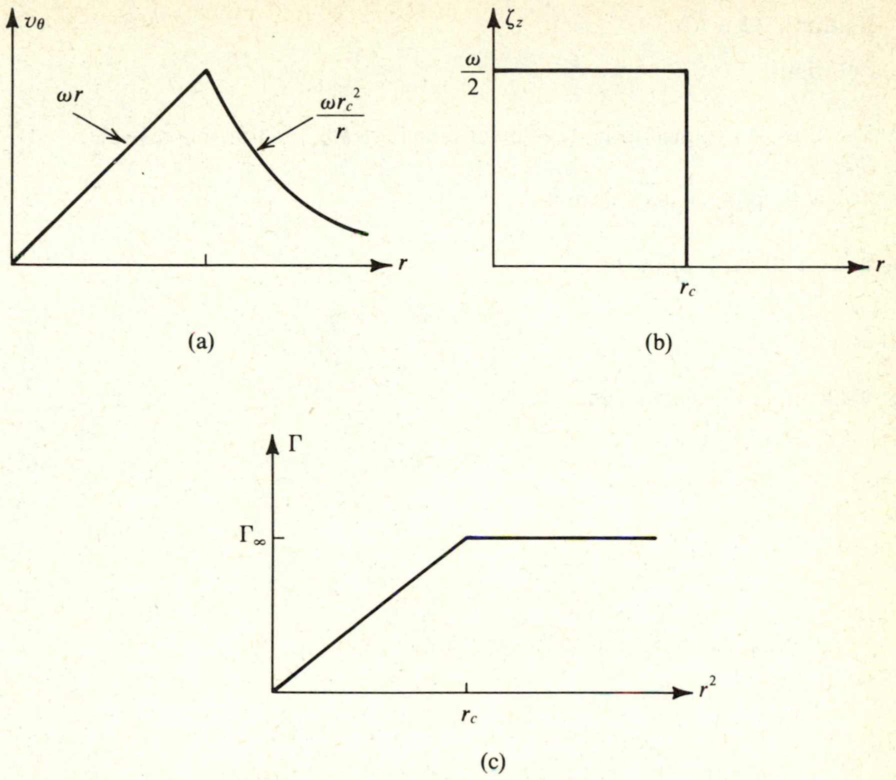

Figure 12.8 *Radial distribution of (a) tangential velocity, (b) vorticity, and (c) circulation.*

Example 12.5 *(Con't.)*

Calculate (a) the radial distribution of vorticity ζ_z about the z-axis, and (b) the radial distribution of circulation.

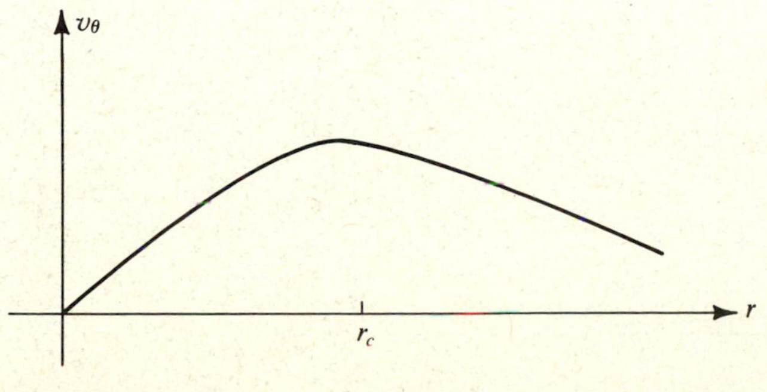

Figure E12.5a

Example 12.5 *(Con't.)*

Solution:

Step 1.

The flow is assumed to be two-dimensional, steady, and incompressible.

Step 3.

The appropriate equations are:

- Definition of vorticity:

$$\zeta_z = \frac{1}{r}\frac{\partial(rv_\theta)}{\partial r} - \frac{1}{r}\frac{\partial v_r}{\partial \theta} \tag{ii}$$

- Definition of circulation:

$$\Gamma = \oint \mathbf{V}\cdot d\mathbf{r} \tag{iii}$$

or

$$\Gamma = \int_A \boldsymbol{\zeta}\cdot d\mathbf{A} \tag{iv}$$

(a) The vorticity ζ_z is calculated from Eq. (ii) as

$$\begin{aligned}\zeta_z &= \frac{v_\theta}{r} + \frac{dv_\theta}{dr} \\ &= \frac{\Gamma_\infty}{2\pi}\left[\frac{1}{r^2} - \frac{\exp(-ar^2)}{r^2} - \frac{1}{r^2}[1 - \exp(-ar^2)] \right. \\ &\quad \left. + \frac{2ar}{r}\exp(-ar^2)\right]\end{aligned} \tag{v}$$

or

$$\zeta_z = 2a\exp(-ar^2) \tag{vi}$$

and is shown plotted in Fig. E12.5b.

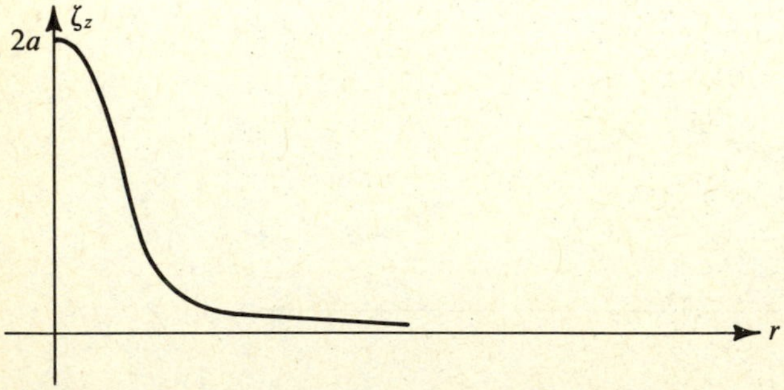

Figure E12.5b

Example 12.5 *(Con't.)*

(b) We can calculate the circulation in two ways:

(1) By formula:

Using Eq. (iii) we find,

$$\begin{aligned}\Gamma &= 2\pi r v_\theta \\ &= 2\pi r\left\{\frac{\Gamma_\infty}{2\pi r}[1 - \exp(-ar^2)]\right\} \\ &= \Gamma_\infty [1 - \exp(-ar^2)]\end{aligned} \tag{vii}$$

(2) By integration:

The results of the vorticity of Eq. (vi) and the circulation vorticity-relationship of Eq. (iv) give

$$\Gamma = \int_o^{2\pi}\int_o^r \zeta_z \, r dr d\theta \tag{viii}$$

or

$$\begin{aligned}\Gamma &= 2\pi \int_o^r 2a \exp(-ar^2)\, r dr \\ &= 2\pi [1 - \exp(-ar^2)]\end{aligned} \tag{ix}$$

The value of the circulation Γ as given by Eqs. (vii) and (ix) is shown in Fig. E12.5c.

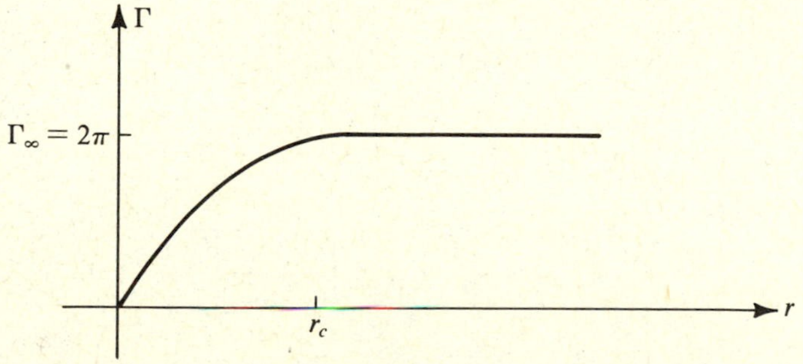

Figure E12.5c

We usually designate $\Gamma_\infty = 2\pi$ as the potential circulation, and we define r_c as the core radius. For $r > r_c$, the vortex is a potential vortex. For $r \leq r_c$, the vortex possesses a vorticity ζ_z and thus the flow is not potential flow.

This completes the solution.

Example 12.6
Given the velocity field $\mathbf{V} = 3xy\mathbf{i} - 3yx\mathbf{j}$, evaluate the circulation Γ around a rectangular path shown in Fig. E12.6.

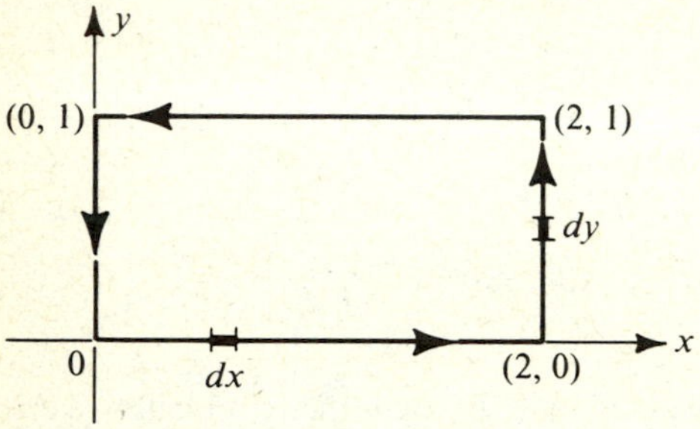

Figure E12.6

Solution:
Step 1.
The flow is assumed to be steady, two-dimensional, and incompressible.
Step 3.
The appropriate equation is:

- Definition of circulation:

$$\Gamma = \oint \mathbf{V} \cdot d\mathbf{r} \tag{i}$$

Substituting the given velocity field into the line integral expression for the circulation of Eq. (i) gives

$$\Gamma = \oint 3xy\,(\mathbf{i} - \mathbf{j}) \cdot (\mathbf{i}\,dx + \mathbf{j}\,dy) \tag{ii}$$

$$= 3\int_o^2 xy\,dx\bigg|_{y=0} - 3\int_o^1 xy\,dy\bigg|_{x=2}$$

$$+ 3\int_2^o xy\,dx\bigg|_{y=1} - 3\int_1^o xy\,dy\bigg|_{x=0} \tag{iii}$$

$$= 3\left[-2\int_o^1 y\,dy - \int_o^2 x\,dx\right] \tag{iv}$$

$$= 3(-1 - 2) = -9 \tag{v}$$

The circulation is clockwise and the flow is rotational since the vorticity is nonzero.

This completes the solution.

12.5.4 Doublet

Consider a source and a sink, each of strength q, located at the points A and B, as shown in Fig. 12.9. The location of the source is $z_o = -ae^{i\alpha}$, and the location of the sink is $z_o = ae^{i\alpha}$. The complex potential $\Omega(z)$ for the source-sink shown in Fig. 12.9 is obtained from Eq. (12.34):

$$\Omega(z) = \frac{q}{2\pi}[\ln(z + ae^{i\alpha}) - \ln(z - ae^{i\alpha})] \tag{12.61}$$

Now let the distance a be very small, as would be the case if the source and sink were close together. We can rearrange the terms in Eq. (12.61) so that

$$\begin{aligned}\Omega(z) &= \frac{q}{2\pi}\left\{\ln\left[z\left(1 + \frac{ae^{i\alpha}}{z}\right)\right] - \ln\left[z\left(1 - \frac{ae^{i\alpha}}{z}\right)\right]\right\} \\ &= \frac{q}{2\pi}\left[\ln\left(1 + \frac{ae^{i\alpha}}{z}\right) - \ln\left(1 - \frac{ae^{i\alpha}}{z}\right)\right] \\ &= \frac{q}{\pi}\left[\frac{ae^{i\alpha}}{z} + \frac{a^3e^{3i\alpha}}{3z^3} + \ldots\right]\end{aligned} \tag{12.62}$$

For very small a and $z \neq 0$, the most significant term is the first term of the power series in the above Eq. (12.62), so that in the limit as $a \to 0$, we can approximate Eq. (12.62) by the expression

$$\Omega(z) = \frac{qae^{i\alpha}}{\pi z} \tag{12.63}$$

We have placed the source and sink close to the origin in Fig. 12.9. If we place the source and sink close to some space point z_o, then the complex potential for the doublet would be

$$\boxed{\Omega(z) = \frac{qae^{i\alpha}}{\pi(z - z_o)}} \tag{12.64}$$

A summary of the complex potentials for elementary potential flows is presented in Table 12.1. The velocity potential and stream function are given in both Cartesian and polar forms.

12.6 Conservation of Circulation

On page 607 we stated that for an irrotational flow, the circulation must be either constant or zero. Since the circulation measures the total swirl, we mean that at any elevation and any radius, the rotation is the same: a constant. Thus, the circulation at

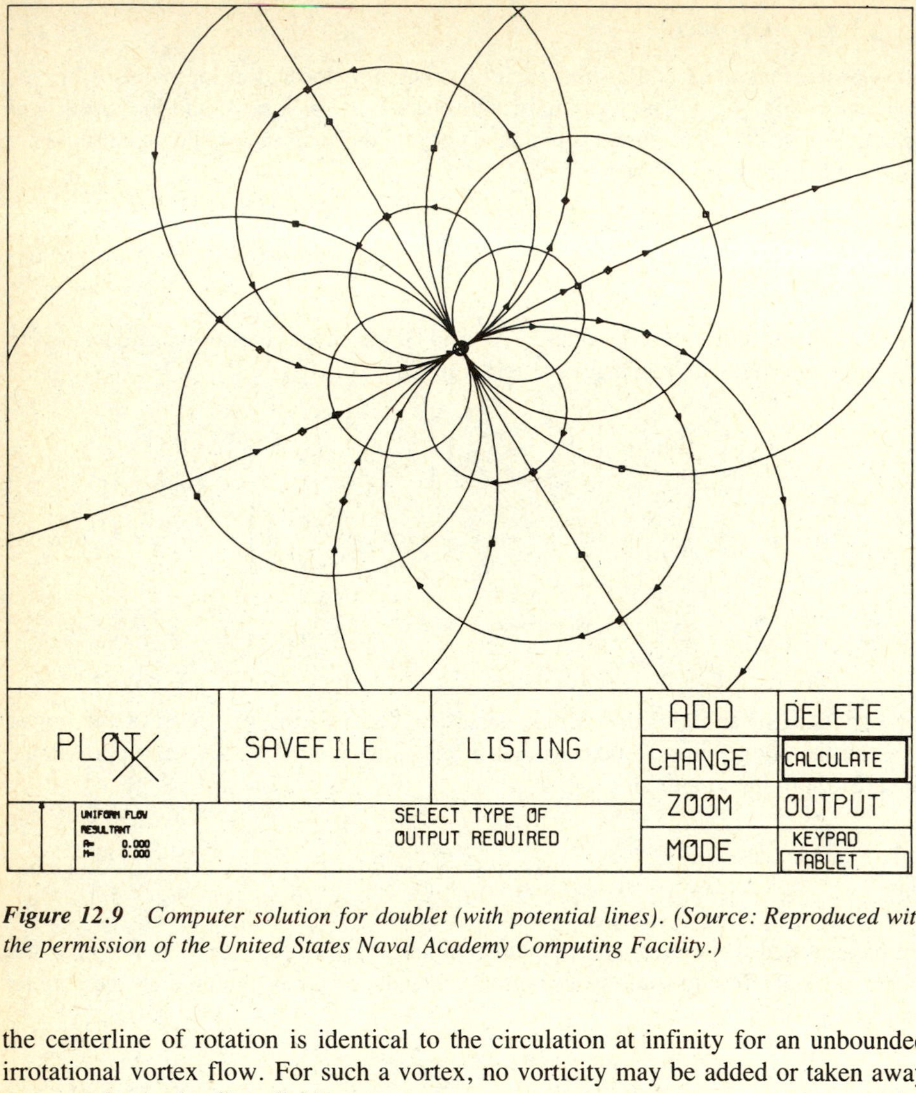

Figure 12.9 *Computer solution for doublet (with potential lines). (Source: Reproduced with the permission of the United States Naval Academy Computing Facility.)*

the centerline of rotation is identical to the circulation at infinity for an unbounded irrotational vortex flow. For such a vortex, no vorticity may be added or taken away at any point in the flow field.

Let us consider the circulation in a circuit moving with an inviscid fluid and subject to conservative external forces. To prove that the circulation is constant we start with the definition of circulation. Taking the total time derivative of both sides of Eq. (12.53), we obtain

$$\begin{aligned}\frac{d\Gamma}{dt} &= \oint \frac{d\mathbf{V}}{dt}\cdot d\mathbf{r} + \oint \mathbf{V}\cdot\frac{d\mathbf{r}}{dt} \\ &= \oint \frac{d\mathbf{V}}{dt}\cdot d\mathbf{r} \\ &= \oint \mathbf{a}\cdot d\mathbf{r}\end{aligned} \tag{12.65}$$

Table 12.1 *Table of Expressions for Elementary Flows*

Elementary Flow	Ω Complex Potential ($z_o = 0$)	$\phi(x, y)$	$\phi(r, \theta)$	$\psi(x, y)$	$\psi(r, \theta)$	Streamlines
1. Uniform flow	$Uze^{-i\alpha}$	$U(x \cos \alpha + y \sin \alpha)$	$Ur \cos (\theta - \alpha)$	$U(y \cos \alpha - x \sin \alpha)$	$Ur \sin (\theta - \alpha)$	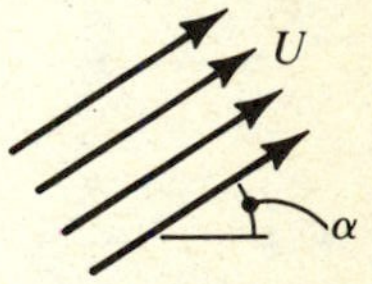
2. Source	$\frac{q}{2\pi} \ln z$	$\frac{q}{4\pi} \ln (x^2 + y^2)$	$\frac{q}{2\pi} \ln r$	$\frac{q}{2\pi} \tan^{-1} \left(\frac{y}{x}\right)$	$\frac{q}{2\pi} \theta$	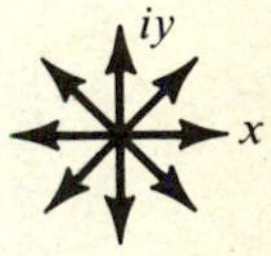
3. Vortex	$\frac{-i\Gamma_\infty}{2\pi} \ln z$	$\frac{\Gamma_\infty}{2\pi} \tan^{-1} \left(\frac{y}{x}\right)$	$\frac{\Gamma_\infty}{2\pi} \theta$	$-\frac{\Gamma_\infty}{4\pi} \ln (x^2 + y^2)$	$-\frac{\Gamma_\infty}{2\pi} \ln r$	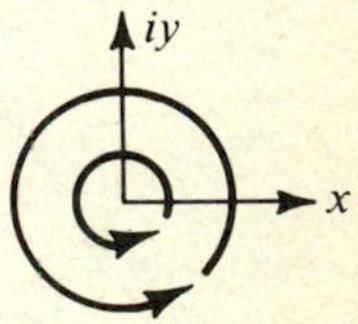
4. Doublet	$\frac{qae^{i\alpha}}{\pi z}$	$\frac{qax \cos \alpha}{\pi(x^2 + y^2)}$	$\frac{qa \cos (\alpha - \theta)}{\pi r}$	$-\frac{qay \sin \alpha}{\pi(x^2 + y^2)}$	$\frac{qa \sin (\alpha - \theta)}{\pi r}$	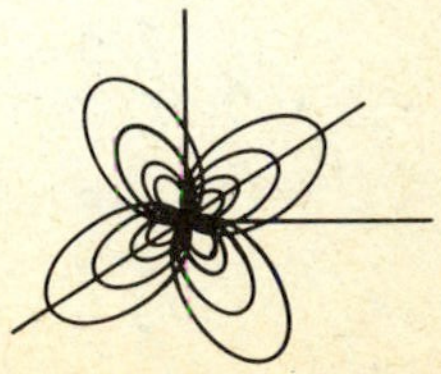

since

$$\oint \mathbf{V}\cdot d\mathbf{V} = \oint \frac{1}{2}\, d\,(V)^2 = 0 \qquad (12.66)$$

because a closed path integral of an exact differential is always zero. Using the Stokes theorem, we introduce the curl of the acceleration **a**:

$$\frac{d\Gamma}{dt} = \oint \mathbf{a}\cdot d\mathbf{r} = \int_A \mathbf{e}_n\cdot(\nabla \times \mathbf{a})\, dA \qquad (12.67)$$

Next let the flow be *ideal*. The acceleration **a** is expressed in terms of the acceleration due to gravity, and the pressure gradient by Euler's equation

$$\mathbf{a} = \frac{\partial \mathbf{V}}{\partial t} + (\mathbf{V}\cdot\nabla)\,\mathbf{V} = \mathbf{g} - \frac{1}{\rho}\nabla p \qquad (4.110)$$

Expressing the gravitational force per unit mass **g** in terms of the gradient of the potential energy Ω

$$\mathbf{g} = -\nabla\Omega \qquad (12.68)$$

we can write the acceleration **a** in terms of a gradient as

$$\mathbf{a} = -\nabla\left(\Omega + \frac{p}{\rho}\right) \qquad (12.69)$$

Since the curl of a gradient always vanishes, then

$$\nabla \times \mathbf{a} = -\nabla \times \nabla\left(\Omega + \frac{p}{\rho}\right) = 0 \qquad (12.70)$$

so that according to Eqs. (12.67) and (12.70), the total rate of change of circulation is zero:

$$\frac{d\Gamma}{dt} = 0 \qquad (12.71)$$

The only way to satisfy Eq. (12.71) is that the circulation be either constant or zero. Such flows are called *circulation-preserving flows* and obey Helmholtz' third law: ''In an inviscid fluid subject to conservative external forces, the circulation in any circuit moving with the fluid is preserved.''

We have shown that the circulation does not vary in an irrotational flow. This is rather nice, because now we can determine, for example, the maximum circulation of an ideal tornado (the maximum value being that circulation for a potential vortex flow), or the circulation over a wing, or the circulation of the earth's atmosphere.

12.7 *Equation of the Body*

We stated in Chap. 8 that the equation of the streamlines is found by setting the stream function ψ equal to a constant, and then evaluating one of the independent variables that appears in the expression for the stream function in terms of the second independent variable. The result will give a mapping of a streamline onto the coordinate plane. Since the complex variable Ω has been defined as

$$\Omega = \phi + i\psi \qquad (12.7)$$

its complex conjugate is

$$\overline{\Omega} = \phi - i\psi \qquad (12.72)$$

Multiplying Eq. (12.7) by the negative of the imaginary number $(-i)$ and Eq. (12.72) by the imaginary number i, then adding the results, gives

$$\psi = \frac{i}{2}(\overline{\Omega} - \Omega) \qquad (12.73)$$

where ψ is to be treated as a constant.

Consider those flows generated by $\Omega = z^m$. The equation of the streamlines are then $\bar{z}^m - z^m = -2\psi i$. Figure 12.10 shows some typical streamlines for various values of m. An expression for an equipotential line is given in the following example.

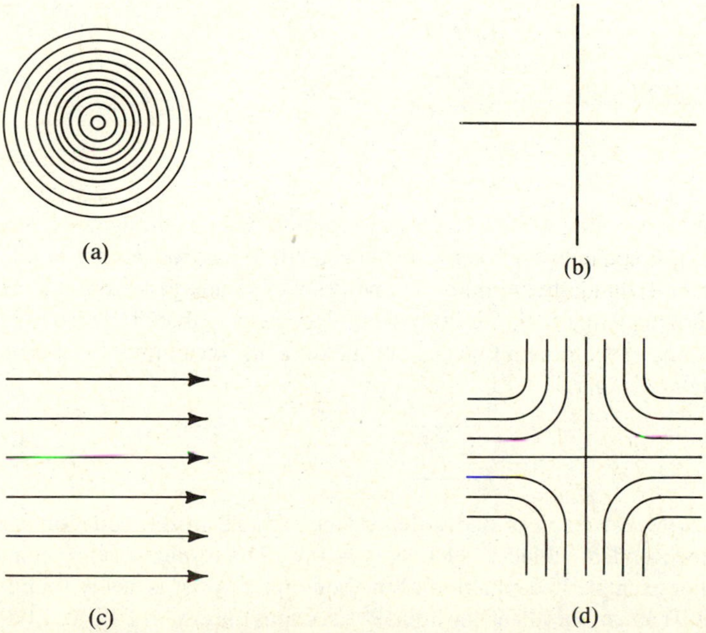

Figure 12.10 *Streamlines for $\Omega = z^m$. (a) $m = -1$ (vortex flow). (b) $m = 0$ (statics). (c) $m = +1$ (uniform flow). (d) $m = +2$ (flow in a corner).*

Example 12.7
Express the velocity potential ϕ in terms of the complete potential Ω.

Solution:
The velocity potential ϕ is related to the complex potential Ω by

$$\Omega = \phi + i\psi \tag{12.7}$$

The conjugate of the complex potential is

$$\overline{\Omega} = \phi - i\psi \tag{i}$$

Adding Eqs. (12.7) and (i) eliminates the stream function ψ so that

$$\phi = \frac{1}{2}(\Omega + \overline{\Omega}) \tag{ii}$$

is the velocity potential expressed in terms of the complex potential. Letting $\Omega = z^m$, we can plot equipotential lines for the flows shown in Fig. 12.1. It is just as easy to sketch lines that intersect the streamlines at 90°.

This completes the solution.

The *zero streamline* can be found by setting the stream function ψ equal to zero, such that from Equation (12.73) the equation of the zero streamline is

$$\Omega = \overline{\Omega} \tag{12.74}$$

and the zero equipotential line is

$$\Omega = -\overline{\Omega} \tag{12.75}$$

The equation for the zero value of the stream function ψ will be designated the fundamental or dividing streamline. We have shown that no flow can cross a streamline. We can thus interpret the dividing streamline as a moving solid boundary, representative of a surface of fluid moving past a fictitious body whose geometry is that of the moving boundary. The zero value of the stream function, or the dividing streamline, is viewed as *the equation of the body*; that is,

$$\boxed{\psi = 0} \tag{12.76}$$

In using Eq. (12.76), we express one independent variable in the equation in terms of the second independent variable, such as $y = y(x)$. The resultant expression can then be plotted. For example, the equation of the body for $\Omega = z^2$ is easily found to be $x = 0$, $y = 0$. Thus any or all of the four 90° sections shown in Fig. 12.10d will create the resultant streamline flow.

12.8 Blasius' Theorem for Forces

The hydrodynamic forces on an object placed in an irrotational incompressible flow field can be found from a theorem developed by H. Blasius in 1910. Consider an arbitrary body of area A shown in Fig. 12.11. When the body lies in a fluid that is flowing with a velocity $\mathbf{V}$, then hydrodynamic forces X, Y, in the x-, y-directions, respectively, act on the body. The components of the elemental force on the surface element $d\mathbf{A}$ are

$$dX = -p \sin \theta \, dA \tag{12.77}$$

$$dY = p \cos \theta \, dA \tag{12.78}$$

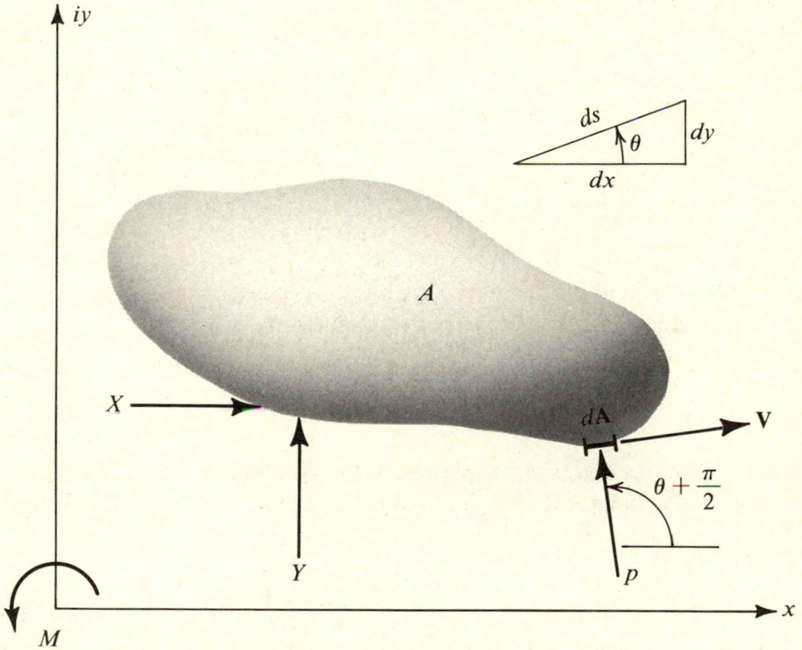

Figure 12.11 *Geometry for forces X, Y, and moment M on a body in a flow field.*

For an irrotational steady two-dimensional flow with no body forces, Bernoulli's equation (4.126) becomes

$$p = \text{const.} - \frac{\rho}{2} \tilde{V}^2 \tag{12.79}$$

The constant is evaluated using far field conditions. For example,

$$\text{const.} = p_\infty + \frac{\rho}{2} U_\infty^2$$

Substitution of Eq. (12.79) into Eqs. (12.77) and (12.78) results in

$$dX = -\left(\text{const.} - \frac{\rho}{2}\tilde{V}^2\right) \sin\theta \, dA \tag{12.80}$$

$$dY = \left(\text{const.} - \frac{\rho}{2}\tilde{V}^2\right) \cos\theta \, dA \tag{12.81}$$

If complex variables are now introduced, the elementary complex force $d\overline{F}$ is defined as

$$d\overline{F} = dX - i\,dY = \left(-\text{const.} + \frac{\rho}{2}\tilde{V}^2\right)(ie^{-i\theta})\,dA \tag{12.82}$$

Note the negative sign before the imaginary part of the force. Integration of Eq. (12.82) over the wetted surface of the body results in

$$\overline{F} = X - iY = \oint_A (-\text{const.})\, ie^{-i\theta}\, dA + \frac{\rho}{2}\oint_A \tilde{V}^2 ie^{-i\theta}\, dA \tag{12.83}$$

The surface area dA is related to the complex variable z as follows: from the geometry of Fig. 12.11, we will treat a section of the body so that

$$dA = ds \tag{12.84}$$

(where the thickness of the section is assumed to be unity). Working from the definition of the complex variable z of Eq. (12.11), and using the geometry of Fig. 12.11, we obtain

$$\begin{aligned} dz &= dx + idy \\ &= ds\cos\theta + ids\sin\theta = ds\, e^{i\theta} \end{aligned} \tag{12.85}$$

Substituting Eqs. (12.84) and (12.85) into our elemental force expression of Eq. (12.83), we can show that the first integral expression involving the constant leaves us with

$$\begin{aligned} \overline{F} = X - iY &= \frac{\rho}{2}\oint \tilde{V}^2\, ie^{-i\theta}\, ds \\ &= \frac{\rho}{2}\oint \tilde{V}^2\, ie^{-2i\theta}\, e^{i\theta}\, ds \\ &= \frac{i\rho}{2}\oint (\tilde{V}\, e^{-i\theta})^2\, dz \end{aligned} \tag{12.86}$$

Recalling that the relationship of the complex velocity, Eq. (12.19), is

$$u - iv = \tilde{V} e^{-i\theta} = \frac{d\Omega}{dz} \tag{12.87}$$

we transform Eq. (12.86) into:

$$\boxed{\bar{F} = X - iY = \frac{i\rho}{2} \oint \left(\frac{d\Omega}{dz}\right)^2 dz} \tag{12.88}$$

This result is known as *Blasius' theorem*. The beautiful aspect of this expression is that we can now evaluate it once and for all using the calculus of residues. We can show that the residue of a closed integral is

$$\oint f(z)\, dz = 2\pi i\, a_{-1} \tag{12.89}$$

so that if we *arbitrarily* identify the integrand of Eq. (12.88) as a Laurent series expanded about the origin

$$\left(\frac{d\Omega}{dz}\right)^2 \equiv a_n z^n + \ldots + a_0 + \frac{a_{-1}}{z} + \frac{a_{-2}}{z^2} + \ldots + \frac{a_{-n}}{z^n} \tag{12.90}$$

then the only term in the integrand that will give a nonzero value for the integral will be a term of the form a_{-1}/z. All the other terms in the series expression of Eq. (12.90) will be zero once they are substituted into the integral of Eq. (12.88) and evaluated. Thus the complex force becomes simply and neatly

$$\boxed{\bar{F} = X - iY = -\pi\rho a_{-1}} \tag{12.91}$$

If a_{-1} is real, then there will only be a force X, and if a_{-1} is imaginary there will only be a force Y. We shall demonstrate how to use Blasius' theorem in Sec. 12.10.*

12.9 Various Complex Potentials $\Omega(z)$ and Corresponding Physical Flows

Table 12.2 presents a few potential flows and their corresponding expressions for the complex velocity potential $\Omega(z)$. Knowing the complex potential Ω for these elementary flows, we can construct more complicated flows by the principle of superposition which will be discussed in the following section.

12.10 Combined Flows

In Sec. 12.9 various complex potentials are given for a variety of elementary flow configurations. Each complex potential is a unique solution of Laplace's equation

*Blasius' theorem for moments produced by hydrodynamic forces is $M = \text{Re}\,[-\,i\pi\rho\, a_{-2}]$ meaning that only the real part of the expression is to be considered.

Table 12.2 *Various Complex Potentials and Corresponding Flows*

Complex Potential Ω $\Omega = \phi + i\psi$	**Configuration**	**Flow Geometry**
$Uze^{-i\alpha}$	Uniform flow in the direction α	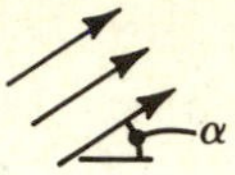
$\dfrac{q}{2\pi}\ln(z - z_o)$	Source located at z_o	
$\dfrac{-i\Gamma_\infty}{2\pi}\ln(z - z_o)$	Vortex located at z_o	
$\dfrac{qae^{i\alpha}}{\pi(z - z_o)}$	Doublet located at z_o	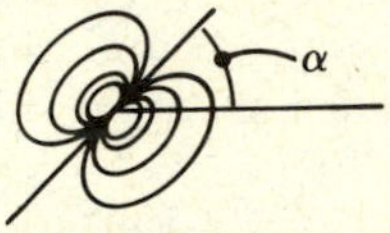
Az^n	Flow in a corner of angle π/n	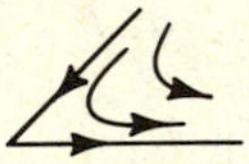
$\dfrac{q}{2\pi}\ln\sinh \pi z/a$	Source at the center of a channel	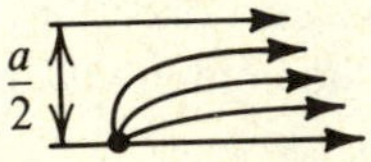
$Uz + \dfrac{q}{2\pi}\ln z$	Flow about a half body	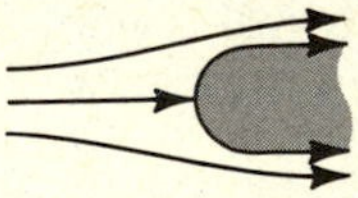
$U\left(z + \dfrac{b^2}{z}\right) + i\dfrac{\Gamma_\infty}{2\pi}\ln z$	Flow about a cylinder with circulation	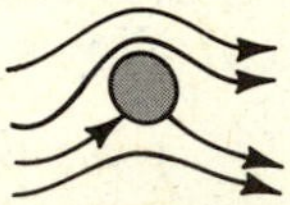
$Uz + \dfrac{q}{2\pi}\ln\dfrac{z + a}{z - a}$	Flow about a Rankin oval	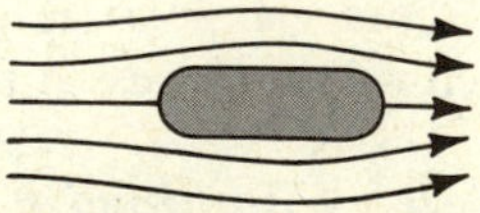
$\dfrac{i\Gamma_\infty}{2\pi}\ln\dfrac{z + a}{z - a}$	Line vortex near a wall	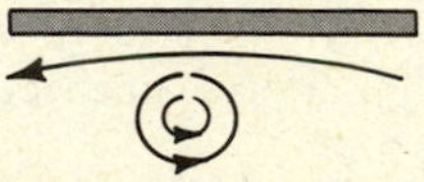

$\nabla^2 \phi = 0$, and therefore satisfies the continuity equation that expresses the conservation of mass in differential form. We note that some of the expressions of the complex potential $\Omega(z)$ in Sec. 12.9 are simply the sum of two or more of the elementary expressions. For instance, the complex potential for uniform flow over a semi-infinite half body is the sum of the complex potential for *uniform flow* and the complex potential for a *source*. What this means is that the "trick" of using a source in a uniform flow allows us to view the zero streamline as that streamline which moves past the boundary of a semi-infinite half body with perfect slip. Disregarding the "no-slip" boundary condition enables us to use fundamental solutions of the Laplace equation that "simulate" a particular irrotational flow.

This section will develop expressions for the complex potential for a few combined flows some of which are given in Table 12.2. We shall first consider the superposition principle, then the complex potential for uniform flow and a source, and conclude with the complex potential for uniform flow past a cylinder with and without circulation.

12.10.1 Principle of Superposition

The governing equation of motion for an irrotational flow is a second-order *linear* partial differential equation which we have identified as Laplace's equation. That the equation is linear enables us to make use of an important property: various solutions of Laplace's equation may be *superimposed* (added or subtracted). Thus more complicated flows can be constructed by addition of and/or subtraction of the fundamental flows treated in Sec. 12.5, or any of those presented in Table 12.2.

Example 12.8

If ϕ_1 and ϕ_2 represent two different solutions of Laplace's equation, show that

$$\phi_3 = \phi_1 \pm \phi_2 \tag{i}$$

also satisfies Laplace's equation, and that

$$\phi_4 = \phi_1 \phi_2 \tag{ii}$$

does not satisfy Laplace's equation.

Solution:

Substituting the velocity potential ϕ_3 given by Eq. (i) into Laplace's equation (12.2) gives

$$\begin{aligned} \frac{\partial^2 \phi_3}{\partial x^2} + \frac{\partial^2 \phi_3}{\partial y^2} + \frac{\partial^2 \phi_3}{\partial z^2} &= \frac{\partial^2 \phi_1}{\partial x^2} + \frac{\partial^2 \phi_1}{\partial y^2} + \frac{\partial^2 \phi_1}{\partial z^2} \\ &+ \frac{\partial^2 \phi_2}{\partial x^2} + \frac{\partial^2 \phi_2}{\partial y^2} + \frac{\partial^2 \phi_2}{\partial z^2} = 0 \end{aligned} \tag{iii}$$

since $\nabla^2 \phi_1 = 0$ and $\nabla^2 \phi_2 = 0$.

Example 12.8 *(Con't.)*

Consider next the velocity potential ϕ_4. From Eq. (ii), since

$$\frac{\partial \phi_4}{\partial x} = \phi_1 \frac{\partial \phi_2}{\partial x} + \phi_2 \frac{\partial \phi_1}{\partial x} \tag{iv}$$

and

$$\frac{\partial^2 \phi_4}{\partial x^2} = \phi_1 \frac{\partial^2 \phi_2}{\partial x^2} + \phi_2 \frac{\partial^2 \phi_1}{\partial x^2} + 2\frac{\partial \phi_1}{\partial x}\frac{\partial \phi_2}{\partial x} \tag{v}$$

with similar expression for $\dfrac{\partial^2 \phi_4}{\partial y^2}$ and $\dfrac{\partial^2 \phi_4}{\partial z^2}$ then

$$\begin{aligned}\frac{\partial^2 \phi_4}{\partial x^2} + \frac{\partial^2 \phi_4}{\partial y^2} + \frac{\partial^2 \phi_4}{\partial z^2} &= \phi_1 \left(\frac{\partial^2 \phi_2}{\partial x^2} + \frac{\partial^2 \phi_2}{\partial y^2} + \frac{\partial^2 \phi_2}{\partial z^2}\right) \\ &+ \phi_2 \left(\frac{\partial^2 \phi_1}{\partial x^2} + \frac{\partial^2 \phi_1}{\partial y^2} + \frac{\partial^2 \phi_1}{\partial z^2}\right) \\ &+ 2\left(\frac{\partial \phi_1}{\partial x}\frac{\partial \phi_2}{\partial x} + \frac{\partial \phi_1}{\partial y}\frac{\partial \phi_2}{\partial y} + \frac{\partial \phi_1}{\partial z}\frac{\partial \phi_2}{\partial z}\right)\end{aligned} \tag{vi}$$

But $\nabla^2\phi_1 = 0$ and $\nabla^2\phi_2 = 0$ by Eq. (12.2), so that Eq. (vi) becomes

$$\nabla^2 \phi_4 = 2\left(\frac{\partial \phi_1}{\partial x}\frac{\partial \phi_2}{\partial x} + \frac{\partial \phi_1}{\partial y}\frac{\partial \phi_2}{\partial y} + \frac{\partial \phi_1}{\partial z}\frac{\partial \phi_2}{\partial z}\right) \tag{vii}$$

which is not necessarily zero. It is important to remember we must always satisfy Laplace's equation for potential flow since mass must always be conserved.

This completes the solution.

12.10.2 Flow about a Half-Body

Consider combining a source of strength $2\pi q$ located at the origin of the coordinate system ($z_o = 0$) together with a uniform flow in the x-direction. The streamlines of each of the two flows are shown in Fig. 12.12. We obtain the complex potential for the two flows from Eqs. (12.26) and (12.33):

$$\Omega(z) = U\,z + \frac{q}{2\pi} \ln z \tag{12.92}$$

From Eq. (12.19) we evaluate the complex velocity as

$$u - iv = \frac{d\Omega}{dz} = U + \frac{q/2\pi}{z} \tag{12.93}$$

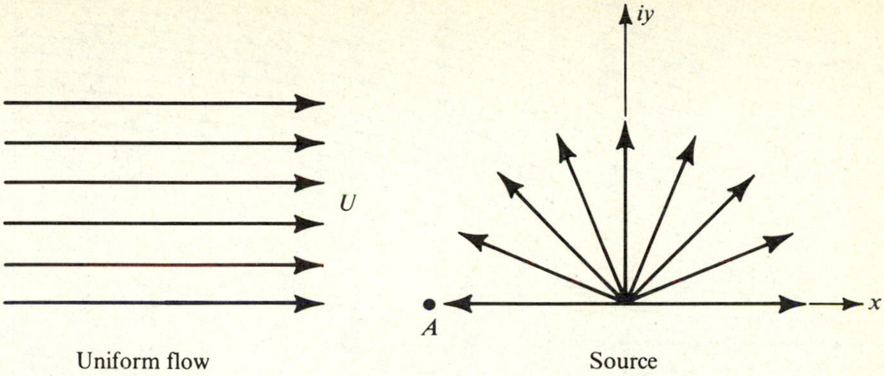

Figure 12.12 *Streamlines for uniform flow and a source.*

We locate the stagnation point by setting $d\Omega/dz$ equal to zero, and then using Eq. (12.93)

$$z = x + iy = -q/2\pi U \tag{12.94}$$

The above result states that the stagnation point lies at a point A shown in Fig. 12.12 (where A is located at $x = -q/2\pi U$ and $y = 0$). At this point A the fluid from the uniform flow exactly meets the fluid from the source and comes to rest. At all other points in the flow field, the fluid has a finite velocity.

The streamlines for the resultant flow can easily be found by identifying the imaginary part of the complex potential given by the Eq. (12.92)

$$\psi = U y + \frac{q}{2\pi} \tan^{-1} (y/x) \tag{12.95}$$

Shown in Fig. 12.13, the streamlines are found by setting Eq. (12.95) equal to different constants.

The *dividing streamline* is also shown in Fig. 12.13. It is the streamline that passes through the stagnation point and divides the flow into two regions. We can imagine the dividing streamline being replaced by a *solid wall*, where flow next to this solid wall can slip perfectly past it. We find the *equation of the dividing* streamline by setting

$$\psi = 0 \tag{12.96}$$

and expressing y as a function of x: from Eqs. (12.95) and (12.96) we obtain

$$\frac{1}{x} = -\frac{1}{y} \tan \left(\frac{2\pi U}{q} y \right) \tag{12.97}$$

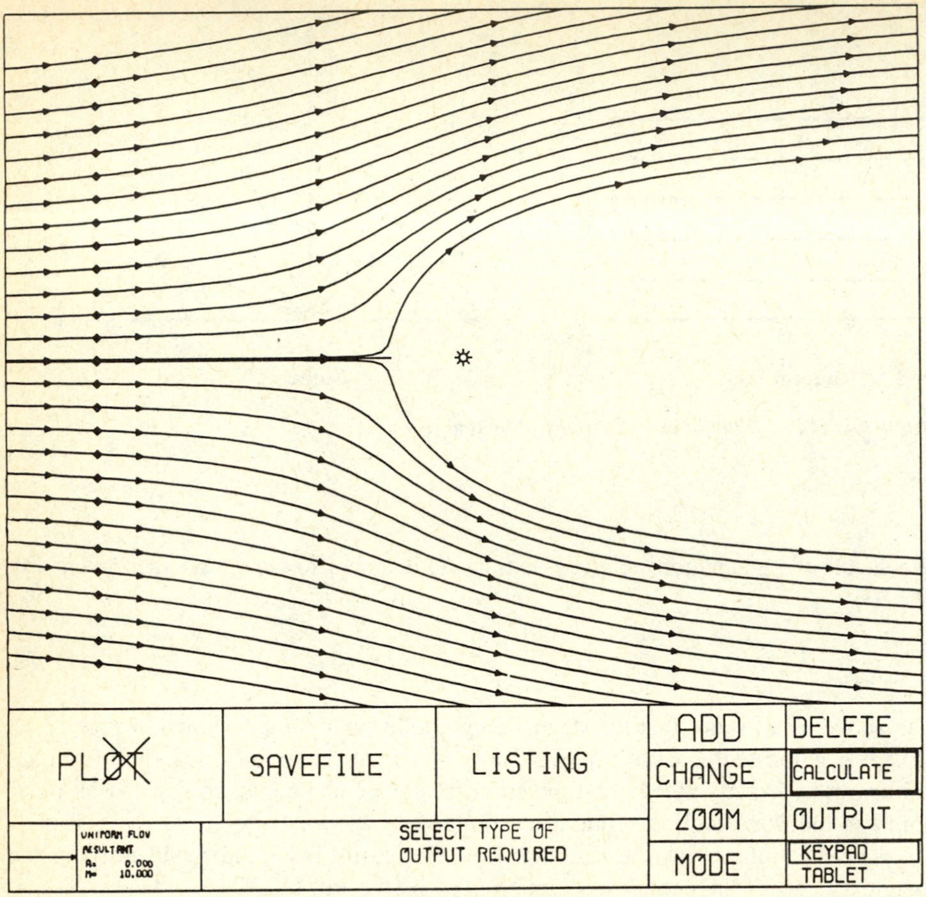

Figure 12.13 *Streamlines for uniform flow past a source. (Source: Reproduced with the permission of the United States Naval Academy Computing Facility.)*

as the equation of the half-body. The width h of the half-body is

$$h = \frac{q}{U} \tag{12.98}$$

obtained by setting x equal to infinity and solving for the value of y in Eq. (12.97).

Example 12.9

A source of flow rate $q(\text{ft}^2/\text{s})$ exists in a uniform flow of velocity U in the positive x-direction. Calculate the force X on the body generated by the equation of the body using the integral momentum theorem, given the pressure p_∞ and density ρ far removed from the body.

Example 12.9 *(Con't.)*

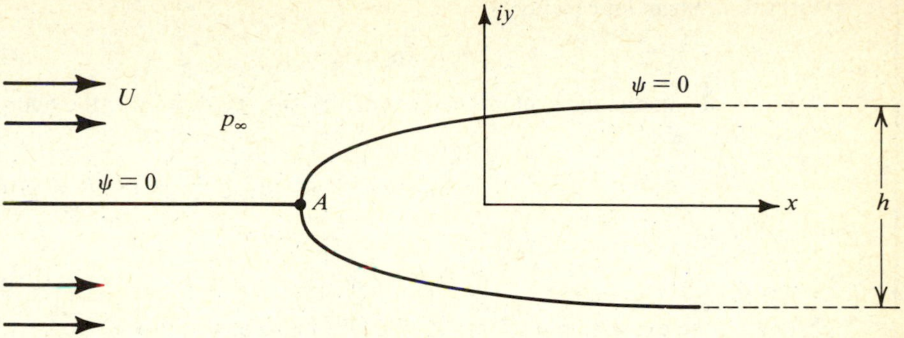

Figure E12.9a

Solution:
Step 1.
The flow is assumed to be steady, incompressible, and potential.
Step 2.
Draw the control volume.

A control volume is first established to identify all the existing external forces.

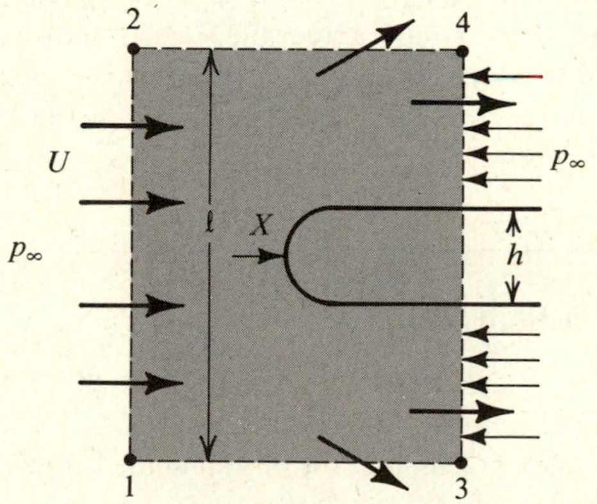

Figure E12.9b

From continuity the flow through surfaces 1–2 and 3–4 is

$$Ul = U(l - h) \tag{i}$$

Equation (i) shows that more flow is entering than leaving the control volume. Thus, the amount of fluid through the surfaces 2–4 and 1–3 of the control volume is Uh, if we expect to preserve the conservation of mass. The total momentum flow through the surfaces 2–4 and 1–3 is then $\rho U^2 h$.
Step 3.
Write the appropriate equation:

Example 12.9 *(Con't.)*

- Integral momentum equation:

$$\Sigma F_x = \rho Q \Delta V_x \tag{ii}$$

From the integral momentum theorem of Eq. (ii), we determine the hydrodynamic force X:

$$X + p_\infty l - p_\infty(l - h) = -\rho U^2 l + \rho U^2 (l - h) + \rho U^2 h \tag{iii}$$

or simplifying

$$X = -p_\infty h \tag{iv}$$

Substituting the expression for h of Eq. (12.98) into Eq. (iii) gives

$$X = -\frac{p_\infty q}{U} \tag{v}$$

as the drag force obtained from integral momentum principles. But according to Blasius' theorem of Eq. (12.88), along with the complex potential $\Omega(z)$ of Eq. (12.92), the complex force $\overline{F}$ is $-\rho Uq$, or

$$X = -\rho Uq \tag{vi}$$

The results of Eqs. (v) and (vi) are similar, differing by a factor of 2. The drag force X as calculated from potential theory is the drag due to pressure. Since potential theory disregards any friction drag force, the fluid is allowed to slip past the body effortlessly. But there is no such restriction in using the integral momentum theorem. Which is the correct result? That obtained from the integral momentum equation is more nearly correct.

This completes the solution.

12.10.3 Uniform Flow Past a Source and a Sink

Consider a source at $z_o = -a$, and a sink at $z_o = +a$. Let the strength of the source equal the strength of the sink. Let uniform flow of magnitude U flow in the positive x-direction. We obtain the complex potential Ω for this combined flow using Eqs. (12.26), (12.33), and (12.34):

$$\begin{aligned}\Omega(z) &= Uz + \frac{q}{2\pi}\ln(z + a) - \frac{q}{2\pi}\ln(z - a) \\ &= Uz + \frac{q}{2\pi}\ln\frac{(z + a)}{(z - a)}\end{aligned} \tag{12.99}$$

The complex velocity is therefore

$$u - iv = \frac{d\Omega}{dz} = U + \frac{q}{2\pi}\left(\frac{1}{z + a}\right)\left[\frac{(z - a) - (z + a)}{(z - a)}\right]$$

$$= U - \frac{q}{\pi} \frac{a}{z^2 - a^2} \tag{12.100}$$

$$= U - \frac{qa}{\pi[(x^2 - y^2 - a^2) + 2ixy]}$$

At the stagnation point, the velocity components do not exist, $u = v = 0$, so that from Eq. (12.100) we find the stagnation point to be located at $y = 0$. At the location $y = 0$, the x location for stagnation is

$$x = \pm \sqrt{\frac{qa}{U\pi} + a^2} \tag{12.101}$$

From the imaginary part of Eq. (12.99) we obtain the stream function ψ

$$\psi = Uy + \frac{q}{2\pi} \tan^{-1} \left(\frac{2ya}{a^2 - x^2 - y^2} \right) \tag{12.102}$$

We can now calculate the streamlines for specific values of a, q, and U by assigning various values to ψ, then calculating x from assorted given values of y. Figure 12.14 shows typical streamlines. The zero streamline generates a figure which is popularly called a *Rankine body*.

Example 12.10

Consider a Rankine body that has a fineness ratio $2b/l$ such that $l = 2a$, making the body extremely slender. Given that $b = q/U$, and the pressure far removed from the body is p_∞, find the pressure at the central section on the body in terms of density ρ, velocity U, pressure p_∞, and lengths l and b.

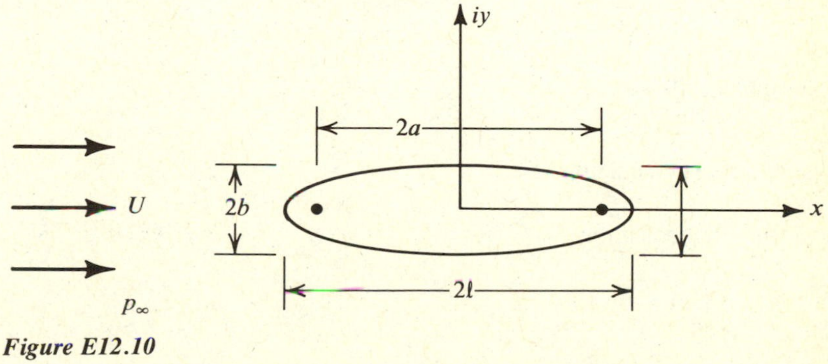

Figure E12.10

Solution:

Step 1.

The flow is assumed to be steady, incompressible, and potential.

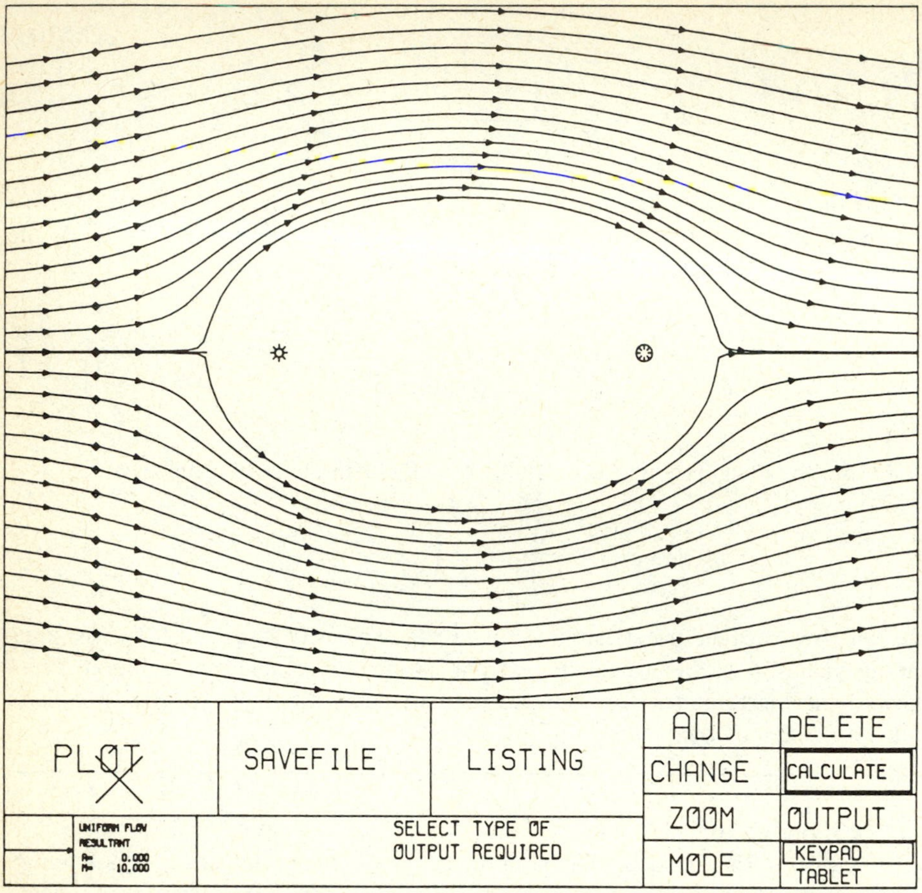

Figure 12.14 *Streamlines about a Rankine body. (Source: Reproduced with the permission of the United States Naval Academy Computing Facility.)*

Example 12.10—*(Con't.)*

Step 3.
The appropriate equations are:

- Bernoulli's equation:

$$\frac{p}{\gamma} + \frac{\tilde{V}^2}{2g} + z = \text{const.} \tag{i}$$

- Complex velocity field:

$$\frac{d\Omega}{dz} = U - \frac{qa}{\pi[(x^2 - y^2 - a^2) + 2ixy]} \tag{ii}$$

Example 12.10 *(Con't.)*

The pressure p is found using Bernoulli's equation (i):

$$p = p_\infty + \tfrac{1}{2}\,\rho(U^2 - \tilde{V}^2) \tag{iii}$$

where

$$\tilde{V}^2 = u^2 \tag{iv}$$

at $x = 0$, and

$$u = U - \frac{qa\,(x^2 - y^2 - a^2)}{\pi[(x^2 - y^2 - a^2)^2 + 4x^2y^2]} \tag{v}$$

from Eq. (ii). At $x = 0$ and $y = b$, the x-component of velocity u is

$$u\,(0, b) = U + \frac{qa}{\pi(b^2 + a^2)} \tag{vi}$$

Substituting Eqs. (iv) and (vi) into Eq. (iii) gives

$$p = p_\infty - \frac{\rho qa}{\pi(b^2 + a^2)}\left[2U + \frac{qa}{\pi(b^2 + a^2)}\right] \tag{vii}$$

Since we were given

$$b = q/U \tag{viii}$$

Eq. (vii) can be expressed as

$$p = p_\infty - \frac{\rho abU^2}{\pi(b^2 + a^2)}\left[2 + \frac{ab}{\pi(b^2 + a^2)}\right] \tag{ix}$$

But for very slender bodies, we can neglect the term $ab/\pi(a^2 + b^2)$ in comparison with the integer 2, so that

$$p \cong p_\infty - \frac{2\rho abU^2}{\pi(a^2 + b^2)} \tag{x}$$

Since $2a \cong l$, and $a^2 >> b^2$, a further simplification gives

$$p = p_\infty - \frac{2\rho U^2}{\pi}\left(\frac{2b}{l}\right) \tag{xi}$$

Equation (xi) is a fair approximation for the minimum pressure on slender bodies of slenderness ratio $(2b/l)$.

This completes the solution.

12.10.4 Uniform Flow Past a Doublet: Flow Past a Cylinder

Let a doublet be located at the origin ($z_o = 0$) and consider uniform flow moving in the positive x-direction. Equations (12.26) and (12.64) give the complex potential $\Omega(z)$ for this confined flow

$$\Omega(z) = Uz + \frac{Ub^2}{z} \tag{12.103}$$

where

$$b^2 \equiv \frac{qa}{\pi U} \tag{12.104}$$

We obtain the dividing streamline by setting $\psi = 0$ and solving for $y = y(x)$. The stream function is the imaginary part of the complex potential of Eq. (12.103):

$$\psi = Uy\left(1 - \frac{b^2}{x^2 + y^2}\right) \tag{12.105}$$

The dividing streamline therefore consists of the line $y = 0$ and the circle $r^2 = b^2$, that is to say, a circle of radius b and that part of the x-axis which lies outside it (as shown in Fig. 12.15) since by definition two streamlines may not intersect. Thus, uniform flow past a doublet has a zero streamline that is a cylinder of radius b lying in a uniform flow.

The stagnation points of the flow are found by setting the complex velocity to zero and solving for the value of the independent variables. From the equation of the stagnation point, Eq. (12.20),

$$u - iv = \frac{d\Omega}{dz} = U - \frac{Ub^2}{z^2} = 0 \tag{12.106}$$

such that the stagnation point is found at

$$z = \pm b \tag{12.107}$$

The x, y locations of stagnation are

$$x = \pm b, \; y = 0 \tag{12.108}$$

These are shown as the points A and B in Fig. 12.15.

To calculate the flow speed on the cylinder (the cylindrical surface is defined as $z = be^{i\theta}$), we evaluate the velocity components u, v, and set the independent variable r equal to b. From the previous result given by Eq. (12.106), the fluid velocity distribution on the cylinder is quite simply

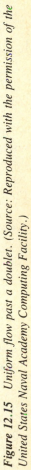

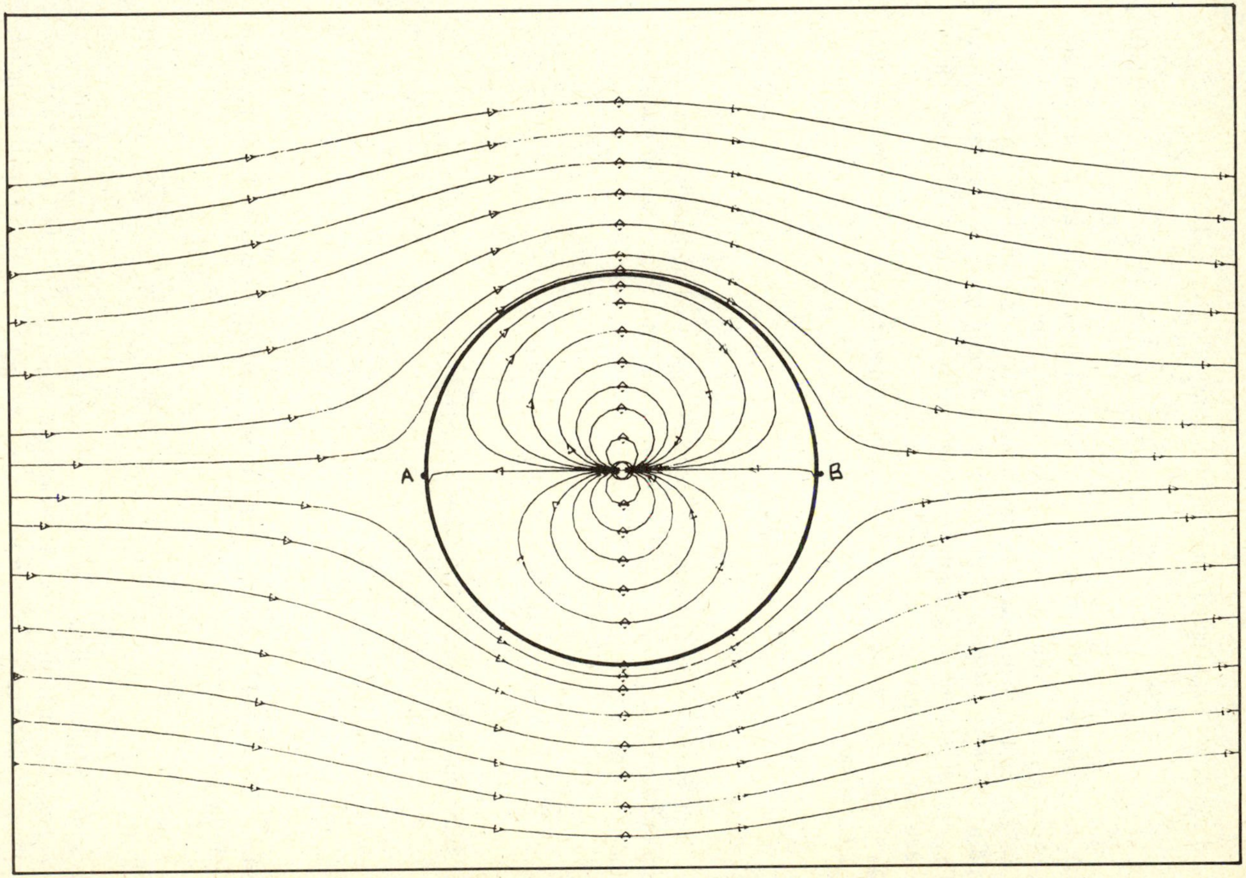

Figure 12.15 Uniform flow past a doublet. (Source: Reproduced with the permission of the United States Naval Academy Computing Facility.)

$$u - iv = U - \frac{Ub^2}{z^2}\bigg|_{r=b} = U(1 - e^{-2i\theta}) \tag{12.109}$$

The speed can now be calculated using the relationship of Eq. (12.21):

$$V = 2U \sin\theta \tag{12.110}$$

Note that the speed is maximum at $\theta = \pm\pi/2$, with the value $2U$ at the top and bottom of the cylinder. Thus, if we consider an infinitely long cylinder whose axis lies in the horizontal plane and is perpendicular to the uniform flow, the velocity measured at the top or bottom of the cylinder is exactly twice that of the uniform flow. This results in a pressure less than ambient pressure. We can now see the reason why objects rise by blowing over them.

To calculate the pressure distribution on the cylinder, we use Bernoulli's equation (4.126) for two-dimensional flow:

$$\frac{p}{\rho} + 2U^2 \sin^2\theta = \frac{p_\infty}{\rho} + \frac{1}{2}U^2 \tag{12.111}$$

where p_∞ is the pressure at a distance sufficiently far removed so that the cylinder will not affect the flow in any way. Rearranging terms, we find the pressure difference on the cylindrical surface to be

$$p - p_\infty = \frac{1}{2}\rho U^2 (1 - 4\sin^2\theta) \tag{12.112}$$

Recalling our definition of the pressure coefficient

$$C_p = \frac{p - p_\infty}{\frac{1}{2}\rho U^2} \tag{7.46}$$

we substitute Eq. (12.112) into Eq. (7.46) to obtain the distribution of the pressure coefficient on the cylindrical surface:

$$C_p = 1 - 4\sin^2\theta \tag{12.113}$$

which is shown plotted in Fig. 12.16. Note the large region on the cylinder that has a negative pressure difference, a phenomenon that helps explain why convertible tops always billow out in a manner similar to the envelope of the pressure coefficient shown in Fig. 12.16. We see the pressure coefficient is unity at $\theta = 0, \pi$, i.e., at the two stagnation points. We should recall that this is where the pressure is a maximum. At $\theta = \pm 30°$, the pressure coefficient is zero. The minimum pressure coefficient occurs at $\theta = \pi/2, 3\pi/2$, where the fluid velocity is a maximum.

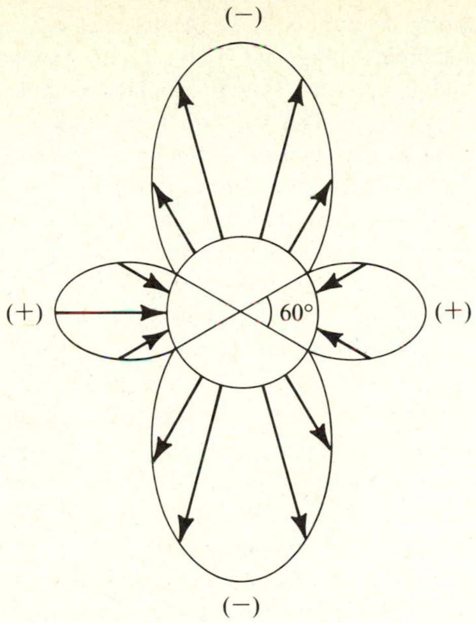

Figure 12.16 *Pressure distribution on a cylinder due to uniform flow.*

Sometimes it is of interest to calculate the pressure coefficient at distances other than on the body. For example, in the present problem, consider the pressure coefficient $C_p\,(r, \theta)$ for $r > b$. Using Eqs. (12.106), (4.112), and (7.46), we find the pressure coefficient for any value of $r \geqslant b$ and for any θ to be

$$C_p(r, \theta) = \frac{b^2}{r^2}\left(2 \cos 2\theta - \frac{b^2}{r^2}\right) \tag{12.114}$$

Comparing the two expressions for the pressure coefficient as given by Eqs. (12.113) and (12.114), we note that the latter is less than the former as can easily be verified by letting $r \to \infty$ in Eq. (12.114).

We find the force on the cylinder due to a uniform potential flow by applying Blasius' theorem of Eq. (12.88). Comparison of the square of the complex potential of Eq. (12.106) with the Laurent series expression of Eq. (12.90) shows that there is no term of the form $(a_{-1})/z$, so that the value of the residue $a_{-1} = 0$ exists for uniform flow past a cylinder. The x- and y-components of the force per unit width are found from Eq. (12.91):

$$\overline{F} = X - iY = 0 \tag{12.115}$$

which means that the x-component of the force per unit width is zero and the y-component of the force per unit width is likewise zero. Thus, *no forces act on the cylinder for irrotational uniform flow over a cylinder.*

We have treated the cylinder in a uniform irrotational flow. Any other body of profile symmetry which can be made up of sources, sinks, and doublets, either singly or in line distribution, will result when placed in a uniform flow in a squared complex velocity expression having no term of the form a_{-1}/z. Thus all similar bodies have a zero hydrodynamic force acting on them. This is known as *d'Alembert's paradox* (previously mentioned in Chap. 5). The paradox states there is no lift or drag for any body that has central symmetry. Anyone can easily verify by experiment that a force does arise when a body moves through a fluid. In the late nineteenth century George Stokes formulated his famous result for the drag on a sphere moving through fluid that gave a plausible explanation for forces on bodies moving through *real* fluids. Later, Helmholtz treated the forces on finite bodies moving in steady *ideal* fluids, using the notion of vortex motions.

Drag is zero (i.e., $X = 0$) not only because the flow is irrotational, but also because the sum of all stresses on the central symmetric body is zero so that the friction drag is zero. The body has central symmetry so that the pressure drag is also zero.

Some authors like to define a *potential flow as being both inviscid* ($\nu = 0$) *and irrotational* ($\zeta = 0$). Some fluids, such as air, have extremely small values of kinematic viscosity, and their motion is very nearly irrotational even in a region close to a solid boundary. Here the effect of fluid friction near the solid boundary cannot be neglected. This viscous effect which causes velocity gradients, and hence stresses in the "near" region of the solid boundary, is of tremendous importance in evaluating the drag force, as we shall later discover.

12.10.5 Uniform Flow Past a Cylinder with Circulation

In the previous section we learned that the mathematical model for uniform flow past a cylinder was given by

$$\Omega(z) = Uz + \frac{Ub^2}{z} \tag{12.103}$$

where the doublet was located at the origin of the coordinate system and the free-steam velocity U was in the positive x-direction. We found there was no lift or drag on the cylinder. What do we have to do to the flow to obtain a force? We shall now find out.

Consider a clockwise rotating potential circulation Γ_∞ about the cylinder. The complex potential $\Omega(z)$ for uniform flow past a cylinder with circulation is obtained from Eqs. (12.26), (12.60), and (12.64) resulting in

$$\Omega(z) = Uz + \frac{Ub^2}{z} + \frac{i\Gamma_\infty}{2\pi}\ln z \tag{12.116}$$

The velocity potential ϕ and stream function ψ are the real and imaginary parts of Eq. (12.116), respectively:

$$\phi(x, y) = Ux\left(1 + \frac{b^2}{x^2 + y^2}\right) - \frac{\Gamma_\infty}{2\pi}\tan^{-1}\frac{y}{x} \tag{12.117}$$

$$\psi(x, y) = Uy\left(1 - \frac{b^2}{x^2 + y^2}\right) + \frac{\Gamma_\infty}{4\pi}\ln(x^2 + y^2) \tag{12.118}$$

expressed in Cartesian coordinates, and

$$\phi(r, \theta) = Ur\cos\theta\left(1 + \frac{b^2}{r^2}\right) - \frac{\Gamma_\infty}{2\pi}\theta \tag{12.119}$$

$$\psi(r, \theta) = Ur\sin\theta\left(1 - \frac{b^2}{r^2}\right) + \frac{\Gamma_\infty}{2\pi}\ln r \tag{12.120}$$

expressed in polar coordinates. We evaluate the radial and tangential velocity components of the fluid as

$$v_r = \frac{\partial\phi}{\partial r} = U\cos\theta\left(1 - \frac{b^2}{r^2}\right) \tag{12.121}$$

$$v_\theta = \frac{1}{r}\frac{\partial\phi}{\partial\theta} = -U\sin\theta\left(1 + \frac{b^2}{r^2}\right) - \frac{\Gamma_\infty}{2\pi r} \tag{12.122}$$

For stagnation, *both* the radial and tangential velocity components must vanish. Setting the radial velocity $v_r = 0$, we obtain

$$r = \pm b, \theta = \pm\pi/2 \tag{12.123}$$

and setting the circumferential velocity $v_\theta = 0$, we obtain

$$\theta = \sin^{-1}\left(\frac{-\Gamma_\infty}{4\pi U b}\right) \tag{12.124}$$

We note that there are two stagnation points, A and B, which are shown in Fig. 12.17. Their locations are given by Eq. (12.124). When $\sin^{-1}(\pm 1)$, then points A and B coincide at the location $(0, -b)$. Quite obviously, the streamline pattern depends strongly on the magnitude of the circulation Γ_∞.

We would expect that as the circulation increases, the stagnation points A and B would move further downward along the cylinder until they reach their maximum at $\theta = 3\pi/2$. Subsequently a further increase of circulation will not allow the streamlines to cross. We would also expect that as the circulation increases, the flow over the top of the cylinder would increase, and the streamlines would pack in even more tightly.

The streamlines shown in Fig. 12.17i are symmetric about the cylinder since no circulation is present. Here the stagnation points are at the leading and trailing edges of the cylinder. As the circulation Γ increases, both stagnation points move downward as shown in Fig. 12.17ii, until they coincide in Fig. 12.17iii at $\theta = 3\pi/2$ for $\Gamma =$

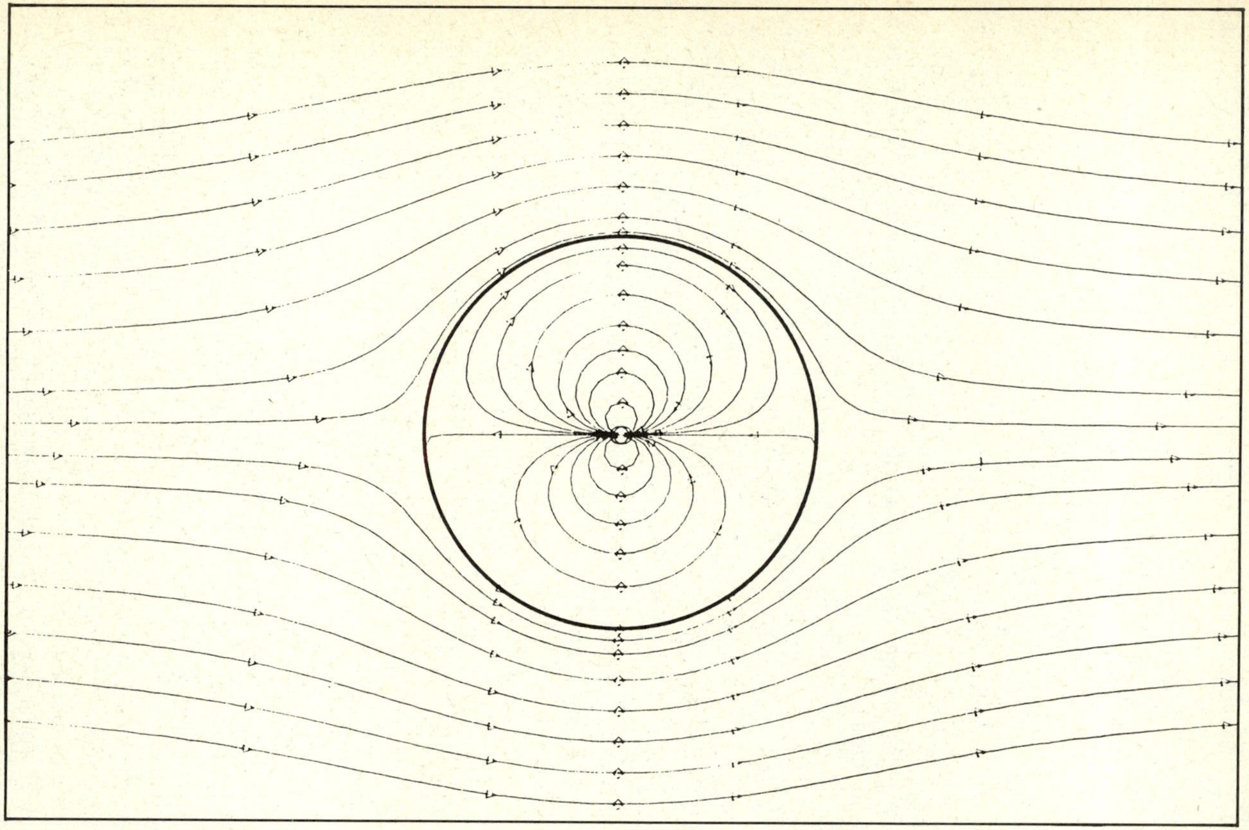

(i)

Figure 12.17 *(i) Flow about a cylinder with no circulation.*

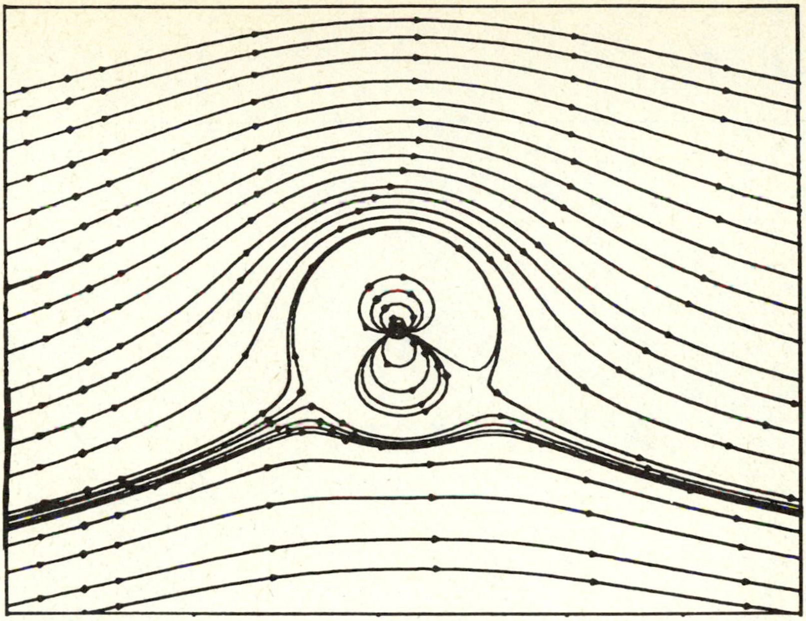

(ii)

Figure 12.17 (*Con't.*) *(ii) Flow net about a cylinder with circulation, $\Gamma/aU<4\pi$.*

$4\pi b\ U$. At values of $\Gamma > 4\ \pi b\ U$, the velocity is nonzero everywhere on the cylinder. As shown in Fig. 12.17iv, the stagnation point has moved away from the cylinder directly below $\theta = 3\pi/2$. Thus a portion of the flow simply recirculates around the cylinder.

If we wish to calculate the force per unit width on the cylinder, we should use the Blasius expression

$$\overline{F} = X - iY = \frac{i\rho}{2} \oint \left(\frac{d\Omega}{dz}\right)^2 dz \tag{12.88}$$

Substituting the complex potential $\Omega(z)$ of Eq. (12.116) into Eq. (12.88) gives

$$X - iY = \frac{i\rho}{2} \oint \left[U\left(1 - \frac{b^2}{z^2}\right) + \frac{i\Gamma_\infty}{2\pi z}\right]^2 dz \tag{12.125}$$

The residue a_{-1} is easily found as the coefficient of the term $1/z$ in the integrand of Eq. (12.125), so that the complex force becomes

$$\begin{aligned} X - iY &= \frac{i\rho}{2}(2\pi i)\left(\frac{iU\Gamma_\infty}{\pi}\right) \\ &= -\,i\rho U\Gamma_\infty \end{aligned} \tag{12.126}$$

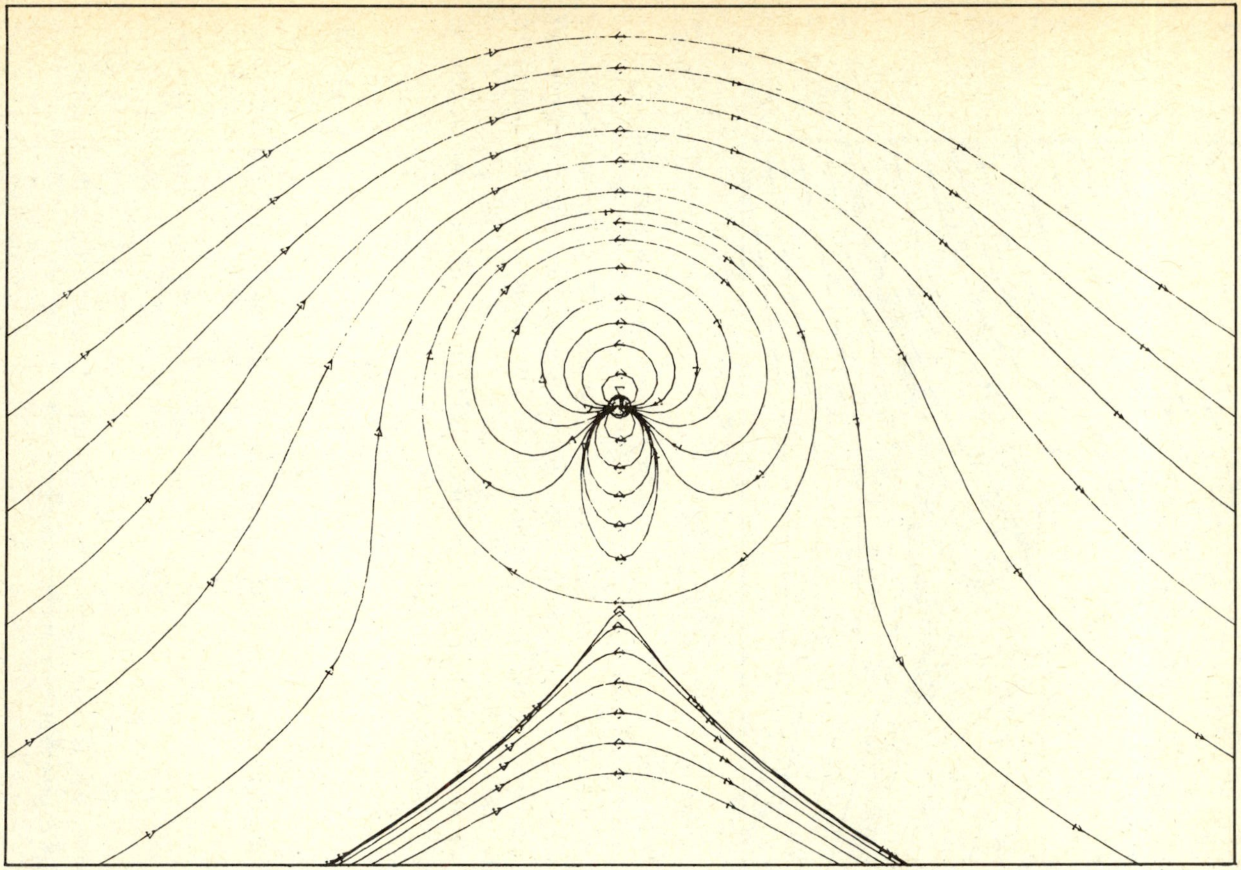

(iii)

Figure 12.17 *(Con't.)* *(iii) Flow net about a cylinder with circulation,* $\Gamma/aU = 4\pi$.

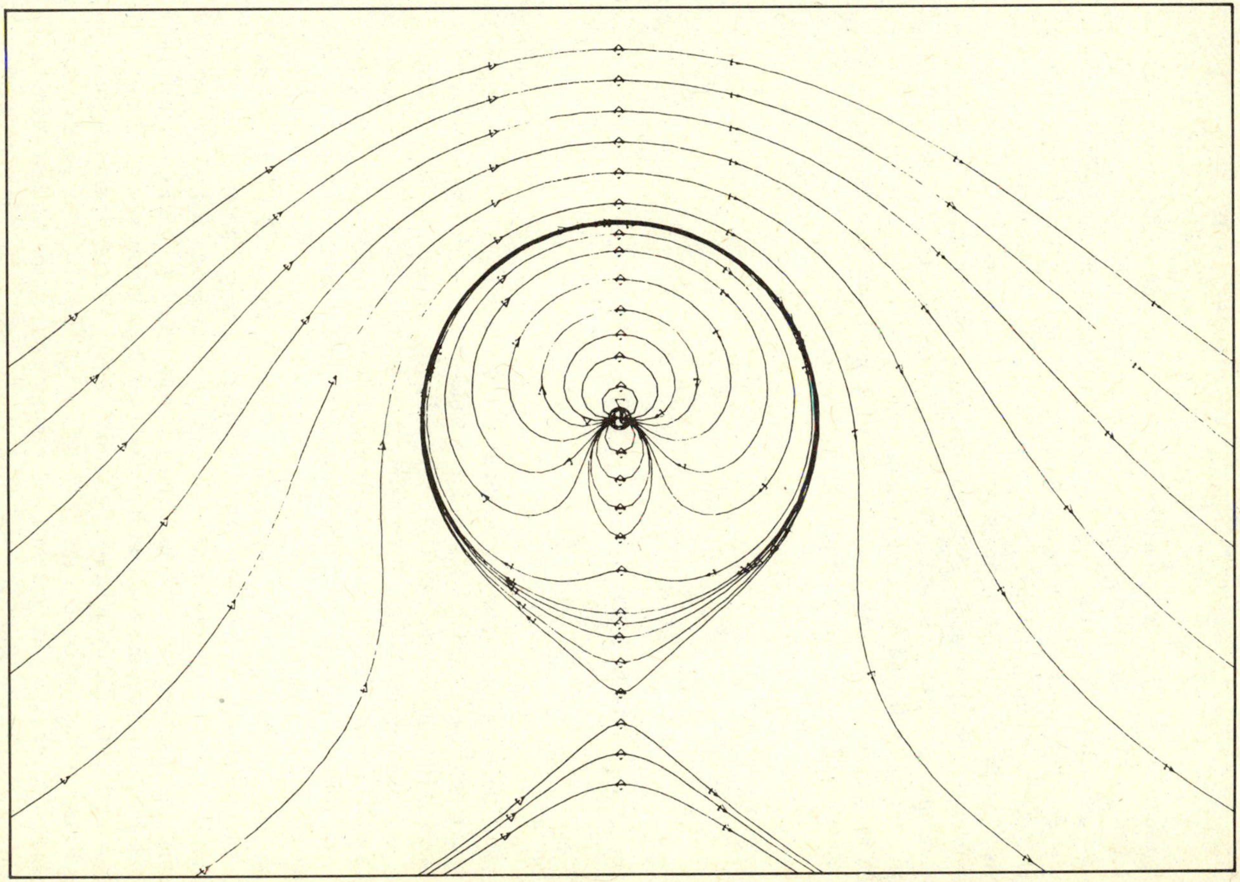

(iv)

Figure 12.17 *(Con't.)*   *(iv) Flow net about a cylinder with circulation,* $\Gamma/aU > 4\pi$.

Comparing the real and imaginary parts of the two sides of Eq. (12.125) results in

$$X = 0 \tag{12.127}$$

$$Y = \rho U \Gamma_\infty \tag{12.128}$$

which represents the x- and y-components of the force per unit width, respectively. Thus, we get positive lift if the circultation Γ_∞ is clockwise and negative lift if the circulation Γ_∞ is counterclockwise. This result is identified as the Kutta-Zhukowskii lift theorem: in two-dimensional incompressible steady irrotational flow about a boundary of any centrally symmetric shape, the lift per unit width always equals the product of the density ρ times the velocity U times the potential circulation Γ_∞. The circulation originates from circular flow, and the velocity originates from translatory flow. Many practical examples come to mind where such flows create a force: slicing a golfball, english on a billiard ball, and the curve on a baseball.

Some marvelous experiments illustrating the lift on bodies with circulation are given by Rogers [12.3].* Three are especially interesting.

DEMONSTRATION 12.1

Purpose:

To illustrate lift on a rotating body

Description:

Approximately three feet of ribbon is wrapped around the central portion of an empty paper tube of toilet paper. The tube is placed on a table so that the ribbon will unwind from the bottom. Holding the free end of the ribbon slightly below the table, the ribbon is yanked. The sudden unwinding of the ribbon promotes an angular velocity, and moving the ribbon forward gives the tube a forward speed. When done correctly, the tube will rise, indicating lift, and if the rotation of the tube is great enough, the tube will go into a vertical loop, like that shown in Fig. 12.18a.

A similar demonstration again uses a paper cylinder. Figure 12.18b shows the set-up. A wooden dowel b is the axis of the cylinder c that is centrally located by wooden insert a. Two 2 ft lengths of string are wound around each end of the dowel with one free end attached to a wooden hanger f at locations g. A third string h is attached to keep the cylinder in a fixed position. When string h is removed, the cylinder falls and starts to rotate. The cylinder has uniform flow plus a vortex around it which, according to theory, will produce a lift. The lift in this case is a side force on the cylinder that moves the cylinder in a direction perpendicular to the vertical plane.

The last demonstration illustrates the slice and hook of golf. The apparatus is shown in Fig. 12.18c. Two sections of sandpaper are formed making a 100° angle that permits a ping pong ball to rest. Launching the ball at various speeds and angles

*Prof. C. Wu of the USNA described similar set-ups at a ASEE meeting in 1974.

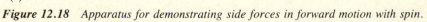

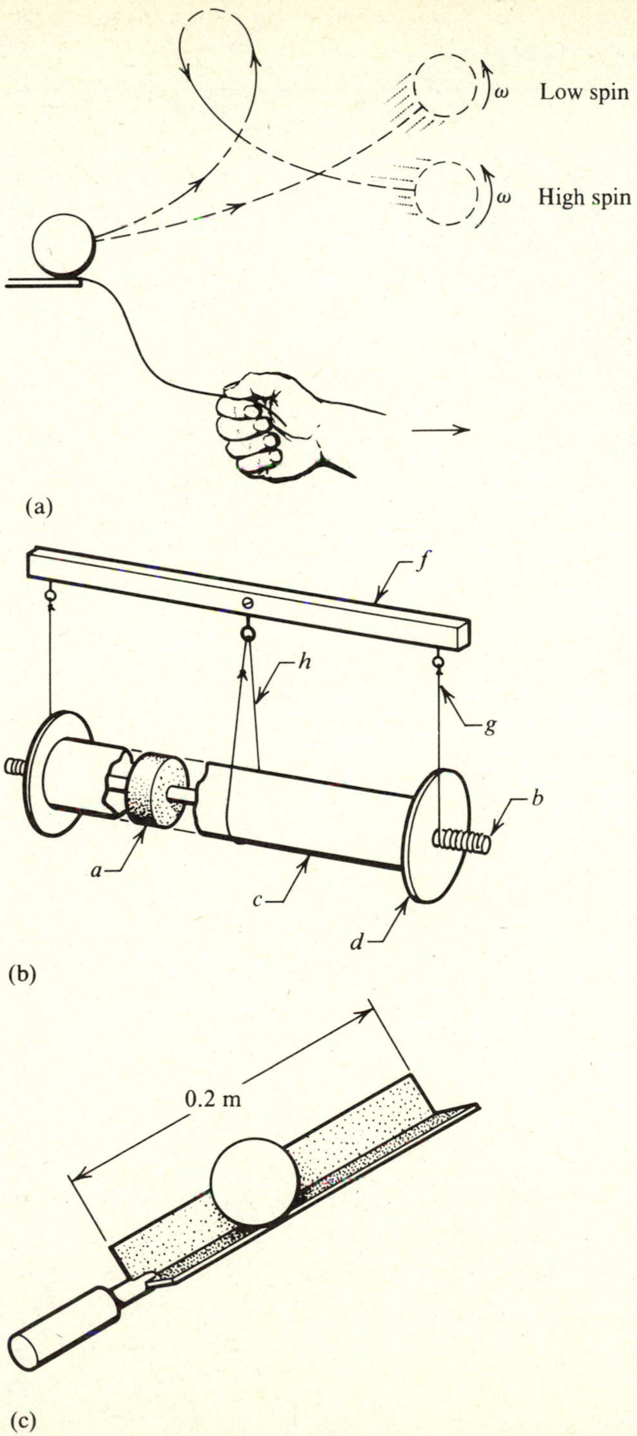

Figure 12.18 *Apparatus for demonstrating side forces in forward motion with spin.*

create different motions illustrating how spin combined with forward motion creates side forces that result in a cycloid path.

Example 12.11

A uniform velocity U flows past a cylinder of radius a, that is rotating with angular speed ω. The resultant lift force is $4\rho U^2 a$. (a) Find the magnitude and sense of rotation of the angular speed ω. (b) Find the angle θ measured from the horizontal x-axis to the stagnation point A or B. (c) Find the pressure p at the point $r = 2a$, $\theta = 0$.

Solution:

Step 1.

The flow is assumed to be steady, incompressible, and potential.

Step 3.

The appropriate equations are:

- Circulation of Eq. (12.57):

$$\Gamma_\infty = 2\pi\omega r_c^2 \tag{i}$$

- Lift per unit width on a cylinder with a vortex:

$$Y = -\rho U\Gamma \tag{ii}$$

- Location of stagnation:

$$\theta = \sin^{-1}\left(\frac{-\Gamma_\infty}{4\pi Ub}\right) \tag{iii}$$

- Bernoulli's equation:

$$\frac{p}{\gamma} + \frac{\tilde{V}^2}{2g} + z = \text{const.} \tag{iv}$$

(a) For solid body rotation ω of the cylinder of radius a, we evaluate the potential circulation Γ_∞ using Eq. (i):

$$\Gamma_\infty = 2\pi a^2\omega \tag{v}$$

Substituting the given lift force and Eq. (v) into the expression for the lift of Eq. (ii) gives

$$4\rho U^2 a = -2\pi\rho a^2 U\omega \tag{vi}$$

Solving for the angular speed, we obtain

$$\omega = -\frac{2U}{\pi a} \tag{vii}$$

with the sense of rotation being clockwise.

Example 12.11 *(Con't.)*

(b) The angle θ is obtained from Eq. (iii):

$$\sin\theta = -\frac{\Gamma_\infty}{4\pi U a} \tag{viii}$$

Substituting the circulation of Eq. (v) and the angular speed of Eq. (vii) into Eq. (viii) gives

$$\sin\theta = \left(-\frac{1}{\pi}\right) \tag{ix}$$

or

$$\theta = -18^\circ\, 34' \tag{x}$$

(c) We evaluate the pressure p from Bernoulli's equation (iv):

$$p = p_\infty + \frac{1}{2}\rho\,(U^2 - v_r^2 - v_\theta^2) \tag{xi}$$

where v_r and v_θ come from Eqs. (12.121) and (12.122), respectively:

$$p = p_\infty + \frac{1}{2}\rho\left[\frac{2a^2U^2}{r^2}(\cos^2\theta - \sin^2\theta) - \frac{U^2a^4}{r^4} - \frac{\Gamma_\infty^2}{4\pi^2r^2} + \frac{U\Gamma_\infty}{\pi r}\left(1 + \frac{a^2}{r^2}\right)\sin\theta\right] \tag{xii}$$

At $r = a$, $\theta = 0°$, the pressure is

$$p = p_\infty - \frac{1}{32}\rho U^2\left(7 - \frac{\Gamma_\infty^2}{\pi^2a^2U^2}\right) \tag{xiii}$$

From Eqs. (v) and (vii), the circulation is

$$\Gamma_\infty = -4Ua \tag{xiv}$$

such that the pressure p becomes

$$p = p_\infty + 0.28786\rho U^2 \tag{xv}$$

A practical application of this example is the propulsion system of the ship Flettner. Two upright rotating cylinders mounted on the deck of the ship provided a pressure distribution over the cylinders resulting in a forward thrust.

This completes the solution.

12.11 Lift and Drag

This section is devoted to a discussion of the Kutta-Zhukovskii result that has been presented in the previous section. We showed that lift is directly proportional to the density of the fluid, the velocity of the body moving through an undisturbed fluid at rest, and the potential circulation over the body. The physical mechanism for the circulation must now be presented to explain the origin of the vortex flow. A brief discussion of the effect of circulation follows, concluding with a brief exposition on images. The method of images is extremely useful in obtaining a mathematical model for a body placed near a solid boundary in a potential flow field.

12.11.1 The Phenomenon of Lift

Consider a body with the shape of an airfoil having constant cross section and an infinite span. Such an airfoil is called a *two-dimensional airfoil* since the analysis excludes the effects of fuselage and wing tip. Let the wing be straight, as against swept back, and all sections at the same angle-of-attack α. Thus, any cross-sectional area of the wing is representative of the whole wing.

Assume the fluid flow to be potential. The resultant streamlines would then be somewhat like those shown in Fig. 12.19. In Fig. 12.19 there are two stagnation points, A, B. For streamlines very close to the bottom of the airfoil, the fluid particles must pass the sharp rear edge with exceedingly large velocities. For the streamline at the bottom surface, the fluid must have an infinite velocity at the trailing edge, and then drop off to zero at point B. We know that physically, nature does not allow infinite velocities or negative infinite pressures. But if the flow is potential, then conceptually infinite velocities and pressures can exist. We want to make velocities everywhere finite even though the flow is potential. For a symmetric airfoil at zero angle-of-attack in a potential flow, we would expect no forces to exist, and the pressure distribution over the top surface to equal that over the lower surface. We can show this by mapping the geometry of an airfoil onto the circumference of a cylinder. We have already shown that there is no force on a cylinder with zero circulation. We have also shown in Sec. 12.10.5 that a lift force does exist when there is circulation. If we consider circulation in a clockwise direction around the foil, then from the Kutta-Zhukovskii lift theorem, Eq. (12.128), the lift is

$$L = \rho U \Gamma_\infty \tag{12.129}$$

where Γ_∞ is the potential circulation given by Eq. (12.53). Thus, to calculate the lift, we must know the circulation. The circulation depends upon the angle-of-attack, the flow velocity, and the geometric shape of the airfoil.

Let us first consider the actual physics of the flow in order to comprehend fully the phenomenon of lift. Suppose the airfoil is at rest in a stationary real fluid. Let the flow of velocity U past the wing be uniform. At the instant motion starts the flow is potential, and we notice experimentally an extremely large velocity near the trailing edge with streamlines like that shown in Fig. 12.19a. Next, the velocity in the region between the trailing edge and stagnation point B decreases very slowly to zero; thus the pressure increases, as predicted by Bernoulli's equation. The flow close to the surface is proceeding against the positive pressure gradient (which is called an *adverse pressure gradient*). This resistance is quickly overcome by nature through the action of the shear stresses. The stagnation point B moves to the trailing edge, as shown in Fig. 12.19c. At this time a single vortex is created, as pictured in Fig. 12.20.

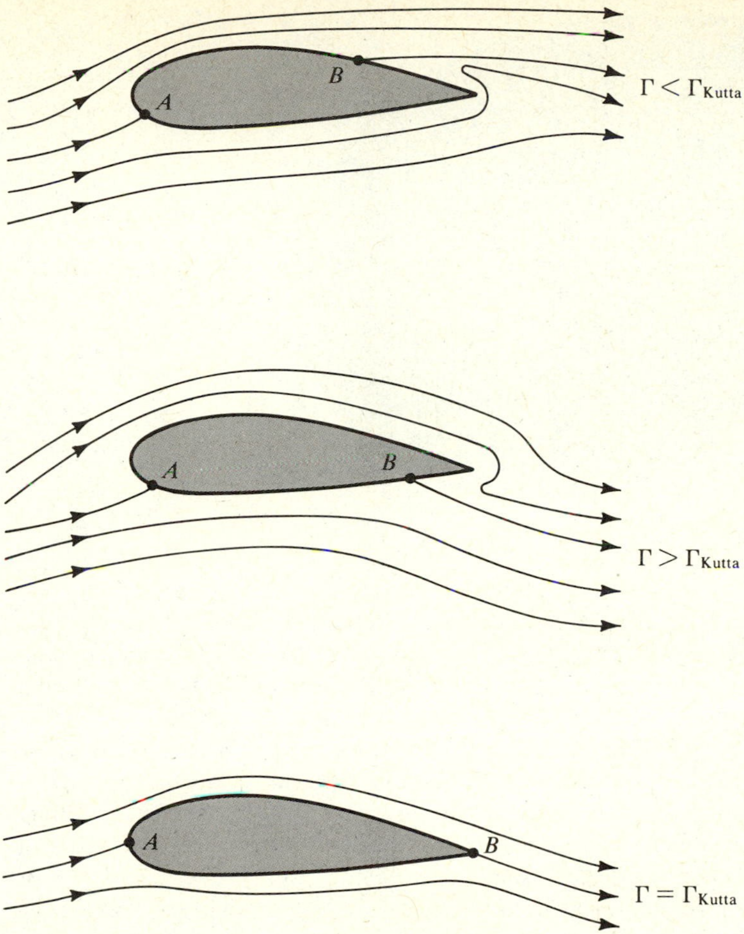

Figure 12.19 *Potential flow about a two-dimensional airfoil at small angle-of-attack* α.

Consider the region C. For a fluid initially at rest, the circulation of region C *must* be zero. According to Kelvin's theorem, Eq. (12.71), the circulation of an irrotational flow is conserved for all times, so that if it is initially zero, then it must forever be zero so long as the flow is irrotational. If the circulation is zero in C, then it must be zero in region C' and in region C'', since C' and C'' are within region C. C'', however, has a vortex with a circulation. Then for the circulation in C to be zero, there must be a circulation equal in magnitude but opposite in direction to the circulation of region C'' in the region $C - C''$, which is the region C'. Though it is not shown, the region C' happens to be the region of the wing section.

Proceeding along with the flow development, the strength of the initial vortex grows rapidly until the velocity on the upper surface of the trailing edge precisely equals the velocity on the lower surface of the trailing edge. At this time in the development, the stagnation point has moved from point B in Fig. 12.19a to the trailing edge of the airfoil, point B in Fig. 12.19c.

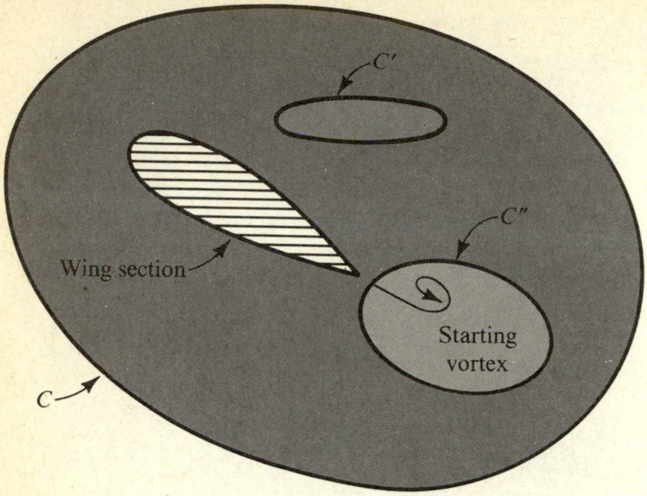

Figure 12.20 *The starting vortex.*

In 1909 S. Chaplygin discussed Zhukovskii's result and hypothesized that "out of the infinite number of theoretically possible flows past a profile with a sharp trailing edge (parts *a, b,* and *c* of Fig. 12.19), the flow that actually occurs is the one with a finite velocity at the trailing edge." This statement is identified as the *Zhukovskii-Chaplygin condition*. We therefore pose the following question: What is the strength of the clockwise moving irrotational vortex that can move the stagnation point *B* of Fig. 12.19a or 12.19b to the trailing edge? The answer is a body with a sharp trailing edge which is moving through a fluid. It will create about itself a circulation of sufficient strength to hold the rear stagnation point at the trailing edge. If the velocity *U*, or angle-of-attack, is increased, another vortex is shed off the trailing edge having the same direction of rotation as the starting vortex. If the velocity, or angle-of-attack, is decreased, a vortex, having an opposite direction of rotation as the starting vortex, is shed. If the wing is suddenly accelerated from rest and then stopped suddenly, two vortices of equal strength but opposite direction are shed off the trailing edge.

Once the initial vortex of Fig. 12.20 has reached its proper strength, then it will shed from the wing and move downstream; the circulation around the wing, however, will remain there until the flow ceases. When the flow ceases, the circulation will move off the wing downstream, just like the starting vortex only opposite in rotation, slowly diminish in strength, because of viscosity, and finally become absorbed in the flow. If the flow is ideal, then the vortex goes to infinity.

The *actual measured* lift *L* is somewhat smaller than that predicted by the Kutta-Zhukovskii expression of Eq. (12.129). It is smaller because of shear stresses in the boundary layer, i.e., that region where vorticity exists. As the flow moves further downstream over the wing, the thickness of the boundary layer increases. That is, the effect of the fluid being a real fluid near a solid body becomes more important as we move downstream. The greater the thickness of the boundary layer, the further away from the body lies the potential flow, which in turn reduces the angle-of-attack, because of the increased size in the suction side of the geometry for an equivalent potential wing section. This in turn results in a smaller lift.

Once again, shape of the body is important in achieving desired results—so important that an entire family of airfoil shapes has been identified. Some of the members of the family are shown in Fig. 12.21. As stated earlier in Chap. 7, the four digit numbering system is important:

- First integer: Maximum value of the mean-line ordinate in percent chord
- Second integer: Distance from leading edge to the location of maximum camber in tenths of the chord
- Last two integers: Section thickness in percent of the chord

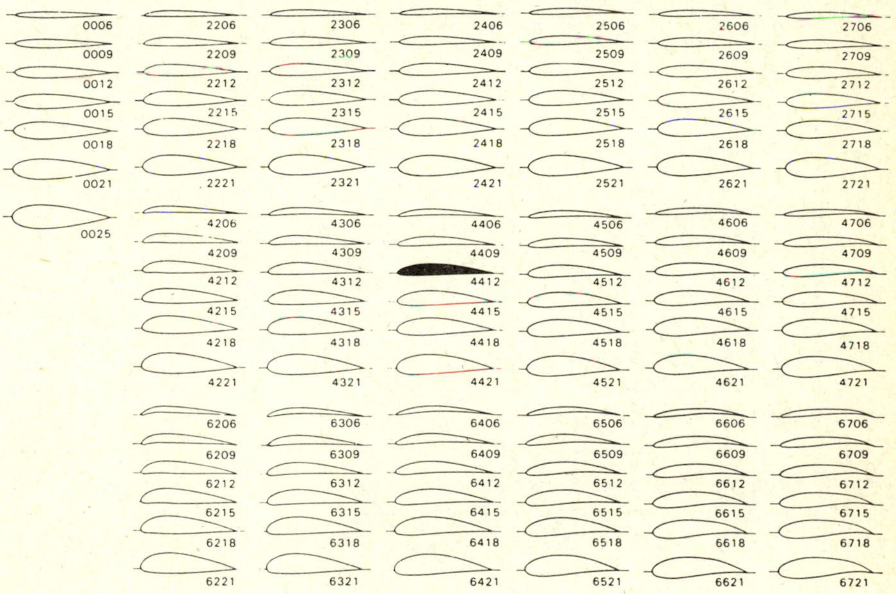

Figure 12.21 *Family of NACA wing sections. (Source: Courtesy of NASA.)*

For example, wing section 4412 has a 4% camber at 0.4 of the chord from the leading edge and is 12% thick.

Each airfoil section has a composite history of its lift, drag, and moment, collected in Ref. 12.4. Since we have just presented the potential theory for lift, let us look at the results for the NACA 4412 airfoil.

Figure 12.22 shows the variation of the coefficient of lift, C_L, versus angle-of-attack. Lift increases almost linearly with angle-of-attack, α, up to a maximum value. A further increase in α would result in a decrease in lift. The angle-of-attack that gives maximum lift is called *stall*. Stall is a condition where the flow *separates*, and a *dead region* (stagnation region) is formed. At the stall angle, the circulation theory would, of course, be no longer applicable.

Sta.	Up'r.	L'w'r.
0	—	0
1.25	2.44	−1.43
2.5	3.39	−1.95
5.0	4.73	−2.49
7.5	5.76	−2.74
10	6.59	−2.86
15	7.89	−2.88
20	8.80	−2.74
25	9.41	−2.50
30	9.76	−2.26
40	9.80	−1.80
60	8.14	−1.00
70	6.69	−0.65
80	4.89	−0.39
90	2.71	−0.22
95	1.47	−0.16
100	(0.13)	(−0.13)
100	—	0

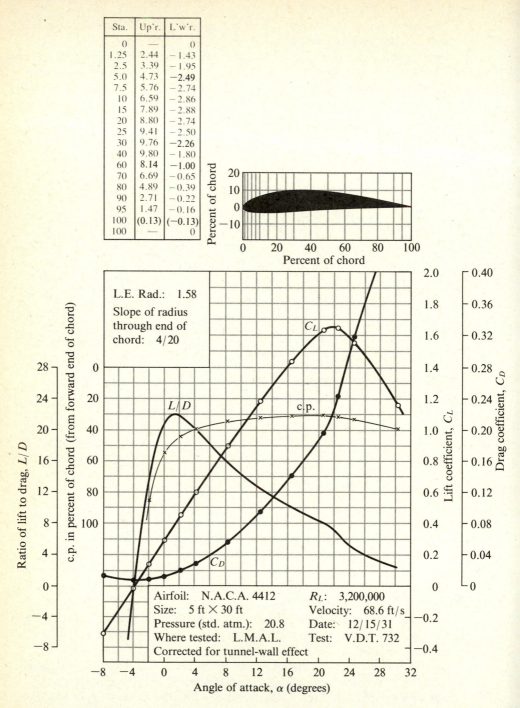

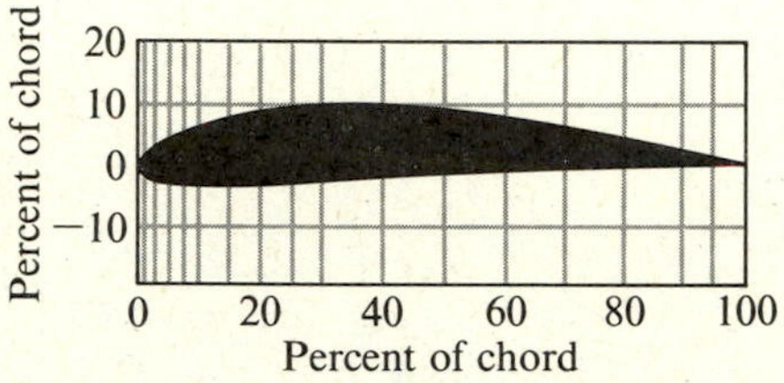

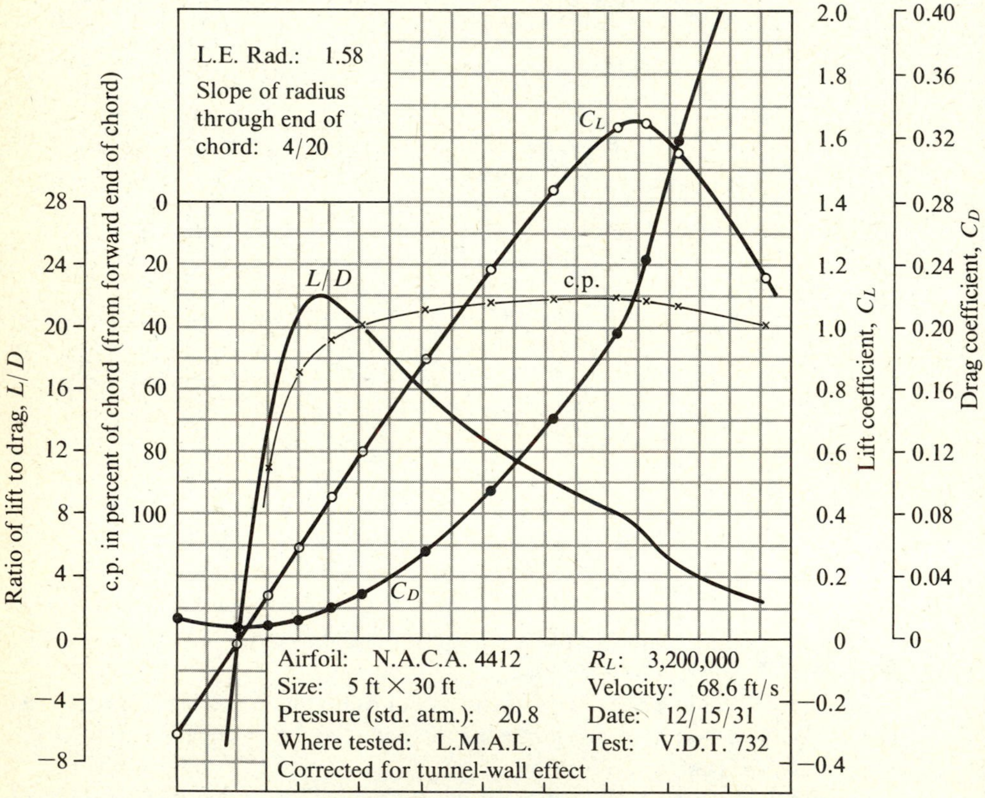

Figure 12.22 *C_L, C_D, C_M variations versus angle-of-attack for NACA 4412 airfoil. (Source: Adapted from I. H. Abbott and A. E. von Doenhoff,* Theory of Wing Sections, *© 1949, Dover Publications, Inc. Used with the permission of the publisher.)*

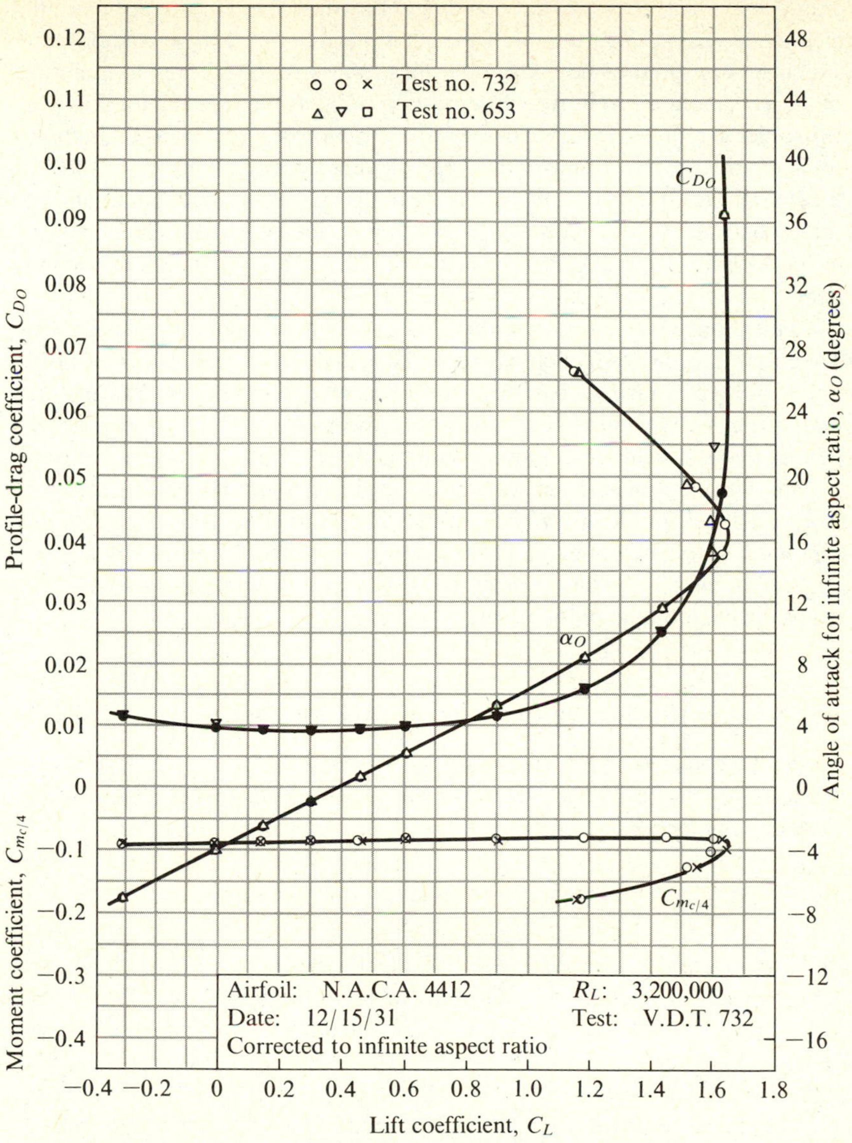

***Figure 12.22** (Con't.)*

12.11.2 The Phenomenon of Drag

Figure 12.22 shows another force, and for completeness, it should be discussed briefly. The other component of the force is the drag. Drag is that force which lies in the direction of the approach velocity. The total drag consists of two parts: drag due to pressure (considering that component of the pressure in the direction of the velocity

U), and drag due to shear stress (or friction drag). Pressure drag largely occurs in the *wake*, which is the region downstream of the *separation point*. The separation point is defined as that point on the body where the shear stress is zero. Thus, to minimize pressure drag, we should try to move the separation point as close to the trailing edge as possible in order to reduce the region of the wake. Friction drag is kept to a minimum, on the other hand, by careful shaping of the wing's geometry, notably its upper surface, and reducing the surface roughness. We will discuss all this in some detail in Chap. 14.

The drag coefficient, C_D, is seen in Fig. 12.22 to increase nonlinearly with an increase of angle-of-attack, rising sharply past stall. Notice that its minimum value is approximately at $\alpha = -2°$, largely because of the wing's camber.

The aforementioned physical description is sufficient to account for potential wing theory. Except for drag, it closely approximates the actual ''real'' fluid flow past two-dimensional wings. We have to introduce the concept of circulation, and in turn, concentrated vorticity. We must recall that any finite portion of a fluid in motion, once it is irrotational, continues to move irrotationally thereafter, since the vorticity is zero to begin with, and there is no way to introduce vorticity: with one exception. Vortex sheets can arise from *boundaries* of immersed solid bodies. However, Kelvin's theorems are *not violated* by vortex sheets arising from boundaries of wetted surfaces since the conservation of angular momentum is preserved. What is required in the analysis is a proper interpretation of results.

12.11.3 Some Illustrative Projects on Lift and Drag

Illustrating the concepts of lift and drag can be fun and the do-it-yourself projects below—described so that even the beginner can appreciate them—can be used as lively classroom demonstrations.

Purpose

1. An article in *Scientific American*, vol. 193 (Oct. 1955), pp. 124–134, uses a Hele-Shaw apparatus to study the dynamics of air and water flows. The article shows how to build a homemade smoke tunnel and flexible *airplane models* that allow one to study certain aeroelastic effects.
2. A second *Scientific American* article, vol. 220 (April 1969), pp. 130–136, offers a detailed explanation of the elements of diamond, Malay, French war, and jib *kites*, along with a nice description of the aerodynamics of flying kites. It presents an excellent description of the construction of kites, with explanations of forces and optimum design configurations.
3. *Scientific American*, vol. 195 (Aug. 1956), pp. 128–138, gives a physical explanation of the wind and water forces on *sailboats* with emphasis on their effect on a boat's speed. The article includes excellent descriptions of good sailboat design for the novice, and it gives suggestions for experiments to test sail coefficients, centerboard designs, and locations, plus racing strategy.

4. One of the clearest explanations of the anatomy of a bird's wing is presented in *Scientific American*, vol. 186 (April 1952), pp. 24–29, which discusses the aerodynamics of *bird flight*, is given showing how it is related to the airfoil and propeller. The article proposes a physical explanation of bird flight, relating the action to that of an airplane wing moving through the air.
5. The lift and drag on a *spinning ball* falling in a wind tunnel is a great demonstration for serious golfers, tennis players, and baseball enthusiasts. Details of the simple apparatus are presented in *Bull. Mech. Engrg. Educ.*, vol. 3, pp. 9–11. Additional reading material on the explanation of forces on a spinning ball is Coxon's article, *Bull. Mech. Engrg. Educ.*, vol. 4, pp. 89–92. Appropriate equations of motion are given in this article.
6. A novel description of a do-it-yourself apparatus to test airplanes in a bathtub can be found in *Scientific American* (April 1954), pp. 100–106. This is a fascinating project that gives great insight into the principles of flight.

12.12 Method of Images (or A Way to Create Straight Boundaries)

The method of images is an extremely useful mathematical tool to simulate a fluid boundary, or two bodies in close proximity. The method was introduced by Kelvin for use in electricity and later used by Helmholtz and Stokes in fluid dynamics.

The method of images can be stated as follows: "Whenever a rigid planar boundary separates two flow regions, the flow inside the boundary must reflect the image of all singularities appearing outside the boundary." The key word in the method is the word "boundary." For a potential flow we require the streamline next to the boundary to possess the geometry identical to the boundary. Since the streamline is determined from the appropriate fundamental solutions that model the flow, then the only way to create a straight boundary is to duplicate the fundamental solution, locating its origin symmetrically from the boundary.

Using the superposition principle of potential theory, we can use the method of images to synthesize flows with particular boundary conditions.

Example 12.12

Consider an irrotational vortex of strength $\Gamma_\infty/2\pi$ located a distance h above a wall. Find the stream function for a point P in the flow field.

Solution:

Step 1.

The flow is assumed to be steady, incompressible, and potential.

Step 3.

The appropriate governing equation is:

Example 12.12 *(Con't.)*

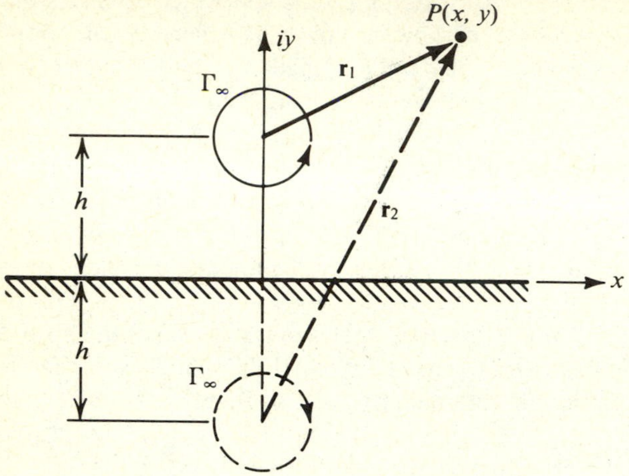

Figure E12.12

- Complex velocity potential for a vortex:

$$\Omega(z) = -\frac{i\Gamma_\infty}{2\pi} \ln (z - z_o) \tag{i}$$

The wall in Fig. E12.12 is straight and solid, so that the flow is linear and cannot cross the boundary. For potential flow, such a wall is a streamline. This is an important fact.

The stream function for a vortex whose origin is at $y = h$ is, from Eq. (i),

$$\psi = -\frac{\Gamma_\infty}{4\pi} \ln r_1 \tag{ii}$$

where

$$r_1 = x^2 + (y - h)^2 \tag{iii}$$

At the wall, $y = 0$, the stream function is to be constant. From Eqs. (ii) and (iii) the stream function

$$\psi(x, 0) = -\frac{\Gamma_\infty}{4\pi} \ln (x^2 + h^2)$$

We see that ψ is highly nonlinear rather than the desired constant which represents our wall. Thus, we must *find* a function to *nullify* this nonlinearity, thereby

Example 12.12 *(Con't.)*

making it a constant. To do this, consider a vortex of the same strength $\Gamma_\infty/2\pi$ but moving in the *opposite* direction (clockwise), with origin located at $y = -h$ as shown in Fig. E12.12. Even though this vortex does not exist physically, we still introduce it so as to create the straight wall at $y = 0$.

The stream function for the two vortices are

$$\psi = -\frac{\Gamma_\infty}{4\pi}\ln r_1 + \frac{\Gamma_\infty}{4\pi}\ln r_2 \tag{iv}$$

where

$$r_2 = [x^2 + (y + h)^2] \tag{v}$$

Once again let us consider the wall. At $y = 0$, Eq. (iv) becomes

$$\psi = \frac{\Gamma_\infty}{4\pi}\ln\frac{x^2 + h^2}{x^2 + h^2} = 0 \tag{vi}$$

which tells us that at the wall our streamline is not only a constant, but is the zero streamline, or dividing streamline.

This completes the solution.

12.13 Potential and Stream Functions in Real Fluids

In this chapter we have been treating potential flows exclusively. We should briefly mention fluids different from incompressible potential flows so that similarities and dissimilarities may be observed, and so that we can make our own observation of what is and what is not significant.

12.13.1 Compressible Fluids

In comparing the differential form of the two-dimensional steady compressible continuity equation with the two-dimensional incompressible continuity equation,

$$\frac{\partial u}{\partial x} + \frac{\partial v}{\partial y} = 0, \quad (\rho = \text{const.}) \tag{12.130}$$

$$\frac{\partial(\rho u)}{\partial x} + \frac{\partial(\rho v)}{\partial y} = 0, \quad (\rho \neq \text{const.}) \tag{12.131}$$

we see that Eq. (12.130) is satisfied when

$$u = \frac{\partial\psi}{\partial y}, \quad v = -\frac{\partial\psi}{\partial x} \tag{12.132}$$

and Eq. (12.132) is satisfied if

$$\rho u = \rho_o \frac{\partial \psi}{\partial y}, \qquad \rho v = -\rho_o \frac{\partial \psi}{\partial x} \tag{12.133}$$

where ρ_o is some reference density, assumed constant. Thus, the vorticity equation for a compressible irrotational flow will *not* be Laplace's equation as it was for the incompressible case. Also, the velocity potential ϕ, *though it exists for compressible flow*, will not solve Laplace's equation either. The equation for compressible flow in terms of the velocity potential can be shown to be

$$(M^2 - 1)\frac{\partial^2 \phi}{\partial x^2} + \frac{\partial^2 \phi}{\partial y^2} = 0 \tag{12.134}$$

where $(M^2 - 1)$ is constant and measures the compressibility of the flow. We note that Eq. (12.134) holds only for two-dimensional flow. It is of hyperbolic differential form and is the classical wave equation found in physics. What make it a difficult equation to solve are the boundary conditions. Note that it blows up when $M = 1$. We have an entire new class of problems when faced with compressible fluid flows. We shall discuss this in Chap. 15.

12.13.2 Viscous Flows

For two-dimensional shear flows, though the stream function exists, the velocity potential need not since vorticity can be finite. The velocity potential ϕ exists solely because the flow is irrotational.

Viscous flows are important in regions near bodies and in those boundaries moving at velocities different from the central portion of the flow. We usually use Prandtl's idea of dividing the flow into two regions: a region close to a body or boundary where viscous effects are important in sustaining vorticity, and a region removed from the body where the flow may be considered potential flow even though the fluid is real because the fluid's vorticity is negligible. The former region is handled theoretically by what is popularly called boundary layer theory (see Chap. 14). The stream function ψ will be shown to be of tremendous value as it reduces the number of dependent variables by one.

12.13.3 Rotational Flows Being Ideal

As pointed out in Chap. 2, an ideal fluid is one which cannot sustain shear. We stated that for the shear stress to be zero, one of two conditions must exist: either the viscosity is zero (inviscid flow), or certain velocity gradients are zero. It is quite possible, however, that we can have vorticity and yet have no shear stress. This is illustrated as follows:

Let the flow be two-dimensional with a given velocity gradient relationship $\partial v/\partial x = -(\partial u/\partial y)$. Typical distribution of velocity where this can occur is for $u = ax^{n-1}y^{m+1}$, and

$$v = \frac{-a(m+1)}{n} x^n y^m$$

Then, the z-component of vorticity ζ_z is

$$\begin{aligned} \zeta_z &\equiv \frac{\partial v}{\partial x} - \frac{\partial u}{\partial y} \\ &= 2\frac{\partial v}{\partial x} \end{aligned} \tag{12.135}$$

and is nonzero for $n \neq 0$. The shear stress p_{xy} is defined as

$$p_{xy} \equiv \mu\left(\frac{\partial u}{\partial y} + \frac{\partial v}{\partial x}\right)$$

such that in this illlustration

$$p_{xy} = 0 \tag{12.136}$$

Thus, the shear stress does not exist but vorticity does. We say then *the flow is ideal but rotational.*

A number of actual rotational flows *appear* ideal. Many porous flows, and the classic Hele-Shaw flow are two examples. Hele-Shaw showed in 1899 that a two-dimensional flow between two closely spaced flat plates exhibited an ideal flow pattern. G. Stokes later gave a formal proof of this seemingly contradictory situation.

12.14 Comparison of Potential Theory with Experiment

Potential theory certainly is a simple and straightforward mathematical tool for modeling a fairly large number of fluid situations. Given a particular flow situation, we would probably try to apply this theory before applying any other in order to gain some insight into the problem. We know we cannot use the theory close to the surface of a body, since the slip condition would be violated. Yet, we would hope that regions exist near the body surface where potential theory can give fairly accurate results when compared to carefully measured experimental data. So we ask, how accurate is potential theory when applied to the surface of bodies?

Since we have applied potential theory to uniform flow past a cylinder, suppose we now compare the pressure difference along the surface of a cylinder as calculated from potential theory with experiment. Let p_o be the upstream ambient pressure. Figure

12.23 shows the distribution of pressure differences ($p - p_o$) normalized with respect to the *dynamic pressure* $\frac{1}{2}\rho U^2$ along the upper and lower surface of a cylinder. The Reynolds number R_D is a measure of the viscosity of the flow. Hence, for an inviscid fluid, R_D equals infinity, and represents the potential flow solution. The significance of an increased value of R_D means that the viscosity has decreased or the velocity U has increased or both. We see from the figure that as velocity U gets larger and larger, the measured pressure over the surface of the cylinder agrees closely with that predicted by potential theory. We also see that the poorest agreement between theory and experiment occurs at $\theta = 120°, 240°$. This is caused by separation (where the shear stress is zero) and will be discussed in later chapters.

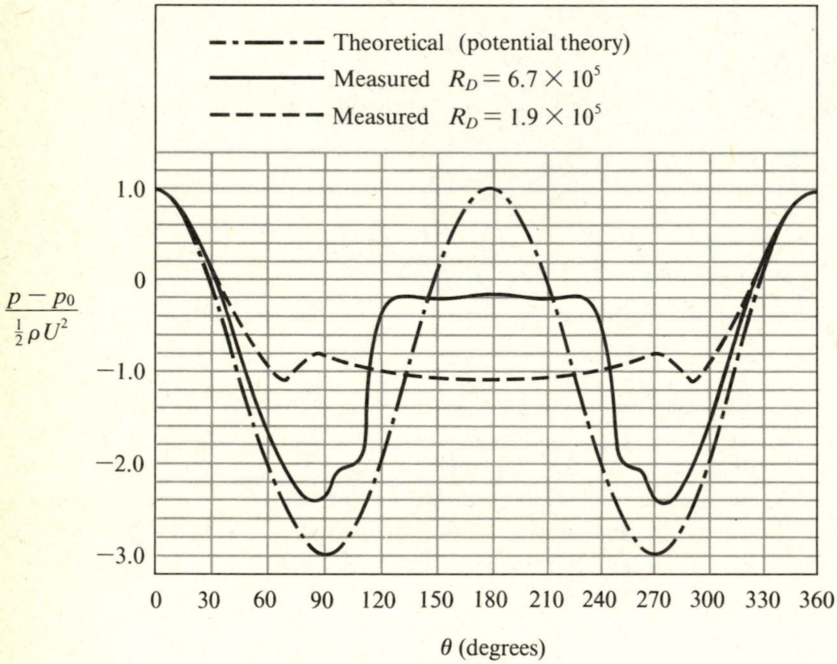

Figure 12.23 *Pressure distribution on a cylinder. (Source: S. Goldstein,* Modern Developments in Fluid Dynamics, *Vol. II, Oxford Engineering Sciences Series, Oxford University Press, England, © 1938.)*

Figure 12.24 shows the pressure distribution across a wing section *as calculated by potential theory* compared to measured data obtained in a wind tunnel. The comparison is good except near the downstream section of the wing. This exception is also caused by separation. Whenever the flow separates, as in the case of a real fluid flowing over bodies that have decelerated flow, potential theory cannot predict results well at all, as borne out in Fig. 12.24a. Thus potential flow is quite useful even when treating real fluids when the flow is accelerating past the body. Only in the decelerated zone will results compare unfavorably with the precise results.

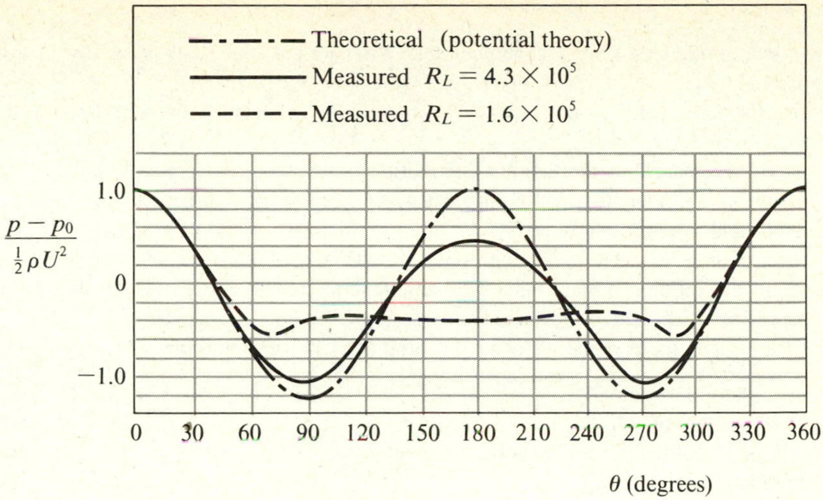

(a)

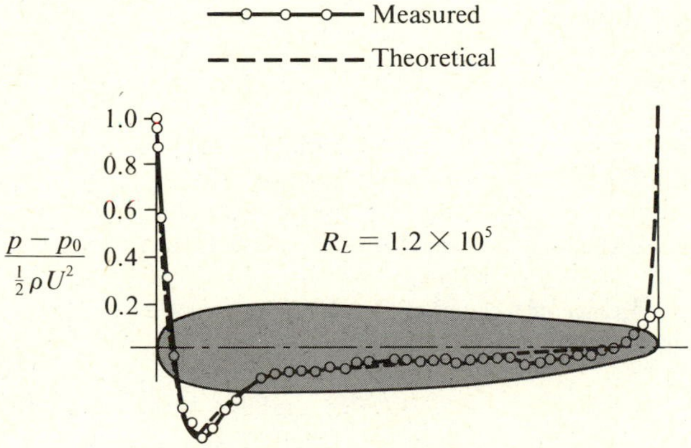

(b)

Figure 12.24 *Pressure distribution on a wing section. (Source: S. Goldstein,* Modern Developments in Fluid Dynamics, *Vol. II, Oxford Engineering Sciences Series, Oxford University Press, England, © 1938.)*

References

12.1 Robertson, James M., *Hydrodynamics in Theory and Application*, Prentice-Hall, Englewood Cliffs, N.J., 1965.

12.2 Roache, P. J., *Computational Fluid Dynamics*, Hermosa Publishers, Albuquerque, N.M., 1972.

12.3 Rogers, Eric M., *Physics for the Inquiring Mind*, Princeton Univ. Press, Princeton, N.J., 1960 (see Part 1).

12.4 Abbott, I. H., and von Doenhoff, A. E., *Theory of Wing Sections*, Dover Publications, New York, 1959.

Study Questions

12.1 What is the principle of linear superposition? What are the conditions necessary for its application?

12.2 Do shear stresses exist in potential flows? Do velocity gradients exist, and if so, what are their forms?

12.3 Why is it necessary to determine the distribution of singularities on a boundary?

12.4 Define (a) complex number, (b) complex variable, (c) complex function, (d) complex potential, and (e) complex velocity.

12.5 If $d\Omega/dz = u - iv$, what is the expression for $d\Omega/dz$ in terms of v_r and v_θ?

12.6 What are the difference and similarities among circulation, vorticity, and tangential velocity?

12.7 Show that given the complex velocity potential $\Omega(z)$, we can find the

1. Velocity potential ϕ
2. Stream function ψ
3. Equation of the streamlines
4. Equation of the body
5. Velocity components u, v, v_r, v_θ
6. Stagnation points
7. Speed
8. Pressure distribution
9. Force field
10. Moment about the origin

12.8 Describe the flow inside a cylinder with uniform flow and circulation Γ_∞.

12.9 Tabulate the stream function ψ and sketch the flow field for: (a) uniform flow in the positive y-direction, (b) a vortex with negative rotation, (c) a source, (d) a source and a sink of equal strength on the y-axis, and (e) a doublet and vortex with clockwise motion.

12.10 What is the significance of the starting vortex? Where does it go? How is it formed? Does it exist in an ideal flow?

Problems

12.1 Starting with $\nabla^2\phi = 0$ in Cartesian coordinates, derive Eq. (12.4).

12.2 Prove that the velocity potential is constant in an irrotational region whose boundary is where ϕ equal constant.

12.3 Given the velocity field

$$\mathbf{V} = e^y[(z \cos x)\mathbf{i} + (z \sin x)\mathbf{j} + (\sin x)\mathbf{k}]$$

(a) find the velocity potential $\phi(x, y, z)$ as well as $\phi(r, \theta, z)$, and (b) show whether the flow is irrotational or not.

12.4 An irrotational flow of velocity U moves past a two-dimensional 90° corner. Using the technique shown in Example 12.12, find the velocity potential ϕ in Cartesian coordinates.

12.5 Consider a body moving through a fluid at a velocity U in the positive x-direction. Given the equation of the boundary surface of the body as $(x - Ut)^2 + y^2/4 + z^2/9$, obtain the kinematic boundary condition of the moving object.

12.6 Starting with

$$\int_{\forall} (\nabla\phi)\cdot(\nabla\phi)\, d\forall$$

show it is equal to

$$\int_A \phi(\nabla\phi)\cdot d\mathbf{A} - \int_{\forall} \phi\nabla^2\phi \, d\forall$$

and thus is equal to

$$\int_A \phi \frac{\partial \phi}{\partial n} \, dA$$

for incompressible flows.

12.7 We know for incompressible potential flows that $\nabla^2\phi = 0$. What is the appropriate differential equation for irrotational flows in terms of the stream function ψ?

12.8 If $\Omega = \phi + i\psi$ and is pure real when $y = c$, show that $\psi = 0$ when $y = c$.

12.9 If $f(z) = 3z + 2iz^2$, what are $f(\bar{z})$ and $\bar{f}(z)$?

12.10 Express the continuity equation in terms of $\Omega(x, y)$, $\Omega(r, \theta)$, and $\Omega(z)$.

12.11 Sketch the streamlines represented by $\Omega = z^2$ and show that the speed is everywhere proportional to the distance from the origin.

12.12 Derive the expression $d\Omega/dz = (v_r - iv_\theta)e^{-i\theta}$ for the complex velocity in polar coordinates.

12.13 What are the radial and tangential velocity components for uniform flow of magnitude U inclined at angle α to the x-axis.

12.14 Draw the streamlines for the flow $\Omega = \frac{1}{2}Ua^3/z^2$.

12.15 Given $\Omega = \sqrt{z^2 - 1}$, show that the $\psi = 1$ streamline is $y = x/\sqrt{1 + x^2}$.

12.16 Find (a) the six roots of $-8i$, (b) the three roots of $-i$, (c) the three roots of $+i$, and (d) the three roots of -1.

12.17 If $F(z) = \sqrt{xy} + ix + y^2/ix$, what is $f(2 - 3i)$?

12.18 What is the value of $\sinh(-1 + 2i)$?

12.19 Given

$$\phi = \frac{x}{x^2 + y^2}, \quad \psi = \frac{-y}{x^2 + y^2}$$

determine (a) $\Omega(x, y)$, (b) $\Omega(z)$, (c) the complex velocity, (d) the speed, and (e) the x and y location of the stagnation point.

12.20 Given

$$v_r = \frac{\cos\theta}{r}, \quad v_\theta = r^2$$

determine the complex velocity $d\Omega/dz$ and then the complex potential $\Omega(z)$. What are ϕ and ψ?

12.21 Find the four roots of $-8i$.

12.22 If $f(z) = xy + i(x^2 - y^2)$, what is $f(-1 + 2i)$?

12.23 Find the value of $\cos(1 + zi)$.

12.24 Evaluate the integral

$$\oint \frac{dz}{(z - a)^{m+1}}$$

around a circle of radius r with center at r equal a, and m is an integer. (Refer to Fig. P12.24.)

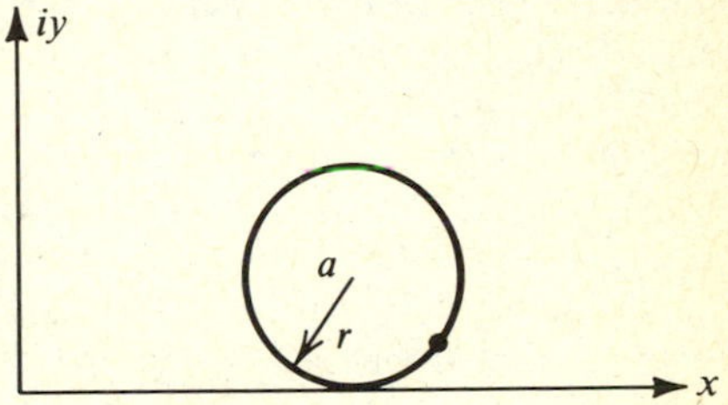

Figure P12.24

12.25 What is the value of

$$\oint \frac{z^3 + 1}{z^2 - 1} \, dz$$

if the integration is about a circle of unit radius with center at (a) $z = 1$, and (b) $z = -1$?

12.26 Given $\Omega = x^2 + y^2$, determine if $\Omega(z)$ is analytic.

12.27 Given $\phi = x + \frac{1}{2}\ln(x^2 + y^2)$, calculate (a) $\psi(x, y)$, (b) $\psi(r, \theta)$, (c) the complex potential $\Omega(z)$, (d) the velocity components u, v, v_r, and v_θ, (e) the accelerations a_x, a_y, a_r, and a_θ, (f) the x, y location of the stagnation points, (g) the r, θ locations of the stagnation points, and (h) the vorticity ζ_z distribution.

12.28 Given $\phi = 2xy$, calculate items a–h of Prob. 12.27.

12.29 Given $\phi = A\theta$, calculate items a–h of Prob. 12.27.

12.30 Given $\phi = (A\cos\theta)/r$, calculate items a–h in Prob. 12.27.

12.31 If $u = Ax$, $v = -Ay$, what is $\Omega(z)$?

12.32 The x, y, and z components of fluid velocity in a flow are given as $u = 3x^2 + 2$,

$v = y + 4xz$, and $w = 2xy + 3z$. (a) Determine if the flow is potential or not, and (b) determine the circulation along a curve formed by straight lines joining the points (1, 0, 0), (0, 1, 0), and (0, 0, 1). Let the direction of traverse be from the point (1, 0, 0) to (0, 1, 0), to (0, 0, 1) and back to (1, 0, 0).

12.33 A source is located at $x = z_o$. Another source is located at $z = z_1$. If $z_o = a$, and $z_1 = ib$, where a and b are both real, determine the complex velocity potential for the two sources, the complex velocity, and where the stagnation point is located.

12.34 A source is located at the origin of a coordinate system. Determine the x, y components of velocity as functions of x, y.

12.35 A uniform flow of magnitude V is flowing in the negative y direction. (a) What is $\phi(x, y)$, $\psi(x, y)$ for this flow? (b) Also, what are v_r and v_θ for this flow?

12.36 Draw the $\psi = 0$, $\psi = \pm 1$ streamlines for a doublet given that $q = 3\pi$, $a = 1/3$ and $\alpha = \pi/2$.

12.37 Prove that the circulation around a doublet is zero.

12.38 Consider $\phi = Ar^{\pi/\alpha} \cos(\pi\theta/\alpha)$, $\psi = Ar^{\pi/\alpha} \sin(\pi\theta/\alpha)$. (a) Prove that the flow is uniform flow for $\alpha = \pi$. (b) When $\alpha = \pi/2$, draw the flow net, and show that this represents flow in a corner. (c) If $A = 1$, what is the velocity of flow along a horizontal wall at 10 ft from the corner?

12.39 Write the expression for the complex velocity potential of a source and a vortex located at (a) $z_o = a$, (b) $z_o = ai$, (c) $z_o = -a$, and (d) $z_o = -ai$ (where a is pure real).

12.40 Given the velocity field $\mathbf{V} = 2x^2y\mathbf{i} - 2xy^2\mathbf{j}$, find the circulation around the rectangular path given in Fig. P12.40.

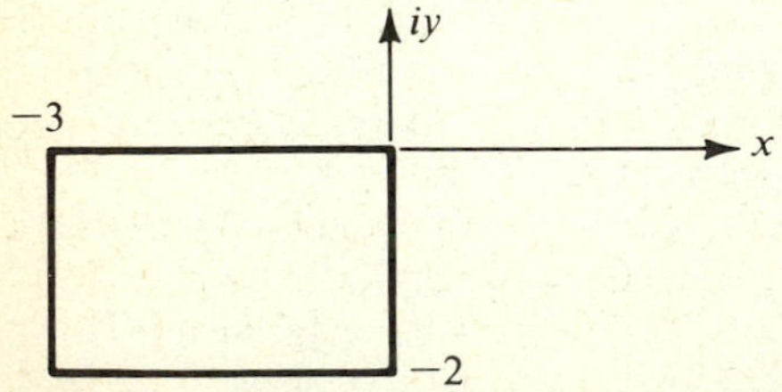

Figure P12.40

12.41 Given $\Omega(z) = (A + iB) \ln z$, (a) plot the zero streamline, (b) find the circulation Γ in a circle of radius a, and (c) find an expression for the tangential velocity $v_\theta(r, \theta)$.

12.42 Construct a flow net for potential flow along two planes that intersect at 135° angle.

12.43 If the circumferential velocity of a tornado is approximated by $v_\theta = \omega/r$, where ω is a constant and represents the local angular speed of the tornado, calculate the value of the circulation and show that it is constant.

12.44 Consider the complex potential $\Omega = (1 + i)z^2$. (a) Sketch the flow net. (b) Determine the complex velocity. (c) Determine the equation of the body. (d) Locate the stagnation points. (e) What is the complex force?

12.45 Consider a circular flow defined by $v_\theta = U(r/a)$, $v_r = 0$. Such a flow is called solid body rotation. Show that this flow is not a potential flow, and calculate the vorticity distribution throughout it. Does the stream function exist for this flow, and if so, what is the equation of the streamline? Is the flow circulation preserving?

12.46 Given $\Omega = z^{1/2} + Uz$, determine (a) if the flow is irrotational, (b) the velocity components u, v, (c) the x and y components of the hydrodynamic force, and (d) the hydrodynamic moment.

12.47 Find the complex potential Ω for uniform flow past two sources of equal strength $q/2\pi$ located at $\pm ia$, respectively.

12.48 Obtain the complex potential for a Kelvin oval; where the vortex pair are located at distance ia on the imaginary axis.

12.49 Given the complex potential for a line vortex near a wall

$$\Omega(z) = \frac{i\Gamma_\infty}{2\pi} \ln \frac{z + a}{z - a}$$

(a) show how the wall condition is satisfied, and (b) what are the velocity component expressions along the wall?

12.50 Consider a point source in a uniform flow of velocity U. Show the difference in pressure $p - p_\infty$ is

$$\tfrac{1}{2}\rho U^2\left[\frac{2\sin\theta}{\theta} - \left(\frac{\sin\theta}{\theta}\right)^2\right]$$

on the surface of the zero streamline.

12.51 Consider uniform flow of velocity U past an infinitely long cylinder of radius a. Show the velocity and pressure variation with radial distance are

$$v_\theta = -U\left[\left(\frac{a}{r}\right)^2 + 1\right]$$

and

$$p - p_\infty = -\tfrac{1}{2}\rho U^2\left[\left(\frac{a}{r}\right)^4 + 2\left(\frac{a}{r}\right)^2\right]$$

respectively, at $\theta = 90°$.

12.52 For the flow in Prob. 12.51, show that the vorticity anywhere in the flow is zero, and that the circulation around any closed path concentric to the cylinder is also zero.

12.53 A source flow is superimposed on a free-vortex flow. Show the streamlines of the combined flow as spirals, and plot typical streamlines for a source with strength $q = 50$ ft²/s, a vortex with circulation $\Gamma = 50\pi$ ft²/s, and with $\psi_{\text{vortex}} = 0$ at $r = 2$ ft.

12.54 Calculate the velocity components u, v for a source in a uniform flow of velocity U. Sketch a few of the streamlines.

12.55 A source of strength $q = \pi$ ft²/s is immersed in a uniform flow of 10 ft/s. Find the distance between the stagnation point, the flow, and the center of the source.

12.56 Water spins into the drain of a bathtub with a circumferential velocity approximately like that of an irrotational vortex. Assuming that the vertical and radial velocity components are negligible, compared to the circumferential velocity, use Bernoulli's equation to find the shape of the free-surface of the flow.

12.57 In Prob. 12.56, consider the drain suddenly closed. The resultant flow is a Rankine vortex. Show that the pressure variation is given as

$$\frac{\partial p}{\partial r} = v_\theta^2/r \qquad \text{and} \qquad \frac{\partial p}{\partial z} = \rho g$$

when

$$v_r = w = 0$$

12.58 Calculate the lift per unit length on a 1 ft diameter cylinder with circulation having a single stagnation point at $\theta = 270°$. The cylinder is in a flow having a velocity of 10 ft/s and a density of 2 slug/ft³.

12.59 Find the source strength, q, and distance between the source and an equal sink for a uniform flow of 10 ft/s around a Rankine oval body 4 ft long and 2 ft thick.

12.60 Prove that $\Omega^2 = z$ is the complex velocity potential bounded by two confocal and coaxial parabolic cylinders.

12.61 As an approximation so that we can simulate tornadoes, suppose we consider a single-celled tornado that behaves as a columnar vortex. Let the columnar vortex travel at a uniform velocity to simulate a tornado moving with a storm. Outside of the tornado core of radius r_c, the flow is assumed potential with a circulation Γ_∞. Let station 1 be some reference station in the irrotational field where the barometric pressure is 33 ft of water, the wind speed is 5 ft/s at a distance of 600 ft from the centerline of the columnar vortex that is moving with an average speed of 15 ft/s. Calculate (a) the ambient circulation Γ_∞ at station 1, (b) the tangential velocity at a 50 ft radius, and (c) the pressure at a 50 ft radius.

12.62 In Prob. 12.61, if the core radius is 25 ft, calculate (a) the tangential velocity, and (b) the pressure at the core radius.

12.63 In Prob. 12.61, if a house has a 100 ft² frontal area, what is the normal force on the frontal area if the tornado's center passes within 25 ft of it?

12.64 Using the transformation $\Omega = z + iz$, find and plot points in the Ω-plane corresponding to (a) $z = 1$, (b) $z = i$, (c) $z = 1 + i$, and (d) $z = 2 + i$. Show that the line $y = 1$ in the z-plane maps into the line $\psi = \phi + 2$ in the Ω-plane.

12.65 If $\Omega = z^2$, find ϕ and ψ.

12.66 Find ϕ and ψ if (a) $\Omega = \ln z$, (b) $\Omega = e^{iz}$, (c) $\Omega = \cosh z$, and (d) $\Omega = \bar{z}$.

12.67 Sketch the two families of curves $\phi = c$, $\psi = k$ for $\Omega = z^2$.

12.68 Given $\Omega = 1/z$, show that (a) $\phi = x/(x^2 + y^2)$, $\psi = -y/(x^2 + y^2)$, (b) the straight lines $y = kx$ map into the straight lines $\psi = -k\phi$, and (c) circles $x^2 + y^2 = r^2$ map into circle $\phi^2 + \psi^2 = 1/r^2$.

12.69 Consider the transformation $\Omega = x^{3/2}$, and sketch the transformation in the z-plane. Use the transformation to describe the flow of air in the neighborhood of the corner of a tall building. (Hint: Use polar coordinates.)

12.70 A cylinder moves through water of density 2 slug/ft³ and at a uniform velocity of 10 ft/s. Calculate the actual pressure at (a) the stagnation point, (b) 30°, (c) 90°, (d) 150°, and (e) 180° if the stagnation point is 100 ft below the water's free-surface.

12.71 Show that the continuity equation for steady, two-dimensional, compressible, potential flow can be expressed as

$$\left(1 - \frac{u^2}{c^2}\right)\frac{\partial^2\phi}{\partial x^2} - 2\frac{uv}{c^2}\frac{\partial^2\phi}{\partial x\partial y} + \left(1 - \frac{v^2}{c^2}\right)\frac{\partial^2\phi}{\partial y^2} = 0$$

12.72 Show that the continuity equation for compressible potential flow can be written

$$\frac{1}{c^2}\frac{\partial^2\phi}{\partial t^2} + \frac{2}{c^2}\mathbf{V}\cdot\frac{\partial \mathbf{V}}{\partial t} = \nabla^2\phi - \frac{1}{c^2}\mathbf{V}\cdot[(\mathbf{V}\cdot\nabla)\mathbf{V}]$$

13 Open-Channel Flow

13.1 Introduction

Examples of free-surface flow abound in nature, and conjure up traditional romantic images of babbling brooks and wide lazy rivers, of white-water rapids, and wind-lashed storms at sea. More pragmatically, ''open channel'' defines the flow in irrigation ditches, flood-control flumes, aquaducts, and water-treatment plants. Obviously a vital aspect of fluid dynamics, with wide applications, open-channel flow differs from anything we have studied thus far. We must now pay attention, as we have not before, to such factors as variations in liquid depth, variable cross-sectional areas, and varying free-surface shapes. Try rafting down the Colorado river *without* attending to the water's depth or to the river's cross-sectional area—it will be an upsetting experience indeed.

We shall first treat an incompressible channel fluid flow which has both variable depth and cross-sectional area. Both the continuity and energy equations are useful in describing the flow process. The results we shall obtain are of the same form as some compressible flows that will be examined in Chap. 15. In fact, we often use free-surface experiments to visualize certain compressible flow phenomena since the governing (I.F.) equations are similar: for example, the normal shock wave of compressible aerodynamic flow is equivalent to the hydraulic jump in channel flows, and bow waves behave much like oblique shock waves.

13.2 Steady Open-Channel Flow

Open-channel flows can be considered irrotational or rotational, inviscid or real, steady or unsteady, laminar, transitional, or turbulent. In addition, the flow may be either uniform or nonuniform. The character of free-surface flows is therefore classified as follows:

- Uniform flow: No variation in cross-sectional area.
- Nonuniform flow: Liquid depth varies and free-surface is nonparallel to the channel floor because of changes in the channel floor.
- Laminar flow: Very shallow depth of flow down an inclined plane. Here viscous effects are important.

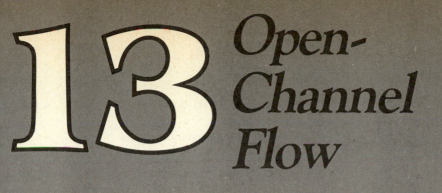

- Irrotational flow: Very deep free-surface flow, where effects of vorticity are ignored.
- Retarded flow: Fluid depth increasing in the downstream direction (also called decelerated flow)
- Accelerated flow: Depth decreasing in the downstream direction.

Consider a steady nonuniform flow, as shown in Fig. 13.1. Let the x-axis be in the direction of flow so that the velocity components v and w are zero (or negligibly small). Thus, the velocity component $u = V$. Let us assume that the flow is *not* turbulent, and that the pressure distribution is determined by the hydrostatic relationship $p = \gamma h$. Let section 1 of Fig. 13.1 be located an infinitesimal distance dx from section 2. The slope of the channel floor is denoted by m. The head loss due to friction in the elemental length is dh_f. The (I.F.) energy equation is written between stations 1 and 2 as

$$h_1 + mdx + \frac{\tilde{V}_1^2}{2g} = h_2 + \frac{\tilde{V}_2^2}{2g} + dh_f \tag{13.1}$$

or in differential form

$$m - \frac{dh}{dx} = \tfrac{1}{2}g\frac{d(\tilde{V}^2)}{dx} + \frac{dh_f}{dx} \tag{13.2}$$

We next express the change in kinetic energy in terms of the geometric changes of

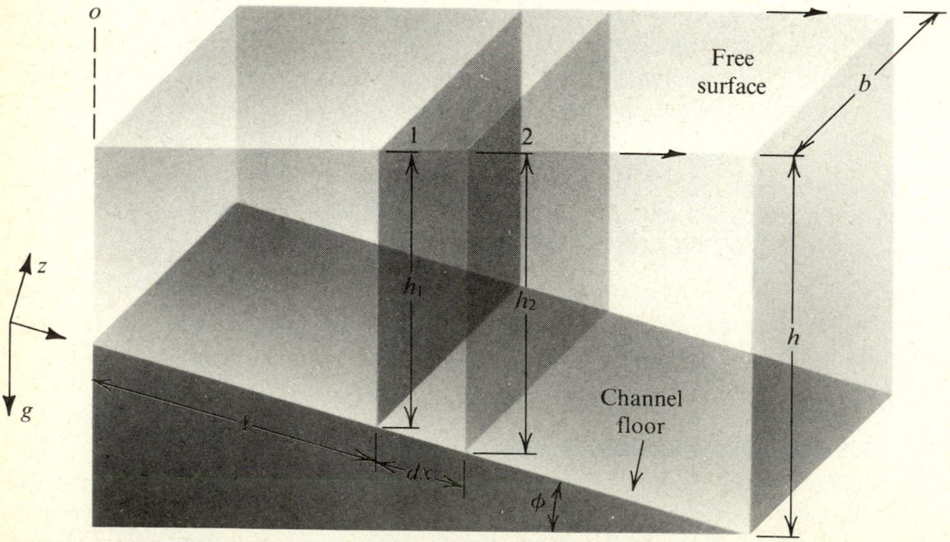

Figure 13.1 Geometry for free-surface flow.

the channel. Since $\tilde{V} = Q/A$, and

$$A = bh \tag{13.3}$$

from Fig. 13.1, then

$$\frac{1}{2g}\frac{d(\tilde{V}^2)}{dx} = -\frac{Q^2}{gA^3}\left(\frac{\partial A}{\partial x} + b\frac{dh}{dx}\right) \tag{13.4}$$

We note that the expression Q^2b/gA^3 is dimensionless. We shall define the Froude number as

$$F_r = \frac{Q}{A}\sqrt{\frac{b}{gA}} \tag{13.5}$$

Next, we define a velocity coefficient C as

$$C = \frac{\tilde{V}}{R_h m} = \tilde{V}\frac{l}{h_f R_h} \tag{13.6}$$

where R_h is the hydraulic radius defined by

$$R_h = \frac{A}{P} \tag{13.7}$$

and P is the perimeter of the wetted channel. Equation (13.6) is called the Chézy formula, after the French engineer, A. Chézy, who conducted numerous experiments in hydraulics in the mid-eighteenth century. The velocity coefficient C is called the Chézy coefficient.

Substituting Eqs. (13.4)–(13.6) into the differential form of the energy Eq. (13.2) yields

$$\frac{dh}{dx} = \frac{m - (Q^2/C^2A^2R_h)\,[1 - (C^2R_h/gA)\,(\partial A/\partial x)]}{1 - F_r^2} \tag{13.8}$$

which predicts the behavior of a nonuniform but steady flow in an open channel of varying width and depth. We must be cautious in using Eq. (13.8). When $F_r = 1$, we observe that $dh/dx = \infty$. This condition is the case of a hydraulic jump: the slope is not vertical, but very steep. We shall discuss the hydraulic jump in Sec. 13.3.2.

Equation (13.8) is the governing equation for incompressible flow in channels having a free-surface. It is a relatively simple expression, and we shall devote the major portion of this chapter to applying it to a variety of flows.

13.2.1 Flow Classification

One initial result is readily observable from Eq. (13.8): the sign of dh/dx depends upon the quantity $(1 - F_r^2)$ in the denominator. We shall define three flow regions. They are

$$F_r = 1.0, \text{ critical flow}$$

$$F_r < 1.0, \text{ subcritical flow}$$

$$F_r > 1.0, \text{ supercritical flow}$$

where the Froude number is either given by Eq. (13.5) or

$$F_r = \frac{\tilde{V}}{\sqrt{gh}} \tag{13.9}$$

This is an important dimensionless parameter in describing free-surface flows. A subcritical flow is a slow flow, or what we call streaming flow as illustrated in Fig. 13.2. A supercritical flow is a rapid flow or the shooting flow of Fig. 13.2. We shall discuss these regions in greater detail on page 678.

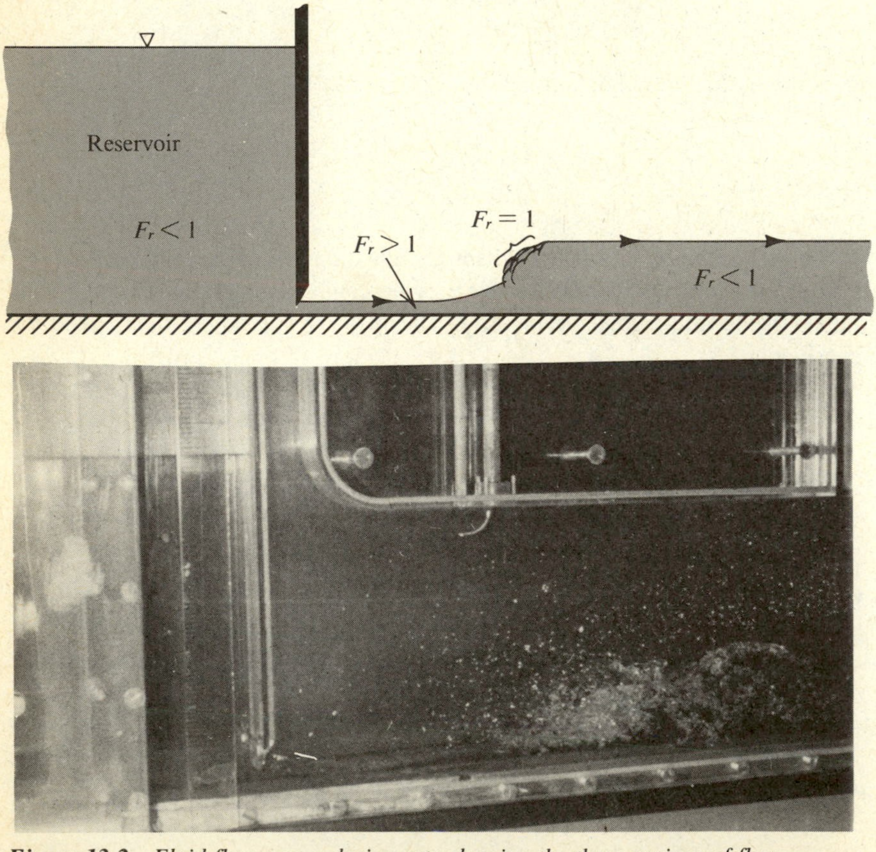

Figure 13.2 *Fluid flow past a device gate showing the three regions of flow.*

13.2.2 Uniform Open-Channel Flow

A uniform open-channel flow is defined as one where the depth h and area A are constant (see Fig. 13.3). Thus, from Eq. (13.8) we obtain the channel slope

$$m = \frac{Q^2}{C^2 A^2 R_h} \tag{13.10}$$

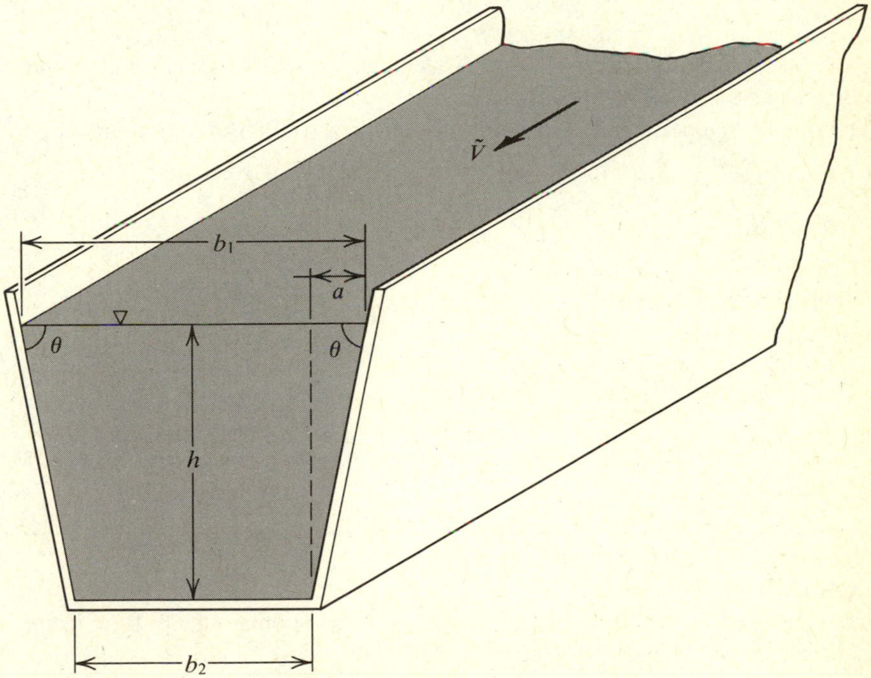

Figure 13.3 *Uniform flow in an open channel.*

The equation of open-channel flow is now vastly simplified. Solutions can be classified into three categories.

- *Category I*. Given the geometry of the channel and the slope of the channel floor m, determine the flow rate Q.
- *Category II*. Given the geometry of the channel, the slope of the channel floor m, and flow rate Q, calculate the fluid depth h.
- *Category III*. Given the flow rate Q and slope of the channel floor, determine the depth h and width b of the channel.

Before we can discuss applications of these three categories we need to set up the parameters for the channel calculations. Questions that need answers are (a) how do we evaluate the Chézy coefficient C, and (b) how do we set up the geometry of the channel?

Evaluating the Chézy Coefficient C

The Chézy coefficient is a function of the channel's roughness and geometry. Because of the complex nature of the flow in the channel, no analytical closed form solution exists for C, so we are left with a collection of empirical relationships. One of the more popular equations is the Manning formula:

$$C = \frac{c}{n} R_h^{1/6} \tag{13.11}$$

where c is the conversion factor (1.0 for SI units and 1.486 for USCS units), and n is the dimensionless roughness coefficient.

The roughness coefficient is related to the channel wall absolute roughness ϵ by

$$n = \frac{\epsilon^{0.2}}{19.6} \tag{13.12a}$$

if the dimension of ϵ is in meters and

$$n = \frac{\epsilon^{0.2}}{15.45} \tag{13.12b}$$

if the dimension of ϵ is in feet. Table 13.1 presents the roughness coefficient n for various channel surfaces.

Example 13.1

Obtain an expression for the Chézy coefficient for laminar steady flow in the uniform inclined channel shown in Fig. E13.1a.

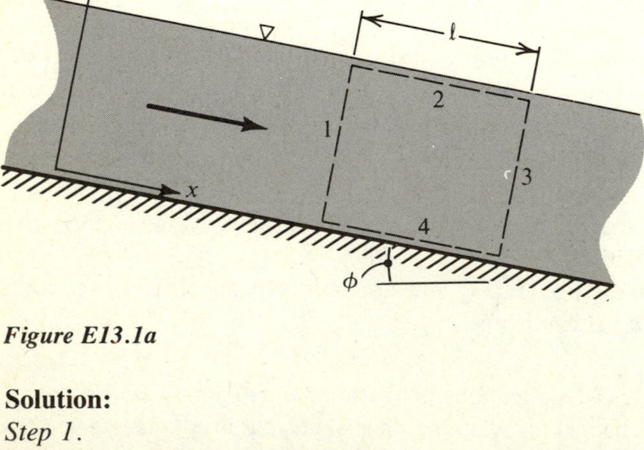

Figure E13.1a

Solution:

Step 1.

Analysis of the physical problem:

Table 13.1 Roughness Coefficient n *for Various Channel Surfaces*

Channel	n
Smooth surface: reinforced cement with plaster, planed boards, metallic surfaces, enameled surfaces	0.011
Slightly rough surfaces: reinforced cement with plaster in fair condition, rusted metallic surfaces, asphalt surfaces	0.011–0.013
Medium rough surfaces: plaster with no reinforcement, serious rust metallic surfaces, stone masonry, average-good brickwork	0.013–0.015
Concrete channel	0.014
Average brickwork surface	0.015
Concrete channels with rough surfaces; rough bricks, sides of canals of rough unplaned boards, slime covered channels, clean finished stone masonry	0.015–0.018
Good ashlar masonry	0.017
Channels coated with thick layer of silt	0.018
Rough old broken brickwork, clay, and well maintained loam channels, canals with sandy bottom, smooth rock canal surfaces	0.018–0.020
Average ashlar masonry, large smooth stones	0.020
Earthen channels in good condition having silt deposits and no serious erosion, surface gravel of 50 mm diameter or less, rough masonry, canals cut through rock	0.020–0.023
Average repaired earth canal	0.025
Earth channels with large stones, roads, plowed and furrowed earth	0.023–0.027
Poorly maintained earth canal	0.03
Gravel canal in poor condition, broken pavement, rough stone canals	0.027–0.031
Canals with many boulders and water foliage, rivers	0.035
Channels in bad need of repair	0.040

Example 13.1 (Con't.)

Since the flow is uniform and steady, there is no acceleration. The forces on the control volume are pressure forces on faces 1 and 3, a shear force on face 4, and a body force due to the weight of the fluid.

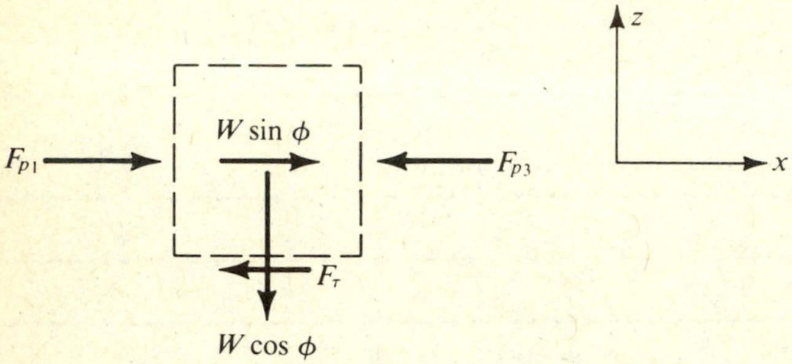

Figure E13.1b

Step 2.
Apply Newton's second law of motion, the hydrostatic relationships for planar surfaces, and the shear stress for pipe flow:

$$F_{p_1} - F_{p_3} + W \sin \phi - F_\tau = 0 \tag{i}$$

or

$$\gamma \bar{h}_1 A_1 - \gamma \bar{h}_3 A_3 + \gamma A l \sin \phi - p_{zx} b l = 0 \tag{ii}$$

Since the areas $A_1 = A_3$, $\bar{h}_1 = \bar{h}_3$, and $p_{zx} = \rho f \tilde{V}^2/8$, where f is the friction factor and is equal to $64/R_D$ for laminar flow, Eq. (ii) becomes

$$\tilde{V} = \sqrt{\frac{gR_D}{8}\left(\frac{A}{b}\sin \phi\right)} \tag{iii}$$

Equating the velocity of Eq. (iii) with the Chézy-velocity expression of Eq. (13.6), we obtain

$$C = \sqrt{\frac{g}{8} R_D} = 2.006\sqrt{R_D} \tag{iv}$$

which serves as the Chézy coefficient for laminar steady flows in a uniform channel for USCS units.

This completes the solution.

Evaluating the Channel Geometry
In applying the governing equation for uniform free-surface flow, Eq. (13.10), we need expressions for the hydraulic radius R_h and the wetted area A. The geometry of the channel for uniform flow is shown in Fig. 13.3, where

$$\tan\theta = \frac{h}{a} \tag{13.13}$$

The appropriate flow area A is found to be

$$A = b_2h + h^2 \cot\theta \tag{13.14}$$

and the wetted perimeter P is

$$\begin{aligned} P &= b_2 + 2h\sqrt{1 + \cot^2\theta} \\ &= b_2 + 2h \csc\theta \end{aligned} \tag{13.15}$$

From Eq. (13.7) we obtain the hydraulic radius R_h as

$$R_h = h(b_2 + h\cot\theta)/(b_2 + 2h\csc\theta) \tag{13.16}$$

The Most Efficient Channel Profile
We are now in a position to obtain the minimum wetted perimeter that will allow the most efficient channel cross-sectional area for minimum flow resistance. The technique is straightforward: maximize the hydraulic radius for a given area. The results are

$$R_h = \frac{h}{2} \tag{13.17}$$

$$P = 4h\csc\theta - 2h\cot\theta \tag{13.18}$$

$$A = \frac{h^2(2 - \cos\theta)}{\sin\phi} \tag{13.19}$$

Notice that the hydraulic radius is independent of the channel angle θ. Equation (13.17) states that the area for uniform flow is most efficient when the hydraulic radius of the channel has the value of one-half the fluid depth.

Example 13.2
Consider a poorly maintained earthen channel having a floor width $b_2 = 4.0$ m, a water depth $h = 3$ m, a 45° side wall ($\theta = 45°$), and a channel slope of 0.001. Calculate the average velocity in the channel and the volume rate of flow. Determine whether erosion will take place if the maximum permissible flow of water in an earthen channel is 1.1 m/s.

Solution:
This a Category I problem.
Step 1.
Determine the geometric properties of the channel.

Example 13.2 (Con't.)

(i) The wetted perimeter P is given by Eq. (13.15) as

$$P = b_2 + 2h \csc \theta$$

$$= 4.0 + 2(3) \csc 45° \qquad \text{(i)}$$

$$= 12.485 \text{ m}$$

(ii) The cross-sectional area is given by Eq. (13.14):

$$A = b_2 h + h^2 \cot \theta$$

$$= (4)(3) + (3)^2 \cot 45° \qquad \text{(ii)}$$

$$= 21 \text{ m}^2$$

(iii) The hydraulic radius is given by Eq. (13.16):

$$R_h = h(b_2 + h \cot \theta)/(b_2 + 2h \csc \theta)$$

$$= (3)[4 + (3) \cot 45°]/[4 + 2(3) \csc 45°] \qquad \text{(iii)}$$

$$= 1.682 \text{ m}$$

Step 2.
Determine the Chézy coefficient C.

(iv) We evaluate the Chézy coefficient from Eq. (13.11), and the Manning roughness coefficient from Table 13.1:

$$C = \frac{R_h^{1/6}}{m}$$

$$= \frac{(1.682)^{1/6}}{0.03} \qquad \text{(iv)}$$

$$= 36.351$$

Step 3.
Determine the volume rate of flow, Q.

We obtain the volume rate of flow from Eq. (13.10):

$$Q = CA\sqrt{mR_h}$$

$$= (36.351)(21)\sqrt{(0.001)(1.682)} \qquad \text{(v)}$$

$$= 31.31 \text{ m}^3/\text{s}$$

Thus, the average velocity through the channel is 1.49 m/s which is greater than the maximum permissible velocity of 1.1 m/s. We can be certain that erosion will take place under these conditions.

This completes the solution.

13.2.3 Specific Energy

One of the most useful parameters for describing the nature of open-channel flow is the energy grade line that was presented in Sec. 5.5.5:

$$\text{E.G.L.} = z + \frac{p}{\gamma} + \frac{\tilde{V}^2}{2g} \tag{13.20}$$

which measures the amount of energy per unit weight that a liquid has flowing past a given effective cross-sectional area like that shown in Fig. 13.4. Letting the zero value of the specific energy be measured from the reference datum line, the specific energy of the stream is

$$\text{E.G.L.} = z_1 + h + \frac{\tilde{V}^2}{2g} \tag{13.21}$$

where the pressure at the free-surface is assumed to be atmospheric.

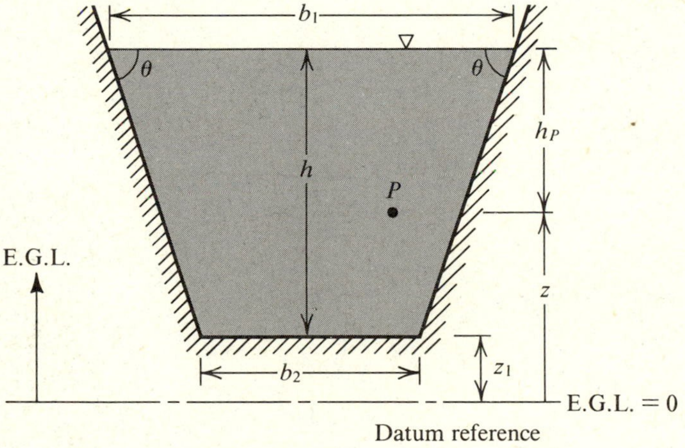

Figure 13.4 *Geometry for energy grade line.*

Specific Energy and Critical Depth for Free-Surface Flow in a Uniform Flat Bed Channel

For convenience, let us treat a free-surface flow in a channel where $\phi = 0$ such that the channel floor is horizontal. It would then be convenient to locate our reference datum at the bottom of the channel so that

$$\text{E.G.L.} = h + \frac{\tilde{V}^2}{2g} \tag{13.22}$$

To make the flow more interesting, let us place a hump on the channel floor as shown in Fig. 13.5. In addition, let the channel cross-sectional area be rectangular; i.e., $b_1 = b_2 = b$. At the free-surface we find

$$\frac{\tilde{V}_1^2}{2g} + h_1 = \frac{\tilde{V}^2}{2g} + h + z \tag{13.23}$$

where z is the elevation of the hump.

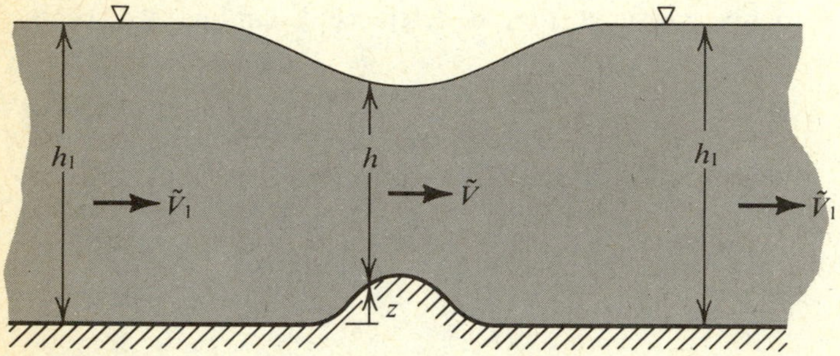

Figure 13.5 Free-surface flow past a hump.

Substituting the expression for the average velocity $\tilde{V}$

$$\tilde{V} = \frac{Q}{hb} \tag{13.24}$$

into Bernoulli's equation (13.21) yields

$$h + \left(\frac{Q^2}{2gb^2}\right)\frac{1}{h^2} = h_1 - z + \frac{Q^2}{2gb^2b_1^2} \tag{13.25}$$

Thus given the quantities b, h_1, z, and Q, we can plot the profile depth h of free-surface shapes.

To accomplish this, we shall express the E.G.L. as

$$\text{E.G.L.} = h + \left(\frac{Q^2}{2gb^2}\right)\frac{1}{h^2} \tag{13.26}$$

Note that the E.G.L. is a constant and is a very simple quantity to evaluate if given h_1, x, b, and Q. Using Eq. (13.26), we obtain an interesting nonlinear curve which is shown in Fig. 13.6a. Notice there is a critical value of depth, h_{crit}. We can evaluate this depth by noting that it occurs at the minimum value of the specific energy. Using

Eq. (13.26), we find with the aid of the calculus that the minimum value of the energy exist when

$$h_{\text{crit}} = \left(\frac{Q^2}{gb^2}\right)^{1/3} \tag{13.27}$$

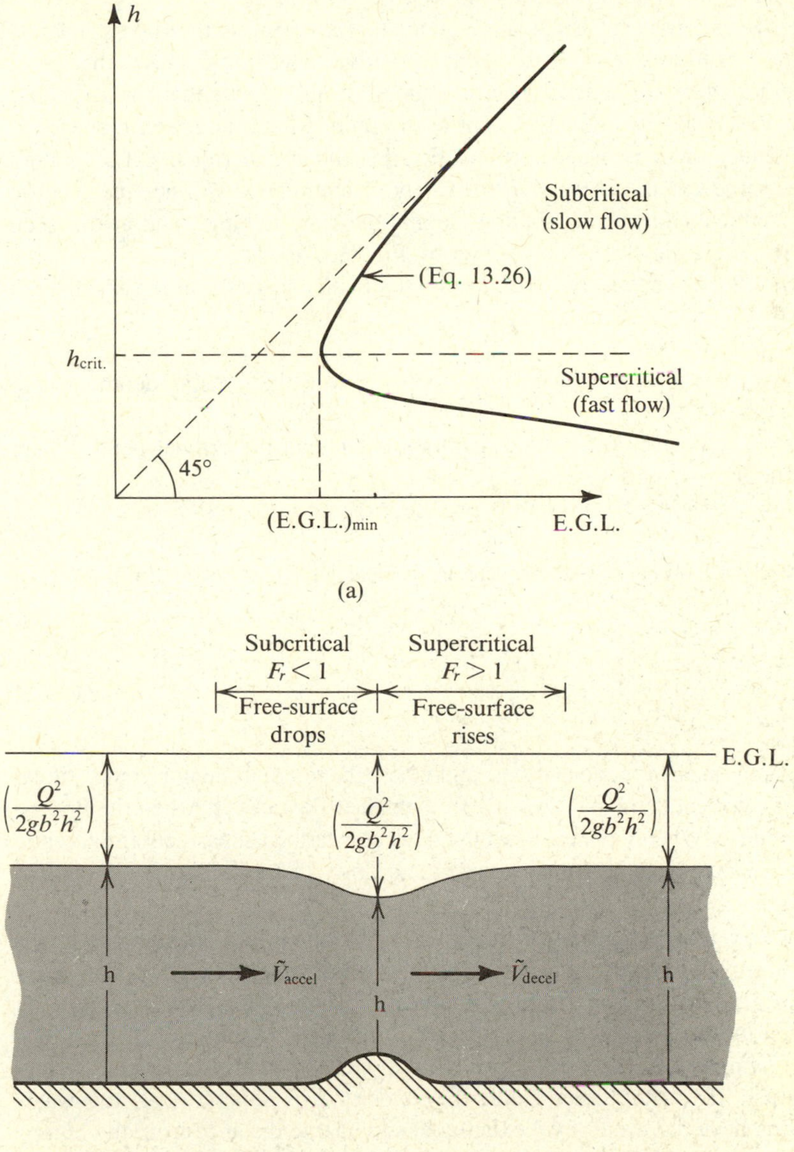

(b)

Figure 13.6 *Variation of the energy grade line with channel depth.*

so that the minimum E.G.L. is evaluated as

$$(\text{E.G.L.})_{\min} = \frac{3}{2}\left(\frac{Q^2}{gb^2}\right)^{1/3} = \frac{3}{2}h_{\text{crit}} \tag{13.28}$$

This is obtained by substituting Eq. (13.27) into Eq. (13.26).

The specific energy is always positive, since the last term on the right-hand side of Eq. (13.26) and the value of h are always positive. The figure shows that for a particular depth h there is a unique specific energy. It also shows that for a specific energy, there are always two depths except at the point where the specific energy is a minimum. Thus, there is a supercritical E.G.L. and a subcritical E.G.L., both possessing the same specific energy but for vastly different depths. Since the specific energies are identical, then at the shallow depths the flow is rapid, and at the deep depth the flow is very slow. This is verified by Eq. (13.26).

The flow in the channel is thus seen in Fig. 13.6a to consist of three very distinct regions:

- Supercritical flow occurring for fluid depths below the critical depth, where $\tilde{V}^2/2g$ is large
- Subcritical flow occurring for fluid depths greater than the critical depth where $\tilde{V}^2/2g$ is small
- Critical flow where the fluid depth is precisely the critical depth

Using the (I.F.) continuity expression and the critical depth expression of Eq. (13.27), we find the result

$$\left(\frac{\tilde{V}^2}{gh}\right)_{\text{crit}} = 1 \tag{13.29}$$

which is recognized as the Froude number, and where the reference length is the critical depth. Thus, the critical depth of flow in a free-channel flow occurs when the Froude number equals unity. When $\tilde{V}^2/gh < 1$, we have what is termed subcritical flow, while for $\tilde{V}^2/gh > 1$ we have what is termed supercritical flow. These regions are illustrated in Fig. 13.6b.

Thus, the elevation of the free-surface flow at the hump in a channel can either be higher or lower than the free-surface at some large distance upstream or downstream from the hump, depending upon the flow velocity $\tilde{V}$. Thus, if the flow velocity $\tilde{V}$ is subcritical, then the free-surface will result in a depression as shown in Fig. 13.6b, since the maximum decrease in water depth will be greater than the maximum change in floor elevation. If, on the other hand, the velocity $\tilde{V}$ is supercritical, the water surface will rise above the water level existing at some large distance from the hump.

For nonrectangular cross sections in the channel geometry, the specific energy E.G.L. is given by

$$\text{E.G.L.} = h + \left(\frac{Q^2}{2g\,A^2}\right) \tag{13.30}$$

where A is the cross-sectional area of the flow given by Eq. (13.14). To find the critical depth, we minimize Eq. (13.30) so that

$$\frac{Q^2}{gA^3}\frac{dA}{dh} = 1 \qquad (13.31)$$

The relationship between A and h is obtained from geometry:

$$b_1\,dh = dA \qquad (13.32)$$

where b_1 is the width of the cross-sectional area at the free-surface. Working with Eq. (13.31) and (13.32), we find

$$\left(\frac{Q^2}{g}\right)\left(\frac{b_1}{A^3}\right)_{\min} = 1 \qquad (13.33)$$

Eliminating the volume rate of flow Q between Eqs. (13.30) and (13.33) results in a minimum specific energy expression

$$(\text{E.G.L})_{\min} = \left(h + \frac{A}{2b_1}\right)_{\min} \qquad (13.34)$$

Comparing the minimum energy expression of Eq. (13.34) with Bernoulli's equation (13.23) for open-channel flow discloses that the minimum specific energy occurs when the velocity head $\tilde{V}^2/2g$ is one-half the depth $(A/b_1)_{\min}$.

Example 13.3

A uniform rectangular smooth surface channel of reinforced cement with plaster has water flowing at a rate of 30 ft^3/s. Determine (a) its critical depth, (b) the nature of the flow if the depth $h = 6$ ft and the width is 10 ft, and (c) the channel slope m.

Solution:

Step 1.

The appropriate equations are:

- Critical depth:

$$h_{\text{crit}} = \left(\frac{Q^2}{gb^2}\right)^{1/3} \qquad \text{(i)}$$

- Slope:

$$m = \frac{Q^2}{C^2A^2R_h} \qquad \text{(ii)}$$

Example 13.3 (Con't.)

(a) From Eq. (i), we find

$$h_{\text{crit}} = \left[\frac{(30)^2}{(32.2)(10)^2}\right]^{1/3} = 0.65 \text{ ft} \tag{iii}$$

(b) Since the critical depth h_{crit} is much less than the actual depth of 6 ft, the flow is subcritical and thus moving slowly.

(c) The slope m is given by Eq. (ii). The Chézy coefficient C is obtained from Eq. (13.11):

$$C = \frac{1.49}{n}\left(\frac{A}{P}\right)^{1/6} \tag{iv}$$

$$= \frac{1.49}{0.011}(22)^{1/6}$$

$$= 226.74 \tag{v}$$

where the value of the Manning roughness coefficient is found using Table 13.1. Substituting Eq. (v) into Eq. (ii) results in

$$m = \frac{(30)^2}{(226.74)^2(60)^2(1.875)}$$

or

$$m = 2.6 \times 10^{-6} \tag{vi}$$

This completes the solution.

13.3 Surge Waves and the Hydraulic Jump

There are two types of surge waves in free-surface channel flow: *positive* and *negative*. Both are unsteady flow phenomena. The type of surge wave depends upon the channel depth variation. For instance, a negative surge wave occurs upstream from a sluice gate that is being opened; or a negative surge wave can occur downstream in the channel from a gate that is being closed, as shown in Fig. 13.7. The symbol c denotes the velocity of the wave.

A positive surge wave, on the other hand, results from an increase in depth (as against the negative surge wave caused by a decrease in depth). Figure 13.8 shows a positive surge wave. A piston lies on the channel floor. When the piston moves either upstream or downstream, a surge will form. Downstream of the surge, the fluid is at rest and at a depth z. Let the fluid at the piston face have a velocity $\delta\tilde{V}$ and a depth $z + \delta z$. The surge will move at a velocity c different from $\delta\tilde{V}$.

We shall consider flows relative to the wave.* The continuity equation gives

$$(c - \delta\tilde{V})(z + \delta z) = cz$$

*The author is indebted to C. J. M. Lee [13.1] for this development.

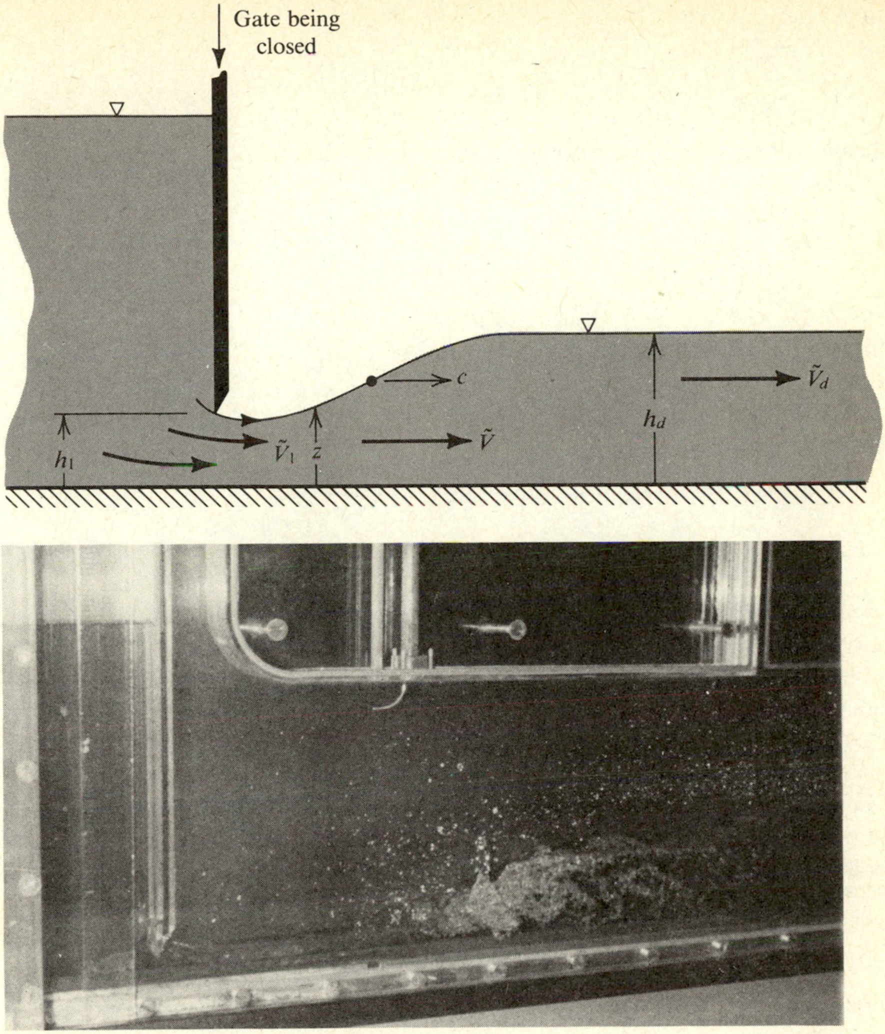

Figure 13.7 Negative surge wave during gate closure.

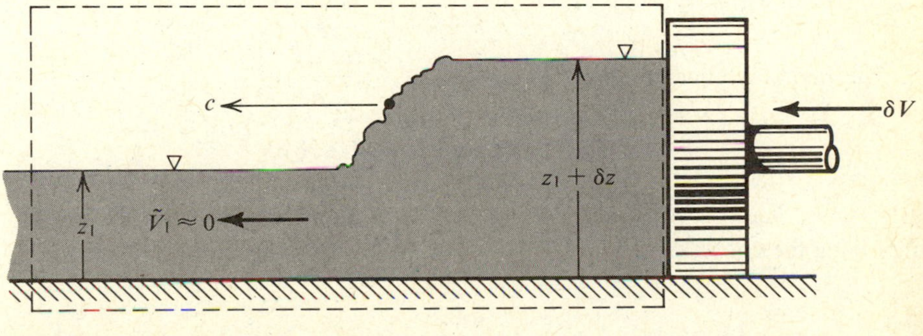

Figure 13.8 Positive surge wave due to a piston.

or simplifying,

$$\delta \tilde{V}(z + \delta z) = c\delta z \tag{13.35}$$

Application of the integral momentum theorem to the control volume sketched in Fig. 13.8 gives

$$\frac{\gamma}{2}[z_1^2 - (z_1 + \delta z)^2] = \rho z_1 c[(c - \delta \tilde{V}) - c] \tag{13.36}$$

or simplifying, after assuming $(\delta z)^2 \approx 0$,

$$g\delta z = c\delta \tilde{V} \tag{13.37}$$

Elimination of δV from Eqs. (13.35) and (13.37) gives

$$c^2 = g(z_1 + \delta z) \tag{13.38}$$

The wave velocity c is thus approximately $\sqrt{gz_1}$. Note that if there exists a flow velocity $\tilde{V}_1 \neq 0$, then the wave speed is $-\tilde{V}_1 + \sqrt{g(z_1 + \delta z)}$.

We can now apply a Froude number analysis similar to that presented in the previous section. Letting $F_r^2 = \tilde{V}^2/gz$, if $F_r < 1$, then $\tilde{V} < c$ and *disturbances in the flow can be propagated upstream*. If $F_r > 1$, then $\tilde{V} > c$ and disturbances in the downstream cannot travel upstream as small waves.

Let us now discuss flows in channels that have a change in width δb and a change in the bed level δz as shown in Fig. 13.9. The analysis for such an open-channel flow is not at all difficult. From the (I.F.) continuity equation,

$$\tilde{V}zb = Q \tag{13.39}$$

Differentiating the above equation produces

$$\frac{db}{b} + \frac{dz}{z} + \frac{d\tilde{V}}{\tilde{V}} = 0 \tag{13.40}$$

The Bernoulli equation gives

$$z + \frac{\tilde{V}^2}{2g} = z + \delta z + a + \frac{(\tilde{V} + \delta \tilde{V})^2}{2g} \tag{13.41}$$

Simplifying the above expression gives

$$dz + a + \frac{\tilde{V}d\tilde{V}}{g} = 0 \tag{13.42}$$

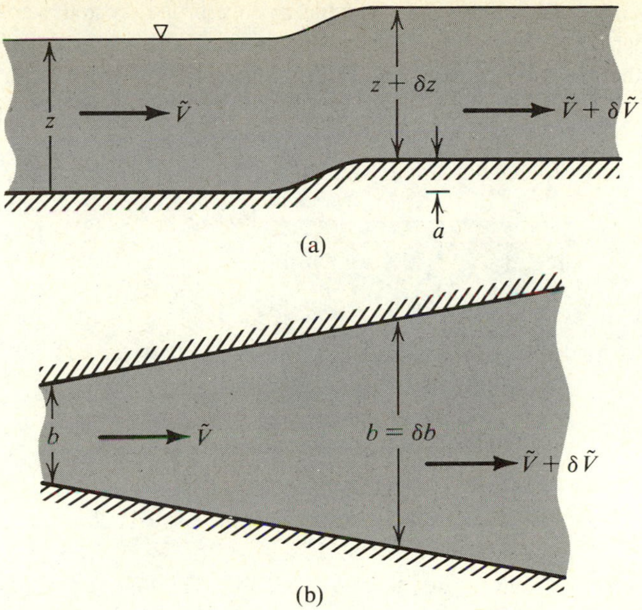

Figure 13.9 *Open channel of variable width and depth. (a) Side view. (b) Top view.*

Introducing the Froude number and eliminating dz between Eqs. (13.40) and (13.42) gives

$$\frac{d\tilde{V}}{\tilde{V}}(1 - F_r^2) + \frac{db}{b} - \frac{a}{z} = 0 \tag{13.43}$$

Consider changes in width and bed level. When $F_r < 1$, the flow is slow. An increase in width results in a decrease in velocity, and an increase in bed level results in an increase in velocity. Let us reverse the condition. When $F_r > 1$ the flow is fast, and all the results in the slow case are just the opposite.

Next, let us eliminate $\delta\tilde{V}$ in Eq. (13.43):

$$\frac{dz}{z}(1 - F_r^2) = F_r^2\frac{db}{b} - \frac{a}{z} \tag{13.44}$$

Once again, let us consider each effect separately. For $F_r < 1$, an increase in width causes an increase in depth, and an increase in bed level causes a *decrease* in depth. For $F_r > 1$, the results are reversed.

13.3.1 Open-Channel Flow Past a Broad-Crested Weir

The above results can be applied to channels containing either a Venturi flume (convergent-divergent section) or a broad-crested weir (an elevated bed similar to that shown in Fig. 13.10). Consider water in an open channel with a broad-crested weir with a constant upstream depth z_1. Downstream of the contraction, a sluice gate can be manipulated to control the downstream depth z_3. With the gate closed, there is no flow. This is represented by line *a* in Fig. 13.10.

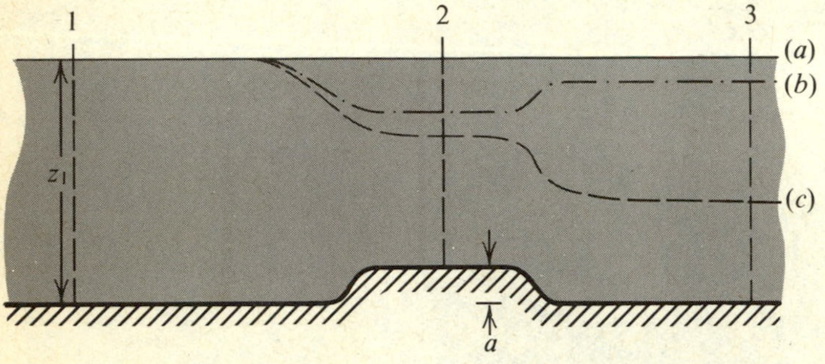

Figure 13.10 *Flow lines for broad-crested weir open-channel flow.*

At some small opening in the sluice gate, the flow is slow. Thus $F_r < 1$. The bed level increases at the throat (and if it is a Venturi flume the width also decreases), both effects individually resulting in a decrease in water depth. This is shown as curve *b*. Here, the Froude number is greatest at the throat since $\tilde{V}$ increases and z decreases for $F_r < 1$ in a contracting section.

As long as the F_{r_2} is less than unity, the velocity will continue to increase as the sluice gate is being opened. Once $F_r = 1$, the velocity reaches a maximum value. Conditions at the throat are termed critical, and $\tilde{V}_2 = \sqrt{gz_2}$.

If we neglect frictional losses (due to turbulent stresses), then Bernoulli's equation gives us

$$\left(z + \frac{\tilde{V}^2}{2g}\right)_1 = \text{E.G.L.}_1 = \left(z + a + \frac{\tilde{V}^2}{2g}\right)_2 \qquad (13.45)$$
$$= a + \frac{3}{2} z_2$$

Solving for the water height at station 2 shows

$$z_2 = \frac{2}{3}(\text{E.G.L.}_1 - a) \qquad (13.46)$$

Thus, for a weir height $a > 0$, the water height at the throat can be appreciably lower than the upstream water depth.

Once the flow in the throat is critical, the divergence in the downstream section will further accelerate the flow, producing the fast flow in the downstream channel depicted by curve *c*. This result is often seen in river beds that have rapids. The rapids are caused by the stream contracting and moving over rocks which act as localized weirs.

13.3.2 The Hydraulic Jump

The diagram given by Fig. 13.10 can also be used to help explain the beautiful hydraulic jump. We have been discussing different types of surges. One special surge occurs when water flowing at a rapid flow and shallow depth suddenly and dramatically leaps to a much greater height and continues to flow at this new deeper depth, but with a much lower velocity. Figure 13.11 shows this surge, i.e., the hydraulic jump in all its turbulent mixing.

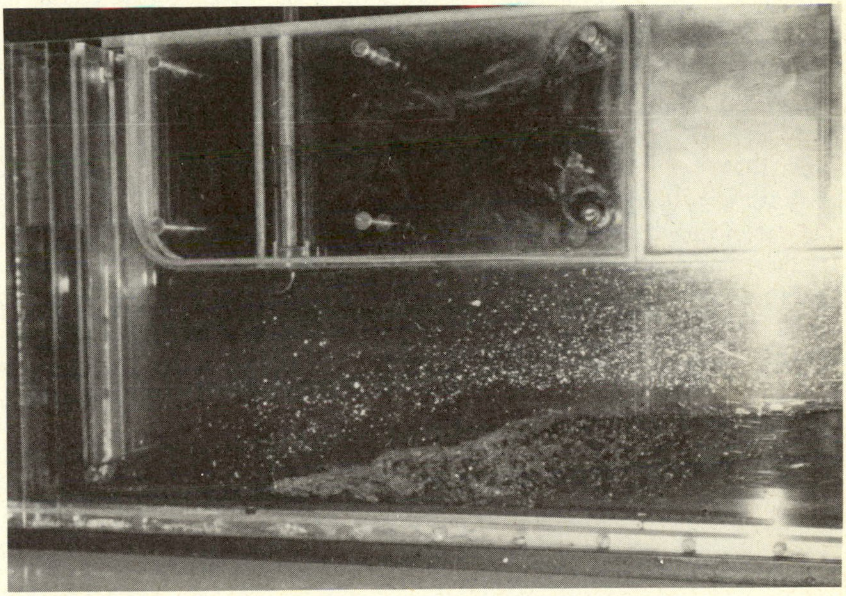

Figure 13.11 *The hydraulic jump.*

The stream depth z_1 before the jump must be less than the critical depth h_{crit}. After the jump, the depth z_2 must be greater than h_{crit}, as shown in Fig. 13.6a. The difference between z_1 and z_2 represents the height of the jump.

Consider the illustration of the jump given in Fig. 13.8. Working with Eqs. (13.35) and (13.36), we obtain

$$\delta \tilde{V} = c \frac{z_2 - z_1}{z_2} \tag{13.47}$$

and

$$\frac{\gamma}{2}(z_2^2 - z_1^2) = \rho z_1 c \delta \tilde{V} \tag{13.48}$$

respectively. We eliminate $\delta \tilde{V}$ from Eqs. (13.47) and (13.48) to obtain

$$c = \left[g\frac{z_2}{2}\left(1 + \frac{z_2}{z_1}\right)\right]^{1/2} \tag{13.49}$$

which is identical to Eq. (13.38) for $z_1 \approx z_2$. The height of the hydraulic jump z_2 is now readily obtained:

$$z_2 = -\frac{z_1}{2} + \sqrt{\left(\frac{z_1}{2}\right)^2 + \frac{2\tilde{V}_1^2 z_1}{g}} \tag{13.50}$$

Note that for $z_2 > z_1$, i.e., for the hydraulic jump to exist, the kinetic energy $\tilde{V}_1^2/2g > 7z_1/16$.

Example 13.4

Water flows in a free-surface uniform channel at a depth $z_1 = 2.5$ m. If the volume rate of flow per unit width is 13 m^2/s, calculate (a) the average velocity $\tilde{V}_1$, (b) the Froude number F_{r_1}, (c) the depth z_2, (d) the average velocity $\tilde{V}_2$, (e) the Froude number F_{r_2}, and (f) the speed of the surge c.

Solution:

(a) The velocity upstream of the surge is

$$\tilde{V}_1 = \frac{Q}{z_1 b} = \frac{13}{2.5} = 5.2 \text{ m/s} \tag{i}$$

(b) The Froude number F_{r_1} is

$$F_{r_1} = \frac{\tilde{V}_1}{\sqrt{gz_1}} = \frac{13}{\sqrt{(9.81)(2.5)}} = 2.625 \tag{ii}$$

(c) From Eq. (13.50), we find the depth of the water downstream of the surge:

$$z_2 = -\frac{z_1}{2} + \left[\left(\frac{z_1}{2}\right)^2 + \frac{2\tilde{V}_1^2 z_1}{g}\right]^{1/2} = -\frac{2.5}{2} + \left[(1.25)^2 + \frac{2(5.2)^2(2.5)}{9.81}\right]^{1/2} = 2.667 \text{ m} \tag{iii}$$

Example 13.4 *(Con't.)*

(d) The velocity downstream of the surge is

$$\tilde{V}_2 = \tilde{V}_1 \left(\frac{z_1}{z_2}\right)$$

$$= (5.2)\left(\frac{2.5}{2.667}\right)$$

$$= 4.874 \text{ m/s} \qquad \text{(iv)}$$

(e) The Froude number, F_{r_2}, is

$$F_{r_2} = \frac{\tilde{V}_2}{\sqrt{gz_2}}$$

$$= \frac{4.874}{\sqrt{(9.81)(2.667)}}$$

$$= 0.953 \qquad \text{(v)}$$

which is subcritical. Thus what we have in this problem is a weak hydraulic jump.

(f) We obtain the speed of the surge from Eq. (13.49):

$$c = \left[\frac{gz_2}{2}\left(1 + \frac{z_2}{z_1}\right)\right]^{1/2}$$

$$= \left[\frac{(9.81)(2.667)}{2}\left(1 + \frac{2.667}{2.5}\right)\right]^{1/2}$$

$$= 5.2 \text{ m/s} \qquad \text{(vi)}$$

which indicates that the surge will be stationary because it is equal in magnitude to the upstream velocity $\tilde{V}_1$.

This completes the solution.

13.4 Flows Past Sharp-Crested Weirs

Weirs are walls or obstructions which block the flow of water. They are extremely important in the design of hydraulic and civil engineering projects involving flood control and water heads in reservoirs and hydroelectric plants. They are also used in measuring the rate of flow in channels.

Figure 13.12 illustrates what can take place when a fluid flow in a channel moves past a sharp-crested weir. The flow accelerates as it approaches the weir, the free-surface deforms and follows the trajectory of the fluid. The nappe is the sheet of water falling from the crest of the weir. In almost every flow past a weir, a region on the weir is exposed to an air cavity. Sometimes the cavities are so large we can stand in them if the velocity is high enough or if H is large, or if weirs are high, as in the case of waterfalls.

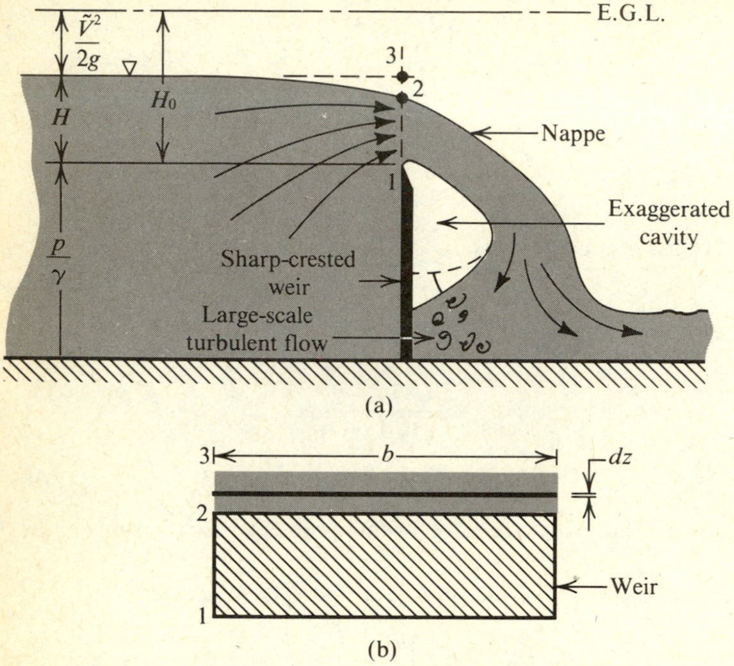

Figure 13.12 *Flow past a sharp-crested weir. (a) Side view. (b) Front view.*

Let us examine a way in which we can evaluate the rate of flow past a sharp-crested weir. We begin by defining the elemental control area at the weir. Figure 13.12a treats an elemental area of width b and height dz, such that the elemental volume rate of flow dQ through the control area dA is found using Bernoulli's equation:

$$dQ = C_d \, dA_o \sqrt{2gz} \tag{13.51}$$

where the differential area of the orifice is $dA_o = bdz$, and C_d defined as the ratio of the actual velocity V_a to the ideal velocity V_i

$$C_d = \frac{V_a}{V_i} \tag{13.52}$$

is the discharge coefficient. Integrating from stations 1 to 2 to obtain the volume rate of flow past the weir gives

$$\begin{aligned} Q &= C_d \, b \sqrt{2g} \int_{H_{o_1}}^{H_{o_2}} \sqrt{z} \, dz \\ &= \frac{2}{3} C_d \, b \sqrt{2g} \left(H_{o_2}^{3/2} - H_{o_1}^{3/2} \right) \end{aligned} \tag{13.53}$$

where the head H_o is seen from Fig. 13.12a to be

$$H_o = H + \frac{\tilde{V}^2}{2g} \tag{13.54}$$

The kinetic energy per weight of fluid can be neglected provided that we take the appropriate value of the discharge coefficient C_d.

A great many weirs do not extend their width to the channel width b. This will create what is called *end contractions*. Since no theoretical solution has been established for this problem, we use empirical corrections to the flow rate of Eq. (13.53). If we define an empirical discharge coefficient as $\overline{C}_d$, then the volume rate of flow through narrow sharp-crested weirs is expressed as

$$Q = \frac{2}{3}\overline{C}_d\, b\sqrt{2g}\, H_o^{3/2} \tag{13.55}$$

where a range of values of $\overline{C}_d$ are given in Ref. 13.2 for different weirs. For example, for a rectangular weir with vertical walls shown in Fig. 13.13a, $\overline{C}_d = 0.63$. For the triangular weir with a 90° base angle shown in Fig. 13.13b, $\overline{C}_d = 0.237$. In SI units, the volume rate of flow per unit width is

$$\frac{Q}{b} = 1.86\, H_o^{3/2} \tag{13.56}$$

for the rectangular narrow weir, and

$$\frac{Q}{b} = 0.7\, H_o^{3/2} \tag{13.57}$$

SI UNITS

for a 90° narrow triangular-shaped weir.

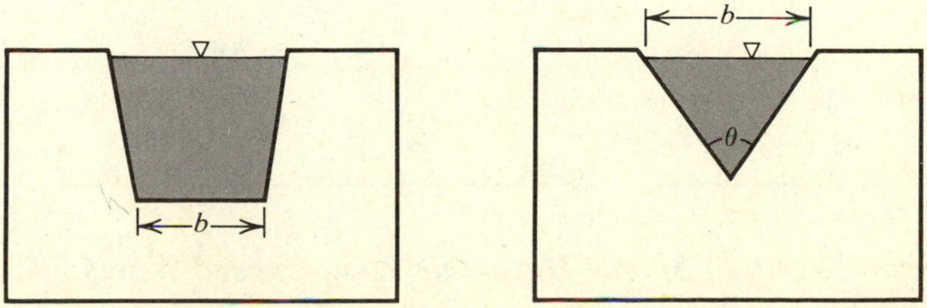

(a) (b)

Figure 13.13 Narrow weirs. (a) Rectangular weir. (b) Triangular weir.

Example 13.5

A weir is placed in a horizontal uniform channel that is 3 m wide and 2 m high. Determine the rate of flow of water if the head above the weir's crest is 0.5 m, and the upstream average flow velocity is 1.0 m/s for (a) a narrow rectangular weir, and (b) a narrow 90° triangular weir.

Solution:

Step 1.

The appropriate equations are:

- Head equation:

$$H_o = H + \tilde{V}^2/2g \tag{i}$$

- Weir equation:

$$Q = \frac{2}{3}\overline{C}_d\, b\, \sqrt{2g}\, H_o^{3/2} \tag{ii}$$

From Eq. (i), the head H_o is

$$H_o = \left(0.5 + \frac{1}{2 \times 9.81}\right) = 0.551 \text{ m} \tag{iii}$$

Substituting Eq. (iii) into the weir Eq. (ii) results in

$$\begin{aligned} Q &= \frac{2}{3}\overline{C}_d\,(3)\,\sqrt{2(9.81)}\,(0.551)^{3/2} \\ &= 3.623\,\overline{C}_d \end{aligned} \tag{iv}$$

(a) For the rectangular weir, $\overline{C}_d = 0.63$. Thus

$$\begin{aligned} Q &= (3.623)(0.63) \\ &= 2.28 \text{ m}^3\text{/s} \end{aligned} \tag{v}$$

(b) For the 90° triangular weir, $\overline{C}_d = 0.237$. Thus

$$\begin{aligned} Q &= (3.623)(0.237) \\ &= 0.86 \text{ m}^3\text{/s} \end{aligned} \tag{vi}$$

Obviously, more flow will pass through the rectangular weir than the triangular weir—more than 2½ times as much.

This completes the solution.

13.5 *Linear Theory of Simple Harmonic Long-Crested Waves of Small Amplitude*

We discussed in the preceding two sections the velocity of surges in translational and rotational flows. Another familiar phenomenon associated with open-channel flows, that of waves on the surface of liquids also deserves some discussion. Waves are produced by disturbances. For example, an object inserted into a fluid must displace the volume of fluid occupied by the object. The result is a displacement of fluid that moves radially outward creating a wave. Secondary waves usually follow behind the primary wave as a result of numerous factors, e.g., the splash resulting in numerous disturbances from water droplets. Waves can also be created by wind shear, objects moving through the water, and rapid shallow flows, to name but a few causes.

This section presents a theory that predicts the wave shape, wave speed c, fluid particle velocities, and the pressure field resulting from *long-crested waves of small amplitude*.

We first turn our attention to what is known as the *free-surface boundary condition*. A free-surface is a surface (it need not be planar) which divides fluids of two different densities (water and air, for example). The lighter fluid transmits a uniform pressure over the free-surface, such that the pressure of the denser fluid equals that of the lighter fluid on the surface owing to the stress boundary condition. This constancy of pressure leads to what is known as the kinematic *free-surface boundary condition.*

In Fig. 13.14, let $\eta\ (x, t)$ be the vertical distance from the equilibrium free-surface state to the disturbed free-surface shape. The gravitational potential energy per weight of fluid is expressed in terms of the free-surface distance $z = \eta$ by expressing Bernoulli's equation for irrotational flow, Eq. (4.131), as

$$\frac{p}{\rho} + \frac{\tilde{V}^2}{2} + g\eta + \frac{\partial\phi}{\partial t} = c_1\ (t) \tag{13.58}$$

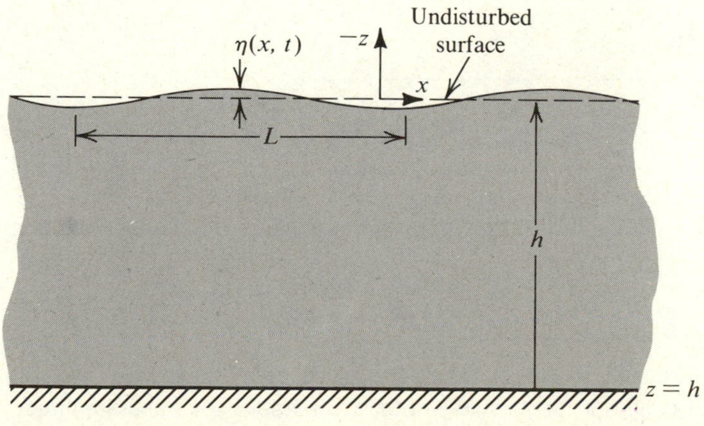

Figure 13.14 *Cross section of an elementary wave.*

At the surface, we assume that the fluid velocity components are small such that the velocity squared term $\tilde{V}^2/2$ is negligible compared to the magnitude of the other terms in Eq. (13.58). Also, since the pressure is constant at the surface, it can be included with the constant $c_1(t)$ in the equation without loss of generality, so that Bernoulli's equation can be expressed as

$$\eta = -\frac{1}{g}\frac{\partial \phi}{\partial t} + c_2(t) \tag{13.59}$$

Consider for a moment the vertical component of velocity w. At the free-surface, it is known from kinematics, and from our expression for the gravitational potential energy per weight of fluid, that

$$w\bigg|_{\text{surface}} = \frac{dz}{dt}\bigg|_{\text{surface}} = \frac{\partial \eta}{\partial t}\bigg|_{z=\eta} \tag{13.60}$$

Recalling that

$$w = \frac{\partial \phi}{\partial z}\bigg|_{z=\eta} \tag{13.61}$$

for irrotational flow, we obtain from Eqs. (13.60) and (13.61)

$$\frac{\partial \eta}{\partial t}\bigg|_{z=\eta} = \frac{\partial \phi}{\partial z}\bigg|_{z=\eta} \tag{13.62}$$

Substituting Eq. (13.59) into Eq. (13.62) yields

$$\frac{\partial^2 \phi}{\partial t} + g\frac{\partial \phi}{\partial z} = \frac{\partial}{\partial t}[g\ c_2(t)] \tag{13.63}$$

For $c_2\ (t) = 0$, we obtain

$$\boxed{\frac{\partial^2 \phi}{\partial t^2} + g\frac{\partial \phi}{\partial z} = 0, \text{ at } z = \eta} \tag{13.64}$$

which is the free-surface condition resulting from constancy of pressure.

There are a number of solutions of Eq. (13.64). For instance, an obvious solution is

$$\phi(\eta, t) = C \exp\,(i\sqrt{gB}t) \exp B\eta \tag{13.65}$$

where C and B are constants, values of which are to be determined from a given problem. Note that this condition exists solely at the free-surface for small amplitude waves. It does not satisfy Laplace's equation, nor should it, necessarily, as we have not invoked the conservation of mass in this discussion.

Consider the cross section of a free-surface flow that has a wave traveling along the free-surface as shown in Fig. 13.14. We define the equation of the surface wave as

$$\eta(x, t) = a \sin\,(\alpha x - \beta t) \tag{13.66}$$

where a is the maximum wave amplitude measured from trough to crest, α is the radian wave number and equals $2\pi/L$, where L is the wavelength measured from trough to trough. It also equals $2\pi f/c$, where f is the frequency of the wave and c is the wave speed. The radial angular frequency is denoted by β and equals $2\pi/T$ or $2\pi f$. Thus the wave speed c can be expressed as

$$c = \frac{\beta}{\alpha} \tag{13.67}$$

For ideal frictionless waves (i.e., potential waves), we have an irrotational motion such that a velocity potential ϕ can exist. Laplace's equation, $\nabla^2\phi = 0$, is the appropriate form for the (D.F.) continuity equation. The motion of the wave is assumed to be two-dimensional and unsteady; hence $\phi = \phi(x, z, t)$. Let the velocity potential solution be of separable form:

$$\phi(x, z, t) = Z(z) \cos (\alpha x - \beta t) \tag{13.68}$$

Substituting this expression into Laplace's equation, and employing the technique of separation of variables, results in one of the equations being the ordinary differential equation:

$$\frac{d^2Z}{dz^2} - \alpha^2 Z = 0 \tag{13.69}$$

Equation (13.69) has the general solution

$$Z(z) = A \exp (\alpha z) + B \exp (-\alpha z) \tag{13.70}$$

Substituting Eq. (13.70) back into the velocity potential expression of Eq. (13.68) gives

$$\phi(x, z, t) = [A \exp (\alpha z) + B \exp (-\alpha z)] \cos (\alpha x - \beta t) \tag{13.71}$$

Now we have to determine the values of the arbitrary constants of integration A and B. We examine the illustration of Fig. 13.14 and note that the normal velocity component w vanishes at the wall, $z = h$:

$$w = \left.\frac{\partial \phi}{\partial z}\right|_{z=h} = 0 \tag{13.72}$$

Substituting the velocity potential expression of Eq. (13.71) into the boundary condition given above results in

$$A \exp (\alpha h) = B \exp (-\alpha h) \tag{13.73}$$

so that

$$B = A \exp(2\alpha h) \tag{13.74}$$

Substituting this value of B back into the velocity potential expression of Eq. (13.71) results in

$$\phi(x, z, t) = A \exp(\alpha h)[\exp(\alpha)(h - z) + \exp(-\alpha)(h - z)] \times \cos(\alpha x - \beta t) \tag{13.75}$$

The following transcendental relationship exists:

$$\exp(\alpha)(h - z) + \exp(-\alpha)(h - z) = 2 \cosh \alpha (h - z) \tag{13.76}$$

If we define

$$C = 2 A \exp(\alpha h) \tag{13.77}$$

and substitute the above two relationships, Eqs. (13.76) and (13.77), into the velocity potential of Eq. (13.75), we will obtain

$$\phi(x, z, t) = C \cosh[\alpha(h - z)] \cos(\alpha x - \beta t) \tag{13.78}$$

(Note that a similar result can be obtained using the stream function $\psi(x, z, t)$ and the vorticity equation $\nabla^2\psi = 0$: $\psi(x, z, t) = C \sinh[\alpha(h - z)] \sin(\alpha x - \beta t)$.)

We are now in a position to calculate the wave velocity. Substituting our solution for the velocity potential $\phi(x, z, t)$ of Eq. (13.78) into Eq. (13.64), we obtain

$$-\frac{\beta^2 C}{g} \cosh(\alpha h) \cos(\alpha x - \beta t) = -\alpha C \sinh(\alpha h) \cdot \cos(\alpha x - \beta t) \tag{13.79}$$

Clearing terms in the above expression yields

$$\frac{\beta^2}{\alpha^2} = \frac{g}{\alpha} \tanh(\alpha h) \tag{13.80}$$

Thus the wave velocity c is

$$c = \sqrt{\frac{g}{\alpha} \tanh(\alpha h)}$$

or

$$c = \sqrt{\frac{gL}{2\pi} \tanh\left(2\pi \frac{h}{L}\right)} \tag{13.81}$$

For deep water, $2h/L \geqslant 1$ so that $\tanh(2\pi h/L) \approx 1$. Thus the speed of an elementary long-crested small amplitude wave in deep water is

$$c = \sqrt{\frac{gL}{2\pi}} \tag{13.82}$$

For shallow water, the quantity $h/L \ll 1$ so that $\tanh 2\pi h/L \approx 2\pi h/L$. The speed of a long-crested small amplitude wave in shallow water is seen to be

$$c = \sqrt{gh} \tag{13.83}$$

which is equivalent to our result in Eqs. (13.38) and (13.49).

The last step in the solution is to evaluate the arbitrary constant C in the velocity potential expression given by Eq. (13.78). From the definition of the free-surface shape η given by Eq. (13.66) and the relationship between η and the velocity potential ϕ given by Eq. (13.59), we obtain

$$a \sin(\alpha x - \beta t) = +\frac{\beta C}{g} \cosh[\alpha(h - z)]\bigg|_{z=0} \cdot [+\sin(\alpha x - \beta t)] \tag{13.84}$$

Evaluating the constant C from the above equation gives

$$C = \frac{ga}{\beta \cosh(\alpha h)} \tag{13.85}$$

We thus have arrived at a complete solution for the velocity potential of simple harmonic long-crested waves of small amplitude:

$$\phi(x, z, t) = \frac{ga}{\beta} \frac{\cosh \alpha(h - z)}{\cosh(\alpha h)} \cos(\alpha x - \beta t) \tag{13.86}$$

Knowing the velocity potential ϕ for a particular flow problem, the fluid flow velocity components u, w, and the pressure p can easily be found.

Example 13.6
For deep water waves, find the speed of the wave c and wavelength L given that the period of the wave is 3 seconds.

Solution:
For deep-water waves, Eq. (13.80) gives the speed c of the wave as

$$c = \sqrt{\frac{gL}{2\pi}} \tag{i}$$

Example 13.6 *(Con't.)*

But, we also know that

$$c = \frac{L}{T} \tag{ii}$$

Substituting Eq. (ii) into Eq. (i) gives

$$c = \frac{gT}{2\pi} = 5.13\ T = 15.39 \text{ ft/s} \tag{iii}$$

as the wave speed. The wavelength L is then

$$\begin{aligned} L &= cT \\ &= 15.39 \times 3 \\ &= 46.17 \text{ ft} \end{aligned} \tag{iv}$$

This completes the solution.

Example 13.7

(a) Find the location and the maximum value of the x- and z-components of velocity, u and w, respectively, in terms of wave height a and period T. (b) What is the expression for the x- and z-components of velocity for deep water?

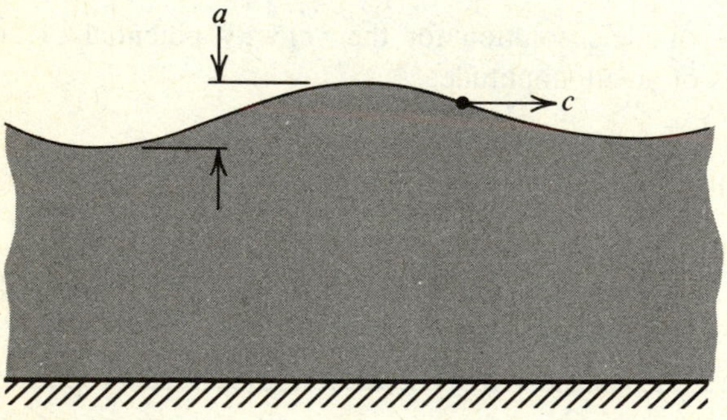

Figure E13.7

Solution:

(a) We obtain the x-component of velocity u by substituting the velocity potential ϕ of Eq. (13.86) into the velocity $\mathbf{V}$ and velocity potential expression $\mathbf{V} = \nabla\phi$, so that

$$u = \frac{\partial \phi}{\partial x} = -\frac{ga\alpha}{\beta}\frac{\cosh \alpha(h - z)}{\cosh (\alpha h)} \sin (\alpha x - \beta t) \tag{i}$$

Example 13.7 *(Con't.)*

Expressed in terms of wave height and period T, the velocity components are

$$u = -\frac{gaT}{L}\frac{\cosh 2\pi(h - z)/L}{\cosh 2\pi\, h/L}\sin 2\pi\left(\frac{x}{L} - \frac{t}{T}\right) \qquad \text{(ii)}$$

$$w = \frac{\partial\phi}{\partial z} = -\frac{ga\alpha}{\beta}\frac{\sinh \alpha(h - z)}{\cosh(\alpha h)}\cos(\alpha x - \beta t) \qquad \text{(iii)}$$

or in terms of wave height a and period T

$$w = -\frac{gaT}{L}\frac{\sinh 2\pi(h - z)/L}{\cosh 2\pi\, h/L}\cos 2\pi\left(\frac{x}{L} - \frac{t}{T}\right) \qquad \text{(iv)}$$

Consider the velocity expressions of Eqs. (ii) and (iv). The velocities are maximum when z is zero, since for $z \neq 0$, the arguments of the hyperbolic sine and cosine are smaller than for the arguments when $z = 0$. For $z = 0$,

$$u_{\max} = -\frac{gaT}{L}\sin 2\pi\left(\frac{x}{L} - \frac{t}{T}\right) \qquad \text{(v)}$$

and

$$w_{\max} = -\frac{gaT}{L}\tanh\frac{2\pi h}{L}\cos 2\pi\left(\frac{x}{L} - \frac{t}{T}\right) \qquad \text{(vi)}$$

(b) For deep water waves ($h/L > 0.5$), the two hyperbolic functions are similar. Thus

$$\cosh\frac{2\pi(h - z)}{L} \approx \sinh\frac{2\pi(h - z)}{L} \qquad \text{(vii)}$$

or

$$\frac{\cosh 2\pi(h - z)/L}{\cosh 2\pi\, h/L}$$

$$= \frac{\exp(2\pi h/L)\exp(-2\pi z/L) + \exp(-2\pi h/L)\exp(2\pi z/L)}{\exp(2\pi h/L) + \exp(-2\pi h/L)} \qquad \text{(viii)}$$

$$\approx \exp\frac{-2\pi z}{L} \qquad \text{(ix)}$$

Using the relationships of Eqs. (ix) and (vii) in the x- and z-component velocities,

$$u = -\frac{gaT}{L}\exp\frac{-2\pi z}{L}\sin 2\pi\left(\frac{x}{L} - \frac{t}{T}\right) \qquad \text{(x)}$$

$$w = -\frac{gaT}{L}\exp\frac{-2\pi z}{L}\cos 2\pi\left(\frac{x}{L} - \frac{t}{T}\right) \qquad \text{(xi)}$$

Example 13.7 *(Con't.)*

Note the exponential decay of the fluid particle velocities for waves in deep water. Note in particular, that the velocities of Eqs. (x) and (xi) are less than 5% of the surface value at half a wavelength down, i.e., at $z = L/2$, and are less than 0.2% at $z = L$. This helps explain why submariners seldom get seasick while submerged at deep depths.

This completes the solution.

This introductory study of free-surface flow comprises certain important flows where gravitational forces have been predominant (as opposed to viscous or compressible forces). We should recall that, in the formulation of the problem, the Froude number played a key role in describing the nature of the flow, just as the Euler number (or the various pressure coefficients) did for potential flow. In the remaining two chapters, we shall be concerned with the role Reynolds number plays for viscous flow and Mach number for compressible flows. It will soon be apparent why we made a rather extensive study of dimensional analysis before we started to analyze various important types of flows.

References

13.1 Lee, C. J. M., "Rationalization in Fluid Mechanics with Particular Reference to Open Channels," *Bull. Mech. Engrg. Educ.*, vol. 4, pp 293–296, Pergamon, Oxford, 1965.

13.2 *Fluid Meters*, 6th ed., American Society of Mechanical Engineers, New York, 1971.

Study Questions

13.1 Show that the hydraulic grade line and the free-surface coincide in open channel flow.

13.2 What is a surge? How does a surge differ from a wave?

13.3 How is a wave's velocity different from the velocity of a fluid particle?

13.4 What are the conditions for a hydraulic jump?

13.5 Is the hydraulic grade line parallel to the liquid surface for nonuniform flow and/or unsteady flow? Explain.

13.6 Can we use $R_D \leqslant 2400$ for laminar flow in an open channel? Explain.

13.7 What is the difference between the Chézy coefficient and the velocity coefficient C_v?

13.8 What is the difference between positive and negative waves, traveling and standing waves, surface and internal waves?

13.9 Give the speed of an elementary wave in a still liquid of shallow depth and one of deep depth.

13.10 Give the speed of an elementary wave in a uniform moving liquid of shallow depth and deep depth.

Problems

13.1 For laminar steady flow in an open channel, as shown in Fig. P13.1, show that the vertical distribution of velocity u is parabolic in z.

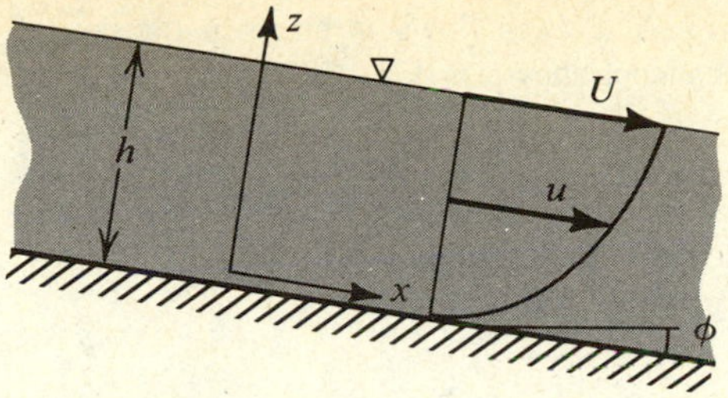

Figure P13.1

13.2 Show that the range of Reynolds number for open channel flow to be laminar is $R_l = \tilde{V}R_h/\nu \leqslant 600$, and for transitional flow $600 < \tilde{V}R_h/\nu < 2000$.

13.3 In reference in Prob. 13.2, show that for steady laminar flow in an open channel the volume flow rate Q cannot be greater than $600\, b\nu$.

13.4 Consider steady laminar flow of water at 20°C in a channel inclined at 10° with respect to the horizon. What is the depth of water h if the Reynolds number is 600?

13.5 Consider an open channel where the hydraulic radius $R_h = 2$ m, and the cross-sectional area $A = 25\ \text{m}^2$. What is the average velocity $\tilde{V}$ if the slope $m = 0.01$ and the channel is structured out of average brickwork?

13.6 Show that the average shear stress along the surfaces of a rectangular open channel is

$$(\tilde{p}_{zx})_o = mR_h\gamma$$

where m is the slope of the channel. (See Fig. P13.6.)

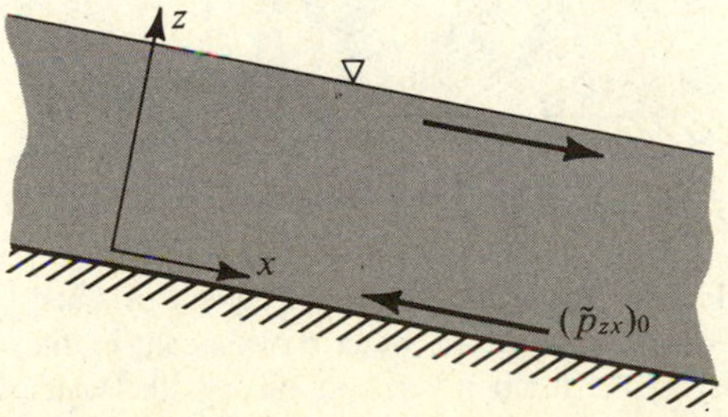

Figure P13.6

13.7 In conjunction with Prob. 13.6, calculate the average shear stress on a rectangular channel's surface if the channel is 5 m wide and 2 m deep and possesses a slope of 1 m in every 1000 m length (see Fig. P13.7).

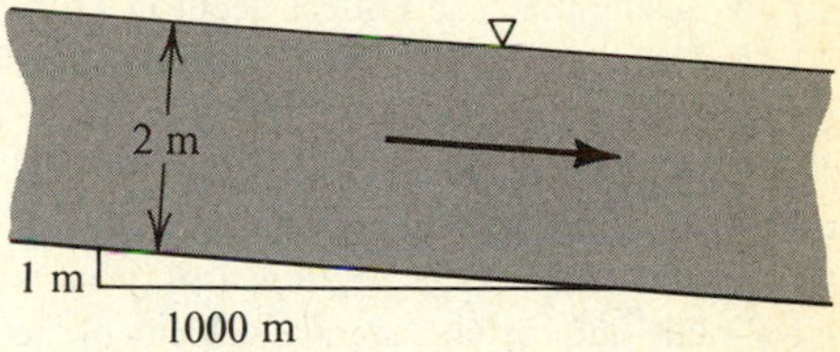

Figure P13.7

13.8 Derive a relationship between friction factor f and the Manning roughness coefficient n, $f = 116\, n^2/R_h^{1/3}$. What assumptions are necessary to obtain this relationship?

13.9 In reference to the channel of Fig. 13.3, suppose the width b_2 of the trapezoidal channel is $2h$. For a slope $\theta = 10°$, calculate (a) the hydraulic radius R_h, and (b) the depth h if the channel has a flow rate of 100 ft³/s and is constructed out of large smooth stones.

13.10 In reference to the channel of Fig. 13.3, suppose the width b_2 of the trapezoidal channel equals the depth h. For a slope of 1 m/1000 m, calculate (a) the hydraulic radius R_h, and (b) the depth h if the channel has a flow rate of 50 m³/s and is a poorly maintained earth canal. Let $\theta = 60°$.

13.11 A rectangular channel 10,000 ft long is constructed out of average brickwork. It has a slope of 1 ft/100 ft through which 50 ft³/s of water flows. If there are 6 bricks in a 1 ft² area, how many bricks are needed to cover the wetted area?

13.12 What is the best cross-sectional area and the best depth to move 1000 m³/s in an open channel where the slope $m = 0.003$, and the roughness coefficient $n = 0.020$?

13.13 Calculate the flow rate in a 2 m wide rectangular channel with a slope of 1 m in 3000 m if the channel is lined with a thick layer of silt and the water is 1 m deep.

13.14 Hydraulic engineers need to model the flow in a rectangular channel that is 2 m wide, 1 m deep, and has a slope of 0.0005. They measure the flow rate and find it is 5 m^3/s. What is the surface of the channel?

13.15 Calculate the depth h of a channel flow having a specific energy E.G.L. = 10 ft for a flow rate of 200 ft^3/s if the channel is 10 ft wide.

13.16 Determine the critical depth in Prob. 13.15. Draw the E.G.L. versus depth h curve for this flow.

13.17 Calculate the critical depth for 100 m^3/s of water through a rectangular channel of width 5 m. If the water depth is 4 m, what is the Froude number and average velocity of the water?

13.18 What is the most efficient geometry for a trapezoidal channel with many boulders and water foliage that transports 500 ft^3/s where the angle of the sloping sides are 63.43° (Fig. P13.18).

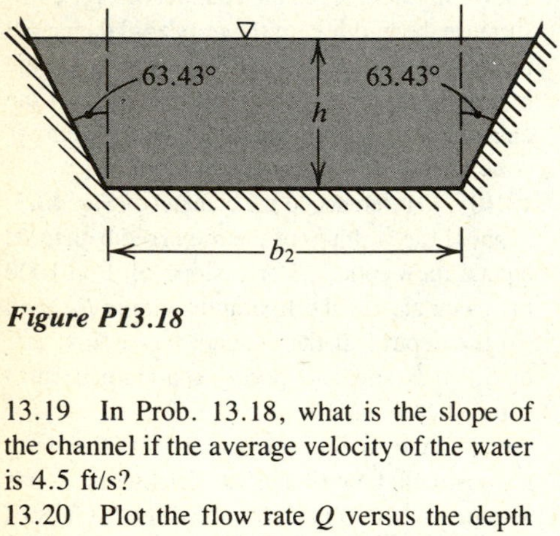

Figure P13.18

13.19 In Prob. 13.18, what is the slope of the channel if the average velocity of the water is 4.5 ft/s?

13.20 Plot the flow rate Q versus the depth h for a constant specific energy E.G.L., and show the region of $F_r > 1$, $F_r < 1$ and where the critical condition exists.

13.21 A 0.5 m diameter drainage pipe is to be laid to handle 1 m^3/s when it is half full. (a) What slope should the pipe be laid if the pipe is constructed of reinforced cement with plaster? (b) Does the slope need to be changed if drainage fills the pipe?

13.22 A pipe lined with copper rests on a slope m = 0.001. It is designed to transport valuable petroleum minerals at a volume rate of flow of 5 ft^3/s at 75% full. Determine the necessary pipe diameter.

13.23 Water flows in an open trapezoidal channel shown in Fig. P13.23 at a depth of 4 m. The channel floor is 5 m wide and has sloping sides at 45° angles. Calculate (a) the slope m of the channel, and (b) the critical depths for a volume rate of flow of 100 m^3/s, given n = 0.025.

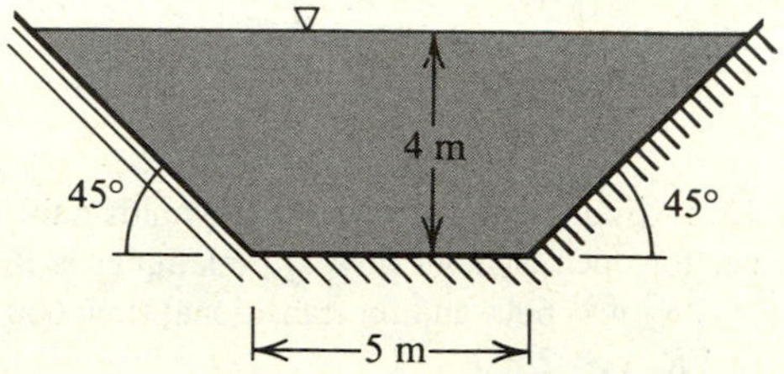

Figure P13.23

13.24 Water flows in a rectangular aqueduct that is 10 m wide. (a) Determine the specific energy E.G.L. for a volume rate of flow of 10 m^3/s if the depth is 1 m. (b) Determine whether the channel flow is subcritical or supercritical.

13.25 As shown in Fig. P13.25, a sluice gate allows water to flow from a reservoir into a 1 m wide channel. If the reservoir is 10 m deep, and the loss coefficients are C_c = 0.85, C_v = 0.97, and C_d = 0.82, calculate the volume of fluid flowing through the gate in 1 minute if the gate is 1 m off the floor. Assume gate width is 1 m.

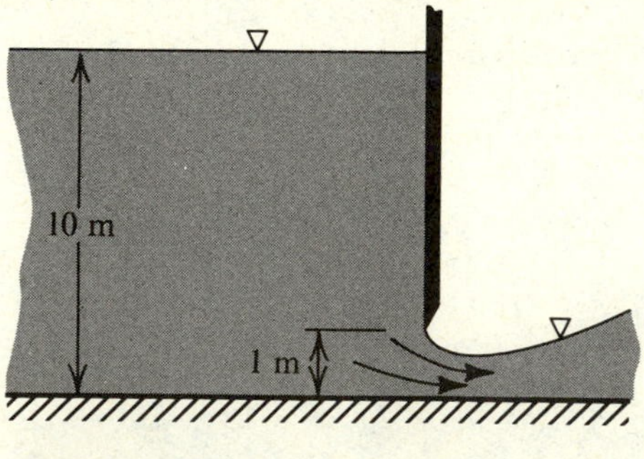

Figure P13.25

13.26 Water in a flume 1 ft wide flows at 5 ft^3/s. The depth of water downstream of the hydraulic jump is 2 ft. (a) What is the water depth upstream of the jump, and (b) what is the loss of energy, h_f?

13.27 Water in a flume 0.5 m wide flows at a rate of 1 m^3/s. The depth of the water upstream of the hydraulic jump is 0.1 m. (a) What is the water depth downstream of the jump, and (b) what is the loss of energy, h_f?

13.28 A rectangular weir is placed in an open channel that is 6 ft high and 6 ft wide. Calculate the volume rate of flow of water if the surface of the water is 1 ft above the top of the weir and the average velocity upstream of the weir is 0.5 ft/s. Assume $\overline{C}_d = 0.63$.

13.29 Some hydraulic engineers use the relationship

$$Q = \frac{2}{3}\overline{C}_d\, b\sqrt{2g}\,H_o^{3/2}$$

of Eq. (13.55) where $\overline{C}_d = 0.577$ for a broad-crested weir. Estimate the volume rate of flow per unit width of a weir in a horizontal channel that is 15 ft wide and 6 ft high, given the elevation H_o is 0.75 ft for the case (a) the weir is sharp crested, and (b) the weir is broad crested.

13.30 A wave propagates along the surface of a basin of water of depth h. At the basin floor, a sensitive pressure gauge detects the pressure fluctuation. Show that the pressure records

$$p = \gamma h + \frac{\eta}{\cosh \dfrac{2\pi h}{L}}$$

13.31 Consider shallow water waves. Show that if $h/L \rightarrow 0$, then Eq. (13.64) can be written as

$$\frac{1}{g}\frac{\partial^2\phi}{\partial t^2} = h\frac{\partial^2\phi}{\partial x^2}$$

13.32 A wave of wavelength L and maximum wave amplitude a travels along the surface of a deep water basin. (a) Find the equation of the free-surface $\eta = \eta(x, t)$ for the case $a \ll L$.

13.33 In Prob. 13.32 show that the behavior of the fluid particles below the free-surface move out in circular paths.

13.34 A wave of 0.5 m amplitude has a wavelength of 50 m. (a) What is the speed of the wave? (b) What is the speed of a fluid particle on the surface of the water?

14 Boundary Layer Flows

14.1 Introduction

All fluids have viscosity, that stickiness we are familiar with when we touch honey or molasses. Just as we pick up a delicious film when we dip into honey, so an airplane moving through air is sheathed with a film of air that is forever dragged along on its flight. A submarine is cloaked with a similar mantle, only thicker because of the submarine's slower speed in a fluid more viscous than air. There is nothing at all mysterious about this mantle: we define it as the region where shear stresses exist. Outside this layer, therefore, the shear stresses will be zero.

Take some oil and pour it on a flat horizontal board. Then lift the board slowly to a near vertical position and study the resultant flow of oil. You will see the frictional stresses that act whenever a viscous fluid moves over a surface. The stresses work their way from the boundary into the flow to retard it. Thus we would expect a loss of energy from the flow. To maintain a given flow, we must resupply energy, most often from some upstream source like a pump.

Prandtl showed that the differences in velocity of the fluid particles between the object and a moving fluid were confined to a region he called *Grenzschicht*, or *boundary layer*. Outside of this layer, the fluid flows as if there were no frictional resistance. Within the layer the distribution of velocity is similar to that shown in Fig. 14.1. Figure 14.1a is a typical laminar flow velocity profile, Fig. 14.1b is a typical turbulent flow velocity profile, Fig. 14.1c depicts conditions for separation, and Fig. 14.1d shows a profile of the separated region and the wake. The region between the dash-line and the wing surface represents the boundary layer.

The boundary layer is defined as the region where the shear stresses are finite. It is the shear stress on the surface that produces the friction drag force* on the body. The drag force due to friction is quite simple to calculate if the flow is laminar. It is not so simple, however, to calculate the drag force for turbulent flow.

We know that the fluid flow of a continuous medium is fully described by the Navier-Stokes equation. Only a few explicit solutions of these equations are known. In many cases it is extremely difficult even to determine approximations. We want to consider the following very simple problem: can we solve the Navier-Stokes equation for the two-dimensional steady flow of an incompressible fluid around a simply con-

*The friction drag is only one aspect of the total drag on an object. For an infinitesimally thin flat plate at zero angle-of-attack, the friction drag is the total drag. For thick bodies, we have to consider other forms of drag such as pressure (force) drag. In supersonic flow, the drag due to normal stresses is called wave drag.

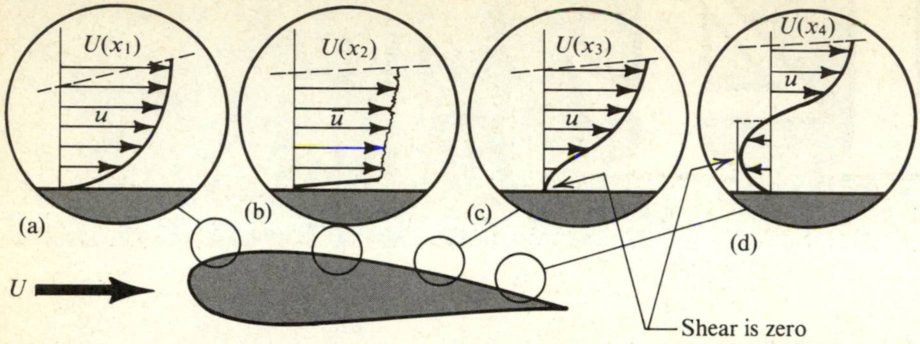

Figure 14.1 Velocity distributions in the boundary layer over an airfoil section. (Source: Lectures on viscous flows by Professor T. Sarpkaya. Reproduced here by written permission of Professor T. Sarpkaya.)

nected finite body which has a smooth continuous boundary? To solve this problem Prandtl simplified the Navier-Stokes equation by omitting certain terms that were small when compared to other terms. The result of his analysis, the *boundary layer* equations, are simpler than the Navier-Stokes equation, and need to be discussed in some detail.

14.1.1 Reynolds' Experiment

In order to gain some comprehension of what transpires in the transition from laminar to turbulent flow, we construct the classical experiment performed by Osborne Reynolds. Reynolds was the first to distinguish laminar from turbulent flow, showing that the fluid often passes from one into the other, just as a river moves from quiet pools to white rapids back to languid flows. Reynolds conducted his experiments on the flow of water through a glass tube in a manner shown in Fig. 14.2.

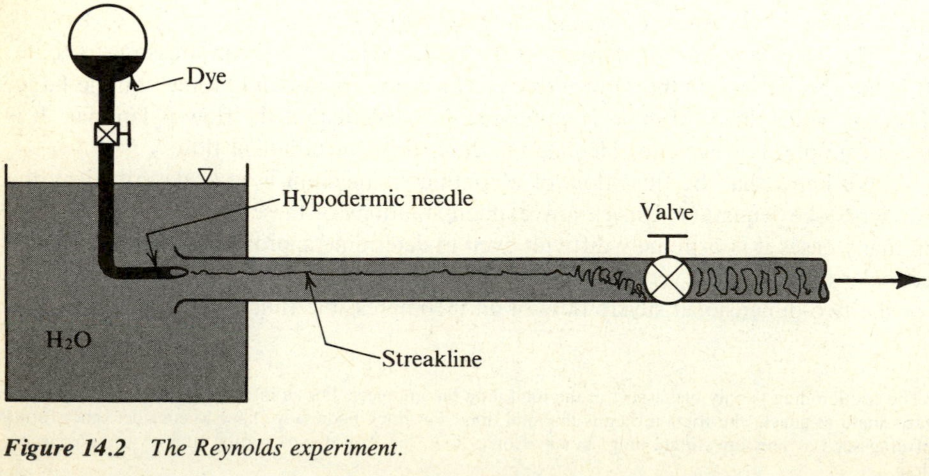

Figure 14.2 The Reynolds experiment.

A glass tube is mounted horizontally with one end placed in a large reservoir and the other end open to the atmosphere. The flow rate through the tube is controlled by a valve. A smooth bell mouth entrance is provided at the entrance to the tube along with a dye jet arrangement to allow a fine stream of dye to be injected into the stream and enter the tube with minimal disturbances. For low velocity flows the dye streakline moves, generating a straight line through the tube in a steady stable manner. Such a behavior is characteristic of a *laminar flow* through a pipe. Now, if the flow rate is gradually increased, the dye streakline will begin to waver until suddenly it bursts. The streakline is diffused downstream of the burst, indicating that the fluid and the dye are mixing. When this condition is reached, the flow has changed to *turbulent flow*, the chief characteristic of which is a violent interchange of macroscopic momentum that completely disrupts the orderly movement observed when the flow was laminar. (The photographs of Fig. 11.6 show this clearly.) By carefully controlling the experiment, Reynolds was able to obtain a value of Reynolds number $R_D = 12{,}000$ before turbulence was seen to occur. Many experimentalists since Reynolds have reproduced his results. Some have been able to maintain laminar flow up to Reynolds number 50,000. These upper critical Reynolds numbers have no practical significance in engineering design, since in the majority of cases, we cannot isolate the fluid flow from disturbances (such as external vibrations and inherent flow turbulence) that cause the flow to trip from laminar into turbulent flow.

On the other hand, the lower critical Reynolds number is of practical utility in fluid dynamics design. It is the highest value of the Reynolds number that can be tolerated and still allow the external disturbances (which might precipitate turbulence) to be dampened out. Even though transition occurs in the pipe between $R_D = 2300$ and $R_D = 4000$, we will assume in our calculations the lower value of $R_D = 2300$ to be the upper limit of laminar flow in a pipe.

In the laminar flow regime the pressure losses are directly proportional to the average velocity, whereas in turbulent flow the losses are proportional to the velocity to a power varying from 1.7 to 2.0. All this takes place in an internal flow. An analogous phenomenon occurs in boundary layer flows.

14.2 The Boundary Layer Concept

In 1904 Prandtl originated the concept of a boundary layer. He hypothesized that for fluids having relatively small viscosity, the effect of internal friction in the fluid is significant only in a narrow region surrounding solid boundaries or bodies where the fluid adheres to the boundary. Thus, close to a body in a region called the boundary layer is where shear stresses exert an increasingly larger effect on the fluid as we move toward the solid boundary, because of the increased velocity gradient $\partial u/\partial y$ as $y \rightarrow 0$. But outside the boundary layer where the effect of the shear stresses on the flow is small compared to values inside the boundary layer (since the velocity gradient $\partial u/\partial y$ is negligible), the fluid particles experience no vorticity, and the flow is similar to potential flow. Hence, the "surface" at the boundary layer is a rather fictitious surface dividing a rotational and irrotational flow. Note that *fluid can pass through this "surface" of boundary layer.*

The classical procedure that illustrates many of the characteristics of the boundary layer is to place a semi-infinite flat plate in a uniform stream of infinite extent. Such a geometry is shown in Fig. 14.3.

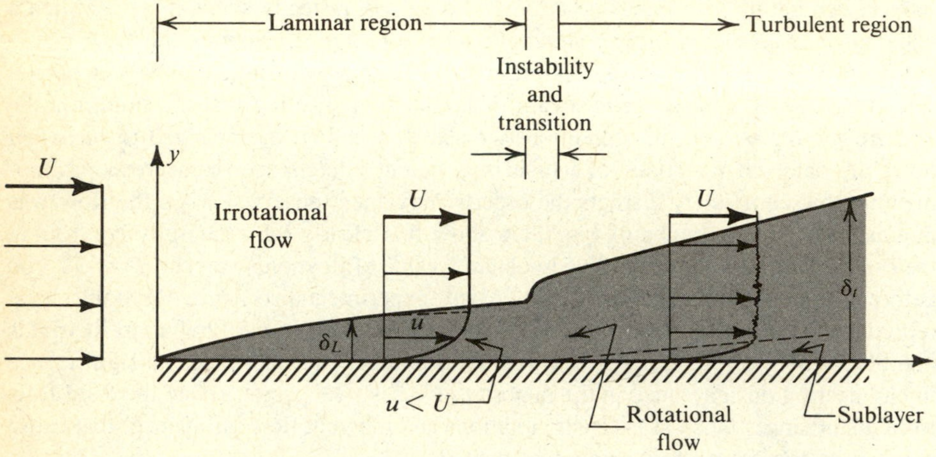

Figure 14.3 *Detail of flow over a flat plate.*

Experimental observations indicate that the fluid on the plate has zero velocity. As the vertical distance increases, so does the velocity, which attains very nearly the free stream velocity a short distance, δ, away from the plate. As with Reynolds' pipe flow experiment, the boundary layer has a region where the flow can become turbulent, as indicated in Fig. 14.3. (Experiments indicate that the flow in the boundary layer on a flat plate becomes turbulent at a Reynolds number $R_x = Ux/\nu$ between 5×10^5 and 10^7.)

Figure 14.3 shows three distinct regions of flow: laminar, transitional, and turbulent flow. To see how a fluid parcel behaves in these three zones, consider a rectangular fluid parcel following the streamline in Fig. 14.4. Outside the boundary layer, the parcel moves in pure translation since the stresses on the bottom and top faces are identical. In the laminar boundary layer, the fluid parcel follows a rectilinear path, but, since unequal shear stresses cause the upper face to move faster than the lower face the parcel deforms, and we observe both translation and rotation. In the transition region, the path changes from rectilinear to curvilinear. Oscillations build up, and the path becomes unstable. When the parcel reaches the turbulent region, the path is undefined and three-dimensional, and the parcel rotates unpredictably.

The boundary layer thickness of each region is different, and so one of the first tasks before us is to estimate the thickness of the laminar region. This can be accomplished quite simply by considering the boundary layer as that region where the viscous force per unit volume is of the same order of magnitude as the inertial force per unit volume. The thickness of the boundary layer is denoted by δ, and is in the y-direction. Since we are concerned solely with the boundary layer of the flow, the order of

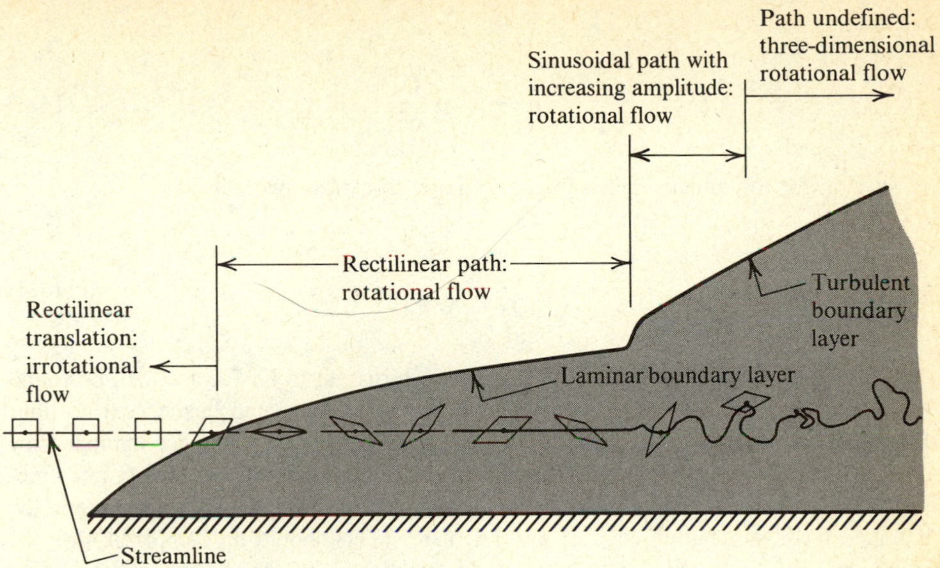

Figure 14.4 *Behavior of a fluid particle traveling along a streamline through a boundary layer along a flat plate.*

magnitude of y is δ. The largest value of the velocity component u is the free-stream velocity U. The inertial force per unit volume is $\rho u(\partial u/\partial x)$, and for laminar flow the viscous force per unit volume is

$$\frac{\partial p_{xy}}{\partial y} = \mu \frac{\partial^2 u}{\partial y^2}$$

if the flow is parallel. Making the following estimations on orders of magnitude:

$$u \sim U, \; y \sim \delta \tag{14.1}$$

$$\frac{\partial u}{\partial x} \sim \frac{U}{x}$$

$$\frac{\partial u}{\partial y} \sim \frac{U}{\delta}$$

$$\frac{\partial^2 u}{\partial y^2} \sim \frac{U}{\delta^2} \tag{14.2}$$

and setting

$$\frac{\partial u}{\partial x} = \mu \frac{\partial^2 u}{\partial y^2}$$

we get

$$\rho \frac{U^2}{x} \sim \mu \frac{U}{\delta^2} \tag{14.3}$$

or solving for the dimensionless boundary layer thickness, we get

$$\frac{\delta}{x} \sim \sqrt{\frac{\mu}{\rho U x}} \sim \frac{1}{\sqrt{R_x}} = \frac{k}{\sqrt{R_x}} \tag{14.4}$$

The constant of proportionality k of Eq. (14.4) turns out to be 5.0 and will be found by solving the laminar boundary layer equation to be presented in the next section. Examination of Eq. (14.4) shows that the boundary layer thickness for laminar flow over a flat plate is inversely proportional to the square root of the density and free-stream velocity, and proportional to the square root of the dynamic viscosity and the distance from the leading edge of the flat plate. (This should be compared with the result obtained for the suddenly accelerated flat plate, Eq. (9.24).) Thus, at the leading edge of the flat plate, we expect and obtain zero thickness. Also, the more viscous a fluid is, the thicker the boundary layer. Thus at supersonic and hypersonic speeds, imagine the tremendous *magnitude* of the shear stresses that move the fluid from zero velocity to the velocity of the body. And these stresses must act in an infinitesimally thin region no thicker than a hair filament.

14.3 Prandtl's Boundary Layer Equations

Let us now discuss the approximate equations of motion for a two-dimensional laminar boundary layer. We make this approximation by estimating the order of magnitude of each term in the Navier-Stokes equations. The result will give us the Prandtl boundary layer equations, first published by Prandtl [1.29] in 1904. The following assumptions are made:

1. The motion is two-dimensional, and lies in a horizontal plane with the y-axis normal to the plate as shown in Fig. 14.3: thus gravitational effects are neglected.
2. The flow is laminar within the boundary layer.
3. The boundary layer thickness is much smaller than the length of the plate L, i.e., $\delta^* = \delta/L \ll 1$.
4. The flow is steady.

We start the discussion with the x and y components of the dimensionless Navier-Stokes equations as given by Eq. (7.26). We recall that our dimensionless independent variables are $x^* = x/L$, $y^* = y/L$, and $z^* = z/L$, and are based on a characteristic length L that is constant. Our dimensionless dependent variables are $u^* = u/U$, $v^* = v/U$, $w^* = w/U$, and $p^* = p/\rho U^2$, and are based on a characteristic velocity U

that is constant. Using the above dimensionless variables, we reduce the dimensionless Navier-Stokes equation (7.26) to

$$u^* \frac{\partial u^*}{\partial x^*} + v^* \frac{\partial u^*}{\partial y^*} = -\frac{\partial p^*}{\partial x^*} + \frac{1}{R}\left(\frac{\partial^2 u^*}{\partial x^{*2}} + \frac{\partial^2 u^*}{\partial y^{*2}}\right) \tag{14.5}$$

$$u^* \frac{\partial v^*}{\partial x^*} + v^* \frac{\partial v^*}{\partial y^*} = -\frac{\partial p^*}{\partial y^*} + \frac{1}{R}\left(\frac{\partial^2 v^*}{\partial x^{*2}} + \frac{\partial^2 v^*}{\partial y^{*2}}\right) \tag{14.6}$$

and the dimensionless continuity equation is

$$\frac{\partial u^*}{\partial x^*} + \frac{\partial v^*}{\partial y^*} = 0 \tag{14.7}$$

Use of Eq. (14.1) reveals that the orders of magnitude of $\partial u^*/\partial x^*$ and $\partial^2 u^*/\partial x^{*2}$ are 1, which means that $\partial v^*/\partial y^*$ must have an order of magnitude of 1 in order to satisfy continuity. Now, since v^* equals zero on the boundary, then $v^* = \int_o^{\delta^*} 1\, dy = \delta^*$. Thus the dimensionless y-component velocity has an order of magnitude δ^*. Also $\partial v^*/\partial x^*$ and $\partial^2 v^*/\partial x^{*2}$ have orders of magnitude δ^*. From Eq. (14.2) $\partial u^*/\partial y^*$ and $\partial^2 u^*/\partial y^{*2}$ have orders of magnitude of $1/\delta^*$ and $1/\delta^{*2}$, respectively. The dimensional pressure gradient $\partial p/\partial x$ is assumed to be known in advance from Bernoulli's equation applied to the outer inviscid flow

$$\frac{\partial p}{\partial x} = \frac{dp}{dx} = -\rho U \frac{dU}{dx} \tag{14.8}$$

The distribution of $U(x)$ along a surface is known from the inviscid analysis described in Chap. 12. Thus $\partial p^*/\partial x^*$ is retained since its order of magnitude is 1.

If we insert these orders of magnitude into Eqs. (14.5), (14.6), and (14.7), we obtain

$$u^* \frac{\partial u^*}{\partial x^*} + v^* \frac{\partial u^*}{\partial y^*} = -\frac{\partial p^*}{\partial x^*} + \frac{1}{R}\left(\frac{\partial^2 u^*}{\partial y^{*2}}\right) \tag{14.9}$$

$$\frac{\partial p^*}{\partial y^*} = 0(\delta^*) \cong 0 \tag{14.10}$$

Equation (14.10) states that the pressure across the boundary layer does not change. The pressure is impressed on the boundary layer, and its value is determined by hydrodynamic considerations. This is all true only if the flow does not separate, and it will not separate if the flow is past a flat plate with no wall transpiration. Transforming back to the dimensional variables $(u, v, p; x, y)$, we obtain the Prandtl boundary layer equations

$$u \frac{\partial u}{\partial x} + v \frac{\partial u}{\partial y} \approx U \frac{dU}{dx} + \nu \frac{\partial^2 u}{\partial y^2} \tag{14.11}^*$$

$$\frac{\partial p}{\partial y} = 0 \tag{14.12}$$

$$\frac{\partial u}{\partial x} + \frac{\partial v}{\partial y} = 0 \tag{14.13}$$

subject to the following necessary and sufficient boundary conditions:

- No slip at the wall: $u = v = 0$ at $y = 0$ (14.14a)
- Patching: $u \to U$ as $y \to \delta$ (14.14b)

We solve the Prandtl boundary layer equations for $u(x, y)$ and $v(x, y)$ with $U(x)$ known from the outer inviscid flow analysis. The equations are solved by starting at the leading edge of the body and moving downstream to the separation point.**

Note that the remaining momentum Eq. (14.11) is still nonlinear. However, it does allow the no-slip boundary condition to be satisfied which constitutes a significant improvement over potential flow analysis in the solution of real fluid flow problems. The Prandtl boundary layer equations are thus a simplification of the Navier-Stokes equations. They can be regarded as *asymptotic equations of the Navier-Stokes equations in the limit of vanishing viscosity*.

Example 14.1

Using the Prandtl boundary layer Eq. (14.11), show that the velocity profile for a laminar flow past a flat plate has an infinite radius of curvature on the surface of the plate.

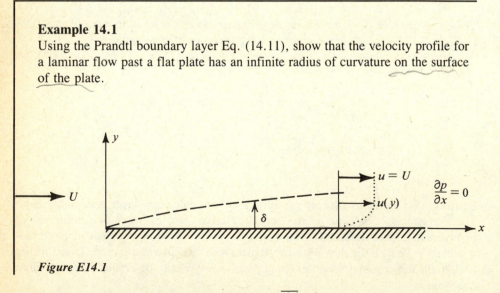

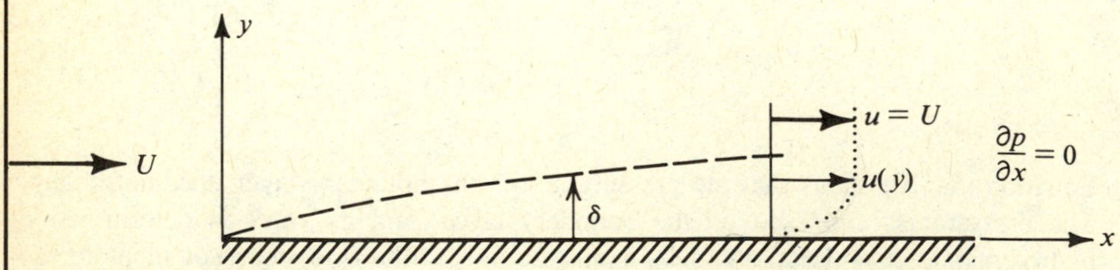

Figure E14.1

*For turbulent flow we add the turbulent acceleration $-[\partial(\overline{u'v'})/\partial y]$.

**See Ref. 14.1 for details of the mathematical analysis.

Example 14.1 *(Con't.)*

Solution:

The radius of curvature ρ of the distribution of velocity $u(y)$ is that used in the calculus:

$$\rho = \frac{[1 + (du/dy)^2]^{3/2}}{|d^2u/dy^2|} \tag{i}$$

The boundary conditions at the surface of the flat plate are

$$u = v = 0 \text{ at } y = 0 \tag{ii}$$

Substituting the above boundary conditions into the Prandtl boundary layer Eq. (14.11) yields

$$u\frac{\partial u}{\partial x} + v\frac{\partial u}{\partial y} = -\frac{1}{\rho}\frac{\partial p}{\partial x} + \nu\frac{\partial^2 u}{\partial y^2} \tag{iii}$$

resulting in

$$\frac{\partial^2 u}{\partial y^2} = 0 \text{ at } y = 0 \tag{iv}$$

Substituting the gradient of the shear stress of Eq. (iv) into the expression for the radius curvature of Eq. (i) gives

$$\rho = \infty \tag{v}$$

which means that very close to the surface of the plate, the velocity is linear and the shear stress is constant.

This completes the solution.

Example 14.2

Reduce the Prandtl boundary layer equations to a simpler form than that given by Eqs. (14.11)–(14.13) for (a) flow over a flat plate, (b) the case $p_{yx} = c_1$ (a constant), and (c) the case where $v \propto \nu$. (d) Solve the Prandtl boundary layer equations for the special case $v = \nu$ and where the pressure gradient $\partial p/\partial x$ is zero.

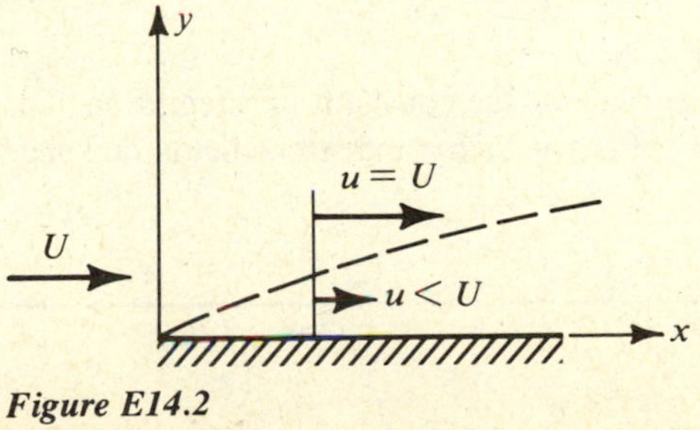

Figure E14.2

Example 14.2 *(Con't.)*

Solution:

(a) For flow past a flat plate, $\partial p/\partial x = 0$, Eq. (14.11) reduces to

$$u \frac{\partial u}{\partial x} + v \frac{\partial u}{\partial y} = \nu \frac{\partial^2 u}{\partial y^2} \tag{i}$$

Equation (i) is the partial differential equation H. Blasius solved for his Ph.D. dissertation in Göttingen, Germany [14.2].

(b) For the constant shear stress case, Eq. (14.11) reduces to

$$u \frac{\partial u}{\partial x} + \frac{c_1}{\mu} v = -\frac{1}{\rho} \frac{\partial p}{\partial x} \tag{ii}$$

which can be altered to yield

$$p + \frac{1}{2} \rho u^2 = -\frac{c_1}{\nu} \int v dx \tag{iii}$$

Thus the total pressure can be determined if we know how the y component of velocity v varies in the flow.

(c) For the case $v \propto \nu$, the Prandtl boundary layer equations reduce to

$$\frac{d^2 u}{dy^2} - a \frac{du}{dy} = -\frac{1}{\mu} \frac{dp}{dx} \tag{iv}$$

where a is a constant of proportionality. We note that the left-hand side is a function of y, and the right-hand side is a function of x, which signifies both the left- and right-hand side terms are constant.

(d) Setting dp/dx to zero, Eq. (iv) reduces to

$$\frac{d(du/dy)}{du/dy} = a dy \tag{v}$$

Integrating Eq. (v) and taking antilogs of both sides yields

$$\frac{du}{dy} = c_1 \exp(ay) \tag{vi}$$

Integrating Eq. (vi) yields

$$u = \frac{c_1}{a} \exp(ay) + c_2 \tag{vii}$$

We need two boundary conditions to evaluate the constants of integration. These would stem from known conditions of the velocity and stress fields on specific boundaries that define the flow.

This completes the solution.

14.4 Blasius Solution for Laminar Boundary Layer Flow over a Flat Plate

H. Blasius [14.2] in 1908 first treated the incompressible steady flow over a flat plate as an explicit solution of the Prandtl equations. Note that *Blasius does not solve the Navier-Stokes equation*. The *practical* success of the Blasius solution of boundary layer theory was staggering. Almost everyone overlooked the fact that the mathematical basis of the boundary layer theory was vague, and for over 50 years fundamental mathematical questions could not be answered. There was no sound mathematical connection between Prandtl's equations and the Navier-Stokes equations. For example, to show that equations are self-sufficient, we must establish existence, uniqueness, and the well-posedness of a solution. But no one could prove any convergence of the numerical approximation; no one could define errors or even their bound set. So let us look at precisely what Blasius did accomplish.

The classical problem Blasius considered was a two-dimensional steady, incompressible flow over a flat plate at zero angle of incidence with respect to the uniform oncoming stream of velocity U. The fluid extends to infinity in all directions from the plate. The physical problem is shown in Fig. 14.3.

Blasius wanted to determine the velocity field solely within the boundary layer, the boundary layer thickness δ, the shear stress distribution on the plate, and the drag force on the plate.

Since the plate is flat and has negligible thickness and the stream velocity U is uniform, the pressure gradient $\partial p/\partial x$ must vanish. [This can be seen from Euler's equation (4.110).] The resultant Prandtl boundary layer equation from Eq. (14.11) is then

$$u\frac{\partial u}{\partial x} + v\frac{\partial u}{\partial y} = \nu\frac{\partial^2 u}{\partial y^2} \tag{14.15}$$

and the continuity equation is

$$\frac{\partial u}{\partial x} + \frac{\partial v}{\partial y} = 0 \tag{14.13}$$

The problem is thus well posed: two equations, two unknowns with boundary conditions given by Eq. (14.14).

To solve this problem, Blasius first sought to reduce the number of variables. The continuity equation can be satisfied by the stream function ψ as

$$u = \frac{\partial \psi}{\partial y} \tag{14.16}$$

$$v = -\frac{\partial \psi}{\partial x} \tag{14.17}$$

Substituting the stream function-velocity relationships into Prandtl's boundary layer Eq. (14.15) yields

$$\frac{\partial\psi}{\partial y}\frac{\partial^2\psi}{\partial x\partial y} - \frac{\partial\psi}{\partial x}\frac{\partial^2\psi}{\partial y^2} = \nu\frac{\partial^3\psi}{\partial y^3} \tag{14.18}$$

which is a third-order, nonlinear, partial differential equation with now a single unknown function, ψ.

A large class of boundary layer problems is solved by transforming the partial differential equation into an ordinary differential equation. Recall that this technique allowed a solution of the accelerated flat plate problem. If, in fact, *our partial differential equation can be transformed into an ordinary differential equation then the dependent variable must be a function of a single variable*, i.e.,

$$\begin{aligned} u &= UF(\eta) \\ v &= UG(\eta) \end{aligned} \tag{14.19}$$

where $F(\eta)$ and $G(\eta)$ are dimensionless similarity functions and η a dimensionless independent similarity variable.

Blasius defined a similarity variable η as

$$\eta = \frac{y}{\delta_o} \tag{14.20}$$

where δ_o is proportional to the boundary layer thickness δ and is a function of x only. This means geometrically that all velocity profiles are similar.

If we set $\psi = 0$ on the plate, which is equivalent to $\eta = 0$, then the stream function ψ is evaluated from continuity as

$$\psi = \int_o^y u dy \tag{14.21}$$

Blasius expressed the above integrand in terms of the similarity function with the result

$$\psi = U\int_o^\eta F(\eta)\delta_o d\eta = U\delta_o\int_1^\eta F(\eta)d\eta \tag{14.22}$$

or

$$\psi = U\delta_o f(\eta) \tag{14.23}$$

where

$$f(\eta) = \frac{dF}{d\eta} = F' \tag{14.24}$$

From the chain rule of the calculus, we obtain from Eq. (14.23)

$$-v = \frac{\partial\psi}{\partial x} = U\delta_o' f - U\delta_o' f'\eta \tag{14.25a}$$

$$u = \frac{\partial\psi}{\partial y} = Uf' \tag{14.25b}$$

$$\frac{\partial u}{\partial y} = \frac{\partial^2\psi}{\partial y^2} = \frac{Uf''}{\delta_o} \tag{14.25c}$$

$$\frac{\partial u}{\partial x} = \frac{\partial^2\psi}{\partial x\partial y} = -\frac{U\delta_o'}{\delta_o}\eta f'' \tag{14.25d}$$

$$\frac{\partial^2 u}{\partial y^2} = \frac{\partial^3\psi}{\partial y^3} = \frac{Uf'''}{\delta_o^2} \tag{14.25e}$$

where the symbols δ_o', f' are defined as

$$\delta_o' = \frac{d\delta_o}{dx} \tag{14.26}$$

$$f' = \frac{df}{d\eta} \tag{14.27}$$

Substituting the above relationships for the similarity variables into the transformed Prandtl layer Eq. (14.18) gives

$$f''' + \frac{U}{\nu}\delta_o\delta_o' ff'' = 0 \tag{14.28}$$

Since η and x are independent variables for similar profiles, then the coefficient $(U/\nu)\,\delta_o\delta_o'$ must be a constant, i.e.,

$$\frac{U}{\nu}\delta_o\frac{d\delta_o}{dx} = C \tag{14.29}$$

Integrating Eq. (14.29) results in

$$\frac{U}{\nu}\frac{\delta_o^2}{2} = Cx + C_1 \tag{14.30}$$

Choosing a coordinate origin such that $\delta_o = 0$ at $x = 0$ results in $C_1 = 0$, and thus from Eq. (14.30), the boundary layer thickness δ_o is found to be

$$\delta_o = \sqrt{2C\frac{x\nu}{U}} \tag{14.31}$$

Since the constant C can be any value, Blasius conveniently chose $C = ½$, such that the similarity variable η was defined as

$$\eta = y\sqrt{\frac{U}{\nu x}} \tag{14.32}$$

The Prandtl boundary layer equation for uniform flow over a flat plate becomes now an ordinary nonlinear differential equation:

$$\boxed{\frac{d^3f}{d\eta^3} + \frac{1}{2}f\frac{d^2f}{d\eta^2} = 0} \tag{14.33}$$

The transformed boundary conditions are

$$\left.\begin{aligned}\frac{df}{d\eta} &= 0\\ f &= 0\end{aligned}\right\} \text{for } \eta = 0 \tag{14.34}$$

from $u(x, 0) = v(x, 0) = 0$, *respectively, and*

$$\frac{df}{d\eta} \rightarrow 1 \text{ as } \eta \rightarrow \infty \tag{14.35}$$

from $u(x, y) \rightarrow U(x)$ as $y \rightarrow \delta$. Since we do not know $\delta(x)$ at this point of development, we modify the condition at $y = \delta$ to be $u(x, \infty) = U(x)$.

To solve the boundary value problem thus posed, Blasius assumed a series for the similarity function f in power series form

$$f(\eta) = A_o + A_1\eta + \frac{A_2}{2!}\eta^2 + \frac{A_3}{3!}\eta^3 + \ldots \frac{A_m}{m!}\eta^m \tag{14.36}$$

where the A_i are undetermined coefficients to be determined from the boundary conditions of Eqs. (14.34) and (14.35). Since there are only three independent boundary conditions, only three coefficients of A_i can be found, and all other coefficients must be related to these three coefficients. It is quite simple to show that the first two boundary conditions given by Eqs. (14.34) require that the coefficients $A_o = A_1 = 0$.

Substituting the power series expression for the similarity function f of Eq. (14.36) into Prandtl's Eq. (14.33), Blasius obtained after some rather simple algebra the series expression

$$2A_3 + 2A_4\eta + (A_2^2 + 2A_5)\frac{\eta^2}{2!} + (4A_2A_3 + 2A_6)\frac{\eta^3}{3!} + \ldots = 0 \quad (14.37)$$

Since this last expression must be true at all points in the boundary layer, all the above coefficients must vanish. Thus

$$A_3 = A_4 = A_6 = A_7 = A_9 = A_{10} = \ldots = 0 \quad (14.38)$$

and

$$A_5 = -\frac{1}{2}A_2^2, \qquad A_8 = -\frac{11}{2}A_2, \qquad A_5 = \frac{11}{4}A_2^3 \quad (14.39)$$

A recursion formula can be generated so that the series for the similarity function f can be rewritten as

$$f(\eta) = \sum_{m=0}^{\infty}\left(-\frac{1}{2}\right)^m \frac{A_2^{m+1}\,C_m}{(3m + 2)!}\,\eta^{3m+2} \quad (14.40)$$

where $C_o = 1, C_1 = 1, C_2 = 11, C_3 = 375, C_4 = 27{,}897, C_5 = 3{,}817{,}137, \ldots$.

A_2 is the only unknown coefficient in the power series expression and must be determined from the last boundary condition of Eq. (14.35). We do this by setting $f = \Phi(\eta)$ when $A_2 = 1$, i.e.,

$$\Phi(\eta) = \frac{\eta^2}{2} - \frac{\eta^5}{2\cdot 5!} + \frac{C_2\eta^8}{4\cdot 8!} - \frac{C_3\eta^{11}}{8\cdot 11!} + \frac{C_4\eta^{14}}{16\cdot 14!} - \ldots \quad (14.41)$$

This leads to the result

$$f(\eta) = A_2^{1/3}\,\Phi(A_2^{1/3}\eta) \quad (14.42)$$

Thus, from the boundary condition at $\eta = \infty$,

$$\lim_{\eta\to\infty} f'(\eta) = A_2^{2/3}\lim_{\eta\to\infty}\Phi'(A_2^{1/3}\eta) = A_2^{2/3}\lim_{\eta\to\infty}\Phi'(\eta) = 1$$

and therefore

$$A_2 = \lim_{\eta\to\infty}\left[\frac{1}{\Phi'(\eta)}\right]^{3/2} \quad (14.43)$$

In principle then, A_2 can be determined by computing $\Phi'(\eta)$ from its series for successively increasing values of η until a sufficiently constant asymptotic value is obtained. For large values of η, however, convergence of the series is slow and we prefer to compute $\Phi'(\eta)$ by numerical integration of the differential equation. In any case, we find that $A_2 = 0.33206$.

This completes the solution for the similarity function $f(\eta)$. It is now possible to compute all other properties of the fluid. The computer program* below enables us to obtain values of the dimensionless velocity distribution $f(\eta)$, the dimensionless shear stress $f'(\eta)$, and the rate of growth of the dimensionless shear stress $f''(\eta)$. The language is Fortran.

```
      IMPLICIT REAL*8 (A-H, O-Z)
      WRITE(6,1005)
1005  FORMAT(1X,'INSERT NUMBER OF TRIAL VALUES OF
      F2PRIME(0) TO BE RUN']
      READ(5,2005)NORUNS
2005  FORMAT(I2)
      N = 0
5000  ETA = 0.0D0
      NCNT = 0
      F = 0.0D0
      DETA = 0.005D0
      F1 = 0.0D0
      WRITE(6,1000)
1000  FORMAT(///,1X,'INSERT GUESSED VALUE OF F2PRIME
      EVALUATED AT ZERO')
      READ(5,2000)F2
2000  FORMAT(F20.0)
      WRITE(6,1001)F2
1001  FORMAT(1X,////,1X,'F2PRIME(0) =',F20.14,///)
      WRITE(6,1007)
1007  FORMAT(7X,'ETA',10X,'F',11X,'FPRIME',8X,
      'F2PRIME')
      WRITE(6,2001)ETA,F,F1,F2
 200  F3 = -0.50D0*(F*F2)
      NCNT = NCNT + 1
      F4 = -0.5D0*(F1*F2 + F*F3)
      F5 = -0.5D0)*(F2*F2 + 2.0D0*F1*F3 + F*F4)
      F = F + F1*DETA + F2*DETA**2 + F3*DETA**3
      F1 = F1 + F2*DETA + F3*DETA**2 + F4*DETA**3
      F2 = F2 + F3*DETA + F4*DETA**2 + F5*DETA**3
      ETA = ETA + DETA
      IF(ETA.GT.6.0D0)GO TO 2
      IND = NCNT - (NCNT/40)*40
      IF(IND.NE.0)GO TO 200
      WRITE(6,2001)ETA,F,F1,F2
2001  FORMAT(1X,F10.5,3F15.10)
      GO TO 200
```

*From lectures on viscous flows by Professor T. Sarpkaya. Reproduced here by written permission of Professor T. Sarpkaya.

```
2 CONTINUE
  N = N + 1
  IF(N.LT.NORUNS)GO TO 5000
  STOP
  END
```

The results from the computer program are shown in Table 14.1, and the velocity component profiles are shown in Fig. 14.5.

Table 14.1 Numerical Solution of Flow Along a Flat Plate

η	$f(\eta)$	$f'(\eta)$	$f''(\eta)$
0	0	0	0.33206
0.2	0.00681	0.06641	0.33198
0.4	0.02689	0.13276	0.33145
0.6	0.06023	0.19892	0.33003
0.8	0.10676	0.26468	0.32731
1.0	0.16638	0.32971	0.32288
1.2	0.23890	0.39366	0.31641
1.4	0.32406	0.45608	0.30763
1.6	0.42151	0.51650	0.29634
1.8	0.53079	0.57440	0.28259
2.0	0.65135	0.62929	0.26636
2.2	0.78254	0.68071	0.24793
2.4	0.92362	0.72825	0.22766
2.6	1.07379	0.77158	0.20602
2.8	1.23216	0.81049	0.18359
3.0	1.39786	0.84489	0.16098
3.2	1.56998	0.87480	0.13880
3.4	1.74763	0.90037	0.11760
3.6	1.92997	0.92184	0.09787
3.8	2.11621	0.93954	0.07996
4.0	2.30565	0.95388	0.06413
4.2	2.49763	0.96527	0.05047
4.4	2.69164	0.97415	0.03896
4.6	2.88719	0.98093	0.02950
4.8	3.08393	0.98603	0.02190
5.0	3.28553	0.98977	0.01595
5.2	3.47978	0.99247	0.01139
5.4	3.67848	0.99438	0.00797
5.6	3.87750	0.99571	0.00547
5.8	4.07674	0.99661	0.00368
6.0	4.27613	0.99721	0.00243
7.0	5.279	0.9999	—
8.0	6.279	1.000	—

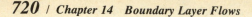

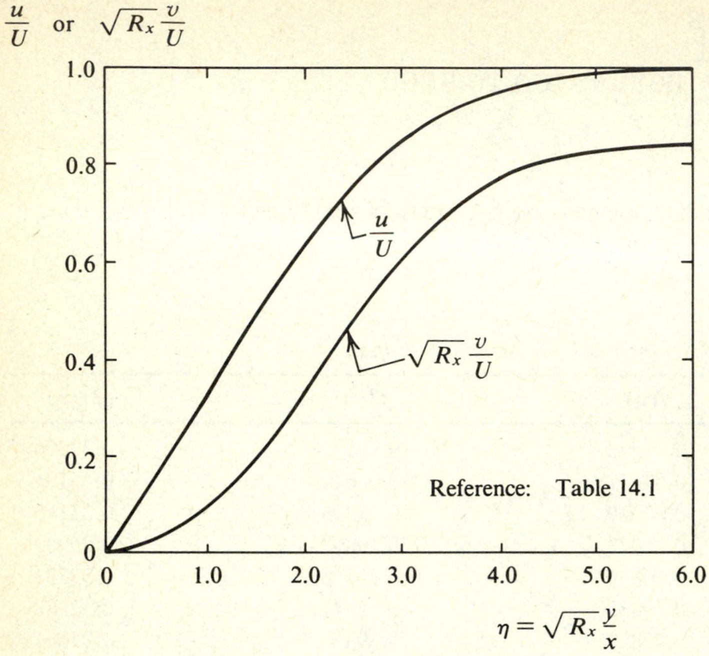

Figure 14.5 *Velocity profiles for flow in boundary layer past a flat plate: laminar case.*

Note that the dimensionless x component of velocity u/U is seen to be quite linear near the wall. In fact, at the wall, since the velocities $u = v = 0$, we find using Eq. (14.15) that $\partial^2 u/\partial y^2 = 0$. Another interesting observation is that the normal velocity component v does not vanish for large values of y.

Consider the value of η where the velocity component u is 99% the value of the free-stream velocity U. Using Table 14.1, we observe that this would result in $\eta \simeq 5.0$ when $f'(\eta) \simeq 0.99$. Accepting this as a good approximation, and using this criteria for the thickness of a laminar boundary layer, we substitute this value into Eq. (14.32) and obtain

$$\boxed{\delta \approx 5.0 \sqrt{\frac{\nu x}{U}}} \tag{14.44}$$

This then represents the Blasius value of the boundary layer thickness.

The wall shear stress, $(p_{xy})_o$, stems from the Newtonian definition

$$(p_{xy})_o = \mu \left.\frac{\partial u}{\partial y}\right|_{y=0} \tag{14.45}$$

where

$$\frac{\partial u}{\partial y} = \frac{Uf''(\eta)}{\delta_o} = \frac{Uf''(\eta)\sqrt{R_x}}{x} \tag{14.46}$$

where the local Reynolds number is $R_x = Ux/\nu$, so that the shear stress on the flat plate is

$$(p_{xy})_o = \mu U f''(0)\,\frac{\sqrt{R_x}}{x} = \frac{1}{2}\,\rho U^2\,\frac{2A_2}{\sqrt{R_x}} \tag{14.47}$$

where $A_2 = 0.33206$. Note that the wall shear stress is infinite at the leading edge and decreases as $x^{-1/2}$, which shows that Prandtl's similarity hypothesis and the boundary layer concepts do not hold at the stagnation point. The same can be said of the local skin friction; i.e., the dimensionless local skin friction coefficient c_f is evaluated as

$$\boxed{c_f = \frac{(p_{xy})_o}{\frac{1}{2}\rho U^2} = \frac{0.664}{\sqrt{R_x}}} \tag{14.48}$$

The drag force D per unit width for one side of the plate of length l is

$$D = \int_o^l (p_{xy})_o\,dx = \frac{1}{2}\,\rho U^2 \int_o^l \frac{2A_2 dx}{\sqrt{R_x}} \tag{14.49}$$

where the total skin friction drag coefficient C_{D_f} is defined as

$$C_{D_f} = \frac{D}{\frac{1}{2}\,\rho U^2 l} = \frac{1}{l}\int_o^l \frac{2A_2 dx}{R_x^{1/2}} \tag{14.50}$$

The integral in Eq. (14.50) is quite easy to evaluate:

$$C_{D_f} = \frac{2A_2}{R_l}\int_o^{R_l} R_x^{-1/2}\,dR_x = \frac{4A_2}{\sqrt{R_l}}$$

Since $A_2 = 0.33206$, then the total skin friction coefficient becomes

$$\boxed{C_{D_f} = \frac{1.328}{\sqrt{R_l}}} \tag{14.51}$$

Caution: In using Eq. (14.51) to find the drag on a flat plate, remember that it is valid *for one wetted side* of the plate only. These theoretical results, which we owe primarily to Blasius, closely match experimental results.

Example 14.3

Determine the distance downstream from the bow of a ship moving at 7.59 knots relative to still water at which the boundary layer becomes turbulent. Also, determine the boundary layer thickness at this point and the total friction drag

Example 14.3 *(Con't.)*

coefficient for this portion of the surface of the ship. Let the maximum value of Reynolds number $R_x = 5 \times 10^5$ be where the flow ceases to be laminar, such that the flow is turbulent for $R_x > 5 \times 10^5$. Let the kinematic viscosity of water be 1.21×10^{-5} ft²/s.

Solution:

Step 1.

Determine the region where the flow is laminar.

Given that the length along the flat plate measured from the bow is

$$x = \frac{5 \times 10^5 \nu}{U} = \frac{(5 \times 10^5)(1.21 \times 10^{-5})}{12.8} = \underline{0.47 \text{ ft}} \tag{i}$$

means that the boundary layer becomes turbulent at a distance of only 0.47 ft downstream from the bow. Thus for $x > 0.47$ ft, our solution will be invalid.

Step 2.

Calculate δ.

The boundary layer thickness δ is calculated from Eq. (14.44)

$$\delta\Big|_{x=0.47\text{ ft}} = 5\sqrt{\frac{\nu x}{U}}\Bigg|_{x=0.47\text{ ft}}$$

or

$$\delta = 3.33 \times 10^{-3} \text{ ft} \tag{ii}$$

Step 3.

Calculate C_{D_f}.

The total friction drag coefficient is evaluated from Eq. (14.51)

$$C_{D_f} = \frac{1.328}{\sqrt{R_l}} = \frac{1.328}{\sqrt{5 \times 10^5}} = 1.88 \times 10^{-3} \tag{iii}$$

This completes the solution.

14.5 Boundary Layer Thicknesses of Displacement and Momentum

Another meaningful measure of the boundary layer thickness is the *displacement thickness* δ^* [not to be confused with the symbol for the dimensionless boundary layer thickness defined on p. 709 (Fig. 14.6a)]. The displacement thickness is a measure of that distance by which the external potential flow is displaced outward as a consequence of the decrease in velocity in the boundary layer. The decrease in volume flow due to the influence of friction is $\int_o^\delta (U - u)\, dy$, so that for a displacement thickness δ^*, we have from the continuity equation

$$U\delta^* = \int_o^\delta (U - u)\, dy$$

or

$$\delta^* = \int_o^\delta \left(1 - \frac{u}{U}\right) dy \tag{14.52}$$

A straightforward substitution using the Blasius solution gives

$$\delta^* = \frac{1.73x}{\sqrt{R_x}} \tag{14.53}$$

and being a displacement in the direction normal to the plate represents a displacement of the potential flow due to a loss of fluid momentum in the boundary layer.

A third important boundary layer thickness is the *momentum* thickness θ (Fig. 14.6b). It is associated with the decrease in momentum flux due to friction as compared with a potential flow. Since the decrease in momentum flux is

$$\rho \int_o^\delta u(U - u)\, dy$$

we shall define

$$\rho U^2 \theta = \rho \int_o^\delta u(U - u)\, dy$$

or

$$\theta = \int_o^\delta \frac{u}{U}\left(1 - \frac{u}{U}\right) dy \tag{14.54}$$

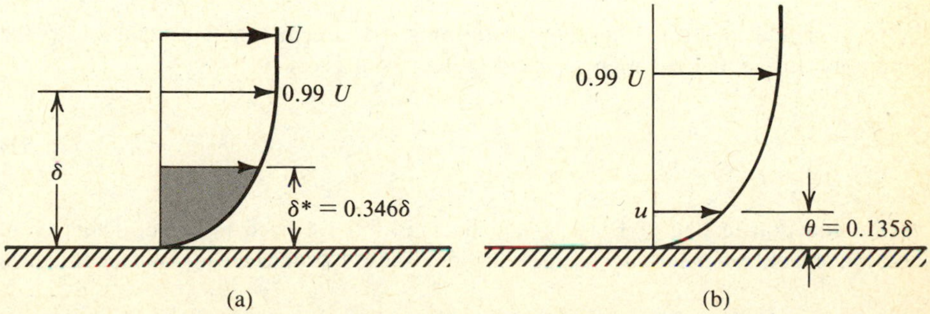

Figure 14.6 *(a) Displacement thickness. (b) Momentum thickness.*

as the momentum thickness. Notice that it has the dimensions of length. We can evaluate the integral using Blasius' results to obtain

$$\theta = \frac{0.664x}{\sqrt{R_x}} \tag{14.55}$$

which is a measure of the loss of momentum due to viscosity. Comparing Eq. (14.44) with Eq. (14.55), we observe that the momentum thickness θ is 13½% as thick as the boundary layer thickness δ.

In Sec. 14.7, application of the momentum principle will lead us into an important use of the momentum thickness.

In conclusion, we point out once again that near the leading edge of the plate the boundary layer theory *ceases to apply*, since there the Prandtl assumption

$$\left|\frac{\partial^2 u}{\partial x^2}\right| \ll \left|\frac{\partial^2 u}{\partial y^2}\right|$$

is not satisfied. The Prandtl boundary layer theory applies *only for large* Reynolds number $R_x = Ux/\nu$, where x is measured *from* the leading edge, so that $x \neq 0$.

14.6 Prandtl's Boundary Layer Theory from the Viewpoint of a Mathematician

The Blasius solution is a special case of a more general class of boundary layer flows which lead to similar velocity profiles. This more general solution, discovered by Falkner and Skan, is appropriate for flows in which the free-stream potential flow along the plate is given by $U = cx^m$. This means that the pressure term in the Prandtl boundary layer equation becomes

$$\frac{dp}{dx} = -\rho U \frac{dU}{dx} \tag{14.8}$$

$$= -\rho c^2 m x^{2m-1} \tag{14.56}$$

Performing the identical steps in transforming the stream function equation to the similarity differential equation we used in Sec. 14.4 gives

$$f''' + \frac{m+1}{2} ff'' - mf'^2 + m = 0 \tag{14.57}$$

as the equation in lieu of Blasius' equation (14.33) we need to solve. There is an enormous family of solutions* to this equation. These solutions are of great significance

*For a detailed discussion of the solution of the boundary layer Eq. (14.57) see Sec. 36.5 by D. F. Rogers in *The Introduction to the Dynamics of Incompressible Fluid Flow*, vol. 5, by R. Granger [14.3]. Listed for at least two reasons. First, in addition to flow along a flat plate, it gives the flow near a forward stagnation point; and second, it shows the effects of pressure gradients on the velocity profile, which is of special interest near a separation point. It is beyond the scope of this text to solve the Falkner-Skan equation for various values of m [14.3]. However, since the pressure gradient is related to m by Eq. (14.56), it is important that we understand the effect of the pressure gradient on the boundary layer flow. For curved surfaces, the pressure outside the boundary layer will vary with a distance measured along the wall, and the pressure gradient for $y > \delta$ may be calculated from potential flow theory. Consider the two types of curved surfaces shown in Figs. 14.7 and 14.8.

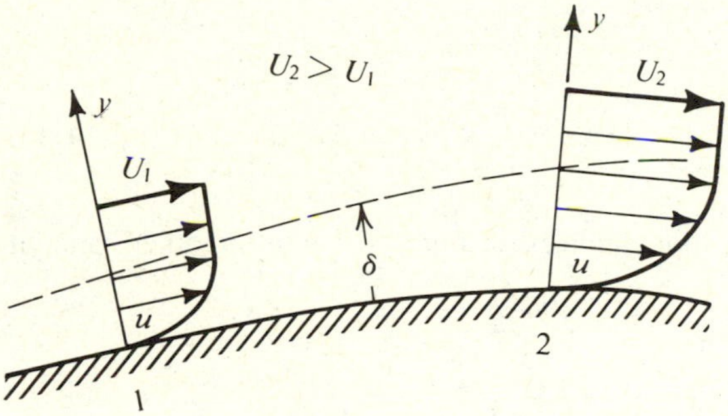

Figure 14.7 *Accelerating laminar boundary layer flow.*

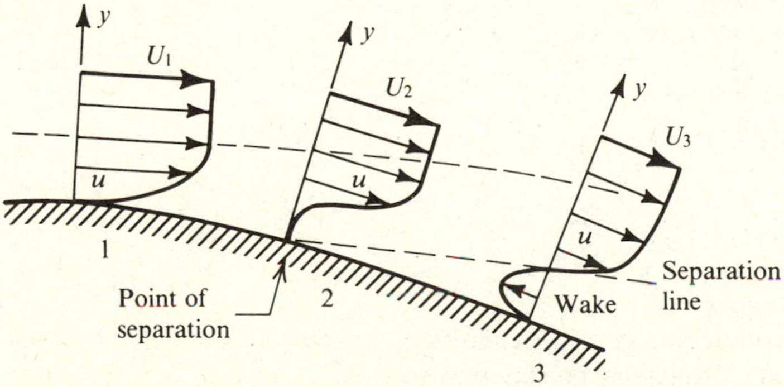

Figure 14.8 *Decelerating laminar boundary layer flow.*

are values of f, f', and f'' for a complete range of η for various values of m compiled from the works of numerous investigators. Further tables (as well as additional references) relating to boundary layer equations are given by M.T.A.C., 7:179, 1953, and N.B.S. project (velocity distribution in boundary layers) under "Projects and Publications of the Applied Mathematics Division," April to June, 1958, p. 32.

For the flow situation shown in Fig. 14.7, $dp/dx < 0$. Such flows are said to flow under a *favorable pressure gradient*, a type of pressure gradient that causes the fluid to accelerate.

Figure 14.8 shows a case of an *adverse pressure gradient*, i.e., the pressure causes the fluid to decelerate. The effect of the pressure gradient can be approximated if we analyze the motion close to the wall. By assuming the convective acceleration at this point within the boundary layer to be zero, we have the pressure acceleration in balance with the viscous deceleration:

$$\frac{\partial p}{\partial x} = \mu \frac{\partial^2 u}{\partial y^2} \tag{14.58}$$

From the preceding relationship, we see that if there is a favorable pressure gradient, then

$$\frac{\partial p}{\partial x} < 0, \quad \text{then} \quad \frac{\partial^2 u}{\partial y^2} < 0 \tag{14.59}$$

which is indicated by the flow situation depicted in Fig. 14.7. On the other hand, if there is an adverse pressure gradient

$$\frac{\partial p}{\partial x} > 0, \quad \text{then} \quad \frac{\partial^2 u}{\partial y^2} > 0 \tag{14.60}$$

indicating the possibility that a reverse flow can occur as shown in Fig. 14.8. We should realize that for both laminar and turbulent flow, a reverse flow need not occur.

At the point where $\partial p/\partial x = 0$, the velocity profile has an inflection point and we say that separation occurs.

14.6.1 Separation

The separation point is defined as

$$\left.\frac{du}{dy}\right|_{y=0} = 0$$

or where the wall shear stress is zero.* The physical explanation is that in the region downstream of the minimum pressure point (which need not be at the point of maximum thickness of the body), the pressure increases. Since the pressure drop across the boundary layer is assumed to be negligible, then this pressure drop (called an *adverse gradient*) would be experienced along the wetted surface of the body. Both the adverse pressure gradient and the shear stress cause the flow to decelerate rapidly in the

*Good only for two-dimensional flows. In three-dimensional flows, the definition of the separation line is exceedingly complex and subject to debate.

boundary layer until the flow decelerates to zero velocity, at which point the viscous forces would be zero. But at this point, the adverse pressure continues to exist, and so downstream of this stagnation point (i.e., separation point) the flow could act in the direction of the adverse pressure gradient, namely in the reversed direction resulting in a backward flow.

The actual point of separation has not been determined analytically. Experimentally, the separation point on a circular cylinder is close to the minimum pressure point for laminar flow. Actually, the separation point depends upon both Reynolds number and surface roughness. Separation is the engineer's nightmare, less science than art. It cannot be disregarded just because it is difficult to determine, for within its behavior, it can drastically change drag, pressure recovery, shock wave configurations, dissipation of energy, and effects of heat transfer.

14.6.2 On Prandtl's Boundary Layer Equations

There are a number of questions that must be raised and answered now in order to place any value on Prandtl's boundary layer equations.

The questions we must ask are:

1. Is the boundary condition satisfied at infinity?
2. Are there special explicit "simple" solutions to the Prandtl boundary layer equations?
3. Is there at least one solution to the boundary layer equations?
4. Are Prandtl's boundary layer equations singular from a mathematical point of view, and are they well posed?
5. Can we prove that for $\nu \rightarrow 0$, the solution of the Prandtl boundary layer equation asymptotically corresponds to a solution of the Navier-Stokes equation?
6. Can we establish general statements about the velocity profile of u without integration?
7. What methods are available for solving the Prandtl boundary layer equations?

The drawbacks to the material presented in this chapter are that (a) no exact solutions are available for comparison of the results, (b) the Blasius boundary layer solution cannot be obtained in closed form, and (c) the outer flow ($u = U$) is too trivial and does not illustrate the concept of matching the outer limit of the inner solution (boundary layer) to the inner limit of the outer solution (potential theory). Some authors evade this difficulty by demonstrating mathematical properties of a boundary layer on some artificial example of an equation that has little or nothing to do with fluid dynamics.

The Prandtl boundary layer equations are not exact forms of the Navier-Stokes equations, which are exact. They are solely approximations, but they work. At least they give results that are good as the best experimental results.*

*W. H. Mason's "Boundary Layer Analysis Methods" PAK *4, Aerodynamic Calculation Methods for Programmable Calculators & Personal Computers with Programs for the TI-59, AEROCAL, Box 799, Huntington, NY 11743, 1981, supplements the theoretical analysis with actual numerical results to develop engineering skills. The material is organized in workbook fashion.

14.7 Integral Momentum Principles

The boundary layer equations developed thus far, and hence their solutions, are applicable only to laminar flow within the boundary layer. Experimental observation shows that the boundary layer becomes unstable at a Reynolds number based on the distance along the surface of $(Ux/\nu)|_{cr} = 5 \times 10^5$. The numerical Example 14.3 indicated that the flow of water at 7.59 knots over a flat ship surface resulted in a laminar boundary layer for only 0.47 ft downstream from the bow. For a ship whose length is 470 ft this shows that the laminar boundary layer represents only 0.1% of the total length of the ship.

If we are to employ boundary layer concepts in real engineering designs, we need to devise approximate methods that would in many cases quickly lead to an answer even if the accuracy is 7 to 10% of the answer obtained by the exact solution. Von Kármán and Pohlhausen devised a simplified method by *satisfying only the boundary conditions of the boundary layer flow rather than satisfying Prandtl's differential equations for each and every particle within the boundary layer*. Their procedure was as follows. They assumed a velocity profile which satisfies the boundary conditions on the surface and at the edge of the boundary, and applied the integral momentum principle to a control volume that includes the boundary layer. They then showed how to calculate the growth of boundary thickness, separation, and distribution of skin friction drag on a simple continuous body.

In the following sections of this chapter we will formulate the integral momentum method of analysis. This method applies to turbulent as well as laminar boundary layer flows. We shall examine its use in considerable detail.

14.7.1 Momentum Principle for Boundary Layer Analysis

Consider applying the integral momentum principle of fluid dynamics to the control volume shown and defined in Fig. 14.9. The analysis is still restricted to steady, two-dimensional, incompressible flow. The flow within the boundary layer is, however, not restricted to being either laminar or turbulent.

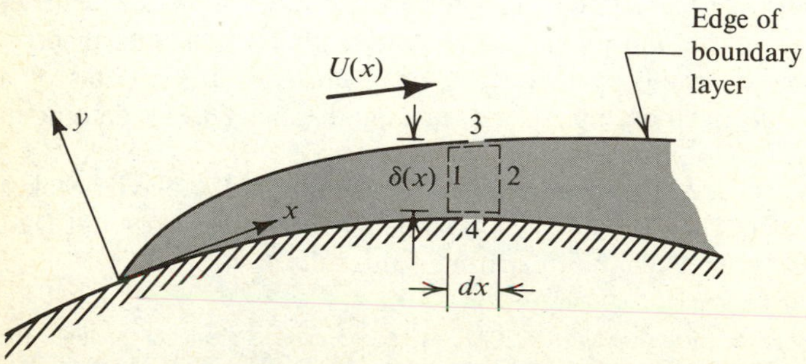

Figure 14.9 Control volume in a laminar boundary layer over a curved surface.

Let

$p = p(x)$ ≡ the pressure on the edge and within the boundary layer, determined from potential flow analysis

$U = U(x)$ ≡ the velocity at the edge of the boundary layer, determined from potential flow analysis

$u = u(x, y)$ ≡ the velocity profile within the boundary layer, an unknown function *to be approximated by satisfying boundary conditions at both the surface and at the edge of the boundary layer* and whose shape can be estimated from experimental observations

y ≡ coordinate normal to surface

x ≡ coordinate tangent to surface

$\delta = \delta(x)$ ≡ the boundary layer thickness, an unknown function, in the direction of y

The mass flow rate, $\dot{M}$, through face 1, the upstream face of the control volume of Fig. 14.9, is

$$\dot{M}_1 = \int_o^\delta \rho u dy \tag{14.61}$$

and that through face 2 is

$$\dot{M}_2 = \dot{M}_1 + \frac{\partial \dot{M}_1}{\partial x} dx$$

or

$$= \dot{M}_1 + \frac{\partial}{\partial x}\left(\int_o^\delta \rho u dy\right) dx \tag{14.62}$$

From continuity, and the fact that the boundary is a material impenetrable surface,

$$\dot{M}_1 + \dot{M}_3 = \dot{M}_2 \tag{14.63}$$

The mass rate through face 3 becomes, from Eqs. (14.61) and (14.62),

$$\dot{M}_3 = \frac{\partial}{\partial x}\left(\int_o^\delta \rho u dy\right) dx \tag{14.64}$$

The x component of the integral momentum equation is

$$\sum F_x = \int_A \rho u \mathbf{V} \cdot d\mathbf{A} \tag{14.65}$$

Let the net surface forces be identified as

$$\sum F_x = F_{1x} + F_{2x} + F_{3x} + F_{4x}$$

where

$$F_{1x} = p\delta$$

$$F_{2x} = -\left(F_{1x} + \frac{dF_{1x}}{dx}dx\right) = -\left[p\delta + \frac{d(p\delta)}{dx}dx\right]$$

$$F_{3x} = p\frac{d\delta}{dx}dx$$

$$F_{4x} = -(p_{xy})_o\, dx$$

and

$$\sum F_x = -(p_{xy})_o\, dx - \delta\frac{dp}{dx}dx \qquad (14.66)$$

The partial derivative operators have been replaced by ordinary derivatives, since the functions involved are functions of a single variable x.

The momentum flux, MF, into the control volume of Fig. 14.9 through face 1 is

$$MF_{1x} = \int_o^\delta \rho u^2 dy \qquad (14.67)$$

and that out of the control volume through face 2 is

$$\begin{aligned} MF_{2x} &= MF_{1x} + \frac{\partial}{\partial x}(MF_{1x})\, dx \\ &= \int_o^\delta \rho u^2 dy + \frac{\partial}{\partial x}\left(\int_o^\delta \rho u^2 dy\right) dx \end{aligned} \qquad (14.68)$$

The momentum flux into the control volume through face 3 is

$$MF_{3x} = \dot{M}_3 U$$

or with Eq. (14.64),

$$MF_{3x} = U\frac{\partial}{\partial x}\left(\int_o^\delta \rho u dy\right) dx \qquad (14.69)$$

The net efflux of x momentum is

$$\int_A \rho u \mathbf{V} \cdot d\mathbf{A} = MF_{2x} - MF_{1x} - MF_{3x}$$

or

$$= \frac{\partial}{\partial x}\left(\int_o^\delta \rho u^2 dy\right) dx - U\frac{\partial}{\partial x}\left(\int_o^\delta \rho u dy\right) dx \tag{14.70}$$

Equating the results of Eqs. (14.66) and (14.70), dividing through by dx, and rearranging terms yields

$$(p_{xy})_o = -\delta\frac{dp}{dx} - \frac{\partial}{\partial x}\int_o^\delta \rho u^2 dy + U\frac{\partial}{\partial x}\int_o^\delta \rho u dy \tag{14.71}$$

as the expression for the surface shear stress.

This last equation can be rearranged into a more useful form by relating the pressure p and velocity U at the edge of the boundary layer, using Bernoulli's equation, and noting that

$$U\frac{\partial}{\partial x}\int_o^\delta \rho u dy = \frac{\partial}{\partial x}\int_o^\delta \rho U u dy - \frac{dU}{dx}\int_o^\delta \rho u dy \tag{14.72}$$

From Bernoulli's equation we have our familiar two-dimensional result

$$p = p_s - \frac{1}{2}\rho U^2$$

or

$$\frac{\partial p}{\partial x} = -\rho U\frac{dU}{dx}$$

where p_s is the stagnation pressure.

By substituting these two last results into Eq. (14.71) we get

$$(p_{xy})_o = \delta\rho U\frac{dU}{dx} - \frac{d}{dx}\int_o^\delta \rho u^2 dy + \frac{d}{dx}\int_o^\delta \rho U u dy - \frac{dU}{dx}\int_o^\delta \rho u dy \tag{14.73}$$

where the differential operators are total differentials: all functions depend solely on x.

The second and third terms in Eq. (14.73) can be combined to give

$$\frac{d}{dx}\int_o^\delta \rho u(U - u)\,dy = \frac{d(\rho U^2\theta)}{dx} \tag{14.74}$$

where θ is the momentum thickness, defined in Sec. 14.5.

The first and last term in Eq. (14.73) can also be combined:

$$\frac{dU}{dx}\left(U\rho\delta - \int_o^\delta \rho u dy\right) = \frac{dU}{dx}\left(\int_o^\delta \rho U dy - \int_o^\delta \rho u dy\right)$$

or

$$\frac{dU}{dx}\int_o^\delta \rho(U - u)\,dy = \rho U\delta^*\frac{dU}{dx} \tag{14.75}$$

where δ^* is the displacement thickness also defined in Sec. 14.5.

Putting everything altogether, we substitute Eqs. (14.74) and (14.75) into (14.73) to give

$$\boxed{(p_{xy})_o = \frac{d}{dx}(\rho U^2\theta) + \rho U\delta^*\frac{dU}{dx}} \tag{14.76}$$

which is called the *von Kármán-Pohlhausen integral momentum equation*. Note that where U equals a constant, the rate of change of momentum thickness determines the shear stress just as the rate of change of momentum determines the forces acting on a body.

14.7.2 Method of Solution of the von Kármán-Pohlhausen Integral Momentum Equation*

The integral momentum equation (14.76) is quite general and holds for laminar as well as turbulent flow provided that the velocity distribution $u(x, y)$ is known. However, it should be noted that in turbulent boundary layers, the wall shear stress $(p_{xy})_o$ cannot be computed from the constitutive equations for a Newtonian fluid but must be derived from turbulent boundary layer analysis. By inspection, Eq. (14.76) does not appear to be of any use so long as the velocity profile $u(x, y)$ in the boundary layer is unknown. If, however, instead of integration of the Prandtl boundary layer equations, we are content with an approximate equation for the velocity distribution which satisfies all or most boundary conditions, then Eq. (14.76) can be integrated. The result of the integration of Eq. (14.76) results in an ordinary differential equation for the boundary layer thickness $\delta(x)$ from which we can approximate the most important characteristics of the boundary layer.

*A complete computer solution to the Pohlhausen laminar boundary layer equation can be found in an article by John S. Milne in the *Bull. Mech. Engng.*, vol. 9, pp. 99–104, 1970.

The procedure used to solve Eq. (14.76) is as follows:

Step 1.

Given a specific boundary geometry, calculate the velocity at the edge of the boundary layer $U(x)$ using potential flow theory. Since the boundary layer thickness is very small, at least up to the point of separation, we can use the boundary of the body in calculating $U(x)$.

Step 2.

Assume a reasonable velocity profile shape for the flow inside the boundary layer. That is, assume u to be some reasonable algebraic or transcendental function of y. Experimental observations play an important role here. The velocity distribution will of course be different for laminar and turbulent flows. In either case, choose the velocity profile to satisfy *all* the boundary conditions.

Step 3.

If the boundary layer flow is laminar, we use Newton's viscosity law to replace $(p_{xy})_o$ by $\mu(\partial u/\partial y)_{y=0}$. If the boundary layer is turbulent we replace $(p_{xy})_o$ by the *turbulent shear stress*. The turbulent stresses and turbulent concepts will be discussed in sections that follow.

Step 4.

By substituting the above information into Eq. (14.76), we reduce the momentum integral equation to an ordinary differential equation for $\delta(x)$, which is of separable form and can thus be integrated in a straightforward manner. Using the expression for $\delta(x)$, we can compute most of the important characteristics of the boundary layer.

For *laminar* boundary layers, Pohlhausen devised an approximate method for solving Eq. (14.76). In this method we choose a velocity profile in the quartic form

$$\frac{u}{U} = f(\eta) = a_0 + a_1\eta + a_2\eta^2 + a_3\eta^3 + a_4\eta^4 \tag{14.77}$$

where $\eta = y/\delta(x)$, and evaluate the coefficients a_i satisfying the following boundary conditions:

$$\begin{aligned} &\text{At } y = 0, \quad u = 0 \\ &\qquad \nu\frac{\partial^2 u}{\partial y^2} = \frac{1}{\rho}\frac{dp}{dx} = -U\frac{dU}{dx} \\ &\text{At } y = \delta, \quad u = U \\ &\qquad \frac{\partial u}{\partial y} = 0 \\ &\qquad \frac{\partial^2 u}{\partial y^2} = 0 \end{aligned} \tag{14.78}$$

Note that these boundary conditions are all satisfied by the Blasius solution. The last two conditions, namely $\partial u/\partial y = 0$, $\partial^2 u/\partial y^2 = 0$ at $y = \delta$ mean that the boundary layer profile smoothly joins the outer flow, and that there is no inflection at this juncture.

By substituting the above boundary conditions into Eq. (14.77), we find that

$$
\begin{aligned}
a_0 &= 0 \\
a_1 &= 2 + \frac{\lambda}{6} \\
a_2 &= -\frac{\lambda}{2} \\
a_3 &= -2 + \frac{\lambda}{2} \\
a_4 &= 1 - \frac{\lambda}{6}
\end{aligned}
\tag{14.79}
$$

where

$$\lambda = \frac{\delta^2}{\nu}\frac{dU}{dx} = -\frac{dp}{dx}\frac{\delta}{\mu U/\delta} \tag{14.80}$$

is called the shape factor. We can see that λ is the ratio of the pressure force to the viscous force. It is restricted to a range $-12 \leq \lambda \leq 12$. For example $\lambda = 0$ corresponds to a zero pressure gradient, or the flat plate case. With these values of the coefficients, the velocity profile becomes

$$\frac{u}{U} = f(\eta) = F(\eta) + \lambda G(\eta) \tag{14.81}$$

where

$$F(\eta) = 1 - (1 - \eta)^3(1 + \eta) \tag{14.82}$$

and

$$G(\eta) = \frac{1}{6}\eta(1 - \eta)^3 \tag{14.83}$$

The F and G functions as given by Eqs. (14.82) and (14.83) are plotted versus η in Fig. 14.10.

The velocity distribution constitutes a one-parameter family of curves in terms of the shape factor λ. The velocity profile u/U as given by Eq. (14.81) is shown in Fig. 14.11 for a variety of shape factors. The result for the flat plate, $\lambda = 0$, compares favorably with Blasius' result. The velocity profiles $\lambda > 0$ correspond to a decreasing

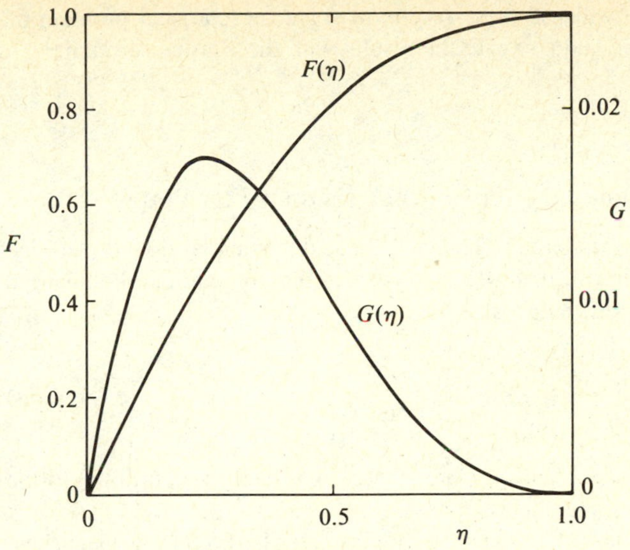

Figure 14.10 *F(η) and G(η) functions.*

pressure and vice versa. Thus the von Kármán-Pohlhausen method is valuable in calculating flows about thick objects. We need to know only the pressure gradient and boundary layer thickness; then we can evaluate the shape factor $\lambda = \lambda(x)$, and thus determine the velocity component u.

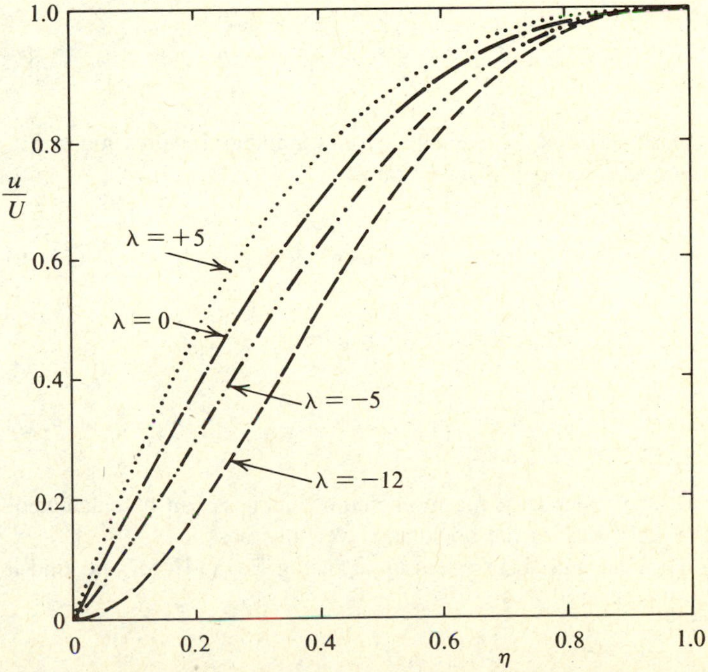

Figure 14.11 *Velocity profiles for a variety of shape factors.*

In the next section we will use these results to solve the classical problem of laminar flow over a flat plate and compare the results with the Blasius solution.

14.7.3 Laminar Boundary Layer Analysis on a Flat Plate

Consider uniform flow parallel to a flat plate. Since the potential flow at the edge of the boundary layer is uniform, then $dU/dx = 0$, and the von Kármán-Pohlhausen momentum integral Eq. (14.76) reduces to

$$(p_{xy})_o = \rho U^2 \frac{d\theta}{dx} \tag{14.84}$$

which states that the shear stress on the plate surface is directly proportional to the axial rate of growth of the momentum thickness.

By definition, the boundary layer shape factor λ as given by Eq. (14.80) vanishes. The vanishing of the shape factor λ *means that the shape of the velocity profiles within the boundary layer is identical as one moves along the length* x *of the plate. It also indicates that the boundary layer cannot separate from the boundary.*

From the assumed power series expression for the velocity component u and the boundary conditions given by Eqs. (14.77) and (14.78), respectively, Pohlhausen's velocity profile approximation becomes

$$\frac{u}{U} = f(\eta) = 2\eta - 2\eta^3 + \eta^4 \tag{14.85}$$

The momentum thickness θ can now be evaluated from the integral expression of Eq. (14.74), along with the above velocity distribution as

$$\frac{\theta}{\delta} = \int_o^1 (2\eta - 4\eta^2 - 2\eta^3 + 9\eta^4 - 4\eta^5 - 4\eta^6 + 4\eta^7 - \eta^8)\, d\eta \tag{14.86}$$

so that upon integration, the momentum thickness becomes

$$\theta = \frac{37}{315}\delta \tag{14.87}$$

Knowing the boundary layer thickness, the momentum thickness can be calculated. Note that it is approximately 10% of the boundary layer thickness.

We compute the laminar wall shear stress $(p_{xy})_o$ using Eq. (14.45), and find it to be

$$(p_{xy})_o = 2\frac{\mu U}{\delta} \tag{14.88}$$

The first-order ordinary differential equation for the boundary layer thickness can now be generated by substituting Eqs. (14.87) and (14.88) into Eq. (14.84) with the result that

$$2\frac{\mu U}{\delta} = \frac{37}{315}\rho U^2 \frac{d\delta}{dx}$$

Separating variables and simplifying, we obtain a simple ordinary differential equation:

$$\delta\, d\delta = \frac{630}{37}\frac{\nu}{U}dx$$

This can be integrated to give

$$\frac{\delta^2}{2} = \frac{630}{37}\frac{\nu}{U}x$$

(Note that the boundary thickness vanishes at $x = 0$.) Thus the boundary layer thickness δ, as approximated by the von Kármán-Pohlhausen series, is

$$\delta = 5.83\sqrt{\frac{\nu x}{U}} \tag{14.89}$$

and agrees fairly well with the solution obtained by Blasius. The mathematical simplicity of the Pohlhausen technique compared to Blasius' solution is considerable. As was pointed out earlier, *any reasonable velocity profile can be used to approximate the exact distribution*. In order to gain familiarity with the von Kármán integral momentum method, we should calculate the various parameters using the velocity profiles as given in Table 14.2, and fill in the blank spaces with values. See Example 14.4.

Table 14.2 *Comparison of Pohlhausen's Results with Blasius' Results*

Velocity profile	$\delta\Big/\sqrt{\frac{\nu x}{U}}$	δ^*/δ	θ/δ	$\frac{\theta\sqrt{R_x}}{x}$	$C_{D_f}\left(\frac{Ul}{\nu}\right)^{1/2}$
Blasius solution	5.0	0.346	0.1348	0.664	1.328
Pohlhausen solution $u = U(a_o + a_1\eta + a_2\eta^2 + a_3\eta^3 + a_4\eta^4)$	5.83	$\frac{3}{8}$	$\frac{37}{315}$	0.686	1.372
$u = U\sin\left(\frac{\pi}{2}\eta\right)$					
$u = U(a_o + a_1\eta)$					
$u = U(a_o + a_1\eta + a_2\eta^2 + a_3\eta^3)$					

Example 14.4
Nikuradse obtained experimental data for laminar flow over a flat plate placed at zero angle-of-attack. Examination of his observed and calculated measurements suggested

$$\frac{u}{U} = a\left(\frac{y}{\delta}\right) + b\left(\frac{y}{\delta}\right)^3 \tag{i}$$

as a good approximation to his data. Using Eq. (i), (a) evaluate the coefficients a, b; (b) obtain expressions for the boundary layer thickness δ, displacement thickness δ^*, momentum thickness θ, and shear stress $(p_{xy})_o$ on the plate's surface and compare the results with Blasius; and (c) evaluate the total skin friction drag coefficient C_{D_f}.

Solution:
Step 1.
Determine the velocity distribution.

The velocity distribution given by Eq. (i) must satisfy the boundary conditions that $u = 0$ at $y = 0$, $u = U$ at $y = \delta$, and $\partial u/\partial y = 0$ at $y = \delta$. The no-slip boundary condition is already satisfied by Eq. (i). At $y = \delta$, we obtain

$$a + b = 1$$

$$a + 3b = 0$$

so that $b = -\frac{1}{2}$, $a = \frac{3}{2}$. Thus Eq. (i) becomes

$$\frac{u}{U} = \frac{3}{2}\left(\frac{y}{\delta}\right) - \frac{1}{2}\left(\frac{y}{\delta}\right)^3 \tag{ii}$$

Step 2.
Write the governing equations of motion.

The von Kármán momentum integral equation for laminar flow over a flat plate is, from Eq. (14.76),

$$(p_{xy})_o = \rho U^2 \frac{d\theta}{dx} \tag{iii}$$

Step 3.
Obtain expressions for momentum thickness wall shear stress:

$$\theta = \int_o^\delta \frac{u}{U}\left(1 - \frac{u}{U}\right) dy \tag{iv}$$

and

$$(p_{xy})_o = \mu\left(\frac{\partial u}{\partial y}\right)_{y=0} \tag{v}$$

Step 4.
Reduce governing equations to their most basic form.

Example 14.4 *(Con't.)*

Substitute Eqs. (iv) and (v) into Eq. (iii):

$$\mu\left(\frac{\partial u}{\partial y}\right)_{y=0} = \rho U^2 \frac{\partial}{\partial x}\int_o^\delta \frac{u}{U}\left(1 - \frac{u}{U}\right) dy \qquad \text{(vi)}$$

Substitute the assumed velocity distribution, Eq. (ii), into Eq. (vi):

$$\frac{3\mu U}{2\delta} = \rho U^2 \frac{\partial}{\partial x}\int_o^\delta \left[\frac{3}{2}\left(\frac{y}{\delta}\right) - \frac{1}{2}\left(\frac{y}{\delta}\right)^3\right] \cdot \left[1 - \frac{3}{2}\left(\frac{y}{\delta}\right) + \frac{1}{2}\left(\frac{y}{\delta}\right)^3\right] dy$$

$$= \rho U^2 \frac{\partial}{\partial x}\left[\delta\left(\frac{3}{4} - \frac{9}{12} + \frac{3}{20} - \frac{1}{8} + \frac{3}{20} - \frac{1}{28}\right)\right]$$

or

$$\frac{3\mu U}{2\delta} = \frac{39\rho U^2}{280}\frac{d\delta}{dx} \qquad \text{(vii)}$$

Step 5.
Solve the differential equation for δ.

Separating the variables in Eq. (vii),

$$\frac{140}{13}\mu dx = \rho U \delta d\delta$$

and integrating yields

$$\delta = \sqrt{\frac{280\nu x}{13U} + c}$$

The constant of integration c is zero since at $x = 0$, the boundary layer has zero thickness. Thus

$$\delta = \frac{4.64x}{\sqrt{R_x}} \qquad \text{(viii)}$$

which is about 6% below the Blasius solution. Experimental errors can be as much as 7%, so this is an acceptable theoretical estimate.

Step 6.
Evaluate θ and δ*.

The momentum thickness θ is found using the results in Eqs. (vii) and (viii) with the result

$$\theta = \frac{0.64x}{\sqrt{R_x}} \qquad \text{(ix)}$$

Example 14.4 *(Con't.)*

The displacement thickness δ^* is evaluated using Eqs. (14.75) and (i):

$$\delta^* = \int_o^\delta \left(1 - \frac{u}{U}\right) dy$$

$$= \int_o^\delta \left[1 - \frac{3}{2}\left(\frac{y}{\delta}\right) + \frac{1}{2}\left(\frac{y}{\delta}\right)^3\right] dy \qquad \text{(x)}$$

$$= \frac{3}{8}\delta$$

which is larger than the Pohlhausen value of $\frac{3}{10}\delta$. Substituting the boundary layer thickness of Eq. (viii) into Eq. (x) gives

$$\delta^* = \frac{1.73x}{\sqrt{R_x}} \qquad \text{(xi)}$$

which is approximately 1% above the Blasius exact value.

Step 7.

Evaluate the wall shear stress.

The shear stress $(p_{xy})_o$ at the plate surface is obtained from Eqs. (vii) and (viii) with the result

$$(p_{xy})_o = \mu U \frac{3\sqrt{R_x}}{(2)(4.64)x}$$

$$= 0.323\mu \frac{U\sqrt{R_x}}{x}, \quad x \neq 0 \qquad \text{(xii)}$$

which is approximately 3% below Blasius' exact solution for the wall shear stress. Note that the shear stress falls off as $1/\sqrt{x}$ as is the case for any assumed velocity distribution that satisfies the boundary conditions.

Step 8.

Evaluate the drag coefficient.

The total skin friction drag coefficient C_{D_f} is evaluated from the drag force D

$$C_{D_f} = \frac{D}{\frac{1}{2}\rho U^2 A} = \frac{0.323\mu U \int_o^l \frac{\sqrt{R_x}}{x} dx}{\frac{1}{2}\rho U^2 l}$$

$$= \frac{1.292}{\sqrt{R_l}} \qquad \text{(xiii)}$$

which is approximately 3% below the Blasius exact value.

This completes the solution.

Example 14.5

Water of kinematic viscosity $\nu = 1 \times 10^{-5}$ ft²/s is flowing steadily over a smooth flat plate at zero angle-of-attack with a velocity of 5 ft/s. The length l of the plate is 1 ft. Calculate (a) the thickness of the boundary layer at 6 in. from the leading edge, (b) the boundary layer rate of growth at 6 in. from the leading edge, and (c) the total drag coefficient on both sides of the plate.

Solution:

Step 1.

Calculate Reynolds number to check if the flow is laminar.

The Reynolds number for the flow is

$$R_x = \frac{Ux}{\nu} = 500{,}000x \tag{i}$$

so that the flow is laminar for $0 \leq x \leq 12$ in.

Step 2.

Calculate boundary layer thickness.

(a) The boundary layer thickness δ is calculated from Blasius' expression

$$\delta = \frac{5x}{\sqrt{R_x}} \tag{ii}$$

so that at 6 in. from the leading edge

$$\delta = \frac{30}{\sqrt{25 \times 10^4}} = 0.06 \text{ in.} \tag{iii}$$

or nearly 1/16 in. thick.

(b) The boundary layer rate of growth is $d\delta/dx$ and from Blasius' expression for the boundary layer thickness

$$\delta = 5\sqrt{\frac{\nu x}{U}} \tag{iv}$$

then

$$\frac{d\delta}{dx} = \frac{5}{2}\sqrt{\frac{\nu}{Ux}} \tag{v}$$

so that at 6 in. from the leading edge

$$\frac{d\delta}{dx} = \frac{5}{2R_x}\frac{1}{2} = \frac{2.5}{\sqrt{5 \times 10^5}} = 3.54 \times 10^{-3} \tag{vi}$$

which means the boundary layer is growing very slowly.

Example 14.5 *(Con't.)*

Step 3.
Calculate the drag coefficient.
(c) The total drag coefficient C_{D_f} is obtained from the Blasius expression

$$C_{D_f} = \frac{1.328}{(R_l)^{1/2}}$$

$$= \frac{1.328}{\sqrt{5 \times 10^5}} \qquad \text{(vii)}$$

$$= 0.00187$$

This completes the solution.

This section probably should have been entitled "Impractical Methods of Solving the Boundary Layer Equations." Though the methods are all accurate, they will not likely be of any practical use in themselves. We do not wish to spend unnecessary time performing detailed computations, and in many practical situations less accurate solutions are acceptable. Such approximate solutions are, apparently, much in vogue. The Pohlhausen method appears to give good results in regions of favorable pressure gradient. With no pressure gradient there is about a 3% error. The method is rapidly less accurate when the pressure gradient is large and unfavorable. In particular, it cannot accurately predict boundary layer separation (about 30% in error).

There are two general criticisms of the Pohlhausen method. The first arises from the arbitrary choice of which boundary conditions shall be satisfied by the approximate velocity-profile that is selected. Why four? Why not a cubic or a quintic? In any case, would the method be less or more accurate? There are many good arguments regarding the higher-order tangencies at $y = \delta$. The other criticism is that the second derivative of the velocity appears explicitly in the formulation. If the velocity is given analytically, it does not matter, but if obtained by experiment, the scatter would make it difficult to obtain even a rough estimate of the second derivative. Two improved methods are Thwaites's method and Stratford's method, as described in Ref. 14.3.

14.8 Mechanics of Boundary Layer Transition

One of the more interesting problems in fluid mechanics is the physical mechanism of transition from laminar to turbulent flow. The problem evolves about the generation of both steady and unsteady vorticity near a body, its subsequent molecular diffusion, its kinematic and dynamic convection and redistribution downstream, and the resulting feedback on the velocity and pressure fields near the body. We can perhaps appreciate the complexity of the transition problem by examining the behavior of a real flow past a cylinder. We will start off with an extremely low Reynolds number ($R_D < 3$), and increase it in increments beyond 3.5×10^6.

Figure 14.12a shows a few streamlines for very low Reynolds number, so low that the flow is almost ideal. If it were truly ideal, then the streamlines would be symmetrical. This case has large lateral vorticity penetration including an axial boundary influence due to the shear stresses.

Figure 14.12b shows typical streamlines and velocity distribution for 3–5 $< R_D$ $<$ 30–40. As the flow moves over the cylinder, the fluid deforms, rotates, and, because of the relatively high velocity (high compared to Fig. 14.12a), shears and forms a standing vortex. A similar behavior occurs over the lower surface, with a standing vortex rotating in the opposite direction to the one above it.

Figure 14.12c shows the case where an imminent instability can arise. Near the cylinder, the flow is symmetric and the twin vortices are attached and well behaved. The flow is laminar and stable, but the wake has a slight instability due to the increased velocity.

Figure 14.12d shows the streamlines for 80–90 $< R_D <$ 150–300, the range for von Kármán vortex shedding. (We can often detect this condition by standing near telephone wires and hearing the noticeable "singing" of wires.) The twin vortices can no longer stay attached, but move off the cylinder in alternate fashion. This is a classic laminar instability, the vortices creating eddies that sweep downstream much like the eddies from an oar.

Figure 14.12e shows the subcritical case which exists when 150–300 $< R_D <$ $1 \times 10^5 - 1.3 \times 10^5$. There is a pronounced laminar-transitional-turbulent zone, much as in the Reynolds pipe experiment. Turbulence progresses forward with a trend toward three-dimensionality. This is the zone where the flow is extremely sensitive to free-stream turbulence and surface roughness.

Figure 14.12f shows the results of the streamlines for $1 \times 10^5 - 1.3 \times 10^5 <$ $R_D < 3.5 \times 10^6$. This is the critical and postcritical stage. The wake has narrowed, corresponding to local reattachments with laminar bubbles. There are timewise and spanwise separation patterns, with little to no periodicity in the behavior of the flow.

Figure 14.12g is the transcritical stage and exists for $3.5 \times 10^6 < R_D$. The transition to turbulence occurs on the front face of the cylinder at $\theta < 80°$. The wake is fully turbulent, though some periodicity can be noted. The separation has moved from near the top to a point on the cylinder's aft side, thereby reducing the area of pressure drag.

An understanding of the transitional flow processes will help in practical problems either by improving procedures for predicting positions or for determining methods of advancing or retarding the transition position.

Earlier, in Chap. 11, we discussed the Reynolds' experiment; in particular we briefly suggested a sequence of events that takes place once the instability commences. In the following sections, we shall identify each event.

14.8.1 The Nonlinear Region

Transitional flow consists of six events, as shown in Fig. 14.13. Let us consider them one at a time.

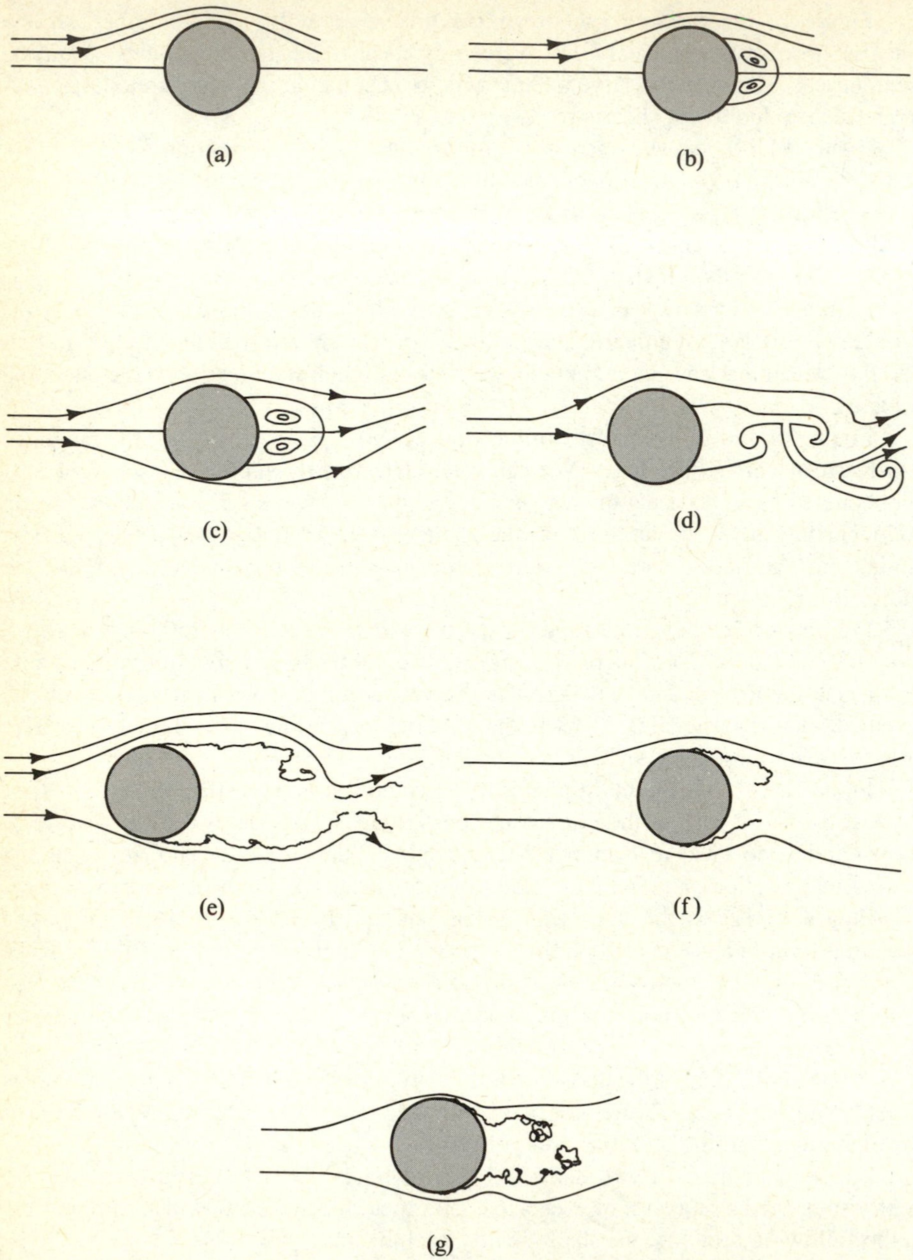

Figure 14.12 *Characteristics of flow past a cylinder. (a) Very low* R_D*. (b)* $3 - 5 < R_D < 30 - 40$*. (c)* $30 - 40 < R_D < 80 - 90$*. (d) von Kármán stage,* $80 - 90 < R < 150 - 300$*. (e) Subcritical,* $150 - 300 < R_D < 1 \times 10^5 - 13 \times 10^5$*. (f) Critical,* $1 \times 10^5 - 1.3 \times 10^5 < R_D < 35 \times 10^6$*. (g) Transcritical,* $3.5 \times 10^6 < R_D$*. (Source: Reproduced with the permission of the American Society of Mechanical Engineers, from ''Flow Around Circular Cylinder—A Kaleidoscope of Challenging Fluid Phenomena'' by M. V. Morkovin,* Symposium on Fully Separated Flows, *1964.)*

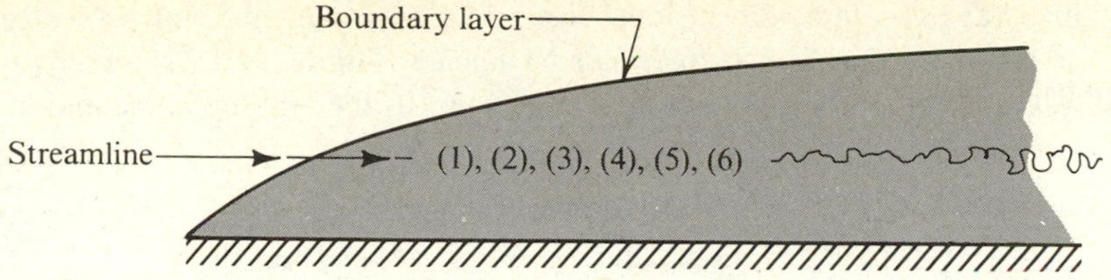

Figure 14.13 *Sequence of events in the nonlinear region leading to turbulence.*

Region of Instability to Small Wavy Disturbances, (1)

Figure 14.14a is a photograph of a laminar flow over a flat plate aligned with a flow where $R_l = 20{,}000$. At $R_l = 100{,}000$ a two-dimensional wave appears (see Fig. 14.14b), made visible by injecting dye in water. In 1921 Tollmien and Schlichting [14.4] predicted that waves would form and grow in the boundary layer. Near two-dimensional flow is achieved.

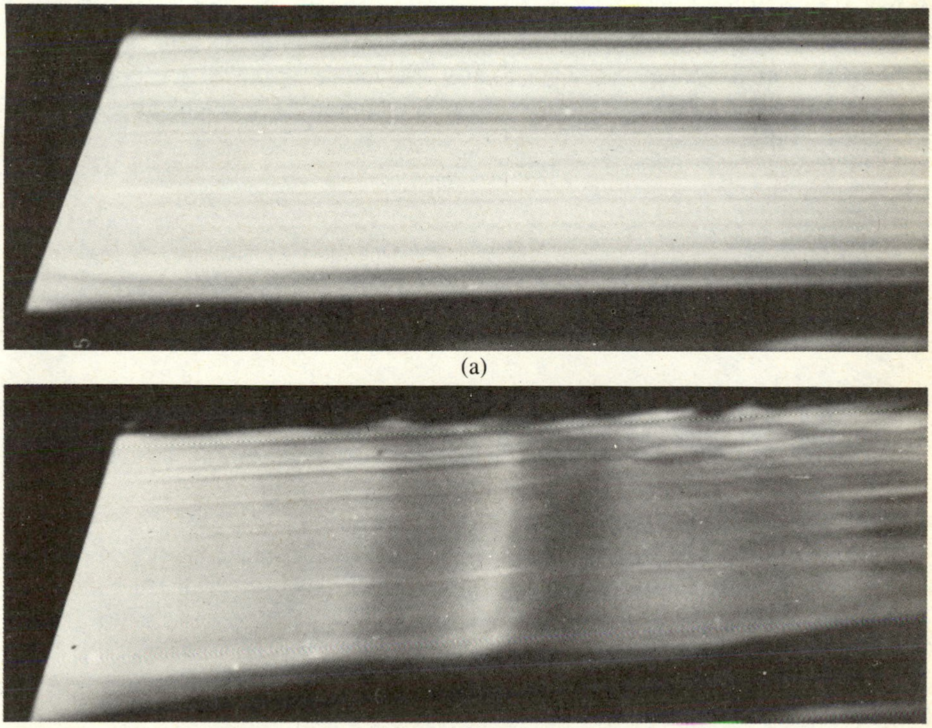

(a)

(b)

Figure 14.14 *Instability of the boundary layer on a plate. (Source: H. Werlé and ONERA, in* An Album of Fluid Motion, *M. Van Dyke, Parabolic Press.)*

Region of Three-Dimensional Wave Amplification, (2)

Minor irregularities in the free-stream or in the upstream boundary layer can give rise to a rate of wave growth which varies with spanwise position, thus leading an initially two-dimensional wave to a three-dimensional form. In many cases, the flow is nearly periodic in the spanwise direction (u may not be linear). Figure 14.15 shows two-dimensional waves at the left becoming three-dimensional as they roll up in the middle of the photograph.

Figure 14.15 *Region of three-dimensional wave amplification. (Source: Paper 12 by F. X. Wortmann in AGARD Conference Proceedings No. 224, 1977. Used with the permission of the author and publisher.)*

Peak-Valley Development with Streamwise-Vortex System, (3)
As the generated three-dimensional wave progresses downstream, the boundary layer flow develops into a more pronounced three-dimensional structure with an associated streamwise vortex system (see Fig. 14.16). At certain spanwise locations, called "peaks," the velocity fluctuations develop rather strongly. The neighboring locations are called "valleys."

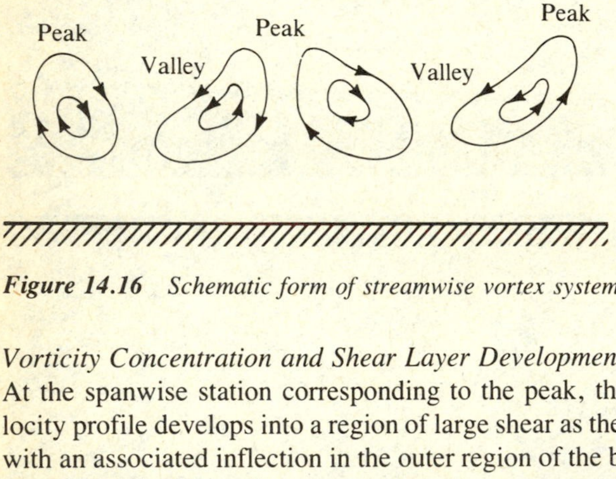

Figure 14.16 *Schematic form of streamwise vortex system for peak-valley development.*

Vorticity Concentration and Shear Layer Development, (4)
At the spanwise station corresponding to the peak, the instantaneous streamwise velocity profile develops into a region of large shear as the downstream distance increases with an associated inflection in the outer region of the boundary layer (see Fig. 14.17). This inflectional profile appears and disappears once after each cycle of the basic wave.
Breakdown, (5)
When the instantaneous velocity profile has a sufficiently large shear in the outer region of the boundary layer, a velocity fluctuation develops from the shear layer at a much higher frequency than that of the basic wave. This phenomenon is called breakdown. Klebanoff et al. [14.5] called these high-frequency fluctuations *hairpin eddies*. Though difficult to isolate, a hairpin eddy can be recognized at approximately the center of Fig. 14.18.

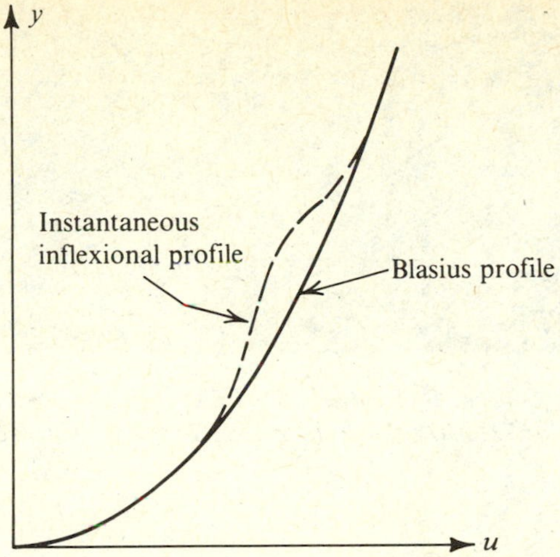

Figure 14.17 *Blasius and instantaneous profiles.*

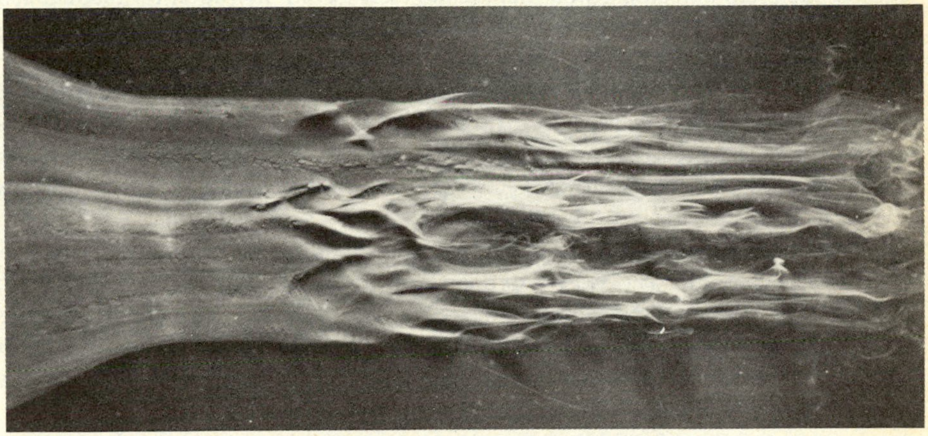

Figure 14.18 *Natural transition on a slightly inclined plate. At the same Reynolds number of 100,000 but 1° angle of attack, transition to turbulence occurs on the plate. (Source: H. Werlé, ONERA photograph in* An Album of Fluid Motion, *M. Van Dyke, Parabolic Press.)*

Turbulent-Spot Development, (6)

The "hairpin eddies" travel at a speed greater than the wave speed of the primary wave. As they travel downstream at a fraction of the free-stream speed, the eddies spread spanwise and toward the wall in the form of turbulent spots. Each spot grows almost linearly with distance while moving downstream. The turbulent spots grow until they encompass the whole boundary layer flow. This is considered to be the main region of transition and is pictured in Fig. 14.19.

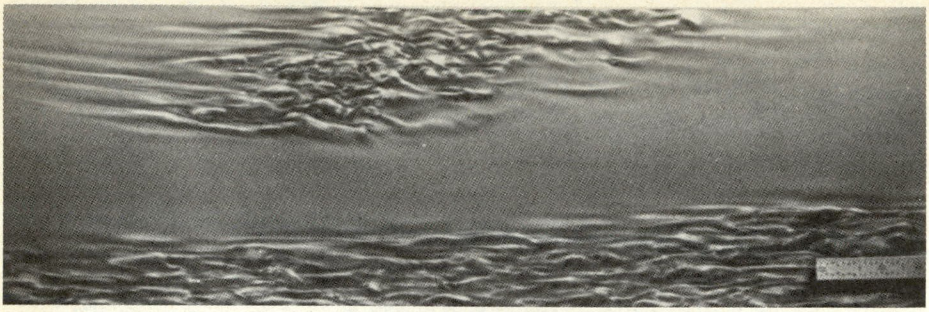

Figure 14.19 *Turbulent spot development. (Source: B. Cantwell, D. Coles, P. Dimotakis,* Journal of Fluid Mechanics, *vol. 87, pp. 641–672, 1978. With the permission of Cambridge University Press.)*

14.8.2 Salient Aspects of Transition

The following is a partial listing of the salient features of transitional flow. We shall be interested in learning what fluid parameters govern transition for internal and external flows.

Transition in Pipe Flows

The critical value at which transition occurs in pipe flow is $R_{cr} = 2300$. The actual value depends upon disturbances in the flow. Some experiments have shown the critical Reynolds number to reach as high as 40,000. The precise upper bound is not known, but the lower bound appears to be $R_{cr} = 2300$. Below this value, the flow remains laminar even when subjected to strong disturbances.

Transition is always accompanied by a change in the law of resistance. In laminar flow, the pressure gradient $\partial p/\partial x$ is proportional to the average flow velocity $\tilde{V}$. In turbulent flow, $\partial p/\partial x$ is proportional to the mean flow velocity. This partially explains the two different slopes in the Moody diagram of friction factor f versus Reynolds number. On log-log paper, the slope is a negative 45° for laminar flow. For completely turbulent flow, the slope is also negative but is less steep than the laminar flow curve.

For $2300 \leq R_D \leq 2600$, the flow alternates randomly between being laminar and partially turbulent. Near the centerline, the flow is more laminar than turbulent, whereas near the wall the flow is more nearly turbulent than laminar.

Transition in Flows over Bodies

Transition for boundary layer flow over bodies is dependent upon the pressure distribution, nature of the wall, wall roughness, and the nature of disturbances in the main flow. We shall discuss each briefly.

Pressure Affects Transition

If we assume that disturbances exist in the primary flow, then in the region where the pressure decreases (or velocity increases), transition is retarded. For instance, consider a flow of Reynolds number R_l. At a location x on the body, λ_1 is known, where

$$\lambda_1 \propto U_1 \frac{dU_1}{dx}$$

Let us vary the geometry by decreasing λ to λ_2, where $\lambda_2 < \lambda_1$, which is at a location where the velocity U_2 is less than the velocity U_1. Then the previously stable region for the R_{l_1} flow may be in an unstable region because λ_2 has decreased sufficiently.

Reynolds Number Affects Transition

An *increase* of velocity from a laminar flow to turbulent flow can bring about instability. Thus an increase of Reynolds number for a fixed λ enhances transition; or, stated alternatively, transition moves upstream. Therefore transition will occur earlier.

Curvature Affects Transition

The curvature of the body can be a *destabilizing* factor in fluid flow. Figure 14.20 shows two types of curvature and the resulting stabilizing effect.

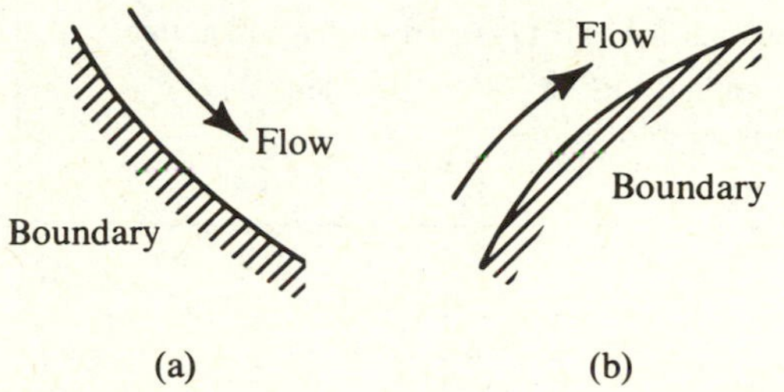

Figure 14.20 *Curvature effects on transition. (a) Destabilizing curvature. (b) Stabilizing curvature.*

Roughness Affects Transition

The absolute roughness ϵ affects the likelihood of transition and also increases the level of turbulence in the same manner as was discussed in turbulent pipe flow, Sec. 11.6. (See the discussion on turbulent pipe flow.)

Suction of the Fluid in the Boundary Layer Retards Transition

If we plot the log R_{cr} versus a shape parameter H (defined as the ratio of the displacement thickness δ^* to the momentum thickness θ)

$$H = \frac{\delta^*}{\theta} \tag{14.90}$$

we can observe the effect of suction by comparing the two curves in Fig. 14.21. The curves were obtained using a fairly broad range of assumed velocity profiles, ranging from the classic Blasius profile through sixth-order polynomials. Though there is a large scattering of data for $H > 2.4$, the results compare favorably. The Blasius solution is simply the $H \simeq 2.6$ case. The ordinate value for the Blasius solution is obtained as

$$\log \frac{U\delta^*}{\nu} = \log(1.71 \times \sqrt{2300}) \simeq 2.6 \tag{14.91}$$

The curves show the critical Reynolds number for various suction conditions as given by the shape factor H. With suction we can go to a slightly higher Reynolds number before transition.

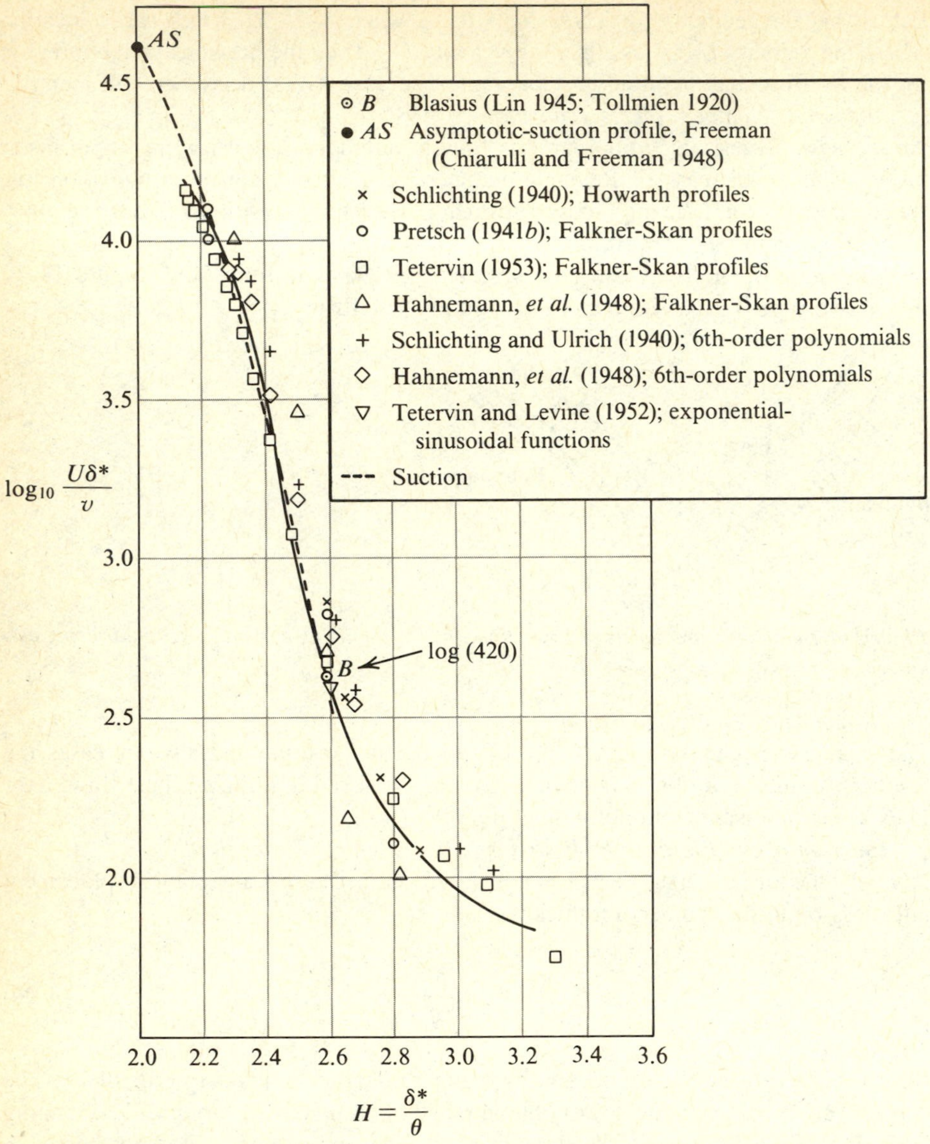

Figure 14.21 *Critical Reynolds number as a function of the shape parameter H. (Source: J. H. Stuart in L. Rosenhead, Ed.,* Laminar Boundary Layers, *Oxford University Press, England, © 1963.)*

Initial Turbulence or Disturbances Affect Transition

The recent efforts of Elder [14.6], Mochizuki [14.7], and Tani et al. [14.8] shed light on the effect of finite three-dimensional disturbances on transition. In order to appreciate the significance of their findings, let us first recall that in the pioneering work of Schubauer and Skramstad [14.9] they carefully delineated the neutral boundary beyond which small disturbances created by an oscillating ribbon could amplify. The neutral, or minimum, critical Reynolds number, $R_{\delta*}$, obtained was 410 for the flat-plate with zero pressure gradient, a value considerably lower than that computed earlier in the equally pioneering work by Schlichting [14.10]. It turns out, however, that the limiting conditions are particularly sensitive to the detailed assumptions of the theory, so that later refinements led Shen [14.11] to a value of 420 (see Example 14.6). Based on the distance from the leading edge, x, the value corresponding to that of Schubauer, Skramstad, and Shen, is approximately 60,000; whereas Elder uses the older value of 112,000. Elder, working with disturbances created by sparks discharging in a flat-plate boundary layer, comes to the rather strong conclusion that whenever disturbances* exceed 18 ± 2.5% of the free-stream velocity U, breakdown of the layer ensues for all R_x between 2×10^4 and 10^6, even for R_x values as much as three times smaller than the critical Tollmein-Schlichting (TS) value. The magnitude of the maximum root means that square fluctuations just before breakdown, which were observed in the unstable regime by Klebanoff et al. [14.5], reached 16% of U, which is in general agreement with the maximum values obtained by others such as Kovasznay et al. [14.12].

We should note that u' fluctuations exceeding 12% of maximum mean velocity in developed turbulent layers have not been observed in flat plate boundary layer or pipe or channel flows, so that a fluctuation of $0.18U$ represents a very intense disturbance indeed. Such intensities are unlikely to occur "naturally" and may be difficult to trigger artificially without recourse to sparks. Even for nurtured free-stream turbulence, the measure u'/U seldom exceeds 3–4% in a test section.

We are probably safe in generalizing that in the vast majority of cases, some TS amplification is needed for transition, and that the latter will not be complete across the span of a two-dimensional flow until some distance downstream of the TS critical point. Thus spherical roughness and even sparks apparently do not trigger large disturbances in their immediate wake.

Heat Affects Transition

Heat transferred from the boundary layer into the wall of the body is a stabilizing effect, whereas heat added from the body to the fluid in the boundary layer is destabilizing, as indicated somewhat in Fig. 14.22. (Notice the intense vortex filament.)

14.8.3 Instability versus Transition

We have stated that we cannot have transition without first having instability. The instability must grow from zero amplitude to some maximum, and the process must be continuous. Just as instability and transition differ, so too do the point of transition and the point of instability differ. Tollmein found for a flat plate at zero angle-of-attack that the point of instability was at $R_l = 420$, whereas the point of transition occurred at $R_l = 950$. The distance between the point of instability and the point of transition depends upon the degree of amplification. The point of instability must always lie upstream of the observed point of transition. The point of instability coincides with the point of minimum pressure for $10^6 \leq R_l \leq 10^7$.

*The disturbances are measured by a velocity fluctuation in the x-direction.

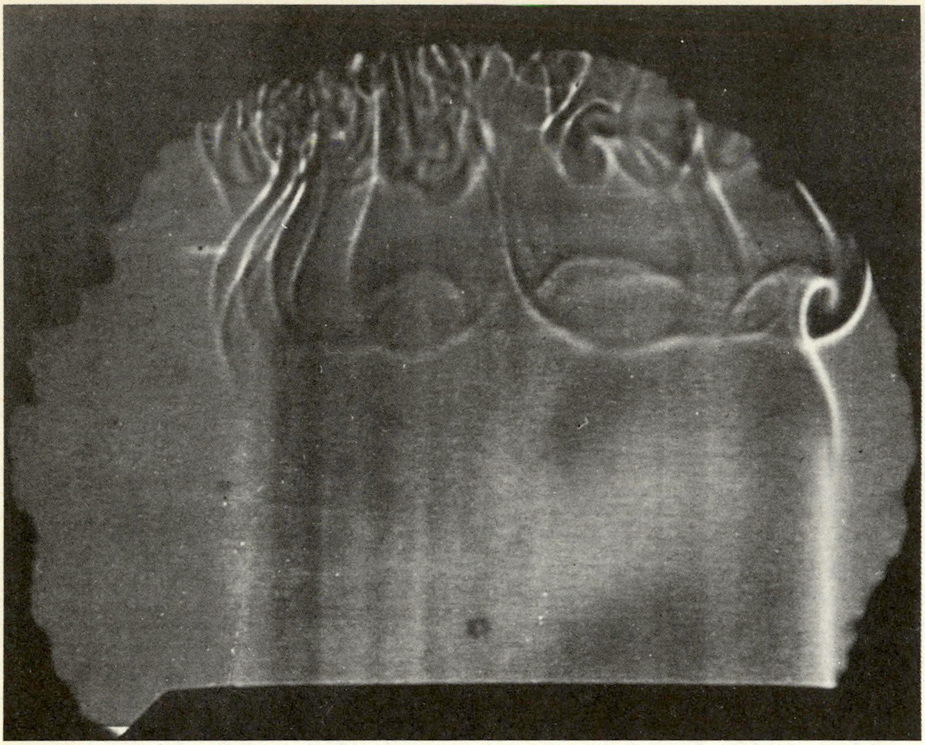

Figure 14.22 *Heat producing longitudinal vortices similar to Gortlër vortices (author unknown).*

Determining the Position of the Instability Point for a Body

The procedure we shall use in determining the position of the point of instability is as follows:

Step 1.

Given the body geometry, determine the pressure gradient using potential theory.

Step 2.

Knowing the pressure gradient, determine the shape factor λ, where

$$\lambda = \frac{\delta^2}{\nu} \frac{dU(x)}{dx} \tag{14.80}$$

Step 3.

Using Pohlhausen's method, calculate the displacement thickness δ^*:

$$\delta^* = \frac{1}{U(x)} \int_o^\delta (U - u)\, dy \tag{14.92}$$

where

$$\frac{u}{U} = F(\eta) + \lambda G(\eta) \qquad (14.81)$$

See Fig. 14.10.

Step 4.

Satisfy the condition for calculating the point of instability by

$$\frac{U(x)\delta^*}{\nu} = \left(\frac{U(x)\delta^*}{\nu}\right)_{cr} \qquad (14.93)$$

The critical Reynolds number is related to the pressure gradient λ by Fig. 14.23.

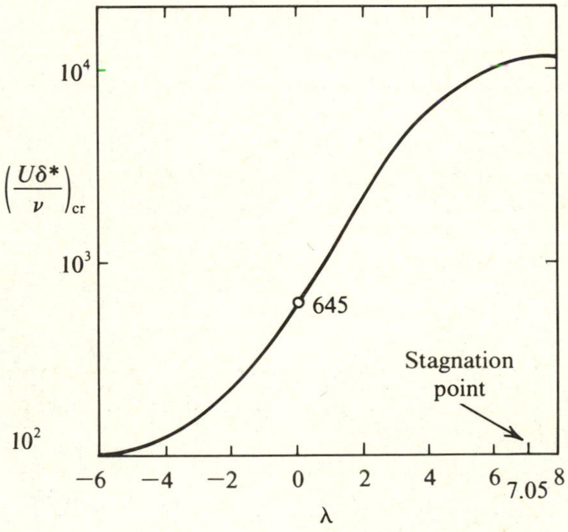

Figure 14.23 *The critical Reynolds number of boundary layer profiles with pressure gradient. (Source:* Boundary Layer Theory *by H. Schlichting. Copyright © 1960 McGraw-Hill Book Company. Used with the permission of McGraw-Hill Book Company.)*

Step 5.

For a given Reynolds number $R_l = Ul/\nu$, calculate $U(x)\delta^*/\nu$ and obtain the results similar to those shown in Fig. 14.24.

Step 6.

Superimpose the already existing $U(x)\delta^*/\nu$ versus λ curve of Fig. 14.23 onto the curves of Fig. 14.24. *Where the curves intersect on the points* x *on the body is where instability occurs.*

Example 14.6 illustrates the procedure.

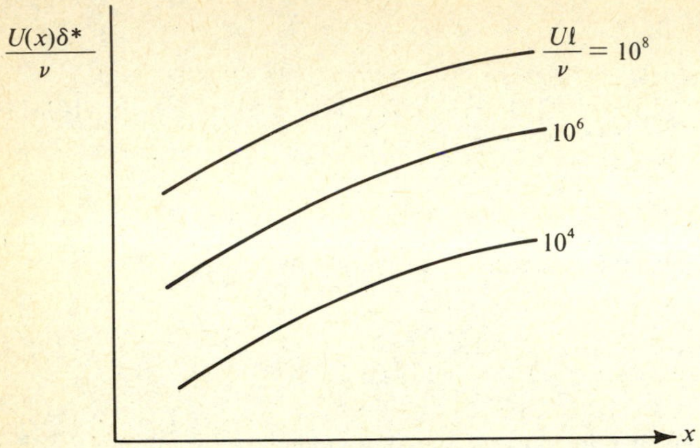

Figure 14.24 Curves of $U(x)\delta^*/\nu$ versus x for different Reynolds number $R_l = Ul/\nu$.

Example 14.6

Calculate the point of instability and the wave length λ_w for laminar boundary layer over a flat plate given a flow velocity U of 10 ft/s and kinematic viscosity $\nu = 1 \times 10^{-5}$ ft²/s.

Solution:

Step 1.

Since the body is a flat plate, the pressure gradient is zero.

Step 2.

Using Eq. (14.80) the shape factor

$$\lambda = 0 \tag{i}$$

Step 3.

The displacement thickness for a flat plate is

$$\delta^* = \frac{1.73x}{\sqrt{R_x}} \tag{ii}$$

The point of instability is that location in the free-stream direction x where the smallest value of Reynolds number exists. Here we find a single disturbance. Thus from Fig. 14.25, the curve C_i equals zero is the neutral disturbance curve for the Blasius flow. The region inside the curve corresponds to unstable disturbances, and that region outside the curve contains stable flow. The minimum value of Reynolds number is thus, from Fig. 14.25,

$$\frac{U\delta^*}{\nu} = 420 \tag{iii}$$

and represents the instability Reynolds number.

Example 14.6 *(Con't.)*

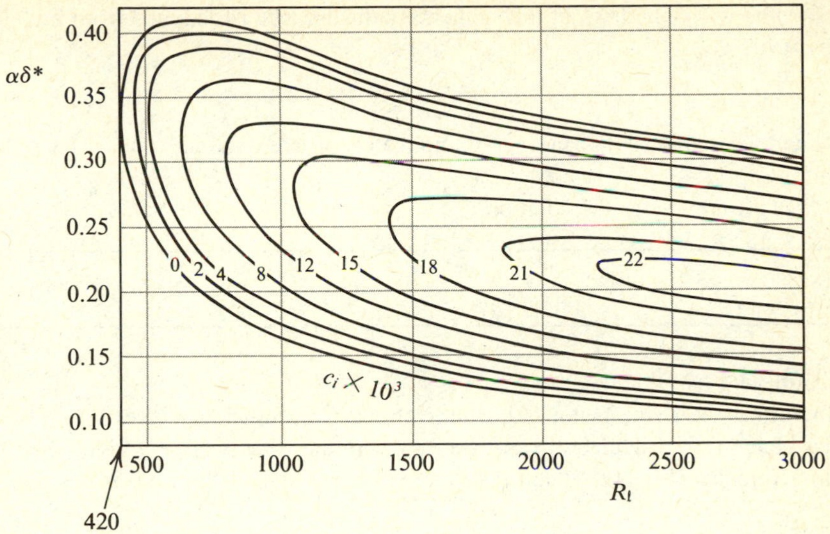

Figure 14.25 *Curves of neutral stability for incompressible flows. (Source: S. F. Shen,* Journal of the Aeronautical Sciences, *vol. 21, 1954. Copyright American Institute of Aeronautics and Astronautics.)*

We substitute Eq. (ii) into Eq. (iii) to obtain

$$x = \left(\frac{420}{1.73}\right)^2 \frac{\nu}{U} \qquad \text{(iv)}$$
$$= 0.059 \text{ ft}$$

We should compare this value of the point of instability with that value obtained by use of the shape factor H in the next Example 14.7.

The wavelength λ_w is associated with α of Fig. 14.25 by

$$\lambda_w = \frac{2\pi}{\alpha} \qquad \text{(v)}$$

At the critical Reynolds number $R_{cr} = 420$, Fig. 14.25 shows

$$\alpha\delta^* = 0.35 \qquad \text{(vi)}$$

Substituting Eq. (iii) into Eq. (vi) gives

$$\alpha = \frac{0.35U}{420\nu}$$
$$= \frac{0.35 \times 10}{420 \times 10^{-5}} \qquad \text{(vii)}$$
$$= 833.3$$

Example 14.6 *(Con't.)*

Thus the wavelength λ_w of the elementary disturbance is obtained from Eqs. (v) and (vii) as

$$\lambda_w = 7.54 \times 10^{-3} \text{ ft} \qquad \text{(viii)}$$

Thus, what looked like a very complicated problem turned out to be rather simple.

This completes the solution.

Example 14.7

Show that the shape factor H has the value 2.6 for boundary layer flow over a flat plate. Then calculate the position x where the flow is critical given that the flow velocity U is 10 ft/s and the kinematic viscosity of the fluid is 1×10^{-5} ft^2/s.

Solution:

The shape factor H is defined as

$$H = \frac{\delta^*}{\theta} \qquad \text{(i)}$$

where δ^* and θ are the displacement and momentum thickness, respectively. Their value is determined from the exact results of Eqs. (14.53) and (14.55):

$$\delta^* = \frac{1.73x}{\sqrt{R_x}} \qquad \text{(ii)}$$

$$\theta = \frac{0.664x}{\sqrt{R_x}} \qquad \text{(iii)}$$

Substituting Eqs. (ii) and (iii) into Eq. (i) gives

$$H = \frac{1.73}{0.664} = 2.605 \qquad \text{(iv)}$$

which agrees with the value given in Fig. 14.22.

The location on the flat plate where the flow is critical is obtained using Fig. 14.22. Since

$$\log R_{cr} = 2.605 \qquad \text{(v)}$$

and

$$\log\left(\frac{U\delta^*}{\nu}\right) = 2.605 \qquad \text{(vi)}$$

Example 14.7 *(Con't.)*

(where the displacement thickness δ^* is given by Eq. (ii)). Substituting Eq. (ii) into Eq. (vi) gives

$$\log(1.73\sqrt{R_x}) = 2.605 \qquad \text{(vii)}$$

For a flow velocity of 10 ft/s and a value of kinematic viscosity $\nu = 1 \times 10^{-5}$ $\text{ft}^2\text{/s}$, we obtain

$$\log(1730\sqrt{x_{cr}}) = 2.605 \qquad \text{(viii)}$$

using Eq. (vii). Solving for the critical location x gives

$$x_{cr} = \left(\frac{403}{1730}\right)^2 = 0.054 \text{ ft} = 0.65 \text{ in.} \qquad \text{(ix)}$$

Thus, at a distance of 0.65 in. from the plate's leading edge, the flow is critical.
This completes the solution.

14.9 Turbulent Boundary Layers

Before we delve into some of the formalism of turbulent boundary layers, we should point out a few of the similarities and dissimilarities between turbulent flow past a flat plate and turbulent flow through a circular pipe.

In Chaps. 10 and 11, we stated that viscosity plays an important role at the pipe entrance; that at some distance L downstream of the entrance, the boundary layers meet as shown in Fig. 14.26a. For turbulent flow over a flat plate, the boundary layer starts out as laminar flow at the leading edge, and then, depending upon many factors, the flow turns into transitional flow and very shortly thereafter turns into turbulent flow. The turbulent boundary continues to grow in thickness, as shown in Fig. 14.26b, with a small region below it called a viscous sublayer. In this sublayer, the flow is well behaved, just as in the laminar boundary layer.

The velocity profiles in the turbulent region for pipe flow and flow past a flat plate are similar, but not identical, when plotted using certain dimensionless parameters. Furthermore, the method of analysis for the two flows is identical. Both have power-law relationships, and both utilize the integral momentum equation.

Some of the dissimilarities are as follows. The pipe is pressurized by a pumping action, whereas for a flat plate the pressure is largely hydrostatic. Although roughness causes the pressure of both flows to drop, the relative roughness ϵ/D for turbulent flow in rough pipes is constant as long as the absolute roughness does not change and the pipe diameter is constant; while for flow over a flat plate, the relative roughness is replaced by ϵ/δ, and thus the relative roughness decreases along the plate for constant absolute roughness since the boundary layer thickness δ increases with distance. Thus we may find the plate becoming smooth if it is sufficiently long.

The complexity of the turbulent motion can be appreciated by examining Fig. 14.27. Tiny oil droplets were introduced into the laminar boundary layer close to the wind

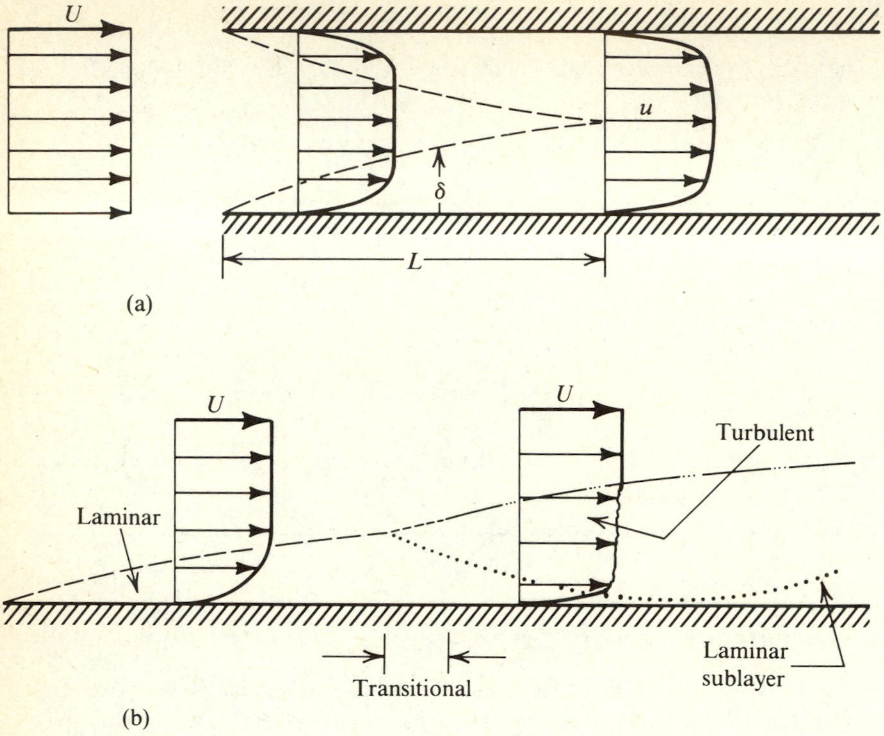

Figure 14.26 (a) Turbulent pipe flow. (b) Turbulent flow past a flat plate.

tunnel test section floor. A vertical sheet of light visualizes the flow pattern where $R_\delta = U\delta/\nu = 4000$.

In order to better describe a turbulent boundary layer, we partition it into an *inner layer* and an *outer layer*. The inner layer is broken down into a *viscous sublayer* and a *log-law region*. It is useful to divide the viscous sublayer into a *linear sublayer* and a *buffer layer*. The outer layer is divided into an *outer-law region* and a *viscous superlayer*, where the outer layer is approximately 80% of the turbulent boundary layer thickness. Table 14.3 presents a summary of characteristics of a turbulent boundary layer based on the contribution of H. Fernholz [14.13]. It is important that the characteristic length L and velocity U be defined for each region.

14.9.1 The Inner Layer, $0 \leq y/\delta \leq 0.2$

We assume two-dimensional incompressible flow past a flat plate in a region close to the surface $y/\delta \leq 0.2$ where the total shear stress approximately equals the wall shear stress $(p_{xy})_o$.

Linear Sublayer, $0 \leq u_* y/\nu < 3$

Figure 14.28 shows the turbulent and laminar shear stress and the mean flow velocity $\bar{u}$ for turbulent flow along a smooth wall. Very close to the wall, the viscous stress far outweighs the turbulent stress, because of the very large velocity gradient $d\bar{u}/dy$. The turbulent stresses are practically nonexistent because of the lack of eddies for

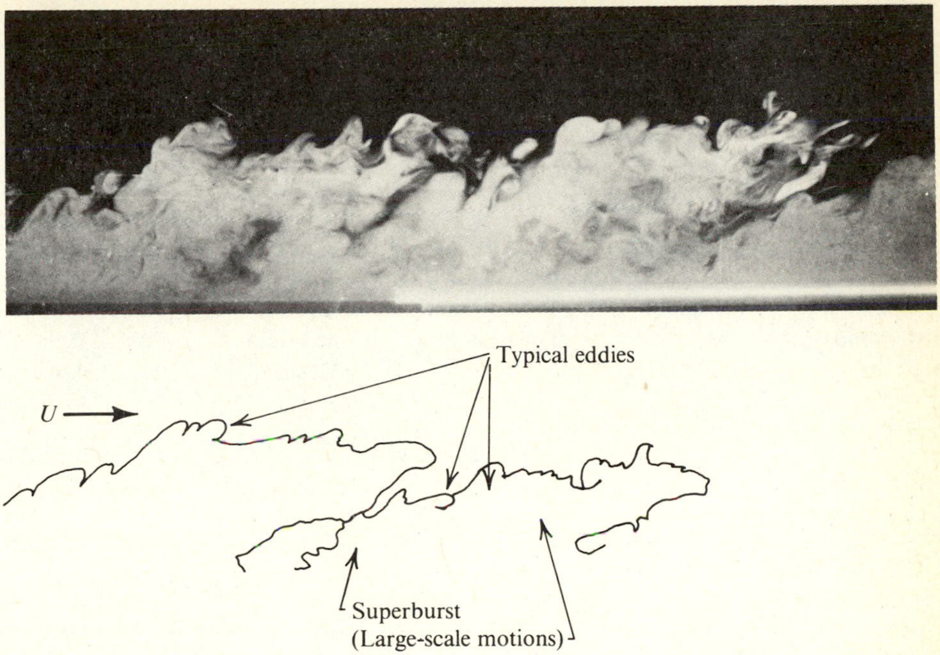

Figure 14.27 *Turbulent boundary layer past a wall. (Source: R. E. Falco,* Physics of Fluids, *vol. 20, pp. S124–S132, 1977. Used with the permission of the American Institute of Physics.)*

turbulent mixing. Within this very small layer, which we call the linear sublayer, we use Eq. (11.27) to obtain

$$p_{xy} = (p_{xy})_{\text{lam}} \approx \mu \frac{d\bar{u}}{dy} \tag{14.94}$$

where the shear stress at the wall $(p_{xy})_o$ is not necessarily the same as if it were a truly laminar flow. Integrating Eq. (14.94), we obtain

$$u^+ = \frac{\bar{u}}{u_*} = \frac{u_* y}{\nu} = y^+, \qquad y^+ < 3 \tag{14.95}$$

and hence the name "linear sublayer." Equation (14.95) is shown plotted in Fig. 14.29. the triggering mechanisms and phenomenology of the flow in the linear sublayer is given by Kline et al. [14.14]. Experimental evidence of Eckelmann [14.15] reveals that the pressure gradient has negligible effect on the flow in the sublayer.

Buffer Layer, $3 < u_* y/\nu < 40$

The buffer zone in the larger part of the viscous sublayer is a region where both viscous and turbulent shear stresses are important. Thus we cannot use either Eq. (14.95) or (11.37). Reichardt [14.16] suggested an empirical relationship

$$u^+ = 5 \ln y^+ - 3.05 \tag{14.96}$$

Table 14.3 Summary of Turbulent Boundary Layer Characteristics

	Inner Layer		**Outer Layer**	
	Viscous Sublayer	*Log-Law Region*	*Outer-Law Region*	*Viscous Superlayer*
Flow Condition	Intermittent turbulent	Fully turbulent	Fully turbulent, part rotational, part irrotational	Laminar-like
Dominating Mechanism	Dissipation	Production of turbulent energy	Extraction of kinetic mean flow energy by Reynolds stress gradient	Transfer of vorticity to the irrotational free stream
Dependence on Viscosity	Dependent	Independent	Independent	Dependent
Characteristic Length L	$\frac{\nu}{u_*}$	$\frac{\nu}{u_*}$	δ	$\frac{\nu}{V_e}$
Thickness within Boundary Layer	$0 < y < \frac{40\nu}{u_*}$	$\frac{40\nu}{u_*} < y < 0.2\delta$	$0.2\delta < y/\delta$	In the region $y > 0.4\delta$
Characteristic Velocity	u_*	u_*	u_*	V_e
Dominant Shear Stress	$\overline{u'v'} < \mu\frac{\partial\bar{u}}{\partial y}$	$\overline{u'v'} > \mu\frac{\partial\bar{u}}{\partial y}$		

Source: Adapted from *Topics in Applied Physics,* vol. 12, "Turbulence," Springer-Verlag, 1976. Used with the permission of the publisher and the author of Chapter 2, H.-H. Fernholz.

for the velocity, and this result is shown plotted in Fig. 14.29. No simple relationship for the mean velocity in this layer can be derived. The region is complicated by the production of turbulence connected to the bursts observed in the sublayer.

Logarithmic Law Region, $40 < u_*y/\nu < 0.2\ \delta u_*/\nu$

Ludwieg and Tillmann [14.17] studied the flow in the inner region where the viscous stress is negligible in comparison to the turbulent shear stress. A dimensional analysis* gives

$$\frac{\partial\bar{u}}{\partial y} = \frac{u_*}{y} f\left(\frac{u_* y}{\nu}\right) \tag{14.97}$$

*See P. Bradshaw [11.1], p. 34.

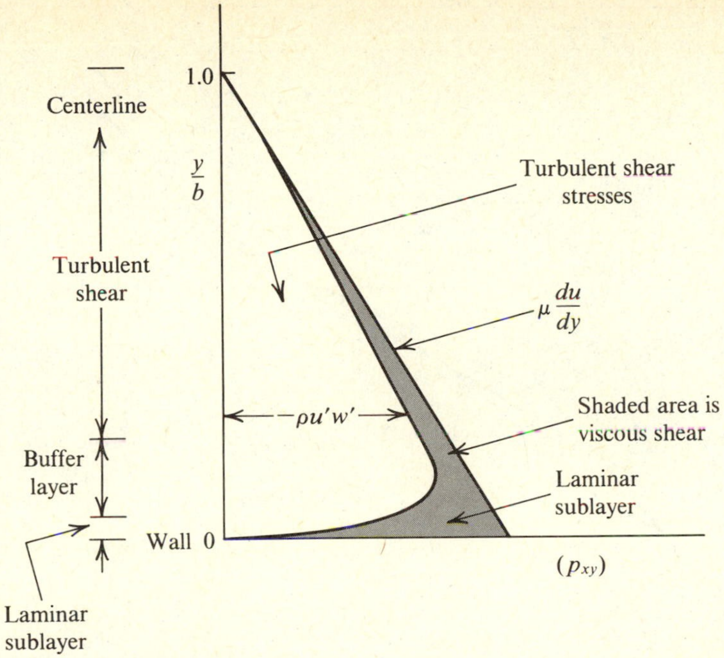

Figure 14.28 *Shear stress distribution for turbulent flow past a wall. (Source: Lectures on viscous flows by Professor T. Sarpkaya. Reproduced here by written permission of Professor T. Sarpkaya.*

Integrating Eq. (14.97), we get

$$u^+ = \frac{1}{\kappa} \ln y^+ + C \tag{14.98}$$

where experimentally we find $\kappa \approx 0.41$ and $C = 5.0$. This is one of the most important equations for turbulent boundary layers, since all measurements for skin friction drag are based on it. It is also appropriate to apply it to curved bodies, i.e., surfaces with pressure gradients.

14.9.2 The Outer Layer, $0.2 \leq y/\delta \leq 1.0$

If we carefully study the turbulent boundary layer in Fig. 14.27, we can see a strong interaction between an irregular unsteady rotational flow moving outwards and an irrotational regular flow. The enlarged free-surface of the turbulent boundary layer is due to the large eddies in the outer layer. These eddies are the major contributor to the turbulent energy that is associated with the velocity fluctuations u', v'. There is still doubt as to how these eddies arise. They could originate from the interfacial instability near the superlayer (see Fig. 14.30) or from the effect of the bursts from the inner region. We thus divide the flow in this outer layer into a rotational and irrotational flow, the boundary being intermittent. If the mean velocity is averaged

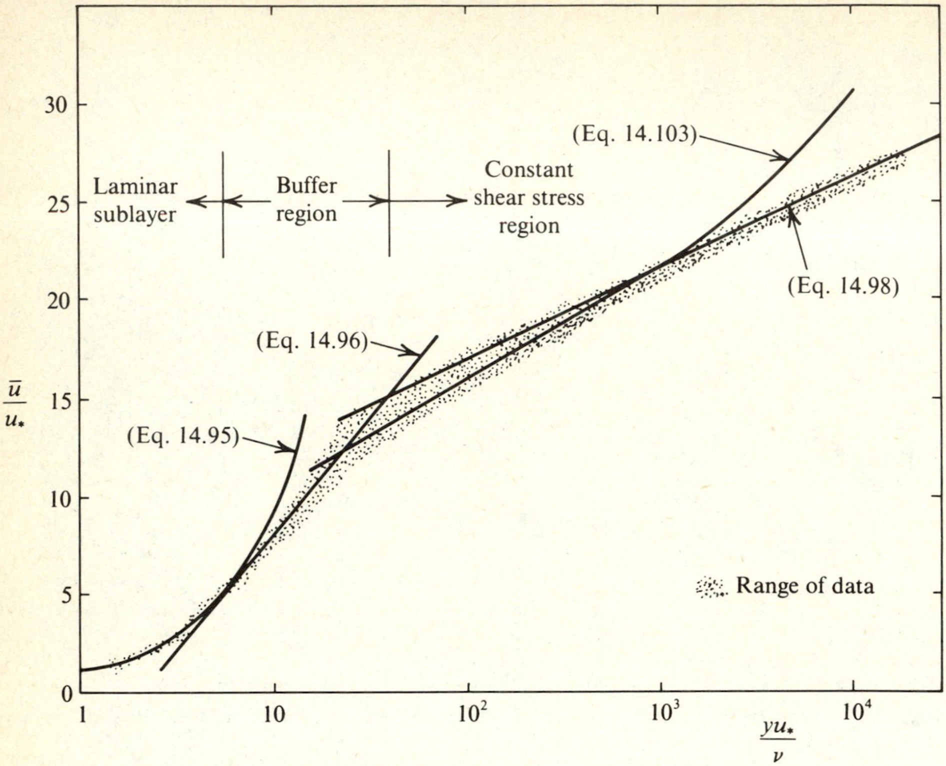

Figure 14.29 *Turbulent velocity distribution past a wall.*

over the rotational and irrotational regions of the outer layer, we find it convenient to express the velocity distribution in terms of the velocity defect $U - \bar{u}$, such that from Rotta [14.18],

$$\frac{U - \bar{u}}{u_*} = -\frac{1}{\kappa} \ln\left(\frac{y}{\delta}\right) + \frac{B}{\kappa}\left[2 - w\left(\frac{y}{\delta}\right)\right] \tag{14.99}$$

where $w(y/\delta)$ is Coles' wake function

$$w\left(\frac{y}{\delta}\right) = 39\left(\frac{y}{\delta}\right)^3 - 125\left(\frac{y}{\delta}\right)^4 + 183\left(\frac{y}{\delta}\right)^5 - 133\left(\frac{y}{\delta}\right)^6 + 38\left(\frac{y}{\delta}\right)^7 \tag{14.100}$$

and B is a constant, dependent upon the value of the constant pressure flow. Notice that $w(1) = 2$ such that $\bar{u} = U$ at the turbulent boundary layer $y = \delta$.

The Viscous Superlayer

Figure 14.30 shows a thin interface at the outer edge of the boundary layer. Corrsin and Kistler [14.19] postulated that a corrugated layer separates the turbulent flow from

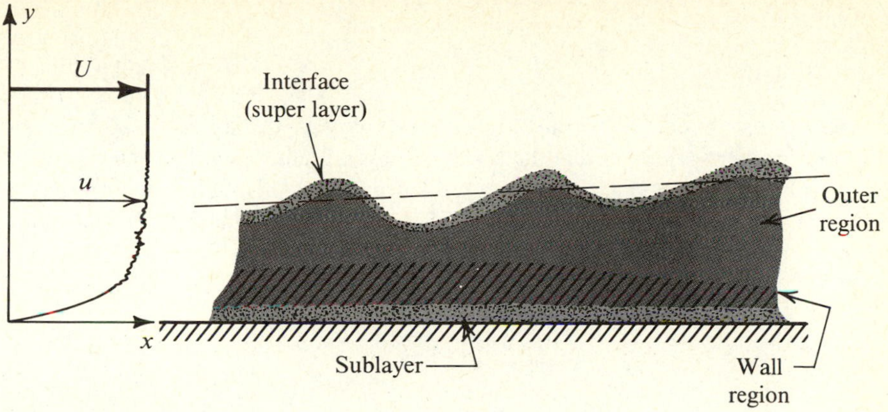

Figure 14.30 *The four regions of the turbulent boundary layer.*

the outer layer where only irrotational velocity fluctuations are observed. Its thinness is maintained by the random stretching of vortex tubes (see Sec. 11.3.1). The mean velocity component normal to the mean position of the interface is the entrainment velocity V_e, and is the characteristic velocity in the superlayer. The characteristic length L is defined as ν/V_e. Applying the conservation of mass to a control volume bounded by $y = 0$ and $y = \delta$, $x = x_o$ to $x = x_o + \Delta x$ yields

$$\rho V_e = \frac{d}{dx}\int_o^\delta \rho\bar{u}\,dy \tag{14.101}$$

14.9.3 Fully Turbulent Boundary Layer Flow

There is an obvious weakness in the mathematical modeling of the velocity for turbulent flow past a flat plate, as can be seen in the results presented in Fig. 14.29. First, it appears that we must use a partitioning of turbulent flow past a wall. Second, each region has its own pet equation for the mean velocity. It seems unnecessary to partition any flow field when we know that nature is a continuous phenomenon, despite being often random and unpredictable. Some of the equations in this section have discontinuities in the derivatives, and this should not be permitted in modeling nature. The answer we should seek lies in trying to unify these different results in a proper expression for either the eddy viscosity or mixing length. For instance, some theorists assume an eddy viscosity of exponential form and can therefore combine the various layers together in the inner region. Some have tried using certain tricks. The reader is urged to read the article "Turbulent Boundary Layer" by L. Kovasznay [14.20], to obtain a greater appreciation of what is involved here.

One of the more important quantities we shall soon be interested in calculating is the friction drag coefficient C_{D_f} for a flat plate. Von Kármán pointed out that the drag coefficient can be calculated from Eqs. (14.49), (14.50), and (14.76) as

$$C_{D_f} = \frac{2}{L}\int_o^\delta \frac{\overline{u}}{U}\left(1 - \frac{\overline{u}}{U}\right) dy \tag{14.102}$$

The velocity distribution $\overline{u}/U$ in a turbulent boundary layer has been shown to be logarithmic in every region except the linear sublayer. Rather than become entangled with logarithmic integrations, Prandtl suggested that for fully turbulent flow past a smooth wall the mean velocity distribution in the region where turbulent shear dominates is

$$\overline{u} = 8.74u_* \left(\frac{yu_*}{\nu}\right)^{1/7} \tag{14.103}$$

so that the shear stress at the wall $(p_{xy})_o$ becomes

$$(p_{xy})_o = 0.0225\ \rho\overline{u}^2 \left(\frac{\nu}{y\overline{u}}\right)^{1/4} \tag{14.104}$$

The usefulness of Eqs. (14.103) and (14.104) requires that $10^5 < R_x < 10^7$, where $R_x = Ux/\nu$. Let us assume that the mean velocity distribution $\overline{u}$ and the shear stress at the wall $(p_{xy})_o$ hold equally as well throughout a fully developed turbulent boundary layer over a flat plate that is held parallel to the ambient flow. Let us further assume that the mean velocity distribution is modified to read:

$$\boxed{\frac{\overline{u}}{U} = \left(\frac{y}{\delta}\right)^{1/7}} \tag{14.105}$$

where δ is now the turbulent boundary layer thickness. The rationale behind using results of turbulent pipe flow for turbulent boundary layer flow over a flat plate is to consider the thickness of the boundary layer to be equivalent to the radius of the pipe, such that the maximum velocity at the center of the pipe is analogous to the free flow velocity U. Furthermore, the velocity distribution as given by Eq. (14.105) is a fair approximation to fully turbulent flow in a region where turbulent shear dominates, as shown by Fig. 14.31.

Substituting the velocity and shear stress expressions of Eqs. (14.105) and (14.104), respectively, into the von Kármán momentum integral Eq. (14.16) results in the expression

$$0.0225\ \rho U^2 \left(\frac{\nu}{U\delta}\right)^{1/4} = \rho U^2 \left\{\frac{d}{dx}\int_o^\delta \left(\frac{y}{\delta}\right)^{1/7}\left[1 - \left(\frac{y}{\delta}\right)^{1/7}\right] dy\right\} \tag{14.106}$$

Performing the indicated integration, we obtain a first-order differential equation of a form similar to what we had for laminar flow:

$$0.0225 \left(\frac{\nu}{U\delta}\right)^{1/4} = \frac{7}{72}\frac{d\delta}{dx} \tag{14.107}$$

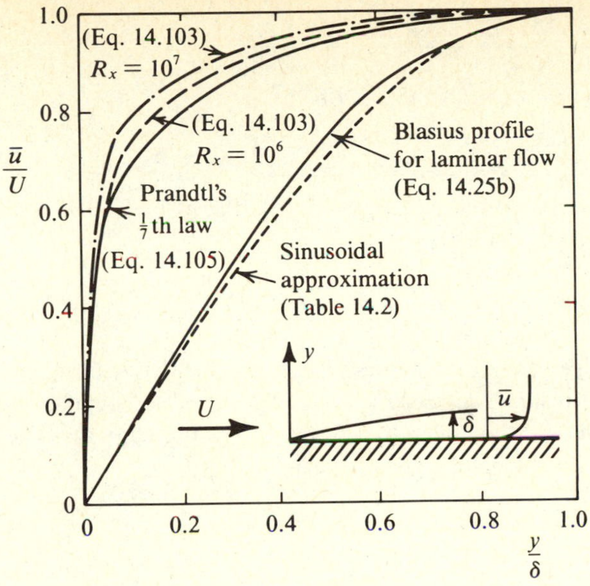

Figure 14.31 *Comparison of laminar and turbulent velocity distributions past a flat plate.*

This equation is separable and, integrated once again, the turbulent boundary layer thickness δ becomes

$$\boxed{\delta = 0.37x\left(\frac{\nu}{Ux}\right)^{1/5}} \qquad (14.108)$$

We note that the turbulent boundary layer grows as $x^{4/5}$, whereas we learned earlier that the laminar boundary layer grows as $x^{1/2}$. Thus the boundary layer for laminar flow does not grow as fast as the turbulent boundary layer, as would be expected.

The drag force on a flat plate of width b due to a fully turbulent flow can be found by integrating the shear stress $(p_{xy})_o$ of Eq. (14.104):

$$D = 2b\int_o^l (p_{xy})_o\,dx = 0.045\,b\rho U^2\left(\frac{\nu}{U}\right)^{1/4}\int_o^l \frac{dx}{\delta^{1/4}} \qquad (14.109)$$

Substituting the expression for the turbulent boundary layer thickness δ of Eq. (14.108) into the integrand of the drag force of Eq. (14.109) yields, after integration, the drag force

$$D = 0.072\,\rho U^2 bl\left(\frac{\nu}{Ul}\right)^{1/5} \qquad (14.110)$$

The drag coefficient for a smooth flat plate is then

$$C_{D_f} = \frac{D}{\frac{1}{2}\rho U^2(2bl)} = 0.072\, R_l^{-0.2} \tag{14.111}$$

where R_l is the Reynolds number based on the length l of the flat plate. *Experiment shows*, however, that a more realistic value is

$$C_{D_f} = 0.074\, R_l^{-0.2}, \qquad 5 \times 10^5 < R_l < 10^7 \tag{14.112}$$

Equation (14.111) was derived assuming the flat plate to be completely turbulent over its entire length. If the flat plate has a sharp leading edge, then its boundary is not completely turbulent. A portion of it is laminar from the leading edge to some downstream position. For this case, Prandtl suggested using

$$C_{D_f} = 0.074\, R_l^{-0.2} - A\, R_l^{-1}, \qquad 5 \times 10^5 < R_l < 10^7 \tag{14.113}$$

where A has various values depending on the value of R_l (see Fig. 14.32 for plots of C_{D_f} versus Reynolds number for three different values of A).

We should be extremely cautious in using Eq. (14.112). Recent LDV experimental measurements reveal that it is $\pm 25\%$ inaccurate. A better expression for the velocity distribution is the logarithmic expression of Eq. (11.37). Schlichting used Nikuradse's expression in the von Kármán momentum integral expression and, in a manner similar to that which we used to obtain the drag coefficient of Eq. (14.111), he obtained the semiempirical equation

$$C_{D_f} = \frac{0.455}{(\log R_l)^{2.58}} \qquad R_l > 10^7 \tag{14.114}$$

Once again, *to allow for the flow being laminar at the front edge of the flat plate*, we can modify Eq. (14.114) to read

$$C_{D_f} = \frac{0.455}{(\log R_l)^{2.58}} - \frac{A}{R_l} \tag{14.115}$$

For instance A equals 1700 corresponds to transition where $Ux/\nu = 5 \times 10^5$. Other values of A are available for different critical Reynolds numbers, and are presented in Table 14.4.

Figure 14.32 presents a summary of the friction drag coefficients C_{D_f} for turbulent flows over a flat plate. A great deal of experimental data is scattered in the lower turbulent Reynolds number range, indicating the great difficulty in obtaining consistent results.

All that we have presented has been for the smooth plate. Figure 14.33 depicts the problem that arises when the plate is rough. As we move downstream on the plate,

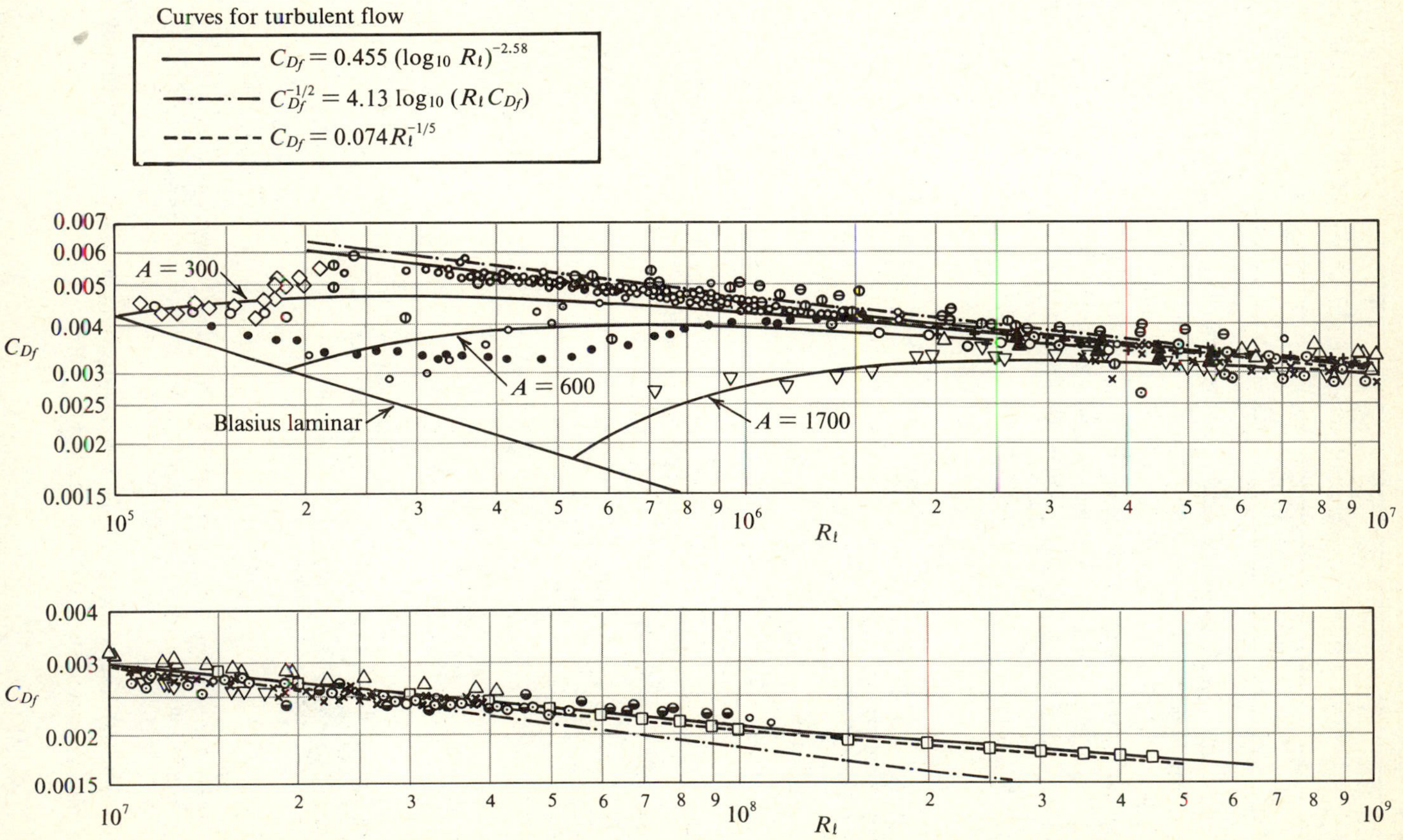

Figure 14.32 *Friction drag coefficient versus Reynolds number for flow over a flat place. (Source: S. Goldstein,* Modern Developments in Fluid Dynamics, *Vol. II, Oxford Engineering Science Series, Oxford University Press, England, © 1938.)*

Table 14.4 Values of Transition Constant A in Eq. (14.113)

$R_{cr} = \dfrac{Ux_{cr}}{\nu}$	3×10^5	4×10^5	5×10^5	6×10^5	10^6
A	1060	1400	1700	2080	3340

the wall shear stress $(p_{xy})_o$ decreases and the thickness of the viscous sublayer increases. There are three zones: a completely rough zone, a transition zone, and a hydraulically smooth zone. The friction drag will depend upon the density distribution of the protrusions, and on their height and shape. If they protrude far enough into the flow (case b or c), they produce vortices that will add to the turbulence in the region. As in pipe flow, the drag will not depend upon Reynolds number but on these elements of roughness.

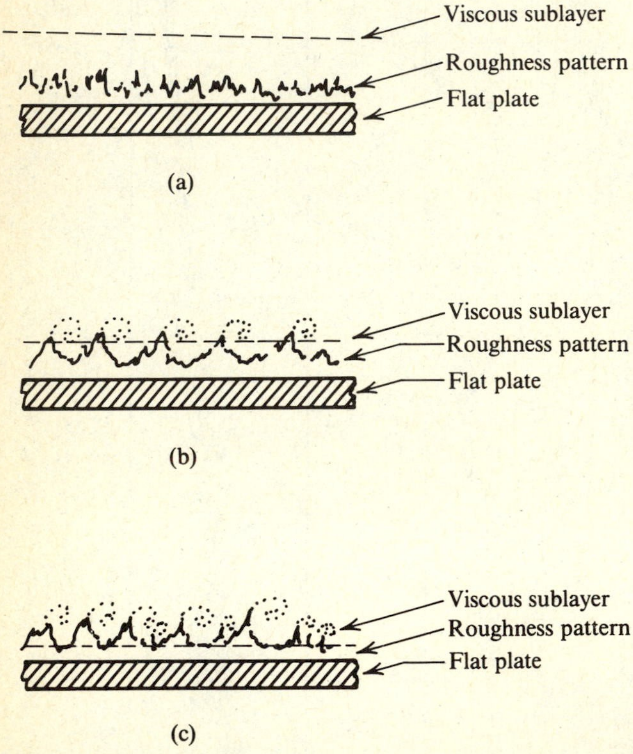

Figure 14.33 *Types of flow over rough plate. (a) Hydraulically smooth zone,* $0 < (\epsilon u_*/\nu) < 5$. *(b) Transition zone,* $5 \leq (\epsilon u_*/\nu) \leq 70$. *(c) Completely rough zone,* $(\epsilon u_*/\nu) > 70$.

Experimental results show that the velocity distribution can be represented by a logarithmic expression such that the drag coefficient is

$$C_{D_f} = \left(A + 1.62 \log \frac{l}{\epsilon}\right)^a \tag{14.116}$$

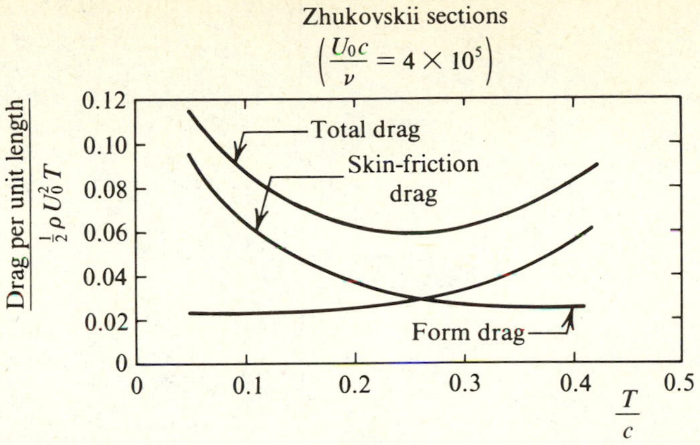

Figure 14.36 *Drag per unit length versus thickness ratio T/c for $R_C = 4 \times 10^5$ (based on frontal area). (Source: S. Goldstein,* Modern Developments in Fluid Dynamics, *Vol. II, Oxford Engineering Science Series, Oxford University Press, England, © 1938.)*

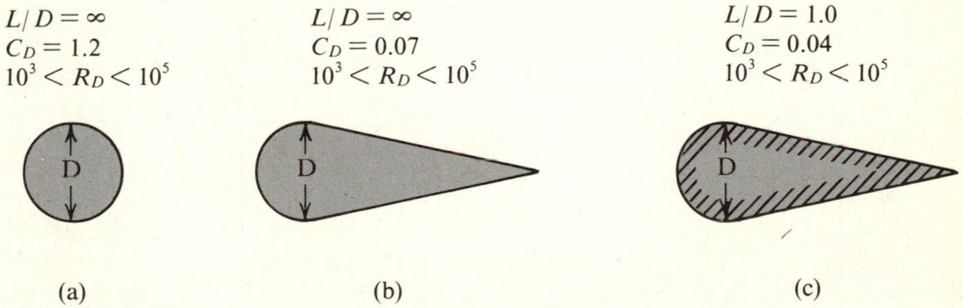

Figure 14.37 *Effect of shape on C_D. (a) Bluff body: a cylinder. (b) 2-D streamlined body. (c) 3-D streamlined body.*

Figure 14.38 shows the general nature of the change in C_D for bodies of variously shaped sections such as the airfoil, oval, and ellipse. Streamlining (shaping the afterbody) reduces the drag. (Compare the lenticular body *g* drag curve with the strut body *h* drag curve.) Additional streamlining can even cut these drags in half.

Figure 14.39 gives the drag coefficient for a variety of bluff bodies. Such information would be useful in calculating the drag force on buildings, water towers, and smoke stacks. The shape that gives the greatest drag is seen to be the square cylinder, which means skyscrapers have one of the worst aerodynamic shapes.

At moderate Reynolds number, a bluff body has a drag composed principally of pressure drag caused by the pressure drop fore and aft of the body. A bluff body that has an afterbody will have a small pressure drag because of the narrow wake, but an increased friction drag because of the greater surface area. The total drag, however, will be less for the attached afterbody object than the bluff body. Hence C_D is much smaller for streamlined bodies at high and moderate Reynolds number than for bluff bodies of the same projected frontal area.

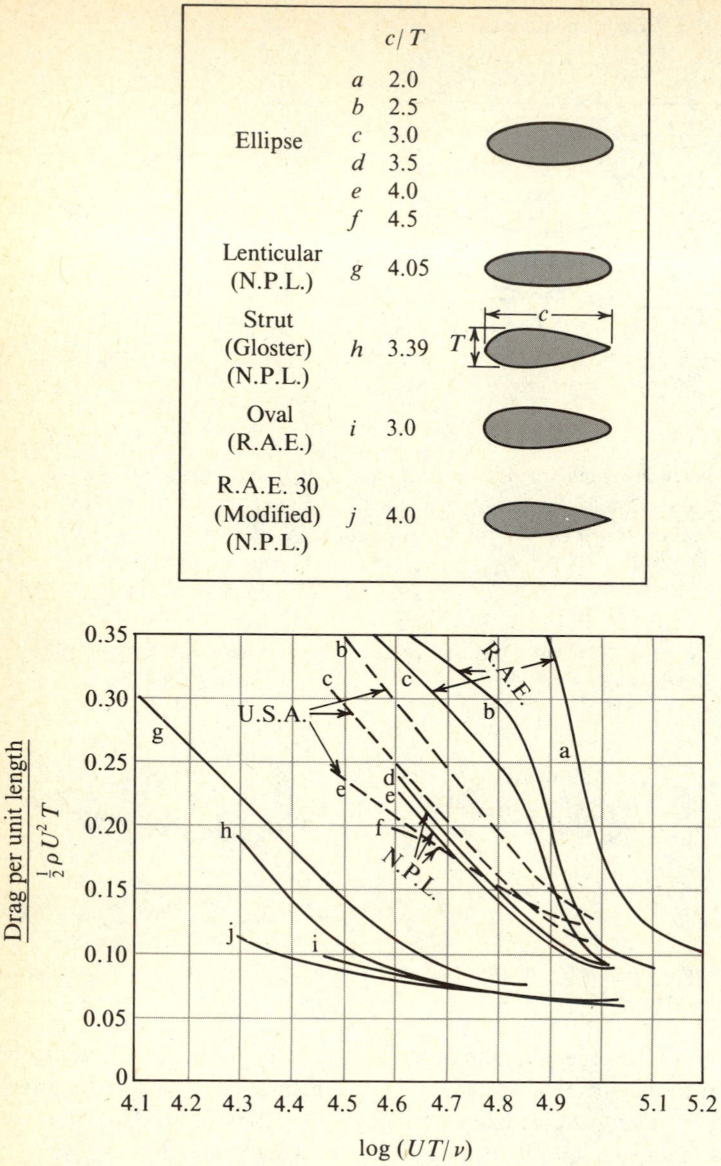

Figure 14.38 *C_D versus R_l for various cylindrical surfaces. (Source: S. Goldstein,* Modern Developments in Fluid Dynamics, *Vol. II, Oxford Engineering Sciences Series, Oxford University Press, England, © 1938.)*

14.10.4 Effect of Roughness on the Drag of Airfoil Shapes

Now that we have seen that the streamlined airfoil is the best shape for minimum drag, we can ask whether any other factor besides shape affects the drag force on an airfoil? The answer lies in the degree of surface roughness. The influence of surface

Object	C_D	R_D
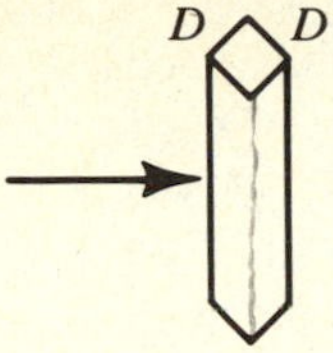 Square cylinder	2.0	$\frac{UD}{\nu} = 3.5 \times 10^4$
 Hollow hemisphere	1.42	$10^4 < \frac{UD}{\nu} < 10^6$
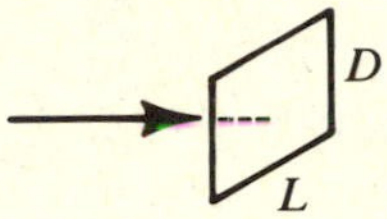 Normal flat plate	L/D 1 1.18 5 1.2 10 1.3 20 1.5 30 1.6 ∞ 1.95	$\frac{UD}{\nu} > 10^3$
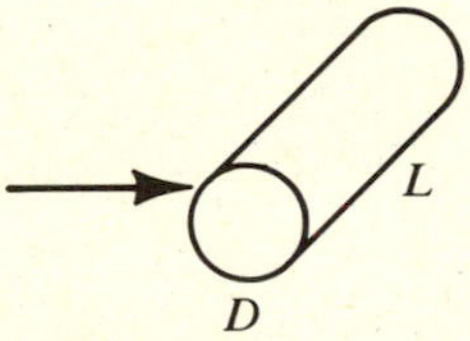 Circular cylinder	L/D 1 0.63 5 0.8 10 0.83 20 0.93 30 1.0 ∞ 1.2	$10^3 < \frac{UD}{\nu} < 10^5$
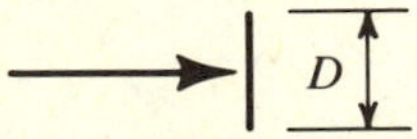 Normal circular disk	1.17	$\frac{UD}{\nu} > 10^3$
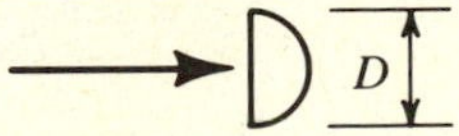 Solid hemisphere	1.17	$10^4 < \frac{UD}{\nu} < 10^6$
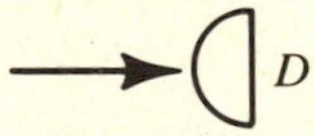 Solid hemisphere	0.42	$10^4 < \frac{UD}{\nu} < 10^6$
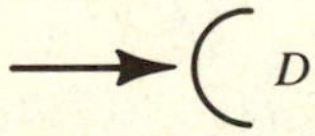 Hollow hemisphere	0.34	$10^4 < \frac{UD}{\nu} < 10^6$
 Sphere	$24 R_D$ 0.47 0.2	$\frac{UD}{\nu} < 1$ $10^3 < \frac{UD}{\nu} < 3 \times 10^5$ $\frac{UD}{\nu} > 3 \times 10^5$
 Flat plate	$1.328/\sqrt{R_L}$ $0.074\, R_L^{-0.2}$	Laminar Turbulent, $R_L \leqslant 10^7$

Figure 14.39 *C_D for various bluff bodies.*

roughness on airfoil drag is particularly noticeable at high Reynolds number, a result we found in pipe flows. This can best be illustrated by the following test described in Ref. 14.3. An NACA 0012 symmetrical foil's surface was roughened by painting it with a paste made by mixing carborundum powder with a suitable lacquer. Two grades of carborundum were used: the average size of the particles of the first grade (FF) was about 0.001 in. and that of the second (FFF) about 0.0004 in. Curves obtained from the tests are shown in Fig. 14.40 for a Reynolds number defined as $R_c = Uc/\nu$. Examining Fig. 14.40, we see that roughening causes a much greater drag once the flow reaches a Reynolds number near 10^6. For $R_c < 10^{5.7}$, roughness appears to play no significant role in increasing drag. Thus, there is some experimental evidence that *small degrees of roughness significantly increase drag at high Reynolds numbers.*

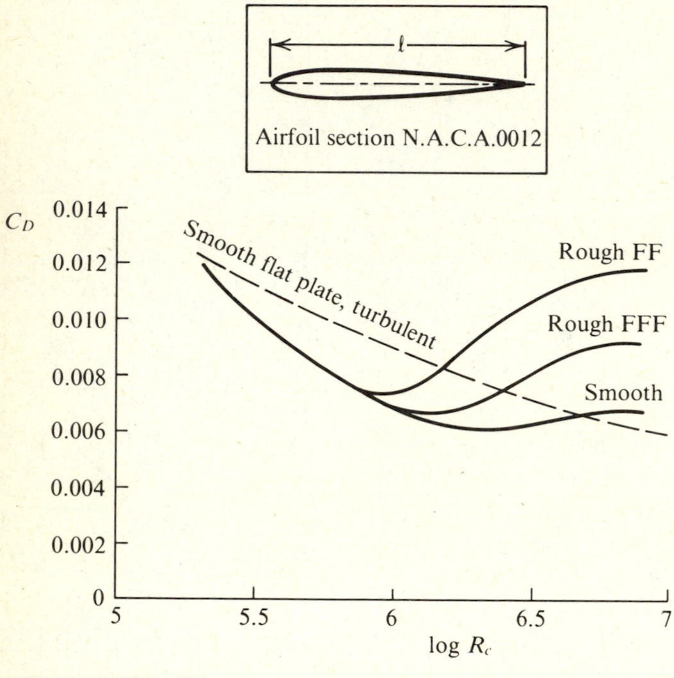

Figure 14.40 *Effect of roughness on drag. (Source: S. Goldstein,* Modern Developments in Fluid Dynamics, *Vol. II, Oxford Engineering Sciences Series, Oxford University Press, England, © 1938.)*

14.10.5 Aspects of Design for Minimum Drag

We have previously stated that the total drag on a body that is completely immersed in a fluid is composed of pressure (or form) drag and friction (or surface) drag. Examples of bodies totally immersed in a fluid are airplanes, rockets, submarines, torpedoes, and spacecraft. Bodies that are immersed in more than one fluid are hydrofoils and surface vessels, where additional components of drag such as surface tension forces and wave drag must be included in calculating the total drag.

We have demonstrated that friction drag depends not only upon the roughness of the wetted surface, but also on the type of flow, i.e., laminar, transitional, or turbulent flows. We have shown that in a laminar boundary layer, the friction drag is far less in magnitude than in a turbulent boundary layer, so that we want to maintain a laminar boundary layer along the body for as long as possible so that the instability point can be postponed as far downstream as possible. Thus, the aerodynamicist designs laminar airfoils that promote laminar boundary layer flow: he does this by moving the chordwise station of maximum thickness as far rearward as possible, where the point of minimum pressure (which is located at the point of maximum thickness) exists. Earlier in Sec. 14.8.3, we showed that the transition point occurs close to the point of minimum pressure.

As we indicated in Sec. 14.10.3, we can minimize the pressure drag by streamlining the body so as to minimize the wake region. Since the wake region is downstream of the separation point, we move the separation point as far downstream as possible. Thus, as we have stated, a slender body with a tapering section rearward has far less pressure drag than the same slender body with a blunt rearward section.

The position of the separation point for curved surfaces cannot be precisely determined analytically. The separation point depends on whether the flow is laminar, transitional, or turbulent. The separation point for laminar flow is approximately in the region of the largest body cross section. By increasing the velocity (and therefore the Reynolds number), the laminar flow may become turbulent upstream of the separation point. When this occurs, the separation point moves downstream, resulting in the smaller wake.

Form drag can be of significant magnitude even in the absence of separation. It tends to increase as the fineness ratio, (axial length/maximum diameter), is decreased to a certain limit.

14.10.6 Applications

Drag on a Cylinder

Consider flow past a circular cylinder of infinite axial length placed normally to the ambient flow velocity U and in the horizontal plane. At low Reynolds number, $R_D < 1$, the flow is termed creeping motion with no eddies or vortices existing downstream of the cylinder. For $5 < R_D < 50$, the laminar flow separates on the cylinder and encloses a stable vortex attached to the cylinder followed by a laminar wake of low vorticity. At $60 < R_D < 5000$, the oscillatory wake increases in amplitude and begins to roll up into distinct vortex patterns. The sequences are shown in Fig. 14.41. Because of the growing amplitude of the vortices, the wake is now no longer stable. The vortices alternate and detach, moving downstream in the flow. This phenomenon is called the *Kármán vortex street*. For $R_D > 5000$, there is no laminar vortex shedding. Throughout the period of vortex shedding, $60 < R_D < 5000$, and the drag coefficient $C_D \approx 1.2$.

For $R_D > 5 \times 10^3$, the wake behind the cylinder can no longer sustain the laminar shed vortices: it is completely turbulent. The flow in the boundary layer around the cylinder from the stagnation point to the separation point is laminar up to a Reynolds number $R_D = 5 \times 10^5$. From the point where $R_D = 5 \times 10^3$, to the point where

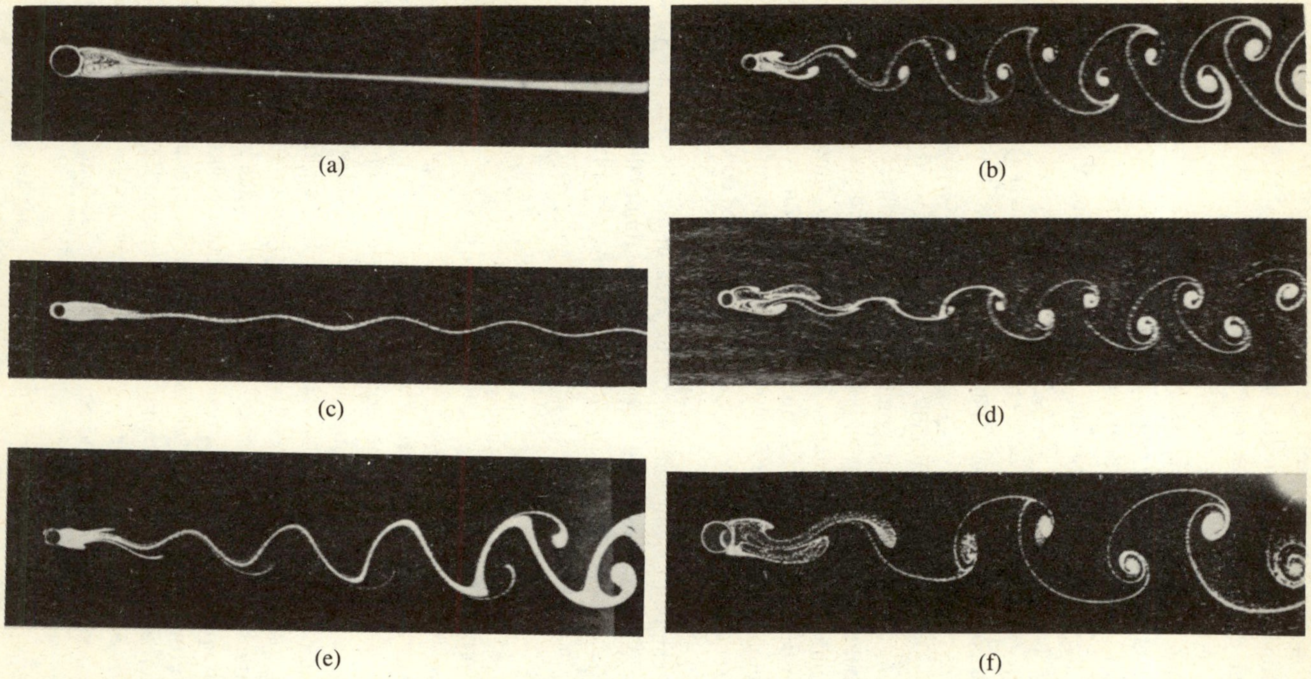

Figure 14.41 *Streaklines in the wake behind a circular cylinder in a stream of oil, (a)* $R_D = 32$*. (b)* $R_D = 73$*. (c)* $R_D = 55$*. (d)* $R_D = 102$*. (e)* $R_D = 65$*. (f)* $R_D = 161$*. (Source: F. Homann,* Forschg. Ing.-Wes., *vol. 7, pp. 1–10, 1936. Used with the permission of the publisher VDI-Verlag GMBH.)*

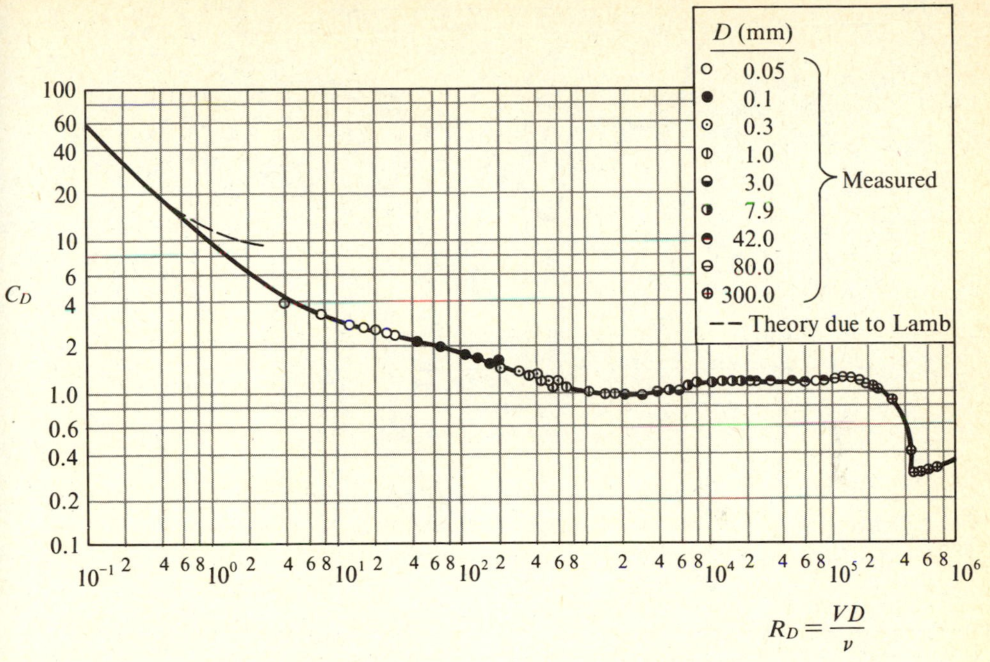

Figure 14.42 *Drag coefficient for an infinitely long cylinder. (Source:* Boundary Layer Theory *by H. Schlichting. Copyright © 1960 McGraw-Hill Book Company. Used with the permission of McGraw-Hill Book Company.)*

$R_D = 5 \times 10^5$, the drag coefficient C_D is nearly constant at a value of 1.2. This is shown in Fig. 14.42.

At $R_D = 5 \times 10^5$, we note in Fig. 14.42 a sharp decrease in the drag coefficient C_D. This results from a delay of the separation point from $\theta = 83°$ to $50° < \theta < 60°$ on the body (we have already discussed the mechanism of this earlier in this chapter). We stated in the section on factors affecting transition that surface roughness is important. The size of the roughness element on a cylinder is seen in Fig. 14.43 to have a pronounced effect on the drag coefficient, particularly around R_{cr}. An increase of roughness has the effect of decreasing the critical Reynolds number, and thus enhancing turbulence, which in turn would delay separation. Thus, *a practical way to reduce wake drag would be to roughen the surface of the cylinder*. This helps explain why a golf ball is dimpled, and why vortex generators are attached on the leading edge of an aircraft's wing.

The intensity of skin friction, i.e., the shear stress $(p_{xy})_o$ on a circular cylinder, is presented in Fig. 14.44. The figure shows a maximum shear stress $(p_{xy})_o$ at $\theta = 60°$. Beyond this angle, the intensity falls off rapidly to zero, which occurs at the separation point on the cylinder. Its location, we recall, is a function of Reynolds number. The minimum values of $(p_{xy})_o$ correspond to transition from laminar to turbulent flow in the boundary layer.

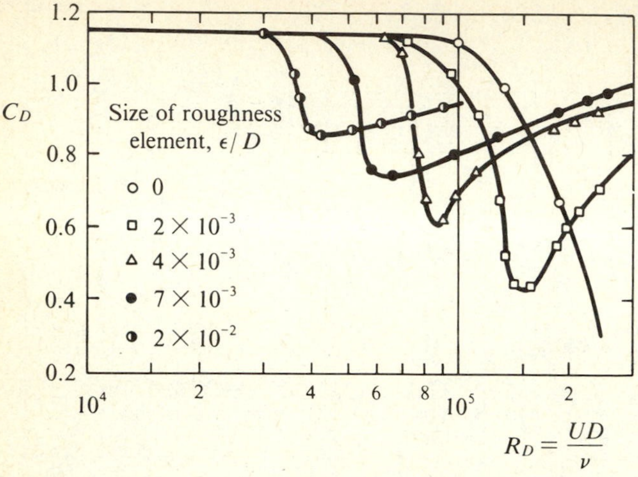

Figure 14.43 *The effects of turbulence and surface roughness on the drag of cylinders. (Source: The Effects of Turbulence and Surface Roughness on the Drag of Circular Cylinders'' by A. Faye and J. H. Warsap, ARC R&M 1283, 1930. Used with permission.)*

Example 14.11

The periscope of a nuclear-powered submarine is a streamlined strut. The optical parts of the periscope are housed in a cylinder 1 ft in diameter. (a) What should the chord c of the strut be for minimum drag if the thickness of the metal strut is 0.5 in. with a length of 20 ft? (b) Determine the power saved using the streamlined strut as against using a cylinder if the velocity $U = 0.923$ ft/s, a kinematic viscosity $\nu = 1 \times 10^{-5}$ ft²/s, and density $\rho = 1.935$ slug/ft³.

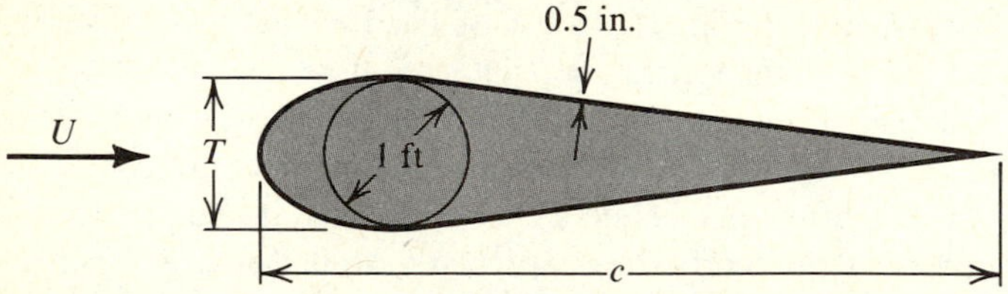

Figure E14.11

Solution:

(a) Using Fig. 14.36, we find that minimum drag occurs at

$$\frac{T}{c} = 0.25 \tag{i}$$

where the maximum thickness T is 12 in. + 1 in. = 13 in. Thus, the chord length c of the strut is, from Eq. (i),

$$c = \frac{13}{0.25} = 52 \text{ in.} \tag{ii}$$

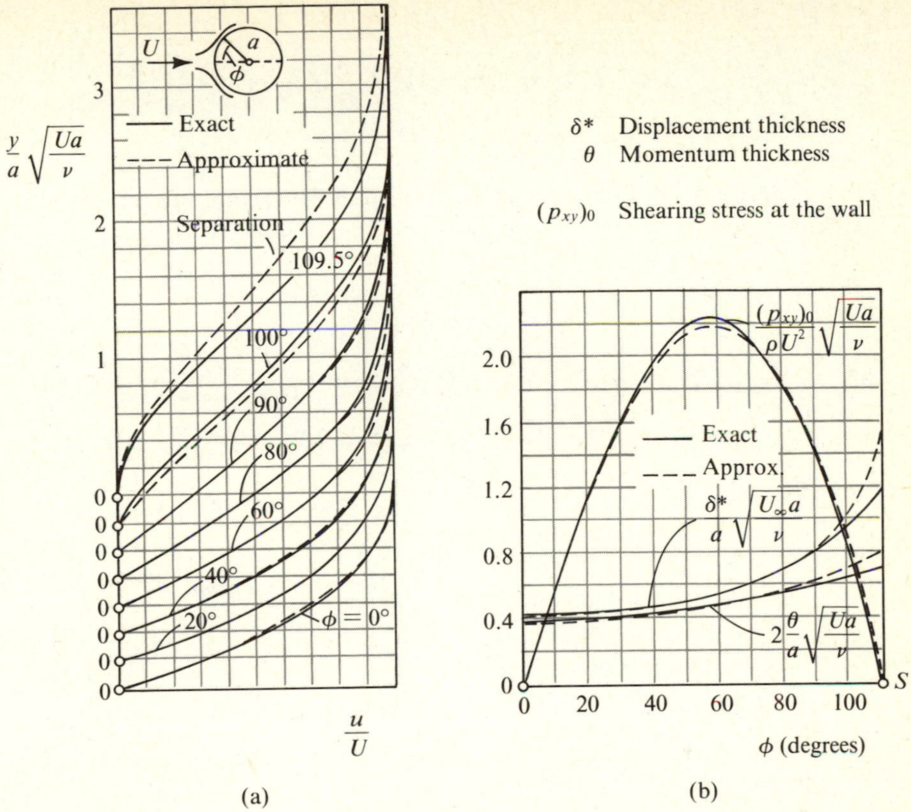

Figure 14.44 *Shear stress $(p_{xy})_o$ on the surface of a cylinder. (a) Comparison of Pohlhausen's approximate solution with the exact solution for the case of a circular cylinder. (b) Comparison of Pohlhausen's approximate solution with the exact solution for the case of a circular cylinder. (Source:* Boundary Layer Theory *by H. Schlichting. Copyright © 1960 McGraw-Hill Book Company. Used with the permission of McGraw-Hill Book Company.)*

Example 14.11 *(Con't.)*

(b) The Reynolds number for the flow past the strut is

$$R_c = \frac{Uc}{\nu} = \frac{0.923 \times 52}{12 \times 1 \times 10^{-5}} = 4 \times 10^5 \qquad \text{(iii)}$$

The drag coefficient C_D for the strut, based on the Reynolds number of Eq. (iii) and value of thickness ratio of Eq. (i), is obtained from Fig. 14.36 as

$$C_{D_{st}} = 0.06 \qquad \text{(iv)}$$

The drag force D of the strut is then

$$D_{st} = ½\rho U^2 A_f C_D \qquad \text{(v)}$$

Example 14.11 *(Con't.)*

$$D_{st} = \tfrac{1}{2}(1.935)(0.923)^2 \frac{13}{12}(20)(0.06)$$

$$= 1.072 \text{ lbf} \qquad \text{(vi)}$$

The drag coefficient for a cylinder is found in Fig. 14.42. For a Reynolds number $R_D = 3.69 \times 10^5$, we obtain

$$C_{D_{cyl}} = 0.8 \qquad \text{(vii)}$$

so that the drag force on the cylinder would be

$$D_{cyl} = \tfrac{1}{2}\rho U^2 A_f C_D$$

$$= \tfrac{1}{2}(1.935)(0.923)^2 (1)(20)(0.8) \qquad \text{(viii)}$$

$$= 13.188 \text{ lbf}$$

The horsepower saved in streamlining the cylinder is

$$\text{hp} = \frac{(D_{cyl} - D_{st})}{550} \times U = \frac{(13.188 - 1.072)(0.923)}{550}$$

$$= 0.02 \text{ hp} \qquad \text{(ix)}$$

or the power saved is 0.02 hp.

This completes the solution.

Drag on a Sphere

Stokes [14.22] predicted the drag on a sphere of diameter D at very low Reynolds number as

$$C_D = \frac{\text{Drag}}{\tfrac{1}{2}\rho U^2\left(\dfrac{\pi}{4}D^2\right)} = \frac{24}{R_D} \qquad (14.119)$$

In the derivation of Eq. (14.119), the inertial forces were assumed to be negligible in magnitude compared to the viscous forces. Using the above expression we calculate the drag force D_s on a sphere as

$$D_s = 6\pi\mu U a \qquad (14.120)$$

A number of investigators have attempted to derive a relationship for the drag on a sphere which includes the inertial forces. For instance Oseen [14.23] obtained an expression

$$C_D = \frac{24}{R_D} + 4.5 \qquad (14.121)$$

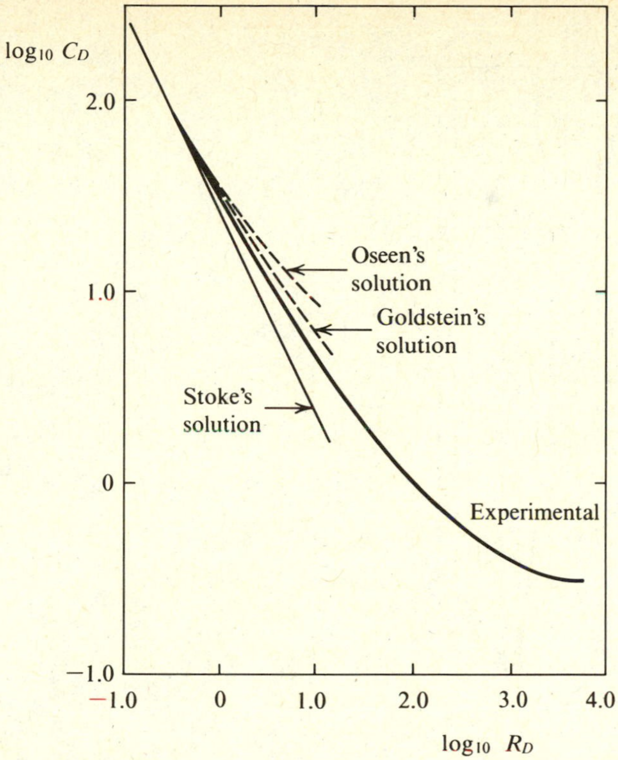

Figure 14.45 *C_D for a sphere based on different mathematical models compared with experimental results.*

Equation (14.119) represents a *lower limit* and Eq. (14.121) represents an *upper limit* for the drag coefficient on a sphere. These two limits are shown plotted in Fig. 14.45.

Goldstein [14.24] refined Oseen's results and obtained a series formula for the drag on a sphere:

$$C_D = \frac{24}{R_D}\left(1 + \frac{3}{16}R_D - \frac{19}{1280}R_D^2 + \frac{71}{20480}R_D^3 - \frac{30179}{34406400}R_D^4 + \frac{122519}{550502400}R_D^5 - \dots\right) \tag{14.122}$$

Goldstein's result is also shown plotted in Fig. 14.45. It is seen to be in better agreement with experiments than Stokes' or Oseen's for large values of Reynolds number.

Experimental determinations of the drag coefficient C_D for a sphere are presented in Fig. 14.46. The curve for the drag coefficient is similar to the C_D curve for the cylinder of Fig. 14.42. For $R_D < 20$, the drag coefficient for the sphere is greater than for the cylinder, but for $R_D > 20$, the reverse is true. Explain why this is true.

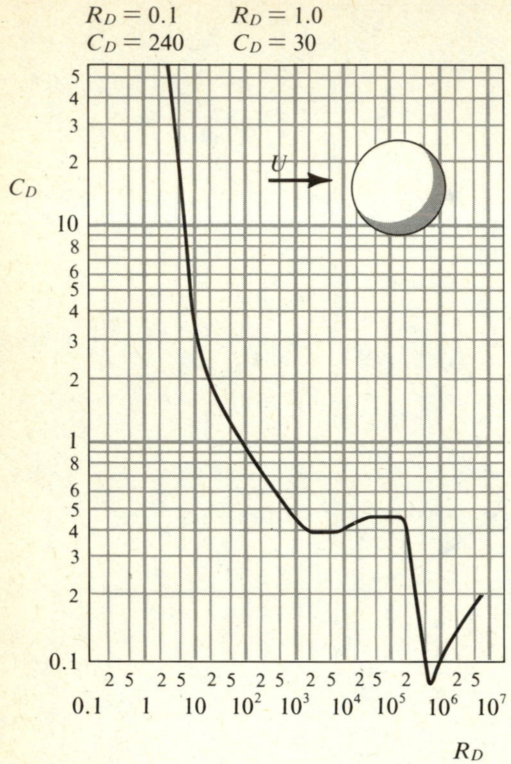

Figure 14.46 C_D *versus Reynolds number for the sphere.*

Example 14.12

Consider a sphere 1 ft in diameter dropped into fresh water at temperature 60°F and specific weight 62.4 lbf/ft^3. Calculate the terminal velocity of the sphere in the water, if the sphere is twice as heavy as an equal volume of water.

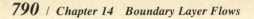

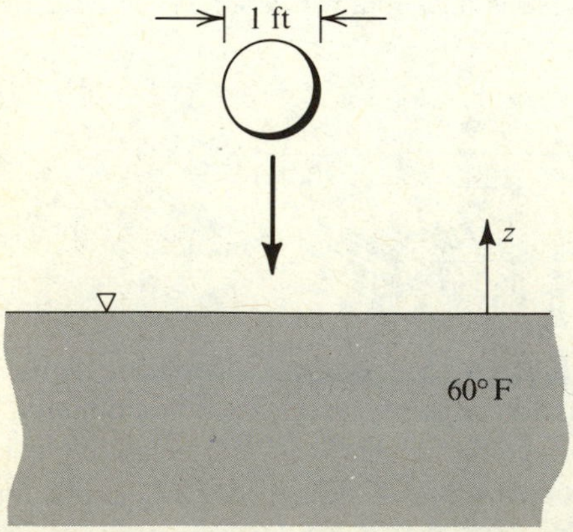

Figure E14.12

Example 14.12 *(Con't.)*

Solution:
At terminal velocity, the deceleration is zero such that the drag force exactly balances the external forces of weight and buoyancy. Thus, at terminal velocity

$$\Sigma F_z = 0 \qquad \text{(i)}$$

so that

$$\begin{aligned} D_s &= \gamma_s \forall - \gamma\forall \\ &= 2 \times 62.4 \frac{\pi}{6} - 62.4 \frac{\pi}{6} = 32.6 \text{ lbf} \end{aligned} \qquad \text{(ii)}$$

The drag force D_s is related to the drag coefficient by

$$\begin{aligned} D_s &= \frac{1}{2} \rho U^2 A C_D \\ &= \frac{1}{2} (1.935) (U^2) \left(\frac{\pi}{4}\right) (C_D) \end{aligned} \qquad \text{(iii)}$$

Substituting Eq. (iii) into Eq. (ii) gives

$$U^2 C_D = 42.8 \text{ ft}^2/\text{s}^2 \qquad \text{(iv)}$$

The Reynolds number R_D for the flow is

$$R_D = \frac{UD}{\nu} = 0.82 \times 10^5 U \qquad \text{(v)}$$

Next we turn to Fig. 14.46 to iterate on the velocity U to obtain the correct value of C_D.

(i) First iteration: Let $U = 10$ ft/s, then

$$R_D = 8.2 \times 10^5$$

$$C_D = 0.15$$

$$C_D = 0.428 \text{ (from Eq. (iv))}$$

(ii) Second iteration: Let $U = 15$ ft/s, then

$$R_D = 12.3 \times 10^5$$

$$C_D = 0.2 \text{ (from Fig. 14.39)}$$

$$C_D = 0.19 \text{ (from Eq. (iv))}$$

Thus, for a velocity of approximately 15 ft/s, the inertial force vanishes, and the drag force is in balance with the weight and buoyancy forces. At this velocity, the sphere has reached terminal velocity.

This completes the solution.

References

14.1 Roache, P. J., *Computational Fluid Dynamics*, Hermosa, Albuquerque, N.M., 1972.

14.2 Blasius, H., "Grenzschlichten in Flussigkeiten mit kleiner Reiburg," *Z. Math. Physics*, 56: 1–37, 1908.

14.3 Granger, R. A., *The Introduction to the Dynamics of Incompressible Fluid Flow*, vol. 5, U.S. Naval Academy, Annapolis, Md., 1975.

14.4 Tollmein, W., "Ein allgemeines Kriterium der Instabilität laminarer Geschwindigkeitsverteilungen," *Nachr. Ges. Wiss. Göttingen, Math. Phys. Kl., Fachgruppe* I, 1: 79–114, 1935.

14.5 Klebanoff, P. S. Tidstrom, K. D., and Sargent, L. M., "The Three-Dimensional Nature of Boundary Layer Instability," *J. Fluid Mech.,* 12: 1–34, 1962.

14.6 Elder, J. W., "An Experimental Investigation of Turbulent Spots and Breakdown to Turbulence," *J. Fluid Mech.,* 9: 235–246, 1960.

14.7 Mochizuki, M., *J. Phys. Soc. Japan*, 16: 5, 995–1008, 1961.

14.8 Tani, I., et al., *Aero Res. Inst., Tokyo*, Rept. 375, Nov. 1962.

14.9 Schubauer, G. B., and Skramstad, H. K., "Laminar Boundary Layer Oscillations and Stability of Laminar Flow," National Bureau of Standards Research Paper, 1972. (See also *J. Aero. Sci.,* 14: 69, 1947.)

14.10 Schlichting, H., *Boundary Layer Theory*, McGraw-Hill, New York, 1960.

14.11 Shen, S. F., "Calculated Amplified Oscillations in Plane Poiseuille and Blasius Flows," *J. Aero. Sci.*, 21: 62–64, 1954.

14.12 Kovasznay, L. S. G., and Komoda, H., *Proc. of the 1962 Heat Transfer Fluid Mech. Inst.,* Stanford, 1–6, 1962.

14.13 Fernholz, H.-H., "Turbulence," (Chap. 2) *External Flows*, P. Bradshaw, Ed., Springer-Verlag, Berlin, 1976.

14.14 Kline, S. J., Reynolds, W. C., Schraub, F. A., and Runstadler, P. W., "The Structure of Turbulent Boundary Layers," *J. Fluid Mech.*, 30: 741, 1967.

14.15 Eckelmann, H., "The Structure of the Viscous Sublayer and the Adjacent Region in a Turbulent Channel Flow," *J. Fluid Mech.,* 65: 439, 1974.

14.16 Reichardt, H., *z. a angew. Math. Mech.*, 18: 358, 1938. (Trans. as *NACA TM* 1313, 1951.)

14.17 Ludwieg, H., and Tillmann, W.; *Ingr.-Arch.*, 17: 288, 1949; also translated as *NACA TM 1285*.

14.18 Rotta, J. C., "Fluid Mechanics of Internal Flow," G. Sovran, Ed., Elsevier, Amsterdam, 1967.

14.19 Corrsin, S., and Kistler, A. L., "Free Stream Boundaries of Turbulent Flows," *NACA TR 1244,* 1955.

14.20 Kovasznay, L. S. G., "The Turbulent Boundary Layer," *Ann. Rev. of Fluid Mech.*, 3: 95–112, 1970.

14.21 Hoerner, S. F., "Fluid-Dynamic Drag," published by author, 1965.

14.22 Stokes, George, *Trans. Camb. Phil. Soc.,* 8, 1845.

14.23 Oseen, C. W., *Ark. f. mat. astr. och. fys.,* 6: 29, 1910.

14.24 Goldstein, S., *Proc. Roy. Soc.*, A, 123: 225–235, 1929.

Study Questions

14.1 What assumptions are imposed in obtaining the Prandtl boundary layer equation from the Navier-Stokes equations? Explain why the equation is not exact.

14.2 Does $p_{xy} = -\mu\zeta_z$ for two-dimensional real fluid flow? Discuss.

14.3 Why isn't the upper bound on the integral expression taken to infinity rather than δ?
14.4 What are the similarities and dissimilarities between turbulent flow through a pipe and over a flat plate?
14.5 What is u_*? How does it differ from $\bar{u}$, u', and u?
14.6 Discuss the shear stress distribution in Fig. 14.28.
14.7 Explain the family of curves shown in Fig. 14.25. Where is the flow stable and where is it unstable? What is the value of C_i for a flat plate? What is the physical meaning of C_i?
14.8 What is a decelerating layer? Where does it occur? What does it signify? What can take place in the layer?
14.9 What is a shape factor? What is the significance of the curve for shape factor versus Reynolds number?
14.10 What does an inflection point in the velocity distribution mean? Explain.

Problems

14.1 Express Prandtl's boundary layer equation for laminar flow over a flat surface if v has the same form as u; i.e., $v \propto u$. Express $u = u(x, y)$.

14.2 If $u \neq u(x)$ for laminar flow, evaluate the pressure distribution from Prandtl's boundary layer equation given $v \neq v(x, z, t)$ nor a constant.

14.3 Using Blasius' results for laminar flow past a flat plate, plot the shear stress distribution as a function of η.

14.4 Discuss the value of the shear stress at $x = 0$, i.e., $p_{xy}(0, y)$ for laminar flow over a flat plate. What does this suggest about some claiming that the solutions are exact?

14.5 A submarine lies dead in the water at a depth of 1000 ft below the ocean's surface. What is the thickness of the boundary layer 10 m from the bow if the ocean's current is 5 cm/s? Assume the ocean's temperature is 10°C. (See Fig. P14.5.)

14.6 In Prob. 14.5, calculate the drag force over the deck of the submarine if it is 50 m long and 7 m wide.

14.7 In Prob. 14.5, calculate the displacement and momentum thickness.

14.8 Estimate the friction drag on an aircraft's wing that has a total span of 50 m and an average chord of 3 m moving through air at a Mach number 0.3. Let the pressure and temperature of the air be 83 kPa and 18°C, respectively (Fig. P14.8). Assume flow is turbulent and wing is approximated by a flat plate.

14.9 Using the velocity distribution $u = a + b \sin cy$, calculate the laminar boundary layer thickness $\delta = \delta(x)$ over a flat plate.

14.10 Investigate the use of the velocity distribution $u = a + b \cos cy$ to calculate the laminar boundary layer thickness $\delta = \delta(x)$ over a flat plate.

14.11 Calculate the boundary layer thickness $\delta(x)$ given $u = U[a + b(y/\delta) + c(y/\delta)^2]$ for laminar flow over a flat plate.

Figure P14.5

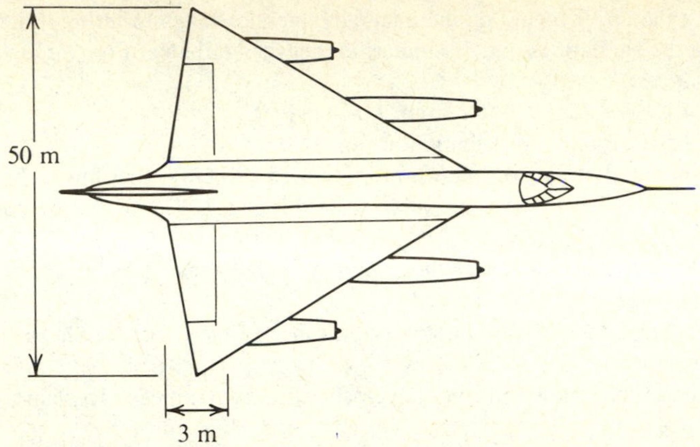

Figure P14.8

14.12 Calculate the displacement and momentum thickness, the shear stress on a flat plate, and the drag coefficient for the flow given by Prob. 14.11.

14.13 Compare the drag coefficients C_D of the velocity distributions given by Prob. 14.9 with that given by Prob. 14.10 and the results of Blasius.

14.14 A cube of surface area 2 cm by 2 cm floats on an oil surface. What forces of the wind are necessary to move the block with a velocity 0.05 cm/s if the viscosity is 8.1×10^{-3} Pa·s, and density 860 kg/m^3 (Fig. P14.14)?

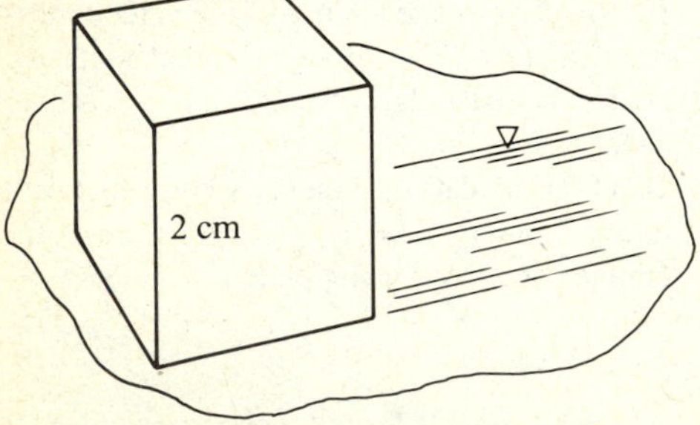

Figure P14.14

14.15 A storm of wind velocity 20 m/s strikes a flat smooth roof of surface area 7 m × 7 m. Calculate the friction drag on the roof if the air temperature is 27°C (Fig. P14.15).

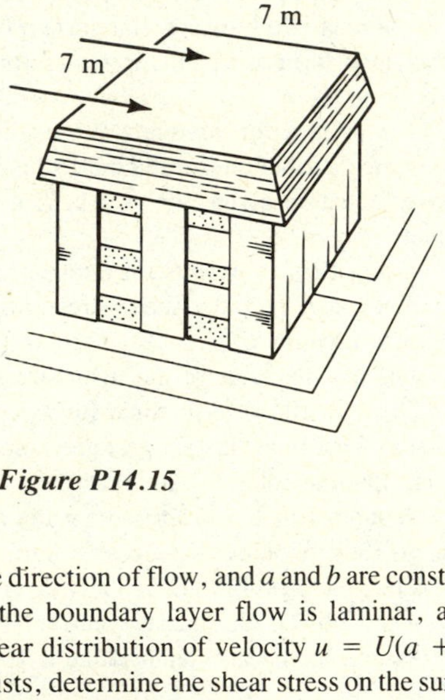

Figure P14.15

14.16 Air at temperature 115°F moves over a linear sand dune whose surface produces an ambient velocity $U = a + bx$, where x is in the direction of flow, and a and b are constants. If the boundary layer flow is laminar, and a linear distribution of velocity $u = U(a + by)$ exists, determine the shear stress on the surface of the sand (Fig. P14.16).

14.17 Calculate the turbulent boundary layer thickness δ if $u = U(y/\delta)^{1/8}$ and Prandtl's shear stress $(p_{xy})_o$ of Eq. (14.106).

14.18 As shown in Fig. P14.18, the top of a bus is a smooth flat surface 38 ft long and 10 ft wide and moves at uniform velocity 10 ft/s through a still atmosphere of 70°F. Calculate the boundary layer thickness at the trailing edge of the bus, the shear stress halfway along on the bus, and the power required to move the upper surface through the air, assuming the flow is laminar.

Figure P14.16

Figure P14.18

14.19 Examine the possibility that there are other solutions of Prandtl's boundary layer equation than the Blasius solution. For example, look for a solution of the form $\psi = Ug(x)f(\eta)$, where $\eta = g/g(x)$ and where $u = Uf'(\eta)$. What forms of $g(x)$ are possible?

14.20 Wind blows over a blacktop road at a velocity 100 km/h. At what length over the road is the turbulent boundary layer thickness 6 mm given the pressure and temperature of air is 10 kPa and 30°C, respectively?

14.21 A small plane flies through air at 200 ft/s, and requires only 8 hp to overcome friction and move through the air (Fig. P14.21). Determine the coefficient of skin friction drag if the projected area of the wing's surface is 8 ft^2 and the air temperature is 70°F.

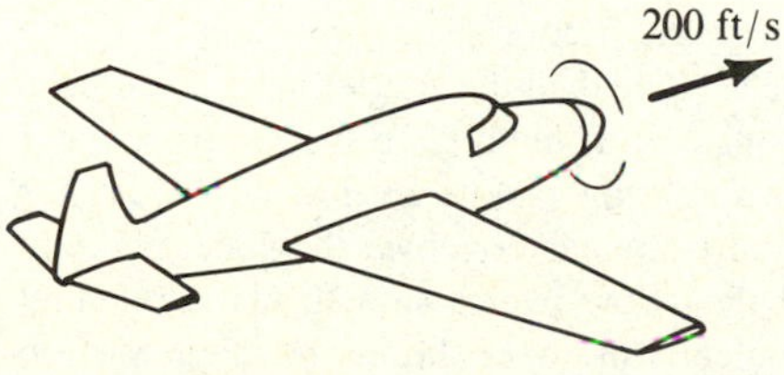

Figure P14.21

14.22 A flat bottom scow that is 68 ft long and 30 ft wide moves through the Seine at 5 ft/s. Calculate the skin friction drag if the water's temperature is 55°F.

14.23 Water moves over a hydrofoil at a uniform velocity of 20 ft/s. At 0.5 ft downstream of the foil's leading edge, what is the boundary layer's displacement and momentum thicknesses, and the wall shear stress if the water temperature is 70°F? Let the flow be laminar. (Refer to Fig. P14.23.)

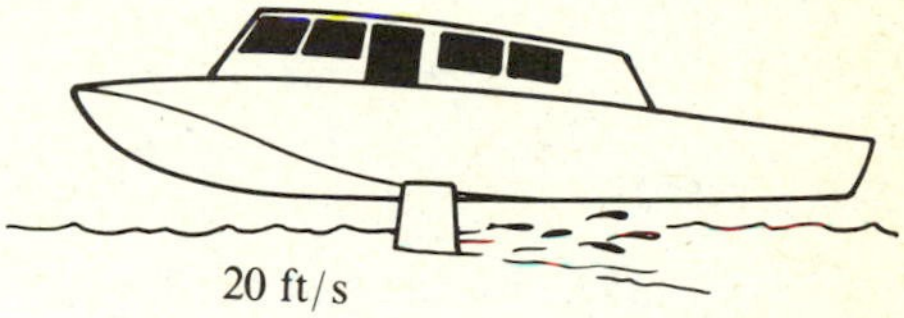

Figure P14.23

14.24 Water flows past a 60 ft long section of a smooth flat ship's hull at 12 ft/s. At 23 ft downstream of the leading edge, what are the x- and y-components of velocity at distances of 0.05 and 5 in. from the hull? Let $\nu = 1.08 \times 10^{-5}$ ft^2/s.

14.25 Calculate the drag force on a flat plate 5 m long and 3 m wide placed in a water tunnel at standard room temperature conditions. The water moves past the plate at a velocity of 50 m/s with the following conditions: (a) the boundary layer is laminar over the entire portion of the plate, (b) transition occurs at $R_x = 5 \times 10^5$, (c) the boundary layer is turbulent over the entire surface of the smooth plate, and (d) the boundary layer is turbulent over the entire surface of a rough plate, where $U\epsilon/\nu = 10^5$.

14.26 Determine the boundary layer thickness δ, the shear stress distribution on the surface $(p_{xy})_o$, and the total drag on one surface of a flat aircraft wing of average chord L assuming the flow is laminar over the entire wing.

Compare the results by assuming a velocity profile (a) $u = U(a_0 + a_1\eta)$, (b) $u = U \sin (\pi/2)\eta$, and (c) $u = U(a_0 + a_1\eta + a_2\eta^2 + a_3\eta^3)$. Determine the thickness of the boundary layer δ at the trailing edge of the wing if $U = 7$ ft/s, $L = 2$ ft, and the wing span is 3 ft. Assume the fluid is air at a temperature of 72°F (Fig. P14.26).

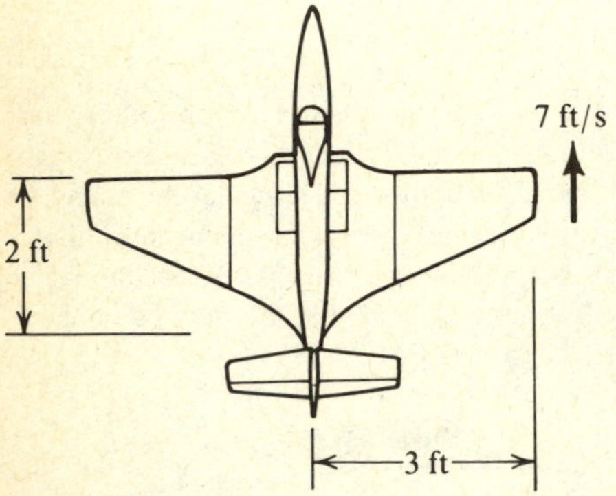

Figure P14.26

14.27 Determine the ratios of the displacement and momentum thicknesses to the boundary layer thickness when the velocity profile is represented by $u = U \sin(\pi/2)\eta$.

14.28 By substitution, show that when Pohlhausen's boundary conditions are used to determine the coefficients, Eq. (14.77) becomes

$$\frac{u}{U} = 2\eta - 2\eta^3 + \eta^4 + \frac{1}{6}\frac{\delta^2}{\nu}\frac{d\grave{U}}{dx}\eta(1-\eta)^3$$

What is the value of the plate's shear stress $(p_{xy})_o$?

14.29 Approximate solutions to the boundary layer equations may be obtained from the momentum integral equation using the approximate velocity profile

$$\frac{u}{U} = a + b \sin \frac{\pi}{2}\eta + c \sin^2 \frac{\pi}{2}\eta$$

Indicate how the coefficients a, b, and c may be obtained for a flat plate.

14.30 Assuming a cubic profile for the velocity distribution over a flat plate determine the ratio δ/x using the von Kármán momentum integral equation for a zero pressure gradient. What is the percentage error of your results as compared to Blasius' solution?

14.31 Consider a flat plate 15 ft wide and 20 ft long. Determine the drag on the top surface of the plate, assuming a cubic profile for the velocity distribution. Assume a laminar boundary layer.

14.32 Compute the total drag force on one side of a 38 ft long, 9 ft wide, flat smooth ship's hull in fully turbulent flow over the entire length of the ship's wetted surface. Let the ship move through water at 55°F and move at a velocity of 20 ft/s. Use (a) the power-law expression for the velocity distribution and (b) the logarithmic expression for the velocity distribution.

14.33 For the plate and velocity of Prob. 14.32, calculate the drag on one side of the plate assuming laminar flow over the entire surface. Compare the drag coefficient with that given by experiment in Fig. 14.32.

14.34 What is the boundary layer thickness and shear stress on the flat plate 1.5 ft from the leading edge assuming turbulent flow over the entire length of the plate? Assume the fluid is air at 70°F and 1 atm, moving at a velocity of 70 ft/s.

14.35 A destroyer with wetted dimensions of 220 ft by 67 ft moves through fresh water at 58°F. Estimate the thrust necessary to overcome skin friction drag if the ship has a velocity of 20 knots. Assume fully turbulent flow at $R_l < 10^7$.

14.36 A flat plate moves through air at 60°F at a velocity of 100 ft/s, and requires 10 hp to accomplish this. Given an area of 10 ft^2, determine the friction drag coefficient.

14.37 Calculate the friction drag force on a flat plate 30 ft long × 15 ft wide placed in a high speed tow tank at a velocity of 43 ft/s for (a) fully laminar flow over the plate, (b) fully turbulent flow over a smooth plate, (c) fully turbulent flow over the entire rough surface given that $U\epsilon/\nu = 10^4$, and (d) that transition occurs at a critical Reynolds number of 5×10^5. Compare results.

14.38 Consider the laminar sublayer of a turbulent boundary layer flow in a two-dimensional flow, as illustrated by Fig. P14.38. Show that at the surface of the curved body, the velocity profile is concave upwards in flows with a favorable pressure gradient ($\partial p/\partial x < 0$), whereas it is concave downwards for flow with an unfavorable pressure gradient ($\partial p/\partial x > 0$).

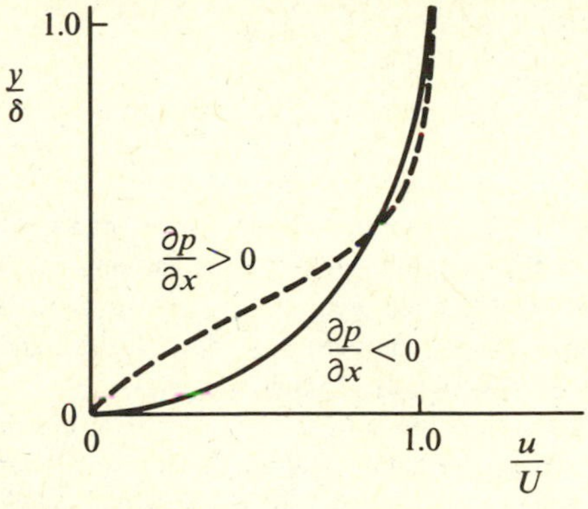

Figure P14.38

14.39 Let a turbulent boundary layer flow be from the leading edge to the trailing edge of a smooth flat plate. The shape of the velocity profile is similar at all positions along the plate. If the wall shear stress $(p_{xy})_o$ is inversely proportional to x^n, where n is a positive constant, show that the turbulent boundary layer thickness obeys the power-law $\delta \propto x^{1-n}$.

14.40 Explain why certain wing profiles at certain angles-of-attack have lower drag when the boundary layer is turbulent than when it is laminar.

14.41 A signpost 1 ft wide by 2 ft long is exposed parallel to the air that flows with a uniform velocity of 15 ft/s. Determine the drag coefficients (a) when the 1 ft edge is parallel to the wind, and (b) when the 2 ft edge is parallel to the wind, given $\nu = 1.6 \times 10^{-4}$ ft²/s and $R_{cr} = 5 \times 10^5$ (see Fig. P14.41).

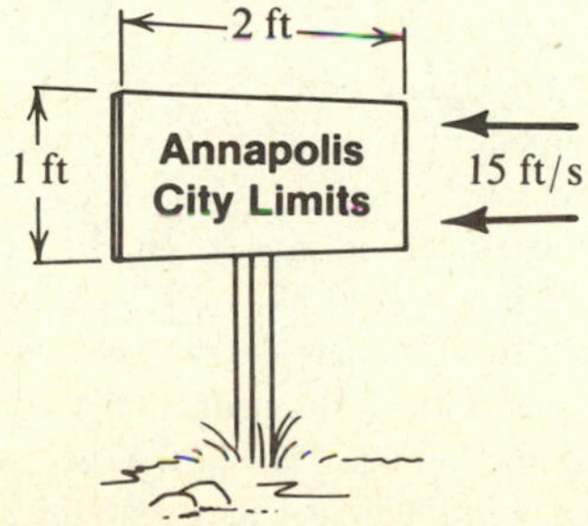

Figure P14.41

14.42 For laminar flow over a flat plate, determine an expression for u/U at the location $y = \delta^*$, and show that the expression is independent of path x and viscosity ν. What is the value of u for $U = 5$ m/s?

14.43 Energy thickness is sometimes used to evaluate the loss of energy of flow in a laminar boundary layer. Defining the energy thickness as δ_E where

$$\delta_E = \int_o^\delta \frac{u}{U}\left(1 - \frac{u^2}{U^2}\right) dy$$

calculate its value assuming (a) Blasius solution, (b) a linear velocity profile, and (c) a cubic velocity profile.

14.44 Using Pohlhausen's approximation for the laminar velocity distribution over a flat plate, Eq. (14.85), calculate (a) the displacement thickness, (b) local skin friction coefficient, (c) drag, and (d) total skin friction drag coefficient, and compare results with Blasius' results.

14.45 As shown in Fig. P14.45, a stream of cars move at constant speed 24 m/s bumper to bumper along a flat stretch of straight road in an otherwise stationary atmosphere at temperatures 30°C. (a) Find the shear stress at $x = 100$ m. (b) What is the velocity of the air if one stood 0.3 m from the edge of the cars 10 seconds after the first car passed?

14.46 Wind blows over a straight stretch of flat sidewalk at a velocity of 1 m/s, where the earth's boundary layer thickness is 0.5 m. What is the velocity of the wind at 5 cm and 10 cm assuming (a) laminar flow and (b) turbulent flow (assume Prandtl's one-seventh law)?

14.47 Calculate the point where instability arises for laminar boundary layer flow over a flat plate of velocity U equal 7 ft/s, and kinematic viscosity ν equal 1×10^{-5} ft²/s. What is the displacement thickness δ^* and momentum thickness θ at this location?

14.48 Consider laminar boundary layer flow over a cylinder of radius 1 m. If the water temperature is 20°C and has a flow velocity of 10 cm/s, determine the position x where the flow is critical.

14.49 Consider turbulent boundary layer flow over a flat plate of length 10 m. Let the water

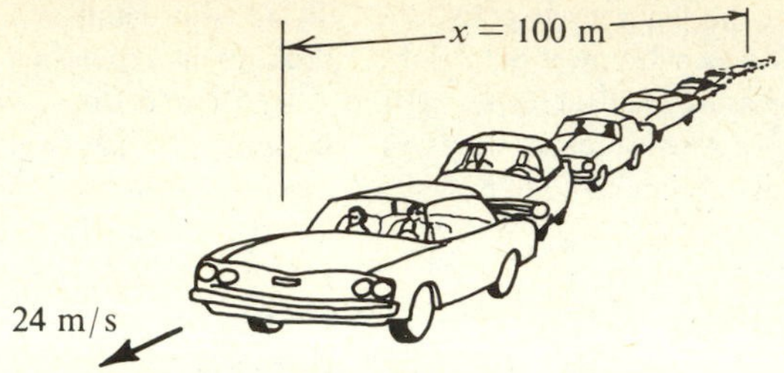

Figure P14.45

temperature be 20°C and the flow have a free-stream velocity of 6 m/s. If the critical position is 0.07 m, calculate the drag force per unit width over the plate.

14.50 Estimate the frictional drag on a plate 10 ft long and 20 ft wide placed in a wind tunnel with air flowing past it at 150 ft/s for ν = 0.00015 ft²/s and for (a) a completely laminar boundary layer, (b) transition at R_{cr} = 5×10^5, and (c) turbulent boundary layer over the entire surface. Assume the plate is smooth. What conclusions can be made from all three answers?

14.51 A flat plate, 3 ft × 20 ft, is held in standard air of density ρ = 0.002377 slug/ft³ moving at 40 ft/s. The plate is first held (long edge) parallel to the flow, then perpendicular. What is the ratio of the drag for the parallel divided by the drag for the perpendicular orientation? Assume turbulent flow (Fig. P14.51).

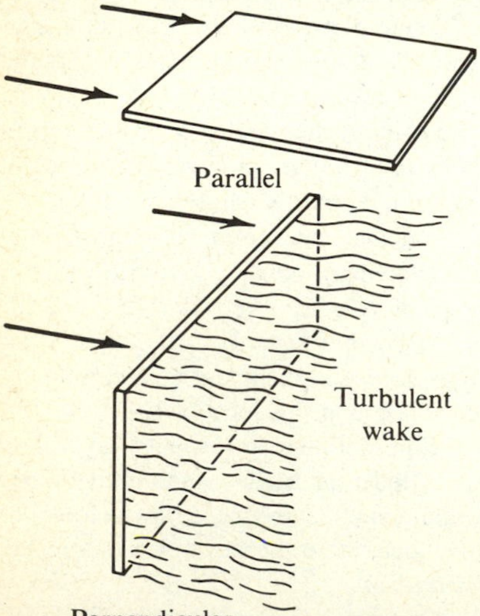

Figure P14.51

14.52 Estimate the frictional resistance of a 600 ft long ship in seawater moving at 10 ft/s, if its wetted area is 30×10^3 ft² and the water has the properties γ = 64 lbf/ft³, and $\mu = 2.391 \times 10^{-5}$ slug/ft·s.

14.53 We are told that a sphere falling in water has a Reynolds number of 0.3. (a) What is the drag coefficient, and (b) what is the drag per unit diameter if $\mu = 2.39 \times 10^{-5}$ slug/ft·s and ρ = 2 slug/ft³?

14.54 Referring to Fig. P14.54, calculate the force per running foot on a vertical fence 6 ft high in a 10 ft/s average wind normal to the fence, given ρ = 0.002378 slug/ft³ and $\mu = 0.0373 \times 10^{-5}$ slug/ft·s.

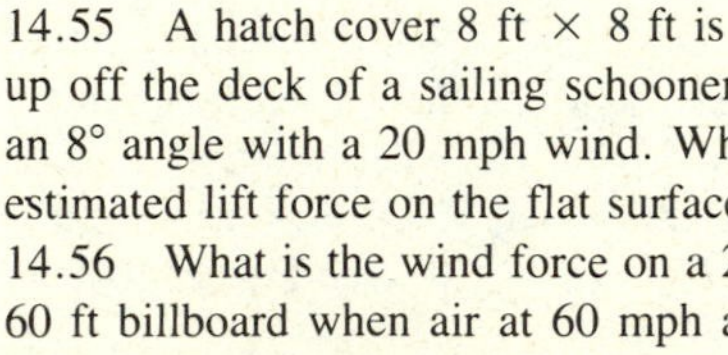

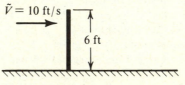

Figure P14.54

14.55 A hatch cover 8 ft × 8 ft is propped up off the deck of a sailing schooner making an 8° angle with a 20 mph wind. What is the estimated lift force on the flat surface?

14.56 What is the wind force on a 20 in. × 60 ft billboard when air at 60 mph and 60°F blows normal to it?

14.57 A thin flat square plate, 3 ft on a side, is held submerged in a parallel flow of oil whose free-stream velocity is 3 ft/s. The kinematic viscosity of the oil is 8×10^{-4} ft²/s and its specific gravity is 0.89. (a) Calculate the total drag D. (b) Calculate the thickness of the boundary layer at the trailing edge. (c) Why is it safe to assume laminar flow? (d) What is the diameter of a cylinder for the same drag force?

14.58 A long, thin, round, hollow rod (closed at both ends), as shown in Fig. P14.58, is dropped from an airplane. The rod falls through the air such that the skin friction is the only appreciable source of drag. Treat the surface area as you would the area of a smooth flat plate. The rod weighs 4 lbf and is 25 ft long. The terminal velocity of the rod at sea level and standard conditions is 400 ft/s. (a) What is the drag force at that velocity? (b) Write an expression for the Reynolds number and indicate to what each term of the expression refers. (c) What is the skin friction coefficient? (d) Calculate the diameter of the rod. (e) If the same rod were to fall horizontally (approximating a 2-D cylinder), what would be the terminal velocity, assuming $C_D \approx 1$? Assume $\nu = 1.7 \times 10^{-4}$ ft²/s.

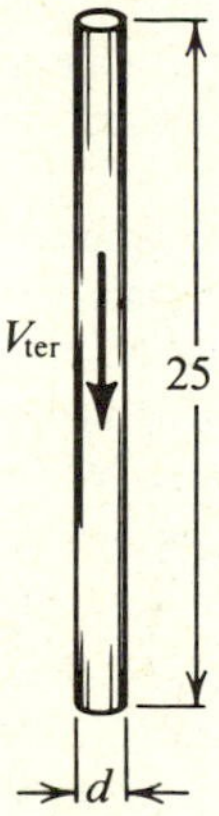

Figure P14.58

14.59 A student is considering purchasing a van as he cannot afford a home. He recognizes that since vans are not streamlined, their wind resistance is high and therefore the gas mileage is low. This student never goes anywhere at less than 60 mph. The van has the following dimensions: front area is 6 ft wide by 5 ft high; top and bottom are 6 ft wide by 12 ft long; sides are 5 ft high and 12 ft long. (a) Estimate the pressure drag. (b) Estimate the total skin friction on the sides, top, and bottom of the vehicle in standard air. Assume a turbulent boundary layer. (c) Estimate the total drag on the vehicle. (d) Calculate the drag coefficient. (e) Estimate the power required for steady horizontal motion exclusive of rolling friction. (f) By what percent is the power required increased or decreased if he takes the van to Daytona Beach where the barometer is 29.5 in. Hg and the temperature is 95°F?

14.60 A balloon 50 ft long and 10 ft in diameter is shaped like a cylinder. Its deflated weight is 100 lbf. It is filled with hydrogen of density 0.0001626 slug/ft³. Calculate the terminal velocity of the balloon at low altitude where the air's density is 0.002335 slug/ft³, assuming the wind's relative velocity is zero (Fig. P14.60).

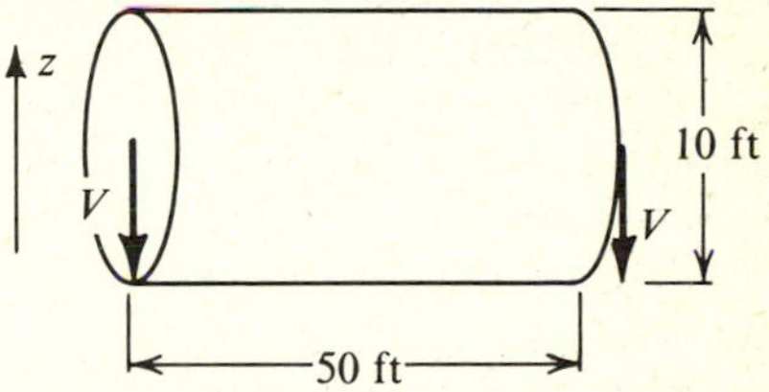

Figure P14.60

14.61 In a TV shampoo commercial, a pearl is dropped into the shampoo and reaches the terminal velocity of 0.15 ft/s. If the density of the shampoo is 2.9 slug/ft³, the diameter and weight of the pearl are 0.025 ft and 0.002 lbf, respectively, find the dynamic viscosity of the shampoo.

14.62 A spherical weather balloon 10 ft in diameter and filled with helium is being towed by a cable attached to a truck moving at 15 mph. The weight of the balloon's material (not including the helium) is 3 lbf. The atmospheric pressure and temperature are 2116 psf and 68°F, respectively. (a) What is the buoyancy force on the balloon? (b) What vertical force is required to hold the balloon? (c) What is the drag force acting on the balloon? (d) What is the tension in the cable (Fig. P14.62)?

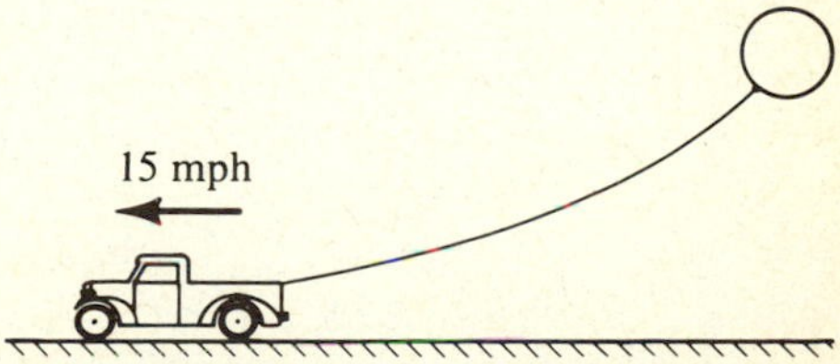

Figure P14.62

14.63 An automobile has a drag coefficient of 0.45 based on a frontal area of 27 ft^2. Determine the drag force on the car and the horsepower required to move it through standard air at a velocity of 55 mph.

14.64 A boat 3 m long with average width of 1 m moves through fresh water at 20°C with a velocity of 2 m/s. Let the hull have an equivalent sand roughness of 0.031 cm. Using flat plate theory, determine the friction drag on the hull assuming the laminar boundary layer becomes turbulent at a Reynolds number of 4×10^5.

14.65 An 800 ft long by 80 ft wide Navy vessel cruises at 35 mph in 50°F seawater. Using flat plate theory, find (a) the total shear force on the ship assuming the boundary layer is completely turbulent and the hull is smooth, (b) the total shear force assuming a laminar boundary layer up to 4×10^5 value of Reynolds number.

14.66 An iron ball of 4 in. diameter has a mass of 0.04 slug, rests on a bed of gunk that has a density of 2 slug/ft^3. After one year, the ball has moved 9 ft into the gunk. Assuming the gunk is Newtonian, determine the viscosity of the gunk and the Reynolds number of the ball's motion. Assume net weight equals weight.

14.67 The curved portion of a submarine may be approximated by a flat plate of 200 ft length and 20 ft diameter. Calculate the friction drag on the submarine that cruises at 30 mph in 48°F seawater. Assume the roughness of the submarine's surface is 0.005 ft, and the Reynolds number transition is 5×10^5.

14.68 A hawk flies at 35 mph and weighs 7 lbf. The span of the wing is 6 ft and has an average 9 in. chord. Estimate the profile drag coefficient of the hawk assuming its aerodynamic characteristics are identical to an airfoil.

14.69 A 700 ft long ship moves at 40 mph through 55°F seawater using 78,000 shaft horsepower supplied to the propellers. The wetted hull area is 50,000 ft^2 and has an equivalent sand roughness of 0.0025 ft. Assuming that the propeller's thrust times the ship's speed is 60% of the shaft horsepower, calculate the frictional as well as residual drag.

14.70 A free-falling parachutist weighing 170 lbf has a terminal velocity of 118 mph when his body lies horizontal moving through air at 1 atm and 60°F. If the man's frontal area is 8.5 ft^2, calculate the drag coefficient.

14.71 An airplane is to cruise at 250 mph at 14,000 ft. It's fuselage length is 72 ft and mean fuselage diameter of 12 ft. Determine the frictional drag force on the fuselage if the turbulent boundary layer on the fuselage is assumed to be equivalent to that over a flat plate having the same length and width.

15 One-Dimensional Compressible Flow

15.1 Introduction

Up to this point we have concentrated on fluids whose density is constant throughout the flow. Most liquid flows fall into this category, provided that we observe the criteria for incompressible flows as described in Sec. 4.6. With gases, compressibility is more likely. This can be seen by careful examination of the relationship $\rho \propto p/T$. For example, the temperature of the perfect gas in a free-floating piston-cylinder arrangement rises when heat is added to the confined gas resulting in decreased density caused by the pressure remaining invariant.

Since density and temperature are variables in compressible flows, the number of unknowns has increased from two (velocity $\mathbf{V}$ and pressure p) to four. Thus where we once had two unknowns requiring two governing equations (continuity and linear momentum), we now require four equations: continuity, linear momentum, energy, and the equation of state.

The physics of the problem has also changed. Compressing the fluid can create shock waves. Every time there is an explosion, or an airplane moves fast enough or a fluid accelerates through passages at velocities in multiples of the speed of sound, shock waves form and affect the flow's behavior. Shock waves are of enormous interest to the mechanical and aeronautical engineer, the astronomer, the nuclear physicist, and the chemical engineer.

We are also interested in another type of wave. When we clap our hands or drop a pebble into a pond, we create a sound wave. A shock wave is to a sound wave what a tsunami is to a ripple in a pond. Sound waves are caused by *small* pressure disturbances radiating from their source at the speed of sound. The increase or decrease is only 1/100,000 of the ambient pressure. The sound wave disturbs the fluid particles, moving them a small distance with a velocity a tiny fraction of the acoustic velocity.

With shock waves on the other hand, we are treating pressure changes often greater than the ambient pressure. For example, an automobile tire blowout or a cannon being fired involves pressures greater than atmospheric. The detonation of an atomic bomb in 1951 sounded as if I were standing with my ear glued to the muzzle of a 105 howitzer, yet I was three miles from ground zero.

Shock waves and sound waves can also be created from sudden pressure reduction. In the case of the 1951 atomic bomb for example, there were two shock waves: one from the explosion of the bomb, and one following it a short time later from the implosion. They were of nearly the same decibel level. When a light bulb breaks, the sudden pressure reduction creates a sound wave. Reduced pressure can create rarefaction waves, whereas increased pressure creates compression waves.

When discussing shock waves, we shall note temperature changes. Temperature will increase across a compression shock; temperature decreases when air expands. The speed of sound is also influenced by temperature, such that the speed of sound at low altitudes is greater than at high altitudes.

Shock waves are also formed by electrical discharges. By the time a thunderclap has reached our ear, most of the energy that created the compression of the air molecules has dissipated, resulting in the transformation of the shock wave into a sound wave, i.e., the transformation of a large pressure to a very small pressure increase. Shock waves such as those from the sonic boom, thermonuclear detonations, or volcanic eruptions, can flatten forests and houses. A sudden increase of only 0.5 psia can turn window glass into lethal shrapnels moving at 170 mph. (We can easily verify such velocities by Bernoulli's equation.) Shock waves are to be respected and deserve our consideration.

This chapter serves only as an introduction into the exciting world of compressible fluid flow. We cannot treat *unsteady* compressible flow, a topic of vital importance in the design of commercial and military aircraft, particularly in stability analyses like flutter and divergence of wings. Nor can we treat three-dimensional steady compressible flow, a subject that is important in the behavior of thick tapered bodies moving at high speeds. We shall treat instead steady one-dimensional compressible flow, leaving all other topics in steady flow to such fine texts as Anderson [15.1] and Shapiro [15.2]. For unsteady compressible flow, the text by Bisplinghoff, Ashley, and Halfman [15.3] is recommended.

15.2 The Description of a Perfect Gas

The principle properties of a gas are pressure, temperature, and density. In fact, if the only work mode of a particular flow is simple compressible work, i.e., if

$$_1W_2 = -\int_{\forall_1}^{\forall_2} p \, d\forall \tag{15.1}$$

then it can be proved that only two independent intrinsic properties are required to fix the state of the gas.* All the other intrinsic properties will be found from either tabulated data in the gas tables of Ref. 15.4, or from equations that we shall develop in this section.

There are many functional relationships among the thermodynamic variables of pressure, temperature, and density. One such relationship is

$$\frac{p}{\rho RT} = 1 + a_1(T)\rho + a_2(T)\rho^2 + \ldots \tag{15.2}$$

*The state postulate is based on the substance being homogeneous and single-phase.

which is useful when the pressure is less than 500 psi and the temperature is reasonably high. As a first approximation to Eq. (15.2), let us take $a_i(T) = 0$, ($i = 1, 2, \ldots$), so that

$$p = \rho RT \tag{2.3}$$

The above equation is recognized as the equation of state of a *perfect gas*, where the gas constant R is defined by Eq. (2.4). Some typical values of R for a variety of gases are presented in Table 15.1.

Table 15.1 *Values of the Gas Constant R*

Gas	R (ft·lbf/lbm°R)	R (bar·m³/kg K)
Argon	38.683	0.00208
Helium	385.960	0.02077
Hydrogen	765.610	0.04120
Nitrogen	55.163	0.00297
Oxygen	48.281	0.00260
Air	53.34	0.00287
Carbon monoxide	35.106	0.00297
Methane	96.322	0.00518
Acetylene	59.332	0.00319
Ethylene	55.080	0.00296
Ethane	51.380	0.00277
Propylene	36.716	0.00198
Propane	35.042	0.00189
n-Butane	26.583	0.00143
n-Pentane	21.414	0.00115
n-Octane	13.527	0.00073

A perfect gas is a substance which has a specific internal energy i that Joule proved is solely a function of temperature:

$$i = i(T) \tag{15.3}$$

Equation (15.3) is valid for *all* thermodynamic processes of a perfect gas, be it isobaric, isothermal, or whatever.

The change in internal energy Δi for any finite process is related to a temperature change through the specific heat:

$$\Delta i = \int_{T_1}^{T_2} C_v \, dT \tag{15.4}$$

where C_v is the specific heat at constant volume. Equation (15.4) is *not* restricted to constant volume processes, but is applicable in *any* process for which $i = i(T)$.

Equation (15.4) can be evaluated if we know how the specific heat C_v varies with temperature for a particular gas. Figure 15.1 shows some typical variations of $C_v(T)$ for air, nitrogen, and oxygen, all at low pressures. Obviously, at high pressures (several hundred psi), the curves would be different from those shown, since $C_v = C_v(p, T)$.

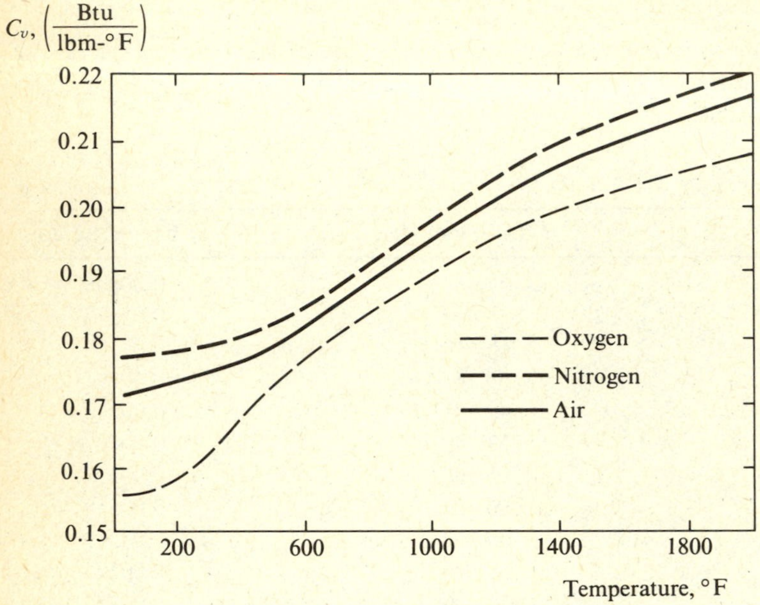

Figure 15.1 *C_v versus temperature for three gases at low pressures. (Source: Thermodynamics, 3rd ed., by K. Wark. Copyright © 1983 McGraw-Hill Book Company. Used with the permission of McGraw-Hill Book Company.)*

Enthalpy, h, is defined as

$$h = i + p/\rho \tag{15.5}$$

and though it has the dimension of energy per unit mass, it does not possess any physical interpretation. It is merely a defined property which is the sum of internal energy and p/ρ. It is *not* a specific form of energy. It is very useful in control volume analysis of flow problems, however, since the work per unit mass transported across the control surface per unit mass flux can be expressed as p/ρ. Comparing Eqs. (15.3) and (2.3) with Eq. (15.5), we see that enthalpy is solely a function of temperature at low pressures. The change of specific enthalpy for a perfect gas is

$$\Delta h = \int_{T_1}^{T_2} C_p \, dT \tag{15.6}$$

where C_p is the specific heat at constant pressure. Equation (15.6) holds for *all* processes, as does C_v, not solely for an isobaric process.

A special relationship exists between the two specific heats C_p and C_v that can be derived by differentiating Eq. (15.6) and using the differential forms of Eqs. (15.4) and (15.6):

$$C_p - C_v = R \tag{15.7}$$

Thus, knowing how C_v varies with temperature for a specific gas, we can obtain the behavior of $C_p(T)$.

We shall define the average specific heats as

$$C_{v_o} \doteq \frac{1}{\Delta T} \int_{T_1}^{T_2} C_v \, dT \tag{15.8}$$

$$C_{p_o} \doteq \frac{1}{\Delta T} \int_{T_1}^{T_2} C_p \, dT \tag{15.9}$$

and the specific heat ratio k as

$$k \doteq \frac{C_{p_o}}{C_{v_o}} \tag{2.11}$$

The specific heat ratio is an important thermostatic property, particularly useful in isentropic processes of compressible aerodynamic flows. Its value for monatomic gases is 1.67, and 1.4 for many of the common diatomic gases. Note that the specific heat ratio is independent of temperature. From Eqs. (15.7) and (2.11), we find that

$$C_{v_o} = \frac{R}{k - 1} \tag{15.10a}$$

$$C_{p_o} = \frac{Rk}{k - 1} \tag{15.10b}$$

Table 15.2 *Properties of Gases at Standard Conditions*

Gas	Molecular Weight	C_{po}^*	C_{vo}^*	k
Air	29	0.24	0.171	1.4
Carbon monoxide	28	0.249	0.178	1.4
Helium	4	1.25	0.753	1.66
Hydrogen	2.02	3.43	2.43	1.4
Nitrogen	28	0.248	0.177	1.4
Oxygen	32	0.219	0.157	1.4
Steam	18	0.444	0.335	1.33

*Units of specific heats are Btu/lbm·°R.

Source: *Thermodynamics*, 3rd ed., by K. Wark. Copyright © 1983 McGraw-Hill Book Company. Used with the permission of McGraw-Hill Book Company.

For air, we shall use the following property values (unless otherwise stated):

$$C_{v_o} = 718 \text{ m}^2/\text{s}^2\cdot\text{K} = 4293 \text{ ft}^2/\text{s}^2\cdot°\text{R} \tag{15.11a}$$

$$C_{p_o} = 1005 \text{ m}^2/\text{s}^2\cdot\text{K} = 6010 \text{ ft}^2/\text{s}^2\cdot°\text{R} \tag{15.11b}$$

Example 15.1

Find an expression for the ratio of specific heats k in terms of the number of degrees of freedom n of molecules of a perfect gas given that the energy of molecular translation $i_t = nRT/2$ for a monatomic gas, $i_r = RT$ for a diatomic gas, and $i_r = 3RT/2$ for a polyatomic gas.

Solution:

The specific heat ratio k is given by Eq. (2.11)

$$k = \frac{C_{p_o}}{C_{v_o}} \tag{i}$$

where C_{p_o} is the specific heat coefficient at constant pressure given by

$$C_{p_o} = \frac{dh}{dT} \tag{ii}$$

C_{vo} is the specific heat coefficient at constant volume given by

$$C_{v_o} = \frac{di}{dT} \tag{iii}$$

and the enthalpy h is defined as

$$h = i + RT \tag{iv}$$

for a perfect gas. For a single molecule of gas

$$i = i_t + i_r = \frac{nRT}{2} \tag{v}$$

such that from Eq. (iii)

$$C_{v_o} = \frac{nR}{2} \tag{vi}$$

and from Eqs. (ii) and (iv),

$$\begin{aligned} C_{p_o} &= \frac{d}{dT}\left(\frac{nRT}{2} + RT\right) \\ &= \left(\frac{n+2}{2}\right)R \end{aligned} \tag{vii}$$

Example 15.1 *(Con't.)*

Substituting Eqs. (vii) and (iii) into Eq. (i) gives

$$k = \frac{n+2}{n} \tag{viii}$$

for a monatomic gas,

$$k = \frac{n+4}{n+2} \tag{ix}$$

for a diatomic gas, and

$$k = \frac{n+5}{n+3} \tag{x}$$

for a polyatomic gas.

This completes the solution.

15.3 The Second Law of Thermodynamics

The second law of thermodynamics determines the direction of change of equilibrium for a particular compressible fluid flow. It deals with the *quality* of energy. Recall that the first law deals with the *quantity* of energy: the first law is an energy balance statement of how energy can be transformed within and through the system, whereas the second law states whether such a transformation can actually take place. The manner in which the latter can be accomplished is through examination of the change in the disorder of the system. Entropy S is the property that directly measures disorder, or chaos, of the fluid flow. It is related to energy by the second law which states

$$dS \geq \frac{đQ}{T} \tag{15.12}$$

Recall that $đQ$ is a small increment of heat transfer. The symbol $đ$ signifies that the heat added in a process depends exclusively upon that process, not on the initial or final states. It is therefore not an exact differential. Note that the product of an inexact differential $đQ$ and an integrating factor $(1/T)$ may be an exact differential dS.

For an internally reversible process, integration of Eq. (15.12) yields

$$\Delta S = \int_1^2 \frac{đQ}{T} \tag{15.13}$$

For an adiabatic reversible process $đQ = 0$, so that $S_2 = S_1$. Hence the entropy S is a constant. We call such a process isentropic and denote it by the symbol Ⓢ.

15.4 Equations of a Process

A process is a continuous succession of states. It exists whenever there is a fluid flow. We customarily know conditions at some stations; perhaps an upstream or downstream station, at an inlet or exit, or a stagnation point. Thus, in most instances, one state of the flow is known. The properties of gas in steady flow will not change unless some outside factor such as a boundary, or a transfer of energy as heat, or some piece of machinery like a compressor, turbine, or nozzle causes a change in the properties.

When we consider the changes that occur during a process, it is apparent that some relationship among the various properties is necessary. One such relationship applicable for perfect gases is the general polytropic equation

$$\left(\frac{p}{\rho^n}\right)_1 = \left(\frac{p}{\rho^n}\right)_2 = \text{const.} \tag{15.14}$$

where n is a parameter that has a specific value depending upon the gas and the process. Some typical processes are defined as follows:

- $n = 0$: isobaric process
- $n = 1$: isothermal process
- $n = -\infty$: isometric process
- $n = k$: isentropic process

Consider a compressible flow of a fluid of constant specific heats. Let the flow be confined in a control volume of fixed mass, such as the piston-cylinder arrangement of Fig. 15.2. The energy equation for the closed system is derived from Eq. (5.91) as

$${}_1Q_2 + {}_1W_2 = M(i_2 - i_1) \tag{15.15}$$

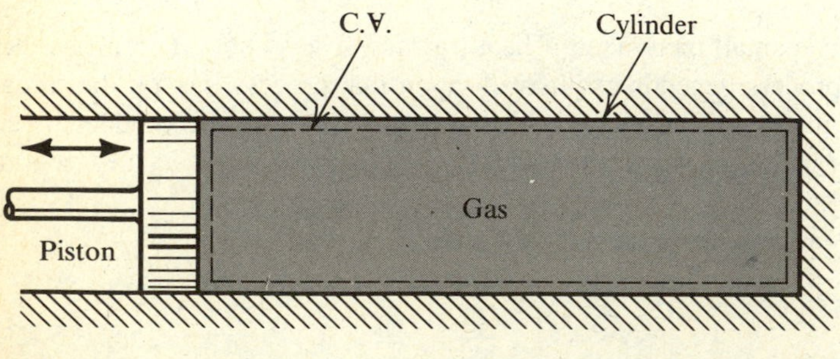

Figure 15.2 *Closed system compressible flow. First law of thermodynamics: ${}_1Q_2 + {}_1W_2 = I_2 - I_1$. Work mode: ${}_1W_2 = -\int p\,d(1|\rho)$. Internal energy: $I = Mi$. Second law of thermodynamics: $S_2 - S_1 = \int^2 \frac{đQ}{T}$.*

where the work mode ${}_1W_2$ is due to simple compressible work:

$$ {}_1W_2 = -M\int_1^2 pd\left(\frac{1}{\rho}\right) \tag{15.16} $$

Note that both heat and work are positive energy transfers if added into the system. Thus if the environment does work *on* the system, work is positive. Substitution of Eqs. (15.16), (15.4), and (15.13) into the differential form of the closed system energy Eq. (15.15) yields for a reversible process

$$ TdS - Mpd\left(\frac{1}{\rho}\right) = MC_v\, dT \tag{15.17} $$

Assuming the gas is a perfect gas, integration of Eq. (15.17) yields

$$ \frac{p_2}{p_1} = \left(\frac{\rho_2}{\rho_1}\right)^k \exp{(s_2 - s_1)/C_v} \tag{15.18} $$

Notice that if the process is isentropic,

$$ \frac{p_2}{p_1} = \left(\frac{\rho_2}{\rho_1}\right)^k \tag{15.19} $$

which is sometimes referred to and identified as the isentropic equation of state for constant specific heats of a perfect gas.

Use of the perfect gas equation of state with Eq. (15.19) gives two other forms that are equally important:

$$ \frac{T_2}{T_1} = \left(\frac{p_2}{p_1}\right)^{(k-1)/k} \tag{15.20} $$

and

$$ \frac{T_2}{T_1} = \left(\frac{\rho_2}{\rho_1}\right)^{k-1} \tag{15.21} $$

Example 15.2

Air flows through a channel where the inlet pressure and density are 300 psia and 1.2 lbm/ft^3, respectively. Calculate (a) the density ρ_2, (b) the temperature T_1, (c) the change in entropy Δs if $p_2 = 25$ psia and $T_2 = 240$°F, and (d) the change in enthalpy Δh.

Example 15.2 *(Con't.)*

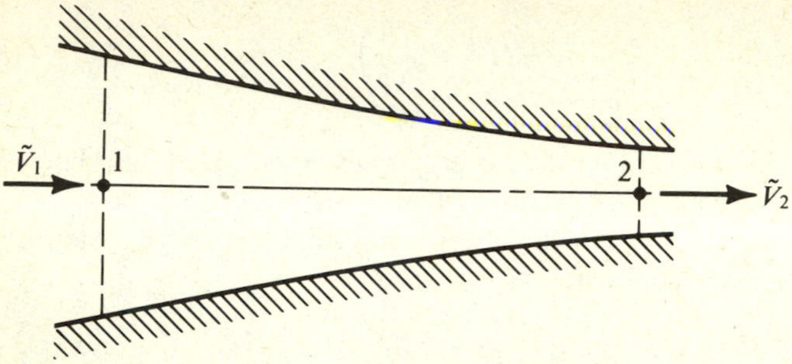

Figure E15.2

Solution:

(a) The density ρ_2 is found from the equation of state of a perfect gas given by Eq. (2.3):

$$\rho_2 = \frac{p_2}{RT_2}$$

$$= \frac{(25)(144)}{(1717)(700)} \tag{i}$$

$$= 0.003 \text{ slug/ft}^3$$

(b) The temperature of air at the inlet is also solved by the equation of state of a perfect gas:

$$T_1 = \frac{p_1}{R\rho_1}$$

$$= \frac{(300)(144)}{(1717)(1.2/32.2)} \tag{ii}$$

$$= 657°\text{R}$$

(c) The change of entropy is evaluated from Eq. (15.18):

$$s_2 - s_1 = C_{v_o} \ln\frac{p_2}{p_1} - C_{p_o} \ln\frac{\rho_2}{\rho_1} \tag{iii}$$

Substituting the known values of pressure, density, specific heats, and the result of Eq. (i) into Eq. (iii) gives

$$s_2 - s_1 = 4293 \ln\frac{25}{300} - 6010 \ln\frac{0.003}{0.0373}$$

$$= -10667.7 + 15147.5 \tag{iv}$$

$$= 4479.8 \text{ ft}^2/(\text{s}^2\cdot°\text{R})$$

Example 15.2 *(Con't.)*

which states that the entropy has increased. If there is no transfer of heat or work, the increase of entropy signifies that the process is not isentropic, but has irreversibilities. This is the case of real fluid flows.

(d) The change in enthalpy is an important quantity to calculate as it indicates how the internal energy of air changes in the channel. Assuming constant specific heat C_{p_o} between the heating from 215°F to 240°F, Eq. (15.6) reveals that

$$\begin{aligned} \Delta h &= C_{p_o} \Delta T \\ &= 6010(240 - 215) \qquad \text{(v)} \\ &= 1.5025 \times 10^5 \text{ ft}^2/\text{s}^2 \end{aligned}$$

revealing that there is an increase of enthalpy. The change in internal energy is seen to be

$$\begin{aligned} \Delta i &= C_{v_o} \Delta T \\ &= 4293(240 - 215) \\ &= 1.073 \times 10^5 \text{ ft}^2/\text{s}^2 \end{aligned}$$

This completes the solution.

15.5 *The Compressible Flow Energy Equation*

The one-dimensional steady state form of the energy equation that we shall spend our time discussing originates from Eq. (5.102). It can be expressed in the form

$$ {}_iq_e + {}_iw_e = \left(h + \frac{\tilde{V}^2}{2} + gz\right)_e - \left(h + \frac{\tilde{V}^2}{2} + gz\right)_i \qquad (15.22)$$

with units in terms of energy per unit weight of fluid, or as

$$ {}_i\dot{Q}_e + {}_i\dot{W}_e = \dot{M}\left(\Delta h + \frac{\Delta\tilde{V}^2}{2} + g\Delta z\right) \qquad (15.23)$$

with units in terms of time rate of energy. The work expressions are described in Sec. 5.5.2, and consist of mechanical or shaft work, work due to a change in control volume, and work lost because of shear stresses.

15.6 Problem Solution Technique in Applying the Energy Equation

Following is a suggested format for solving compressible flow problems:

Step 1.

Identify a suitable control volume for the analysis. The choices are either a closed system (fixed fluid mass) or an open system (fluid mass crosses the system boundary). A sketch of the control volume is useful so that boundaries and stations can be identified.

If the system is closed, let 1 denote the initial state, and 2 denote the final state.

If the system is open, let i denote the inlet station and e denote the exit station.

Step 2.

Identify all energy transfers such as heat and work. Let energy coming from the environment into the control volume be positive.

Step 3.

Write the appropriate form of the energy equation. Check to see if the flow is steady and one-dimensional. Also check to see if the control volume is fixed or varying.

Step 4.

Identify the fluid, and tabulate the data by organizing the given values of the properties by stations or state. If the fluid is a perfect gas, determine if the specific heats can be assumed constant.

Step 5.

Identify the process. Is it ⓟ, Ⓣ, ⓟ, ⓢ, or ⓗ?

Step 6.

What mechanical properties are given? Can changes in kinetic and potential energies be neglected?

Step 7.

Complete the solution by applying the (I.F.) energy equation, and check the units of each term in the equations that are used.

15.6.1 Application of the (I.F.) Energy Equation for Compressible Flow

A number of mechanical devices will be encountered in compressible flow. We can fully describe each device by noting certain conditions on the thermodynamic and mechanical properties. These conditions simplify the energy equation.

Nozzles and Diffusers

A nozzle is a device which changes the velocity of a fluid. In the sections to come, we shall speak of three types of nozzle and diffusers: (i) subsonic, (ii) supersonic, and (iii) combinations of both (i) and (ii). Figure 15.3 shows the various types.

For an adiabatic nozzle or diffuser, the one-dimensional steady state (I.F.) energy equation is

$$h_i - h_e = \tfrac{1}{2}(\tilde{V}_e^2 - \tilde{V}_i^2) \tag{15.24}$$

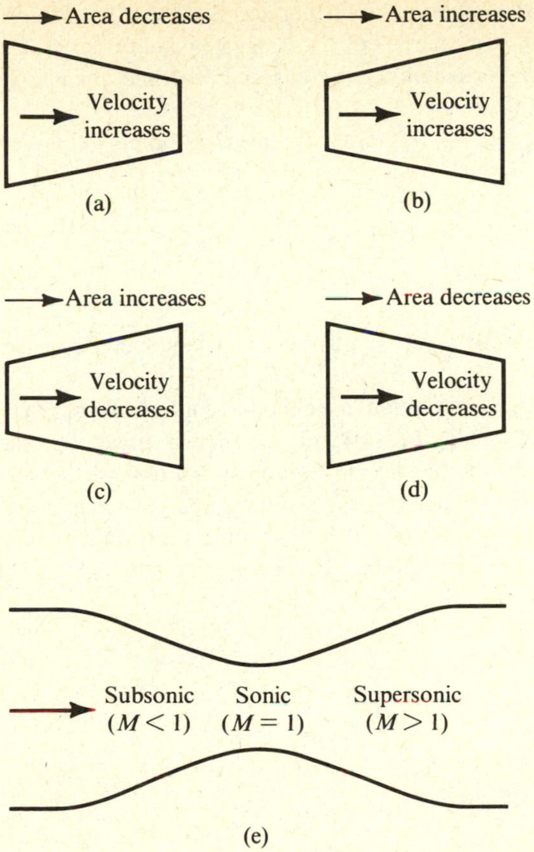

Figure 15.3 *General shapes of nozzles and diffusers for subsonic, transonic, and supersonic flow. (a) Subsonic nozzle. (b) Supersonic nozzle. (c) Subsonic diffuser. (d) Supersonic diffuser. (e) Converging-diverging nozzle.*

where the change of enthalpy is caused by the change in internal energy and displacement work p/ρ. Equation (15.24) can also be expressed as

$$C_{p_o}(T_i - T_e) = \tfrac{1}{2}(\tilde{V}_e^2 - \tilde{V}_i^2) \tag{15.25}$$

for a perfect gas with constant specific heats. Thus for a subsonic nozzle, $\tilde{V}_e > \tilde{V}_i$. From Eq. (15.25) we see that the gas has cooled down.

Compressors, Turbines, and Fans

We know that compressors and fans are machines that perform work on the gas adding energy to the fluid and resulting in an increase of pressure. Fans by and large do not cause a significant increase in pressure (the exit pressure usually being slightly above the inlet pressure), whereas the compressor's exit pressure may range from twice to over ten times the inlet pressure.

Turbines extract energy from the fluid, resulting in a pressure drop. Both turbines and compressors are usually assumed to be adiabatic devices, though each case has to be carefully examined to ascertain the importance of heat transfer effects. In addition, though the velocities at the inlet and exit stations of these devices can be large, the change in their kinetic energies may be insignificant compared with the changes in internal energy and displacement work.

Assuming therefore that $\Delta KE = {}_i\dot{Q}_e = 0$, the steady state (I.F.) energy equation becomes

$$({}_iw_e)_{\text{mech}} = h_e - h_i \tag{15.26a}$$

$$= C_{p_o}(T_e - T_i) \tag{15.26b}$$

the latter expression being appropriate for a perfect gas with constant specific heats.

Throttle

A familiar flow regulator is a throttle: an obstruction in a duct that controls the pressure drop in a line without performing any transfer of work. Figure 10.10b shows a needle valve: one type of a throttle valve. As the needle moves downward toward the seat of the valve, it creates a greater flow resistance thereby resulting in a greater pressure drop across the valve. We would therefore surmise that this would result in a change in the enthalpy. But such is not the case. Examining the steady state adiabatic (I.F.) energy equation, we obtain for the flow in Fig. 15.4

$$h_i = h_e \tag{15.27}$$

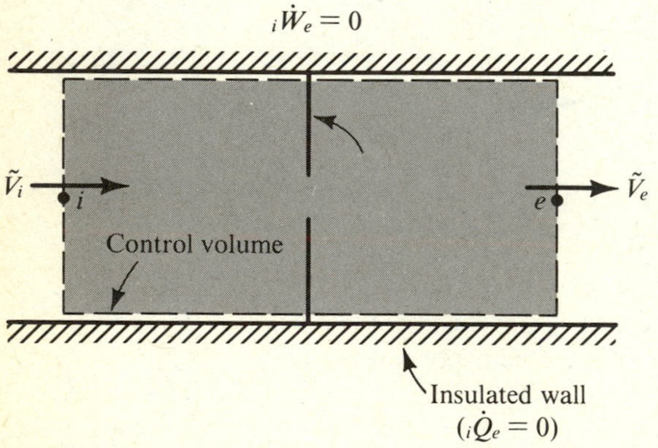

Figure 15.4 *The throttle,* $h_i = h_e$.

One-Dimensional Isentropic Pipe Flow

Another device that is frequently encountered in compressible gas flows is one-dimensional steady flow through a duct of varying cross-sectional area, like that shown in Fig. 15.5. The one-dimensional steady (I.F.) energy equation becomes

$$h_i + \frac{\tilde{V}_i^2}{2} = h_e + \frac{\tilde{V}_e^2}{2} \tag{15.28}$$

Specific categories of the above energy equation can now be treated.

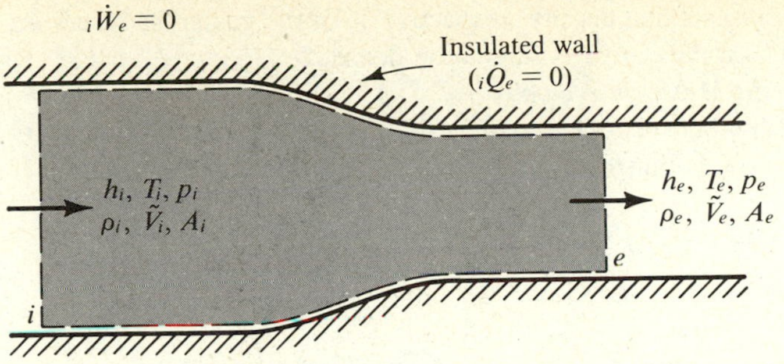

Figure 15.5 *One-dimensional steady isentropic gas flow in a duct*

$$h_i + \frac{\tilde{V}_i^2}{2} = h_e + \frac{\tilde{V}_e^2}{z} \qquad \rho_i \tilde{V}_i A_i = \rho_e \tilde{V}_e A_e$$

Stagnation Conditions

Let us assume that the inlet area A_i is so large that $\tilde{V}_i \approx 0$. The enthalpy of the gas at rest is denoted by h_s and is called the stagnation enthalpy. For this condition. Eq. (15.28) becomes

$$h_s = h_e + \frac{\tilde{V}_e^2}{2} \tag{15.29}$$

If the gas is a perfect gas, substitution of Eqs. (15.5) and (15.10b) into Eq. (15.29) yields

$$\frac{k}{k-1} RT_s = \frac{k}{k-1} RT_e + \frac{\tilde{V}_e^2}{2} \tag{15.30}$$

which indicates that the exit temperature increases with a decrease of velocity $\tilde{V}_e$.

The Maximum Velocity

Equation (15.29) illustrates that the maximum velocity exists when the exit enthalpy is zero. That is

$$\tilde{V}_{e_{\max}} = \sqrt{2h_s} \tag{15.31}$$

Hence the maximum gas velocity is a function of one variable: the enthalpy of the gas at the stagnation condition. We now are in a position to show that the maximum velocity attainable by the gas is independent of the exit pressure (i.e., the pressure where the gas escapes). From Eq. (15.31),

$$\begin{aligned} \tilde{V}_{e_{\max}} &= \sqrt{\frac{2k}{k-1} RT_s} \\ &= \sqrt{\frac{2k}{k-1} \left(\frac{p}{\rho}\right)_s} \end{aligned} \tag{15.32}$$

where p_s and ρ_s are the pressure and density, respectively, of the gas in the reservoir, and are called the stagnation pressure and stagnation density.

Speed of Sound versus Maximum Velocity

We next seek a relationship between the exit velocity and the maximum velocity for an isentropic flow. The flow is illustrated in Fig. 15.5. Starting with Eq. (15.29),

$$\frac{h_e}{h_s} = 1 - \frac{\tilde{V}_e^2}{2h_s}$$

we substitute Eq. (15.30) to obtain

$$\frac{h_e}{h_s} = 1 - \frac{\tilde{V}_e^2}{\tilde{V}_{e_{\max}}^2}$$

or

$$\frac{\tilde{V}_e}{\tilde{V}_{e_{\max}}} = \sqrt{1 - \frac{T_e}{T_s}} \tag{15.33}$$

The speed of sound c has been defined by Eq. (7.40).* If c_s is the speed of sound in a gas at rest, then

$$c_s = \sqrt{kg_cRT_s} \tag{15.34}$$

such that Eq. (15.33) becomes

$$c^2 = \frac{k-1}{2}(\tilde{V}_{e_{\max}}^2 - \tilde{V}^2) \tag{15.35}$$

Notice that the speed of sound c increases with a decrease in the velocity of the gas.

Effect of Mach Number on Isentropic Flows Through a Variable Duct

We are still discussing a compressible flow through the duct illustrated in Fig. 15.5. We seek to find the relationships between changes of pressure, duct cross-sectional area, density, and speed. Such relationships are valuable in analyzing nozzle flows out of rockets, jet engines, turbines, and high-speed wind tunnels.

Let us select the differential control volume of Fig. 15.6. The one-dimensional steady (I.F.) continuity equation for such a flow is

$$\rho\tilde{V}A = \text{const.} \tag{15.36}$$

*See Example 15.4.

$$A_i = A \qquad A + dA = A_e$$

$$\tilde{V}_i = \tilde{V} \longrightarrow \qquad \longrightarrow \tilde{V} + d\tilde{V} = \tilde{V}_e$$

$$p_i = p \longrightarrow \qquad \longleftarrow (p + dp) = p_e$$

Control volume——

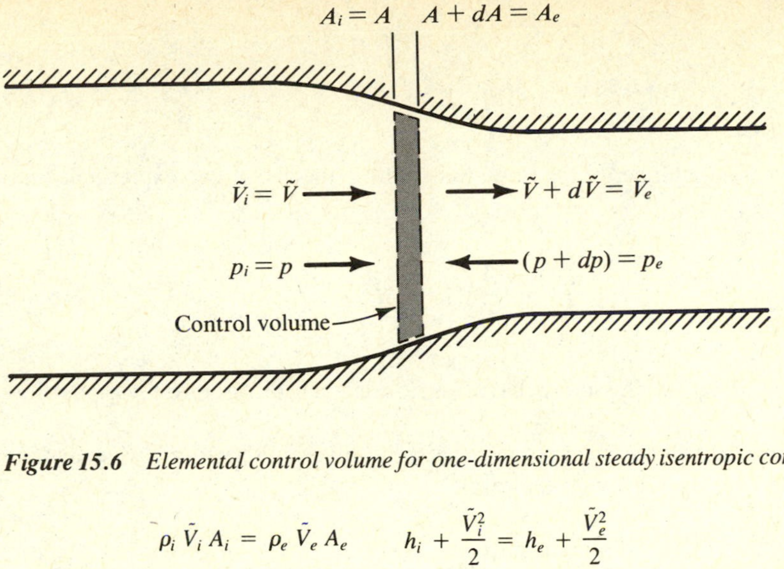

Figure 15.6 *Elemental control volume for one-dimensional steady isentropic compressible flow*

$$\rho_i \, \tilde{V}_i \, A_i = \rho_e \, \tilde{V}_e \, A_e \qquad h_i + \frac{\tilde{V}_i^2}{2} = h_e + \frac{\tilde{V}_e^2}{2}$$

Taking the logarithms of both sides and then differentiating yields

$$\frac{d\rho}{\rho} + \frac{d\tilde{V}}{\tilde{V}} + \frac{dA}{A} = 0 \tag{15.37}$$

Since the process is isentropic, there are no irreversibilities. This means that the flow meets no resistance, or that the flow is ideal. Example 4.10 treats just such a flow. Equation (x) in this example can be expressed as

$$\frac{dp}{\rho} + \frac{d\tilde{V}^2}{2} = 0 \tag{15.38}$$

for the case where there is negligible change in a flow's elevation. Substituting Eq. (15.38) into Eq. (15.37) yields

$$\frac{dA}{A} = \frac{dp}{\rho} \left(\frac{1}{\tilde{V}^2} - \frac{d\rho}{dp} \right) \tag{15.39}$$

This equation can be transformed in terms of the Mach number through the use of the speed of sound c; c was shown to be expressed as

$$c = \sqrt{\frac{dp}{d\rho}} \tag{7.37}$$

or

$$c = \frac{\tilde{V}}{M} \tag{7.36}$$

where M is the Mach number of the flow. Substituting the above two expressions into Eq. (15.39) gives

$$\frac{dA}{dp} = \frac{A}{\rho\tilde{V}^2}(1 - M^2) \tag{15.40}$$

which describes how the pressure behaves in nozzles and diffusers for various flow conditions.

To see how the velocity behaves in converging-diverging sections, we substitute Eq. (15.38) into Eq. (15.40) with the result

$$\frac{dA}{d\tilde{V}} = \frac{A}{\tilde{V}}(M^2 - 1) \tag{15.41}$$

Hence

1. For subsonic flow ($M < 1$), an increase of area results in an increase of pressure $dA/dp > 0$, and a decrease of velocity $dA/d\tilde{V} < 0$.
2. For supersonic flow ($M > 1$), an increase of area results in a decrease of pressure $dA/dp < 0$, and an increase of velocity $dA/d\tilde{V} > 0$.
3. For sonic flow ($M = 1$), the area is a minimum: $dA = 0$.

Thus, a nozzle is a device which can increase the velocity of the flow, whereas a diffuser can decrease the velocity of the flow. We see that subsonic flows cannot be accelerated to velocities greater than the speed of sound c in devices with decreasing areas. This is true despite the pressures that may exist at the inlet. The only way in which the flow can be accelerated from subsonic to supersonic velocities is in a converging-diverging device like that shown in Fig. 15.3.

The Converging Nozzle

Let us return to our discussion of flow out of a reservoir (where the flow is stagnant) into a converging nozzle and subsequently exhausting into a receiver which is another reservoir. Figure 15.7 identifies the three stations. Such a flow can be found in most blowdown channels and compressible flow wind tunnels that are used in testing aircraft models. Station s is at the inlet to the nozzle from which the gas flows. Station 1 is deliberately placed at the nozzle's throat or section of minimum area of the nozzle. Station r is the downstream reservoir station to which the fluid flows. Conditions at s and r govern what transpires in the channel.

Once again, we stipulate that the process is isentropic, such that the (I.F.) energy equation is that given by Eq. (15.30) applied between stations s in the reservoir and station 1 at the throat:

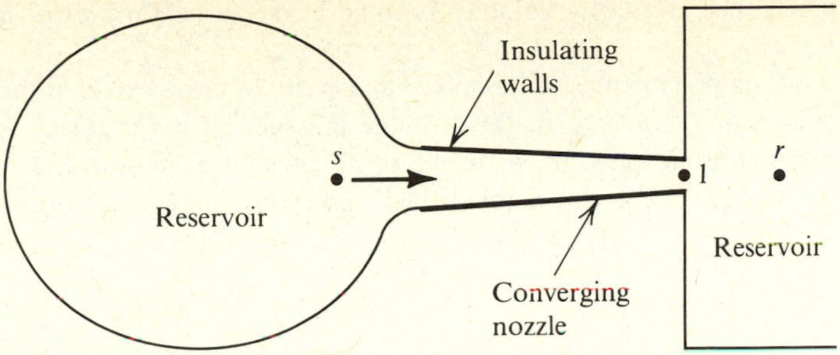

Figure 15.7 *The converging nozzle.*

$$\frac{Rk}{k-1}(T_s - T_1) = \frac{\tilde{V}_1^2}{2} \tag{15.42}$$

Expressing Eq. (15.42) in terms of pressure and density, we write

$$\frac{k}{k-1}\left[\left(\frac{p}{\rho}\right)_s - \left(\frac{p}{\rho}\right)_1\right] = \frac{\tilde{V}_1^2}{2} \tag{15.43}$$

Since the process is isentropic, we can make use of the isentropic relationship for a perfect gas with constant specific heats given by Eq. (15.19) between the stagnation points and throat station 1, and substitute the result into Eq. (15.43):

$$\frac{k}{k-1}\left[p_s\left(\frac{p_1}{p_s}\right)^{1/k} - p_1\right] = \frac{\rho_1\tilde{V}_1^2}{2} \tag{15.44}$$

Let us stipulate that the flow is one where we have sonic conditions at the throat. This is, as we have stated previously, where the maximum velocity exists. Substituting

$$\tilde{V}_1 = c = \sqrt{kg_cRT_1} = \sqrt{\frac{kp_1}{\rho_1}} \tag{15.45}$$

into Eq. (15.44) yields

$$\frac{1}{k-1}\left(p_s^{(k-1)/k}p_1^{1/k} - p_1\right) = \frac{p_1}{2} \tag{15.46}$$

or rearranging terms to obtain the ratio p_1/p_s, we find that

$$\left(\frac{p_1}{p_s}\right)_{cr} = \frac{p^*}{p_s} = \left(\frac{2}{k+1}\right)^{k/(k-1)} \tag{15.47}$$

We label this condition (where the flow velocity is sonic at the throat) the *critical pressure ratio*.

The critical pressure ratio states that if an acoustic velocity of flow exists at the throat of a converging section of a channel, the absolute pressure p^* at the throat is a fixed proportion of the absolute pressure p_s in the upstream tank reservoir and is independent of the pressure p_r in the downstream tank reservoir. For air at normal conditions

$$\frac{p^*}{p_s} = 0.528 \tag{15.48}$$

If the pressure p_r in the downstream tank reservoir is reduced below the value of p^* at the throat necessary to establish the critical pressure ratio, then the pressure in the receiver cannot be transmitted back into the throat of the nozzle, because the fluid in the throat is moving with the same velocity as the velocity of pressure propagation, i.e., the acoustic velocity. The pressure p_1 remains fixed by the critical pressure ratio. On the other hand, if the pressure in the receiver is above that required by the critical pressure ratio, then the fluid velocity at the throat is less than the velocity of sound; the receiver pressure can be telegraphed back into the throat and p^* will equal p_r. In such a case the velocity in the throat may be computed from the (I.F.) energy equation as

$$\tilde{V}_1 = \left[\frac{2Rk}{k-1}(T_s - T_1)\right]^{1/2} \tag{15.49}$$

Example 15.3

Equation (15.47) is the pressure ratio of the critical pressure at the sonic point referenced to the stagnation value p_s. The critical pressure is noted by an asterisk: p^*. Obtain expressions for the critical density ρ^*, critical temperature T^*, critical speed of sound c^*, and velocity $\tilde{V}^*$ in terms of stagnation properties.

Solution:

The critical pressure relationship has been derived and is given by Eq. (15.47):

$$\frac{p^*}{p_s} = \left(\frac{2}{k+1}\right)^{k/(k-1)} \tag{i}$$

Use of the isentropic process Eq. (15.19) with Eq. (i) results in

$$\frac{\rho^*}{\rho_s} = \left(\frac{2}{k+1}\right)^{1/(k-1)} \tag{ii}$$

Substituting the perfect gas relationship

$$\left(\frac{p^*}{\rho^*}\right)\left(\frac{\rho_s}{p_s}\right) = \frac{T^*}{T_s} \tag{iii}$$

Example 15.3 *(Con't.)*

with Eqs. (i) and (ii) results in

$$\frac{T^*}{T_s} = \frac{2}{k + 1} \tag{iv}$$

The critical speed of sound c^* with respect to the speed of sound in a gas at rest is obtained from Eqs. (15.34) and (iv):

$$\frac{c^*}{c_s} = \left(\frac{2}{k + 1}\right)^{1/2} \tag{v}$$

Since the critical velocity $\tilde{V}^*$ is exactly the critical speed of sound by definition ($M^* = 1$), we obtain from Eq. (v)

$$\tilde{V}^* = \left[\left(\frac{2}{k + 1}\right) k g_c R T_s\right]^{1/2} \tag{vi}$$

If the gas is air, Eqs. (i)–(v) yield $p^*/p_s = 0.528$, $\rho^*/\rho_s = 0.634$, $T^*/T_s = 0.833$, and $c^*/c_s = 0.913$.

This completes the solution.

The Convergent-Divergent Nozzle

We have shown that in a contracting nozzle, the flow velocity can never exceed the acoustic velocity no matter what we do to the exhaust pressure in the reservoir. Is there no way to obtain a higher velocity?

There is a way, but it will not alter the relationship we have developed for the converging channel. Let us consider critical conditions existing in a converging nozzle where the pressure at the throat $p_1 = 0.53 p_s$. This will result in sonic conditions existing at the throat. Then we shall add an expanding nozzle to create the converging-diverging nozzle shown in Fig. 15.3.

The (I.F.) continuity equation for such a flow is

$$\dot{M} = (\rho A \tilde{V})_i = (\rho A \tilde{V})_e \tag{15.50}$$

The (I.F.) steady one-dimensional energy Eq. (15.44) is expressed in terms of the average velocity $\tilde{V}$ in the nozzle:

$$\tilde{V} = \left\{\left(\frac{p}{\rho}\right)_s \left(\frac{2k}{k - 1}\right)\left[1 - \left(\frac{p}{p_s}\right)^{(k-1)/k}\right]\right\}^{1/2} \tag{15.51}$$

Substitution of the mass flow rate given by Eq. (15.50) into Eq. (15.51) results in

$$\frac{\dot{M}}{A} = \left\{2 p_s \rho_s \frac{k}{k - 1}\left[\left(\frac{p}{p_s}\right)^{2/k} - \left(\frac{p}{p_s}\right)^{(k+1)/k}\right]\right\}^{1/2} \tag{15.52}$$

We see that there is no flow when $p/p_s = 0$ or when $p/p_s = 1.0$ since $A \rightarrow \infty$. It is also seen that the mass rate is parabolic. Thus, plotting $\dot{M}/A$ versus the pressure ratio p/p_s in Fig. 15.8 shows that a maximum mass rate of flow through a convergent-divergent nozzle exists for a particular pressure ratio. Differentiating $(\dot{M}/A)$ with respect to p/p_s and setting the result equal to zero yields

$$\frac{2}{k}\left(\frac{p}{p_s}\right)^{(2-k)/k} - \left(\frac{k+1}{k}\right)\left(\frac{p}{p_s}\right)^{1/k} = 0$$

or

$$\frac{p}{p_s} = \left[\frac{2}{(k+1)}\right]^{k/(k-1)} \tag{15.53}$$

a result identical to Eq. (15.47). It should, since the critical conditions must exist at the throat.

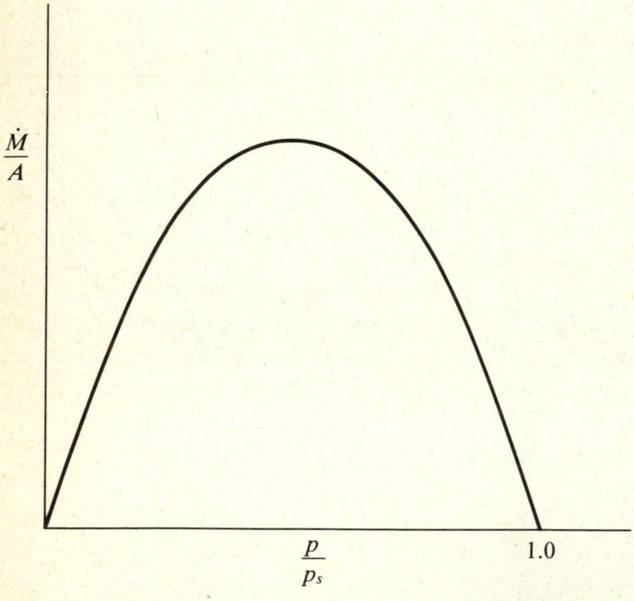

Figure 15.8 *Mass rate of flow per unit area variation with pressure ratio p/p_s.*

Use of the isentropic pressure-temperature relationship of Eq. (15.20) gives

$$\frac{T}{T_s} = \frac{2}{k+1} \tag{15.54}$$

and for a density relationship, we would use Eqs. (15.19) and (15.53) to obtain

$$\frac{\rho}{\rho_s} = \left[\frac{2}{(k+1)}\right]^{1/(k-1)} \tag{15.55}$$

Example 15.4
Show that the velocity corresponding to the maximum rate of flow occurs when sonic velocity is reached in the convergent-divergent passage.

Solution:
Substitution of Eq. (15.53) into the expression for the velocity in a convergent-divergent nozzle, Eq. (15.51), yields

$$\tilde{V} = \left[\left(\frac{2k}{k-1}\right)\left(\frac{p}{\rho}\right)_s\left(1 - \frac{2}{k+1}\right)\right]^{1/2} \tag{i}$$

or

$$\tilde{V} = \left[\left(\frac{2k}{k+1}\right)\left(\frac{p}{\rho}\right)_s\right]^{1/2} \tag{ii}$$

Since the fluid is a perfect gas, then

$$\left(\frac{p}{\rho}\right)_s = g_cRT_s \tag{iii}$$

Substituting Eq. (iii) into Eq. (ii) results in

$$\tilde{V} = \left[\left(\frac{2k}{k+1}\right)(g_cRT_s)\right]^{1/2} \tag{iv}$$

But the maximum rate of flow is when

$$\frac{T}{T_s} = \frac{2}{k+1} \tag{v}$$

Substituting the foregoing result into Eq. (iv) shows

$$\tilde{V} = \sqrt{kg_cRT} \tag{vi}$$

which is the acoustic velocity.
This completes the solution.

Example 15.5
Consider a high-speed compressible gas flow through a convergent-divergent nozzle as shown in Fig. E15.5. If the air enters at a pressure of p_i = 0.5 mPa, and a temperature T_i = 23°C and exits at a pressure p_e = 0.3 mPa, find the

Example 15.5 *(Con't.)*

exit velocity of the air from the nozzle, if the gas has a molecular weight of 44, and $k = 1.4$.*

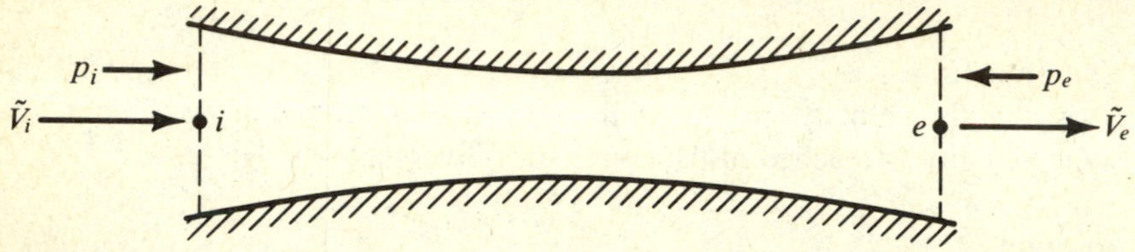

Figure E15.5

Solution:

We first must determine if the pressure ratio p_e/p_i is greater, equal to, or less than the critical pressure ratio. Given that

$$\frac{p_e}{p_i} = \frac{30}{50} = 0.6 \tag{i}$$

we notice that it is greater than $(p_e/p_i)_{cr} = 0.53$, such that we can use the expression for the average velocity given by Eq. (15.51):

$$\tilde{V}_e = \left\{\left(\frac{p_i}{\rho_i}\right)\left(\frac{2k}{k-1}\right)\left[1 - \left(\frac{p_e}{p_i}\right)^{(k-1)/k}\right]\right\}^{1/2} \tag{ii}$$

The density of the gas at the inlet is found from the equation of state for a perfect gas:

$$\rho_i = \frac{p_i}{RT_i} \tag{iii}$$

where

$$R = \frac{8312}{44} = 188.91 \text{ m·N/kg·K} \tag{iv}$$

The given data at station i are substituted into Eq. (iii) with the result

$$\begin{aligned} \rho_1 &= \frac{0.5 \times 10^6 \text{ N/m}^2}{(188.91 \text{ m·N/kg·K})(300 \text{ K})} \\ &= 8.8 \text{ kg/m}^3 \end{aligned} \tag{v}$$

*It should be mentioned that $p_e/p_i > 0.528$ does not ensure that subsonic flow exists throughout the convergent-divergent nozzle. Another possibility is case (v), p. 827.

Example 15.5 *(Con't.)*

The exit velocity then becomes

$$\tilde{V}_e = \left\{\left(\frac{0.5 \times 10^6}{8.8}\right)\left(\frac{2.8}{0.4}\right)[1 - (0.6)^{0.4/1.4}]\right\}^{1/2}$$

$$= (0.054 \times 10^6)^{1/2} \qquad \text{(vi)}$$

$$= 232.4 \text{ m/s}$$

which is subsonic, as we suspected it would be.

This completes the solution.

We have shown from the (I.F.) continuity equation for compressible flow that

$$\frac{dA}{d\bar{V}} = \frac{A}{\bar{V}}(M^2 - 1) \qquad (15.41)$$

It stipulates that sonic conditions exist only at a point where the area is not changing. This occurs only at the throat of a convergent-divergent channel. We also know that the mass rate $\dot{M}$ is always a constant in the nozzle, and therefore $\dot{M}/A$ is a maximum at the point where A is a minimum. This is again at the throat.

Therefore, for compressible fluid flow in a convergent-divergent nozzle, the maximum mass rate of flow occurs when

$$\frac{p_1}{p_s} = \left(\frac{2}{k + 1}\right)^{k/(k-1)}$$

where p_1 is the pressure at the throat.

Figure 15.9 shows the variation of the thermodynamic properties of pressure, temperature, and density, and the variation of velocity through a convergent-divergent nozzle for varying conditions at the reservoir station r. We start the analysis of results by stating that the stagnation pressure p_s is a constant. The following results are important:

1. When $p_r = p_s$, there is no flow through the nozzle. This is obvious from Eq. (15.52).
2. When p_r is slightly less than p_s, the gas will move through the nozzle, as described by curve A. The flow accelerates or decelerates always at subsonic speeds reaching a maximum at the throat. In accordance with the energy equation, as the velocity increases, the pressure will decrease. As pressure decreases, so shall the temperature and density. However, if p_r is not sufficiently less than p_s so that the pressure

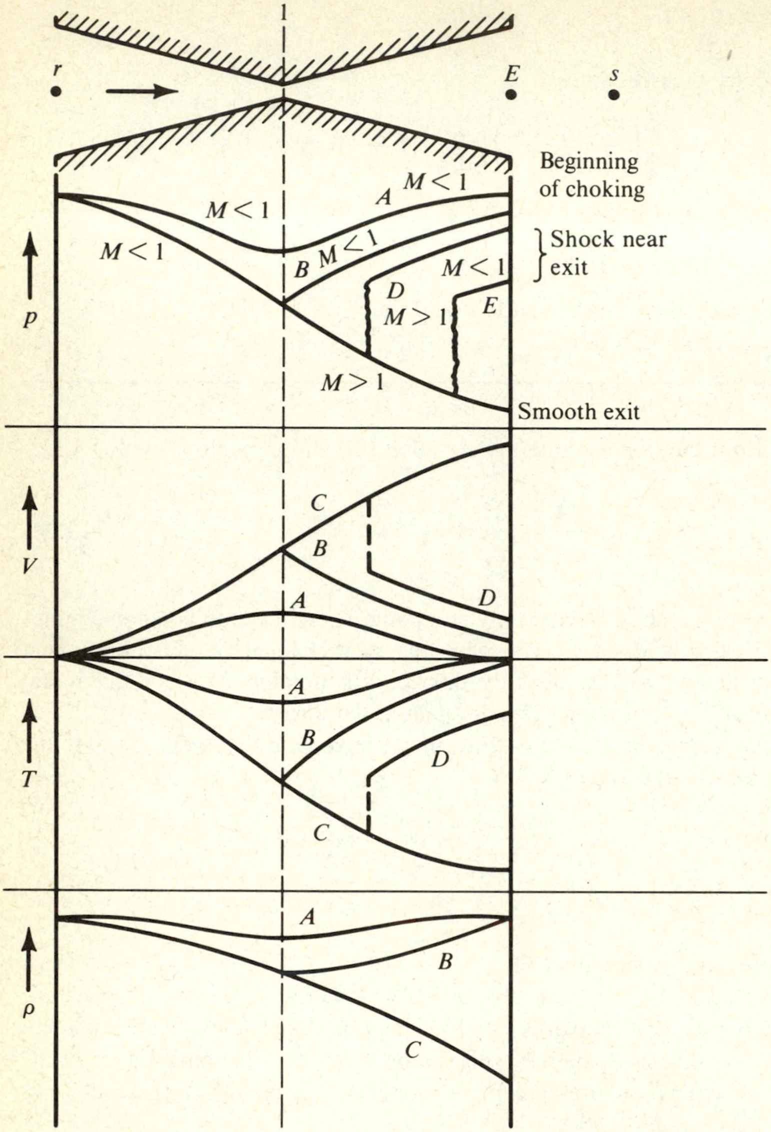

Figure 15.9 Behavior of thermodynamic properties and velocity through a converging-diverging nozzle.

ratio $p_1/p_s > (p/p_s)_{cr}$, sonic velocity will not be obtained in the throat. This approximates flow through a Venturi.

3. If the receiver pressure p_r is further decreased, a point will be reached (depending upon the design of the nozzle) at which sonic velocity will be achieved at the throat. Thus $p_1/p_s = 2/(k + 1)^{k/(k-1)}$. This condition is shown as curve B. Isentropic expansion takes place from point s to the throat 1, and isentropic compression takes place from 1 to the exit. We call this latter result *subsonic recompression*.

As the velocity decreases from the throat to the exit, the pressure increases as expected. So will the temperature increase and so will the density.

4. If the receiver pressure p_r is decreased sufficiently, the pressure distribution in the nozzle will follow curve C. Sonic velocity will exist at the throat and the gas will follow an isentropic expansion from entrance s to exit r. This process is called *full expansion*. The velocity continues to increase throughout the nozzle, which will result in a continuous decrease of pressure. Thus, both the temperature and density will decrease.
5. For receiver pressures p_r which are lower than that necessary for "full recompression" and higher than that for "full expansion," sonic velocity will exist at the throat and $p_1/p_s = 2/(k + 1)^{k/(k-1)}$. As p_r decreases from just below that value necessary for full expansion, progressively greater amounts of expansion and lesser amounts of recompression will occur in the divergent section. However, for the fluid pressure to reach the receiver pressure p_r, a "discontinuity" must take place in the divergent section of the nozzle. This abrupt change is called *compression shock wave*, and the pressure distribution in the divergent section will appear as curves D or E, depending on the magnitude of the receiver pressure.
6. If the receiver pressure is lower than the pressure for curve C, no change will be observed in the nozzle. As stated previously, this is because the flow is supersonic at the exit and pressure disturbances cannot propagate upstream.

We have not discussed what takes place if viscosity is significant. Viscosity is important when shock waves are present, particularly in the interaction of the shock waves with the boundary layer. By and large, viscosity does not play a significant role in the behavior of a pressure wave.

15.7 Normal Shock Waves

Normally, i.e., if things are done slowly, fluids avoid being compressed. It is natural for fluids to deform when subjected to external stresses and to move out of the way of solid objects passing through the medium. But if an object or a disturbance is moving so fast that the fluid particles cannot move out of the way, then the molecular structure of the fluid permits the molecule's relative positions to be altered very easily. The molecules can move together; the gas can be compressed; a shock wave can be formed. We call a shock wave a front of compressed gas molecules. Some call the shock wave a discontinuity. In actuality, the shock is not discontinuous. It may represent an abrupt change, but the properties vary continuously across a shock.

Figure 15.10 presents a normal shock wave in a steady one-dimensional adiabatic flow. Let the upstream conditions be denoted by the subscript 1 and the downstream conditions be denoted by the subscript 2. The gas is assumed to be a perfect gas, without energy loss. We purposely select a control volume of small axial length, large enough to include the normal shock yet small enough so that $A_1 \approx A_2$. We shall find it more convenient to identify the upstream reservoir station s as 01, and the downstream reservoir station r as 02 so that we can use the gas tables of Ref. 15.4. The stagnation temperature T_{01} is equal to the receiver temperature T_{02} so the process is isoenergetic.

The pressure p_{01} must be greater than the receiver pressure p_{02} so that there is a normal shock standing in the channel. Upstream at station 1 the flow is supersonic, whereas downstream at point 2 the flow is subsonic. The flow is isentropic from the stagnation station 01 to station 1, and it is isentropic from station 2 to the receiver station 02. The flow is *not* isentropic from 1 to 2 because of the internal irreversibilities associated with the shock wave.

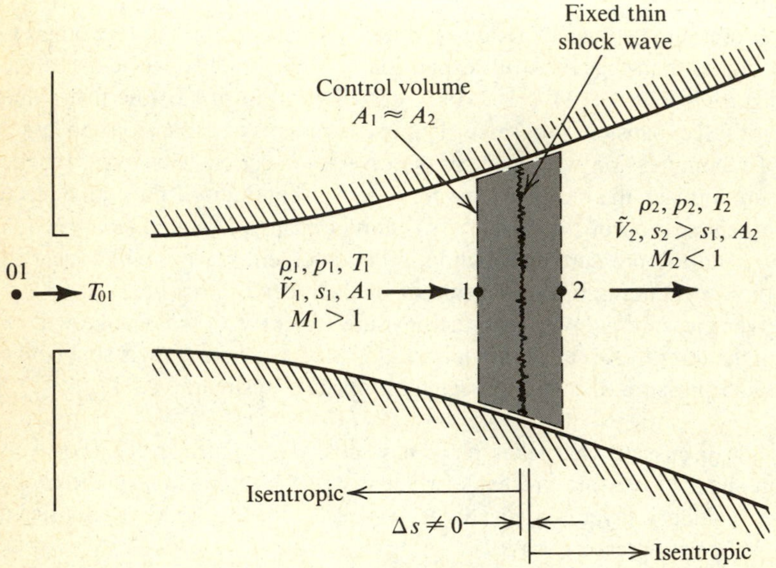

Figure 15.10 *Flow through a normal shock wave,* $T_{01} = T_{02}$, $p_{01} > p_{02}$.

The appropriate (I.F.) governing equations are the

- Continuity equation:

$$\rho_1 \tilde{V}_1 A_1 = \rho_2 \tilde{V}_2 A_2 \tag{15.56}$$

- The linear momentum equation:

$$p_1 A_1 - p_2 A_2 = \rho_1 A_1 \tilde{V}_1 (\tilde{V}_2 - \tilde{V}_1) \tag{15.57}$$

- The energy equation:

$$\frac{Rk}{k-1} T_1 + \frac{\tilde{V}_1^2}{2} = \left(\frac{Rk}{k-1}\right) T_2 + \frac{\tilde{V}_2^2}{2} \tag{15.58}$$

- The equation of state for a perfect gas:

$$\frac{p_1}{\rho_1 R T_1} = \frac{p_2}{\rho_2 R T_2} \tag{15.59}$$

We seek to derive expressions for the changes of the properties (ρ, p, T, and $\tilde{V}$) that occur across the normal shock. Because the shock wave is very thin (see Sec. 15.7.6), the changes that occur take place abruptly and thus we cannot expect thermal equilibrium to exist. It is known that viscosity effects and heat conduction are two very important aspects of heat transfer that contribute to making the shock wave internally irreversible.

We begin by combining the (I.F.) continuity Eq. (15.56) with the (I.F.) energy Eq. (15.58) to give

$$h_1 = h_2 + \frac{\dot{M}^2}{2A^2}\left(\frac{1}{\rho_y^2} - \frac{1}{\rho_x^2}\right) \tag{15.60}$$

which is the equation of the *Fanno line*, illustrated in Fig. 15.11. In Eq. (15.60), $\dot{M}$ is the mass rate and A is the area of the duct shown in Fig. 15.10. Lines of constant mass flow are called Fanno lines. An alternate and more popular way of expressing a Fanno line is to express the (I.F.) continuity Eq. (15.56) in terms of temperature, pressure, and Mach number using the equations for a perfect gas and the speed of sound. We obtain

$$\frac{T_2}{T_1} = \left(\frac{p_2 M_2}{p_1 M_1}\right)^2 \tag{15.61}$$

and combine it with the (I.F.) energy Eq. (15.58) with the result that

$$\frac{p_2}{p_1} = \frac{M_1}{M_2}\sqrt{\frac{2 + (k - 1)M_1^2}{2 + (k - 1)M_2^2}} \tag{15.62}$$

which is tabulated in Tables 42–46 of the gas tables in Ref. 15.4. When Eq. (15.62) is plotted on an h-s diagram (see Fig. 15.12), the upper branch of the Fanno line indicates that the flow is subsonic. As entropy increases, so does the flow velocity. Maximum entropy can occur only at sonic velocity ($M = 1.0$). The effect of friction is thus to accelerate the flow when the flow is subsonic, and decelerate the flow when the flow is supersonic. If friction continues to exist after the entropy has reached a maximum, we say the flow is choked.

Suppose we combine all three governing (I.F.) equations and see what we get. From Eq. (15.58) we obtain with the use of Eq. (15.59)

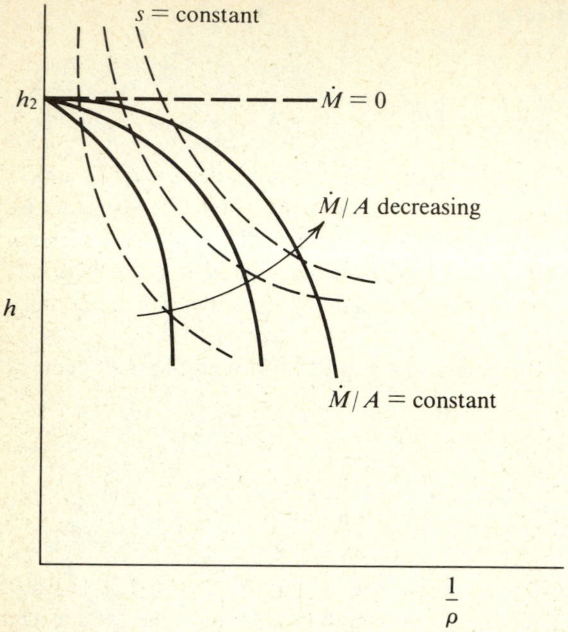

Figure 15.11 *Fanno lines for adiabatic flow of a real fluid in a duct.*

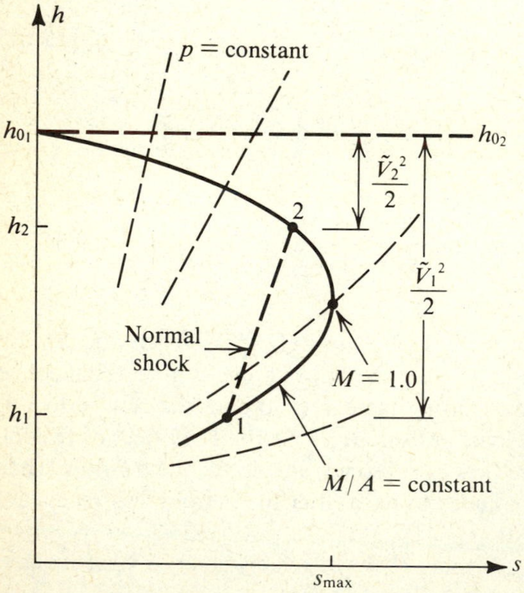

Figure 15.12 *Fanno line for adiabatic flow of a real fluid in a duct on an h-s diagram.*

$$\tilde{V}_2^2 - \tilde{V}_1^2 = \frac{2k}{k-1}\left(\frac{p_1}{\rho_1} - \frac{p_2}{\rho_2}\right) \tag{15.63}$$

The (I.F.) linear momentum equation can be expressed as

$$p_1 - p_2 = \rho_1\tilde{V}_1\tilde{V}_2 - \rho_1\tilde{V}_1^2 \tag{15.64}$$

for a constant diameter control volume. Using the continuity Eq. (15.56), we transform Eq. (15.64) into

$$p_1 + \rho_1\tilde{V}_1^2 = p_2 + \rho_2\tilde{V}_2^2 \tag{15.65}$$

If we divide the left-hand side by $\rho_1\tilde{V}_1$ and the right-hand side by $\rho_2\tilde{V}_2$, and rearrange terms and then multiply the result through by $(\tilde{V}_1 + \tilde{V}_2)$, we obtain

$$(p_1 - p_2)\left(\frac{1}{\rho_1} + \frac{1}{\rho_2}\right) = \tilde{V}_2^2 - \tilde{V}_1^2 \tag{15.66}$$

Substituting this expression for the change in kinetic energy into the energy Eq. (15.63), then expanding and collecting terms results in

$$\frac{p_2}{p_1} = \frac{\rho_1(k-1) - \rho_2(k+1)}{\rho_2(k-1) - \rho_1(k+1)} \tag{15.67}$$

Dividing both numerator and denominator by $\rho_1\,(k-1)$ and rearranging terms, we arrive at the expression for the pressure ratio across a normal shock:

$$\frac{p_2}{p_1} = \left[\left(\frac{\rho_2}{\rho_1}\right)\left(\frac{k+1}{k-1}\right) - 1\right]\Big/\left[\left(\frac{k+1}{k-1}\right) - \frac{\rho_2}{\rho_1}\right] \tag{15.68}$$

Similarly, we can manipulate the density ratios of Eq. (15.68) to obtain

$$\frac{\rho_2}{\rho_1} = \left[\left(\frac{p_2}{p_1}\right)\left(\frac{k+1}{k-1}\right) + 1\right]\Big/\left[\left(\frac{k+1}{k-1}\right) + \left(\frac{p_2}{p_1}\right)\right] = \frac{\tilde{V}_1}{\tilde{V}_2} \tag{15.69}$$

Equations (15.68) and (15.69) are known as the *Rankine-Hugoniot* equations. They are quite different from the isentropic relationships between pressure and density changes for a continuous flow in a channel that we presented on page 814.

The change in entropy across the fixed normal shock is given by Eq. (15.18)

$$s_2 - s_1 = \ln\left[\left(\frac{p_2}{p_1}\right)\left(\frac{\rho_1}{\rho_2}\right)^k\right]^{C_v} \tag{15.70}$$

Substituting the density ratio of Eq. (15.69) into Eq. (15.70), we find the change of entropy across the normal shock as

$$s_2 - s_1 = \ln\left\{\left(\frac{p_2}{p_1}\right)\left[\left(\frac{p_2}{p_1}\right) + \left(\frac{k+1}{k-1}\right)\right]^k \Big/ \left[\left(\frac{p_2}{p_1}\right)\left(\frac{k+1}{k-1}\right) + 1\right]^k\right\}^{C_v} \tag{15.71}$$

Thus, given a particular gas, we can determine the change of entropy by measuring the pressures p_1 and p_2 upstream and downstream of the shock.

Example 15.6

Consider air where $k = 1.4$ and $C_{v_o} = 0.716$ kJ/(kg)(°C) flowing in a tunnel where a normal shock exists. Calculate the change of entropy $s_2 - s_1$ for (a) a pressure ratio $p_2/p_1 = 2.0$, (b) a pressure ratio $p_2/p_1 = 1.0$, and (c) a pressure ratio $p_2/p_1 = ½$.

Solution:

The pressure ratio across a normal shock is given by Eq. (15.71):

$$s_2 - s_1 = \ln\left\{\left(\frac{p_2}{p_1}\right)\left[\left(\frac{p_2}{p_1}\right) + \left(\frac{k+1}{k-1}\right)\right]^k \Big/ \left[\left(\frac{p_2}{p_1}\right)\left(\frac{k+1}{k-1}\right) + 1\right]^k\right\}^{C_v} \tag{i}$$

(a) For a pressure ratio $p_2/p_1 = 2.0$, Eq. (i) becomes

$$\begin{aligned} s_2 - s_1 &= \ln\left[2\left(\frac{8}{13}\right)^{1.4}\right]^{0.716} \\ &= 9.62 \times 10^{-3} \text{ kJ/(kg)(°C)} \end{aligned} \tag{ii}$$

which shows there has been an increase of entropy across the normal shock. Because the increase is rather small, the shock is considered a weak shock.

(b) For a pressure ratio $p_2/p_1 = 1.0$, we would suspect that this is the condition for the beginning of choking. The change of entropy becomes

$$s_2 - s_1 = \ln\left[1\left(\frac{7}{7}\right)^{1.4}\right]^{0.716} = 0 \tag{iii}$$

which states that the flow is isentropic. So far no shock has formed. This flow condition is shown in Fig. 15.9 as curve B.

(c) For a pressure ratio $p_2/p_1 = ½$, Eq. (i) becomes

$$\begin{aligned} s_2 - s_1 &= \ln\left[\tfrac{1}{2}\left(\frac{6.5}{4}\right)^{1.4}\right]^{0.716} \\ &= -\,9.62 \times 10^{-3} \text{ kJ/(kg)(°C)} \end{aligned} \tag{iv}$$

which shows a drop in entropy. Though we can have increases of entropy, we cannot permit a decrease in entropy without an increase of entropy of the surroundings. This is so that the second law of thermodynamics is obeyed.

This completes the solution.

We are now in the position where we can find conditions downstream of the normal shock in terms of such upstream quantities as pressure p_1, density ρ_1, and velocity $\tilde{V}_1$. The basic equations necessary in deriving such a relationship are the (I.F.) continuity Eq. (6.2) and the (I.F.) momentum Eq. (6.5) adapted to the notation of Fig. 15.10:

$$\rho_1 \tilde{V}_1 = \rho_2 \tilde{V}_2 \tag{15.72}$$

$$(p_2 - p_1) A_1 = \rho_1 Q_1 \tilde{V}_1 - \rho_2 Q_2 \tilde{V}_2 \tag{15.73}$$

Substituting Eq. (15.72) into Eq. (15.73) results in

$$p_2 = p_1 + \rho_1 \tilde{V}_1^2 \left(1 - \frac{\tilde{V}_2}{\tilde{V}_1}\right) \tag{15.74}$$

Substituting the Rankine-Hugoniot relationship for the velocity ratio $\tilde{V}_1/\tilde{V}_2$ of Eq. (15.69) into Eq. (15.74) results in

$$p_2 = p_1 + \rho_1 \tilde{V}_1^2 \left[\left(\frac{p_2}{p_1}\right)\left(\frac{2}{k-1}\right) - \left(\frac{2}{k-1}\right)\right] \Big/ \left[\left(\frac{p_2}{p_1}\right)\left(\frac{k+1}{k-1}\right) + 1\right] \tag{15.75}$$

After a bit of algebraic juggling, we obtain

$$p_2 = \frac{1}{k+1}\left[2\rho_1 \tilde{V}_1^2 - p_1 (k-1)\right] \tag{15.76}$$

This equation is often useful in evaluating the pressure downstream of the shock when we know the upstream pressure, density, and velocity.

Example 15.7

Consider pressure upstream of a normal shock being 107.05 kPa. If the temperature of the air at this location is 100°C, calculate the pressure downstream of the shock if the velocity of the gas $\tilde{V}_1$ = 637.8 m/s.

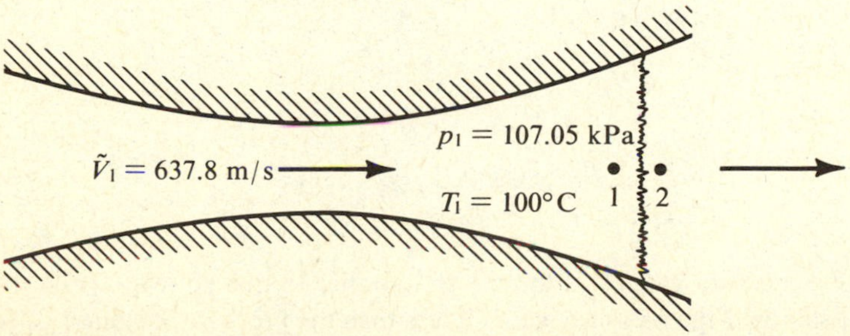

Figure E15.7

Example 15.7 *(Con't.)*

Solution:
The pressure downstream of the shock can be determined from Eq. (15.76):

$$p_2 = \frac{1}{k + 1} [2\rho_1 \tilde{V}_1^2 - p_1 (k - 1)] \tag{i}$$

We first calculate the density of air at station 1:

$$\begin{aligned} \rho_1 &= \frac{p_1}{RT_1} \\ &= \frac{107.05 \text{ kPa}}{\left(0.287 \dfrac{\text{kJ}}{\text{kg·K}}\right)(373 \text{ K})} \\ &= 1 \text{ kg/m}^3 \end{aligned} \tag{ii}$$

Substituting the given data and Eq. (ii) into Eq. (i) results in

$$\begin{aligned} p_2 &= \frac{1}{2.4} [2(1)(637.8)^2 - 107.05 (0.4) \times 10^3] \\ &= 321.15 \text{ kPa} \end{aligned} \tag{iii}$$

Thus, the pressure has tripled, which means the entropy will have increased, according to Eq. (15.71).

This completes the solution.

15.7.1 Mach Number Relationships for a Normal Shock

It is often convenient to determine pressure, temperature, or density rise in terms of the initial Mach number upstream of the shock. Using the relationships that $M = \tilde{V}/c$ and $c = \sqrt{kp/\rho}$, we transform the pressure ratio p_2/p_1 of Eq. (15.76) as

$$\frac{p_2}{p_1} = \frac{1}{k + 1} [2 M_1^2 k - (k - 1)] \tag{15.77}$$

such that

$$\frac{p_2 - p_1}{p_1} = \frac{2k}{k + 1} (M_1^2 - 1) \tag{15.78}$$

Suppose we have pressure gauges in the nozzle, similar to that shown in Fig. 15.13. We note that at station 1 the pressure is much less than the pressure at station 2, which means that the flow is supersonic ($M_1 > 0$) since the pressure difference is positive.

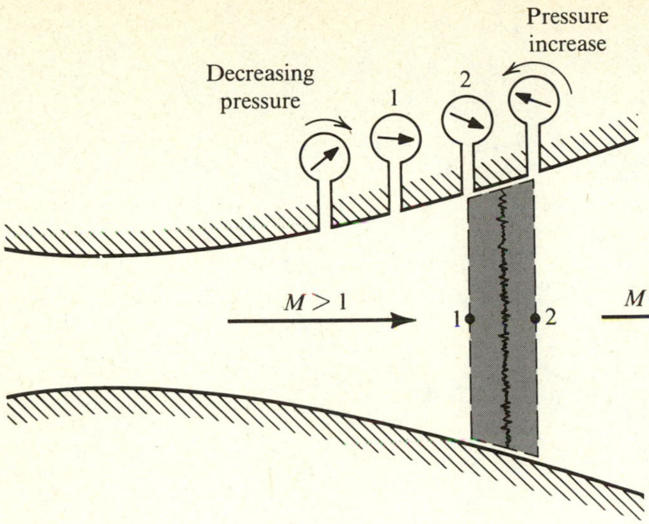

Figure 15.13 *Pressure distribution in a supersonic nozzle.*

Let us see how the downstream Mach number M_2 enters the picture. It is best to start with the (I.F.) momentum Eq. (15.73):

$$p_2 - p_1 = \rho_1 \tilde{V}_1^2 - \rho_2 \tilde{V}_2^2 \tag{15.79}$$

which states that the change in total pressure is precisely balanced by the change in dynamic pressure. Substituting the relationship for the speed

$$\tilde{V}^2 = kM^2 \frac{p}{\rho} \tag{15.80}$$

into Eq. (15.79) results in the *Rayleigh* equation

$$\frac{p_2}{p_1} = \frac{(1 + k\,M_1^2)}{(1 + k\,M_2^2)} \tag{15.81}$$

which is tabulated in Table 36 of the gas tables of Ref. 15.4. Equation (15.81) can be plotted on an enthalpy-entropy diagram as shown in Fig. 15.14. The *Rayleigh line* represents the loci of Mach number. As the temperature along the $M < 1$ line increases, the Mach number increases up to $1/\sqrt{k}$ for $h_{\max}$, then increases continuously as the temperature increases as the velocity decreases. Heat addition from friction enables the velocity of the flow to approach the speed of sound. It is important to notice that according to the second law (that entropy must increase when heat is added to the flow), it is not possible to pass from supersonic flow to subsonic flow. This can only be accomplished by cooling.

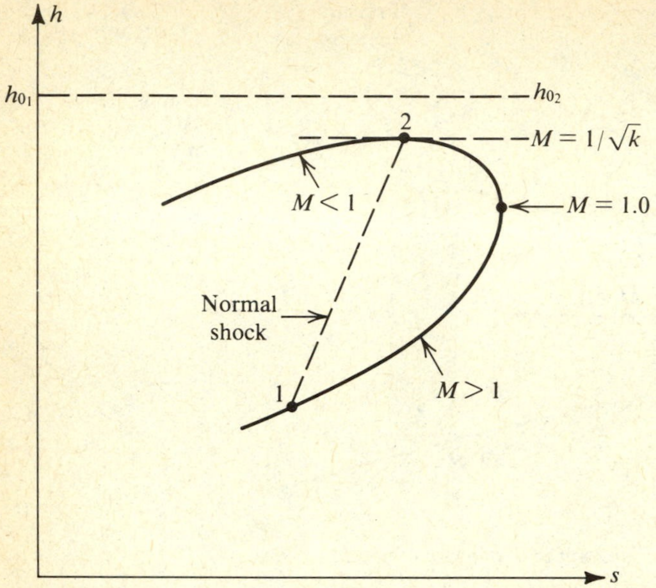

Figure 15.14 *Rayleigh line: enthalpy-entropy diagram for adiabatic flow in a duct with friction.*

Equating the two pressure ratios of Eqs. (15.77) and (15.81) gives a relationship for the downstream Mach number M_2 in terms of the upstream Mach number M_1:

$$M_2 = \left[\frac{2 + M_1^2 (k - 1)}{1 - k + 2k M_1^2}\right]^{1/2} \tag{15.82}$$

For air,

$$M_2 = \left(\frac{2 + 0.4 M_1^2}{2.8 M_1^2 - 0.4}\right)^{1/2}$$

Note that if M_1 is sonic, then so is M_2. If $M_1 > 1$, then $M_2 < 1$. Thus within the thickness of nanometers, the shock wave slows the flow from supersonic speeds to subsonic ones, resulting in an increase of pressure. Figure 15.15 shows the behavior of the downstream Mach number based upon the upstream Mach number.

Another important relationship we need to present is the functional dependency of the density ratio ρ_2/ρ_1 on the upstream Mach number M_1. Substitution of the pressure ratio Eq. (15.77) into Eq. (15.69) results in

$$\frac{\rho_2}{\rho_1} = \frac{M_1^2 (k + 1)}{M_1^2 (k - 1) + 2} \tag{15.83}$$

The temperature ratio is obtained from Eqs. (15.58) and (15.83) along with the equation of a perfect gas such that

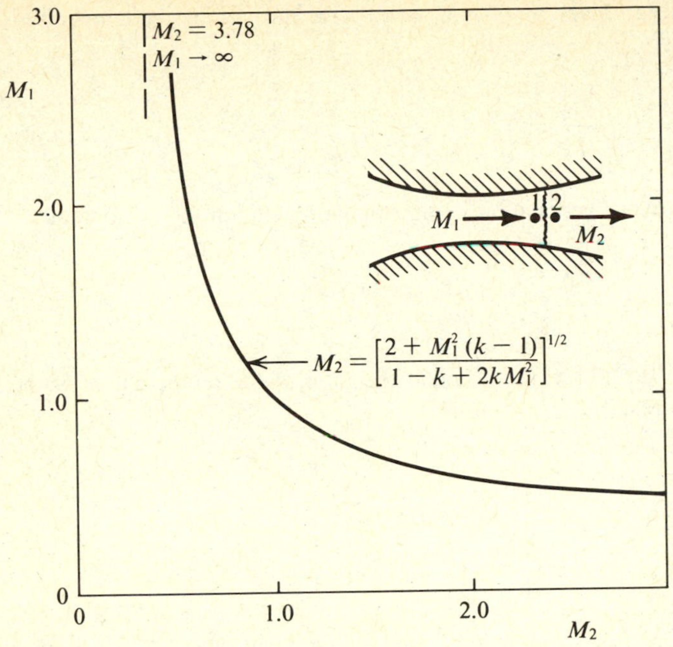

Figure 15.15 M_2 *versus* M_1 *for air in a supersonic nozzle.*

$$\frac{T_2}{T_1} = \frac{(1 + kM_1^2)}{(1 + kM_2^2)} \cdot \left[\frac{2 + (k - 1)M_1^2}{M_1^2 (k + 1)}\right] \tag{15.84}$$

or

$$\frac{T_2}{T_1} = \frac{[2 + M_1^2 (k - 1)] (2kM_1^2 - k + 1)}{(k + 1)^2 M_1^2} \tag{15.85}$$

15.7.2 Mach Number Relationships for Stagnation Conditions in Isentropic Nozzles

The gas tables of Ref. 15.4 contain relationships for the pressure, density, and temperature in terms of stagnation values for an *isentropic* flow of a perfect gas and for normal shock relations. We usually know upstream conditions, such as pressure p_1, temperature T_1, and velocity $\tilde{V}_1$. Parameters we often like to evaluate are the stagnation pressure, density, and temperature, as well as downstream values including the receiver pressure p_{02}.

We start the formulation with the (I.F.) energy Eq. (15.29) between the upstream station 1 and the stagnation station 01:

$$T_{01} = T_1 + \frac{k - 1}{2k} \tilde{V}_1^2$$

We transform the velocity to Mach number M_1 and divide through by the temperature T_1 to obtain

$$\frac{T_{01}}{T_1} = 1 + \frac{k - 1}{2} M_1^2 \qquad (15.86)$$

Similarly between the downstream station 2 and stagnation station 02, we obtain

$$\frac{T_{02}}{T_2} = 1 + \frac{k - 1}{2} M_2^2 \qquad (15.87)$$

Dividing Eq. (15.86) by Eq. (15.87), we obtain the ratio of the temperatures of the two stagnation temperatures:

$$\frac{T_{01}}{T_{02}} = \frac{T_1\left[1 + \dfrac{(k - 1)}{2} M_1^2\right]}{T_2\left[1 + \dfrac{(k - 1)}{2} M_2^2\right]} \qquad (15.88)$$

The stagnation pressure ratio across the normal shock is obtained by substituting the isentropic relationship of Eq. (15.20) into Eq. (15.86) such that

$$\frac{p_{01}}{p_1} = \left(1 + \frac{k - 1}{2} M_1^2\right)^{k/(k-1)} \qquad (15.89)$$

and then combining this equation with the static pressure ratio across the shock given by Eq. (15.81) with the result

$$\frac{p_{02}}{p_{01}} = \left[\frac{(k + 1)M_1^2}{2 + (k - 1)M_1^2}\right]^{k/(k-1)} \cdot \left[\frac{k + 1}{2kM_1^2 + 1 - k}\right]^{1/(k-1)} \qquad (15.90)$$

Notice that as $M_1 \rightarrow \infty$, the pressure ratio $p_{02}/p_{01} \rightarrow 0$ whereas $p_2/p_1 \rightarrow \infty$.

Values of T/T_0, p/p_o, and ρ/ρ_o are given in Tables 30 to 35 of the gas tables [15.4] for a particular value of k. Table C.1 of the Appendix has been generated from Table 30 of the gas tables for a perfect gas where $k = 1.4$.

To gain some insight into the energy transfer involved in going across the shock, we turn to a T-s diagram. The upstream station 1 has a given pressure p_1 and temperature T_1 such that the entropy s_1 can be determined by the gas tables [15.4]. The stagnation pressure p_{01} and temperature T_{01} have the same entropy s that exists at station 1 (since the process is isentropic) where the increase in temperature ΔT is obtained by the energy equation and is proportional to the kinetic energy:

$$T_{01} - T_1 = \left(\frac{k - 1}{k}\right) \tilde{V}_1^2 \qquad (15.91)$$

We note from Fig. 15.16 that across the shock the temperature and the pressure both rise and the kinetic energy decreases, which results in the flow decelerating.

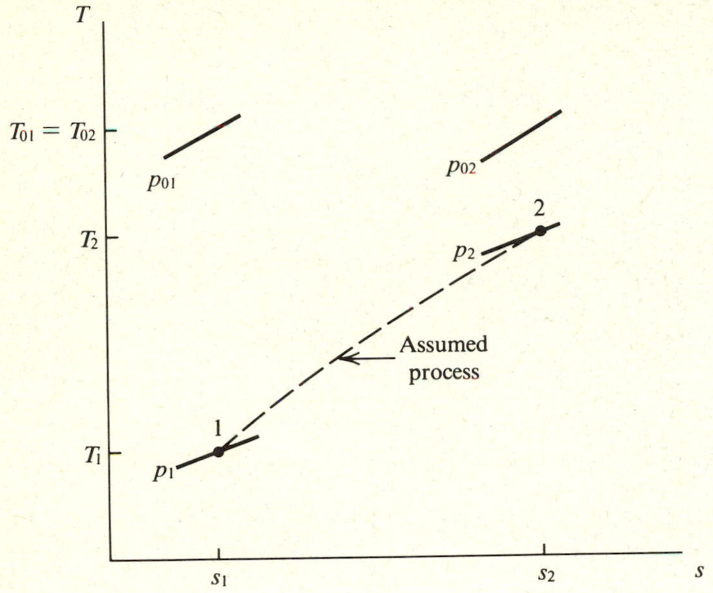

Figure 15.16 *T-s diagram for a normal shock.*

15.7.3 Mass Rate through an Isentropic Nozzle

The previous section presented a number of thermodynamic ratios based on stagnation properties. These relationships are used in isentropic compressible flows in ducts and pipes, as well as in nonisentropic flows, and where the local stagnation properties are evaluated using actual local values of the thermodynamic properties.

One of the more important relationships used in compressible flow analysis is the mass rate of fuel or exhaust gases through nozzles. In order to obtain such a relationship, we will need an expression for the area ratio A/A^*. We begin with the (I.F.) compressible continuity equation:

$$\frac{\dot{M}}{A} = \rho\tilde{V} \qquad (15.92)$$

$$= \frac{p\tilde{V}}{RT}$$

Using the fact that the speed of sound is $c = \sqrt{kg_cRT}$, we write Eq. (15.92) as

$$\frac{\dot{M}}{A} = pM\sqrt{\frac{kg_c}{RT}}$$
$$= pp_o\frac{M}{p_o}\sqrt{\frac{kg_c}{R}}\cdot\sqrt{\frac{T_o}{T}}\cdot\sqrt{\frac{1}{T_o}} \tag{15.93}$$

Substituting Eqs. (15.86), (15.87), and (15.89) into Eq. (15.93) gives

$$\frac{\dot{M}}{A} = \sqrt{\frac{kg_c}{RT_o}}\,p_oM\left(1 + \frac{k-1}{2}M^2\right)^{(k+1)/2(1-k)} \tag{15.94}$$

At the throat, $M = 1$ and $A = A^*$, so that

$$\frac{\dot{M}}{A^*} = \sqrt{\frac{kg_c}{RT_o}}\,p_o\left(\frac{k+1}{2}\right)^{(k+1)/2(1-k)} \tag{15.95}$$

The ratio A/A^* is thus found by combining Eqs. (15.94) and (15.95) with the result

$$\frac{A}{A^*} = \frac{1}{M}\left[\frac{2 + (k-1)M^2}{k+1}\right]^{(k+1)/2(k-1)} \tag{15.96}$$

We tabulate Eq. (15.96) in Table C.1 of the Appendix for a range of Mach numbers and for $k = 1.4$.

The above area ratio can now aid us in designing a nozzle for a missile. Suppose the design requires a missile to function in the atmosphere where the pressure is known. This is our exit pressure p_e. Suppose we know the missile's chamber pressure (p_o) and chamber temperature (T_o), values critical in achieving a certain thrust. We can now design the throat and exit areas, the latter from the (I.F.) linear momentum equation, and the former from Eq. (15.80) or Table C.1.

15.7.4 Location of a Normal Shock in a Nozzle

Consider a perfect gas flowing through a converging-diverging nozzle of Fig. 15.17. Flow conditions are such that a normal shock exists downstream of the throat yet upstream of the exit e of the nozzle. Given the exit and throat areas of the nozzle, A_e and A_T, respectively, and the upstream stagnation pressure p_{01} and exit pressure p_e, the problem is to predict the location of the normal shock in the tunnel. That is, we wish to find A_x at which a normal shock will occur for a particular value of pressure slightly downstream of the shock.

The solution of this problem depends on two important factors:

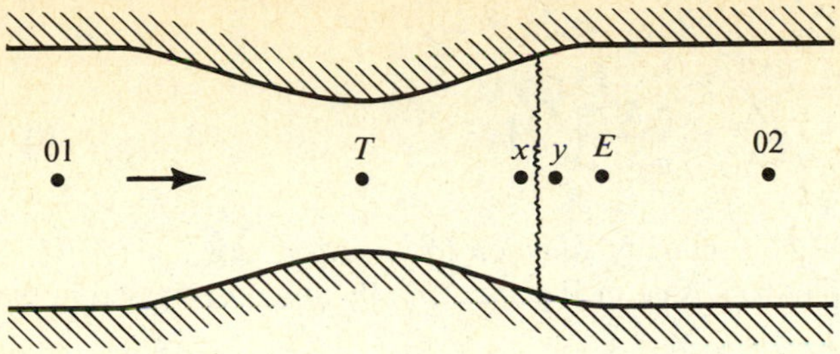

Figure 15.17 *Location of a normal shock in a nozzle.*

- The product

$$\left(\frac{A}{A^*}\right)\left(\frac{p}{p_o}\right) \tag{15.97}$$

values of which are obtained from the gas tables in Appendix C, and

- The area relationship for nonisentropic flow from the throat across the shock to the exit

$$\frac{A_e}{A_T} = \left(\frac{p_{01}}{p_{02}}\right)\left(\frac{A_e}{A_e^*}\right) \tag{15.98}$$

Before we continue with the procedure used in calculating A_x, it is worthwhile deriving Eq. (15.98). We start by setting $A_T = A_1$ and $A_e = A_2$, since it is convenient to use these stations for easy referral to the isentropic relationships among pressures and temperatures.

The (I.F.) continuity equation states that

$$\frac{A_2}{A_1} = \frac{\rho_1 \tilde{V}_1}{\rho_2 \tilde{V}_2} \tag{15.99}$$

Since

$$\frac{\rho_1}{\rho_2} = \frac{p_1 T_1}{p_2 T_2} \tag{15.100}$$

and

$$\frac{\tilde{V}_1}{\tilde{V}_2} = \frac{M_1 c_1}{M_2 c_2} \tag{15.101}$$

they can be combined with Eq. (15.99) to yield

$$\frac{A_2}{A_1} = \frac{p_1 M_1}{p_2 M_2}\left(\frac{T_2}{T_1}\right)^{1/2} \tag{15.102}$$

which is an alternate form of the continuity equation for a perfect gas.

We next multiply the right-hand side of Eq. (15.102) by unity, expressed in the form

$$\left(\frac{p_{01}}{p_{01}}\right)\left(\frac{p_{02}}{p_{02}}\right)\left[\left(\frac{T_{01}}{T_{01}}\right)\left(\frac{T_{02}}{T_{02}}\right)\right]^{1/2}$$

with the result

$$\frac{A_2}{A_1} = \left(\frac{p_1}{p_{01}}\right)\left(\frac{p_{02}}{p_2}\right)\left(\frac{p_{01}}{p_{02}}\right)\cdot\frac{M_1}{M_2}\left[\left(\frac{T_1}{T_{01}}\right)\left(\frac{T_{02}}{T_2}\right)\right]^{1/2} \tag{15.103}$$

since the flow is isoenergetic ($T_{01} = T_{02}$). While the flow is nonisentropic from 1 to 2, it *is* isentropic from 01 to 1, and from 2 to 02, though at different entropy levels. Therefore, substituting the isentropic pressure-temperature relationship of Eq. (15.20) into the continuity Eq. (15.103) gives

$$\frac{A_2}{A_1} = \frac{p_{01}}{p_{02}}\frac{M_1}{M_2}\left(\frac{T_1/T_{01}}{T_2/T_{02}}\right)^{(k+1)/2(k-1)} \tag{15.104}$$

The relationship between stagnation temperature T_o and a temperature T for an isentropic process is given by the (I.F.) energy Eqs. (15.86) and (15.87). Substituting the temperature ratio T/T_o into Eq. (15.104) yields

$$\frac{A_2}{A_1} = \frac{p_{01}}{p_{02}}\frac{M_1}{M_2}\left[\frac{2 + (k - 1)M_2^2}{2 + (k - 1)M_1^2}\right]^{(k+1)/2(k-1)} \tag{15.105}$$

The area ratio A/A^* is given by Eq. (15.96), so that by arrangement

$$\frac{A_2/A_2^*}{A_1/A_1^*} = \frac{M_1}{M_2}\left[\frac{2 + (k - 1)M_2^2}{2 + (k - 1)M_1^2}\right]^{(k+1)/2(k-1)} \tag{15.106}$$

Comparison of Eq. (15.106) with Eq. (15.105) shows

$$\frac{A_2}{A_1} = \frac{p_{01}}{p_{02}}\frac{A_2/A_2^*}{A_1/A_1^*} \tag{15.107}$$

Now we can go back to our former notation where station 1 is the throat T and station 2 is the exit e. Noting that $A_T/A_T^* = 1$, Eq. (15.107) gives us

$$\frac{A_e}{A_T} = \frac{p_{01}}{p_{02}}\left(\frac{A_e}{A_e^*}\right) \tag{15.98}$$

It is often more convenient to use the form

$$\left(\frac{A_e}{A_T}\right)\left(\frac{p_e}{p_{01}}\right) = \left(\frac{p_e}{p_{02}}\right)\left(\frac{A_e}{A_e^*}\right) \tag{15.108}$$

since the left-hand side is determined from given data, and the right-hand side is evaluated with the assistance of the tables of Appendix C. Note that if $p_{02} = p_{0y}$ and $p_{01} = p_{0T} = p_{0x}$ since the flow is isentropic, then

$$\frac{p_{02}}{p_{01}} = \frac{p_{0y}}{p_{0x}} \tag{15.109}$$

Example 15.8

Consider the flow of air in the nozzle shown in Fig. E15.8. If the air expands from 222 psia at $T = 500°F$ to a back pressure of 150 psia, find the Mach number where the normal shock exists and the corresponding area.

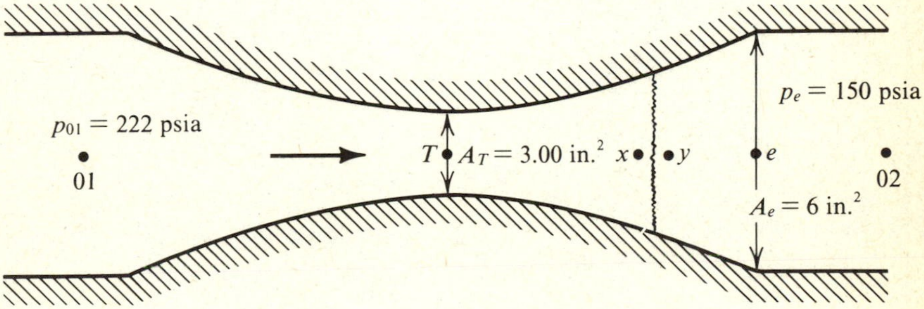

Figure E15.8

Solution:

To find the Mach number M_x, we need to find the pressure ratio p_{0y}/p_{0x}. From Eq. (15.109), we begin with

$$\frac{p_{0y}}{p_{0x}} = \frac{p_{02}}{p_{0T}} \tag{i}$$

where $p_{0T} = p_{01} = 222$ psia. Using Eq. (15.108), we find

$$\left(\frac{p_e}{p_{02}}\right)\left(\frac{A_e}{A_e^*}\right) = \left(\frac{A_e}{A_T}\right)\left(\frac{p_e}{p_{01}}\right) \tag{ii}$$

$$= \left(\frac{6}{3}\right)\left(\frac{150}{222}\right) = 1.35$$

Example 15.8 *(Con't.)*

Entering the last column of Table 30 of the gas tables [15.4] gives $p_e/p_{02} = 0.885$. Thus

$$p_{02} = \frac{p_e}{0.885} = \frac{150}{0.885} = 170.5 \text{ psia} \qquad \text{(iii)}$$

Substituting Eq. (iii) into Eq. (i) gives

$$\frac{p_{0y}}{p_{0x}} = \frac{170.5}{222} = 0.768 \qquad \text{(iv)}$$

Entering Table C.2 of Appendix C, with the above value we obtain

$$M_x = 1.90 \qquad \text{(v)}$$

as the Mach number slightly upstream of the normal shock.

To find the location of the normal shock, we enter Table C.1 of Appendix C with $M = 1.90$ to get

$$\frac{A}{A^*} = 1.5552 \qquad \text{(vi)}$$

Therefore

$$A_x = A_x^* (1.5552) = (3)(1.5552) = 4.67 \text{ in.}^2 \qquad \text{(vii)}$$

With this area, we can identify where that area exists in the nozzle, which is where the normal shock stands.

This completes the solution.

15.7.5 The Prandtl Relation

Another important condition across the normal shock is the *Prandtl* relation:

$$\bar{V}_1 \bar{V}_2 = \frac{\bar{V}_1^2}{M_1^2} \frac{[2 + (k - 1)M_1^2]}{k + 1} \qquad (15.110)$$

Using Eq. (15.86), we express the product of the flow velocities in terms of the stagnation temperature:

$$\tilde{V}_1\tilde{V}_2 = 2\,c_1^2\left(\frac{T_{01}/T_1}{k+1}\right) \tag{15.111}$$

$$= \frac{2c_0^2}{k+1}$$

$$= (c^*)^2 \tag{15.112}$$

where c^* is the critical speed of sound. Thus if the velocity of the gas in front of the normal shock $\tilde{V}_1$ is greater than the speed of sound c_1, then the velocity of the gas behind the shock is less than the corresponding speed of sound, which is exactly what we have already shown in the previous section.

15.7.6 Thickness of the Normal Shock

How important is the thickness of a shock wave? How can it be defined? Actually the two questions are the same. To answer these questions let us look into the physics of the shock wave. We know a gas is composed of molecules plus empty space that in a microscopic analysis is far from being homogeneous in composition. The gas molecules move rapidly about, colliding into other molecules and surfaces. When we speak of the density of the gas, we are talking about the average number of molecules of the gas per volume. This in effect is the same thing as describing the mean free path that any one molecule can travel before colliding with another. Temperature is the measure of the average energy of the molecules, and pressure is the force per unit area exerted by the molecule as a result of the density of the gas and the energy it contains. In order to create a shock, the gas must be compressed. This can only be accomplished if a motion has a speed so large that the gas molecules cannot move out of the way. Thus, the mean free path of the molecules is shortened. This is accomplished in layers, or waves. As one wave overtakes another, a pressure build-up or pressure pulse is created which grows larger and larger until it becomes a shock wave wherein the mean free path is now very small, resulting in a rise in temperature. The rise in temperature is a manifestation of the rise in the total energy of each molecule, principally the energy of translation. The molecular energy of rotation and vibration has also increased.

The shock wave has a thickness caused by the ''packing'' of the molecules. The ''packing'' or density tries to spread out evenly, its distribution governed by the viscosity of the gas. We shall assume that in the region of the shock the shear stress p_{yx} is of the same magnitude as the normal stress. Let t denote the thickness of a shock, and p the normal stress, as shown in Fig. 15.18. As a first estimate,

$$\Delta p \simeq p_{yx} \tag{15.113}$$

or

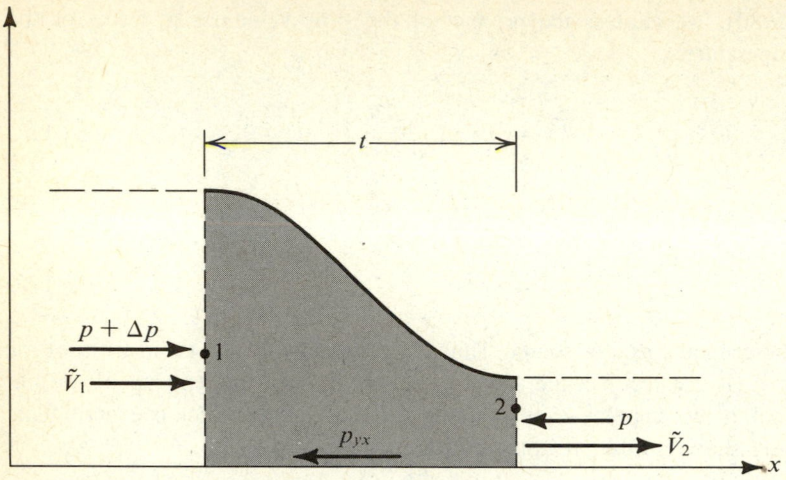

Figure 15.18 Thickness of a normal shock.

$$p_2 - p_1 = \mu \frac{(\tilde{V}_1 - \tilde{V}_2)}{t} \tag{15.114}$$

Substituting Eq. (15.74) into Eq. (15.114) results in

$$t = \frac{\nu}{\tilde{V}_1} \tag{15.115}$$

For example, the viscosity of air at 1 atm and 100°F is $\nu = 1.8 \times 10^{-4}$ ft^2/s. Assuming a gas flow of 2000 ft/s, the thickness of the shock is 9.0×10^{-8} ft, which is a very narrow region. Because it is so very thin, many consider it a surface of discontinuity.

15.8 Isothermal Gas Flow in a Pipe

In today's energy technology, vast quantities of gas are transported enormous distances through pipelines. One of the more important parameters used in designing pipelines is the weight discharge $\dot{W}$ of the gas. Figure 15.19 shows a long pipe of average pipe diameter D and length L containing a gas of specific weight γ.

Pipelines are designed to keep the flow velocity as low as possible to avoid pressure losses due to friction. Thus, though the gas is a compressible fluid, the flow is considered incompressible. We shall assume that the process is isothermal, and the pipe sufficiently long so that the only losses are major friction losses due to pressure drop.

We start the analysis by expressing the (I.F.) energy Eq. (6.11) in elemental form:

$$\frac{dp}{\gamma} + \frac{\tilde{V}\, d\tilde{V}}{g} = dh_f \tag{15.116}$$

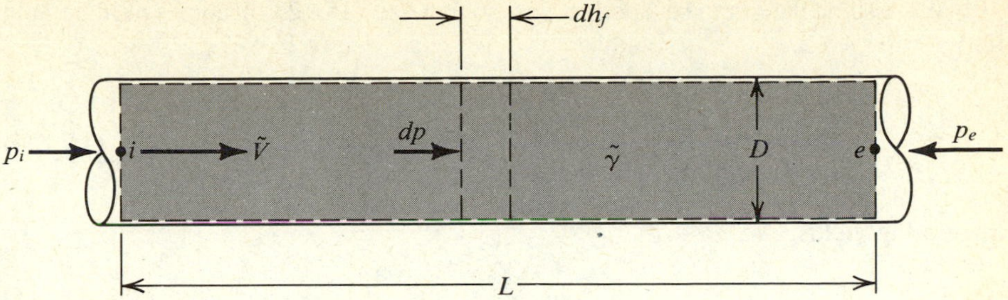

Figure 15.19 Long pipe carrying a gas.

where the elemental friction loss is obtained from Eq. (10.30) as

$$dh_f = f\frac{\tilde{V}^2}{2g}\frac{dL}{D} \tag{15.117}$$

Combining Eqs. (15.116) and (15.117) results in

$$\frac{dp}{\gamma} + \frac{\tilde{V}^2}{2g}\left(2\frac{d\tilde{V}}{\tilde{V}} - f\frac{dL}{D}\right) = 0 \tag{15.118}$$

The (D.F.) continuity Eq. (6.1) for steady flow states

$$d(\gamma\tilde{V}) = 0 \tag{15.119}$$

or

$$\frac{d\tilde{V}}{\tilde{V}} = -\frac{d\gamma}{\gamma} \tag{15.120}$$

Substituting Eq. (15.120) into Eq. (15.118) results in

$$\frac{dp}{\gamma} - \frac{\tilde{V}^2}{g}\left(\frac{d\gamma}{\gamma} + \frac{f}{2}\frac{dL}{D}\right) = 0 \tag{15.121}$$

We obtain the weight rate of flow $\dot{W}$ from Eq. (6.3) as

$$\dot{W} = \gamma\tilde{V}A \tag{15.122}$$

so that the energy Eq. (15.121) can be expressed as

$$\gamma dp - \frac{\dot{W}^2}{A^2 g}\left(\frac{d\gamma}{\gamma} + \frac{f}{2}\frac{dL}{D}\right) = 0 \tag{15.123}$$

We assume the gas to be a perfect gas, so that Eq. (15.123) becomes of integrable form:

$$\frac{gpdp}{RT} - \frac{\dot{W}^2}{A^2g}\left(\frac{d\gamma}{\gamma} + \frac{f}{2}\frac{dL}{D}\right) = 0 \tag{15.124}$$

Integrating yields

$$\frac{gp^2}{2RT} - \frac{\dot{W}^2}{A^2g}\ln\gamma - \frac{\dot{W}^2 fL}{2A^2gD} = c \tag{15.125}$$

The constant c is evaluated using the conditions that at the inlet, $L = 0$, $p = p_i$, and $\gamma = \gamma_i$, whereas at the exit $p = p_e$ and $\gamma = \gamma_e$. Thus Eq. (15.125) becomes

$$p_i^2 - p_e^2 = \frac{2\dot{W}^2RT}{g^2A^2}\left(\ln\frac{\gamma i}{\gamma_e} + \frac{f}{2}\frac{L}{D}\right) \tag{15.126}$$

Solving for the weight rate of flow $\dot{W}$ yields

$$\dot{W} = cD^2\left[\frac{p_i^2 - p_e^2}{RT\left(\ln\frac{\gamma i}{\gamma_e} + f\frac{L}{D}\right)}\right]^{1/2} \tag{15.127}$$

where $c = 17.885$ ft/s^2 for USCS units and $c = 5.449$ m/s^2 for SI units. If the pipe is fairly short, so that there is negligible change in the density, then Eq. (15.127) becomes

$$\dot{W} = cD^{5/2}\left(\frac{p_i^2 - p_e^2}{fRTL}\right)^{1/2} \tag{15.128}$$

Example 15.9
Gas flows in a pipeline that is fairly short; only 3 km long. The pressure at the inlet is 500 kPa and at the exit is 110 kPa. Determine the weight rate of flow of the gas given that the pipe diameter is 0.1 m in diameter and the gas temperature and gas constant are 300 K and 1000 m^2/s^2·K, respectively. Assume that the flow is at the Reynolds number $R_D = 2000$.

Solution:
The weight discharge rate is given by Eq. (15.128) for a short pipe:

$$\dot{W} = 5.449\ D^{5/2}\left(\frac{p_1^2 - p_e^2}{fRTL}\right)^{1/2} \tag{i}$$

Example 15.9 *(Con't.)*

The friction factor f can be evaluated knowing the flow is at the Reynolds number of 2000, so that

$$f = \frac{64}{R_D} = \frac{64}{2000} = 0.032 \tag{ii}$$

Substituting the given data and Eq. (ii) into Eq. (i) results in

$$\dot{W} = 5.449\,(0.1)^{5/2}\left[\frac{(500)^2 - (110)^2}{(0.032)(1000)(300)(3\times 10^3)}\right]^{1/2} = 0.05 \text{ N/s} \tag{iii}$$

This completes the solution.

15.9 Other Types of Shock Waves

When there is supersonic flow through a nozzle or channel, or when there is supersonic flow past bodies at rest, the thermodynamic properties of pressure, density, and temperature, as well as the velocity of the gas, can change abruptly, as we have noted in the previous sections. The abrupt change takes place in a common shock, sometimes termed a condensation discontinuity. In the former sections, we have presented normal shock waves in supersonic channel flows. Now we shall examine normal shock waves for supersonic external flows.

Let us suppose that a source of disturbances is introduced at an arbitrary point O in a supersonic external flow of gas as shown in Fig. 15.20. The disturbance sends out waves in all directions with the velocity of sound c.

For a gas at rest, the disturbances from point O form concentric circles. In a moving gas, the sound wave from point O propagates through the gas with the velocity of sound c, while the wave center is carried by the flow at a velocity $u > c$, where u is supersonic and is the velocity of the flow. At time t_1, the sound wave has traveled a distance ct_1, while its center has traveled a greater distance ut_1. At time t_2, the wave has traveled a distance ct_2, while its center is now at a distance ut_2, creating an envelope of sound waves popularly termed a *Mach cone*. The angle α (see Fig. 15.20) of the Mach cone satisfies the relationship

$$\sin\alpha = \pm\frac{ct}{ut} = \pm\frac{1}{M} \tag{15.129}$$

Since the disturbances will be confined within the conical region, we would suspect that there are quite different behaviors of the gas inside and outside the Mach cone.

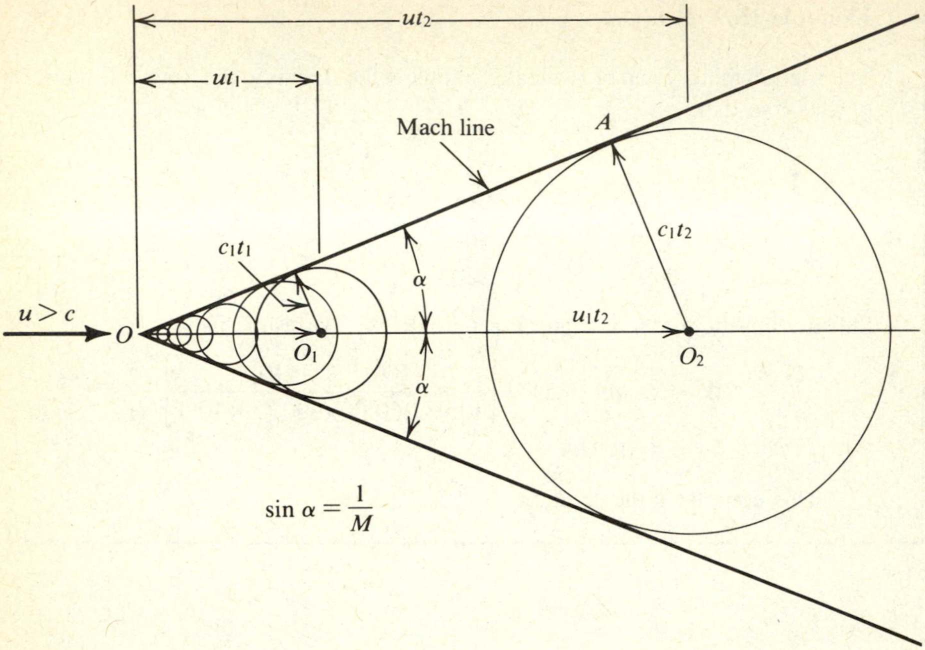

Figure 15.20 *Propagation of a disturbance in supersonic flow.*

Figure 15.21 is a photograph of a detached normal shock as a result of supersonic flow past a blunt object. Notice that the shock wave is bowed, with the leading section orthogonal to the flow. Within the shock, the flow decelerates from supersonic to subsonic flow with a significant rise in temperature, hence a rise in entropy resulting in a large thermal energy that is convected through the gas. (The detached shock is a complicated flow and is discussed by Shapiro [15.2].)

A simpler compression shock is shown in Fig. 15.22. Supersonic flow moves past a sharp object of vertex angle 2θ. When the flow reaches the object, it will deflect through an angle θ. There are two prominent observations we can note about Fig. 15.22. The first is the existence of a shock wave. The second is that the shock wave is attached to the body. There is a minimum value of supersonic flow where the attached shock becomes detached from the sharp object. This is not attributed to the normal component of the Mach number dropping below its sonic value, as might be suspected. Actually, there is no simple physical explanation why this occurs. Obviously some fundamental law of nature is being violated.

With the above as background, let us return to our one-dimensional analysis for channel flow and set up the physical situation leading to oblique shock waves. Figure 15.9 illustrates how the shock wave can be moved in the nozzle by adjusting the back pressure. Certainly there exists a value of back pressure that permits the normal shock to stand at the exit *e* of the nozzle. By further decreasing the pressure, an interesting development occurs. If the back pressure is decreased slightly below that pressure where the normal shock stands at the exit, the shock has to move out of the exit, yet be attached at the exit walls, creating what is termed an oblique shock (see Fig. 15.23).

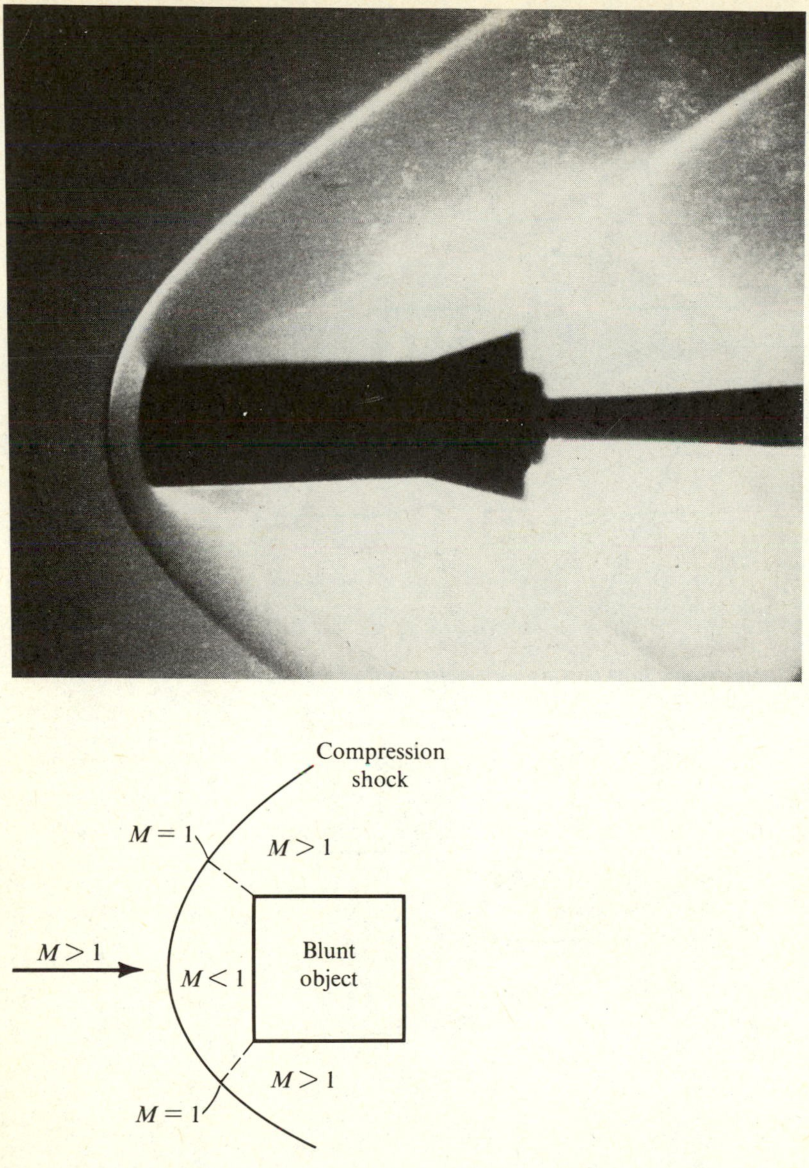

Figure 15.21 *Supersonic flow past a blunt object obtained by Schlieren photography.*

Consider the velocity components of Fig. 15.24 for a two-dimensional oblique shock inclined at an angle α with respect to the horizontal. The velocity upstream of the shock is denoted by $\tilde{V}_1$ and has components $\tilde{V}_{1_n}$ and $\tilde{V}_{1_t}$, denoting normal and tangential velocity, respectively. When the gas passes through the shock, the velocity $\tilde{V}_2$ changes in a unique manner: the tangential component $\tilde{V}_{2_t} = \tilde{V}_{1_t}$, resulting in the normal component $\tilde{V}_{2_n}$ being subsonic and less than $\tilde{V}_{1_n}$. Thus, only the normal com-

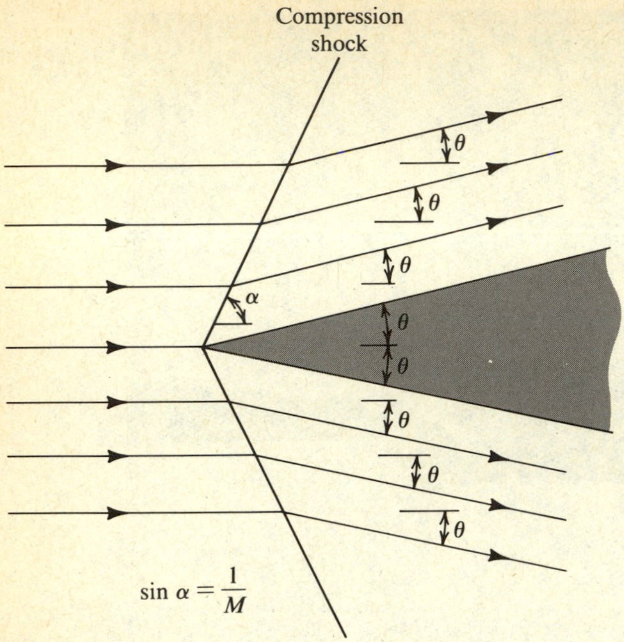

Figure 15.22 Supersonic flow past a sharp object, $\sin \alpha = 1/M$.

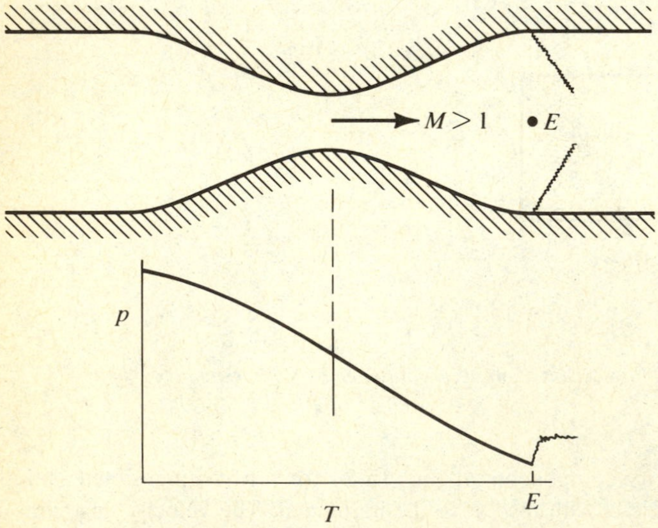

Figure 15.23 Oblique shock at nozzle exit and pressure variation.

ponent of the velocity is affected by the shock. From the geometry of Fig. 15.24, we note that

$$\tilde{V}_{1_n} = \tilde{V}_1 \sin \alpha \tag{15.130}$$

$$\tilde{V}_{2_n} = \tilde{V}_2 \sin \beta \tag{15.131}$$

$$\tilde{V}_{1_t} = \tilde{V}_{2_t} = \tilde{V}_1 \cos \alpha = \tilde{V}_2 \cos \beta \tag{15.132}$$

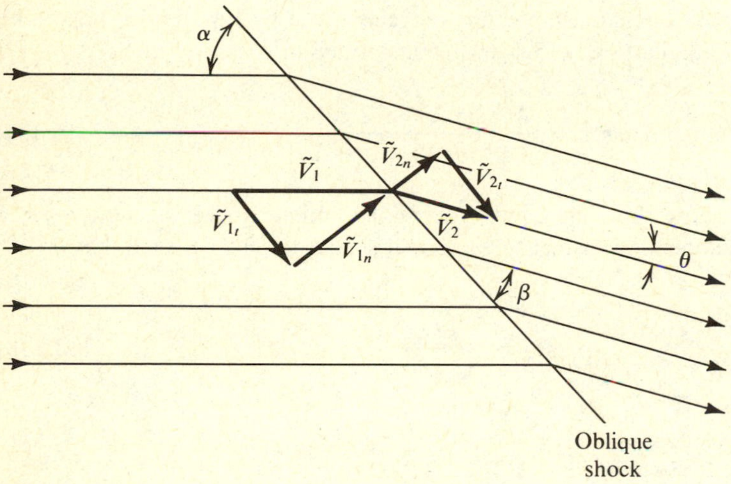

Figure 15.24 *Velocity polygons of oblique shock. α = angle of oblique shock with respect to $\tilde{V}_1$. θ = turning angle of $\tilde{V}_2$. β = angle of oblique shock with respect to $\tilde{V}_2$.*

Since $M_1 = \tilde{V}_1/c_1$ and $M_2 = \tilde{V}_2/c_2$, we find using Eq. (15.76) that

$$\frac{p_2}{p_1} = \frac{1}{k+1}\left[1 - k + 2k(M_1 \sin \alpha)^2\right] \tag{15.133}$$

and from Eq. (15.83) that

$$\frac{\rho_2}{\rho_1} = \frac{\tilde{V}_{2_n}}{\tilde{V}_{1_n}} = \frac{1}{k+1}\left(k - 1 + \frac{2}{M_1^2 \sin^2 \alpha}\right) = \frac{\tan \beta}{\tan \alpha} \tag{15.134}$$

It is quite simple to obtain

$$M_2 \sin \beta = \left[\frac{2 + (k-1)\, M_1^2 \sin^2 \alpha}{1 - k + 2k\, M_1^2 \sin^2 \alpha}\right]^{1/2} \tag{15.135}$$

from Eq. (15.82), which is equivalent to the velocity orthogonal to the shock in the region behind it.

Now we are in a position to determine the turning angle θ of the flow through the oblique shock. The turning angle θ is obtained from the geometry of Fig. 15.24:

$$\tan\theta = \tan(\alpha - \beta) = \frac{\tan\alpha - \tan\beta}{1 + \tan\alpha\tan\beta} \tag{15.136}$$

Using the trigonometric relationship of the tangents given by Eq. (15.134) and a bit of arithmetic manipulation, Eq. (15.136) is transformed to

$$\tan\theta = \frac{M_1^2 \sin 2\alpha - 2\cot\alpha}{2 + M_1^2(k + \cos 2\alpha)} \tag{15.137}$$

Figure 15.25 shows how air deflects across an oblique shock of angle α for a variety of Mach numbers. Note that for a normal shock, $\tan\theta = 0$ such that

$$M_1 = \sqrt{\frac{2\cot\alpha}{\sin 2\alpha}} \tag{15.138}$$

or

$$\alpha = \sin^{-1}\left(\frac{1}{M_1}\right) \tag{15.139}$$

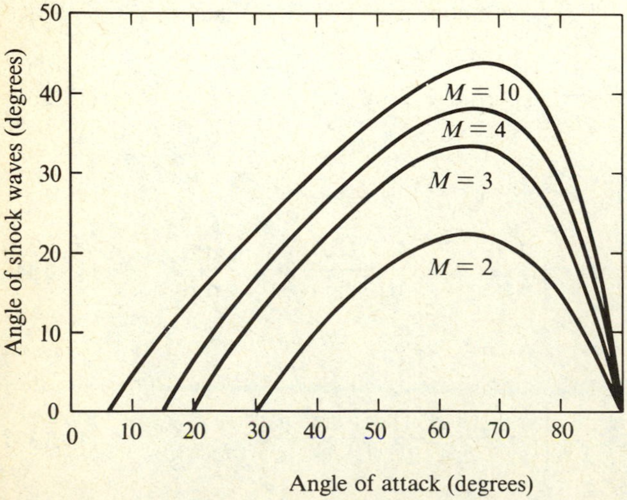

Figure 15.25 *Flow deflection θ versus oblique shock angle α for various Mach numbers.*

which was obtained in an earlier section. The value of α where there is no flow deflection is shown in Fig. 15.25 as the intersection of the Mach number curve with the abscissa. The case of very high Mach numbers reveals that

$$\tan\theta \simeq \frac{\sin 2\alpha}{k + \cos 2\alpha} \tag{15.140}$$

or that θ is nearly linear with 2α.

For the case of three-dimensional supersonic flows, the shock surface becomes a cone rather than the plane shock wave surface illustrated in Fig. 15.24. Like the oblique two-dimensional shock, the flow may be supersonic or subsonic behind the shock depending on the angle α.

Example 15.10

Consider an aircraft traveling through air at $M_1 = 3.0$. If the pressure of the air is 11.5 psia and the oblique wave angle is 40°, calculate (a) the downstream pressure p_2, (b) the downstream Mach number M_2, and (c) the turning angle θ of the flow.

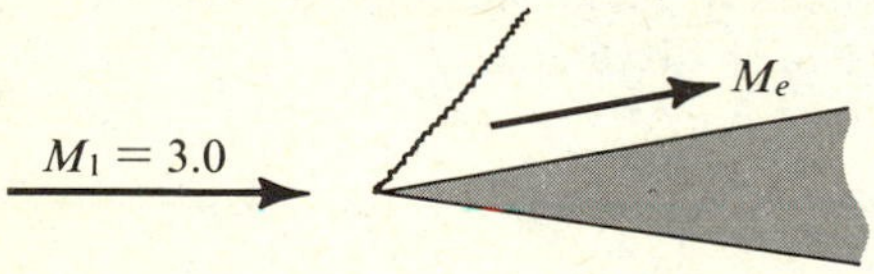

Figure E15.10

Solution:

(a) The downstream pressure p_2 is obtained from Eq. (15.133) as

$$p_2 = \frac{p_1}{k+1}[1 - k + 2k(M_1 \sin\alpha)^2]$$

$$= \frac{11.5}{2.4}[1 - 1.4 + 2.8(3 \times \sin 40°)^2] \tag{i}$$

$$= 47.97 \text{ psia}$$

(b) The downstream Mach number M_2 can be obtained from Eq. (15.135) provided that we know β. The angle β is found using Eq. (15.134):

$$\tan\beta = \frac{\tan\alpha}{k+1}\left(k - 1 + \frac{2}{M_1^2 \sin^2\alpha}\right)$$

$$= \frac{\tan 40°}{2.4}\left[0.4 + \frac{2}{(3 \sin 40°)^2}\right] \tag{ii}$$

$$= 0.5024$$

Example 15.10 *(Con't.)*

Thus,

$$\beta \approx 27° \tag{iii}$$

Substituting Eq. (iii) into Eq. (15.135) yields

$$M_2 \sin 27° = \left[\frac{2 + 0.4\,(3 \sin 40°)^2}{-\,0.4 + 2.8\,(3 \sin 40°)^2}\right]^{1/2}$$
$$= 0.59$$

or

$$M_2 = 1.3 \tag{iv}$$

(c) The turning angle of the flow θ is obtained through the application of Eq. (15.137):

$$\tan\theta = \frac{9 \sin 80° - 2 \cot 40°}{2 + 9(1.4 + \cos 80°)} = 0.4 \tag{v}$$

Thus the turning angle θ is approximately 22°, which agrees with the value obtained directly from Fig. 15.25 for $\alpha = 40°$ and $M = 3.0$.

This completes the solution.

15.10 *Drag Coefficient C_D* for Compressible Flow

One of the finest compilations and evaluations on fluid dynamic drag on bodies in compressible flows is S. Hoerner's opus [15.5]. To try to equal or improve upon it would be improper. It is also beyond the scope of this text to delve into the theory. Thus this section comprises merely a brief summary.

The theory of compressible flow drag is usually partitioned into four regimes: subsonic compressible, transonic (including sonic), supersonic, and hypersonic. The results we shall present for each regime will be based on experimental measurements, rather than theory. Expressions for the drag coefficient C_D will be based on a curve fit of the experimental data.

15.10.1 Subsonic Compressible Drag Coefficients

The drag coefficient on objects placed in subsonic compressible flow is largely caused by the stagnation pressure on the object's forward face. The base pressure and surface friction drag are rather small and are not affected by compressibility to any significant degree. Figure 15.26 presents the drag coefficient for a few familiar objects in subsonic flow. The flat plate results of curve (a) have a drag coefficient composed of 0.85 due to stagnation pressure and 1.13 due to base pressure. When the Mach number approaches 0.8 the drag is greatly influenced by compressibility. The blunt body of revolution has some stagnation forward drag like the flat plate but a greatly reduced base pressure. The drag coefficient for the blunt cylinder follows rather closely the drag coefficient for the 12% thick wedge and blunt body of revolution. Only by streamlining will the drag coefficient drop to very low values. In all cases, the drag coefficient rises dramatically as the transonic regime is approached.

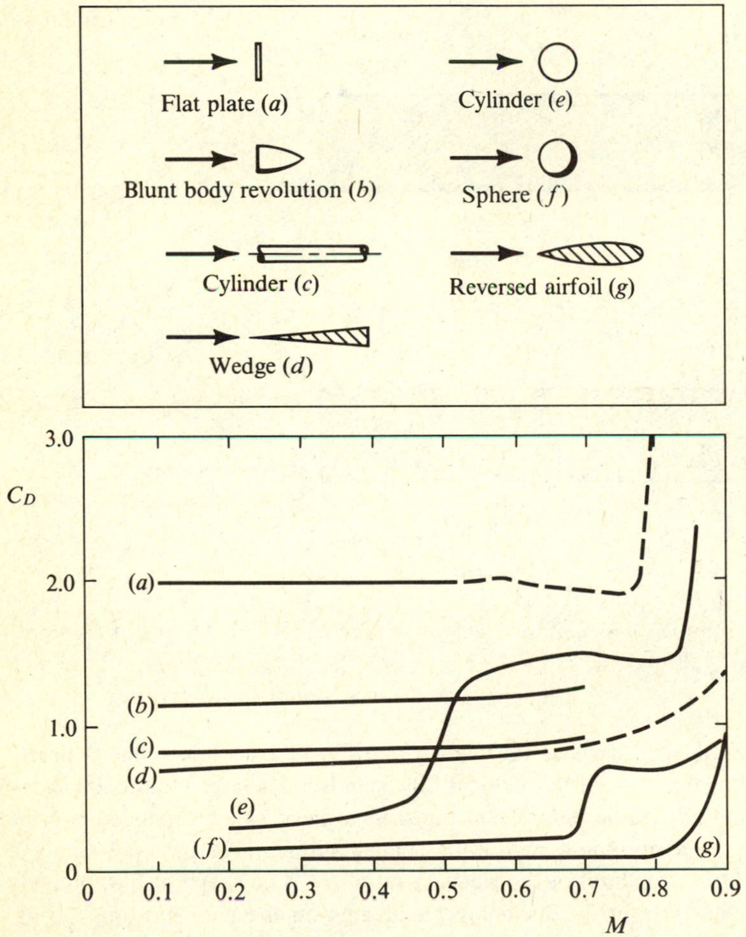

Figure 15.26 *Drag coefficients of various bodies in subsonic compressible flow.*

15.10.2 Transonic and Supersonic Drag Coefficients

Transonic flows cover the range $0.9 \leq M \leq 1.5$, and are quite different from the subsonic or supersonic flows. The equations of fluid motion are vastly dissimilar among the three flow regimes, resulting in entirely different formulations and different methods of solution. Figure 15.27 illustrates the flow patterns of transonic and supersonic flow past a cylindrical body in the region of the base. We note the existence of expansion waves and shocks, turbulence, and a very complicated viscous wake.

A dead region aft of the body base will cause a backflow from the conelike compression shock that is subsonic mixing, highly turbulent and diffusive from the minimum area to the shear layer. This viscous mixing and suction contribute to the base drag.

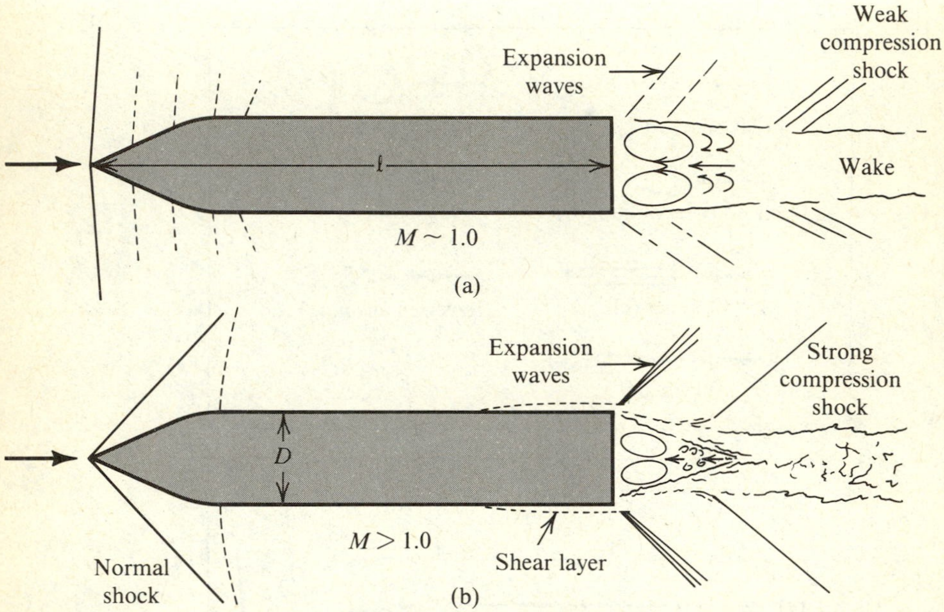

Figure 15.27 *The transonic and supersonic flow patterns on a cylindrical body. (a) Transonic. (b) Supersonic.*

The existence of a dead space creates a flow that is a backflow into a nearly conical shape. At the terminus of this conical flow stands a conical compression shock wave. The base and expansion pressure of the conical dead space depends upon the flow pattern through the minimum area of the conical surface. The base pressure and thus base drag is also a function of the fineness ratio l/D. The shape of the forebody has a bearing on the flow in the shear layer and thus on the parasitic drag. Thus, because of a less viscous forebody drag, a slender ogive-shaped body has a greater base pressure drag coefficient than the cone-shaped forebody.

Figure 15.28 presents the drag coefficients for a cylinder, sphere, and two streamlined cylinders with blunt-ended afterbodies in transonic and supersonic flow. We see that a maximum drag on the more streamlined cylinder occurs at $M = 1.0$, whereas for the other three bodies, the maximum drag coefficient occurs in the supersonic regime. Once again we see that streamlining reduces drag significantly.

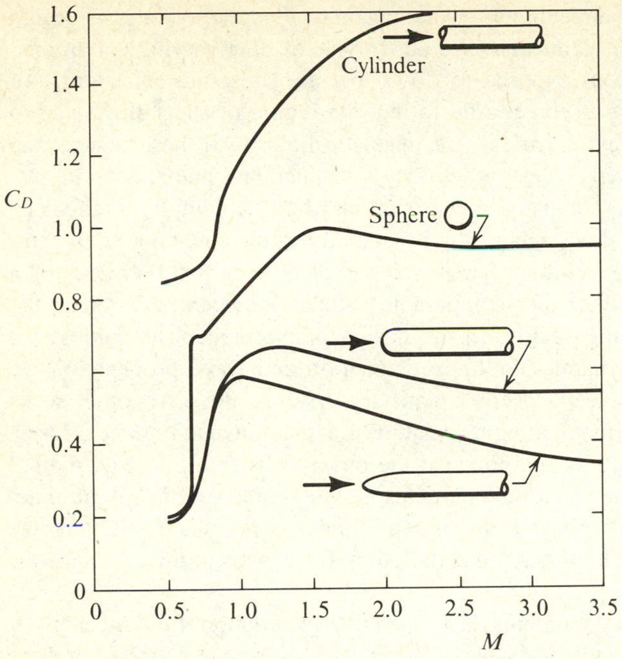

Figure 15.28 C_D *versus M of various bodies in transonic and supersonic flows.*

15.11 Closure

We have presented a few topics in compressible one-dimensional steady flow to whet the mental palate of the reader. Because of the enormity of compressible flow topics, only an introduction can suffice in this text. References 15.6 to 15.12 are provided at the end of this chapter, with a comment on why they were selected, but such references serve only to invite the student to examine the books for an overview of what lies ahead.

The study of compressible flow perhaps illustrates better than any other topic how incredibly complex and far reaching fluid mechanics can become. In future vehicles, speeds shall be attained that boggle the imagination. Special designs will be required for unique atmospheres, unusual missions, and other assorted criteria that govern aircraft shape, flexibility, and stability. The interplay among the mass distribution, elastic properties of the aircraft, and the aerodynamic forces for complex structures using exotic materials demand in complicated high-speed computing routines to analyze the aircraft's deflection, mode shapes, flutter speed, and flutter frequencies—all requisite for stable performance. What does all that mean? Wings, tails, and fuselage are all elastic, and when subjected to time-varying pressure distributions, they deflect, and thus vibrate with various amplitudes and frequencies. As the structure deforms, the shape changes, and the pressure and shear stresses vary, which in turn results in new loads and new deflections—which again change the loading. A design engineer must try to control a vicious never-ending cycle of time-varying conditions. Add to these conditions regions exposed to flows that are subsonic, supersonic, or transonic, or combinations of all three. Add to that the region of shear flow, and to that add the effects of heat and electrical and magnetic fields. All these factors and many more must be considered in designing today's manned and unmanned aircraft. No one person, not even one company, can design and build a complex high-speed aircraft, involving as it does thousands of highly skilled technicians, enormous computers, and elaborate testing facilities. No wonder it costs over $10,000,000 for a single sophisticated jet aircraft. Other comparable complexities can be listed in the design of submarines or racing yachts. In the design of thermonuclear bombs, we require sophisticated fluid dynamics to study detonation and wave propagation. In studying the behavior of galaxies and other planetary systems, the astronomer seeks mathematical models to describe dilating thermal masses that move in complex vortex motions. To unlock the secrets of the universe the physicist is trying to find mathematical models for the behavior of atomic particles moving at the speed of light under magnetic and electrical fields. Physicians and biofluid engineers are studying the behavior of cells in the blood, to help find the cures for cancer, arthritis, diabetes, and other afflictions.

These varied and complex problems have one thing in common: fluid mechanics. Whether a fluid be incompressible, or steady, or inviscid, or even newtonian, or none of these, it is still a fluid. It is still 96% of what we are, where we came from, of what we breathe, and what we eat. Inescapably, fluid mechanics is one of the single most important subjects in our study of the physical universe.

References

15.1 Anderson, John D., *Modern Compressible Flow; with Historical Perspective*, McGraw-Hill, New York, 1982.

15.2 Shapiro, A., *The Dynamics and Thermodynamics of Compressible Fluid Flow*, 2 vols., Ronald, New York, 1953.

15.3 Bisplinghoff, Ashley, and Halfman, *Aeroelasticity*, Addison-Wesley, Reading, Mass., 1955.

15.4 Keenan, J. H., and Kaye, J., *Gas Tables*, Wiley, New York, 1948.

15.5 Hoerner, S. F., *Fluid-Dynamic Drag*, published by the author, 1965.

Additional References

15.6 Ackeret, J., "Air Forces on Airfoils Moving Faster than Sound Velocity," *NACA TM 317*, 1925.
One of the earliest definitive papers on steady aerodynamic forces on thin airfoils at supersonic speeds.

15.7 Ames Research Staff, "Equations, Tables, and Charts for Compressible Flow," *NACA TR 1135*, 1953.
A must for students of compressible flow. It contains a compilation of all requisite data and relationships for compressible flow analyses.

15.8 Courant, R., and Friedrichs, K. O., *Supersonic Flow and Shock Waves*, Interscience, New York, 1948.
One of the great mathematical treatments of steady compressible flow that is a required reference for any theoretical investigation into shock waves.

15.9 Hayes, W. D., and Probstein, R. F., *Hypersonic Flow Theory*, vol. 1, Academic, New York, 1966.
The most influential text on hypersonic flow. Its delightful style and elegant format make it the best reference on the subject.

15.10 Liepmann, H. W., and Roshko, A., *Elements of Gasdynamics*, Wiley, New York, 1957.
Probably the most widely accepted undergraduate text in compressible flow written by two of the most respected professors in their field. It is a book hard to beat for clarity and completeness.

15.11 Reynolds, W. C., and Perkins, H. C., *Engineering Thermodynamics*, McGraw-Hill, New York, 1977.
There are many fine thermodynamic texts on the market, but this still has to stand out as one of the best. What Reynolds did for thermodynamics is what Goldstein did for mechanics.

15.12 von Mises, R., *Mathematical Theory of Compressible Fluid Flow*, Academic, New York, 1958.
The classic mathematical presentation of compressible flow. Rather demanding to read unless one is well grounded in mathematical analysis, but required reading to gain insight into a great mind.

Study Questions

15.1 Define one-dimensional steady compressible flow. What are the dependent and independent variables? What are the (I.F.) and (D.F.) of the governing equations?

15.2 What is a sound wave? How is it formed? How does the wave propagate? Why is it called a sound wave? What is the differential equation describing its motion?

15.3 What is a shock wave? How does it differ from a sound wave? What is the velocity of a shock wave? How is a shock wave formed?

15.4 What is a rarefaction wave? How does it differ from a compression wave? What is the velocity of a rarefaction wave?

15.5 If simple compressible work is defined by Eq. (15.1) for a simple compressible substance, what is

$$\int_{p1}^{p2} \forall \, dp \; ?$$

15.6 Discuss the application of the equation of state of a perfect gas for gases at temperatures of 1000°F. Can Eq. (2.3) be applied to steam? Explain.

15.7 Give the units of the gas constant R for SI and USCS systems. How is R related to the universal gas constant and the molecular weight?

15.8 How does entropy change for a perfect gas? Is it solely a function of temperature? Can the entropy of a gas change if there is no heat transfer? Explain.

15.9 What is the effect of adding heat to compressible flow? How does it affect the pressure, temperature, density, and velocity of a subsonic flow?

15.10 What is the best geometric shape for minimum drag in (a) incompressible flow, (b) subsonic compressible flow, (c) supersonic flow, and (d) hypersonic flow?

Problems

15.1 Nitrogen is maintained at a temperature of 47°C and pressure 2.2 bar. Calculate the density if the gas is a perfect gas.

15.2 What is the mass of air in the cabin of an aircraft 15 ft × 70 ft × 9 ft if the cabin pressure is 14.8 psia and the temperature is 75°F? Assume air is a perfect gas.

15.3 One kg of air is heated in a reservoir from 300 K to 500 K. Assuming air is a perfect gas, calculate the change of enthalpy (a) assuming $C_p = 1.002$ and 1.031 kJ/kg·K, respectively, at the two temperatures, and (b) that $C_p = 27.43 + 0.0062T - 0.9 \times 10^{-6} T^2$. Assume the pressure is small.

15.4 Low pressure air at 1 bar and 300 K is compressed to 5 bar and 510 K. The enthalpies at these two temperatures are 300.19 and 513.32 kJ/kg, respectively. If the power input to the air is 10 kW, and there is a loss of heat from the compressed air of 10 kJ/kg during the compression, calculate the mass flow rate. Assume inviscid flow, and negligible changes in kinetic and potential energies.

15.5 Hydrogen's specific heat at constant pressure is sometimes given as $C_p = 5.76 + 5.78 \times 10^{-4} T + 20 T^{-1/2}$ (Btu/lb·mol·°R) where T is in degrees Rankine. Calculate the change in entropy per lbm of hydrogen when it is heated from a temperature 500°R to 3000°R resulting in a pressure drop from 25 psia to 20 psia. The molecular weight of hydrogen is 2.016.

15.6 Calculate the change in entropy of 10 slugs of steam given that $p_1 = 20$ psia, $T_1 = 800$°R, $p_2 = 50$ psia, and $T_2 = 550$°R.

15.7 Calculate the change of enthalpy of 10 kg of helium if it is compressed from $p_1 = 110$ kPa and $T_1 = 300$ K to a final state of $p_2 = 700$ kPa and $T_2 = 400$ K.

15.8 Oxygen flows through a piping system in a hospital where $p_1 = 200$ psia and $\rho_1 = 1.23$ lbm/ft^3 to a reservoir where $p_2 = 25$ psia and $T_2 = 70$°F. Calculate (a) the temperature T_1, (b) the density ρ_2, (c) the change in enthalpy Δh, and (d) the change in entropy.

15.9 The specific heats for helium are found from Table 15.2 to be $C_{p_o} = 1.25$ Btu/lbm·°R and $C_{v_o} = 0.753$ Btu/lbm·°R. Find the value of the gas constant R and specific heat ratio k, and compare it with values given in Tables 15.1 and 15.2, respectively.

15.10 Derive the isentropic equation of state p/ρ^k = const. from the (D.F.) energy Eq. (4.152) for steady isentropic flow.

15.11 Steam enters a turbine at 700°R and expands through a pressure ratio of 1:9 (Fig. P15.11). If the expansion is internally reversible and the turbine is insulated so that there is negligible transfer of thermal energy, calculate the change in enthalpy.

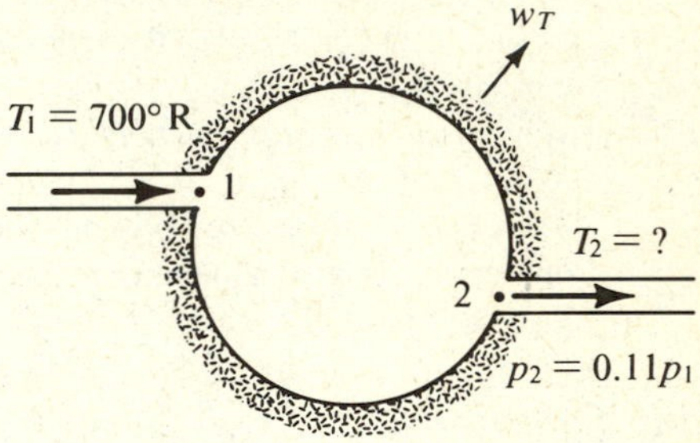

Figure P15.11

15.12 Show that the change of entropy for an incompressible fluid is $\Delta s = c \ln (T_2/T_1)$ if the specific heat of the incompressible substance $c = C_{v_o} = C_{p_o}$.

15.13 Show that the change of enthalpy for an isentropic process is

$$h_2 - h_1 = C_{p_o} T_1 \left[\left(\frac{\rho_2}{\rho_1} \right)^{k-1} - 1 \right]$$

15.14 Consider 0.2 slugs of air moving through a mechanical device that increases the pressure from 14 psia to 30 psia. Let the temperature $T_1 = 100$°F. (a) Calculate the change in density $\Delta\rho$. (b) Determine the process equation. (c) Determine the work done by or on the air. (d) What is the change of entropy? (e) Determine if heat transfer occurred in the compression process and the amount transferred.

15.15 Calculate the speed of sound of a perfect gas for an isothermal process if the temperature of the gas is 17°C.

15.16 Calculate the speed of sound in hydrogen at a temperature of 0°F, and compare it with the speed of sound in air at the same temperature.

15.17 Calculate the speed of sound in water at 1034 bar and at a temperature of 93.33°C, and compare it with the speed of sound in water at 1 bar and at the same temperature of 93.33°C, where the density is 2 slug/ft^3.

15.18 A 747 commercial airliner is cruising at 50,000 ft altitude at $M = 0.75$. Calculate the ground speed of the airplane.

15.19 A sailor drops a wrench on the deck of a submerged submarine. A second submarine picks up the disturbance 0.7 seconds later. How far apart are the two subs? (Assume $\rho = 2.00$ slug/ft^3, $p = 1$ bar, and $T = 20$°C.)

15.20 Consider a wave propagating through air at 40°F forming a Mach cone of angle $\alpha = 60°$. Calculate (a) the Mach number of the disturbance, (b) the speed of sound, and (c) the velocity of the disturbance.

15.21 Air is flowing out of a reservoir where the pressure and temperature are 150 psia and 120°F, respectively. The air moves through a nozzle and into a receiver reservoir. (a) Calculate the temperature T_1 of the stagnation reservoir if the receiver pressure is 100 psia. (b) Calculate the velocity $\tilde{V}_1$, density ρ_1, and flow rate $\dot{M}$ out of a 0.01 ft^2 exit area. (Refer to Fig. P15.21.)

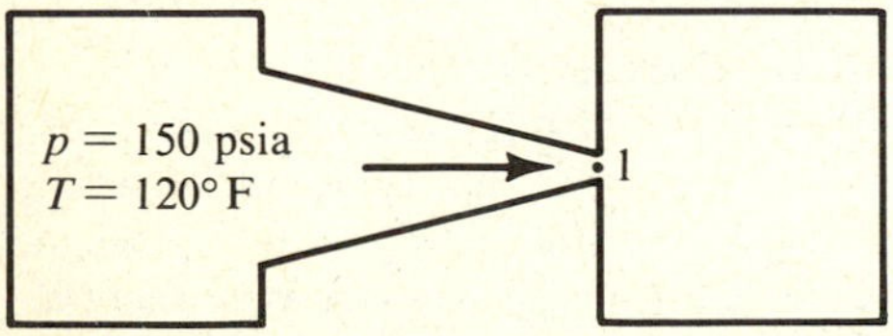

Figure P15.21

15.22 Repeat Prob. 15.20 where the receiver is open to the atmosphere such that the receiver pressure is 14.7 psia. Calculate (a) the exit temperature, (b) the velocity $\tilde{V}_1$, (c) density ρ_1, and (d) the flow rate $\dot{M}$ of the 0.01 ft^2 nozzle.

15.23 Consider a tunnel that draws air in from the atmosphere at $T = 70$°F by a vacuum tank at the exit of the tunnel. At station 1 the pressure is 5 psia, and the cross-sectional area is 2 ft^2. Calculate the Mach number and mass flow rate at this station assuming isentropic flow (Fig. P15.23).

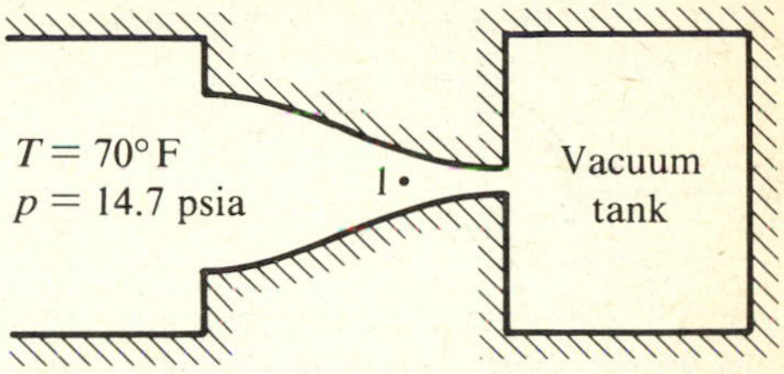

Figure P15.23

15.24 Air at an absolute pressure of 1000 kPa and temperature 20°C moves through a converging nozzle that has an exit area of 1×10^{-3} m^2. (a) Calculate the exit pressure and exit temperature if the receiver pressure is 500 kPa. (b) Calculate the velocity at the receiver station. (c) Calculate the density of air at the receiver station.

15.25 Suppose a diverging section is attached to the nozzle in the above Prob. 15.23. Let the throat have an area of 0.01 m^2, and the exit have an area 0.04 m^2. Calculate the exit pressure, the exit temperature, the exit Mach number, and the mass rate $\dot{M}$ for a total expansion in the nozzle (Fig. P15.25).

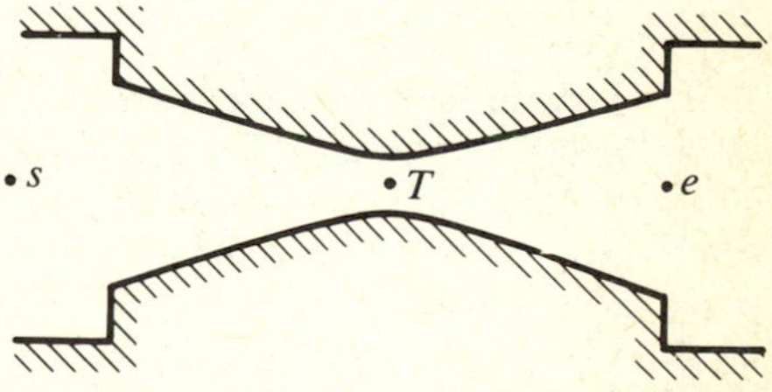

Figure P15.25

15.26 Air flows through a converging-diverging nozzle where there is no transfer of energy with the surroundings. At a station 1 in the nozzle, the pressure and temperature are found to be 30 psia and 600°R, respectively, and $M_1 = 0.7$. Refer to Fig. P15.26. (a) Calculate the stagnation thermodynamic properties of pressure, temperature, and density. (b) Calculate the velocity of the gas $\tilde{V}_1$.(c) Calculate the maximum velocity $\tilde{V}_{max}$. (d) Calculate the sonic velocity $\tilde{V}^*$.

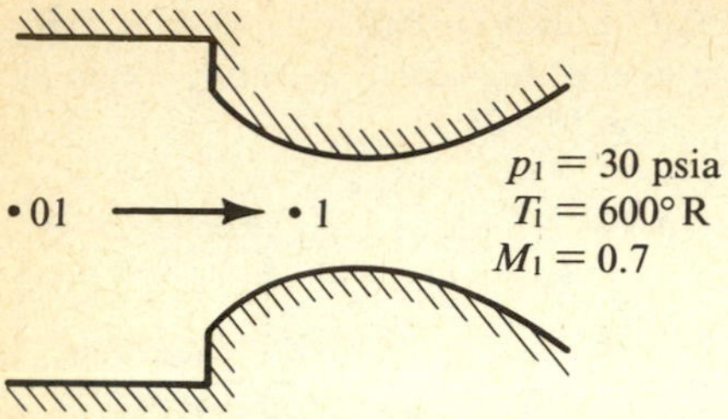

Figure P15.26

15.27 Show that the dimensionless term $-\nabla^* p^*$ of the dimensionless Navier-Stokes equation (7.26) can be expressed as

$$-\frac{1}{kg}\nabla^*\left(\frac{1}{M^2}\right)$$

15.28 As illustrated in Fig. P15.28, nitrogen at 65°F flows through a normal shock at an upstream velocity $\bar{V}_1 = 1300$ ft/s where $p_1 = 12$ psia. Calculate the pressure p_2 and density ρ_2 downstream of the shock and the Mach number M_1.

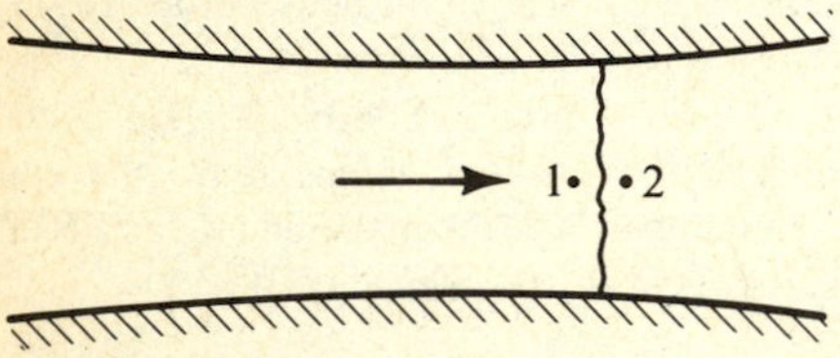

Figure P15.28

15.29 Oxygen flows from the upstream reservoir through a converging-diverging nozzle, as shown in Fig. P15.29. Given the throat and exit areas as 0.01 ft^2 and 0.04 ft^2, respectively, and the upstream reservoir thermodynamic properties $p_{01} = 41$ psia, $T_{01} = 75°$F, find the exit pressure necessary to fix the normal shock wave at a point in the diffuser where $A_1 = 0.03$ ft^2.

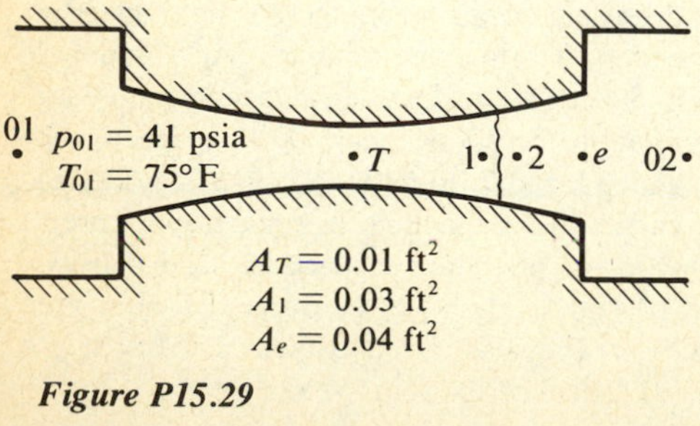

Figure P15.29

15.30 Consider the shock wave in Prob. 15.29 where $p_1 = 35$ psia and p_{01} has been increased to 80 psia. Calculate the Mach numbers M_1 and M_2 and the pressure p_2.

15.31 Air in the reservoir of Prob. 15.29 is open to the atmosphere at a pressure $p_{01} = 14.7$ psia and $T_{01} = 75°$F. Upstream of the shock, $M_1 = 2.5$. (a) Find T_1, $\bar{V}_1$, p_1, and ρ_1. (b) Calculate M_2, $\bar{V}_2$, ρ_2, p_2, and T_2. (c) Determine p_{02} and T_{02} in the reservoir tank. Solve the values using equations in the text and not the tables.

15.32 Steam originates from a boiler at a stagnation pressure of 100 psia and stagnation temperature of 1000°R. It moves through a pipe isentropically into a nozzle where it expands to an exit pressure of 25 psia. Determine the areas of the throat A_T and exit A_e if there is a mass rate of 50,000 lbm/hr. (This problem must use steam tables or the Mollier chart for steam, both of which are not in this text but are in most thermodynamics texts.)

15.33 Consider a perfect gas in a reservoir that is at a stagnation pressure p_s. The volume of the reservoir is $\forall$ and the mass of the gas in the reservoir is M. Derive the expression for the maximum velocity of the gas leaving the tank

$$\bar{V}_{max} = \left[\frac{2k\forall p_s}{(k-1)M}\right]^{1/2}$$

15.34 Consider the shock tube shown in Fig. P15.34. A metal diaphragm separates a region of high pressure from a region of low pressure where $p_2 = 5$ psia and $T_2 = 70°$F. When the diaphragm is ruptured, a shock wave is formed. Calculate the velocity behind the shock if the shock wave is moving at $M = 2.5$.

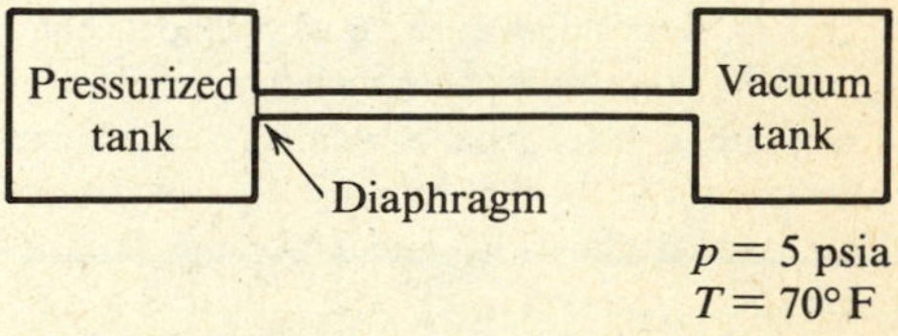

Figure P15.34

15.35 Consider a frictional steady adiabatic flow of a perfect gas through a duct. Let the upstream conditions p_1, ρ_1, and $\tilde{V}_1$ be fixed. (a) Plot a Mollier diagram of enthalpy versus entropy for constant mass flow. (b) Where is the entropy a maximum? (c) What are the conditions of the enthalpy for the flow to the subsonic and supersonic? (d) Show the isobaric lines for p_{01} and p_{02}, and the line for h_{01} and h_{02}. (e) Identify where the kinetic energies $\tilde{V}_1^2/2$ and $\tilde{V}_2^2/2$ are.

15.36 Air moves through a shock wave such that its downstream Mach number is M_2 = 0.5. (a) Calculate the downstream value of p_2, T_2, ρ_2, and $\tilde{V}_2$ and the upstream Mach number M_1, given p_1 = 10 psia and T_1 = 30°F. (b) Estimate the major loss h_f due to the shock wave.

15.37 As Fig. P15.37 shows, a missile has a propulsion system that develops 100,000 lbf at design conditions of 14.7 psia and 60°F. The combustion chamber pressure and temperature are 1000 psia and 10,000°R, respectively. Calculate (a) the areas of the throat A_T and exit A_e, and (b) the Mach number at the nozzle exit if the exhaust gases have an average specific heat ratio of 1.3, R = 52 ft-lbf/lbm°R, and the flow is isentropic.

15.38 A pitot tube is often used to determine the velocity in supersonic flow. Because it is blunt, a detached shock wave somewhat like that shown in Fig. P15.38 exists. Let the pressure be 20 psia at the stagnation point s, and the temperature T_s = 800°R. Calculate the pressure p_1 and velocity $\tilde{V}_1$ upstream of the normal shock portion of the wave if M_1 = 3.0, and p_2 = 15 psia.

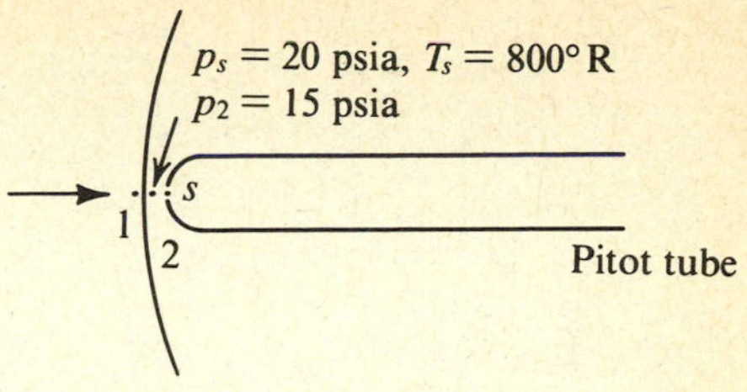

Figure P15.38

15.39 As shown in Fig. P15.39, an F-16 supersonic fighter aircraft is traveling at a Mach number of 2.2 where the atmospheric pressure is 50 kPa. At the inlet of the jet engine stands a normal shock. Assuming isentropic flow everywhere except across the shock, find the pressure and Mach number of the air leaving the jet engine at e if A_e/A_2 = 4.

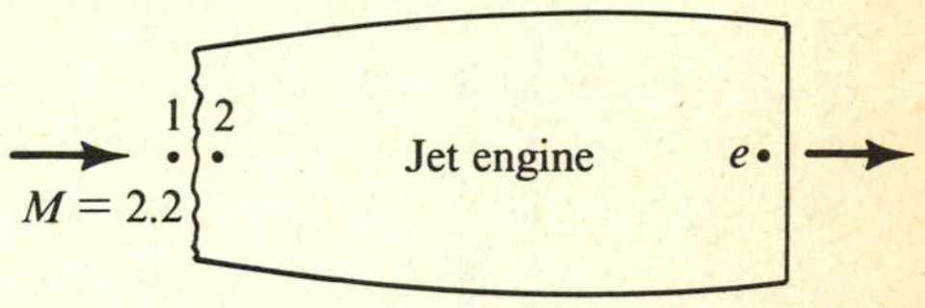

Figure P15.39

15.40 Helium at 200°F is flowing in a duct of constant cross-sectional area of 0.025 ft^2 at a velocity of 500 ft/s. A Venturi is placed in the duct. Calculate the minimum diameter of the Venturi that prevents choked flow.

15.41 Consider the flow of propane gas through a pipe entering a factory through a 80° bend. The conditions at the inlet are p_i = 50 psia, T_i = 80°F, $\tilde{V}_i$ = 400 ft/s through an area A_i = 3 in.^{2}. The exit condition of the pipe is 20 psia across an area A_e = 4 in.^{2}. Calculate the restraining force **R** on the pipe between the inlet and exit station given k = 1.4 (see Fig. P15.41). (Note, if $M_e > 1$, the exit pressure and exit area will have to be determined for M_e = 1.0.)

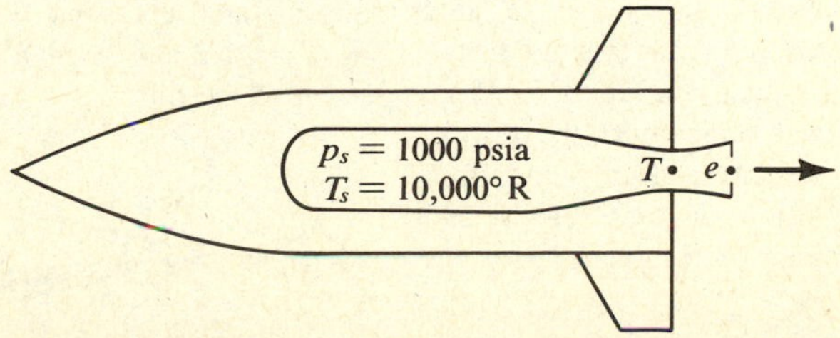

Figure P15.37

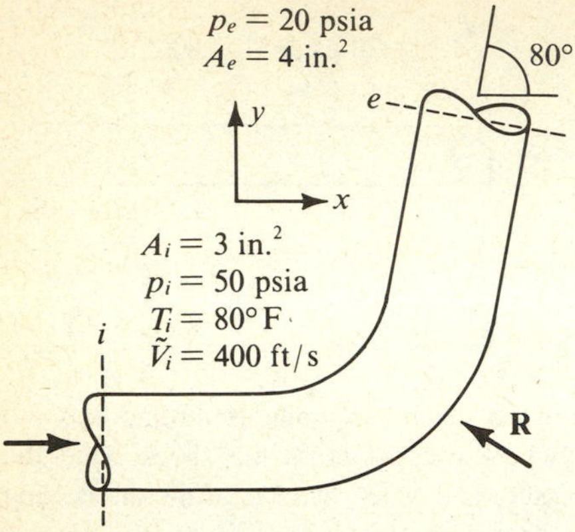

Figure P15.41

15.42 Derive the differential equation of a one-dimensional steady flow of a compressible gas through a duct of varying diameter

$$(1 - M^2)\frac{dp}{dx} = \rho\frac{\tilde{V}^2}{A}\frac{dA}{dx}$$

illustrated in Fig. P15.42.

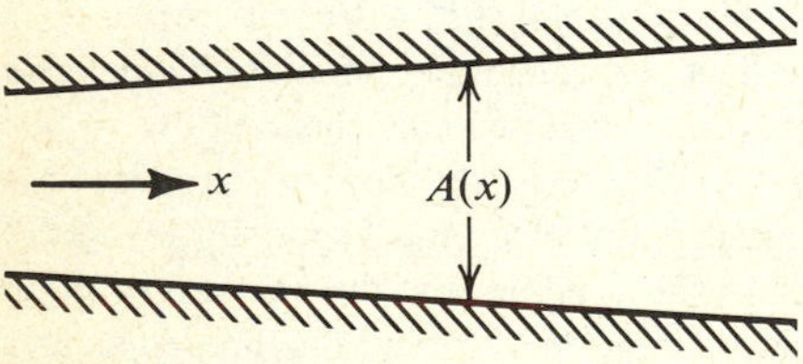

Figure P15.42

15.43 Air moves through a converging-diverging nozzle that has a throat area $A_T = 0.02$ ft^2 and exit area $A_e = 0.05$ ft^2. Attached to the exit of the nozzle is a reservoir. Air is drawn into the nozzle from the atmosphere at $p = 14.7$ psia and $T = 70$°F. Look at Fig. P15.43 and answer the following. (a) What is the pressure in the reservoir tank so that a normal shock exists at the nozzle exit? (b) What is the velocity and pressure at the throat? (c) What is the velocity and pressure before the shock wave? (d) What is the velocity and pressure downstream of the shock? Use gas tables.

15.44 Consider the supersonic flow past the oblique shock wave of Fig. 15.24. If the Mach

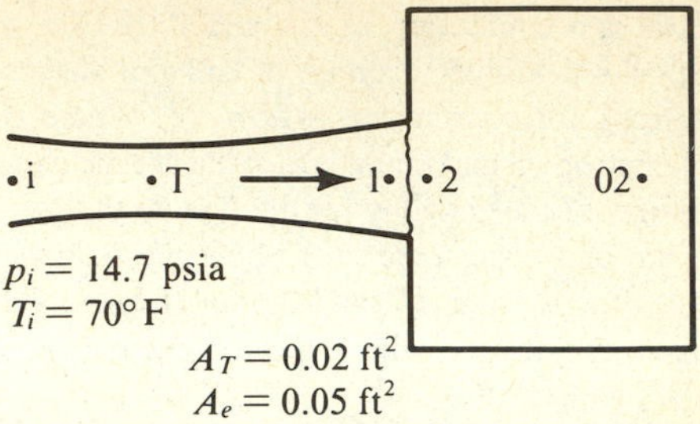

Figure P15.43

number of the approaching flow is 2.0, and the air temperature is 0°F, calculate the Mach number of the flow downstream of the shock if the angle of the shock is 30°.

15.45 An airplane is flying supersonically at 22,875 m altitude, as Fig. P15.45 illustrates. When it is 50,000 m past a house, the sonic boom breaks the window glass. Calculate the velocity of the aircraft.

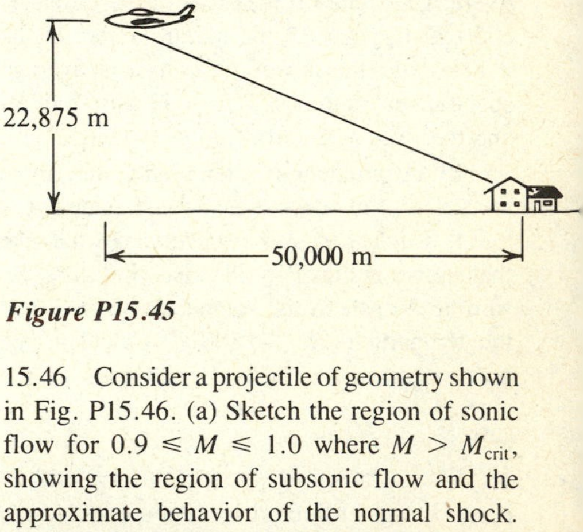

Figure P15.45

15.46 Consider a projectile of geometry shown in Fig. P15.46. (a) Sketch the region of sonic flow for $0.9 \leq M \leq 1.0$ where $M > M_{\text{crit}}$, showing the region of subsonic flow and the approximate behavior of the normal shock. (b) Sketch the regions of subsonic and supersonic flow for $M = 1.01$ and $M = 1.04$ showing the shock waves. (c) Sketch the region of subsonic and supersonic flow, and the shocks for $M = 1.2$ and $M = 2.0$. Discuss the behavior of the flow.

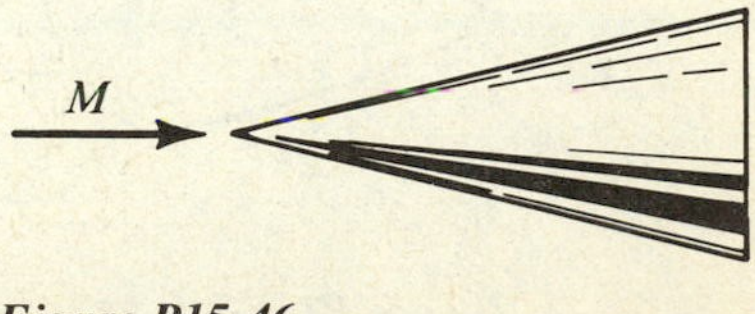

Figure P15.46

15.47 Air at $M = 3.0$ is forced to flow past a sharp wedge of semi vertex angle of 10° forcing the flow to follow the geometry of the wedge. An oblique shock forms. Calculate the angle of the wave α, the shock angle β, and the Mach number behind the shock M_2. (See Fig. P15.47.)

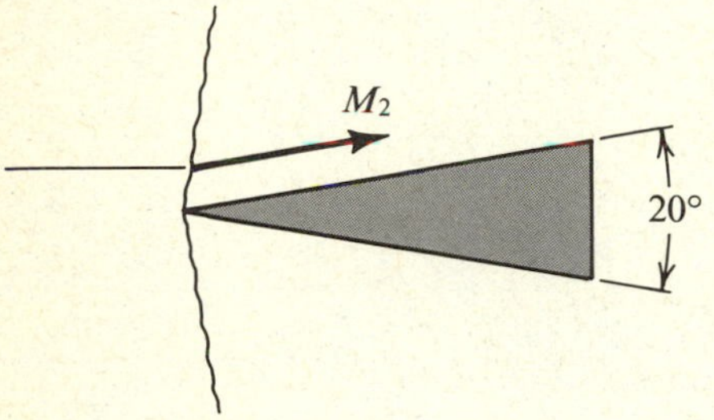

Figure P15.47

15.48 In Fig. 15.24, let $M_1 = 3.0$ and $M_2 = 2.0$ for the Mach numbers before and after the oblique shock. If $T_1 = 0$°F, calculate the angle α between the ambient flow and the oblique shock.

15.49 Air flows from a reservoir into a pipe of diameter 0.5 ft. At a station 1 in the pipe, the pressure and temperature are measured and found to be 50 psia and 70°F. (a) Calculate the pressure, temperature, and velocity at a point 2 where sonic conditions exist if $\bar{V}_1 = 400$ ft/s. (b) Calculate the heat transfer involved from the reservoir to where sonic conditions exist.

15.50 Derive the relationship

$$\bar{V}_2/\bar{V}_1 = \rho_1/\rho_2 = \frac{k - 1 + 2/M_1^2}{k + 1}$$

starting with the (I.F.) continuity equation

$$\rho_1\bar{V}_1 = \rho_2\bar{V}_2,$$

$$p_1 + \rho_1\bar{V}_1^2 = p_2 + \rho_2\bar{V}_2^2$$

and

$$\bar{V}_1^2 + \frac{2k}{k-1}\left(\frac{p_1}{\rho_1}\right) = \bar{V}_2^2 + \frac{2k}{k-1}\left(\frac{p_2}{\rho_2}\right)$$

15.51 Derive the one-dimensional wave equation in spherical coordinates

$$\frac{\partial^2 u}{\partial t^2} = \frac{c^2}{r^2 \sin\theta}\frac{\partial}{\partial r}\left(r^2 \sin\theta \frac{\partial u}{\partial r}\right)$$

where u is the radial velocity. Start with the (D.F.) one-dimensional energy equation $(\partial u/\partial t) + u\nabla u = -(1/\rho)\nabla p$ and the (D.F.) continuity equation $(\partial \rho/\partial t) + \nabla(\rho u) = 0$.

15.52 Consider a perfect gas flowing steadily through a duct of variable cross sections (Fig. P15.52). Derive an expression for the heat per unit mass of the flow in terms of stagnation temperatures T_{01} and T_{02}.

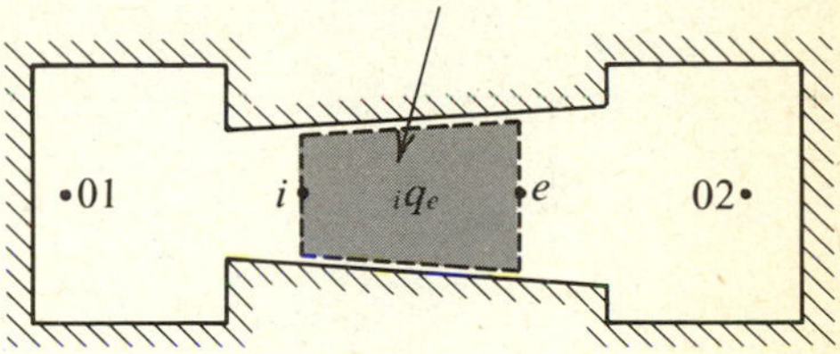

Figure P15.52

15.53 Using Fig. P15.52, derive the relationship

$$\frac{T_o}{T_o^*} = \frac{(k+1)M^2[2 + (k-1)M^2]}{(1 + kM^2)^2}$$

starting with

$$\frac{T_1}{T_2} = \left(\frac{M_1}{M_2}\right)^2\left(\frac{1 + kM_2^2}{1 + kM_1^2}\right)^2$$

15.54 Derive the expression

$$\frac{p}{p_{01}} = 1 - \frac{k-1}{2}M_o^2$$

where $M_o = \bar{V}/c_o$ starting with the (I.F.) energy equation $h + (\bar{V}^2/2) = h_{01}$.

15.55 Prove that

$$M^* = \left[\frac{(k+1)M^2}{2 + (k-1)M^2}\right]^{1/2}$$

15.56 Consider steady compressible isentropic flow past a two-dimensional wing shown in Fig. P15.56. Derive the expression

$$\frac{p}{p_\infty} = \left[1 + \frac{k-1}{2} M_\infty^2 \left(1 - \frac{V^2}{U^2}\right)\right]^{k/(k-1)}$$

where $V^2 = u^2 + v^2 + w^2$, and U is the undisturbed velocity of the wing.

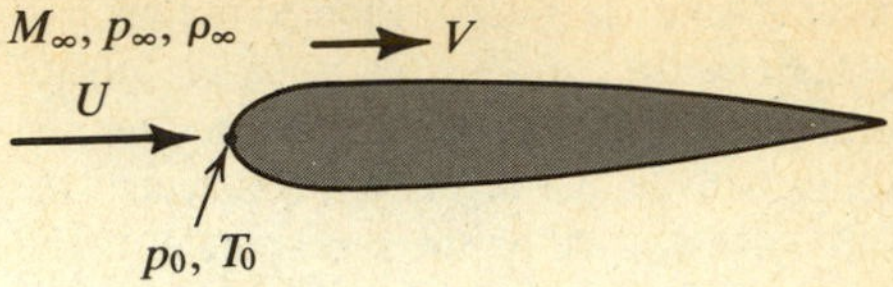

Figure P15.56

A Complex Variables

The results of complex variables that are applicable to fluid mechanics are summarized below.

A.1 Complex Numbers

A complex number is written in the form $a + ib$, where a and b are real scalars, and i is the imaginary unit defined by $i = \sqrt{-1}$. The quantity $x + iy$ is called a *complex independent variable* where x is called the real part and y is a real variable and together with i it is called the imaginary part. This complex independent variable $x + iy$ is defined by the complex variable z:

$$z = x + iy \tag{A.1}$$

Any two complex numbers $(a + ib)$ and $(c + id)$ are said to be equal if and only if the real and imaginary parts of the first number are equal, respectively, to the real and imaginary parts of the second. Thus, for $z = 0$, two conditions must be met: *the real part must be zero, and the imaginary part must be zero.*

A.2 de Moivre's Theorem

A power series for the exponential e^x can be written as

$$e^x = 1 + \frac{x}{1!} + \frac{x^2}{2!} + \frac{x^3}{3!} + \ldots + \frac{x^m}{m!} \tag{A.2}$$

If $x = i\theta$, then

$$e^{i\theta} = \cos\theta + i\sin\theta \tag{A.3}$$

Also,

$$\begin{aligned} re^{i\theta} &= r\cos\theta + ir\sin\theta \\ &= x + iy \\ &= z \end{aligned} \tag{A.4}$$

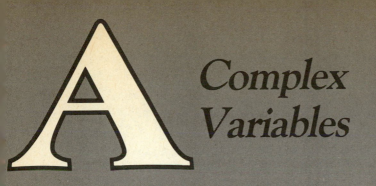

Note that

$$\frac{1}{z} = \frac{e^{-i\theta}}{r} \tag{A.5}$$

or

$$z^{1/n} = r^{1/n}\left[\cos\left(\frac{\theta + 2k\pi}{n}\right) + i\sin\left(\frac{\theta + 2k\pi}{n}\right)\right] \tag{A.6}$$

A.3 Some Useful Definitions

Some useful relationships that often appear in fluid mechanics are shown as follows.

- Modulus of z:

$$\begin{aligned} r &= (x^2 + y^2)^{1/2} \\ &= |z| \end{aligned} \tag{A.7}$$

- Argument of z:

$$\arg z = \theta \tag{A.8}$$

- Conjugate of z:

$$\bar{z} = x - iy = re^{-i\theta} \tag{A.9}$$

$$z\bar{z} = r^2 \tag{A.10}$$

- Logarithms:

$$\log z = \log(r) + i\theta \tag{A.11}$$

$$= 0.5\log(x^2 + y^2) + i\tan^{-1}\left(\frac{y}{x}\right) \tag{A.12}$$

- Hyperbolic functions:

$$\cosh\theta = \frac{e^{\theta} + e^{-\theta}}{2} = \cos i\theta \tag{A.13}$$

$$\sinh\theta = \frac{e^{\theta} - e^{-\theta}}{2} = -i\sin i\theta \tag{A.14}$$

A.4 Regular Function

A function $f(z)$ is regular (or analytic) if df/dz exists at the point and the neighborhood of z_o. The value of df/dz can be calculated in any direction, that is, $df/dz = \partial f/\partial x = \partial f/\partial y$.

A.5 Singular Points

Given a function $f(z)$, a point z_o is said to be a *singularity*, or a singular point, of $f(z)$ if in every neighborhood of z_o there are regular points of $f(z)$, whereas z_o itself is not a regular point of $f(z)$.

A.6 Taylor Series

If $f(z)$ is regular at all points with a circle of radius less than the radius of convergence ρ of the series $\Sigma a_n(z - z_o)^n$, then $f(z)$ can be expressed as

$$\begin{aligned} f(z) &= f(z_o) + f'(z_o)(z - z_o) + f''(z_o)\frac{(z - z_o)^2}{2!} + \dots \\ &\quad + f^k(z_o)\frac{(z - z_o)^k}{k!} + \dots \qquad \text{(A.15)} \\ &\equiv \sum_o^\infty f^k(z_o)\frac{(z - z_o)^k}{k!} \end{aligned}$$

The case $z_o = 0$ is the Maclaurin's series.

A.7 Laurent Series

If $f(z)$ is regular at all points within two circles of radius r_o and r_1, where $0 < r_1 < r_o$, then for every z in the region

$$f(z) = \sum_{-\infty}^{\infty} a_n(z - z_o)^n \qquad \text{(A.16)}$$

where

$$a_n = \frac{1}{2\pi i}\oint_C \frac{f(z)\,dz}{(z - z_o)^{n+1}}, \quad n = \dots, -1, 0, 1, \dots \qquad \text{(A.17)}$$

For the special case where $r_1 \rightarrow 0$, the series transforms to the Taylor series.

A.8 Cauchy-Goursat Theorem

If $f(z)$ is regular throughout a simply connected open region R and C is any simple closed path in R, then

$$\oint_C f(z)\,dz = 0 \tag{A.18}$$

A.9 Cauchy's Integral Formula

If $f(z)$ is regular throughout a simply connected open region R, z_o is any point in R, and C is any simple closed path in R with z_o on its interior, then

$$f(z) = \frac{1}{2\pi i} \oint_C \frac{f(z)\,dz}{z - z_o} \tag{A.19}$$

If $f(z)$ is regular at a point z_o, then $f^n(z_o)$ exists and is given by the formula

$$f^n(z_o) = \frac{n!}{2\pi i} \oint_C \frac{f(z)\,dz}{(z - z_o)^{n+1}}, \quad \text{for} \quad n = 0,1, \ldots \tag{A.20}$$

A.10 Residue

If a function $f(z)$ is regular in a deleted neighborhood of a point z_o and C is a simple closed path in that neighborhood with z_o as its interior, we speak of the value of

$$\frac{1}{2\pi i} \oint_C f(z)\,dz$$

as the *residue* of $f(z)$ at z_o. This residue equals the coefficient a_{-1} of $1/(z - z_o)$ in the Laurent expansion of $f(z)$ in the neighborhood of z_o; it is, of course, independent of C.

A.11 Residue Theorem

If $f(z)$ is regular in a simply connected open region R_1 except possibly at a finite number of points $z_1, z_2, \ldots, z_n$, and C is a simple closed path in R with these points on its interior, then

$$\oint_C f(z)\,dz = 2\pi i \sum \text{res}_j[f(z)] \tag{A.21}$$

where $\text{res}_j[f(z)]$ means the residue of $f(z)$ at the point z_j, $j = 1, 2, \ldots, n$.

B Vectors

The results of vector analysis that are applicable to fluid mechanics are summarized here. In the following formulas, ϕ is any scalar function and **A**, **B**, **C**, and **D** are any vector functions.

B.1 Vector Products

$$\mathbf{A}\cdot\mathbf{B} = \mathbf{B}\cdot\mathbf{A} \tag{B.1}$$

$$\mathbf{A}\times\mathbf{B} = -\mathbf{B}\times\mathbf{A} \tag{B.2}$$

$$\mathbf{A}\cdot(\mathbf{B}\times\mathbf{C}) = \mathbf{B}\cdot(\mathbf{C}\times\mathbf{A}) = \mathbf{C}\cdot(\mathbf{A}\times\mathbf{B}) = \begin{vmatrix} A_x & A_y & A_z \\ B_x & B_y & B_z \\ C_x & C_y & C_z \end{vmatrix} \tag{B.3}$$

$$\mathbf{A}\times(\mathbf{B}\times\mathbf{C}) = \mathbf{B}(\mathbf{A}\cdot\mathbf{C}) - \mathbf{C}(\mathbf{A}\cdot\mathbf{B}) \tag{B.4}$$

$$(\mathbf{A}\times\mathbf{B})\cdot(\mathbf{C}\times\mathbf{D}) = (\mathbf{A}\cdot\mathbf{C})(\mathbf{B}\cdot\mathbf{D}) - (\mathbf{B}\cdot\mathbf{C})(\mathbf{A}\cdot\mathbf{D}) \tag{B.5}$$

B.2 Differentiation with Respect to a Scalar

Given $\mathbf{A} = \mathbf{A}(u)$ and $\mathbf{B} = \mathbf{B}(u)$, then

$$\frac{d}{du}(\mathbf{A}+\mathbf{B}) = \frac{d\mathbf{A}}{du} + \frac{d\mathbf{B}}{du} \tag{B.6}$$

$$\frac{d}{du}(\phi\mathbf{A}) = \phi\frac{d\mathbf{A}}{du} + \mathbf{A}\frac{d\phi}{du} \tag{B.7}$$

$$\frac{d}{du}(\mathbf{A}\cdot\mathbf{B}) = \mathbf{A}\cdot\frac{d\mathbf{B}}{du} + \mathbf{B}\cdot\frac{d\mathbf{A}}{du} \tag{B.8}$$

$$\frac{d}{du}(\mathbf{A}\times\mathbf{B}) = \mathbf{A}\times\frac{d\mathbf{B}}{du} + \frac{d\mathbf{A}}{du}\times\mathbf{B} \tag{B.9}$$

B.3 Formulas of Partial Differentiation

- Gradient:

$$\nabla\phi = \mathbf{i}\frac{\partial\phi}{\partial x} + \mathbf{j}\frac{\partial\phi}{\partial y} + \mathbf{k}\frac{\partial\phi}{\partial z} \tag{B.10}$$

$$\nabla\phi = \mathbf{e}_r\frac{\partial\phi}{\partial r} + \mathbf{e}_\theta\frac{1}{r}\frac{\partial\phi}{\partial\theta} + \mathbf{k}\frac{\partial\phi}{\partial z} \tag{B.11}$$

- Divergence:

$$\nabla\cdot\mathbf{A} = \frac{\partial A_x}{\partial x} + \frac{\partial A_y}{\partial y} + \frac{\partial A_z}{\partial z} \tag{B.12}$$

$$= \frac{1}{r}\frac{\partial(rA_r)}{\partial r} + \frac{1}{r}\frac{\partial A_\theta}{\partial\theta} + \frac{\partial A_z}{\partial z} \tag{B.13}$$

- Curl:

$$\nabla\times\mathbf{A} = \begin{vmatrix} \mathbf{i} & \mathbf{j} & \mathbf{k} \\ \dfrac{\partial}{\partial x} & \dfrac{\partial}{\partial y} & \dfrac{\partial}{\partial z} \\ A_x & A_y & A_z \end{vmatrix} \tag{B.14}$$

$$= \begin{vmatrix} \dfrac{\mathbf{e}_r}{r} & \mathbf{e}_\theta & \dfrac{\mathbf{k}}{r} \\ \dfrac{\partial}{\partial r} & \dfrac{\partial}{\partial\theta} & \dfrac{\partial}{\partial z} \\ A_r & rA_\theta & A_z \end{vmatrix} \tag{B.15}$$

- Laplacian:

$$\nabla^2\phi = \frac{\partial^2\phi}{\partial x^2} + \frac{\partial^2\phi}{\partial y^2} + \frac{\partial^2\phi}{\partial z^2} \tag{B.16}$$

$$= \frac{1}{r}\frac{\partial}{\partial r}\left(r\frac{\partial\phi}{\partial r}\right) + \frac{1}{r^2}\frac{\partial^2\phi}{\partial\theta^2} + \frac{\partial^2\phi}{\partial z^2} \tag{B.17}$$

- Given $f(u_1, u_2, \ldots, u_n)$, (u_r are scalar fields):

$$\nabla f = \sum_{r=1}^{r=n}\frac{\partial f}{\partial u_r}\nabla u_r \tag{B.18}$$

- Let f and g be scalar fields:

$$\nabla(fg) = g\nabla f + f\nabla g \tag{B.19}$$

$$(\mathbf{A}\cdot\nabla)f = \mathbf{A}\cdot\nabla f \tag{B.20}$$

$$(\mathbf{A}\times\nabla)f = \mathbf{A}\times(\nabla f) \tag{B.21}$$

$$\nabla\cdot\sum_n \mathbf{A}_n = \sum_n \nabla\cdot\mathbf{A}_n \tag{B.22}$$

$$\nabla\times\sum_n \mathbf{A}_n = \sum_n \nabla\times\mathbf{A}_n \tag{B.23}$$

$$(\mathbf{A}\times\nabla)\cdot\mathbf{B} = \mathbf{A}\cdot(\nabla\times\mathbf{B}) \tag{B.24}$$

$$\nabla\cdot\phi\mathbf{A} = \phi(\nabla\cdot\mathbf{A}) + \mathbf{A}\cdot\nabla\phi \tag{B.25}$$

$$\nabla\times\phi\mathbf{A} = \phi(\nabla\times\mathbf{A}) - \mathbf{A}\times(\nabla\phi) \tag{B.26}$$

$$\nabla\cdot(\mathbf{A}\times\mathbf{B}) = \mathbf{B}\cdot(\nabla\times\mathbf{A}) - \mathbf{A}\cdot(\nabla\times\mathbf{B}) \tag{B.27}$$

$$\nabla\times(\mathbf{A}\times\mathbf{B}) = \mathbf{A}(\nabla\cdot\mathbf{B}) + (\mathbf{B}\cdot\nabla)\,\mathbf{A} - \mathbf{B}(\nabla\cdot\mathbf{A}) - (\mathbf{A}\cdot\nabla)\mathbf{B} \tag{B.28}$$

$$\begin{aligned}\nabla(\mathbf{A}\cdot\mathbf{B}) = {} & (\mathbf{A}\cdot\nabla)\mathbf{B} + (\mathbf{B}\cdot\nabla)\mathbf{A} + \\ & \mathbf{A}\times(\nabla\times\mathbf{B}) + \mathbf{B}\times(\nabla\times\mathbf{A})\end{aligned} \tag{B.29}$$

$$(\mathbf{A}\cdot\nabla)\mathbf{A} = \frac{1}{2}\nabla A^2 - \mathbf{A}\times(\nabla\times\mathbf{A}) \tag{B.30}$$

$$\nabla\times(\nabla\phi) = 0 \tag{B.31}$$

$$\nabla\cdot(\nabla\times\mathbf{A}) = 0 \tag{B.32}$$

$$\nabla\times(\nabla\times\mathbf{A}) = \nabla(\nabla\cdot\mathbf{A}) - \nabla^2\mathbf{A} \tag{B.33}$$

- Divergence of a position vector:

$$\nabla\cdot\mathbf{r} = 3 \tag{B.34}$$

- Curl of a position vector:

$$\nabla\times\mathbf{r} = 0 \tag{B.35}$$

$$(\mathbf{A}\cdot\nabla)r = \mathbf{A} \tag{B.36}$$

B.4 Formulas of Integration

In the following theorems, $\forall$ is any volume, **A** is the surface which encloses the volume, and $\boldsymbol{l}$ is a line. **F** is any vector function.

- Gauss' theorem (divergence theorem):

$$\int_A \mathbf{F}\cdot\mathbf{e}_n \, dA = \int_{\forall} \nabla\cdot\mathbf{F} \, d\forall \tag{B.37}$$

- Stokes' theorem:

$$\oint \mathbf{F}\cdot d\mathbf{l} = \int_A (\nabla \times \mathbf{F})\cdot\mathbf{e}_n \, dA \tag{B.38}$$

C Gas Tables

Table C.1 *One-Dimensional Isentropic Compressible Flow Functions for a Perfect Gas with Constant Specific Heat and Molecular Weight and k = 1.4**

M	M^*	$\frac{A}{A^*}$	$\frac{p}{p_o}$	$\frac{\rho}{\rho_o}$	$\frac{T}{T_o}$	$\left(\frac{A}{A^*}\right)\left(\frac{p}{p_o}\right)$
0	0	∞	1.00000	1.000000	1.00000	∞
.10	.10943	5.8218	.99303	.99502	.99800	5.7812
.20	.21822	2.9635	.97250	.98027	.99206	2.8820
.30	.32572	2.0351	.93947	.95638	.98232	1.9119
.40	.43133	1.5901	.89562	.92428	.96899	1.4241
.50	.53452	1.3398	.84302	.88517	.95238	1.12951
.60	.63480	1.1882	.78400	.84045	.93284	.93155
.70	.73179	1.09437	.72092	.79158	.91075	.78896
.80	.82514	1.03823	.65602	.74000	.88652	.68110
.90	.91460	1.00886	.59126	.68704	.86058	.59650
1.00	1.00000	1.00000	.52828	.63394	.83333	.52828
1.10	1.08124	1.00793	.46835	.58169	.80515	.47206
1.20	1.1583	1.03044	.41238	.53114	.77640	.42493
1.30	1.2311	1.06631	.36092	.48291	.74738	.38484
1.40	1.2999	1.1149	.31424	.43742	.71839	.35036
1.50	1.3646	1.1762	.27240	.39498	.68965	.32039
1.60	1.4254	1.2502	.23527	.35573	.66138	.29414
1.70	1.4825	1.3376	.20259	.31969	.63372	.27099
1.80	1.5360	1.4390	.17404	.28682	.60680	.25044
1.90	1.5861	1.5552	.14924	.25699	.58072	.23211
2.00	1.6330	1.6875	.12780	.23005	.55556	.21567
2.10	1.6769	1.8369	.10935	.20580	.53135	.20087
2.20	1.7179	2.0050	.09352	.18405	.50813	.18751
2.30	1.7563	2.1931	.07997	.16458	.48591	.17539
2.40	1.7922	2.4031	.06840	.14720	.46468	.16437
2.50	1.8258	2.6367	.05853	.13169	.44444	.15432
2.60	1.8572	2.8960	.05012	.11787	.42517	.14513
2.70	1.8865	3.1830	.04295	.10557	.40684	.13671
2.80	1.9140	3.5001	.03685	.09462	.38941	.12897
2.90	1.9398	3.8498	.03165	.08489	.37286	.12185
3.00	1.9640	4.2346	.02722	.07623	.35714	.11528

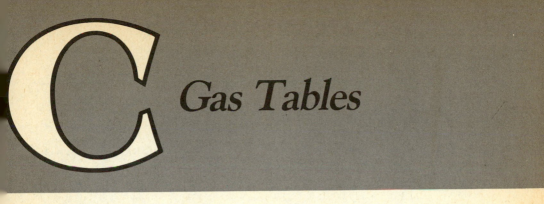

Table C.1 (Con't.)

M	M^*	$\frac{A}{A^*}$	$\frac{p}{p_o}$	$\frac{\rho}{\rho_o}$	$\frac{T}{T_o}$	$\left(\frac{A}{A^*}\right)\left(\frac{p}{p_o}\right)$
3.50	2.0642	6.7896	.01311	.04523	.28986	.08902
4.00	2.1381	10.719	.00658	.02766	.23810	.07059
4.50	2.1936	16.562	.00346	.01745	.19802	.05723
5.00	2.2361	25.000	$189(10)^{-5}$	.01134	.16667	.04725
6.00	2.2953	53.180	$633(10)^{-6}$	.00519	.12195	.03368
7.00	2.3333	104.143	$242(10)^{-6}$	.00261	.09259	.02516
8.00	2.3591	190.109	$102(10)^{-6}$	.00141	.07246	.01947
9.00	2.3772	327.189	$474(10)^{-7}$	.000815	.05814	.01550
10.00	2.3904	535.938	$236(10)^{-7}$	.000495	.04762	.01263
∞	2.4495	∞	0	0	0	0

Source: Abridged from Table 30 of *Gas Tables*, by J. Keenan and J. Kaye, copyright © 1948 John Wiley & Sons, Inc. Publishers. Reprinted by permission of John Wiley & Sons, Inc.

Table C.2 *One-Dimensional Normal Shock Functions for a Perfect Gas with Constant Specific Heat and Molecular Weight and k = 1.4**

M_x	M_y	$\frac{p_y}{p_x}$	$\frac{\rho_y}{\rho_x}$	$\frac{T_y}{T_x}$	$\frac{p_{o2}}{p_{o1}}$	$\frac{p_{o2}}{p_x}$
1.00	1.00000	1.00000	1.00000	1.00000	1.00000	1.8929
1.10	0.91177	1.2450	1.1691	1.06494	0.99892	2.1328
1.20	0.84217	1.5133	1.3416	1.1280	0.99280	2.4075
1.30	0.78596	1.8050	1.5157	1.1909	0.97935	2.7135
1.40	0.73971	2.1200	1.6896	1.2547	0.95819	3.0493
1.50	0.70109	2.4583	1.8621	1.3202	0.92978	3.4133
1.60	0.66844	2.8201	2.0317	1.3880	0.89520	3.8049
1.70	0.64055	3.2050	2.1977	1.4583	0.85573	4.2238
1.80	0.61650	3.6133	2.3592	1.5316	0.81268	4.6695
1.90	0.59562	4.0450	3.5157	1.6079	0.76735	5.1417
2.00	0.57735	4.5000	2.6666	1.6875	0.72088	5.6405
2.10	0.56128	4.9784	2.8119	1.7704	0.67422	6.1655
2.20	0.54706	5.4800	2.9512	1.8569	0.62812	6.7163
2.30	0.53441	6.0050	3.0846	1.9468	0.58331	7.2937
2.40	0.52312	6.5533	3.2119	2.0403	0.54015	7.8969
2.50	0.51299	7.1250	3.3333	2.1375	0.49902	8.5262
2.60	0.50387	7.7200	3.4489	2.2383	0.46012	9.1813
2.70	0.49563	8.3383	3.5590	2.3529	0.42359	9.8625
2.80	0.48817	8.9800	3.6635	2.4512	0.38946	10.569
2.90	0.48138	9.6450	3.7629	2.5632	0.34773	11.302
3.00	0.47519	10.333	3.8571	2.6790	0.32834	12.061

Table C.2 *(Con't.)*

M_x	M_y	$\frac{p_y}{p_x}$	$\frac{\rho_y}{\rho_x}$	$\frac{T_y}{T_x}$	$\frac{p_{o2}}{p_{o1}}$	$\frac{p_{o2}}{p_x}$
4.00	0.43496	18.500	4.5714	4.0469	0.13876	21.068
5.00	0.41523	29.000	5.0000	5.8000	0.06172	32.654
10.00	0.38757	116.50	5.7143	20.388	0.00304	129.217
∞	0.37796	∞	6.000	∞	0	∞

Source: Abridged from Table 48 of *Gas Tables*, by J. Keenan and J. Kaye, copyright © 1948 John Wiley & Sons, Inc. Publishers. Reprinted by permission of John Wiley & Sons, Inc.

Index